VERTEBRATE LIFE
ELEVENTH EDITION

VERTEBRATE LIFE

ELEVENTH EDITION

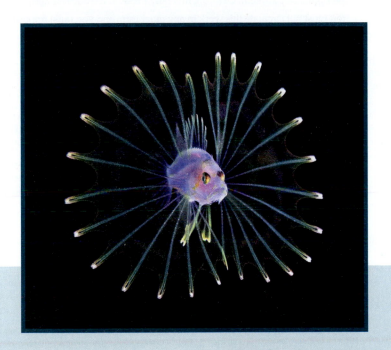

F. HARVEY POUGH
Professor Emeritus, Rochester Institute
of Technology

WILLIAM E. BEMIS
Professor of Ecology and Evolutionary Biology and
Faculty Curator of Ichthyology, Cornell University

BETTY MCGUIRE
Senior Lecturer, Cornell University

CHRISTINE M. JANIS
Professor Emerita of Biology, Brown University,
USA, Honorary Professor of Earth Sciences,
University of Bristol, UK

Chapter 18 "Avemetatarsalia and the Origin of Dinosauria"
and Chapter 19 "Theropods and the Origin of Birds"
by Emanuel Tschopp, Universität Hamburg, Germany

Chapter 24 "Primate Evolution and the Emergence of Humans"
by Sergi López-Torres, Institute of Evolutionary Biology, Faculty of Biology,
Biological and Chemical Research Centre, University of Warsaw, Poland

Art Development by William E. Bemis

SINAUER ASSOCIATES
NEW YORK   OXFORD
OXFORD UNIVERSITY PRESS

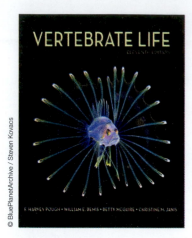

© BluePlanetArchive / Steven Kovacs

On the cover:
Lionfish larva, *Pterois* sp., photographed during blackwater drift dive in open ocean at 20–40 feet with bottom at 700 plus feet below, Palm Beach, Florida, USA, Atlantic Ocean.

Oxford University Press is a department of the University of Oxford. It furthers the University's objective of excellence in research, scholarship, and education by publishing worldwide. Oxford is a registered trade mark of Oxford University Press in the UK and certain other countries.

Published in the United States of America by Oxford University Press
198 Madison Avenue, New York, NY 10016, United States of America

© 2023, 2019 Oxford University Press
Sinauer Associates is an imprint of Oxford University Press.

For titles covered by Section 112 of the US Higher Education Opportunity Act, please visit www.oup.com/us/he for the latest information about pricing and alternate formats.

Address editorial correspondence to:

Sinauer Associates
23 Plumtree Road
Sunderland, MA 01375 USA

ACCESSIBLE COLOR CONTENT   Every opportunity has been taken to ensure that the content herein is fully accessible to those who have difficulty perceiving color. Exceptions are cases where the colors provided are expressly required because of the purpose of the illustration.

NOTICE OF TRADEMARKS   Throughout this book trademark names may have been used, and in some instances, depicted. In lieu of appending the trademark symbol to each occurrence, the authors and publisher state that these trademarks are used in an editorial fashion, to the benefit of the trademark owners, and with no intent to infringe upon the trademarks. Every effort has been made to determine and contact copyright holders. In the case of any omissions, the publisher will be pleased to make suitable acknowledgment in future editions.

Library of Congress Cataloging-in-Publication Data

Names: Pough, F. Harvey, author. | Bemis, William E., author. | McGuire, Betty, author. | Janis, Christine M., author.

Title: Vertebrate life / F. Harvey Pough, William E. Bemis, Betty McGuire, Christine M Janis

Description: Eleventh edition. | New York, NY : Oxford University Press, [2023] | Includes bibliographical references and index. | Summary: "Vertebrate Life distills the necessary information from vertebrate anatomy, physiology, ecology, and behavioral studies and then helps students see important connections across levels of biological scale. Students come to understand how organisms function effectively in their environments and how lineages of organisms change through evolutionary time. Processing complex detailed information about expansive phylogenies and diverse anatomies can be difficult for even the most motivated students, and Vertebrate Life addresses this challenge by combining appropriately-detailed, clearly-written text with outstanding phylogenies and figures, making it a thorough and engaging reference for students and instructors alike. The text's impressive illustration program helps students visualize complex concepts, allowing them to parse difficult anatomical information. The 11th edition has an upgraded illustration program with several new and revised figures, including layered figures presented in the new enhanced eBook. The text's integrative approach to vertebrate biology addresses the need to engage students in a challenging subject through the inclusion of material about extant organisms and also by linking the fossil record to the present day. The text further addresses this need by highlighting real-world applications, particularly as they relate to the conservation and protection of biodiversity in the modern world. The comprehensive and up-to-date coverage in Pough gives instructors the freedom of choice when structuring their vertebrate biology courses-while remaining an indispensable guide for students"-- Provided by publisher.

Identifiers: LCCN 2021052848 (print) | LCCN 2021052849 (ebook) | ISBN 9780197558621 (paperback) | ISBN 9780197564899 (ebook)

Subjects: LCSH: Vertebrates--Textbooks. | Vertebrates, Fossil--Textbooks.

Classification: LCC QL605 .P68 2023 (print) | LCC QL605 (ebook) | DDC 596--dc23/eng/20211028

LC record available at https://lccn.loc.gov/2021052848

LC ebook record available at https://lccn.loc.gov/2021052849

9 8 7 6 5 4 3 2 1

Printed in the United States of America

Brief Contents

Contents

Chapter 10

Geography and Ecology of the Mesozoic   223

Chapter 11

Living on Land   237

Chapter 12

Lissamphibians   259

Chapter 13

Synapsids and Sauropsids: Two Ways of Living on the Land   291

Preface

In the fall of 1972, I started my college career with a course that changed my life. That course was called "The Vertebrates" and the professor who taught it was Dr. F. Harvey Pough, the lead author of this book. I learned so much in that class and became so fascinated with vertebrates that I have studied, written, and taught about them ever since. Today I teach The Vertebrates in the same classroom where I took the course as a student 50 years ago.

When Harvey invited my wife, Betty McGuire, and me to join the author team for the eleventh edition of *Vertebrate Life*, the book that Harvey and his original coauthors first published in 1979, we were thrilled.

The approach of this eleventh edition traces back to the course that I took so long ago. Then, as now, the goal was to integrate morphology, physiology, ecology, and behavior in a phylogenetic context to provide an organism-level understanding of vertebrates and their evolution. Back then, it was a little easier to contemplate such an undertaking, for only about half as many vertebrate species were known in 1972. As of this writing there are more than 70,000 species of vertebrates, and scientists describe several hundred new species each year.

When the first edition of *Vertebrate Life* was published in 1979, the types of phylogenetic rigor and methods of organizing phylogenetic information that we use today were in their infancy and no one had access to the molecular tools that we now take for granted. The many subsequent editions of *Vertebrate Life* progressively incorporated modern phylogenetics, more behavior, and more ecology, as well as new insights from paleontology and Earth history, and more recently from evolutionary developmental biology, or evo-devo. Especially and always, *Vertebrate Life* reflects a deep appreciation and respect for vertebrate diversity and conservation.

For this new edition, we focused first on phylogenetic content. We oriented the trees horizontally to improve readability, providing bracketed phylogenetic classifications to the right. We incorporated the dagger symbol (†) to indicate extinct taxa, a helpful convention in a field that requires constant comparisons between extinct and extant taxa. For most groups, we provide anatomical synapomorphies that students can see in the specimens that they study in course laboratories. We also discuss and incorporate molecular phylogenetic interpretations.

We worked to eliminate nonmonophyletic groups in the text and figures with the goal of encouraging clear thinking about phylogenetic trees and classifications. I always emphasize to my students that all trees are hypotheses, many are wrong, and all are subject to revision. But I also tell them that trees and phylogenetic classifications are the most useful and concise way to organize what we know about organisms and that their challenge as biologists is to think critically about the information that supports or falsifies a particular branching pattern. This prepares students to take the next step of using trees to explore anatomy, behavior, ecology, and other aspects of vertebrate life.

We are delighted by the return of Sergi López-Torres who authored the chapter on primates and the evolution of humans in the previous edition, and we welcome Emanuel Tschopp, the author of two chapters on dinosaurs in this edition. These are extraordinarily active fields of research and Sergi and Emanuel bring the depth and breadth of knowledge needed to describe these increasingly complex topics.

The eleventh edition includes more than 475 figures composed of nearly 2,000 individual photographs and illustrations, of which more than 1,100 are new. We critically evaluated every existing figure, redrew many, replaced others, and worked to find the best photographs to highlight vertebrate diversity. We developed new and informative part labels that appear directly on the figures, providing "instant captions" to orient students and serve instructors who depend on illustrations for teaching. Graphic balloons call attention to key points in many figures. Our goal has been to develop fun and informative visual interpretations to help students think about vertebrate evolution. We are indebted to Elizabeth Morales, Will Sillin, Sabrina Shih, as well as hundreds of amazing photographers and illustrators from around the world who graciously shared their work through Creative Commons licensing. All uncredited color illustrations of vertebrates are based on images by William E. Bemis and William B. Sillin, rendered by Sabrina Shih.

New to this edition are Learning Objectives presented at the start of each section within a chapter to highlight key take-away concepts for the section. Other specific improvements include:

- Development of background materials for the study of vertebrate evolution, exemplified by Figure 1.2, and the many new time trees and cladograms throughout the book

- Many updated examples using key fossil vertebrates to explore new phylogenetic interpretations and explain how particular anatomical features, such as the mammalian middle ear, evolved (see Figure 22.11)

- Reorganization of key topics within chapters to improve continuity and connections among chapters

along with new cross-referencing to link topics between chapters

- Enhanced coverage and reorganization of chapters on sharks (Chapter 6), ray-finned fishes (Chapter 7), sarcopterygians (Chapter 8), early tetrapods, lissamphibians, and amniotes (Chapter 9), dinosaurs (Chapters 18 and 19), synapsids (Chapter 22), therians (Chapter 23), and primates (Chapter 24)

- Consolidation and reorganization of chapters to better highlight the role of thermoregulation and comparative physiology in vertebrate evolution (Chapter 14)

- Incorporation of new behavioral and ecological information to showcase vertebrate lives;

- Updated Discussion Questions at the end of each chapter

- A list of the sources that we consulted in preparing this edition, as well as many sources from earlier editions, that are available on the book's web page: https://learninglink.oup.com/access/content/pough-11e

We hope that you will find the eleventh edition of *Vertebrate Life* an inspiring introduction to the biology of these remarkable and enduringly fascinating animals.

William E. Bemis
Freeville, NY
January 2022

Sources for extant species diversity and common and scientific names

We relied upon these sources for the numbers of extant species and their common and scientific names.

- Eschmeyer's Catalog of Fishes: https://www.calacademy.org/scientists/projects/eschmeyers-catalog-of-fishes

- FishBase: http://www.fishbase.se/search.php

- AmphibiaWeb: https://amphibiaweb.org/

- The Reptile Database: http://www.reptile-database.org/

- IOC World Birds list: http://www.worldbirdnames.org/

- Avibase: http://avibase.bsc-eoc.org/checklist.jsp

- Mammal Species of the World: https://www.departments.bucknell.edu/biology/resources/msw3/

- ASM Mammal Diversity Database: https://mammaldiversity.org/

- IUCN Red List of threatened species: http://www.iucnredlist.org/

- Time Tree of Life: http://www.timetree.org

Acknowledgments

Expert librarians are essential to any scholarly undertaking, and we are fortunate to have had the outstanding assistance of Adwoa Boateng and Morna Hilderbrand (Wallace Library, Rochester Institute of Technology) and Michael Cook, Mel Jensen, Erica Johns, Matthew Kibbee, and Matt Ryan (Mann Library, Cornell University).

We are grateful to the many colleagues who answered questions and provided data and photographs.

Todd Ahern
Iván Trinidad Ahumada-Carrillo
Warren Allmon
Robin Andrews
José Carlos Arenas-Monroy
Iulia Bădescu
Carole Baldwin
Matthew G. Baron
Andy Bass
Aaron Bauer
Katherine Bemis
Michael Benton
Maureen Bickley
Ron Blakey
Dee Blakey
Adwoa Boateng
Dmitry Bogdanov
Nanci Bompey
Richard Bornemann
Doug Boyer
Cinthia Brasileiro
Julie Brigham-Grette
Ralf Britz
William Brown
Christopher W. Brown
David Brown
Larry Buckley
Ken Catania
Ignacio Cerda
James Cerda
Zhao Chuang
Brian Choo
Kaitlyn Cisz
René Clark
Sara Clark
Chip Clark
Richard Cloutier
Michael Coates
Rafael Camargo Consolmagno
Heath Cook
Andrew Corso
Cristina Cox Fernandes
Marty Crump
Julius Csotonyi

Nina Cummings
Myrna C. Cureg
Ted Daeschler
Christine Dahlin
Charles Dardia
Fábio Perin de Sá
Matteo De Stefano
Diego Delso
Didier Descouens
Dominique Didier
Casey Dillman
Jignasu Dolia
Bernard Dupont
Hugo Dutel
Devin Edmonds
Liam Elward
Robert Espinoza
Allyson Evans
Jan Factor
Stacy Farina
Marty Feldner
Jane Fenelon
Danté Fenolio
Daniel Field
Will Flaxington
Claire Hope Fox
Ethan France
Gabriela Franzoi Dri
Stella Freund
Juergen Freund
Jeff Gage
Mark Gautreau
Aline Ghilardi
Pamela Gill
Phillip Gingerich
Terry Grande
Lance Grande
Kate Bruder Greenman
Chris Griffin
Nathan Grigg
Célio Haddad
James Harding
David Haring
Scott Hartman
Danielle Haulsee
Christophe Hendrickx
Anthony Herrel

Eric Hilton
Rob Holmes
Tom Hübner
Elaine Hyder
Robert Hynes
Philippe Janvier
Carlos Jared
Mike Johnson
Dave Johnson
Jean Joss
Dale Kalina
Johanne Kerr
John Klausmeyer
Phillip Krzeminski
James Kuether
Jeffrey Lang
Björn Lardner
George Lauder
Michel Laurin
Mark Leckie
Gustavo Lecuona
Mike Letnic
Brenna Levine
John Long
Tyler Lyson
John Maisey
David Martill
Jeffrey Martz
Brad Maryan
Frank Mazzotti
Amy McCune
Jennifer McElwain
Kevin McGraw
John McPhee
Rita Mehta
Robert Mendyk
Stefan Merker
Axel Meyer
Jeffrey Milisen
Tetsuto Miyake
Carrie Mongle
Gustavo Monroy-Becerril
Suzi Moore
Josh More
Josh Moyer
Bruce Mundy
Andy Murch

Laura Murch
Darren Naish
Anna Nekaris
Sterling Nesbitt
Janna Nichols
Andreas Nöllert
Ai Nonaka
John Nyakatura
Patrick O'Connor
Kevin Padian
Charles Peterson
Neil Pezzoni
Fritz Pfeil
Ted Pietsch
Graciela Piñeiro
Teresa Porri
Yin Qi
Malcolm Ramsay
Ernesto Raya
Jane Reeves
Roger Repp
Aaron Rice
Alan Richmond
Jennifer Rieser
Ross Robertson
Vanya Rohwer
Jim Rorabaugh
Ken Rosenberg
Robert Rothman
Tim Rowe
Michael J. Ryan
Ariadne Sabbag
Kalyan Singh Sajwan
Alan Savitzky
Susan Schalbe
Rob Schell
Torsten Scheyer
Nalani Schnell
Rainer Schoch
Alexander Schreiber
Katlin Schroeder
Hans-Peter Schultze
Chris Scotese
Anusha Shankar
Charles Sharp
Jeff Shaw
Wade Sherbrooke

Sabrina Shih	Todd Stailey	Michael Tuccinardi	Grahame Webb
Rick Shine	Frank Steinmann	Susan Turner	Qiwei Wei
John Sibbick	Hans-Dieter Sues	Peter Uetz	Kentwood Wells
Greg Sievert	Nobu Tamura	Jan van der Ploeg	Rom Whitaker
Mary Silcox	Wawan Tarniwan	Wayne Van Devender	Fred Wierum
Will Sillin	Glenn Tattersall	Merlijn van Weerd	Mark Wilson
Roberto Sindaco	Robert Thomson	Audrey Vinton	Marc Wisniak
Miranda Smith	Ian Tibbets	Laurie Vitt	Mark Witton
Scott A. Smith	Angelica Torices	Richard Vogt	Joost Woltering
Leo Smith	Luke Tornabene	Sebastian Voigt	Dandan Wu
Jake Socha	Kathy Townsend	Gunter Wagner	Stephen Zozaya
Timothy Sosa	Ray Troll	Amber Walker-Bolton	

We are also grateful to these organizations for assistance:

Canadian Museum of Nature
Cornell Laboratory of Ornithology
Cornell University Museum of Vertebrates
Field Museum of Natural History
Fossil Butte National Monument
Miguasha National Park
Museum of the Earth

National Systematics Lab
National Museum of Natural History
NOAA Office of Exploration and Research
The Natural History Museum
United States Fish and Wildlife Service
United States National Park Service

The authors are only the visible tip of the iceberg that is a textbook

Mere words cannot express the gratitude and admiration we feel for the people with whom we have worked at Sinauer Associates division of Oxford University Press. Their encouragement and support enabled us to create an entirely new book: Jason Noe (Acquisitions Editor) for supporting and encouraging such an extensive revision; Joan Gemme (Production Manager), who in addition to managing production also provided critical feedback to improve figure concepts, accessibility, and design; Martha Lorantos (Lead Production Editor) and Carol Wigg (Copy and Production Editor), whose skill, experience, and extraordinary patience eased us through writing, rewriting, editing, paging, proofing, and glossary preparation; Beth Roberge (Book Designer and Production Specialist), for her beautiful book design, super-efficient page layout and production, and patience with our many figure revisions; Mark Siddall (Photo Researcher), who located obscure photographs for us and, when all else failed, photographed the needed material himself; Michele Beckta (Permissions Supervisor) and Cailen Swain (Permissions Coordinator), whose expertise in copyright law made it possible for us to manage the nearly 1,200 new permissions needed for the book; Peter Lacey (Digital Resource Development Editor), whose expertise allowed us to include creative elements in the eBook; Grant Hackett (Indexer), whose sure hand with the index made complex topics accessible; and Sarah A. D'Arienzo (Editorial Assistant) for helping us to keep communications working.

About the Authors

Harvey F. Pough is Professor Emeritus at Rochester Institute of Technology.

William E. Bemis is Professor of Ecology and Evolutionary Biology at Cornell University, and Faculty Curator of Fishes for the Cornell University Museum of Vertebrates (CUMV).

Betty McGuire is Senior Lecturer in Ecology and Evolutionary Biology at Cornell University.

Christine M. Janis is Professor Emerita of Biology, Brown University, USA, Honorary Professor of Earth Sciences, University of Bristol, UK.

Digital Resources for Vertebrate Life, Eleventh Edition

For the instructor

(Available at learninglink.oup.com/access/pough11e) Instructors using *Vertebrate Life*, Eleventh Edition, have access to a wide variety of resources to aid in course planning, lecture development, and student engagement. Content includes:

- *News Links:* A continually-updated collection of vertebrate biology news links.
- *Video Guide:* A curated collection of video links for further exploration of specific key topics within each chapter.
- *Image PowerPoints*: The complete set of figures from each chapter.
- *Layered Figures:* Selected key figures throughout the textbook are prepared as step-by-step and animated presentations that build the figure one piece at a time.
- *Active Learning Exercises:* Engaging exercises for each chapter are provided as Word documents.
- *Answers to End-of-Chapter Discussion Questions:* Answers to all the end-of-chapter Discussion Questions are provided as Word documents.
- *Complete References:* The complete citations for the literature referenced through the textbook.

For the student

Enhanced e-book

(ISBN 9780197564899)

Ideal for self-study, the *Vertebrate Life*, Eleventh Edition, enhanced e-book delivers extensive digital resources in a format independent from any courseware or learning management system platform. The enhanced e-book is available through leading higher-education e-book vendors and includes the following student resources:

- *Learning Objectives* outline the important take-aways of every major section
- *Layered Figures* let students work through complex figures step-by-step
- *Discussion Questions* following each chapter to help foster engagement with the material
- *Flashcards* help students learn and review the many new terms introduced in the textbook
- *Complete References* to the research literature cited in the textbook

© William E. Bemis

1

Diversity, Classification, and Evolution of Vertebrates

A student curator cares for vertebrate specimens

Evolution is central to biology. It is the overarching principle that helps us understand how living organisms work and allows us to organize their diversity. Evolutionary organization is at the core of the field of phylogenetic systematics, an approach that uses both morphological and molecular data to interpret the evolutionary history of extant (living) and extinct species. This book is specifically about the evolution of vertebrates, from their early forms to their current diversity, habitats, interactions, and means of survival.

This chapter reviews the extant vertebrate groups, summarizes phylogenetic systematics, and discusses genetic mechanisms and environmental events that shaped the evolution and biology of the vertebrates.

1.1 The Vertebrate Story

LEARNING OBJECTIVES

1.1.1 Summarize the diversity of living species in different vertebrate groups and understand how they are classified based on their phylogeny.

1.1.2 Explain the historical origin of binominal nomenclature and why we still rely on it.

1.1.3 Understand terms used to describe the organization of phylogenetic trees.

1.1.4 Differentiate the concepts of basal versus derived taxa.

Say the word "animal" and most people picture a vertebrate. Vertebrates are abundant and conspicuous parts of people's experience of the natural world. They are also remarkably diverse—more than 70,000 extant species ranging in size from fishes weighing as little as 0.1 gram to whales weighing over 100,000 kilograms (**Figure 1.1**). Vertebrates live in virtually all of Earth's habitats. Bizarre fishes, some with mouths so large they can swallow prey bigger than their own bodies, live in the depths of the sea, sometimes luring prey with glowing lights. Some 15 kilometers above these fishes, migrating birds fly over the peaks of the Himalayas.

But extant species are only a small proportion of the species of vertebrates that have existed. For each of the more than 70,000 extant species, there may be hundreds of extinct species, and some of these have no counterparts among extant forms. For example, the non-avian dinosaurs that dominated Earth for 180 million years are so entirely different from extant animals that it is hard to reconstruct the lives they led. Even mammals were once more diverse than they are now. The Pleistocene saw giants of many kinds, such as ground sloths as big as modern rhinoceroses and raccoons as large as bears. The number of species of terrestrial vertebrates probably reached its maximum in the middle Miocene, between 14 and 12 million years ago, and has been declining since then.

The diversity of extant and extinct vertebrates makes organizing them into a coherent system of classification that takes their evolutionary history into account an extraordinarily difficult task. Initially, classifying species was treated like an office filing system: Each species was placed in a pigeonhole marked with its name, and when all species were in their pigeonholes, the diversity of vertebrates would have been encompassed. This approach to classification was satisfactory as long as species were regarded as static and immutable; once a species was placed in the filing system, it was there to stay. Acceptance of the fact that species evolve made that kind of classification inadequate.

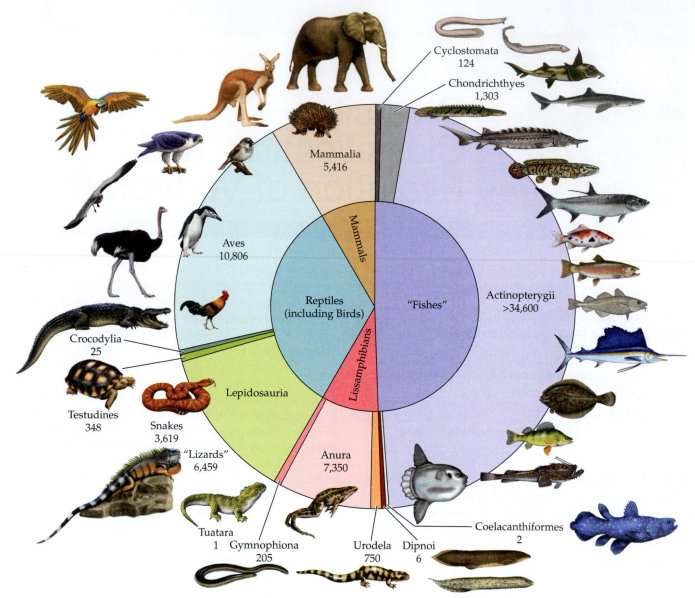

Figure 1.1 **Diversity of extant vertebrates.** Areas in the pie chart correspond to approximate numbers of extant species in each group as of 2021; numbers change as new species are described or existing species become extinct. Common names are in the center circle, with formal names for the groups shown in the outer circle. The two major extant lineages of vertebrates are Actinopterygii (ray-finned fishes) and Sarcopterygii (see the classification brackets in Figure 1.2), each of which includes about 35,000 extant species. (Some illustrations after K. Liem et al. 2001. *Functional Anatomy of the Vertebrates*, 3rd ed. Cengage/Harcourt College: Belmont, CA.)

Today's phylogenies are expected to express evolutionary relationships among species. Ideally, a classification system should not only attach a label to each species but also encode information about evolutionary relationships between that species and other species. Modern phylogenetic systematics—the naming and evolutionary classification of organisms—generates testable hypotheses about evolution, as we will describe in the next three sections. The first step, however, remains attaching a label—a unique name—to every species.

Binominal nomenclature

Our system of naming species is pre-Darwinian. It traces back to methods established by the naturalists of the 17th and 18th centuries, especially those of Carl von Linné, the Swedish naturalist better known by his Latin pen name, Carolus Linnaeus. The Linnaean system names species using **binomial nomenclature**: two names, a genus name and a species epithet. Species are grouped in hierarchical categories (family, order, class, and so on). These categories are called **taxa** (singular *taxon*), and the discipline of naming organisms is **taxonomy** (Greek *taxo*, "to arrange"; *nomos*, "order"). Goals for scientific names are threefold: uniqueness (i.e., no two species can share the same name); universality (i.e., everyone agrees to use the same name); and stability (i.e., a species name cannot be changed once it is properly named).

The scientific naming of species became standardized with the publication of the 10th edition of Linnaeus's monumental work *Systema Naturae* (*The System of Nature*) in 1758. The work was published in 12 editions between 1735 and 1766, each edition improving—and expanding on—the previous ones. The names that Linnaeus used in the 10th edition represent the formal start of biological nomenclature; any names published before 1758 are not accepted as scientific names (i.e., they are invalid). Linnaeus attempted to give an identifying binomial name to every known species of plant and animal. Familiar examples include *Homo sapiens* for human beings (Latin *homo*, "human" and *sapiens*, "wise"), *Passer domesticus* for the house sparrow (Latin *passer*, "sparrow" and *domesticus*, "belonging to the house"), and *Canis familiaris* for the domestic dog (Latin *canis*, "dog" and *familiaris*, "domestic").

Why use Latin? Latin was the early universal language of European scholars and scientists. It provided a uniform usage that scientists, regardless of their native language, continue to recognize worldwide. The same species may have different colloquial names, even in the same language. For example, *Felis concolor* (Latin, "uniformly colored cat") is known in various parts of North America as the cougar, puma, mountain lion, American panther, painter, and catamount. In Central and South America it is called león Colorado, león de montaña, pantera, onça vermelha, onça parda, yagua pytá, and suçuarana. But biologists of all nationalities recognize the name *Felis concolor* as referring to this specific kind of cat.

If stability of names is a goal of nomenclature, then why do species names seem to change so often? There are several reasons. First, upon further study, a systematist may decide that what was thought to be one species turns out to be two or more species. In other cases, a species thought to be distinct is found to be the same as another species; or that a species originally placed in one genus belongs in another genus; or that the species name was already in use for another organism (often an insect).

Another reason for instability of names is the principal of priority, which means that the first person who recognizes and names a species in an appropriate publication is credited as the author of that species name; any other names that someone else subsequently applies to that species are invalid. This happens less frequently now than it did 200 years ago, because methods of communication and dissemination of scientific names are much faster today, but it still happens as systematists study and revise taxa. Even though changes in the names of familiar species may seem inconvenient, they represent important scientific progress, which is why we work to keep scientific names used in this book up to date.

Extant vertebrate groups

Figure 1.2 shows extant vertebrate groups arranged in a **phylogeny**, or **phylogenetic tree** (or just "tree," for convenience). It is important to understand this figure because we use this design to organize the trees throughout this book. The **root** (sometimes called the **base**) is at the upper left in our design (it could have been placed at the bottom or top, which would change the tree's appearance but not the information it contains). From its root, the tree branches—splits in two—at a **node**. Splits at subsequent nodes, in an earliest-to-most-recent time progression, lead to all the branches in the tree. Each branch ends in a **terminal taxon,** the group named at the tip of the branch. The first branch split leads to two terminal taxa, hagfishes (Myxiniformes) and lampreys (Petromyzontiformes). Together, hagfishes and lampreys are within the **classification bracket** labeled Cyclostomata. All groups in Figure 1.2 are nested within the classification bracket Vertebrata ("nested" in this case means that all of the smaller classification brackets are included within the large bracket labeled Vertebrata).

The second node of Figure 1.2 branches out to eventually encompass all extant vertebrates that are *not* cyclostomes; these comprise the classification bracket Gnathostomata, or jawed vertebrates. Morphological characters indicated at each node along the "trunk" of the tree give some of the major evidence that supports this particular phylogeny. For example, the character "Four limbs with digits" characterizes the classification bracket Tetrapoda; the character "Amniotic egg" characterizes Amniota; and so forth. We often use the term **basal** when referring to taxa such as Cyclostomata that branch closer to the root of a particular tree. Taxa that branch after Cyclostomata, from cartilaginous fishes to placental mammals at the bottom of the figure, are said to be **derived** relative to hagfishes.

The arrangement of branches and taxa in a tree are collectively termed the tree's **topology**. Note that simply *rotating* a branch does not change the tree's topology. For example, if you swap the positions of hagfishes and lampreys with each other, then they will still be connected to the same branch, and will still be enclosed in the classification bracket Cyclostomata. However, branch *rearrangement*—changing the positions of nodes—changes the tree's topology and the resulting phylogenetic classification. For example, if you found convincing evidence that the branch leading to lampreys should connect not with hagfishes but with the branch that leads to chimaeras, that would yield a new topology and change the classification—and you could probably count on publishing your results in an extremely prestigious journal, because such a dramatic rearrangement would be extremely noteworthy.

You can consider the relationships in Figure 1.2 as you study the overview of extant vertebrates on pages 5–6.

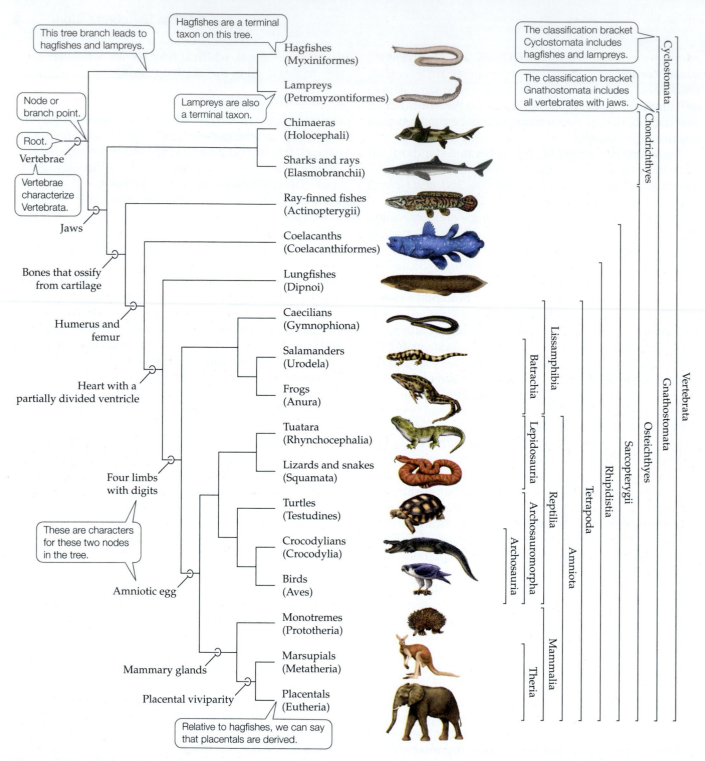

Figure 1.2 Phylogenetic relationships of extant vertebrate groups. Tree of extant vertebrate groups summarizing the phylogeny and classification used in this book. Branches of the tree lead to 18 terminal taxa, all of which are covered in this book. Nodes on the left indicate major characters that distinguish clades. Classification brackets are shown on the right. (Some illustrations after K. Liem et al. 2001. *Functional Anatomy of the Vertebrates*, 3rd ed. Cengage/Harcourt College: Belmont, CA.)

1.2 Phylogenetic Systematics

LEARNING OBJECTIVES

1.2.1 Understand the significance of monophyly.
1.2.2 Define the terms used in phylogenetic systematics and understand their significance in evolution.

Once we've given species their names, how do we decide where each one fits on our tree? All methods of classifying organisms are based on similarities shared by the included species, but some similarities are more significant than others. For example, nearly all terrestrial vertebrates have paired limbs with digits, but only a few vertebrate groups have mammary glands. Consequently, knowing that two species have mammary glands tells you more about the closeness of their relationship than does knowing they have paired limbs. A way to assess the relative importance of different characteristics was

Extant Vertebrate Groups

The topology of the tree in Figure 1.2 is well supported by modern morphological, molecular, and genomic research. Names for the nested classification brackets follow the most currently accepted concepts for these taxa. Here we briefly treat all terminal taxa in Figure 1.2 and their higher classification with the goal of familiarizing you with names used throughout this book.

All animals on the tree belong to **Vertebrata**, the vertebrates, and are distinguished by the presence of vertebral elements or vertebrae: serially arranged structures that form a spinal column.

Hagfishes and lampreys: Cyclostomata   Named for their round mouth openings (Greek *cyclos*, "round"; *stoma*, "mouth"), hagfishes and lampreys lack jaws; members of both groups are elongate, limbless, scaleless, and lack bone. Hagfishes (Myxiniformes) are highly specialized for life as marine scavengers. They live burrowed in deep-water muddy seafloors. Typical adult lampreys (Petromyzontiformes) feed parasitically on blood and tissue fluids; they hatch from eggs laid in rivers or creeks, spend several years as filter-feeding larvae, migrate to sea to mature, and return to freshwater to spawn.

All extant vertebrates more derived than Cyclostomata belong to **Gnathostomata** (Greek *gnathos*, "jaw"), a group united by the presence of jaws.

Sharks, rays, and chimaeras: Chondrichthyes   Chondrichthyes (Greek *chondros*, "gristle"; *ichthys*, "fish") refers to the cartilaginous skeletons of these fishes. Sharks have captured human imagination since antiquity, and have a reputation for ferocity that all but a few shark species would have difficulty living up to. Some are small (< 25 cm), while the largest species, the whale shark (*Rhincodon typus*), grows to 17 m and filter feeds on plankton. Most rays are dorsoventrally flattened bottom feeders that swim using undulations of their broad pectoral fins. Chimaeras, also known as ratfishes or rabbit fishes, are a small but ancient group of marine fishes that forage on the seafloor and feed on hard-shelled prey such as crustaceans and mollusks.

All extant gnathostomes more derived than Chondrichthyes belong to **Osteichthyes** (Greek *osteon*, "bone"), a group united by presence of bones that ossify from cartilages (endochondral bone; see Section 2.5).

Ray-finned fishes: Actinopterygii   Most of the animals we think of as fishes belong to Actinopterygii (Greek *aktin* "ray"; *pteryg* "fin"), a group with nearly 35,000 extant species living in nearly every imaginable aquatic habitat, from rivers, ponds, and lakes to coral reefs and deep ocean trenches, to subzero waters of the Antarctic Ocean. Some survive temperatures in desert pools that would kill any other vertebrate. Most actinopterygians are higher bony fishes, or teleosts (Greek, *teleios* "complete"; *osteon*, "bone"), a group that spans tarpons, herrings, carps, catfishes, and trout to perches, swordfishes, tunas, flatfishes, and giant ocean sunfishes. Our understanding of teleost diversity continues to expand: on average, about 400 new species are described annually, and many thousands of new species await discovery.

All extant osteichthyans more derived than Actinopterygii belong to **Sarcopterygii**, a group united by presence of a single bone, the humerus or femur, as the only skeletal element at the base of the paired fins or limbs. Muscles associated with these bones are the source of the name

Sarcopterygii (Greek *sarcos*, "flesh"; *pteron* "fin"). From a phylogenetic standpoint, Sarcopterygii includes all extant groups of terrestrial vertebrates—lissamphibians, reptiles (including birds), and mammals, which together are about 35,000 species. In a broadly comparative sense, about half of the living vertebrates are actinopterygians; most of the rest are sarcopterygians.

Coelacanths: Coelacanthiformes   The two extant species of Coelacanthiformes (Greek *koilos*, "hollow"; akantha "spine") occur in moderately deep waters (~200 m) off the east coast of Africa and Indonesia. They are relicts of a small group that has persisted since the Devonian Era.

All extant sarcopterygians more derived than Coelacanthiformes belong to **Rhipidistia**. This group is characterized by a heart with a partially divided ventricle, which helps to separate oxygenated from deoxygenated blood flowing through the heart.

Lungfishes: Dipnoi   Six extant species of lungfishes or Dipnoi (Greek *di*, "double"; *pnoe*, "breath") live in South America, Africa, and Australia. They can extract oxygen from water with gills and from air with lungs. Well represented in the Devonian, lungfishes are the closest living relatives of land vertebrates.

All extant rhipidistians more derived than Dipnoi are terrestrial vertebrates belonging to **Tetrapoda**, a group characterized by the presence of four limbs with digits. An extensive series of fossils allows us to trace the transformation of characters in fish-like sarcopterygians to those of tetrapods.

Caecilians, salamanders, and frogs: Lissamphibia   Popularly known as amphibians (Greek *amphis*, "double"; *bios*, "life"), most species of lissamphibians have biphasic life histories, which typically include an aquatic larval form (larva for caecilians and salamanders; tadpole for frogs) and a terrestrial adult. We infer that such an amphibious life history also characterized the earliest Devonian tetrapods, and that it was retained in the branch of tetrapods leading to lissamphibians. Although we often think of lissamphibians as tied to water for reproduction, many species lack aquatic larvae and spend their entire lives on land.

Lissamphibians are characterized by smooth skin without scales (Greek *lissos*, "smooth") that is important in the exchange of water, ions, and gases with their environment. Caecilians (Gymnophiona) are legless aquatic or burrowing animals; salamanders (Urodela) are elongate, mostly terrestrial, and usually have four legs; frogs (Anura) are short-bodied, with large heads and large hindlegs used for walking, jumping, and climbing. Much evidence shows that salamanders and frogs are more closely related to each other than they are to caecilians, and they are placed together in **Batrachia**.

All extant tetrapods more derived than Lissamphibia belong to **Amniota**, a group characterized by the presence of an amniotic egg with an amnion, chorion, and allantois. These three new membranes develop from the embryo's body and are not made by the reproductive tract of the

(Continued)

Extant Vertebrate Groups (continued)

mother (this contrasts with egg membranes, egg cases, or eggshells, which are secreted by the mother's oviduct). The innermost extraembryonic membrane, the amnion, encloses the embryo in fluid. The other two membranes serve roles in gas exchange and protection. Most amniotes are terrestrial, but there are secondarily aquatic species such as sea turtles and whales.

Amniota has two extant clades: **Reptilia** and **Mammalia**. Reptilia includes the lepidosaurs, turtles, crocodylians, and birds.

Tuatara, lizards, and snakes: Lepidosauria   Lepidosaurs (Greek *lepsis*, "scale"; *sauros*, "lizard") have scaly skins and share many skeletal characters. The tuatara (*Sphenodon punctatus*, a stocky lizardlike animal found only on some offshore islands of New Zealand) is the sole living remnant of a more diverse Mesozoic lineage. In contrast, **Squamata** (lizards and snakes) is now at its peak diversity.

Remaining groups of extant reptiles belong to **Archosauromorpha**, represented today by turtles, crocodylians, and birds.

Turtles: Testudines   The ~350 species of Testudines (Latin *testudo,* "turtle") belong to a distinctive and immediately recognizable group. The shell that encloses the body of a turtle does not occur in other vertebrates, and unique morphological modifications associated with the shell make turtles extremely peculiar animals. They are, for example, the only vertebrates with the shoulders (pectoral girdle) and hips (pelvic girdle) enclosed by the ribs. Their distinctiveness makes them something of an evolutionary enigma, and over the years systematists have placed turtles in several different locations on the tree. Our placement in Figure 1.2 is based chiefly on molecular phylogenetic interpretation.

All extant archosauromorphs more derived than Testudines belong to **Archosauria**, represented today by crocodylians and birds.

Alligators and crocodiles: Crocodylia   These impressive animals draw their name from the Latin word for them (*crocodilus*). Extant crocodylians are semiaquatic predators with long snouts with large pointed teeth. They range from dwarf crocodiles and caimans (1.5 m long) to the saltwater crocodile (*Crocodylus porosus*), which can grow to 6 m. The skin contains many bones (osteoderms; Greek *osteon*, "bone"; *derma*, "skin") that form in their scales and provide a kind of armor plating. Crocodylians provide parental care for eggs and young.

Birds: Aves   Birds (Latin *avis*, "bird") are dinosaurs that evolved flight in the Mesozoic. Discoveries of fossil archosaurs that lack wings but have feathers ("feathered dinosaurs") show that feathers evolved long before flight. The first feathers were almost certainly used in courtship displays and for insulation, and their modifications in birds (as airfoils and for streamlining) were secondary events. This disparity illustrates an important principle: The function of a trait in an extant species is not necessarily the same as that trait's function when it first appeared. In other words, current utility is not the same as evolutionary origin.

The last amniote group on the tree is Mammalia, all of which feed their young with milk produced by female mammary glands.

Mammals: Mammalia   Extant mammals (Latin *mamma*, "teat") include about 5,400 species in three groups: monotremes (**Prototheria**), marsupials (**Metatheria**), and placentals (**Eutheria**). Monotremes—platypus and echidnas—occur today in Australia, and one species of echidna reaches New Guinea. Monotreme young hatch from eggs. Today, marsupials dominate the mammalian fauna only in Australia, although more than 100 species occur in South and Central America, and one, the Virginia opossum (*Didelphis virginiana*), lives in North America. Placentals are by far the largest extant group of mammals in terms of species (most are rodents or bats), but the name "placentals" is misleading, because both marsupials and placentals have placentas, which are structures that transfer nutrients from the mother to the embryo and remove waste products of an embryo's metabolism. Because of the shared presence of a placenta, we group Metatheria and Eutheria together as **Theria**.

developed in the mid-20th century by Willi Hennig, who introduced a method of hypothesizing evolutionary relationships called **phylogenetic systematics** (Greek *phylon*, "tribe"; *genesis*, "origin").

A core concept of phylogenetic systematics is the recognition of groups of organisms at different hierarchical levels within an **evolutionary lineage,** or **clade.** Clades must have a single evolutionary origin—that is, they must be **monophyletic** (Greek *mono*, "one" or "single"), and include all the descendants of a **common ancestor.** Members within monophyletic groups are linked by nested sets of characters that trace the evolutionary history of the group. These principles—monophyly and the inclusion of all descendants—underpin the discipline of **cladistics.**

Hennig's approach embodied the concept that monophyletic groups can be identified only on the basis of **shared derived characters**—that is, characters that have the same evolutionary origin (i.e., are homologous) and that differ from an ancestral condition (are derived). A derived character is called an **apomorphy** (Greek *apo*, "away from"; *morphe*, "form"; thus, "away from the ancestral condition"). A shared derived character is a **synapomorphy** (Greek *syn*, "together"). In our example from the previous paragraph, mammary glands are a synapomorphy for Mammalia (note that this is shown as a character for Mammalia in Figure 1.2).

Of course, organisms within a clade can also share characters that they have inherited unchanged from their ancestors. These are called **plesiomorphies** (Greek *plesios,* "near," in the sense of "similar to the ancestor"). The vertebral column of terrestrial vertebrates, for example, was inherited from sarcopterygian fishes, and, by itself, the mere presence of a vertebral column cannot tell us anything about evolutionary relationships within tetrapods. The same character can be either plesiomorphic or apomorphic, depending on the level within the phylogeny. A vertebral column is a plesiomorphic character of vertebrates, so it provides no information about evolutionary relationships of vertebrates to one another, but it is a synapomorphy of vertebrates when compared with nonvertebrate chordates.

1.3 Applying Phylogenetic Criteria

LEARNING OBJECTIVES

1.3.1 Explain what an outgroup is and understand the use of outgroups in proposing a phylogeny.

1.3.2 Apply principles of phylogenetic systematics to interpret phylogenetic trees.

1.3.3 Explain how morphological and molecular characters are used to propose phylogenies.

1.3.4 Explain what the dagger (†) symbol means and why it is important to use it.

Although the conceptual basis of phylogenetic systematics is straightforward, applying the criteria can become complicated. Scientists analyze character states and apply phylogenetic criteria to propose hypotheses about evolutionary relationships among organisms. To proceed, however, we need to address a central issue of phylogenetic systematics: How do scientists know which character state is ancestral (plesiomorphic) and which is derived (apomorphic)? That is, how can we determine the direction (polarity) of evolutionary transformation of the characters? The basic method is to compare characters in the group of interest, the **ingroup**, to those of an **outgroup**, a reference group or groups that, although known to be related to the organisms we are studying, is less closely related to any member of the ingroup than the ingroup members are to each other. For example, sarcopterygian fishes such as coelacanths and lungfishes are appropriate outgroups for studying characters of terrestrial vertebrates.

Evaluating possible phylogenies

To illustrate tree thinking and phylogenetic classifications, consider **Figure 1.3**, which shows the possible phylogenies of three taxa identified as 1, 2, and 3. Note that when we are comparing information about three taxa, there are always three possible resolved phylogenies. Each of the resolutions specifies that two of the taxa are more closely related to each other than they are to the third taxon.

To make this hypothetical example more concrete, we'll consider three characters: the number of toes on the front feet, the skin covering (scales or no scales), and the presence or absence of a tail. In the ancestral character

state—based on the outgroup—there are five toes on the front feet, and in the derived state there are four toes; the tail is present in the ancestral state and absent in the derived state; the ancestral state is scaly skin, and the derived state is a lack of scales (Figure 1.3A). This information can be presented as a data matrix in which the ancestral state in indicated by 0, the derived state by 1:

	Five front toes	Tail	Scales
Outgroup	0	0	0
Taxon 1	0	0	1
Taxon 2	0	1	0
Taxon 3	1	1	0

How can we use the information in this data matrix to decipher the evolutionary relationships of the three groups of animals? Notice that the derived number of four toes occurs only in taxon 3, the derived tail condition (absent) is found in taxa 2 and 3, and the derived loss of scales occurs only in taxon 1. *Any change in a structure is an unlikely event, so the most plausible phylogeny is the one requiring the fewest changes.* The **most parsimonious phylogeny** (i.e., the branching sequence requiring the fewest number of changes) is represented by Figure 1.3B, where only three changes are needed to produce the current distribution of character states. In the evolution of taxon 1, scales are lost; in the evolution of the lineage that includes taxa 2 and 3, the tail is lost; and in the evolution of taxon 3, a toe is lost. The phylogenies in Figures 1.3C and D are possible, but they require tail loss to occur independently twice (i.e., four evolutionary changes rather than three), so they are less parsimonious than the first phylogeny.

Like any scientific hypothesis, a phylogeny is constantly tested as new character data or taxa are discovered. If it fails the test, it is falsified; that is, it is rejected, and a different hypothesis (a different branching sequence) is proposed. The process of testing phylogenetic hypotheses and replacing those that are falsified is a continuous one, and changes in the phylogenies in successive editions of this book show where new information has generated new hypotheses.

Molecules and morphology

Initially, scientists classified organisms according to their morphology, and morphology is still paramount in taxonomy. We will cover characters that define all vertebrates in Chapter 2, and subsequent chapters will describe the various and sometimes unique ways these characters have become adapted among members of the different groups. In the last half-century, however, we have also gained access to a vast amount of molecular and genetic data. Such data have become an important source of information and have led to many revisions in phylogenies.

In the 1960s, American anthropologist Vincent Sarich pioneered the use of molecular character information by using immunological comparison of blood serum

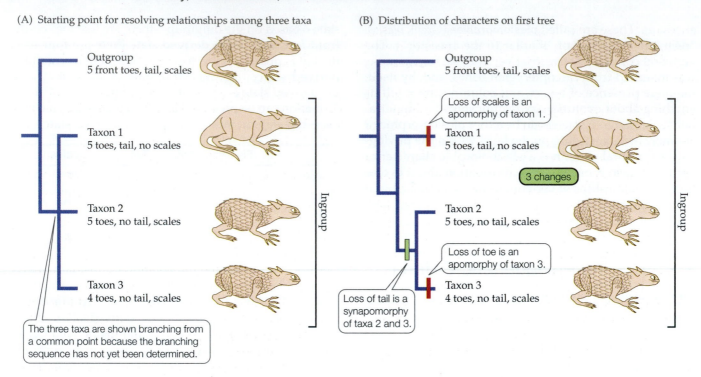

(A) Starting point for resolving relationships among three taxa

Outgroup
5 front toes, tail, scales

Taxon 1
5 toes, tail, no scales

Taxon 2
5 toes, no tail, scales

Taxon 3
4 toes, no tail, scales

Ingroup

The three taxa are shown branching from a common point because the branching sequence has not yet been determined.

(B) Distribution of characters on first tree

Outgroup
5 front toes, tail, scales

Loss of scales is an apomorphy of taxon 1.

Taxon 1
5 toes, tail, no scales

3 changes

Taxon 2
5 toes, no tail, scales

Loss of toe is an apomorphy of taxon 3.

Taxon 3
4 toes, no tail, scales

Loss of tail is a synapomorphy of taxa 2 and 3.

Ingroup

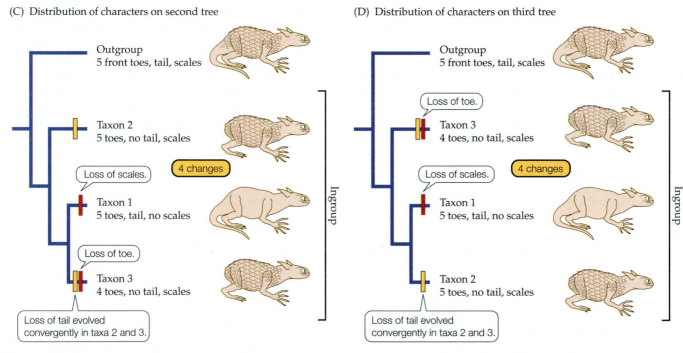

(C) Distribution of characters on second tree

Outgroup
5 front toes, tail, scales

Taxon 2
5 toes, no tail, scales

Loss of scales.

4 changes

Taxon 1
5 toes, tail, no scales

Loss of toe.

Taxon 3
4 toes, no tail, scales

Loss of tail evolved convergently in taxa 2 and 3.

Ingroup

(D) Distribution of characters on third tree

Outgroup
5 front toes, tail, scales

Loss of toe.

Taxon 3
4 toes, no tail, scales

Loss of scales.

4 changes

Taxon 1
5 toes, tail, no scales

Taxon 2
5 toes, no tail, scales

Loss of tail evolved convergently in taxa 2 and 3.

Ingroup

Figure 1.3 Resolving phylogenetic relationships. (A) An outgroup is used to determine the best branching sequence for the evolution of the ingroup (taxa 1, 2, 3). The outgroup has the ancestral (plesiomorphic) character states, which are five toes on the front feet, a tail, and a scaly body covering. (B–D) Three possible evolutionary trees with different arrangements of the ingroup taxa. Bars identify derived evolutionary changes, or apomorphies. The green bar in (B) indicates a shared derived character, or synapomorphy, of the lineage that includes taxa 2 and 3. Orange bars in (C) and (D) indicate two independent evolutionary losses of the tail; we say that tail loss in the taxa on these two trees occurred convergently. Because phylogeny (B) requires only three changes in characters, we say it is more parsimonious than either (C) or (D) and thus is most likely to reflect evolutionary history.

albumins to determine that chimpanzees are the apes most closely related to humans. In the 1970s, amino acid sequences of proteins and base sequences of mitochondrial and nuclear DNA were added to the repertoire of molecular phylogeny. That same decade saw the development of computer algorithms that can rapidly sort and arrange large numbers of characters in a phylogeny. More recently, we have entered the era of genomic-level comparisons, which may help to confirm or refute earlier phylogenetic hypotheses. It is still early days, however, and it will be fascinating to see what happens as we learn more.

Any set of characters can produce multiple possible phylogenies, and the number of possible trees grows exponentially as the number of taxa and characters increase, putting it beyond one person's ability to interpret all possible trees. Algorithms developed since the 1960s use various methods (e.g., parsimony, maximum likelihood, Bayesian inference) to discover and identify the most plausible sequences of changes. Ever more powerful computers made it possible for anybody to use these algorithms. With the establishment of public-access repositories such as GenBank in 1982, Morpho-Bank in 2005, and MorphoSource in 2012, vast quantities of both genetic and morphological data became available on the Internet. As a result, phylogenetic systematics moved from humans analyzing a few characters to computer-based analyses of huge data sets (although analyses of the largest phylogenetic data sets can require weeks of processing time).

Molecular characters came to be regarded as superior to morphological characters for evaluating phylogenies. The ease with which large quantities of molecular data can be obtained contributed to this perception, as did the assumption that molecular data would be free from some of the problems of interpretation associated with morphological data. More recently, however, molecular phylogenies have lost some of their luster. The assumptions that all types of molecular data are equally useful for phylogenetics and that more data always produce better results are being questioned. Some molecular characters are appropriate for looking at recent divergences, while others are better at resolving older splits; mixing the two can produce conflicting results.

Differences of that sort are usually easy to see in morphological data, but harder to see in molecular data. For example, coat color of mammals can be useful for distinguishing among different species within a genus, but it is not useful at higher taxonomic levels. Different patterns of striping distinguish different species of zebras, but stripes cannot be used as a character to ally zebras with tigers. Molecular characters are more difficult to assess in this fashion, because the functional significance of differences in base sequences or amino acids is rarely known—most molecular and even genomic characters remain black boxes as far as function is concerned. Furthermore, fossils seldom yield molecules, and when they do the molecules are often degraded, so molecular characters can rarely be used to determine the relationships of fossil organisms to one another or to extant species. This is a critical shortcoming. A morphology-based phylogeny can be drastically rearranged by the addition of a single fossil that alters the polarity of morphological characters, but molecular phylogenies cannot be tested in this manner.

The problem of dating

Dating the time that lineages separated is another weakness of molecular phylogenies, because dates are based on assumptions about the rates at which mutations occur and must be anchored by reference to an estimated age of a group based on its fossil record. The age of a fossil can almost always be determined—provided that you have a fossil. One difficulty with fossils is the incompleteness of the fossil record. There are gaps, sometimes millions of years long, in the record of many taxa. For example, we have fossil lampreys from the Late Devonian (~360 Ma[1]), the Late Carboniferous (~300 Ma), and the Early Cretaceous (~145 Ma), but nothing between those dates.

Considering the difficulties with both molecular and morphological phylogenies, it's remarkable that they mostly agree about branching patterns. Disagreements often center on dates of divergence. For example, molecular phylogenies indicate that the extant lineages of lissamphibians diverged in the Late Carboniferous (~315–300 Ma), whereas the earliest fossils indicate that the divergence occurred in the Late Permian (~260–255 Ma). However, molecular and morphological methods sometimes agree; molecular evidence indicates that humans separated from their common ancestor with chimpanzees about 6.6 Ma, and this date fits well with the earliest fossil in the human lineage, †*Ardipithecus* (5.8 Ma). There are, of course, some particularly stubborn groups, in the sense that they have been difficult to place whether you study molecules or morphology. Turtles are a good example of this, and their placement on the tree in Figure 1.2 as the sister taxon of archosaurs is just one of many possible arrangements proposed in recent decades.

The best information sometimes comes from combining molecular and morphological data. Studies that include extant and extinct organisms often employ the technique of **molecular scaffolding**: the extant taxa are placed in their phylogenetic position by the relationships established by the molecular data, and then morphological data are used to integrate the fossil taxa with the extant taxa.

Dagger (†) convention adopted in this book

Fossil vertebrates offer information about morphology, evolutionary relationships, and paleoenvironments, but it is not always easy to integrate fossils into phylogenetic analyses. Fossils are usually incomplete. Even so-called whole-body fossils that preserve the skeleton in a lifelike position and sometimes include evidence of soft tissues can never allow us to see and evaluate all the types of characters that we can study in a living vertebrate. Thus, we will always know less about vertebrates known only from fossils. This is important to remember, because it colors our ideas about the quality of the information

---

[1]Ma is the abbreviation for mega-anna, "millions of years" or, when referring to a specific date range, "million years ago."
Analogously, ka = kilo-anna (thousand years, thousand years ago), and Ga = giga-anna (billion years, billion years ago).

available for our evaluation. In this book, then, we indicate wholly extinct taxa with a dagger symbol (†) before the taxon name. The dagger symbol helps to distinguish extinct from extant taxa and reminds us about the lack of information for extinct taxa. This convention is particularly important for extinct higher taxa (higher taxa are genera and any group names, such as family names, that include genera), which often prove difficult to place with confidence on trees of extant vertebrates.

1.4 Using Phylogenetic Trees

LEARNING OBJECTIVES

1.4.1 Use the concept of an extant phylogenetic bracket to make inferences about extinct taxa.

1.4.2 Understand why paraphyly is problematic.

1.4.3 Explain the difference between crown groups and stem groups.

Phylogenetic systematics is based on the assumption that organisms in a lineage share a common heritage that accounts for their similarities. Because of that common heritage, we can use phylogenetic trees to ask questions about evolution. By examining the origin and significance of characters of extant animals, we can make inferences about the biology of extinct species. For example, some fossilized dinosaur nests contain remains of partly grown baby dinosaurs, suggesting that at least some extinct dinosaurs cared for their young. Is that a plausible inference? Obviously there is no direct way to determine what sort of care extinct dinosaurs provided to their eggs and young. The tree in **Figure 1.4** provides an indirect way to approach the question by examining the closest living relatives of extinct dinosaurs, which are crocodylians and birds.

Extant phylogenetic brackets

Take a look at the phylogenetic relationship of crocodylians, non-avian dinosaurs, and birds in Figure 1.4. The intermediate lineages in the phylogenetic tree (†pterosaurs and non-avian dinosaurs) are extinct, so we cannot observe their reproductive behavior. But crocodylians are the extant sister taxon of birds, and together, crocodylians and birds form what is called an **extant phylogenetic bracket**. We know that both crocodylians and birds provide parental care. Looking at extant representatives of more distantly related lineages (outgroups), we see that parental care is not universal among lepidosaurs, or turtles. The most parsimonious explanation of the occurrence of parental care in both crocodylians and birds is that it evolved after archosaurs separated from turtles but before the lineages leading to crocodylians and birds separated from each other. We cannot "prove" that parental care did not evolve independently in crocodylians and birds, but a single origin of parental care is more likely than two separate origins. Thus, the most parsimonious hypothesis is that parental care is a derived character of the evolutionary lineage containing crocodylians + birds. That means we are probably correct when we interpret fossil evidence as showing that non-avian dinosaurs did indeed exhibit parental care.

Figure 1.4 also provides an example of the important term **sister group** (or **sister taxon**), which refers to the monophyletic lineage most closely related to the monophyletic lineage being discussed. Here the lineage that includes crocodylians + †phytosaurs is the sister group of ornithodires. Similarly, †pterosaurs are the sister group of dinosaurs, †ornithischians are the sister group of saurischians (including birds), and †*Tyrannosaurus rex* on this tree is the sister group of birds.

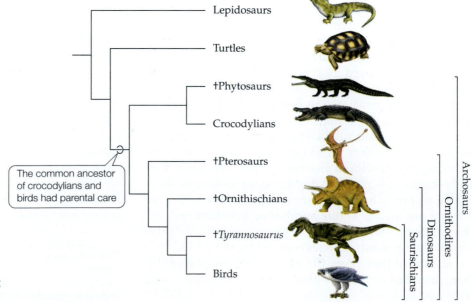

Figure 1.4 Using a phylogenetic tree to make inferences about behavior. Did extinct dinosaurs provide parental care? The tree shows currently accepted relationships among Archosauria, a group that includes several extinct lineages as well as extant crocodylians and birds. Crocodylians and birds (together, the extant phylogenetic bracket labelled archosaurs) display extensive parental care of eggs and young, but turtles do not. The most parsimonious explanation is that parental care evolved in the archosaur lineage after its separation from turtles but before the separation of crocodylians from ornithodires. (†Phytosaur by Jeff Martz; all others after K. Liem et al. 2001. *Functional Anatomy of the Vertebrates*, 3rd ed. Cengage/Harcourt College: Belmont, CA.)

The common ancestor of crocodylians and birds had parental care

Paraphyly

Figure 1.4 also reveals how phylogenetics has made talking about certain groups of animals more complicated than it used to be. Suppose you wanted to refer to just the two branches in this tree that are popularly known as "dinosaurs"—†ornithischians and extinct saurischians, such as †*Tyrannosaurus rex*. What do you call them? If you call them "dinosaurs," you're not being phylogenetically correct, because Dinosauria as a lineage includes birds. So when you say "dinosaurs," you are including †ornithischians + all saurischians, which in a phylogenetic sense includes birds, even though any 7-year-old would understand that you mean to restrict the conversation to extinct Mesozoic animals.

In fact, there is no technically correct name in phylogenetic systematics for just those animals popularly known as "dinosaurs." That's because cladists recognize only monophyletic groups (see Section 1.2), and a monophyletic group includes an ancestral form and all of its descendants. If birds are omitted from Dinosauria, then the group no longer includes all the descendants of the common ancestor. The lineage †ornithischians + saurischians *minus* birds is **paraphyletic** (Greek *para*, "beside" or "near," meaning a taxon that includes the common ancestor and some, but not all, of its descendants).

The only technically correct way of referring to the animals popularly known as dinosaurs is to call them "non-avian dinosaurs," and you will find that term and other examples of paraphyletic groups in this book. Sometimes even this construction does not work because there is no appropriate name for the part of the lineage you want to distinguish. In that situation, we use quotation marks (e.g., "†ostracoderms") to indicate that the group is paraphyletic, or we refer to individual taxa (usually genera) when we are talking about fossil vertebrates.

We try to avoid naming or referring to paraphyletic groups in this book, but sometimes it is convenient to do so, particularly when the name is in common usage. For example, "Fishes" in Figure 1.1 is not monophyletic because it does not include all of the descendants of a single common ancestor. Still, the term "fishes" no doubt calls to mind the image of an aquatic animal that uses gills to breathe water. Sometimes such a mental picture is worth a thousand words, and we use quotation marks as the standard convention for referring to such groups.

Crown and stem groups

Even though phylogenetic methods allow us to decipher evolutionary pathways and construct phylogenetic classifications for extant groups, difficulties arise when we try to find names for groups that include fossils. The derived characters found in extant groups did not all evolve at the same time. On the contrary, derived characters typically evolve in a stepwise or mosaic fashion. Extant members of a group have all of the derived characters of that group because that is how we define the group today, but as we move backward through time to fossils that represent outgroups to the extant species, we encounter forms that have a mosaic of ancestral and derived characters.

The farther back in time we go, the fewer derived characters the fossils may have. What can we call the parts of a lineage that contain these fossils? They cannot have the same name as the extant group because they lack some of the derived characters that we use to define the group. The solution lies in recognizing crown groups and stem groups. **Crown groups** have all of the derived character states found in extant species. Not all members of a crown group have to be extant; an extinct species with all of the derived characters is a member of the crown group. **Stem groups** are those extinct forms that lack some of the derived characters. Put another way, stem groups are fossils with some derived characters, and crown groups contain extant species plus fossils that have all the derived characters of the extant group. Stem groups are paraphyletic because they do not contain all the descendants of the ancestor of the stem group.

1.5 Genetic Mechanisms of Evolutionary Change

LEARNING OBJECTIVES

1.5.1 Describe the various lines of research that converged to become the Modern Synthesis.

1.5.2 Understand the significance of developmental biology in current evolutionary studies.

1.5.3 Explain the difference between "survival of the fittest" and "arrival of the fittest."

1.5.4 Describe an example of developmental gene regulation and its effect on a phenotypic character.

1.5.5 Define heterochrony, heterotopy, and heterometry and explain how each can produce phenotypic change.

"Descent with modification" is the phrase that Charles Darwin used to describe evolution. He drew his evidence from the animals and plants he encountered during his voyage aboard the HMS *Beagle* (1831–1836) and from his familiarity with selective breeding of domestic animals. Darwin emphasized the roles of natural selection and sexual selection as the mechanisms of evolution, although the basis of the traits he described was a mystery at the time. Gregor Mendel's meticulous and remarkable work documenting evidence of particulate inheritance (i.e., separate inheritance of specific traits, such as yellow versus green and smooth versus wrinkled peas) was published in 1866, but was not widely read at the time and was largely forgotten for more than 40 years. The rediscovery of Mendel's work in the early 20th century resulted in its extension and application to Darwin's ideas and led to the firm establishment of genes as the basis of heritable traits.

The 1930s, 40s, and early 50s saw the blending of Darwin's ideas on natural selection, Mendelian genetics, paleontology, and quantitative population biology into a comprehensive view of evolution known as the **Modern Synthesis**. Molecular research in the 1950s and

60s established DNA as the gene's self-replicating repository of the information that produces life's essential proteins, followed by the elucidation of the genetic code, and eliminated any doubt that evolution is the force that drives diversity.

The last three decades of the 20th century saw exponential increases in our knowledge of molecular biology, and in the 1980s and 90s this vast new understanding transformed the study of embryos into the modern field of developmental biology. Embryological studies were not a significant element of the Modern Synthesis, but the ever expanding findings of developmental biology were quickly seen to be integral to evolution, giving rise to the important field of **evolutionary developmental biology** (colloquially known as **evo-devo**). Current evolutionary approaches blend the insights and methods of all data sources and methods, and the new science of genome editing using CRISPR/Cas9 promises even more insights into the links between genotypes and phenotypes.

Phenotypes and fitness

Among many other findings, the Modern Synthesis established that most genes are polymorphic; that is, they have two or more **alleles** (forms of a gene that differ in their DNA base sequence). The phenotype, or physical form, of an organism is determined by its genotype—that is, by its particular combinations of alleles—and natural selection acts on phenotypes via differential survival and reproduction. Genetic mutations that result in new alleles occur randomly, and heritable allelic variation produced by mutation is the raw material of evolution.

An axiom of evolution is that *it is populations, not individuals, that evolve.* Although the phrase "survival of the fittest" may conjure images of individual combat, in fact evolutionary success is measured in terms of reproductive success: some phenotypes leave more descendants than others. The Modern Synthesis quantified this concept, showing that the frequency of the alleles that produce those phenotypes increases from one generation to the next. **Darwinian fitness** is a shorthand term that refers to the genetic contribution of a specific genotype to succeeding generations relative to the average contribution of all the genotypes in that generation. **Positive selection** is revealed by an increase in the frequency of a genetic trait in successive generations. Among humans, for example, we know that positive selection has increased the frequencies of genotypes associated with traits that are controlled by hundreds of genes, including adult height (there is positive selection for taller individuals in most human populations), adult female hip size, and infant head circumference (both of which affect maternal and infant mortality).

Developmental regulatory genes

Natural selection is a series of compromises because it is possible only to tinker with what is already present, not to redesign structures from scratch. The Modern Synthesis in combination with natural selection sought to explain the *survival* of the fittest, but it did not account for the *arrival*

of the fittest—that is, the origin of the phenotypic variation on which natural selection acts. For this, we need to consider genetic mechanisms of changes in embryonic development and the effects of these changes on phenotype. Evolutionary developmental biology emphasizes interactions of developmental regulatory genes that are arranged in hierarchical networks.

Groups of cells can release molecules (paracrine factors) that diffuse to neighboring cells and tissues, where they trigger **signal cascades** that result in transcription factors (proteins) that bind to DNA and can activate or repress (silence) gene expression. Changing gene expression in one group of cells can in turn produce factors that change gene expression in their neighboring cells, sometimes feeding back to influence genes in the cells that produced the first set of paracrine factors.

Genes are often grouped in **gene families** that produce multiple structurally related forms of the same protein, and regulatory genes from a relatively small number of families control a host of developmental processes. These **developmental regulatory genes** can be expressed in different parts of the body, affect diverse aspects of the phenotype, and can interact. For example, there are at least 23 bone morphogenetic proteins (BMPs). Originally identified by their ability to induce bone formation, BMPs also regulate cell division, differentiation, cell migration, and apoptosis (programmed cell death). Sonic hedgehog (Shh), a member of the Hedgehog family, participates in determining the left–right body axis, the proximal–distal axis of limbs, and the formation of feathers, among other processes. Gremlin, a member of the deadenylating nuclease (DAN) family, blocks the action of BMP and works with Shh to regulate limb growth.

The *Runx2* gene encodes a factor that regulates the transcription of genes associated with the formation of bone, including expression of BMPs. Acting through BMPs, Runx2 stimulates development of bone by inducing the formation of osteoblasts (bone-forming cells) and by delaying the conversion of osteoblasts to mature bone cells (osteocytes) that no longer form new bone. Changes in the timing and extent of expression of Runx2 and BMPs have profound effects on the phenotype of a developing embryo because early formation of osteoblasts or delayed conversion to osteocytes allows more bone to develop.

Alleles of *Runx2* differ in the relative number of glutamine and alanine residues they contain; alleles with higher glutamine/alanine ratios are expressed more strongly, resulting in more active synthesis of bone. The effect of this variation in the glutamine/alanine ratio can be seen in changes in the heads of bull terriers that were produced by selective breeding between 1931 and 2004 (**Figure 1.5**). Human cleidocranial dysplasia (*kleis*, "clavicle"; *kranion*, "skull"; *dysplasia*, "abnormal form") is also the result of *Runx2* alleles that differ in glutamine/alanine ratio. Analysis of the glutamine/alanine ratio of *Runx2* alleles in 30 species of carnivores showed that the glutamine/alanine ratio of the *Runx2* alleles characteristic of each species correlates with the snout length of that species (**Figure 1.6**).

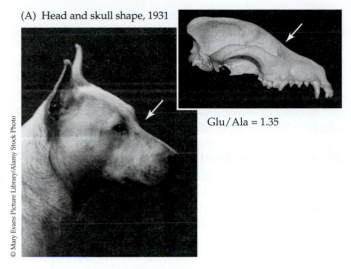

(A) Head and skull shape, 1931

Glu/Ala = 1.35

© Mary Evans Picture Library/Alamy Stock Photo

(B) Present

Glu/Ala = 1.46

© Luril Konoval/123RF

Figure 1.5 Rapid evolution of skull shape in bull terriers. Selective breeding changed the head shape of bull terriers. (A) In the early 20th century, the inflection point between the snout and the cranium (arrow) was at the level of the eyes, and the snout was horizontal. (B) In modern bull terriers, the inflection point has shifted behind the eyes, and the snout slopes downward. Selective breeding favored alleles of *Runx2* with high glutamine/alanine ratios that cause strong expression of *Bmp*. Stronger expression of *Bmp* is associated with accelerated bone growth leading to the sharply angled snout. (Skull images from J.W. Fondon et al. 2004. *PNAS* 101: 18058–18063, © National Academy of Sciences USA.)

Three kinds of change in the expression of developmental genes can produce phenotypic variation that is subject to natural selection: the *time* during development that a gene is expressed, the *place* it is expressed, and *how strongly* it is expressed (i.e., the amount of protein synthesized).

Heterochrony   Heterochrony (Greek *heteros*, "different"; *chronos*, "time") refers to changes in the timing of gene expression during development. Heterochrony can involve the length of time during which a gene is expressed during development, as in the case of expression of the *Runx2* gene in bull terriers, or the time at which one gene is expressed relative to expression of other genes. Heterochrony can occur at any stage of development, and can produce phenotypic changes in morphology, physiology, or behavior.

Body proportions of most vertebrates change substantially between infancy and maturity. For example, the head of a human infant is about 25% of its total body length, whereas the head of an adult is only about 13% of total body length. The cranium of an infant is large in proportion to its trunk, its snout is short, and its eyes are large in proportion to its head. The body proportions of infants explain why the rounded heads, big eyes, and short snouts

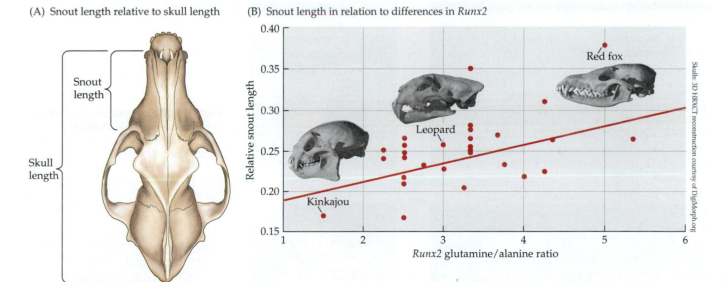

(A) Snout length relative to skull length

Snout length

Skull length

(B) Snout length in relation to differences in *Runx2*

Red fox

Leopard

Kinkajou

Relative snout length

Runx2 glutamine/alanine ratio

Skulls: 3D HRXCT reconstruction courtesy of DigiMorph.org

Figure 1.6 Changes in the snouts of carnivores. Increased glutamine/alanine ratios in *Runx2* among 30 species of Carnivora are associated with lengthening of the snout. (A) Relative snout length is expressed as the ratio of the snout length to the skull length. (B) Relative snout length is proportional to the glutamine/alanine ratio of each species' *Runx2* gene. (After K.E. Sears et al. 2007. *Evol. Dev.* 9: 555–565.)

of puppies and kittens are so appealing to humans. Larval and juvenile fishes also have proportionally larger heads and eyes than adult fish.

Because body proportions change during development, stopping development early produces an adult that retains body proportions that are characteristic of juveniles of its species, a phenomenon called **paedomorphosis** (Greek *pais*, "child"; *morph*, "form"). Paedomorphosis is a widespread form of phenotypic change among vertebrates, and the paedomorphic nature of the heads of birds compared with those of non-avian dinosaurs clearly reveals the role of heterochrony in the evolution of birds.

Once again, comparisons of the two extant groups of archosaurs—crocodylians and birds—provides evidence of developmental changes. Juvenile alligators may not be as cute as puppies, but they do have rounded heads, big eyes, and short snouts (**Figure 1.7A**). Differential growth of the cranium, eyes, and snout during maturation leads to a very different morphology in adult alligators, which have flat heads, small eyes, and long snouts (**Figure 1.7B**). In contrast, the heads of juvenile and adult ostriches are much more alike (**Figure 1.7C,D**). Heterochrony—in this case early truncation of head development—has left adult ostriches with head proportions much like those of juveniles.

Heterotopy   A change in the physical location of a gene's expression is called **heterotopy** ("different place") and can lead to dramatic phenotypic changes. During embryonic development, the fingers and toes of vertebrates are initially connected by a web of skin (remnants of this interdigital webbing are visible at the bases of your fingers and toes). Most of the webbing is lost before birth by apoptosis—except in web-footed aquatic vertebrates such as ducks, which retain interdigital webbing as adults. Embryonic ducks and chickens both have interdigital webbing, but chickens lose the webbing before hatching while ducks retain it. Apoptosis is initiated when *Bmp2*, *Bmp4*, and *Bmp7* are expressed in the webbing of both chickens and ducks. This BMP-induced apoptosis removes the webbing between the toes of chickens, but the feet of ducks also express the Gremlin protein, a BMP inhibitor that prevents apoptosis, thus leaving the webbing in place.

Webbing between the digits in the wings of bats is also preserved by inhibition of BMP. In this case, both Gremlin and a second inhibitor of BMP, fibroblast growth factor 8 (Fgf8), are expressed in the interdigital webbing, leaving the webbing intact to form the bat's wings. Additional examples of heterotopic changes include the reduction of limbs seen in many lepidosaurs.

Heterometry   **Heterometry** ("different measure") refers to a change in the *amount* of a gene product. A heterometric change in the production of BMP4 is responsible for one of the classic examples in evolution, beak evolution in Darwin's finches. Found in the Galápagos and Cocos islands, this radiation of about 15 species of ground-finches (genus *Geospiza*), descended from a single ancestral species from South America, provides one of the best-studied examples of adaptive radiation and natural selection.

These birds forage for seeds on the ground, and their beak morphology is correlated with the kinds of seeds each species selects (**Figure 1.8**). The large ground-finch (*G. magnirostris*) uses its massive beak to crack hard-shelled seeds but the small ground-finch (*G. fuliginosa*)

(A) Juvenile alligator

(B) Adult alligator

(C) Juvenile ostrich

(D) Adult ostrich

Figure 1.7 Heterochrony results in paedomorphosis in birds. Paedomorphosis (retention of juvenile traits) in extant birds can be seen by comparing alligators and ostriches. (A) A juvenile alligator has a short snout, rounded cranium, and large eyes relative to the rest of the head. (B) The head of an adult alligator has quite different proportions: the snout is long, the cranium is flat, and the eyes are small in proportion to the size of the head. (C) A juvenile ostrich also has a short snout, rounded cranium, and large eyes. (D) These proportions are little changed in the adult ostrich as a result of heterochrony—specifically, the early truncation of head development.

(A) Beaks of three *Geospiza*

(B) *Bmp4*, stage 26   (C) *Bmp4*, stage 29

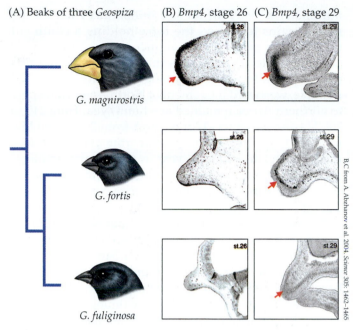

G. magnirostris

G. fortis

G. fuliginosa

B,C from A. Abzhanov et al. 2004. *Science* 305: 1462–1465

Figure 1.8 Heterometry, heterochrony, and heterotopy affect beak shape. (A) Three species of Galápagos ground-finches (*Geospiza*) show different beak phenotypes. (B,C) Dark areas (arrows) in the histological sections of developing beaks indicate the location and intensity of *Bmp4* expression. (B) At developmental stage 26, *Bmp4* is strongly expressed in *G. magnirostris*, somewhat expressed in *G. fortis*, and not expressed in *G. fuliginosa*. (C) At developmental stage 29, *Bmp4* expression is intense in *G. magnirostris*. It is also expressed, but limited to smaller areas, in *G. fortis* and *G. fuliginosa*.

picks up small, soft seeds with its pointed beak. The medium ground-finch (*G. fortis*) has a beak intermediate between those of the other two species and feeds on a broader range of seeds but is less effective than either specialist at opening soft-shelled or hard-shelled seeds. The different beak phenotypes of these three species result from differences in the expression of *Bmp4* during development of the upper beak. Early onset and a high level of expression of the *Bmp4* gene produces the heavy beak of *G. magnirostris*, whereas progressively later and weaker expression leads to the smaller beaks of *G. fortis* and *G. fuliginosa*.

1.6 Epigenetic Effects

LEARNING OBJECTIVES

1.6.1 Know how an epigenetic effect differs from a genetic effect.

1.6.2 Explain what is meant by an "intragenerational epigenetic effect," an "intergenerational epigenetic effect," and a "transgenerational epigenetic effect."

Some phenotypic variation results from modification of gene expression via **epigenetic effects** (Greek *epi*, "above, outside of, around"). Epigenetic mechanisms modify the expression of genes without changing the DNA sequence of the gene. A wide variety of factors, including diet, stress, temperature, and chemicals in the environment, can modify the behavior, physiology, or morphology of a developing organism, a phenomenon called **phenotypic plasticity**.

For example, chemicals released into the water by predators of tadpoles lead to epigenetic changes in the tadpoles' body form and behavior that improve their chances of escaping predation (**Figure 1.9**). This is an example of an **intragenerational epigenetic effect** ($F_0 \rightarrow F_0$). That is, the phenotypic changes are limited to those individuals exposed to the stimulus; their offspring do not inherit the modified

(A) Predators absent

(B) Chemicals from larval dragonflies

Broad tailfin

(C) Larval salamanders present

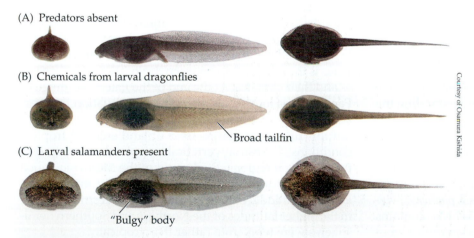

"Bulgy" body

(D) Larval dragonfly   (E) Larval salamander

© Robert Hermo/Alamy Stock Photo

Courtesy of Osamura Kishida

Figure 1.9 Phenotypic plasticity of tadpoles. (A–C) Head-on, lateral, and dorsal views of tadpoles of the Hokaido brown frog (*Rana pirica*) show epigenetic changes produced by exposure to predators. (A) Tadpoles raised in the absence of predators have streamlined bodies with low tailfins. (B) Tadpoles exposed to chemicals released by dragonfly larvae have streamlined bodies and broad tailfins. (C) Tadpoles living in aquaria with larval salamanders develop bulgy bodies and broad tailfins. (D) Dragonfly larvae seize and dismember resting tadpoles. Broad tails (photo B) may increase acceleration during swimming, allowing tadpoles to evade these predators. (E) Salamander larvae swallow tadpoles whole, and thus may be unable to swallow a tadpole with a bulgy body (photo C).

phenotypes. Many examples of intragenerational epigenetic effects have been described, including development of a cannibal morph of lissamphibian larvae and the role of incubation temperature in determining the sex of a variety of reptiles, topics that are discussed in later chapters.

Intergenerational ($F_0 \rightarrow F_1$) and **transgenerational** ($F_0 \rightarrow F_n$) epigenetic effects have been difficult to document in free-living animals, but laboratory studies have shown that epigenetic effects can extend across generations. For example, exposing adult stickleback fishes to predators modified the behavioral responses of the sticklebacks' offspring to predators. Moreover, the behavioral change seems to depend on whether it was the mother or the father that was exposed to predators, and on the sex of the offspring. Sons, but not daughters, of predator-exposed fathers were *less* cautious than sons of control fathers, whereas both sons and daughters of predator-exposed mothers were *more* cautious than offspring of control mothers. Studies of laboratory mice have documented behavioral changes resulting from postnatal stress in the mother that extend to the F_2 generation (grandchildren).

1.7 Earth History and Vertebrate Evolution

LEARNING OBJECTIVES

1.7.1   Know the difference between eons, eras, periods, and epochs.

1.7.2   Describe the pattern of continental movements from the Cambrian to the present.

1.7.3   Explain three components of continental drift that have influenced the evolution of vertebrates.

Since their origin in the early Paleozoic, vertebrates have been evolving in a world that has changed enormously and repeatedly. These changes have affected vertebrate evolution both directly and indirectly. Understanding the sequence of changes in the locations of continents and the significance of those locations for climates and interchange of faunas is central to understanding the vertebrate story. The history of Earth spans three geological **eons**: the Archean, Proterozoic, and Phanerozoic. Only the Phanerozoic, the eon of visible life (Greek *phanero*, "visible"; *zoon*, "animal") that began about 541 Ma, contains vertebrate life. The Phanerozoic is divided into three geological **eras**: Paleozoic (Greek *palaios*, "ancient"), Mesozoic (Greek *mesos*, "middle"), and Cenozoic (Greek *kainos*, "recent"), which we describe in Chapters 5, 10, and 20, respectively. These eras are divided into **periods**, which can be further subdivided in a variety of ways, such as the **epochs** within the Cenozoic era (see Appendix).

Movements of landmasses—**continental drift**—have been a feature of Earth's history at least since the Proterozoic, and the course of vertebrate evolution has been shaped by continental movements. The rate of continental drift remains low (movement of only about 1

millimeter a year) for tens of millennia. Then, as the strain begins to exceed the force holding a continent together, the rate of drift can increase to 20 mm/year (about the rate at which fingernails grow), causing rifts to develop and widen in a geologically brief period. For example, westward movement of North America relative to northern Africa remained at ~1 mm/year from 240 to 200 Ma, increased to ~12 mm/year from 200 to 190 Ma, increased again to ~20 mm/year from 190 to 180 Ma, then fell back to ~10 mm/year. A still greater acceleration, reaching 40 mm/year, accompanied the separation of southern Africa from South America between 128 and 120 Ma.

The continents are still drifting. Indeed, Australia is moving north so rapidly that land-based latitudinal coordinates established in 1994 are now 1.5 m out of register with global coordinates determined by GPS satellites. Because the movements are so complex, the sequence, the varied directions, and the precise timing of the changes are difficult to summarize. When the movements are viewed broadly, however, a simple pattern unfolds during vertebrate history: continental fragmentation during the Cambrian, coalescence by the Devonian, and a return to fragmentation by the Late Cretaceous.

During that coalescence some 300 Ma, the continents combined to form a single landmass known as Pangaea, which was the birthplace of terrestrial vertebrates. Persisting and drifting northward as an entity, this huge continent began to break apart about 150 Ma. Its separation occurred in two stages: first into Laurasia in the north and Gondwana in the south, and then into a series of units that have drifted to become the continents we know today.

The complex movements of the continents through time had major effects on the evolution of vertebrates. Most obvious is the relationship between the location of landmasses and their climates. At the end of the Paleozoic, much of Pangaea was located on the Equator, a situation that persisted through the middle of the Mesozoic; thus, throughout this time frame large areas of land enjoyed tropical conditions. Terrestrial vertebrates evolved and spread in these tropical regions. By the end of the Mesozoic, much of Earth's landmass had moved out of equatorial regions, and by the middle of the Cenozoic most terrestrial climates in the higher latitudes of the Northern and Southern hemispheres were temperate rather than tropical.

A less obvious effect of the position of continents on terrestrial climates comes from changes in patterns of oceanic circulation. For example, the Arctic Ocean is now largely isolated from the other oceans and does not receive warm water via currents flowing from more equatorial regions. High latitudes are cold because they receive less solar radiation than do areas closer to the Equator, and the Arctic basin does not receive enough warm water to offset the lack of solar radiation. As a result, the Arctic Ocean has an extensive covering of ice, and cold climates extend well southward across the continents.

The Atlantic Meridional Overturning Circulation (AMOC) drives the Gulf Stream as it brings warm water north and east near the surface of the Atlantic Ocean and transports cold water south at greater depth, eventually to warm, rise to the surface, and move north again in the Gulf Stream. Heat transported by the AMOC is responsible for the relatively warm climates of northern North America and northern Europe. Sudden, drastic weakenings of the AMOC during the middle of the last ice age (~35 ka) were associated with abrupt cooling events at intervals of about 1.5 ka. Icebergs released from the ice sheet covering Canada melted and diluted seawater in the North Atlantic, making it less dense so that it did not sink to form the southward flow of deep, cold water that drives the AMOC.

Another factor that influences climates is the relative levels of the continents and the seas. At some periods in Earth's history, most recently in the late Mesozoic and again in the first part of the Cenozoic, shallow seas flooded large parts of the continents. These **epicontinental seas** extended across the middle of North America and the middle of Eurasia during the Cretaceous and early Cenozoic, forming barriers between the eastern and western portions of those landmasses.

Water absorbs heat as air temperature rises, then releases that heat as air temperature falls. Thus, areas of land bordering large bodies of water have maritime climates—they do not get very hot in summer or very cold in winter, and they are usually moist because water that evaporates from the sea falls as rain on the land. Continental climates, which characterize areas far from the sea, are usually dry with cold winters and hot summers. The draining of the epicontinental seas at the end of the Cretaceous probably contributed to the demise of the non-avian dinosaurs as climates in the Northern Hemisphere became more continental.

On a continental scale, advances and retreats of glaciers throughout the Pleistocene caused homogeneous habitats to split and merge repeatedly, isolating populations of widespread species and leading to the evolution of new species. For example, during glacial maxima, when much of the world's water was trapped in ice, much of the continental shelf of North America was exposed as land and the ranges of many terrestrial and freshwater vertebrates were forced southward into glacial refugia. As the ice retreated, recolonization of more northern environments provided many opportunities for diversification.

In addition to changing climates, continental drift has formed and broken land connections between the continents. Isolation of different lineages on different landmasses has produced dramatic examples of the independent evolution of similar types of organisms, an evolutionary phenomenon known as **convergent evolution**. Many groups of mammals on separate continents independently evolved superficially similar forms during the mid-Cenozoic, a time when Earth's continents reached their greatest separation during the history of vertebrates. For example, jerboas are small, desert-dwelling saltatorial (ricocheting, which means bipedal jumping) rodents from Africa and the Middle East that strongly resemble kangaroo rats living in southwestern deserts of North America (**Figure 1.10**).

Much of evolutionary history appears to depend on whether a particular lineage was in the right place at the right time. This random element of evolution is assuming increasing prominence as more detailed information about the times of extinction of old groups and radiation of new groups suggests that competitive replacement of one group by another is not the usual mechanism of large-scale evolutionary change. As a result, movements of continents and their effects on climates and the isolation or dispersal of animals are taking an increasingly central role in our understanding of vertebrate evolution.

(A) Hairy-footed jerboa, *Dipus sagitta*

(B) Merriam's kangaroo rat, *Dipodomys merriami*

Svyatoslav Knyazev CC BY-SA 4.0

Nick Bonzey CC BY-SA 4.0

Figure 1.10 Convergent evolution of ricocheting rodents. (A) Jerboas (family Dipodidae) are found in deserts from North Africa through the Middle East into Asia. (B) Kangaroo rats (family Heteromyidae) inhabit the deserts in southwestern North America. Although the lineages separated in the Late Cretaceous (~75 Ma), jerboas and kangaroo rats are similar in appearance, ecology, and behavior.

Summary

1.1 The Vertebrate Story

There are more than 70,000 extant species of vertebrates. Vertebrates live in virtually all of Earth's habitats and range in size from 0.1 g to more than 100,000 kg.

Some key events and characters in vertebrate history include the evolution of jaws (Gnathostomata), bones that ossify from cartilage (Osteichthyes), the humerus and femur (Sarcopterygii), a heart with a partially divided ventricle (Rhipidistia), four limbs with digits (Tetrapoda), the amniotic egg (Amniota), mammary glands (Mammalia), and placental viviparity (Theria).

Scientists name species using a system from the 18th century developed by Linnaeus for his great work, *Systema Naturae*. Every named species has a Latinized binomial consisting of a genus name that can encompass closely related species and a second identifier known as the species epithet that is unique to that species (e.g., humans = *Homo sapiens*).

1.2 Phylogenetic Systematics

Phylogenetic systematics produces branching evolutionary diagrams or phylogenetic trees showing changes in characters. New monophyletic groups at each branch point can be named, producing a nested series of named lineages.

Phylogenetic systematics groups organisms in evolutionary lineages on the basis of synapomorphies, which are homologous characters shared by the organisms in a group that differ from those of an ancestor.

1.3 Applying Phylogenetic Criteria

Parsimony is one basis for identifying the most likely sequence of evolutionary changes; any change is an unlikely event, so the phylogeny that requires the fewest changes to account for the observed distribution of characters is favored.

Today many phylogenies combine molecular and morphological characters. The branching patterns generally agree. When they disagree, the discrepancy often lies in estimated dates of divergence of lineages.

1.4 Using Phylogenetic Trees

An extant phylogenetic bracket allows us to draw inferences about characters of the enclosed extinct lineages when direct evidence is lacking.

It is sometimes convenient to refer to groups that are not phylogenetically correct in that they do not include an ancestral group and all of its descendants. Such a group is referred to as paraphyletic.

Crown groups are composed of extant or extinct species that have all of the derived characters of the lineage. Stem groups include extinct taxa that lack some derived characters of other taxa in the lineage but are nonetheless more closely related to the taxa in the crown group than they are to taxa in other lineages.

1.5 Genetic Mechanisms of Evolutionary Change

In the 19th century, Charles Darwin characterized evolution as "descent with modification" and invoked natural selection and sexual selection. Gregor Mendel described and documented the particulate nature of genetic inheritance.

During the first half of the 20th century, the Modern Synthesis combined the perspectives of Darwinian selection, Mendelian genetics, and quantitative population biology to explain "survival of the fittest" as the result of differential reproduction of phenotypes and consequent changes in allele frequencies within populations.

In the late 20th century, molecular genetics and embryology came together in the field of evolutionary developmental biology ("evo-devo") to show how genes acting early in embryonic development can produce profound phenotypic changes and provide novel raw material for the action of selection (the "arrival of the fittest").

1.6 Epigenetic Effects

Epigenetic mechanisms modify the expression of genes without changing the DNA sequence of the gene.

Intragenerational epigenetic effects result from changes in gene expression initiated by external factors such as diet, temperature, or the presence of predators. Intragenerational changes are limited to the individuals exposed to the stimulus and are not transmitted to their offspring ($F_0 \rightarrow F_0$).

Intergenerational (parent to children, $F_0 \rightarrow F_1$) and transgenerational (parent to grandchildren, $F_0 \rightarrow F_2$) epigenetic effects have been demonstrated in laboratory studies.

1.7 Earth History and Vertebrate Evolution

At the time vertebrates first evolved, the continental landmasses were scattered across the globe. They coalesced into one enormous continent, Pangaea, about 300 Ma, which began to fragment again about 150 Ma.

This pattern of fragmentation, coalescence, and fragmentation has isolated and renewed contacts of major groups of vertebrates, producing the biogeographic distributions of vertebrates today.

Discussion Questions

1.1 Why is the nested set of names in Figure 1.2 referred to as a phylogenetic classification? What would happen to the classification if new data changed the arrangement of the branches on the tree?

1.2 Why doesn't phylogenetic systematics have a fixed number of hierarchical categories like the Linnaean system?

1.3 Why does phylogenetic systematics represent evolution more clearly than pre-cladistic methods?

1.4 What inference can you draw from Figure 1.4 about parental care by †pterosaurs? What about †phytosaurs?

1.5 Suppose that you have firm evidence that †phytosaurs did not exhibit parental care. What would be the most parsimonious hypothesis about the appearance or disappearance of parental care in the archosaur lineage?

1.6 What problems might the types of convergent evolution seen between jerboas and kangaroo rats pose for phylogenetic analyses?

1.7 Today, molecular, genomic, and developmental data play crucial roles in examining evolutionary relationships among species and lineages. What difficulties arise when we try to apply these tools to the phylogenetic analysis of fossil vertebrates?

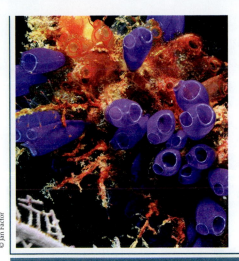

Bluebell tunicate, *Clavelina puertosecensis*

© Jan Factor

What Is a Vertebrate?

Vertebrates are a diverse and fascinating group of animals. Because we are vertebrates ourselves, that statement may seem chauvinistic, but vertebrates are remarkable in comparison with most other animal groups. Vertebrates are **chordates**, members of phylum **Chordata**. Only arthropods (Arthropoda, which includes insects, crustaceans, and spiders) rival vertebrates in diversity of forms and habitats. And it is only among mollusks (Mollusca) that we find animals such as octopuses and squids that approach the large sizes of some vertebrates and that have a capacity for complex learning.

The relationship of chordates to other animal phyla is revealed by anatomical, physiological, biochemical, and developmental characters. In this chapter, we first discuss evolutionary relationships among vertebrates and other members of the animal kingdom. We then describe some characteristic structures of vertebrates and present an overview of the organ systems that make vertebrates functional animals. These fundamentals will help you appreciate changes that occurred during vertebrate evolution and will allow you to trace homologies between basal and more derived vertebrates.

2.1 Vertebrates in Relation to Other Animals

LEARNING OBJECTIVES

2.1.1 Differentiate between diploblasty and triploblasty and describe the features that distinguish bilaterian animals.

2.1.2 Compare Protostomia and Deuterostomia with respect to their diversity.

2.1.3 Specify the developmental feature that is synapomorphic for Deuterostomia.

2.1.4 Describe phylogenetic relationships among the major clades of Ambulacraria.

The animal kingdom, or **Metazoa**, contains more than 30 phyla, of which Chordata is one (**Figure 2.1**). All metazoans are multicellular heterotrophs (feeding on other organisms). They are also motile (capable of movement) for at least part of their life cycle (many are motile as larvae but sessile as adults). Sponges, the most basal metazoans, differentiate from a single layer of cells. Cnidarians are **diploblastic**, having two embryonic germ layers (Latin *germen*, "bud"): ectoderm that becomes the cells of the outer body, and endoderm that differentiates and lines the gut. All other phyla in Figure 2.1 have a third germ layer, mesoderm, that contributes to many organ systems and makes these organisms **triploblastic** (see Section 2.4). The triploblastic phyla comprise **Bilateria** (Latin, "two sides"). At some point in their life cycle, whether as larvae or adults or both, all bilaterians have a body plan with two sides that are mirror images of each other (i.e., they are bilateral).

Current molecular data place the origin of Bilateria late in the Proterozoic, probably between 688 and 596 Ma. Fossils of undisputed bilaterians first occur ~553 Ma, at the end of the Proterozoic. Most bilaterian phyla first appear in the geological record during the early Cambrian (~541 Ma), and the term "Cambrian Explosion" has been used to describe this phenomenon. The apparent suddenness of this diversification probably relates to two artifacts of the fossil record: (1) the frequency with which hard parts, such as shells, are preserved; and (2) a few fossil localities with exceptional preservation that record soft-bodied animals, which are usually absent from the fossil record.

There are two major divisions within Bilateria: **Protostomia** (Greek *pro*, "earlier"; *stoma*, "mouth") and **Deuterostomia** (Greek *deuteros*, "second"). This division was originally based on embryonic features, including development of the mouth and anus during **gastrulation**.

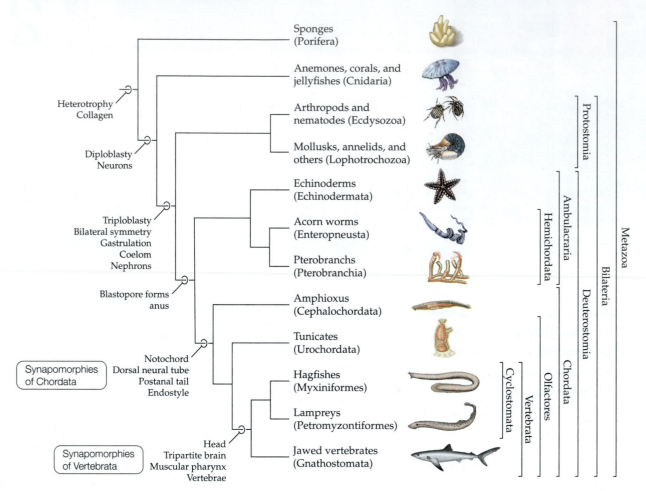

Figure 2.1 Simplified phylogeny of Metazoa. This tree emphasizes Deuterostomia, the clade that includes Vertebrata.

During gastrulation, undifferentiated cells of the early embryo move into three germ layers and the body axes are defined. The original embryonic gut opening is the **blastopore** (Greek *poros*, "small opening"), and a second opening develops before gastrulation concludes. The classic view is that the blastopore becomes the mouth of protostomes and the second opening becomes the anus, whereas in deuterostomes the second opening becomes the mouth and the blastopore becomes the anus.

Protostomia comprises more than 90% of bilaterian phyla and is divided into two main groups: Ecdysozoa, including arthropods and nematodes, and Lophotrochozoa, including mollusks and annelids, such as earthworms.

Deuterostomia is a much smaller grouping, but it is our focus here, since it includes Chordata (and thus Vertebrata). In addition to Chordata, there are two other deuterostome phyla: Echinodermata (starfishes, sea urchins, and similar animals) and Hemichordata (acorn worms and pterobranchs, which uniquely share a three-part body plan consisting of a proboscis, collar, and trunk). Fossils and molecular data indicate that the common ancestor of extant deuterostomes was probably a small, marine, wormlike, free-swimming filter-feeder.

Adult echinoderms lack a distinct head and have pentaradial (fivefold and circular) symmetry. Based on appearance alone, it seems unlikely that echinoderms would be related to vertebrates, which have distinct head (anterior) and tail (posterior) ends, bilateral limbs, and bilateral (left–right) symmetry as adults. However, echinoderm *larvae* are bilaterally symmetrical, and a relationship between echinoderms and vertebrates is firmly established by developmental and molecular information. Diversity of extinct echinoderms was greater than that of extant forms, and some Paleozoic echinoderms were bilaterally symmetrical, meaning that the fivefold symmetry of adult echinoderms is a derived character.

Molecular characters unite echinoderms and hemichordates as **Ambulacraria** (Latin *ambulacrum*, "alley," in reference to the ambulacral grooves for the tube-feet of echinoderms). Hemichordates were long considered to be the sister group of chordates because both groups have a **pharynx** (Latin *pharynx*, "throat") with **pharyngeal slits** that open to the outside. However, a pharynx with slits now is considered to be a feature of the common ancestor of all deuterostomes, a view supported by a shared cluster of genes that are present in all deuterostomes and are expressed during pharyngeal development. Thus, a pharynx with slits is not a shared derived character uniting chordates and hemichordates.

2.2 Characteristics of Chordates

LEARNING OBJECTIVES

2.2.1 Name the shared derived features for chordates and explain their significance.

2.2.2 List the shared derived features for cephalochordates and for urochordates.

2.2.3 Explain why ascidean tunicates, despite their dramatic metamorphosis from larva to adult, are chordates.

2.2.4 Categorize the evidence (morphological, developmental, molecular, or behavioral) indicating urochordates are the sister group of vertebrates.

Chordates are united by four synapomorphies—shared derived features—that occur in all members of the group at some point in their life cycle:

- A *notochord*, a dorsal (Latin *dorsum*, "back") stiffening rod that gives phylum Chordata its name
- A dorsal *neural tube*
- A muscular *postanal tail* (i.e., a tail that extends beyond the anus)
- An *endostyle*, a ciliated, glandular groove on the floor of the pharynx

The first three of these features allow larvae to swim using a muscular tail. The **notochord** (Greek *notos*, "back"; *chorde*, "string") serves not only as a stiffening rod but also as an attachment site for segmental muscles that power swimming. In most vertebrates, the notochord is transient, replaced during development by the vertebral column and remaining in adults only as a portion of the intervertebral discs. The neural tube coordinates muscle activity needed for swimming and, in vertebrates, develops into the spinal cord and brain, collectively known as the central nervous system. Muscles of the **postanal tail** provide more power for swimming than other mechanisms, such as cilia. In addition to secreting mucus to trap food particles during filter feeding, the **endostyle** takes up iodine like the vertebrate thyroid gland.

Other features have been debated as chordate characters. For example, all chordates pump blood, but not all chordates have a definitive heart. Cephalochordates pump blood using a simple enlargement of the main ventral blood vessel consisting of a single layer of contractile myocardial cells. Gene expression as this enlargement develops is similar to that in the development of the heart in vertebrates.

Chordate origins and evolution

Compared with other metazoans, the internal organization of chordates appears to be "upside down." Annelids such as earthworms, for example, have a nerve cord on the ventral (Latin *venter*, "belly") side and heart and primary blood vessel on the dorsal side; in chordates, these positions are reversed. In the early 19th century, the French naturalist Étienne Geoffroy Saint-Hilaire proposed an "inversion hypothesis" to explain this observation. His idea fell into disrepute but has been revitalized in recent years by evolutionary developmental biology. We have learned that the early embryos of chordates express genes on the dorsal side that are expressed on the ventral side in nonchordates. This heterotopic event (see Section 1.5) seems to have happened at the evolutionary origin of chordates.

Although the notochord is unique to chordates, gene expression patterns suggest similarities with the annelid axochord. Likewise, genomic studies identify a possible homolog of the endostyle in the epibranchial ridge of hemichordates.

As mentioned at the end of Section 2.1, gene expression patterns during development show that pharyngeal slits are homologous across deuterostomes to the extent that they represent outpocketings of the pharynx that perforate to the exterior. Although pharyngeal slits are not found in extant echinoderms, they may have been present in some Paleozoic echinoderms and are present in their living sister group, the hemichordates.

Vertebrates have pharyngeal arches between pharyngeal pouches containing neural, muscular, circulatory, and skeletal tissues derived from neural crest cells. Neural crest is unique to vertebrates, but the skeletal supports of the pharynx of other deuterostomes are cartilages based on fibrillar collagen as in vertebrates. Neural crest and pharyngeal arches will be described in more detail in Section 2.4.

Extant nonvertebrate chordates

Extant nonvertebrate chordates are small marine animals in two clades: lancelets (**Cephalochordata**) and tunicates (**Urochordata**). Cephalochordata refers to the notochord's extension to the anterior tip of the body (Greek *kephale*, "head"). Urochordata refers to the position of the notochord in the larval tail (Greek *oura*, "tail"). The anterior tip of the vertebrate notochord ends midway through the head region and extends posteriorly to the tip of the tail. Other nonvertebrate chordate groups existed in the past, but such soft-bodied animals are rarely preserved as fossils.

Cephalochordates The 29 species of lancelets, also known as amphioxus (Greek *amphis*, "double"; *oxys*, "sharp"), are superficially fishlike marine animals usually less than 5 cm long (**Figure 2.2A**). Lancelets share with vertebrates some anatomical features that are absent from tunicates. However, molecular analyses place tunicates as the sister group of vertebrates. Structures such as segmental muscles along the body and a tail fin might be basal chordate characters that have been lost in tunicates. In addition to their unique notochord, lancelets have a ring of oral cirri that prevents coarse particles from entering the mouth (**Figure 2.2B**).

Lancelets swim using **myomeres** (Greek *mys* "muscle"; *mere* "segment"). Myomeres are segmental blocks of skeletal muscle fibers arranged along both sides of the body and separated by sheets of connective tissue termed myosepta. Sequential contraction of myomeres bends the

(A) A lancelet, *Branchiostoma lanceolatum*

Courtesy of Arthur Anker

(B) Anatomical features of lancelets

Segmented,   1. Dorsal neural tube   2. Notochord   3. Postanal
V-shaped   tail
myomeres

Oral cirri   Atriopore   Anus
                                                      Gut
4. Endostyle Pharyngeal slits Gonads

Figure 2.2 Cephalochordates.   (A) Adult lancelets are burrowing, sedentary animals that are widely distributed in shallow marine waters of the continental shelves. (B) The four synapomorphies lancelets share with all chordates are numbered in this schematic view of a generalized lancelet.

(A) An appendicularian tunicate, *Oikopleura dioica*

3. Postanal tail with segmental muscles

Gut   1. Dorsal neural tube   2. Notochord

Photo by R. Rudolph

4. Endostyle Pharynx Incurrent siphon   Excurrent siphon and anus

(B) An adult ascidian   (C) Anatomy of an adult ascidian
tunicate, *Ciona savigny*

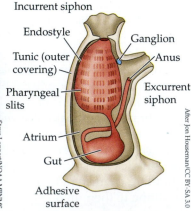

Incurrent siphon

Endostyle   Ganglion

Tunic (outer   Anus
covering)

Pharyngeal   Excurrent
slits   siphon

Atrium

Gut

Adhesive
surface

Steve Lonhart/NOAA MBNMS

After Jon Houseman/CC BY-SA 3.0

Figure 2.3 Urochordates.   (A) Appendicularian tunicates are free-swimming throughout their lives and retain all four chordate synapomorphies (numbered). (B) The larvae of ascidian tunicates metamorphose into sedentary adults attached to the substrate. These adult tunicates do not retain the structural similarity to cephalochordates or vertebrates seen in their larval stages. (C) Schematic view of a generalized adult ascidian. (A from N.H. Patel. 2004. *Nature* 431: 28–29.)

body from side to side, resulting in forward or backward swimming. The notochord acts as an incompressible elastic rod that prevents the body from shortening when the myomeres contract.

Lancelets have a cerebral vesicle, a thickening at the anterior end of the neural tube that expresses genes also expressed in portions of the vertebrate brain. These studies reinforce the growing understanding that differences in how, when, and where genes are expressed (see Section 1.5) are as important as differences in which genes are present.

Urochordates   Tunicate synapomorphies include a tadpole larva, an outer covering known as a tunic, and heartbeat reversal. Appendicularian tunicates, also known as larvaceans, are motile, filter-feeding animals of the marine pelagic realm that retain all four chordate characters throughout life (**Figure 2.3A**). The more commonly studied ascidian tunicates have motile but nonfeeding larvae that metamorphose to become sessile adults (**Figure 2.3B**), in the process losing the postanal tail and notochord and reducing the neural tube to a ganglion (bundle of nerve cells). The tunic is made primarily from cellulose, a structural carbohydrate otherwise seen only in plants. Tunicates appear to have acquired the ability to synthesize cellulose by horizontal gene transfer from bacteria. They filter food particles from the water with a basketlike perforated pharynx and capture them in secretions from the endostyle (**Figure 2.3C**). Blood is pumped toward the pharynx and, after a short pause, in the reverse direction, toward the intestine. The function of this reversal is unknown.

Molecular analyses place tunicates and vertebrates as sister taxa (see Figure 2.1), in a group that was given the unfortunate name of **olfactores** based on the mistaken interpretation that tunicates, certain fossil echinoderms, and vertebrates share characters of the olfactory system. In addition to a heart, tunicates share some derived features with vertebrates, such as cells that may be homologous with vertebrate neural crest cells (see Section 2.4). The simple morphology of adult tunicates is considered to be highly derived rather than primitive, and tunicates have a greatly reduced genome in comparison with other chordates. This phylogenetic interpretation implies that tunicates lost features such as myomeres.

2.3 What Distinguishes a Vertebrate?

LEARNING OBJECTIVES

2.3.1   Describe how the head and brain of vertebrates differ from those of nonvertebrate chordates.

2.3.2   Explain how differences in structure of the respiratory and digestive tracts of nonvertebrate chordates and vertebrates allowed the evolution of active, large-bodied vertebrates.

2.3.3   Identify the key embryonic tissue of vertebrates that is responsible for many of the differences between nonvertebrate chordates and vertebrates.

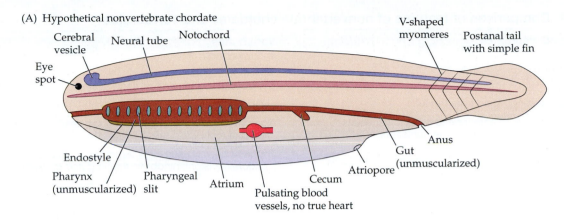

(A) Hypothetical nonvertebrate chordate

Cerebral vesicle · Neural tube · Notochord · V-shaped myomeres · Postanal tail with simple fin · Eye spot · Anus · Gut (unmuscularized) · Endostyle · Atriopore · Pharynx (unmuscularized) · Pharyngeal slit · Atrium · Cecum · Pulsating blood vessels, no true heart

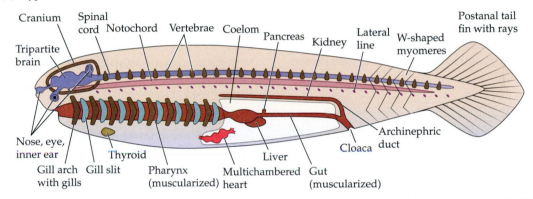

(B) Hypothetical ancestral vertebrate

Cranium · Spinal cord · Notochord · Vertebrae · Coelom · Pancreas · Kidney · Lateral line · W-shaped myomeres · Postanal tail fin with rays · Tripartite brain · Nose, eye, inner ear · Thyroid · Gill arch with gills · Gill slit · Pharynx (muscularized) · Multichambered heart · Liver · Gut (muscularized) · Cloaca · Archinephric duct

Figure 2.4 Comparing structures of nonvertebrates and vertebrates. A hypothetical nonvertebrate chordate (A) is shown in comparison with a hypothetical vertebrate (B). The neural tube of vertebrates expands in the head to form the tripartite brain; posteriorly, it is the spinal cord. Vertebrae are segmental skeletal elements flanking the spinal cord. The pharyngeal region of vertebrates is muscularized and has gill arches that support gills. Other differences include presence in vertebrates of organs such as the liver and pancreas, kidneys and associated ducts, and a multichambered heart. Myomeres, segmental blocks of muscle used for locomotion, are W-shaped in vertebrates. The cloaca of vertebrates is the common exit for the digestive, urinary, and reproductive systems.

Vertebrates take their name from **vertebrae**, serially arranged bones that make up the **vertebral column**, or backbone. The earliest known vertebrates, from Cambrian sediments, have segmented dorsal structures interpreted as vertebrae. **Figure 2.4** compares a generalized, hypothetical nonvertebrate chordate with a hypothetical ancestral vertebrate; **Table 2.1** compares features of extant nonvertebrate chordates and vertebrates.

Vertebral structure varies, but all vertebrates have segmentally organized elements composed of cartilage or bone that protect the spinal cord (see Figure 2.4B). In **gnathostomes**, the jawed vertebrates, vertebral elements known as **centra** (singular centrum) typically form in close association with the embryonic notochord, which is largely replaced by centra in most adult vertebrates.

Vertebrates have additional unique characters, including:

- *Cranium.* The cranium, or braincase, is a bony, cartilaginous, or fibrous structure surrounding the brain.
- *Head, sense organs, and brain.* Although many animals have heads, the vertebrate head is notable for its prominence and its array of complex sense organs, including the nose, eyes, and ears. The embryonic vertebrate brain is tripartite, having a forebrain, midbrain, and hindbrain to integrate sensory information.

- *Complex endocrine organs.* Endocrine glands, such as the thyroid, produce hormones that regulate many body functions.
- *Muscularized gut tube.* Vertebrates have muscles in the wall of the gut tube for efficient processing of large amounts of food.
- *Multichambered heart.* Vertebrate circulation is powered by a multichambered heart that distributes respiratory gases and nutrients to all cells of the body.
- *Mineralized tissues.* Vertebrates deposit minerals (primarily calcium compounds) in tissues, creating rigid structures such as calcified cartilage and bone and exposed surfaces with different degrees of resistance to abrasion including enamel, enameloid, dentine, and cementum.
- *Gills derived from endoderm.* Vertebrates have gills derived from the embryonic germ layer endoderm (see Section 2.4) for efficient respiration.

Table 2.1 Comparison of features of nonvertebrate chordates and vertebrates

Nonvertebrate chordates	Vertebrates
Head and brain	
No cranium	Cranium surrounds and protects brain
Head poorly distinguished from trunk	Head distinct from trunk
Notochord extends to anterior end of head (lancelets) or is restricted to the tail of larvae (tunicates)	Notochord starts midway through the head region and extends to tip of embryonic tail
Simple "brain"	Tripartite brain for sensory integration: Forebrain (olfaction); midbrain (vision); hindbrain (hearing and balance)
Simple sense organs and no cranial nerves	Specialized sense organs concentrated in the head served by cranial nerves
Neural tube gives rise to neurons	Neural tube, neural crest, and neurogenic placodes give rise to neurons
Endocrine control	
Limited endocrine integration of body functions	Complex endocrine system with many glands to integrate body functions
Pharynx and respiration	
Pharynx is not muscularized; water moved by ciliary action	Muscular pharynx moves water by pumping for feeding and respiration
No gills	Gills derived from endoderm
Feeding and digestion	
No teeth	Teeth composed of keratin in cyclostomes and dentine and enamel in gnathostomes
Gut not muscularized; food moved by ciliary action	Muscular contractions (peristalsis) move food
Limited differentiation of gut tube	Differentiated digestive organs and regionalized gut tube
Intracellular digestion: Food particles are taken into cells lining the gut and digested	Extracellular digestion: Food particles are digested in the gut and breakdown products are absorbed across gut wall
Heart and circulation	
Few capillaries	Extensive capillary networks
Blood moved by vessel contractions (lancelets); single-chambered heart with heartbeat reversal (tunicates)	Multichambered heart pumps blood
No neural control of the heart	Neural control of rate and force of cardiac contractions
No respiratory pigment or red blood cells; O_2 and CO_2 transported in solution	Red blood cells contain hemoglobin, which binds O_2 and CO_2 and aids their transport and delivery to tissues
Osmoregulation	
Body fluid osmolal concentration and ion composition same as external environment	Body fluid osmolal concentration and ion composition differ from external environment (except hagfishes)
Support and locomotion	
Mineralized tissues absent	Mineralized tissues present in osteognathostomes, including calcified cartilage, bone, dentine, enamel, enameloid
Acellular cartilage	Cellular cartilage
Vertebral elements absent	Vertebral elements present
V-shaped myomeres	W-shaped myomeres
True fins absent	Median fins in cyclostomes + gnathostomes; paired fins or limbs in gnathostomes

Together, these features give vertebrates far more powerful and efficient feeding, respiration, and locomotion than nonvertebrate chordates. Vertebrates are large, active animals. The evolution of almost all of the organ systems that support their large body sizes and activity is linked to embryonic development, and notably to the neural crest, a unique embryonic tissue that contributes to the formation of many vertebrate organ systems.

2.4 Vertebrate Embryonic Development

LEARNING OBJECTIVES

2.4.1   Describe the three embryonic germ layers and their fates.

2.4.2   Explain how compartments of the coelom differ in number and location among adult sharks, salamanders, and cats.

2.4.3   Explain the functions of neural crest and its significance in vertebrate evolution.

2.4.4   Summarize similarities and differences between neural crest and neurogenic placodes.

Studies of embryonic development and its physical, functional, and evolutionary constraints and opportunities provide clues about ancestral conditions and homologies between structures in different groups.

Development of the body

Bodies of all bilaterians form from three embryonic germ layers. The fates of these germ layers are largely conserved throughout vertebrate evolution.

- **Ectoderm**, the outermost layer, forms the superficial layers of skin (epidermis), linings of the most anterior and most posterior parts of the digestive tract, and most of the nervous system (Greek *ectos*, "outside"; *derm*, "skin").

- **Endoderm**, the innermost layer, forms the rest of the digestive tract lining, as well as glands associated with the gut, including the liver and pancreas (Greek *endos*, "within"). It also lines portions of the urinary system and forms respiratory surfaces of gills and lungs. Taste buds and the thyroid, parathyroid, and thymus glands develop from endoderm. Germ cells, which give rise to gametes, migrate from the endoderm into the progenitors of the gonads.

- **Mesoderm**, the middle layer, forms muscles, skeleton (including the notochord, vertebrae, skull, limb girdles, and limb bones), deeper layers of the skin (dermis), connective tissues, and the circulatory and urogenital systems (including the heart, kidneys, portions of the gonads and urogenital ducts; Greek *mesos*, "middle").

Within the mesoderm is the body cavity or **coelom** (Greek *koilia*, "cavity") that contains the internal organs of the body, such as the heart and gut tube (see Figure 2.4B). The coelom becomes subdivided in vertebrates and differs across them, being relatively simple in sharks and more complicated in mammals (**Figure 2.5**). In sharks, it is divisible into the **pericardial cavity**, which contains the heart, and the **peritoneal cavity**, which contains the viscera (gastrointestinal tract, liver, pancreas, and gallbladder). The two cavities are separated by the transverse septum, a partition between the heart and liver, as shown in the adult shark in Figure 2.5A. Note that the heart and surrounding pericardial cavity are located beneath

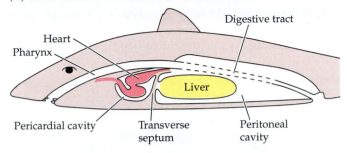

(A) Coelom and its two subdivisions in a shark

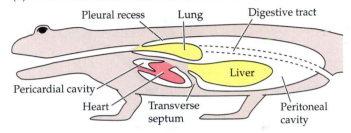

(B) Coelom and its two subdivisions in a salamander

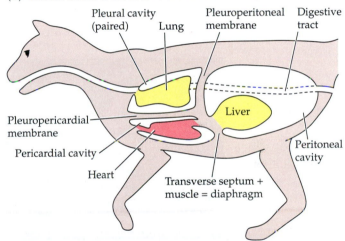

(C) Coelom and its four subdivisions in a cat

Figure 2.5 Divisions in the vertebrate coelom.
(A) A shark's coelom has a pericardial cavity containing the heart and a peritoneal cavity containing abdominal organs. (B) Like the shark, a salamander has a pericardial cavity and a peritoneal cavity. However, the salamander heart is located more caudally than the shark's, and portions of the peritoneal cavity containing the lungs (pleural recesses) lie directly above the pericardial cavity. (C) A cat has a pericardial cavity, paired pleural cavities containing the lungs, and a peritoneal cavity. (After K. Liem et al. 2001. *Functional Anatomy of the Vertebrates*, 3rd ed. Cengage/Harcourt College: Belmont, CA.)

the pharynx in the anterior portion of the body and the peritoneal cavity lies posteriorly. The same arrangement characterizes embryonic tetrapods, but during development the heart moves caudally to a position beneath the lungs; this condition can be seen in the adult salamander (Figure 2.5B). In the adult salamander, the anterior portion of the peritoneal cavity lies above the pericardial cavity and the areas around the lungs are called pleural recesses. In derived tetrapods such as mammals, the paired pleural recesses develop into cavities separate from the rest of the peritoneal cavity (Figure 2.5C). Thus, in adulthood, mammals have a pericardial cavity, paired pleural cavities, and a peritoneal cavity (i.e., four separate compartments rather than the two compartments of sharks and salamanders). The separate pleural cavities allow the lungs to expand and contract without interfering with other organs. Also, the muscular diaphragm of mammals develops to help separate the pericardial and paired pleural cavities from the peritoneal cavity; it functions in respiration. Coelomic cavities are lined by thin sheets of mesoderm referred to as **peritoneum** (in the peritoneal and pleural cavities) and **pericardium** (on the surface of the heart and lining the pericardial cavity). The gut is suspended in the peritoneal cavity by folds of peritoneum called **mesenteries**.

Postcranial portions of the vertebrate body develop in a segmented fashion from anterior to posterior, with each segment having an initial component of spinal nerves, major blood vessels, progenitors of bone and muscles, and other internal structures. This segmentation can be seen in the arrangement of human vertebrae and ribs. **Figure 2.6A** shows how the developing mesoderm is divided into somitic, intermediate, and lateral plate mesoderm.

- **Somites** are segmented blocks of mesoderm that form on either side of the neural tube (**Figure 2.6B**). Derivatives of **somitic mesoderm** include the segmentally arranged vertebrae, dermis, and skeletal muscles. Dorsally forming skeletal muscles and dermis later grow ventrally and are innervated by the voluntary nervous system.

- Derivatives of **intermediate mesoderm** are the kidney tubules and ducts, gonads (except for the cells that give rise to gametes, which come from endoderm), and gonadal ducts.

- Derivatives of the **lateral plate mesoderm** are principally the viscera, smooth muscle lining the gut, and cardiac muscle of the heart, which are innervated by the autonomic, or involuntary, nervous system.

Development of the pharyngeal region

The term **pharyngeal arch** encompasses the segmental structures of the vertebrate pharynx that include internal skeletal components, associated muscles, nerves, and blood vessels best seen in a **pharyngula**, an early developmental stage common to vertebrate embryos. Pharyngeal arches are delimited on the outside by grooves called pharyngeal clefts (**Figure 2.7**). On the inside of the pharynx, a series of outpocketings known as pharyngeal pouches separate the arches from each other. Derivatives of the pharyngeal arches are strongly conserved throughout vertebrate history (**Table 2.2**).

Development of the brain

Initially represented by a thickening on the dorsal surface of the embryo known as the **neural plate**, the nervous system undergoes early and dramatic development to form the **neural tube**. The neural tube is the primary

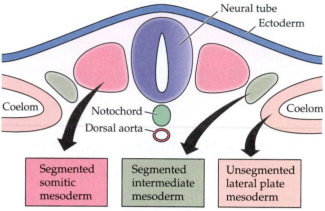

(A) Cross section through trunk of a developing amniote

Neural tube
Ectoderm
Coelom
Notochord
Dorsal aorta
Coelom

Segmented somitic mesoderm

Segmented intermediate mesoderm

Unsegmented lateral plate mesoderm

(B) Dorsal view of developing somites

Anterior

Broken edge of peeled ectoderm

Neural tube

Somites

Somitic mesoderm rounding into somites

Courtesy of Kathryn Tosney

Posterior

Figure 2.6 Major divisions of mesoderm.   (A) Segmented somitic mesoderm forms somites that contribute to segmented structures such as vertebrae and unsegmented structures such as endothelial cells that line the vessels of the circulatory system. Segmented intermediate mesoderm gives rise to segmented structures such as kidney tubules as well as kidney and reproductive ducts and portions of the gonads. Unsegmented lateral plate mesoderm contributes to smooth muscle layers of the gut and muscles of the ventral body wall. (B) When surface ectoderm is peeled away, scanning electron microscopy reveals somites forming adjacent to the neural tube. Development proceeds from the anterior end of the embryo toward the posterior, where somites are just beginning to form. (A, after S.F. Gilbert and M.J.F. Barresi. 2016. *Developmental Biology*, 11th ed. Oxford University Press/Sinauer: Sunderland, MA)

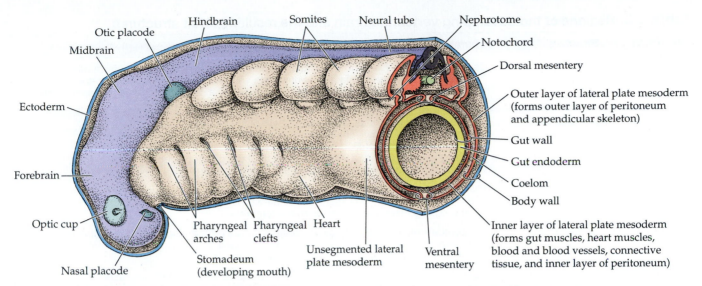

Figure 2.7 Development of the pharyngeal region. Three-dimensional view of a portion of a generalized vertebrate pharyngula, the embryonic stage when pharyngeal pouches appear. The ectoderm is stripped off the left side to show pharyngeal development and segmentation of the mesoderm in the trunk region. (After J.S. Kingsley. 1912. *Comparative Anatomy of Vertebrates.* Philadelphia, P. Blakiston's Son & Co.)

neurogenic (neuron-generating) tissue of vertebrates. By the early embryonic stage shown in Figure 2.7, the three primary brain regions have differentiated (**Table 2.3**). The **forebrain** is associated with the olfactory system and in most vertebrates becomes an integrative and associative area. Information from the eyes is projected to centers in the **midbrain**. The **hindbrain** receives input from many senses, including balance, hearing, taste, and touch. It also coordinates outgoing voluntary and involuntary motor activities, including respiration and circulation.

Other neurogenic tissues of vertebrates

Many differences between vertebrates and other chordates relate to the appearance of two new embryonic tissues, neural crest and neurogenic placodes, which contribute to formation of the nervous system and sense organs.

Table 2.2 Fates of the pharyngeal arches of gnathostomes

Arch number and name	Skeletal and muscle derivatives (sharks)	Cranial nerve (sharks and mammals)	Aortic arch (mammals)	Skeletal and muscle derivatives (mammals)[a]
Arch 1: Mandibular arch	Upper and lower jaws and jaw-closing muscles	V (trigeminal)	Remnant forms maxillary artery	Incus and malleus in middle ear and muscles of mastication
Arch 2: Hyoid arch	Hyomandibula, ceratohyal, and associated muscles	VII (facial)	Stapedial artery	Stapes, dorsal portions of hyoid, and muscles of facial expression
Arch 3: Carotid arch	1st gill arch and associated muscles	IX (glossopharyngeal)	Carotid artery	Ventral portions of hyoid and stylopharyngeus muscle
Arch 4	2nd gill arch and associated muscles	X (vagus)	Systemic aorta	Thyroid cartilage and extrinsic laryngeal muscles
Arch 5	3rd gill arch and associated muscles	X (vagus)	Lost in adults	None
Arch 6	4th gill arch and associated muscles	X (vagus)	Pulmonary artery	Arytenoid cartilages and intrinsic laryngeal muscles

[a]Only major muscles are included.

Table 2.3 Regions of the developing vertebrate brain and the resulting adult structures

Region	Structure	Characteristics
Forebrain	Telencephalon	Develops in association with the olfactory system and coordinates inputs from other sensory modalities. The area for olfaction is unique to vertebrates. In derived vertebrates the telencephalon becomes enlarged and is known as the cerebrum or cerebral hemispheres, the area responsible for associative processing of information
	Diencephalon	The pineal organ, a median dorsal outgrowth of the diencephalon, regulates circadian (daily) rhythms based on light The floor of the diencephalon (location of the hypothalamus) and the adenohypophysis (anterior pituitary gland, a ventral outgrowth of the diencephalon) form the primary center for neural-hormonal coordination and integration
Midbrain	Mesencephalon	The tectum (dorsal portion) receives sensory information (visual, auditory, and touch). The tegmentum (ventral portion) is the pathway for incoming sensory information and outgoing responses to and from the forebrain
Hindbrain	Metencephalon	The cerebellum, a dorsal outgrowth, receives sensory information from the vestibular system and coordinates and regulates motor activities (whether reflexive, such as maintenance of posture, or directed, such as escape movements)
	Myelencephalon	Controls involuntary (autonomic) functions such as respiration and circulation

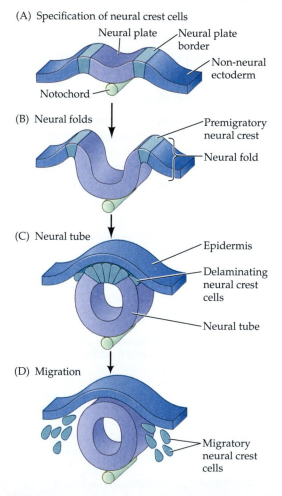

(A) Specification of neural crest cells

Neural plate

Neural plate border

Non-neural ectoderm

Notochord

(B) Neural folds

Premigratory neural crest

Neural fold

(C) Neural tube

Epidermis

Delaminating neural crest cells

Neural tube

(D) Migration

Migratory neural crest cells

Figure 2.8 Formation and migration of neural crest cells. (A) Neural crest cells are specified at the border of the neural plate. (B) As the neural plate folds to form the neural tube, the premigratory neural crest cells are carried to the top of the fold. (C, D) As the neural tube closes, neural crest cells delaminate (separate from the epidermal sheet) and migrate. (After S.F. Gilbert and M.J.F. Barresi. 2016. *Developmental Biology*, 11th ed. Oxford University Press/Sinauer: Sunderland, MA.)

Neural crest Key features of **neural crest** cells are their migratory ability and multipotency—they move to distant sites in the body where they differentiate into many different cell types, resulting in even more cell types than are formed by mesoderm (**Figure 2.8**). Neural crest occurs only in vertebrates and is perhaps the most important innovation in the evolutionary origin of the vertebrate body plan. Because of this, many scientists regard neural crest as a fourth germ layer, in addition to ectoderm, mesoderm, and endoderm.

Neural crest develops in both the head and trunk and contributes to many unique vertebrate structures, including cranial skeletal elements, cranial nerves and sense organs, cranial muscles, adrenal glands, pigment cells in the skin, secretory cells of the gut, and smooth muscle tissue of the blood vessels. Gnathostomes (jawed vertebrates) have even more features that form from neural crest, including the sympathetic nervous system and dentine found in teeth. Neural crest cells interact extensively with other tissues to form these structures.

Lancelet larvae express a gene during early development that may be homologous with a vertebrate gene active in neural crest formation, and similar genes are expressed in some tunicates. Cells resembling migratory neural crest cells occur in larvae of some tunicate species, where they differentiate into pigment cells. Thus, neural crest probably had antecedents among nonvertebrate chordates.

Neurogenic placodes Embryonic vertebrates uniquely have thickenings of the ectoderm in the anterior portion of the head, called **neurogenic placodes**, that give rise to nerves and sensory receptors of the nose, ear, and other sensory systems (see Figure 2.7). Like neural crest, cells of neurogenic placodes migrate. Some placodal cells migrate caudally to contribute to the trunk lateral lines of fishes and lissamphibians and to the cranial nerve that innervates them. Other placodal thickenings in the embryonic ectoderm are not neurogenic, giving rise to structures such as feathers and hair.

2.5 Vertebrate Tissues

LEARNING OBJECTIVES

2.5.1 Categorize the four types of tissues of vertebrates.
2.5.2 Differentiate dermal bone, endochondral bone, and perichondral bone.

At the whole-animal level, increased body size and increased activity levels distinguish vertebrates from nonvertebrate chordates. Because of their relatively larger sizes, vertebrates need efficient, specialized tissues and organ systems to carry out processes accomplished by diffusion or cilia in smaller animals. The origin of vertebrates probably was linked to changes from filter feeding to active predation. These changes, including a muscular pharynx, cranial nerves, and more complex sense organs, are largely based on tissues derived from neural crest. A muscularized, multichambered heart also supports higher levels of activity and oxygen transport in the circulatory system.

Here and in Section 2.6 we introduce general aspects of vertebrate anatomy and function. Evolutionary changes and further specializations of these structures in the different vertebrate taxa are described in later chapters.

Adult tissue types

Vertebrate bodies are made of four types of tissues:

1. *Epithelial tissues* consist of sheets of tightly connected cells that form boundaries between the inside and outside of the body, such as the skin, and between compartments within the body.

2. *Muscular tissues* are made of cells containing the filamentous proteins actin and myosin, which together cause muscle cell contraction to exert forces.

3. *Neural tissues* include neurons, which are the cells that transmit information via electrical and chemical signals; and glial cells, which support the neurons and perform many functions, including formation of myelin sheaths that enhance conduction speeds of neural impulses.

4. *Connective tissues* provide structural support, protection, and strength. Connective tissues include not only bone and cartilage that form the skeleton but also tendons, ligaments, adipose tissue, and blood.

Mineralized tissues

Mineralized connective tissues are typically composed of cells that secrete a proteinaceous tissue matrix, usually composed of collagen and crystals of calcium hydroxyapatite (**Table 2.4**). Hydroxyapatite crystals often align on the matrix of collagenous fibers in layers with alternating directions, much like the structure of plywood. This combination of cells, fibers, and minerals gives bone a latticework structure that combines strength with relative lightness. Mineralized tissues readily fossilize and supply most of the information we have about extinct vertebrates. Thus,

we discuss mineralized tissues throughout this book as we trace vertebrate evolutionary history.

Bone   Vertebrates have three main types of bone: dermal, endochondral, and perichondral.

- **Dermal bone** forms in the dermis of the skin by a process called **intramembranous ossification**; such bones lack a cartilaginous precursor. The precursor tissue in embryos is called mesenchyme, and it consists of cells surrounded by an extensive extracellular matrix. During intramembranous ossification, mesenchyme cells transform into osteocytes (bone cells) that secrete additional extracellular matrix specialized to bind with bone minerals.

- **Endochondral bone** is made up of osteocytes that form within a cartilaginous precursor deeper within the body. In bony fishes and tetrapods, endochondral bone forms the internal skeleton.

- **Perichondral bone** is like dermal bone in some respects and like endochondral bone in others. It forms deep within the body, like endochondral bone, but within the perichondral membrane surrounding an individual cartilage or bone, and thus has some aspects of dermal bone.

Genes expressed during development of endochondral bone are absent from cyclostomes and were secondarily lost in chondrichthyans (cartilaginous fishes), but the presence of bone in extinct jawless fishes indicates that bone is an ancestral character for gnathostomes.

Teeth   Toothlike components called odontodes were part of the dermal armor of early fossil vertebrates, and recent studies indicate that odontodes were the primary elements from which the oral teeth of gnathostomes evolved. Gnathostome teeth are harder than bone and more resistant to wear because they are composed of

Table 2.4 Mineralized tissues of vertebrates

Tissue	Occurrence	Approximate mineral content
Enamel	Teeth, some fish scales, and armor of early vertebrates	96%
Enameloid	Teeth and scales of some extant and fossil fishes	96%
Dentine	Teeth, some fish scales, and armor of early vertebrates	90%
Bone	Internal skeleton and dermal skeleton of bony fishes and tetrapods, as well as external structures such as antlers	70%
Calcified cartilage	Internal skeleton of chondrichthyans; mineral typically organized as prisms and sometimes referred to as prismatic calcification	70%
Cementum	Mammalian teeth; helps fasten teeth into sockets	45%

dentine and enamel, both of which are more mineralized than bone (see Table 2.4).

An oral tooth of a gnathostome typically forms at the interface between ectoderm and a dental papilla derived from neural crest. Enamel-forming cells (ameloblasts) develop from ectoderm; the neural crest-derived dental papilla forms the dentine-forming cells (odontoblasts); and reciprocal developmental interaction leads to tooth formation.

2.6 Vertebrate Organ Systems

LEARNING OBJECTIVES

2.6.1 List and characterize the organ systems of vertebrates.

2.6.2 Describe the three components of the cranial skeleton.

2.6.3 Explain why it is important to think beyond the five sensory systems of mammals.

An animal is more than the sum of its parts, and a key level of organization lies in integrated combinations of tissues to form organs and organ systems. In this section, we briefly describe the general characteristics of the vertebrate organ systems listed in **Table 2.5**. Subsequent chapters will describe modifications in these systems that evolved in particular taxa as vertebrates reached their current extraordinary diversity.

Integumentary system

The integumentary system consists of skin and its derivatives, including dermal bone, teeth, glands (mucus, oil, sweat, wax-producing, and mammary), and so-called "appendages" of the skin (scales, feathers, hair, horns, antlers, and hooves). The main functions of skin and its derivatives are protection, regulation of body temperature, sensation, and communication.

Vertebrate skin consists of two layers, the **epidermis** (outer layer) and **dermis** (inner layer); the basement membrane between them consists of extracellular material secreted by both epidermal and dermal cells and serves as the foundation on which the epithelial cells of the epidermis are organized (**Figure 2.9**). The epidermis is relatively thin and derived from ectoderm. It is constantly renewed by cell division in its most basal layer, the stratum germinativum. The skin of non-amniotes typically has a layer of mucus secreted by goblet cells. The epidermal cells proper, especially those of terrestrial vertebrates, synthesize keratin (Greek *keras*, "horn"), an insoluble protein that ultimately fills the cells of the outermost layer of epidermis, the stratum corneum (Latin *cornu*, "horn"), before they are shed at the skin's surface.

The dermis is unique to vertebrates and is much thicker than the epidermis. Derived from mesoderm and neural crest, the dermis contains connective tissue, nerves, and blood vessels. Beneath the skin is the **hypodermis**, a loose connective tissue layer that anchors the skin to underlying muscles and organs. Derived from mesoderm and also unique to vertebrates, the hypodermis protects internal organs and serves as a site for fat storage in some vertebrates, including birds and mammals.

Skeletal system

Bones and joints make up the skeletal system. Functions of this organ system include protection of internal organs, movement, storage of fat and minerals, and production of formed elements of the blood (red blood cells, white blood cells, and platelets). Fat is stored in yellow bone marrow,

Table 2.5 Vertebrate organ systems

Integumentary system	Skin (epidermis and dermis) and its derivatives, including dermal bone, teeth, claws, scales, feathers, hair, and some exocrine glands (mucus, oil, sweat, wax-producing, and mammary)
Skeletal system	Bones and joints, organized as part of the axial skeleton (cranial skeleton, vertebrae, ribs, sternum, and median fins) or appendicular skeleton (pectoral and pelvic girdles and associated fin or limb bones)
Muscular system	Three types of muscles that control either voluntary movements (skeletal muscle) or involuntary movements (cardiac muscle and smooth muscle)
Nervous system	Cells and organs that regulate and integrate information from the internal and external environments. Includes the brain and spinal cord, as well as sensory organs that receive signals for chemosensation (taste and smell), mechanosensation (touch and pressure), vision, and hearing. Some aquatic vertebrates can perceive electrical impulses generated by muscle contractions of other organisms.
Endocrine system	Ductless glands that regulate and coordinate internal body functions by secreting hormones that interact with specific target cells
Respiratory system	Organs involved in ventilation, the movement of water or air across surfaces such as those of the gills, skin, or lungs, where respiratory gases are exchanged
Circulatory system	Heart pumps blood through a closed system of vessels that transports oxygen and nutrients to cells and removes carbon dioxide and other metabolic wastes
Digestive system	Organs (mouth, pharynx, esophagus, stomach, and small and large intestines) that take in complex food compounds and break them down into small molecules that are absorbed across the gut wall for transport to cells
Excretory system	Kidneys filter nitrogenous waste from the blood and regulate salt and water balance
Reproductive system	Gonads (testes in males and ovaries in females) and their ducts. Oviducts in some female vertebrates enlarge to form paired uteri or fuse to form a single uterus where fertilized eggs develop. Some male vertebrates have an intromittent organ used for internal fertilization

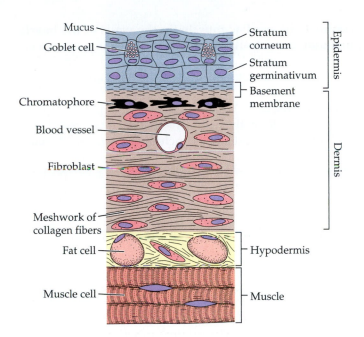

Figure 2.9 Simplified schematic of vertebrate skin.
In the epidermis, cells of the stratum germinativum multiply mitotically, with some cells moving toward the surface where they are sloughed off (in terrestrial vertebrates these cells become filled with keratin as they move upward to the surface and die). Goblet cells in the epidermis produce mucus. Chromatophores are pigment cells derived from neural crest. Fibroblasts are connective tissue cells that produce fibers of the extracellular matrix of the dermis. (After K. Liem et al. 2001. *Functional Anatomy of the Vertebrates*, 3rd ed. Cengage/Harcourt College: Belmont, CA.)

and blood cells and platelets are formed from stem cells in red bone marrow.

Structures of the vertebrate internal skeleton include the vertebrae, chondrocranium, and gill arch skeleton (**Figure 2.10A**). We infer that both neural arches (which are components of the vertebrae) and elements that support the caudal and other median fins were present in early Paleozoic vertebrates. Subsequently, vertebrates evolved a dermal skeleton of external plates and scales that protected soft tissues of the body. Many dermal skeletal elements persisted throughout vertebrate history. For example, much of our skull is dermal bone.

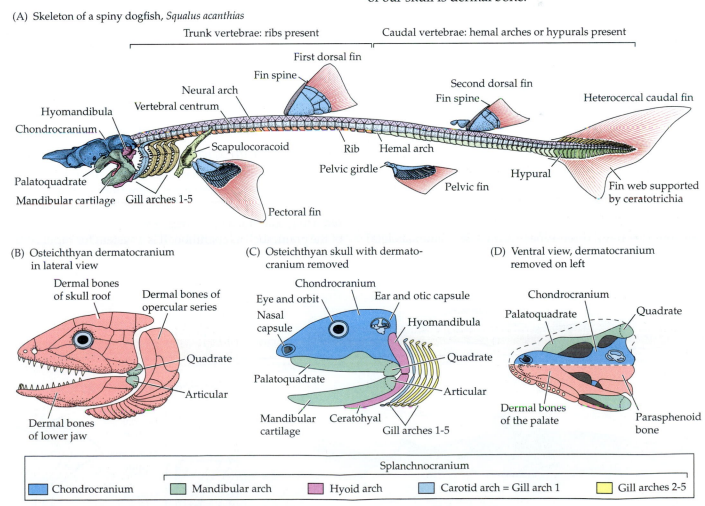

Figure 2.10 Components of the skeleton. (A) Skeleton of a shark. (B–D) Cranium of an idealized osteichthyan based on the bowfin, *Amia calva*. (B) Lateral view of dermatocranium showing dermal bones and jaw joint between quadrate and articular bones. (C) Lateral view with dermatocranium removed to show chondrocranium and splanchnocranium. (D) Ventral view with left side of dermatocranium removed. (A after S.G. Gilbert. 1973. *Pictorial Anatomy of the Dogfish*. U. Washington Press. B–D after K. Liem et al. 2001. *Functional Anatomy of the Vertebrates*, 3rd ed. Cengage/Harcourt College: Belmont, CA.)

In addition to considering the types of bone, ossification patterns, and position within the body, another way to think about organization of the vertebrate skeleton is to recognize its components: the **axial skeleton** (cranial skeleton, vertebrae, ribs, sternum, and median fins) and the **appendicular skeleton** (pectoral and pelvic girdles and associated fin or limb bones).

The cranial skeleton of most extant vertebrates forms from three components (**Figure 2.10B–D**):

1. **Dermatocranium** (Greek *derma*, "skin"; Latin *cranium*, "skull"), which forms by intramembranous ossification in the dermis of the skin and functions to protect the brain, anchor teeth, and provide attachment sites for muscles. Anterior regions of the dermatocranium form from neural crest and posterior regions form from mesoderm. Bones of the dermatocranium often cover cartilage or bones derived from the chondrocranium and splanchnocranium.

2. **Chondrocranium** (Greek *chondros*, "gristle"), which forms in part from neural crest and in part from mesoderm and functions to protect the brain, nose, and inner ear.

3. **Splanchnocranium** (Greek *splanchnon*, "viscera"), which forms from neural crest and contributes to the upper and lower jaws as well as the gill arch skeleton and functions in feeding and respiration.

Components of the splanchnocranium are known by many names: visceral arches, gill arches, pharyngeal arches, and branchial arches. In adult gnathostomes, the first two arches are the mandibular and hyoid, and they form jaws and jaw supports, respectively. In adult fishes, arches 3–7 bear gill tissue, and we call these **gill arches**.

Muscular system

Vertebrates have three types of muscle tissue: skeletal muscle (attached to the skeleton; voluntary control of contraction), cardiac muscle (in the walls of the heart; involuntary control), and smooth muscle (in walls of visceral organs and blood vessels; involuntary control). We focus here on skeletal muscles of the head and trunk.

Cranial muscles   Vertebrates have two main groups of skeletal muscles in the head. The first are those associated with the pharyngeal arches, including the jaws. Muscles of the mandibular arch form the jaw-closing muscles of all gnathostomes. In fishes, muscles of the more posterior arches power gill ventilation, but in adult tetrapods, which have lost gill ventilation, these muscles evolved new functions (see Table 2.2). The second group of skeletal muscles in the head are the extrinsic eye muscles, which rotate the eyeball.

Axial muscles   Axial musculature consists of segmental myomeres folded in three dimensions so that each extends anteriorly and posteriorly to span several body segments (**Figure 2.11**). The original function of myomeres was to produce lateral undulations of the body for swimming. Myomeres of nonvertebrate chordates are V-shaped whereas those of vertebrates are W-shaped (see Figure 2.4). The segmental pattern of axial muscles is clearly visible in fishes; you may have observed it when a piece of cooked fish flakes apart into zigzag blocks, each block representing a myomere. The segmental pattern is less obvious in tetrapods, but can be observed in the six-pack abdomen of a body builder, where each ridge represents a segment of the rectus abdominis muscle.

Nervous system and sense organs

The nervous system regulates and integrates information from both the external and internal environments to control motor, sensory, and automatic body functions (those not requiring conscious thought), as well as higher functions of the brain, such as cognition. It is a system for rapid communication that works via both electrical signals (action

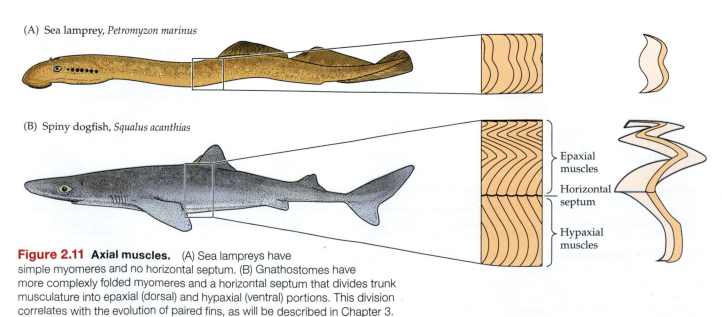

(A) Sea lamprey, *Petromyzon marinus*

(B) Spiny dogfish, *Squalus acanthias*

Epaxial muscles

Horizontal septum

Hypaxial muscles

Figure 2.11 Axial muscles.   (A) Sea lampreys have simple myomeres and no horizontal septum. (B) Gnathostomes have more complexly folded myomeres and a horizontal septum that divides trunk musculature into epaxial (dorsal) and hypaxial (ventral) portions. This division correlates with the evolution of paired fins, as will be described in Chapter 3.

potentials) and chemical signals (neurotransmitters). We focus on the sense organs of vertebrates because these organs often reveal much about the sensory world of an animal. Sense organs contain receptor cells that transduce environmental signals, such as light or sound, into action potentials that travel via sensory neurons to the brain.

Because we are mammals, we think of vertebrates as having five senses—taste, touch, sight, smell, and hearing—but this list does not include all senses of extant vertebrates. Many groups have complex, multicellular sense organs formed from neurogenic placodes tuned to the sensory stimuli important to those species.

Chemosensation   The ability to perceive and react to chemical signals (molecules) in the external environment is the most universal of sensory abilities—plants, fungi, animals, and even bacteria all have this vital and often extraordinary ability (think of sharks that can detect small amounts of blood in the open ocean). Among vertebrates, taste (gustation) and smell (olfaction) involve detection of dissolved molecules by specialized receptors. We humans think of taste and smell as interlinked; for example, our sense of taste is poorer if our sense of smell is blocked when we have a cold. However, these two senses operate at different distances and have different pathways in the brain. Odorants are detected at a distance, and the sensations are received in the forebrain. Tastant molecules are perceived upon direct contact, and the sensations are received initially in the hindbrain.

Vision   A vertebrate's eye consists of a cornea, lens, iris, and retina. The retina, which develops as an outgrowth of the brain, has three main layers: photoreceptors, which are the light-sensitive cells, interneurons, and output neurons. Rods are photoreceptors used to see in low light levels, such as dusk. Cones are color-sensitive photoreceptors that require higher light levels. Both rods and cones contain opsin photopigments. Color vision using different opsins appears to be a plesiomorphic feature of vertebrates.

Electroreception   The capacity to perceive electrical impulses generated by muscle contractions of other organisms is a form of distance reception that works only in water. Electroreception was probably an important feature of early vertebrates, and it is widespread among extant fishes and aquatic lissamphibians but has been lost in amniotes. Electroreception secondarily evolved in monotreme mammals such as the platypus, but the mechanism and innervation of the system differ from those of fishes. Some fishes not only detect weak electric fields but also produce electrical discharges to sense their environment and communicate with other individuals. Still others, such as electric eels, produce strong electrical discharges that can subdue prey and deter predators.

Mechanoreception   All vertebrates have sensory cells or organs that respond to mechanical forces; touch receptors in the skin are a familiar example. A more specialized class of mechanoreceptors are **hair cells**; found in all vertebrates, hair cells detect motion in fluids. The fluid can be outside the body, such as water surrounding a fish, or within the body, such as fluid within the inner ear. They are called hair cells because they have hairlike microscopic specializations on the cell surface. When these hairlike projections are deflected, the cell detects the direction of fluid motion. Hair cells function in many sensory systems, including the vestibular system and hearing (see below), and in the lateral line system of fishes and aquatic lissamphibians, which detects water movements (see Chapter 4).

Vestibular system   The inner ear is enclosed within the braincase (**Figure 2.12.A**). The **vestibular system** is the component of the inner ear that detects changes in an animal's position in space using hair cells and fluid (endolymph) contained within its chambers and semicircular canals (**Figure 2.12B**). Chambers of the inner ear, the utriculus and sacculus, house sensory patches called maculae that contain tiny crystals of calcium carbonate resting in a jellylike substance on hair cells. Macular sensations tell the animal which way is up and detect linear acceleration

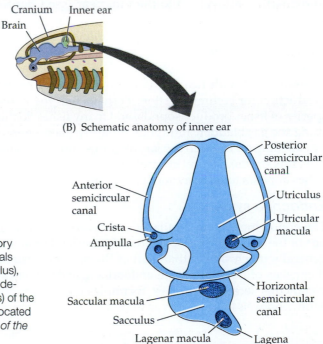

(A) Head of hypothetical vertebrate

(B) Schematic anatomy of inner ear

Figure 2.12 **Schematic interpretation of the inner ear.** The inner ear, located within the braincase (A), serves vestibular and auditory senses. An idealized gnathostome inner ear (B) has three semicircular canals (anterior, posterior, and horizontal) and two chambers (utriculus and sacculus), allowing the animal to detect motions in three dimensions. Sensations are detected by hair cells in small sensory patches (cristae) in ampullae (swellings) of the semicircular canals, and by hair cells in larger sensory patches (maculae) located within the chambers. (Modified from Liem et al. 2001. *Functional Anatomy of the Vertebrates*, 3rd ed. Cengage/Harcourt College: Belmont, CA.)

(movement in a straight line). At the base of each semi-circular canal is a swelling (ampulla), which contains a smaller sensory patch called a crista. Cristae also have hair cells embedded in the same jellylike substance found in the sensory areas of the sacculus and utriculus. Cristae detect angular acceleration (such as head rotation) by monitoring the displacement of endolymph within the semicircular canals during motion.

Gnathostomes have three semicircular canals—anterior, posterior, and horizontal—that allow detection of motion in three dimensions (see Figure 2.12B). Cyclostomes differ, with hagfishes interpreted as having only one semicircular canal and lampreys two. Recent work suggests that the vestibular system of lampreys is more complicated than previously believed and that it functions as effectively as that of gnathostomes.

We often fail to realize the importance of our own vestibular sense because we usually depend on vision to determine our position. We can sometimes be fooled, however, as when sitting in a stationary train or car and thinking that we are moving, only to realize from the lack of input from our vestibular system that it is the vehicle next to us that is moving.

Hearing   The inner ear of fishes and tetrapods is also used to detect sound waves. The lagena is an extension off the sacculus that is specialized for sound reception in different groups of vertebrates (see Figure 2.12B); in mammals, for example, the lagena evolved into an elongated, snail-shaped cochlea. Sound waves transmitted to the cochlea create waves of compression that pass through the endolymph, stimulating auditory hair cells.

The ways in which sound waves reach the inner ear differ across vertebrates. An elegant example in some fishes is the Weberian apparatus, a chain of small bones that connects the gas bladder (which is also important in regulating buoyancy) to the inner ear (see Chapter 7). Mammals have a large external ear, or pinna, and a chain of three small bones in the middle ear that transmit vibrations from the ear drum to the inner ear (see Chapter 22).

Associated with hearing is the evolution of sound production for communication. Perhaps half the living species of fishes produce sounds, and many fishes do this using the gas bladder. Tetrapods typically make sounds by expelling air through the larynx (frogs and mammals) or syrinx (birds). Neural control of sound production in fishes appears to be homologous with that in tetrapods.

Endocrine system

The endocrine system regulates and coordinates activities of the body, but the speed of information transfer is generally much slower than that of the nervous system. It consists of many separate **endocrine glands**, ductless glands that secrete hormones. Examples include the adenohypophysis (also known as the anterior pituitary) and the thyroid and adrenal glands. Organs that contain some endocrine tissue but have additional functions also are part of the endocrine system, including parts of the digestive system (stomach, small intestine, and pancreas), as well as reproductive organs (ovaries and testes).

Endocrine cells secrete hormones into the surrounding fluid where they typically diffuse into the bloodstream for transport throughout the body (some are transported in other fluids, such as cerebrospinal fluid or lymph). Although hormones contact virtually all cells, they influence only target cells that have receptors for specific hormones. Hormones can initiate relatively short-term changes in behavior and physiology (for example, the fight-or-flight response mediated by the adrenal glands) or longer-term changes, such as those associated with development and growth (mediated by the adenohypophysis and the thyroid gland).

Endocrine glands differ from **exocrine glands**, which secrete products such as mucus, oil, sweat, wax, milk, saliva, or digestive enzymes into ducts leading to the skin, mouth, or digestive tract. Mucus, oil, sweat, wax-producing, and mammary glands are part of the integumentary system mentioned earlier. Salivary glands, the liver, and exocrine glands of the stomach, intestines, and pancreas are part of the digestive system.

Respiratory system

Getting oxygen to cells requires both ventilation and circulation. **Ventilation** refers to movement of the respiratory medium (water or air) across a respiratory surface through which gases are exchanged via diffusion with the blood. Respiratory surfaces such as those of the gills, skin, or lungs are thin, moist, highly vascularized, and characterized by large surface areas. One or more pumps are needed to move the respiratory medium over the respiratory surface; examples of pumps used by vertebrates include the mouth and oral cavity, rib cage, and diaphragm. Fishes typically use movements of the oral cavity to draw water into the mouth and pump it out over the gills. Mammals use movements of the rib cage and diaphragm to move air in and out of the lungs.

Circulatory system

Circulation is movement of blood through vessels to the vicinity of cells where diffusion of gases again occurs. In all vertebrates, the pump that powers circulation is the heart.

Blood is a fluid connective tissue that is pumped through blood vessels. It is composed of liquid plasma, red blood cells (erythrocytes, which contain the iron-rich protein hemoglobin used to transport respiratory gases), several different types of white blood cells (leukocytes), and specialized cells (or cell-like structures) that promote clotting of blood (such as platelets in mammals). Blood transports oxygen and nutrients to cells, removes carbon dioxide and other metabolic waste products, and

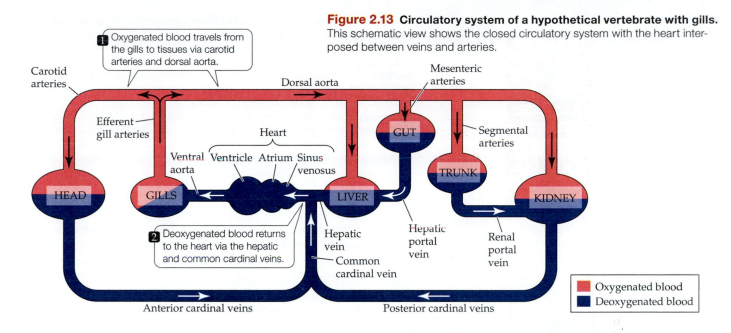

Figure 2.13 **Circulatory system of a hypothetical vertebrate with gills.** This schematic view shows the closed circulatory system with the heart interposed between veins and arteries.

stabilizes the internal environment. It also carries hormones from their sites of release to target tissues and plays essential roles in immunity.

The vertebrate circulatory system and the pump that powers it—the heart—offer an interesting look at some important and unique vertebrate synapomorphies and their evolution.

Closed circulation Vertebrates have **closed circulatory systems**—arteries and veins are connected by capillaries so that blood always stays within vessels. **Arteries** carry blood away from the heart, and **veins** return blood to the heart. Blood pressure is higher in arteries than in veins, and arterial walls have a layer of smooth muscle that is absent from veins.

The dorsal aorta is flanked by paired cardinal veins that return blood to the heart. Anterior cardinal veins (draining the head) and posterior cardinal veins (draining the body) unite on each side in common cardinal veins that enter a common chamber known as the sinus venosus before entering the atrium of the heart (**Figure 2.13**). In lungfishes and tetrapods, the posterior cardinal veins are functionally replaced by a single median vessel, the posterior vena cava.

Interposed between the smallest arteries (arterioles) and smallest veins (venules) are **capillaries**, which are sites of exchange between blood and tissues. Capillaries pass close to every cell and their walls are only one cell layer thick, facilitating diffusion. Collectively, capillaries provide an enormous surface area for exchange of gases, nutrients, and waste products. Arteriovenous anastomoses connect arterioles directly to venules, allowing blood to bypass a capillary bed, and normally only a fraction of the capillaries in a tissue have blood flowing through them.

Blood vessels that connect two capillary beds are called **portal vessels**. For example, the hepatic portal vein, present in all vertebrates, connects capillary beds of the gut and liver (see Figure 2.13). Substances absorbed from the gut are transported directly to the liver, where toxins are rendered harmless and some nutrients are processed or stored. Most vertebrates also have a renal portal system connecting veins returning from the tail and posterior region of the trunk to the kidneys. The renal portal system is not well developed in jawless vertebrates and has been lost in adult mammals.

The vertebrate heart The heart develops as a muscular tube that folds on itself and becomes constricted into sequential chambers: **sinus venosus**, **atrium**, and **ventricle** (see Figure 2.13). The **conus arteriosus** is added in gnathostomes. Therian mammals—marsupials and placentals—incorporate the sinus venosus into the wall of the right atrium as the **sinoatrial node**, which controls heartbeat; in these mammals the conus arteriosus is subsumed into the base of the aorta. The heart wall is composed of a thick layer of cardiac muscle—the myocardium—with a thin outer epicardium and inner endocardium.

In basal gnathostomes such as sharks, the sinus venosus is a thin-walled sac with few cardiac muscle fibers. Blood moves from the sinus venosus into the atrium, which has valves at each end to prevent backflow. The ventricle has thick muscular walls with an intrinsic pulsatile rhythm that can be speeded up or slowed down by the nervous system. Ventricular contraction forces blood into the ventral aorta. Researchers think that, originally, six pairs of aortic arches branched from the ventral aorta to supply the gills (**Figure 2.14**). Oxygenated blood flows from the gills to the paired dorsal aortae above the gills.

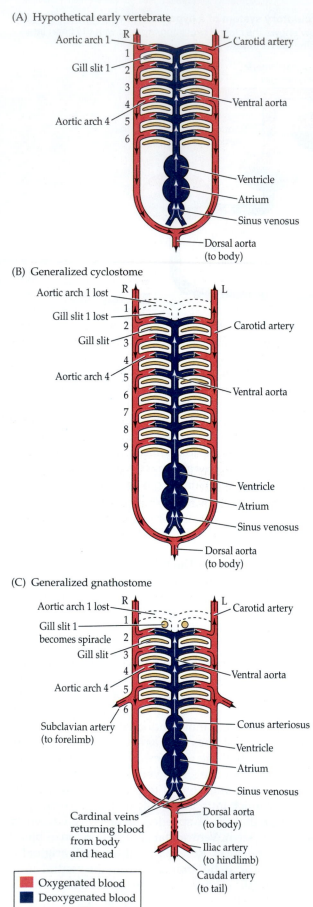

(A) Hypothetical early vertebrate

Aortic arch 1
Gill slit 1
Aortic arch 4
Carotid artery
Ventral aorta
Ventricle
Atrium
Sinus venosus
Dorsal aorta
(to body)

(B) Generalized cyclostome

Aortic arch 1 lost
Gill slit 1 lost
Gill slit
Aortic arch 4
Carotid artery
Ventral aorta
Ventricle
Atrium
Sinus venosus
Dorsal aorta
(to body)

(C) Generalized gnathostome

Aortic arch 1 lost
Gill slit 1 becomes spiracle
Gill slit
Aortic arch 4
Subclavian artery (to forelimb)
Carotid artery
Ventral aorta
Conus arteriosus
Ventricle
Atrium
Sinus venosus
Cardinal veins returning blood from body and head
Dorsal aorta (to body)
Iliac artery (to hindlimb)
Caudal artery (to tail)

■ Oxygenated blood
■ Deoxygenated blood

Figure 2.14 Heart and aortic arches of vertebrates viewed from the ventral side. (A) Early vertebrates probably had six pairs of aortic arches, as do the embryos of extant vertebrates; aortic arch 1 is lost in the adults of all extant vertebrates. (B) Cyclostomes have additional posterior aortic arches to accommodate more gill openings. In the generalized cyclostome shown there are nine aortic arches. (C) Gnathostomes have subclavian and iliac arteries to supply forelimbs and hindlimbs. The conus arteriosus between the ventricle and ventral aorta dampens the pulsatile component of the blood flow. Some people have referred to the conus arteriosus as a "fourth" heart chamber, but this confuses the generalized gnathostome condition with the four-chambered heart of archosaurs and mammals.

Vessels from the most anterior aortic arch run forward to the head as carotid arteries. Posterior to the gills, the paired dorsal aortae unite to carry blood posteriorly.

Different groups of vertebrates exhibit fascinating adaptations of the heart and circulation, such as the evolution of four-chambered hearts in crocodylians, birds, and mammals. These differences are described in later chapters.

Digestive system

Feeding and digestion are accomplished by the digestive system. **Feeding** refers to taking food into the mouth, mechanically processing it using teeth and tongue, moving it through the pharynx, and swallowing it into the esophagus. **Digestion** refers to the breakdown of complex compounds into small molecules that can be absorbed across the wall of the gut and transported to tissues. In all vertebrates, food moves through the gut by involuntary rhythmic muscular contractions called peristalsis, but basal vertebrates lack the regional differentiation of the gut seen in mammals.

In land vertebrates, digestion begins in the mouth with breakdown of food by enzymes in saliva or venom and continues in the stomach and small intestine. Enzymes break down carbohydrates and proteins, and lipids are emulsified for ready absorption. Nutrient absorption occurs primarily in the small intestine. The large intestine, or colon, absorbs water and inorganic ions. In most vertebrates, feces are eliminated through the cloaca, which is a shared exit for digestive, urinary, and reproductive systems. In therian mammals, the cloaca becomes divided during development into separate urogenital and rectal portions, and feces are eliminated directly from the posteriormost portion of the large intestine, termed the rectum.

Excretory and reproductive systems

Although functions of the excretory and reproductive systems differ, both develop from intermediate mesoderm (see Figure 2.6). The principal organs of the excretory system are the kidneys, and the system is also known as the renal system (Latin *renes*, "kidney").

Kidneys develop from a portion of the intermediate mesoderm called nephrotome (**Figure 2.15**). They are initially segmental, whereas gonads (ovaries in females, testes in males) are not. Kidneys and gonads develop and

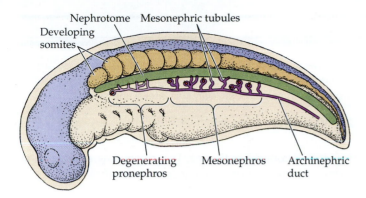

Nephrotome   Mesonephric tubules
Developing somites
Degenerating pronephros   Mesonephros   Archinephric duct

Figure 2.15 Early kidney development. Vertebrate embryos initially develop a pronephros drained by the archinephric duct. The pronephros degenerates and the mesonephros develops posterior to it. In amniote embryos, the familiar compact, bean-shaped kidney known as a metanephros subsequently develops (not shown in this schematic).

lie behind the peritoneum on the dorsal body wall; only in some therian mammals do testes descend into a scrotum.

Kidneys   Kidneys dispose of waste products (primarily nitrogenous waste from protein metabolism) and regulate the body's water and minerals, especially sodium, chloride, calcium, magnesium, and potassium, as well as bicarbonate and phosphate ions. Kidneys of tetrapods perform almost all these functions, but the gills and skin of fishes and lissamphibians also play important roles.

The functional units of kidneys are microscopic structures called **nephrons**. Vertebrate kidneys work by ultrafiltration: high blood pressure forces water, ions, and small molecules through tiny gaps in capillary walls. Nonvertebrate chordates lack true kidneys, although lancelets have excretory cells known as solenocytes that may be homologous with portions of the vertebrate nephron.

Gonads   Gonads are single in cyclostomes but paired in gnathostomes. Although structural tissues of gonads derive from mesoderm, the progenitor cells of gametes (eggs and sperm), known as germ cells, originate from endoderm. Early in development, these progenitor cells migrate through the dorsal mesentery to reach the gonads. As mentioned when we described the endocrine system, gonads also produce hormones (for example, ovaries produce estrogen and testes produce testosterone).

Lampreys lack a special duct for passage of gametes to the outside; rather, sperm or eggs erupt from the gonad and move through the coelom to pores that open to the outside. In gnathostomes, gametes are transported via specialized ducts. In males, sperm are released into the archinephric ducts, which originally formed to drain the kidneys. In females, eggs released into the coelom are drawn into the oviduct. In many vertebrates, oviducts produce substances associated with the egg, such as an eggshell. Oviducts can enlarge to form paired uteri or fuse to form a single uterus in which fertilized eggs develop.

Some vertebrates lay eggs that develop outside the body (**oviparity**), while others retain eggs within the mother's body until embryonic development is complete and young are born (**viviparity**). Shelled eggs must be fertilized in the oviduct before the shell forms. Viviparous vertebrates and many vertebrates that lay shelled eggs have an intromittent organ—such as the pelvic claspers of sharks or the penis of mammals—by which sperm are transferred into a female's reproductive tract.

Summary

2.1 Vertebrates in Relation to Other Animals

Deuterostomia and Protostomia are the two subdivisions of Bilateria (animals with bilateral symmetry).

Bilaterian bodies develop from three germ layers (ectoderm, endoderm, and mesoderm) that become organized during gastrulation.

Vertebrates belong to Chordata, which along with echinoderms and hemichordates comprise Deuterostomia. All nonchordate deuterostomes are small marine animals.

2.2 Characteristics of Chordates

Chordates have four unique characteristics.

- All have a notochord, which gives the group its name. The notochord is a dorsal stiffening rod to which muscles attach for lateral undulation. It is a transient structure in most vertebrates, largely replaced during development by vertebrae.

- Chordates also have a dorsal neural tube, a muscular postanal tail, and an endostyle at some point in development.

In addition to vertebrates, Chordata contains two groups of small, filter-feeding marine animals.

- Cephalochordata contains the lancelets, also known as amphioxus. The notochord extends from the tip of the head to the tip of the tail.

- Urochordata contains the tunicates. Most adult tunicates are sessile and have lost three of the four identifying chordate characteristics that were present in their larvae (the endostyle remains).

(Continued)

Summary *(continued)*

Although cephalochordates look more like vertebrates than do tunicates, molecular data show that tunicates and vertebrates are sister groups, united in Olfactores.

2.3 What Distinguishes a Vertebrate?

Vertebrates are defined by vertebrae and other features.

- A cranium, which is a bony, cartilaginous, or fibrous structure surrounding the brain.

- A prominent head containing a muscular pharynx, complex sense organs, and a tripartite brain.

- A muscular multichambered heart, complex endocrine organs, mineralized tissues, and gills derived from endoderm.

Early vertebrates evolved an active, predaceous lifestyle and larger body sizes than nonvertebrate chordates.

2.4 Vertebrate Embryonic Development

Embryonic development is important for understanding adult anatomy. Like all bilaterians, vertebrates develop from three germ layers.

- Ectoderm forms epidermis, sense organs, and nerves.

- Mesoderm consists of segmented somitic mesoderm, intermediate mesoderm, and unsegmented lateral plate mesoderm. Somites give rise to the body's skeletal muscles, dermis of the skin, and vertebral column. Intermediate mesoderm forms kidneys, ducts, and portions of gonads. Lateral plate mesoderm forms smooth muscle of the gut and components of the circulatory system including the heart. In vertebrates with limbs, it also forms limb bones.

- Endoderm gives rise to the lining of the gut, gills, lungs, major digestive organs, and much of the circulatory system. Germ cells that will form gametes also come from endoderm.

Vertebrates have a coelom (body cavity) that can have two compartments (pericardial cavity and peritoneal cavity, as in adult sharks or salamanders) or four compartments (pericardial cavity, paired pleural cavities, and peritoneal cavity, as in adult mammals).

Pharyngeal development, with its component pharyngeal arches, pouches, and clefts, is strongly conserved within vertebrates. Each pharyngeal arch has characteristic skeletal elements, nerves, muscles, and blood supply. The fates of each arch can be traced throughout the vertebrate groups, from sharks to mammals.

Neural crest, unique to vertebrates and considered by many scientists to be a fourth germ layer, forms pigment cells, nerves, endocrine organs, and many structures in the head including teeth.

Neurogenic placodes that form in the head of embryos are also unique to vertebrates and give rise to complex sense organs such as the nose, ear, and lateral line system.

2.5 Vertebrate Tissues

The vertebrate body is composed of epithelial, muscular, neural, and connective tissues. These tissue types have specific characteristics and functional properties that make them unique to vertebrates.

Mineralized connective tissues are particularly important in the study of the vertebrate fossil record. These include bone, mineralized cartilage, dentine, enamel, enameloid, and cementum.

Bone occurs in the skin (dermal bone) and in the internal skeleton, where it surrounds cartilaginous skeletal elements (perichondral bone) or replaces cartilage (endochondral bone). Among extant vertebrates, endochondral bone occurs only in bony fishes and tetrapods.

Teeth are composed of dentine and enamel (or enameloid), which are more mineralized than bone and thus harder.

2.6 Vertebrate Organ Systems

The integumentary system consists of skin and its derivatives, such as dermal bone, teeth, glands, and "appendages" of the skin, including scales, feathers, and hair. The main functions of the integumentary system are protection, regulation of body temperature, sensation, and communication. Skin has an upper layer, the epidermis, and a lower layer, the dermis.

The gnathostome skeleton includes the axial skeleton (skull, vertebral column, ribs, and sternum) and the appendicular skeleton (limbs and limb girdles). The original axial skeleton was the notochord. Vertebrates added vertebral elements and, in gnathostomes, ribs.

The dermal skeleton of bony fishes and tetrapods includes many bones of the skull, such as the cranial roofing bones, facial bones, and lower jaw.

Axial musculature consists of complexly folded segmental myomeres, which are W-shaped in vertebrates rather than V-shaped as in cephalochordates.

Vertebrates have two systems for internal communication, the nervous system and endocrine system.

- The nervous system includes specialized sense organs and transmits information rapidly.

- The endocrine system consists of specialized glands (and some tissues) that secrete hormones, which then travel to target cells throughout the body. Speed of information transfer by the endocrine system is generally much slower than that by the nervous system.

Summary *(continued)*

Smell and taste are chemosenses that function at different distances and have different receptor organs innervated by different nerves that project to different brain regions.

Vision depends on the retina of the eye, which is an outgrowth of the brain containing photosensory rods and cones.

Some fishes have electroreceptors that allow them to perceive electrical stimuli generated by other animals. Electroreception secondarily evolved in monotreme mammals.

Mechanoreceptors sense touch, pressure, vibrations, and sound waves. These include touch receptors in the skin, as well as hair cells in the inner ear that function in balance, body orientation, and hearing. For some aquatic vertebrates, hair cells of the lateral line system allow the detection of vibrations in the water.

Getting oxygen to cells requires ventilation and circulation, provided by the respiratory and circulatory systems, respectively.

- Ventilation is movement of a respiratory medium across a respiratory surface through which gases are exchanged via diffusion with the blood.

- Circulation is movement of blood through vessels to the vicinity of cells where diffusion of gases again occurs. Blood is a fluid connective tissue composed of plasma, red and white blood cells, and specialized cells (or cell-like structures) that promote clotting.

- Vertebrates have a closed circulatory system, with arteries and veins connected by capillaries and powered by the pumping action of the heart.

- The heart is a muscular tube that during development becomes folded and divided into chambers with valves in between them.

Vertebrates feed in diverse ways, break down their food in the digestive tract, and excrete some wastes via kidneys, which develop in close association with gonads.

Digestion begins with the mouth and continues with the stomach and small and large intestines. Food moves down the gut via muscular contractions. Basal vertebrates lack the degree of gut regionalization seen in mammals.

Vertebrate kidneys (the main organs of the excretory system) and the gonads of the reproductive system form from intermediate mesoderm.

- Kidneys are composed of nephrons and dispose of waste products; they function via ultrafiltration and require high blood pressure.

- Gonads release gametes. In male gnathostomes, each testis shares a duct with the kidney. Females have a separate gonadal duct, the oviduct.

Vertebrates may lay eggs or retain fertilized eggs within the mother's body. In the latter case (and also with shelled eggs), internal fertilization is essential, and males may have an intromittent organ.

Discussion Questions

2.1 Suppose new molecular data showed that cephalochordates and vertebrates are sister taxa. What difference would this make to our interpretations about early chordates?

2.2 Would evidence of complex sense organs in a fossil nonvertebrate chordate suggest a close relationship with vertebrates? What vertebrate features would have to be present?

2.3 Which neurogenic placodes of embryonic fishes are absent from amniote embryos?

2.4 Explain the different processes of ossification. How does the type of ossification relate to the position of a bone in the body?

2.5 Describe the three components of a mammalian skull. Include in your answer position in the body, type of ossification, functions of each component, and embryonic sources.

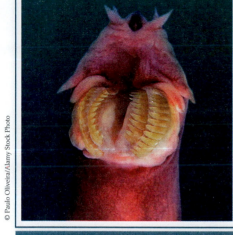

© Paulo Oliveira/Alamy Stock Photo

3

Jawless Vertebrates and the Origin of Gnathostomes

Hagfish jaws open horizontally

Early jawless vertebrates represented an advance over their filter-feeding ancestors. Their most conspicuous novel features included an anatomically distinct head with complex sense organs and a tripartite brain enclosed by a cartilaginous cranium (braincase). Newly evolved pharyngeal muscles powered the new gill arch skeleton to draw water into the mouth and over newly evolved gills for respiration.

Early vertebrates were more active than nonvertebrate chordates, and many of them had external armor formed by dermal bone and other mineralized tissues. The evolution of jaws represents a further increase in complexity that allowed even greater activity and the evolution of predation.

No fossils directly document the origin of jaws, but studies of embryonic development and functional morphology help us understand some aspects of jaw evolution. Jaws may have evolved originally to improve the effectiveness of gill ventilation before later modification to seize and hold prey.

Today the jawed vertebrates, or **gnathostomes** (Greek *gnathos*, "jaw"; *stoma*, "mouth") encompass all extant vertebrate species other than lampreys and hagfishes (together the **cyclostomes**; Greek *cyclos*, "round"). In this chapter, we trace early steps in the radiation of vertebrates that began more than 500 million years ago. We discuss Paleozoic jawless vertebrates and the two extant jawless groups, as well as jawed fishes that did not survive the Paleozoic.

3.1 Earliest Evidence of Vertebrates

LEARNING OBJECTIVES

3.1.1 Describe features of Cambrian vertebrates.

3.1.2 Name the fundamental feature that unites members of Osteognathostomata.

3.1.3 Compare the evidence for a freshwater versus a marine origin for vertebrates.

Until recently, the most ancient evidence of jawless vertebrates consisted of fragments of dermal armor from the Ordovician, ~480 Ma. Complete fossils of both jawless and jawed vertebrates are known from the late Silurian, ~425 Ma. Discoveries of soft-bodied animals from the Chengjiang Fauna of the early Cambrian of China now extend the vertebrate fossil record back into the Cambrian, ~525 Ma. An example is †*Haikouichthys* (probably synonymous with †*Myllokunmingia*), a fish-shaped animal about 3 cm long known from hundreds of well-preserved individuals (**Figure 3.1A**). A slightly younger animal, †*Metaspriggina*, occurs in the middle Cambrian Burgess Shale, about 508 Ma.

Evidence of a notochord marks these animals as chordates, and presence of a cranium and paired sensory structures mark them as vertebrates, because these structures form from neural crest and neurogenic placodes (see Section 2.4). These animals also had other vertebrate features: a dorsal fin and a ribbonlike ventral fin (but without fin rays); six or seven gill pouches with filamentous gills and a branchial skeleton; W-shaped myomeres (rather than V-shaped; see Figure 2.4); and segmental structures flanking the notochord interpreted as vertebral rudiments. Cambrian vertebrates lack bone and features that would convincingly unite them with either cyclostomes or gnathostomes.

The earliest complete vertebrate fossils are torpedo-shaped †arandaspids from the Late Ordovician, a time of great radiation and diversification in the wake of large-scale extinctions at the end of the Cambrian. Fossil †arandaspids are small (12–35 cm long; **Figure 3.1B**) and have been found in rocks in Bolivia, Australia, North America, and Arabia.

(A) Cambrian vertebrate, †*Haikouichthys*

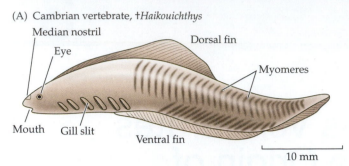

(B) Ordovician vertebrate, †*Astraspis*

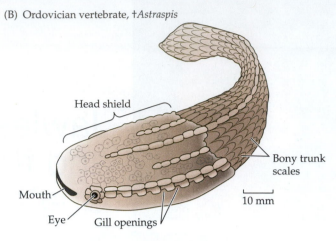

Figure 3.1 Two of the earliest vertebrates. (A) Discoveries of organisms such as the tiny, soft-bodied †*Haikouichthys* in the early Cambrian Chengjiang Fauna of China extended the presumed origin of vertebrates back to about 525 Ma. (B) An †arandaspid, †*Astraspis*, from Ordovician deposits in North America. The head and gill regions of †*Astraspis* were encased by many small, close-fitting, polygonal bony plates that formed a head shield with sensory canals, two openings between the eyes interpreted as paired nostrils (not shown here), bony plates around the eye, and as many as eight gill openings on each side of the head. (A after D-G. Shu et al. 1999. *Nature.* 402: 42–46; B after D.K. Elliott. 1987. *Science* 237: 190–192.)

Enigmas: †*Conodonts and* †*Tullimonstrum*

Spinelike or comblike microfossils (<1 mm) known as **conodont elements** are widespread and abundant in marine deposits from the late Cambrian to the Late Triassic (**Figure 3.2A**). Discovery of impressions of a more complete †conodont animal with features such as a notochord, myomeres, and large eyes, led to their interpretation as vertebrates (**Figure 3.2B**), but some doubt remains. Conodont elements are arranged within the animal's pharynx in a complex apparatus. Our current interpretation is that the mineralized tissue of conodont elements evolved convergently with hard tissues of other vertebrates.

Another enigmatic fossil sometimes interpreted as a vertebrate is †*Tullimonstrum* ("Tully monster") named for Francis Tully, an amateur fossil collector. Tully collected the first known specimen in 1955 from the late Carboniferous Mazon Creek formation in central Illinois. (Mazon Creek is a **Lagerstätte**, a locality in which fossils exhibit extraordinary preservation and, typically, taxonomic diversity.) No fossils similar to the soft-bodied †*Tullimonstrum* are known, and its affinities to other animals stumped original investigators. Although many researchers do not consider †*Tullimonstrum* a vertebrate, it has been interpreted as a cyclostome (**Figure 3.2C**).

(A) Isolated conodont elements

From Jarochowska, E. and A. Munnecke. 2016. *Geol. J.* 51: 683–703

(B) Reconstruction of †conodont animal

© Bill Parsons

(C) Reconstruction of †*Tullimonstrum* as a vertebrate

From V.E. McCoy et al. 2016. *Nature* 532: 496–499.
Illustration courtesy of Sean McMahon/Yale University

Figure 3.2 †Conodonts and †*Tullimonstrum*. (A) Mineralized conodont elements, assumed to have been used in feeding, are abundant in Paleozoic deposits. (B) Reconstruction of a †conodont animal, †*Idiognathodus*. Complex arrangements of conodont elements lined the animal's pharynx. (C) Although its status is debated, some consider the soft-bodied †*Tullimonstrum*, here reconstructed as a cyclostome, to be a vertebrate. (A from E. Jarochowska and A. Munnecke. 2016. *Geol J* 51: 683–703; C from V.E. McCoy et al. 2016. *Nature* 532: 496–499.)

Early mineralized tissues

Cambrian vertebrates and extant cyclostomes lack mineralized tissues. Mineralized tissues composed of hydroxyapatite represent a major evolutionary advance within vertebrates (see Section 2.5), and we group all vertebrates with bone in **Osteognathostomata** (**Figures 3.3** and **3.4**).

Odontodes are small, toothlike elements in the skin consisting of projections of dentine from a bony base covered with an outer layer of enamel, like the placoid scales of extant sharks. They are basic units of mineralized tissue in early osteognathostomes, and contributed to dermal armor. Odontodes are similar to gnathostome teeth; they are composed of two different tissues, dentine and enameloid, that combine to yield a hard, wear-resistant structure.

The large plates and head shields of early osteognathostomes consisted of aggregates of odontodes and underlying dermal bone known as aspidin, which lacked

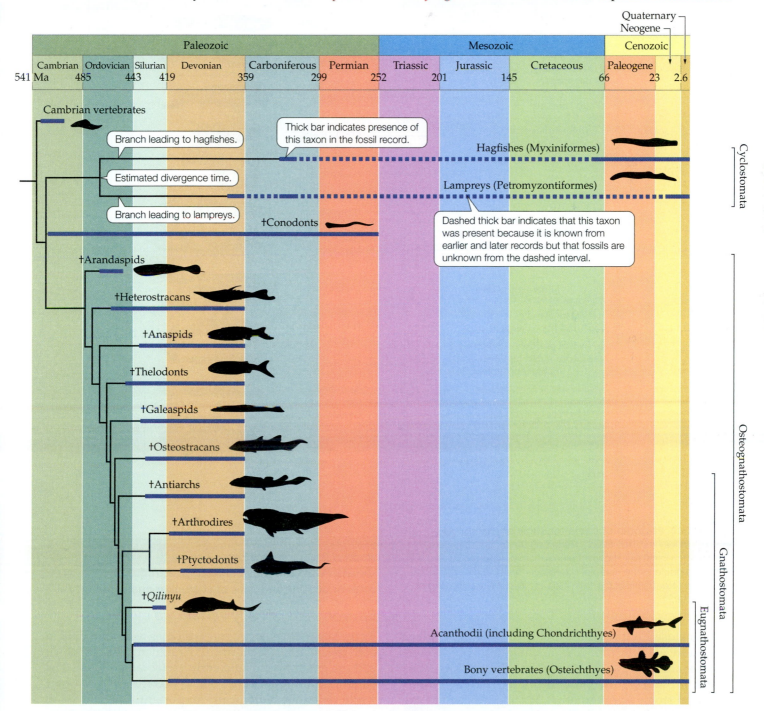

Figure 3.3 Time tree of vertebrates. Narrow lines show phylogenetic relationships and estimated divergence times. Divergence times are based on molecular phylogenetic interpretation and often predate the first known occurrences of fossils. Wide horizontal bars mark occurrences of taxa in the fossil record. Wide dashed bars indicate that a taxon is present because it is known from earlier and later records but that no fossils are known from the dashed interval. Daggers (†) indicate extinct groups. (After K. Liem et al. 2001. *Functional Anatomy of the Vertebrates*, 3rd ed. Cengage/Harcourt College: Belmont, CA.)

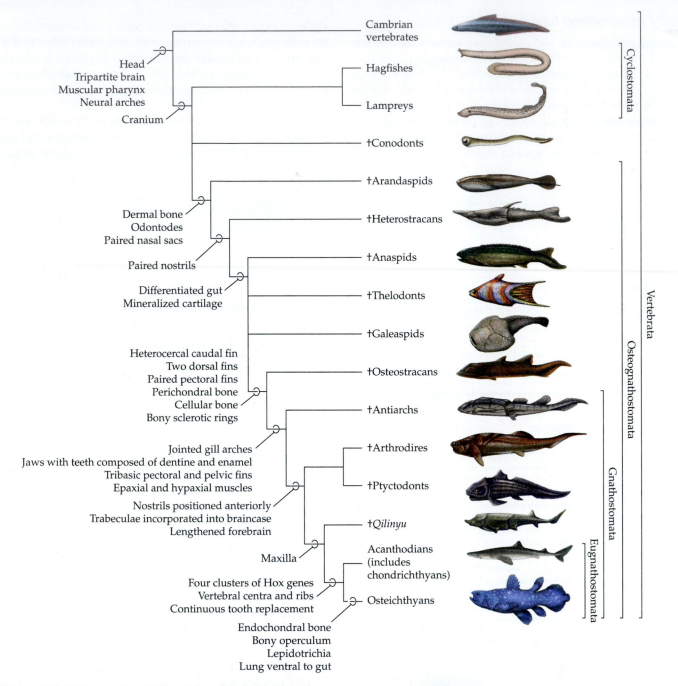

Figure 3.4 Phylogenetic relationships of vertebrates including key extinct taxa. Taxa and their shared characters (synapomorphies) are shown. Taken together, Figures 3.3 and 3.4 provide a guide to the material covered in this chapter and will be referenced frequently in the text. (See credit details at the end of this chapter.)

cells enclosed within the bone (cellular bone evolved later, and we interpret it as a synapomorphy of †osteostracans and gnathostomes; see Figure 3.4). Bony armor provided protection and may also have functioned in storage or regulation of minerals such as calcium and phosphorus (our bones serve these functions). Speculation that the armor insulated electroreceptive organs has been discredited.

Toothlike structures are found within the pharynx of some early osteognathostomes, but these may represent internal scales not homologous with the teeth of gnathostomes. Basal gnathostomes had teeth on tooth plates associated with the jaws consisting of multiple cusps with cores of dentine and caps of enamel or enameloid.

Environment of early vertebrate evolution

By the late Silurian, osteognathostomes were abundant in both freshwater and marine environments. But in which of these environments did early vertebrates evolve? The vertebrate kidney is very good at excreting excess water while retaining biologically important molecules and ions. That is what a freshwater fish must do: because its body fluids are continuously diluted by osmotic inflow of water, a freshwater fish must excrete that water to regulate its internal environment (see Chapter 4). Thus, kidney physiology suggests that vertebrates evolved in fresh water. Despite the logic of that inference, however, a marine origin of vertebrates is now widely accepted. Osmoregulation is complex, and fishes use cells in gills as well as the kidney to control their internal fluid concentrations. Probably the structure of the kidney is merely fortuitously suited to life in fresh water.

Three lines of evidence support the hypothesis of a marine origin of vertebrates:

1. The earliest vertebrate fossils are from marine sediments.

2. Invertebrate deuterostomes are exclusively marine, and have body fluids with approximately the same osmolal concentration as their surroundings.

3. Hagfishes have concentrated body fluids like surrounding seawater, and this probably represents the original vertebrate condition.

3.2 Cyclostomes: Extant Jawless Vertebrates

LEARNING OBJECTIVES

3.2.1 List five derived features of cyclostomes.

3.2.2 Name a derived feature and three reduced or rudimentary sensory systems of hagfishes.

3.2.3 Compare feeding and respiration in larval lampreys and adults of parasitic species of lampreys.

Hagfishes and lampreys, together grouped as **Cyclostomata** (Greek, *cyclos*, "circle"; *stoma*, "mouth") lack jaws, paired fins, specialized reproductive ducts, and mineralized tissues. In other respects, such as cranial anatomy, cyclostomes are highly derived in ways different from extant gnathostomes. The fossil record of cyclostomes is sparse because they are soft-bodied.

Evolutionary relationships of hagfishes and lampreys to each other and to early osteognathostomes have been controversial because their unusual anatomy and phylogenetically uninformative fossil record make direct comparisons difficult. However, molecular and genomic studies, as well as newly discovered anatomical synapomorphies confirm that hagfishes and lampreys are sister taxa. Much is still to be learned about relationships of cyclostomes to Paleozoic groups, but here we interpret cyclostomes as the sister group of osteognathostomes (see Figure 3.4).

Characters of cyclostomes

Hagfishes and lampreys share several anatomical features, but it is sometimes difficult to determine whether these are plesiomorphic for vertebrates or synapomorphies of cyclostomes. Key features of cyclostomes that differentiate them from gnathostomes include:

1. A single median nostril.

2. Gills supported by an unarticulated gill arch skeleton known as a branchial basket (**Figure 3.5A**); in gnathostomes, the gill arch skeleton has joints between the elements (**Figure 3.5B**).

(A) Lateral view of cranial skeleton of sea lamprey, *Petromyzon*

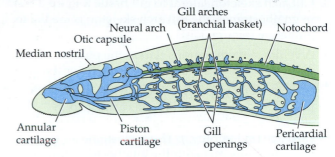

(B) Lateral view of cranial skeleton of spiny dogfish, *Squalus*

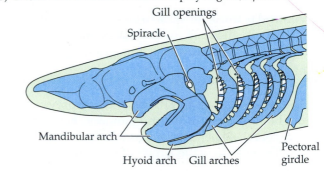

(C) Frontal section through pharynx of sea lamprey, *Petromyzon*

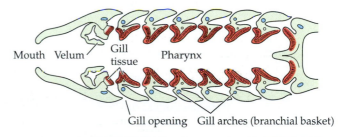

(D) Frontal section through pharynx of spiny dogfish, *Squalus*

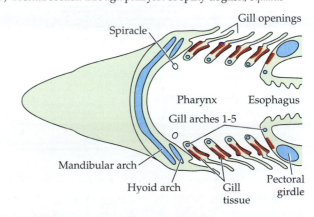

Figure 3.5 Gill arch skeleton and gills of cyclostomes and gnathostomes. Both the conditions (A,B) and positions (C,D) of these structures differ between the two clades. (A) In cyclostomes, the gill arch skeleton is not articulated, instead forming a branchial basket. (B) In gnathostomes, the gill arch skeleton is articulated, with well-developed joints between the elements. (C) The branchial basket of cyclostomes is lateral to gill tissue. (D) The gill arch skeleton of gnathostomes (mandibular arch, hyoid arch, and gill arches 1–5) is medial to gill tissue. (A,C,D after K. Liem et al. 2001. *Functional Anatomy of the Vertebrates*, 3rd ed. Cengage/Harcourt College: Belmont, CA.)

3. Gill arch skeleton lateral to gill tissue (**Figure 3.5C**); in gnathostomes, the gill arch skeleton is medial to gill tissue (**Figure 3.5D**).

4. A velum, a membranous flap in the pharynx moved by muscles that pumps water over the gills.

5. A tongue that bears keratinous teeth and is supported by a prominent lingual cartilage plus accessory cartilages (in hagfishes) and the piston cartilage (in lampreys). The cyclostome tongue is not homologous with the gnathostome tongue because its muscles are innervated by cranial nerve V rather than XII.

6. Periocular trunk muscles (trunk muscles that extend into the head around the eyes).

7. An immune system that is different from that of gnathostomes.

The notochord of hagfishes and lampreys persists throughout life. Hagfishes and lampreys have rudimentary vertebral elements (see Figure 2.4B). Lampreys have these elements along the length of the body dorsal to the notochord (interpreted as equivalent to neural arches) and hagfishes have elements below the notochord in the tail (interpreted as equivalent to hemal arches). We think that the common ancestor of cyclostomes and osteognathostomes probably had rudimentary vertebral elements both dorsal and ventral to the notochord and that cyclostomes simplified this condition. Although hagfishes and lampreys lack jaws, they have several cranial cartilages that do not have obvious homologues in gnathostomes.

The single median nostril and a single nasal sac of cyclostomes is unlike the paired nostrils and paired nasal sacs of gnathostomes. Early fossil osteognathostomes also had a single median nostril, so the condition is presumably plesiomorphic for vertebrates.

Hagfishes: Myxiniformes

There are about 75 species of hagfishes in two major genera, *Eptatretus* and *Myxine*. Adults are elongated, scaleless, pinkish to purple in color, and about 0.5 m long (**Figure 3.6A**). All are marine, with a nearly worldwide distribution except for polar regions. They live in deep, cold-water habitats and are scavengers of the sea floor, drawn to carcasses by their sense of smell.

Structural characteristics   The single median nostril and the mouth are surrounded by three pairs of barbels (also known as tentacles) that are moved by muscles when a hagfish searches for food (**Figure 3.6B**). The keratinized teeth are arranged in longitudinal (front to back) rows on paired horizontal cartilaginous supports that open and

(A) External anatomy

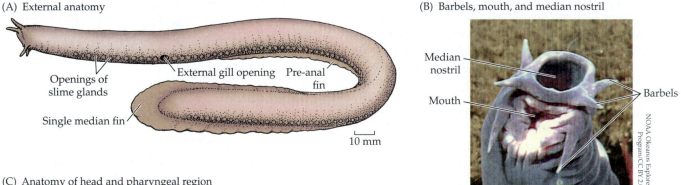

(B) Barbels, mouth, and median nostril

NOAA Okeanos Explorer Program/CC BY 2.0

(C) Anatomy of head and pharyngeal region

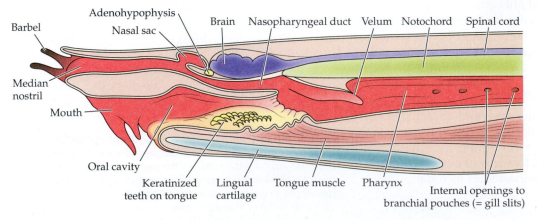

Figure 3.6 Hagfish anatomy   (A) Lateral view of a hagfish. (B) Barbels, mouth, and single median nostril of a hagfish. (C) Sagittal section through head and pharynx of a hagfish. The single median nostril opens into the nasopharyngeal duct, which joins with the oral cavity at the level of the velum to form the pharynx. The single nasal sac and adenohypophysis are adjacent to each other, ventral to the forebrain. (A,C after F.H. Martini. 1998. *Sci Am* 279: 70–75.)

close in the horizontal plane, unlike the jaws of gnathostomes that open and close in the vertical plane. The median nostril of hagfishes leads from its opening in the front of the head to the pharynx and is used to take in water for both olfaction and gill ventilation (**Figure 3.6C**). The number of gill openings on each side of the body varies among genera, from a single opening for all gill pouches up to 15 separate openings, one for each gill pouch. The pharynx and gill slits are posterior to the head, perhaps associated with the feeding behavior of burrowing into carcasses. Hagfishes have a single median fin that extends from the trunk around the tail as the caudal fin and continues anterior to the cloaca as a pre-anal fin; they are the only extant vertebrates with a pre-anal fin, but some early fossil osteognathostomes also have a pre-anal fin.

Many hagfish features appear to be reduced or rudimentary, including: small, skin-covered, non-image-forming eyes lacking eye muscles to move them; absence of electroreception; absence of a well-developed lateral line system; a single semicircular canal in the inner ear (all other extant vertebrates have at least two); persistence of a pronephric kidney in adults (see Figure 2.15); absence of autonomic innervation of the heart (from the vagus nerve, X); and presence of accessory hearts in posterior vessels of the circulatory system. These conditions are either plesiomorphic for vertebrates, secondarily simplified or lost in hagfishes, or apomorphies unique to hagfishes.

One shared derived character of hagfishes is the presence of **slime glands** on each side of the body that open via 90–200 pores in the skin and secrete enormous quantities of mucus and long proteinaceous threads. The threads, coiled within thread cells and released when the cells rupture, unravel on contact with seawater and interact with released mucus to produce a mass of slime that can choke and deter gnathostome predators. If the hagfish becomes entrapped in its own slime, it makes a knot in its body and scrapes the slime off.

Osmoregulation and ion regulation   Hagfishes are unique among vertebrates in maintaining the nonvertebrate chordate condition of body fluids that have the same osmotic concentration as seawater. Hagfishes may represent the original condition; lampreys and gnathostomes convergently evolved the ability to regulate internal concentrations of ions. However, unlike nonvertebrate chordates, hagfishes regulate the divalent ions calcium (Ca^{2+}) and magnesium (Mg^{2+}) and maintain an internal ion concentration different from that of seawater. They also regulate acid–base balance. Hagfish slime may excrete monovalent ions (e.g., sodium and chloride, Na^+ and Cl^-) and divalent ions (e.g., Ca^{2+}).

Feeding   Lacking image-forming eyes, a lateral line system, and electroreception, hagfishes locate food using olfactory cues (chemoreception; see Section 2.6). An experimental study showed that Atlantic hagfish are more attracted to traps baited with dead fish than to traps baited with dead clams or crabs, possibly because fishes

have more lipids in their tissues. Hagfishes typically feed on dead or dying vertebrates, burying their heads in the prey's body; their low metabolic rates allow hagfishes to tolerate the resulting hypoxia. After teeth engage the food, hagfishes can tie a knot in their body and pass it forward, brace it against the food, and tear off the flesh. Baited underwater remote video cameras document that hagfish can catch and kill smaller fish.

Reproduction   Female hagfishes can outnumber males by 100 to 1; the reasons for this strange sex ratio are unknown. Some species are hermaphroditic, having both male and female gonads. Oval yolky eggs about 1 cm long are encased in a tough, clear covering and secured to the seafloor by hooks. Development of eggs is very slow, and there is no larval stage.

Fossil hagfishes   There is a large gap in the fossil record of hagfishes (see Figure 3.3). †*Myxinikela* and †*Myxineidus* are hagfishes from the late Carboniferous Mazon Creek Lagerstätte; †*Gilpichthys* from the same formation may also be a hagfish. The recently described †*Tethymyxine* from Cretaceous sites in Lebanon allowed recognition and definition of a new synapomorphy for cyclostomes: the presence of periocular trunk muscles.

Hagfishes and humans   Strangely enough, hagfishes have economic importance for humans. Almost all so-called eelskin leather products (for example, wallets and belts) are made from hagfish skin. Worldwide demands for this leather and for specialty food markets have decimated hagfish populations in Asian waters and in some sites in North America. However, we still have very little knowledge of hagfish ecology, or virtually any of the other information needed for management of commercially exploited populations.

Lampreys: Petromyzontiformes

There are about 40 extant species of lampreys in two major genera, *Petromyzon* and *Lampetra*. Most adult lampreys parasitize other fishes, although some small freshwater species have nonfeeding adults. Adults of different species range in length from 10 cm to 1 m. Lampreys occur in northern and southern temperate latitudes.

Structural characteristics   Shared derived characters of lampreys include seven pairs of gill pouches that open to the outside in a slanted line just behind the head; and a round mouth located at the bottom of the buccal funnel (**Figure 3.7A,B**). Evolution of the buccal funnel displaced the single median nostril to the top of the head. The nostril opens into the nasohypophyseal duct, which leads to the single median nasal sac and ends blindly in the nasohypophyseal pouch (**Figure 3.7C**). The nostril does not connect to the oral cavity or pharynx as it does in hagfishes. The nasohypophyseal pouch acts like a pipette bulb to move water in and out to tidally irrigate the nasal sac. The nasohypophyseal duct passes close to

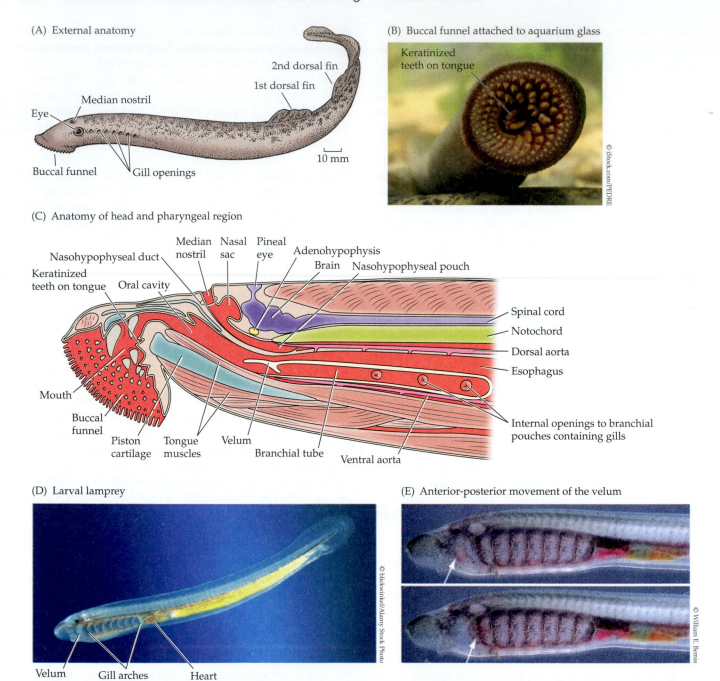

(A) External anatomy

2nd dorsal fin

1st dorsal fin

Median nostril

Eye

Buccal funnel   Gill openings

10 mm

(B) Buccal funnel attached to aquarium glass

Keratinized teeth on tongue

© iStock.com/PEDRE

(C) Anatomy of head and pharyngeal region

Median   Nasal   Pineal
nostril   sac   eye   Adenohypophysis

Nasohypophyseal duct   Brain   Nasohypophyseal pouch

Keratinized teeth on tongue   Oral cavity

Spinal cord

Notochord

Dorsal aorta

Esophagus

Mouth

Buccal funnel

Piston cartilage   Tongue muscles   Velum   Branchial tube   Ventral aorta

Internal openings to branchial pouches containing gills

(D) Larval lamprey

© blickwinkel/Alamy Stock Photo

Velum   Gill arches (branchial basket)   Heart

(E) Anterior-posterior movement of the velum

© William E. Bemis

Figure 3.7 Lamprey anatomy.   (A) Lateral view of an adult lamprey. (B) Lamprey mouth surrounded by the buccal funnel with its keratinous teeth. Teeth in the center of the funnel are on the tongue. (C) Sagittal section through head and pharynx. The oral cavity opens into the esophagus, leading to the gut. In adult lampreys, the pharynx is a blind-ending structure known as the branchial tube with paired internal gill openings leading to gills. (D) Lamprey larvae (also known as ammocoetes larvae) have a fleshy oral hood, tiny eyes, and large pharynx used for filter feeding. (E) When the velum moves from its anterior position to a more posterior position (arrows), it pumps water into the pharynx and out over the gills. There can be several pumping cycles per second.

the adenohypophysis, which develops in embryos as an outpocketing of the duct (adults lose this open connection with the duct). The large, laterally placed eyes of lampreys are well developed, with good color vision; the **pineal eye** (a light-sensitive structure homologous with the pineal gland of mammals) underlies a pale spot in the skin just posterior to the median nostril. Lampreys uniquely have a piston cartilage that supports the tongue.

Feeding   Lamprey larvae are filter feeders. They use their velum to pump water into the pharynx and out over the gills (**Figure 3.7D,E**), while the endostyle produces mucus that entraps microscopic organisms and organic particles. The endostyle of the larva develops into the thyroid gland in the adult. Adults of parasitic species attach to the body of another vertebrate (usually a larger bony fish) by suction. The tongue has keratinous teeth.

Muscles span the dorsal and ventral sides of the piston cartilage and, by alternate contractions, move the keratinous teeth against the skin of the host to rasp a shallow, seeping wound. The inner surface of the buccal funnel is studded with keratinous teeth, and an oral gland secretes an anticoagulant to prevent the host's blood from clotting. Feeding is probably continuous while a lamprey is attached to its host.

Most of a parasitic adult lamprey's diet consists of body fluids sucked from fishes. The digestive tract is straight and simple, as one would expect for an animal with a diet as rich and easily digested as blood and tissue fluids. Lampreys generally do not kill their hosts, but they leave them weakened and with an open wound. Lampreys feed on several species of whales and porpoises in addition to sharks and ray-finned fishes. Swimmers in the Great Lakes, after having been in the water long enough for their skin temperature to drop, have reported attempts by lampreys to attach to their bodies.

Respiration and osmoregulation   Larval lampreys and adults of nonparasitic species employ **flow-through ventilation**: they draw water into the mouth and pump it out over the gills. Adult parasitic lampreys, however, spend much of their time with their buccal funnel affixed to hosts to feed, so they cannot ventilate their gills in a flow-through fashion. Instead, they use **tidal ventilation** to draw water in and expel it out through the gill openings. The velum prevents water from flowing out of the branchial tube into the mouth. This mode of ventilation is not very efficient at oxygen extraction, but a compromise required by the specialized parasitic mode of feeding.

As in jawed fishes, chloride-transporting cells in the gills and well developed kidneys regulate ions, water, and nitrogenous wastes, as well as overall concentration of body fluids, allowing lampreys to live in a wide range of salinities.

Reproduction   The most familiar lampreys, such as *Petromyzon marinus*, are anadromous or potamodromous—that is, adults live in the ocean (**anadromy**) or large lakes (**potamodromy**) and ascend rivers and streams to breed. Some nonparasitic freshwater lampreys do all of their feeding as larvae, with adults acting solely as a reproductive stage. In the late spring, sea lampreys make extensive migrations related to reproduction and are commonly observed at fish passage stations at the bases of dams.

Female lampreys produce thousands of small eggs about 1 mm in diameter that lack specialized coverings such as those of hagfish eggs. The name Petromyzon means "stone sucker," and mating pairs of lampreys use their buccal funnels to suck onto and move rocks to construct a nest, known as a redd, typically in a shallow stream or river. The arrangement of rocks creates a turbulent flow of water through the redd, which oxygenates the eggs. The female attaches to an upstream rock in the nest to lay her eggs and the male wraps around her, fertilizing eggs as they are extruded.

Lamprey larvae look so different from adults that they were originally described as a distinct genus, *Ammocoetes*, a name that we now use as a common name for lamprey larvae (**Figure 3.7D**). Ammocoetes are wormlike with a large, fleshy oral hood, which will transform into the adult buccal funnel. The eyes are hidden beneath the skin. Ammocoetes burrow into soft mud of backwaters, where they spend 3–7 years as sedentary filter feeders before metamorphosing into adults. Adults usually live one to two years and die after breeding once, the reproductive strategy known as **semelparity**. Semelparity contrasts with **iteroparity**, the reproductive strategy characterized by multiple reproductive episodes over a lifetime.

Fossil lampreys   Like hagfishes, lampreys have major gaps in their fossil record (see Figure 3.3). Lampreys are known from the Late Devonian (†*Priscomyzon*, a short-bodied form from South Africa); the late Carboniferous (†*Hardistiella* from Montana, †*Mayomyzon* and †*Pipiscius* from Mazon Creek); and the Early Cretaceous (†*Mesomyzon* from southern China, the first fossil lamprey known from a freshwater deposit). Like most extant forms, all fossil lampreys appear to have been specialized parasites.

Recent discovery of a hatchling-to-adult growth series for †*Priscomyzon* shows that, unlike extant lampreys, †*Priscomyzon* did not have ammocoetes larvae. Instead, their larvae resembled adults and were macrophagous (eating relatively large prey), not filter feeders. Other Paleozoic lampreys also have macrophagous larvae. These findings call into question the long-held interpretation that the common ancestor of cyclostomes and osteognathostomes had filter-feeding larvae.

Lampreys and humans   In North America, humans and lampreys have been at odds for the past century. Historically, sea lampreys moved up the St. Lawrence River as far as Lake Ontario, but Niagara Falls prevented them from moving westward into the other Great Lakes. That changed with construction and subsequent expansions of the Welland Canal between 1829 and 1932. From the 1920s to the 1950s, the range of sea lampreys expanded across the entire Great Lakes basin, which impacted populations of economically important fishes. Chemical lampricides as well as electrical barriers and mechanical weirs (low dams) at the mouths of spawning streams have been used to control Great Lakes lamprey populations.

3.3 Jawless Osteognathostomes

LEARNING OBJECTIVES

3.3.1 Describe how Paleozoic osteognathostomes are similar to and different from extant cyclostomes.

3.3.2 Differentiate between hypocercal and heterocercal caudal fins.

Diverse vertebrate fossils from the Silurian and Devonian periods have dermal bone but lack jaws (see Figure 3.4).

Those without jaws have often been referred to as "†ostracoderms," but this group is not monophyletic (see Section 1.2) because some jawless fishes are more closely related to gnathostomes than are others. Early osteognathostomes had large plates of dermal bone forming a head shield, as in †*Astraspis* (see Figure 3.1B). Most early osteognathostomes were small, but some reached lengths of 50 cm. Although they lacked jaws, many had plates around a circular mouth and probably ate small, soft-bodied prey. As in extant cyclostomes, well-formed vertebrae were absent, and the notochord was the main postcranial axial support.

We can trace the evolution of some key derived features within early osteognathostomes (see Figure 3.4). For

(A) A †heterostracan, †*Pteraspis*

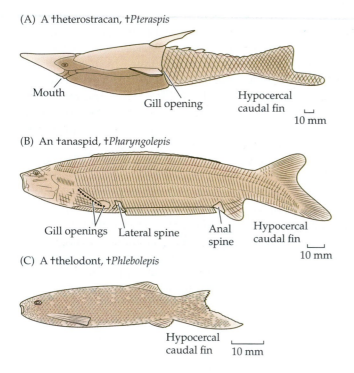

(B) An †anaspid, †*Pharyngolepis*

(C) A †thelodont, †*Phlebolepis*

(D) An †osteostracan, †*Hemicyclaspis*

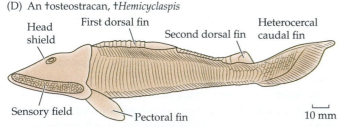

Figure 3.8 Diversity of jawless osteognathostomes.
(A) †*Pteraspis*, a late Silurian †heterostracan presumed to be pelagic because of its body shape and hypocercal caudal fin. Other †heterostracans were dorsoventrally flattened, and presumably benthic. All †heterostracans had a single gill opening. (B) †*Pharyngolepis*, an †anaspid from the late Silurian with fine scales, multiple gill openings, and lateral and anal spines. (C) †*Phlebolepis*, a †thelodont, showing pattern of small scales covering the body. (D) †*Hemicyclaspis*, an †osteostracan from the Late Devonian. †Osteostracans had paired pectoral fins, two dorsal fins, and a heterocercal caudal fin. The multiple gill openings on the ventral surface of the head are not shown. (A–C after J.A. Moy-Thomas and R.S. Miles. 1971. *Paleozoic Fishes*, Chapman & Hall, London; D after E.A. Stensio. 1932. *The Cephalaspids of Great Britain*. Br. Mus. [Nat. Hist.], London.)

example, paired nostrils were present in †heterostracans such as †*Pteraspis* and the lower lobe of its caudal fin was larger than the upper lobe to form a **hypocercal caudal fin** (**Figure 3.8A**). Regionalization of the gut is evident in fossil †anaspids such as †*Pharyngolepis* (**Figure 3.8B**), and in the †thelodont †*Phlebolepis* (**Figure 3.8C**).

†Osteostracans share several characters with gnathostomes. For example, †*Hemicyclaspis* had a **heterocercal caudal fin** in which the upper lobe is larger than the lower lobe and the vertebral column bends upward to support it (**Figure 3.8D**). †Osteostracans had first (= anterior) and second (= posterior) dorsal fins, and some had an anal fin posterior to the cloaca. Like gnathostomes, †osteostracans had paired pectoral fins, but pelvic fins are not known in †osteostracans. Other features shared with gnathostomes include perichondral bone, cellular bone, and bony, sclerotic rings around the eyes (see Figure 3.4).

Jawless forms coexisted with gnathostomes during the late Silurian and for most of the Devonian—about 50 million years—so it is unlikely that the origin and radiation of gnathostomes directly caused their extinction. Extinctions of jawless forms at the end of the Devonian occurred simultaneously with mass extinctions among many marine invertebrate groups, as well as extinctions of many early groups of gnathostomes (see Figure 3.3).

3.4 Gnathostome Body Plan

LEARNING OBJECTIVES

3.4.1 Explain the relevance of Hox genes to the early anatomical diversity of gnathostomes.

3.4.2 Describe derived skeletal and muscular features of gnathostomes that relate to ventilation, feeding, and locomotion.

3.4.3 Compare gnathostomes to cyclostomes with respect to presence of myelin, number of semicircular canals in the ear, and placement of nostril(s).

Gnathostomes have joints between skeletal elements of the gill arches, jaws with true teeth (composed of dentine and enamel), two dorsal fins, two sets of tribasic paired fins (pectoral and pelvic), and segmental W-shaped axial muscles divided by a horizontal septum into epaxial and hypaxial muscles (**Figure 3.9A**).

As you can see in Figure 3.3, gnathostomes are known from the early Silurian onward, and by the early Devonian they were already phylogenetically and anatomically diverse. This may relate to duplications of the Hox gene complex. **Hox genes** occur in clusters arranged in the genome in the same linear sequence as the structures they control along the anterior–posterior axis of the body. The gene at the anterior end of a cluster is first to be expressed, and this occurs in the head region. Expression of the subsequent genes proceeds posteriorly, with the gene in the posteriormost position expressed at the posteriormost tip of the body. All animals, with the possible exception of sponges, have Hox genes (they were first identified in fruit flies), but vertebrates have more Hox

(A) Derived features of an idealized gnathostome

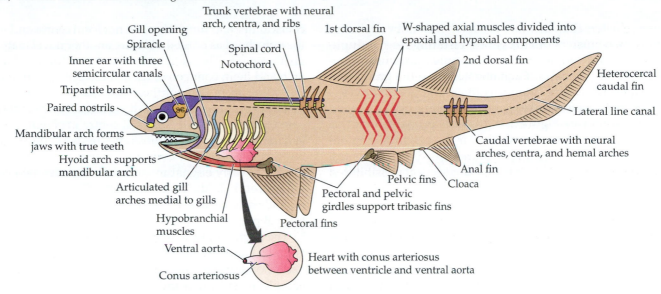

- Gill opening
- Spiracle
- Inner ear with three semicircular canals
- Tripartite brain
- Paired nostrils
- Mandibular arch forms jaws with true teeth
- Hyoid arch supports mandibular arch
- Articulated gill arches medial to gills
- Hypobranchial muscles
- Ventral aorta
- Conus arteriosus
- Trunk vertebrae with neural arch, centra, and ribs
- Spinal cord
- Notochord
- 1st dorsal fin
- W-shaped axial muscles divided into epaxial and hypaxial components
- 2nd dorsal fin
- Heterocercal caudal fin
- Lateral line canal
- Caudal vertebrae with neural arches, centra, and hemal arches
- Anal fin
- Cloaca
- Pelvic fins
- Pectoral and pelvic girdles support tribasic fins
- Pectoral fins
- Heart with conus arteriosus between ventricle and ventral aorta

(B) Cranium, jaws, hyoid arch, and gill arches of an idealized gnathostome

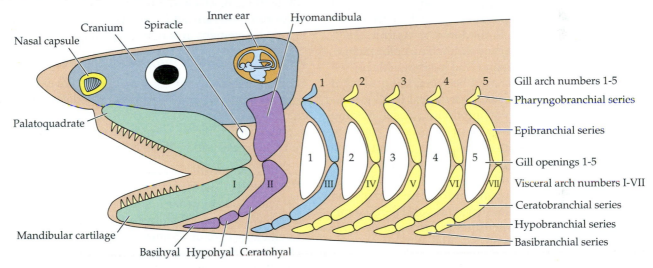

- Nasal capsule
- Cranium
- Spiracle
- Inner ear
- Hyomandibula
- Palatoquadrate
- Mandibular cartilage
- Basihyal   Hypohyal   Ceratohyal
- Gill arch numbers 1–5
- Pharyngobranchial series
- Epibranchial series
- Gill openings 1–5
- Visceral arch numbers I–VII
- Ceratobranchial series
- Hypobranchial series
- Basibranchial series

(C) Pectoral girdle and tribasic pectoral fin of an actinopterygian

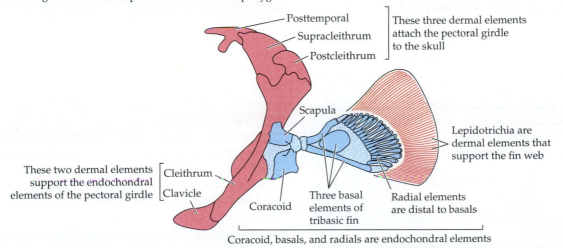

- Posttemporal
- Supracleithrum
- Postcleithrum
- These three dermal elements attach the pectoral girdle to the skull
- Scapula
- Lepidotrichia are dermal elements that support the fin web
- These two dermal elements support the endochondral elements of the pectoral girdle
  - Cleithrum
  - Clavicle
- Coracoid
- Three basal elements of tribasic fin
- Radial elements are distal to basals
- Coracoid, basals, and radials are endochondral elements

Figure 3.9 Basic characters of gnathostomes.
(A) Schematic diagram of an idealized gnathostome showing derived features. Epaxial and hypaxial components of the trunk muscles are divided by the horizontal septum. The trunk lateral line canal lies in the plane of this septum. (B) Schematic diagram of the jaws (mandibular arch; green), hyoid arch (purple), carotid arch (= gill arch 1; blue), and gill arches 2–5 (yellow) of an idealized gnathostome loosely based on a shark. (C) Pectoral girdle and tribasic pectoral fin of *Polypterus*, a basal actinopterygian, showing contributions of endochondral and dermal bones. The series of three basal elements of the pectoral fin of gnathostomes is reduced to a single element, called the humerus, in sarcopterygians. (B,C after K. Liem et al. 2001. *Functional Anatomy of the Vertebrates*, 3rd ed. Cengage/Harcourt College: Belmont, CA.)

gene clusters than do other groups. All extant vertebrates have two clusters of Hox genes, resulting from a duplication of the single Hox gene complex present in nonvertebrate chordates. **Eugnathostomata**—the group that includes the extant Chondrichthyes and Osteichthyes—have four clusters of Hox genes resulting from a second duplication (see Figure 3.4). It seems probable that this second duplication occurred near the origin of gnathostomes and that it resulted in more refined genetic control of development, leading to new anatomical possibilities.

Gnathostome skeletons

In common with †osteostracans, gnathostomes have cellular bone and perichondral bone (among gnathostomes, endochondral bone is known only in bony vertebrates; see Figure 3.4). Other features shared with †osteostracans include pectoral fins, two dorsal fins with internal skeletal supports, and a heterocercal caudal fin.

On each side of the gnathostome body, a typical gill arch has five skeletal components with articulations between them (**Figure 3.9B**). From dorsal to ventral, skeletal components are the pharyngobranchial, epibranchial, ceratobranchial, hypobranchial, and basibranchial (often a single basibranchial is shared between left and right sides of the body). In teleosts, pharyngeal teeth associated with these elements play important roles in feeding adaptations, as we will describe in Chapter 7. Anterior to the typical gill arches, the hyoid arch consists of the hyomandibula, ceratohyal, hypohyal, and basihyal (typically shared between left and right sides of the body). The hyomandibula functions in jaw support and projection in chondrichthyans. In actinopterygians and fishlike sarcopterygians, it becomes modified to form a large component of the suspensorium (a structure composed of endochondral and dermal bones from which the lower jaw is suspended). In tetrapods, the hyomandibula is greatly reduced in size and functions in hearing.

The articulated gill arches of gnathostomes allow **double-pump gill ventilation** in which water moves between muscularized chambers in the head. The mouth and pharyngeal cavity together form one chamber. Water is sucked into the mouth during inhalation and then forcefully pumped across respiratory surfaces of the gills before entering the second muscularized chamber or chambers. Muscles span the joints of the gill arch skeleton to power changes in the volume of the pharynx during respiration and feeding. The net result is that large volumes of water can be moved across the gills to enhance exchange of respiratory gasses or ions as needed to maintain homeostasis (relatively steady internal physical and chemical conditions).

Vertebrae of most eugnathostomes have a **centrum** surrounding the notochord in addition to neural arches along the length of the vertebral column, ribs in the trunk region, and hemal arches in the caudal region (see Figure 3.9A). In adults, centra form bony rings attached to the neural and hemal arches, replacing the notochord completely, although a few bony vertebrates (such as sturgeons,

coelacanths, and lungfishes) do not form centra and retain the notochord as adults). Ribs are another novel feature of gnathostomes. They articulate with the centrum and neural arch and lie in connective tissue between the myomeres, providing anchorage for those muscles.

The evolutionary history of vertebral centra is complicated, and we are not certain if centra evolved once as a synapomorphy of eugnathostomes (as we show in Figure 3.9A), or whether they evolved independently in different groups of early eugnathostomes. Well-developed centra are not found in the earliest acanthodians and bony fishes, and early tetrapods also retained centra composed of several portions.

The pectoral girdle of gnathostomes consists of endochondral and dermal components (**Figure 3.9C**). Some endochondral elements remain cartilaginous, while others complete ossification and form bones such as the scapula, coracoid, and the three basal elements. The condition of having three basal elements in the pectoral fin is retained in chondrichthyans and actinopterygians, but modified to a single element in sarcopterygians (see Section 7.1).

What about soft anatomical features?

Because fossilized soft tissues are rare, we cannot know with certainty whether certain derived soft anatomical characters are unique to gnathostomes or whether they evolved earlier. But we can make some inferences, summarized below and in **Table 3.1**.

- Insulating sheaths of **myelin** surround motor nerves of all extant gnathostomes and increase the speed of transmission of nerve impulses. Cyclostomes lack myelin. We can infer that myelin sheaths were present in some basal gnathostomes such as †*Romundina* because motor nerves supplying eye muscles, preserved as tracts within the skull, were much longer than those in jawless osteognathostomes—too long to adequately transmit impulses without myelination.

- Gnathostomes have a **cerebellum**, a hindbrain structure that functions in motor coordination; the cerebellum is poorly developed in cyclostomes. Impressions on the inner surface of the dermal head shield reveal the presence of a cerebellum in early osteognathostomes.

- Gnathostomes have a small chamber anterior to the ventricle of the heart, known as the **conus arteriosus** (see Figure 3.9A) not found in cyclostomes. The conus arteriosus acts as an elastic reservoir that helps to smooth pulsatile blood flow produced by ventricular contractions.

- Extant gnathostomes have an acid-secreting stomach, but cyclostomes and nonvertebrate chordates do not. It is unclear whether a stomach was present in the common ancestor of cyclostomes and osteognathostomes. Some fossil †anaspids and †thelodonts preserve impressions interpreted as a stomach, and it is tempting to regard this as evidence of the evolution

Table 3.1 Comparison of features of cyclostomes and eugnathostomes

Cyclostomes	Eugnathostomes
Head and brain	
Branchial basket without articulations between cartilages	Articulations between skeletal elements of the gill arches
No jaws	Jaws
Rudimentary cerebellum	Cerebellum
One (hagfishes) or two (lampreys) semicircular canals	Three semicircular canals
No myelin	Myelin
Single median nostril (monorhiny)	Paired nostrils (diplorhiny)
Short forebrain	Elongated forebrain
Pharynx and respiration	
Gill arch skeleton lateral to gill tissue	Gill arch skeleton medial to gill tissue
Feeding and digestion	
Keratinous teeth	True teeth composed of dentine and enamel
No stomach	Acid-secreting stomach
Heart and circulation	
No conus arteriosus between ventricle and ventral aorta	Conus arteriosus between ventricle and ventral aorta
Osmoregulation	
Same osmolal concentration as seawater (hagfishes); osmoregulation (lampreys)	Osmoregulation
Support and locomotion	
No horizontal septum	Horizontal septum divides epaxial and hypaxial trunk muscles
Pre-anal fin (hagfishes); two dorsal fins (lampreys)	No pre-anal fin; two dorsal fins
No paired pectoral or pelvic fins	Paired pectoral and pelvic fins
Straight caudal fin	Heterocercal caudal fin
Periocular trunk muscles	No periocular trunk muscles
No cellular or perichondral bone	Cellular and perichondral bone
No centrum	Vertebrae have centrum with ribs in trunk region and hemal arches in caudal region
Reproduction	
No specialized reproductive ducts	Specialized reproductive ducts lead to the cloaca, a common passage for urogenital products and feces

of early feeding specializations within osteognathostomes. However, it is equally possible that cyclostomes lost the stomach during the evolution of their specialized diets.

- Gnathostome gonads have specialized ducts leading to the cloaca; cyclostomes lack equivalent ducts (see Section 2.6).

- Gnathostomes have three semicircular canals in the inner ear (anterior, posterior, and horizontal semicircular canals; see Figure 2.12). Lampreys have only two semicircular canals (interpreted as homologues of the anterior and posterior semicircular canals) and hagfishes only have one semicircular canal.

- Gnathostomes have paired nasal sacs and nostrils, a condition termed **diplorhiny**. Cyclostomes have a single median nostril and nasal sac, a condition termed **monorhiny**.

Together, these gnathostome features allowed evolution of new feeding behaviors, including the ability to grasp prey and exploit many food sources, more powerful swimming and sophisticated maneuvering, as well as much larger body sizes.

3.5 Origin of Jaws

LEARNING OBJECTIVES

3.5.1 Explain how extant cyclostome and gnathostome embryos provide information about the evolutionary origin of jaws.

3.5.2 Explain how the position of the developing nose relates to the evolution of diplorhiny and jaws.

3.5.3 Describe the ventilation hypothesis for the selective value of jaws.

No fossils document the origin of jaws, so we draw inferences about jaw evolution by studying jaw development in embryos of extant forms. Our knowledge is increasing as we learn more about gene expression during early development of both cyclostomes and gnathostomes.

Figure 3.10A shows the head of a pharyngula (the embryonic stage shown in Figure 2.7) and depicts eight idealized **developmental domains** that give rise to structures in the head. The premandibular domain is innervated by the first division of the trigeminal nerve, V_1. Hox genes are not expressed in the premandibular domain, nor does

this domain give rise to features associated with more posterior domains of the pharynx, such as branchiomeric muscles, aortic arches, or gill tissue. The mandibular arch domain, which also lacks Hox gene expression, gives rise to the upper and lower lips of lampreys and to the upper and lower jaws of gnathostomes; these structures are innervated by the second and third branches of the trigeminal nerve, V_2 and V_3. The spiracular pouch forms the spiracle of early gnathostomes such as sharks and

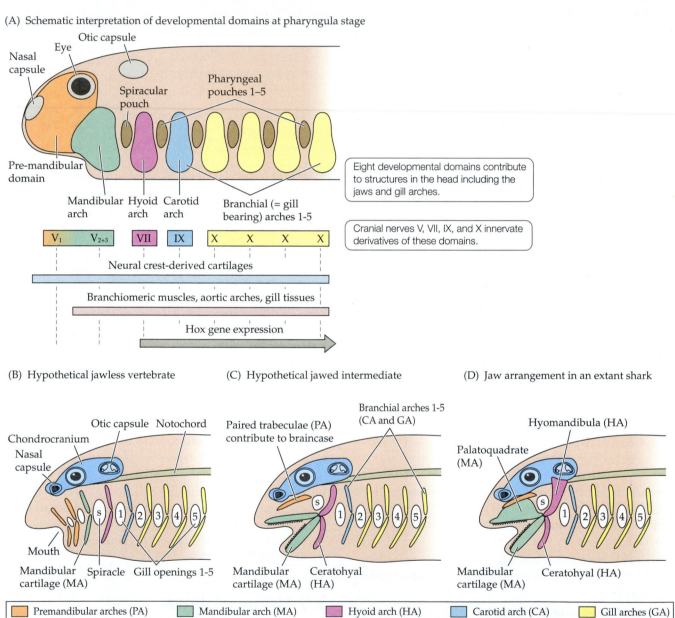

(A) Schematic interpretation of developmental domains at pharyngula stage

Eight developmental domains contribute to structures in the head including the jaws and gill arches.

Cranial nerves V, VII, IX, and X innervate derivatives of these domains.

(B) Hypothetical jawless vertebrate

(C) Hypothetical jawed intermediate

(D) Jaw arrangement in an extant shark

| Premandibular arches (PA) | Mandibular arch (MA) | Hyoid arch (HA) | Carotid arch (CA) | Gill arches (GA) |

Figure 3.10 Development and evolution of jaws.
(A) Idealized pharyngula showing eight domains of the premandibular region, mandibular arch, hyoid arch, and five gill-bearing arches separated by pouches. The premandibular domain forms cartilages and bones derived from neural crest, but it lacks segmentally organized muscles and blood vessels found in the mandibular, hyoid, and gill arches. The sensory nerve for the premandibular region is a branch of the trigeminal nerve, V_1; other portions of the trigeminal nerve supply the mandibular arch itself, V_2 and V_3. (B) Nineteenth century hypothesis of the ancestral jawless condition in which the mandibular arch bore functional gills and the spiracle was a large,

fully functional gill slit; there is no direct evidence that such a form ever existed. (C) Hypothetical intermediate form in which jaws are present but not closely associated with the braincase; the hyoid arch supports the jaws. Paired trabeculae, derived from premandibular arches, support an elongated braincase. (D) Idealized gnathostome condition based on an adult shark. The mandibular arch forms the jaws, the hyoid arch supports the jaws, and the spiracular pouch forms a spiracle with a small gill. (A after T. Miyashita. 2016. *Biol Rev* 91: 611–657. © 2015 Cambridge Philosophical Society; B–D after K. Liem et al. 2001. *Functional Anatomy of the Vertebrates*, 3rd ed. Cengage/Harcourt College: Belmont, CA.)

eustachian tube of mammals. Hox gene expression occurs in the hyoid arch domain, which gives rise to the skeleton and muscles of the hyoid arch, innervated by the facial nerve, VII. The segmental arrangement of structures that develop from the five posterior arch domains suggests that there once was a complete series of gill arches and gill slits, as shown in the schematic interpretation of hypothetical jawless vertebrate in **Figure 3.10B**.

Hypotheses of jaw origins

Gnathostome jaws develop from the mandibular arch, and the hyoid arch supports the jaws (**Figure 3.10C,D**). It was long thought that the mandibular arch bore gills in jawless ancestors, just like the more posterior gill-bearing arches, and that the spiracle between the mandibular and hyoid arches is a remnant of a once much larger gill slit (see Figure 3.10B). However, neither fossils nor embryos provide evidence that the mandibular arch ever supported gills or that the spiracle ever formed a large gill slit. Given the absence of Hox gene expression and trigeminal innervation of the anterior part of the head, how, when, and why did the mandibular arch evolve to form the jaws of gnathostomes?

Importance of the nose

A key difference between monorhiny in cyclostomes and diplorhiny in gnathostomes relates to the position of the olfactory system relative to the adenohypophysis—the anterior portion of the pituitary gland—a median structure in all vertebrates. The olfactory system and adenohypophysis develop in the premandibular domain of the head. The single median nasohypophyseal plate of cyclostomes gives rise to both the median nasal sac and the adenohypophysis (**Figure 3.11A**). In contrast, eugnathostomes have three separate placodes: one adenohypophyseal placode that gives rise to the median adenohypophysis, and paired nasal placodes that give rise to the nasal sacs (**Figure 3.11B**).

Anatomical differences in the olfactory system dictate size and shape of the head anterior to the mandibular arch as well as the size of the forebrain. In cyclostomes, the single nasal capsule (the part of the cranium surrounding the nasal sac) overlies the forebrain, so the forebrain is correspondingly short. In gnathostomes, however, the paired nasal capsules are positioned anterior to the adenohypophysis, leaving a space between the placodes into which mandibular neural crest can migrate to form the upper jaw. Correspondingly, the separation between the adenohypophysis and nasal capsules allows lengthening of the forebrain. To support the lengthened forebrain from beneath, a pair of cartilages called trabeculae contribute to the braincase (see Figure 3.10C,D).

The cyclostome condition of monorhiny thus precludes development of an upper jaw. We infer that monorhiny is the basal vertebrate condition because early osteognathostomes, such as †galeaspids, also had a single nostril that served paired nasal sacs (**Figure 3.12**). Evolution of the jaw became possible only after the change from a single

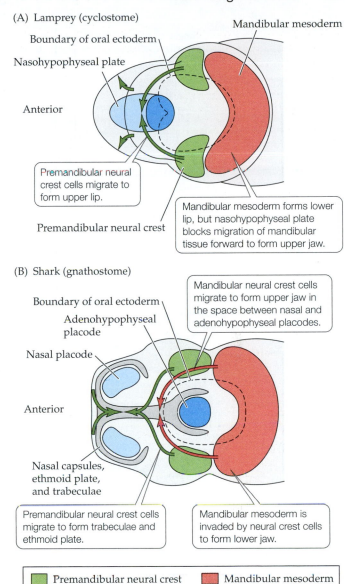

Figure 3.11 Development of premandibular region and mandibular arch in lampreys and gnathostomes. (A) Idealized ventral view of developing lamprey embryo. (B) Idealized ventral view of developing gnathostome embryo at a similar stage. During lamprey embryogenesis, the single median nasohypophyseal plate gives rise to both the nose and adenohypophysis. Premandibular mesoderm migrates beneath the nasohypophyseal plate to form the upper lip. In gnathostomes, the adenohypophysis develops from a separate adenohypophyseal placode and the paired nasal placodes are located anteriorly on the sides of the head. Neural crest cells from the mandibular arch migrate into the space between the adenohypophyseal placode and the nasal placodes to form the upper jaw. Premandibular neural crest cells grow forward to form the trabeculae and ethmoid plate. (After S. Kuratani. 2004. *J Anat* 205: 335–347. © 2004 Anatomical Society of Great Britain and Ireland.)

nasal sac to laterally displaced paired nasal sacs, a change made possible by new patterns of tissue interactions and gene expression.

Selective value of jaws

Ideas about evolutionary events that drove these profound shifts in head development and the evolution of

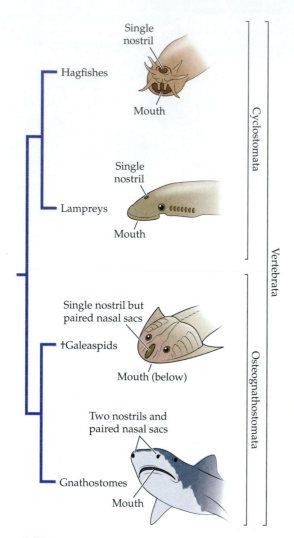

Figure 3.12 Number of nostrils differs in cyclostomes and gnathostomes. Cyclostomes have a single nostril and a single nasal sac above the forebrain. It is thought that this condition limited the possibility for jaw formation and expansion of the forebrain. Early jawless osteognathostomes such as the †galeaspid *Shuyu* also lacked an expanded forebrain but had paired nasal sacs (accessed through a single nostril) separated from the adenohypophysis (based on an open duct in fossils that has been interpreted as the hypophyseal duct). In gnathostomes, paired nasal sacs are separated from the adenohypophysis; this separation may have allowed evolution of jaws and an expanded forebrain. (After Z. Gai et al. 2011. *Nature* 476: 324–327.)

jaws remain speculative. We do not know whether any early osteognathostomes had a mandibular arch modified into cyclostome-like cartilages and velar bars, but the velum and velar pumping seen in cyclostomes are absent in extant gnathostomes. The presence of bony oral plates in some early osteognathostomes suggests the presence of premandibular cartilages (as in cyclostomes), and a toothed biting structure could have been formed from these. So why modify mandibular cartilages into a biting apparatus?

The **ventilation hypothesis** proposes that jaws would have allowed early gnathostomes to suck larger volumes of water into the oral cavity, close the mouth, and then forcefully pump water out over the gills, thus improving respiration and gas exchange. Like the double-pump system of gill ventilation in extant jawed fishes, the ability to open the mouth and expand the oral cavity also would have improved suction feeding.

Two differences yet to be explained are (1) how the gill arch skeleton shifted in position from lateral to the gills as in cyclostomes to medial to the gills in gnathostomes; and (2) why the gill arches of cyclostomes lack articulations between skeletal elements as in gnathostomes (see Figure 3.5).

3.6 Origin of Paired Appendages

LEARNING OBJECTIVES

3.6.1 Explain how development of trunk muscles differs between cyclostomes and gnathostomes.

3.6.2 Define lateral somitic frontier.

3.6.3 Differentiate median and paired fins with respect to locomotor functions such as control of roll, yaw, and pitch.

†Osteostracans had paired pectoral fins, but paired pelvic fins first appeared in early gnathostomes such as †antiarchs. Pectoral fins of †antiarchs were attached to the posterior corner of the skull by dermal bones, so the head and pectoral girdle could not move independently. (Not until the origin of tetrapods does a true mobile neck region between the pectoral girdle and head evolve.) The fin skeleton and fin girdles are components of the internal skeleton, although because the pectoral fin was originally connected to the skull, it retains close connections with dermal bones; our clavicle is an example. Pectoral and pelvic fins in extant chondrichthyans and basal bony vertebrates are **tribasic**, meaning that three main elements within the fin, known as basal cartilages or bones, articulate with the limb girdle (see Figure 3.9C). Developmental and molecular studies support anatomical interpretations that tribasic paired fins are the ancestral gnathostome condition.

Fin development and the lateral somitic frontier

Unsegmented **lateral plate mesoderm** (see Figure 2.6) has two distinct layers that define the coelom in early and intermediate developmental stages of vertebrate embryos (**Figure 3.13A,B**). The outer layer, adjacent to ectoderm, is the somatic layer of the lateral plate mesoderm; the inner layer, adjacent to endoderm of the gut, is the splanchnic layer. From this general condition early in embryogenesis, cyclostomes and gnathostomes exhibit different developmental paths related to two derived features of gnathostomes: (1) presence of paired fins; and (2) presence of a **horizontal septum** in the trunk of gnathostomes, dividing the trunk muscles into epaxial and hypaxial portions (see Figure 2.11).

Cyclostome trunk muscles develop from somites, and unsegmented lateral plate mesoderm contributes only to

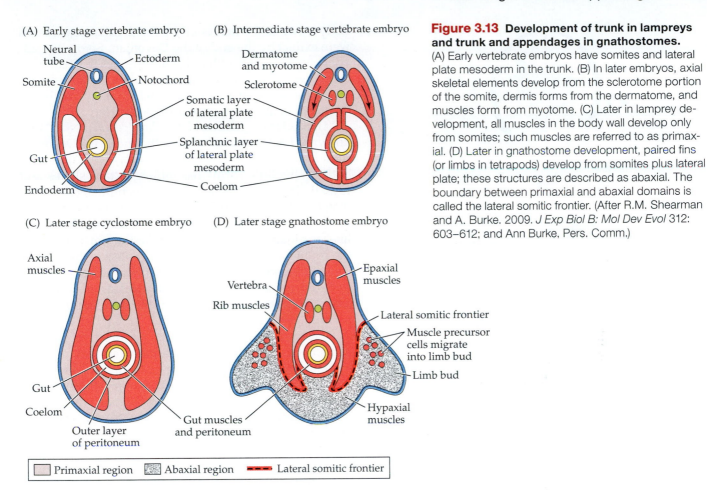

Figure 3.13 Development of trunk in lampreys and trunk and appendages in gnathostomes. (A) Early vertebrate embryos have somites and lateral plate mesoderm in the trunk. (B) In later embryos, axial skeletal elements develop from the sclerotome portion of the somite, dermis forms from the dermatome, and muscles form from myotome. (C) Later in lamprey development, all muscles in the body wall develop only from somites; such muscles are referred to as primaxial. (D) Later in gnathostome development, paired fins (or limbs in tetrapods) develop from somites plus lateral plate; these structures are described as abaxial. The boundary between primaxial and abaxial domains is called the lateral somitic frontier. (After R.M. Shearman and A. Burke. 2009. *J Exp Biol B: Mol Dev Evol* 312: 603–612; and Ann Burke, Pers. Comm.)

the peritoneal lining of the coelom and muscles of the gut (**Figure 3.13C**). The somatic layer of lateral plate mesoderm exists only in early development, and is eliminated as the myotome (the part of the somite from which muscles develop) expands ventrally, effectively separating cyclostome ectoderm from lateral plate mesoderm. This is reflected in the myomeres of hagfishes and lampreys: as shown in Figure 2.11, no horizontal septum is present.

Limb development in gnathostomes disrupts the cyclostome pattern because appendages (fins or limbs) develop as outgrowths of lateral plate mesoderm plus the adjacent ectoderm. As shown in **Figure 3.13D**, the gnathostome trunk consists of:

1. A primaxial region (somitic mesoderm only) that forms structures near the dorsal midline of the body, including the epaxial muscles.

2. An abaxial region (mixed somitic and lateral plate mesoderm), away from the dorsal midline of the body, that forms hypaxial muscles and components of the fins or limbs.

The boundary between the primaxial and abaxial regions is the **lateral somitic frontier.**

Loosely, you can think of the lateral somitic frontier as the horizontal septum that divides the epaxial and hypaxial musculature seen in the trunk of a shark or

bony fish (see Figure 2.11). However, the lateral somitic frontier is dynamic and not a simple dorsal–ventral division of the body, because it is deflected by the ventrally extending ribs and intercostal muscles that form from somites. Embryonic patterning genes appear to be regulated independently in the primaxial and abaxial regions, yet there are many structural similarities between fully developed primaxial fins (e.g., dorsal fins) and abaxial paired fins. We infer that the gene expression program for dorsal fins, which evolved early in vertebrate history, has been co-opted to control development of the paired fins of gnathostomes.

Advantages of fins

Guiding a body in three-dimensional space is complicated. Fins act as hydrofoils, applying pressure to the surrounding water. Because water is practically incompressible, force applied by a fin to the water produces a reaction force in the opposite direction. Reaction forces can be resolved into vectors, and the vector parallel to the body axis produces forward thrust. A caudal fin increases the area of the tail, allowing greater thrust and permitting rapid acceleration. Quick adjustments of body position are especially important for active predatory fishes, including Devonian gnathostomes. Median fins (dorsal, anal, and caudal) control **roll** (rotation around the body axis) and **yaw**

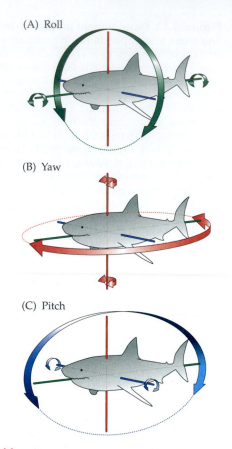

(A) Roll

(B) Yaw

(C) Pitch

Figure 3.14 Fins stabilize a fish in three dimensions.
Median fins (dorsal, anal, and caudal) control roll (A), which is
rotation around the body axis, and yaw (B), which is a swing to
the right or left. (C) Paired pectoral and pelvic fins can limit pitch,
which is vertical tilt.

(swings to the right or left; **Figure 3.14**). Paired pectoral
and pelvic fins can limit **pitch** (vertical tilt), act as brakes,
or become specialized to provide thrust, as in the enlarged
pectoral fins of skates and rays.

Fins have nonlocomotor functions as well. Spiny fins
are used in defense and, when combined with glandular
secretions, may become systems to inject venom. Colorful
fins send visual signals to potential mates, rivals, and
predators. In some fishes, such as guppies, fins also serve
as intromittent organs.

3.7 Extinct Paleozoic Jawed Fishes

LEARNING OBJECTIVES

3.7.1 Describe the structure, fossil record, and habitats of
"†placoderms."

3.7.2 Explain the evolutionary importance of †*Materpiscis*,
†*Qilinyu*, and †*Entelognathus*.

As seen in Figure 3.3, many successive diversifications
and extinctions of vertebrates occurred during the Phanerozoic. †Conodonts flourished in the Ordovician but
disappeared in the Permian. Jawless osteognathostomes
reached their greatest diversity in the late Silurian and

Early Devonian, whereas groups such as †osteostracans
were extinct by the end of the Devonian. Before considering extant groups of basal gnathostomes in Chapters 6
and 7, we turn to the fascinating Paleozoic forms commonly known as "†placoderms," which, as currently understood, do not form a monophyletic group.

"†Placoderms" (Greek, *plax*, "plate"; *derma*, "skin") are
known from the early Silurian to the end of the Devonian
from localities all over the world. Many were flattened,
bottom-dwelling forms resembling extant skates and
rays. Body sizes ranged from a few centimeters to more
than 6 m, the size of the extant white shark, *Carcharodon
carcharias*. Early "†placoderms" were marine, but some
lineages adapted to freshwater and estuarine habitats. As
the common name "†placoderm" implies, a thick, bony
shield covered the anterior one third of the body (**Figure
3.15**). These plates formed a head shield, as in more basal
osteognathostomes, and a trunk shield connected to the
pectoral girdle and fin.

Here we briefly treat †antiarchs, †arthrodires, †ptycto-
donts, and the so-called †maxillate placoderms exemplified
by †*Entelognathus* and †*Qilinyu* (see Figures 3.3 and 3.4).

†Antiarchs occur in many Devonian localities, including the Escuminac Bay formation in Quebec, where many
individuals of †*Bothriolepis* have been found. The large
plates of its trunk shield are shown in **Figure 3.15A**. Its
pectoral fin extends from the girdle as a series of jointed
segments covered by dermal bone.

†Arthrodires (**Figure 3.15B–D**) had a joint between the
head and trunk shields, enabling the head to be lifted
independently of the body; this feature is the source of
the group's name (Greek *arthron*, "hinge"; *deira*, "neck").
Some exceptionally preserved fossils show muscles spanning the joint; we infer that these were used to elevate
and depress the head. Underneath this armor, the internal
skeleton resembles that of extant cartilaginous fishes.

Fossils of the appropriately named †*Materpiscis* (Latin
mater, "mother"; *piscis*, "fish"), a †ptyctodontid from
the Late Devonian of Australia, preserve embryos and a
structure interpreted as an umbilical cord. This evidence
of viviparity matches the observation that both basal
and derived "†placoderms" had claspers resembling the
pelvic claspers extant male chondrichthyans use for internal fertilization. However, "†placoderm" claspers are
not homologous with those of extant chondrichthyans
because they were separate structures behind the pelvic
fins. We infer from evidence for viviparity and claspers
that "†placoderms" had internal fertilization and perhaps
complex reproductive behaviors similar to those of extant
chondrichthyans.

The so-called "†maxillate placoderms" are known from
the late Silurian of China. These forms had tooth-bearing
dermal bones around the mouth that have been homolo-
gized with those of gnathostomes. For example, †*Qilinyu*
had a maxilla and a premaxilla in its upper jaw, and
†*Entelognathus* had these bones as well as a dentary in
its lower jaw.

(A) An †antiarch, †*Bothriolepis*

© The Natural History Museum/Alamy Stock Photo

(B) An †arthrodire, †*Dunkelosteus*

James St. John/CC BY 2.0

(C) Jaws of an †arthrodire, †*Coccosteus*

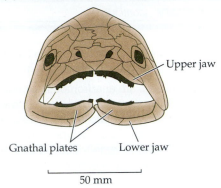

Upper jaw

Gnathal plates

Lower jaw

50 mm

50 mm

(D) Head and trunk shield of an †arthrodire, †*Coccosteus*

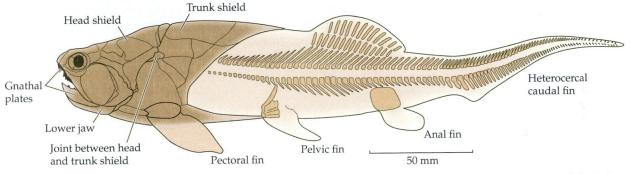

Head shield

Trunk shield

Gnathal plates

Lower jaw

Joint between head and trunk shield

Pectoral fin

Pelvic fin

Anal fin

Heterocercal caudal fin

50 mm

Figure 3.15 **Paleozoic "†Placoderms."** (A) Dorsal view of head and trunk shields of †*Bothriolepis* (Middle to Late Devonian), an †antiarch. (B) Fossilized head and trunk shields of †*Dunkleosteus* (Late Devonian), an 8-m-long †arthrodire closely related to eugnathostomes. (C) Jaws of †*Coccosteus* (Middle Devonian), an †arthrodire. (D) Reconstruction of head and trunk shield of †*Coccosteus*. (C,D after J.A. Moy-Thomas and R.S. Miles. 1971. *Paleozoic Fishes*, Chapman & Hall, London.)

Summary

3.1 Earliest Evidence of Vertebrates

Vertebrates are known from the start of the Paleozoic. Fossils of soft-bodied vertebrates occur in early Cambrian rocks. The earliest known vertebrates with bone, Osteognathostomata, are from the Late Ordovician.

†Conodonts have toothlike structures that are not homologous with mineralized tissues in osteognathostomes.

The earliest evidence of bone is in odontodes of the dermal head shield of osteognathostomes. Their mineralized tissues include acellular bone, dentine, and enameloid and may have functioned as mineral storage sites as well as armor.

Some early osteognathostomes had toothlike structures within the pharynx, but true teeth are a synapomorphy of gnathostomes.

Early vertebrates evolved in marine environments, although they rapidly radiated in freshwater habitats.

3.2 Cyclostomes: Extant Jawless Vertebrates

The extant jawless vertebrates, hagfishes and lampreys, are cyclostomes. They lack paired fins, jaws, and paired nostrils. However, they evolved complex head structures different from gnathostomes, including a cartilage that supports a muscular tongue that is not homologous with the tongue of gnathostomes.

Cyclostomes lack mineralized tissues but have keratinized teeth. Their vertebrae consist of rudiments interpreted as homologues of neural and hemal arches. The gill arch skeleton forms a branchial basket of unjointed cartilage that lies lateral to gill tissue.

Hagfishes are marine. Uniquely among extant vertebrates, their body fluids have the same osmotic concentration as seawater. Their sense organs are less specialized than those of other vertebrates, and they have a primitive kidney.

Hagfishes usually feed on dead or dying fishes, although they also can capture small prey. To deter predators, hagfishes secrete slime from a series of paired slime glands along the length of the body.

Hagfish skin is used to make "eelskin" leather products. Some populations have been overfished, in part because

(Continued)

Summary *(continued)*

we know so little about their life history that it has been difficult to develop management strategies.

Many lampreys are anadromous, breeding and spending larval life in freshwater but migrating to the ocean after they metamorphose into adults. Larval lampreys, known as ammocoetes, are filter feeders. Adults of some species are parasites on other vertebrates, usually bony fishes, sucking blood by affixing themselves to a host via their buccal funnel. While attached to a host, a lamprey uses tidal ventilation of its gills.

Lampreys have colonized the North American Great Lakes and reduced stocks of economically important fishes; control measures are controversial.

3.3 Jawless Osteognathostomes

Armored, jawless osteognathostomes radiated in the Ordovician. Ordovician osteognathostomes such as †arandaspids had head shields composed of dermal bone.

†Osteostracans share many features with gnathostomes, including a heterocercal caudal fin, two dorsal fins, paired pectoral fins, perichondral bone, cellular bone, and bones in the connective tissue surrounding the eyeball.

Many jawless osteognathostomes lived alongside gnathostomes in the Silurian and Devonian but disappeared in the Late Devonian extinctions that also claimed several lineages of gnathostomes and marine invertebrates.

3.4 Gnathostome Body Plan

Gnathostomes have many derived features in addition to jaws and two sets of tribasic paired fins. Duplications of Hox genes in the lineage leading to gnathostomes may be linked to their anatomical complexity, including:

- Articulated gill arches medial to gill tissue, which is ventilated using a double-pump mechanism

- Myelinated nerves

- Three semicircular canals

- Conus arteriosus in the heart

- Specialized ducts connecting gonads to the cloaca

These and other features allowed further increases in activity and complexity related to active predation.

3.5 Origin of Jaws

Jaws evolved from the mandibular arch. In contrast to older hypotheses about jaw origins, the mandibular arch may not have been part of the branchial series—that is, it never supported gill tissue. Because there are no intermediate fossil forms our inferences are based on developmental studies of extant cyclostomes and gnathostomes.

During early development, cyclostomes have a single median nasohypophyseal plate that gives rise to the single nasal sac and adenohypophysis. Extant gnathostomes have two nasal placodes, one on each side of the head, separated from the single median adenohypophyseal placode. This difference has been linked to the evolutionary origin of gnathostomes.

The transition from monorhiny (single nasal sac and nostril) to diplorhiny (paired nasal sacs and nostrils) can be seen in the fossil record. †Galeaspids represent a stage in which there is a single nostril but two nasal sacs. Basal gnathostomes show a second stage in which the nostrils are separate but not located anteriorly on the head. More derived "†placoderms" show a third stage, in which the nostrils are anterior, the forebrain is longer, and trabecular cartilages contribute to the braincase to support the forebrain.

A selective advantage of jaws is that a mouth that can be closed increases the force with which water can be pumped over the gills. The articulated skeleton of the gill arches medial to the gill tissue proper is also part of this system of improved gill ventilation of gnathostomes.

3.6 Origin of Paired Appendages

Paired fins have internal skeletal supports, and dermal bones contribute to the pectoral girdle.

†Osteostracans have paired pectoral fins and basal gnathostomes have paired pectoral and pelvic fins. Paired fins have a tribasic internal skeleton of cartilage or bone that articulates with limb girdles. The pectoral girdle was initially connected to dermal bones of the head shield, and dermal bone contributes to the pectoral girdle of more derived bony vertebrates (e.g., the human clavicle is a dermal bone).

Paired fins are important for maneuvering, braking, and in some species, propulsion; they also have non-locomotory functions, such as defense, communication, and reproduction.

Evolution of paired fins required changes in early development. All body muscles of cyclostomes derive from embryonic somites, which are segmented portions of the mesoderm. In gnathostomes, the somatic layer of the lateral plate mesoderm contributes to fin and hypaxial muscles. The lateral somitic frontier marks the division between structures derived from somites (primaxial) and structures derived from a mixture of somites and lateral plate mesoderm (abaxial). This can be most easily visualized by comparing the myomeres of cyclostomes and gnathostomes and noting how the horizontal septum divides gnathostome myomeres into epaxial and hypaxial portions.

Summary (continued)

3.7 Extinct Paleozoic Jawed Fishes

"†Placoderms" occur in rocks from the early Silurian through the end of the Devonian. They were structurally diverse and do not form a monophyletic group. Some had a unique hinge between the head and trunk shields, and some had structures interpreted as claspers for internal fertilization. Some fossils preserve evidence of developing young in the body cavity, supporting the inference that at least some basal gnathostomes were viviparous.

†"Maxillate placoderms" such as †*Entelognathus* had tooth-bearing dermal bones surrounding the mouth, homologous with the premaxilla, maxilla, and dentary bones of osteichthyans (including humans).

Discussion Questions

3.1 Did jawed vertebrates outcompete the jawless ones? Does the fossil record provide any evidence?

3.2 How did the realization that †conodonts were vertebrates change our ideas about the pattern of vertebrate evolution?

3.3 Cyclostomes are often thought of as "primitive" because they lack jaws. In what ways are their heads specialized and different from those of gnathostomes?

3.4 Ventilating gills by taking water in through the gill openings instead of the mouth is not very efficient. Why, then, does a parasitic adult lamprey ventilate its gills this way?

3.5 We usually think of biting as the original function of jaws. What might have been a different original use of jaws, and what is the evidence for this?

3.6 A recently discovered †ptyctodont placoderm has a developing embryo inside its body. In the absence of such direct evidence, would we have been able to speculate that placoderms had internal fertilization?

4

Living in Water

School of eeltail catfish, *Plotosus lineatus*

4.1 **Aquatic Environments**

4.2 **Sensory World of Aquatic Vertebrates**

4.3 **Maintaining an Internal Environment**

4.4 **Osmoregulation in Different Environments**

Although life evolved in water and the earliest vertebrates were aquatic, the physical properties of water create some difficulties for animals. To live successfully in water, a vertebrate must adjust its buoyancy to remain at a specific depth and must force its way through a dense medium to pursue prey or escape predators. Heat flows rapidly between an animal and surrounding water, and it is difficult for an aquatic vertebrate to maintain a body temperature different from water temperature. This phenomenon was dramatically illustrated when the *Titanic* sank in the frigid waters of the North Atlantic (estimated to be about −2°C). Most victims died from extreme hypothermia rather than by drowning. (Hypothermia is defined as a decrease in body temperature from 37°C to below 35°C; death occurs when body temperature drops below 28°C.)

Ions and water molecules move readily between water and an animal's internal body fluids, so maintaining a stable internal environment can be difficult. On the plus side, CO_2 and ammonia are extremely soluble in water. Consequently, CO_2 can easily be lost through the gills or skin to the surrounding water, making disposal of nitrogenous (nitrogen-containing) waste products such as ammonia much easier in aquatic environments than on land. Oxygen concentration, however, is much lower in water than it is in air, and water is denser than air, so more energy is needed to move water across a respiratory surface.

In this chapter, we consider how aquatic vertebrates evolved solutions to some of the physical challenges of life in water. Our focus is fishes and lissamphibians, but to make general points about differences in sensory systems in water and on land, we also make some comparisons to tetrapods such as cetaceans that secondarily returned to water. Evolutionary diversification of teleosts resulted in an enormous array of sizes and ways of life in freshwater and marine environments. Aquatic lissamphibians include some highly specialized forms such as the crab-eating frog that experiences twice-daily changes in salinity.

4.1 Aquatic Environments

LEARNING OBJECTIVES

4.1.1 Compare fresh water and air with respect to density, heat capacity, oxygen content, and oxygen diffusion rate.

4.1.2 Describe the structure of teleost gills.

4.1.3 Define countercurrent flow.

4.1.4 Explain when lungs evolved and why they might have been advantageous.

4.1.5 Compare buoyancy regulation in actinopterygians and chondrichthyans.

Water covers 73% of Earth's surface. Most water is held in ocean basins, which are populated everywhere except the deepest ocean trenches by vertebrates. Freshwater lakes and rivers hold a negligible amount of water—about 0.01%, much less than the amount of water tied up in the atmosphere, ice, and groundwater—but freshwater habitats are exceedingly rich biologically; nearly 40% of the more than 33,000 species of ray-finned fishes (Actinopterygii) live in fresh water.

Both water and air are fluid at biologically relevant temperatures and pressures, but their different physical properties make them dramatically different environments in which to move (**Table 4.1**). In air, for example, gravity is an important force acting on a moving animal, but fluid resistance to movement (i.e., air resistance) is trivial for all but the fastest birds. In water, the opposite relationship holds: gravity is negligible, but fluid resistance to movement is a factor with which all but the smallest aquatic vertebrates must contend. Because the physical properties of water constrain the evolution of

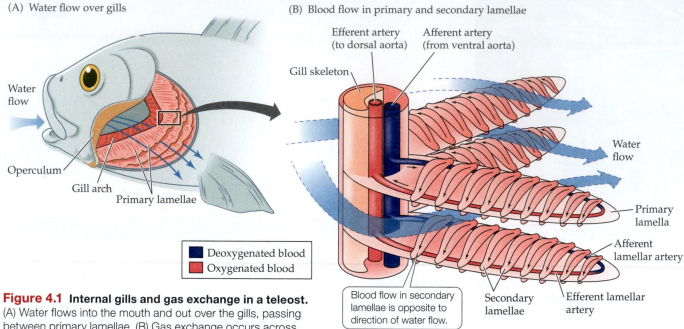

(A) Water flow over gills

(B) Blood flow in primary and secondary lamellae

Efferent artery (to dorsal aorta)
Afferent artery (from ventral aorta)
Gill skeleton
Water flow
Operculum
Gill arch
Primary lamellae

■ Deoxygenated blood
■ Oxygenated blood

Water flow
Primary lamella
Afferent lamellar artery

Blood flow in secondary lamellae is opposite to direction of water flow.
Secondary lamellae
Efferent lamellar artery

Figure 4.1 **Internal gills and gas exchange in a teleost.** (A) Water flows into the mouth and out over the gills, passing between primary lamellae. (B) Gas exchange occurs across the surfaces of microscopic secondary lamellae. (A after D.W. Townsend. 2012. *Oceanography and Marine Biology*, Oxford University Press/Sinauer: Sunderland, MA.)

some body shapes, most aquatic vertebrates are streamlined as adults. Each major clade of aquatic vertebrates met the various environmental challenges in different but related ways. Here we consider breathing and regulating buoyancy in water.

Obtaining oxygen from water using gills

Most aquatic vertebrates have external or internal **gills**: vascularized structures where oxygen, carbon dioxide, and other substances are exchanged between the body and surrounding water. Some larval fishes and larval and adult lissamphibians have external gills, which are filamentous processes attached to the sides of the head. Internal gills consist of **primary lamellae** (sometimes known as **gill filaments**) attached to the walls of gill pouches in hagfishes and lampreys and to gill arches in other fishes. We start our coverage with teleost gills.

In teleosts, the gills separate the buccopharyngeal cavity (= the oral cavity and pharynx) from the opercular cavity. On each side of the head, the opercular cavity is covered by an **operculum** (plural, *opercula*) composed of bones and associated soft tissues (**Figure 4.1A**). Unlike the tidal (in–out) airflow that ventilates the lungs of mammals, flow of water across fish gills is unidirectional—in through the mouth, over the gills, and out through the opercular opening. Flaps just inside the mouth and at the margins of the opercula of teleosts act as valves to prevent backflow. Two columns of primary lamellae extend from each gill arch. As water leaves the buccopharyngeal cavity, it passes over primary lamellae. Gas exchange takes place at the **secondary lamellae**, microscopic projections from the primary lamellae (**Figure 4.1.B**).

The pumping action of the mouth and pharyngeal region, called **buccopharyngeal pumping**, creates a positive pressure across the gills so that the respiratory current is only slightly interrupted during each pumping

Table 4.1 Comparison of physical properties of fresh water and air at 20°C

Property[a]	Comparison
Density	Water is ~833 times more dense than air
Dynamic viscosity	Water is ~55 times more viscous than air
Heat capacity	It takes ~3,500 times more heat to raise a unit volume of water 1°C than it does air
Heat conductivity	Heat moves through water ~25 times faster than through air
Oxygen content	A unit volume of air has ~35 times as much oxygen as a unit volume of water
Oxygen diffusion rate	Oxygen diffuses nearly 8,500 times faster in air than in water
Velocity of sound	Sounds travels 4.3 times faster in water than in air
Refractive index	Refractive index of water = 1.33 (nearly the same as the cornea of the eye). Refractive index of air = 1.0

[a]Most of these properties change with temperature and atmospheric pressure, and some are affected by presence of solutes.

cycle. Some filter-feeding fishes and many pelagic fishes—including certain sharks, mackerel, tunas, and swordfishes—have reduced or even lost the ability to pump water across the gills. They create a respiratory current by swimming with their mouth open, a method known as **ram ventilation**, and they must swim continuously. Many other fishes rely on buccopharyngeal pumping when at rest and switch to ram ventilation when swimming.

The arrangement of blood vessels in gills maximizes oxygen exchange. Each primary lamella has two vessels: an afferent lamellar artery running from the gill arch to the tip, and an efferent lamellar artery returning blood to the arch (**Figure 4.2A**). Each secondary lamella connects the afferent and efferent lamellar arteries. The direction of blood flow through the secondary lamellae is opposite to the direction of water flow. This arrangement, known as **countercurrent exchange**, assures that as much oxygen as possible diffuses from water into blood (**Figure 4.2B,C**).

Obtaining oxygen from air using lungs and other respiratory structures

Although most fishes depend on gills to extract oxygen from water, fishes that live in water with low oxygen levels cannot obtain enough oxygen via gills alone. These fishes obtain additional oxygen either via lungs or accessory respiratory structures known as accessory air-breathing organs (ABOs).

Accessory ABOs used to take up oxygen from air include enlarged lips that can be extended just above the water surface and a variety of internal structures into which air is gulped. Labyrinth fishes of tropical Asia (including many species featured in pet stores, such as Siamese fighting fish and gouramis, **Figure 4.3A**) have a vascularized labyrinth organ in the posterodorsal part of the opercular chamber. Air is sucked into the mouth and transferred to the chamber surrounding the labyrinth organ, where gas exchange takes place. The walking catfish, *Clarias batrachus* (**Figure 4.3B**), a highly invasive species that can move over land between ponds, extracts oxygen from air using arborescent organs in a posterior extension of the opercular chamber. Bronze corydoras catfishes, *Corydoras aeneus* (**Figure 4.3C**), often kept in home aquariums, are intestinal breathers: they dash to the surface, inhale a bubble of air, and swallow it. Gas exchange takes place in the posterior portion of the intestine and gas is expelled out the anus. Corydoras catfishes and many other air-breathing fishes facultatively switch oxygen uptake from their gills to accessory respiratory structures when oxygen in the water decreases. Others, like electric "eels" (*Electrophorus*, which are not eels but rather South American knifefishes; see Section 4.2) are obligatory air breathers. The gills alone cannot meet their respiratory needs, even if the water is saturated with oxygen, and they drown if they cannot reach the surface to breathe air.

(A) Water flow in relation to blood flow

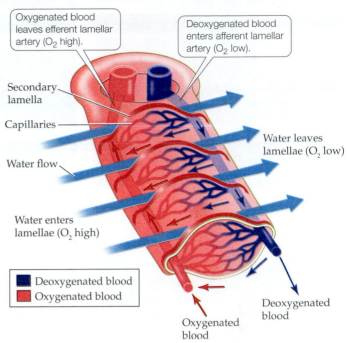

Oxygenated blood leaves efferent lamellar artery (O_2 high).

Deoxygenated blood enters afferent lamellar artery (O_2 low).

Secondary lamella

Capillaries

Water flow

Water leaves lamellae (O_2 low)

Water enters lamellae (O_2 high)

■ Deoxygenated blood
■ Oxygenated blood

Oxygenated blood

Deoxygenated blood

(B) With countercurrent flow

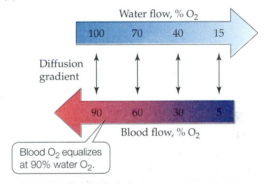

Water flow, % O_2

| 100 | 70 | 40 | 15 |

Diffusion gradient

| 90 | 60 | 30 | 5 |

Blood flow, % O_2

Blood O_2 equalizes at 90% water O_2.

(C) With concurrent flow

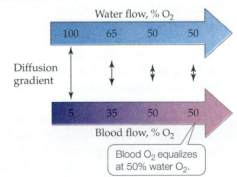

Water flow, % O_2

| 100 | 65 | 50 | 50 |

Diffusion gradient

| 5 | 35 | 50 | 50 |

Blood flow, % O_2

Blood O_2 equalizes at 50% water O_2.

Figure 4.2 Countercurrent exchange in gills of a teleost. (A) Direction of water flow across the gills is opposite to the flow of blood through secondary lamellae. (B) Countercurrent flow maintains a difference in oxygen concentration (a diffusion gradient) between the water and the blood along the full length of a secondary lamella, resulting in a high oxygen concentration (90%) in blood leaving the gills. (C) Were water and blood to flow in the same direction (concurrent flow), the diffusion gradient would initially be high but would decrease as the concentration of oxygen equalizes at 50% in both water and the blood. (A after D.W. Townsend. 2012. *Oceanography and Marine Biology*, Oxford University Press/Sinauer: Sunderland, MA.)

(A) Dwarf gourami, *Colisa lalia*

Courtesy of Quatermass

(B) Walking catfish, *Clarias batrachus*

© Svietlieisha Olena/Shutterstock.com

Figure 4.3 Teleosts with accessory air-breathing organs. (A) Gouramis and their relatives have a specialized labyrinth organ in the posterodorsal portion of the opercular chamber. (B) The walking catfish has arborescent organs in the posterior portion of its opercular chamber, which evolved independently from the labyrinth organ of gouramis. (C) Catfishes in the genus *Corydoras* are intestinal breathers.

(C) Bronze corydoras, *Corydoras aeneus*

Courtesy of Quatermass

We think of lungs as respiratory structures used by terrestrial vertebrates, as indeed they are, but lungs first evolved in fishes and preceded the origin of tetrapods by millions of years. Lungs develop embryonically as outpocketings (evaginations) of the pharyngeal region, originating from either its ventral or dorsal surface. Lungs of bichirs (basal actinopterygian fishes from Africa), lungfishes, and tetrapods develop from the ventral surface of the pharynx, whereas lungs and gas bladders of gars and teleosts (derived actinopterygians) develop from the dorsal surface. Gas exchange organs require large surface areas, and many have ridges or pockets in their walls. Such alveolar lungs occur in gars, lungfishes, and tetrapods, but bichirs have non-alveolar lungs.

For many years it was assumed that lungs evolved in fishes living in stagnant, oxygen-depleted water, where gulping oxygen-rich air would supplement oxygen uptake by the gills. However, although some lungfishes live in stagnant, anoxic environments, other air-breathing fishes such as the bowfin (*Amia calva*) live in oxygen-rich waters. An alternative explanation is that lungs first evolved in active fishes that lived in well-oxygenated waters, and that the additional oxygen from lung ventilation supplied heart muscle rather than general tissues of the body.

Adjusting buoyancy

Holding a bubble of air inside the body changes the buoyancy of an aquatic vertebrate, and bony fishes use lungs and gas bladders to regulate their position in the water. Air-breathing aquatic tetrapods (whales, dolphins, seals, and penguins, for example) can adjust their buoyancy by altering the volume of air in their lungs when they dive.

Actinopterygians Many actinopterygians are neutrally buoyant (i.e., they have the same density as water),

meaning they do not have to expend energy swimming to maintain their vertical position in the water column. The only movements they make when at rest is backpedalling with the pectoral fins to counteract the forward thrust produced by water as it is ejected from the gills, and a gentle undulation of the caudal fin to keep them level in the water. Fishes capable of hovering in the water like this usually have a well-developed **gas bladder** (sometimes called a swim bladder), a sac containing gas that helps regulate buoyancy.

Gas bladders lie in the body cavity just beneath the vertebral column, occupying about 5% of body volume in marine teleosts and 7% in freshwater teleosts (see Figure 4.4). The difference in volume corresponds to different water densities: because saltwater is denser than fresh water, it provides greater buoyancy, so a smaller gas bladder is sufficient. Guanine crystals in the wall of the gas bladder make it impermeable to oxygen and give it a silvery appearance.

Neutral buoyancy produced by a gas bladder works as long as a fish remains at one depth, but if it swims vertically up or down, the surrounding water exerts hydrostatic pressure on the bladder, which in turn changes the bladder's volume. Water pressure increases as a fish dives deeper, compressing the gas bladder and reducing buoyancy. When the fish swims toward the surface, water pressure decreases, the gas bladder expands, and the fish becomes more buoyant. Thus, to maintain neutral buoyancy, a fish adjusts the volume of gas in its bladder as it changes depth.

Most actinopterygians regulate volume of the gas bladder by adding gas when they swim down and removing gas when they swim up. Basal teleosts such as eels, herrings, anchovies, minnows, and salmon retain a connection—the pneumatic duct—between the gut and gas bladder (**Figure 4.4A**). These fishes are called **physostomous** (Greek *physeter*, "bellows"; *stoma*, "mouth"). Because they have a connection between the gut and the gas bladder, they can gulp air at the surface to fill the bladder and burp gas out to reduce bladder volume.

Adult teleosts from more derived clades lack a pneumatic duct, a condition called **physoclistous** (Greek *cleistos*, "closed"). Physoclistous teleosts increase volume of the gas bladder by secreting gas from the blood into the bladder and decrease volume by absorbing gas from the bladder and releasing it at the gills. Physoclists have a gas gland located in the anterior ventral floor of the gas bladder (**Figure 4.4B**). Underlying the gas gland is an area with many capillaries arranged to give countercurrent flow of blood entering and leaving the area. This structure, known as a **rete mirabile** (Latin, "wonderful net"), moves oxygen from the blood into the gas bladder. It is remarkably effective at doing this, even when the pressure of oxygen in the bladder is many times higher than its pressure in blood. For example, a fish living at a depth of 1,000 m (a typical depth at the edges of a continental shelf) experiences surrounding pressures of 100 atmospheres (1 atmosphere per 10 m). This means the fish must build up the partial pressure of oxygen in the gas gland to more than 100 atmospheres in order to cause gas to diffuse into the gas bladder.

The gas gland secretes oxygen by releasing lactic acid and carbon dioxide, which acidify blood in the rete mirabile. Acidification causes hemoglobin to release oxygen into solution. The Bohr effect reduces hemoglobin's affinity for oxygen when pH is low (**Figure 4.4C**). The Root effect, a property of teleost hemoglobin, not only reduces hemoglobin's affinity for oxygen at low pH, but also reduces the maximum amount of oxygen that hemoglobin can bind (**Figure 4.4D**). Because of the anatomy of the rete mirabile, which folds back on itself in a countercurrent multiplier arrangement, oxygen released from the hemoglobin accumulates and is retained within the gland until its pressure exceeds the oxygen pressure in the gas bladder. At this point, oxygen diffuses into the gas bladder, increasing its volume.

Physoclistous fishes have no connection between the gas bladder and gut, so they cannot burp to release excess gas from the bladder. Instead, physoclists open a sphincter muscle to allow gas to enter a region called the ovale, located in the posterior dorsal region of the gas bladder (see Figure 4.4B). The ovale contains a capillary bed, and the high internal pressure of oxygen in the gas bladder causes oxygen to diffuse into the blood when the ovale sphincter is opened.

Gas bladder adaptations and loss   Many deep-sea teleosts have deposits of lightweight oil or fat in the gas bladder, and others have reduced or lost the gas bladder entirely and have lipids distributed throughout the body. These lipids provide static lift. Because a smaller volume of the bladder contains gas, the amount of secretion required for a given change in depth is smaller. Teleosts that migrate over large vertical distances depend more on lipids, such as wax esters, than on gas for buoyancy, whereas close relatives that do not undertake extensive vertical migrations depend more on gas. Finally, fishes like tunas that swim fast and change depth quickly have lost the gas bladder, as have many other teleosts that live exclusively either near the surface or on the bottom.

Cartilaginous fishes   Sharks, rays, and chimaeras do not have a gas bladder. Instead, midwater chondrichthyans use the liver to create neutral buoyancy. A shark's liver has

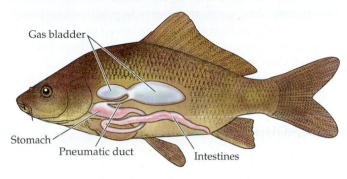

(A) Gas bladder and pneumatic duct in a physostomous fish

Gas bladder

Stomach

Pneumatic duct

Intestines

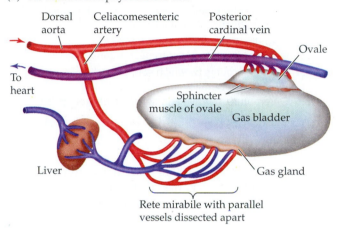

(B) Gas bladder in a physoclistous fish

Dorsal aorta   Celiacomesenteric artery   Posterior cardinal vein

Ovale

To heart

Sphincter muscle of ovale

Gas bladder

Liver

Gas gland

Rete mirabile with parallel vessels dissected apart

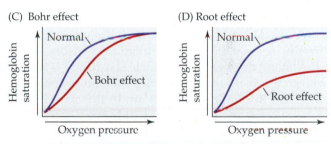

(C) Bohr effect

Hemoglobin saturation

Normal

Bohr effect

Oxygen pressure

(D) Root effect

Hemoglobin saturation

Normal

Root effect

Oxygen pressure

Figure 4.4 Gas bladders of teleosts.   (A) In a physostomous fish, the gas bladder retains a connection to the gut via the pneumatic duct. Physostomous fishes gulp air into the gas bladder and burp air out to adjust the bladder's volume. (B) The gas bladder of adult physoclistous fishes lacks a connection to the gut; oxygen is added via the gas gland and removed via the ovale. (C) At low pH, the Bohr effect increases oxygen delivery to tissues by about 25%. (D) At low pH, the Root effect increases oxygen delivery to tissues by about 50%. (A after K. Liem et al. 2001. *Functional Anatomy of the Vertebrates*, 3rd ed. Cengage/Harcourt College: Belmont, CA.)

a high oil content and is lighter than water (0.95 g/mL), and the liver can make up as much as 25% of total body weight. Not surprisingly, the liver cells of bottom-dwelling sharks such as the nurse shark, *Ginglymostoma cirratum*, have fewer and smaller oil vacuoles, and these sharks are negatively buoyant. The sand tiger shark, *Carcharias taurus*, gulps air into its stomach and has a correspondingly smaller liver than other sharks of its size.

In chondrichthyans, nitrogen-containing compounds contribute to buoyancy because urea and trimethylamine oxide (TMAO) in the blood and muscle tissue are less dense than an equal volume of water. Chloride ions are also lighter than water and provide positive buoyancy, whereas sodium ions and protein molecules are denser than water. The net effect of these solutes is to provide positive buoyancy for chondrichthyans.

4.2 Sensory World of Aquatic Vertebrates

LEARNING OBJECTIVES

4.2.1 Compare how images are focused on the retinas of fishes and terrestrial mammals.

4.2.2 State whether fishes and terrestrial vertebrates couple olfaction and respiration.

4.2.3 Describe the function of the lateral line system and the structure of neuromast organs.

4.2.4 List three routes by which sound reaches the inner ear of fishes.

4.2.5 Differentiate passive and active electroreception.

Water's properties (see Table 4.1) influence the behaviors of fishes and other aquatic vertebrates. Light is absorbed by water molecules and scattered by suspended particles. Objects more than 100 m distant are invisible even in the clearest water, whereas distance vision is virtually unlimited in clear air. Fishes supplement vision with other senses, some of which can operate only in water. The most important of these aquatic senses is mechanical and consists of detecting water flow via the lateral line system. Because water is dense and viscous, even small currents in the water can stimulate the sensory organs of the lateral line system. Electrical sensitivity is another important sensory mode that depends on properties of water and does not operate in air. In this case it is the electrical conductivity of water that is key.

Vision

Vertebrates generally have well-developed eyes, but the way an image is focused on the retina differs in terrestrial and aquatic animals. Air has an index of refraction of 1.00, and light rays bend as they pass through a boundary between air and a medium with a different refractive index. The amount of bending is proportional to the difference in indices of refraction, and corneas of the eyes of both terrestrial and aquatic vertebrates have an index of refraction of about 1.37. Water has a refractive index of 1.33, and the bending of light as it passes between air and water causes underwater objects to appear closer than they really are.

The difference in refractive indices of air and the corneas of vertebrate eyes means light is bent as it passes through the air–cornea interface, and the cornea plays a substantial role in focusing an image on the retina of a terrestrial vertebrate. This relationship does not hold in water, however, because the refractive index of the cornea is too close to that of water for the cornea to have much light-bending effect. That is why your vision is blurred underwater; a diving mask restores the air–cornea interface and allows you to see more clearly.

The lens plays the major role in focusing light on the retina of an aquatic vertebrate, and fishes have spherical lenses with high refractive indices. The entire lens is moved toward or away from the retina to focus images of objects at different distances from the fish. Terrestrial vertebrates have flatter lenses, and in birds and mammals, muscles in the eye change the shape of the lens to focus images, rather than moving the entire lens forward and backward as in fishes. Flatter lenses work for terrestrial vertebrates because the cornea contributes to bending light at the air–cornea interface. Aquatic mammals such as whales have spherical lenses like those of fishes.

(A) Olfactory system of sea lamprey, *Petromyzon marinus*

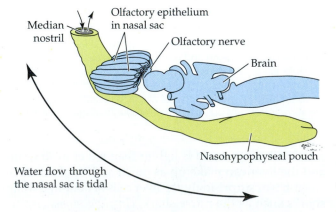

(B) Olfactory system on left side of American eel, *Anguilla rostrata*

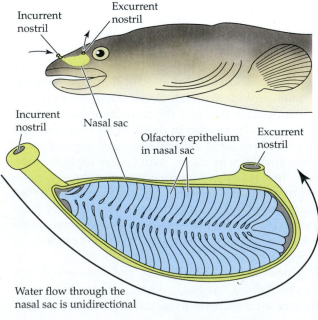

Figure 4.5 Olfactory systems of fishes. (A) Olfactory system of sea lamprey, *Petromyzon marinus*, illustrating tidal water flow over the olfactory epithelium powered by contractions of the nasohypophyseal pouch. (B) Lateral view of the head of American eel, *Anguilla rostrata*, showing incurrent and excurrent nostrils and one-way flow of water over the olfactory epithelium. (After K. Liem et al. 2001. *Functional Anatomy of the Vertebrates*, 3rd ed. Cengage/Harcourt College: Belmont, CA.)

Chemosensation: Olfaction and taste

The olfactory organs of fishes detect chemicals dissolved in water. Although land vertebrates couple olfaction and respiration (see Section 11.6), most fishes do not couple these two functions. Recall from Chapter 3 that sea lampreys have a single median nostril on top of the head that leads to the nasal sac and ends blindly in the nasohypophyseal pouch. The nasohypophyseal pouch moves water in and out tidally, irrigating the nasal sac where odorants are detected (**Figure 4.5A**). Jawed fishes such as eels have paired nasal sacs. Associated with each nasal sac is an anterior (incurrent) and posterior (excurrent) nostril. Thus, on each side of the head, water flows into the incurrent nostril, over the olfactory epithelium in the nasal sac, and out through the excurrent nostril (**Figure 4.5B**). In both lampreys and eels, water drawn in for olfaction does not pass over the gills.

Olfaction is well developed in some fishes, such as sharks and salmon that can detect odors at concentrations of less than 1 part per billion. Sharks compare the time of arrival of an odor stimulus on the left and right sides of the head to locate the odor source. Homeward-migrating salmon locate their natal stream from astonishing distances using the stream's chemical signature that was permanently imprinted when they were juveniles. Plugging the nostrils of salmon destroys their ability to home.

Fishes have taste buds in the mouth and around the head and anterior fins. Some, such as catfishes, have taste buds in the skin over the entire body, with the highest concentrations in the barbels surrounding the mouth.

Detecting water displacement

In fishes and aquatic lissamphibians, clusters of sensory hair cells form **neuromast organs** (see Section 2.6 for a description of hair cells). Neuromast organs are distributed in different configurations: within canals below the skin, exposed in epidermal depressions, or on the skin surface. Many fishes have multiple arrangements. Gnathostomes have a series of canals on the head (cranial or cephalic canals) and a trunk canal (or canals) that pass along each side of the body to the tail (**Figure 4.6A,B**). This receptor system is the **lateral line system**.

Lateral line systems are found only in aquatic vertebrates because air is not dense enough to stimulate neuromast organs. Lissamphibian larvae have lateral line systems, and permanently aquatic lissamphibians retain lateral lines throughout life. Terrestrial lissamphibians lose their lateral line system when they metamorphose

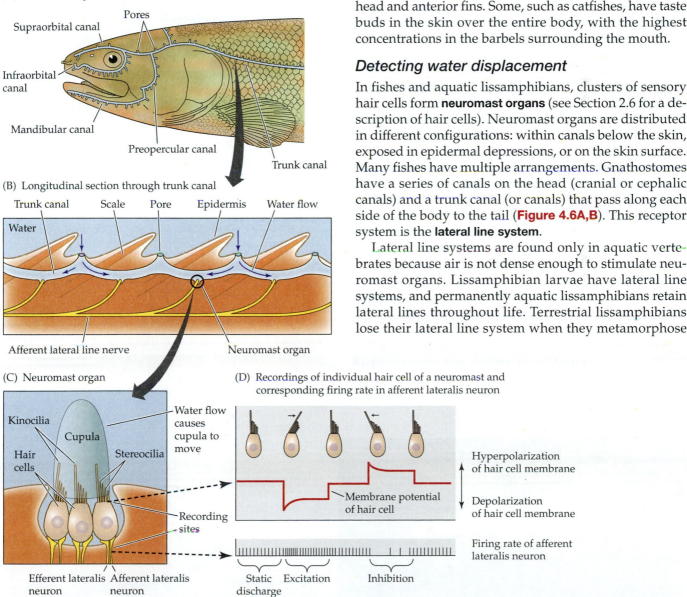

Figure 4.6 Lateral line systems. (A) Neuromast organs of bony fishes, such as the bowfin, *Amia calva*, are in lateral line canals on the head and in a canal that runs down each side of the trunk. The cranial lateral line system includes supraorbital, infraorbital, preopercular, and mandibular canals. (B) Canals open to the surface via pores, and water movements around the fish create water flow within the canals that can be sensed by neuromast organs.

(C) Each hair cell of a neuromast organ has a kinocilium placed asymmetrically in a cluster of stereocilia, all of which are embedded in a gelatinous cupula. As water flows in the canals, it displaces the cupula, bending the kinocilia. (D) Bending a kinocilium in one direction increases a hair cell's firing rate (excitation) and bending in the other direction decreases firing rate (inhibition). (See credit details at the end of this chapter.)

into adults. Amniotes do not have lateral line systems, having lost the neuromast organs, and lack the nerves serving the lateral line system and the brain regions that process lateral line information in non-amniotes. Terrestrial amniotes that secondarily returned to water, such as sea snakes, sea turtles, and whales, do not have lateral line systems.

Hair cells of neuromast organs have a kinocilium (a single, long, modified cilium) placed asymmetrically in a cluster of stereocilia (modified microvilli that are shorter than the kinocilium; **Figure 4.6C**). The kinocilium and stereocilia are embedded in a gelatinous substance, the **cupula** (Latin, "small tub"). Hair cells of neuromast organs generate a constant base rate of nerve impulses (static discharge; **Figure 4.6D**). Displacement of the cupula by water movement changes the base rate: bending the cupula toward the kinocilium increases the rate of impulses (excitation), whereas bending the cupula away from the kinocilium decreases the rate (inhibition). Afferent lateralis neurons carry impulses from hair cells to the brain. Efferent lateralis neurons extend from the brain to the hair cells, where they can modulate activity to suppress hair cell responses to the animal's own movements.

Within a lateral line canal, hair cells are arranged in sets. In one set, kinocilia might be located in the forward-facing part of a hair cell (i.e., on the side closest to the head of the fish), whereas in the other set the kinocilia might be located in the posterior part of a hair cell (i.e., on the side closest to the tail). This arrangement enables a fish to detect directional displacement of kinocilia, and from this sense the direction of water flow (from head to tail, for example).

Surface-feeding fishes provide vivid examples of the lateral line system functioning under natural conditions. The striped panchax (*Aplocheilus lineatus*), a type of killifish, finds insects on the water surface by detecting surface waves created by prey movements (**Figure 4.7A**). Panchax have an upturned mouth and an array of 18 very large neuromast organs arranged in six lines on top of the head (**Figure 4.7B**). Each neuromast has a unique

receptive field, and the fish can detect differences in arrival time of surface waves at different neuromast organs, which can be used to precisely locate prey at the surface, even in compete darkness.

The large numbers of neuromast organs on the heads of some fishes might be important for sensing vortex trails in the wakes of adjacent fishes in a school. Many fishes that form extremely dense schools (such as herring) lack neuromast organs along the trunk but retain neuromast organs in canals on the head. The cephalic neuromast organs sense water motion near the head to detect turbulence (caused by other fish in the school) into which the fish is swimming. Reduction of trunk neuromast organs reduces noise from turbulence to the side of the fish.

Hearing and equilibrium

Hair cells also function in hearing, where they respond to sound waves. Sound travels about four times faster in water than in air, and for any given frequency the wavelength in water is about four times longer than in air. Sound generally travels farther in the open sea than in terrestrial environments for two reasons:

1. Sound rarely has an unobstructed path on land—for example, solid objects reflect sound waves, and vegetation absorbs sound energy. The open sea, in contrast, provides an unobstructed path for sound waves.

2. In air, sound energy obeys the inverse square law: as sound propagates from a source, the energy spreads over larger and larger areas. Doubling the distance from the source reduces the energy by a factor of four. In the open ocean, sound waves can reflect off thermoclines (interfaces between water masses with different temperatures) and remain within a sound channel rather than spreading outward. As a result, sound energy can travel many kilometers in the open ocean, allowing whales to communicate over vast distances.

(A) Striped panchax killifish, *Aplocheilus lineatus*

(B) Dorsal view of neuromasts

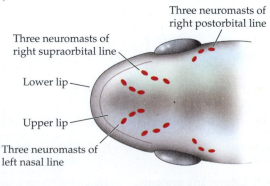

Three neuromasts of right postorbital line

Three neuromasts of right supraorbital line

Lower lip

Upper lip

Three neuromasts of left nasal line

Figure 4.7 **A specialization of the lateral line system for detecting prey in darkness.** (A) The striped panchax, *Aplocheilus lineatus*, detects surface movements of prey in darkness. (B) To do this it uses its 18 large neuromast organs on the dorsal surface of the head. (B after J.S. Schwarz et al. 2011. *J Exp Biol* 214: 1857–1866.)

Sound reaches and stimulates auditory hair cells in the inner ear by several routes. In chondrichthyans and osteichthyans, sound travels through tissues of the skin and skull to reach the inner ear. In osteichthyans with gas bladders, sound travels through tissues to the gas bladder, which then vibrates in resonant frequency with the impinging sound; if the gas bladder has contacts with the inner ear, then the vibrations can be transmitted to it. Such systems allow the detection of higher frequency sound waves. Finally, in some osteichthyans, such as catfishes, minnows, and goldfishes, vibrations that reach the gas bladder are transmitted to the inner ear by a series of small bones called Weberian ossicles. Fishes with Weberian ossicles are even more sensitive to high-frequency sounds than fishes relying on other routes of sound transmission to the ear (see Section 7.4).

Like all vertebrates, fishes have an inner ear with a vestibular system that detects changes in speed and direction of motion and allows them to distinguish up from down (see Section 2.6 and Figure 2.12).

Electroreception and electrogenesis

Unlike air, water conducts electricity. Electroreception is the ability to detect external electric fields using receptors that convert electric currents into nerve impulses. Electroreception can be either passive or active, with the former being more common than the latter.

Passive electroreception   Passive electroreception is the ability to detect weak electric fields generated by muscle contractions of aquatic prey. All muscle activity generates electric potential: motor neurons produce extremely brief changes in electric potential, and muscular contraction generates changes of longer duration. In addition, a steady potential issues from an aquatic organism due to its chemical imbalance with surrounding water. Animals with passive electroreceptive abilities detect prey using electroreceptors called **ampullary organs**, or **ampullae**.

Passive electroreception is universal among chondrichthyans, who use it to detect and capture prey. Their ampullary organs, known as **ampullae of Lorenzini**, are on the head and are most concentrated just in front of the mouth where they help direct the final strike at a prey animal (**Figure 4.8A,B**). Rays have ampullary organs on the pectoral fins as well as on the head. An ampullary organ consists of a subcutaneous canal with receptor cells located in its slightly enlarged base; the receptor cells are modified hair cells that have a single kinocilium and no stereocilia. The canal opens to the skin surface by a pore and is filled with an electrically conductive gel (**Figure 4.8C**). The walls of the canal are nonconductive. Because the canal runs for some distance beneath the skin surface, the receptor cells can detect differences in electric potential between the tissue in which they lie and the distant pore opening. Thus, ampullary organs can detect electric fields, which are differences in electric potential in space.

(A) Ampullae of Lorenzini are concentrated on snout

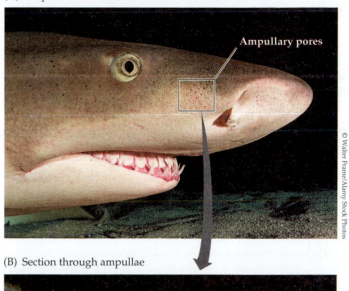

Ampullary pores

© Walter Frame/Alamy Stock Photos

(B) Section through ampullae

Skin surface

Ampullae   Lateral line canal

© William E. Bemis

(C) Anatomy of an ampullary organ

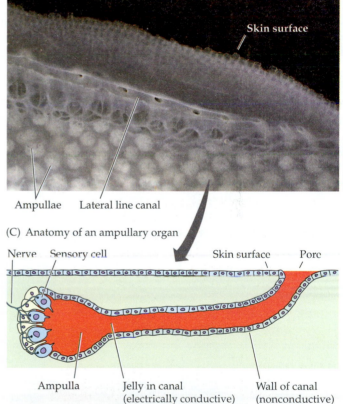

Nerve   Sensory cell   Skin surface   Pore

Ampulla   Jelly in canal   Wall of canal
                 (electrically conductive)   (nonconductive)

Figure 4.8 Ampullae of Lorenzini. (A) Ampullae of Lorenzini are electroreceptive organs clustered in fields on the head of chondrichthyans. Pores marking openings of ampullae are clearly visible on the snout of this lemon shark, *Negaprion brevirostris*. (B) Section through snout of the spiny dogfish, *Squalus acanthias*, showing supraorbital lateral line canal and clusters of ampullary organs beneath the skin surface. (C) Anatomy of single ampullary organ. (See credit details at the end of this chapter.)

Ampullae of Lorenzini act like voltmeters, measuring differences in electric potentials across the body surface. These organs are remarkably sensitive, with thresholds lower than 0.01 microvolt per centimeter, a level of detection achieved by only the best voltmeters. A shark can accurately

(A) Shark can locate live prey under sand

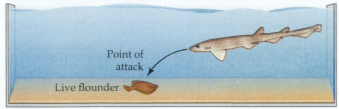

Point of attack

Live flounder

(B) Shark preferentially attacks electrical cue over olfactory cue

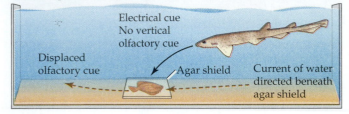

Electrical cue
No vertical olfactory cue

Displaced olfactory cue

Agar shield

Current of water directed beneath agar shield

(C) When no electrical cue, shark attacks olfactory cue

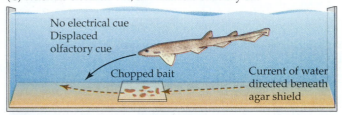

No electrical cue
Displaced olfactory cue

Chopped bait

Current of water directed beneath agar shield

(D) Without electrical or olfactory cues, shark does not detect prey

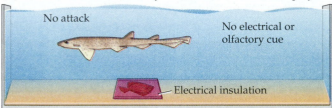

No attack

No electrical or olfactory cue

Electrical insulation

(E) Shark attacks electrical cue from inanimate electrodes

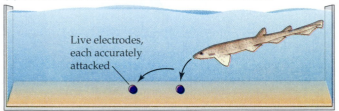

Live electrodes, each accurately attacked

(F) Shark preferentially attacks electrodes over dead bait

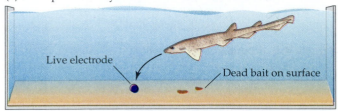

Live electrode

Dead bait on surface

Figure 4.9 Passive electroreception and prey detection in sharks. In a classic series of experiments, Adrianus Kalmijn tested the ability of the small-spotted catshark, *Scyliorhinus canicula*, to locate hidden prey. (A) Flounders normally rest on the bottom, covered by a layer of sand. The shark had no difficulty locating the flounder; was it using electroreception, olfaction, or both sensory modalities? (B) The flounder was covered by an agar shield that allowed electric impulses but not olfactory cues to pass through it. A current of water flowing under the shield and over the flounder displaced olfactory cues downstream. The shark directed its attack at the flounder, showing that it was responding to electrical activity of the flounder's muscles (used to ventilate its gills). (C) When the flounder was replaced by pieces of dead fish that emitted an odor but no electrical activity, the shark swam past the bait without detecting it, but responded when it reached the odor plume. (D) When the flounder was covered with insulation that blocked the electric signal *and* with odor-blocking agar, the shark was unable to detect the fish. (E) The shark attacked electrodes emitting an electric signal duplicating the activity of a flounder's respiration. (F) The shark preferentially attacked a "live" electrode over pieces of dead bait on the surface, showing it responds more strongly to electrical cues than to olfactory and visual cues. (A–E after A.J. Kalmijn. 1971. *J Exp Biol* 55: 371–383.)

feature of coelacanth anatomy for more than 350 million years. Lungfishes and basal actinopterygians such as the North American paddlefish, *Polyodon spathula*, have electroreceptors similar to those of chondrichthyans, although with much shorter canals. (This is because the electrical resistance of the skin is greater in fresh water than in saltwater; freshwater chondrichthyans show the same pattern of short canals.) Paddlefishes are named for their paddle-shaped rostrum, which is essentially an antenna with as many as 70,000 electroreceptive ampullae, more than known in any other fish. Paddlefishes use their network of ampullae to detect electrical fields generated by swarms of *Daphnia* and other crustaceans on which they filter-feed.

Active electroreception Animals with active electroreception have electric organs (modified muscle or neuronal tissue) that generate an electric organ discharge (EOD). Distortions in the electric field generated by the discharge, caused for example by a nearby fish or physical feature of the environment, are sensed by tuberous electroreceptors (also called tuberous organs). Active electroreception is sometimes called active electrolocation.

African elephantfishes (family Mormyridae), the aba aba (*Gymnarchus niloticus*; family Gymnarchidae), and South American knifefishes (Gymnotiformes) are freshwater teleosts that use active electroreception. Their body creates a dipole during EOD, with the positive pole near the head and the negative pole near the tail. The EOD of some species is a continuous, nearly sinusoidal wave, whereas other species produce individual pulses. These fishes are only weakly electric, with EOD amplitudes less than 1 volt. Sensory cells of tuberous electroreceptors lack a kinocilium and instead have stereocilia. They evolved by modification of neuromasts and are not homologous to the ampullary organs of chondrichthyans. EODs are used

locate and attack a live flatfish buried under the sand using these differences in electric potential (**Figure 4.9**).

Coelacanths in the genus *Latimeria* also have passive electroreception. They use a specialized rostral organ in the snout that is anatomically different from the typical ampullary organs of chondrichthyans and that has been a

for orientation, detection of prey and predators, species identification, and electrocommunication (i.e., tuberous organs detect EODs of nearby fishes). In addition to tuberous organs, these fishes possess ampullary organs, which aid in prey detection via passive electroreception.

Other electrogenic fishes   A few electric fishes can generate EODs strong enough to stun prey and deter predators; others produce weak EODs to facilitate communication. **Figure 4.10** summarizes the types of electrogenic fishes, where they occur, and the locations of their electric organs.

Few marine fishes have electric organs. Among chondrichthyans, only the torpedoes and electric rays (Torpediniformes) and some skates (Rajiformes) are electrogenic, and among marine teleosts, only the stargazers (Uranoscopidae) produce electric discharges. The electric organs of torpedo rays are modified gill muscles; those of skates are modified tail muscles; and those of stargazers are modified eye muscles.

The three species of South American electric "eels" (Gymnotiformes) live in fresh water. The electric organ in these fishes runs nearly the length of the body and sums output from thousands of specialized muscle cells, collectively known as electroplax, to produce EODs over 650 V; researchers recorded a discharge from one of the newly described species at 860 V. Despite the amplitude of EODs, the shock administered by an electric eel depends on location of the animal receiving the shock relative to the positive and negative ends of the eel's dipole. Michael Faraday, the famed English pioneer in the study of electricity, reported in 1839 that he felt only a weak shock when he placed one or both hands at the same location on the body of an electric eel, but that the shock was very strong when he moved his hands close to the poles (**Figure 4.11**). Electric catfishes from Africa independently evolved the ability to produce powerful voltages using modified trunk muscles.

Other electroreceptive vertebrates   Electroreception is not restricted to fishes. Monotremes (platypus and echidnas) have electroreceptors on their bill or beak. Electroreceptors are much more numerous on the bill of a platypus than on the beak of an echidna (which makes sense, since platypus are semiaquatic and echidnas are terrestrial), and there is behavioral evidence that platypus use electroreception to detect prey under water. It is possible that echidnas use electroreception when searching for prey in moist soil. Electroreceptors also have been described in another mammal, the Guiana dolphin (*Sotalia guianensis*); this species may use electroreception to detect prey at short distances and rely on echolocation to detect distant prey.

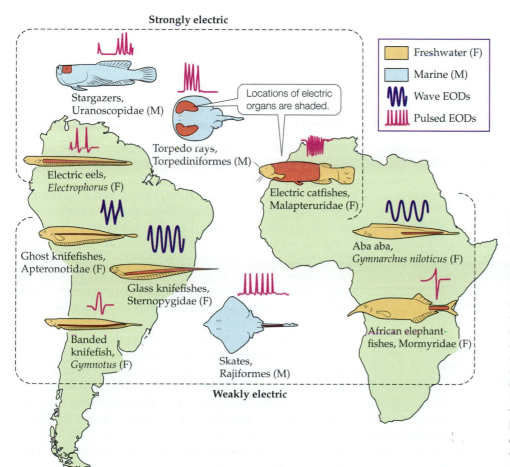

Figure 4.10 Biogeography and types of electric fishes. Freshwater electric fishes occur in the rivers and lakes of South America and Africa (gold images on continents); marine species live in oceans around the world (blue images). Locations of electric organs within the fishes' bodies are shaded. All marine species generate pulsed electric organ discharges (EODs; readings with sharp peaks, shown in red). Some freshwater species produce this same type of pulsed EOD, while others generate more even wavelengths (wave EODs, shown in blue). (After P. Moller. 1995. *Electric Fishes: History and Behavior.* Chapman & Hall, London; and M.V.L. Bennett. 1970. *Annu Rev Physiol* 32: 471–528.)

(A) Electric fields in the water around an electric eel

(B) After an electric eel seizes prey

(C) Electric eel attacking model alligator head

LEDs installed in the model reveal intensity of the eel's electrical discharges.

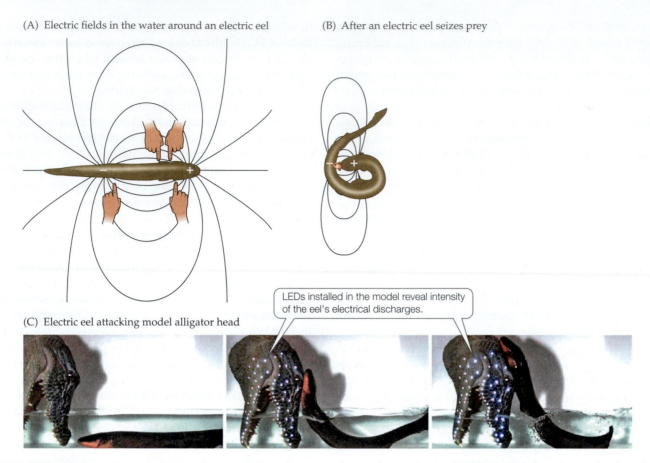

Figure 4.11 Intensity of shock delivered by electric eels.
Electric "eels" can generate electric potentials above 650 V. Intensity of the shock depends on position of the recipient relative to the eel's positive and negative poles. (A) Black lines show the distribution of electric potential in the water around an electric eel. Touching the eel produces only a small shock when both hands are in nearly the same place along the eel's body but moving the hands closer to each pole increases the intensity of the shock. (B) After an eel seizes a fish, it bends its body, placing the prey between its positive and negative poles, thus increasing the voltage. Repeated electrical discharges overstimulate the prey's muscles, leading to muscle fatigue that renders it unable to escape. (C) Video images show an electric eel attacking a partially submerged predator by leaping from the water and pressing its body against it. (Here the "predator" is a model of an alligator head equipped with LED lights that respond to the eel's electrical discharges.) By leaping higher from the water, the eel increases the voltage that the predator receives, as shown by the greater intensity of the lights in the rightmost panel. (A,B after K.C. Catania. 2015. *Curr Biol* 25: 2889–2898 with permission from Elsevier; C from K.C. Catania. 2016. *PNAS* 113: 6979–1984, courtesy of Kenneth C. Catania.)

4.3 Maintaining an Internal Environment

LEARNING OBJECTIVES

4.3.1 Explain why it is difficult for an aquatic vertebrate to maintain a body temperature different from water temperature.

4.3.2 Differentiate ammonotely and ureotely.

4.3.3 Describe structure and function of the glomerulus of a nephron.

4.3.4 Explain the importance of osmoregulation.

4.3.5 Understand and apply the terms isosmolal, hyposmolal, and hyperosmolal.

Physiological processes are temperature-sensitive. Rates of chemical reactions underlying an animal's physiology typically increase with temperature. Reactions in a metabolic pathway usually have different temperature sensitivities, so a change in temperature can mean that too much or too little substrate is produced to sustain the next reaction in the pathway. Limiting the range of temperatures that an organism experiences is easier in aquatic environments because of the thermal stability of water, which has a heat capacity 3,500 times that of air (see Table 4.1). Thus, stability of water temperature simplifies the task of maintaining a constant body temperature, as long as the body temperature the animal needs to maintain is the same as the temperature of the surrounding water. An aquatic animal has a hard time maintaining a body temperature different from water temperature, however, because water conducts heat 25 times faster than does air. Heat flows out of the body if an animal is warmer than the surrounding water and into the body if the animal is cooler. But thermoregulation is not the only challenge to homeostasis when living in water. Others include coping with metabolic wastes and maintaining water and salt balance of body fluids.

Nitrogenous wastes and kidneys

Excretion is the removal of waste products of metabolism. Carbohydrates and fats are composed of carbon, hydrogen, and oxygen, and the waste products from their metabolism—carbon dioxide and water—are easily voided. Proteins and nucleic acids are another matter, because they contain nitrogen. Most nitrogenous waste comes from the breakdown of proteins. When protein is metabolized, nitrogen is enzymatically reduced to ammonia through a process called deamination. Deamination occurs primarily in the liver and involves removal of an amino group (NH_2), which then picks up a hydrogen ion to form ammonia (NH_3). Ammonia is soluble in water and diffuses readily, but it is also extremely toxic. Rapid excretion of ammonia, whether as NH_3 or in a modified form, is crucial.

Differences in the form in which nitrogenous wastes are excreted reflect their toxicity and cost of production. Most bony fishes and aquatic lissamphibians eliminate nitrogen as a mixture of ammonia and urea. Excretion of nitrogenous wastes primarily as ammonia is **ammonotely**, excretion primarily as urea is **ureotely**.

- *Ammonotely*: Bony fishes are primarily ammonotelic and excrete ammonia through the skin and gills as well as in urine produced by the kidneys. Ammonia is produced by deamination of proteins and no metabolic energy is needed to produce it.

- *Ureotely*: Aquatic lissamphibians are ammonotelic when water is plentiful but can shift to ureotely during dehydration. Urea is synthesized from ammonia in an enzymatic process that occurs primarily in the liver. Urea synthesis requires more energy than does ammonia production, but urea is less toxic and can be concentrated in urine, thus conserving water.

Vertebrates rely on kidneys composed of **nephrons** to remove nitrogenous wastes from the blood. For scale, a human typically has about one million nephrons in each kidney. Each nephron consists of a **glomerular capsule** and **renal tubule**. First, blood is filtered at the glomerulus, a tuft of capillaries unique to vertebrates and enclosed within the glomerular capsule. Arterial blood pressure forces water and small molecules from the blood in the glomerulus into the glomerular capsule to form an ultrafiltrate; blood cells and larger molecules cannot pass through the glomerular filter. Blood flow is regulated to adjust glomerular filtration rate (often reported as GFR in reports from medical laboratories). As the ultrafiltrate moves along the renal tubule, it is processed to recover essential metabolites (e.g., glucose, amino acids) and to remove wastes. Fluid that remains after this processing is urine, which contains nitrogenous wastes, excess ions, foreign substances such as pesticides, and varying amounts of water. We will discuss kidneys in more detail in Chapters 11 and 13.

Osmoregulation

An organism can be described as a leaky bag of dirty water. That is not an elegant description, but it accurately identifies two important characteristics: (1) a living animal contains organic and inorganic substances dissolved in water; and (2) this fluid is enclosed by a permeable body surface. Exchanges of matter and energy with the environment are essential to survival.

Osmoregulation refers to the process of maintaining both water and salt balance to prevent body fluids from becoming too concentrated or too dilute. The bodies of most vertebrates are 70–80% water, so the internal chemical reactions that release energy and synthesize new chemical compounds take place in an aqueous environment. Body fluids of vertebrates contain a complex mixture of ions and other solutes. Some ions are cofactors that control rates of metabolic processes; others help regulate pH, stability of cell membranes, or electrical activity of nerves and muscles. Metabolic substrates and products must move from sites of synthesis to sites of use. Almost everything that happens in vertebrate tissues involves water, so maintaining concentrations of water and solutes within narrow limits is vital.

Water may sound like an ideal place to live for an animal that itself is mostly water, but an aquatic environment can be too much of a good thing. Saltwater vertebrates must prevent the water in their bodies from being sucked out into seawater, but freshwater vertebrates—especially fishes and lissamphibians—face the threat of being flooded with water that flows into them from their environment. **Water balance** is the state when equal amounts of water are entering and leaving an animal's body. Proper water balance prevents overhydration (too much water) and dehydration (too little water).

Regulation of ions and body fluids

Small molecules, collectively termed solutes, are dissolved in body fluids such as blood plasma. Salt ions, urea, and some small carbohydrate molecules are solutes that influence osmoregulation because their presence lowers the potential activity of water. In the process of **osmosis**, water moves from areas of high water potential to areas of lower water potential. This means that water flows from a dilute solution (one with a high water potential) across membranes to a more concentrated solution (with a lower water potential).[1]

Three terms—isosmolal, hyposmolal, and hyperosmolal—describe the concentration of an aquatic vertebrate's body fluids relative to the surrounding water (**Table 4.2**). Seawater has a solute concentration of ~1,050

---

[1] More water means a higher water potential, and when solutes are present, they occupy some of the volume of the solution, so the presence of *more* solutes (higher osmolality) means there is *less* water (lower water potential).

Table 4.2 Osmolality of vertebrate blood

Environment	Osmolality[a] (mmol/kg)	Na+	Cl-	Urea	TMAO[b]
Seawater	1,050	475	550	—	—
Fresh water	<10	5	5	—	—
Type of vertebrate					
Marine					
Hagfishes	1,035	485	515	—	—
Teleosts	350	150	135	—	—
Spiny dogfish (shark)	1,080	255	240	440	70
Coelacanths	930	200	190	380	120
Freshwater					
Teleosts	300	140	125	—	—
Lungfishes	240	100	45	—	—
Aquatic lissamphibians	200	100	90	—	—
Euryhaline					
Bull shark					
in seawater	1,070	290	295	370	45
in fresh water	640	210	200	190	20
Crab-eating frog adult					
in 80% seawater	830	250	225	350	—
in fresh water	290	125	100	40	—
Crab-eating frog tadpole					
in seawater	525	265	255	—	—
in fresh water	270	125	100	—	—
Terrestrial					
Terrestrial lissamphibians	250	125	90	—	—
Birds	350	160	140	—	—
Mammals	340	160	110	—	—

[a] Concentrations are expressed in millimoles of solute per kilogram of water and have been rounded to the nearest 5 units. Note that osmolality is in reference to the weight of a solution, whereas osmolarity (not used here) is in reference to volume.

[b] TMAO, trimethylamine oxide

millimoles per kilogram (mmol/kg) of water. Body fluids of hagfishes are in osmotic equilibrium with seawater, so hagfishes are **isosmolal** relative to seawater. Body fluid concentrations of marine teleosts are about 350 mmol/kg, a value much lower than the surrounding seawater. Therefore, water flows outward from a marine teleost's body to the sea (i.e., from a region of high water potential to a region of lower water potential), making marine teleosts **hyposmolal** (lower solute concentrations than the surrounding water). Sharks and other cartilaginous fishes retain urea and other nitrogen-containing compounds, raising the osmolality of their blood to slightly above that of seawater. Water flows from the sea into their bodies, and they are described as **hyperosmolal** (higher solute concentrations than the surrounding water). Freshwater fishes, including many teleosts and lungfishes, and freshwater lissamphibians have a higher content of solutes (ranging from 200 to 300 mmol/kg)

than the surrounding water (<10 mmol/kg), and thus are hyperosmolal.

Monovalent ions such as sodium (Na+) and chloride (Cl-) can diffuse through an animal's surface membranes (such as gills), so both water balance and salt balance of a marine teleost are constantly threatened by outflow of water and inflow of ions. In contrast, freshwater vertebrates must cope with inflow of water and outflow of ions.

4.4 Osmoregulation in Different Environments

LEARNING OBJECTIVES

4.4.1 Explain the difference between stenohaline and euryhaline fishes.

4.4.2 Describe osmoregulation of marine cartilaginous fishes.

4.4.3 Explain why marine teleosts drink seawater.

4.4.4 Describe the osmoregulatory challenges faced by freshwater teleosts.

4.4.5 Contrast anadromy and catadromy.

Aquatic vertebrates live in the sea, in estuaries and tidal rivers, and in freshwater streams, rivers, ponds, and lakes. As we saw in the previous section, water and ions diffuse into or out of the body depending on the habitat. Most fishes are **stenohaline** (Greek *stenos*, "narrow"; *hals*, "salt"), meaning that they inhabit *either* seawater *or* fresh water and tolerate only modest changes in salinity. Because stenohaline fishes remain in one osmotic environment, the magnitude and direction of the osmotic gradients to which they are exposed are stable. Some fishes, however, migrate between fresh water and seawater and tolerate large changes in salinity. These fishes are **euryhaline** (Greek *eurys*, "wide"), and water and salt gradients are reversed as they move from one habitat to the other.

Here we examine some details of excretion and osmoregulation in several major groups of fishes summarized in **Figure 4.12**. These fishes are stenohaline and live in either seawater or fresh water. We also briefly cover freshwater lissamphibians before moving to euryhaline fishes and lissamphibians.

Marine cartilaginous fishes and coelacanths

Cartilaginous fishes minimize osmotic flow by maintaining internal concentration of their body fluids close to that of seawater. To do this, they retain nitrogen-containing compounds, primarily urea and trimethylamine oxide (TMAO), that increase blood osmolality. The body fluids of spiny dogfish, for example, are usually slightly hyperosmolal to seawater (see Table 4.2). As a result, dogfish gain water by osmosis across the gills and thus do not need to drink seawater (see Figure 4.12A). This constant influx of water allows high filtration rates and rapid elimination of metabolic wastes from the blood as it passes through the kidneys; spiny dogfishes excrete copious amounts of urine. Urea is very soluble and diffuses through most biological membranes, which could be problematic for fishes that retain urea to remain hyperosmotic to their surroundings. The gills of cartilaginous fishes are nearly impermeable to urea, however, and the kidney tubules actively reabsorb it.

Because sharks have internal sodium and chloride ion concentrations that are low relative to seawater, these ions enter body fluids across the gills. However, the gills of cartilaginous fishes have low permeabilities to these ions relative to those of marine teleosts. Cartilaginous fishes generally do not have salt-excreting cells in the gills. Rather, they achieve ion balance by secreting a salty fluid from the rectal gland (= digitiform gland) into the digestive tract (see Figure 4.12A). Cartilaginous fishes that inhabit fresh water have reduced rectal glands and retain less urea in comparison to marine species.

Coelacanths also minimize osmotic flow by maintaining the internal concentration of their body fluids close to that

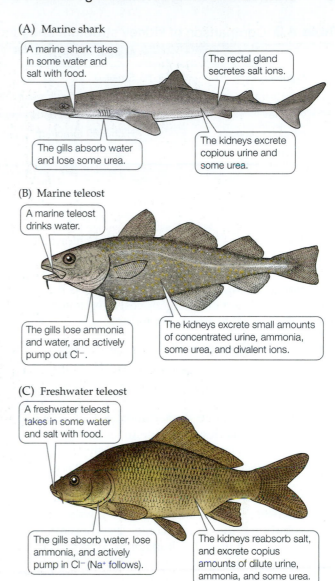

(A) Marine shark

A marine shark takes in some water and salt with food.

The rectal gland secretes salt ions.

The gills absorb water and lose some urea.

The kidneys excrete copious urine and some urea.

(B) Marine teleost

A marine teleost drinks water.

The gills lose ammonia and water, and actively pump out Cl⁻.

The kidneys excrete small amounts of concentrated urine, ammonia, some urea, and divalent ions.

(C) Freshwater teleost

A freshwater teleost takes in some water and salt with food.

The gills absorb water, lose ammonia, and actively pump in Cl⁻ (Na⁺ follows).

The kidneys reabsorb salt, and excrete copius amounts of dilute urine, ammonia, and some urea.

Figure 4.12 Osmoregulation and excretion by marine and freshwater fishes.   The main sites of exchange of water, salts, and nitrogenous wastes in marine and freshwater fishes. (A) A marine shark (spiny dogfish). (B) A marine teleost (cod). (C) A freshwater teleost (carp). (B,C after K. Liem et al. 2001. *Functional Anatomy of the Vertebrates*, 3rd ed. Cengage/Harcourt College: Belmont, CA.)

of seawater. Like cartilaginous fishes, they retain urea and TMAO, thus increasing blood osmolality (see Table 4.2).

Marine teleosts

Recall that marine teleosts are hyposmolal relative to surrounding seawater and in constant danger of dehydration. To compensate for osmotic dehydration, marine teleosts do something unusual: they drink seawater (see Figure 4.12B). The glomeruli of marine teleosts are small and glomerular filtration rate is low (**Table 4.3**). As a result, little urine is formed, so the amount of water lost by this route is reduced. The urine is concentrated.

Many marine teleosts drink more than 25% of their body weight each day and absorb 80% of this ingested water. Drinking seawater to compensate for osmotic

Table 4.3 Comparison of kidney structure and function in marine and freshwater teleosts

Measure	Marine teleosts	Freshwater teleosts
Size of glomeruli	Small	Large
Glomerular filtration rate	Low (1% of body weight/day)	High (10% of body weight/day)
Urine volume	Small (< 1% of body weight/day)	Large (7% of body weight/day)
Urine concentration	High (300 mmol/kg)	Low (20 mmol/kg)
Urine/plasma concentration ratio	High (1/1.1)	Low (1/15)

water loss increases influx of sodium and chloride ions. To compensate for this salt load, chloride cells in the gills actively pump chloride ions outward against a large concentration gradient. Active transport of ions is metabolically expensive, and the energetic cost of osmoregulation may be 25% of daily energy expenditure.

Freshwater teleosts and lissamphibians

As noted in Section 4.3, the osmoregulatory challenges for freshwater teleosts and lissamphibians are water gain by osmosis and ion loss by diffusion. To compensate for the influx of water, the large glomeruli of freshwater teleosts produce a copious flow of dilute urine (see Table 4.3). To conserve salts, the renal tubules of freshwater teleosts reabsorb ions (see Figure 4.12C). Reabsorption is an active process, requiring energy.

Because the distal portions of the renal tubules of freshwater teleosts are impermeable to water, water remains in the tubule and the urine becomes less concentrated as ions are removed from it. Ultimately, the urine becomes hyposmolal to blood. In this way, water absorbed across the gills is removed and ions are conserved. Nonetheless, some ions are lost in urine, in addition to those lost by diffusion across the gills. Salts from food compensate for some ion loss, and teleosts have ionocytes in their gills, specialized cells that take up chloride ions from water. The chloride ions are moved by active transport against a concentration gradient, requiring energy. Sodium ions also enter through the gill tissue, passively following chloride ions (see Figure 4.12C). A freshwater teleost does not drink water because osmotic water movement is already providing more water than the fish needs—drinking would only increase the amount of water it would have to excrete via the kidneys.

Like freshwater fishes, aquatic lissamphibians lose ions by diffusion and do not drink because the osmotic influx of water more than meets their needs. Cells in lissamphibian skin actively take up ions from surrounding water. Acidity inhibits the active transport of ions in both lissamphibians and fishes, and inability to maintain internal ion concentrations can cause death in habitats made acidic by either acid rain or drainage from mine wastes.

Euryhaline vertebrates

Many species of teleosts are euryhaline, moving between fresh water and the sea as part of their life history.

Anadromous species such as the Atlantic salmon (*Salmo salar*) breed in fresh water, migrate to the sea as juveniles, and return as adults to fresh water to spawn. **Catadromous** species such as the American eel (*Anguilla rostrata*) follow the opposite pattern: they breed in the sea, migrate to fresh water as juveniles, and return to the sea as adults. Some teleosts live in estuaries, where salt concentrations of the water rise and fall with the tides. In all of these situations, teleosts maintain stable internal osmotic and ion concentrations despite changes in their environments.

About 50 species of cartilaginous fishes are euryhaline. The bull shark, *Carcharhinus leucas*, readily enters rivers and may venture thousands of kilometers from the sea. It has been recorded from Indiana (in the Ohio River) and from Illinois (in the Mississippi River), and a land-locked population lives in Lake Nicaragua. In seawater, bull sharks retain high levels of urea and TMAO, but in fresh water their blood levels of these compounds decline (see Table 4.2).

Although most lissamphibians live in fresh water or terrestrial habitats, nearly 150 species occur in saline coastal or inland habitats; four such euryhaline species are shown in **Figure 4.13**. The Southeast Asian crab-eating frog, *Fejervarya cancrivora* (Figure 4.13A), illustrates how these animals adjust to fluctuating environments.

Adult frogs inhabit intertidal mudflats and must tolerate exposure to 75% seawater at each high tide. Full-strength seawater has an ion content of 1,050 mmol/kg (also expressed as a salinity of 35 parts per thousand, or ppt), meaning that *F. cancrivora* tolerates water up to about 788 mmol/kg (26 ppt). During seawater exposure, the frog allows its blood ion concentrations to rise, thus reducing the ionic gradient. In addition, ammonia removed from proteins is rapidly converted to urea, which is released into the blood. Blood urea concentration quickly rises from 40 to 350 mmol/kg, and the frog becomes hyperosmolal to surrounding water. In this sense, *F. cancrivora* acts like a shark and absorbs water osmotically. Frog skin is permeable to urea, however, so the urea that these frogs synthesize is rapidly lost. To compensate for this loss, activity of their urea-synthesizing enzymes is very high. The genetic bases of these adaptations have been studied, and there appears to have been rapid evolutionary change in genes associated with ion transport.

(A) Crab-eating frog, *Fejervarya cancrivora*

Bernard Dupont CC BY-SA 2.0

(B) Green toad, *Bufotes viridis*

Greg Schechter/CC BY 2.0

(C) California slender salamander, *Batrachoseps attenuatus*

Marshal Hedin CC BY-SA 2.0

(D) Smooth newt, *Lissotriton vulgaris*

gailhampshire/CC BY 2.0

Figure 4.13 Salinity-tolerant lissamphibians. (A) The crab-eating frog can tolerate nearly full strength seawater (35 parts per thousand, ppt). (B) The green toad breeds in brackish pools measured at 20 ppt. (C) Adults of the California slender salamander live on beaches under driftwood, where they experience 17 ppt based on laboratory measurements. (D) On the coast of the Baltic Sea, the smooth newt breeds in brackish tidal pools measured at 17 ppt.

Unlike adult frogs, *F. cancrivora* tadpoles are trapped in pools on the mudflats and exposed to high salt concentrations when water evaporates from the pools. Like most tadpoles, they lack urea-synthesizing enzymes until late in development. Thus, tadpoles of crab-eating frogs use a different method of osmoregulation: they have salt-excreting cells in their gills that actively pump ions outward even as ions diffuse inward. This active pumping does not match the rate at which ions diffuse inward, however, and as a result the osmolality of the tadpoles' blood doubles during low tide.

Summary

4.1 The Aquatic Environment

Fishes exchange oxygen and carbon dioxide at gills and regulate buoyancy via a gas bladder or lipids.

Water is both dense and viscous, and in fishes the respiratory flow of water is typically unidirectional, rather than tidal (in–out). Buccopharyngeal pumping draws water into the mouth and expels it across internal gills. A countercurrent flow of blood in the gills and water maximizes transfer of oxygen from water to blood.

Some fishes supplement gill respiration with air-breathing, using structures that range from enlarged lips to labyrinth organs to lungs.

Cartilaginous fishes have a large, oil-filled liver that makes the body lighter than water to provide buoyancy.

Most bony fishes actively adjust their buoyancy when changing depth. Teleosts have a gas bladder to which they add or release gas. There are two types of gas bladders: physostomous bladders retain a connection to the gut, whereas physoclistous bladders do not.

4.2 Sensory World of Aquatic Vertebrates

Aquatic vertebrates have spherical eye lenses; the lenses of terrestrial vertebrates are flatter. In fishes, the entire lens moves toward or away from the retina to focus images. In mammals and birds, muscles change the shape of the lens to focus images.

The olfactory organs of fishes detect chemicals dissolved in water. Most fishes do not couple olfaction with respiration.

Unlike terrestrial vertebrates, taste buds of fishes are widely distributed in the mouth and on the body.

The density of water allows detection of small water currents, and fishes and aquatic lissamphibians have mechanical sensors (neuromasts) on the head and trunk in the lateral line system.

Sound reaches and stimulates auditory hair cells in the inner ear of fishes by several routes, including through tissues of the skin and skull. Sound travels about four times faster in water than in air, and in the open sea sound can travel very long distances.

(Continued)

Summary *(continued)*

Like other vertebrates, fishes have a vestibular system in their inner ear that detects changes in speed and direction of motion.

Water conducts electricity, and electroreception is the ability to detect external electric fields using specialized receptors that convert electric currents into action potentials.

Electroreception can be either passive or active.

- Passive electroreception is the ability to detect weak electric fields generated by muscle movements of aquatic prey. Ampullary organs serve as electroreceptors.

- Vertebrates with active electroreception have electric organs that generate an electric organ discharge. Distortions in the generated electric field are sensed by electroreceptors called tuberous organs and allow electrolocation and electrocommunication.

4.3 Maintaining an Internal Environment

Temperatures of aquatic environments are more stable than those of terrestrial environments because the heat capacity of water is greater than that of air.

It is hard for an aquatic vertebrate to maintain a body temperature different from that of surrounding water because heat conduction is higher in water than in air.

Disposal of nitrogenous wastes is easier in aquatic environments than on land.

Ammonia and urea are the nitrogenous wastes usually excreted by aquatic vertebrates.

- Ammonia, which is soluble in water, is the product of deamination, so it requires no added energy for synthesis. However, it is highly toxic and must be excreted quickly. Ammonotely is characteristic of fishes.

- Urea is less toxic than ammonia, but its synthesis requires energy. Ureotely is characteristic of lissamphibians when dehydrated.

Vertebrate kidneys regulate salt and water balance. Within kidneys, nephrons are the sites of urine production and processing.

Osmoregulation refers to the process of maintaining both water and salt balance to prevent body fluids from becoming either too concentrated or too dilute.

Most vertebrates are 70–80% water, so cellular and tissue processes take place in aqueous solutions.

Physiological stability depends on regulating the concentration of solutes in body fluids. The ease with which water and many ions move in and out of aquatic organisms makes such regulation challenging.

Ions move from areas of high concentration to areas of low concentration (i.e., down a concentration gradient).

Water moves from areas of high water potential (i.e., dilute solutions) to areas of low water potential (i.e., concentrated solutions).

Hagfish body fluids are in osmotic equilibrium with seawater; their fluids are isosmolal to seawater.

Body fluid concentrations of marine teleosts are hyposmolal, having lower solute concentrations than the surrounding water.

Cartilaginous fishes retain nitrogen-containing compounds, raising the osmolality of their blood to slightly above that of seawater; their fluids are hyperosmolal, having higher solute concentrations than the surrounding water.

4.4 Osmoregulation in Different Environments

Most freshwater and marine fishes are stenohaline, tolerating only modest changes in salinity. Euryhaline fishes tolerate large changes in salinity and can move between fresh water and seawater.

Because cartilaginous fishes maintain internal concentrations of solutes slightly above those of seawater, water flows inward. Cartilaginous fishes produce copious amounts of urine and excrete excess sodium and chloride ions via the rectal gland.

Marine teleosts have body fluids less concentrated than the surrounding water; consequently, they lose water by osmosis and gain sodium and chloride ions by diffusion. They drink seawater and actively excrete chloride ions through the gills. Their kidneys have small glomeruli that produce small amounts of concentrated urine.

Body fluids of freshwater teleosts are more concentrated than the water surrounding them; as a result, they lose sodium and chloride ions by diffusion and gain water by osmosis. To compensate for loss of ions, their kidneys reabsorb ions and cells in the gills pump in chloride ions (sodium ions follow). To cope with water gain, they have kidneys with large glomeruli that produce a large volume of urine (which is dilute due to the conservation of ions by the kidneys).

Euryhaline fishes live in estuaries or migrate between fresh and saltwater for reproduction. Anadromous fishes live in saltwater and breed in fresh water; catadromous fishes live in fresh water and breed in saltwater.

Approximately 150 species of lissamphibians are euryhaline, with evolutionary adaptations that allow them to live in tidal flats and other environments with fluctuating salinities.

Discussion Questions

4.1 When deep-sea fishes are pulled quickly to the surface, they often emerge with their gas bladder protruding from their mouth.

   a. Why does this happen, and what does it tell you about whether these deep-sea fishes are physoclistous or physostomous?

   b. Why could you have determined whether deep-sea fishes are physostomous or physoclistous simply by considering how air enters the gas bladder of a fish?

4.2 Explain how you can tell whether a vertebrate sees in either water or air by simply examining the shape of its lens.

4.3 Some species of frogs have conspicuous lateral lines. What does this tell you about their ecology?

4.4 While visiting an aquarium, you notice a killifish hovering slightly below the surface. What specializations does this fish have to catch prey on the surface? Can it do this in darkness?

4.5 What osmoregulatory challenges are faced by marine and freshwater teleosts, and how do sizes of glomeruli, glomerular filtration rate, and urine volume and concentration differ in teleosts living in these two different habitats?

Figure credits
Figure 4.6: A after K. Liem et al. 2001. *Functional Anatomy of the Vertebrates*, 3rd ed. Cengage/Harcourt College: Belmont, CA; C,D after A. Flock in P.H. Cahn [Ed.] 1967. *Lateral Line Detectors*, Indiana Univ. Press, Bloomington.
Figure 4.8: C after K. Liem et al. 2001. *Functional Anatomy of the Vertebrates*, 3rd ed. Cengage/Harcourt College: Belmont, CA.

© turtix/Shutterstock.com

5

Geography and Ecology of the Paleozoic

The Grand Canyon: 1.5 Ga of Earth's history

5

.1 Deep Time
5.2 Continental Geography
5.3 Paleozoic Climates
5.4 Paleozoic Ecosystems
5.5 Extinctions

Many sources of physical evidence indicate that Earth formed about 4.6 billion years ago, and that life, in the form of self-replicating unicellular organisms, first evolved about 4 billion years ago. Were we to transpose 4.6 billion years onto the common 12-month, 365-day calendar, the earliest life would appear in the first week of March. However, vertebrates (and indeed almost all animal lineages) are known only from the last 500 million years or so. In our calendar analogy, vertebrates show up in the third week of November, and humans do not show up until December 31.

In this chapter, we briefly overview geological time and the physical and environmental changes that Earth's shifting continents and changing climates have undergone in 4 billion years. We then describe some specific changes that occurred during the Paleozoic, the first of the three eras of the Phanerozoic eon; the Mesozoic and Cenozoic eras are treated in later chapters. We consider the evolving nature of vertebrate communities in the context of continental drift, climate change, and many associated changes in flora and fauna. We close with a discussion of extinctions and the massive extinction event that ended the Paleozoic (and very nearly all multicellular life).

5.1 Deep Time

LEARNING OBJECTIVES

5.1.1 Grasp the concept of deep time and why it is intrinsic to understanding life on Earth.

5.1.2 Become familiar with the major time divisions of Earth history as described in this section and the Appendix.

5.1.3 In general terms, describe life on Earth prior to the appearance of vertebrates.

Throughout this book, one of our goals is to help you develop a sense of **deep time**, a phrase and concept coined by author John McPhee to help us think about the inconceivably long spans of time that have passed since the first multicellular organisms arose, the first plants and animals appeared, the first fishes evolved, the last giant dinosaurs dominated terrestrial landscapes, or even the time the earliest humans lived. The calendar analogy we used in the introduction is a simplified illustration, but to grasp the sense of change, we suggest a thought experiment.

A thought experiment in deep time begins by thinking of a natural place that you are familiar with. It might be a pond, a stand of trees, a rock outcrop on a mountain, or a windswept beach. Perhaps birds are flying overhead, you can smell salt water or flowers in the air, and you can hear noises made by other, unseen animals. Now imagine this place 100 years ago... then 1,000... then 500,000... then a million. Would what you are seeing, hearing, or smelling today have been the same a million years ago? What evidence would you need to have in order to answer that question? Now cast your imagination even farther back: 50 million, 100 million, now half a billion years. How can a person, with a life span measured in decades, begin to comprehend such expanses of time and the uncountable physical and biological changes that have taken place on Earth? Developing a sense of deep time (and knowing a few critical dates as well as the sequence of the geological periods) is essential for anyone who wants to understand biology, for the long and ongoing story of life's evolution builds on a basic knowledge of Earth's geological history.

Geologists divide the history of Earth into units of time lasting tens of millions to hundreds of millions of years (**Figure 5.1** and the Appendix). Eons are the longest intervals we will consider. Each eon contains several eras, each era contains several periods, and some periods contain several epochs. Vertebrates are known from the **Phanerozoic eon** (Greek *phaneros*, "visible"; *zoon*, "animal"), which began 541 million years ago and includes the **Paleozoic era** (Greek *palaios*, "ancient"), **Mesozoic era** (Greek *mesos*, "middle"), and **Cenozoic era** (Greek

Eons: Duration ~0.5 billion to ~2 billion years.
We live in the Phanerozoic ("visible animals") eon.

Eras: Durations up to several hundred million years.
We live in the Cenozoic ("recent animals") era.

Periods: Duration millions to ~100 million years.
We live in the Quaternary period.

Epochs: Duration thousands to tens of millions of years.
We live in the Holocene epoch.

Figure 5.1 Hierarchy of geological time units. Each unit encompasses all those listed beneath it. Complete names and durations of geological timeframes are detailed in the Appendix. These designations are used throughout the book and their names and order of occurrence should become familiar to you. Note that the words "era," "period," and "epoch" often are not appended; we typically speak, for example, of the "Jurassic," not the "Jurassic period."

kainos, "recent"). At least 99% of all fossil species are from Phanerozoic rocks, although the oldest known fossils are about 3.4 billion years old.

The Precambrian world

The time before the Phanerozoic is informally referred to as the **Precambrian**, and it represents the vast majority of Earth's 4.6-billion-year history (**Figure 5.2A**). The Precambrian has two formal divisions: the Archean (Greek *archaios*, "ancient"), commencing with the cooling of Earth's crust and formation of continents about 4 billion years ago; and the Proterozoic (Greek *pro-*, a prefix meaning "earlier" or "before"), which began about 2.5 billion years ago. The lifeless period between the formation of Earth 4.6 billion years ago and the start of the Archean 4 billion years ago is informally known as the "Hadean," and no rocks are preserved from this time. "Hades" is the classical term for Hell, reflecting the notion that the Earth of this time would have been hellish.

A globe of the Archean world would look rather like portions of today's South Pacific Ocean, with small volcanic islands separated by immense tracts of ocean. Life in the Archean consisted of microbial mats (**Figure 5.2B**), and organisms more complex than unicellular prokaryotes are not known before the Proterozoic. The start of the Proterozoic (~2.5 billion years ago) was marked by the appearance of the large continental blocks seen today.

Substantial levels of atmospheric oxygen were not present until about 2.2 billion years ago—some 300 hundred million years into the Proterozoic. This rise was a result of the evolution of photosynthesis among unicellular cyanobacteria and the ensuing release of O_2, a by-product of photosynthesis. Atmospheric O_2 did not approach present-day levels until late in the Proterozoic, about 700 Ma. The evolution of eukaryotic

(A) Key events during the Precambrian

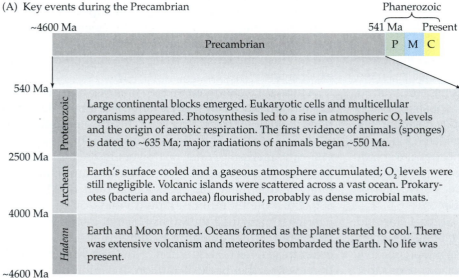

Earth and Moon formed. Oceans formed as the planet started to cool. There was extensive volcanism and meteorites bombarded the Earth. No life was present.

(B) Microbial mats at volcanic hot springs in Yellowstone National Park

James St. John/CC BY 2.0.

Figure 5.2 The Precambrian accounts for most of Earth's geologic history. (A) Of the 4.6 billion years since the formation of Earth, proliferation of multicellular and structurally complex animals occurred only during the Phanerozoic ("visible animals") eon, which is divided into the Paleozoic (P), Mesozoic (M), and Cenozoic (C) eras. Time prior to the Phanerozoic is referred to as the Precambrian. (B) Life during most of the Precambrian consisted of unicellular prokaryotes (bacteria and archaea). Much of this life may have taken the form of dense microbial mats, as seen today in certain harsh environments such as the volcanic hot springs of Yellowstone National Park.

Figure 5.3 Key events during the Paleozoic. The first period of the Paleozoic era—the Cambrian—saw a burst of evolutionary radiations. By the end of the Cambrian, representatives of all extant major animal taxa, including vertebrates, were present. The Paleozoic ended with a mass extinction that eliminated many taxa, especially marine species.

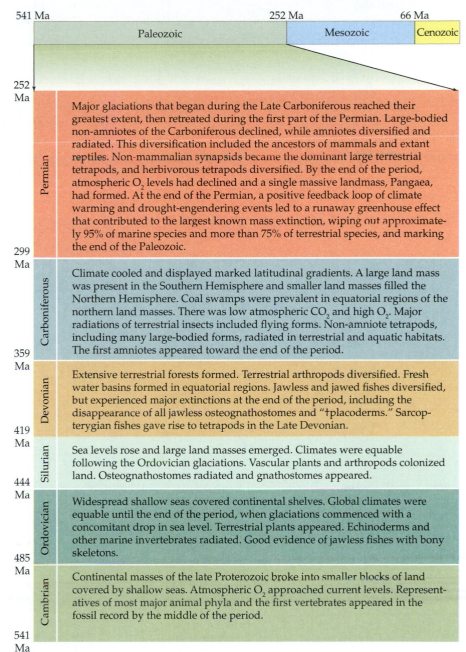

organisms, which depend on O_2 for respiration, commenced about 1.8 billion years ago. There is some evidence of early animals (sponges) at about 635 Ma, following a second rise in levels of O_2. However, metazoans did not start their major radiation until near the end of the Proterozoic, some 550 Ma; the oldest vertebrates known are from early in the Phanerozoic, in early Paleozoic rock formations.

The Paleozoic

The first of the three Phanerozoic eras was the Paleozoic, beginning with the Cambrian about 541 Ma and extending through the end of the Permian, about 252 Ma (**Figure 5.3**). The Paleozoic was very different from today's world. The continents were in different places (Section 5.2) and climates were different (Section 5.3). Initially most Paleozoic life was aquatic—there was little structurally complex life on land. By the Early Devonian, however, terrestrial environments supported a substantial diversity of plants and invertebrates, setting the stage for the first terrestrial vertebrates (the earliest tetrapods), which appeared in the Late Devonian (see Section 5.4 and Chapter 8).

5.2 Continental Geography

LEARNING OBJECTIVES

5.2.1 Describe some examples of how changing continental positions influenced the history of life on Earth.

5.2.2 Describe differences in the positions of the continents at the start and close of the Paleozoic, and compare them to continental positions today.

Continental positions have changed during Earth's history, causing global climate to vary. Today the continents are widely separated from one another, and the main continental landmasses are in the Northern Hemisphere—two conditions that did not exist during most of vertebrate evolution.

Continental drift and plate tectonics

Our understanding of Earth's dynamic nature and variable climate over geological time is fairly recent. Patterns of global climate, including such features as ice at the poles, continental glaciations, and the directions of major winds and ocean currents, are influenced by the positions of the continents. The observation that continents have moved substantially during geological time, termed **continental drift**, was explained in the late 1960s as the model of **plate tectonics**. Oceanographic research showed how the great oceans opened by the spreading of the seafloor via movements of tectonic plates underlying the continents. These movements are responsible for the sequence of fragmentation, coalescence, and re-fragmentation of the continents that has occurred during Earth's history. Plants

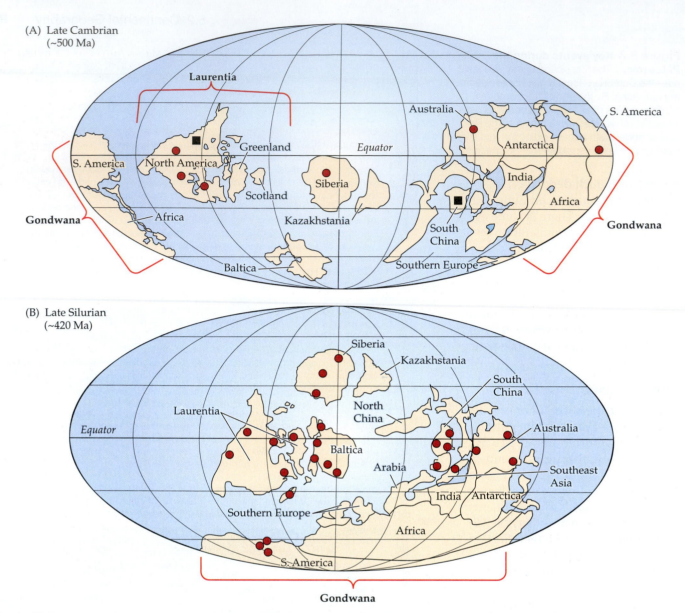

Figure 5.4 **Location of continental blocks in the early Paleozoic.** (A) Late Cambrian. Black squares indicate locations of fossil vertebrates from the Cambrian; red dots mark the locations of Ordovician vertebrate fossils. Baltica consisted of northeastern Europe plus part of western Asia. (B) Late Silurian. Baltica has combined with western Europe and parts of northeastern North America. Dots indicate fossil vertebrate localities. (After C.R. Scotese and W.S. McKerrow. 1990. In *Palaeozoic Palaeogeography and Biogeography, Geological Society Memoir No.* 12. W.S. McKerrow and C.R. Scotese [eds.], pp.1-21. The Geological Society: London.)

and animals were carried along as continents slowly drifted, collided, and separated (see Section 1.7).

As continents collided, terrestrial floras and faunas that had evolved in isolation mixed and populations of marine organisms were separated. A recent (in geological terms) example of this phenomenon is the joining of North and South America by a land bridge about 3 Ma. The faunas and floras of the two continents mingled, which is why we now have opossums (of South American origin) as far north as southern Canada and some species of deer (of North American origin) in Argentina. In contrast, marine organisms originally found in the sea between North and South America were separated by the rise of the Central American isthmus, and with the passage of time the separated populations on the Atlantic and Pacific sides of the land bridge have become increasingly different from each other.

Shifting continents of the Paleozoic

The world of the early Paleozoic contained at least six major continental blocks (**Figure 5.4**). **Laurentia** included most of present-day North America plus Greenland and Scotland. **Gondwana** (sometimes called Gondwanaland) included most of the land that is now in the Southern Hemisphere—the continents of South America, Africa, Antarctica, and Australia—as well as southern portions of Asia and Europe. Four smaller blocks contained other parts of continents now in the Northern Hemisphere.

In the late Cambrian (~500 Ma), Gondwana and Laurentia straddled the Equator, and the north–south orientations of the modern continents within Gondwana were different than they are today—Africa and South America appear to be upside down (see Figure 5.4A). By the late Silurian (~420 Ma), the eastern portion of Gondwana was over the South Pole, and Africa and South America were in orientations similar to those that they occupy today, but were entirely south of the Equator (see Figure 5.4B).

From the Devonian through the Permian, the continents drifted closer to each other (**Figure 5.5A**). The continental blocks that make up parts of modern North America, Greenland, and western Europe came into proximity along the Equator by the Late Devonian, forming a continental land mass known as **Euramerica**. With the later addition of Siberia, these blocks formed a northern supercontinent known as **Laurasia**. Most of Gondwana was in the far south, overlying the South Pole. The Tethys Sea, which separated Gondwana from Laurasia, did not close completely until the Late Carboniferous, when Africa moved northward, eventually colliding with the east coast of North America (**Figure 5.5B**). By the Permian, most of the continental surface was united in a single supercontinent, **Pangaea** (sometimes spelled Pangea). At its maximum extent, the landmass of Pangaea covered 36% of Earth's surface, compared to 27% for the present arrangement of continents. Pangaea persisted for 160 Ma, from the mid-Carboniferous to the mid-Jurassic.

Shifting continents and changing climates

Continental positions affect the flow of ocean currents, and because these currents transport vast quantities of heat, changes in their flow affect climates worldwide. For example, the breakup and northward movements of continents during the late Mesozoic and Cenozoic led first to the isolation of Antarctica and the formation of the Antarctic ice cap (~45 Ma) and eventually to the isolation of the Arctic Ocean, with the formation of Arctic ice some 5 Ma. The presence of this ice cap influences global climate conditions in many ways, and Earth today is colder and drier than it was earlier in the Cenozoic.

5.3 Paleozoic Climates

LEARNING OBJECTIVES

5.3.1 Describe the major global climate changes seen in the Paleozoic.

5.3.2 Explain how the evolution of land plants resulted in changes to Earth's climate.

5.3.3 Describe how the composition of atmospheric gases varied during the Paleozoic, and how these are related to changes in climate.

During the early Paleozoic, sea levels were at or near all-time highs for the Phanerozoic. Atmospheric carbon dioxide levels were also very high—at least 10 times today's level—while oxygen levels were

Figure 5.5 Location of continental blocks later in the Paleozoic. (A) Late Devonian. Laurentia and Baltica (i.e., the continental areas including present-day North America, Greenland, and northern Europe) joined to form the landmass Euramerica, which straddled the Equator. An arm of the Tethys Sea extends westward between Gondwana and the northern continents. (B) Late Carboniferous. This map illustrates an early stage of the supercontinent Pangaea. The location and extent of continental glaciation are shown by dashed lines and arrows radiating outward from the South Pole. Dots indicate localities of fossil tetrapods. Heavy gray shading shows the extent of the forests that formed coal beds. (A,B after C.K. Seyfert and L.A. Sirkin, 1979. *Earth History and Plate Tectonics. An Introduction to Historical Geology.* Harper & Row: New York.)

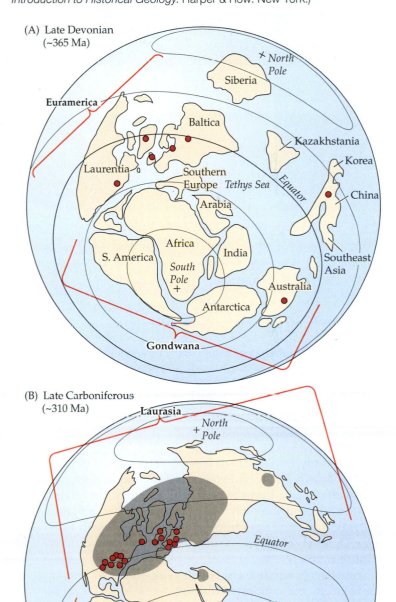

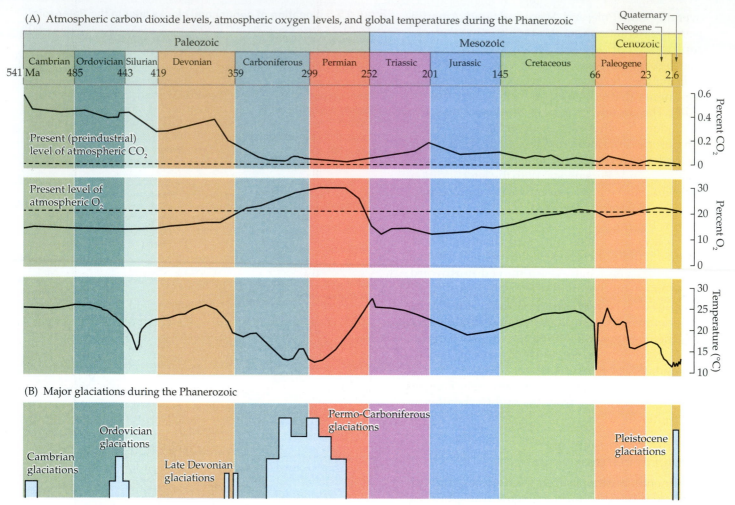

(A) Atmospheric carbon dioxide levels, atmospheric oxygen levels, and global temperatures during the Phanerozoic

(B) Major glaciations during the Phanerozoic

Figure 5.6 Climate has changed many times during Earth's history. (A) Air temperatures correlate with atmospheric levels of CO_2 and O_2. These conditions led to warm climates in the earlier part of the Paleozoic, but plummeting levels of atmospheric CO_2 in the late Paleozoic brought cold temperatures and the formation of ice caps. (B) Major glaciations occurred several times during the era, most extensively from the Late Carboniferous through the Early Permian. These charts combine information from multiple sources. (See end of chapter for details.)

low, probably two-thirds or less of today's level of 21% (**Figure 5.6A**). The high levels of CO_2 created a greenhouse effect, and Earth experienced hot, dry terrestrial climates during the early Paleozoic. However, a major glaciation in the Late Ordovician combined with falling CO_2 levels resulted in cooler and moister conditions (**Figure 5.6B**).

A relatively equable climate developed by the end of the Silurian and continued through the Middle Devonian. Glaciation recurred in the Late Devonian. From the mid-Carboniferous until the Middle Permian, ice sheets covered much of Gondwana. The waxing and waning of glaciers created oscillations in sea level, resulting in cyclical formation of coal deposits from buried plant material ("carboniferous" means "coal-bearing"). The climate over Pangaea was fairly uniform in the Early Carboniferous, but glaciations led to significant regional differences in flora in the Late Carboniferous and Early Permian. Most vertebrates lived in equatorial regions during this time.

The covering of Earth by terrestrial vegetation during the Devonian led to profound changes in climate and geological processes because plants are ecosystem engineers, affecting water cycles and key biogeochemical cycles. Soils formed as plant roots penetrated and broke apart underlying rocks, and the roots' organic secretions plus decomposition of dead plant material created acids that dissolved minerals from the rocks. These processes of mechanical and chemical weathering increased the flow of dissolved phosphorous and calcium ions into the ocean. There, the phosphorous acted as a nutrient, fueling plankton blooms and creating anoxic waters that may have caused some marine extinctions. Calcium ions (Ca^{2+}) in the ocean reacted with atmospheric CO_2 to form carbonates such as $CaCO_3$, which then precipitated out of the water column to form massive limestone deposits. The spread of land plants and their burial in coal-forming deposits also sequestered carbon as coal ("sequestered" means that the carbon is tied up in coal and cannot directly exchange with CO_2 in the atmosphere).

Because so much carbon became sequestered in carbonate rocks and coal, atmospheric CO_2 levels dropped sharply from the Late Devonian through the Late Carboniferous, and by the middle Permian were at about the same level as today. (Even with the recent anthropogenic increase, atmospheric CO_2 concentrations today are lower than they were for most of the Phanerozoic.) The reverse greenhouse effect of reduced atmospheric CO_2 probably caused extensive Permo-Carboniferous glaciations (see Figure 5.6B).

The spread of land plants led not only to decreased levels of atmospheric CO_2 but also to increases in atmospheric O_2, which reached present-day levels (~21%) early in the Carboniferous and peaked at ~30% in the Permian.

Average temperatures started to climb after the Permo-Carboniferous glaciations, and eventually reached an extreme of about 30°C. In the latest Permian, equatorial areas were mainly deserts with seasonal average temperatures as high as 45°C, far above the thermal tolerance of many organisms. Vertebrate fossils from the latest Permian are known mainly from higher latitudes, probably because equatorial regions were essentially uninhabitable.

5.4 Paleozoic Ecosystems

LEARNING OBJECTIVES

5.4.1 Describe what early land plants were like, and how they are related to other organisms and to present-day plants.

5.4.2 Describe changes in land plant communities during the Paleozoic.

5.4.3 Describe the radiation of terrestrial invertebrates during the Paleozoic.

5.4.4 Explain how tetrapod communities changed during the later Paleozoic

Climate is an important characteristic of **ecosystems**, which are communities of interacting organisms and their abiotic environment. Ecosystems are complex, including biogeochemical cycles and myriad microbial communities as well as multicellular plants, fungi, and animals. Although the focus of this book is vertebrates and their evolutionary history, this section will also consider (albeit briefly) plants and invertebrates of the Paleozoic. Plants are important because they create structural habitats and are the base of terrestrial food webs, and many vertebrates eat insects and other invertebrates that rely on plants.

Aquatic life

All early Paleozoic vertebrates were marine. Fragmentary remains from the latest Cambrian and Early Ordovician represent jawless vertebrates such as †conodonts (see Section 3.1) and basal osteognathostomes (see Section 3.3). Basal osteognathostomes (both jawless and jawed) diversified in the Silurian, and acanthodians and osteichthyans appeared toward the end of that period. Many of these early Paleozoic fishes were heavily armored with plates or scales made of dermal bone.

The Devonian is popularly known as the "Age of Fishes." Jawless forms such as †heterostracans and †osteostracans remained prominent in the Early Devonian. Basal gnathostomes, including groups such as †antiarchs, †arthrodires, and †ptyctodonts (previously referred to as "†placoderms," a non-monophyletic group; see Chapter 3), were the most diverse of the Devonian vertebrates, but by the end of the period only eugnathostomes survived. Acanthodians (including chondrichthyans; see Chapter 6) and osteichthyans commenced major radiations in the Devonian, with both freshwater and marine forms.

Chondrichthyans were especially diverse during the Carboniferous, and basal members of the lineages that led to extant sharks and rays appeared by the end of the Paleozoic. Devonian actinopterygians, the group that includes about half of extant vertebrate species, were heavily armored; more modern forms appeared by the end of the Carboniferous. Sarcopterygians, which make up most of the other half of extant vertebrate diversity, radiated in the Devonian and gave rise to tetrapods by the end of the period. By the Middle Permian, however, lungfishes and coelacanths were the only remaining fishlike sarcopterygians. We trace the evolution of Devonian tetrapodomorphs, the lineage that includes some Devonian fishlike sarcopterygians as well as tetrapods, in Chapter 8.

During the Carboniferous and Permian, many aquatic non-amniote tetrapods inhabited fresh waters but the only Paleozoic amniotes that evolved secondarily aquatic forms were †mesosaurs, small reptiles of the Permian.

Terrestrial flora

Photosynthesizing bacteria (cyanobacteria, once called blue-green algae) probably have existed in wet terrestrial habitats since the late Archean, around 2.7 billion years ago. We have geochemical evidence for a land-covering of microbes in the late Proterozoic, around 850 Ma, and green algae may also have colonized the land around this time, with their photosynthetic activity resulting in increased levels of atmospheric O_2. Complex terrestrial ecosystems, with evidence of plants and associated fungi and animals (arthropods), are known from the early Silurian, but may date back to the Ordovician. Such landscapes would have looked bleak by our standards—barren, with a few kinds of low-growing, water-dependent plants in moist areas.

Cambrian rivers were wide, without stable banks or associated muddy floodplains. The subsequent explosive evolution of land plants transformed the physical landscape. Plant roots helped stabilize the land, and, by the Devonian, banks of meandering rivers were fertile habitats for both plants and animals.

Extant land plants include mosses, liverworts, hornworts, and **tracheophytes**, the clade that encompasses all other extant land plants. Mosses, liverworts, and hornworts lack internal channels to conduct food and water through their tissues, and are thus low-growing. Tracheophytes have vascular channels (the source of their common name, vascular plants; "tracheophyte" loosely

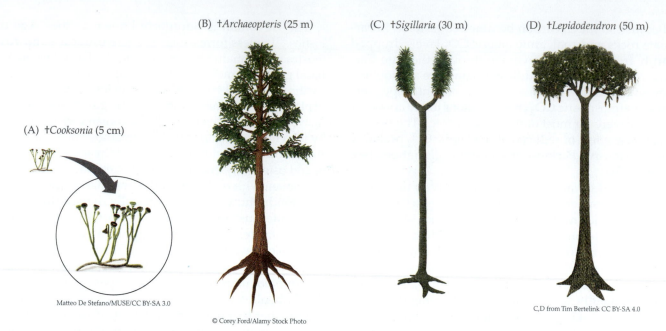

(A) †*Cooksonia* (5 cm)

(B) †*Archaeopteris* (25 m)

(C) †*Sigillaria* (30 m)

(D) †*Lepidodendron* (50 m)

Matteo De Stefano/MUSE/CC BY-SA 3.0

© Corey Ford/Alamy Stock Photo

C,D from Tim Bertelink CC BY-SA 4.0

Figure 5.7 **Carboniferous vascular plants.** (A) †*Cooksonia* was an early vascular plant. Shown is the sporophyte generation with spore-bearing capsules (sporangia) on unbranched stalks. (B) With branches emerging from reinforced joints on the trunk, †*Archaeopteris* was more treelike than other Early Carboniferous plants. (C) †*Sigillaria* and (D) †*Lepidodendron* were lycopods.

translates to "plants with pipes," derived from the Greek term for the human windpipe) and can grow much taller than mosses, liverworts, or hornworts.

Some Paleozoic tracheophytes reproduced by spores, as do the extant lycopods (club mosses), horsetails, and ferns. The vast majority of extant tracheophytes reproduce using seeds, an evolutionary innovation that took place during the Devonian. With more than 300,000 extant species, and tens of thousands of species known as fossils, it is far beyond the scope of this book to trace the evolutionary history of tracheophytes, but we can briefly survey how tracheophyte diversification changed terrestrial habitats and paved the way for the evolution of new types of terrestrial vertebrates.

The early tracheophyte †*Cooksonia*, first known from the mid-Silurian, reached heights of only a few centimeters (**Figure 5.7A**). Subsequent evolution of lignin, a complex polymer that makes vascular plants stiff and woody, enabled greater heights, and vascular plants increased in size over the course of the Devonian, eventually attaining the stature of trees. They also developed rooting structures that probably had mutualistic associations with mycorrhizal fungi, accelerating weathering and creating soils rich in organic matter. Later, during the Carboniferous, vegetation was sufficiently luxuriant that dead plant parts accumulated, forming the coal deposits that modern humans have exploited for industrial purposes. These coals often contain charcoal, documenting that occasional forest fires were also part of these ecosystems.

As terrestrial ecosystems increased in complexity and diversity, taller vegetation produced wet microclimates

at ground level—a step toward the variation in terrestrial microclimates that plants and animals exploit today. Treelike forms evolved independently within several ancient plant clades, and stratified forest communities consisted of plants of different heights. Plants with large leaves first appeared in the Late Devonian, a time that saw the development and spread of forests dominated by †*Archaeopteris*, an early seed plant with a trunk up to 1 m in diameter, a complex and deep rooting system, and height as much as 25 m (**Figure 5.7B**). Ferns, horsetails, and lycopods such as †*Sigillaria* and †*Lepidodendron* (**Figure 5.7C,D**) dominated Early Carboniferous forests. †*Lepidodendron* reached heights of 50 m, tall even by the standards of extant trees, but its extant relatives, the clubmosses, are small ground plants. Carboniferous terrestrial landscapes would have looked superficially like current-day scenes, although the types of plants were completely different.

By the end of the Devonian, land plants had evolved most of the major innovations for life on land except fruits and flowers. Together, mosses, liverworts, hornworts, and tracheophytes such as lycopods and early seed plants made up the **Paleophytic Flora** (**Figure 5.8**). Fundamental changes occurred near the end of the Carboniferous, when ecosystems previously dominated by spore-producing plants were replaced by seed-producing ones, yielding the **Mesophytic Flora**. This floral change, known as the **Carboniferous Rainforest Collapse**, coincided with low levels of atmospheric CO_2, aridification, glaciation, and low sea levels. Plants that reproduce using spores need moist habitats for reproduction. In contrast, seed plants, such as **gymnosperms** (extant conifers and cycads), can reproduce without water, which allowed them to spread into drier upland habitats.

Mesophytic forests were dominated by tree ferns, had diverse vines and ground cover, and included early conifers and cycads. This floral change had a profound effect

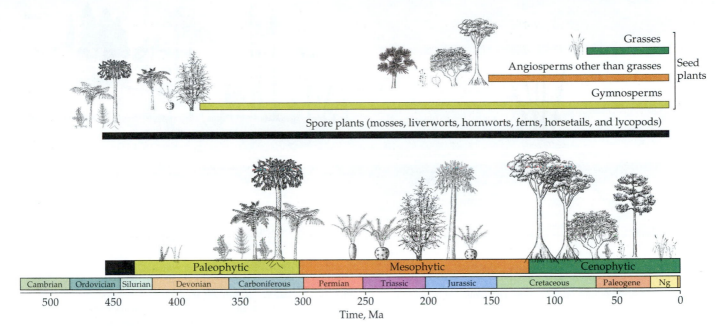

Figure 5.8 **Evolution of land plants during the Phanerozoic.**
The three evolutionary floras, Paleophytic, Mesophytic, and Ceno-
phytic, do not correspond exactly with the Paleozoic, Mesozoic, and
Cenozoic geological eras. (Modified from J.C. McElwain. 2018. *Annu
Rev Plant Biol* 69: 761–787, with permission.)

on the fauna: the first herbivorous tetrapods appeared
at this time, and herbivorous insects radiated. Diversification of herbivorous tetrapods in turn had profound
effects on land plants.

By the Permian, forests typical of drier environments
characterized terrestrial habitats worldwide. These forests
were dominated by conifers and contained many deciduous plants, in contrast to the earlier evergreen forests. By
the Late Permian, highly productive forests in Antarctica reflected the return to an ice-free Earth. **Angiosperms**,
the flower- and fruit-producing seed plants (including
grasses) that predominate today, did not evolve until the
Mesozoic. They form the basis of the **Cenophytic Flora**
(see Figure 5.8). Grasses evolved late in the Mesozoic and
diversified in the Cenozoic. As terrestrial environments
became cooler and drier during the Cenozoic, grasslands
expanded, leading to the diversification of many groups
of mammals.

Terrestrial fauna

In addition to vertebrates, three other animal clades also
diversified on land: arthropods (arachnids and insects),
mollusks (slugs and snails), and annelids (earthworms).
Terrestrial arthropods are known from the late Silurian,
but the first evidence of terrestrial mollusks and annelids
is in the Carboniferous.

Three arthropod lineages independently colonized
land during the Paleozoic. Although molecular data show
that these events took place during the Cambrian and
fossil arthropod tracks are known from that period, fossil myriapods (millipedes and centipedes) and arachnids
(spiders, scorpions, and mites) are first known from the
late Silurian. Springtails and primitive flightless insects

appear in the Early Devonian record. Fossils of terrestrial
crustaceans such as isopods (woodlice) and amphipods
(land hoppers) are not known until the Cretaceous. Early
arthropod colonizers probably ate organic debris from
dead and dying plants, as well as each other.

Flying insects were present in the Early Carboniferous, and herbivorous insects appeared near the end of the
Early Carboniferous. Giant dragonflies (one species had
a wingspan of 63 cm) flew through the air, and predatory
†arthropleurids (millipede-like arthropods) were nearly
2 m long. Arthropod groups appearing in the Permian
that survive to the present include hemipterans (bugs), coleopterans (beetles), and forms resembling but not closely
related to mosquitoes.

The earliest tetrapods are known from Late Devonian
rocks (see Chapter 8). The transition from the Carboniferous to the Permian was marked by increasingly arid
terrestrial habitats that favored the radiation of amniotes,
which do not need water for reproduction. Molecular data
place the split between lissamphibians and amniotes in
the Early Carboniferous, and fragmentary fossils from this
time hint that amniotes might have been present. However, the first definitive amniote fossils date from the Late
Carboniferous. Soon afterward, amniotes diverged into
two major lineages, synapsids and sauropsids (see Chapter
13). The synapsid lineage, which eventually led to mammals, was the more diverse during the Paleozoic. The
sauropsid lineage led to reptiles (including birds), which
became preeminent during the Mesozoic. Even today, reptiles have a far greater diversity than mammals.

By the Early Carboniferous an established fauna of
terrestrial tetrapods consisted of animals ranging in size
from a few centimeters to more than 2 m (**Figure 5.9**),
almost all of them carnivores. Aquatic non-amniotes,
such as the aquatic apex predators †*Crassigyrinus* and
†*Loxomma* lived in large bodies of open water, while several

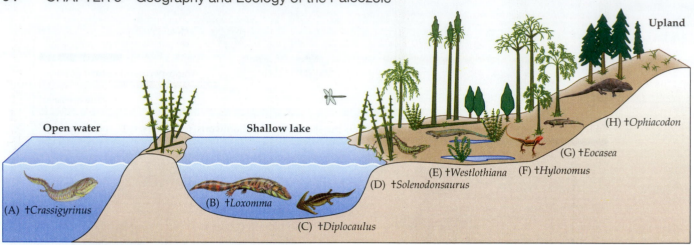

Upland

Open water

Shallow lake

(H) †*Ophiacodon*

(G) †*Eocasea*

(F) †*Hylonomus*

(E) †*Westlothiana*

(D) †*Solenodonsaurus*

(A) †*Crassigyrinus*

(B) †*Loxomma*

(C) †*Diplocaulus*

Figure 5.9 **A Carboniferous scene.** Reconstruction of a hypothetical Late Carboniferous lake and its surrounding landscape. Plants and animals are not shown to scale. (A) †*Crassigyrinus*, an apex aquatic carnivore, probably lived in large bodies of water; ~2 m. (B) †*Loxomma*, a slightly smaller aquatic carnivore, lived in swamps; ~1.5 m. (C) †*Diplocaulus*, a genus of aquatic carnivores with several species ranging from 30 cm to 1 m. Flaps of skin may have extended from the sides of the head to the trunk, creating a hydrofoil that provided lift in flowing water. (D) †*Solenodonsaurus*, a medium-size terrestrial carnivore; ~45 cm. (E) †*Westlothiana*, a terrestrial predator; ~20 cm. (F) †*Hylonomus*, a basal amniote, was a small terrestrial insectivore; ~20 cm. (G) †*Eocasea*, the earliest known terrestrial herbivore; ~20 cm. (H) †*Ophiacodon*, a terrestrial apex carnivore; ~2 m. (Landscape after M. Benton. 2014. *Vertebrate Palaeontology*, 4th Edition, Wiley-Blackwell.)

species of †*Diplocaulus* lived in shallower pools. Other non-amniotes, such as †*Solenodonsaurus* and †*Westlothiana* were terrestrial predators. Carboniferous amniotes included †*Hylonomus*, a terrestrial insectivore, and two synapsids: †*Ophiacodon*, a terrestrial apex predator, and †*Eocasea martini*, the first herbivorous tetrapod.

†*Eocasea* was small (20 cm) and had pointed teeth, but derived †caseids of the Permian were large (up to 5.5 m) with blunt teeth and capacious rib cages, housing the large guts required to digest vegetation. (Multicellular animals cannot digest cellulose, the structural material of plants, and must rely on symbiotic microorganisms in the gut to ferment plant material.) Herbivore diversity—mainly large-bodied synapsids and †parareptiles (an extinct group of sauropsids)—reached a peak in the Late Permian. Following the end-Permian extinctions (see Section 5.5), new types of herbivorous tetrapods evolved over the course of the Triassic.

Late Carboniferous and Early Permian terrestrial faunas showed less geographic variation in composition than terrestrial faunas of earlier times, probably reflecting greater connectivity between communities following the Carboniferous Rainforest Collapse. During the Early Permian, diversity of non-amniote tetrapods declined, while that of amniotes increased. Several amniote lineages gave rise to small, insectivorous predators rather like extant lizards. Larger synapsids (up to 1.5 m long), such as the sail-backed †*Dimetrodon*, were apex terrestrial predators.

By the Late Permian, terrestrial communities had started to resemble those of today, with diverse herbivorous vertebrates supporting a smaller variety of carnivorous forms, although the kinds of plants and animals in Permian ecosystems were almost entirely different from those of today.

5.5 Extinctions

LEARNING OBJECTIVES

5.5.1 Describe the two major mass extinctions in the Paleozoic.

5.5.2 Explain why major mass extinction events have been used as milestones to mark the boundaries between eras (e.g., the Paleozoic–Mesozoic boundary).

5.5.3 Describe the climatic events and other upheavals that led to the end-Permian extinctions.

Extinctions occurred constantly throughout the Phanerozoic, including several mass extinctions that greatly impacted the evolutionary history of vertebrates. Based chiefly on the marine fossil record of invertebrates (because it is far more extensive than the record of terrestrial animals), paleontologists recognize five major mass extinctions and many minor ones. There is no sharp distinction between "major" and "minor" mass extinctions. A major mass extinction results in sharply increased extinction rates for many groups of organisms in many habitats. For example, the end-Permian, or Permo-Triassic, mass extinction marks the boundary between the Paleozoic and Mesozoic and is estimated to have eliminated 95% of marine species and 70% of terrestrial species. Earlier in the Permian, 15 clades of tetrapods (primarily non-amniote tetrapods and basal synapsids) disappeared in a relatively minor mass extinction known as Olson's extinction.

Recovery from a mass extinction is an equally critical evolutionary event that is measured by the time it takes to recover the level of diversity that existed prior to the extinction. Levels of diversity can be difficult to estimate because the taxa after an extinction event are typically different from the ones present prior to the event, but some patterns are apparent. Recovery from the

(A) †*Lystrosaurus*

(B) Fossilized mummy of †*Lystrosaurus*

(C) Artist's reconstruction showing carcass prior to fossilization

DataBase Center for Life Science (DBCLS)/CC BY 4.0

B.C from R.M.H. Smith and J. Botha-Brink 2014. *Palaeoecology* 396: 99–118

15 cm

Figure 5.10 **Vertebrate fossils show evidence of end-Permian drought conditions.**
(A) †*Lystrosaurus murrayi*, an end-Permian synapsid relatively close to the ancestry of mammals. With a body length of ~0.5 m, this was one of the smallest species in the genus. (B) †*Lystrosaurus* fossil from the latest Permian of the Karoo basin of South Africa, preserved in a lakebed. (C) This animal died and was mummified during a drought and the mummy was subsequently buried in lake sediments. Mummification preserved the skeleton intact and with the bones articulated in the position assumed soon after death. Fossilized, mummified skin has been found on many skeletons preserved in the Karoo at about the Permo-Triassic boundary.

end-Permian mass extinction took some 5 million years, and may have taken as long as 10 million years. Recovery from other mass extinctions has been faster.

Several minor extinctions occurred during the Cambrian, but a major mass extinction of marine invertebrates occurred at the end of the Ordovician. The vertebrate fossil record from the Ordovician is too scanty to know how these extinctions affected vertebrates. Additional minor extinctions occurred in the Silurian and the Devonian, followed by two major mass extinctions during the Late Devonian. An estimated 44% of vertebrate lineages became extinct at this time, including jawless osteognathostomes, †antiarchs, †arthrodires, †ptyctodonts, and many families of sarcopterygians including early tetrapods. The Devonian extinctions may have been related to an increase in UV-B radiation following thinning of the ozone layer associated with a period of global warming. At least one study has suggested that a supernova explosion contributed to this ozone depletion. The end-Devonian extinctions resulted in restructuring of marine vertebrate communities and the rise to dominance of chondrichthyans and actinopterygians.

The greatest extinction of the Paleozoic—indeed, the greatest extinction of all time—occurred at the end of the Permian, 252 Ma. Approximately 57% of marine invertebrate families and 95% of all marine species, including 12 families of fishes, disappeared. Among tetrapods, four of 11 non-amniote families and 17 of 32 amniote families disappeared, with especially heavy losses among

synapsids. Indeed, the end-Permian extinction events ended the dominance of synapsids as the large terrestrial vertebrates, a role they would not play again until the evolution of large mammals in the Cenozoic. Terrestrial sauropsids—including the giant dinosaurs that so captivate our imaginations—would dominate Mesozoic fauna of the "Age of Reptiles."

The end-Permian extinction occurred in pulses extending over about 180,000 years, so it is unlikely that it was caused by a single catastrophic event like an asteroid impact. A more likely explanation lies in massive volcanic eruptions in Siberia. In a period of less than 1 million years, enough molten lava was released to cover an area half the size of the United States. Carbon isotope changes in boundary sediments indicate that massive amounts of the greenhouse gases CO_2 and methane (CH_4) were released, probably from the interaction of volcanic magma with oil shales.

Such a rapid change in atmospheric composition would have produced profound and rapid global warming, acid rain on land, and ocean acidification. Charcoal- and soot-bearing sediments are evidence of extensive wildfires, indicating habitat destruction and terrestrial ecosystem collapse. Tetrapod distributions shifted away from the Equator and toward the poles, presumably to escape extreme temperatures. The fossil record of end-Permian synapsids, with many specimens mummified before preservation, suggests that they succumbed during droughts (**Figure 5.10**).

Thus, the Permo-Triassic boundary was marked by a **runaway greenhouse effect**, a positive feedback loop of events resulting in planetary warming that disrupted global environmental mechanisms for hundreds of thousands of years and caused the extinction of almost all multicellular life on Earth.

Summary

5.1 Deep Time

Earth's history is divided into eons, eras, periods, and epochs. The vast majority of this history, extending from Earth's formation 4,600 Ma to 541 Ma, is referred to as the Precambrian.

The Precambrian consists of a lifeless time informally known as the "Hadean," followed by two eons, the Archean and the Proterozoic. The Archean supported only unicellular life forms. Multicellular organisms first appeared in the Proterozoic.

The Phanerozoic eon includes only the last 541 million years, but encompasses almost all the history of multicellular life. The Phanerozoic is divided into three eras, the Paleozoic, Mesozoic, and Cenozoic.

5.2 Continental Geography

Continents move over geological time, and their positions affect how heat is transferred via oceanic circulation, determining whether there is ice at the poles or continental glaciation.

Continental movements affected the flora and fauna, at different times resulting in isolation or mingling of organisms.

Continents were initially widely separated, but by the end of the Paleozoic they had come together, forming the supercontinent Pangaea.

5.3 Paleozoic Climates

In addition to the climatic effects of shifting continents, Earth's climate is affected by changing concentrations of atmospheric gases such as O_2, CO_2, and CH_4.

The Paleozoic saw extreme changes in atmospheric CO_2 levels. Early in the Paleozoic a CO_2 level 10 times that of today resulted in a greenhouse climate.

A major glaciation in the Late Ordovician cooled Earth, resulting in suitable conditions for the invasion of terrestrial life. A second major glaciation occurred from the mid-Carboniferous to the mid-Permian, accompanied by low levels of atmospheric CO_2 and high levels of atmospheric O_2.

The later Permian saw a return to an ice-free world. At the end of the Permian the equatorial regions were so hot and dry as to be uninhabitable. The poles were free of permanent ice caps until the mid-Cenozoic, and there were no major glaciations until the late Cenozoic ice ages, which are still in progress.

Climate changes were influenced by the covering of Earth by vegetation. Plants affected water cycles and biogeochemical cycles, notably by the sequestration of carbon in coals and carbonate rocks such as limestone, which lowered atmospheric CO_2 levels and ultimately cooled the planet.

5.4 Paleozoic Ecosystems

Marine animals diversified explosively in the early Cambrian, with the appearance of all the major invertebrate phyla as well as the first vertebrates.

- The jawless †conodonts appeared in the mid-Cambrian and survived until the end of the Paleozoic. Jawless armored osteognathostomes were prominent in the Ordovician, but all were extinct by the end of the Devonian.

- Gnathostomes diversified in the Silurian and Devonian. Cartilaginous and bony fishes started to radiate in the Devonian, but many lineages of sarcopterygian fishes were extinct by the Middle Permian.

Life on land was initially sparse. Land plants probably originated in the Cambrian and diversified in the Ordovician.

The Paleophytic Flora that characterized most of the Paleozoic consisted of both nonvascular and vascular plants. Nonvascular plants are represented today by mosses, liverworts, and hornworts. Vascular plants (tracheophytes), including all of the dominant extant plant groups, have internal channels for the conduction of food and water and thus can grow taller than nonvascular plants.

- Early low-growing land plants are known from the Silurian, and terrestrial floral ecosystems rapidly became established.

- During the Devonian and throughout most of the Carboniferous, the predominant global vegetation was swampy rainforest. Ferns, horsetails, and lycopods dominated the flora, along with some early seed plants.

- During the Carboniferous, oscillations in sea level resulted in the formation of coal beds from buried vegetation.

A profound shift in terrestrial vegetation occurred near the end of the Carboniferous with the transition to the Mesophytic Flora. Earlier ecosystems collapsed in most of the world and were replaced by more complex forests characteristic of seasonally dry environments. Permian forests were dominated by seed-bearing plants such as conifers and contained many deciduous plants.

Conifers and cycads (gymnosperms) appeared near the end of the Carboniferous, but the flowering seed plants, or angiosperms, that dominate the modern Cenophytic Flora were unknown during the Paleozoic.

Land plants formed the food base for terrestrial faunal ecosystems based on terrestrial invertebrates.

Three groups of arthropods colonized land during the Paleozoic. Myriapods (centipedes and millipedes) and

Summary *(continued)*

arachnids (spiders, scorpions, and mites) appeared in the late Silurian. Fossils of springtails and primitive flightless insects are known from the Early Devonian, although all may have originated in the Cambrian.

Flying insects are known from the Early Carboniferous. Slugs, snails, and earthworms also first appeared in the Carboniferous. Radiations of herbivorous insects correlate with the shift from Paleophytic to Mesophytic flora near the end of the Carboniferous.

Vertebrates did not evolve terrestrial forms until the Late Devonian. Many early tetrapods were aquatic, but terrestrial communities of tetrapods were established early in the Carboniferous.

- Non-amniote tetrapods (terrestrial and semiaquatic) remained diverse through the Carboniferous. They began to decline in the Early Permian, and amniotes diversified.

- The earliest amniotes are known from the Late Carboniferous. Soon after their first appearance,

amniotes diversified in two lineages: synapsids (mammals) and sauropsids (reptiles, including birds).

Vertebrate communities of the Late Permian resembled those of today, but with different players: herbivores were mainly synapsids, insectivores were small sauropsids, and the principal predators were large synapsids.

5.5 Extinctions

Several major mass extinctions occurred during the Paleozoic, with the largest at the end of the Paleozoic era, 252 Ma.

- The Ordovician extinction affected mainly marine invertebrates, but vertebrate diversity was greatly reduced by the extinction at the end of the Devonian.

- The end-Permian or Permo-Triassic extinction was the greatest in Earth's history. Volcanic eruptions resulted in a runaway greenhouse effect, with global warming and ocean acidification adversely affecting all life.

Discussion Questions

5.1 We think of the presence of ice caps at the poles as the "normal" condition, but in fact this situation is less common over Earth's geological history than are ice-free polar regions. How does continental movement relate to this?

5.2 What other aspects of Earth's history (besides continental movement) might make the world colder, with ice at the poles?

5.3 What effect did the Late Carboniferous/Early Permian climate changes have on global vegetation, and how did this affect vertebrate evolution?

5.5 What profound change happened to the community structure of tetrapods at the end of the Carboniferous that led to a change for the entire terrestrial vertebrate community?

5.6 Several groups of invertebrates colonized the land before vertebrates did. What is a possible reason for this? Did things have to happen this way?

Figure credits
Figure 5.6: J. Hansen et al. 2013. *Environ. Res. Lett.* 8: 011006; L.E. Lisiecki and M.E. Raymo. 2005. *Paleooceanog Paleoclimatol.* 20: 1; S.A. Marcott et al. 2013. *Science* 339: 1198-1201; J.C. McElwain 2018. *Annu Rev Plant Biol* 69: 761–787; D.L. Royer et al. 2004. CO2 as a primary driver of Phanerozoic climate. *GSA Today*; v. 14; no. 3, doi: 10.1130/1052-5173; C.R. Scotese. 2015. PALEOMAP Project. https://www.researchgate.net/publication/275277369_Some_Thoughts_on_Global_Climate_Change_The_Transition_for_Icehouse_to_Hothouse_Conditions; Zachos, J. C. et al. 2008. *Nature* 451: 279–283.

© iStock.com/ELizabethHoffmann

6

Origin and Radiation of Chondrichthyans

As you saw in Figure 1.2, Gnathostomata includes two crown groups: cartilaginous fishes (Chondrichthyes) and bony vertebrates (Osteichthyes). From tracing the evolutionary history of features such as the maxilla (a dermal bone that forms part of the upper jaw), we know that characters of gnathostomes evolved mosaically rather than all at once. Although gnathostomes have vertebral centra, ribs, and continuous tooth replacement, we have yet to completely understand the sequence In which these synapomorphies evolved.

In the Silurian and Early Devonian, Paleozoic eugnathostomes rapidly diversified into an array of body forms, some known by remarkably complete fossils and others by tantalizing fragments. But even relatively complete specimens of Devonian eugnathostomes have puzzled generations of vertebrate paleontologists because of difficulties in assessing their relationships to either Chondrichthyes or Osteichthyes. New phylogenetic approaches, new preparation methods for fossils, and especially the discovery of taxa with new combinations of characters are shedding light on stem taxa for both chondrichthyans and osteichthyans. There are still many uncertainties, and the phylogenetic interpretations of chondrichthyans presented in **Figures 6.1** and **6.2** will undoubtedly change as we learn more about this fascinating group.

This chapter begins with a discussion of stem chondrichthyans (long referred to as spiny sharks or acanthodians) and an overview of Chondrichthyes. We then focus on extant chondrichthyans—Chimaeriformes (chimaeras) and Neoselachii (sharks and rays)—and close with a discussion of the ecological implications of declining populations.

6.1 Acanthodii

LEARNING OBJECTIVES

6.1.1 Explain the phylogenetic position of acanthodians.

6.1.2 Describe the "Wonder Block" and why it exemplifies the "good fossil effect."

6.1.3 Explain the significance of †*Doliodus problematicus*.

For more than a century, paleontologists have struggled to interpret many taxa of small Paleozoic fishes that had stout spines anterior to well-developed dorsal, anal, and paired fins (**Figure 6.3A,B**). In addition to these spines, some had up to six pairs of prepelvic spines, and, in exceptionally well-preserved specimens, each can be seen to support a small fin. Known as spiny sharks or **acanthodians** (Greek *acanthos*, "spine"), these fishes had dermal bones in the skull, a partially ossified internal skeleton, and a heterocercal caudal fin (one in which the vertebral column turns upward into the upper lobe of the fin). Many had denticles covering the body like the placoid scales of living chondrichthyans. Were acanthodians shark relatives? Early osteichthyans? Or something in between?

In recent years, the phylogenetic position of acanthodians has been clarified: they are stem chondrichthyans, which means that acanthodians are not extinct and that Acanthodii includes Chondrichthyes (see Figure 6.2). This new understanding has been driven in part by discoveries of exceptional fossils such as the "Wonder Block" from the Man on the Hill locality, an Early Devonian deposit in the Northwest Territories of Canada (**Figure 6.3C**). This single block preserves two articulated specimens (i.e., fossils with their bones connected) of the stem chondrichthyan †*Brochoadmones* that show prepelvic spines supporting small fins (**Figure 6.3D**), as well as specimens of other stem chondrichthyans, including †*Ischnacanthus*,

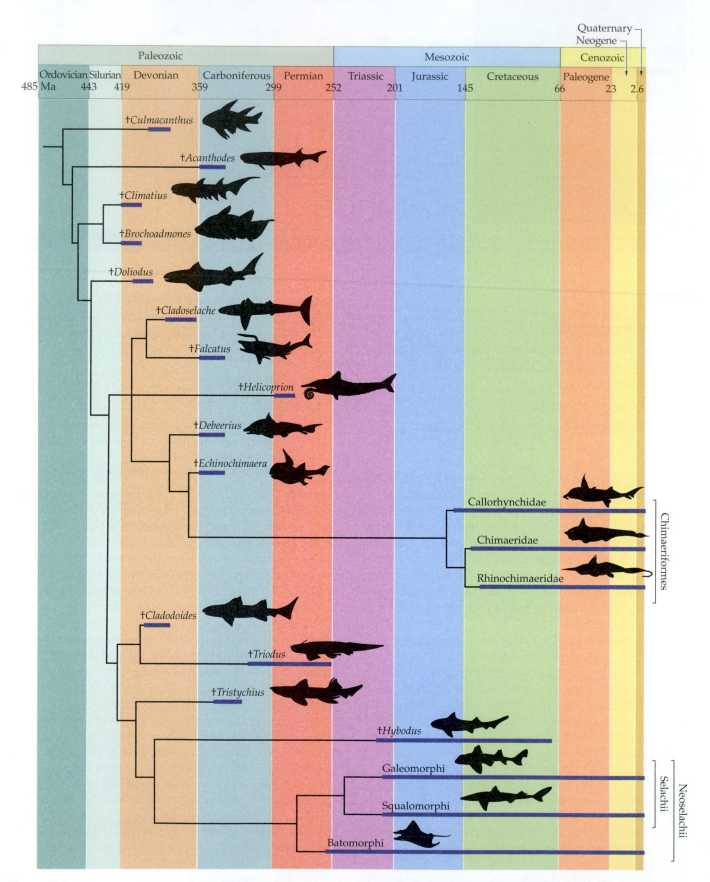

Figure 6.1 Time tree of Acanthodii and Chondrichthyes.
Occurrences of acanthodians in the fossil record are indicated by blue bars. Narrow lines show phylogenetic relationships. Daggers (†) indicate extinct taxa. (Based on multiple sources including J.A. Moy-Thomas and R.A. Miles. 1971. *Palaeozoic Fishes*. Chapman & Hall, London; J.S. Nelson, T. C. Grande and M. V. H. Wilson. 2016. *Fishes of the World*. 5th edition. John Wiley & Sons, Inc.: Hoboken, N.J.; S. Turner et al. 1983. *Mem. Ass. Australas. Palaeontols*. 1: 51–65; J.R. Dick. 1978. *Trans Royal Soc Edinburgh* 70: 63–108.)

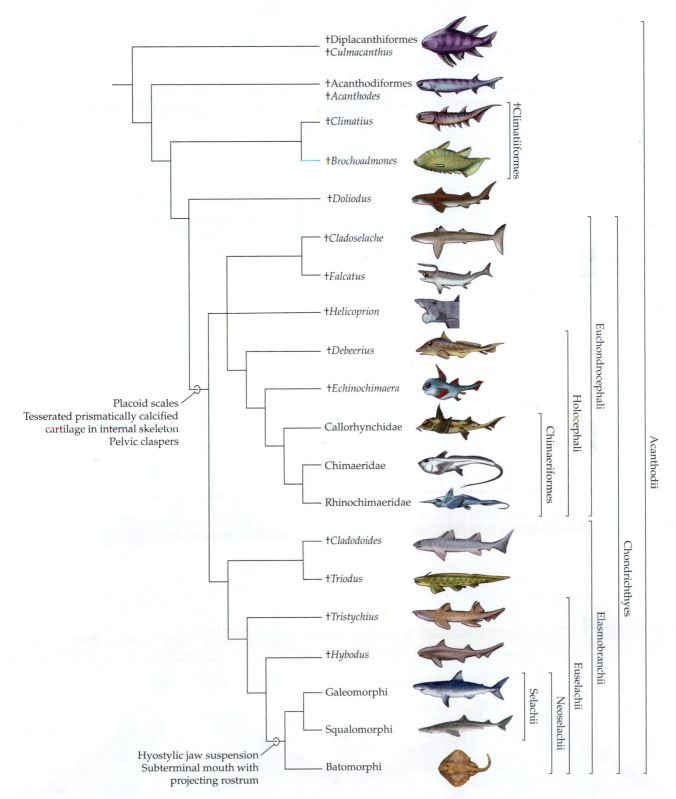

Figure 6.2 Phylogeny of Acanthodii and Chondrichthyes. Taxa and characters are shown. Daggers (†) indicate extinct taxa. (Based on M.I. Coates et al. 2018. *Proc R Soc B* 285: 20172418. Silhouettes based on multiple sources including J.A. Moy-Thomas and R.A. Miles. 1971. *Palaeozoic Fishes*. Chapman & Hall, London; J.S. Nelson et al. 2016. *Fishes of the World*. 5th edition. John Wiley & Sons, Inc.: Hoboken, N.J.; S. Turner, J.A. Long. 1983. *Mem. Ass. Australas. Palaeontols.* 1: 51–65; J.R. Dick. 1978. *Trans Royal Soc Edinburgh* 70: 63–108.)

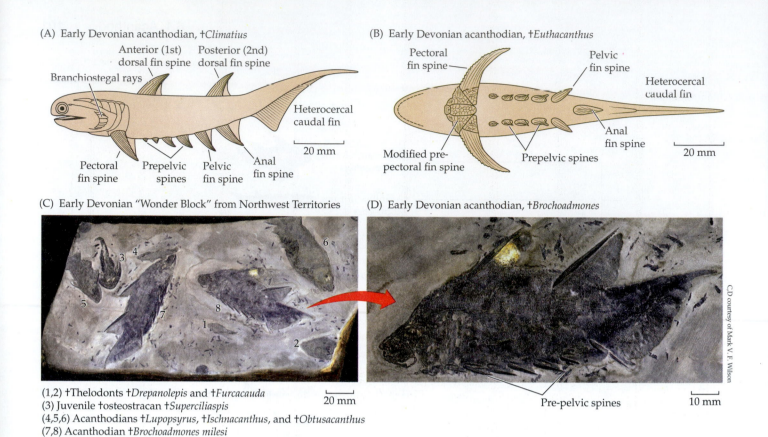

(A) Early Devonian acanthodian, †*Climatius*

Branchiostegal rays
Anterior (1st) dorsal fin spine
Posterior (2nd) dorsal fin spine
Heterocercal caudal fin
Pectoral fin spine
Prepelvic spines
Pelvic fin spine
Anal fin spine
20 mm

(B) Early Devonian acanthodian, †*Euthacanthus*

Pectoral fin spine
Pelvic fin spine
Heterocercal caudal fin
Anal fin spine
Prepelvic spines
Modified pre-pectoral fin spine
20 mm

(C) Early Devonian "Wonder Block" from Northwest Territories

(1,2) †Thelodonts †*Drepanolepis* and †*Furcacauda*
(3) Juvenile †osteostracan †*Superciliaspis*
(4,5,6) Acanthodians †*Lupopsyrus*, †*Ischnacanthus*, and †*Obtusacanthus*
(7,8) Acanthodian †*Brochoadmones milesi*

20 mm

(D) Early Devonian acanthodian, †*Brochoadmones*

Pre-pelvic spines
10 mm

C,D courtesy of Mark V. F. Wilson

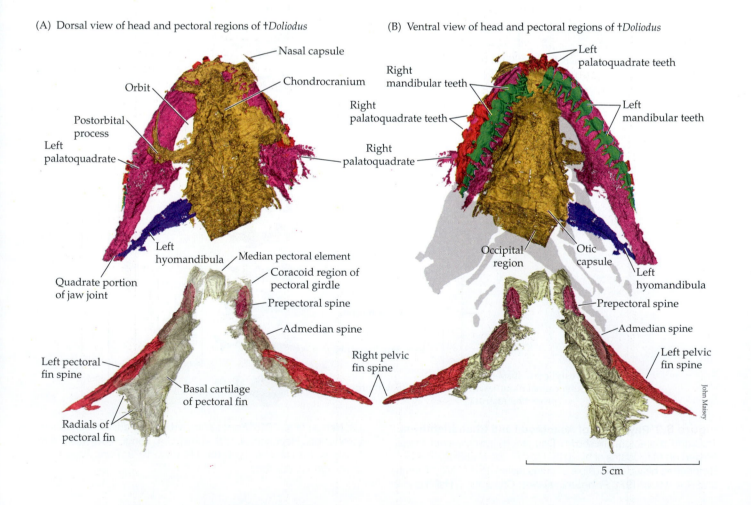

(A) Dorsal view of head and pectoral regions of †*Doliodus*

Nasal capsule
Orbit
Chondrocranium
Postorbital process
Left palatoquadrate
Quadrate portion of jaw joint
Left hyomandibula
Median pectoral element
Coracoid region of pectoral girdle
Prepectoral spine
Admedian spine
Left pectoral fin spine
Basal cartilage of pectoral fin
Radials of pectoral fin

(B) Ventral view of head and pectoral regions of †*Doliodus*

Left palatoquadrate teeth
Right mandibular teeth
Right palatoquadrate teeth
Left mandibular teeth
Right palatoquadrate
Occipital region
Otic capsule
Left hyomandibula
Prepectoral spine
Admedian spine
Right pelvic fin spine
Left pelvic fin spine

5 cm

John Maisey

◄ **Figure 6.3** Early Devonian acanthodians. (A) Lateral view of †*Climatius* showing paired pre-pelvic spines. (B) Ventral view of †*Euthacanthus*. (C) The Early Devonian "Wonder Block" from the Man on the Hill Locality in the Northwest Territories, Canada, contains eight vertebrate fossils representing seven different taxa. (D) Details of †*Brochoadmones*. (A,B after J.A. Moy-Thomas and R.A. Miles. 1971. In *Palaeozoic Fishes*. Chapman & Hall, London; C,D modified from G.F. Hanke and M.V.H. Wilson. 2006. *J Vert Paleontol* 26: 526–537.)

†*Lupopsyrus* and †*Obtusacanthus*. Amazingly, the "Wonder Block" also includes fossils of six other Early Devonian fishes, including jawless forms such as †thelodonts and an †osteostracan. In addition to showing that fish communities in the Early Devonian were remarkably diverse, the "Wonder Block" exemplifies the **"good fossil effect"**: spectacularly well-preserved fossils that drive new ideas and interpretations.

At the base of the acanthodian tree is †Diplacanthiformes, including taxa such as the deep-bodied †*Culmacanthus* from the Middle Devonian of Australia. Next is †Acanthodiformes, which includes species of †*Acanthodes* known from relatively well preserved specimens from the Permian of Germany (see Figure 6.2). Their body shape somewhat resembles that of the unrelated extant anchovies and, like anchovies, †*Acanthodes* had long gill rakers for filter feeding. All †Acanthodiformes were extinct by the end of the Paleozoic. The clade that includes †*Climatius* and †*Brochoadmones* contained many other taxa, but none of these survived the mass extinction at the end of the Devonian (see Figure 6.1).

An important transitional fossil between basal acanthodians and chondrichthyans is †*Doliodus problematicus* from the Early Devonian of Canada. Its teeth are organized like those of extant sharks, but, like more basal acanthodians, it retained paired prepectoral and prepelvic fin spines not found in chondrichthyans (**Figure 6.4**). Paleoichthyologist John Maisey has referred to †*Doliodus* as "sharkyopteryx" because it has a mixture of characters that illuminate the transition to chondrichthyans, just as †*Archaeopteryx* illuminates the transition to birds (see Section 19.2).

◄ **Figure 6.4** A unique acanthodian. Reconstructions based on computed tomography (CT) of the head and pectoral region of the only known specimen of †*Doliodus problematicus* from the Early Devonian of Canada. Until the 1990s, the only ways to study fossil specimens were by preparing them either with mechanical tools such as needles and miniature sandblasters, or with chemical methods using acids. Such methods risk damaging or destroying a unique fossil, but CT scanning makes it possible to study the anatomy of specimens such as †*Doliodus* without that risk. (Original figure by John Maisey from J. G. Maisey et al 2019. In *Evolution and Development of Fishes*, pp. 87–109. Used with permission of Cambridge University Press.)

6.2 Chondrichthyes

LEARNING OBJECTIVES

6.2.1 Compare number of opercular openings in chimaeras to number of gill openings in elasmobranchs.

6.2.2 Dispel the myth that †megalodon (†*Carcharocles megalodon*) was closely related to the extant white shark (*Carcharodon carcharias*).

6.2.3 Describe how the skeleton of chondrichthyans differs in composition from that of osteichthyans.

6.2.4 Explain tooth replacement in elasmobranchs by contrasting locations of functional and replacement teeth.

6.2.5 State the structural differences and functional implications of holostylic and hyostylic jaw suspension.

6.2.6 Describe the type of fertilization universal among chondrichthyans and the male structures that make this possible.

Chondrichthyes includes diverse forms in two extant radiations, Euchondrocephali and Elasmobranchii. Fossils and molecular dating place the split between euchondrocephalians and elasmobranchs before the start of the Devonian (see Figures 6.1 and 6.2).

- Within Euchondrocephali, Holocephali includes the extant Chimaeriformes (ratfishes and allies, commonly known as chimaeras). Chimaeriforms have separate right and left opercular openings (**Figure 6.5A**).

- Within Elasmobranchii, Neoselachii includes the extant Selachii (sharks) and Batomorphi (rays). The name Elasmobranchii (Greek *elasma*, "plate"; *branchia*, "gills") refers to the location of the gills on plates of tissue that extend to the body surface, creating multiple gill openings on each side of the head. Typically there are five gill openings (**Figure 6.5B**), although some elasmobranchs have six or even seven.

Historically, many authors interpreted extant chondrichthyans as representative of primitive gnathostomes, but the many unique and remarkable specializations of chondrichthyans do not make them good models to interpret the anatomy and biology of Early Devonian gnathostomes.

There are problems interpreting early but incomplete fossils that resemble extant chondrichthyans. For example, †*Mongolepis* and related Ordovician forms are examples of "scale taxa" because they are known only from fossilized scales. Although these scales resemble those of extant chondrichthyans, they predate the earliest definitive chondrichthyan fossils by about 50 Ma, and we cannot confidently assign them to Chondrichthyes without additional information.

Habitats and diversity

Paleozoic chondrichthyan fossils occur in rocks that formed in freshwater, estuarine, and marine environments. All extant chimaeras and most elasmobranchs are

(A) Opercular openings on each side in a chimaera

(B) Spiracle and five gill openings on each side in a typical elasmobranch

Figure 6.5 Differences in opercular openings of chimaeriforms and gill openings of elasmobranchs. (A) Chimaeriforms have a single opercular opening on each side of the head (example shown is a spotted ratfish, *Hydrolagus colliei*). (B) Elasmobranchs typically have a spiracle and five gill openings on each side of the head, exemplified in this brownbanded bamboo shark, *Chiloscyllium punctatum*.

marine, but more than 50 species of elasmobranchs live in estuarine or freshwater habitats. Several species of sharks, sawfishes, skates, and stingrays routinely use estuaries, and the bull shark, *Carcharhinus leucas*, can live in fresh waters far from the ocean. One family of South American stingrays, Potamotrygonidae, occurs only in fresh water.

Hundreds of species of chondrichthyans are known from fossils, but in most cases the only material consists of fossilized teeth or vertebral centra. For example, large teeth of †*Carcharocles megalodon* (commonly known simply as †megalodon, which is Greek for "big tooth") are found in Miocene and Pliocene deposits around the world. No complete specimens of †megalodon are known, but they may have reached body lengths of 15 m; length estimates vary depending on the method of calculation. Like the extant (and cinematically renowned) white shark *Carcharodon carcharias*, †megalodon preyed on marine mammals, as shown by fossil whale vertebrae from the same deposits with distinctive bite marks that match †megalodon's large teeth.

Formerly, some workers believed that †megalodon was closely related to white sharks, but recent work confirms that †megalodon belongs to a family that is sister to the family Lamnidae, which includes the extant porbeagle, salmon, mako, and white sharks. It is a myth that †megalodon gave rise directly to any extant forms, but that has not stopped television programmers and movie makers from selling the myth. (The similar generic names †*Carcharocles* and *Carcharodon* probably do not help people understand the distinction.)

There are about 50 extant species of chimaeras, 550 species of sharks, and 700 species of skates and rays. Species discovery of both fossil and extant chondrichthyans continues at a rapid pace: more than 200 new species of extant elasmobranchs have been described in the last decade, many from deepwater marine habitats.

Placoid scales

Basal gnathostomes such as †*Qilinyu* (see Section 3.7) and basal acanthodians such as †*Brochoadmones* had large plates of dermal bone (see Figure 3.3D). Chondrichthyans lost those plates but retained scales composed of dentine and enamel in the skin (portions of tissue at the bases of scales and teeth have been interpreted as bone). Like oral teeth, these unique **placoid scales** develop from a dental papilla and have a pulp cavity (**Figure 6.6A**). There are two theories about the relationship between placoid scales and oral teeth. One is the "outside in" theory, in which teeth are interpreted as having evolved from external placoid scales that were associated with the oral region. An alternative theory, appropriately called "inside out," proposes that teeth originated first in the pharynx and only subsequently became distributed on the skin as placoid scales.

Placoid scales are remarkably diverse and serve many functions across the extant species of chondrichthyans. For example, the small, closely spaced scales of fast-swimming sharks reduce drag during swimming (**Figure 6.6B**). Large placoid scales of bullhead sharks (*Heterodontus*) help to resist abrasion from the rocky reefs they inhabit. Spaces between the small scales of lantern sharks (*Etmopterus*) reveal bioluminescent skin between the scales (see Section 6.5).

Cartilaginous skeleton

Chondrichthyan skeletons are mineralized with the same mineral—calcium hydroxyapatite—that osteichthyans use in bones. Chondrichthyans, however, deposit the

(A) Section through skin and placoid scale of an elasmobranch

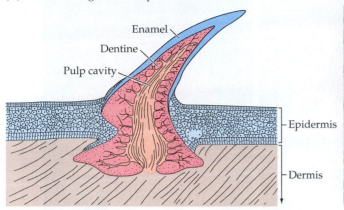

(B) Surface view of placoid scales of bonnethead, *Sphyrna tiburo*

Courtesy George V. Lauder

Figure 6.6 Microstructure of placoid scales. (A) Placoid scales are composed of a core of dentine (surrounding the pulp cavity) and an outer layer of enamel. (B) Scanning electron micrograph shows a surface view of placoid scales of a bonnethead. The animal's anterior is to the left. (A after K. Liem et al. 2001. *Functional Anatomy of the Vertebrates*, 3rd ed. Cengage/Harcourt College: Belmont, CA.)

mineral differently, forming **calcified cartilage**. Vertebral centra of sharks have dense internal calcification, whereas surfaces of other skeletal elements such as the jaws have a crust of calcification in the form of small prisms known as **tesserae** that can be seen in radiographs (see Figure 6.7A).

Genome sequencing of the Australian ghost shark (*Callorhinchus milii*) offers insight into the loss of bone in the internal skeleton of chondrichthyans. The gene family that controls conversion of cartilage into bone in osteichthyans, including humans, is absent in *Callorhinchus*. Experimental studies that knocked out these genes in zebrafish reduced their ability to make bone. While we do not understand the original selective value (if any) of the loss of this gene family, both cartilage and bone perform well in compression tests (cartilage performs less well in tension tests). Loss of the ability to form bone in the internal skeleton did not impose major constraints on breathing, feeding, or locomotion. But bone, particularly the cellular bone typical of early osteichthyans and most descendant clades (including mammals) is more metabolically active and more easily remodeled and repaired than is cartilage. The evolutionary significance of these and other differences between the skeletons of chondrichthyans and osteichthyans remain fascinating and much-debated topics in the history of vertebrates.

The chondrocranium (braincase) of chondrichthyans has a solid cartilaginous roof that protects the brain. This is in contrast to the dermal bones (frontal, parietal, and others) that cover the braincase of osteichthyans (see Figure 2.10).

Teeth and tooth plates

Tooth shapes of extant sharks range from broad, triangular teeth of white sharks (**Figure 6.7A**) to needlelike points in frilled sharks (**Figure 6.7B**), to the **heterodont dentition** of horn sharks in the genus *Heterodontus* (Greek *heteros*,

"different"; *odous*, "tooth"), in which anterior teeth are specialized to puncture prey and the broad posterior molariform teeth are used for crushing (**Figure 6.7C**). Shark teeth form in **tooth whorls**, and are continually replaced as they wear and are lost. Among extant vertebrates, tooth replacement by whorls is retained only by elasmobranchs. Each functional tooth is attached to a ligamentous band that courses down the inside of the jaw cartilage deep below the fleshy lining of the mouth. Developing teeth are aligned in a series behind the functional tooth or teeth, and together the developing and functional teeth are known as a **tooth file** (sometime known as a tooth family). Tooth replacement is often rapid. Young sharks can replace teeth about every 8 days, although rates of replacement vary with the species, age, and general health of the shark, as well as with environmental factors such as water temperature.

Sharks that chase down, capture, and cut into vertebrate prey often have serrated teeth. A typical example is the characteristically shaped teeth of tiger sharks, *Galeocerdo cuvier*, which resemble an old-fashioned pocket can opener (**Figure 6.7D**). These sharks also have smaller primary and secondary serrations on the cutting edges of the teeth (**Figure 6.7E**). Together, the can-opener, primary, and secondary serrations concentrate forces on prey at three different scales, allowing the teeth to cut through prey or process chunks of food of different sizes.

Tooth shapes tell us about roles in three feeding functions: tearing, cutting, and crushing. Biomechanical analyses by Lisa Whitenack and her colleagues found many examples of convergent evolution of tooth shapes within sharks; most teeth tested served at least two of these three functions, and almost half could perform all three. The flexible attachment of a tooth to its ligament allows some movement and rotation of the tooth as force is applied during a bite, which may facilitate multiple tooth functions. In addition, each tooth acts in concert with adjacent teeth: their bases overlap so that forces on one tooth may be transmitted to adjacent teeth.

Early fossil acanthodians and basal osteichthyans had tooth whorls, so it is likely that tooth whorls are a

(A) Three views of functional and replacement teeth in the upper jaw of a white shark, *Carcharodon carcharias*

Tesserae on labial surface of upper jaw

Tesserae on lingual surface of upper jaw

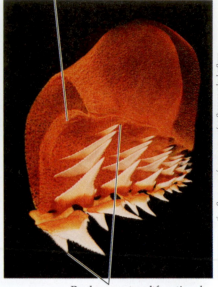

Functional teeth

Functional teeth   Replacement teeth

Replacement and functional teeth together form a tooth file

(B) Teeth in the lower jaw of a frilled shark, *Chlamydoselachus*

(C) Teeth in the lower jaw of a horn shark, *Heterodontus*

Crushing teeth

Puncturing teeth

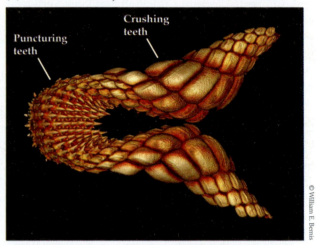

(D) Simple pocket can opener compared with tooth of a tiger shark, *Galeocerdo cuvier*

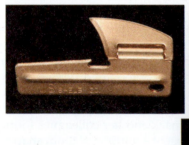

(E) Primary and secondary serrations

Primary serration   Secondary serrations

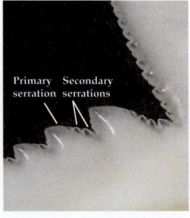

Figure 6.7 **Shark teeth and tooth replacement.** (A) Three views of the upper jaw of a white shark, reconstructed from a CT dataset to show position of the functional and replacement teeth in relation to the upper jaw. Also visible are bright orange crusts of mineralized tesserae on the labial (facing the lip) and lingual (facing the tongue) surfaces of the jaw cartilage. The internal cartilaginous tissue does not mineralize, which is why the cartilage appears hollow in these reconstructions. (B) Rows of functional teeth in the lower jaw of a frilled shark. (C) Teeth in the lower jaw of a horn shark showing sharp puncturing teeth anteriorly and large crushing teeth posteriorly. (D) Comparison of a pocket can opener to the tooth of a tiger shark. Note the prominent cusp in the center of the tooth. (E) Close-up of the tiger shark tooth showing primary and secondary serrations.

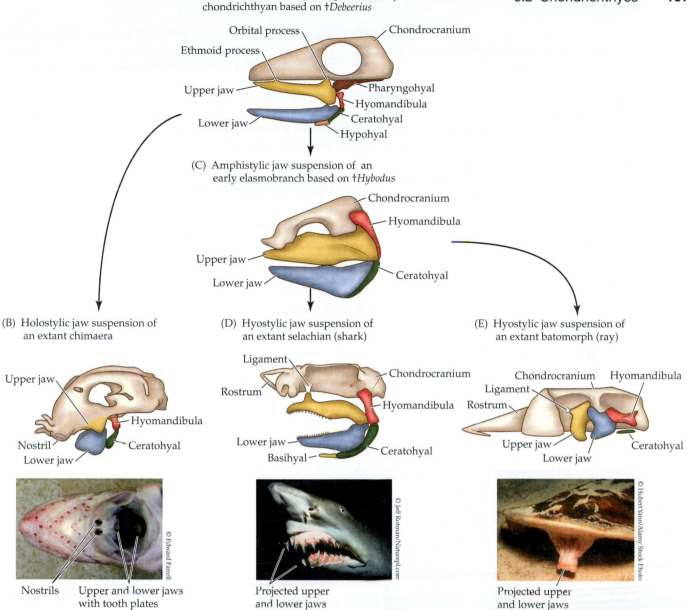

(A) Autodiastylic jaw suspensiopn of an early chondrichthyan based on †*Debeerius*

Orbital process
Ethmoid process
Chondrocranium
Upper jaw
Lower jaw
Pharyngohyal
Hyomandibula
Ceratohyal
Hypohyal

(C) Amphistylic jaw suspension of an early elasmobranch based on †*Hybodus*

Chondrocranium
Hyomandibula
Upper jaw
Lower jaw
Ceratohyal

(B) Holostylic jaw suspension of an extant chimaera

Upper jaw
Hyomandibula
Nostril
Ceratohyal
Lower jaw

Nostrils   Upper and lower jaws with tooth plates

(D) Hyostylic jaw suspension of an extant selachian (shark)

Ligament
Rostrum
Chondrocranium
Hyomandibula
Lower jaw
Basihyal
Ceratohyal

Projected upper and lower jaws

(E) Hyostylic jaw suspension of an extant batomorph (ray)

Chondrocranium   Hyomandibula
Ligament
Rostrum
Upper jaw
Ceratohyal
Lower jaw

Projected upper and lower jaws

Figure 6.8 Evolution of jaw suspension in chondrichthyans. (A) A basal chondrichthyan, †*Debeerius*, with autodiastylic jaw suspension. (B) Chimaeras have a holostylic articulation; the photograph shows a ventral view of the head with the mouth open. (C) The early elasmobranch †*Hybodus* has an amphistylic jaw suspension. (D,E) Neoselachians have a hyostylic jaw suspension in which the upper jaw articulates with the mobile hyomandibula, and its anterior attachment is a flexible ligament. Photographs illustrate projection of jaws in sharks and rays (A,B,C,E after C.D. Wilga. 2002. *Biol J Linn Soc* 75: 483–502. © The Linnean Society of London; D after E.S. Goodrich. 1909. *A Treatise on Zoology*, Part IX, R. Lankester, ed. London, Adam and Charles Black Publishers.)

synapomorphy of eugnathostomes. Basal euchondrocephalians also had tooth whorls (see Figure 6.10), but derived forms such as chimaeras evolved three pairs of tooth plates used to crush food. As material wears away on the surfaces of a chimaeriform tooth plate, new tissue is added from beneath so that the tooth plate itself is never completely lost.

Jaws and jaw suspension

As shown in **Figure 6.8A**, early chondrichthyans such as †*Debeerius* had an **autodiastylic jaw suspension** (Greek *auto*, "self"; *dia*, "across"; *stylos*, "pillar") in which ethmoid and orbital processes on the upper jaw attached it to the chondrocranium, preventing the jaw from moving

in relation to the chondrocranium. The hyoid arch was not involved in jaw suspension, and consisted of four skeletal elements (pharyngohyal, hyomandibula, ceratohyal, and hypohyal) serially homologous with those of the typical gill arches posterior to it (see Figure 3.9). From this condition, holocephalans evolved a **holostylic jaw articulation**, in which the upper jaw is fused to the chondrocranium (**Figure 6.8B**). The fused upper jaw suspends the short lower jaw, which pivots like a nutcracker to crush prey between tooth plates. This fused condition is the source of the name Holocephali (Greek *holos*, "whole, entire"; *kephale*, "head").

In contrast, early elasmobranchs such as †*Hybodus* had an **amphistylic jaw suspension** (Greek *amphis*, "double"), in which the anterior end of the upper jaw articulated directly with the chondrocranium and the posterior end was supported by the hyomandibula (**Figure 6.8C**). (Note the absence of the pharyngohyal, which could represent the loss of the pharyngohyal or its fusion with the hyomandibula; opinions differ).

Neoselachians (extant sharks and rays) have a **hyostylic jaw suspension**. The upper jaw articulates with the chondrocranium via the large and mobile hyomandibula, and flexible ligaments connect the anterior part of the upper jaw to the chondrocranium (**Figure 6.8D,E**). Hyostyly allows the upper jaw of neoselachians to be projected and retracted during feeding.

Internal fertilization and claspers

Internal fertilization is universal in chondrichthyans. Males have a pair of **pelvic claspers**, one associated with each pelvic fin, supported by a jointed skeleton. Each clasper has a muscular subcutaneous siphon sac extending anteriorly beneath the skin of the pelvic fins (**Figure 6.9A**). Siphon sacs have a secretory lining, and the male pumps them up with seawater before copulation. The male uses one clasper during copulation (**Figure 6.9B**). During copulation, the clasper bends 90° to the long axis of the male's body, so that the dorsal groove on the clasper lies under the cloacal papilla from which sperm are emitted. This flexed clasper is inserted into the female's cloaca and locked there by barbs near the clasper's tip. Simultaneously, the siphon sac contracts and sperm are ejaculated into the clasper's groove. Fluid from the siphon sac washes sperm down the groove into the female's cloaca, and the sperm swim into the female's oviducts.

Distinctive soft tissue and physiological features

In comparison to extant osteichthyans, extant chondrichthyans have many distinctive features of internal anatomy and physiology, notably a large, lipid-filled liver that renders them neutrally buoyant. They retain nitrogenous compounds that raise their internal osmotic concentrations, and ketones are primary metabolic substrates of their skeletal and cardiac muscles. Several of these features are described in Chapter 4.

6.3 Euchondrocephali and Chimaeriformes

LEARNING OBJECTIVES

6.3.1 Describe unique features of †*Cladoselache*, †*Damocles*, and †*Helicoprion*.

6.3.2 Explain the name chimaera.

6.3.3 Identify the reproductive mode of extant chimaeras.

6.3.4 State the function of the cephalic clasper found in male chimaeras.

†*Cladoselache*, known from well-preserved specimens from the Devonian Cleveland Shale and long interpreted as an elasmobranch, is a euchondrocephalian. It was about 2 m long, with large fins and a large terminal mouth (opening at the tip of the head, with upper and lower jaws of the same length). †*Cladoselache* was probably a pelagic predator, engulfing its prey whole or slashing them with daggerlike teeth.

†Stethacanthid euchondrocephalians from the Early Carboniferous showed remarkable sexual dimorphism. The first dorsal fin and spine of males in this family were elaborated into a structure called a spine-brush complex that was probably used in courtship. In two genera, †*Orestiacanthus* and †*Stethacanthus*, this structure projected upward and ended in a blunt surface covered with spines. Male †*Damocles* and †*Falcatus* had a swordlike spine that projected forward parallel to the top of the head (see Figure 6.2). A unique fossil preserves a pair of †*Falcatus* that appear to have died in a precopulatory courtship position, with the female apparently grasping the male's dorsal spine in her jaws.

†*Helicoprion* (a euchondrocephalian of uncertain phylogenetic affinity) is known for its bizarre spiral tooth whorl. There have been many weird and wonderful reconstructions of †*Helicoprion* (**Figure 6.10A**), but a detailed biomechanical analysis shows that the tooth

(A) Male gripping female's right pectoral fin in his jaws

© Tony Wu/Minden Pictures

(B) Male inserting right clasper into female's cloaca

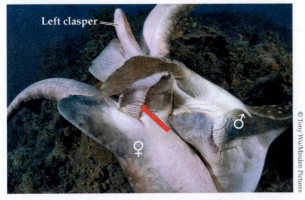

Left clasper

© Tony Wu/Minden Pictures

Figure 6.9 Copulation of whitespotted bamboo shark, *Chiloscyllium plagiosum*. (A) Male gripping female's right pectoral fin in his jaws. Siphon sacs beneath the male's pelvic fins, visible here as a bulge (arrow), have been inflated with seawater. (B) Male inserting right clasper into the female's cloaca (arrow). The siphon sacs have nearly been emptied.

(A) Some hypothetical reconstructions of †*Helicoprion*

(B) Current interpretation of †*Helicoprion* places tooth whorl at symphysis of left and right mandibular cartilages

© Ray Troll

Figure 6.10 Tooth whorl of †*Helicoprion*. (A) Because †*Helicoprion* was initially known only from fossil tooth whorls, there have been many fanciful reconstructions. (B) Current interpretation of the position of tooth whorl in life. (C) Reconstruction of tooth whorl of †*Helicoprion* in relation to a hypothetical chondrocranium. (C after J.B. Ramsay et al. 2015. *J Morphol* 276: 47–64.)

(C) Reconstruction of jaws with hypothetical chondrocranium

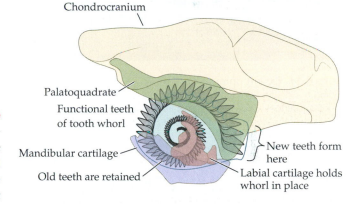

Chondrocranium

Palatoquadrate
Functional teeth of tooth whorl
Mandibular cartilage
Old teeth are retained
New teeth form here
Labial cartilage holds whorl in place

whorl was contained entirely within the mouth (**Figure 6.10B**). The new model interprets the whorl at the mandibular symphysis (where the left and right sides of the lower jaw meet; **Figure 6.10C**). †*Helicoprion* probably fed on soft-bodied prey, using the tooth whorl rather like the stationary blade of a circular saw to trap, pierce, and cut its prey. As teeth wore out at the front of the saw, new teeth in the whorl moved into place at the back of the saw.

†*Debeerius* is a well-preserved early member of Holocephali (see Figure 6.2). Other Paleozoic holocephalans are known only by partial skeletons or tooth plates, so we have little knowledge about their external morphology. Most lineages of holocephalans were extinct by the end of the Paleozoic.

Biology of extant Chimaeriformes

The extant families of Chimaeriformes—Callorhinchidae, Chimaeridae, and Rhinochimaeridae—diverged in the Late Jurassic and Early Cretaceous and today contain 56 species (**Figure 6.11**). Callorhinchids have a heterocercal caudal fin, but the tail is long and whiplike in the other two families. Chimaeras swim using lateral undulations of the body and powerful movements of large pectoral fins. The common names chimaera, rabbitfish, or ratfish reflect their seeming combination of different animals. The head of some species resembles a caricature of a rabbit, with big eyes and broad tooth plates that look like a rabbit's teeth.

Many species of chimaeras live at depths of 500 m or more, and range across entire ocean basins. Males and females may spend most of the year separately and come together only during an annual inshore spawning migration. Some chimaeras have the same life-history characteristics that make sharks vulnerable to overexploitation—they take 10–12 years to reach maturity, reproduce once a year, and produce only a few young (see Section 6.6). All extant chimaeras are oviparous, the reproductive mode in which a mother deposits eggs that develop outside her body. However, fossil evidence suggests that some extinct species were viviparous, a reproductive mode in which offspring are retained, protected, and nourished within the female's body.

Plownose and longnose chimaeras have rostra that are densely studded with lateral line mechanoreceptors and ampullary electroreceptors, as are the blunt snouts of shortnose chimaeras. These receptors are especially dense around the mouth, which is relatively small and faces downward. Chimaeras prefer soft substrates, feeding on invertebrates and small fishes that live on the seafloor. Although they consume soft-bodied organisms, including anemones and jellyfishes, their tooth plates can crush hard-bodied prey such as crabs.

(A) Australian ghost shark, *Callorhinchus milii*

(B) Egg cases of Australian ghost shark, *Callorhinchus milii*

(C) Ratfish, *Chimaera sp.*

(D) Longnose chimaera, *Rhinochimaera sp.*

Figure 6.11 **Extant chimaeras.** (A) An Australian ghost shark, showing the distinctive rostrum that gives this genus its common name of plownose chimaera. (B) All chimaeras are oviparous. This photo shows the unusual egg cases of the Australian ghost shark. (C) A chimaera, showing why this group has the common name of rabbitfishes, based on large eyes and head shape. (D) Most longnose chimaeras live at depths of 2,000 m or more.

Many chimaeras have a venom gland associated with a mobile dorsal spine. The spine can be erected when the chimaera is attacked, and a predator that is stabbed in the mouth might well decide to release the chimaera and seek less noxious prey. Males of some species of chimaeras have a spine-encrusted **cephalic clasper** used in courtship. Oddly, although the clasper is on the top of the head, it is controlled by muscles that also move the lower jaw. During copulation, the cephalic clasper is believed to fit into a corresponding hollow on the female's head.

6.4 Elasmobranchii, Euselachii, and Neoselachii

LEARNING OBJECTIVES

6.4.1 Distinguish between Elasmobranchii, Euselachii, and Neoselachii.

6.4.2 Identify a key difference in the pectoral fins of selachians and batomorphs.

6.4.3 Name the two extant groups of Selachii.

6.4.4 Compare body shapes, appearance of the caudal fin, and reproductive modes of skates and rays.

Elasmobranchii has a long fossil record marked by some strikingly well-preserved Paleozoic and Mesozoic taxa (see Figures 6.1 and 6.2). For example, fossils of †*Cladodoides wildungensis*, the earliest known elasmobranch, occur in upper Devonian limestones in Germany. The mouth of †*Cladodoides* is terminal. †Xenacanths such as the Carboniferous †*Triodus* were freshwater bottom-dwellers with prominent dorsal spines, robust fins, and heavily calcified skeletons that may have decreased their buoyancy; some had elongated, eel-like bodies.

More derived sharks belong to Euselachii (Greek *eu*, "true"; Latin *selachos*, "cartilaginous fish"). †*Tristychius* is a Carboniferous euselachian known from many specimens found near Edinburgh, Scotland. As in other gnathostomes, its pectoral fins are tribasic, meaning that they have three skeletal elements, termed basals, that support the base of the fin where it attaches to the body (**Figure 6.12**). Unlike the earlier elasmobranchs, however, the base of the euselachian pectoral fin where it meets the body is narrow. This narrow-based condition allows rotation of the pectoral fin for greater control of locomotion; such fins also have smaller radial cartilages, meaning that most of the fin is supported by flexible fin rays called **ceratotrichia** (Greek *kerat*, "horn"; *trich*, "hair"; see Figure 6.12).

The remarkable three-dimensional preservation of the skeletons of †*Tristychius* allows detailed functional anatomical analyses of its feeding system. Like extant nurse sharks

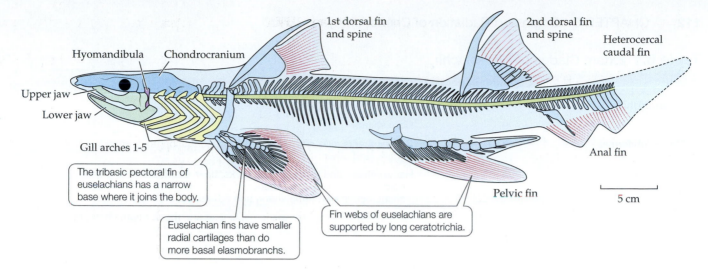

Upper jaw

Lower jaw

Hyomandibula   Chondrocranium

1st dorsal fin
and spine

2nd dorsal fin
and spine

Heterocercal
caudal fin

Gill arches 1-5

The tribasic pectoral fin of
euselachians has a narrow
base where it joins the body.

Euselachian fins have smaller
radial cartilages than do
more basal elasmobranchs.

Fin webs of euselachians are
supported by long ceratotrichia.

Pelvic fin

Anal fin

5 cm

Figure 6.12 Skeleton of †*Tristychius*, a Carboniferous euselachian. Reconstruction showing major components of the skeleton, including well-preserved chondrocranium, jaws, and gill arches. Dashed outline indicates the hypothetical shape of the caudal fin. Important characters of euselachians include the narrow base of the tribasic pectoral fin and support of fin webs by long ceratotrichia rather than radial cartilages. (Redrawn from J.R.F. Dick. 1978. *Trans R Soc Edinburgh* 70: 63–109. Reproduced with permission.)

(Ginglymostomatidae), †*Tristychius* was a suction-feeding specialist, with a broad head and stout hyomandibula.

†*Hybodus* is a well-known Mesozoic euselachian that retained a terminal mouth. Complete skeletons 2 m long have been found. †*Hybodus* had heterodont dentition: anterior teeth had sharp cusps and may have been used for piercing, holding, and slashing soft foods, while posterior teeth were stout and blunt and may have been used to crush hard-bodied prey such as crabs.

The 13 extant elasmobranch clades belong to Neoselachii (Greek *neos*, "new"). Selachii includes nine extant groups of sharks, and Batomorphi includes four extant groups of skates and rays (**Figure 6.13**; **Table 6.1**). The mobile upper jaw of neoselachians is attached to the chondrocranium by ligaments and supported by the hyomandibula (hyostyly; see Figure 6.8D,E).

Most sharks are torpedo-shaped with 5–7 gill openings on each side of the head. Batomorphs are dorsoventrally flattened forms with the eyes and spiracle on the dorsal surface of the body and five pairs of gill openings on the ventral surface of the head. When a batomorph rests on the bottom, it draws water in through the spiracles and pumps it out through the gill slits.

A key difference between selachians and batomorphs relates to

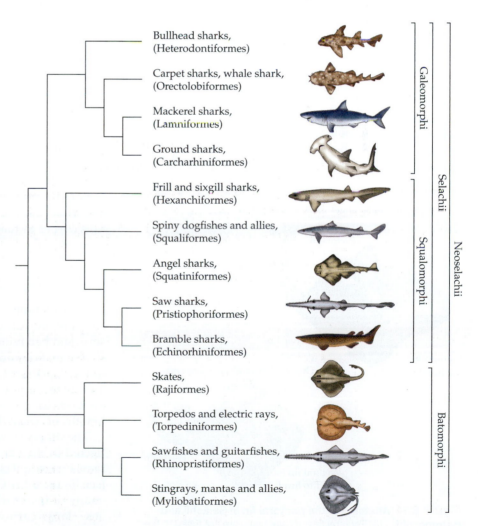

Bullhead sharks,
(Heterodontiformes)

Carpet sharks, whale shark,
(Orectolobiformes)

Mackerel sharks,
(Lamniformes)

Ground sharks,
(Carcharhiniformes)

Frill and sixgill sharks,
(Hexanchiformes)

Spiny dogfishes and allies,
(Squaliformes)

Angel sharks,
(Squatiniformes)

Saw sharks,
(Pristiophoriformes)

Bramble sharks,
(Echinorhiniformes)

Skates,
(Rajiformes)

Torpedos and electric rays,
(Torpediniformes)

Sawfishes and guitarfishes,
(Rhinopristiformes)

Stingrays, mantas and allies,
(Myliobatiformes)

Galeomorphi

Selachii

Squalomorphi

Neoselachii

Batomorphi

Figure 6.13 Phylogenetic relationships of the 13 extant clades of Neoselachii. Both anatomical and current molecular phylogenetic characters support this interpretation.

Table 6.1 Extant Clades of Neoselachii

Clade	Number of extant species	Examples
Galeomorphi		
Heterodontiformes	9	Horn sharks
Orectolobiformes	45	Carpet sharks, epaulette sharks, wobbegongs, bamboo sharks, whale shark, and others
Carcharhiniformes	296	Hammerhead and bonnethead sharks, tiger shark, bull shark, smooth dogfish, catsharks, swellsharks, blue shark, lemon shark, whitetip reef shark, oceanic whitetip shark, and others
Lamniformes	16	Goblin shark, sand tiger shark, basking shark, megamouth shark, thresher sharks, mako sharks, salmon shark, porbeagle, white shark, and others
Hexanchiformes	7	Frilled sharks, sixgill sharks, and sevengill sharks
Squalomorphi		
Squaliformes	164	Spiny dogfishes, gulper sharks, lantern sharks, Greenland shark, cookie-cutter sharks, and others
Echinorhiniformes	2	Bramble sharks
Pristiophoriformes	10	Sawsharks and sixgill sawsharks
Squatiniformes	25	Angel sharks
Batomorphi		
Rajiformes	304	Skates and others
Rhinopristiformes	74	Sawfishes and guitarfishes
Torpediniformes	73	Electric rays and torpedoes
Myliobatiformes	252	Stingrays, eagle rays, cownose rays, manta rays, and others
TOTAL	**1277**	

their pectoral fins. For example, angel sharks (*Squatina*) superficially resemble rays, but their pectoral fins are not fused with the head (**Figure 6.14A**). In contrast, the pectoral fins of batomorphs fuse to the sides of the head during ontogeny (**Figure 6.14B**). Batomorphs swim with their enlarged pectoral fins; one derived group of rays, Myliobatiformes, which includes cownose, eagle, devil, and manta rays, evolved specialized cephalic lobes of the pectoral fin that are used in feeding.

Batomorph fossils are known from the Late Permian, and molecular data place the split between sharks and rays earlier in that period. The earliest fossil selachians come from the Triassic. Most of the extant selachian clades were present by the end of the Cretaceous.

Selachii: Sharks

Throughout their evolutionary history, sharks have been consummate carnivores. Derived locomotor, trophic, sensory, and behavioral characteristics produced predatory sharks that dominate the top levels of marine food webs today. Shark sizes range from the tiny dwarf lanternshark and spined pygmy shark, deepwater forms that are about 22–25 cm long as adults, to whale sharks, which can attain lengths of nearly 19 m.

The diverse environments inhabited by sharks are reflected in the variety of their body forms. Selachii contains two extant clades, Galeomorphi and Squalomorphi (see Figure 6.13). Galeomorphi (**Figure 6.15A–D**) includes many of the sharks featured on television and in movies—large carnivores such as the white shark, mako sharks (*Isurus*), and ground sharks such as hammerheads and bonnetheads (*Sphyrna*). The enormous whale shark (*Rhincodon typus*) and the smaller basking shark

(A) Dorsal view of an angel shark, *Squatina*

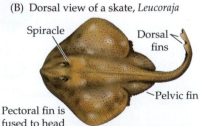

Spiracle
Pectoral fin   Caudal fin
Dorsal fins
Pelvic fin
Pectoral fin is not fused to head

(B) Dorsal view of a skate, *Leucoraja*

Spiracle   Dorsal fins
Pelvic fin
Pectoral fin is fused to head

Figure 6.14 **Anatomy of the pectoral fin in Selachii and Batomorphi.** (A) Pectoral fins of selachians are not fused to the head. (B) Pectoral fins of adult batomorphs are fused to the head. (After K. Liem et al. 2001. *Functional Anatomy of the Vertebrates*, 3rd ed. Cengage/Harcourt College: Belmont, CA.)

(A) Galapagos bullhead shark, *Heterodontus quoyi*

(B) Whale shark, *Rhincodon typus*

(C) White shark, *Carcharodon carcharias*

(D) Great hammerhead shark, *Sphyrna mokarran*

(E) Spiny dogfish, *Squalus acanthias*

(F) Bluntnose sixgill shark, *Hexanchus griseus*

(G) Longnose sawshark, *Pristiophorus cirratus*

(H) Japanese angel shark, *Squatina japonica*

Figure 6.15 Examples of extant sharks. **(A–D) Galeomorph sharks.** (A) The Galapagos bullhead shark reaches about 1 m in length, making it a relatively small member of Galeomorphi. (B) The whale shark filter feeds on plankton and can attain lengths of 18.8 m, making it the largest extant species of fish. (C) The iconic white shark reaches 6.4 m in length and specializes on mammalian prey such as seals. (D) The peculiarly shaped head of the scalloped hammerhead consists of left and right cephalofoils supported by cartilage. The eyes, nostrils, lateral line, and electroreceptors are spread out, increasing directional sensitivity for vision, olfaction, mechanoreception, and electroreception. **(E–H) Squalomorph sharks.** (E) The spiny dogfish, one of the most abundant species of selachians, has a defensive spine anterior to each of its two dorsal fins. (E) The bluntnose sixgill shark is one of the world's largest predators at 3.1 to 4.2 m long; most sharks have five gill slits whereas this species has six. It specializes on eating other sharks. (G) The rostrum of the longnose sawshark makes up more than one-fourth of the shark's total length (1.25 m). The rostrum is studded with "teeth" (modified placoid scales) and has a pair of sensory barbels at its midpoint. (H) The Japanese angel shark (~1.5 m) is an ambush predator that lies buried on the bottom during the day. Needle-sharp teeth pointing backward into the mouth prevent prey from escaping once caught. Angel sharks are superficially raylike, but their pectoral fins are not fused to the head.

(*Cetorhinus maximus*; up to 15 m) are also galeomorphs. Not all galeomorphs are large, however: horn sharks (*Heterodontus*) reach lengths of only 1.5 m.

Squalomorphs (**Figure 6.15E–H**) are generally smaller than galeomorphs. The spiny dogfish (*Squalus acanthias*), the most abundant species of shark, reaches 1.5 m; its many smaller relatives include the green lanternshark (*Etmopterus virens*; 0.23 m) and the cookie-cutter shark (*Isistius brasiliensis*; 0.5 m). The much larger sixgill sharks (*Hexanchus griseus*; up to 5 m) eat other sharks and are among the largest carnivores on Earth. The long rostrum of sawsharks (*Pristiophorus*) has a series of enlarged placoid scales that form a sawlike blade (this structure is convergently like the saw of sawfishes, which are batomorphs). Angel sharks (*Squatina*) are sit-and-wait ambush predators that reach lengths of 2 m.

Batomorphi: Skates and rays

Batomorphi includes about 700 extant species of skates, rays, guitarfishes, and sawfishes. Most of these are skates (Rajiformes; about 300 species) and rays (Myliobatiformes; about 250 species). Electric rays and torpedoes

(A) Little skate, *Leucoraja erinacea*

(B) Spotted eagle ray, *Aeotobatis narinari*

(C) Marbled torpedo ray, *Torpedo marmorata*

(D) Giant freshwater whipray, *Urogymnus polylepis*

(E) Common sawfish, *Pristis pristis*

(F) Cownose ray, *Rhinoptera bonasus*

(Torpediniformes; about 70 species) are noteworthy for electroplax, a tissue derived from modified gill arch muscles that can produce powerful electric currents to stun prey and deter predators (see Section 4.2). Rhino-pristiformes (about 70 species) includes guitarfishes and sawfishes (see Table 6.1).

Body forms of batomorphs are diverse (**Figure 6.16**). In general, skates have a circular body outline with a long rostrum, whereas rays (which include the largest extant batomorphs) are diamond-shaped. Most skates have a thick tail with two dorsal fins that lack spines. Most rays have a long, whiplike tail, many with modified venomous barbs; when a ray whips its spine against a predator, the barb penetrates its cutaneous covering that contains the venom-producing tissue, and venom is carried into the wound by the barb. Skates are oviparous, with eggs that are protected by stiff egg cases, whereas rays are viviparous.

The characteristic suite of specializations in bato-morphs relates to bottom-dwelling, durophagous habits (Latin *durus*, "hard"; Greek *phago*, "eat"). The flattened teeth form a pavement-like dentition used to crush hard-bodied prey. The mouth is often extremely pro-trusible (see Figure 6.8E), providing powerful suction to dislodge shelled invertebrates from the substrate. Skates and rays have diverse adaptations of their jaws and teeth, and many similar-appearing morphologies have evolved independently. Notably, different species of Myliobati-formes invaded pelagic and freshwater environments (see Figure 6.16B,D), and some evolved filter feeding.

Most skates are ambush hunters, resting on the sea-floor and covering themselves with a thin layer of sand. They spend hours partially buried and nearly invisible except for their prominent eyes, surveying their sur-roundings and lunging at prey that come too close. Tor-pedo rays (Torpediniformes) are ambush feeders: during the day, they surge upward from the sand, enclose prey in their pectoral fins, and emit an electrical discharge to stun it. At night, torpedo rays hover in the water column 1–2 m above the bottom and drop on a fish that passes beneath them, cupping it in their pectoral fins while they stun it.

◀ **Figure 6.16 Examples of extant skates and rays.** (A) Like all male chondrichthyans, this little skate has prominent claspers behind its pelvic fins. (B) The spotted eagle ray (~3 m) is a wide-ranging predator of mollusks and crustaceans that flies through the water using its powerful pectoral fins. (C) The marbled torpedo ray is electrogenic, producing pulses of electric discharges from modi-fied gill muscles during prey capture and defense against predators. (D) The giant freshwater whipray, found in rivers and estuaries from the Mekong River in southeast Asia to northern Australia, reaches 2.5 m in length and weights of 600 kg. (E) Like other sawfishes (Pristidae), the common sawfish (~7.5 m) has a "toothed" rostrum that resembles those of sawsharks (Pristiophoridae) but it lacks barbels. ((F) The cownose ray (~1 m), common along the east coast of the United States, uses broad, flat teeth (inset) to crush mollusks and crustaceans.

6.5 Biology of Neoselachii

LEARNING OBJECTIVES

6.5.1 Identify the sequence of sensory systems used by sharks to detect and attack prey.

6.5.2 Contrast bioluminescence with biofluorescence.

6.5.3 Explain why hypoxia can develop in marine lagoons and how epaulette sharks respond.

6.5.4 Describe endothermal heterothermy in mackerel sharks and its functional significance.

6.5.5 Relate caudal fin shapes of sharks to habitats and swimming behaviors.

6.5.6 Distinguish lecithotrophy from matrotrophy and why they represent extremes in a continuum.

6.5.7 Evaluate evidence that manta rays pass the Mirror Self-Recognition (MSR) Test.

6.5.8 Discuss evidence that sand tiger sharks display fission-fusion behavior.

In this section, we examine selected topics in the biology of extant Neoselachii, beginning with feeding mecha-nisms and sensory systems used in hunting.

Feeding

Sharks employ three basic mechanisms of prey capture, often in combination. **Suction feeding** draws prey into the mouth. Engulfing prey with an open mouth, known as **ram feeding**, is effective with smaller prey items. **Filter feeding** is a special case in which prey are much smaller than the shark. **Biting** consists of opening the mouth and closing it on large prey, usually tearing out a chunk of flesh.

Filter feeding and gigantism have evolved at least four times within Neoselachii. The largest sharks—the whale shark, basking shark, and megamouth shark (*Megachasma pelagios*; up to 7 m)—are filter feeders. Whale and bask-ing sharks feed at or near the surface, whereas the mega-mouth shark forages at depths down to 600 m. The largest batomorphs—manta and devil rays (*Mobula*)—are also filter feeders. These highly specialized rays swim with winglike motions of the pectoral fins, filtering plankton from the water as they go.

Sharks as predators   Sharks are versatile and effective predators. White sharks swim back and forth parallel to the shore near seal or sea lion breeding sites, intercepting them as they come and go from rookeries. On sunny days, white sharks place the sun behind them as they approach, making it difficult for prey to detect them.

White sharks kill their mammalian prey by exsanguina-tion—that is, bleeding them to death. Sharks can hold a seal in the jaws until it is dead, but they repeatedly seize and re-lease sea lions until the prey dies from blood loss, probably because sea lions, unlike seals, have powerful front flippers that are effective in defense. These behavioral observations support a hypothesis that emerged from biomechanical analyses of shark teeth: perhaps a main factor influencing tooth anatomy is the severity of damage a bite can inflict.

Although sharks are formidable predators, they do not turn down an easy meal. Tiger sharks (*Galeocerdo cuvier*) assemble near Raine Island (off the coast of Queensland, Australia) during the nesting season for green sea turtles (*Chelonia mydas*) to scavenge dead turtles and prey on weak individuals.

Shark sensory systems   Sharks have been described as "swimming noses" so acute is their sense of smell, which can detect odors at concentrations of less than 1 part per billion. Sharks compare the time of arrival of an odor stimulus on the left and right sides of the head to locate the odor's source. Their vision at low light intensities is especially well developed, because many sharks feed at dusk or at depths in the sea where little sunlight penetrates. This sensitivity is due to a rod-rich retina and cells with numerous platelike crystals of guanine located just behind the retina, in the eye's choroid layer. Collectively called the **tapetum lucidum**, the shiny crystals of guanine in these cells act like mirrors to reflect light back through the retina and increase the chance that photons will be absorbed. (Many nocturnal animals, including domestic cats, have a tapetum lucidum, which accounts for the eyeshine seen when light shines on their eyes in the dark.)

In addition to smell and vision, sharks have mechanoreceptors, including an exquisitely sensitive lateral line system that detects water movements, such as those in the turbulent wake of a swimming fish. Hair cells in the ears detect low-frequency sounds, such as vibrations produced by a struggling fish (20 hz to 100 hz; see Section 2.6).

Ampullary electroreceptive organs, also known as ampullae of Lorenzini, passively detect electrical signals. The ampullae are concentrated on the underside of the snout immediately anterior to the mouth (see Figure 4.8). A projecting rostrum is a derived character of neoselachians, and electroreception may have been an important sensory modality in their ecological and evolutionary success.

Sensory systems are closely linked to the distinctive head shape of hammerhead sharks (*Sphyrna*; see Figure 6.15D). Left and right extensions of the braincase termed cephalofoils support expanded sensory systems, including ampullae of Lorenzini and a uniquely enlarged olfactory system used for tracking chemical cues from prey. The eyes are at the ends of the cephalofoils, allowing exceptional depth perception and extensive fields of view.

Hunting behavior of sharks   Prey send out a variety of signals that a shark might detect, but the sequence in which a shark employs different senses varies with distance from the prey (**Figure 6.17**). Olfaction often alerts a

(A) Cues from prey

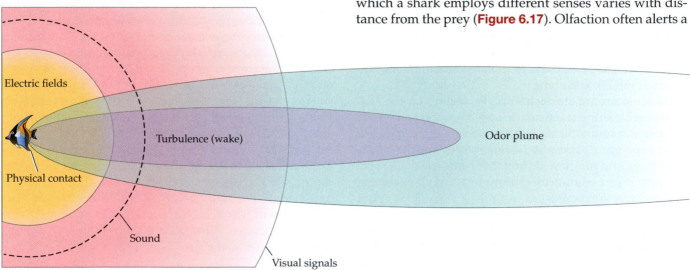

(B) Sequence of sensory systems used to detect and capture prey

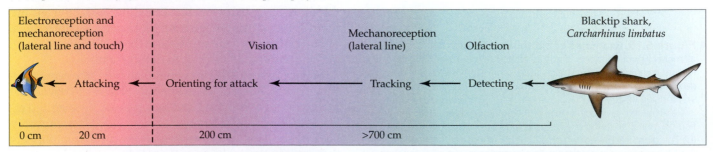

Figure 6.17 Sharks integrate sensory information when hunting.   (A) Potential prey generate many signals that can be detected by a shark downstream. (B) During the day, a blacktip shark, *Carcharhinus limbatus*, may initially detect and track prey using olfaction and then employ mechanoreception (via the lateral line sense) to detect the prey's wake. As it approaches prey, the

shark switches to vision to orient for attack. At close range, its snout blocks its view, so the attack is adjusted using turbulence detected by the shark's lateral line system. Finally, the shark switches to electroreception and touch to capture the prey. (After J.M. Gardiner et al. 2014. *PLOS One* 9: e93036. CC-BY 4.0. https://creativecommons.org/licenses/by/4.0/)

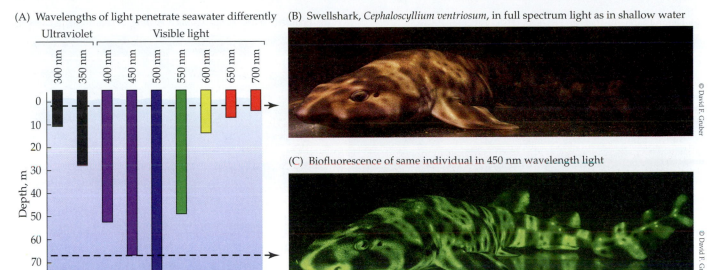

(A) Wavelengths of light penetrate seawater differently

(B) Swellshark, *Cephaloscyllium ventriosum*, in full spectrum light as in shallow water

(C) Biofluorescence of same individual in 450 nm wavelength light

Figure 6.18 Biofluorescence. (A) Different wavelengths of light penetrate to different depths in the ocean with blue light (500 nm wavelength) penetrating deeper than other colors. (B) In surface light, the skin of the swellshark appears as mottled brown colors. (C) At depth, the skin absorbs blue light and re-emits it as fluorescent green light. This vivid pattern may be involved in courtship and mating.

shark to potential prey, especially wounded individuals, from great distances. Next, mechanoreception allows the shark to track the wake of its prey via the lateral line sense. Once the shark closes in on its prey, vision becomes a primary mode of detection. If the prey is easily recognized, the shark may proceed directly to attack. In the last moments before contact, some sharks draw an opaque eyelid (the nictitating membrane) across each eye to protect it. As the shark opens its mouth, electroreception guides the final attack.

Sharks can treat unfamiliar prey somewhat differently, as studies aimed at developing shark deterrents have discovered. A circling shark may suddenly turn and rush toward unknown prey. Instead of opening its jaws to attack, however, the shark bumps and scrapes the surface of the potential prey with its rostrum. Opinions differ on whether this is an attempt to determine texture through mechanoreception (in this case, touch), to make a quick electrosensory appraisal, or to use the rough placoid scales to abrade the prey's surface and release fresh olfactory cues.

Bioluminescence and biofluorescence

Some small, deepwater squaliform sharks known as lantern sharks are **bioluminescent**, meaning that they emit light. The organs responsible for bioluminescence are known as **photophores**, and their production of light is believed to be controlled by hormones. Photophores on the ventral surface of the body may camouflage the shark by counterillumination—that is, by emitting light matching the intensity of down-welling light from the water surface, thereby concealing the shark from predators looking upward. Photophores on the sides of the body or on a dorsal spine may be used for intraspecific recognition.

A different phenomenon is **biofluorescence**, in which the skin absorbs light of one wavelength and re-emits it at a different wavelength. Light of different wavelengths within the visual spectrum penetrates seawater differently, with blue light (450 to 500 nm wavelengths) penetrating to depths between ~65 and 75 m (**Figure 6.18A**). Biofluorescence has evolved at least three times in chondrichthyans. When exposed to light in blue wavelengths, the shark's skin reradiates it as green light, enhancing color patterns in the skin (**Figure 6.18B,C**). These patterns, which are both species-specific and sexually dimorphic, may play roles in intraspecific communication.

Hypoxia and the epaulette shark

Heron Island is a small coral cay at the southern end of the Great Barrier Reef in Australia (**Figure 6.19A**). It is surrounded by a large, fringing platform coral reef. The shallow lagoon created by the reef is home to the epaulette shark, *Hemiscyllium ocellatum* (**Figure 6.19B**). Epaulette sharks, named for the black spot on each shoulder, grow to about 1 m in length and are probably most famous for their behavior of walking on the bottom using their paired fins and a diagonal gait that resembles salamander locomotion (see Figure 11.5). However, they also tolerate extreme hypoxic (low oxygen) conditions, which they periodically encounter due to tide cycles in the Heron Island Lagoon. About every two weeks, extremely low tides leave the water in the lagoon out of contact with water outside of the reef crest. Respiration by coral polyps and other animals in the lagoon quickly depletes the oxygen content of the water, and this is exacerbated at night, when photosynthetic organisms keep respiring but stop releasing oxygen. All this leads to a depletion of as much as 80% of the water's oxygen.

When faced with such hypoxic conditions, most aquatic vertebrates with gills either move to waters with higher oxygen content or gulp air from the surface, but the epaulette shark uses a different mechanism. As oxygen levels in the surrounding water decline, the shark initially uses

(A) Heron Island, Australia and its lagoon

(B) Epaulette shark, *Hemiscyllium ocellatum,* in Heron Island Lagoon

(C) Section through reef crest and lagoon

At low tides, the reef crest limits influx of oxygenated seawater into the lagoon.

Respiration of organisms depletes oxygen in lagoon during extreme low tides.

Heron Island

Epaulette sharks tolerate hypoxia by decreasing metabolic activity.

Figure 6.19 Hypoxia tolerance in the epaulette shark. (A) Aerial view of Heron Island, Australia and its large lagoon surrounded by a fringing reef. (B) An epaulette shark, *Hemiscyllium ocellatum,* showing one of its two large shoulder spots from which its name derives. (C) At low tides, the lagoon can be cut off from the surrounding waters by the reef crest. When this happens, oxygen levels in the water of the lagoon can decrease by 80% due to respiration by aquatic animals. Epaulette sharks respond by suppressing their metabolism until fresh oxygenated water returns. (C after G.E. Nillson and G.M.C. Renshaw. 2004. *J Exp Biol* 207, Pt 18: 3131–3139.)

anaerobic respiration (glycolysis) to maintain as much ATP production as possible, then initiates metabolic suppression like the hibernating mammals described in Section 14.3.

Endothermal heterothermy

Mackerel sharks (Lamnidae), including the shortfin mako (*Isurus oxyrinchus*) and white shark (*Carcharodon carcharias;* **Figure 6.20**), can maintain different temperatures in different parts of the body. These large, fast-swimming marine predators have nearly global distributions, typically preying on bony fishes, other sharks, sea turtles, and marine mammals.

Countercurrent heat exchange in the rete mirabilia (dense networks of capillaries; see Figure 4.4) in the muscles captures heat produced by muscular contractions as a lamnid shark swims, raising muscle temperatures above water temperature, a physiological process known as endothermy (Greek *endos,* "within"; *thermos,* "heat"; see Chapter 14). A second set of rete mirabilia in the head carries cold blood from the gills through a venous sinus filled with blood warmed by the swimming muscles, keeping eye temperature above water temperature. As a result, swimming muscles and the eyes are warmer than other parts of the shark's body, a condition called **endothermal heterothermy** (also known

Figure 6.20 Muscles of lamnid sharks are warmer than water temperatures. Muscles of lamnid sharks can be 10°C warmer than the surrounding water. Species that live in colder waters, such as the porbeagle *Lamna nasus* and salmon shark *L. ditropis,* can maintain greater temperature differentials than species from warmer water, such as the white shark, *Carcharodon carcharias,* and mako sharks, *Isurus oxyrinchus* and *I. paucus.* (Graph after F.G. Carey et al. 1985. *Mem So Calif Acad Sci* 9: 92–108.)

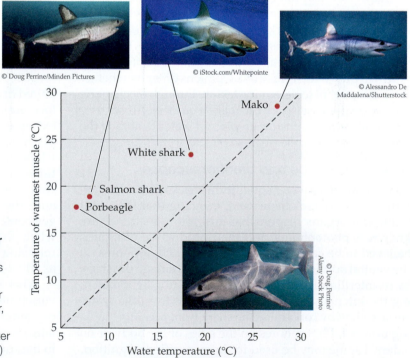

(A) Caudal skeleton of hammerhead shark, *Sphyrna*

Upper lobe

Subterminal lobe

© William E. Bemis

Ceratotrichia

Lower lobe

10 cm

(B) Elements of the caudal skeleton

Neural arches and
supraneurals

Caudal vertebrae
(110 total)

Hemal arches
and hemal spines

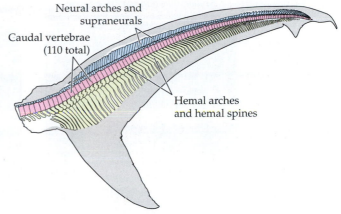

(C) Caudal fin of white shark, *Carcharodon carcharias*

Pelagic habitats: Tall and forked
caudal fin with nearly equally sized
upper and lower lobes for
high-speed swimming and long
distance migrations.

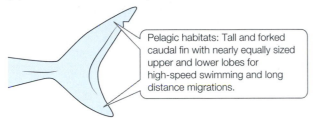

(D) Caudal fin of sand tiger shark, *Carcharias taurus*

Pelagic habitats: Less vertical
caudal fin with smaller lower lobe
for constant swimming and shorter
migrations.

(E) Caudal fin of nurse shark, *Ginglymostoma cirratum*

Benthic and reef associated
habitats: Shallow upturn into upper
lobe and small lower lobe of caudal
fin for slow swimming and short
migrations.

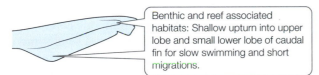

(F) Caudal fin of spiny dogfish, *Squalus acanthias*

Epibenthic habitats: Upturned
upper lobe and small lower lobe for
swimming close to the bottom and
in the water column; the spiny
dogfish is highly migratory.

Figure 6.21 Heterocercal tails of selachians. (A) Caudal fin of a hammerhead shark showing upper, subterminal, and lower lobes. Fin rays (ceratotrichia) support the fin web. (B) Elements of the caudal skeleton of the specimen shown in (A). (C–F) The caudal fins of different shark species are adapted to different habitats and swimming abilities. Pelagic habitats include upper levels of the open sea; benthic habitats include the seafloor; epibenthic refers to areas above but near the bottom. (C–F based on K.S. Thomson and D.E. Simanek, 1977. *Am Zool* 17: 343–354. By permission of Oxford Univ. Press.)

as regional endothermy). Retaining heat in the muscles may increase power output, and hence swimming speed, allowing lamnid sharks to forage in deep, cold waters and to range far north and south into cold seas. Keeping the eyes warm probably increases their ability to detect fast-moving prey in dim light.

Swimming

Extant sharks have a strong, helically wound layer of collagen fibers firmly attached to thick myosepta that run between muscle segments and attach to the vertebral column. This "elastic jacket" with anchors to the axial skeleton appears to store and release energy during the lateral undulations used for swimming. Muscles pulling on the skin probably contribute more to swimming than do muscles pulling on the axial skeleton.

Most chondrichthyans have a heterocercal caudal fin in which the vertebral column angles up into the upper lobe of the fin (**Figure 6.21A,B**). Many sharks have more than 100 caudal vertebrae, and thresher sharks (*Alopias*)

can have more than 250 vertebrae in their caudal fin, which extends about half the length of the body. Thresher sharks swim into a school of fish such as sardines and whip their elongated caudal fin. The fin swings as much as 180 degrees in less than one-third of a second to physically hit prey and stun others with the shock wave created by the movement.

The external shape of the caudal fin offers clues about swimming capabilities and habitats. Pelagic sharks typically have a tall, crescent-shaped caudal fin with large upper and lower lobes used for high speed swimming and long ocean migrations (**Figure 6.21C,D**). Sharks that live in closer association with the bottom often have a less tall caudal fin, with smaller lower lobes (**Figure 6.21E**). Spiny dogfish and other Squalidae have a characteristically shaped caudal fin for epibenthic swimming (above but near the bottom) and long migrations (**Figure 6.21F**).

The functional value of the heterocercal tail lies in its flexibility and control of shape, made possible by intrinsic musculature. When it undulates from side to side, the fin

(A) Lesser spotted dogfish,
Scyliorhinus canicula, in egg case

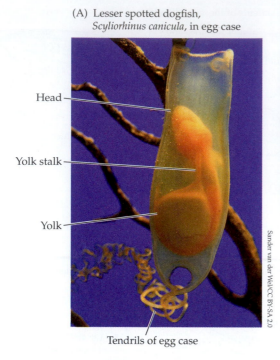

Head

Yolk stalk

Yolk

Sander van der Wel/CC BY-SA 2.0

Tendrils of egg case

(B) Little skate, *Leucoraja erinacea*, removed from its egg case

Mouth   Developing pectoral fin

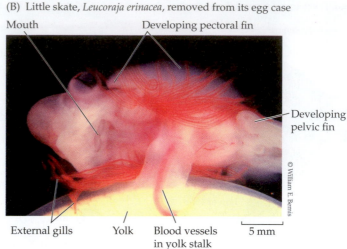

Developing
pelvic fin

© William E. Bemis

External gills   Yolk   Blood vessels   5 mm
                                      in yolk stalk

Figure 6.22 Oviparity. (A) Lesser spotted dogfish developing in an egg case. Tendrils on egg cases entangle vegetation or egg cases can be wedged into protected sites on the substrate. (B) Developing little skate and its yolk in ventral view. The developing pectoral fin will eventually fuse to the head below the eyes. External gills are resorbed later in development.

twists so that the flexible lower lobe trails behind the stiff upper one. This distribution of force produces forward and upward thrust that can lift a fish from a resting position or counteract its tendency to sink as it swims.

Reproduction

Many neoselachians are oviparous. Because extant chimaeriforms are also oviparous, we interpret that oviparity was the ancestral mode of reproduction of chondrichthyans. Female neoselachians have paired ovaries, oviducts, and uteri.

Lecithotrophy (Greek *lekithos,* "yolk"; *trophos,* "one who feeds") describes the pattern in which yolk supplies all nourishment for developing offspring. Oviparous neoselachians are lecithotrophic, and most produce eggs with large yolks (the size of a chicken egg yolk or larger). A specialized region near the anterior end of the oviduct, the nidamental or shell gland, secretes a proteinaceous case around the fertilized egg. Egg cases typically have tendrils or extensions that help them stay attached to rocks or other substrate materials (**Figure 6.22A**). Development within the egg case can last from several weeks to 15 months. During this time, the yolk stalk connects the developing young to its yolk. Blood vessels in the yolk stalk transfer nutrients from the yolk to the young (**Figure 6.22B**). Respiratory gases are exchanged with the water in the egg case using filamentous external gills; these gills are resorbed as development proceeds and internal gills and the muscles that move the gill arches form. Small openings in the egg case allow water to flow through, aided by tail movements. The young hatch from the egg case as miniature replicas of adults.

Some lecithotrophic neoselachians are viviparous. For example, female spiny dogfish, *Squalus acanthias,* exemplify **nonplacental viviparity:** they complete their development inside the mother's uteri, relying only on yolk. After ovulation, two or three eggs are typically enclosed within a thin, proteinaceous shell termed a candle (**Figure 6.23A**). There may be more than one candle in a uterus, yielding litter sizes of 2–14 pups). Development proceeds within the candle inside the uteri for about 12 months before young hatch from the candle. They then may spend another 12 months in utero, eventually to be born as miniature adults. For populations of *Squalus acanthias* along the east coast of the United States, this 2-year gestation period takes place while the mother makes two round-trip migrations between summer feeding grounds in the Gulf of Maine and overwintering habitats off the coast of North Carolina, near Cape Hatteras.

Matrotrophy (Latin *mater,* "mother") refers to the pattern in which the reproductive tract of the mother supplies additional nutrients during development beyond the nutrients supplied in the yolk. For example, **placental viviparity** is a common reproductive pattern of Carcharhiniformes such as the smooth dogfish, *Mustelus canis* (**Figure 6.23B**). In this pattern, a placental attachment allows efficient transfer of nutrients from the mother to the young, allowing shorter gestation periods than in nonplacental species.

Lecithotrophy and matrotrophy are ends of a continuum; many sharks (and other vertebrates) use a combination, with developing young receiving some nourishment from yolk and some from the reproductive tract of the mother. For example, in lamniform sharks such as the sand tiger shark, *Carcharias taurus,* developing young gain nutrients in addition to their own yolk by eating eggs (oophagy) or siblings (uterine cannibalism; this was discovered when a shark biologist reached into the uterus

(A) Nonplacental viviparity of spiny dogfish, *Squalus acanthias*

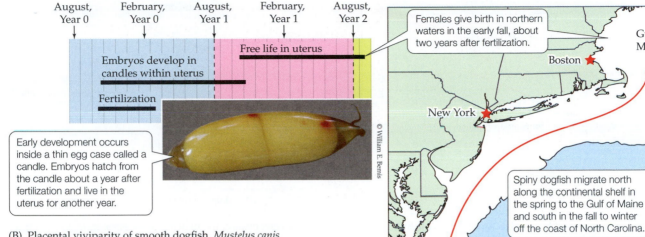

Females give birth in northern waters in the early fall, about two years after fertilization.

Gulf of Maine

Boston

New York

Spiny dogfish migrate north along the continental shelf in the spring to the Gulf of Maine and south in the fall to winter off the coast of North Carolina.

North Carolina

Cape Hatteras

Early development occurs inside a thin egg case called a candle. Embryos hatch from the candle about a year after fertilization and live in the uterus for another year.

© William E. Bemis

(B) Placental viviparity of smooth dogfish, *Mustelus canis*

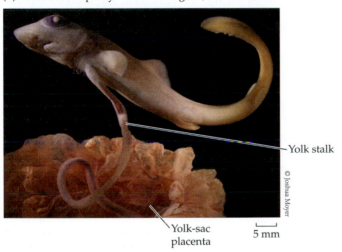

Yolk stalk

Yolk-sac placenta

5 mm

© Joshua Moyer

Figure 6.23 Viviparity. (A) Reproduction of spiny dogfish in the western North Atlantic. Developing spiny dogfish spend approximately one year inside a candle (a thin egg case) in the mother's uterus and a second year as pups inside the uterus. During the two years that a litter of pups is developing, the mother makes two round-trip migrations along the east coast of the United States. Red stars indicate locations for context. (B) Yolk sac placenta and developing pup of a smooth dogfish. In contrast to the spiny dogfish (Squaliformes), developing smooth dogfish (Carcharhiniformes) have a placental attachment to the mother's uterus. Although the litter sizes for the two species are similar, the gestation period of the smooth dogfish is about half that of the spiny dogfish. (A, concept for seasonal data based on T.S. Jones and K.I. Ugland, 2001. *Fishery Bull* 99: 685–690.)

of a pregnant sand tiger shark and was bitten by a 23-cm pup). In the early stages of stingray development, the egg yolk nourishes the embryos; when the yolk is exhausted, the embryos consume a nutrient secretion called uterine milk that is produced by the walls of the uterus.

Elasmobranch brains

Because a large animal has more sensory, motor, and interneurons than a small animal, large vertebrates have larger brains than do small vertebrates. In **Figure 6.24A**, polygons outline values for brain size in relation to body weight for different vertebrate groups. The polygon for chondrichthyans overlaps those of birds and mammals, meaning that at least some chondrichthyans have relatively large brains compared across all vertebrate groups. If we compare brain sizes within chondrichthyans (**Figure 6.24B**), then manta and devil rays (*Mobula*) and the tiger shark (*Galeocerdo cuvier*) are near the apex of the polygon, indicating that they have the largest brains relative to body weight among chondrichthyans studied to date.

The relationship between intelligence, behavior, and brain-to-body weight ratios is not easy to tease out from the simple relationships shown in Figure 6.24. But there

are some intriguing recent findings concerning giant oceanic manta rays (formerly *Manta birostris*, now *Mobula birostris*) and the Mirror Self-Recognition (MSR) test. The MSR test assesses whether animals are capable of self-recognition and (potentially) self-awareness, traits associated with species that exhibit large brain-to-body weight ratios and complex social behaviors. The MSR test typically involves placing an odorless visual mark on an animal and observing the animal's behavior when it sees itself in a mirror.

When two giant manta rays were exposed to a mirror mounted in their aquarium, they showed great interest in their reflections, repeatedly circling, moving their cephalic fins, and blowing bubbles. They also checked body parts that would not be visible without the mirror, such as their ventral sides. These behaviors were displayed either more frequently or only when the mirror was present. The mantas did not respond aggressively to their mirror images, and changes in white markings on the head,

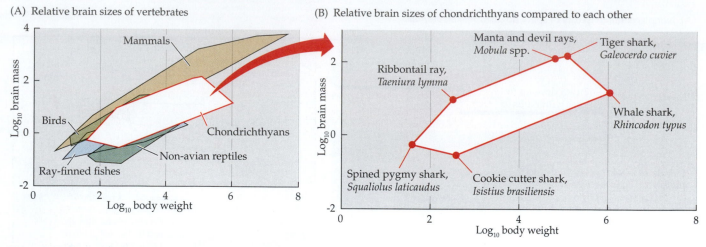

(A) Relative brain sizes of vertebrates

(B) Relative brain sizes of chondrichthyans compared to each other

Figure 6.24 Polygons of brain size and body weight in vertebrates. Neurobiologists express the ratio between brain size and body weight using logarithmic axes because vertebrates span such a great range of body sizes. (A) Brain sizes of chondrichthyans overlap those of other vertebrates, including birds and mammals. (B) Among chondrichthyans, tiger sharks and giant rays have the largest brains. (Based on K.E. Yopak, 2012. *J Fish Biol* 80: 1968–2023.)

which typically occur during encounters with unfamiliar mantas, did not occur.

Although these findings suggest that giant mantas can recognize themselves, the visual mark component of the MSR test could not be done because of difficulty placing marks on mantas. Subsequent video recordings of free-living mantas by other researchers revealed that some of the behaviors interpreted as self-directed in the aquarium study also occurred during intraspecific social interactions, raising the possibility that such behaviors are social and not an indication of self-recognition. However, further research into the cognitive abilities and self-awareness of mantas and other chondrichthyans with large brains may reveal more similarities to complex mammalian social behaviors, a topic we turn to next.

Social networks and migration in sand tiger sharks

Not all sharks lead solitary lives; some are members of social groups. As in other social animals, group-living sharks likely experience costs, such as intraspecific competition for food, as well as benefits, such as enhanced acquisition and defense of food. In some species of group-living sharks, individuals exhibit complex social behaviors, which might be predicted given the relatively large brain-to-body weight ratios of many chondrichthyans (see Figure 6.24). One such behavior, described as fission–fusion, involves changes in group size in relation to changing costs and benefits of group living. For example, groups may increase in size when food resources are abundant and intraspecific competition for food is therefore low (fusion), and break into smaller groups when food becomes scarce and intraspecific competition is high (fission).

This appears to be the case for the sand tiger shark, *Carcharias taurus*, as revealed by tagging data from two males, Sand Tiger 1 (ST1) and Sand Tiger 2 (ST2), summarized as a "social network" in **Figure 6.25**. Upon their initial capture in August of 2012 in Delaware Bay, each male received an acoustic transceiver that not only emitted a uniquely coded ping, but also received and archived pings from nearby conspecifics and other species of fishes with compatible transmitters. The two males were tracked for almost a year as they carried out their annual migration along the east coast of the United States, from Delaware Bay to Cape Hatteras, North Carolina and back again. Upon their return to Delaware Bay in 2013, ST1 and ST2 were recaptured, their transceivers removed, and detection records examined.

September, when food was plentiful in Delaware Bay, was a time of fusion, indicated by the high degree of overlap in the social network graphs of ST1 and ST2 (see Figure 6.25A). December was another time of fusion, this time around the Carolinas, possibly related to constriction of the continental shelf and presence of shipwrecks in that area, which attract sharks and prey species.

In contrast, April was a time of fission, when ST1 and ST2 moved away from conspecifics they previously encountered and entered a more solitary phase. Fission is indicated by the few encounters with sand tigers for both ST1 and ST2 in April, as shown in Figure 6.25A. Numbers of friends, best friends, mutual friends, and acquaintances for ST1 and ST2 are shown in Figure 6.25B.

The social network data from ST1 and ST2 demonstrate that sand tiger sharks exhibit fission–fusion behavior, previously thought to be displayed primarily by mammals. The data also have implications for conservation, because changes in sizes of groups could influence population estimates of sand tigers and affect conservation management decisions.

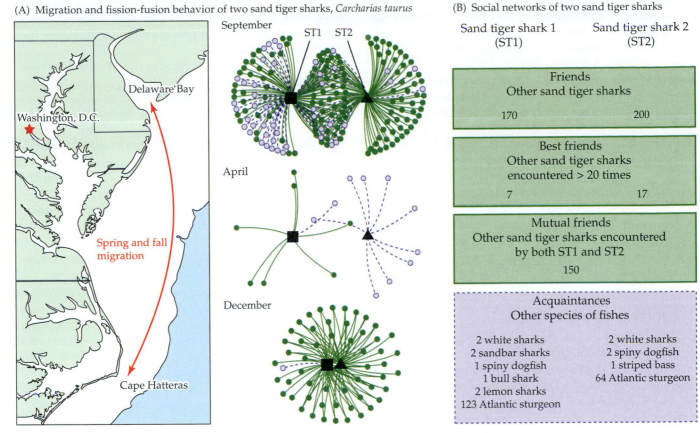

Figure 6.25 Social networks of two sand tiger sharks.
(A) Along the east coast of the United States, sand tiger sharks migrate from wintering areas off the coast of Cape Hatteras, North Carolina, to summer feeding grounds in Delaware Bay. Tagging data from two individuals, ST1 and ST2, show increases in sizes of their social networks (fusion) in September and December, and decreases in sizes of social networks (fission) in April. (B) A summary of the social networks of the two tagged sand tiger sharks. (Based on D.E. Haulsee et al. 2016. *Sci Reports* 6: 34087. CC BY 4.0.)

6.6 Declining Elasmobranch Populations

LEARNING OBJECTIVES

6.6.1 Discuss the plight of sawfishes and list major threats to shark populations caused by humans.

6.6.2 Describe characteristics of chondrichthyans that may limit recovery from population declines.

6.6.3 Explain how decreases in large predatory sharks affect organisms at lower trophic levels.

6.6.4 Distinguish target-based and limit-based policies to protect elasmobranchs.

The 2021.1 version of the International Union for the Conservation of Nature (IUCN) Red List classified 36% of elasmobranch species as Critically Endangered, Endangered, Vulnerable, or Near Threatened. About 70% of species of large predatory sharks fall into one of these IUCN risk categories, and some species have been particularly hard-hit; for example, over the last 50 years, hammerhead shark populations have declined by anywhere from 76% to more than 99% in the Atlantic, Pacific, and Mediterranean basins.

The plight of smaller sharks also has caused international concern among biologists and fisheries managers. A 2020 study of 371 reefs in 58 countries found that sharks are functionally extinct on 20% of them. Reefs in the Dominican Republic, the French West Indies, Kenya, Vietnam, the Windward Dutch Antilles, and Qatar had few to no sharks. Reef sharks were more commonly observed in some other countries, including Australia, Federated States of Micronesia, French Polynesia, and the United States, where there are bans on all shark fishing or science-based management plans to limit shark overfishing.

In this section, we consider a case history of decline, protection, and slow recovery of sawfishes. Next, we examine threats, vulnerabilities, and ecological impacts of population declines. We end with a discussion of policies designed to protect chondrichthyans.

Conservation and sawfishes

The IUCN regards all seven species of sawfishes (Pristidae; see Figure 6.16E) as highly endangered. Sawfishes are the only elasmobranchs included in the Convention on International Trade in Endangered Species (CITES)

Appendix 1, a list of about 1,200 species worldwide that are at critical risk of extinction, and which signatory nations agree to ban from commercial trade.

Problems for sawfishes are manifold. Some fisheries still target sawfishes for meat, fins, and the saw, which was unfortunately a popular curio in the past. Also because of their saw, they are highly susceptible to entanglement in fishing nets targeting other species. Finally, sawfishes live in coastal waters in close contact with human populations, where they are susceptible to habitat loss and degradation. On the East Coast of the United States, the common sawfish *Pristis pristis* once ranged from New York to Texas but is now restricted to peninsular Florida. It was the first marine fish to be included on the United States Endangered Species List, making it illegal to catch, harm, kill, or harass a sawfish in United States waters. There are some hopeful signs of recovery off the southwest coast of Florida, but it will take generations for populations to rebuild, and occupation of the former range is extremely unlikely.

Threats to chondrichthyans

Fishing causes most declines in shark populations. People eat sharks, although they often do not realize it. In England, spiny dogfish is fried for fish and chips, and populations of that species in European waters are very low. Shark steaks have been marketed as an alternative to swordfish in American supermarkets, and as a result mako sharks (genus *Isurus*) are now protected in the Exclusive Economic Zone of the United States (EEZ; the zone within 200 nautical miles of the coastline of any country that borders the ocean). Shark fins are harvested to make shark fin soup. Because the fins are the most valuable part of the shark, "finners" cut the fins from live sharks and throw the rest of the animal, often still alive, back into the sea. Sharks are caught as bycatch in fisheries that target other fishes, and most die before they are returned to the sea.

Many species of sharks and rays extensively use coastal habitats where they interact with people. Coastal development and pollution of areas that sharks use as pupping grounds and nurseries, higher ocean temperatures due to global warming, and increased silt loads in the water column that can make it difficult for sharks to see and avoid people swimming represent new challenges for sharks. Globally, the rare encounters with sharks that result in bites or fatalities are tracked by the International Shark Attack File. In 2019 there were 64 unprovoked shark attacks on humans, including one fatality in the Bahamas and one at Réunion Island in the Indian Ocean. Forty-one of the 62 nonfatal shark attacks occurred in the United States, but this may be an artifact of better reporting.

Because of safety concerns, drum nets are sometimes deployed at beaches to prevent shark–human interactions. Unfortunately, such drum nets are nonselective and kill thousands of sharks. Improved safety awareness at beaches offers a better long-term solution for both people and sharks. For example, when a tagged white shark is detected off a beach on Cape Cod, people are notified to stay out of the water. Tagged sharks can even be tracked using cell phone apps. Techniques that analyze eDNA (molecules of DNA shed into the environment where they are collected by water sampling and analyzed using DNA sequencing) have been able to detect white sharks along the California coast. As technology matures, it may be possible to rapidly detect sharks and other species to provide real-time information on a large scale.

	Spiny dogfish *Squalus acanthias*	Greenland shark *Somniosus microcephalus*	White shark *Carcharodon carcharias*	Scalloped hammerhead *Sphyrna lewini*
Age to maturity (years)	♂ 6–14, ♀ 10–12	~150	♂ 9–10, ♀ 12–14	♂ 4–10, ♀ 4–15
Size at maturity	♂ 65 cm, ♀ 85 cm	~4 m	♂ 3.5–4.1 m, ♀ 4.0–4.3 m	♂ 1.4–2.8 m, ♀ 1.5–3.0 m
Lifespan (years)	75	400	36	35
Reproductive mode	Viviparous, nonplacental	Viviparous, nonplacental	Viviparous, nonplacental	Viviparous, placental
Litter size	2–14 pups	10 pups	10–14 pups	15–23 pups
Reproductive frequency	Biennial	Biennial?	Biennial?	(?)
Gestation period (months)	18–24	(?)	>12	9–10

Figure 6.26 Life-history characteristics of sharks. Late maturation, small litter sizes, and biennial or longer reproductive cycles make sharks vulnerable to overexploitation. For some species, we don't even know the frequency with which individuals reproduce; this is a key life history characteristic necessary for informed management and conservation. (Modified from P. Klimley 1999. *Am Sci* 87: 488–491.)

(A) Predation by large sharks controls populations of mesopredators

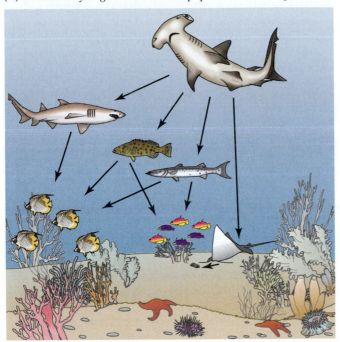

(B) Without large sharks, populations of mesopredators increase

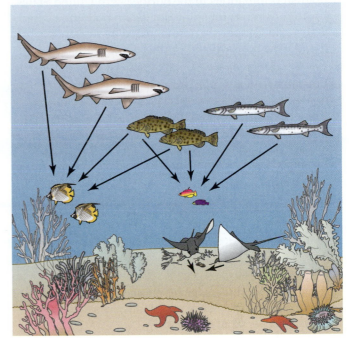

Figure 6.27 Top-down control by apex-predator sharks. Predator–prey relationships are shown by arrows. (A) Predation by large sharks controls population sizes of mesopredators (smaller sharks, predatory bony fishes such as groupers and barracudas, and mollusk-eating rays). (B) When apex predators are removed, mesopredator populations increase and eat more of their prey species (such as smaller fishes), causing prey populations to decline. Habitat disturbance may increase (e.g., from more rays making feeding pits in their search for buried mollusks), and ecosystem services (e.g., cleaning services provided by small fishes) may be lost, leading to increased parasite loads.

Vulnerabilities of chondrichthyans

Migratory behavior places some chondrichthyans at risk. Although many species of sharks spend most of their lives in restricted areas, others migrate hundreds of kilometers along a coast, and some pelagic species travel enormous distances. For example, white sharks travel thousands of kilometers across ocean basins, moving between favored feeding and breeding areas. Nations that protect sharks within the waters of their EEZ are powerless to protect sharks that migrate across national boundaries, and pelagic sharks swimming in the open ocean are beyond any nation's jurisdiction.

The ecology and life-history characteristics of chondrichthyans are poorly suited for a world dominated by humans and make it difficult for shark populations to recover from declines. Sharks grow slowly, mature late, and have only a few young at a time (**Figure 6.26**). The Greenland shark (*Somniosus microcephalus*) is an extreme example, living for at least 400 years (and perhaps longer) and requiring more than 150 years to reach sexual maturity. Females of many viviparous species reproduce only every other year because they must accumulate enough energy to nourish developing young (see Section 6.5).

Ecological impacts of shark population declines

Large predatory sharks such as the white shark, tiger shark, and great hammerhead are **apex predators** that eat species smaller than themselves. In contrast, most coastal and reef-dwelling sharks are **mesopredators**, sharing a trophic level with large bony fishes such as groupers.

Apex predators often exert top-down control on mesopredator populations, stabilizing the structure of the communities in which they occur. As populations of large sharks have crashed due to overfishing, populations of mesopredators have increased. Without top-down control, mesopredator populations could increase to the point where they decimate populations of their prey (**Figure 6.27**). Ecological effects of declining shark populations thus extend far beyond the sharks.

Policies to protect sharks

With public support, we may be able to develop better policies to conserve sharks globally, but it will be extremely challenging because the losses are already so great. Policies to protect elasmobranchs generally fall into these two categories:

- Target-based policies focus on sustainable exploitation of individual species. This approach has shown promise for small, fast-growing species of sharks that occur in waters of developed nations that have the fisheries management infrastructure necessary to plan and implement regulations. In the United States, policies implemented by NOAA's National Marine Fisheries Service led to successful recovery of spiny dogfish populations in the northeastern United States. In 2016, fisheries for spiny dogfish in both the United States and Canada were recognized as sustainable by the Marine Stewardship Council Certification.

- Limit-based policies ban all exploitation of sharks. Fin bans, for example, prohibit sale or purchase of shark fins from any species of shark. Although that phrase is pleasing, the reality is less happy. Fin bans allow sharks to be caught and sold if the fins are not sold separately from the rest of the shark. Protected areas where all fishing is prohibited and

shark sanctuaries where only sharks are protected are approaches to a limit-based policy. Although regulations are often poorly enforced, protected areas have been effective in rebuilding some populations of bony fishes and may prove helpful for chondrichthyans.

There is one small but brighter note in the generally dark prospects for sharks. Viewing sharks and rays in the wild has become a tourist attraction in many parts of the world and may give living sharks economic value rivaling or exceeding that of dead sharks.

Summary

6.1 Acanthodii

Basal members of Acanthodii, also known as spiny sharks, are small Paleozoic fishes with stout spines anterior to dorsal, anal, and paired fins.

The phylogenetic position of acanthodians, unclear for a long time, was clarified by fossil discoveries confirming that spiny sharks are stem chondrichthyans, meaning that Acanthodii includes Chondrichthyes and thus is not extinct.

†*Doliodus problematicus* from the Early Devonian of Canada is an important transitional fossil between basal acanthodians and chondrichthyans.

6.2 Chondrichthyes

Chondrichthyes includes two extant radiations: Euchondrocephali and Elasmobranchii.

- Within Euchondrocephali, Holocephali includes extant Chimaeriformes (ratfishes and allies, commonly known as chimaeras). Chimaeriforms have separate right and left opercular openings.

- Within Elasmobranchii, Euselachii includes some Paleozoic forms and extant Neoselachii. Neoselachii has two groups: Selachii (sharks) and Batomorphi (skates and rays). Elasmobranchs have multiple gill openings (typically five) on each side of the head.

Paleozoic chondrichthyans occur in rocks that formed in freshwater, estuarine, and marine environments, but all extant chimaeras and most elasmobranchs are marine.

There are about 1,300 extant species of chondrichthyans, including about 50 species of chimaeras, 550 species of sharks, and 700 species of skates and rays.

Chondrichthyans have a cartilaginous internal skeleton, mineralized with calcium hydroxyapatite.

The chondrocranium of chondrichthyans has a solid cartilaginous roof to protect the brain. This is because chondrichthyans lack the dermal bones that cover the trough-shaped chondrocranium of osteichthyans.

Chondrichthyans have individual teeth in whorls or tooth plates. Shark teeth are continually replaced as they wear

and are lost. Chimaeras have three pairs of tooth plates used to crush food. As material wears away on the surfaces of a tooth plate, new tissue is added from beneath so that the tooth plate itself is never completely lost.

Jaw suspension differs between holocephalans and neoselachians.

- Holocephalans have a holostylic jaw articulation, in which the upper jaw is fused to the chondrocranium.

- Neoselachians have a hyostylic jaw suspension. The upper jaw articulates with the chondrocranium via the large and mobile hyomandibula, and flexible ligaments connect the anterior part of the upper jaw to the chondrocranium. Hyostyly allows the upper jaw to be projected and retracted during feeding.

Internal fertilization is universal among chondrichthyans. During copulation, males insert one of their paired pelvic claspers into the cloaca of a female to transfer sperm.

Extant chondrichthyans have a large, lipid-filled liver that makes them neutrally buoyant. They also retain nitrogenous compounds that raise their internal osmotic concentrations.

6.3 Euchondrocephali and Chimaeriformes

Holocephali are first known from the Middle Devonian.

- †*Cladoselache*, historically interpreted as an elasmobranch, is a euchondrocephalian.

- †Stethacanthids from the Early Carboniferous showed sexual dimorphism, with the first dorsal fin and spine of males elaborated into a spine-brush complex.

- †*Helicoprion* is known for its bizarre spiraling tooth whorl.

Extant chimaeras are distinctive, mostly deep-water chondrichthyans. Males of some species have a cephalic clasper used in courtship. All extant species are oviparous.

The common names chimaera, rabbitfish, or ratfish reflect the seeming combination of different animals that make up chimaeras.

Summary (continued)

6.4 Elasmobranchii, Euselachii, and Neoselachii

Some strikingly well-preserved Paleozoic and Mesozoic members of Elasmobranchii include †*Cladodoides wildungensis*, the earliest known elasmobranch, and †xenacanths such as the Carboniferous †*Triodus*. †*Tristychius* is a Carboniferous euselachian specialized for suction feeding.

†*Hybodus* is a well-known Mesozoic genus that retained a terminal mouth and had teeth of different shapes in different regions of the jaw, a condition known as heterodonty.

The 13 extant clades of elasmobranchs are the Neoselachii. Within Neoselachii is Selachii, which includes 9 clades of sharks belonging to either Galeomorphi or Squalomorphi; and Batomorphi, which includes 4 clades of skates and rays.

Most sharks are torpedo-shaped; their pectoral fins are not fused to the head.

Batomorphs are dorsoventrally flattened forms whose pectoral fins are fused to the head.

6.5 Biology of Neoselachii

Sharks feed in several ways, including ram feeding, suction feeding, biting off chunks, and filter feeding.

The typical sequence of sensory modalities used by sharks to detect and capture prey is as follows: olfaction; mechanoreception; vision; electroreception; and touch.

The skin of some species of sharks is bioluminescent, emitting light. In other sharks and some rays, the skin is biofluorescent, absorbing blue light and reradiating it as green light.

Epaulette sharks in the Heron Island lagoon, Australia, face periodic low-oxygen conditions due to tide cycles and have adapted to extreme hypoxia by using anaerobic respiration and metabolic depression.

Some species of mackerel sharks can maintain different temperatures in different parts of the body, a condition called endothermal heterothermy.

An elastic jacket of collagen fibers and a heterocercal caudal fin facilitate swimming in sharks.

Oviparity is the ancestral form of reproduction for chondrichthyans, but viviparity has evolved independently in several lineages.

Oviparous species are lecithotrophic, meaning the energy for development is provided entirely by the egg yolk. Viviparous chondrichthyans can exhibit a combination of lecithotrophy and matrotrophy, with the embryo receiving some nourishment from yolk and some from the mother's reproductive tract.

Some chondrichthyans have large brains and complex behaviors (fission–fusion and possibly self-recognition), with brain-to-body weight ratios that overlap those of birds and mammals.

6.6 Declining Elasmobranch Populations

The IUCN Red List classified 36% of elasmobranch species as Critically Endangered, Endangered, Vulnerable, or Near Threatened.

Life-history characteristics of neoselachians make them vulnerable to overfishing. Sharks are particularly slow to mature, produce few young in a reproductive season, and may breed every other year.

Because large sharks are apex predators, reductions in their populations can reduce predation on mesopredators, with potentially disruptive effects for marine ecosystems.

Policies to protect elasmobranchs include target-based policies that focus on sustainable exploitation of individual species and limit-based policies that ban all exploitation of sharks.

Discussion Questions

6.1 *Chondrichthyes* means "cartilaginous fishes." Was bone lost during the evolutionary history of Chondrichthyes, or do they retain a more ancient type of skeleton that evolved before bone? What types of evidence from the fossil record, comparative anatomy, and genetics substantiate your answer?

6.2 Are members of the Paleozoic genus †*Cladoselache* sharks? Why or why not?

6.3 A hyostylic jaw suspension in chondrichthyans allows projection of the upper jaw independent of the lower jaw. Evolutionary biologists have assumed that the ability to project the upper jaw is advantageous for feeding—in other words, that it is an adaptive derived character. An alternative hypothesis is that a hyostylic jaw suspension is a neutral ancestral character (i.e., one that is neither advantageous nor disadvantageous). These two hypotheses (i.e., hyostylic jaw is advantageous versus hyostylic jaw is neutral) generate different predictions about the phylogenetic distribution of hyostyly. What are those predictions, and which one is supported by the phylogenetic distribution of hyostyly?

© Alex Mustard/Nature Picture Library/Alamy Stock Photo

Origin of Osteichthyes and Radiation of Actinopterygians

Striped mackerel, *Rastrelliger kanagurta*

We now turn to the bony fishes, Osteichthyes (Greek *osteon*, "bone"; *ichthys*, "fish"). As we learned in Chapter 1, all extant vertebrate groups other than cyclostomes and chondrichthyans are osteichthyans (see Figure 1.2). Each of the two clades of Osteichthyes—Sarcopterygii, which encompasses the tetrapods, and Actinopterygii, the ray-finned fishes—includes more than 34,000 extant species, with the edge going to Actinopterygii. In other words, there are more species of actinopterygians than all tetrapod groups combined; as land-dwelling tetrapods, humans seem often to overlook the diversity of this remarkable clade.

Most species of actinopterygians belong to the subclade Teleostei (higher bony fishes), and a central fact about this group is its diversity. Our discovery of that diversity continues at a far more rapid rate than for other vertebrate groups. Since the year 2000, ichthyologists have described about 7,500 new species of teleosts. (Think about that number for a minute because it means that more than 20% of the species of teleosts have been discovered in your lifetime). Of these new species, more than 60% come from freshwaters in Asia, Africa, and South America. This is for several reasons, including the sheer physical sizes of tropical freshwater systems such as the Amazon and Congo river basins, difficulties in sampling freshwater fishes, the remoteness of many freshwater habitats, and physical barriers that separate watersheds and promote allopatric speciation (allopatric speciation can occur when there are barriers to gene flow within a population, such as a ridge dividing two watersheds). As the number of ichthyologists looking for new species of fishes continues to increase, new species are being discovered all of the time.

Until recently, the vastness of Earth's waters and their uncountable numbers of species helped lull humans into a sense of "boundless bounty," there for the harvesting. But the industrial world has brought increasing destruction and pollution of resources that once seemed endless, and—the ongoing discovery of new species notwithstanding—many species of fishes are threatened. Because of the enormous diversity of bony fishes, different groups face different anthropogenic threats and require different conservation solutions. Dams on large rivers impact spawning migrations of many fishes. Overfishing and bycatch (a term that refers to incidental capture of species that are not targeted by fisheries) threaten many marine species. Invasive freshwater fishes such as carps cause habitat changes that have impacted freshwater ecosystems across the world. And, everywhere, rapidly warming waters are affecting aquatic ecosystems and threatening many species. We will address these and other conservation issues throughout the chapter as we discuss individual taxa.

We begin this chapter with a discussion on the origin and characters of bony fishes and some key differences between actinopterygians and sarcopterygians before turning to actinopterygian evolution, diversity, swimming, reproduction, and ecology. We describe some specializations of actinopterygians, especially teleosts, for life in Earth's waters from small, still freshwater ponds to coral reefs and deep ocean habitats. Our understanding of marine teleosts living in remote and previously inaccessible environments is rapidly expanding through the use of scuba, crewed submersibles, and remotely operated vehicles (ROVs).

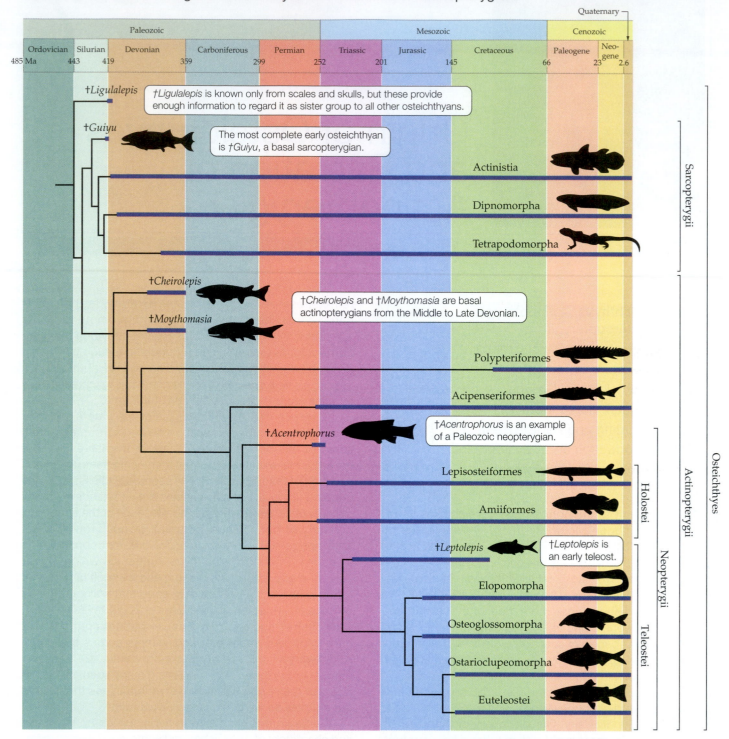

Figure 7.1 **Time tree for osteichthyans.** Time tree of well supported phylogenetic relationships of bony fishes. The split between sarcopterygians (coelacanths, lungfishes and tetrapods) and actinopterygians (ray-finned fishes, including teleosts) occurred early in the Silurian. (See credit details at the end of this chapter.)

7.1 Osteichthyes, Actinopterygii, and Sarcopterygii

LEARNING OBJECTIVES

7.1.1 Summarize the synapomorphies of Osteichthyes.

7.1.2 Explain how tribasic paired fins became modified differently in Actinopterygii and Sarcopterygii.

7.1.3 Understand how mobility of skull components of actinopterygians and sarcopterygians differ.

7.1.4 Discuss the different evolutionary fates of lungs and hearts of actinopterygians and sarcopterygians.

As you can see from the timeline in **Figure 7.1**, one of the earliest osteichthyans, †*Ligulalepis*, is known from the late Silurian. Although †*Ligulalepis* is known only from scales and a fossil braincase, recent workers interpret it as the sister group of all other Osteichthyes, including the two

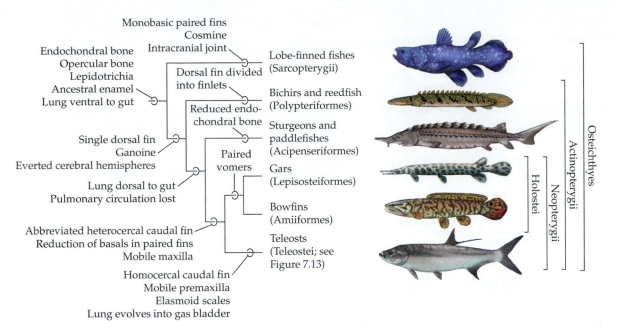

Figure 7.2 Phylogenetic relationships of extant osteichthyans. Sarcopterygii, the lobe-finned fishes, includes tetrapods, as we describe in Chapter 8. Most of the ~34,000 species of actinopterygians belong to the neopterygian group Teleostei (see Figures 7.13 and 7.22 and Tables 7.2 and 7.4). Total species diversity of the four extant groups of non-teleostean actinopterygians is low (see Table 7.1). (Illustrations after K. Liem et al. 2001. *Functional Anatomy of the Vertebrates*, 3rd ed. Cengage/Harcourt College: Belmont, CA.)

large extant clades—Sarcopterygii (*sarkos*, "flesh"; *pteron*, "wing, fin"), which includes eight extant species of fishlike sarcopterygians, and all tetrapods (groups that we will cover extensively in Chapter 8). Actinopterygii (Greek *aktis*, "ray") includes the vast majority of fishlike osteichthyans (**Figure 7.2**; **Table 7.1**). The split between sarcopterygians and actinopterygians occurred in the Silurian, and actinopterygians rapidly diversified during the Devonian.

†*Guiyu oneiros* (Mandarin, "ghost fish") is a small fish from the late Silurian (~ 425 Ma) of China, here interpreted as a sarcopterygian (**Figure 7.3A**). It is among the most complete early osteichthyans yet discovered (although its caudal fin is unknown). Like earlier eugnathostomes, it had two dorsal fins. In contrast to sarcopterygians, actinopterygians have a single dorsal fin, as in the Devonian species reconstructed in **Figures 7.3B** and **7.3C**.

Osteichthyan characters

All osteichthyans have an internal skeleton with ossified endochondral bone in addition to dermal and perichondral bone seen in more basal gnathostomes. Many

Table 7.1 Extant Actinopterygii shown as terminal taxa in Figure 7.2

	Number of extant families	Number of extant species	Habitats	Distribution
Bichirs and reedfishes (Polypteriformes)	1	12	Freshwater	Africa
Sturgeons and paddlefishes (Acipenseriformes)	2	25	Freshwater, anadromous	Northern Hemisphere
Gars (Lepisosteiformes)	1	7	Freshwater, marine nearshore	North America, Central America, Cuba
Bowfins (Amiiformes)	1	1	Freshwater	North America
Teleosts (Teleostei)	>99% of extant actinopterygians*			

Compiled from information in J.S. Nelson et al. 2016. *Fishes of the World*, 5th Ed. Hoboken NJ: Wiley; R. Fricke et al., eds. 2021. *Eschmeyer's Catalog of Fishes: Genera, Species, References*. Online, California Academy of Sciences; R. Froese and D. Pauly, eds. 2021 FishBase. www.fishbase.org.

*See Table 7.2 and Figure 7.13.

(A) Silurian sarcopterygian, †*Guiyu*

First dorsal fin   Second dorsal fin

Diphycercal caudal fin (hypothetical)

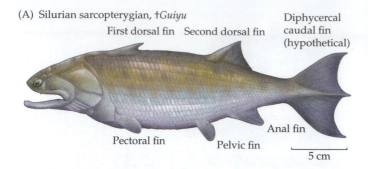

Pectoral fin   Pelvic fin   Anal fin

5 cm

(B) Devonian actinopterygian, †*Cheirolepis*

Maxilla sutured to cheek bones   Small trunk scales with ridges   Single dorsal fin   Heterocercal caudal fin

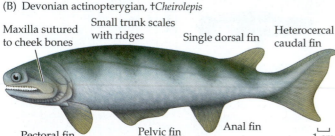

Pectoral fin   Pelvic fin   Anal fin

1 cm

(C) Devonian actinopterygian, †*Moythomasia*

Maxilla sutured to cheek bones   Thick, rhombic trunk scales   Single dorsal fin   Heterocercal caudal fin

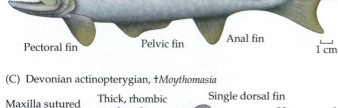
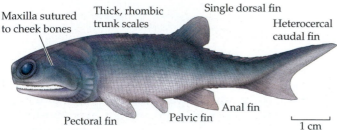

Pectoral fin   Pelvic fin   Anal fin

1 cm

Figure 7.3 Early osteichthyans. (A) Late Silurian †*Guiyu oneiros* from Qujing, Yunnan is the oldest known whole-body fossil of an osteichthyan. It was about 30 cm long, and retained the two dorsal fins characteristic of more basal taxa. Its scales were rhomboid and covered with ganoine; like all osteichthyans, its fin webs were supported by lepidotrichia. The caudal fin is unknown in the specimen, but is reconstructed as a three-lobed diphycercal caudal fin based on fossils of other early bony fishes. (B) †*Cheirolepis canadensis* is known from Quebec's Miguasha Lagerstätte. It has the single dorsal fin characteristic of basal actinopterygians and small, triangular body scales. The heterocercal caudal fin is typical of Paleozoic actinopterygians. (C) †*Moythomasia*, a well-known genus of Devonian actinopterygians from Europe and Australia. Like †*Guiyu*, †*Moythomasia* had rhomboid, ganoine-covered scales. (A after M. Zhu et al. 2009. *Nature* 458: 469–474; B after J.A. Moy-Thomas and R.S. Miles. 1971. *Paleozoic Fishes.* London: Chapman & Hall; C after J.A. Moy-Thomas and R.S. Miles. 1971. *Paleozoic Fishes.* London: Chapman & Hall.)

osteichthyan synapomorphies are identifiable in extant basal actinopterygians. These include:

- A gas-containing lung ventral to and derived from the embryonic gut used for breathing (see Section 4.1).

- Many dermal bones in the head, including the **opercular bone** (or opercle), which covers the gills and aids in gill ventilation. Individual teeth attach to other dermal bones, including the **premaxilla**, **maxilla**, and **dentary**, that surround the mouth (**Figure 7.4A**).

- Dermal bones such as the vomer and parasphenoid form the palate and bear teeth in many forms.

- Dermal bones of the **supracleithral series** extend posteriorly from the skull and attach to the **cleithrum**, a dermal bone of the pectoral girdle.

- Fanlike dermal bones called **branchiostegal rays** form the floor of the gill chamber. Their movements help maintain unidirectional flow of water over the gills for respiration (see Section 4.1).

- Fin rays with **lepidotrichia**, signet-ring shaped dermal bones derived by evolutionary modification of scales (**Figure 7.4B**).

- A shiny, enamel-like tissue, referred to as ancestral enamel, on scales and dermal bones. This tissue was

an evolutionary precursor to true enamel, which occurs in sarcopterygians, and also to **ganoine**, which is found in actinopterygians. These updated interpretations are based on developmental genetic studies of proteins involved in ganoine and enamel formation. To understand what this ancestral enamel was like, it is useful to compare the scales of actinopterygians and sarcopterygians. Paleozoic actinopterygians such as †*Moythomasia* and a few extant lineages such as gars have thick **ganoid scales** (bony, rhomboid-shaped scales covered with a shiny layer of ganoine; **Figure 7.4B,C**). In contrast, basal Devonian sarcopterygians had **cosmoid scales** composed of a dentine-like tissue (not bone as in actinopterygians) and covered with a true enamel surface (not ganoine as in actinopterygians). Within a cosmoid scale or a surface coating of cosmine on a dermal bone is an internal systems of canals that open to the surface via pore-canal openings (**Figure 7.4D**).

Fin adaptations

At the start of this chapter, you probably noticed the same root word in the names of both clades: Greek *pteron*, used for both "fin" and "wing." Actinopterygians are thus "ray fins" and sarcopterygians are "flesh fins." As these names suggest, different fin specializations distinguish the two groups.

The pectoral fin of basal osteichthyans resembles that of chondrichthyans (see Figure 6.12). It has a row of three skeletal elements, called basals (metapterygium, mesopterygium, and propterygium) that articulate with the endochondral shoulder girdle (scapulocoracoid), an arrangement called a **tribasic fin**. Distal to the basals is a row of metapterygial and mesopterygial radials, followed by distal radials, to which attach the bases of the fin rays that support the fin web (**Figure 7.5A**). Basals and radials are endochondral elements that typically ossify to form bone. The characteristic lepidotrichia of

(A) Dermatocranium of a bowfin, *Amia calva*

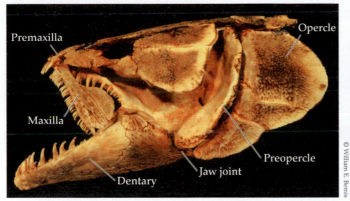

(C) Posterior edge of ganoid scale of a longnose gar, *Lepisosteus osseus*

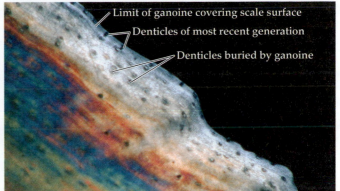

(B) Caudal fin of a longnose gar, *Lepisosteus osseus*

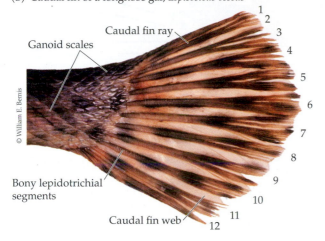

(D) Cosmoid scales of a Devonian lungfish, †*Chirodipterus australis*

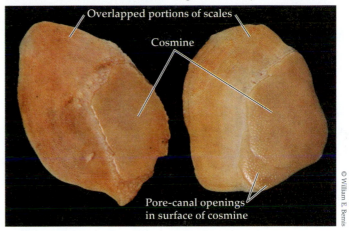

Figure 7.4 Some osteichthyan characters. (A) The derma-tocranium of a bowfin, *Amia calva*, showing examples of bones shared by actinopterygians and sarcopterygians. (B) The caudal fin of a longnose gar has 12 fin rays reinforced by bony lepidotrichial segments that support the caudal fin web. (C) When viewed with polarized light, the posterior edge of a ganoid scale shows the enamel-like tissue ganoine and generations of buried denticles of ganoine (growth rings) closer to the center of the scale. (D) Cosmoid scales of a Devonian lungfish, †*Chirodipterus*. No extant vertebrates have cosmine.

osteichthyans, however, are rings of dermal bone that make the fin rays look segmented (see Figure 7.4B). (Recall that the ceratotrichia that support the fins of chondrichthyans are unsegmented and made of an elastic, keratinlike protein.)

Some extant actinopterygians, such as bichirs and paddlefishes, retain ancestral features of osteichthyan paired fins, but teleosts (which are more derived) lost the basal elements (**Figure 7.5B**), meaning that the paired fin of a teleost is composed mostly of fin rays. It also means that a teleost's fins can be collapsed against the side of the body, contributing to a streamlined profile. Actinopterygian fin rays are also unique in having a bilaminar structure—two halves combining to make a single ray, allowing muscles to curve it to change the fin's shape. Also, because a fin consists of flexible fin rays supporting a thin fin web, it can be spread or collapsed like a handheld fan. This ability to change a fin's surface area allows it to perform different functions in swimming.

In contrast to actinopterygians, sarcopterygian fins have a single basal element, as in the Australian lungfish,

Neoceratodus forsteri (**Figure 7.5C**); this single basal element is homologous to an actinopterygian metapterygoid. As noted in Figure 7.2, this **monobasic fin** is synapomorphic for sarcopterygians. Your humerus is homologous to the basal in a lungfish fin and, like the muscles of your shoulder joint, muscles at the base of the lungfish's fin power its rotation, flexion, and extension. A skeletal axis of mesomeres extends centrally down the fin of the lungfish, flanked by postaxial (posterior edge of fin) and preaxial (anterior edge) radials. These radials bear fin rays, which in the paired fins of lungfishes have little ability to be spread or collapsed. Paired fins of more derived sarcopterygians and tetrapodomorphs have reduced numbers of radials, particularly on the postaxial side of the fin (**Figure 7.5D**). The mesomeres and radials along the preaxial side of the fin have a one-to-two pattern in which the single basal metapterygium supports mesomere 1 + one radial, and mesomere 2 supports mesomere 3 + one radial (**Figure 7.5E**). (This pattern is the same as the humerus in your arm and ulna + radius in your forearm; we discuss other homologies of wrist and ankle bones in Section 8.6.)

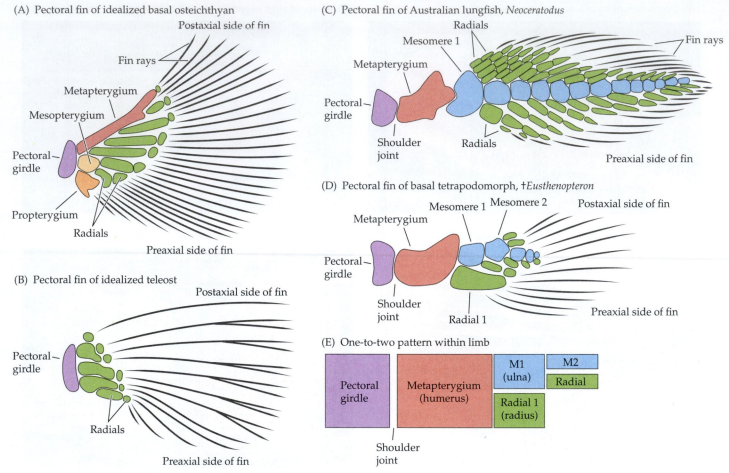

Figure 7.5 Pectoral fins of osteichthyans. Drawings are of a left pectoral fin, with the animal facing left and the preaxial (leading) edge of the fin downward. (A) Basal osteichthyan fin type with three basal skeletal elements: metapterygium, mesopterygium, and propterygium. (B) Pectoral fin of a teleost (a derived actinopterygian) showing loss of basal elements. (C) Monobasic fin of a sarcopterygian, the extant Australian lungfish *Neoceratodus* (see Figure 8.10). Of the three basal elements shown in (A), *Neoceratodus* retains only the metapterygium. The fin axis is formed by mesomeres; radial skeletal elements branch off to both the preaxial and postaxial sides, and fin muscles attach to the skeleton to form a fleshy lobe that moves the fin. (D) Basal tetrapodomorphs such as †*Eusthenopteron* (see Figure 8.3B) had an asymmetrical fin skeleton with most radials branching from the preaxial edge so that the mesomeres and radials formed a "one-bone two-bone" pattern, foreshadowing the pattern seen in tetrapod limbs (see Section 8.5).

Other differences between actinopterygians and sarcopterygians

Other ways in which actinopterygians and sarcopterygians differ concern their skulls. Basal actinopterygians such as polypterids retain a complete dermal skull roof and have little mobility of their upper jaw, but more derived actinopterygians have reduced the dermal bones of the cheek region and have mobile premaxillae and maxillae (see Figure 7.2). In contrast, jaw mobility of sarcopterygians is partially derived from movement between the anterior and posterior portions of the braincase. This intracranial joint is a synapomorphy of sarcopterygians (see Figure 7.2).

Forebrains of actinopterygians are everted (that is, they develop by eversion, in which the cerebral hemispheres fold outward). Sarcopterygians retain inversion, the ancestral osteichthyan form of brain development in which the cerebral hemispheres fold inward on themselves during growth. As sarcopterygians, we retain the inverted pattern of forebrain development.

We hypothesize that the osteichthyan ancestor of actinopterygians and sarcopterygians had a pulmonary circulation and a heart with a single atrium (**Figure 7.6A**). Within Actinopterygii, teleosts secondarily lost the pulmonary circulation because the gas bladder took on a role in buoyancy regulation rather than respiration (**Figure 7.6B**). Within Sarcopterygii, the atrium became divided and the pulmonary veins delivered oxygenated blood from the lungs to the left atrium, allowing better separation of oxygenated and deoxygenated blood flow through the heart and aortic arches (**Figure 7.6C**).

Although the earliest sarcopterygians had gills and fins supported by fin rays, one subgroup ultimately left aquatic environments for land, where they lost their gills and fin rays and radiated into the many groups of tetrapods. The pectoral and pelvic fins of sarcopterygians are homologous to the forelimbs and hindlimbs, respectively, of the tetrapods that will be the subject of the remaining chapters of this book. But before turning to tetrapods, we will examine the astonishing diversity of actinopterygians.

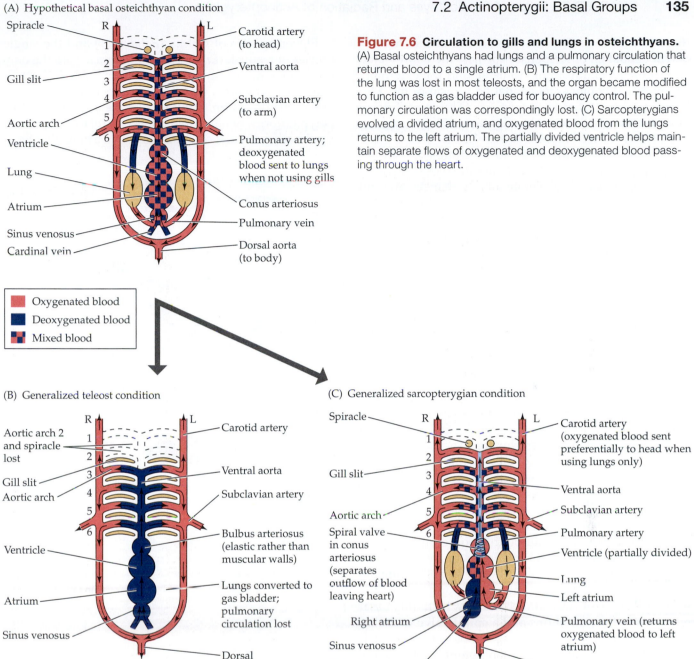

(A) Hypothetical basal osteichthyan condition

Spiracle
R L
1
2
Gill slit
3
4
5
Aortic arch
6
Ventricle
Lung
Atrium
Sinus venosus
Cardinal vein

Carotid artery (to head)
Ventral aorta
Subclavian artery (to arm)
Pulmonary artery; deoxygenated blood sent to lungs when not using gills
Conus arteriosus
Pulmonary vein
Dorsal aorta (to body)

Oxygenated blood
Deoxygenated blood
Mixed blood

Figure 7.6 Circulation to gills and lungs in osteichthyans. (A) Basal osteichthyans had lungs and a pulmonary circulation that returned blood to a single atrium. (B) The respiratory function of the lung was lost in most teleosts, and the organ became modified to function as a gas bladder used for buoyancy control. The pulmonary circulation was correspondingly lost. (C) Sarcopterygians evolved a divided atrium, and oxygenated blood from the lungs returns to the left atrium. The partially divided ventricle helps maintain separate flows of oxygenated and deoxygenated blood passing through the heart.

(B) Generalized teleost condition

R L
Aortic arch 2 and spiracle lost
1
2
3
Gill slit
4
Aortic arch
5
6
Ventricle
Atrium
Sinus venosus

Carotid artery
Ventral aorta
Subclavian artery
Bulbus arteriosus (elastic rather than muscular walls)
Lungs converted to gas bladder; pulmonary circulation lost
Dorsal aorta

(C) Generalized sarcopterygian condition

Spiracle
R L
1
2
Gill slit
3
4
Aortic arch
5
Spiral valve in conus arteriosus (separates outflow of blood leaving heart)
6
Right atrium
Sinus venosus
Posterior vena cava

Carotid artery (oxygenated blood sent preferentially to head when using lungs only)
Ventral aorta
Subclavian artery
Pulmonary artery
Ventricle (partially divided)
Lung
Left atrium
Pulmonary vein (returns oxygenated blood to left atrium)
Dorsal aorta

7.2 Actinopterygii: Basal Groups

LEARNING OBJECTIVES

7.2.1 Contrast the extant diversity of non-teleostean actinopterygians with teleosts.

7.2.2 Describe anatomical differences between heterocercal and homocercal caudal fins.

7.2.3 Explain why sturgeons and paddlefishes are at high risk of extinction.

7.2.4 Compare the current and prehistoric diversity of Amiiformes and suggest possible reasons for differences.

Paleozoic actinopterygians were mostly small (5–25 cm) with a single dorsal fin, a heterocercal caudal fin, and thick, bony ganoid scales such as those of †*Moythomasia*

(see Figure 7.3C). Its paired fins retained three basal elements like the idealized form in Figure 7.5A. More than 300 extinct genera of actinopterygians, historically referred to as "†paleonisciforms," lived from the Devonian to the Jurassic, and lifetimes of work will be required to sort out their phylogenetic relationships. Precursors of three small extant lineages, Polypteriformes, Acipenseriformes, and Holostei, diverged in the Paleozoic (see Figure 7.1). A fourth lineage, Teleostei, appeared in the mid-Triassic, and today the teleosts are by far the most diverse, most familiar, and most commercially important actinopterygians, as we will describe in Sections 7.3, 7.4, and 7.5.

A general and usually visible distinction among these four lineages is the internal and external structure of their caudal fins. The sharply asymmetrical **hetero-cercal caudal fin** of Acipenseriformes, the abbreviated heterocercal caudal fin of Holostei, and the externally symmetrical **homocercal caudal fin** typical of most teleosts are illustrated in **Figure 7.7**. (The caudal fin of Polypteriformes, the most basal extant actinopterygians, is diphycercal; see Figure 7.8B.)

We can easily characterize some groups of fishes as "freshwater" or "marine," but fishes in other groups live in estuaries, where they tolerate salinity changes, and still others are diadromous, meaning that they migrate between freshwater and marine habitats and have distinct life stages in both. Diadromy is the general term for such movements, which are related to feeding and reproduction, but there are two types of diadromy. Anadromous fishes spawn in freshwater but grow to adulthood in the ocean, while catadromous fishes grow to adulthood in freshwater but spawn in the ocean. We characterize an order as "deep sea" if adults routinely live in waters deeper than 300 m and "nearshore" if its adult members routinely use shallow coastal waters. Many freshwater fishes have relatively circumscribed distributions, but marine fishes often have global or near-global distributions in waters of appropriate temperatures. Other orders contain both freshwater and marine species and have global distribution in continental freshwaters and global marine distributions. For groups covered in this chapter, these categories are listed in Tables 7.1, 7.2, and 7.4.

Polypteriformes

Polypteriformes is the living sister group of Acipenseriformes + Neopterygii. Although Neopterygians (comprising the groups Holostei and Teleostei; see Figure 7.2) largely replaced more basal actinopterygians during the early Mesozoic, extant polypteriforms and acipenseriforms retain many ancestral characters, including basal elements in the paired fins and a spiral valve in the intestine, that offer clues about the biology of earlier actinopterygians. The oldest fossil polypterids are from the mid-Cretaceous, but a related Triassic group, †Scanilepidiformes, has been recognized as stem polypterids. Thus, the group is ancient, although molecular clock data indicate a Miocene origin for the extant crown polypterids.

Eleven species of bichirs (*Polypterus*) and the single species of reedfish (*Erpetoichthys calabaricus*) are the only extant genera of Polypteriformes (see Table 7.1). A diagnostic character is the series of dorsal finlets instead of a single dorsal fin (**Figure 7.8**). Like sarcopterygians, polypteriforms retain paired lungs ventral to the gut with a pulmonary circulation, which is why we think that these

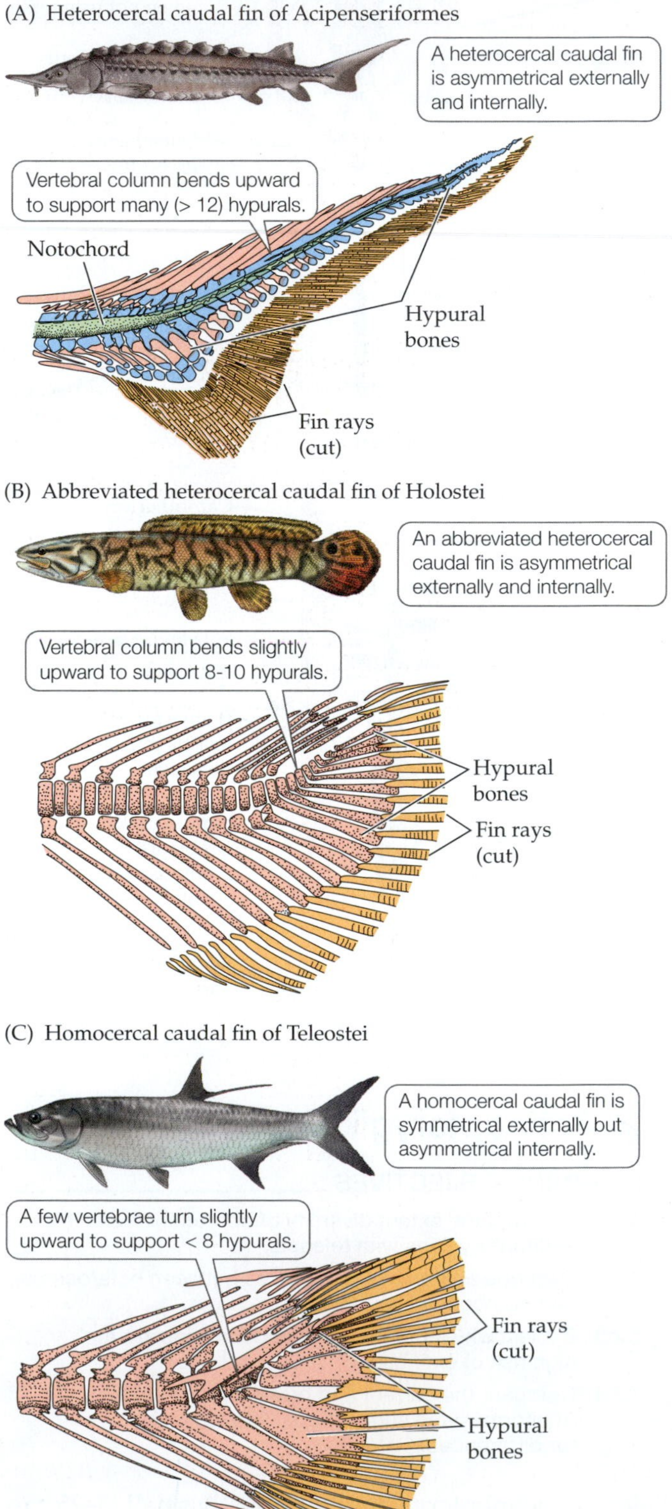

(A) Heterocercal caudal fin of Acipenseriformes

A heterocercal caudal fin is asymmetrical externally and internally.

Vertebral column bends upward to support many (> 12) hypurals.

Notochord

Hypural bones

Fin rays (cut)

(B) Abbreviated heterocercal caudal fin of Holostei

An abbreviated heterocercal caudal fin is asymmetrical externally and internally.

Vertebral column bends slightly upward to support 8-10 hypurals.

Hypural bones

Fin rays (cut)

(C) Homocercal caudal fin of Teleostei

A homocercal caudal fin is symmetrical externally but asymmetrical internally.

A few vertebrae turn slightly upward to support < 8 hypurals.

Fin rays (cut)

Hypural bones

Figure 7.7 Evolution of the caudal fin. (A) Like sharks (see Figure 6.21), sturgeons and paddlefishes (Acipenseriformes) retain an asymmetrical heterocercal caudal fin in which the vertebral column extends into the upper lobe and the fin rays attach to modified hemal arches called hypural bones. (B,C) Neopterygian caudal fins. (B) The abbreviated heterocercal caudal fin of bowfins and gars (Holostei) has fewer vertebrae and fewer hypural bones and remains asymmetrical externally. (C) Teleosts typically have a homocercal caudal fin supported by few vertebrae. Even though its skeleton is asymmetrical (i.e., the vertebral column bends upward into the caudal fin), the caudal fin is symmetrical externally. (Insets after K. Liem et al. 2001. *Functional Anatomy of the Vertebrates*, 3rd ed. Cengage/Harcourt College: Belmont, CA.)

(A) Saddled bichir, *Polypterus endlicheri*

© Koen van Uitert

(B) External anatomy of bichir

Incurrent nostril

Dorsal fin divided into finlets

Diphycercal caudal fin

Pectoral fin   Ganoid scales   Pelvic fin   Anal fin

Figure 7.8 Polypteriformes. (A) Eleven species of bichirs (genus *Polypterus*) and a single species of reedfish are the only extant polypteriforms. (B) External anatomy of bichir. Instead of a dorsal fin, a distinct row of finlets characterizes polypteriforms. The diphycercal caudal fin is symmetrically expanded and the vertebrae extend to its tip, a condition shared with sarcopterygians. (B after K. Liem et al. 2001. *Functional Anatomy of the Vertebrates*, 3rd ed. Cengage/Harcourt College: Belmont, CA.)

two traits were present in the common ancestor of actinopterygians and sarcopterygians. All 12 extant species of polypteriforms are from Africa, where they live in freshwater habitats and use lungs to supplement gill respiration. Ganoine covers their thick, rhomboid-shaped scales. Bichirs are heavy-bodied ambush predators that eat other fishes. The reedfish (also known as the ropefish) is slimmer and more eel-like, moving sinuously through thick vegetation and feeding primarily on snails and invertebrates.

Acipenseriformes

Acipenseriformes (sturgeons and paddlefishes) share with Neopterygii a single lung or gas bladder dorsal to the gut with reduced or absent pulmonary circulation. Some endochondral elements of acipenseriforms partially ossify, but most remain cartilaginous, and a large notochord is retained throughout life. The heterocercal caudal fin is strongly asymmetrical both externally and internally (see Figure 7.7A). Many vertebrae turn upward into the dorsal lobe of an acipenseriform's caudal fin, where they support a series of hypural bones (Greek *hypo*, "below"; *uro*, "tail") that, in turn, support the caudal fin rays.

The upper jaw of acipenseriforms attaches to the cranium by ligaments, so there is no bony cheek as in †*Cheirolepis* or *Polypterus*, and the jaws are highly mobile. Stem acipenseriforms occur in the Late Jurassic, and the group has an important Cretaceous record.

Acipenseridae The 25 extant species of sturgeons (Acipenseridae) are large (1–6 m), active, demersal (bottom-dwelling) fishes, all native to the Northern

Hemisphere (see Table 7.1). Their protrusible jaws make sturgeons effective suction feeders—they can slurp prey from the substrate by projecting the upper and lower jaws (**Figure 7.9A**). Sturgeons have five rows of enlarged armor-like scutes along the body. All lay eggs in freshwater, but some species are anadromous, leaving their natal rivers to forage in the relatively shallow waters of estuaries and continental shelves (sturgeons do not routinely cross great ocean basins). A tagging study of the green sturgeon, *Acipenser medirostris*, documented movements greater than 4,000 km per year between spawning rivers, estuaries, and along the coast of western North America. The beluga sturgeon (*Huso huso*) of eastern Europe and Asia is among the world's largest fishes, growing to >8 m and weights of 1,300 kg. The smaller shovelnose sturgeons (**Figure 7.9B**) live in the Mississippi River basin and other rivers of North America.

Because they depend on large rivers for spawning, all species of sturgeons are at risk of extinction. The three species of Asian sturgeons in the genus *Pseudoscaphirhynchus*, small sturgeons native to the Aral Sea basin, are an example. The Aral Sea was at one time the fourth largest lake in the world, but water diversion during the Soviet era caused a cascade of environmental problems resulting in desertification of much of its former area. All three species of *Pseudoscaphirhynchus* are critically endangered, and one may in fact be extinct.

Polyodontidae Paddlefishes (Polyodontidae) uniquely have an array of star-shaped dermal bones (stellate bones) supporting an elongated and flattened rostrum shaped like a paddle. Spaces between the arms of these stars are filled with tens of thousands of ampullary organs used to detect minute electric fields of prey (see Section 4.2). There is a rich fossil history of Cretaceous and Cenozoic paddlefishes, including the Eocene Green River Paddlefish (**Figure 7.9C**). The American paddlefish (*Polyodon spathula*; **Figure 7.9D**) occurs in the Mississippi River drainage from western New York to central Montana and from Canada to Louisiana. This species is a filter feeder that uses elongate gill rakers to capture planktonic crustaceans such as *Daphnia*. Fossil species of *Polyodon* from the Cretaceous Hell Creek Formation have identical gill rakers.

The Chinese paddlefish (*Psephurus gladius*) occurred in the Changjiang (Yangtze) River valley (**Figure 7.9E**). One of the largest freshwater fishes in the world, it was declared extinct in 2020, a victim of dam construction that limited its spawning migrations. Like sturgeons, the Chinese paddlefish used its protrusible mouth to feed on small fishes and crustaceans.

Neopterygii: Holostei

Species in the Permian genus †*Acentrophorus* (**Figure 7.10A**) are among the oldest known members of Neopterygii, a group that includes extant Holostei and Teleostei (see Figure 7.2). Neopterygians have an abbreviated heterocercal caudal fin (see Figure 7.7B). We describe it as

(A) Sterlet, *Acipenser ruthenus*, showing projected mouth

(B) Shovelnose sturgeon, *Scaphirhynchus platorhynchus*

(C) Green River paddlefish, †*Crossopholis magnicaudatus*

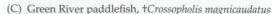

(D) American paddlefish, *Polyodon spathula*

(E) Chinese paddlefish, †*Psephurus gladius*

Figure 7.9 Acipenseriformes.   (A) Sterlet, showing projected mouth and four chemosensory barbels in front of the mouth, features that characterize sturgeons. (B) Shovelnose sturgeons are native to the Mississippi and other American rivers. (C) Fossil paddlefish from the Eocene Green River Formation (~48.5 Ma) in southwestern Wyoming. Superbly preserved specimens from this fossil Lagerstätten offer insights into many groups of fishes (see Section 20.2). (D) American paddlefish, *Polyodon spathula* filter feed for *Daphnia* and other freshwater zooplankton. (E) The Chinese paddlefish of the Yangtze River was among the largest freshwater fishes; it was declared extinct in 2020.

abbreviated because relatively few vertebrae turn up into the caudal fin and it supports only nine hypural bones instead of the dozens in an acipenseriform. Neopterygians also have reduced the basal elements in the paired fins and have a mobile maxilla (mobile because its posterior end is not sutured into other bones of the cheek).

Holostei includes many extinct Mesozoic taxa and two extant groups, Lepisosteiformes and Amiiformes.

Lepisosteiformes   Lepisosteiformes, the gars, includes seven extant species in two genera, *Lepisosteus* (**Figure 7.10B**) and *Atractosteus*. The alligator gar, *A. osseus*, reaches lengths of 4 m. Gars live in freshwater habitats in North and Central America and Cuba and use their vascularized gas bladder as a lung. They feed on other fishes, which are taken unaware when a seemingly lethargic and well-camouflaged gar eases alongside them and, with a sideways flip of its body, grasps its prey with needlelike teeth. Species of both extant genera occur in the Eocene Green River formation of Wyoming. The interlocking, multilayered ganoid scales of gars (see Figure 7.4B,C) resemble those of Paleozoic and Mesozoic actinopterygians, and alligators are the only natural predators able to cope with the thick armor of an adult gar.

Amiiformes   The single extant species of Amiiformes is the bowfin, *Amia calva*, the only living member of a group that, during the Mesozoic, also occurred in Europe, Asia, Africa, and South America. Named for its bow-shaped dorsal fin, *Amia calva* lives in weed-filled lakes and slow-flowing rivers from Quebec to Florida and Texas (**Figure 7.10C**). The bowfin is an excellent model for understanding many aspects of osteichthyan evolution, particularly skeletal anatomy. It reaches lengths of about 1 m, and eats almost any organism smaller than itself. Its scales lack ganoine, are much thinner than those of gars,

(A) Reconstruction of †*Acentrophorus*

Thick, rhombic trunk scales   Single dorsal fin   Abbreviated heterocercal caudal fin

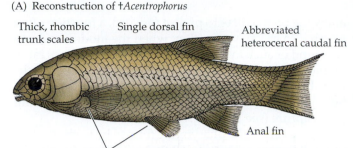

Anal fin

Pectoral and pelvic fins separated from each other

Figure 7.10 An early neopterygian and extant Holostei. Neopterygii includes the extant Holostei (seven species of gars and the bowfin, *Amia calva*) as well as teleosts (Teleostei). (A) †*Acentrophorus*, a neopterygian from the Permian, was 10–15 cm long. The reconstruction shows thick, rhombic scales and an abbreviated heterocercal caudal fin. (B) Gars have multilayered ganoid scales that interlock, protecting adult gars from most natural predators except alligators. (C) The caudal ocellus, or eyespot, of a bowfin confuses predators and causes them to attack the wrong end of the fish. (A after E.L. Gill, 1923. *Proc Zool Soc Lond* 1923: 19–40.)

(B) Longnose gar, *Lepisosteus osseus*
Rhombic, ganoid scales   Abbreviated heterocercal caudal fin

Courtesy of Chris Crippen

(C) North American bowfin, *Amia calva*
Bow-shaped dorsal fin   Ocellus at base of abbreviated heterocercal caudal fin

Pectoral and pelvic fins separated from each other

Ryan Hagarty/USFWS

and are composed of a single layer of bone, as are those of teleosts. Males have a dark tail spot (ocellus) and build and guard nests during the breeding season.

Neopterygii: Teleostei

As we noted in the introduction to this chapter, there are a lot of fish in the sea, and most of them are teleosts. More than 99% of the extant species of ray-finned fishes are teleosts (Greek *teleos*, "entire"; *osteon*, "bone"), which means there are many more teleost species than there are species of birds and mammals combined. We devote Sections 7.3–7.5 to the characters, phylogeny, and diversity of teleosts, as well as describing some recent advances in understanding their evolutionary history.

7.3 Characters of Teleostei

LEARNING OBJECTIVES

7.3.1 Understand the implications of pelagic eggs.

7.3.2 Describe evolutionary changes at the base of teleost diversity related to improved swimming performance.

7.3.3 Explain the evolutionary importance of changes in the cheek bones, maxilla, premaxilla, and pharyngeal jaws of teleosts.

Among several synapomorphies of teleosts is the presence of a homocercal caudal fin, as described in Figure 7.7. Such a fin appears symmetrical on the outside of

the fish—that is, its upper and lower lobes have similar shapes. Internally, however, the skeleton of a homocercal caudal fin retains vertebrae that bend upward to support fewer and larger hypural bones than those of an abbreviated heterocercal caudal fin. Long fin rays make a teleost's homocercal caudal fin extremely flexible.

Teleost scales lack ganoine and are composed of a much thinner layer of bone than those of polypterids or gars. Much of each scale is embedded in the dermis, but the posterior one-third of the scale is exposed, meaning that you can see it when you look at the external surface of a fish. The exposed parts overlap like the shingles on a roof, a pattern described as imbricated (**Figure 7.11A**). Together, thinner scales and imbrication make a teleost's body far more flexible during swimming than that of a polypterid or gar. Most basal teleosts have round, **cycloid scales** (**Figure 7.11B**). As a defensive mechanism, some, such as herrings, easily shed scales when attacked; such scales are described as deciduous. More derived teleosts often have **ctenoid scales** (Greek *cteni*, comb), named for the small, comblike, bony projections on the free margin of the scale (**Figure 7.11C**).

The cheek bones of teleosts are reduced relative to those of more basal actinopterygians, and this allows the suspensorium (the hyomandibula and several other skeletal elements that together suspend the lower jaw from the skull) to swing laterally when the mouth opens, flaring the oral cavity and rapidly increasing its volume to suck

(A) Section through skin and thin, imbricated scales of a teleost

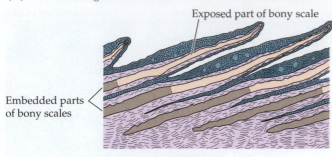

(B) Surface view of a cycloid scale of a teleost

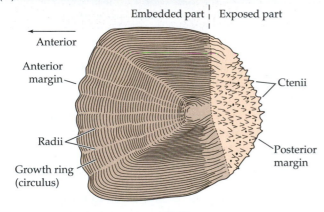

(C) Surface view of a ctenoid scale of a teleost

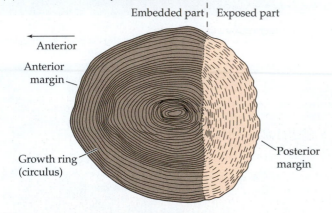

Figure 7.11 Anatomy of teleost scales. (A) Teleosts have thin, imbricated (overlapping) scales like the shingles on a roof. Thinner scales and imbrication give a teleost's body flexibility while swimming. (B) Most basal teleosts have rounded cycloid scales. (C) Derived teleosts such as acanthopterygians often have ctenoid scales, named for the cteni (tiny bony projections) on their exposed posterior edges. Growth rings (circuli) on scales can be counted to estimate a fish's age. (See credit details at the end of this chapter.)

prey into the mouth. The open cheek also allows space for larger jaw closing muscles to bulge when biting prey.

The term **oral jaws** refers to the upper and lower jaws and associated skeletal elements that surround and define the mouth; these elements may or may not have teeth. Teleosts that gather food from the water surface have upward-facing mouths; mouths of bottom feeders face downward. Teleosts have a mobile premaxilla, and many teleost groups evolved further specializations that

allow the premaxillae to project forward, a function often referred to as jaw protrusion, when the mouth opens to surround or suck in prey. The maxilla also becomes more mobile in teleosts, and, in the most derived forms, it lacks teeth because its primary function is not to process prey, but to support and define the shape of the mouth during prey capture (see Figure 7.12B).

Tooth plates in the pharynx are formed by teeth attached to small bones associated with the gill arches. Originally, pharyngeal tooth plates were not very mobile, a condition retained today in bowfin and gars. In teleosts, however, the tooth plates became consolidated, fused to one another and to gill arch elements above and below the esophagus, and subsequently attained great mobility as **pharyngeal jaws**. Newly differentiated muscles associated with pharyngeal jaws allowed new functions such as crushing that often differ from the stabbing or capture functions of oral jaws. With so many separate systems to work with, it is little wonder that some of the most extensive adaptive radiations among teleosts occurred in groups such as cichlids that have protrusible oral jaws and separately mobile pharyngeal jaws with diverse types of pharyngeal teeth.

Many evolutionary advances in teleosts relate to improved swimming performance (see Section 7.6). Thinner, less massive scales and the resulting increase in body flexibility link to new possibilities for fast swimming to catch prey and escape predators. Teleosts use gas bladders for buoyancy regulation (and air-breathing in some), which allows a teleost to swim horizontally without using its paired fins to stay up in the water column. No longer responsible for generating lift, the skeleton of the paired fins of teleosts became simplified. For example, loss of basal elements and radials attached directly to the limb girdles mean that the visible external part of the fin consists primarily of fin rays (see Figure 7.5B). The paired fins of teleosts are used during social signaling, courtship, sound production, walking, sensing the environment, sucking onto the substrate, and gliding over the surface of the water, as well as turning and braking during swimming.

Pectoral fins of holosteans such as the bowfin and of basal teleosts are located ventrally, about one-third of the way along the body, but the pelvic fins are positioned farther posteriorly in an abdominal position, leaving a wide gap separating the bases of the pectoral and pelvic fins. This arrangement changed during teleost evolution. The base of the pectoral fin of derived teleosts is located on the side of body, behind the operculum, and the base of the pelvic fin is located directly ventral to it (the thoracic position of the pelvic fin) or, in some cases, even anterior to the pectoral fin (the jugular position).

Another innovation of teleosts was the evolution of **pelagic eggs** (Greek *pélagos*, "open sea") only a few millimeters in diameter. These eggs float in the water column because each contains a small oil droplet, and, when they hatch, the resulting larvae typically stay in the water

(A) †*Leptolepis knorri*, a generalized Jurassic teleost

Homocercal caudal fin

10 mm

(B) Teleosts have a mobile premaxilla and maxilla

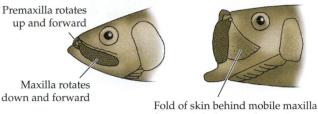

Premaxilla rotates
up and forward

Maxilla rotates
down and forward

Fold of skin behind mobile maxilla
forms a tubular mouth opening

(C) †*Leedsichthys problematicus*, a giant filter-feeding Jurassic teleost

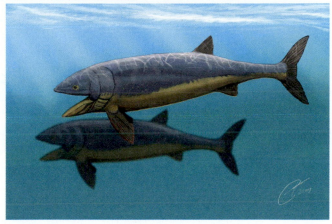

1 m

(D) †*Xiphactinus audax*, a predatory Cretaceous teleost

Figure 7.12 Mesozoic teleosts. (A) †*Leptolepis* and related forms lived in freshwater and marine habitats from the Late Triassic to the Early Cretaceous. (B) Teleosts have a derived jaw structure with enlarged mobile maxillae that rotate when the jaws opened; a fold of skin closes the gaps behind the bones, forming a tubular mouth opening that increased the effectiveness of prey capture. (C) Reconstruction of †*Leedsichthys problematicus*, a filter-feeding Jurassic teleost that occupied the niche currently occupied by baleen whales. †*Leedsichthys* is the largest fish ever discovered, reaching lengths estimated at 30 m. (D) †*Xiphactinis audax*, a predatory Cretaceous teleost related to extant osteoglosso-morphs, reached lengths more than 3 m. Many specimens have large prey in their stomachs.

column as pelagic larvae. There are many implications of this reproductive innovation, including the possibility of spawning millions of eggs and larval dispersal across great expanses of water. Although not all teleosts have pelagic eggs and larvae, many highly diverse subgroups do, and broadcasting such small eggs into the environment does not occur in any other vertebrates.

By the Jurassic, the remarkable radiation of teleosts was well underway, with forms such as †*Leptolepis* common in many localities (**Figure 7.12A**). During mouth opening, the mobile maxilla of early teleosts supported a fold of skin that limited water inflow to directly in front of the mouth (**Figure 7.12B**). Giant filter-feeding teleosts such as †*Leedsichthyes* (**Figure 7.12C**), estimated to have reached lengths of 15 m or more and the largest actinopterygians in history, filled the ecological role in Late Jurassic seas that today is occupied by baleen whales. Noteworthy Cretaceous teleosts included predators such as †*Xiphactinis* (**Figure 7.12D**), which reached lengths greater than 4 m in the epicontinental sea that once covered Kansas. Specimens of †*Xiphactinis* with whole prey almost 2 m long inside the body attest to their voracious appetite.

7.4 Teleostei: Basal Groups

LEARNING OBJECTIVES

7.4.1 State a synapomorphy for each of the following groups: Elopomorpha, Ostariophysi, and Otophysi.

7.4.2 Distinguish true eels from electric eels.

7.4.3 Describe factors affecting the conservation status of Atlantic cod.

Because of their great diversity, it has been far more challenging to develop robust phylogenetic hypotheses for teleost fishes than for other groups of vertebrates. Conservatively, there are about 470 extant families of teleosts belonging to about 65 orders. Ichthyologists are reasonably confident about the monophyly of many of these families and orders, but this is always subject to change—and in fact has changed frequently for some groups.

Using characters derived from the skeleton and other organ systems, ichthyologists sorted out many phylogenetic problems, but other questions remained, particularly about evolutionary relationships of teleost subgroups that radiated very rapidly after the end-Cretaceous mass extinction (see Section 10.4). In the last 5 years, analyses of very large datasets of DNA sequence data resulted in new phylogenetic hypotheses for some of these groups, and

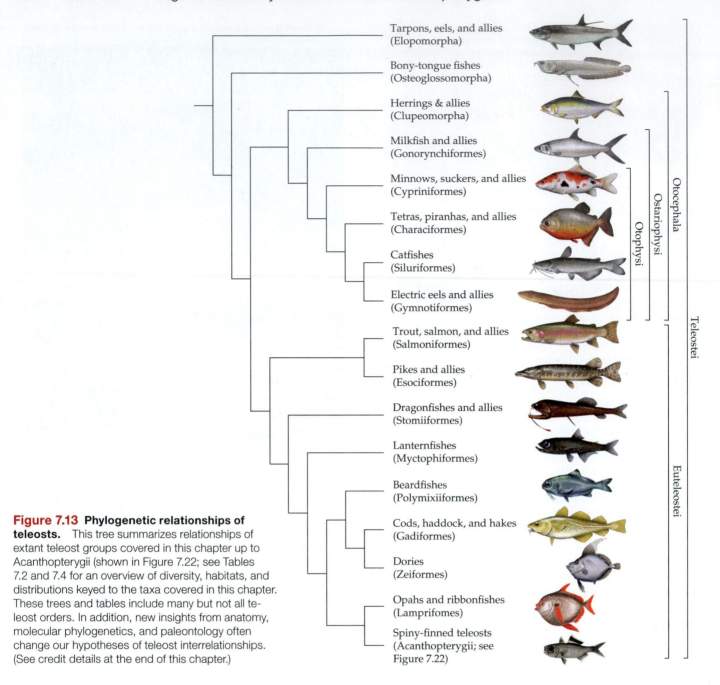

Figure 7.13 Phylogenetic relationships of teleosts. This tree summarizes relationships of extant teleost groups covered in this chapter up to Acanthopterygii (shown in Figure 7.22; see Tables 7.2 and 7.4 for an overview of diversity, habitats, and distributions keyed to the taxa covered in this chapter. These trees and tables include many but not all teleost orders. In addition, new insights from anatomy, molecular phylogenetics, and paleontology often change our hypotheses of teleost interrelationships. (See credit details at the end of this chapter.)

schemes based on morphological and molecular data sets are coming together to show us many surprising things about the course of teleost evolution. We are approaching a better consensus about major patterns in teleost evolution while also identifying the remaining challenges.

The tree in **Figure 7.13** shows the relationships of basal teleost groups, from tarpons, eels, and allies (Elopomorpha) through opah and ribbonfishes (Lampriformes). **Table 7.2** provides an overview of the diversity, habitats, and distributions of these groups. Section 7.5 picks up at Lampriformes to explore the diversity of the spiny-finned teleosts (Acanthopterygii), whose relationships are shown in Figure 7.22.

Elopomorpha

Elopomorphs (Greek *elopos*, "a marine fish") are known as fossils from the Late Jurassic to Recent; extant species include tarpons, bonefishes, ladyfishes, and more than 800 true eels. All but a few species are marine, and all share a specialized larva known as a leptocephalus (Greek *leptos*, "small"; *kephale*, "head"). Leptocephalus larvae of some species spend a long time drifting near the ocean surface, where they can be widely dispersed by currents (**Figure 7.14A**). We are only now beginning to understand the diversity of specializations of leptocephali of different species of elopomorphs.

Table 7.2 Extant Teleostei shown as terminal taxa in Figure 7.13

	Number of extant families	Number of extant species	Habitats	Distribution
Tarpons, ladyfishes, true eels (Elopomorpha)	24	~1,300	Marine, nearshore	Global warm waters
Bony-tongue fishes (Osteoglossomorpha)	5	~300	Freshwater	South America, Africa, Southeast Asia, Australia
Herrings, sardines, anchovies (Clupeomorpha)	5	~500	Freshwater, marine, anadromous	Global
Milkfishes, beaked sandfishes (Gonorynchiformes)	3	~40	Freshwater and marine	Southeast Asia, Africa, Indo-Pacific
Minnows, suckers, and allies (Cypriniformes)	13	~4,600	Freshwater	North America, Europe, Asia
Tetras, piranhas, and allies (Characiformes)	24	~2,250	Freshwater	Africa, South, Central, and North America
Catfishes (Siluriformes)	40	~4,600	Freshwater and marine	Global
Electric eels and allies (Gymnotiformes)	5	~230	Freshwater	Neotropics
Trout, salmon, and allies (Salmoniformes)	1	~225	Freshwater and anadromous	Northern Hemisphere
Pikes and allies (Esociformes)	2	12	Freshwater	Northern hemisphere
Dragonfishes (Stomiiformes)	5	~460	Marine, ranging to deep sea	Global
Lanternfishes (Myctophiformes)	2	~260	Marine, ranging to deep sea	Global
Beardfishes (Polymixiiformes)	1	~12	Marine, ranging to deep sea	Atlantic, Indian, and western Pacific oceans
Cods, haddock, and hakes (Gadiformes)	13	~625	Freshwater and marine, ranging to deep sea	Global
Dories (Zeiformes)	6	~35	Marine, ranging to deep sea	Global
Opahs and ribbonfishes (Lampriformes)	6	~25	Marine, pelagic ranging to deep sea	Global
Spiny-finned teleosts (Acanthopterygii)	Most teleost species*			

Compiled from information in J.S. Nelson et al. 2016. *Fishes of the World*, 5th Ed. Hoboken NJ: Wiley; R. Fricke et al., eds. 2021. *Eschmeyer's Catalog of Fishes: Genera, Species, References*. Online, California Academy of Sciences; R. Froese and D. Pauly, eds. 2021 FishBase. www.fishbase.org.

*See Table 7.4 and Figure 7.22.

Tarpons (**Figure 7.14B**) and bonefishes are fast-swimming, predatory fishes that inhabit shallow coastal waters and reefs in the tropics and subtropics. They are prized by sport fishers—even though not eaten as food—because they are exciting to catch.

Like other eels, the American eel, *Anguilla rostrata* (**Figure 7.14C**), has an unusual life history. After growing to sexual maturity (which can take up to 25 years for females) in rivers, lakes, and ponds of eastern North America, these catadromous eels swim downriver, enter the ocean, and migrate to the Sargasso Sea, an area in the central North Atlantic between the Azores and the West Indies. Here they are thought to spawn at great depths. Eggs and newly hatched larvae float to the surface and drift in the currents. Larval life continues until the larvae reach continental margins, where they transform into miniature eels and ascend rivers to feed and mature.

The European eel, *Anguilla anguilla*, spawns in the same part of the Atlantic as its American kin, although it may choose a somewhat more northeasterly part of the Sargasso Sea, and perhaps a different (but also deep) depth at which to spawn. Its larvae stay in the clockwise currents of the North Atlantic (principally the Gulf Stream) and

(A) Leptocephalus larvae of ladyfish and moray eel

© Danté Fenolio

(B) Atlantic tarpon, *Megalops atlanticus*

© Anthony Aneese Totah Jr/Dreamstime.com

(C) American eel, *Anguilla rostrata*

Courtesy of Sean Landsman

(D) Pharyngeal jaws of a moray eel

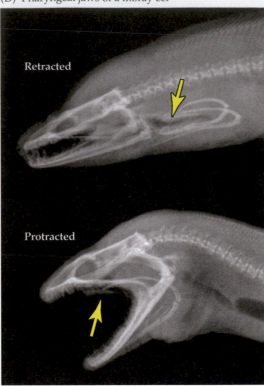

Retracted

Protracted

Courtesy of Rita Mehta, UC Santa Cruz

Figure 7.14 Elopomorpha. (A) Elongate, flattened, and leaf-shaped leptocephalus larvae characterize elopomorphs. Shown are the larvae of an eel (above) and bonefish (below). (B) Nicknamed "silver king," the Atlantic tarpon is prized by sport fishers, which has led to the implementation of strict catch-and-release programs in many places. (C) The American eel is catadromous. Adults spawn in the Sargasso Sea, and their leptocephalus larvae drift in ocean currents before coming inshore to ascend rivers. (D) X-ray images of a moray eel compare its pharyngeal jaws in retracted and protracted positions. A moray seizes and holds prey with its teeth, then advances its pharyngeal jaws into its mouth to grasp the prey. The oral jaws release their hold, and retraction of the pharyngeal jaws drags the food into the throat.

ride to Europe before entering river mouths to migrate upstream and mature.

Moray eels are famous reef predators. They will attack humans if disturbed, and their bite can be vicious. Their feeding behavior is enhanced by a pharyngeal jaw that advances and retracts to draw prey into the throat (**Figure 7.14D**).

Osteoglossomorpha

Osteoglossomorpha means "bony tongue" (Greek *glossa*, "tongue"), and the presence of a specialized tongue and a slicing surface formed by the tongue and parasphenoid (a "tongue parasphenoid bite") have been regarded as synapomorphic for the group. These are complex characters, and details differ among osteoglossomorphs, but there is abundant evidence that the group is monophyletic.

The predatory South American silver arowana, *Osteoglossum bicirrhosum*, feeds on fishes, amphibians, and insects at the water surface, approaching from below and engulfing prey in their upward-directed mouths

(**Figure 7.15A**). The pirarucu, *Arapaima gigas*, an even larger Amazonian predator, is among the largest freshwater teleosts (**Figure 7.15B**). Before intense fishing reduced their populations, pirarucus reached lengths of at least 3 m, and perhaps as much as 4.5 m. The African bonytongue, *Heterotis niloticus*, is distantly related to the South American forms, and other species of bonytongues occur in southeast Asia and Australia. All of these extant osteoglossomorphs live in freshwater and cannot cross large ocean basins. These groups diversified in freshwaters after the continents separated in the Mesozoic. Their disjunct biogeographic distributions may be the result of independent invasions of freshwater habitats on different continents by now-extinct marine ancestors.

Osteoglossomorpha also includes about 200 species of freshwater elephantfishes (Mormyridae) in Africa. An example is *Gnathonemus petersii*, a small bottom feeder that uses weak electric discharges to communicate with conspecifics and detect objects in its murky environment (**Figure 7.15C,D**).

(A) Silver arowana, *Osteoglossum bicirrhosum*

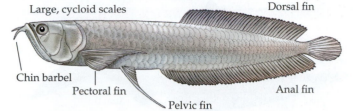

Large, cycloid scales

Dorsal fin

Chin barbel

Pectoral fin

Pelvic fin

Anal fin

(B) Pirarucu, *Arapaima gigas*

© Todd Stailey, Tennesse Aquarium

(C) Peters' elephantfish, *Gnathonemus petersii*

Hippocampus Bildarchiv

(D) Electric fields of elephantfish

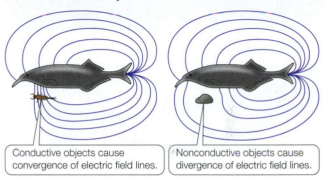

Conductive objects cause convergence of electric field lines.

Nonconductive objects cause divergence of electric field lines.

Figure 7.15 Osteoglossomorpha. (A) The silver arowana of South America has a distinctive lower jaw and upward-directed mouth that work like a trap door as it feeds on prey at the water's surface. (B) The pirarucu, one of the largest freshwater fishes, is a commonly exhibited Amazonian fish. (C) Peter's elephantfish is a weakly electric mormyrid from sub-Saharan Africa. (D) Electric discharges establish a three-dimensional electric field around the elephantfish, allowing it to detect objects in its environment at night or in murky water. (A after K. Liem et al. 2001. *Functional Anatomy of the Vertebrates*, 3rd ed. Cengage/Harcourt College: Belmont, CA; D after W. Heiligenberg. 1977. *Principles of Electrolocation and Jamming Avoidance in Electric Fish*. Berlin: Springer-Verlag.)

Otocephala

Otocephala includes two very speciose extant lineages: Clupeomorpha (herrings, shads, sardines, and anchovies) and Ostariophysi (milkfish and Otophysi). The subgroup Otophysi contains the minnows, carps, and suckers; tetras and piranhas; catfishes; and the electric eels and South American knifefishes (see Figure 7.13).

Clupeomorpha Clupeomorphs (**Figure 7.16A,B**) are silvery, mostly marine teleosts that feed on plankton. They are of great commercial importance and critical components of marine food webs. Several species are anadromous, migrating from seawater to freshwater to spawn. Although depleted by dams, pollution, and fishing, millions of American shad (*Alosa sapidissima*, **Figure 7.16B**) still migrate from the North Atlantic into rivers along the east coast of North America, and there are efforts to conserve and rebuild their populations. Adult shad spawn in the rivers, and the juveniles live for a summer in freshwater before heading out to sea in the fall.

Ostariophysi The presence of an alarm substance in the skin, known as *Schreckstoff* (German, "fright substance"), is a synapomorphy of Ostariophysi. It is released into the water when the skin is damaged and causes nearby conspecifics and even other ostariophysan species to rush for cover or to form tighter schools. Ostariophysans represent about 80% of freshwater species of fishes and about 30% of all extant fishes. Many have protrusible oral jaws and pharyngeal teeth in the throat to process a range of food items. Many species have fin spines or armor for protection, and the skin typically has glands that produce substances used in olfactory communication.

Gonorynchiformes is a small order of ostariophysans with a fossil record including the Eocene Green River Formation of Wyoming (**Figure 7.16C**). An extant relative, the milkfish (*Chanos chanos*, **Figure 7.16D**) lives in marine and estuarine environments of the Pacific Ocean; it is a target of sport fishers and important in aquaculture.

Ostariophysi: Otophysi Otophysi is a large and important subgroup within Ostariophysi whose members share the presence of a **Weberian apparatus**, a series of modified vertebral elements that connects the gas bladder with the inner ear (**Figure 7.17A**). Using the gas bladder as

(A) School of sardines, *Sardina pilchardus*

Photo by Koki Okagawa

(B) American shad, *Alosa sapidissima*

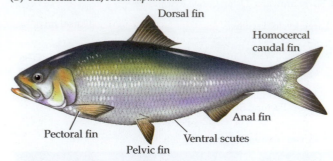

Dorsal fin

Homocercal caudal fin

Anal fin

Pectoral fin

Ventral scutes

Pelvic fin

(C) Green River gonorynchid, †*Notogoneus osculus*

© Lance Grande, Field Museum

(D) Catch and release fishing for milkfish, *Chanos chanos*

© Blue Safari Seychelles

Figure 7.16   Clupeomorpha and Gonorynchiformes. Some species in these two groups are of great commercial importance. (A) Sardines school during a "sardine run" off the coast of South Africa. (B) Shad are another group of fishes that exhibit schooling behavior. The anadromous American shad is endangered by overfishing at sea and pollution of the North American rivers in which it spawns. (C) Gonorynchiformes is a small order of ostariophysans with a fossil record that includes the Eocene Green River gonorynchid. Extant gonorynchids live in continental shelf environments. (D) An extant gonorynchid, the milkfish, is an important food and game fish of the Indo-Pacific. (B after K. Liem et al. 2001. *Functional Anatomy of the Vertebrates*, 3rd ed. Cengage/Harcourt College: Belmont, CA.)

(A) Weberian apparatus of otophysans

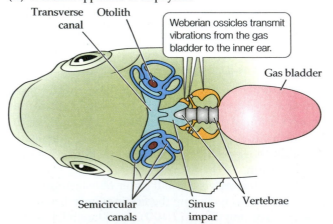

Transverse canal   Otolith

Weberian ossicles transmit vibrations from the gas bladder to the inner ear.

Gas bladder

Semicircular canals   Sinus impar   Vertebrae

(B) Kinethmoid bone, jaws, and jaw protrusion of cypriniforms

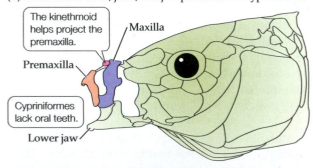

The kinethmoid helps project the premaxilla.

Maxilla

Premaxilla

Cypriniformes lack oral teeth.

Lower jaw

(C) Cypriniform nuptial tubercles

Nuptial tubercles form on the skin of the head during the breeding season.

© Mark Gautreau

Figure 7.17   Characters of Otophysi and Cypriniformes. (A) Otophysans have Weberian ossicles, a series of small, modified vertebral bones used to transmit sound vibrations from the gas bladder to the inner ear. (B) Cypriniforms uniquely have a kinethmoid bone, which allows the premaxilla to project forward during feeding. (C) During the breeding season, cypriniforms such as the fallfish, *Semotilus corporalis*, develop nuptial tubercles on the skin that function as contact organs to help pairs stay close to each other during spawning. (A after K. von Frisch. 1938. *Nature* 141: 8–11; B after W.K. Gregory. 1933. *Trans Am Philosoph Soc* 23: Article 2.)

a receiver for sound and the chain of bones as conductors, the Weberian apparatus enhances the hearing sensitivity of these fishes, particularly for higher frequency sounds that other fishes cannot so easily detect. As you can see in Figure 7.13, this subgroup contains many familiar and commercially important groups, including Cypriniformes (carps, minnows, suckers, chubs, loaches, and many others); Characiformes (tetras, piranhas, and allies); Siluriformes (catfishes); and Gymnotiformes (electric eels and South American knifefishes).

All but a few species of Cypriniformes live in freshwater. Cypriniforms are native to North America, Europe, Asia, and Africa, but were absent from South America and Australia until non-native cypriniforms such as carp were widely introduced. They lack oral teeth but evolved a unique bone, the kinethmoid, that allows the premaxilla to be projected forward to form a tube used for suction feeding (**Figure 7.17B**). Male cypriniforms develop nuptial tubercles on the skin during the breeding season that serve as contact organs during spawning (**Figure 7.17C**).

Commercially and ecologically important cypriniforms include several species of Asian carps, which are often introduced for food or aquaculture and are invasive pests in many parts of the world. Goldfish and minnows (**Figure**

7.18A) are other familiar cypriniforms. The zebrafish, *Danio rerio*, is a common aquarium fish native to freshwaters of southern Asia. It has become a preeminent model for studies of vertebrate genetics and development because it is small, easy to maintain, and breeds readily in captivity.

In contrast to cypriniforms, characiforms such as tetras and (especially) piranhas have noteworthy teeth. Characiforms radiated extensively in South America, where many of the most important Amazonian food fishes are tetras, and in Africa, where they co-occur with cypriniforms. Characiforms familiar from the aquarium-keeping hobby include the cardinal tetra (**Figure 7.18B**). Large aquariums often exhibit carnivorous piranhas (**Figure 7.18C**) and the closely related pacus (**Figure 7.18D**), herbivores famous for their grinding teeth, which they use to crack Brazil nuts.

There are about 35 extant families of catfishes (Siluriformes). They have scaleless skins, sensory barbels that superficially resemble a cat's whiskers (**Figure 7.19A**), and a spine at the leading edge of the dorsal and pectoral fins. In some species, the spine is covered by toxin-producing tissue. Pairs of barbels on the upper and lower jaws are covered with taste buds, allowing catfishes to live in murky or dark water environments. Most catfishes are predators, and many possess electroreceptive organs.

(A) Blue shiner minnow, *Cyprinella caerulea*

(B) Cardinal tetra, *Paracheirodon axelrodi*

(C) Red-bellied piranha, *Pygocentrus nattereri*

(D) Pacu, *Collosoma macropomum*

Figure 7.18 Cypriniformes and Characiformes. (A) "Minnow" is the common name for cyprinid fishes, a group that also includes carp and goldfish. The blue shiner minnow occurs in eastern North America. (B–D) Characiforms radiated extensively in South America and Africa and are noted for their teeth. (B) Cardinal tetra from upper reaches of the Orinoco and Negro rivers in South America. (C) The red-bellied piranha occurs throughout the Amazon basin, and it is one of the carnivorous species that give piranhas their fearsome (but largely unwarranted) reputation. (D) The herbivorous pacu is closely related to piranhas. They swim into flooded forests to feed on fruits (including Brazil nuts) and are seed dispersers for Amazonian trees.

(A) Barbels of an ictalurid catfish

(B) Electric eel, *Electrophorus electricus*

Courtesy of Brian Gratwicke

(C) Glass knifefish, *Eigenmannia virescens*

Ictiología Universidad Católica de Oriente/CC BY 2.0

© Aleron Val/Shutterstock.com

Figure 7.19 Siluriformes and Gymnotiformes. Catfishes and South American knifefishes are sister taxa. (A) Channel catfish showing sets of barbels. The resemblance of siluriform barbels to a cat's whiskers is the source of the common name. (B) Electric eels are gymnotiforms from South America (genus *Electrophorus*, not to be confused with true eels in Elopomorpha). They grow to lengths of 2.5 m and use powerful electrical discharges to immobilize prey and defend against predators. (C) The glass knifefish uses weak electric discharges for orientation and communication.

We know of thousands of species of catfishes, and more are discovered each year. This is because most catfishes live in inland freshwaters, where speciation can happen when bodies of water become isolated from each other. Also, until recently, many interior freshwaters have been difficult to reach to survey for fishes. Such a diverse group not surprisingly displays some fascinating adaptive specializations, including:

- Marine catfishes (Ariidae) are mouthbrooders; males carry marble-sized eggs in their mouths until they hatch.

- Some armored catfishes (family Loricariidae) have large, sucker-shaped mouths used to eat biofilms or even wood (the gut microbiome of wood-eating catfishes helps them process cellulose).

- Marine eel catfishes (Plotosidae) form beautiful schools on coral reefs. This family also includes the freshwater catfishes of Australia, which has no other native freshwater catfishes because it has been isolated from other continents for so long.

- Walking catfish (Clariidae) from Southeast Asia use a specialized arborescent organ located in the gill cavity to exchange gases with air. They can move over land, and are an invasive species in Florida, where their ability to move over wet land has allowed range expansion into aquaculture facilities, where they prey on farmed species.

- Electric catfishes of Africa (Malapteruridae) are sausage-shaped and can generate powerful electric fields (~ 350 V) for defense and prey capture (as opposed to sensory electroreception, known in many catfishes). Known to the ancient Egyptians, an electric catfish figures on the Narmer Pallette, which is about 5,000 years old and one of the earliest known examples of hieroglyphic writing.

The ability to generate electric discharges is also found in a different group of otophysans, Gymnotiformes, from South America (see Figure 4.10). Gymnotiformes is considered to be the sister group to Siluriformes (see Figure 7.13), although electrogeneration probably evolved independently in the electric catfishes. Electric eels (genus *Electrophorus*; **Figure 7.19B**) should not be confused with "true" eels, which are elopomorphs (see Figure 7.14). The three species of *Electrophorus* all are capable of generating strong electric discharges (up to 640 V; see Section 4.2 and Figure 4.11). Gymnotiformes also includes the weakly electric South American knifefishes (**Figure 7.19C**).

Basal euteleosts

Teleostei accounts for the vast majority of the ~34,000 extant actinopterygian species, and of these more than 20,000 species belong to Euteleostei (Greek *eu*, "true"), a group that began its diversification in the Cretaceous. In turn, most euteleosts belong to Acanthopterygii, and we describe these "spiny fin" fishes in Section 7.5.

The first branch of euteleosts leads to salmon and trout (Salmoniformes) and pikes and pickerels (Esociformes; see Figure 7.13). Salmonids are important commercial and game fishes (**Figure 7.20A**). They are native to the Northern Hemisphere but have been widely introduced and cultivated for aquaculture in many parts of the world. Anadromous salmonids spend their adult lives at sea, making epic journeys up freshwater rivers to spawn. Esociforms such as muskellunges (**Figure 7.20B**) are ambush predators that eat other vertebrates, relying on cryptic colors and patterns to conceal them until a prey fish or frog is close enough to be captured by a sudden lunge powered by fast-start locomotion (see Section 7.6).

(A) Brown trout, *Salmo trutta*

Courtesy of Chris Crippen

(B) Muskellunge, *Esox masquinongy*

Courtesy of Chris Crippen

(C) Tanned bristlemouth, *Cyclothone pallida*

NOAA Office of Exploration and Research/CC BY

(D) Atlantic cod, *Gadus morhua*

© Paulo Oliveira/Alamy Stock Photo

(E) Oarfish, *Regalecus*

© Planetfelicity/Dreamstime.com

Figure 7.20 **Salmoniformes, Esociformes, Stomiiformes, Gadiformes, Lampriformes.** These five groups of basal teleosts (see Figure 7.13) include many commercially and ecologically important species (A) Some populations of the widely introduced brown trout are anadromous, as are many other species of salmonids. (B) Esociforms such as the freshwater muskellunge are ambush predators. (C) Taken together, the 13 species of the stomiiform genus *Cyclothone* are the most abundant vertebrates, with populations estimated to number in the trillions or even quadrillions. (D) Populations of Atlantic cod (Gadiformes), long a staple of commercial fishing, are threatened today by overfishing and global climate change. (F) Oarfish (Lampriformes) reach lengths of 8 m and may be the basis for legends of sea serpents.

Dragonfishes (Stomiiformes) are fishes of the open ocean, where they play critical ecological roles because of the sheer number of individuals. Trillions—possibly quadrillions—of individuals of different species of bristlemouths (stomiiforms in the family Gonostomatidae; **Figure 7.20C**) are swimming in the deep sea at this moment, making them by far the most abundant vertebrates. Another midwater group, the lanternfishes (Myctophiformes) is also extremely abundant. Named for the photophores (light-producing organs) on their bodies, lanternfishes swim upward in the water column at night to feed and return to deeper depths during the day, thus transferring nutrients within the ocean.

There are several Cretaceous genera and 10 extant species of beardfishes (Polymixiiformes). The phylogenetic relationships of these marine euteleosts have been unstable for many decades and have attracted much attention from fish systematists. Figure 7.13 shows beardfishes as the sister group of cods and dories based on one current interpretation. Named for a pair of long barbels extending from the hyoid, they live in close association with the bottom at depths between 100 and 800 m. For such a phylogenetically interesting group, remarkably little is known about their reproductive biology and behavior.

Cods and allies (Gadiformes) are among the most important wild-caught commercial fishes (**Figure 7.20D**). Sadly, a legacy of overfishing and warming ocean waters have reduced populations of Atlantic cod to a fraction of their former abundance. Despite nearly 30 years of intensive regulations and restrictions, cod populations off the northeast coast of the United States have not rebounded, and it is unclear whether they can ever recover to former levels because the water temperature has increased so greatly. A small group of deep-bodied marine fishes known as dories (Zeiformes; see Figure 7.13) are related to Gadiformes based on molecular phylogenetic research.

Opah and ribbonfishes (Lampriformes) are iconic marine pelagic fishes. For example, the oarfish (*Regalecus*) is famous for achieving giant sizes and inspiring stories of sea serpents (**Figure 7.20E**).

7.5 Teleostei: Acanthopterygii

LEARNING OBJECTIVES

7.5.1 Identify three anatomical characteristics of Acanthopterygii.

7.5.2 Explain why some groups within Percomorpha recognized using molecular approaches are not well supported by anatomical characters.

7.5.3 Relate the major events of flatfish metamorphosis to the unusual behaviors and abilities of adult flatfishes.

7.5.4 Assess some functional and evolutionary consequences of the low vertebral number of Tetraodontiformes.

Spiny-finned teleosts—Acanthopterygii (Greek, *acanthos*, "spine"; *pteron*, "fin")—have stiff, bony fin spines in the dorsal, anal, and pelvic fins (**Figure 7.21A**). Spines support the first part of the dorsal fin, while flexible fin rays support the soft part. The maxilla lacks teeth and supports the premaxilla during jaw projection. Unlike basal euteleosts, the pectoral fin is on the side of the body, directly behind the operculum, and the pectoral girdle is vertical (**Table 7.3**). Pelvic fins of acanthopterygians are far forward on the body (in the thoracic position) and the pelvic girdle is closely associated with the pectoral girdle. The premaxillae have elongated ascending processes that can slide forward to project the oral jaws (**Figure 7.21B**). An open pneumatic duct connecting the gut tube to the gas bladder is lost, and the scales are ctenoid.

Acanthopterygians began their rapid diversification in the Late Cretaceous, and most extant groups were present by the Eocene. Marine taxa explosively diversified during the first 10 million years of the Cenozoic, which appears

(A) Fins of acanthopterygians

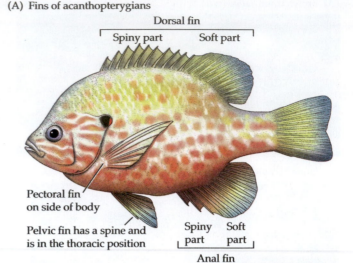

(B) Jaws of acanthopterygians

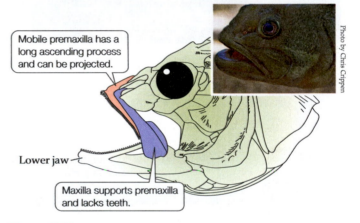

Mobile premaxilla has a long ascending process and can be projected.

Lower jaw

Maxilla supports premaxilla and lacks teeth.

Figure 7.21 Characters of spiny-finned teleosts, Acanthopterygii. (A) Acanthopterygians have bony fin spines in their dorsal, anal, and pelvic fins. (B) Long, ascending processes of the premaxillary bones allow the upper jaw to be protruded during feeding. (B after W.K. Gregory. 1933. *Trans Am Philosoph Soc* 23: Article 2.)

Table 7.3 Some key structural differences between basal euteleosts and acanthopterygians

Structure	Generalized basal euteleost (e.g., rainbow trout, Salmoniformes)[a]	Generalized acanthopterygian (e.g., yellow perch, Perciformes)
Bony spines in fins	Absent	Present in dorsal, anal, and pelvic fins
Dorsal fins	One dorsal fin supported by flexible fin rays plus an adipose fin[b]	Spines support the spiny portion of the dorsal fin and flexible fin rays support the soft portion of the dorsal fin
Pectoral fin base and girdle	Ventral and horizontal	Lateral and vertical
Pelvic fin position	Abdominal and widely separated from the pectoral girdle and fin	Thoracic and closely associated with the pectoral girdle and fin
Bones of upper jaw; teeth	Short premaxilla and long maxilla; both bear teeth	Premaxilla and maxilla; only the premaxilla bears teeth
Gas bladder	Pneumatic duct present (physostomous)	Pneumatic duct absent (physoclistous)
Scales	Cycloid	Ctenoid

After J.S. Nelson et al. 2016. *Fishes of the World*, 5th Ed. Hoboken NJ: Wiley.

[a]These characters are not synapomorphies of euteleosts, but are useful to compare with percomorphs.

[b]An adipose fin is present in some but not all basal euteleosts. It is not regarded as homologous to the second (soft) dorsal fin of percomorphs.

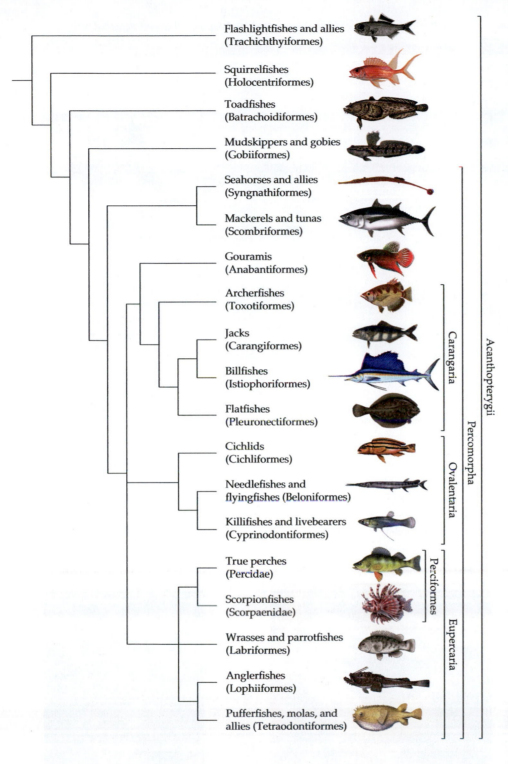

Figure 7.22 Phylogenetic relationships of Acanthopterygii. This tree, a continuation of Figure 7.13, summarizes relationships of acanthopterygian taxa covered in this chapter (also see Table 7.4). As noted earlier, these summary trees and tables include many but not all teleost taxa, and new insights from anatomy, molecular phylogenetics, and paleontology often change our hypotheses of teleost interrelationships. (Based on K. Liem et al. 2001. *Functional Anatomy of the Vertebrates*, 3rd ed. Cengage/Harcourt College: Belmont, CA; J. T. Williams et al. 2010. *PLOS ONE* 5: e10676/CC0; Vlad Butsky/CC 2.0; D.S. Jordan. 1907. *Fishes*. Henry Holt and Co: New York, courtesy of Freshwater and Marine Bank at University of Washington; and, other public domain sources.)

use a large photophore beneath the eye like a flashlight (**Figure 7.23A**). They can turn the flashlight on and off (there are different mechanisms for this in different groups of flashlightfishes), using the light in prey detection, predator avoidance, and schooling. Squirrelfishes (Holocentriformes) are nocturnal reef fishes known for sound production and detection (they make a sound like a squirrel). Another famous group of light- and sound-producing fishes are the toadfishes and midshipmen (Batrachoidiformes; **Figure 7.23B**). Male midshipmen (*Porichthys* spp.) call females to lay eggs in the nests they have made, and then guard the eggs until they hatch. Their choruses can be so loud that people have sometimes mistaken them for human-generated noises, as in a famous case where houseboat residents in California thought the U.S. Navy was conducting tests under their boats.

More than 2,200 species of gobies and allies (Gobiiformes) inhabit shallow nearshore marine habitats in tropical and temperate regions. A few species live in freshwater, particularly on islands such as Hawaii, where four species of gobies are among the only native freshwater fishes. Mudskippers are gobies famous for their ability to move and forage on mudflats at low tide; they also have elaborate courtship displays and defend territories (**Figure 7.23C**).

in part to be linked to the availability of many new marine habitats, including reefs formed by stony corals similar to those of today.

Many of the groups shown in **Figure 7.22** and listed in **Table 7.4** are incredibly diverse, in some cases containing hundreds to thousands of species, so it is impractical to offer more than a brief overview here.

Basal acanthopterygians

The basalmost acanthopterygian lineage, Trachichthyiformes, includes flashlightfishes and allies. Flashlightfishes

(A) Flashlight fish, *Photoblepharon palpebratum*

(C) Territorial behavior of male bluespotted mudskippers, *Boleophthalmus boddarti*

(B) Oyster toadfish, *Opsanus tau*

Figure 7.23 Some basal acanthopterygians. (A) Flashlight-fishes (Trachichthyiformes) have a bright photophore underneath the eye that can be turned on or off like a flashlight. Different genera evolved different mechanisms to do this. (B) Toadfishes (Batrachoidiformes) are sluggish, with powerful jaws and well-developed sonic muscles used to vibrate the gas bladder to produce sounds. (C) Highly territorial mudskippers (Gobiiformes) can move across mudflats at low tide and defend their portions of the mudflat.

Percomorpha

All remaining acanthopterygians belong to Percomorpha (see Figure 7.22). Recent molecular phylogenetic research has upended many conventional ideas about evolutionary relationships among percomorphs. Because they radiated so rapidly in the late Cretaceous and Paleogene, the divergences between major groups of percomorphs occurred very quickly, which can complicate molecular phylogenetic interpretations. Some groups within Percomorpha recognized using molecular approaches are well

(A) Striped mackerel, *Rastrelliger kanagurta*

(B) Mating seahorses

(C) Male seahorse giving birth

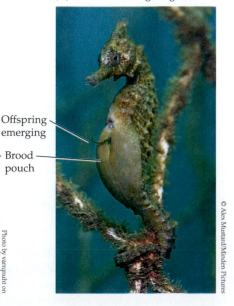

Offspring emerging

Brood pouch

Figure 7.24 Basal percomorphs. A surprising result of recent molecular phylogenetic research is the placement of mackerels and tunas (Scombriformes) with seahorses and allies (Syngnathiformes) near the base of the percomorph tree (see Figure 7.22). (A) The open mouth of a striped mackerel shows the long gill rakers it uses for filter feeding. (B,C) Seahorses (family Syngnathidae) are viviparous, and males brood the eggs. (B) When the eggs of the female seahorse are released, the male fertilizes them as they are moved into his brood pouch. (C) A juvenile emerges tail-first from the brood pouch of a male White's seahorse, *Hippocampus whitei*.

Table 7.4 Extant Acanthopterygii shown as terminal taxa in Figure 7.22

	Number of extant families	Number of extant species	Habitats	Distribution
Flashlightfishes and allies (Trachichthyiformes)	5	~75	Marine, ranging to deep sea	Global
Squirrelfishes and allies (Holocentriformes)	1	~90	Marine, reef associated	Atlantic, Indian, and Pacific oceans
Toadfishes and allies (Batrachoidiformes)	1	~110	Marine and estuarine; benthic	Global
Mudskippers and gobies (Gobiiformes)	8	~2,500	Freshwater, estuarine, and marine	Global
Seahorses and allies (Syngnathiformes)	8	~350	Freshwater, estuarine, and marine	Global
Mackerels and tunas (Scombriformes)	9	~200	Marine; pelagic	Global
Gouramis (Anabantiformes)	4	~230	Freshwater	Africa, southern Asia
Archerfishes (Toxotiformes)	1	7	Marine coastal, estuarine, and freshwater	India, Philippines, Australia, Polynesia
Jacks and allies (Carangiformes)	6	~170	Marine and estuarine	Global
Billfishes (Istiophoriformes)	3	~40	Marine; pelagic	Global
Flatfishes (Pleuronectiformes)	14	~800	Freshwater, estuarine, and marine; benthic	Global
Cichlids (Cichliformes)	1	~2,250	Freshwater	North, Central, and South America, Africa, Asia
Needlefishes and flyingfishes (Beloniformes)	6	~300	Freshwater, estuarine, and marine	Global
Killifishes and livebearers (Cyprinodontiformes)	10	~1,350	Freshwater, estuarine, and marine	Global
True perches and allies (Perciformes)	~100	~3,500	Freshwater, estuarine, and marine	Global
Wrasses and parrotfishes (Labriformes)	3	~700	Marine; many reef associated	Global
Anglerfishes (Lophiiformes)	18	~350	Marine; many deep sea	Global
Pufferfishes, molas, and allies (Tetraodontiformes)	10	~450	Freshwater, estuarine, and marine; many reef associated	Global

Compiled from information in J.S. Nelson et al. 2016. *Fishes of the World*, 5th Ed. Hoboken NJ: Wiley; R. Fricke et al., eds. 2021. *Eschmeyer's Catalog of Fishes: Genera, Species, References*. Online, California Academy of Sciences; R. Froese and D. Pauly, eds. 2021 FishBase. www.fishbase.org.

supported by anatomical characters, but others are not, and there will certainly be many changes to come in our understanding of percomorph interrelationships.

Basal percomorphs   One surprising result of recent molecular phylogenetic research is the placement of mackerels and tunas (Scombriformes) with seahorses and allies (Syngnathiformes) at the base of the percomorph tree (see Figure 7.22). Tunas, with their typically "fishlike" streamlined bodies, are among the fastest swimmers in the sea, whereas seahorses and pipefishes are encased in bony armor, unique in body shape (many people do not even think of them as fish), and move very slowly. Bluefin tuna are pelagic piscivores that chase

down and even hunt prey in packs like some mammalian predators. Mackerels filter feed using elongated gill rakers (**Figure 7.24A**), a mechanism convergently like that of other filter-feeding actinopterygians such as paddlefishes or clupeomorphs. Seahorses (family Syngnathidae) are known for male brooding of eggs (**Figure 7.24B,C**). Populations of many species of seahorses have declined because dried seahorses were once regarded as suitable curios for the souvenir trade (they are both unsuitable and cruel curios). As of 2002, all seahorses in the genus *Hippocampus* are listed in CITES Appendix 2. Eighteen species of seahorses and the closely related pipefishes are listed by the IUCN as vulnerable, endangered, or critically endangered.

Many species of gouramis (Anabantiformes), such as the betta (**Figure 7.25A**), are familiar aquarium fishes. Anabantiforms have an accessory air-breathing structure known as a labyrinth organ located in the posterodorsal portion of the gill cavity (see Section 4.1). This allows them to live in tropical freshwaters that have low concentrations of dissolved oxygen. Anabantiforms such as snakeheads can make terrestrial excursions between bodies of water, making them potentially dangerous invasive species outside their normal ranges in Asia.

Carangaria   One of the most interesting groups recently recognized using molecular phylogenetic tools is Carangaria, a large group of marine fishes including barracudas, archerfishes, jacks, billfishes, and flatfishes. Eight species of archerfishes (Toxotidae) occur in brackish and freshwater

habitats from India to the Philippines, Australia, and Polynesia and are famous for their ability to shoot a precise jet of water to knock insects off plants overhanging the water (**Figure 7.25B**). An archerfish can strike insects as far as 2 m away, despite the difficulty caused by the refraction of light as it passes through the water surface, which makes prey appear to be in a different position from their actual location. They have an amazing ability to discriminate patterns, even recognizing different human faces.

Jacks (Carangiformes) have a characteristic body shape associated with fast and agile swimming (**Figure 7.25C**). The caudal peduncle (the portion of the body immediately in front of the caudal fin) is narrow; the tall caudal fin is semilunate (half-moon shaped), with equally sized upper and lower lobes. Rapid tail beats propel jacks forward as they forage on fast-swimming prey such as other fishes

(A) Breeding bettas and bubble nest

(B) Archerfish shooting water

(C) Greater amberjack, *Seriola dumerili*

(D) Atlantic sailfish, *Istiophorus albicans*, hunting

Figure 7.25 Anabantiformes and Carangaria   (A) Bettas and many gouramis (Anabantiformes) make bubble nests at the surface. They spawn beneath the nest and push fertilized eggs into the bubbles, where oxygen exchange is maximized. (B–D) Carangaria is a large group of mostly marine fishes recently recognized using tools of molecular phylogenetics. (B) With its opercula closed, compression of the buccal cavity shoots a jet of water out the mouth of an archerfish (*Toxotes jaculatrix*; Toxotiformes). Archerfish can hit an insect on an overhanging plant several body lengths away and can accurately predict where the falling prey will hit the water, enabling it to reach it before another fish steals it. (C) Greater amberjack (Carangiformes), a fast-swimming pelagic predator with a forked caudal fin. (D) Billfishes—swordfishes, marlins, and sailfishes—belong to Istiophoriformes. An individual Atlantic sailfish usually attacks prey from either the right or left side, and capture success is higher for attacks from the sailfish's favored side. Hunting groups typically include approximately equal numbers of individuals that attack from right and left, making it difficult for the prey (here, sardines) to determine the direction of the next attack.

and squids. Popular food and game fishes, jacks are common in marine temperate and tropical waters globally.

Billfishes (marlins, swordfishes, and sailfishes; Istiophoriformes) are among the largest and fastest pelagic predators. The bill is an extension of the premaxillae, rounded in marlins and sailfishes and laterally flattened in swordfishes, used as a club to stun prey or, in the case of swordfishes, to slash squid into bite-sized pieces. Sailfishes engage in cooperative hunting (**Figure 7.25D**). Groups of from 6 to 40 sailfish have been observed attacking schools of sardines. Relatively few sardines are

captured outright in these attacks, but large numbers are injured, pursued, and captured.

There are more than 700 species of flatfishes (flounders, halibuts, plaice, soles, hogchokers, and turbots; see Figure 7.22). Both eyes of adult flatfishes are on the same side of the head, and adults lie on one side of the body on the seafloor (**Figure 7.26A**). Some flatfishes are right-eyed (meaning that both eyes are on the right side of the body) while others are left-eyed. Sidedness is usually consistent within a species, but in some 50% of the individuals are right-eyed and 50% are left-eyed.

(A) Flatfish with both eyes on left side of body

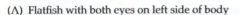

(B) Three stages in eye migration during metamorphosis

(C) Peacock flounder, *Bothus mancus*, changes colors to match the background

(D) Hogchoker, *Trinectes maculatus*, walks on the tips of dorsal and anal fin rays

Figure 7.26 Carangaria: Flatfishes. Flatfishes (Pleuronectiformes) include flounders, halibuts, plaice, soles, hogchokers, and turbots. (A) A flatfish's eyes are on one side of the body (the left side in this individual); the other side lies on the substrate. (B) Three stages in eye migration during metamorphosis of a right-eyed flatfish, the bamboo sole, *Heteromycteris japonicus*. (C) Peacock flounders can rapidly change color to match the background. (D) Bunches of fin rays form "fin feet" in the dorsal and anal fins of the hogchoker, *Trinectes maculatus*. By passing a wave of fin feet posteriorly, the flatfish walks forward on the substrate.

Integrated phylogenetic studies incorporating DNA sequence data, fossils, and morphological studies of extant forms show that the flatfish body form evolved very quickly in a period <4 Ma in the early Paleogene. Flatfishes begin life as normal-looking bilateral pelagic larvae, but as they metamorphose, one eye migrates across the top of the head (**Figure 7.26B**). They rely on crypsis and have remarkable abilities to match their background by changing colors and patterns (**Figure 7.26C**).

A newly discovered feature of some flatfishes is their ability to walk on the bottom on the tips of their dorsal and anal fin rays, which bunch together to form fin feet that move like a wave down the fins as the animal creeps forward on the substrate (**Figure 7.26D**).

Ovalentaria   Ovalentaria is one of the more controversial groups revealed by molecular phylogenetic and genomic studies, because reliable anatomical synapomorphies to support it have yet to be found. Confirmation may require time, or perhaps Ovalentaria will be disassembled by future research. Nevertheless, Ovalentaria as currently understood includes some extremely important fishes (see Figure 7.22 and Table 7.4). For example, cichlids (Cichliformes) occur in South and Central America (and just reach Texas in the United States), Africa, Madagascar, and Asia. There may be as many as 3,000 extant species of cichliforms, although only a much smaller number have been formally described. Some, such as "tilapia" (species of *Coptodon, Oreochromis* and *Sarotherodon*), are important in aquaculture; others, such as angelfishes, discus, and oscars (**Figure 7.27A**) are

popular for home and exhibit aquariums. The explosive radiations of cichlid species in the Great Lakes of eastern Africa have inspired generations of evolutionary biologists who hope to understand how the extensive species diversity of cichlids in these relatively young lakes (such as Lake Victoria, where there were hundreds of species of cichlids until recent human-related extinctions) could possibly have happened.

Also within Ovalentaria are needlefishes, flyingfishes, and allies (Beloniformes). Flyingfishes glide through the air to escape predators using expanded pectoral fins as airfoils (**Figure 7.27B**); some species have expanded pelvic fins as well. To initiate flight, a flyingfish swims rapidly toward the water surface with its fins furled and leaps through the surface at a shallow angle, spreading its fins and beating the water with its caudal fin to gain enough speed to glide. As the glide slows and the fish descends toward the water, it may beat the water with its caudal fin to prolong the glide.

Killifishes and allies belong to Cyprinodontiformes, which means "tooth carps" (Greek, *cyprinus* "carp"; *odontos*, "teeth"). This is an apt but somewhat confusing name, for these small fishes superficially resemble true minnows (Cypriniformes) in body form and behavior—often schooling in shallow waters—yet they have oral teeth, which are absent in cypriniforms, and in fact the two groups are not closely related (compare Figures 7.13 and 7.22). Known for its salinity and temperature tolerance, a classic example of a cyprinodontiform is the mummichog, *Fundulus heteroclitus*, which inhabits tidal creeks and estuaries along the Atlantic coast of North America (**Figure 7.27C**). If stranded on land

(A) Oscar, *Astronotus ocellatus*

(C) Mummichog, *Fundulus heteroclitus*

(D) Swordtail, *Xiphophorus hellerii*

(B) Flyingfish

(E) Male mosquitofish, *Gambusia*

Pectoral fin   Dorsal fin

Caudal fin

Anal fin modified as gonopdium

Pelvic fin

Figure 7.27 Ovalentaria. Ovalentaria is supported by molecular phylogenetic data, but reliable anatomical synapomorphies are lacking and monophyly of this group (see Section 1.2) remains controversial. (A) Cichlids such as the oscar are popular aquarium fishes. (B) Flyingfishes can glide over the water using large pectoral, and, in some enlarged pelvic fins. (C) Some cyprinodontiforms, such as the mummichog, tolerate extreme environmental conditions such as high temperatures and high salinities. (D) Male swordtails have elongated caudal rays, a sexual dimorphism. (E) The anal fin of male mosquitofish forms a gonopodium used for internal fertilization. (E after K. Liem et al. 2001. *Functional Anatomy of the Vertebrates*, 3rd ed. Cengage/Harcourt College: Belmont, CA.)

by the tide, a mummichog flips upright on its belly to visually locate the water, and then flips toward it. Pupfishes (*Cyprinodon* spp.) also tolerate wide ranges of temperature and salinity; the desert pupfishes of southwestern North America can live in waters that reach temperatures lethal for other vertebrates. Still other cyprinodontiforms live in waters that dry up annually, leaving behind eggs that hatch when the water returns.

Livebearers—including guppies, mollies, and swordtails—are cyprinodontiforms familiar from the aquarium hobby (**Figure 7.27D**). Male livebearers have a rod-shaped anal fin, or gonopodium, that they use to inseminate females (**Figure 7.27E**). Young develop inside the female's reproductive tract and are born alive as miniature adults.

Eupercaria: Perciformes   Two hundred years ago, European zoologists viewed freshwater perch (*Perca*) as something of a model for a fish (**Figure 7.28A**). Perch anatomy was depicted in many works, and *Perca* leant its name to many concepts for higher taxa, including Perciformes. Zoologists subsequently referred many different types of higher bony fishes to Perciformes, which as a result became a gigantic paraphyletic group. Ichthyologists are making great strides in breaking up that paraphyletic concept and returning the meaning of Perciformes to a group that can be recognized and readily diagnosed. Thus, the concept of Perciformes as a group within Eupercaria (true perches; see Figure 7.22) differs from phylogenies that are only a few years old, and future research will undoubtedly continue to refine and probably further restrict these phylogenetic concepts.

Another prominent and familiar group within Perciformes is the scorpionfish family (Scorpaenidae), marine teleosts famous for their bright colors, many sharp spines, and toxins. For example, lionfishes (**Figure 7.28B**) have elongated fin spines with venom-producing tissues that can cause a painful reaction in humans, potentially lasting for days. Native to the Pacific and Indian oceans, species of lionfishes have been introduced via the aquarium trade into the Caribbean, where they now pose a significant threat as invasive species because they eat smaller fishes on coral reefs. Their range is rapidly expanding, driving new efforts to control their spread, including commercial harvesting (the toxins are associated with the fins, which are removed before cooking, making lionfish delicious).

Eupercaria: Other groups   Coral reefs would not be the same without wrasses (Labriformes). Colorful, inquisitive, abundant, and fun to watch, forms such as the birdmouth wrasses (**Figure 7.28C**) provide the backdrop for many diving adventures. The group is important to functional anatomists because it lends its name to the type of locomotion based on up and down movements of the pectoral fins, causing the body to move slightly up and down with each stroke, that they typically use to cruise around a reef.

Angling (fishing with a rod, line, and lure) is the source of the common name anglerfishes (Lophiiformes). Anglerfishes have a modified first dorsal fin spine that migrates forward during development to the top of the fish's head, just dorsal to its mouth (**Figure 7.29A**). This spine, or illicium, serves as a "fishing pole," and it supports a flap of skin, the esca, that serves as a lure. Waved in front of the mouth, the lure attracts prey fish, which can be inhaled in milliseconds, movements generally considered to be the fastest feeding movements among vertebrates. For such a seemingly specialized group, there is surprising diversity among anglerfishes, with some forms living in *Sargassum* (a genus of brown algae) floating on the sea surface, others on rocky or coral reefs, and some on muddy bottoms where they camouflage both their body and breathing movements. There are also deep-sea anglerfishes that live in open pelagic waters.

(A) Yellow perch, *Perca flavescens*

(B) Lionfish, *Pterois volitans*

(C) Male green birdmouth wrasse, *Gomphosus*

Robert Colletta/USDA ARS

Michael Gäbler/CC BY 3

Hectonichus/CC BY-SA 3.0

Figure 7.28 Eupercaria: Perciformes and Labriformes. (A) The yellow perch is a common freshwater fish in North America. Many 19th century zoologists considered the freshwater perch as a model, and perch anatomy and physiology were often presented as the standard for advanced teleost fishes. An unfortunate result of this idealization was that many taxa were lumped into in a huge paraphyletic group "Perciformes." The modern taxonomic concept of Perciformes presented here is much narrower. (B) Our current concept of Perciformes includes the scorpionfishes, such as this lionfish, which have elongated, venom-producing fin spines. (C) Wrasses (Labriformes) are common, colorful, and active fishes of coral reef environments.

(A) Anglerfish, *Lophius piscatorius*

The first dorsal spine is the illicium (pole) and the esca (lure) is at its tip.

© Jeff Rotman/Alamy Stock Photo

(B) Queen triggerfish, *Balistes vetula*

Mark Peter & James St. John/CC BY 2.0

(C) Inflated freshwater pufferfish, *Pao* sp.

© Daniel Heuclin/Minden

(D) Ocean sunfish, *Mola mola*

Molids lack a caudal fin and swim using their dorsal and anal fins.

© Mike Johnson/ earthwindow.com

Figure 7.29 Eupercaria: Lophiiformes and Tetraodontiformes. (A) Lophiiforms (anglerfishes) such as *Lophius piscatorius* have a specialized first dorsal fin that serves as a fishing pole tipped with a flap of skin that lures prey fishes near enough to be caught by powerful suction feeding. (B–D) Tetraodontiforms. (B) Most species of triggerfishes are coastal and coral reef dwellers (a few are oceanic). The queen triggerfish is a fairly large (30–60 cm) and aggressive predator that eats sea urchins and other bottom-dwelling invertebrates. (C) Pufferfishes can inflate their body with water (or air if removed from the water, as in this photograph) to deter predators. (D) Ocean sunfishes are the heaviest teleosts, routinely reaching 1,000 kg, with a record of >2,200 kg. They make very fast deep dives to forage on jellyfishes, after which they bask near the surface to warm up.

Pufferfishes and their allies (Tetraodontiformes) are weird and wonderful animals. The 10 extant families include about 450 species, ranging from triggerfishes to puffers and ocean sunfishes (**Figure 7.29B–D**). As a group, tetraodontiforms have relatively short vertebral columns with few vertebrae, which limits the ability of their bodies to bend for typical lateral undulation. New locomotor solutions—such as swimming by flapping the dorsal and anal fins from side to side—evolved as alternatives to lateral undulation. There was also a corresponding increase in defense: if you cannot swim fast, you'd better be able to either: (1) swim precisely so that you can hide in inaccessible cracks; (2) armor yourself so that you cannot be easily eaten; (3) puff up your body to make yourself harder to swallow; or (4) grow to supersize. If all else fails, you can concentrate or make toxins that render you safer from predation. The amazing thing about tetraodontiforms is that they have done all of these things, sometimes convergently in different families. Add to this that pufferfishes are curious, fun to watch, and exhibit behaviors that endear them to people, and you have a group worthy of a lifetime of investigation.

7.6 Swimming and Hydrodynamics

LEARNING OBJECTIVES

7.6.1 Compare the body shapes of fishes that swim continuously and fishes with fast-start performance.

7.6.2 Define undulation and oscillation with respect to fin movements.

7.6.3 Understand the concept of Reynolds numbers as they relate to swimming by adult and larval fishes.

7.6.4 Explain how a fish stays in place in the water column.

In this section, we use simple hydrodynamic principles to explore how fishes generate forward thrust, overcome the drag of turbulence, and steer.

Generating forward thrust

Fishes swim using sequential anterior-to-posterior contractions of the myomeres (muscle segments; see Figure 2.11) along one side of the body and simultaneous relaxation of the corresponding myomeres on the opposite side. Thus, a portion of the body momentarily bends, the bend moves posteriorly, and the fish oscillates from side to side. These **lateral undulations** are most obvious in

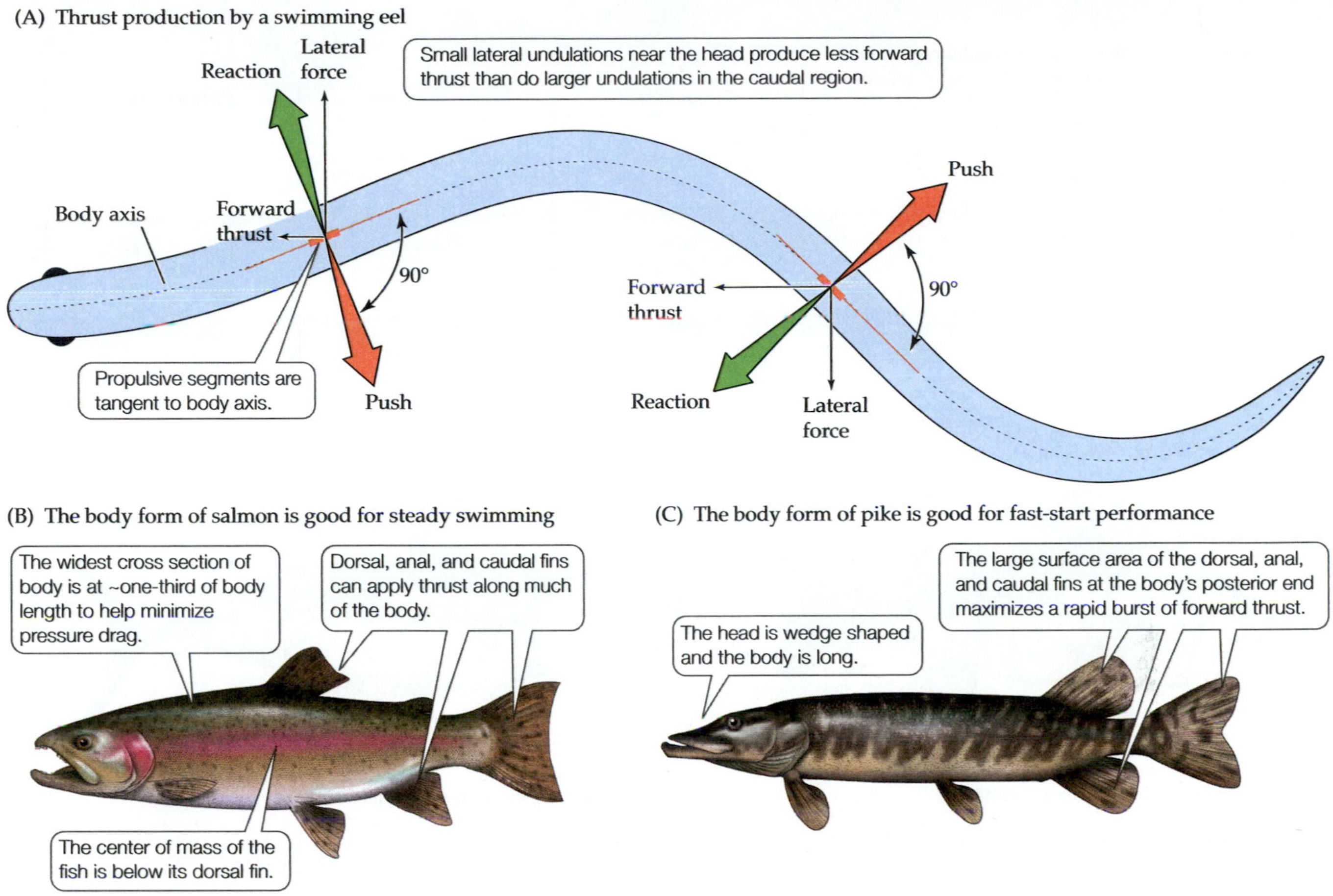

(A) Thrust production by a swimming eel

Lateral
Reaction force

Small lateral undulations near the head produce less forward thrust than do larger undulations in the caudal region.

Push

Body axis

Forward
thrust

90°

Forward
thrust

90°

Propulsive segments are
tangent to body axis.

Push

Reaction

Lateral
force

(B) The body form of salmon is good for steady swimming

The widest cross section of
body is at ~one-third of body
length to help minimize
pressure drag.

Dorsal, anal, and caudal fins
can apply thrust along much
of the body.

The center of mass of the
fish is below its dorsal fin.

(C) The body form of pike is good for fast-start performance

The large surface area of the dorsal, anal,
and caudal fins at the body's posterior end
maximizes a rapid burst of forward thrust.

The head is wedge shaped
and the body is long.

Figure 7.30 Swimming mechanics. (A) Lateral undulations push against the water, creating a reaction force. A vector diagram of the reaction resolves it into forward thrust and lateral force. If a fish swims in a straight line, then its right and left lateral forces of undulation cancel out. Forward thrust is always in the same direction, and thus is additive, so the fish moves forward through the water. (B) The median fins of salmon (Salmoniformes) can apply force to the water along the body axis and thus are good for steady swimming. (C) The median fins of pike (Esociformes) are located near the posterior end of the body and are suited for fast-start locomotion. (See credit details at the end of this chapter.)

elongated fishes such as eels but also occur in powerful swimmers such as tunas.

We can view lateral undulation as a series of propulsive segments along the body, each one tangent to the body axis of the fish (**Figure 7.30A**). As a propulsive segment sweeps through the water, it pushes water away from the body. An equal and opposite reaction results, which can be resolved into vectors, one for forward thrust and the other for lateral force. Smaller undulations near the head of the fish produce relatively less forward thrust compared to broader undulations nearer the caudal region.

The disproportionately greater thrust that can be generated by posterior body segments helps to explain the variety of different body forms of fishes. For example, fishes such as salmon swim steadily and often for great distances during migrations to feed and reproduce, and several aspects of their body form relate to this type of

swimming (**Figure 7.30B**). The median fins (dorsal, anal, and caudal fins) can apply thrust to the water along the length of the body. The dorsal fin is above the center of mass, where it helps to stabilize the direction of swimming. Also, the widest cross section of a salmon's body is about one-third along its length, a position that helps to minimize pressure drag, a topic we explore in the next section. In contrast, pikes are lurking predators that dart out to quickly capture prey such as other fishes or frogs. Their body form exemplifies specializations for **fast-start performance**, such as a tubular, elongate, arrow-shaped body with most of the area of median fins located far posteriorly to maximize forward thrust (**Figure 7.30C**). Such a body form is unsuited for long-distance swimming, but it works very well for dash and grab feeding.

Modes of locomotion

Looking more broadly at body form and swimming ability, we see a continuum of swimming modes (**Figure 7.31A**) that range from fishes that swim using eel-like or anguilliform locomotion involving most of the body length, through salmon (subcarangiform locomotion) and jacks (carangiform locomotion), to tunas and white sharks that use only the caudal peduncle and caudal fin (thunniform locomotion). A boxfish, which cannot laterally undulate its trunk because it is enclosed in bony armor, is limited to motions of the caudal fin, a swimming mode termed ostraciiform locomotion.

(A) Proportion of the trunk and caudal region used for swimming

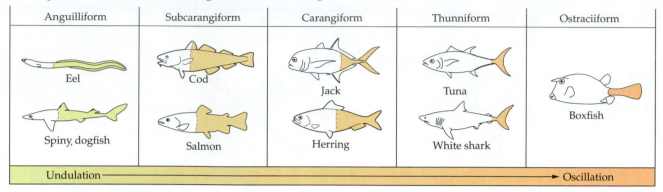

Anguilliform	Subcarangiform	Carangiform	Thunniform	Ostraciiform
Eel	Cod	Jack	Tuna	Boxfish
Spiny dogfish	Salmon	Herring	White shark	

Undulation ⟶ Oscillation

(B) External anatomy of a yellow perch, *Perca flavescens*

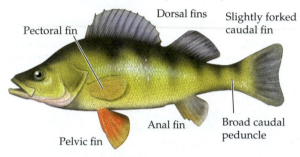

Pectoral fin
Dorsal fins
Slightly forked caudal fin
Anal fin
Broad caudal peduncle
Pelvic fin

(C) External anatomy of a yellowfin tuna, *Thunnus albacares*

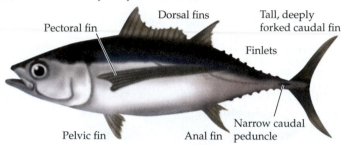

Pectoral fin
Dorsal fins
Tall, deeply forked caudal fin
Finlets
Pelvic fin
Anal fin
Narrow caudal peduncle

(D) Fins used for undulatory swimming

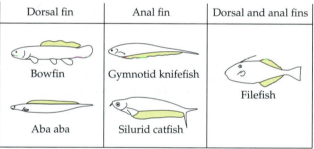

Dorsal fin	Anal fin	Dorsal and anal fins
Bowfin	Gymnotid knifefish	Filefish
Aba aba	Silurid catfish	

(E) Fins used for oscillatory swimming

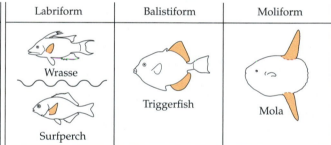

Labriform	Balistiform	Moliform
Wrasse	Triggerfish	Mola
Surfperch		

Figure 7.31 Modes of swimming. (A) Proportions of the trunk and caudal region used for swimming. There is a continuum between undulation, in which the body makes wide excursions from side to side, and oscillation, in which much of the movement occurs in the caudal peduncle and caudal fin. (B) The body form and fins of a yellow perch (Perciformes) make it a locomotor generalist because it can spread or collapse its fins for different types of swimming. (C) The yellowfin tuna (Scombriformes) is among the speediest of fishes. Unlike perches, tunas cannot change the surface area of their tall, deeply forked, crescent shaped caudal fins, making tunas speed specialists. Other specializations for speed include a narrow caudal peduncle and long, stiff dorsal and anal fins that, together with the long pectoral fins, act like feathers on the shaft of an arrow to keep the body moving in a straight line. (D) Many fishes independently evolved undulatory locomotion, in which a propulsive wave passes along a median fin while the body is held in a relatively straight line. (E) Other fishes use oscillations of paired or median fins to propel themselves through the water. In the case of wrasses or surfperches, the body moves up and down in a wave as the fish applies force either upward or downward. (See credit details at the end of this chapter.)

The ability to spread and collapse its median fins allows a yellow perch to swim using several swimming modes, making it a good example of a locomotor generalist (**Figure 7.31B**). Tunas represent apices of power and speed, and have many specializations for fast swimming, such as a narrow caudal peduncle (**Figure 7.31C**). Other swimming specializations include long tendons that connect the body muscles to a stiff, crescent-shaped caudal fin. This allows the fish to rapidly oscillate the caudal fin from side to side, maximizing thrust, while holding more anterior portions of its body relatively straight to minimize drag.

Median and paired fins of other fishes are specialized for different types of undulation (refers to passing a wave along the body or a fin) and oscillation (refers to rapid movements of an entire fin). For example, bowfins and aba-aba can pass an undulatory wave along the length of elongated dorsal fins while holding the body straight in the water column (**Figure 7.31D**). Gymnotid knifefishes and silurid catfishes accomplish the same thing using elongate anal fins, and filefishes use both the dorsal and anal fins. In all of these cases, undulations of median fins allow precise locomotion either forward or backward in order to move more stealthily than when using typical body undulations.

Some fishes rely on oscillations of paired or median fins for swimming (**Figure 7.31E**). Among these are wrasses (see Figure 7.28C), which swim by oscillating the pectoral fins up and down. If you watch a wrasse in a tank or on a reef, then you will see that the body bobs up and down with each stroke of the fins, producing a wavelike pattern as the fish swims. Although triggerfishes (see Figure 7.29B) are closely related to filefishes, trigger-fishes typically oscillate their stiff dorsal and anal fins to provide powerful forward locomotion. Finally, an extreme in oscillatory motion characterizes ocean sunfishes in the family Molidae (see Figure 7.29D). The caudal fin was lost during molid evolution, and their bodies are too short and too inflexible for lateral undulation, so molids use powerful and rapid oscillations of the dorsal and anal fins to achieve great swimming speeds.

Speed and drag

If you have ever watched fishes schooling in a large aquarium exhibit, you may have noticed a curious fact: fishes often school together not by species, but by the lengths of the schooling individuals. There is a strong correlation between a fish's body length and its swimming speed: longer fishes swim faster, fishes of similar lengths tend to have similar swimming speeds, and the ability to keep up with the other fish in a school is paramount in schooling behavior. But the correlation between fish length and swimming speed is not the only question worth considering.

We study many aspects of swimming fishes using tools originally developed for hydrodynamics, the study of fluids in motion. Some of the most important principles of hydrodynamics are intuitive because we experience them in daily life. Understanding these basics can contribute to your ability to watch fishes and to understand more about the challenges they face by living in a moving fluid.

Reynolds numbers   Reynolds numbers are central to many concepts in hydrodynamics and helpful for thinking about fish swimming. Abbreviated Re, a Reynolds number is a dimensionless number that expresses the ratio of inertial forces to viscous forces for an object (such as a fish) moving through a fluid (such as water). The same math applies to liquids flowing through pipes or air flowing around the body of an aircraft. The equation is:

$$Re = \mu L / \nu$$

For our purposes, inertial forces are μL, the product of a fish's speed (μ) and its length (L). A tuna swimming rapidly experiences higher inertial forces than does a guppy swimming slowly. Kinematic viscosity, represented as ν, measures the stickiness (viscosity) of a fluid. Water at 20°C has a kinematic viscosity of about 10^{-6} m^2/s, whereas maple syrup is 150 times more viscous ($\nu = 150 \times 10^{-6}$ m^2/s).

As noted in **Table 7.5**, for a fast-swimming tuna, inertial forces are much greater than viscous forces, so it has a large Reynolds number (Re = 5,500,00). Even our swimming guppy has a relatively large Reynolds number (Re = 5,000). But at small scales and lower speeds, Re is much lower. Thus, viscous forces dominate swimming larval porgies (Re = 110) and herrings (Re = 24). This means the larvae experience water as sticky and difficult to push through. To visualize this, imagine that our guppy is swimming in maple syrup; its Re would decrease from 5,000 to 33, putting it into the range of Reynolds numbers for larvae. This simple thought experiment helps you understand that larval fishes typically operate at very low Reynolds numbers and that their biology must, of necessity, be very different from that of adults.

Displacement and turbulence   We can describe the flow of water around a fish's body using additional concepts from hydrodynamics. As a fish swims and pushes through the water, it displaces a volume of water. Ideally, it would displace the same volume of water as its body and as it swims forward there would be no impact on the smooth flow of water around its body. We can visualize this as smooth lines of flow, or streamlines, that do not break up as they pass around the body. But such an ideal displacement of water around a fish's body is never achieved—a fish always displaces more water than its body volume as it swims. It is hard to push water out of the way, and the more water a swimming fish has to displace, the more energy it uses.

The boundary layer is the layer of water immediately adjacent to the fish's body. It has a velocity of zero (0.0 m/s) closest to the skin of the fish, and it increases up to

Table 7.5 Estimated Reynolds numbers and flow regimes for five swimming fishes

	μ (swimming speed, m/s)	L (body length, m)	ν (kinematic viscosity, m^2/s)[a]	Reynolds number ($\mu L / \nu$)	Flow regime
Yellowfin tuna	5.4	1.0	0.000001	30,000,000	Turbulent
Trout	0.35	0.14	0.000001	49,000	Turbulent
Guppy	0.10	0.05	0.000001	5,000	Turbulent
Larval perciform	0.01	0.011	0.000001	110	Laminar
Larval herring	0.004	0.006	0.000001	24	Laminar

Compiled from information in H.S.H. Yuen. 1966. *Trans. Am. Fisheries Soc.* 95: 2: 203–209; Tytell, E.D. 2010. In G. Taylor, M.S. Triantafyllou, and C. Tropea (Eds.): *Animal Locomotion.* Springer; T. Landgraf et al. 2016. *Bioinspir. Biomim.* 11: 015001; J.M. Leis et al. 2006. *Marine Biol.* 148: 655–669.

[a] The kinematic viscosity given here is for water at 20°C.

the speed of local water flow as a function of kinematic viscosity: objects moving in viscous fluids like maple syrup have thicker boundary layers than objects moving in water. At Reynolds numbers lower than ~2,000, the boundary layer stays attached to the body and the flow lines smoothly separate around the fish, continuing behind it without any turbulence (laminar flow; **Figure 7.32A**). But at higher Reynolds numbers, the boundary layer separates from the body and turbulent vortices form in the water (**Figure 7.32B**). Flow lines do not become smoothed out and regular again until some distance behind the fish, and the region of turbulence behind the fish is its wake.

Drag   The concepts of water displacement and Reynolds numbers help us understand the types of **drag**—backward force opposing forward motion—that a swimming fish experiences. The two types of drag respond differently to speed and body form:

- **Viscous drag** is a function of the kinematic viscosity of water and is caused by friction between the fish's body and the surrounding water at the boundary layer. It is sensitive to the surface area and smoothness of the body. A smooth sphere, a body approximated by many pelagic larval teleosts, has the smallest surface area for a given volume, and thus minimal viscous drag. A long, thin body has relatively higher viscous drag because of its large surface area relative to body mass; scaleless fishes (such as eels) have relatively lower viscous drag than fishes with scales.

- **Pressure drag**, also known as inertial drag, is caused by pressure differences resulting from the fish's displacement of water. It increases as speed increases and is sensitive to the shape and cross-sectional area of the body. A thick body induces high inertial drag because it displaces a large volume of water as it moves forward. Streamlined (teardrop) shapes produce minimum inertial drag when their maximum width, usually near the insertion of the pectoral fin, is about one-fourth their length and is situated about one-third body length along the body.

Fast swimmers such as jacks and tunas limit body undulation insofar as possible to the narrow caudal peduncle and caudal fin. The caudal peduncle is flattened from top to bottom and edged by sharp scales, so it generates minimal pressure drag as it sweeps from side to side. Swordfishes, reputed to be the fastest-swimming fishes, release drag-reducing oil from pores on the head. Caudal fins cause turbulent vortices (whirlpools of swirling water) in the wake that are part of the fish's inertial drag (see Figure 7.32B). The total drag caused by the caudal fin depends on its shape. When the **aspect ratio** of the caudal fin (its dorsal-to-ventral height divided by its anterior-to-posterior length) is high, the amount of thrust produced relative to drag is high. The stiff, deeply forked caudal fin of jacks, mackerels, tunas, and swordfishes have a high aspect ratio and thus contribute to efficient forward motion (see Figure 7.31C). Even the cross

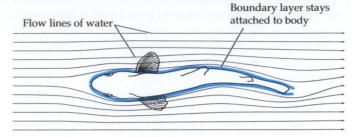

(A) Laminar flow can occur at slow forward speeds

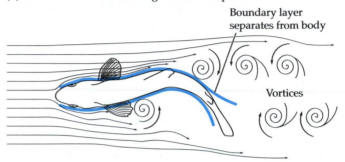

(B) Turbulent flow occurs at higher forward speeds

Figure 7.32 Flow regimes during swimming.   (A) A small fish moving slowly can maintain laminar flow of water over its body. (B) At higher swimming speeds or larger body sizes, flow becomes turbulent, and energy is lost in overcoming pressure drag. Most adult fishes experience this hydrodynamic environment, which is why they have a teardrop (or torpedolike) body shape that minimizes pressure drag. (See credit details at the end of this chapter.)

section of the upper and lower lobes of these caudal fins has a streamlined teardrop shape, further reducing drag and allowing many of these species to swim continuously.

In contrast, the caudal fins of trout, minnows, and perches are not stiffened and typically have low aspect ratios. These fishes bend the body in a swimming mode called subcarangiform (see Figure 7.31A). Subcarangiform swimmers can spread or compress the caudal fin to modify thrust. The caudal peduncle of subcarangiform swimmers is laterally compressed but relatively deep from top to bottom, and because of its depth, it contributes to thrust.

Steering, stopping, and staying in place

When watching fishes, you should always start with the fins because fins are how fishes maneuver through their three-dimensional worlds. Observing swimming fishes in an aquarium or in videos will help you better understand the behaviors described here.

A fish steers by extending fins on one side of its body to generate drag, just as a rower allows an oar to drag in the water to turn a boat. Any combination of fins can be used to steer—extending a pectoral fin produces a gradual turn, whereas extending additional fins increases the drag and sharpens the turn. A fish also uses fins to stop. Pectoral and pelvic fins rotate outward, and the posterior portions of the dorsal, anal, and caudal fins flex, forming surfaces analogous to a parachute to decrease forward momentum.

Watching a fish stay in place in the water is fascinating. Teleosts use the gas bladder to make themselves neutrally buoyant, so you would think a fish could float in the water just as you can, with no movement of its appendages. A moment's observation will disprove that hypothesis, however—a fish staying in place in water actually moves several fins continuously. The pectoral fins maintain a sculling motion, and careful observation will show that the power stroke is directed forward. Looking a bit farther forward, you will see that the opercula are opening and closing as the fish breathes. Each time a fish exhales, water jets backward from beneath the opercula, generating a thrust that would drive the fish forward. The backward sculling movements of the pectoral fins counteract the forward thrust of water leaving the gill chambers.

If you watch carefully, you can see undulatory waves traveling downward along the posterior margin of the caudal fin. A downward wave generates an upward thrust, which should rotate the posterior end of the fish upward; yet the fish does not somersault, but remains horizontal in the water. The position of the gas bladder explains this observation. The volume of gas in the gas bladder can be adjusted to make the fish neutrally buoyant, but the gas bladder is typically anterior to the center of gravity of the fish. As a result, a fish is tail-heavy, and it would float tail-down without the upward thrust produced by the sine wave in its caudal fin.

7.7 Reproduction and Development

LEARNING OBJECTIVES

7.7.1 Compare reproductive strategies of oviparous freshwater and marine species.

7.7.2 Describe viviparity in the families Syngnathidae (see Figure 7.24) and Poeciliidae.

7.7.3 Differentiate gonochorism, protandry, and protogyny.

Actinopterygians show more diversity in reproductive biology than any other vertebrate group (**Figure 7.33**).

- Most species lay eggs, which may be buried in the substrate, deposited in nests, or released to float in the water, and a few species lay eggs out of water (on beaches, for example), but viviparity evolved in several lineages and nutrition of the embryos ranges from lecithotrophy (completely dependent on yolk) to matrotrophy (most of the nutrients come from the mother).

- Parental care ranges from none, through attending eggs in nests and protecting young after they hatch, to mouth brooding (carrying eggs in the mouth during development and allowing the young to flee back into the parent's mouth when danger threatens).

- Even sex determination shows diversity—the sex of most species of teleosts is genetically determined and fixed for life, but some species are **hermaphroditic** (i.e., an individual has functional ovaries and testes, typically at different phases during life), and a few species are **parthenogenetic**, consisting entirely of females.

Diverse modes of reproduction among fishes mean there are exceptions to every generalization. The following discussion focuses on the typical characteristics of the groups of fishes described.

Oviparity

Most ray-finned fishes are oviparous, and freshwater and marine species show contrasting specializations for oviparous reproduction.

Freshwater habitats Actinopterygians that inhabit fresh water generally produce a relatively small number of large, yolk-rich **demersal eggs** (buried in gravel, placed in a nest, or attached to the surface of a rock or plant). Attachment is important because flowing water might carry floating eggs away from suitable habitats. Because eggs stay in one place, parental care is possible, and males of many species guard nests and sometimes young.

Eggs of many freshwater teleosts hatch into young with large yolk sacs that support growth for some time after hatching. When the yolk is exhausted, the young fishes begin eating on their own. After metamorphosis, the juveniles generally resemble their parents.

Marine habitats Marine teleosts generally release large numbers of small, buoyant, transparent eggs that float in the water column. Such pelagic eggs are left to develop and hatch while drifting in the open sea, precluding any possibility of parental care. The small larvae have few yolk reserves remaining after they hatch and begin feeding on microplankton very soon after hatching.

Marine larvae can be specialized for life in the oceanic plankton, feeding and growing while adrift at sea for weeks or months, depending on the species. Larvae are often very different in appearance from adults, and adult forms of many larvae are unknown (see Section 7.8). Larvae eventually metamorphose and settle into the juvenile or adult habitats appropriate for their species. Although they drift with currents, increasing evidence suggests that many larvae actively choose settling habitats, and groups of larvae can stay together until they settle.

The strategy of producing planktonic eggs and larvae exposed to a prolonged and risky pelagic existence appears to be wasteful of gametes. Nevertheless, complex life cycles of this sort are the principal mode of reproduction among marine fishes. Pelagic spawning may offer advantages:

- Predators on eggs may be abundant in the parental habitat but scarce in the pelagic realm.

- Microplankton (bacteria, algae, protozoans, rotifers, and minute crustaceans) are abundant where sufficient nutrients reach sunlit waters. The same winds and currents that transport larvae transport these food items, and both larvae and their food aggregate in convergence zones.

(A) Spawning camouflage groupers, *Epinephelus polyphekadion*

(B) Nesting bluegill sunfish, *Lepomis macrochirus*

(C) Spawning sockeye salmon, *Oncorhynchus nerka*

(D) Male yellowhead jawfish, *Opistognathus aurifrons*

Figure 7.33 **Spawning and parental care by oviparous teleosts.** Reproductive modes of teleosts extend from broadcasting eggs and sperm to extensive parental care. (A) Many marine fishes release eggs and sperm into the sea, where the fertilized eggs hatch into pelagic larvae. Here, spawning groupers are releasing clouds of eggs and sperm. (B) A "nest ring." Male bluegill sunfish clear debris from a wide area to make a nest site. The male attends the nest to defend the eggs against predators, which include other male bluegills, like the one shown in the inset. (C) Female sockeye salmon move streambed pebbles to prepare a nest known as a redd. The female deposits eggs that are fertilized by the male, then moves to the upstream edge of the nest and digs a new nest, covering the eggs in the original nest at the same time. Adult sockeye salmon die after spawning. (D) Males in several groups of teleosts brood eggs in their mouths, as in this yellowhead jawfish, *Opistognathus aurifrons*. Individuals of this species are monogamous; members of a pair live in adjacent burrows, sometimes trading burrows or occupying the same burrow.

Producing floating, current-borne eggs and larvae increases the chances of colonizing all patches of appropriate habitat for adults over a large area. A widely dispersed species is not vulnerable to local environmental changes that could extinguish a species with a restricted geographic distribution.

Early development of ray-finned fishes that have pelagic eggs is usually biphasic, meaning that larval planktonic existence as part of the **ichthyoplankton** is very different from that of an adult. A profound metamorphosis often separates these life stages. After an egg hatches into a larva, the larva has a period of post-hatching development before it starts to feed on its own. It depletes any remaining yolk in its yolk sac, and developmental changes happen quickly, including further development of sense organs, muscles, gills, jaws, and teeth that began before hatching. Ichthyologists often describe these changes in terms of the number of hours or days post-hatching. As

these features develop, most species of marine larval fishes begin to feed on other planktonic organisms. Once they start feeding, growth of such planktotrophic larvae can be remarkably fast. Eventually, the larvae will metamorphose and settle out of the plankton to occupy habitats typical for the adults of that species, such as muddy expanses of the sea floor, rocky or coral reefs, or the water column of the pelagic realm.

Metamorphosis of larval marine fishes is often dramatic, with losses of larval features such as elongate fins, and acquisition of features such as different types of teeth, that are associated with adult life in a very different habitat. Some planktotrophic fish larvae live in the plankton only briefly before they metamorphose and settle out to live in adult habitats such as coral reefs or the relatively shallow waters of the continental shelf (this is typical, for example, for many species of flatfishes; see Figure 7.26B). Others live in the water column for months or potentially even years before settling, such as some species of eels (see Figure 7.14C).

A prolonged period of life as a planktotrophic larva can contribute to **dispersal**, in this case referring to larvae that travel in the open ocean before settling in locations remote from the place where spawning occurred. Because larvae are small, viscous forces dominate their ability to swim (see Section 7.6). Water sticks to the surface of the fish, making it hard to move very far or very fast (it would be like you swimming in maple syrup). Thus larval fishes are dispersed by ocean currents in a process called advection (the transfer of matter by a fluid such as air or, in this case, water). Advection is how larvae of some coral reef fishes are able to reach isolated and widely dispersed reefs. Understanding the ability of planktonic larvae to disperse across vast distances of open ocean helps us predict how the ranges of such species may expand to near-global presence in tropical waters as the oceans warm. For example, some Indo-Pacific reef species are projected to extend their ranges to the Caribbean by the end of the 21st century.

Terrestrial habitats Anyone who has lived near the California shore has probably heard of grunion runs. California grunions (*Leuresthes tenuis*) are small marine fishes that lay their eggs at the top of the tide mark on beaches from March through August. Tides reach their maximum for about two nights on each side of the full moon and the new moon, and this is when grunions spawn, riding a wave ashore. Females squirm tail-first into the wet sand as the wave recedes. One or more males cluster around a female and release sperm as she releases eggs, and then the fishes ride a receding wave back into the sea. Grunion eggs require about 10 days to hatch, and during this period of lower tides the nests are above the reach of waves. At the next high tide cycle—about two weeks after the eggs were laid—the grunion fry hatch and are swept back into the sea.

Even less fishlike are the reproductive behaviors of at least one species of mudskipper (*Periophthalmodon schlosseri*) and the rockhopper blenny (*Andamia tetradactyla*). Mudskippers construct burrows on mudflats. During high tide, mudskippers hide in their flooded burrows, emerging at low tide to forage on the mud (see Figure 7.23C). The burrows contain an upward bend filled with air, even when the burrow entrance is covered by the tide, and the adhesive eggs are deposited on the roof of this chamber, well above the water and exposed only to air. Other mudskipper species also deposit eggs in their burrows, and it is likely that some of those eggs also develop in air.

Rockhopper blennies forage during low tide on wave-splashed mats of algae on rocky coastlines and shelter during high tide in crevices in the rocks. Females deposit eggs in the crevices and males attend the nests, which are entirely out of the water except during the height of the tide or when they are washed by waves during storms.

Viviparity

The widespread occurrence of viviparity among familiar freshwater aquarium fishes in the family Poeciliidae (guppies, mollies, platys, swordtails, and related species) gives the impression that it is a common mode of reproduction for teleosts, but that is not true. Although viviparity evolved in at least 12 lineages of teleosts, only 3% of teleosts are viviparous, and more than one-third of viviparous teleosts are poeciliids. This scarcity of viviparous lineages of teleosts is a bit surprising considering the diversity that teleosts display in other aspects of their reproductive biology.

Intromission is essential for viviparity, and males of many species of viviparous teleosts have modified fins to serve as intromittent organs. As noted above, the anal fin of male poeciliids forms a distinctive long gonopodium (see Figure 7.27E). An unrelated group of teleosts in the family Phallostethidae (the name means "penis chest") have a unique muscular intromittent organ known as the priapium (after Priapus, the Greek fertility god) on the ventral side of the head. The priapium is asymmetrical, with openings for the anus and urinary ducts on one side and the genital opening on the opposite side. The skeleton of the priapium evolved by modifications of the pelvic girdle and fin together with components derived from ribs and possibly portions of the pectoral fins.

The shiner surfperch *Cymatogaster aggregata* is an example of a viviparous marine teleost. Males inseminate females using a thickened anal fin. Females can store sperm for an extended period and, after the eggs are fertilized, gestation lasts for about 6 months. The embryos develop within the ovary; folds in the ovarian wall provide nutrients and gas exchange for the developing young (matrotrophy), which are relatively large when born, appearing as miniature adults (males may even be sexually mature at the time of their birth). Brood sizes are relatively small, ranging from 3 to 36.

Scorpionfishes offer a polar opposite example, with brood sizes for the Bocaccio rockfish *Sebastes paucispinis* being as high as 2.6 million. Like surfperches, gestation occurs in the lumen of the ovary, but the mother supplies very little if any additional nutrients beyond those supplied in the yolk of each egg (lecithotrophy), and the young develop within their egg envelope for most of gestation. Gestation lasts about 1 month, and the tiny young hatch from their egg envelope shortly before birth. Essentially in a late larval state, the hatchlings nonetheless can swim and feed. This extreme life history pattern stands in sharp contrast to other vertebrate livebearers, which usually give birth to a few relatively large, well-developed young.

Syngnathidae (seahorses, sea dragons, and pipefishes) provides another interesting example of viviparity—male gestation. Males of all species in this family carry the fertilized eggs until they hatch, but the details vary. Some species carry the eggs externally, adhering to the male's abdomen or his tail; in other species the eggs are partially enclosed in a pouch that remains open to the external environment. In the most derived species, including the yellow seahorse (*Hippocampus kuda*) of the Indo-Pacific region, the eggs are held in a closed pouch on the male's abdomen. The male and female mate by facing each other and pressing their abdomens together (see Figure 7.24B). During this stage, which is very brief, the eggs are apparently released and pass across the opening of the sperm duct while sperm are released, and the eggs and sperm are carried through the opening of a male's brood pouch, which then closes tightly. The eggs develop within the pouch, and after a gestation period (typically 2–4 weeks), the male expels the young using muscular contractions (see Figure 7.24C). Unfertilized eggs and some of the embryos die during development, and the brooding male reabsorbs nutrients from these tissues.

Sex change in teleosts

Gonochorism is the pattern in which sex is fixed throughout life, a pattern exhibited by the scissortail sergeant-major (*Abudefduf sexfasciatus*; **Figure 7.34A**). This is typical for about 88% of teleost species. Some teleosts, however,

begin life as one sex and change to the other after they are adults. Some even have functional ovaries and testes at the same time. All such sex-changing teleosts are hermaphrodites, but there are important differences in the pattern of hermaphroditism in different teleosts. Underlying these differences in life history is Darwinian fitness—that is, what allocation of time and energy to reproduction yields the highest representation of the alleles of an individual in succeeding generations relative to the alleles of other individuals in the population?

Sequential hermaphroditism describe the condition in which sex change occurs from male to female or from female to male, with the result that different gametes are produced at different times in life. Some teleosts exhibit **protandry** (Greek *prot*, "first"; *andro*, "male"), meaning that individuals start life as males and change to females. Anemonefishes (about 30 species in the genus *Amphiprion*) demonstrate protandrous hermaphroditism (**Figure 7.34B**). The largest individual is always female and the second-largest is male; both are reproductively mature. Smaller individuals, if present, remain sexually immature. If the female dies, then the male changes sex and the largest immature member of the group becomes male.

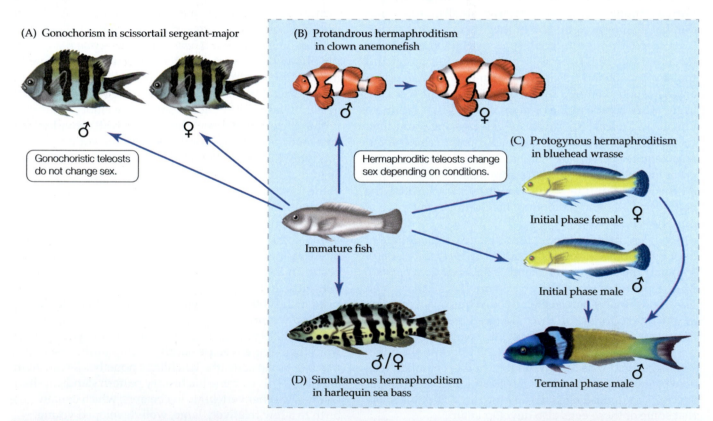

Figure 7.34 Sex determination in teleosts. (A) About 88% of teleosts are gonochoristic, meaning that their sex does not change during life, as shown by the scissortail sergeant-major, *Abudefduf sexfasciatus*. (B) Protandrous hermaphrodites such as the clown anemonefish, *Amphiprion ocellaris*, begin life as males and change to females. (C) Protogynous hermaphroditic fishes such as the bluehead wrasse, *Thalassoma bifasciatum*, begin life as either females or males that mature into an initial phase in which the color pattern on the body is the same for both sexes. Fishes in the initial phase can subsequently transform into terminal phase males, which develop the characteristic blue head of this species. Males and females in the initial phase can reproduce, but the harem-based social system of this species greatly limits the reproductive success of initial-phase males. (D) Simultaneous hermaphrodites such as the harlequin bass, *Serranus tigrinus*, alternate male and female roles during spawning.

The "size advantage hypothesis" proposes that benefits of protandry lie in the relationship between body size and sperm and egg production. Large females produce more eggs than small ones, but even a small male can produce all the sperm needed to fertilize the eggs of a large female. A clown anemone fish can maximize its lifetime reproductive success by starting life as a male and fertilizing the eggs of a large female, and then changing to a female when it has grown to a larger size.

Protogyny (Greek *gyne*, "female") is the reverse of protandry. In protogyny, juveniles mature initially as females and subsequently change sex to become males, as in the bluehead wrasse *Thalassoma bifasciatum* (**Figure 7.34C**). Territorial males defend mating territories, making large body size advantageous. Both males and females start adult life in an initial phase, in which both have a yellow color pattern without the blue head. Some initial-phase individuals (both males and females) subsequently change into terminal-phase males with a blue head, black saddle, and green posterior region.

A terminal-phase male bluehead wrasse defends a territory over a coral head protruding above the general height of the reef, mating with females that come to the coral head to spawn. Initial-phase males also breed, but instead of defending a breeding site they swim in groups and try to intercept females. When a female spawns, any initial-phase males nearby release sperm, but their mating success is lower than that of terminal-phase males. An initial-phase male may have one or two mating opportunities a day, and even then, his sperm must compete with sperm from other initial-phase males in his group. In contrast, a terminal-phase male mates 40–100 times a day, without sperm competition from other males.

The potential reproductive success of a terminal-phase male is greater than that of even the largest female because the male mates so often. Thus it is advantageous for a large individual to be male, and when a terminal-phase male disappears from its territory, the largest female in the group becomes a male. The transition is astonishingly rapid: if the male is artificially removed, the largest female begins to display male behaviors within minutes. Individuals that spawned as females one day have been observed to mate as males on the following day.

A few species of teleosts exhibit **simultaneous hermaphroditism**. For example, an individual harlequin sea bass has functional ovaries and functional testes at the same time and can mate either as a male or as a female (**Figure 7.34D**). This life-history pattern is uncommon, but occurs in about 20 families of teleosts, including many small sea basses. Simultaneous hermaphrodites show two patterns of spawning behavior:

- Egg traders (or parcelers) alternate female and male roles during a single spawning session. One member of the pair releases a portion of its eggs (the "parcel" that gives this mating pattern its name), and the other releases sperm. Then the two individuals switch roles and continue alternating throughout the session, which may last for more than an hour.

- Reciprocating monogamous pairs alternate roles between spawning sessions—that is, the individual that releases eggs in one spawning session releases sperm in the next spawning session, which may occur some days later.

At least one teleost species is a self-fertilizing hermaphrodite. *Kryptolebias marmoratus* is a small fish (7.5 cm) that lives in shallow, brackish waters from Florida through the Caribbean. About 95% of individuals are born as self-fertilizing hermaphrodites that produce homozygous clones of the parent when they reproduce. The remaining 5% are males; females have never been described in this species. In areas with water temperatures below 20°C, about 60% of the hermaphrodites transform to secondary males after 3 or 4 years, but in areas where the water temperature is above 25°C, *K. marmoratus* remain hermaphrodites throughout their lives.

7.8 Ecology of Marine Teleosts

LEARNING OBJECTIVES

7.8.1 Provide examples of materials that might contribute to an extended specimen of a larval teleost.

7.8.2 Define the photic zone and explain key differences in its three subdivisions on a tropical coral reef.

7.8.3 Discuss the relationship between biomass and pelagic life zones.

7.8.4 Describe anatomical and behavioral specializations of bathypelagic teleosts.

More than 15,000 extant species of teleosts live in marine environments ranging from coastal estuaries and coral reefs to the open ocean and its depths. **Demersal** (Latin *demersus*, "to sink") fishes live in close association with substrates ranging from mud, gravel, or rock to coral reefs. As we noted above, species such as oarfishes are pelagic, living their entire lives within the water column and never contacting the substrate (see Figure 7.20E). Although most teleosts move within the water column during their lives (for example, as when flatfishes metamorphose from pelagic larvae to become demersal adults), adults of many species typically live within a relatively narrow depth range. In this section, we consider how direct observations from scuba diving, submersibles, and remotely operated vehicles (ROVs) are changing our understanding of how teleosts use different depth zones of the marine environment.

Black-water diving and larval teleosts

It is twilight in Hawaii, and you are on a boat heading out for a night dive in the open ocean. Below you, the seafloor is thousands of meters away—not the deepest spot in the ocean, but deep enough for you to enter a very different world from that of a reef dive. Your goal is to find and photograph minuscule planktonic animals against the black water of the nighttime depths. This task will push your photographic skills to the limit, but there is potential to make colorful, crisp, and mind-blowing images of some of Earth's least-known vertebrates.

Arriving at the coordinates for your dive, the crew hangs bright lights in the water. Don your scuba gear, grab your camera, focusing lights, and strobes, and descend to your tether 10 m below the boat. As you drift, the bright lights attract small zooplankton, and larger animals such as squid and larval fishes come into view to eat the smaller animals. Welcome to the world of black-water diving.

A thrilling experience in itself, black-water diving offers opportunities to see larval fishes as never before observed. Since the 19th century, studies of larval marine fishes have been based on samples netted at sea. Because fish larvae are small, this means bulk collection of plankton, preservation of samples, and painstaking sorting using a microscope in a shore-based lab to separate larval fishes from the far more abundant crustaceans and other zooplankton. At best, such preserved samples pale in comparison to the colors and shapes of living larval fishes, but until black-water diving, there was no easy way to directly study them alive in the plankton.

Tonight your dive is even more special because you have a second goal: after you photograph a fish larva, you will collect it and preserve it for a permanent natural history collection so it can be studied and compared with other specimens. The important thing is to be able to associate your photograph with the specimen itself, so that the museum can curate and maintain an **extended specimen** that includes much more information about a specimen than was possible in the past.

Color photographs made in the field may reveal new details about larvae, such as their color in life or the extended gut much better seen in living than preserved larval dragonfishes (**Figure 7.35A**). Sometimes black-water photographs record larval behaviors that could never be

(A) Larva of a barbeled dragonfish, *Aristostomias*, photographed in life (left) and after preservation (right)

© Nonaka et al. 2021. *Ichthyology & Herpetology* 109: 138–156

2 mm

(B) Larva of a bigtoothed pomfret, *Brama orcini*, riding a jellyfish (left), which it does using its specialized pelvic fin (right)

© Nonaka et al. 2021. *Ichthyology & Herpetology* 109: 138–156

1 mm

Figure 7.35 Ichthyoplankton and black-water diving. (A) Larval dragonfish photographed in the water (left) and after preservation (right). (B) Larval pomfrets have been photographed riding jellyfishes, which helps to explain why the pelvic fins are modified to form a "saddle."

seen in a preserved sample, such as the jellyfish-riding behavior of the bigtoothed pomfret, which we now can correlate with specializations of the pelvic fin seen in preserved larvae (**Figure 7.35B**).

Larval collection also allows researchers to extract a small DNA sample—usually taken from the lens of the right eye—that can be sequenced and compared with other species using **genetic barcoding**: the use of short sequences of DNA, typically from genes in the mitochondrial genome such as cytochrome *c* oxidase I (COI), to identify species. Barcoding has proved to be remarkably effective for species identification, and it allows researchers to match up odd-looking fish larvae with the often very different looking adults. Sometimes, DNA barcoding shows that a larva is from a species with adults yet unknown to science. For example, the larva in Figure 7.35A belongs to the genus *Aristostomias*, but the larva's DNA does not match any known sequences for an adult *Aristostomias*. Larval species identification using barcoding is growing rapidly as DNA from more species is sequenced and made publicly available via online bioinformatic databases. We are on the cusp of even more powerful identification tools as the mitochondrial genomes of more and more organisms are sequenced and published.

Images of larval fishes in nature made possible by citizen scientists provide new insights into the biology of larval teleosts. For example, people often wondered about odd body forms of preserved larvae—what are those elongated fin rays, bright colors, transparent fins, and weird shapes for? When larvae can be observed alive, interpretation of some of these specializations can be more readily understood, and while we still cannot answer all such questions, there are some intriguing considerations. First, because of their small sizes, larval fishes live in very different flow regimes than do adults, which are large enough and fast enough that overcoming pressure drag is essential for swimming (which is why adult fishes that actively swim typically have fusiform body shapes, see Section 7.6). In contrast, as we also saw in Section 7.6, larval fishes face potentially overwhelming viscous forces making it difficult to swim and much easier to go with the flow. If a larva cannot swim quickly enough to escape a predator, then maybe a transparent body or fins, camouflage to break up the body outline, or long fins with distracting flaps of skin can keep a larva from being eaten. The odd body forms and specializations of larval teleosts might also represent ways to look like something else such as a jellyfish. We are a long way from understanding how such remarkable larval specializations evolved, but there is already enough tantalizing information to show what an exciting area this will be for new research.

The photic zone and its subdivisions

Because of differences in physical environmental factors, including light, pressure, temperature, and substrate, different species of teleosts live at different depths in the ocean. Most of the biomass within the ocean (including fishes) occurs in the top 1000 m of water, which is the depth through which light can penetrate, and consequently is known as the **photic zone**. Below the photic zone, the ocean is perpetually dark except for organisms that emit their own light.

Subdivisions of the photic zone have been the subject of recent study and interpretation made possible by using new technologies. Fishes that live in the shallowest upper waters have been observed since antiquity (Aristotle famously spent two years studying fishes of the Kallonis Lagoon on the Aegean island of Lesvos). Within the last century, it became possible for divers using scuba to reach depths of about 100 m, although the practical working depth limit for scuba observations of fishes is only about 40 m. Coincidentally (and conveniently for divers who study coral reef ecosystems) depths to about 40 m constitute a distinct zone for fishes that has been called the altiphotic fish-faunal zone (**Figure 7.36**). This zone includes most of the depth range within which there is enough light to allow zooxanthellate corals—reef-building corals with symbiotic algae—to flourish.

To directly observe teleosts living in deeper water, you need equipment such as a crewed submersible or camera-equipped ROV. Maturation of these technologies and their application to the study of fishes has allowed recognition of a distinct mesophotic fish-faunal zone, characterized by the decline in reef-building corals (due to decreased light) and increases in soft corals and sponges (see Figure 7.36). Based on direct observations using a submersible at the Caribbean island of Curaçao, we know that many species of teleosts move between the altiphotic and mesophotic zones. Other species inhabit the deeper waters of the rariphotic fish-faunal zone (below ~130 m), which is too deep and too dark for reef-building corals. Importantly, a distinctive fauna of teleosts lives in this zone that differs from those found in deeper waters.

Coral reef fishes

Within the altiphotic zone, the assemblages of organisms associated with coral reefs are the most diverse on Earth. Almost all vertebrates that live on reefs are teleosts—more than 600 species have been documented on a single Indo-Pacific reef. A main reason why reef environments can support such a rich diversity of teleosts is the range of different microhabitats and ways of life that reefs support, with some, such as physical complexity, deriving from the reef structure itself while others are based on anatomical and behavioral differences of the fishes and their prey. Niches available to fishes on a coral reef are also defined by the time of day during which the fish is active, where it hides when it is inactive, what it eats, and how it captures prey.

Reef-building organisms were present throughout most of the Phanerozoic, but reefs based on zooxanthellate

From C.C. Baldwin, et al. 2018. *Scientific Reports* 8: 4920/CC BY 4.0

Figure 7.36   Three reef-fish faunal zones at Curaçao.
Using a crewed submersible to study reef fishes that live at depths below those easily reached using scuba, ichthyologists developed a new classification of reef-fish faunal zones. Note that the depth scale on the figure is not linear but instead indicates major habitat changes at 40 m and 130 m. Scuba is practical for observing fishes only in the top 30–40 m of the water column. The practical working depth limit for small submersibles such as the one used in this study is ~300 m.

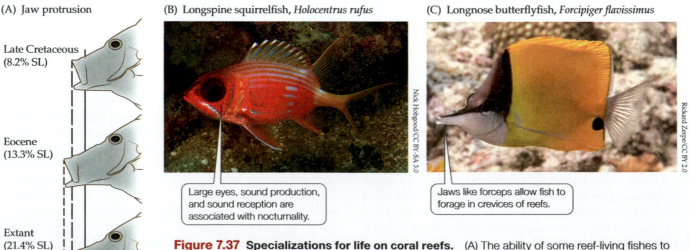

Figure 7.37   Specializations for life on coral reefs.   (A) The ability of some reef-living fishes to protrude their jaws has increased from the Early Cretaceous to the present. Protrusion is measured as a percentage of the standard length (SL) of a fish (the distance from the tip of the snout to the end of the last caudal vertebra). Some extant teleosts can protrude their jaws by more than 20% of their standard length, allowing them to engage in forceful and precise suction feeding. (B,C) Specializations such as large eyes and forceps-like jaws allow different groups of reef fishes to use different parts of the reef at different times. (A after D.R. Bellwood et al. 2015. *Curr Biol* 25: 2696–2700.)

corals like those of today are a Cenozoic phenomenon. Diversification of coral reefs and coral reef habitats co-occurred with rapid diversification of acanthopterygians. Many reef teleosts are feeding generalists that eat diverse invertebrates, but an evolutionary arms race has been occurring on reefs throughout the Cenozoic. For example, the ability of some acanthopterygians to project their jaws has increased since the late Cretaceous (**Figure 7.37A**), allowing capture of ever more elusive prey. In turn, to avoid predation many reef invertebrates evolved nocturnal behaviors and concealment during the day. And in response, some Eocene acanthopterygians evolved large eyes effective at low light intensities. To this day, their extant relatives—squirrelfishes—disperse over the reef at night to feed (**Figure 7.37B**). Other acanthopterygians, such as butterflyfishes, use forceps-like jaws to extract small invertebrates from their daytime hiding places or to snip off coral polyps, bits of sponges, and other exposed reef organisms (**Figure 7.37C**). This mode of predation demands high visual acuity that can be achieved only in the bright light of day, as well as the capacity to maneuver precisely through a complex three-dimensional habitat.

Teleost activity on coral reefs shifts between day and night. At dusk, diurnal fishes seek nighttime refuges, while nocturnal fishes leave their hiding places and replace the diurnal fishes in the water column. The timing of the shift is controlled by light intensity, and the precision with which each species leaves or enters the protective cover of the reef day after day indicates that this is a strongly selected behavioral and ecological characteristic. Correlated with the shift are changes in the soundscape of the reef related to differences in sensory biology. Sound, for example, can be used for communication when light levels are low. Nocturnal species such as squirrelfishes make sounds using specializations of the gas bladder, and squirrelfishes have, in turn, more sensitive hearing. Space, time, and food resources available on a reef are partitioned through these activity patterns, but the growing exposure of coastal habitats to artificial light at night increases the activity of predators and may alter predator–prey relationships.

Pelagic and deep-sea fishes

The life zones of the pelagic realm form the largest habitats on Earth (**Figure 7.38**). Light levels in the **epipelagic** (the top 200 m of the ocean) can support photosynthesis, and, as a result, oceanic biomass is concentrated in the epipelagic and **mesopelagic,** or twilight zone, immediately beneath it. Food webs of the ocean depths largely depend on falling detritus, which can range in size from microscopic remains of plankton to carcasses of large fishes and whales. Below the mesopelagic, the **bathypelagic** is perpetually dark and cold, about 4°C. Many species of teleosts are specialized for life in the bathypelagic, but total biomass is relatively low. Remotely operated vehicles have been used to study teleosts such as snailfishes and flatfishes living on the seafloor at abyssopelagic and hadopelagic depths, but the water column at these depths is very sparsely used by fishes. Increasing pressure eventually limits the maximum depth range of teleosts to about 8,000 m, and no vertebrates have been observed in the deepest part of the ocean at Challenger Deep, 10,900 m below sea level. (Challenger Deep is located southeast of Guam, and was discovered in 1875 by the scientific crew of the HMS *Challenger* during its voyage of marine exploration).

Epipelagic fishes   The sunlit waters of the epipelagic are home to many familiar teleosts, many of which have ranges spanning huge areas of the ocean. Some, such as herrings and mackerels, filter feed on small planktonic organisms. They can be of extreme ecological importance because of the sheer number of individuals; schools of Atlantic herring, *Clupea harengus*, have been estimated to contain several billion individuals. Other pelagic teleosts, such as tunas, are apex predators that chase down and capture smaller fishes and squid within the epipelagic. Still others, such as swordfish or ocean sunfishes, are predators that routinely move between warmer surface waters and great depths to feed. The fastest teleosts, such as tunas, live in the epipelagic, where their streamlined body shapes and powerful muscles allow them to catch fast swimming prey and avoid predators. Correlated with such specializations for speed is the capacity to make long migrations; for example, bluefin tunas are famous for ocean-spanning migrations for food and reproduction.

Vision is a primary sensory modality for epipelagic teleosts, and many have specializations to camouflage themselves or to confuse potential predators. Schooling species such as herrings are often silver, with brightly reflective scales. The reflected light can give predators an impression of transparency. Schooling also can confuse predators and reduce an individual's chances of being caught. Pelagic predators are typically strongly countershaded, meaning that they are darkly colored on the dorsal surface and lightly colored on their ventral surfaces. This makes it difficult for them to be seen from above against the dark background of deeper waters and from below against light streaming down from above.

Physical structure is rare in the pelagic; think of open ocean to the endless horizon. Nevertheless, the water's surface can hold all manner of floating materials—from sargassum (a floating brown algae) to logs washed into the ocean to manmade flotsam—and communities of fishes at many different life stages form in association with it. Some pelagic fishes seek out cover in very specific associations. For example the man-of-war fish, *Nomeus gronovii*, lives in close association with the stinging tentacles of the siphonophore commonly known as a Portuguese man o'war. Relying primarily on agility to avoid being stung,

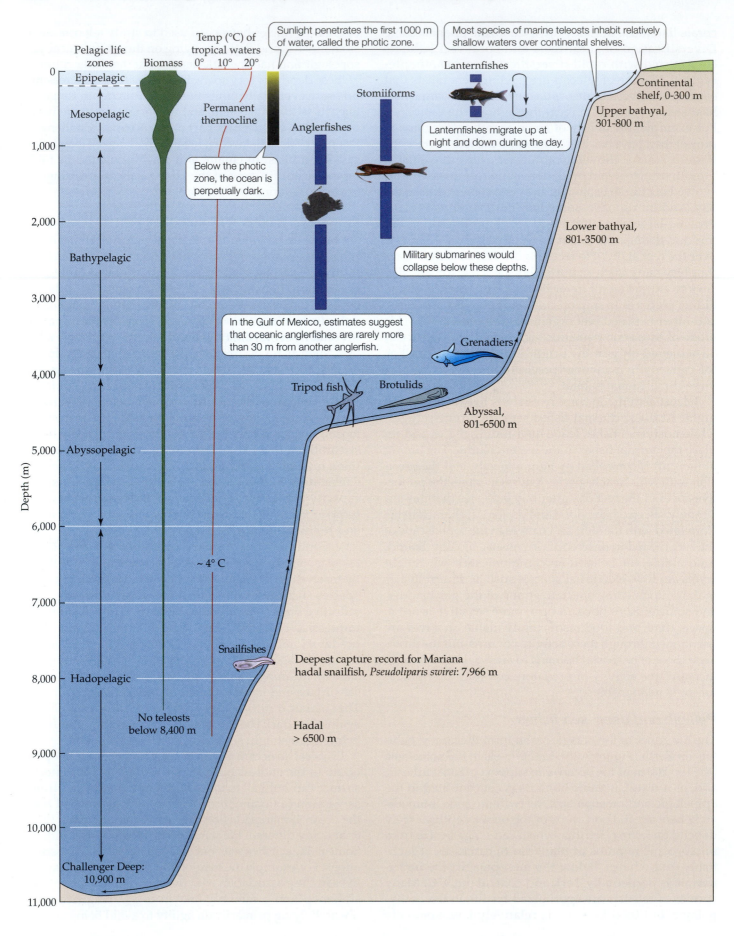

Pelagic life zones

Biomass

Temp (°C) of tropical waters
0°   10°   20°

Sunlight penetrates the first 1000 m of water, called the photic zone.

Most species of marine teleosts inhabit relatively shallow waters over continental shelves.

Lanternfishes

Epipelagic

Mesopelagic

Permanent thermocline

Anglerfishes

Stomiiforms

Continental shelf, 0-300 m

Upper bathyal, 301-800 m

Below the photic zone, the ocean is perpetually dark.

Lanternfishes migrate up at night and down during the day.

Bathypelagic

Lower bathyal, 801-3500 m

Military submarines would collapse below these depths.

In the Gulf of Mexico, estimates suggest that oceanic anglerfishes are rarely more than 30 m from another anglerfish.

Grenadiers

Tripod fish

Brotulids

Abyssal, 801-6500 m

Abyssopelagic

~ 4° C

Depth (m)

Snailfishes

Deepest capture record for Mariana hadal snailfish, *Pseudoliparis swirei*: 7,966 m

Hadal > 6500 m

Hadopelagic

No teleosts below 8,400 m

Challenger Deep: 10,900 m

◄ **Figure 7.38 Pelagic and deep sea life zones.** Zonation by depth characterizes pelagic and deep sea teleosts. Most photosynthesis occurs in the upper 200 m, known as the epipelagic zone, and most of the biomass (indicated by the width of the green "Biomass" bar) is above 1,000 m. Accordingly, most species of teleosts live within the top 1,000 m of water. Water temperature remains relatively constant below 1,000 m. Mesopelagic fishes such as lanternfishes (family Myctophidae) move up in the water column at night to feed where prey is more abundant and return to deeper water during the day (arrows). Deeper depth ranges occupied by adults of other types of teleosts, such as anglerfishes and stomiiforms, are shown with bars. Tripod fishes, brotulids, grenadiers, and snailfishes are deepwater bottom-dwellers that do not make extensive vertical movements once they metamorphose and settle.

the fish benefits from the shade and may eat portions of the non-stinging tentacles of the siphonophore.

Mesopelagic teleosts and daily vertical migrations

Mesopelagic teleosts are often much smaller and much slower than species specialized for life in the epipelagic, but they can be extremely abundant. For example, detection of a "false bottom" at depths between 300 and 500 m confused early sonar operators, particularly because the depth of the false bottom decreased at night. Sonar reflections of the false bottom turned out to be caused by gas bladders of countless small teleosts such as myctophids that move up and down in the water column. Such daily vertical migrations offer one way to cope with food scarcity in the depths: move toward the surface at dusk to enter areas with more food, and descend again near dawn. By rising at dusk, mesopelagic fishes enter a region of higher productivity, where food is more concentrated so they can gather more energy (see Figure 7.38). Shallower water is also warmer, so their metabolic rates increase, and they use energy faster. Daytime descent into cooler waters lowers metabolism and conserves energy.

Bathypelagic teleosts

For a bathypelagic teleost living at depths below 2,000 m, costs in energy and time to migrate to the surface outweigh the energy that would be gained from invading the rich surface waters. Instead of daily migrations, deep-sea fishes evolved specializations that allow adults to permanently live in the depths.

- *Large mouths and stomachs.* If a fish rarely encounters potential prey, then it should have: (1) a mouth large enough to engulf nearly anything it does meet; and (2) a gut that can expand to accommodate nearly any meal. The jaws and teeth of deep-sea fishes are often enormous in proportion to the rest of the body. Many bathypelagic fishes, such as the pelican eel, can be described as a large mouth accompanied by a stomach (**Figure 7.39A**). The unrelated dragonfishes evolved similar specializations, with large teeth, ability to swallow large prey intact, and specialized

biomechanical features such as a flexible neck joint that allows a huge gape for swallowing large prey (**7.39B–D**).

- *Light organs.* Many deep-sea fishes have light-producing photophores arranged on their bodies in species-specific and even sex-specific patterns. The light is produced by symbiotic species of *Photobacterium* and bacteria related to *Vibrio*. Some photophores, such as those of dragonfishes, probably attract prey (**Figure 7.39E,F**). Others may act as signals to potential mates in the darkness of the deep sea. These sources of light are dim and intermittent, however, and the sensitivity of the vertebrate visual system has been pushed to its extreme by bathypelagic fishes.

- *Sensory systems.* Because of the perpetual dark of their habitat, adults of many species of deep-sea fishes have reduced eyes. But many species of freshwater cave fishes that live in perpetual darkness lost eyes altogether, so why didn't eyelessness evolve in deep-sea fishes? The explanation is that planktivorous larvae of deeps-sea fishes typically live in the photic zone, much higher in the water column, where image-forming eyes are essential for catching prey. Thus, the utility of eyes in one life stage prevents their total evolutionary loss in a species. That said, the eyes of some deep-sea fishes are decidedly weird (**Figure 7.39G**). Other fishes that live on bottom of the deep sea rely primarily on touch and chemosensation to detect prey (**Figure 7.39H**).

The life history of ceratioid anglerfishes dramatizes some of the extreme adaptions for life in challenging deep-sea habitats. Adults typically spend their lives in lightless regions below 1,000 m. Fertilized eggs, however, rise to the surface, where they hatch into larvae. The larvae remain mostly in the upper 30 m, where they grow before descending to the lightless region. Descent is accompanied by metamorphic changes that differentiate females and males. During metamorphosis, young females descend to great depths, where they feed and grow slowly, reaching maturity after several years.

Female anglerfishes feed throughout their lives, whereas males feed only during the larval stage. Males have a different future—reproduction, literally by lifelong union. During metamorphosis, males cease eating and begin an extended period of swimming. The olfactory organs of males enlarge at metamorphosis, and the eyes continue to grow. These changes suggest that adolescent males spend a brief free-swimming period finding a female. The journey is precarious, for males must search vast, dark regions for a single female while running a gauntlet of other deep-sea predators.

Having found a female, a male does not want to lose her. He ensures a permanent union by attaching himself as a parasite to the female, biting into her flesh and attaching

(A) The pelican eel, *Eurypharynx pelecanoides*, has an enormous gape

NOAA OKEANOS Explorer Program, 2013 Northeast U.S. Canyons Expedition/CC BY 2.0

(B) Teeth of a viperfish, *Chauliodus sloani*

Prof. Francesco Costa/CC BY-SA 3.0

(C) Xray of a barbeled dragonfish, *Eustomias obscurus*, with a large myctophid in its stomach

© Nalani Schnell and G. David Johnson

(D) Flexible neck of *Grammatostomias dentatus*

Courtesy of Nalani Schnell

1 cm

(E) Light organs on branched barbel of a barbeled dragonfish, *Eustomias fissibarbis*

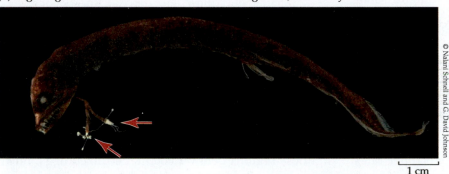

© Nalani Schnell and G. David Johnson

(F) Light organ on cheek of *Photostomias guernei*

NOAA/Ocean Explorer

1 cm

(G) Sleepy eye, *Ipnops murrayi*

Naked retinas

NOAA Okeanos Explorer Program, Gulf of Mexico 2012 Expedition

(H) Fins of the tripod fish, *Bathypterois grallator*, in lateral and anterior views

Pelvic fins   Caudal fin   Pectoral fins with sensory elements

NOAA Okeanos Explorer Program, INDEX-SATAL 2010, NOAA/OER

Figure 7.39 Specializations for life in the deep sea.   (A) The pelican eel can swallow prey larger than itself. (B) Viperfishes and related dragonfishes have large, fang-like teeth. (C) Dragonfishes can swallow large, whole prey. (D) A flexible neck joint (red arow) in a barbeled dragonfish allows the head to pivot upwards, extending its gape. (E) Dragonfishes use light organs on the barbel to attract potential prey. To avoid being detected, they have ultrablack skin that reflects less than 0.5% of light. Because dragonfishes eat luminous prey, they have black stomach walls to limit the amount of light that might otherwise shine through their body wall. (F) Light organs on cheeks of dragonfishes are used to see prey. Some species evolved red-light organs and red-sensitive retinas, allowing them to see prey without the prey detecting them (retinas of most deep sea fishes cannot detect red light). (G) The sleepy eye, *Ipnops murrayi*, has large, naked retinas (i.e., their eye lacks an iris or lens). This specialization increases sensitivity at the expense of directionality; *Ipnops* can determine only that a faint light source is to its left or right. (H) The tripod fish is demersal. It rests on its pelvic and caudal fins (left), while the long, flexible rays of its pectoral fins (right) are directed forward and sense the presence of its prey both by direct contact and by chemoreception.

Figure 7.40 Parasitic males. Males of many species of ceratioid anglerfishes are obligate parasites. Here a 6.2-mm male *Photocorynus spiniceps* has attached to the back of a 46-mm female.

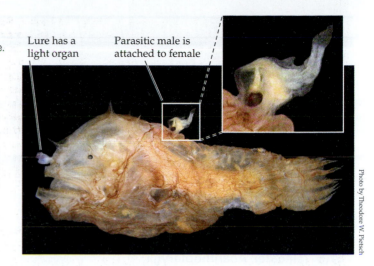

Lure has a light organ

Parasitic male is attached to female

Photo by Theodore W. Pietsch

so firmly that their circulatory systems fuse (**Figure 7.40**). Preparation for this encounter begins during metamorphosis, when the male's teeth degenerate and strong, toothlike bones develop at the tips of the jaws. A male remains attached to the female for life. In this parasitic state, he grows and his testes mature. Monogamy prevails in this pairing, as females usually have only one attached male.

Summary

7.1 Osteichthyes, Actinopterygii, and Sarcopterygii

The two groups of osteichthyans are Actinopterygii (ray-finned fishes) and Sarcopterygii (fleshy-finned fishes and tetrapods). All osteichthyans have an internal skeleton of ossified endochondral bone.

Several osteichthyan synapomorphies identifiable in basal actinopterygians relate to unique dermal bones in the head such as the opercle, branchiostegal rays, and premaxilla, maxilla, and dentary. Other synapomorphies include fin rays with lepidotrichia and ganoine on bones and scales.

Basal actinopterygians retain a complete dermal skull roof and have little upper jaw mobility; derived actinopterygians have a less-complete skull roof and mobile premaxillae and maxillae. Sarcopterygians have an intracranial joint between the anterior and posterior portions of the braincase.

Actinopterygians retain tribasic paired fins from the common ancestor of eugnathostomes; sarcopterygians evolved monobasic paired fins.

Actinopterygian forebrains develop by eversion. Sarcopterygian forebrains retain the ancestral state of development by inversion.

7.2 Actinopterygii: Basal Groups

Three lineages that diverged from the main stem of actinopterygians before the Jurassic are extant: Polypteriformes, Acipenseriformes, and Holostei.

Polypteriformes have paired lungs ventral to the gut and include bichirs and the reedfish. Acipenseriformes have a single dorsal lung or gas bladder, a mostly cartilaginous internal skeleton, and include sturgeon and paddlefishes.

Neopterygii includes extant Holostei and Teleostei. Neopterygians have an abbreviated heterocercal caudal fin, reduction of basals in paired fins, and a mobile maxilla.

Holostei includes gars (Lepisosteiformes), which have rhombic ganoid scales, and bowfins (Amiiformes), with a single extant species, *Amia calva*. Most actinopterygians belong to the more derived Teleostei.

7.3 Characters of Teleostei

Teleosts make up more than 99% of extant species of actinopterygians. The presence of a homocercal caudal fin is synapomorphic for Teleostei.

Scales of teleosts are circular, thin, and lack ganoine.

Teleosts use a gas bladder for buoyancy regulation and have flexible, mobile paired fins; these characteristics are associated with improved swimming performance.

Teleosts have mobile oral and pharyngeal jaws.

7.4 Teleostei: Basal Groups

Elopomorpha includes tarpons, bone fishes, and true eels. Leptocephalus larvae characterize the group.

Osteoglossomorpha (boney-tongue fishes) are freshwater fishes such as arowanas, pirarucu, and African elephant-nose fishes.

Otocephala includes two large extant lineages, Clupeomorpha (herrings, shads, sardines, and anchovies) and Ostariophysi (milkfish; minnows, carps, suckers; tetras and piranhas; catfishes; electric eels and South American knifefishes). An alarm substance (*Schreckstoff*) that is released from damaged skin and the Weberian apparatus used in hearing are synapomorphies of Ostariophysi, which account for about 80% of freshwater fishes.

Euteleostei began diversifying in the Cretaceous and includes more than 20,000 extant species.

The first branch of the euteleost tree leads to the sister groups Salmoniformes (salmon and trout) and Esociformes (pikes and pickerels).

(Continued)

Stomiiformes are deep-sea teleosts. They include dragonfishes and allies, including bristlemouths, whose vast numbers make them the most abundant vertebrates. Myctophiformes is another deep-sea group that includes lanternfishes, which have light-producing organs (photophores).

Gadiformes (cods) and Zeiformes (dories) are sister groups. Atlantic cod were overfished for decades, and despite regulatory efforts, populations off the northeast coast of North America have not rebounded.

Lampriformes (opah and ribbonfishes) include iconic marine pelagic fishes such as the oarfish.

7.5 Teleostei: Acanthopterygii

Acanthopterygii are the spiny-finned teleosts, named for the bony fin spines at the leading edges of their dorsal, anal, and pectoral fins.

Trachichthyiformes (flashlightfishes) are the basalmost lineage, known for the photophore below each eye. Holocentriformes (squirrelfishes) are nocturnal, sound-producing reef fishes. Batrachoidiformes (toadfishes and midshipmen) produce both light and sound.

Gobiiformes (gobies) typically inhabit shallow nearshore marine habitats.

Percomorpha includes all of the remaining acanthopterygians. Rapid radiations during the Paleogene complicate molecular phylogenetic interpretations of relationships among groups of percomorphs.

Scombriformes (mackerels and tunas), some of the fastest marine fishes, are the sister group of Syngnathiformes (seahorses), slow-moving fishes known for male egg brooding. Anabantiformes (gouramis) are popular aquarium fishes with a labyrinth organ.

Carangaria includes Toxotidae (archerfishes), Carangiformes (jacks), which are the sister group of Istiophoriformes (billfishes), pelagic predators whose bill is an extension of premaxillae. Sister to Carangiformes + Istiophoriformes is Pleuronectiformes (flatfishes), known for their background-matching abilities and unique metamorphosis.

Ovalentaria includes Cichliformes (cichlids; popular aquarium fishes and subjects of extensive study for their explosive radiations in lakes of eastern Africa), Beloniformes (flyingfishes; use expanded pectoral fins as airfoils), and Cyprinodontiformes (killifishes). Beloniformes and Cyprinodontiformes are sister groups. Reliable anatomical synapomorphies to support Ovalentaria have yet to be found.

Eupercaria includes Perciformes (perches and scorpionfishes), Labriformes (wrasses), and the sister groups Lophiiformes (anglerfishes, which have a spine and lure used to attract prey) and Tetraodontiformes (puffers). Puffers have a relatively short vertebral column, which limits bending for lateral undulations, which led to the evolution of new locomotor specializations and defense mechanisms.

7.6 Swimming and Hydrodynamics

Fishes swim using sequential anterior-to-posterior contractions of myomeres along one side of the body and simultaneous relaxation of corresponding myomeres on the opposite side. Lateral undulation generates forward thrust.

Based on portion of the body involved, locomotion can be categorized as anguilliform (eel-like; involving most of the body length), subcarangiform, carangiform, thunniform, and ostraciiform (limited to motions of the caudal fin).

In hydrodynamics, Reynolds numbers are dimensionless numbers that express the ratio of inertial forces to viscous forces for objects moving through a fluid. For fishes, the inertial force is the product of the body's length and speed, and the viscous force is the kinematic viscosity (stickiness) of the water the fish is moving through.

A fish always displaces more water than its body volume as it swims: the more water a swimming fish has to displace, the more energy it must use.

Drag is backward force opposing forward motion. Viscous drag is a function of the kinematic viscosity of water and is caused by friction between the fish's body and the surrounding water at the boundary layer. Pressure drag is caused by pressure differences resulting from the fish's displacement of water.

Fishes use fins to steer, stop, and remain in place.

7.7 Reproduction and Development

Most fishes are oviparous. Only 3% of teleosts are viviparous, and most of these are in the family Poeciliidae.

Freshwater teleosts typically produce and care for a relatively small number of large, yolk-rich, demersal eggs; parental care is possible.

Marine teleosts typically release large numbers of small, buoyant, transparent eggs that are left to develop and hatch while drifting in the open sea. A dramatic metamorphosis often separates the larval planktonic stage from the adult. Some marine fishes, such as grunions and blennies, deposit eggs in tidal terrestrial habitats.

In gonochoristic species, sperm and eggs are produced by separate individuals; in hermaphroditic species, sperm and eggs are produced by the same individual, either simultaneously or sequentially. In protandrous species, individuals start life as males and change to females. In protogynous species, individuals start life as females and change to males.

7.8 Ecology of Marine Teleosts

Blackwater diving allows photography and collection of fish larvae that can be deposited in natural history museums as extended specimens. Small DNA samples are sequenced (genetic barcoding) and compared with DNA from other species for identification.

Summary (continued)

Sunlight penetrates the first 1,000 m of water, called the photic zone. Within the photic zone are three fish-faunal zones associated with tropical coral reef ecosystems: the altiphotic, mesophotic, and rariphotic.

Pelagic life zones include (moving from the surface to the ocean floor) epipelagic (top 200 m), mesopelagic (twilight zone), bathypelagic, abyssopelagic, and hadopelagic (no vertebrates recorded by remotely operated vehicles). Most biomass is concentrated in the epipelagic and mesopelagic zones.

- Epipelagic teleosts rely on vision and have specializations to camouflage themselves (countershading) or confuse potential predators and reduce an individual's chance of being caught (schooling).

- Mesopelagic teleosts are often much smaller and slower than epipelagic species. Many make vertical migrations, moving upward at dusk to feed and downward at dawn to conserve energy.

- Bathypelagic teleosts have specializations to live permanently at depth, including large mouths, extensible stomachs, and light organs.

Discussion Questions

7.1 Why do modifications of neopterygian jaws represent an advance in feeding, and how do further modifications seen in the oral jaws of teleosts allow still more specialization?

7.2 In general, freshwater teleosts lay adhesive eggs that remain in one place until they hatch, whereas marine teleosts release free-floating eggs that are carried away by water currents. (There are exceptions to that generalization in both freshwater and marine species.) What can you infer about differences in the biology of newly hatched freshwater and marine teleosts on the basis of that difference in their reproduction?

7.3 Hermaphroditic teleosts exhibit patterns of sex change called protogyny and protandry. Describe the evolutionary principle that underlies those two patterns.

7.4 To synthesize your understanding of actinopterygian diversity, consider four levels of diversification that have occurred since the Devonian. Begin by explaining the initial separation of sarcopterygians and actinopterygians, highlighting key differences between them. Then explain key differences of neopterygians compared to more basal actinopterygians and tie these differences to the fossil record. Summarize the advances that characterize teleosts, and tie these to the fossil record. Finally, characterize acanthopterygians in terms of morphology, fossil record, habitats, and rates of diversification.

7.5 What do the pharyngeal jaws of bony fishes do? How is this different from the actions of the oral jaws (the premaxilla and maxilla)?

Figure credits
Figure 7.1: Based on multiple sources including K. Liem et al. 2001. *Functional Anatomy of the Vertebrates*, 3rd ed. Cengage/Harcourt College: Belmont, CA; J.S. Nelson, T.C. Grande and M.V.H. Wilson. 2016. *Fishes of the World*. 5th ed. John Wiley & Sons, Inc.: Hoboken, N.J.; M. Zhu et al. 2009. *Nature* 458: 469–474.
Figure 7.11: After K. Liem et al. 2001. *Functional Anatomy of the Vertebrates*, 3rd ed. Cengage/Harcourt College: Belmont, CA.
Figure 7.13: Based on multiple sources including K. Liem et al. 2001. *Functional Anatomy of the Vertebrates*, 3rd ed. Cengage/Harcourt College: Belmont, CA; J.S. Nelson, T.C. Grande and M.V.H. Wilson. 2016. *Fishes of the World*. 5th ed. John Wiley & Sons, Inc.: Hoboken, N.J.
Figure 7.30: B,C after K. Liem et al. 2001. *Functional Anatomy of the Vertebrates*, 3rd ed. Cengage/Harcourt College: Belmont, CA.
Figure 7.31: A,B after K. Liem et al. 2001. *Functional Anatomy of the Vertebrates*, 3rd ed. Cengage/Harcourt College: Belmont, CA.
Figure 7.32: After K. Liem et al. 2001. *Functional Anatomy of the Vertebrates*, 3rd ed. Cengage/Harcourt College: Belmont, CA.

© Michael Tuccinardi; Aqua Imports

8

Sarcopterygians and the Origin of Tetrapods

East African lungfish, *Protopterus amphibius*

Three lineages of sarcopterygians were present by the end of the Devonian: coelacanths and their relatives (Actinistia), lungfishes and their relatives (Dipnomorpha), and tetrapodomorphs (Tetrapodomorpha). All three groups are extant (**Figure 8.1**), although today there are only two species of coelacanths and six species of lungfishes, meaning that the vast majority—more than 99.9%—of extant sarcopterygians are tetrapods, most of which are terrestrial. Sometimes we use paraphyletic names such as "fishlike sarcopterygians" for those sarcopterygians that retain a gill cavity and lack legs. Such "fishlike sarcopterygians" reached their greatest diversity in the Paleozoic, primarily in the Devonian and Carboniferous periods, with much lower diversity in the Mesozoic and still lower in the Cenozoic.

In this chapter, we review sarcopterygian characters and groups and describe some of the fascinating fossil sarcopterygians that document transitions in organ systems associated with the origin of tetrapods and the move to land. Finally, we consider some possible advantages of terrestrial activity and the original environment in which the fish–tetrapod transition might have occurred.

8.1 Phylogenetic Concepts of Tetrapoda and Characters for Sarcopterygii

LEARNING OBJECTIVES

8.1.1 Explain the difference between an apomorphy-based definition of Tetrapoda and a crown-group definition of Tetrapoda.

8.1.2 Identify the extant sister group of tetrapodomorphs.

8.1.3 Describe three diagnostic characters of sarcopterygians.

It is important to distinguish clearly between two different concepts of the word Tetrapoda. The first is an older concept: any sarcopterygian with four feet is a tetrapod. It makes sense because that is what the word tetrapod means (Greek *tetra*, "four"; *pod* "feet"). This is an apomorphy-based definition, with the apomorphy being the presence of four feet. But the name Tetrapoda has also been used in a more restricted sense, limited to the clade including extant lissamphibians (caecilians, frogs, and salamanders), amniotes (reptiles, including crocodylians and birds, and mammals), and their most recent common ancestor (MRCA), an animal that might have lived as long ago as the Early Carboniferous. This is a crown-group definition of Tetrapoda, favored by some researchers. In this book we adopt elements of both naming conventions, using the word Tetrapoda to include all sarcopterygians with four legs but using the term Crown Tetrapoda for the clade that includes lissamphibians, amniotes, and their MRCA.

Phylogenetic relationships of coelacanths, lungfishes, and tetrapods were the subject of considerable debate in the 20th century. Both extinct and extant lungfishes have highly specialized skeletons that can be difficult to interpret because they are so different from those of other vertebrates. These and other peculiar features of lungfishes led many 20th-century workers to group coelacanths with tetrapodomorphs based on superficial similarities while placing lungfishes on very different branches of

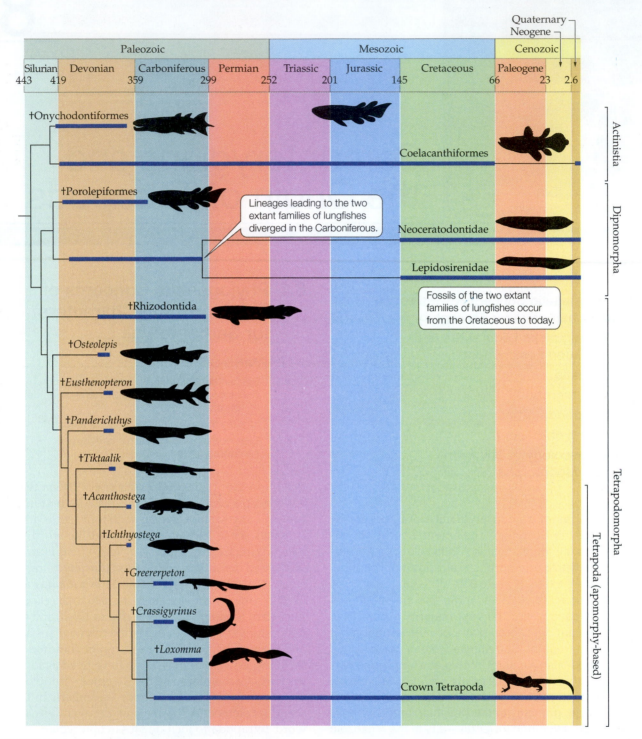

Figure 8.1 Time tree of sarcopterygians. Three clades of sarcopterygians are extant, with more than 99% of the species in Tetrapoda ("four feet"). Molecular phylogenetic studies confirm coelacanths as the most basal extant sarcopterygians, with lungfishes as the extant sister group of tetrapodomorphs. (See credit details at the end of this chapter.)

the vertebrate tree. However, molecular phylogenetic studies confirm coelacanths as the most basal extant sarcopterygians, with lungfishes as the extant sister group of tetrapodomorphs (**Figure 8.2**).

Early and Middle Devonian sarcopterygians were cylindrical fishes ranging from 20 cm to more than 4 m long. They had thick bony scales, two dorsal fins, a heterocercal caudal fin, and share three characters that are diagnostic of sarcopterygians:

1. Presence of the hard tissue **cosmine** on dermal bones and scales (see Figure 7.4D)

2. **Monobasic paired fins** with a scaled, muscular lobe at the base (see Figure 7.5C,D)

3. An **intracranial joint** between the anterior and posterior portions of the braincase

Sarcopterygians from the Devonian and Carboniferous play starring roles in understanding the evolutionary

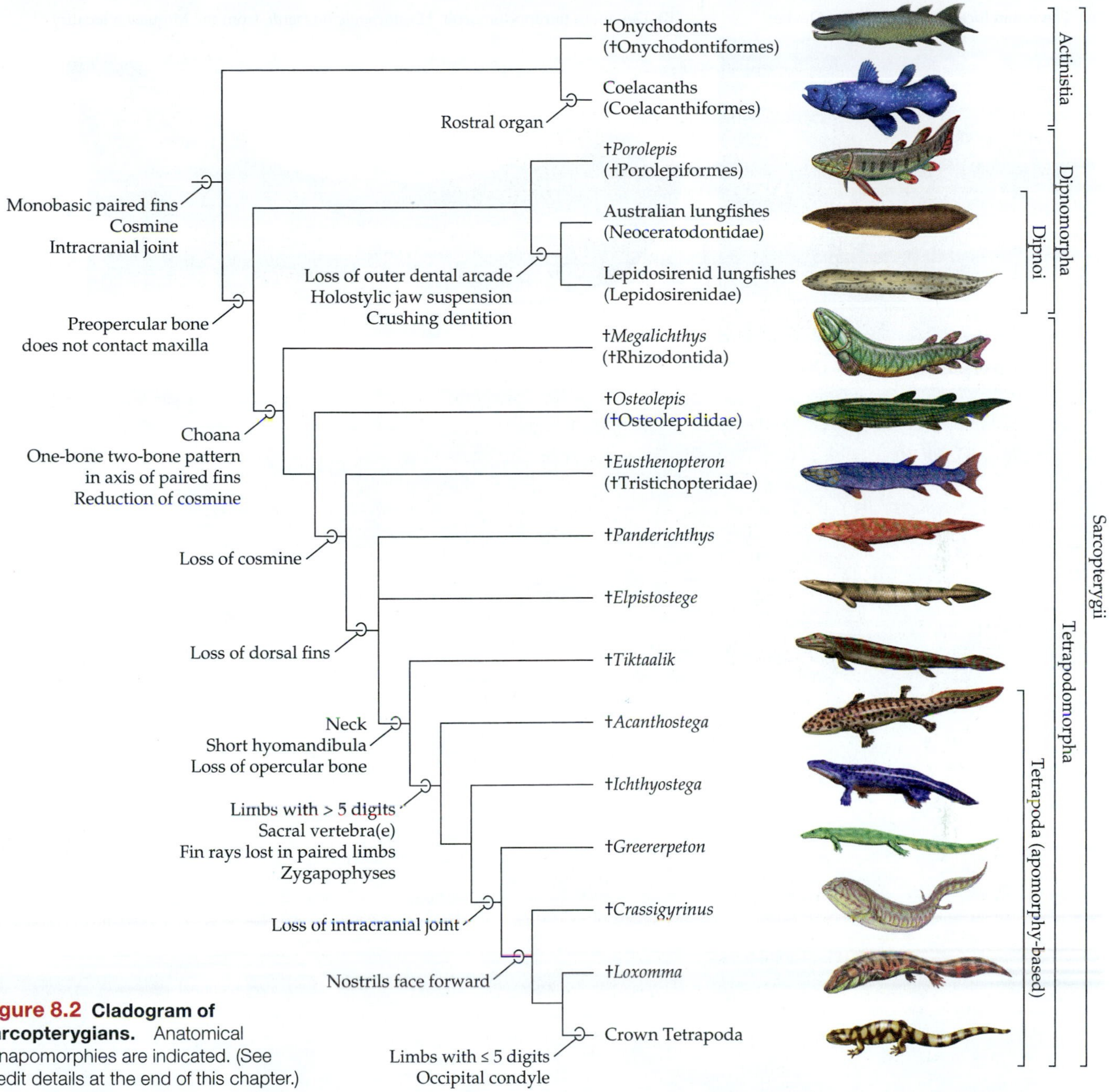
†Onychodonts
(†Onychodontiformes)
Coelacanths
(Coelacanthiformes)
Rostral organ
†*Porolepis*
(†Porolepiformes)
Australian lungfishes
(Neoceratodontidae)
Lepidosirenid lungfishes
(Lepidosirenidae)
Monobasic paired fins
Cosmine
Intracranial joint
Loss of outer dental arcade
Holostylic jaw suspension
Crushing dentition
Preopercular bone
does not contact maxilla
†*Megalichthys*
(†Rhizodontida)
†*Osteolepis*
(†Osteolepididae)
Choana
One-bone two-bone pattern
in axis of paired fins
Reduction of cosmine
†*Eusthenopteron*
(†Tristichopteridae)
Loss of cosmine
†*Panderichthys*
†*Elpistostege*
Loss of dorsal fins
†*Tiktaalik*
Neck
Short hyomandibula
Loss of opercular bone
†*Acanthostega*
Limbs with > 5 digits
Sacral vertebra(e)
Fin rays lost in paired limbs
Zygapophyses
†*Ichthyostega*
†*Greererpeton*
Loss of intracranial joint
†*Crassigyrinus*
Nostrils face forward
†*Loxomma*
Crown Tetrapoda
Limbs with ≤ 5 digits
Occipital condyle
Actinistia
Dipnomorpha
Dipnoi
Sarcopterygii
Tetrapodomorpha
Tetrapoda (apomorphy-based)

Figure 8.2 Cladogram of sarcopterygians. Anatomical synapomorphies are indicated. (See credit details at the end of this chapter.)

origin of tetrapods. The fossils document many of the steps that occurred, including progressive modifications of paired fins, losses and modifications of bones, changes in sensory systems, loss of gills, etc. Their story is exciting because it is not over, and never will be. Insights from newly discovered fossils, new tools for reexamining fossils collected long ago, new approaches in paleoecology, methods in molecular developmental biology, and rigorous application of phylogenetic methodologies (e.g., adding more taxa and more characters to analyses) continually change our picture of how this transition happened. To place the larger story of sarcopterygians and tetrapod origins in context, we start in Miguasha, Quebec, one of the most important localities for Devonian vertebrates yet discovered.

8.2 The Miguasha Lagerstätte and the "Good Fossil Effect"

LEARNING OBJECTIVES

8.2.1 Describe the environmental conditions at Miguasha during the Upper Devonian and explain why these conditions promoted fossilization.

8.2.2 Explain how fossils from Miguasha have influenced core ideas about Devonian vertebrates.

Miguasha National Park is on Canada's Gaspe Peninsula, an 8-hour drive east of Montreal. On the north side of Escuminac Bay, its rocky cliffs extend to the water's edge (**Figure 8.3A**). From those cliffs come many noteworthy fossils, including †antiarchs (e.g., †*Bothriolepis*; see Figure 3.15A), early actinopterygians such as †*Cheirolepis*

(A) Devonian locality at Miguasha, Quebec

(B) Devonian tetrapodomorph, †*Eusthenopteron foordi*, from the Miguasha locality

(C) Paleomap of Euramerica in the Devonian, 375 Ma

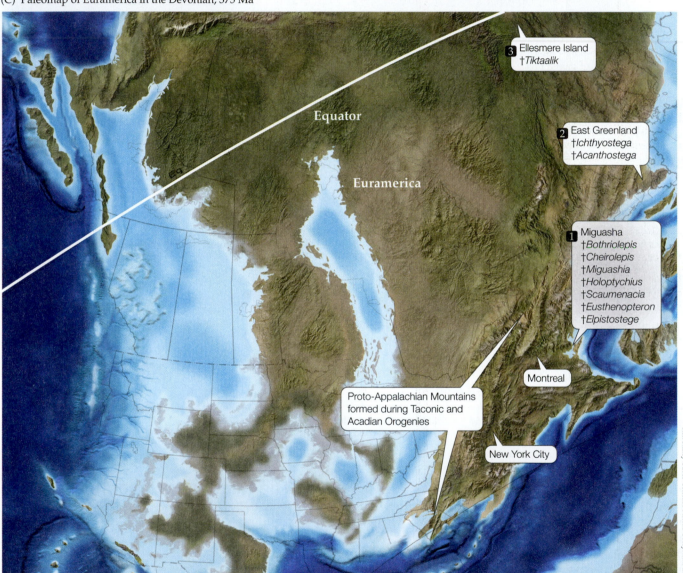

Figure 8.3 Miguasha and other Devonian localities.
(A) The cliff at Miguasha, Quebec, is a Lagerstätte that has yielded some of the most important Devonian fossils. (B) A Devonian tetrapodomorph, †*Eusthenopteron foordi*, from Miguasha. Note the diphycercal caudal fin with three lobes. (C) Paleomap of Euramerica in the Devonian, 375 Ma, indicating approximate locations of three major localities where Devonian tetrapodomorph fossils have been recovered. A series of mountain-building events (orogenies) formed the massive Proto-Appalachian mountain range. Erosion of these mountains produced sediments carried by rivers draining the range, leading to a deltaic environment in which many different organisms were buried and exquisitely preserved as fossils. Although today these localities occur high in Northern Hemisphere latitudes, during the Devonian they were close to the Equator.

(see Figure 7.3B), the oldest well preserved coelacanths, early lungfishes, and one of the best known early tetrapodomorphs, †*Eusthenopteron foordi* (**Figure 8.3B**). There are even well-preserved larval and juvenile specimens of some of these taxa. Declared a Canadian National Park in 1985 and a UNESCO World Heritage Site in 1999, Miguasha preserves a slice in time from the Upper Devonian, about 370 Ma, when the world was very different. At that time, the paleocontinent of Euramerica lay on the Equator (**Figure 8.3C**), so much of the early evolution of sarcopterygians occurred in warm tropical environments. Giant mountains as high as the Himalayas loomed down the eastern spine of North America. Rivers carried enormous volumes of sediments to the oceans, where they were deposited in massive deltas on scales like those of the Mississippi or Ganges river deltas today, rapidly burying dead animals and plants. Rapid burial and high biodiversity in the paleoenvironment of the Miguasha delta meant that many different types of organisms were preserved as fossils, making it among the most famous Lagerstätten (German, "storage places") in the world, preserving exquisitely detailed fossils that document paleodiversity (see Section 6.1).

When Europeans first uncovered Miguasha fossils in the 1840s, the site influenced early ideas about the Devonian world and its organisms. Subsequently, painstaking anatomical studies of fossil sarcopterygians from Miguasha, especially †*Eusthenopteron foordi*, revealed their evolutionary significance for the vertebrate story.

Many other fossils help to tell this story. For example, explorations of localities in East Greenland (location 2 in Figure 8.3C) during the 1930s recovered fossil tetrapods from rocks some 10 million years younger than those at Miguasha. New fossils further documenting the transition of vertebrates to land were discovered in Latvia, and more specimens of key fossil sarcopterygians were collected at other localities in the United States, Australia, and China. Methods for both study and analyses improved. Then in 2006, fossils of a new genus—†*Tiktaalik*—were described from a site on Ellesmere Island (close to the Devonian paleoequator; location 3 in Figure 8.3C). The discovery of †*Tiktaalik* prompted new interpretations and reexaminations of sarcopterygian fossils.

This human historical context for paleontological discoveries is important because each new fossil, particularly if it is a well-preserved specimen, can move the story in a new direction. As in the case of acanthodians in the "Wonder Block" (see Figure 6.3), these sarcopterygian discoveries exemplify the "good fossil effect" because they drove new interpretations. As recently as 2020, a nearly complete fossil of †*Elpistostege* from Miguasha was described, and it has changed ideas about the fin-to-limb transition. Countless new fossils wait to be discovered, making our evolving understanding of the origin of tetrapods among the most exciting stories in vertebrate biology.

8.3 Actinistia

LEARNING OBJECTIVES

8.3.1 Provide a possible explanation for the absence of coelacanths from the Cenozoic fossil record.

8.3.2 Describe key external and internal anatomical features of coelacanths.

Actinistia includes the extinct Devonian group †Onychodontia and a fascinating extant lineage that has persisted since the Devonian, Coelacanthiformes.

†*Onychodontia*

†Onychodonts (Greek *onycho*, "claw"; *odus* "tooth"), sometimes known as dagger-toothed fishes, were large, mostly marine predators. Known from the Devonian of North America, China, Germany, and Australia, †onychodonts had a spiky tooth whorl in the symphysis of the lower jaw (the junction where the right and left halves of the jaw meet in the midline; **Figure 8.4A**). They also had a secondarily symmetrical diphycercal caudal fin in which both the skeleton and external shape of the upper and lower lobes were similar (**Figure 8.4B**). †Onychodonts

(A) Symphysial tooth whorl of a Devonian †onychodont

1 cm

© William E. Bemis, courtesy of Museum of the Earth

(B) Reconstruction of Devonian †onychodont, †*Qingmenodus yui*

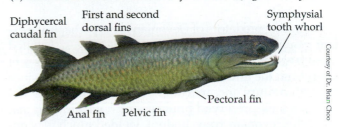

Diphycercal caudal fin | First and second dorsal fins | Symphysial tooth whorl
Pectoral fin
Anal fin | Pelvic fin

Courtesy of Dr. Brian Choo

Figure 8.4 †Onychodonts. †Onychodonts, sometimes called "dagger-toothed fishes," were marine predators of the Devonian. (A) Symphysial tooth whorl of a Middle Devonian †onychodont. These whorls of spiked teeth occurred at the junction (symphysis) of the right and left halves of the lower jaw. (B) †*Qingmenodus yui*, a Devonian †onychodont from Yunnan, China, with a diphycercal caudal fin and a symphysial tooth whorl. In life the animal was about 20 cm long.

(A) External anatomical features of *Latimeria chalumnae*

(B) Intracranial joint in early fetal coelacanth

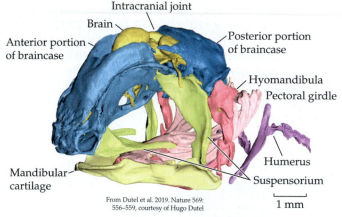

From Dutel et al. 2019. Nature 569: 556–559, courtesy of Hugo Dutel

1 mm

(C) Intracranial joint and basicranial muscle in adult coelacanth

© William E. Bemis

Figure 8.5 **Extant coelacanths.** (A) The better known of the two extant coelacanth species, *Latimeria chalumnae*, is from the West Indian Ocean, where it is known locally as gombessa. (B) Micro-CT reconstruction of skull of a fetal coelacanth shows the development of the intracranial joint, a sarcopterygian character which today is retained only in coelacanths. (C) Contraction of the basicranial muscle spanning the intracranial joint of an adult *Latimeria* depresses the anterior portion of the braincase. The joint is thought to facilitate suction feeding and the generation of bite force. (A,C after K. Liem et al. 2001. *Functional Anatomy of the Vertebrates*, 3rd ed. Cengage/Harcourt College: Belmont, CA.)

coelacanths were extinct. The 1938 discovery of an extant coelacanth, saved for science by South African naturalist and museum curator Marjorie Courtenay-Latimer and subsequently named *Latimeria chalumnae* in her honor, set in motion more than 80 years of research and re-interpretation of coelacanths and their phylogenetic affinities. Adult coelacanths are large (up to 2 m long and weighing 90 kg), and steel blue-gray in color with irregular white or pinkish spots (**Figure 8.5A**). The discovery of *Latimeria* not only made coelacanths among the most widely known fishes, but also inspired poetry, supplied stories for popular books and television programs, and even entered slang (as in "That politician is a coelacanth," one who clings to tired old ideas long after everyone else has moved on).

Since the 1938 discovery of *Latimeria chalumnae*, hundreds of specimens have been caught or observed in waters off South Africa, the Comoros archipelago (where its local name is gombessa), Madagascar, Mozambique, Tanzania, and Kenya. Working at night, fishermen in the Comoros catch coelacanths during summer (December–March in the Southern Hemisphere), usually in 260–300 m of water, near the ocean floor. The islands of the Comoros are volcanic and the bottom drops off with a steep (~45°) slope, making it possible to fish at such depths while remaining within easy distance from and in sight of shore. In Sodwana Bay, South Africa, scuba divers have documented living coelacanths in their natural habitat.

In 1998, a second species of coelacanth, *Latimeria menadoensis*, was discovered in Indonesian waters. DNA sequencing shows that it separated from the East African *L. chalumnae* ~6 Ma. Young individuals of *L. menadoensis* have been observed in their reef habitats by researchers

were lurking predators of Devonian reefs, perhaps like extant moray eels as some paleontologists suggest. Isolated fossil jaws and tooth whorls of large †*Onychodus* (3–4 m long) from the middle Devonian have been known since the 19th century and are common in some localities. Recently, a well-preserved braincase of †*Qingmenodus yui*, an Early Devonian †onychodont, yielded much new anatomical information and confirmed placement of †onychodonts with coelacanths as actinistians.

Coelacanthiformes

Coelacanths, sometimes called "tufted tails," have a long and excellent fossil record beginning in the Devonian. Curiously, their fossil record stops in the Cretaceous, and for nearly a century paleontologists thought that

using remotely operated vehicles. To date, no coelacanths have been held in captivity because none have survived the rapid pressure changes when they are brought up from the depths.

Coelacanth characters   Figure 8.5A shows some key external anatomical features of *Latimeria*, including three paired openings of the rostral organ, a specialized electroreceptive organ in the snout uniquely found in coelacanths; paired fins and second dorsal and anal fins with muscular, lobed bases (note that the first dorsal fin does not have a lobed base); and a diphycercal caudal fin with a terminal tuft.

Latimeria resemble chondrichthyans in retaining urea to maintain a blood osmolality close to that of seawater. This osmoregulatory physiology may be an ancestral gnathostome condition that has been retained in coelacanths, because coelacanths, like chondrichthyans, have a rectal gland that excretes salt. Coelacanths are viviparous and give birth to large, fully formed pups. Internal fertilization must occur, but how copulation happens is unknown because males lack an intromittent organ.

Recent micro-CT studies of fetal coelacanths are revealing new anatomical details, including development of the intracranial joint (**Figure 8.5B**). Coelacanths are the only extant vertebrates with an intracranial joint between the anterior and posterior portions of the braincase (a sarcopterygian character that was lost in tetrapods; see Figure 8.2). This joint is moved by large, paired basicranial muscles innervated by the same cranial nerve as an eye muscle (the lateral rectus muscle, which pulls your eye outward, away from your nose) and is thought to increase capacity for suction feeding and allow generation of large bite forces (**Figure 8.5C**).

The fossil record   Coelacanths continue to fascinate us partly because their external appearance has seemingly changed so little over the last 350 million years, leading to the catchy but somewhat misleading term "living fossil." The term is misleading because, despite anatomical similarities, the biology of early coelacanths may have been completely different from that of living *Latimeria* and we have no way to evaluate this, although genomic studies do show that its protein-coding genes have evolved more slowly than do those of tetrapods.

The oldest well-preserved coelacanths are in the genus †*Miguashia* (**Figure 8.6A**), which had a heterocercal caudal fin (recall that a heterocercal caudal fin is common to †osteostracans + gnathostomes and was retained by the common ancestor of osteichthyans; see Figure 3.4). Carboniferous coelacanths included some peculiar deep-bodied forms with reduced paired fins and a highly modified caudal fin, such as †*Allenypterus*, but most post-Devonian coelacanths have, like *Latimeria*, a diphycercal caudal fin bearing the distinctive tuft at its tip. Some Paleozoic coelacanths lived in shallow fresh waters, but fossil coelacanths from the Mesozoic were largely

(A) Reconstruction of †*Miguashia* with heterocercal caudal fin

DiBgd/CC BY-SA 4.0

(B) †*Mawsonia gigantea*, a giant Cretaceous coelacanth with a tufted, diphycercal caudal fin

Artwork by Joschua Knüppe

Figure 8.6 Fossil coelacanths.   (A) Reconstruction of a Devonian coelacanth, †*Miguashia*, showing heterocercal caudal fin. (B) During the Cretaceous, coelacanths in the genus †*Mawsonia* reached lengths greater than 4 m.

marine, including some gigantic Cretaceous †*Mawsonia* that reached lengths greater than 6 m (**Figure 8.6B**).

Living and fossil coelacanths share many features. In addition to the characteristic hollow fin spines that give the group its name (Greek *koilos*, "hollow"; *akantha*, "spine") and the unique rostral organ, there are skeletal changes such as loss of the maxilla and branchiostegal rays. The lung of extant coelacanths is vestigial, although evidence exists for larger lungs in some fossils, surrounded by calcified plates that perhaps aided ventilation. The fat-filled lung of *Latimeria* serves as a buoyancy organ. Although it has no function in gas exchange, it retains a pulmonary circulation like that of other basal osteichthyans.

Habitat   Extant coelacanths live at depths ranging from ~65 to several hundred meters (deep for humans, but not that deep in the scale of the oceans; see Figure 7.38). These habits may explain the absence of coelacanths from the Cenozoic fossil record, because such marine habitats rarely yield well-preserved fossils. *Latimeria chalumnae* is nocturnal, hiding in underwater lava caves during the day and venturing out at night to dine on small fishes, squids, and octopuses. They do not use their paired fins as props or to walk on the bottom (as once thought), but swim by moving the pectoral and pelvic fins in diagonal pairs, just as tetrapods move their limbs on land.

The peculiar distribution of the two known species of *Latimeria* on opposite sides of the Indian Ocean might reflect poor sampling. It would be difficult and prohibitively expensive to mount a "fishing expedition" to explore all the potential habitats where coelacanths might occur, which could include sea mounts, ridges, and other underwater features across vast areas of ocean. Traditional methods for surveying fishes rely on trawling, which is poorly suited for the types of habitats where coelacanths are known to occur. A case in point is Tanzania, where coelacanths were unknown until 2003, when one was caught using a new type of fishing gear; 22 more individuals were caught in Tanzanian waters within the year. It might be possible to design a new survey of *Latimeria chalumnae* using environmental DNA (eDNA) to sample waters near islands that might provide suitable habitats east of the Comoros (see Section 6.6 for an example of eDNA used to detect white sharks).

Both extant species of *Latimeria* are listed in CITES Appendix 1 (species threatened with extinction; international trade permitted only in exceptional circumstances). The IUCN Red List includes *L. chalumnae* as Critically Endangered and *L. menadoensis* as Vulnerable.

8.4 Dipnomorpha

LEARNING OBJECTIVES

8.4.1 List three synapomorphies of lungfishes.

8.4.2 Describe differences in morphology between extant and Devonian lungfishes and the role of paedomorphosis in generating these differences.

8.4.3 Compare lungs, paired fins, and tooth plates of South American and African lungfishes with those of the Australian lungfish.

Dipnomorpha contains the extinct †Porolepiformes; and Dipnoi, the extant clade that includes six living species of lungfishes (see Figures 8.1 and 8.2).

†Porolepiformes

Once thought to be closely related to tetrapodomorphs, the presence of a long series of mesomeres in the axis of the paired fins flanked by radials on both sides (see Figure 7.5C) as well as cranial features show that †porolepiforms are more closely related to lungfishes. †*Porolepis* is known from Early Devonian rocks in Norway (**Figure 8.7A**). Its name derives from the presence of large pores in the cosmine-coated surfaces of the scales (Greek *poros*, "pore"; *lepis*, "scale"). The function of these pores and the pore–canal system that connects them is still unknown. They did not house electroreceptors, at least not electroreceptors like those of extant vertebrates, which are not closely associated with dermal mineralization even in forms such as paddlefishes (which have more electroreceptors in the soft tissue between stellate bones of the paddle than any other living vertebrate; see Section 7.2). Because of

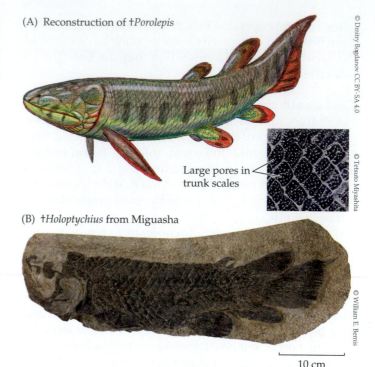

(A) Reconstruction of †*Porolepis*

Large pores in trunk scales

(B) †*Holoptychius* from Miguasha

10 cm

Figure 8.7 †Porolepiforms. (A) Reconstruction of Early Devonian †*Porolepis* from marine deposits in Spitsbergen; it was about 1.5 m long. (B) Cast of a derived †porolepiform, †*Holoptychius jarviki* from Miguasha.

their large teeth, we interpret these fishes as predators of Devonian reefs. They co-occur with other sarcopterygians in many fossil localities, including Miguasha and Ellesmere Island. Other genera of †Porolepiformes include †*Holoptychius* (**Figure 8.7B**), known from whole-body fossils from Miguasha. A dorsoventrally flatted form, †*Laccognathus*, from Ellesmere Island and Latvia, was a benthic sit-and-wait predator.

Dipnoi

The three extant genera of lungfishes—*Neoceratodus* from Australia, *Lepidosiren* from South America, and *Protopterus* from Africa—belong to a group that has persisted for more than 400 million years. Dipnoans (Greek *dis*, "two"; *pneuma*, "breathing"; i.e., they use both gills and lungs) share many synapomorphies, including reduction and loss of the outer dental arcade (the tooth-bearing premaxilla, maxilla, and dentary bones) and fusion of the palatoquadrate to the cranium, a pattern convergently like the holostylic jaw suspension of holocephalans (see Figure 6.8B).

Fossil lungfishes   †*Diabolepis* from the Early Devonian of China is the oldest known dipnoan, and it is a transitional form. It retained remnants of the intracranial joint synapomorphic for sarcopterygians as a group, but lost in more derived dipnoans. †*Diabolepis* had large, rounded teeth on its palate and lower jaw used to crush hard foods.

(A) Dorsal view of olfactory system of lungfish, *Lepidosiren paradoxa*

(B) Lateral view of olfactory system of lungfish, *Lepidosiren paradoxa*

(C) Evolution of nostrils of sarcopterygians

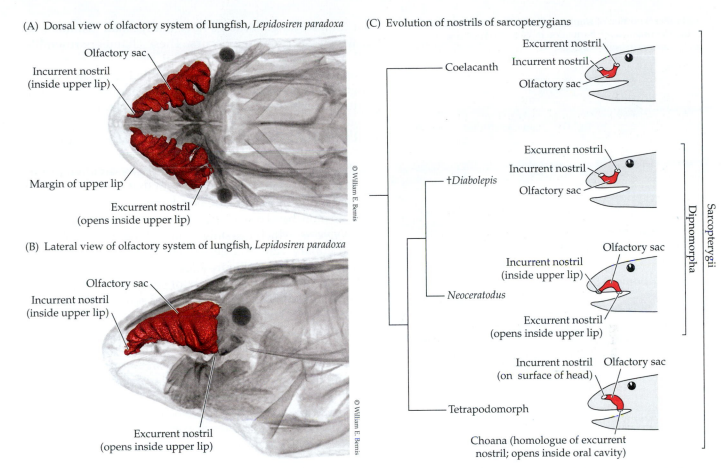

Figure 8.8 Nostrils of Sarcopterygians. (A) Dorsal view of olfactory system of the South American lungfish, *Lepidosiren paradoxa*. The incurrent nostril opens inside the upper lip. The large olfactory sac indicates that olfaction is an important sensory modality for lungfishes. (B) Lateral view showing positions of incurrent and excurrent nostrils inside the upper lip. (C) Evolution of nostrils of sarcopterygians. The opening of the excurrent nostril inside the upper lip of derived lungfishes evolved independently from the choana of tetrapodomorphs. (C after K. Liem et al. 2001. *Functional Anatomy of the Vertebrates*, 3rd ed. Cengage/Harcourt College: Belmont, CA.)

Such **durophagy** (Greek *duros*, "hard"; *phagous*, "eating") has persisted throughout lungfish evolution.

As in other basal gnathostomes, †*Diabolepis* has four openings for its olfactory system—one pair of incurrent nostrils and one pair of excurrent nostrils—on the outside of its snout. This pattern was lost in more derived lungfishes such as *Lepidosiren*, in which the excurrent nostrils open inside the upper lip and are not visible externally (**Figure 8.8A,B**). This means that the position of the excurrent nostril of derived lungfishes evolved independently from the choana of tetrapodomorphs (**Figure 8.8C**).

Well-known Devonian lungfishes include †*Chirodipterus* and †*Dipterus* (**Figure 8.9A,B**). Dentitions of Devonian lungfishes range from simple denticals, to thickened layers of dental tissues that allow the entire palate to function as a single crushing surface, to fan-shaped tooth plates like those of †*Scaumenacia* (**Figure 8.9C**). All post-Devonian lungfishes have generally similar tooth plates, which grow by the fusion of new denticles into the margins of the tooth plates and by the addition of new mineralized tissue from below.

Devonian lungfishes such as †*Dipterus* had two dorsal fins, a heterocercal caudal fin, and an anal fin. (Recall that this is the pattern of median fins typical of basal gnathostomes that sarcopterygians retained.) By the Carboniferous, lungfishes had evolved a distinct body form in which a single continuous fin extends around the posterior third of the body. The caudal fin changed from heterocercal to a symmetrical diphycercal fin, an evolutionary change that occurred convergently within Actinistia, Dipnomorpha, and Tetrapodomorpha. The mosaic of small dermal bones in the skull roof of Devonian dipnoans changed to a pattern of several large elements. Cosmine on the scales and dermal bones was lost and much of the bony internal skeleton typical of Devonian lungfishes remains cartilaginous and never ossifies in extant dipnoans. Such dramatic and correlated changes in morphology appear to be the result of paedomorphosis (Greek *pais*, "child"; *morphe*, "form"), the retention of juvenile characters in adults. For example, early in the development of bony fishes such as *Amia calva*, the initially continuous median fin becomes subdivided into separate dorsal, caudal, and

(A) Skull and portion of trunk of †*Chirodipterus australis*, from the Devonian Gogo Formation in Western Australia

Photo by William E. Bemis, courtesy of The Natural History Museum

(B) Reconstruction of a Devonian lungfish, †*Dipterus*

© Corey A. Ford/123RF

(C) Tooth plates of Devonian lungfish, †*Scaumenacia curta*

© Richard Cloutier

Figure 8.9 Fossil lungfishes. (A) Head and portion of trunk of †*Chirodipterus australis* from the Devonian Gogo formation in Western Australia. (B) Reconstruction of a Devonian lungfish of the genus †*Dipterus*. (C) Palatal view of tooth plates of †*Scaumenacia curta* from Miguasha.

anal fins. Extant dipnoans truncate that and other developmental patterns typical of basal bony fishes.

Many early lungfishes are from marine deposits such as the Gogo Formation of Western Australia, which preserves a Devonian reef environment. Later fossil forms are from freshwater deposits, and all six extant lungfish species live in freshwater. Dipnoans diversified during the Paleozoic. The two lineages that include the extant lungfishes diverged in the Carboniferous.

Structural characters Extant lungfishes share some soft anatomical and physiological features with tetrapods. For example, lungfish lungs have many small chambers in which gas exchange occurs. These chambers resemble but are much larger than typical alveoli of a mammalian lung (the small, grapelike, air-filled structures where gas exchange occurs). Like other vertebrates with lungs, lungfishes secrete detergent-like surfactants onto the gas exchange surface, which reduce surface tension at the interface between gas and body fluids and prevent the chamber from collapsing. Other features lungfishes share with tetrapods include a partially divided atrium in the heart and a pulmonary vein that delivers oxygenated blood to its left side (see Figure 7.6C). This differs from more basal osteichthyans such as the actinopterygian *Polypterus*, which has only a single atrial chamber and a pulmonary vein that delivers oxygenated blood to the sinus venosus (see Figure 7.6A).

Lungfishes possess a mix of sensory systems, some associated with living in water and some with living on land. Typical of fishes, lungfishes have good senses of smell and taste: they have large olfactory organs (see Figure 8.8A,B) and many taste buds in the oral cavity. In addition to the lateral line system on the head and trunk that detects water currents, many ampullary electroreceptors are concentrated on the snout, which may be used to detect prey hidden from view (see Section 4.2). Experimental evidence indicates that lungfishes can detect underwater sounds transduced by air in the lungs, as well as airborne sounds via vibrations reaching the head. Lungfishes also appear to possess a vomeronasal system, an accessory system distinct from the main olfactory system that is found in many terrestrial vertebrates (see Section 11.6).

The Australian lungfish *Neoceratodus forsteri* reaches lengths of 1.5 m and weights of 45 kg (**Figure 8.10A**). They can have long lifespans; one specimen lived at Chicago's Shedd Aquarium for 80 years. They are omnivorous, consuming plant material as well as invertebrates, which they crush with fan-shaped tooth plates. As in Paleozoic dipnoans, *Neoceratodus* retains large, paired fins with muscular bases and radial elements on both sides of the mesomeric axis of the fin (see Figure 7.5C). It uses well-developed gills for gas exchange, and although it can use its single lung to obtain oxygen, it does not depend on air-breathing.

The genome of *Neoceratodus forsteri* is among the largest known for any animal—about 43 Gb (gigabases) in a recent genomic study—and it is by far the largest genome yet sequenced. (*Lepidosiren* and *Protopterus* also have very large amounts of DNA per cell, but have yet to be studied from a genomic perspective.) More than 67% of the *Neoceratodus* genome is composed of repetitive transposable elements, which is the largest percentage of noncoding sequences known for any genome. Because cell size correlates with genome size, *Neoceratodus* has very large cells. Intriguingly, based on histological sections, cell sizes in bones of Devonian lungfishes were much smaller than those of extant lungfishes, suggesting that these extinct species had correspondingly smaller genomes.

(A) Australian lungfish, *Neoceratodus forsteri*

(C) Monitoring lungfish in Brisbane River

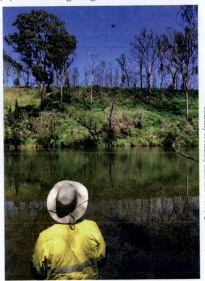

(B) Larval Australian lungfish, *Neoceratodus forsteri*

Courtesy of Helen Sanders

© William E. Bemis

Courtesy of Healthy Land and Water, Queensland, Australia

Figure 8.10 Australian lungfish. *Neoceratodus* is one of three extant genera of lungfishes (the other two are found in Africa and South America, respectively; see Figure 8.11). (A) *Neoceratodus forsteri*, the only extant species of *Neoceratodus*, has lobed paired fins. It is a facultative air-breather that can survive without surfacing and shows no parental care. (B) Larval *Neoceratodus* have a continuous median fin. (C) A researcher manipulates a drone to monitor *Neoceratodus* in the Brisbane River.

Australian lungfishes have complex courtship behaviors that may include male territoriality. Females are selective about the vegetation on which they lay their adhesive eggs, but no parental care (by either parent) has been documented. The jelly-coated eggs, 3 mm in diameter, hatch in 3–4 weeks, but the young are elusive, and nothing is known of their juvenile life in the wild. It is possible to rear eggs in the laboratory, however, which has allowed detailed developmental studies (**Figure 8.10B**).

Neoceratodus forsteri was described in 1871 based on specimens from the Burnett and Mary rivers in southeastern Queensland; its current range includes the Brisbane River, the Enoggera Reservoir, and adjacent fresh waters. Some of its current distribution results from introductions made in the 19th century because of concerns about the species' limited natural range. These rivers run throughout the year, fed by rainfall in the mountains of their drainage basins. Fossils of *Neoceratodus* occur in many other—now much drier—localities across the Australian continent.

The number of wild individuals of *Neoceratodus forsteri* in Queensland is difficult to estimate, but there may be as few as 10,000. It has been listed in CITES Appendix II since 1975, a designation that limits international trade. In 2019, the IUCN listed *N. forsteri* as Endangered based on its severely fragmented range within a small area (328 km²) and habitat alteration due to dams. Ongoing monitoring programs in Queensland assess and work to protect populations of Australian lungfishes (**Figure 8.10C**). For example, recent flooding scoured the riverbeds, killing the aquatic plants on which *Neoceratodus* prefer to lay eggs. Conservation teams have worked to re-plant reaches of rivers where lungfish occur.

South American and African lungfishes In contrast to *Neoceratodus*, both *Lepidosiren* (South America) and *Protopterus* (Africa) have paired lungs, a siphonlike opercular opening, reduced scales, long eel-like bodies, and threadlike paired fins with reduced internal skeletons (**Figure 8.11A,B**). They have comparatively reduced gills, and for decades were termed "obligate air breathers" because laboratory studies suggested that they depend on aerial oxygen. This appears not to be the case in the wild, however, at least not for the marbled lungfish, *P. aethiopicus*, which lives in large lakes in East Africa. Radio telemetric studies of individual marbled lungfish in Lake Baringo, Kenya show that they do not surface to breathe air. The waters of Lake Baringo are well oxygenated, and lungfish can derive all the oxygen they need via their gills.

An often-asked question is: If lungfishes have lungs and can rely on them for oxygen exchange, then why do they retain gills? One answer is that it is easy to lose CO_2 to the surrounding water through gills, and in this sense, African and South American lungfishes can partition gas exchange between lungs and gills. Adult *Protopterus* often retain tiny external gills (used during the larval stage for gas exchange), but if they are functional at all, these gills play an insignificant role in gas exchange in adults.

To air-breathe, a lungfish must travel to the water surface. The frequency of such trips depends in part on water temperature and the animal's activity, with the number of trips ranging from every few minutes to once or twice an hour. Buccal pumping moves air into the lung. A surfacing lungfish first inhales fresh air by depressing the floor of its buccal cavity, then opens the glottis, allowing gas to leave the lung. Next, the mouth closes and the floor of the buccal cavity rises, pumping gas into the lung.

Lepidosiren paradoxa and fossils of its extinct relatives occur in South America. Originally described in 1837 from a tributary of the Amazon, *Lepidosiren* lives in habitats

(A) African lungfish, *Protopterus annectens*

Hippocampus Bildarchiv

(B) South American lungfish, *Lepidosiren paradoxa*

Vassil/CC0 1.0

(C) Amazon habitat of South American lungfish, *Lepidosiren paradoxa*

© Cristina Cox Fernand

Figure 8.11 African and South American lungfishes.
(A) *Protopterus annectens* is one of four species of African lungfishes. Note the siphonlike opercular opening (red arrow) and threadlike pectoral fins. These features are shared with the South American *Lepidosiren*, and differentiate these two genera from *Neoceratodus* (see Figure 8.10). However, new evidence shows that, like *Neoceratodus*, some populations of *Protopterus* are facultative air-breathers. (B) The single species of South American lungfish, *Lepidosiren paradoxa*, is an obligate air-breather. Males guard eggs and young in nests. (C) Amazonian habitat of *Lepidosiren*, near Manaus, Brazil.

areas that flood during the wet season and bake during the dry season—habitats not available to actinopterygians except by immigration during floods. When the floodwaters recede, African lungfishes dig a vertical burrow in the mud with an enlarged chamber at the end and become increasingly lethargic as drying proceeds. Eventually the water in the burrow dries up, and the lungfish in its burrow folds into a U shape with its tail over its eyes. It produces heavy secretions of mucus that then dry, forming a protective envelope around its body; only an opening at its mouth remains for breathing. Such estivation—physiological and behavioral mechanisms that allow an animal to survive extended drought—is an ancient trait of dipnoans, based on the discovery of fossil burrows containing lungfish tooth plates in Carboniferous and Permian deposits in North America and Europe. These fossil taxa, however, are not closely related to *Protopterus*.

African lungfishes are surprisingly powerful swimmers. Their bodies are essentially tubes of muscle and, although they look sedentary, they can move very quickly when necessary. They walk and bound underwater, propelling themselves with ribbonlike pelvic fins. On land they move by planting their head on the ground and pivoting the body around it.

Unlike the fan-shaped tooth plates of *Neoceratodus*, those of *Lepidosiren* and *Protopterus* are self-sharpening. The upper and lower tooth plates closely occlude, with each creating wear that sharpens the other's cutting surfaces. Each tooth plate consists of three regions: (1) a transverse bladelike region anteriorly, which serves to cut prey; (2) a middle region with a grinding surface and a stabbing cusplike peak; and (3) a posterior region with a long cutting blade. Such regionalization along a single tooth plate is unknown in other extant vertebrates, and it allows feeding on both hard and soft foods.

ranging from Peru to the Guianas to Paraguay (**Figure 8.11C**). Despite its broad range, surprisingly little is known about its ecology and behavior. During the breeding season, females lay eggs in a bowl or flask-shaped nest cavity guarded by a male. Males develop vascularized extensions on their pelvic fins, known as pelvic gills. Long thought to function in delivering oxygen from the male's blood to the eggs and young that he protects in the nest, recent work suggests that the pelvic gills have a thickened epidermis and are not well suited for such gas exchange.

There are four species of African lungfishes in the genus *Protopterus* (see Figure 8.11A). They live in many habitats, and together these four species span the African continent except for the very driest regions. Some live in swamps and

8.5 Tetrapodomorpha

LEARNING OBJECTIVES

8.5.1 List three key features of Tetrapodomorpha.

8.5.2 Describe the fishlike and tetrapodlike features of †*Tiktaalik*.

8.5.3 Explain how limb girdles and vertebrae differ between †*Eusthenopteron* and †*Ichthyostega*.

8.5.4 Describe the organization of limb elements typical of both Devonian and extant tetrapods.

8.5.5 List two synapomorphies of Crown Tetrapoda.

Tetrapodomorpha includes Devonian sarcopterygians that retained gills and gill cavities and tetrapods with four limbs (see Figure 8.2). They share the presence of a **choana** (see Figure 8.8C) through which air can be inhaled into the oral cavity through the nose, a one-bone two-bone pattern in the axis of the paired fins (see Figure 7.5D,E), and reduction and eventual loss of cosmine on scales and dermal bones. A series of Devonian and Early Carboniferous fossils helps us understand some key aspects of tetrapod origins. Progress also has been made in interpreting paleoenvironments relevant to this transition.

Basal tetrapodomorphs

Here we use the tree in Figure 8.2 to trace the early history of Tetrapodomorpha. †Megalichthys and its Carboniferous relatives belong to †Rhizodontida, the most basal group of tetrapodomorphs. †Rhizodonts had long, recurved teeth and were among the largest bony fishes of their time, reaching 7 m. One of the most famous †rhizodont fossils is a beautifully preserved right pectoral fin of †Sauripterus, which exemplifies the one-bone two-bone pattern synapomorphic for tetrapodomorphs.

†Osteolepidae (Greek osteon,"bone"; lepis "scale") includes forms such as †Osteolepis macrolepidotus, a shallow-water predator from the Middle Devonian of Northern Scotland. Thick, bony scales cover its stout body. Known since the 1830s, †Osteolepis has contributed much to our knowledge of sarcopterygians. It retains first and second dorsal fins, a heterocercal caudal fin, and an anal fin as well as the intracranial joint. Detailed studies of its skull provided early insight into the evolution of the choana.

†Tristichopterids such as †Eusthenopteron had a near-global distribution during the Devonian. The diphycercal caudal fin had three lobes. Paired, crescent-shaped components of the vertebral centra (intercentra) surrounded the notochord. Large teeth on the palate had irregular, labyrinth-like infoldings of enamel, and thus are known as **labyrinthodont teeth**. The scales of †Eusthenopteron are reduced to bone and lack the cosmine typical of more basal sarcopterygians. †Eusthenopteron foordi from the Miguasha Lagerstätte (see Figure 8.3B) is the best known †tristichopterid, and one of the best known of all Devonian vertebrates.

†Panderichthys (from Latvia) and †Elpistostege (from Miguasha and other localities) share with more derived tetrapodomorphs loss of the dorsal fins. They had greatly reduced caudal fins, a dorsoventrally flattened body, and a flat head with a long snout. Modification of the humerus allowed these basal tetrapodomorphs to prop their heads out of the water; the eyes were on the top of the head (like those of a crocodile), and together these features suggest that they probably lived in shallow water.

†Tiktaalik   The next major advance is the loss of the bones that connected the head to the pectoral girdle, and

the first form to show this is the beautifully preserved holotype[1] of †Tiktaalik roseae (**Figure 8.12A**). Fossils of †Tiktaalik come from Ellesmere Island in the Canadian Arctic (which lay near the Equator during the Devonian; see Figure 8.3C).

†Tiktaalik (pronounced with the accent on the second syllable) means "a large freshwater fish seen in the shallows" in the local Inuktitut language. Skulls and skeletons of †Tiktaalik were discovered in shallow river floodplain deposits from early in the Late Devonian, about 382 Ma, which is about 22 million years before the earliest whole-body fossils of tetrapods.

†Tiktaalik lacks the supracleithral bones that connect the skull to the shoulder girdle in basal osteichthyans. These losses mean that †Tiktaalik had a neck, allowing it to swing its head to the right or left without having to move its pectoral girdle.

In other respects, †Tiktaalik displays a mosaic of fish-like and tetrapodlike characters (**Figure 8.12B**). It remained fishlike in retaining fin rays in the paired fins, the absence of a sacral articulation between the pelvic girdle and vertebral column, lack of an ischium in the pelvis, and a short ulna lacking an olecranon process (to which the triceps tendon attaches in tetrapods). In addition to a neck, its tetrapodlike features include loss of the opercular bone. This loss does not mean that †Tiktaalik lacked gills; rather, it suggests that it had a soft-tissue opercular opening, perhaps like that of Protopterus (see Figure 8.11A, red arrow).

Like early tetrapods, †Tiktaalik had a pelvic girdle similar in size to its pectoral girdle, robust overlapping ribs, a forelimb with wrist bones (at least two are homologues of wrist bones of tetrapods), fingerlike bones distal to the wrist, flexion in the forelimb at the elbow and wrist, and a process on the shoulder girdle interpreted as the origin for strong flexor muscles that attached to the humerus (see Figure 8.12B).

Tetrapods

The earliest known tetrapods come from the Late Devonian, about 360 Ma. In the 1930s, two iconic forms, †Ichthyostega and †Acanthostega, were discovered in East Greenland in riverine sediments. †Ichthyostega was intensively studied in the middle of the 20th century, but only later were good fossils of †Acanthostega recovered and described by Jenny Clack. Both animals ranged from 0.5 to 1.2 m in length and were primarily aquatic rather than terrestrial. Fin rays, which are dermal bones, are absent from the paired limbs of both †Acanthostega and †Ichthyostega, but there are dermal fin rays in the caudal fin (see Figure 8.13).

Devonian tetrapods retained portions of the gill arch skeleton, with grooves on the ceratobranchial bones

---

[1]The holotype is the single specimen that scientists use both to describe the species and to bear the name of the species. Scientists usually select the most perfect specimen in a series as the holotype.

(A) Holotype of †*Tiktaalik roseae*

© William E. Bemis, courtesy of The Museum of the Earth

(B) Interpretive skeletal reconstruction of †*Tiktaalik* highlighting fishlike and tetrapodlike features

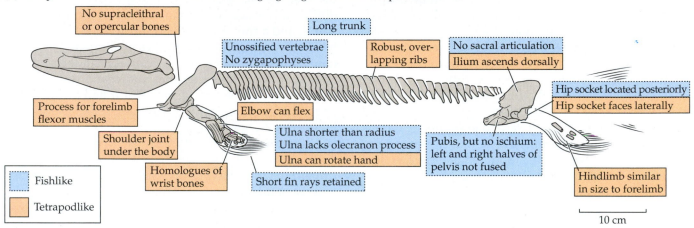

Figure 8.12 An important link. The discovery of the Late Devonian tetrapodomorph fish †*Tiktaalik roseae* was documented in a landmark paper (Daeschler et al. 2006) and opened new perspectives for our understanding of the transition of life to land. (A) Holotype of †*Tiktaalik roseae*. This well-preserved specimen was discovered near what today is Canada's Arctic Circle. (B) Interpretive skeletal reconstruction of †*Tiktaalik* highlighting both its fishlike and its tetrapodlike features. (B based on E.B. Daeschler et al. 2006. *Nature* 440: 757–763; and N.H. Shubin et al. 2013. *PNAS* 11: 893–899.)

indicating the presence of blood vessels supplying functional gills like those of more basal osteichthyans. Although the presence of these gills was originally surprising, subsequent discoveries showed that gills were also present in many Paleozoic tetrapods, especially temnospondyls (the group that includes extant lissamphibians; see Section 9.2).

A flange on the shoulder girdle of †*Acanthostega* that supports the posterior wall of the opercular chamber in bony fishes indicates that it had a soft-tissue operculum covering its gills (**Figure 8.13A**). Although fossils preserving the wrist and hand are not known, the ulna of †*Ichthyostega* has a well-developed olecranon process (the ulnar nerve passes adjacent to your olecranon process where it can be easily bumped, which is why this area is called the funny bone; **Figure 8.13B**). This indicates that it had a

large triceps muscle, which would have allowed †*Ichthyostega* to support the front of its body off the ground and to walk on land. Joints in its paddlelike hindlimbs show that †*Ichthyostega* could not place its feet directly under its body. Lungs are not preserved in the fossils, but we infer that they were present because they are present in lungfishes (the extant sister group of tetrapodomorphs) and in living tetrapods.

It is difficult to know how much time Devonian tetrapods spent in terrestrial activity. Sutures between the skull bones of †*Acanthostega* are like those of tetrapods that feed on land and use the jaws for biting food, but biting also could have occurred underwater. There is evidence that †*Acanthostega* had a juvenile stage of several years, during which limb bones were not ossified, so perhaps these juveniles were entirely aquatic. †*Ichthyostega* had

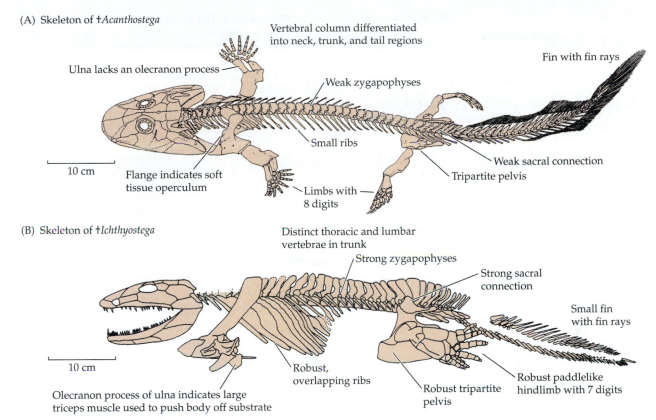

(A) Skeleton of †*Acanthostega*

Vertebral column differentiated into neck, trunk, and tail regions

Ulna lacks an olecranon process

Fin with fin rays

Weak zygapophyses

10 cm

Flange indicates soft tissue operculum

Small ribs

Weak sacral connection

Tripartite pelvis

Limbs with 8 digits

(B) Skeleton of †*Ichthyostega*

Distinct thoracic and lumbar vertebrae in trunk

Strong zygapophyses

Strong sacral connection

Small fin with fin rays

10 cm

Olecranon process of ulna indicates large triceps muscle used to push body off substrate

Robust, overlapping ribs

Robust tripartite pelvis

Robust paddlelike hindlimb with 7 digits

Figure 8.13 Devonian tetrapods. †*Acanthostega* (A) and †*Ichthyostega* (B) are tetrapods from the Late Devonian. Although we cannot be certain how much of their lives either of these animals spent on land versus in the water, it is clear that, by the end of the Devonian, tetrapod diversification and invasion of terrestrial environments were well underway. (A after M.I. Coates, 1996. *Earth Environ Sci Trans R Soc Edinb* 87: 363–421. © 1996 Royal Society of Edinburgh; B after P.E. Ahlberg et al. 2005. *Nature.* 437: 137–140.)

large shoulders (suggestive of terrestrial locomotion), but an ear region specialized for underwater hearing and paddlelike hindfeet specialized for swimming (see Figure 8.13B). In any case, †*Ichthyostega* and †*Acanthostega* differ enough from each other to show that by the end of the Devonian tetrapods occupied different ecological niches, if not different habitats.

We next focus on how the skeletal anatomy of Devonian tetrapods differs from that of Devonian tetrapodomorphs such as †*Tiktaalik* and †*Eusthenopteron*.

Limb girdles, vertebrae, and ribs The fore- and hind-limbs of †*Ichthyostega* and †*Acanthostega* could support more weight and supply more forward propulsion than could the fins of †*Tiktaalik*. Key changes relate to differences in the anatomical relationships of the limb girdles to the axial skeleton. To help you visualize this, **Figure 8.14A** shows the fins and fin girdles of a Devonian tetrapodomorph fish based on †*Eusthenopteron*. Note that the pectoral girdle attaches to the posterior region of the skull and the pelvic girdle does not articulate with the vertebral

(A) Fins and fin girdles of a Devonian tetrapodomorph fish

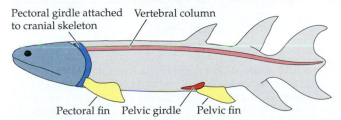

Pectoral girdle attached to cranial skeleton   Vertebral column

Pectoral fin   Pelvic girdle   Pelvic fin

(B) Limbs and limb girdles of a Devonian tetrapod

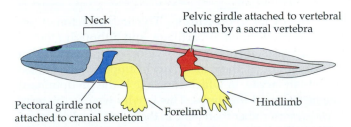

Neck

Pelvic girdle attached to vertebral column by a sacral vertebra

Pectoral girdle not attached to cranial skeleton

Forelimb

Hindlimb

Figure 8.14 Comparisons of fins, limbs, girdles, and axial skeleton in basal tetrapodomorphs and tetrapods. (A) Fins and limb girdles of a tetrapodomorph fish, based on †*Eusthenopteron*. The pectoral girdle attaches to the posterior portion of the skull, so there is no flexible neck. The pelvic girdle is embedded in muscle; it is not attached to the vertebral column. (B) Limbs and limb girdles of a Devonian tetrapod based on †*Acanthostega*. The pectoral girdle does not attach to the skull, which means that there is a flexible neck (cervical) region of the vertebral column. The pelvic girdle attaches to the vertebral column by a sacral vertebra.

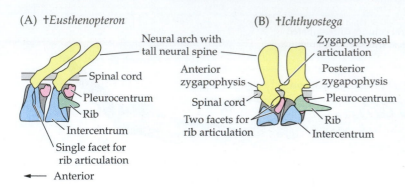

(A) †*Eusthenopteron*

Neural arch with tall neural spine

Spinal cord

Pleurocentrum

Rib

Intercentrum

Single facet for rib articulation

← Anterior

(B) †*Ichthyostega*

Zygapophyseal articulation

Anterior zygapophysis

Posterior zygapophysis

Spinal cord

Pleurocentrum

Two facets for rib articulation

Rib

Intercentrum

Figure 8.15 The vertebral column of tetrapods resists compression, torsion, and the pull of gravity. (A) Vertebrae of †*Eusthenopteron* consisted of a centrum (ossified pleurocentra and intercentra and cartilage) with a neural arch, which protected the spinal cord, and a large neural spine. Ribs had a single articulation with the intercentrum. (B) Vertebrae of Devonian tetrapods such as †*Ichthyostega* retained the multipart centrum and neural arch with a neural spine. Paired anterior and posterior zygapophyses projected from the neural arches and met at articulations between the vertebrae. These zygapophyses interlock each vertebra with two adjacent vertebrae, so that the vertebral column can support the body and resist torsion (twisting) along its length. Each rib articulates with both the intercentrum and neural arch. (See credit details at the end of this chapter.)

column. **Figure 8.14B** shows the limbs and limb girdles based on †*Ichthyostega*. Not only does it lack any attachment of the shoulder girdle to the skull, allowing a distinct neck region as we saw in †*Tiktaalik* (see Figure 8.12B), but the pelvic girdle is attached to the vertebral column via a sacral vertebra, providing additional structural support on land. Also, on each side of the body, the pelvis is tripartite, consisting of an ilium, pubis, and ischium (the ischium is not present in †*Tiktaalik*).

Vertebrae also differ between †*Eusthenopteron* and †*Ichthyostega*. A single vertebra of †*Eusthenopteron* consists of a neural arch with a tall neural spine dorsal to two elements—the pleurocentrum and the intercentrum—that together form the vertebral centrum (**Figure 8.15A**). In life, the gaps between the bony pleurocentrum and intercentrum would have been filled with cartilage. In contrast, the trunk vertebrae of Devonian tetrapods such as †*Ichthyostega* have specialized articulations known as **zygapophyses** (**Figure 8.15B**). Each neural arch has a pair of anterior zygapophyses pointing forward and a pair of posterior zygapophyses pointing posteriorly. They meet at zygapophyseal articulations between adjacent vertebrae and help to strengthen the vertebral column to support the body on land.

There are also differences in the way ribs articulate with vertebrae. Instead of a single articulation with the intercentrum, as in †*Eusthenopteron*, a rib of †*Ichthyostega* articulates with the neural arch as well as the intercentrum, a more robust arrangement that might have aided air-breathing.

Elements of tetrapod limbs The limb skeleton of tetrapods is pre-formed as cartilage, and although most of these skeletal elements undergo endochondral ossification in most tetrapods, there are many examples, such as lissamphibian larvae, in which cartilage is the predominant skeletal tissue of the limbs. This is why we use the term *skeletal element* rather than bone in the following discussion.

The limb elements of Devonian tetrapods preserve the monobasic arrangement that is synapomorphic for Sarcopterygii. They also show the characteristic organization of the limb elements found in extant tetrapods. The first element is the **stylopodium**, which is the humerus in the

forelimb and femur in the hindlimb (**Figure 8.16**). The **zeugopodium** corresponds to the radius and ulna in forelimb and the tibia and fibula in the hindlimb. The **autopodium** refers to the wrist, hand, and digits in the forelimb and the ankle, foot, and digits in the hindlimb. The individual carpal (wrist) and tarsal (ankle) elements are named, but the metacarpals, metatarsals, and phalanges are numbered. Numbering starts on the preaxial (thumb or big toe) side of the autopodium. Devonian tetrapods had more than five digits, and the autopodial portions of the limbs of †*Acanthostega* look more like flippers

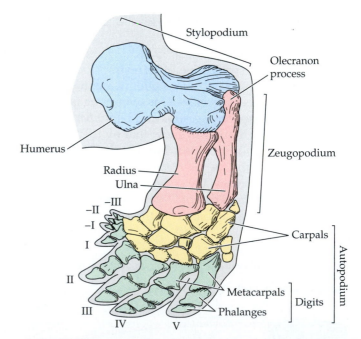

Figure 8.16 Eight-digit forelimb of a Devonian tetrapod. Tetrapod limbs have three regions: the stylopodium, zeugopodium, and autopodium. In a forelimb, the skeletal element of the stylopodium (upper arm) is the humerus. Skeletal elements of the zeugopodium (forearm) are the radius and ulna, and the skeletal elements of the autopodium are the wrist bones (carpals) and digits (metacarpals and phalanges). Comparable regions characterize a hindlimb (not shown). The animal shown has eight digits. The most preaxial digit is −III; digit I in this and other figures corresponds to your thumb. (See credit details at the end of this chapter.)

for maneuvering in shallow water than appendages for walking on land.

Origin of Crown Tetrapoda   Several fossil taxa help us understand the anatomical changes that led to Crown Tetrapoda. The intracranial joint that characterizes Sarcopterygii was lost in tetrapods, as exemplified by the Early Carboniferous †*Greererpeton*. †*Greererpeton* reached lengths of 1.5 m and is interpreted as aquatic, perhaps ecologically like extant giant salamanders (*Andrias*).

†*Crassigyrinus* shares with more crownward elpistostegalians nostrils facing forward (see Figure 8.2). Affectionately known as the "swamp monster," †*Crassigyrinus* comes from the Early Carboniferous of Scotland. It had a huge head and jaws with large teeth; its elongate body reached lengths of ~2 m (see Figure 5.9). Its greatly reduced forelimbs and powerful tail indicate its aquatic habits, for its limbs could not have supported the body on land.

The discovery of †*Loxomma* (also shown in Figure 5.9) in the middle of the 19th century fueled many early ideas about tetrapod evolution. Known only from three well-preserved skulls, its head was broad and flat with needlelike teeth suitable for a diet of fishes. There is no associated postcranial material, but based on the size of its head, it may have reached lengths >4 m.

Synapomorphies of Crown Tetrapoda include (1) an occipital condyle that articulates with the first vertebra, allowing freer movement of the head relative to the trunk, and (2) pentadactyl limbs (limbs with five digits). Further digital reduction evolved in some groups of tetrapods, most notably in Temnospondyli, which have four fingers (see Section 9.1).

8.6 Moving onto Land

LEARNING OBJECTIVES

8.6.1 List possible advantages of terrestrial activity to tetrapodomorphs.

8.6.2 Provide evidence that the tetrapod autopod evolved from postaxial structures of more basal sarcopterygians.

8.6.3 Summarize differences in the functions of hypaxial muscles in fishes and tetrapods.

For many years, we assumed that tetrapods evolved in fresh water, partly because lissamphibians do not tolerate salt water. But many fossil tetrapodomorphs come from marine or estuarine deposits, and the original environment of the fish–tetrapod transition remains unclear. Lungs and digits may have allowed aquatic tetrapodomorphs to make occasional forays onto land. But how did the transition from water to land happen?

One classic story of tetrapod origins posits that the Devonian was a time of seasonal droughts and that tetrapodomorphs used their limbs to crawl to another pond when their pond dried up. In other words, limbs originally aided movement across land. However, it now appears that limbs of early tetrapodomorphs improved locomotion in water and only later became useful for terrestrial locomotion. So the question becomes: What advantages did terrestrial activity offer? Hypotheses (not mutually exclusive) include searching for food, dispersal of juveniles, laying eggs in moist environments, and basking in the sun to elevate body temperature. Another intriguing possibility is that tetrapods evolved in areas that had extreme intertidal exposure during low tides, the idea being that both air-breathing and the ability to move over exposed mudflats were adaptive for intertidal tetrapodomorphs.

How did fins become limbs?

We can understand some aspects of limb evolution by comparing a series of extant and fossil taxa (**Figure 8.17A**). For example, homologues of the bones in a tetrapod's stylopod and zeugopod can be identified in more basal sarcopterygians. But it has been harder to identify and understand the homologues of elements of the tetrapod autopod. A major question has been where did wrist/ankle bones, metacarpals/metatarsals and phalanges (bones of the fingers and toes) come from? We are making progress in answering this question by comparing fin and limb development of extant forms and ever more detailed studies of fossils such as †*Elpistostege*.

Whereas actinopterygians show diverse patterns in endochondral elements of the paired fins, sarcopterygians share a common pattern (see Figure 7.5). There is a single basal element (the metapterygium in Figure 7.5C,D), a series of distal mesomeres running down the axis of the fin, and a series of metapterygial radials, to which the fin rays attach. As dermal bones, fin rays form in the skin, and unlike the rest of the fin skeleton, develop from neural crest cells. But how did tetrapods evolve autopods—wrists, hands, fingers, ankles, feet, and toes?

During development, the autopod of tetrapods is specified by differential expression of the same genes that are present in more basal osteichthyans rather than completely new genetic information. For example, *hoxa13* is expressed throughout the autopod of fishes and tetrapods. Two other regulatory genes, *hoxd13* and *alx4* are expressed on opposite sides of the fin or limb during early stages (**Figure 8.17B,C**). In *Neoceratodus*, the two expression domains specify the preaxial and postaxial sides of the fin (as mentioned above, your thumb is on the preaxial side of your hand, while your little finger is on the postaxial side). But in tetrapods, the expression domain for *hoxd13* expands during development to encompass the portion of the limb bud that will become the hand and fingers. This developmental information supports the interpretation from comparative anatomical studies of living and fossil sarcopterygians: a tetrapod's metapterygial axis (red dashed lines in Figure 8.17B,C) is bent compared to that of *Neoceratodus* or †*Tiktaalik*. This means that components of our hands (feet) and fingers (toes) evolved from postaxial structures in the fin of more basal sarcopterygians.

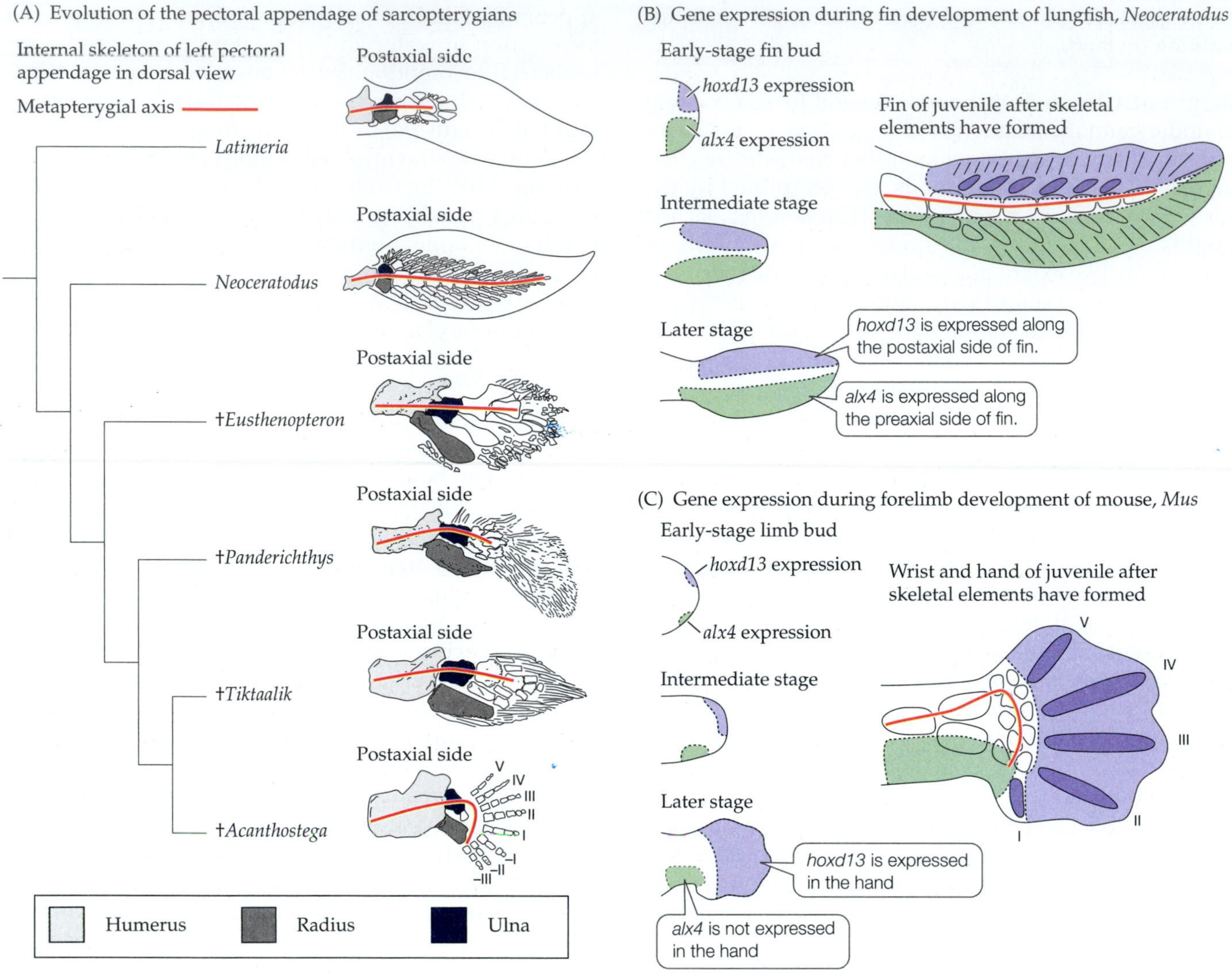

(A) Evolution of the pectoral appendage of sarcopterygians

Internal skeleton of left pectoral appendage in dorsal view

Metapterygial axis ———

Latimeria — Postaxial side

Neoceratodus — Postaxial side

†Eusthenopteron — Postaxial side

†Panderichthys — Postaxial side

†Tiktaalik — Postaxial side

†Acanthostega — Postaxial side

V IV III II I
—I —II —III

Humerus Radius Ulna

(B) Gene expression during fin development of lungfish, Neoceratodus

Early-stage fin bud

hoxd13 expression

alx4 expression

Fin of juvenile after skeletal elements have formed

Intermediate stage

Later stage

hoxd13 is expressed along the postaxial side of fin.

alx4 is expressed along the preaxial side of fin.

(C) Gene expression during forelimb development of mouse, Mus

Early-stage limb bud

hoxd13 expression

alx4 expression

Wrist and hand of juvenile after skeletal elements have formed

V
IV
III
II
I

Intermediate stage

Later stage

hoxd13 is expressed in the hand

alx4 is not expressed in the hand

Figure 8.17 Evolution and development of sarcopterygian fins and limbs. (A) Evolution of the pectoral appendages of sarcopterygians showing how postaxial radial elements contributed to hand and finger bones. (B) Gene expression during fin development of *Neoceratodus*. The *hoxd13* gene is expressed on the postaxial side of the fin bud and *alx4* is expressed on the preaxial side of the fin. (C) Gene expression during forelimb development of *Mus*. Expression patterns of *hoxd13* and *alx4* in the early stages of limb bud development resemble those seen in *Neoceratodus*. As development proceeds, the expression domain of *hoxd13* increases disproportionally, so that by later stages of development it is expressed in the region that will develop into the hand, while *alx4* expression is restricted to the arm and forearm. (B,C modified from J.M. Woltering et al. 2020. *Sci Adv* 6: 34, eabc3510. CC BY 4.0. https://creativecommons.org/licenses/by/4.0/)

In all extant tetrapods, digit formation starts with digit IV (the ring finger on our hands) and proceeds in an arc toward the thumb, with digit V (the "pinky") being added in the opposite direction. The branching of digits from this arc ends with digit I (the thumb), or sooner in animals such as lissamphibians that have lost digit I. If the process of developmental branching continues, however, a polydactylous condition results, and additional digits form beyond the thumb (digits –I, –II, and –III; see Figure 8.16A). This situation sometimes occurs as an abnormality in extant vertebrates but was apparently the normal condition in Late Devonian tetrapods; they have additional digits beyond the thumb. In addition, when digits are lost in more derived tetrapods (whether in limbless lizards, or in mammals and birds specialized for more rapid locomotion), the loss happens in the reverse pattern. Digit I is invariably the first to be lost, followed by digit V and then II; digits III and IV tend to be retained.

Body support and locomotion of early tetrapods

A fish in water is neutrally buoyant, and its vertebral column counteracts compressive forces produced during lateral undulations used for swimming. On land, gravity is a downward force acting on the body, meaning that the trunk experiences torsion (twisting force) during locomotion. The dorsoventrally flattened body of †telpis-tostegalian fishes and early tetrapods amplified these torsional forces on land. In fishes, the epaxial and hypaxial muscles are used only for propulsion, but tetrapod hypaxial muscles evolved modifications to resist torsion and stabilize the trunk. Enlargement of the endochondral limb girdles and ribs also aids trunk stability, and the beginnings of these changes are seen in early tetrapods such as †*Acanthostega* (see Figure 8.13A).

Fins of fishes project horizontally out from the sides of the body. The humerus and femur of early tetrapods also extended horizontally—as do those of extant salamanders and lizards—and the elbows and knees were bent to allow the lower limbs and feet to support the body. This arrangement provides a stable base of support but exerts torsional forces on the elbow and knee joints.

Pectoral fins of †*Tiktaalik* were more ventrally positioned than those of †*Eusthenopteron*, so it could place them under the body. The bony process on its pectoral girdle served as the origin for strong flexor muscles running to the humerus, which would have acted to bring the forelimb forward, elevating the head and anterior part of the body.

Tetrapods use muscles for body support as well as for locomotion, and they are bolstered by internal connective tissue making tetrapod muscles tougher to cut than the flesh of fishes (think of the difference between cutting into a steak versus a slice of fish). Likewise, ribs of Devonian tetrapods such as †*Acanthostega* and especially †*Ichthyostega* are much more robust than the ribs of fishes.

Extant salamanders are highly derived and much smaller than Devonian tetrapods, so they are not good models for the locomotion of early tetrapods. Computed tomography (CT) scans of †*Ichthyostega* suggest that its limbs had only a limited range of motion, moving backward and downward, with little ability to rotate the limb to face the palm forwards or backwards. †*Ichthyostega* probably walked on land using its forelimbs, as does the extant mudskipper (see Figure 7.23C), dragging its belly and tail. The vertebral column of †*Ichthyostega* would have flexed in a dorsoventral plane during this locomotion, and ossified midline elements, perhaps contained within a cartilaginous sternum, would have supported the rib cage.

8.7 Paleoecology of Devonian Tetrapodomorphs

LEARNING OBJECTIVE

8.7.1 List some questions raised by fossil tetrapod trackways regarding the origin of tetrapods.

Our understanding of Devonian tetrapodomorphs is based largely on comparative anatomical study of whole-body fossils such as †*Eusthenopteron*, †*Panderichthys*, and †*Elpistostege* and Late Devonian forms such as †*Ichthyostega* and †*Acanthostega*. But there are Middle Devonian trackways that predate these specimens; for example, footprints in a marine lagoon deposit in Poland predate the earliest whole-body tetrapod fossils by about 22 million years. Some question whether the animal that made the tracks was actually a tetrapod, but uncontested tetrapod tracks are known from deposits in Ireland made only a few million years later. There are also tracks from the Late Devonian of Scotland and Australia, hinting at a much earlier diversity of tetrapods than is currently established.

In addition, Middle Devonian tetrapod trackways occur in deposits that formed far inland, away from the deltaic habitats we classically associate with the origin of tetrapods, and this has spawned some new ideas about early tetrapod evolution. Middle Devonian †telpistostegalians lived in deltaic environments like those preserved at Miguasha, but the inland trackways suggest that tetrapods had already evolved and were living away from deltaic habitats. By the Late Devonian, when the more basal tetrapodomorphs were extinct, the first whole-body fossils of †*Acanthostega* and †*Ichthyostega* show up in deltaic deposits.

Clearly, there is much still to be learned about the origin of tetrapods, including the key questions of exactly when and where it happened. In our current conception, adaptations for life on land evolved in a mosaic fashion over tens of millions of years, making it difficult to draw sharp ecological boundaries between, for example, †*Tiktaalik* and †*Ichthyostega*, despite their morphological differences. One fact is clear: No sooner had tetrapods evolved specializations for living on land than some lineages became more terrestrial while others returned to the water. We will pick up the story of these lineages in Chapter 9.

Summary

8.1 Phylogenetic Concepts of Tetrapoda and Characters for Sarcopterygii

We distinguish Tetrapoda (all sarcopterygians with four legs) from Crown Tetrapoda (extant lissamphibians and amniotes along with their most recent common ancestor).

Molecular phylogenetic studies confirm coelacanths as the most basal extant sarcopterygians, with lungfishes as the extant sister group of tetrapodomorphs.

Sarcopterygians share three diagnostic characters: (1) monobasic paired fins with a scaled, muscular lobe at the base; (2) cosmine on dermal bones and scales; and (3) an intracranial joint between anterior and posterior parts of the braincase.

8.2 The Miguasha Lagerstätte and the "Good Fossil Effect"

Miguasha National Park in Canada protects a Lagerstätte that has produced exquisitely preserved fossils of diverse groups from the Late Devonian.

New fossils that continue to be unearthed in Miguasha as well as other localities have prompted new interpretations and reexaminations of previously discovered sarcopterygian fossils, a phenomenon described as the "good fossil effect."

8.3 Actinistia

Actinistia includes †Onychodontia, an extinct Devonian group, and Coelacanthiformes, the coelacanths, an extant lineage that has persisted since the Devonian.

- †Onychodonts were marine predators with a spiky tooth whorl in the symphysis of the lower jaw and a diphycercal caudal fin.

- Key features of derived coelacanths include: (1) a rostral organ in the snout; (2) paired fins and second dorsal and anal fins with muscular, lobed bases; and (3) a diphycercal caudal fin with terminal tuft.

Coelacanths have a long fossil record beginning in the Devonian, but their fossil record stops in the Cretaceous. For nearly a century, paleontologists thought that coelacanths were extinct, until discovery of an extant coelacanth in 1938 off the east coast of Africa; this species was named *Latimeria chalumnae*. A second species of coelacanth, *L. menadoensis*, was discovered in Indonesian waters in 1998.

8.4 Dipnomorpha

Dipnomorpha contains the extinct †Porolepiformes and the extant Dipnoi, a clade of six living species of lungfishes.

- †*Porolepis* had large pores of unknown function in the cosmine-coated surfaces of the scales.

- Dipnoan synapomorphies include reduction and loss of the outer dental arcade and fusion of the palatoquadrate to the cranium.

The fossil record of lungfishes reveals changes in fins and skull bones, as well as loss of cosmine and lack of ossification in the internal skeleton. These changes probably represent paedomorphosis.

There are three extant genera of lungfishes: (1) *Neoceratodus* from Australia (1 species); (2) *Lepidosiren* from South America (1 species); and (3) *Protopterus* from Africa (4 species).

- The Australian lungfish *Neoceratodus forsteri* retains the large, paired fins with muscular bases characteristic of Paleozoic dipnoans and depends on access to water throughout life. It uses well-developed gills for gas exchange, has a single lung, and does not depend on air-breathing. Its tooth plates are fan-shaped.

- In contrast, both *Lepidosiren paradoxa* and the four species of *Protopterus* have threadlike paired fins with reduced internal skeletons, comparatively reduced gills, and paired lungs; they use both lungs and gills for gas exchange. *Lepidosiren* and *Protopterus* have self-sharpening tooth plates.

8.5 Tetrapodomorpha

Tetrapodomorpha includes Devonian sarcopterygians that retained gills and gill cavities (e.g., †*Eusthenopteron*, †*Panderichthys*, †*Elpistostege*, and †*Tiktaalik*) and Late Devonian tetrapods with four limbs.

Synapomorphies of Tetrapodomorpha include: (1) choana; (2) a one-bone two-bone pattern in the axis of paired fins; and (3) reduction and eventual loss of cosmine on scales and dermal bones.

†*Tiktaalik* displays a mosaic of fishlike and tetrapodlike characters.

†*Ichthyostega* and †*Acanthostega* are some of the earliest known tetrapods from the Late Devonian; both were primarily aquatic rather than terrestrial.

Several key skeletal changes occurred leading to Devonian tetrapods.

- The pectoral girdle lost its connection to the skull.

- The pelvic girdle articulated with the vertebral column via a sacral vertebra.

- Zygapophyses on trunk vertebrae strengthened the vertebral column and supported the body on land.

The limbs of Devonian tetrapods (and extant tetrapods) have three elements: stylopodium, zeugopodium, and autopodium.

Summary *(continued)*

8.6 Moving onto Land

The original environment of the fish–tetrapod transition is unclear. Possible reasons for terrestrial activity include searching for food, dispersal of juveniles, laying eggs in moist environments, and basking in the sun to raise body temperature. For tetrapodomorphs living in intertidal environments, both air-breathing and the ability to move over exposed mudflats might have been advantageous.

Tetrapods lost their dermally derived fin rays and evolved autopods (wrists, hands, fingers, ankles, feet, and toes) from postaxial structures of more basal sarcopterygians.

Changes in skeletal and muscular systems occurred with the move to land.

- In fishes, the epaxial and hypaxial muscles are used only for propulsion, but tetrapod hypaxial muscles evolved modifications to resist torsion and stabilize the trunk. Enlargement of the endochondral limb girdles and ribs also aids trunk stability.

- †*Ichthyostega*'s limbs probably had only a limited range of motion; it might have walked on land using its forelimbs, as do extant mudskippers.

8.7 Paleoecology of Devonian Tetrapodomorphs

Fossil tetrapod trackways raise questions about the timing and location of the origin of tetrapods.

- Middle Devonian tetrapod trackways occur in deposits that formed far inland, away from the deltaic habitats that we classically associate with the origin of tetrapods.

- Fossil trackways from the Late Devonian suggest a much earlier diversity of tetrapods than currently known.

Discussion Questions

8.1 Lungfishes are considered to be more closely related to tetrapods than are coelacanths, but for most of the 20th century, coelacanths were thought to be the sister group of tetrapods. How have new phylogenetic tools and changes in our ways of thinking about phylogenetic relationships reversed the older interpretation? What is it about lungfishes that makes their phylogenetic placement challenging?

8.2 What information about Devonian tetrapods leads to the conclusion that tetrapods evolved in the water rather than on land?

8.3 What is the definition (in phylogenetic terms) of a tetrapodomorph?

8.4 What features of †elpistostegalian fishes lead us to infer that they were shallow-water forms?

8.5 What is the main new piece of information about early tetrapods that contradicts the old "drying pond" hypothesis of the origins of terrestriality?

Figure credits
Figure 8.1: Based on K. Liem et al. 2001. *Functional Anatomy of the Vertebrates*, 3rd ed. Cengage/Harcourt College: Belmont, CA; Brian Choo; Dmitry Bogdanov.
Figure 8.2: Based on K. Liem et al. 2001. *Functional Anatomy of the Vertebrates*, 3rd ed. Cengage/Harcourt College: Belmont, CA; Brian Choo; Dmitry Bogdanov; Nobu Tamura.
Figure 8.15: Based on K. Liem et al. 2001. *Functional Anatomy of the Vertebrates*, 3rd ed. Cengage/Harcourt College: Belmont, CA.
Figure 8.16: Based on K. Liem et al. 2001. *Functional Anatomy of the Vertebrates*, 3rd ed. Cengage/Harcourt College: Belmont, CA.

© Dzulfikri/Shutterstock.com

Origins of Lissamphibia and Amniota

Green treefrogs (*Litoria caerulea*) sometimes eat small mammals.

9.1 **Paleozoic Tetrapods and the Origins of Extant Groups**

9.2 **Characters of Amniotes**

9.3 **Diversification of Amniotes**

For more than 100 million years of the Paleozoic, Crown Tetrapoda, a clade recognized by the presence of pentadactyl limbs and an occipital condyle (see Figure 8.2), radiated into a great variety of aquatic and terrestrial forms. Some lineages were semiaquatic, some secondarily became fully aquatic, and others became increasingly specialized for terrestrial life. During that long span of Earth history, the supercontinent of Pangaea formed, great rainforests of lycopods grew, and climate change plunged Earth into an extended ice age followed by a time of extremely hot and arid environments. As terrestrial vegetation diversified and increased during the Carboniferous, insects radiated, providing a food supply that in turn could support a diverse fauna of tetrapods. Early tetrapods were carnivorous, and no adult lissamphibians eat plants. This is in striking contrast to amniotes, in which herbivory evolved many times.

Like their osteichthyan ancestors, early crown tetrapods laid their eggs in water. We know this because fossilized larvae with gills have been found. Many had a biphasic life history, with an aquatic larval period followed by metamorphosis to terrestrial adults, like most extant lissamphibians (caecilians, frogs, and salamanders; see Chapter 12). But many Carboniferous and Permian tetrapods that had an aquatic larval stage were much larger as adults than any living or extinct lissamphibians, with several groups resembling extant crocodylians in appearance and, presumably, ecology. Evolutionary changes in skulls, vertebrae, limb girdles, limbs, and ankles allow us to trace the diversification of Paleozoic tetrapods and to interpret aspects of their biology.

Evolution of the amniotic egg in the Carboniferous allowed amniotes to escape dependence on water for reproduction. An archetypal amniotic egg, which we describe in Section 9.2, contains the embryo, nutrients in the form of yolk, and extraembryonic membranes contained within a shell, much like a chicken egg. Amniotic eggs are laid on land, not in water, and the shell reduces evaporative water loss. Amniotes also evolved a waterproof skin that limited desiccation. Together, these adaptations allowed amniotes to survive far from wet lowland habitats and to continue to diversify despite the increasing aridity of the late Paleozoic. Eggs and skin are rarely found in association with tetrapod fossils, so we rely on diagnostic skeletal features of amniotes that can be seen in fossils, including an axis vertebra, at least two sacral vertebrae, lateral flanges of the pterygoid bones, and an astragalus bone in the ankle.

We begin this chapter with an overview of Paleozoic tetrapods, focusing on a few taxa that offer insight into the divergence and diversification of lissamphibians and amniotes—that is, the extant groups of tetrapods. We also explore what the fossils can tell us about the biology, ecology, and functional morphology of these animals, using information from fossil skeletons and the types of sites where the fossils have been found. We trace amniote diversification by examining skulls and temporal fenestrae in relation to the evolution of jaw closing muscles. Based on the arrangement of these temporal fenestrae, we distinguish two clades of amniotes, Sauropsida and Synapsida. There have been many more species of sauropsids over the course of Earth history, for this group includes Lepidosauria (tuatara, lizards, and snakes; Chapter 15), turtles (Chapter 16), crocodylians (Chapter 17), and dinosaurs (Chapters 18 and 19), which today are represented by more than 10,000 species of birds (Chapter 21) as well as many extinct groups. Synapsid evolution culminated in the origin and diversification of mammals (Chapters 22, 23, 24).

9.1 Paleozoic Tetrapods and the Origins of Extant Groups

LEARNING OBJECTIVES

9.1.1 Discuss the evidence that caecilians are stereospondyls and batrachians are dissorophoids.

9.1.2 Provide three examples of extinct reptiliomorphs outside Amniota.

9.1.3 Explain the anatomical and phylogenetic importance of the early reptiles †*Captorhinus*, †*Paleothyris*, and †*Hylonomus*.

Figure 9.1 Time tree of tetrapods. Few tetrapod fossils are known from Romer's Gap—an interval from just before the end of the Devonian through the first ~44 Ma of the Carboniferous—because there are few outcrops of terrestrial rocks from that time. This tree shows one of several competing hypotheses about the origin of Lissamphibia from two groups of temnospondyls that diverged in the Carboniferous. Estimated divergence times for extant groups are based on molecular dates, with most divergences predating the earliest known fossils of the groups. (See credit details at the end of this chapter.) ▶

During the early Carboniferous, crown Tetrapoda diversified into clades that ultimately led to Lissamphibia and Amniota. Molecular dating places the split between lissamphibians and amniotes at ~355 Ma, which is ~4 Ma after the end of the Devonian (**Figure 9.1**). We do not know whether there was an explosive radiation of early tetrapods following the end-Devonian extinctions because there are few sites with non-marine sedimentary rocks from the early Carboniferous. This time in earth history is called Romer's Gap, and fossil tetrapods from that time are scarce. Although recent explorations have unearthed early Carboniferous tetrapods in Nova Scotia and Scotland, we still lack a clear and compelling picture of the early evolution of tetrapods.

Our simplified phylogenetic treatment of early tetrapod evolution focuses on two large extant clades: Temnospondyli, which gave rise to lissamphibians; and Reptiliomorpha, which gave rise to amniotes (**Figure 9.2**).

Figure 9.2 Simplified cladogram of tetrapods. Paleozoic taxa of uncertain relationships to extant taxa are omitted from this tree. (See credit details at the end of this chapter.)

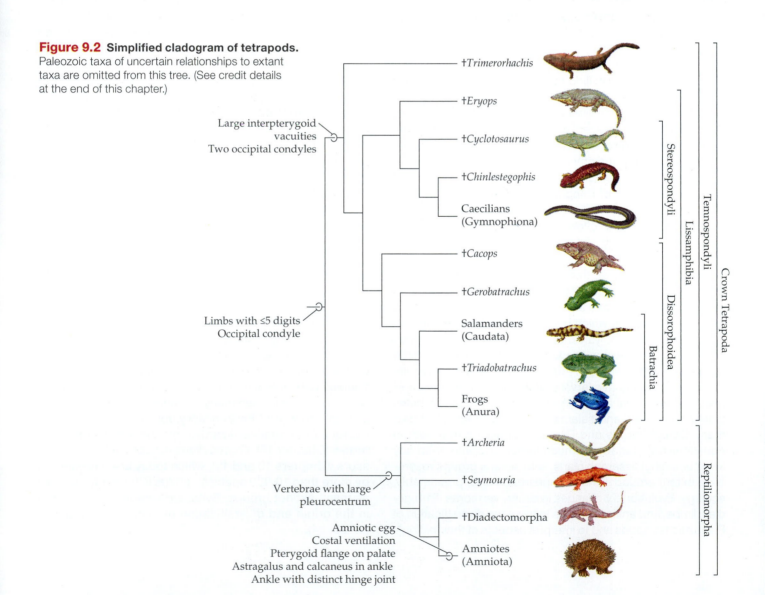

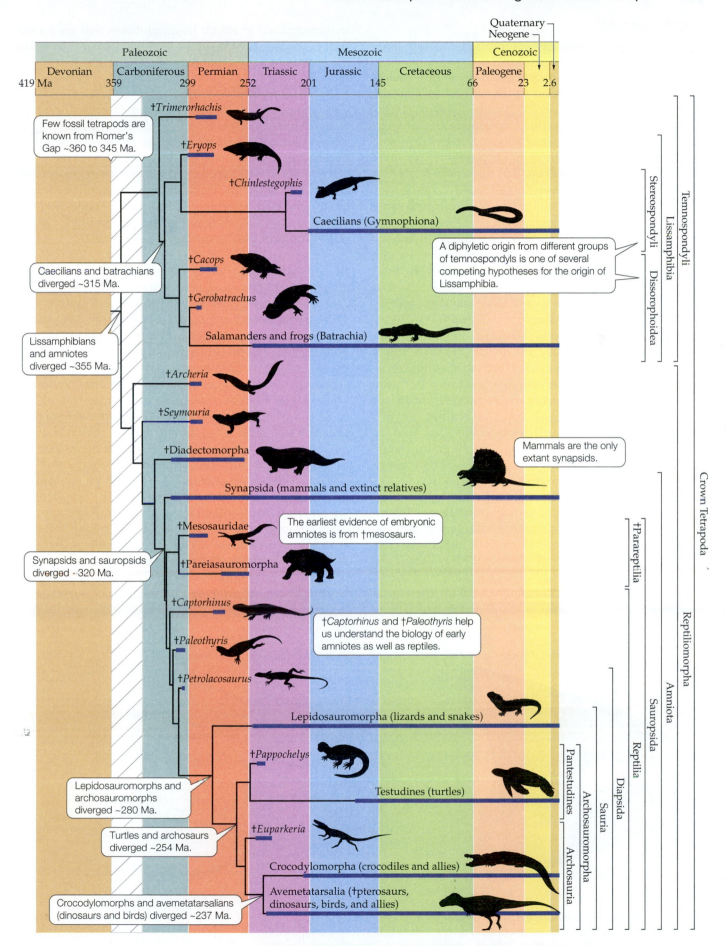

Temnospondyli

Temnospondyli includes Lissamphibia and all tetrapods more closely related to Lissamphibia than to Reptiliomorpha (see Figure 9.2). Soon after their origin in the early Carboniferous, temnospondyls diversified and spread across the world. More than 200 genera of Paleozoic temnospondyls have been described, ranging from a few centimeters in length to the size of alligators. All had aquatic larvae, and most were semiaquatic, but there were also fully aquatic forms and others that were fully terrestrial as adults, returning to water only to breed.

Temnospondyls persisted into the Mesozoic, and some large-bodied species survived to the Cretaceous. There was even a Triassic marine radiation (†Trematosauridae; see Section 10.3). By the start of the Cenozoic, however, the only remaining temnospondyls were the comparatively small lissamphibians—caecilians, salamanders, and frogs—that evolved from Paleozoic temnospondyls and subsequently diversified during the Mesozoic and Cenozoic.

Traditionally, temnospondyls were distinguished from other Paleozoic tetrapods by a vertebral column in which each vertebra consisted of a neural arch that ossified separately from the intercentrum and pleurocentrum (see Figure 8.15). Because the intercentrum and pleurocentrum are not fused, it looks as though each vertebral centrum—which are solid units in groups such as mammals— has been cut into components, and this is the source of the group's name (Greek *temnos*, "cut"; *spondylos*, "vertebrae"). Such a pattern now appears to be plesiomorphic for tetrapods.

A small aquatic Permian tetrapod, †*Trimerorhachis insignis*, is a basal member of Temnospondyli (see Figure 9.2). Known from many specimens collected at localities in the American southwest, †*T. insignis* had a wide, flat head, a short snout, external gills like those of extant salamanders such as *Necturus* (see Figure 12.3), and lateral lines on its skull roofing bones (**Figure 9.3**). The lateral line system functions only in water and is lost during metamorphosis in extant lissamphibians that are terrestrial as adults; retention of the lateral line system by adult †*T. insignis* indicates that they were aquatic.

Origins of Lissamphibia

Over the last 50 years, discoveries of Paleozoic and Mesozoic temnospondyls have provided new evidence that Lissamphibia is a derived group of Temnospondyli. The question is, *which* temnospondyls—stereospondyls or dissorophoids? (Note that we did not put a dagger in front of either of these taxon names, which should be a hint about where we are going.) To answer that question, we consider a few taxa of temnospondyls and phylogenetic implications of their anatomy (see Figure 9.2).

Stereospondyli Paleozoic temnospondyls such as †*Eryops megacephalus* had broad, flat skulls with long snouts and well-ossified postcranial skeletons (**Figure 9.4A**). †*Eryops* was semiaquatic, and its vertebrae had sturdy zygapophyses. Limb girdles and limbs supported its heavy body when it was on land. Current phylogenetic analyses place †*Eryops* as a close outgroup to Stereospondyli.

Like many temnospondyls, †*Cyclotosaurus robustus*, a Triassic form from Europe, had a long snout. The paired nostrils, located far forward in its skull, led to paired choanae that opened into the front of the oral cavity. The palatal view of †*C. robustus* illustrates two features that are probably synapomorphies for Temnospondyli as a whole. The first is the presence of large holes in its palate, termed interpterygoid vacuities (**Figure 9.4B**). These structures are especially wide in temnospondyls compared to other early tetrapods and in life were covered by the oral mucosa, which may have had denticles like the palatal surfaces of some fishes. The second feature is the presence of two occipital condyles. Extant frogs and salamanders have two occipital condyles, but they do not extend as far posteriorly as they do in †*Cyclotosaurus*. However, the two

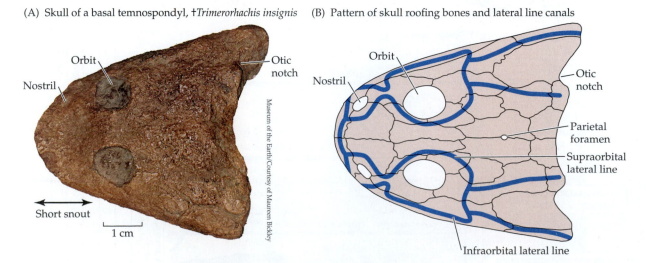

(A) Skull of a basal temnospondyl, †*Trimerorhachis insignis*

Orbit
Nostril
Otic notch
Short snout
1 cm

Museum of the Earth/Courtesy of Maureen Bickley

(B) Pattern of skull roofing bones and lateral line canals

Orbit
Nostril
Otic notch
Parietal foramen
Supraorbital lateral line
Infraorbital lateral line

Figure 9.3 Basal temnospondyls. (A) Forms such as †*Trimerorhachis insignis* were small aquatic animals with flat skulls and lateral line canals (B) in their cranial bones. (B after A.S. Romer. 1966. *Vertebrate Paleontology*. University of Chicago Press: Chicago.)

(A) Skeleton of a Permian stereospondyl, †*Eryops megacephalus*

Large ribs   Robust limb girdles and limbs

Tim Evanson/CC-BY-SA 2.0

Flat, wide head with long snout   5 digits on fore- and hindfeet

Figure 9.4. Paleozoic and Mesozoic stereospondyls.
(A) †*Eryops* lacked a tail fin, had robust limbs and girdles, and a long snout. It is usually interpreted as a semiaquatic predator. (B) Skull of large aquatic †*Cyclotosaurus* showing textured surfaces of cranial roofing bones and large interpterygoid vacuities in the palate. (C) Restoration of †*Cyclotosaurus*, a semiaquatic predator ~4 m long.

occipital condyles of extant caecilians do extend posteriorly behind the skull.

Early temnospondyls had otic notches at the posterior corners of the skull that, at least in some, supported a tympanum (eardrum). A long, tapered stapes transmitted vibrations from the tympanum to the oval window of the inner ear, allowing detection of airborne sounds. †*Cyclotosaurus robustus* had a derived condition in which the otic notch was surrounded by bone to form an otic fenestra across which the tympanum stretched (see Figure 9.4B).

Like other temnospondyls, †*Cyclotosaurus robustus* had large fangs in the front of its mouth. These fangs have a pattern of infolded layers of dentine and enamel described as labyrinthodont. It was formerly thought that labyrinthodont teeth characterized some or all temnospondyls, but we now know that similar teeth evolved independently in other groups, such as gars (Lepisosteidae), and that basal tetrapodomorphs such as †*Eusthenopteron* also had labyrinthodont teeth. Thus, labyrinthodont teeth are not by themselves diagnostic for temnospondyls. †*Cyclotosaurus robustus* also had an outer row of smaller teeth along the margins of the upper and lower jaws, and an inner row of teeth on the palatal bones. In the posterior corners of the palate, large jaw adductor muscles filled the fossae (Latin, *fossa*, "ditch"; in anatomy, a fossa refers to a depression or hollow space; *fossa* is also the root word for fossils). The long jaws would have closed with a rapid snap, and †*Cyclotosaurus* must have been a fearsome predator, for its skull was ~70 cm long (**Figure 9.4C**).

(B) Skull of a large Triassic stereospondyl, †*Cyclotosaurus robustus*, in dorsal (above) and palatal views

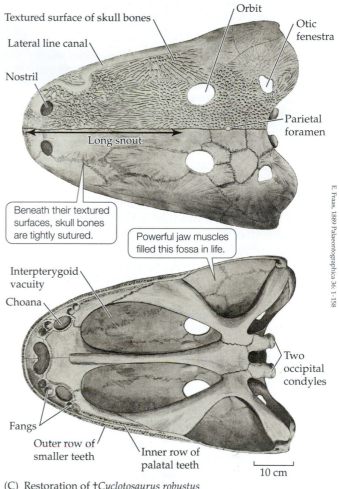

Textured surface of skull bones

Lateral line canal

Nostril

Orbit

Otic fenestra

Parietal foramen

Long snout

Beneath their textured surfaces, skull bones are tightly sutured.

Powerful jaw muscles filled this fossa in life.

Interpterygoid vacuity

Choana

Two occipital condyles

Fangs

Outer row of smaller teeth

Inner row of palatal teeth

10 cm

E. Fraas, 1889 Palaeontographica 36: 1-158

(C) Restoration of †*Cyclotosaurus robustus*

1 m

Based on Hiuppo/CC BY 3.0

Because they lived in lowland swamps, temnospondyls such as †*Cyclotosaurus robustus* have often been compared with extant crocodylians, and some researchers link the demise of large temnospondyls to the radiation of crocodylomorphs during the Triassic. For example, the evolution of the secondary palate in crocodylomorphs strengthened their skulls and allowed more forceful biting than was possible for large temnospondyls. But such speculation does not provide a complete picture, and there may be many other reasons why large temnospondyls did not survive into the Cenozoic.

Figure 9.5 **Interpretation of caecilians as derived stereospondyls.** (A) Restoration of †*Chinlestegophis*, a small Late Triassic fossil interpreted as a close outgroup of extant caecilians, Gymnophiona. (B) Restoration of †*Eocaecilia*, a Jurassic caecilian with tiny limbs. (C) Some characters that support interpretation of caecilians as derived stereospondyls are shown here. Not all workers agree with this interpretation, but it focuses new attention on the evolutionary history of stereospondyls.

(A) †*Chinlestegophis jenkinsi*   (B) †*Eocaecilia micropodia*

(C) Skeletal evidence that †*Chinlestegophis* is the sister group of caecilians

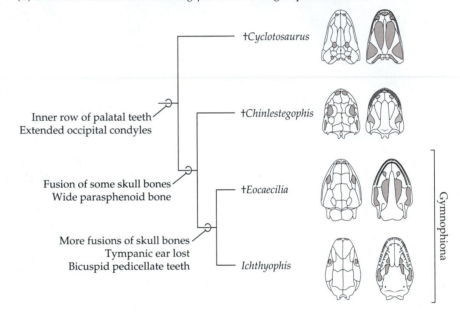

†*Chinlestegophis jenkinsi*, a small stereospondyl from the Triassic of Colorado, was discovered in 2017 (**Figure 9.5A**). It has paired limbs and a skull with a mosaic of features interpreted as intermediate between typical stereospondyls and caecilians, including a Jurassic caecilian, †*Eocaecilia*, which has four tiny legs and relatively large eyes (**Figure 9.5B**); extant caecilians lack legs and have reduced eyes (see Figure 12.8B). If the interpretation of †*Chinlestegophis* as a stem caecilian in **Figure 9.5C** is correct, then Stereospondyli is not extinct because it is represented by extant caecilians.

Dissorophoidea   But what about salamanders and frogs, which together form Batrachia? Here we consider batrachians to be derived Dissorophoidea (see Figure 9.2). Paleozoic dissorophoids are exemplified by the Permian genus †*Cacops* (Greek *kakos*, "ugly"; *ops*, "face"; **Figure 9.6A,B**). †*Cacops* had a sturdy skeleton, a short tail, and a series of osteoderms along its back. Unlike stereospondyls, †*Cacops* had a relatively short snout and large eyes. Its eyes are usually interpreted as bulging above the surface of the head like those of extant frogs or salamanders (see Figure 9.6B). There was a large tympanum on the side of its head, and the stapes/columella connected the tympanum to the inner ear in an arrangement like that of extant frogs.

Another Paleozoic dissorophoid from Texas, †*Gerobatrachus hottoni* (Greek *geros*, "elder"; *batrachus*, "frog"), caused a sensation in 2008 because of its mosaic of froglike and salamanderlike features (**Figure 9.6C,D**). The bones of its broad head are simplified relative to †*Cacops* and are more like those of extant lissamphibians. It has pedicellate teeth (i.e., the tooth crowns attach to a base known as a pedicel), a characteristic of extant lissamphibians. Its caudal vertebrae are like those of salamanders. The press nicknamed †*Gerobatrachus* the "frogamander," and current phylogenetic analyses place it as a very close outgroup of Batrachia (see Figure 9.2).

If caecilians are derived stereospondyls and batrachians are derived dissorophoids, then Lissamphibia as a group has been extant since the Paleozoic. The caecilian lineage would have separated from the batrachian lineage sometime in the Carboniferous, ~315 Ma (see Figure 9.1). There are difficulties with this new phylogenetic hypothesis, however, and not everyone accepts it. For example, caecilians and batrachians have pedicellate teeth but Paleozoic stereospondyls did not, so pedicellate teeth would have had to evolve independently in the caecilian and batrachian lineages. Like all new phylogenetic hypotheses, this one will drive new research, and perhaps its most valuable outcome will be to inspire new studies of the biology and evolution of stereospondyls.

Reptiliomorpha and the origin of amniotes

Reptiliomorpha includes more than 80% of the extant species of tetrapods. All extant reptiliomorphs are amniotes, a subclade of reptiliomorphs characterized by the presence of an amniotic egg and other features we discuss in Section 9.2. The extensive Paleozoic record of reptiliomorphs throws light on important characters found in extant amniotes. Here, we consider three examples: †*Archeria*, †*Seymouria*, and †*Orobates*.

(A) Skeleton of a Permian dissorophoidean, †*Cacops aspidephorus*

Osteoderms

Short tail

Position of tympanum

Parietal foramen

Large orbit

Nostril

Limbs splay out

Short snout

10 cm

Jonathan Chen/CC BY-SA 4.0

(B) Restoration of †*Cacops aspidephorus*

Tympanum

Dmitry Bogdanov/CC BY 3.0

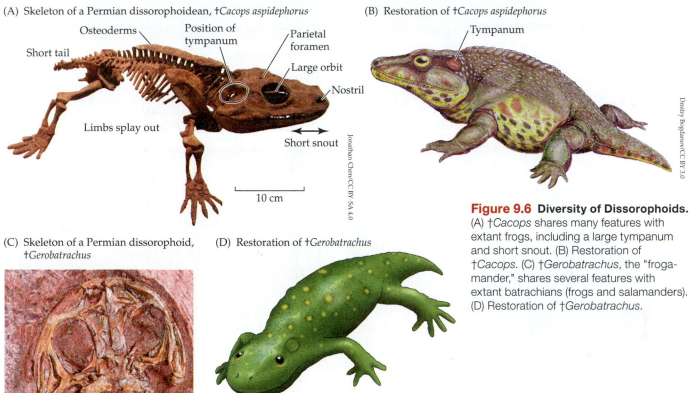

Figure 9.6 **Diversity of Dissorophoids.**
(A) †*Cacops* shares many features with extant frogs, including a large tympanum and short snout. (B) Restoration of †*Cacops*. (C) †*Gerobatrachus*, the "frogamander," shares several features with extant batrachians (frogs and salamanders). (D) Restoration of †*Gerobatrachus*.

(C) Skeleton of a Permian dissorophoid, †*Gerobatrachus*

From Anderson et al 2008; *Nature* 453: 515–518

10 mm

(D) Restoration of †*Gerobatrachus*

The aquatic predator †*Archeria* is known from many complete skeletons from the Early Permian of Texas (**Figure 9.7A**). It shares with other reptiliomorphs features of the vertebrae. In the vertebrae of more basal tetrapods, intercentra are larger than the pleurocentra. In contrast, the intercentra of †*Archeria* are roughly the same size as the pleurocentra (**Figure 9.7B**). The functional significance of this reduction in the size of the intercentrum is unknown, but within amniotes there are further evolutionary reductions and eventual loss of the intercentra.

Many excellent fossils of †*Seymouria*, including whole skeletons, have been collected in Early Permian rocks of North America and Europe. Unlike basal tetrapods and basal temnospondyls, which had relatively flat skulls, the skull of †*Seymouria* is domelike (**Figure 9.7C**). We know that †*Seymouria* reproduced by laying eggs in water, for aquatic larvae with external gills have been found (**Figure 9.7D**).

†*Diadectes* (**Figure 9.7E**) and related taxa from the Late Carboniferous through the Late Permian belong to †Diadectomorpha. The largest members of this group reached 3 m in length and had flattened teeth specialized for herbivory. It is rare to find both fossil trackways and articulated skeletons of the animals that made them, but that is case for †*Orobates pabsti* (**Figure 9.7F,G**). This combination allowed investigators to study details of its locomotion, even building a robot to test the ability of different gaits to reproduce the patterns of the footprints in the trackways. They found that †*Orobates* walked more like a caiman than a salamander, with relatively less side-to-side undulation than is typical for salamanders (see Figure 11.5).

†*Limnoscelis paludis* was a 2-m semiaquatic carnivore of the Late Carboniferous (**Figure 9.7H**). Its long skull with many teeth would have suited it for piscivory. There has been debate about

(A) Basal reptiliomorph, †*Archeria crassidica*

Dmitry Bogdanov/CC BY 3.0

(B) Vertebrae of †*Archeria* have a large intercentrum and a large pleurocentrum

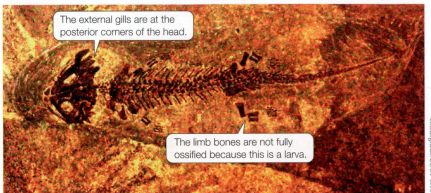

Neural arch with tall neural spine

Anterior zygapophysis — — Posterior zygapophysis

Intercentrum — — Pleurocentrum
— Tuberculum of rib
— Capitulum of rib

(C) Basal reptiliomorph, †*Seymouria*

The domelike skull lacks temporal fenestrae.

© William E. Bemis

(D) Larval †seymouriamorph with external gills

The external gills are at the posterior corners of the head.

The limb bones are not fully ossified because this is a larva.

Ghedoghedo/CC BY-SA 3.0

(E) Large herbivorous †diadectomorph, †*Diadectes sideropelicus*

The skull roof does not have temporal fenestrae.

Ghedoghedo/CC BY-SA 3.0

(F) †*Orobates pabsti* found with associated trackways

J.A. Nyakatura et al. 2015. *PLOS ONE*. doi.org/10.1371/journal.pone.0137284/CC BY 4.0

(G) Trackway assigned to †*Orobates pabsti*

From S. Voigt et al. 2007. *J Vert Paleontol* 27: 553–570

(H) A carnivorous †diadectomorph, †*Limnoscelis paludis*

Dmitry Bogdanov/CC BY-SA 3.0

Figure 9.7 Diversity of basal reptiliomorphs. (A) The Early Permian aquatic reptiliomorph †*Archeria* is known from several articulated skeletons. (B) In the vertebrae of more basal tetrapods, the intercentrum was larger than the pleurocentrum. In †*Archeria*, intercentra and pleurocentra were nearly equal in size; the intercentrum became further reduced in derived reptiliomorphs. (C,D) †*Seymouria* was a terrestrial reptiliomorph with aquatic larvae. (E–H) †Diadectomorpha included large herbivores with specialized teeth for processing vegetation (E); forms such as †*Orobates*, for which articulated fossils (F) have been found together with their tracks (G); and large carnivores such as †*Limnoscelis* (H). (See credit details at the end of this chapter.)

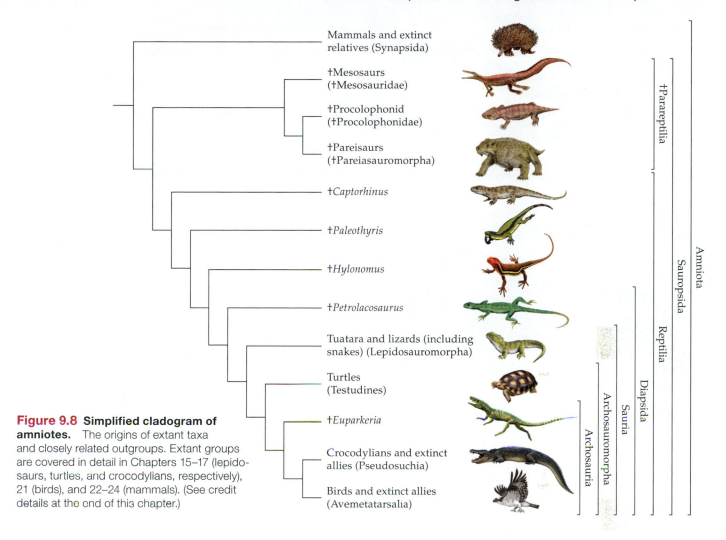

Figure 9.8 Simplified cladogram of amniotes. The origins of extant taxa and closely related outgroups. Extant groups are covered in detail in Chapters 15–17 (lepidosaurs, turtles, and crocodylians, respectively), 21 (birds), and 22–24 (mammals). (See credit details at the end of this chapter.)

how effective it was in catching terrestrial prey, however, for its short limbs would have limited its ability to run. The idea that †*Limnoscelis* was an amniote, and possibly the sister group of Synapsida, has been proposed, but most current analyses find †*Limnoscelis* to be a †diadectomorph.

Some workers have suggested that †diadectomorphs had amniotic eggs or were ovoviviparous because no fossils of aquatic larvae have been discovered. Currently, most authorities place †Diadectomorpha close to but outside of Amniota (see Figure 9.2).

Paleozoic diversification of amniotes

Amniotes were well established by the mid-Carboniferous, although they were at that time a minor part of the terrestrial fauna and did not radiate and diversify until the Permian. It is useful to briefly review the phylogenetic interrelationships of amniotes, shown in **Figure 9.8**, before turning to characters and features of amniotes in Section 9.2.

By the Late Carboniferous, the two extant clades of amniotes, **Synapsida** and **Sauropsida**, had appeared, distinguishable by skeletal differences (see Sections 9.2 and 9.3; in Chapter 13 we describe and compare some

significant physiological differences between the two clades that evolved as they adapted to terrestrial life.) Synapsida includes extant mammals and their many extinct relatives. Carboniferous synapsids were small insectivores, but larger carnivorous and herbivorous synapsids evolved in the Permian. Sauropsida contains †Parareptilia, an extinct lineage of the Late Permian and Early Triassic that included secondarily aquatic forms (†mesosaurs), small terrestrial omnivores (†procolophonids), and large herbivores (†pareiasaurs; **Figure 9.9**). Another sauropsid clade, Reptilia, gave rise to extant reptiles, including birds.

Early reptilians such as †*Captorhinus*, †*Paleothyris*, and †*Hylonomus* were small carnivores and insectivores. †*Captorhinus* lived in the mid-Permian, ~280 Ma. Fossils have been found in many parts of the world, including Texas, Brazil, Zambia, and India. Many examples are extraordinarily well preserved, such as the pair shown in **Figure 9.10A**. Like earlier reptiliomorphs, †*Captorhinus* had a skull with a solid temporal region lacking the temporal fenestration that characterizes more derived sauropsids. The astragalus bone, formed by fusion of three separate ankle bones present in more basal tetrapods, was present and next to a large calcaneus (**Figure 9.10B**).

(A) †*Mesosaurus*, an aquatic piscivore

Nobu Tamura/CC BY-SA 3.0

10 cm

(B) †*Procolophon*, a small omnivore

Nobu Tamura/CC BY-SA 3.0

5 cm

(C) †*Bradysaurus*, a large herbivorous †pareiasaur

Nobu Tamura/CC BY 3.0

50 cm

Figure 9.9 **Diversity of †Parareptilia.** (A) †Mesosaurs from the Early Permian played an important role in the realization that continental drift had occurred: they lived in nearshore environments and could not have crossed ocean basins, yet their fossils occur in both Africa and South America. (B) †*Procolophon* lived in the Permian of South Africa. (C) Reaching 3 m and weights of 600 kg, the herbivorous †*Bradysaurus* was part of a Permian megafauna of herbivores and carnivores that were the largest tetrapods of the Paleozoic.

(A) Two specimens of †*Captorhinus aguti*, a semiaquatic Permian sauropsid

Two sacral vertebrae attach the vertebral column to the ilium.

There is an axis vertebra.

The large astragalus evolved by fusion of three ankle bones.

The calcaneus is next to the astragalus.

The skull roof does not have temporal fenestrae.

10 cm

Didier Descouens/CC BY-SA 4.0

(B) Bones of the left hindlimb

Femur

Fibula

Tibia

Astragalus

Calcaneus

The ankle joint passes between the proximal and distal tarsal bones.

Distal tarsals

Phalanges

Metatarsals

I

II

V

III

IV

Phalangeal formula is 2:3:4:5:3

Figure 9.10 **Amniote characters of the Permian reptile †Captorhinus aguti.** (A) Known from many complete skeletons, †*Captorhinus* demonstrates three skeletal synapomorphies of amniotes including an axis vertebra, an astragalus bone in the ankle, and two sacral vertebrae. There are no temporal fenestrae in its skull (i.e., it is anapsid; see Figure 9.15A), making it a good general model for basal amniotes. (B) Left hindlimb of †*Captorhinus* shows the plane of its ankle joint between the proximal tarsals (astragalus and calcaneus) and distal tarsals. The phalangeal formula is the number of phalangeal bones in each digit. (Count the numbers of phalanges in the fossil shown in panel A to confirm that the phalangeal formula is 2:3:4:5:3.) The pointed terminal phalanges had claws in life.

Fossils of †*Paleothyris acadiana* from ~312 Ma (making it the earliest currently known fossil sauropsid) and the slightly younger †*Hylonomus lyelli* were preserved inside stumps of Late Carboniferous lycopods in Joggins, Nova Scotia. Both genera were small, terrestrial insectivores and provide information about the structure and biology of the earliest amniotes. **Figure 9.11** shows the bones of the skull and lower jaw of †*Paleothyris*. Note that the only openings in the skull roof are the nostril and orbit (i.e., there were no temporal fenestrae; see Section 9.3). †*Paleothyris* is an excellent starting point for understanding evolutionary changes in the arrangements of these bones in amniotes. Many changes in the details of skull roofing bones took place as different lineages evolved, but all amniotes retain the basic pattern seen in †*Paleothyris*; here we introduce key features of that pattern.

There are five series of paired dermal bones in the skull roof (see Figure 9.11A–C):

1. The lateral tooth-bearing series: premaxilla and maxilla
2. The median series: nasal, frontal, parietal, and postparietal
3. The circumorbital series: lacrimal, jugal, prefrontal, postfrontal, and postorbital
4. The temporal series: supratemporal and tabular
5. The cheek series: squamosal and quadratojugal

The lateral view of the outer surface of the lower jaw in Figure 9.11A shows three dermal bones: dentary, angular, and surangular. The dentary bears teeth and makes up about two-thirds of the length of the lower jaw. In the

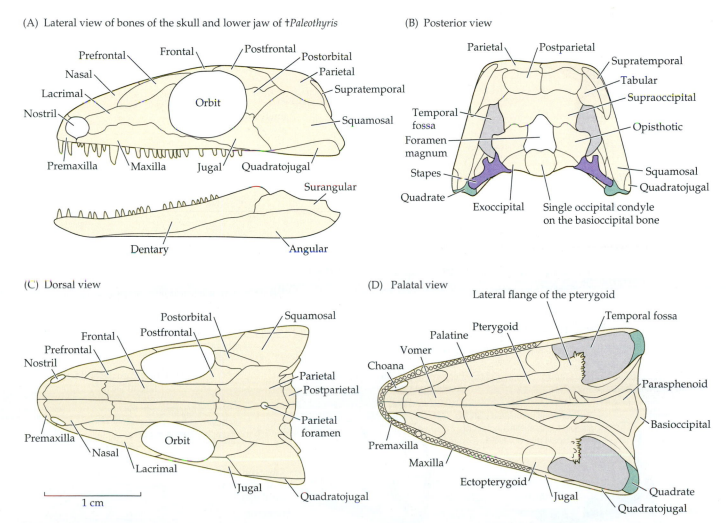

Figure 9.11 †*Paleothyris* **offers a good model for understanding later evolutionary changes in the skull and lower jaw.** (A) †*Paleothyris* retained a solid temporal region (A) and a lower jaw composed of many individual bones. (B) In posterior view, the single occipital condyle (on the basioccipital bone) and the pattern of bones surrounding the foramen magnum can be seen. Also shown are the quadrate bone and the stapes. The quadrate is the upper bone in the jaw joint; the stapes is homologous to the hyomandibula of gnathostomes. (C) Most of the skull roof is composed of long nasal and frontal bones, with parietal bones located behind the orbits and surrounding the parietal foramen (for the parietal, or pineal, eye). (D) The pterygoid bones of amniote palates have lateral flanges that serve as the origin for the pterygoideus muscle (see Figures 9.13 and 9.16). (Based on R.L. Carroll. 1969. *J Paleontol* 43: 151–170.)

posterior view (Figure 9.11B), the median supraoccipital and basioccipital bones, together with paired opisthotic and exoccipital bones, surround the foramen magnum (the opening through which the spinal cord exits the skull). Also seen in Figure 9.11B are the quadrate and the stapes. The quadrate is homologous to the palatoquadrate of an idealized gnathostome (see Figure 3.9B), and as in basal gnathostomes it forms the upper component of the articulation between the upper and lower jaws. The stapes is homologous to the hyomandibula of a basal gnathostome and conducts sound to the inner ear in many tetrapods (see Section 11.6 and Section 22.2).

Dermal bones of the palate of †*Paleothyris* are the paired vomer, palatine, ectopterygoid, and pterygoid

(Figure 9.11D). The pterygoid has a prominent pterygoid flange with a few small teeth (we discuss the role of the pterygoid flange as the origin for the pterygoideus muscle in Section 9.3). Posteriorly, the median parasphenoid and basioccipital bones form the floor of the braincase.

The temporal fossa (Figure 9.11B,D) is the space beneath the skull roof that houses the main jaw closing muscle, the adductor mandibulae. Because †*Paleothyris* does not have any openings in the temporal region of the skull, these muscles would not have been able to bulge very much upon their contraction.

The life restoration of the lizardlike †*Hylonomus* (**Figure 9.12A**) shows the importance of several postcranial skeletal features of amniotes. In that restoration, the animal's

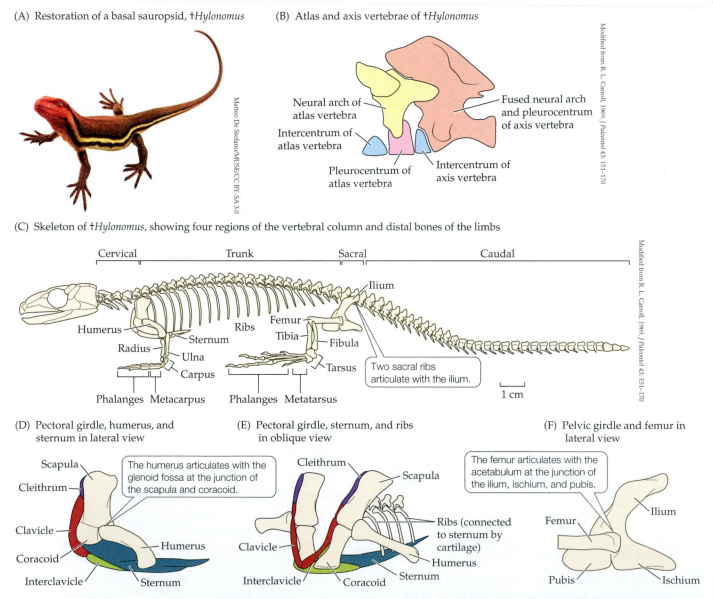

(A) Restoration of a basal sauropsid, †*Hylonomus*

Matteo De Stefano/MUSE/CC BY-SA 3.0

(B) Atlas and axis vertebrae of †*Hylonomus*

Neural arch of atlas vertebra
Intercentrum of atlas vertebra
Pleurocentrum of atlas vertebra
Fused neural arch and pleurocentrum of axis vertebra
Intercentrum of axis vertebra

Modified from R. L. Carroll, 1969. *J Palientol* 43: 151–170

(C) Skeleton of †*Hylonomus*, showing four regions of the vertebral column and distal bones of the limbs

Cervical   Trunk   Sacral   Caudal

Ilium
Humerus
Radius
Ulna
Carpus
Sternum
Ribs
Femur
Tibia
Fibula
Tarsus
Two sacral ribs articulate with the ilium.
Phalanges Metacarpus
Phalanges Metatarsus

1 cm

Modified from R. L. Carroll, 1969. *J Palientol* 43: 151–170

(D) Pectoral girdle, humerus, and sternum in lateral view

Scapula
Cleithrum
Clavicle
Coracoid
Interclavicle
Sternum
Humerus
The humerus articulates with the glenoid fossa at the junction of the scapula and coracoid.

(E) Pectoral girdle, sternum, and ribs in oblique view

Cleithrum
Scapula
Clavicle
Interclavicle
Coracoid
Ribs (connected to sternum by cartilage)
Humerus
Sternum

(F) Pelvic girdle and femur in lateral view

The femur articulates with the acetabulum at the junction of the ilium, ischium, and pubis.
Femur
Ilium
Pubis
Ischium

Figure 9.12 Postcranial features of †*Hylonomus*.
(A) A life restoration of †*Hylonomus* incorporates information based on postcranial osteology. (B) The large axis and atlas vertebrae, a synapomorphy of amniotes, allowed rotation of the head. (C) The vertebral column of †*Hylonomus* has cervical, trunk, sacral, and caudal regions, and, like other amniotes, two sacral ribs that articulate with the pelvic girdle. (D,E) The pectoral (shoulder) girdle consists of endochondral bones (scapula and coracoid) and dermal bones (cleithrum, clavicle, and interclavicle). The interclavicle and some trunk ribs articulate with the sternum. (F) All three bones of the pelvic girdle (ilium, ischium, and pubis) are endochondral derivatives. (See credit details at the end of this chapter.)

head is turned and tilted slightly, movements made possible by the presence of both atlas and axis vertebrae (**Figure 9.12B**). You can see that its limbs are more slender than those of †*Eryops* (see Figure 9.4A), †*Diadectes* (see Figure 9.7E), and †*Captorhinus* (see Figure 9.10A). The long fingers and toes with claws on their tips suggest agility and climbing ability.

†*Hylonomus* and other tetrapods have four regions of the vertebral column: cervical, trunk, sacral, and caudal (**Figure 9.12C**; mammals have a fifth, the lumbar region, composed of trunk vertebrae that lost their ribs; see Figure 11.2). Cervical vertebrae of †*Hylonomus* bear short ribs, and trunk vertebrae bear ribs, several of which articulate with the sternum. Two sacral vertebrae have sacral ribs that articulate with the ilium.

As in basal tetrapods, the pectoral girdle of basal amniotes has both endochondral and dermal bones. In †*Hylonomus*, endochondral bones of the pectoral girdle are the scapula and coracoid; the glenoid fossa where these two bones meet articulates with the head of the humerus (**Figure 9.12D**). Two dermal bones, the cleithrum and clavicle, are incorporated into the pectoral girdle along the anterior border of the scapula, and a new dermal bone, the interclavicle, forms a ventral midline structure beneath the sternum (**Figure 9.12D,E**). The clavicle connects the scapula to the interclavicle and sternum. The general tendency within amniotes is to further reduce the dermal bones of the pectoral girdle. For example, the clavicle is the only dermal bone in your pectoral girdle (see Figure 22.21B on the evolution of the pectoral girdle in synapsids).

Like more basal sarcopterygians, and as in all amniotes, all three paired elements of the pelvic girdle of †*Hylonomus*—the ilium, ischium, and pubis—are endochondral bones. The femur articulates with the pelvic girdle at the acetabulum, where the three pelvic bones meet (**Figure 9.12F**). Further evolutionary changes in pelvic anatomy of amniotes relate to greater use of the hindlimbs for locomotion, and in later chapters we trace key changes in the pelvic girdle of archosaurs (see Figure 18.8) and synapsids (Figure 22.21B).

9.2 Characters of Amniotes

LEARNING OBJECTIVES

9.2.1 Explain the functional significance of each of the four skeletal characters synapomorphic for Amniota.

9.2.2 Describe the functions of the four extraembryonic membranes of an amniotic egg.

9.2.3 Summarize derived features of amniotes associated with the integumentary, excretory, and respiratory systems.

The skeletons, skin, internal organs, and reproductive modes of amniotes differ from those of non-amniotes and, in combination, these derived characters allow amniotes to pursue ways of life that are distinctly different from those of non-amniotes. These characters, the types of information that we can use to understand their phylogenetic origin and distribution, and their functional implications are presented in **Table 9.1**.

Skeletal characters

Four characters of the cranial and postcranial skeleton are synapomorphies of Amniota. The first is the presence of a lateral flange on the pterygoid bones, which can be seen in early amniotes such as †*Paleothyris* (see Figure 9.11D). This flange is the origin for the pterygoideus muscle, a jaw-closing muscle that evolved from the adductor mandibulae of more basal tetrapods and plays many roles in feeding systems of amniotes (**Figure 9.13**). The pterygoid flange and pterygoideus muscle are present in extant synapsids and sauropsids, so this system must have been present in their common amniote ancestor.

The second skeletal synapomorphy of amniotes is modification of the second cervical vertebra to form the axis, which allows head rotation. In addition to its presence in basal amniotes such as †*Hylonomus* (see Figure 9.12B), the axis vertebra is present in extant synapsids and sauropsids, so by phylogenetic inference it is a synapomorphy of amniotes.

Amniotes share the presence of an astragalus bone in the ankle and a mesotarsal joint between the proximal tarsal bones (the astragalus and calcaneus) and the distal tarsals. The latter is mesotarsal in the sense that the joint is between tarsal bones rather than between the tibia and fibula and the tarsal bones. These features can be seen in early amniotes such as †*Captorhinus* (see Figure 9.10B), but there have been many changes in ankle anatomy over the course of amniote evolution, some of which are diagnostic features for amniote groups, as we will describe in Section 9.3.

The fourth skeletal synapomorphy of amniotes is the presence of two or more sacral vertebrae. This is seen in early amniotes such as †*Captorhinus* (see Figure 9.10A) and †*Hylonomus* (see Figure 9.12C). Except for legless forms and a few others with secondary reductions of the hindlimbs (such as whales), no amniotes have fewer than two sacral vertebrae.

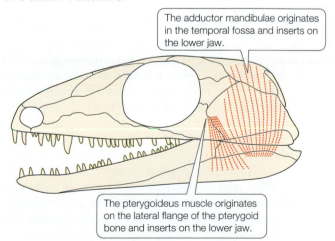

The adductor mandibulae originates in the temporal fossa and inserts on the lower jaw.

The pterygoideus muscle originates on the lateral flange of the pterygoid bone and inserts on the lower jaw.

Figure 9.13 Arrangement of jaw-closing muscles of †*Paleothyris*. Amniotes such as †*Paleothyris* have a pterygoideus muscle that originates on the lateral flange of the pterygoid bone. (Based on R.L. Carroll. 1969. *J Paleontol* 43: 151–170.)

Table 9.1 Derived features of amniotes and their functional significance

Anatomical feature	Visible in fossils?	How to infer presence	Functional significance
Lateral flange of the pterygoid bone	Yes	Can be seen in fossils and confirmed by phylogenetic inference	Jaw adductor muscles become divided into adductor mandibulae and pterygoideus; pterygoideus originates from the pterygoid flange and allows different modes of processing food
Second cervical vertebra modified into an axis	Yes	Can be seen in fossils and confirmed by phylogenetic inference	Greater mobility of head on neck allows changes in feeding and social behavior
Astragalus bone and mesotarsal joint	Yes	Can be seen in fossils and confirmed by phylogenetic inference	Allows foot to be used more like a lever than a holdfast; indicates more limb-based (rather than axial-based) locomotion
More than one sacral vertebra	Yes	Can be seen in fossils and confirmed by phylogenetic inference	Reflects increasing use of hindlimbs in limb-based locomotion
Amniote egg (eggshell and four extraembryonic membranes)	No[a]	Phylogenetic inference	No longer need to return to a body of water for reproduction
Waterproof skin (multilayered keratinous epidermis, thick dermis)	No	Phylogenetic inference	Limits water loss by evaporation and allows greater independence from water
Penis	No	Phylogenetic inference	Used in internal fertilization, which is essential for amniote eggs
Metanephric kidney drained by the ureter	No	Phylogenetic inference	Contributes to improved water retention
Costal aspiration (breathing with ribs)	Not directly	Phylogenetic inference and anatomy of fossils	Allows higher rates of lung ventilation for higher metabolic rates; also allows loss of CO_2 via lungs rather than skin
Trachea with cartilaginous rings (strengthens trachea and prevents it from collapsing)	Not directly	Phylogenetic inference and anatomy of fossils	Long trachea allows a longer neck, which in turn allows changes in feeding and social behavior

[a]Although there are many examples of fossilized amniote eggs in the record (see Figure 18.27 for examples), these fossil eggs occur in deposits that are much younger than the earliest amniote fossils.

The amniotic egg

Like lungfishes, basal tetrapods such as lissamphibians lay their eggs in water and have an aquatic larval stage. In contrast, amniotes produce a shelled **amniotic egg** (think chicken egg) that is laid on land. Embryonic amniotes bypass the aquatic larval stage, do not form functional gills at any stage in development, and exchange gases by diffusion through the eggshell. Aquatic amniotes either come ashore to lay eggs (as in turtles) or give birth to fully-formed young (as in whales). In some amniotes, notably therian mammals (see Chapter 23), evolutionary modifications of the amniotic egg involved loss of the eggshell and changes in the extraembryonic membranes related to viviparity.

A flexible, leathery eggshell is probably the generalized amniote condition, and persists in many extant lizards, snakes, turtles, and all extant monotremes. In some lizards, some turtles, and in all crocodylians and birds, the eggshell contains calcium salts and is rigid. Because fertilization must occur before the eggshell forms around the egg, all amniotes have internal fertilization. To accomplish this, many male amniotes have a penis used to inseminate

females. It is likely that the penis is homologous across amniotes, but functional analogues evolved independently in many fishes as well as in caecilians.

The amniotic egg, exemplified by the chicken egg shown in **Figure 9.14A**, is a remarkable biological structure. The eggshell is the first line of defense against mechanical damage to the embryo, and pores in the eggshell permit exchange of water vapor, oxygen, and carbon dioxide with the surrounding air. Within the shell, albumen (egg white) protects the embryo from mechanical damage and provides a reservoir of water and protein for embryonic development.

The yolk is the energy supply for the developing embryo. Like all vertebrates, the yolk of an amniote is contained within an extraembryonic membrane known as the **yolk sac**. The yolk sac is extraembryonic in the sense that it lies entirely outside the embryo's body (see Figure 9.14A). Nutrients are transported from the yolk sac to the embryo's body via vessels of the circulatory system that develop in the wall of the yolk sac. As yolk is depleted with the growth of the embryo, the size of the yolk sac decreases.

(A) Structure of an amniotic egg

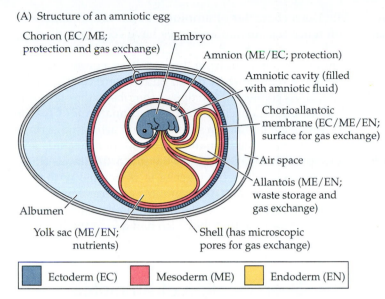

(B) Life restoration of the earliest known amniote embryo, †*Mesosaurus tenuidens*, from the Early Permian of Uruguay

From G. Piñeiro et al. 2012. *Hist Biol* 24: 620–630. Illustration by Gustavo Lecuona

5 mm

Figure 9.14 The amniotic egg. (A) The chicken egg is emblematic of the shelled amniotic egg. Shown are the protective eggshell and albumen, as well as the extraembryonic membranes. Extraembryonic membranes, which derive from the three embryonic germ layers described in Section 2.4, serve for protection, gas exchange, and storage of metabolic waste products. Nutrients are stored in the yolk sac, which shrinks as nutrients are consumed and the embryo grows. (B) Artist's rendition of an embryo of †*Mesosaurus tenuidens*. The Paleozoic specimens on which the rendering is based were found in association with the abdominal regions of larger individuals and represent the only amniote embryos known from that era.

Three additional extraembryonic membranes form in an amniotic egg: the **amnion**, **chorion**, and **allantois** (see Figure 9.14A). The amnion and chorion form initially from folds of ectoderm and mesoderm that grow out from the embryo, rise above it, and eventually fuse to form one membrane, with mesoderm on the outside (the amnion) and a second membrane with mesoderm on the inside (the chorion). The amnion defines the amniotic cavity, which contains amniotic fluid that surrounds and helps to protect a delicate early-stage embryo from mechanical shocks (e.g., if the egg is disturbed, the amniotic fluid dampens physical movements of the embryo). The chorion continues to expand, eventually enclosing the embryo and the other extraembryonic membranes and contacting the inside surface of the eggshell.

The allantois is an outgrowth of the embryonic hindgut and is composed of endoderm on the inside and mesoderm on the outside. The allantois serves two functions: (1) it stores nitrogenous wastes produced by the embryo's metabolism; and (2) it expands to contact the chorion to form and vascularize the chorioallantoic membrane (see Figure 9.14A). The mesodermal layer of the allantois can give rise to blood vessels, thus enabling it to vascularize the chorioallantoic membrane and making it possible to exchange oxygen and carbon dioxide with the environment via pores in the eggshell.

The earliest amniote embryos in the fossil record are of †mesosaurs from an Early Permian Lagerstätte in Uruguay (~286 Ma, at least 34 Ma after the origin of amniotes; see Figure 9.1). A partially articulated embryo of †*Mesosaurus tenuidens*, coiled up like a chicken embryo in its shell, was found in association with the abdominal region of larger individuals (**Figure 9.14B**). An intriguing feature of this specimen is a structure interpreted as an egg tooth—a structure used by extant amniotes with shelled eggs to break through the eggshell during hatching and lost soon after hatching. The eggshell of this specimen of †*M. tenuidens*, however, is unrecognizable, and this led researchers to interpret that †mesosaurs either laid eggs at a very late stage in development or that they were viviparous and gave birth to fully developed young. No other amniote embryos are known from the Paleozoic.

What were the initial evolutionary advantages of extraembryonic membranes? We do not know the answer to this important question. One possibility is that extraembryonic membranes initially appeared during the evolution of viviparity, and that eggshells evolved later. Another is that the membranes originally facilitated gas exchange within the eggshell that provided mechanical support on land, allowing the evolution of larger eggs that produced larger hatchlings. Both possibilities are speculative, but even though we cannot provide a single convincing explanation for the origin of the amniotic egg, its importance in the subsequent evolution of amniotes is clear.

Other soft-tissue characters of amniotes

Phylogenetic inference indicates that amniotes share several soft-tissue characters that, like the amniotic egg and other reproductive characters, typically are not preserved in fossils (see Table 9.1). We do not know whether these characters are unique to amniotes or whether they evolved in reptiliomorphs prior to the origin of amniotes. Nevertheless, it is convenient to describe them here and in the following subsection because they are commonly cited as synapomorphies of amniotes.

Waterproof skin   Amniotes have waterproof skin consisting of a multilayered keratinous epidermis and a thick dermis. Especially important in limiting water loss are hydrophobic lipids in the skin (see Section 11.7). Because both extant synapsids (mammals) and extant sauropsids (lepidosaurs, turtles, crocodylians, and birds) have skin that resists transcutaneous water movement, we can infer that their common amniote ancestor also had waterproof skin.

Skin elaborations   Amniotes have a great variety of skin elaborations—scales, hair, feathers, nails, beaks, and horns—formed from keratin. Scales, hair, and feathers are homologous in the sense that all of them develop from epidermal thickenings, or placodes (note that these epidermal placodes are different from the neurogenic placodes that give rise to sensory structures and nerves; see Section 2.4) and express a similar suite of developmental signaling molecules. Changes in gene expression during development caused the placodes to produce different types of epidermal structures from the same primordium.

Extant amniotes have keratinized skin containing alpha keratin, so we can make the strong phylogenetic inference that alpha keratin was present in the earliest amniotes. Extant sauropsids also have a second, harder type of keratin, beta keratin, that is not found in mammals. Thus, we can use phylogenetic inference to conclude that that the hard, scaly skin of extant sauropsids was not a basal amniote feature.

Metanephric kidneys   As adults, all extant amniotes have metanephric kidneys (see Section 13.4), supporting the strong phylogenetic inference that their common ancestor did, too. Early embryonic amniotes pass through developmental stages in which the kidney consists of simple pronephric tubules, then more complex mesonephric tubules, in both cases drained by the archinephric duct (see Figure 2.15). The metanephric kidney initially develops as the ureter bud, an outgrowth of the archinephric duct that grows into a mass of undifferentiated intermediate mesoderm. The ureter bud induces the mass of intermediate mesoderm to develop kidney tubules, which are then drained via the definitive ureter. The initial pronephric and mesonephric tubules degenerate, and by the time an amniote hatches or is born, only the metanephric kidney functions in waste removal.

Lung ventilation   We regard costal ventilation (moving air into and out of the lungs using movements of the ribs) as a character of amniotes based on phylogenetic inference (i.e., extant amniotes have costal ventilation) and anatomy of fossils. For example, trunk ribs of fossil amniotes extend out from the vertebrae, curve ventrally, and have scars at their distal ends where cartilaginous extensions connected the tips of the ribs to the sternum.

Both phylogenetic inference and anatomy of fossils show that a long trachea supported by cartilaginous rings was present in the first amniotes. Extant synapsids and sauropsids have a long trachea with cartilaginous rings. The presence of such a trachea is associated with relatively long necks, and early amniotes such as †*Hylonomus* had relatively longer necks than more basal tetrapods, so we can assume that they also had a long trachea with cartilaginous rings.

9.3 Diversification of Amniotes

LEARNING OBJECTIVES

9.3.1 Characterize patterns of temporal fenestration and the possible advantages of such openings in the skull.

9.3.2 Describe changes in the ankles of reptiles that characterize lepidosauromorphs, crocodylomorphs, and dinosauromorphs

In this section, we explore the diversification of amniotes by tracing skeletal evolution. First, we describe the evolution of bones and fenestrae in the temporal region of the amniote skull and the differences that define its major subgroups. We then trace the evolution of ankle bones to understand major patterns in the diversification of reptiles.

Temporal fenestration: Synapsids and diapsids

Like more basal tetrapods, reptilomorphs such as †*Seymouria* and †*Diadectes* have a solid, bony skull roof in the temporal region (see Figure 9.7C,E). Such a solid skull roof is also present in basal sauropsids such as †*Captorhinus* (see Figure 9.10) and †*Paleothyris* (see Figure 9.11). By phylogenetic inference, then, the earliest amniotes had a solid skull roof, as in the diagram shown in **Figure 9.15A**. We call this arrangement of the skull roofing bones **anapsid** (Greek *an*, "without"; *apsis*, "a loop," referring to an opening in the side of the skull). Openings in the skull's temporal regions, known as fenestrae (Latin *fenestra*, "window"), evolved within the amniote lineage. The different patterns of **temporal fenestration**—that is, the number and arrangement of fenestrae—are diagnostic of the two extant radiations of amniotes, Synapsida and Diapsida (see Figure 9.17).

In the **synapsid** ("single bar") condition, there is a single lower temporal fenestra on each side of the head (**Figure 9.15B**). Below the fenestra is the lower temporal bar formed by the jugal and quadratojugal bones. Above the fenestra, the postorbital and squamosal bones meet. The synapsid condition characterizes mammals and their extinct relatives in Synapsida (see Chapter 22).

(A) Anapsid condition: solid roof in temporal region

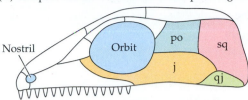

(B) Synapsid condition: lower temporal fenestra

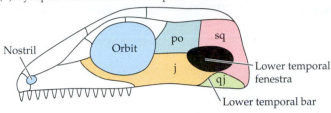

(C) Diapsid condition: lower and upper temporal fenestrae

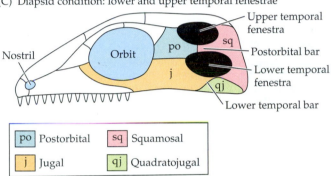

| po | Postorbital | sq | Squamosal |
| j | Jugal | qj | Quadratojugal |

Figure 9.15 Anapsid, synapsid, and diapsid skulls showing bones of the temporal region. The anapsid condition (A) was present in the common ancestor of amniotes. Temporal fenestrae evolved independently in synapsids (B) and diapsids (C). (After A.S. Romer. 1966. *Vertebrate Paleontology*. University of Chicago Press: Chicago.)

In the **diapsid** ("double bar") condition, there are two temporal fenestrae (**Figure 9.15C**). The lower temporal bar (formed by the jugal and quadratojugal bones) is below the lower temporal fenestra. The upper temporal bar (formed by the postorbital and squamosal bones) is between the lower and upper fenestrae. This condition is seen in many sauropsids that comprise the extant group Diapsida (see Figure 9.17)

Reorientation and subdivision of the adductor mandibulae in the more domed skull of basal amniotes may explain convergent evolution of skull fenestration in synapsids and diapsids. The jaw-closing muscles of a typical reptiliomorph were relatively simple and contained beneath the skull roof (**Figure 9.16A**). With the lateral flange of the pterygoid and pterygoideus muscle (see Figure 9.13), evolution of subdivisions of the adductor mandibulae in amniotes had already started. The next step was to further increase the area that would allow origin of the jaw adductor muscles. Even a small opening in the skull roof (for example, in an area where three bones fail to completely meet during development) might allow jaw muscles to expand their origins to the outside surface of the skull (**Figure 9.16B**).

(A) Simple jaw closing muscle of a basal reptiliomorph

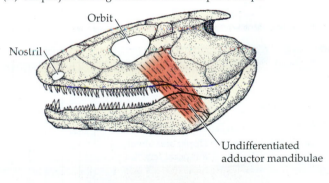

(B) Differentiated jaw closing muscles of a synapsid

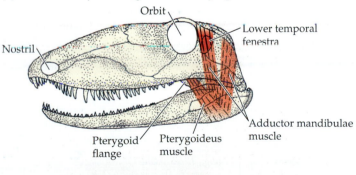

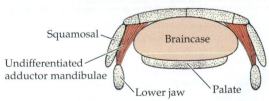

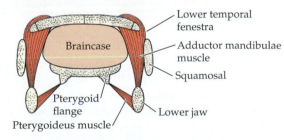

Figure 9.16 Evolution of temporal fenestration is linked to evolution of jaw-closing muscles. (A) Jaw-closing muscles (adductors) were simple and undifferentiated in basal reptiliomorphs. (B) Amniotes share the presence of a pterygoid flange and pterygoideus muscle. Within amniotes, convergent evolution of temporal fenestrae in synapsids and diapsids allowed further differentiation of adductor muscles in the temporal region. (See credit details at the end of this chapter.)

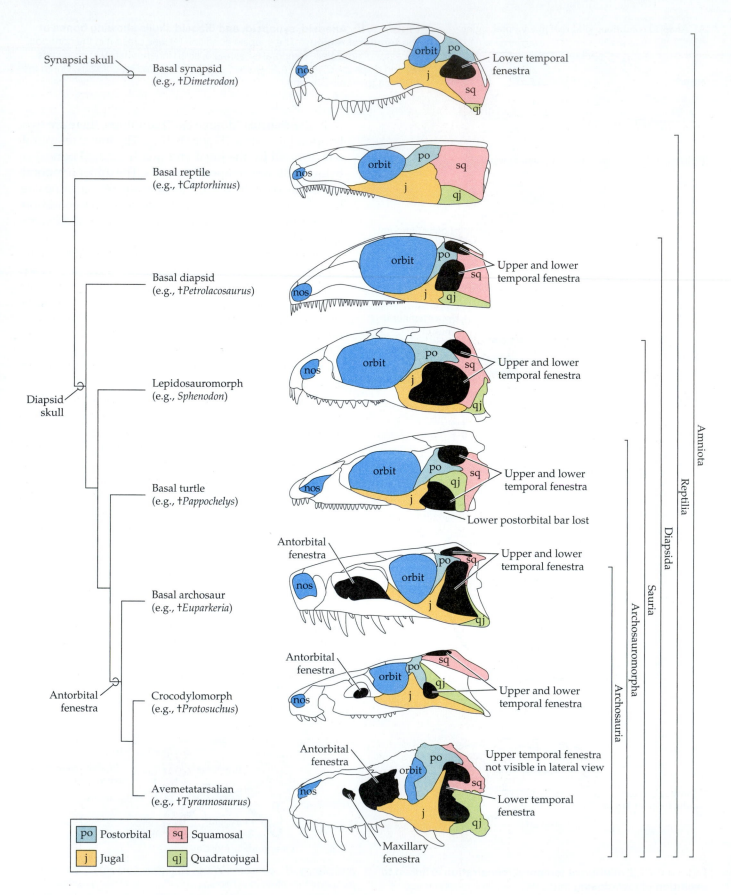

◀ **Figure 9.17 Temporal fenestration of amniotes.** Temporal fenestration evolved twice in amniotes, once in the line leading to Synapsida and once in the line leading to Diapsida. Within diapsids, there have been many changes in the original condition, which is most closely approximated among extant tetrapods by a basal lepidosauromorph, the tuatara (*Sphenodon*). Molecular phylogenetic and paleontological evidence place turtles within Diapsida, even though extant turtles lack temporal fenestrae. Archosaurs share the presence of an additional fenestra in front of the orbit (the antorbital fenestra). In the lineage that leads to extant crocodylomorphs, consolidation of the skull and reduction in the sizes of the fenestrae occurred. In the lineage leading to extant birds, additional cranial fenestrations evolved and the skull became a strutwork of bones. (See credit details at the end of this chapter.)

Evolutionary enlargements of such an opening could have led to larger, more differentiated muscles that allowed more powerful bites.

In the next several paragraphs we describe the evolution of temporal fenestrae among diapsids, illustrated in **Figure 9.17**. The earliest diapsid is †*Petrolacosaurus* from the Late Carboniferous of Kansas. All extant diapsids belong to Sauria, the most basal extant group of which is Lepidosauria (tuatara, lizards, and snakes). The lower temporal bar of diapsids was lost and re-established in the lineage leading to the tuatara (*Sphenodon*), where it braces the upper jaw to withstand the unique anteroposterior shearing movements of the lower jaw (see Section 15.3).

The other extant diapsids are in Archosauromorpha, with turtles as the sister group of Archosauria. The early members of the turtle lineage, such as †*Pappochelys*, retain temporal fenestrae. The anapsid skulls of extant turtles secondarily evolved from that diapsid condition, with extensive loss of bone from the rear of the skull, probably as one of the many anatomical changes that accompanied evolution of the shell (see Table 16.1).

In addition to upper and lower temporal fenestrae, basal archosaurs such as †*Euparkeria* have a large antorbital fenestra thought to be related to reducing the weight of the skull. Further evolutionary changes in the skull occurred in two extant clades of archosaurs. For example, crocodylomorphs reduced the overall sizes of the two temporal fenestrae, and the lineage including extant alligators and crocodiles secondarily lost the antorbital fenestra. Avemetatarsalia is the other extant clade of archosaurs; it includes dinosaurs and extant birds. The skull of avemetatarsalians trended toward more fenestrae with strutlike bones. For example, the skull of †*Tyrannosaurus* has a maxillary fenestra in addition to the antorbital fenestra. Despite its huge size, the skull of this iconic dinosaur looks more like a framework for soft tissues than a solid box (see Figure 19.6C).

Ankle evolution in amniotes

The ankle (tarsus) of basal tetrapods consists of many small bones (**Figure 9.18A**). Movement between the bones is possible, but there is no distinct ankle joint to serve as a plane of bending within the ankle. Axial muscles power locomotion, with the feet functioning mainly as pivot points around which the hindlimb rotates (see Figure 11.5).

In amniotes, three small proximal tarsal bones fuse to form the astragalus (for example, see †*Captorhinus* in Figure 9.10). The astragalus articulates with the tibia, and the calcaneus (the other proximal tarsal bone) articulates with the fibula. A distinct plane of bending, termed a mesotarsal joint because it is in the middle of the ankle, lies between the proximal and distal tarsal bones. This consolidation of bones and the evolution of a distinct mesotarsal joint relate to evolutionary changes in the locomotor role of the hindlimbs. Powerful extensor muscles in an amniote's thigh extend the shank (lower leg), and the distinct bending plane of the mesotarsal joint allows the foot to be used as a lever to push the body forward.

Synapsids inherited an ankle like that of †*Captorhinus*. Over the course of synapsid evolution, the calcaneus of derived synapsids evolved a long process known as the calcaneal heel (**Figure 9.18B**). This process is familiar to us because it serves as the insertion site for the Achilles tendon of the gastrocnemius muscle, a powerful extensor of the foot.

Ankle evolution took different paths in reptiles as shown in Figure 9.18B. Lepidosauromorphs retained a mesotarsal joint, but the astragalus and calcaneus fused to form a single element. Basal turtles such as †*Proganochelys* also have a fused astragalus and calcaneus and a mesotarsal joint. An S-shaped ankle joint passing between the astragalus and fibula and the calcaneus and distal tarsals evolved in basal archosauromorphs such as †*Euparkeria*. A similar arrangement, in which a peg on the astragalus fits into a socket on the calcaneus, occurs in crocodylomorphs. Such an ankle joint allows both extension and rotation of the foot, and helps to explain why crocodiles can crawl, walk, and gallop (see Figure 17.2).

Avemetatarsalians evolved an upright posture and bipedalism, and the plane of motion in the ankle joint became secondarily simplified to a mesotarsal joint between the fused astragalus and calcaneus proximally and distal tarsals fused with metatarsals. The ankle joint functions like a hinge and lacks any ability for rotation but is well suited for fast bipedal running.

(A) Left ankle and foot of basal tetrapod

(B) Ankle evolution in amniotes (left ankle is shown)

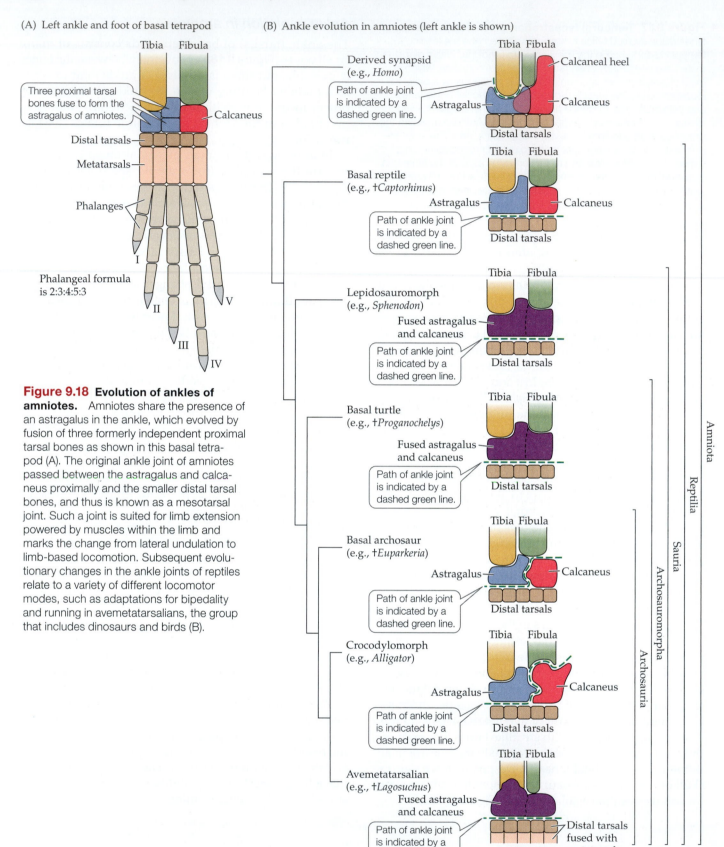

Three proximal tarsal bones fuse to form the astragalus of amniotes.

Phalangeal formula is 2:3:4:5:3

Figure 9.18 Evolution of ankles of amniotes. Amniotes share the presence of an astragalus in the ankle, which evolved by fusion of three formerly independent proximal tarsal bones as shown in this basal tetrapod (A). The original ankle joint of amniotes passed between the astragalus and calcaneus proximally and the smaller distal tarsal bones, and thus is known as a mesotarsal joint. Such a joint is suited for limb extension powered by muscles within the limb and marks the change from lateral undulation to limb-based locomotion. Subsequent evolutionary changes in the ankle joints of reptiles relate to a variety of different locomotor modes, such as adaptations for bipedality and running in avemetatarsalians, the group that includes dinosaurs and birds (B).

Summary

9.1 Paleozoic Tetrapods and the Origins of Extant Groups

In the early Carboniferous, Crown Tetrapoda diversified into clades that ultimately led to Lissamphibia and Amniota.

Temnospondyli and Reptiliomorpha are two large extant clades of tetrapods. Several other types of Paleozoic tetrapods have unknown relationships with extant taxa and are not considered in this chapter.

Temnospondyli includes Lissamphibia and all tetrapods more closely related to Lissamphibia than to Reptiliomorpha. Paleozoic temnospondyls had aquatic larvae, and most adults were semiaquatic; some reached large body sizes. Temnospondyls persisted into the Mesozoic, but by the start of the Cenozoic, the only remaining temnospondyls were the comparatively smaller lissamphibians (caecilians, salamanders, and frogs).

- †*Chinlestegophis jenkinsi*, a small Triassic stereospondyl, has been interpreted as a stem caecilian because its features are intermediate between typical stereospondyls and caecilians. If correct, then Stereospondyli is not extinct because it is represented by extant caecilians.

- Salamanders and frogs together form Batrachia. Batrachians have been interpreted as derived Dissorophoidea. †*Cacops* and †*Gerobatrachus hottoni* are examples of Paleozoic dissorophoids.

This recent phylogenetic hypothesis—that caecilians are derived stereospondyls and batrachians are derived dissorophoids—means that Lissamphibia as a group has been extant since the Paleozoic. It also agrees with the estimated date of ~315 Ma for the split between caecilians and batrachians based on molecular phylogenetic information.

Reptiliomorpha includes more than 80% of extant tetrapod species. All extant reptiliomorphs are amniotes, characterized by presence of an amniotic egg. †*Archeria*, †*Seymouria*, and †*Diadectes* are examples of Paleozoic reptiliomorphs.

By the Late Carboniferous, the two extant clades of amniotes—Synapsida and Sauropsida—had evolved. Synapsida includes mammals and their extinct relatives. Sauropsida contains †Parareptilia and Reptilia.

†Parareptilia, an extinct lineage of the Late Permian and Early Triassic, included forms such as †mesosaurs and †pareiasaurs. Reptilia gave rise to extant reptiles, including birds. †*Captorhinus*, †*Paleothyris*, and †*Hylonomus* are examples of early reptiles.

†*Paleothyris*, a starting point to understand evolutionary changes in the arrangements of skull roofing bones in amniotes, has a solid skull roof (i.e., it is anapsid, meaning that it has no temporal fenestrae). The basic pattern of five series of paired dermal bones in the skull roof of †*Paleothyris* is retained in extant amniotes, despite the many evolutionary changes seen in different groups.

9.2 Characters of Amniotes

Four skeletal characters are synapomorphic for Amniota.

- A lateral flange on the pterygoid bone (the origin for the pterygoideus muscle).

- Modification of the second cervical vertebra into an axis vertebra that allows head rotation.

- An astragalus bone and mesotarsal ankle joint.

- Two or more sacral vertebrae.

Several nonskeletal synapomorphies characterize amniotes, including the amniotic egg.

The shelled amniotic egg is laid on land. Pores of the eggshell permit exchanges of water vapor, oxygen, and carbon dioxide with the surrounding aerial environment. Albumen protects the embryo from mechanical damage and serves as a reservoir of water and protein for embryonic development. Yolk provides nutrition.

The amniotic egg has four extraembryonic membranes: the yolk sac, amnion, chorion, and allantois. The initial evolutionary advantages of the extraembryonic membranes and amniotic egg are unclear but may be related to the evolution of larger eggs and hatchlings.

Fertilization must occur before the eggshell forms around the egg; thus, all amniotes have internal fertilization, and many male amniotes have a penis to inseminate females.

The evolution of viviparity in some amniotes (notably therian mammals) resulted in loss of the eggshell and modifications of the extraembryonic membranes.

In addition to the amniotic egg, derived soft tissue characters of amniotes include waterproof skin, metanephric kidneys drained by ureters, costal (rib) ventilation, and a long trachea supported by cartilaginous rings. Together, these characters allowed amniotes greater independence from moist habitats and the ability to grow to very large body sizes.

9.3 Diversification of Amniotes

Amniotes are classified based on patterns of temporal fenestration. The earliest amniotes had a solid skull roof, the anapsid condition (i.e., no openings in the temporal region). Openings in the skull may have allowed the origins of jaw muscles to expand to the outside surface of the skull. Evolutionary enlargements of such openings might have led to larger, more differentiated muscles, and more powerful bites.

- In the synapsid condition, seen in mammals and their extinct relatives, there is a single lower temporal fenestra on each side of the head.

(Continued)

Summary (continued)

- In the diapsid condition, there are two temporal fenestrae (an upper and lower) on each side of the head.

All extant diapsids are in Sauria, the most basal extant group of which is Lepidosauria (tuatara, lizards, and snakes). Sauria also includes turtles, crocodylians, and birds (and their extinct dinosaur relatives). Evolutionary changes in these lineages can make it challenging to see both temporal openings in extant taxa (for example, the skulls of extant turtles lack temporal openings, but fenestrae can be seen in outgroups of the extant forms).

Ankle evolution in amniotes is phylogenetically informative and relates to new functions of the hindlimbs as levers for fast locomotion.

- Basal amniotes had already consolidated proximal bones of the ankle to form an astragalus and calcaneus.

- Within saurians, variations include fusions of the astragalus and calcaneus; a joint that passes between the astragalus and calcaneus; or the greatly solidified and simple ankle of dinosauromorphs, in which fusions of bones evolved on both sides of the mesotarsal ankle joint.

Discussion Questions

9.1 Most researchers agree that Lissamphibia is a derived group of Temnospondyli. This chapter explored a new phylogenetic hypothesis that caecilians are derived stereospondyls, whereas batrachians are derived dissorophoids. Propose an alternative phylogenetic hypothesis in which Lissamphibia remains a monophyletic group within Temnospondyli. What are some implications of your hypothesis?

9.2 Explain why three taxa of basal sauropsids—†Captorhinus, †Paleothyris, and †Hylonomus—are particularly useful models for interpreting basal amniotes.

9.3 Explain the roles of fossil evidence and phylogenetic inference in assessing skeletal and soft-tissue characters of amniotes.

9.4 Amniocentesis is a prenatal diagnostic procedure in which amniotic fluid is withdrawn from the amniotic cavity. Prior to inserting the needle, physicians use ultrasound imaging to select the safest spot for insertion, since it is important to avoid the fetus, placenta, and umbilical cord. Which two extraembryonic membranes will the needle necessarily pass through?

9.5 Distinguish the pattern of temporal fenestration in synapsids and diapsids, identifying bones that define the temporal openings.

Figure credits
Figure 9.1: †Trimerorhachis: Nobu Tamura/CC BY-SA 4.0; †Eryops: Dmitry Bogdanov/CC BY-SA 3.0; Cacops: Dmitry Bogdanov/CC BY 3.0; Archeria: Dmitry Bogdanov/CC BY 3.0; Seymouria: Nobu Tamura/CC BY-SA 4.0; Diadectes: Dmitry Bogdanov/CC BY-SA 3.0; Dimetrodon: Dmitry Bogdanov/CC BY-SA 3.0; Mesosaurus: Nobu Tamura/CC BY-SA 3.0; Bradysaurus: Nobu Tamura/CC BY 3.0; Captorhinus: Nobu Tamura/CC BY 3.0; Paleothyris: Conty/CC BY-SA 4.0; Petrolacosaurus: Nobu Tamura/CC BY-SA 3.0; Sphenodon: no citation needed; Pappochelys: Rainer Schoch/CC BY-SA 4.0; Chelonia: no citation needed; Euparkeria: Taenadoman/CC BY-SA 3.0; Crocodylus: no citation needed; Tyurannosaurus: Steveoc 86/CC BY-SA 4.0.
Figure 9.2: †Trimerorhachis; Nobu Tamura/CC BY-SA 4.0: †Eryops; Dmitry Bogdanov/CC BY-SA 3.0: Cyclotosaurus; Dmitry Bogdanov/CC BY 3.0: †Chinlestegophis; no citation needed: Caecilian; no citation needed: Cacops; Dmitry Bogdanov/CC BY 3.0: Gerobatrachus; no citation needed: Salamander; no citation needed: Triadobatrachus; Nobu Tamura/CC BY 3.0: Frog; no citation needed: Archeria; Dmitry Bogdanov/CC BY 3.0: Seymouria; Nobu Tamura/CC BY-SA 4.0: Diadectes; Dmitry Bogdanov/CC BY-SA 3.0.
Figure 9.7: B after K. Liem et al. 2001. *Functional Anatomy of the Vertebrates*, 3rd ed. Cengage/Harcourt College: Belmont, CA; G after S. Voigt et al. 2007. *J Vert Paleontol* 27: 553–570.
Figure 9.8: Synapsida: K. Liem et al. 2001. *Functional Anatomy of the Vertebrates*, 3rd ed. Cengage/Harcourt College: Belmont, CA; Mesosaurs: Nobu Tamura/CC BY 3.0; Procolophon: Nobu Tamura/CC BY 3.0; Pereisaurs/Bradyaurus: Nobu Tamura/CC BY 3.0; Captorhinus: Nobu Tamura/CC BY 3.0; Paleothyris: Conty/CC BY-SA 4.0; Hylonomus: Mateo De Stefano/MUSE/CC BY-SA 3.0; Petrolacosaurus: Nobu Tamura/CC BY-SA 3.0; Euparkeria: Taenadoman/CC BY-SA 3.0.
Figure 9.12: B after R. L. Carroll. 1969. *Biol Rev* 44: 393-432; C–F after R.L. Carroll and D.B. Baird. 1972. *Bull Mus Comp Zool* 143: 321–363.
Figure 9.16: Some parts of this figure are based on T.S. Kemp. 1982. *Mammal-Like Reptiles and the Origin of Mammals*. Academic Press: Cambridge, MA and A.S. Romer and L.W. Price. 1940. *Geol Soc Am Special Papers* 28: 1–538.
Figure 9.17: Skull 3 after R. Reisz. 1977. *Science* 196: 1091–1093; Skull 5 after R.R. Schoch and H.-D. Sues. 2016. *Zoology* 119: 159–161; Skull 6 after R. F. Ewer. 1965. *Phil. Trans. R. Soc. Lond. B* 248: 379–435; Skull 7 after S. J. Nesbitt. 2011. *Bull Am Mus Nat Hist* 352: 1–292; Skull 8 After E. Rayfield. 2004. *Proc R Soc Lond B* 271: 1451–1459.

Dan Chure, National Park Service

Students at Dinosaur National Monument, Utah

Geography and Ecology of the Mesozoic

In Mesozoic oceans, neoselachians—sharks, skates, and rays—diversified. Neopterygians, present in the Permian, rapidly diversified in the first half of the Mesozoic, only to be replaced by teleosts after the Jurassic. All extant tetrapod lineages had appeared well before the end of the Mesozoic era, and the radiation of angiosperms (flowering seed plants) during the Cretaceous profoundly impacted the tetrapod fauna. Total diversity of tetrapods increased relatively slowly, with the number of species only doubling or tripling, but there were major changes in the clades present. Lissamphibians evolved and the two major lineages of amniotes diversified:

- Synapsida (the lineage including mammals) radiated during the Permian, leading to both carnivorous and herbivorous therapsids in the Mesozoic.

- Sauropsida diversified in the Carboniferous and led to many groups: lepidosaurs (tuatara, lizards, and snakes), turtles, archosaurs (extant crocodylians and birds as well as the iconic, now-extinct †ornithischian and large †sauropod dinosaurs).

In the Early Triassic, synapsids were the most abundant and diverse tetrapods, but by the mid-Triassic, sauropsids had surpassed them in both abundance and diversity. By the Cenozoic, mammals were the only synapsids.

Synapsids were not part of Mesozoic marine faunas, whereas several lineages of Triassic sauropsids successfully invaded marine habitats. Some (such as †ichthyosaurs and †plesiosaurs) emerged as apex predators; others evolved crushing dentitions. On land, a major mass extinction at the end of the Triassic was followed by dinosaur-dominated faunas of the Jurassic and Cretaceous. Another major mass extinction at the end of the Cretaceous is best known for

the demise of †ornithischians and †sauropods but affected many other groups as well (**Figure 10.1**).

In this chapter, we describe the geography and climates of Mesozoic Earth, terrestrial and marine ecosystems and faunas, and the era's history of extinction events that opened the way for a new faunal balance in the Cenozoic.

10.1 Continental Geography and Climates

LEARNING OBJECTIVES

10.1.1 Describe continental movements during the Mesozoic.

10.1.2 Explain the impact of Pangaea's breakup on terrestrial floras and faunas.

10.1.3 Describe changes in global temperature that occurred during the Mesozoic.

During the Mesozoic, tectonic forces broke up the huge landmass of Pangaea and the continents began to approach their current positions. The planet was warmer than it is today; there were no polar ice caps, and the latitudinal range of climates so familiar to us was absent.

Continental movements

By the Early Mesozoic, Earth's entire land surface had coalesced into a single supercontinent, **Pangaea**, that stretched from pole to pole (**Figure 10.2A**). Fragmentation of Pangaea began in the Jurassic as westward extension of the Tethys Sea separated Pangaea into **Laurasia** in the north and **Gondwana** in the south (**Figure 10.2B**). Laurasia rotated away from other continents, separating North America from its connection with South America and increasing the size of the newly formed Atlantic Ocean.

Continental separation and rotation continued during the Jurassic and into the Cretaceous. By the Late Cretaceous, continents were approaching their current positions. India was still close to Africa, but Madagascar had split from India, while Australia, New Zealand, and New Guinea formed the single landmass of Australasia and were still well south of their present-day positions (**Figure 10.2C**).

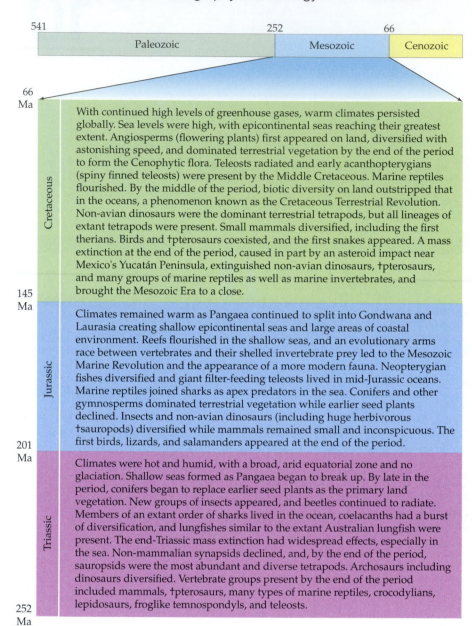

| Paleozoic | Mesozoic | Cenozoic |

Figure 10.1 Key events of the Mesozoic. The Mesozoic was marked by warm climates and large-scale changes in flora and fauna, including the appearance and rapid radiation of angiosperms (flowering plants) and the diversification of herbivores in many terrestrial animal lineages. In the sea, stony corals appeared and coral reefs have been the most diverse marine habitats for the past 240 Ma. The radiations of fishes and crustaceans on coral reefs set the stage for the diversification of marine tetrapods that preyed on those fishes and crustaceans and on each other.

66 Ma

Cretaceous

With continued high levels of greenhouse gases, warm climates persisted globally. Sea levels were high, with epicontinental seas reaching their greatest extent. Angiosperms (flowering plants) first appeared on land, diversified with astonishing speed, and dominated terrestrial vegetation by the end of the period to form the Cenophytic flora. Teleosts radiated and early acanthopterygians (spiny finned teleosts) were present by the Middle Cretaceous. Marine reptiles flourished. By the middle of the period, biotic diversity on land outstripped that in the oceans, a phenomenon known as the Cretaceous Terrestrial Revolution. Non-avian dinosaurs were the dominant terrestrial tetrapods, but all lineages of extant tetrapods were present. Small mammals diversified, including the first therians. Birds and †pterosaurs coexisted, and the first snakes appeared. A mass extinction at the end of the period, caused in part by an asteroid impact near Mexico's Yucatán Peninsula, extinguished non-avian dinosaurs, †pterosaurs, and many groups of marine reptiles as well as marine invertebrates, and brought the Mesozoic Era to a close.

145 Ma

Jurassic

Climates remained warm as Pangaea continued to split into Gondwana and Laurasia creating shallow epicontinental seas and large areas of coastal environment. Reefs flourished in the shallow seas, and an evolutionary arms race between vertebrates and their shelled invertebrate prey led to the Mesozoic Marine Revolution and the appearance of a more modern fauna. Neopterygian fishes diversified and giant filter-feeding teleosts lived in mid-Jurassic oceans. Marine reptiles joined sharks as apex predators in the sea. Conifers and other gymnosperms dominated terrestrial vegetation while earlier plants declined. Insects and non-avian dinosaurs (including huge herbivorous †sauropods) diversified while mammals remained small and inconspicuous. The first birds, lizards, and salamanders appeared at the end of the period.

201 Ma

Triassic

Climates were hot and humid, with a broad, arid equatorial zone and no glaciation. Shallow seas formed as Pangaea began to break up. By late in the period, conifers began to replace earlier seed plants as the primary land vegetation. New groups of insects appeared, and beetles continued to radiate. Members of an extant order of sharks lived in the ocean, coelacanths had a burst of diversification, and lungfishes similar to the extant Australian lungfish were present. The end-Triassic mass extinction had widespread effects, especially in the sea. Non-mammalian synapsids declined, and, by the end of the period, sauropsids were the most abundant and diverse tetrapods. Archosaurs including dinosaurs diversified. Vertebrate groups present by the end of the period included mammals, †pterosaurs, many types of marine reptiles, crocodylians, lepidosaurs, froglike temnospondyls, and teleosts.

252 Ma

Triassic faunas and floras showed some regional differentiation across Pangaea, but there were no oceanic barriers to dispersal. With the breakup of Pangaea, floras and faunas were geographically isolated on continents separated by expanses of ocean and became distinct in different parts of the world. However, in the relatively warm world of the Mesozoic, there was little latitudinal zonation of plants or animals. Cretaceous seas were as much as 200 m above present-day levels, and shallow epicontinental seas extended across vast parts of continents. The Western Interior Seaway that once connected the Gulf of Mexico to the Arctic Ocean reached depths of 760 m, completely flooding regions that are the North American Great Plains of today and separating North America into eastern and western landmasses. Epicontinental seas also split the Eurasian landmass, divided Africa, and made deep incursions into South America and Australia (see Figure 10.2C). Epicontinental seas

receded when sea levels fell near the end of the Cretaceous, changing landscapes as rivers eroded valleys in exposed seabeds.

Climate shifts

Global temperatures spiked at the Permo-Triassic boundary (see Figure 5.6). Early Triassic climates were unstable, with two episodes of ecosystem collapse within the first million years, and at least five greenhouse crises in the first 5 million years. Stressful environmental conditions are indicated by the absences of coal deposits on land and coral reefs in the seas, the latter reflecting anoxia in the ocean.

Global temperatures remained high for most of the Mesozoic, apart from a dip in the mid-Jurassic that lasted into the Early Cretaceous. Carbon dioxide levels peaked around 0.2% at the end of the Triassic and remained constant at around 0.1% (today's level is 0.04% ppm) until they started to fall in the mid-Cretaceous (see Figure 5.6).

During the early part of the Late Triassic, ~234–232 Ma, a strange and transformative climatic event occurred, the **Carnian Pluvial Episode** (Latin *pluvia*, "rain"). During this time (fancifully characterized as "a million years of rain"), the previously arid climate became extremely wet and rainy, then once again became arid. The Carnian Pluvial was accompanied by major faunal turnovers and many extinctions, after which dinosaurs radiated and eventually came to dominate Jurassic faunas.

Early Cretaceous seafloor spreading and continental movements initiated global warming that extended into the Late Cretaceous and caused sea levels to rise about 200 m. Low and middle latitudes were relatively dry until the Late Cretaceous, when coal deposits indicate the presence of swamps.

(A) Triassic

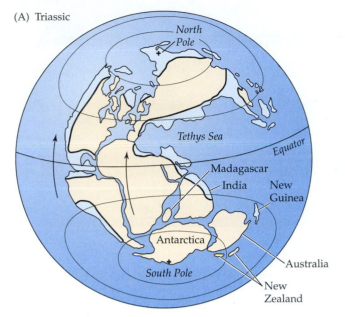

(B) Jurassic

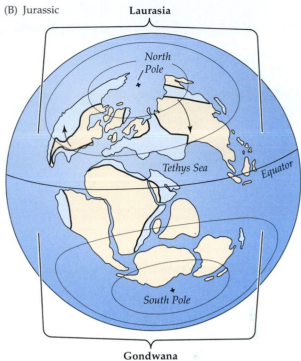

(C) Cretaceous

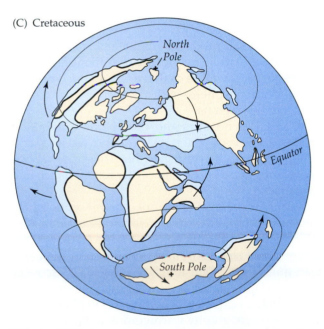

Figure 10.2 Continental blocks in the Mesozoic. Arrows indicate directions of continental movements. (A) In the Early Triassic, the supercontinent of Pangaea had no oceanic barriers to dispersal. (B) During the Jurassic, westward expansion of the Tethys Sea separated Laurasia and Gondwana and the rifts that would create the present-day continents appeared. (C) Continents approached their current positions during the Cretaceous, but vast areas remained covered by shallow epicontinental seas for much of the period. (After C.K. Seyfert and L.A. Sirkin. 1979. *Earth History and Plate Tectonics. An introduction to Historical Geology.* Harper & Row: New York.)

For much of the Cretaceous, temperate forests extended into polar regions, supporting faunas of large †sauropods that would have required temperate or subtropical climates. Global cooling near the end of the Cretaceous is indicated by the loss of plant diversity at higher latitudes, and several cycles of abrupt warming and cooling destabilized marine and terrestrial ecosystems.

10.2 Terrestrial Ecosystems

LEARNING OBJECTIVES

10.2.1 Describe major floral changes of the Mesozoic and the transition from Mesophytic to Cenophytic flora.

10.2.2 Describe major components of the terrestrial vertebrate fauna in each period of the Mesozoic and how these components changed over the era.

10.2.3 Describe how and why latitudinal diversity of tetrapods differs between the Triassic and today.

The massive end-Permian extinctions (see Section 5.5) left an impoverished terrestrial fauna and flora. The first 10 million years of the Triassic were barren, with few forests on land or corals in the sea. As diversity finally began to rebound, however, the flora and fauna of the Mesozoic underwent a series of large-scale changes. By the end of the Triassic, terrestrial ecosystems had a modern form, with a broad base of herbivores supporting several levels of carnivores. An array of Jurassic †ornithischian, †sauropod, and theropod dinosaurs occupied adaptive zones for large vertebrates. By the end of the Cretaceous, all major extant lineages of vertebrates were present, and the flora had come to be dominated by angiosperms, as it is today.

Flora

Triassic vegetation continued the Mesophytic Flora first established in the Carboniferous (see Figure 5.8), including modern gymnosperms such as conifers (pines and other cone-bearing trees), ginkgophytes (relatives of the extant ginkgo), and cycads. Near the end of the Triassic the vegetational understory was composed of ferns, tree ferns, horsetails, and cycads, all of which survive today

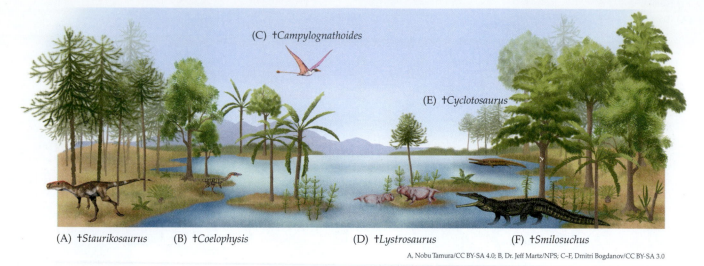

(C) †*Campylognathoides*

(E) †*Cyclotosaurus*

(A) †*Staurikosaurus*   (B) †*Coelophysis*   (D) †*Lystrosaurus*   (F) †*Smilosuchus*

A, Nobu Tamura/CC BY-SA 4.0; B, Dr. Jeff Martz/NPS; C–F, Dmitri Bogdanov/CC BY-SA 3.0

Figure 10.3 **A scene from the Late Triassic.** Archosaurs (sauropsids) include theropod dinosaurs, †pterosaurs, and †phytosaurs, while the diversity of therapsids (synapsids) was greatly reduced after the end-Permian mass extinction event. Temnospondyls (non-amniotes) remain as aquatic and semi-aquatic predators.

but at lower diversities. Tree-size cycads were present, but the canopy was dominated by gymnosperms that provided a new source of food for herbivorous vertebrates, setting the scene for the evolution of long-necked †sauropods that came to prominence in the Jurassic (**Figure 10.3**). Jurassic vegetation had lower overall diversity and leaf morphology indicating that arid environments followed widespread deforestation at the end-Triassic mass extinction (see Section 10.4).

Early Cretaceous vegetation resembled that of the Late Jurassic, but by the Late Cretaceous the Mesophytic Flora had been replaced by the angiosperm-dominated Cenophytic Flora (see Figure 5.8). This rapid radiation of angiosperms contributed to the **Cretaceous Terrestrial Revolution**, the point in Earth's history at which terrestrial diversity outstripped marine diversity. Because many angiosperms depend on insects for pollination, both pollinators and plant-eating insects flourished.

Early angiosperms were small trees, low-growing shrubs, and nonwoody herbs. Disturbances such as fire may have given angiosperms, with their more rapid reproductive cycles, advantages over other kinds of seed plants. Ferns also diversified in the Cretaceous, including lineages that grow on forest floors and in the canopy. Grasses (which are angiosperms) evolved by the latest Cretaceous (see Figure 5.8) but grasslands were not an important feature of landscapes until the Cenozoic. Angiosperm-based tropical forest ecosystems date from the mid-Cretaceous, although equatorial rainforests like those of today are a Cenozoic phenomenon.

Fauna

In contrast to the latitudinal zonation seen today, where species diversity is highest at the Equator, the Early and Middle Triassic continued zonation of the latest Permian, with high diversity in temperate regions and low diversity in arid and probably uninhabitably hot equatorial regions. Large fossil temnospondyls in high-latitude Triassic deposits as well as coal deposits in the Northern and Southern Hemispheres indicate warmer and moister climates in those regions.

Small basal archosaurs and therapsids dominated Early Triassic faunas. Fossil trackways with left and right footprints close together indicate that Triassic tetrapods had more upright postures than earlier forms, which had wider stances.

Early Triassic therapsids included herbivorous †dicynodonts such as †*Lystrosaurus*, a clade that survived the end-Permian mass extinction (see Figure 5.10). Species of †*Lystrosaurus* were extremely abundant in the Early Triassic, and ranged over much of Pangaea. Discovery of fossils of †*Lystrosaurus* in Africa, South America, and Antarctica helped drive early discoveries in plate tectonics and paleobiogeography. Carnivorous and herbivorous therapsids soon became prominent members of the Triassic fauna (see Chapter 22). Although dinosaurs did not diversify until the Late Triassic, fossils of basal dinosaurs occur in the Early Triassic, and skeletons and footprints of close relatives date from ~248–245 Ma.

Low levels of atmospheric oxygen in the Late Permian may have contributed to the end-Permian mass extinction, and continuing low oxygen levels may help explain patterns of faunal turnover starting in the Late Triassic. Archosaurs, especially pseudosuchians (the lineage that includes extant crocodylians) now overshadowed therapsids.

Middle Triassic herbivores, including therapsids, pseudosuchians, and archosaur-related †rhynchosaurs, ranged from 10 to 1,000 kg (goat- to buffalo-size). These generalized browsers would have foraged within a meter of the ground. Arboreal vertebrates were rare; there were no arboreal herbivores, but there were insectivorous sauropsids, including gliding forms.

Crocodylians

Avialans

Lissamphibians   3 meters   Turtles   Lepidosaurs   Mammals

Durbed/CC-BY-SA 3.0

Figure 10.4 **Terrestrial tetrapods of the Late Cretaceous were a mix of the old and the new.** Species present at the Hell Creek and Bug Creek Anthills localities of present-day Montana inhabited wooded, swampy habitats with large streams and some ponds. All extant tetrapod groups—mammals, crocodylians, birds, lepidosaurs, turtles, and lissamphibians—are present at these sites. Dinosaurs and †pterosaurs were only a small portion of the fauna in terms of number of species, but they made up much of the faunal biomass. (See credit details at the end of this chapter.)

The Carnian Pluvial Episode (see Section 10.1) restructured terrestrial communities. †Rhynchosaurs disappeared, but †dicynodonts persisted for another 20 Ma. In the episode's aftermath, dinosaurs diversified worldwide and began to evolve larger body sizes. Tetrapod faunas in the Late Triassic were most diverse at middle latitudes.

Basal mammals (tiny insectivorous forms) first appeared in the Late Triassic, as did other extant groups such as rhynchocephalians (the tuatara, which is the sister group of lizards and snakes), turtles, and crocodyliforms, and now-extinct groups such as †pterosaurs (flying archosaurs). Extant crocodylians are semiaquatic (i.e., they live primarily on land but spend time in water), but their Mesozoic relatives include many terrestrial forms (some of which were herbivorous) as well as marine clades that spent most of their lives in water.

Cockroaches, aphids, moths, flies (dipterans), and stick insects first appeared in the Triassic. Beetles (coleopterans), first known from the Late Carboniferous, further radiated. The expanding insect fauna fed on plants and pollinated others. Changes in insect and plant life in turn affected the fauna, and new terrestrial herbivores characterize the transition to the Jurassic world. Long-necked †eusauropods evolved, reaching gigantic sizes—some with body weights greater than 50,000 kg—in the Jurassic. †Eusauropods could browse tall, small-leaved conifers at treetop height. In turn, they became prey for large carnivorous theropods. †Ornithischians such as †stegosaurs and †ornithopods browsed shorter vegetation growing at 1- to 2-m heights. The large egg clutches of †sauropods and †ornithischians filled Jurassic landscapes with juveniles that were probably the principal prey of carnivores, including diverse terrestrial crocodyliforms.

Most Jurassic mammals were small insectivores, but there were also small omnivores (†multituberculates). Lizards and lissamphibians appeared in the Jurassic, and the stem bird †*Archaeopteryx* is known from the Late Jurassic. New kinds of insects diversified, including termites, fleas, butterflies (lepidopterans), true bugs (hemipterans), and social insects (hymenopterans) including ants, wasps, and hive-forming bees.

New types of herbivorous dinosaurs evolved in the Late Cretaceous, including †hadrosaurs (†duck-billed dinosaurs) and †ceratopsians (†thorned dinosaurs). These were ground-level feeders with grinding teeth and horny beaks that could deal with tough vegetation. Other faunal changes included an early radiation of Aviales (basal birds). The Late Cretaceous also saw the evolution of snakes, modern crocodyliforms, and mammals belonging to all three extant clades (monotremes, metatherians, and eutherians; see Figure 1.2). Most Cretaceous mammals were small—shrew- to mouse-size—and none was larger than a Labrador retriever.

By the Late Cretaceous, terrestrial faunas included all extant tetrapod clades (**Figure 10.4**).

10.3 Marine Ecosystems

LEARNING OBJECTIVES

10.3.1 Characterize the initial radiation of marine tetrapods in the Triassic.

10.3.2 Explain some differences between the Cretaceous Terrestrial Revolution described in Section 10.2 and the Mesozoic Marine Revolution described in this section.

10.3.3 Describe how the diversity of clades with apex predators changed during the Mesozoic.

Increases in marine nutrients derived from weathering of terrestrial rocks led to a burst of animal life in Mesozoic seas. The evolution of planktonic organisms that began in the Late Triassic accelerated through the rest of the Mesozoic, providing the base for food chains that supported radiations of large marine animals.

In response to the evolution of fishes and tetrapods with crushing dentitions, an evolutionary arms race occurred between crushers and the invertebrates that they crushed. As prey organisms evolved thicker shells to avoid predation, predators evolved more powerful dentitions. This arms race contributed to the **Mesozoic Marine Revolution**, yielding marine environments more similar to those of today, with highly diversified invertebrate faunas preyed upon by vertebrates with specialized feeding mechanisms.

Stony (scleractinian) corals appeared and diversified in the Mesozoic following the eradication of horn and tabulate corals in the end-Permian mass extinction. †Rudists, a group of bivalve mollusks related to extant edible clams, were also important reef-building organisms from their appearance in the mid-Jurassic to the end of the Cretaceous.

Reef-building organisms exhibited cycles of diversity and abundance corresponding to mass extinctions and recovery and patterns of continental fractionation and drift. For example, rapid reef evolution was driven by the breakup of Pangaea and the opening of the Tethys Sea, a precursor to the Mediterranean Sea, that separated Laurasia and Gondwana. Large expanses of shallow, low-latitude water flooded the newly formed continental shelves of the Tethys Sea, creating new habitats that drove the evolution of many groups of fishes, including the early acanthopterygians (spiny-finned teleosts; see Section 7.5). Acanthopterygians subsequently diversified and radiated in marine and freshwater environments, leading to about 25% of extant vertebrate species.

Oceans were new territory for tetrapods, and the initial radiation of marine tetrapods in the Triassic gives the impression of a free-for-all: at least 14 clades waded into the sea, with varying degrees of success.

Faunal composition: Apex predators

Four clades of marine tetrapods, all sauropsids, gave rise to apex marine predators. The relative diversity of these clades changed over time, and no more than three clades were ever contemporaneous (**Figure 10.5**). †Ichthyosaurs and †sauropterygians were about equally diverse in the Triassic, but the end-Triassic mass extinction greatly reduced the diversity of both clades, and recovery was slow.

†Thalattosuchian crocodyliforms radiated in the Jurassic. By the Early Cretaceous, †ichthyosaurs, †sauropterygians, and †thalattosuchians had similar levels of diversity. †Thalattosuchians disappeared midway through the Cretaceous, leaving the seas to †ichthyosaurs and †sauropterygians. Apex predators changed yet again in the Late Cretaceous with the extinction of †ichthyosaurs and the radiation of the lizard-related †mosasaurs. †Mosasaurs and †sauropterygians shared the apex predator role until both groups disappeared in the mass extinction at the end of the Cretaceous.

†Ichthyosaurs †Ichthyosaurs existed from the Triassic to the Late Cretaceous but disappeared before the end-Cretaceous mass extinction. Limbs of basal †ichthyosaurs such as †*Sclerocormus* (Figure 10.5A) were modified into flippers, with a loss of distinct digits; the hind flippers were small. They lacked a dorsal fin and their long tails lacked well-developed caudal fins. Derived Triassic †ichthyosaurs such as †*Californosaurus* (Figure 10.5B) had streamlined bodies with small dorsal fins; the now-shorter tail terminated in a vertical caudal fin with an upper lobe formed by stiff tissue and a downward turning of the vertebral column to support the lower lobe of the fin, a condition termed hypocercal (compare with heterocercal caudal fins, shown in Figure 6.21 and Figure 7.7).

The birth position of †ichthyosaurs also changed during the Triassic. Terrestrial tetrapods are born headfirst so they can take their first breath as soon as they emerge, but extant marine tetrapods that give birth in water—cetaceans, dugongs, and manatees—emerge tailfirst so that they can swim to the surface to take their first breath. †*Chaohusaurus*, an Early Triassic †ichthyosaur, retained the head-first birth orientation of its terrestrial ancestors (**Figure 10.6A**), whereas later †ichthyosaurs, such as the Early Jurassic †*Stenopterygius*, were born tailfirst (**Figure 10.6B**).

†Ichthyosaurs increased in size over time; the modal length of Triassic taxa was between 1 and 2 m, and that increased to between 2 and 5 m in the Jurassic. They became progressively more streamlined during the Jurassic; derived forms like †*Excalibosaurus* (Figure 10.5C) evolved fusiform bodies with well-developed dorsal fins and hypocercal tails with caudal fins. A clade of †ichthyosaurs known as †ophthalmosaurids (Greek, "eye lizards") had extremely large eyes (†*Temnodontosaurus*; Figure 10.5D). We know this because the eyeball was supported by a ring of scleral ossicles; similar bony rings occur in the eyes of many extant teleosts, squamates, and birds. †Ophthalmosaurids are thought to have hunted at depths of 500 m or more, detecting light emitted by photophores of their prey, as some species of extant swordfishes do.

Figure 10.5 Four clades of marine tetrapods included large apex predators. ▶ The modal body sizes of all of the clades increased progressively during the Mesozoic. (Proportions based on information in Reeves et al. 2021. *Palaeontology* 64: 31–49. For additional credits, see details at the end of this chapter.)

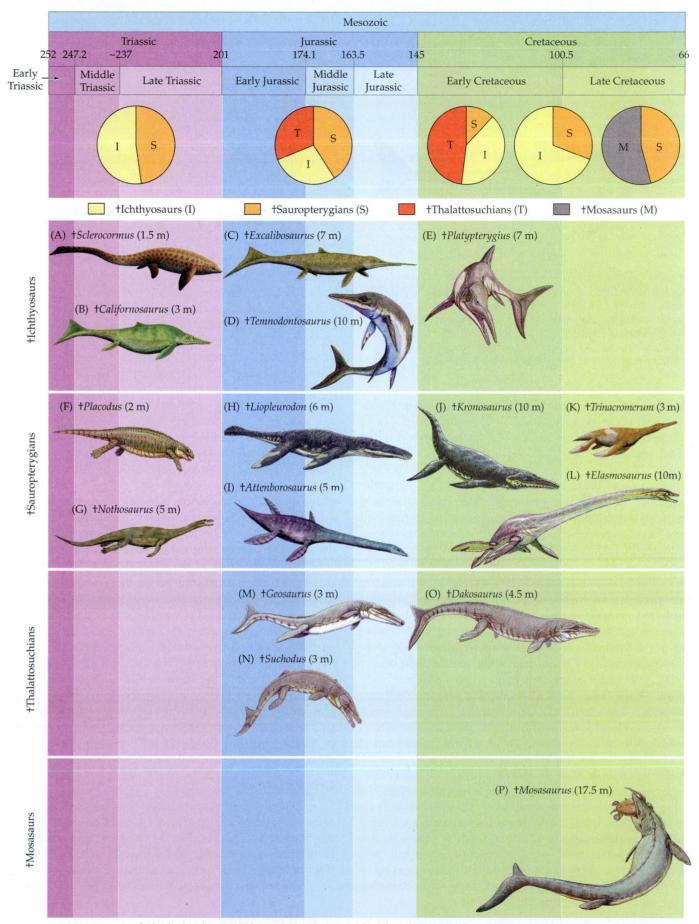

†Ichthyosaurs (I)   †Sauropterygians (S)   †Thalattosuchians (T)   †Mosasaurs (M)

(A) †*Sclerocormus* (1.5 m)

(B) †*Californosaurus* (3 m)

(C) †*Excalibosaurus* (7 m)

(D) †*Temnodontosaurus* (10 m)

(E) †*Platypterygius* (7 m)

(F) †*Placodus* (2 m)

(G) †*Nothosaurus* (5 m)

(H) †*Liopleurodon* (6 m)

(I) †*Attenborosaurus* (5 m)

(J) †*Kronosaurus* (10 m)

(K) †*Trinacromerum* (3 m)

(L) †*Elasmosaurus* (10m)

(M) †*Geosaurus* (3 m)

(N) †*Suchodus* (3 m)

(O) †*Dakosaurus* (4.5 m)

(P) †*Mosasaurus* (17.5 m)

(A) †*Chaohusaurus*, Early Triassic

5 cm

(B) †*Stenopterygius*, Early Jurassic

Note three more developing young in mother's body.

5 cm

Figure 10.6 Birth orientation of †ichthyosaurs.   (A) Early Triassic †*Chaohusaurus* that died while giving birth shows that the young retained the head-first birth orientation of terrestrial tetrapods. (B) Early Jurassic †*Stenopterygius* shows one newborn emerging tail-first, and three others in birth position in the body cavity. (From R. Motani et al. 2014. *PLOS ONE* 9: e88640 /CC BY 4.0.)

By the Middle Cretaceous, †ichthyosaurs were abundant, but no more specialized than Jurassic forms (e.g., †*Platypterygius*.; Figure 10.5E). The clade vanished in the Late Cretaceous.

†Sauropterygians   †Sauropterygians lived from the Triassic to the end-Cretaceous extinction. Basal forms had feet with toes, elongate bodies, long tails without fins, and may have been amphibious, spending most of their time in water but occasionally hauling out on land. †Placodonts were among the least specialized marine tetrapods. Most, like †*Placodus* (Figure 10.5F), were coastal-dwelling, durophagous mollusk-eaters, with protruding anterior teeth that grasped mollusk shells and flat posterior teeth that crushed them. In contrast, more streamlined Triassic †sauropterygians such as †*Nothosaurus* (Figure 10.5G), were pursuit predators of fishes and other marine tetrapods.

Derived †sauropterygians evolved two different body forms—large-headed, short-necked fast-swimming forms commonly known as "†pliosauromorphs," and small-headed, long-necked "†plesiosauromorphs" that used both fore- and hindlimbs as flippers to swim, a form of limb-based locomotion that is unique among vertebrates.

Forms such as †*Liopleurodon* (Figure 10.5H) were pelagic predators capable of attacking large prey, rather like orcas (killer whales) today. Others, such as †*Attenborosaurus* (Figure 10.5I), have no ecological analogs among extant marine tetrapods, and their feeding habits are unclear. Their sharp teeth suggest a diet of fishes, but their necks were not very flexible and bent downward more readily than to the side or upward, suggesting that they may have foraged on the bottom.

Cretaceous †sauropterygians were truly formidable. †*Kronosaurus* (Figure 10.5J) reached 10 m. Its skull was

2.5 m long (25% of its total length), its neck was scarcely detectable (as in whales today), and some of its teeth were 30 cm long.

Short-necked forms with long, narrow snouts, such as †*Trinacromerum* (Figure 10.5K), appeared in the Late Cretaceous. Extreme long-necked forms include †*Elasmosaurus* (Figure 10.5L) with a neck that had ≥72 cervical vertebrae and made up 70% of its total body length. Its proportions are so strange that the paleontologist who reconstructed the first specimen placed the skull at the end of the tail—an error he never lived down.

†Thalattosuchians   †Thalattosuchians, a marine radiation of crocodyliforms, diversified in the Jurassic and had disappeared by the mid-Cretaceous. At least four lineages expanded into marine habitats. Three had crocodilelike body forms and were probably ecologically similar to the extant saltwater crocodile of Australia, a shoreline predator that can make long sea voyages. Derived †thalattosuchians such as †*Geosaurus* (Figure 10.5M) and †*Suchodus* (Figure 10.5N) converged on the body form of extant orcas, although, like modern crocodiles, they swam using lateral undulations of their body and tail (orcas use dorsoventral body undulations). Early Cretaceous †thalattosuchians resembled earlier forms (for example, †*Dakosaurus*; Figure 10.5O). Most disappeared in Early Cretaceous, although fossil teeth from Sicily document the persistence of at least one species until the middle of the period.

†Mosasaurs   †Mosasaurs, a group of giant marine lizards, were the last group of tetrapods to occupy the niche of apex marine predators in the Mesozoic, radiating in the mid-Cretaceous and disappearing in the end-Cretaceous mass extinction. Like †thalattosuchians, Late Cretaceous †mosasaurs resembled modern orcas. They lived in shallow epicontinental seas across North America and Europe. Basal †mosasaurs were probably semiaquatic, as are several extant species of monitor lizards. Their limbs ended in toes and the tail had no fin. Derived taxa such as †*Mosasaurus* (Figure 10.5P) had limbs modified into flippers (but not to the same extent as in †ichthyosaurs and †sauropterygians) and hypocercal tails. Although similarities in body form makes it tempting to think that competition with †mosasaurs led to the extinction of †thalattosuchians, the ~20-million-year gap between these two groups invalidates that hypothesis.

Other clades

Several short-lived clades shared the seas with †ichthyosaurs, †thalattosuchians, and †sauropterygians during the Triassic and Jurassic, and extant lineages (including turtles, snakes, and birds) had appeared by the Cretaceous (**Figure 10.7**). They were smaller than the apex predators of the Mesozoic, and many of them were coastal dwellers.

- **†Hupehsuchians**   †*Hupehsuchus* (Figure 10.7A) was a near-shore predator related to †ichthyosaurs.

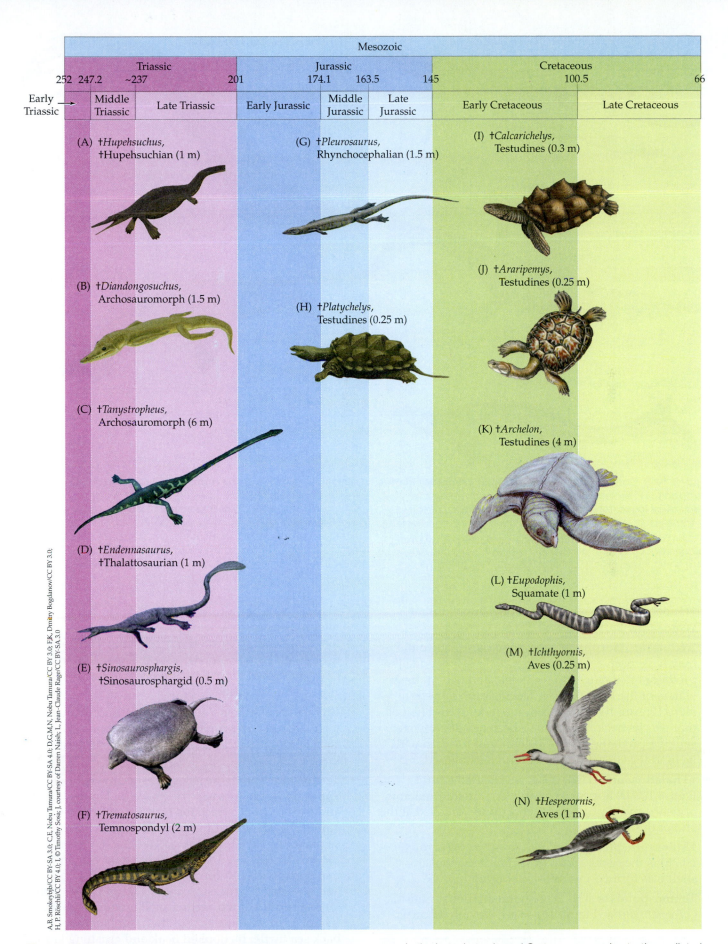

Mesozoic

Triassic			Jurassic			Cretaceous	
252 247.2	~237	201	174.1	163.5	145	100.5	66
Early Triassic →	Middle Triassic	Late Triassic	Early Jurassic	Middle Jurassic	Late Jurassic	Early Cretaceous	Late Cretaceous

(A) †*Hupehsuchus,* †Hupehsuchian (1 m)

(B) †*Diandongosuchus,* Archosauromorph (1.5 m)

(C) †*Tanystropheus,* Archosauromorph (6 m)

(D) †*Endennasaurus,* †Thalattosaurian (1 m)

(E) †*Sinosaurosphargis,* †Sinosaurosphargid (0.5 m)

(F) †*Trematosaurus,* Temnospondyl (2 m)

(G) †*Pleurosaurus,* Rhynchocephalian (1.5 m)

(H) †*Platychelys,* Testudines (0.25 m)

(I) †*Calcarichelys,* Testudines (0.3 m)

(J) †*Araripemys,* Testudines (0.25 m)

(K) †*Archelon,* Testudines (4 m)

(L) †*Eupodophis,* Squamate (1 m)

(M) †*Ichthyornis,* Aves (0.25 m)

(N) †*Hesperornis,* Aves (1 m)

Figure 10.7 Other clades of marine tetrapods. The Triassic saw radiations of clades of tetrapods that never attained large body sizes. None of these initial lineages persisted for long after the end of the Triassic, but during the Jurassic two new clades entered the seas. In the Late Jurassic and Cretaceous, marine turtles radiated widely, there was a small radiation of two-legged marine snakes, and a diversity of toothed birds arose.

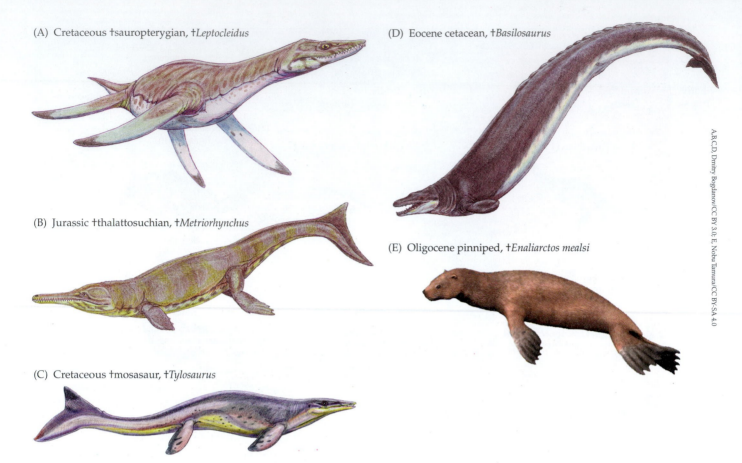

(A) Cretaceous †sauropterygian, †*Leptocleidus*

(B) Jurassic †thalattosuchian, †*Metriorhynchus*

(C) Cretaceous †mosasaur, †*Tylosaurus*

(D) Eocene cetacean, †*Basilosaurus*

(E) Oligocene pinniped, †*Enaliarctos mealsi*

Figure 10.8 **Convergence in body form of predatory marine tetrapods.** Five clades of marine tetrapods converged on similar body forms: (A) †sauropterygians, (B) †thalattosuchians, (C) †mosasaurs, (D) cetaceans, and (E) pinnipeds (flipper-footed marine mammals such as seals). The Eocene cetacean †*Basilosaurus* has a body form and flippers so like those of the marine reptiles that it was initially identified as a reptile, and its name (Greek, "king lizard") preserves that error. †Sauropterygians used their front and hind flippers to swim, a mode of locomotion that is unique to this clade. †Thalattosuchians and †mosasaurs swam with lateral undulations of the body and tail. Cetaceans swim with dorsoventral undulations of the body; pinnipeds use their appendages (either fore- or hindlimbs) for propulsion, along with dorsoventral movements of the body.

†Hupehsuchians persisted for only a few million years in the Early Triassic.

- **Archosauromorphs** †*Diandongosuchus* (Figure 10.7B) may have been a †phytosaur, a clade of crocodilelike predators that reached its greatest diversity in fresh waters of the Late Triassic. The bizarre †*Tanystropheus* (Figure 10.7C) was about 6 m long, and half of that was its neck. It may have been a shallow-water, stalking predator that propelled itself forward with its hind legs to capture fishes and cephalopods.

- **†Thalattosaurs**[1] Thalattosaurs (†*Endennasaurus*; Figure 10.7D) appear to have been pursuit predators of fishes and cephalopods. The group vanished at the end of the Triassic

- **†Sinosaurosphargids** †*Sinosaurosphargis* (Figure 10.7E) is the sister group of †sauropterygians. Its shell was formed by a mosaic of osteoderms.

- **Temnospondyls** The long-snouted †trematosaurids (†*Trematosaurus*; Figure 10.7F) are the only known marine batrachomorphs. Confined to the early and middle Triassic, they were probably shoreline predators somewhat like extant crocodylians. They seized fishes with a sideward swipe of the head and may have ambushed terrestrial animals that came too close to the waterline.

- **Rhynchocephalians** Mesozoic rhynchocephalians were diverse and primarily terrestrial, but a few taxa, such as †*Pleurosaurus* (Figure 10.7G), were coastal marine predators.

- **Testudines** The first turtles in estuarine habitats appeared in the Jurassic (†*Platychelys*; Figure 10.7H) and by the Late Cretaceous turtles comprised nearly half of the diversity of marine tetrapods. Small semiaquatic turtles (†*Calcarichelys* and †*Araripemys*; Figures 10.7I,J) lived in coastal habitats. Streamlined marine turtles with flipperlike limbs lived in bays and on reefs. †*Archelon ischyros* (Figure 10.7K) is the largest turtle known, with a shell length of 4.6 m. Its carapace was like that of the extant leatherback sea turtle. Its hooked beak and crushing jaws suggest that it ate hard-shelled crustaceans and mollusks.

---

[1] †Thalattosaurs are marine reptiles with an unclear phylogenetic relationship to other clades. †Thalattosuchians are marine crocodylomorphs.

- **Squamates** Snakes radiated in the Cretaceous, and many basal snakes retained hindlimbs. A small radiation of snakes with small hindlimbs, such as †*Eupodophis* (Figure 10.7L), occupied marine coastal habitats in the Late Cretaceous.

- **Aves** A radiation of †toothed birds in the Late Cretaceous filled out the roster of marine Mesozoic tetrapods. †*Ichthyornis* (Figure 10.7M) was pigeon-sized, with well-developed wings. Ecologically, it may have been the Mesozoic equivalent of extant gulls. †*Hesperornis* (Figure 10.7N) was a larger (1.8 m long) flightless, foot-propelled diver with highly reduced wings that pursued and captured prey near the sea surface.

By the end of the Cretaceous, all of these iconic Mesozoic marine tetrapods had disappeared. Some marine turtles, snakes, and birds persisted, but tetrapods did not return as apex marine predators until the evolution of cetaceans and pinnipeds in the Eocene (**Figure 10.8**).

10.4 Extinctions

LEARNING OBJECTIVES

10.4.1 Name the Mesozoic extinction events and explain which ones were major mass extinctions.

10.4.2 Explain how volcanic activity can cause mass extinctions.

10.4.3 Describe the volcanic and extraterrestrial events that contributed to the end-Cretaceous mass extinction.

10.4.4 Describe the unequal effects of the end-Cretaceous mass extinction on different vertebrate clades.

Most Mesozoic extinction events resulted from intense episodes of volcanic activity accompanying collisions and splitting of continents as Pangaea formed and later ruptured (see Section 10.1). Although we think of lava flows as deadly accompaniments of volcanic activity, release of vast quantities of gases is the real culprit. The proportions of gases depend on the chemical characteristics of the magma, but SO_2 and CO_2 are always present. SO_2 is directly toxic to plants and animals, and both SO_2 and CO_2 acidify bodies of water. In addition, both gases induce greenhouse climate warming. CO_2 acts directly by trapping infrared radiation, while SO_2 acts indirectly through its effect on atmospheric chemical processes.

Triassic and Jurassic extinctions

The first round of Mesozoic extinctions occurred during the Carnian Pluvial Episode in the Late Triassic (see Section 10.1), which coincided with episodes of volcanism. Although not profound enough to qualify as a mass extinction, much diversity was lost, and in its aftermath dinosaurs gained a foothold in terrestrial ecosystems.

The end-Triassic extinction was one of the "big five" major mass extinctions in Earth's history (see Section 5.5). Eighteen clades of terrestrial and marine tetrapods

became extinct. This extinction coincided with the break-up of Pangaea, and may be related to two episodes of massive volcanism that occurred 20,000 years apart. Atmospheric CO_2 levels trebled from levels at the start of the Triassic to around 0.2% and global temperatures increased by about 4°C.

Another extinction event in the Early Jurassic (the Toarcian Oceanic Anoxic Event, ~183 Ma, characterized by depleted oceanic O_2) resulted in rapid global warming, affected terrestrial plant diversity, and drove extinctions of marine invertebrates. Minor extinctions affecting vertebrates occurred at the end of the Jurassic, principally affecting tetrapods such as dinosaurs, crocodyliforms, †pterosaurs, †sauropterygians, and †ichthyosaurs. About 20% of genera became extinct. Jurassic extinctions may be linked to changes in habitat and sea levels that took place as seas retreated from continental shelves.

Cretaceous extinctions

Two minor extinction events during the Cretaceous (~120 Ma and ~94 Ma) primarily impacted marine life. However, the major mass extinction at the K–Pg (Cretaceous–Paleogene) boundary (66 Ma) saw the extinction of 43% of tetrapod families. Although larger than the end-Triassic extinction, the end-Cretaceous event was not as large as the end-Permian extinction, when almost all vertebrates larger than 10 kg became extinct (see Section 5.5).

End-Cretaceous extinctions were not evenly distributed among clades. †Sauropod, †ornithischian, and most theropod dinosaurs, †sauropterygians, †pterosaurs and 75% of bird families were wiped out. No arboreal lineages of Cretaceous birds survived. However, extinction rates for crocodyliforms and turtles were only twice baseline rates, and those for lissamphibians, lizards, and snakes were indistinguishable from baseline rates. Eutherians suffered somewhat elevated extinction rates, as did †multituberculates and monotremes, but metatherians were profoundly affected, with an 89% loss of diversity.

Extensive extinctions occurred among marine invertebrates, but fishes were only moderately affected. Losses were greatest among large-bodied predators, with the extinction of †hybodont sharks, many kinds of batoids, and several basal neopterygian lineages. The angiosperm-dominated flora, expanding throughout the Late Cretaceous, continued into the Cenozoic relatively undisturbed, although conifers and ferns took a significant hit, especially in the higher latitudes.

The timescale of the K–Pg extinctions has long been debated: were they sudden or gradual? Certainly, some Mesozoic vertebrates became extinct before the end of the Cretaceous. The last †ichthyosaurs disappeared 30 Ma before the K–Pg boundary, declining in two steps during a period of global warming. Likewise, †pterosaurs declined through the latter part of the Cretaceous, and only two clades remained at the end the period. There is debatable evidence for decreases in diversity of †sauropods and †ornithischians earlier than 66 Ma.

There was unquestionably a catastrophic extinction at the very end of the Mesozoic era. The question is, what caused it? Culprits cited most often include massive volcanism, which caused earlier mass extinctions, and an asteroid impact. Good evidence exists for both mechanisms, and they may have acted together in a one-two punch.

The impact of an asteroid about 10 km in diameter has seized our imaginations as "the" cause of the extinction of large dinosaurs. The huge Chicxulub crater on the northern coast of Mexico's Yucatán Peninsula is well established as the impact site. More controversially, the Tanis locality in the Hell Creek Formation of North Dakota, some 3,000 km away, has been interpreted as a record of events in the first few hours after the impact. One hypothesis is that the impact triggered huge standing waves along the Western Interior Seaway that transported aquatic organisms far inland. The jury is still out on this developing interpretation.

The asteroid's impact would certainly have created a cloud of dust and aerosols, as evidenced by the layer of the rare element iridium found in K–Pg boundary sediments globally, including the Tanis locality. This dust cloud reduced the intensity of sunlight reaching Earth's surface, possibly for years, resulting in global cooling, decreased rainfall, and plunging primary productivity. An even more deadly blockage of sunlight might have come from soot released by fires following the impact. Geologic evidence suggests that the world plunged into brief, intense cooling, and climate modeling suggests that

the cold might have been deep enough to cause the demise of many groups of organisms.

Massive volcanic eruptions in the Deccan Traps province of western India began about 66.3 Ma and continued across the K–Pg boundary. These eruptions released enormous quantities of toxic metals as well as CO_2, Cl_2, and SO_2, causing global temperatures to rise and impacting Earth's ecosystems.

Sea levels fell near the end of the Cretaceous, draining epicontinental seas and causing rivers to erode new east–west valleys that could have impeded the north–south migratory movements of large dinosaurs. Rapid cooling near the K–Pg boundary dropped global temperatures 6–8°C below those during the previous 100,000 years. With these events crowding together at the end of the Cretaceous, plus evidence that some vertebrate lineages were already in decline, some researchers speculate that Earth's ecosystems were in trouble generally and that the asteroid impact was simply the final straw.

Debates between proponents of asteroid impact versus volcanism have resembled a tug-of-war for decades, with successive discoveries hailed as proof that one or the other was *the* primary driver of the end-Cretaceous mass extinction. Both sides overlook a biological truism: Great changes rarely if ever have "either/or" causes, and the correct answer to "which one" is usually "both of these, and others as well." A nuanced view is required for phenomena so massive in their effects and yet so curiously selective in terms of which clades perished and which survived.

Summary

10.1 Continental Geography and Climates

By the start of the Mesozoic, the continents had coalesced into the single supercontinent Pangaea.

Separation into a northern continent (Laurasia) and southern continent (Gondwana) began in the Jurassic. By the Late Cretaceous, continents were nearing their present-day positions.

The climate during the Early Triassic was unstable, with periods of lethally high temperatures. Lower latitudes were probably too hot for vertebrates.

The Carnian Pluvial Episode of the Late Triassic represented a switch from an arid climate to a wet one that lasted for a few million years and profoundly affected flora and fauna.

For most of the Mesozoic, polar regions were relatively warm and covered by forests. In contrast, equatorial regions were dry for most of the era and rainforests were absent.

Near the end of the Cretaceous, periods of abrupt warming and cooling disrupted ecosystems on a global scale.

10.2 Terrestrial Ecosystems

Impoverished ecosystems followed the end-Permian extinction, and the first 10 million years of the Triassic was a barren time. Following this, the Mesozoic was a time of major diversification and radiation, leading to large-scale changes in flora and fauna by the end of the era.

Early Triassic plants included ferns, tree ferns, horsetails, cycads, and ginkgophytes. By the end of the Triassic, conifers were the dominant canopy plants.

Angiosperms (flowering plants) first appeared in the Early Cretaceous, heralding the transition between the Mesophytic and Cenophytic Floras. By the Late Cretaceous, angiosperms dominated the vegetation, which had a major effect on the diversity of insects that fed on and pollinated them.

Early Triassic faunas of therapsids and basal archosaurs included herbivores that browsed within 1 m of the ground. These lineages were extinct by the end of the Triassic. Tetrapods were largely absent from arid equatorial regions.

Summary (continued)

By the end of the Triassic, many extant groups had appeared: mammals, lepidosaurs, turtles, and crocodyliforms. †Pterosaurs first appeared and dinosaurs also radiated. These lineages were joined in the Jurassic by lizards, lissamphibians, and stem birds. Except for crocodyliforms, members of these lineages were small in body size.

The largest tetrapods during the Jurassic and Cretaceous were dinosaurs. The Jurassic saw a large radiation of the enormous high-browsing, long-necked †eusauropods and large carnivorous theropods first appeared at this time.

Along with the angiosperm-dominated flora in the Cretaceous, new kinds of insects diversified. New types of dinosaurs also appeared, including herbivorous †hadrosaurs and †ceratopsians and new types of predatory theropods.

10.3 Marine Ecosystems

Nutrients from eroding terrestrial rocks enriched Mesozoic seas, leading to diversification of planktonic organisms that supported marine food webs.

An evolutionary arms race between durophagous vertebrates and their invertebrate prey (mollusks and crustaceans) contributed to the Mesozoic Marine Revolution and establishment of a modern, diversified fauna of marine invertebrates.

Stony corals like those of today, together with an extinct group of bivalves (†rudists) formed marine reef environments. Teleosts diversified in reef habitats, and acanthopterygians (spiny finned teleosts), a clade that includes 25% of the living species of vertebrates, was well established by the end of the era.

As many as 14 clades of tetrapods radiated in marine habitats during the Mesozoic.

Four clades included species that were apex predators: †sauropterygians, †ichthyosaurs, †thalattosuchians, and †mosasaurs, but no more than three of these clades ever coexisted.

Marine tetrapods converged on similar body forms that included limbs and feet modified as flippers and hypocercal tails with the vertebral column bending ventrally to form the lower lobe and an upper lobe formed by stiff tissue. Derived †ichthyosaurs also had a dorsal fin supported by stiff tissue. The body forms of these marine reptiles closely resemble those of sharks and extant cetaceans.

†Sauropterygians persisted through the entire Mesozoic and †ichthyosaurs extended from the Triassic to the Late Cretaceous, disappearing before the end-Cretaceous mass extinction. The diversity of †sauropterygians and †ichthyosaurs was greatly reduced in the end-Triassic mass extinction. †Thalattosuchian crocodylians lived in the Late Jurassic and Early Cretaceous, and the lizard-related †mosasaurs in the Late Cretaceous.

The Late Cretaceous saw the radiation of marine representatives of extant clades: turtles, snakes, and birds.

10.4 Extinctions

Several extinction events related to episodes of volcanism that accompanied the formation and subsequent breakup of Pangaea shaped the history of the Mesozoic.

The Carnian Pluvial Episode resulted in the extinction of many herbivorous tetrapods and set the stage for the radiation of dinosaurs. An end-Triassic mass extinction resulted in establishment of the dinosaur-dominated faunas of the Jurassic and Cretaceous.

Early Jurassic extinctions impacted terrestrial vegetation and marine life. End-Jurassic extinctions were relatively minor and probably related to falling sea levels.

The end-Cretaceous (K–Pg) major mass extinction wiped out †sauropod, †ornithischian, and most theropod clades as well as many other lineages, although some lineages were nearly unscathed. This extinction occurred at a time characterized by changes in sea level, climate instability, and profound volcanism. The impact of an asteroid may have been the final straw.

Discussion Questions

10.1 The Mesozoic often is called the "Age of Dinosaurs." In what way is that phrase misleading when considering how similar—or dissimilar—Mesozoic ecosystems were to those of today?

10.2 Why did so many lineages of tetrapods return to the sea during the Mesozoic?

10.3 Dinosaurs first appeared in the Early Triassic but did not rise to prominence until near the end of the era. What vegetational changes might have benefitted them?

10.4 Several studies indicate a gradual decline in the rate of appearance of new species of dinosaurs toward the end of the Cretaceous. Might this decline have been sufficient to account for their extinction even if there had been no external catastrophic events?

Sharp Photography, sharpphotography.co.uk/CC BY-SA 4.0

Living on Land

Komodo monitor (*Varanus komodoensis*)

Demands of terrestrial life differ from those of aquatic life because air and water have very different physical properties, as we described in Table 4.1. Air is less viscous and less dense than water, so the streamlining in the body forms of fishes is a minor factor for terrestrial tetrapods, whereas a skeleton that supports the body against the pull of gravity is essential. Respiration, too, is different in air and water. In air, the primary gill lamellae of fishes collapse on each other, greatly reducing the surface area available for gas exchange. And, because water is both dense and viscous, pumping water into and out of a closed-ended gas-exchange structure is prohibitively expensive; instead, fishes have flow-through gas exchange. Terrestrial tetrapods need gas-exchange organs that won't collapse, and given the low density and viscosity of air, they can use a tidal flow of air into and out of a saclike lung, although not all do. Muscles assumed new roles in tetrapods, including postural support of the body and ventilation of the lungs. These functions required more differentiated muscles, as did the complicated movements made by tetrapods during locomotion.

In this chapter, we consider changes to body structures and key organ systems associated with living on land, as summarized in **Figure 11.1** and **Table 11.1**. These adaptations include changes to feeding mechanisms, blood circulation, and sensing the world, as well as critical adaptations for conserving water while surrounded by a dry atmosphere. We begin with changes in the skeletal and muscular systems and their roles in support on land.

11.1 Support on Land

LEARNING OBJECTIVES

11.1.1 Describe how gravity influences support for terrestrial tetrapods.

11.1.2 Compare the roles played by axial and appendicular muscles in fishes and terrestrial tetrapods.

11.1.3 Describe two ways in which tetrapods can support greater body weight.

Perhaps the most important difference between living in water and on land is the effect of gravity on support and locomotion. Gravity has little significance for a fish because vertebrate bodies are approximately the same density as water, and hence fish are essentially weightless in water. Gravity is a critically important factor on land, however, and skeletons of terrestrial tetrapods must be able to support the body.

The skeletons of terrestrial tetrapods must be rigid enough not only to resist gravity but also the forces exerted as an animal starts, stops, and turns. Bone remodelling is of great importance for a tetrapod, and the internal structure of bone adjusts continually to changing demands of an animal's life. In humans, for example, intense physical activity results in an increase in bone mass, whereas inactivity (prolonged bed rest or being in outer space) results in loss of bone mass. And, because the skeletons of terrestrial animals experience greater stress than those of aquatic animals, their bones are more likely to break. Bone remodelling allows broken bones to mend.

The vertebrate skeleton has two major subdivisions, the axial skeleton and the appendicular skeleton. We begin with the axial skeleton.

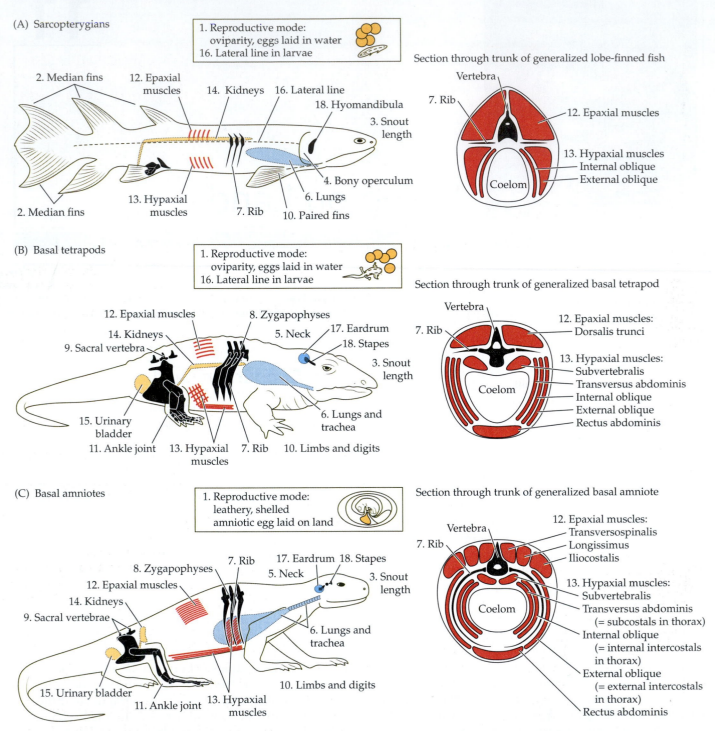

Figure 11.1 Adapting to life on land. The figure compares general body plans of (A) sarcopterygians, (B) basal tetrapods, and (C) basal amniotes, with attention to 18 characters that changed as tetrapods adapted to life on land. Note that many of the characters involve skeletal and muscular changes, as well as changes in the respiratory system and adaptations for conserving water in a nonaqueous environment. Numbered characters are described more fully in Table 11.1 (facing page) and referred to throughout this chapter. (After K. Liem et al. 2001. *Functional Anatomy of the Vertebrates*, 3rd ed. Cengage/Harcourt College: Belmont, CA.)

Axial skeleton

The **axial skeleton** includes the skull, vertebral column, ribs, and sternum (**Figure 11.2**). In all vertebrates, it protects the brain, spinal cord, and vital organs. In terrestrial vertebrates the axial skeleton is the primary source of support and locomotion, providing the attachment sites for muscles that move the head, neck, and back, as well as muscles that act in the shoulder and hip joints to move the limbs of the appendicular skeleton.

Skull The skulls of sarcopterygians have numerous dermal bones, and some but not all of these bones are

Table 11.1 Generalized characters of sarcopterygians, tetrapods, and amniotes[a]

Character[b]	Sarcopterygians (Figure 11.1A)	Basal tetrapods (Figure 11.1B)	Basal amniotes (Figure 11.1C)
1. Reproductive mode	Oviparity; eggs laid in water	Oviparity; eggs laid in water	Shelled amniotic egg laid on land; shell lost in viviparous mammals
2. Median fins supported by skeletal elements[c]	Present	Absent in all post-Devonian tetrapods	Absent
3. Snout length (distance from eyes to anterior tip of head)	Short	Long	Long
4. Bony operculum	Present	Absent	Absent
5. Neck	Absent	Present, allows head to turn separately from rest of body	Present and elongated, allows more mobility and longer trachea
6. Lungs and trachea	Simple lungs; no trachea	Simple lungs; short trachea without cartilaginous rings leads directly to lungs	Complex lungs divided to increase gas-exchange area; long trachea with cartilaginous rings branches into bronchi
7. Ribs	Short and slender	Longer and wider, provide greater body support	Longer; used in lung ventilation
8. Zygapophyses	Absent	Present	Present and well developed
9. Sacral vertebra(e)	Absent	Single sacral vertebra attaches pelvic girdle to vertebral column	Typically 2 (more in some derived groups; fused into synsacrum in birds)
10. Limbs and digits	Paired monobasic fins, no digits	Stout limbs with digits	More-gracile limbs with digits that play important roles in locomotion
11. Ankle joint	Absent	Simple	Distinct mesotarsal joint in middle of ankle
12. Epaxial muscles	Myomeric segments are not differentiated into discrete longitudinal muscles	Single muscle mass spans many vertebrae	Differentiated into 3 muscles that span many vertebrae
13. Hypaxial muscles	Not differentiated	Differentiated into 5 groups	Differentiated into 5 major groups, some with further subdivisions in thorax
14. Kidneys	Elongate, extending the length of the trunk; opisthonephric	Elongate, extending the length of the trunk; opisthonephric	Compact and posterior; metanephric (characterized by large numbers of nephrons)
15. Urinary bladder	Absent	Present	Present; lost in some
16. Lateral line	Present	Present in aquatic larvae and adults; lost during metamorphosis in terrestrial forms	Absent at all life stages
17. Eardrum (tympanum)	Absent	Present in adult frogs, but derived independently from the eardrum of amniotes	Present
18. Stapes (ossicle of middle ear)	Absent (homolog is hyomandibula, a large bone used in jaw support, reduced to stapes in basal tetrapods)	Present and functional in middle ear cavity of adult frogs; transmits vibrations from eardrum to inner ear	Present and functional in middle ear cavity; two additional ossicles (malleus and incus) in mammals

[a] In addition to the comparisons noted for basal tetrapods and basal amniotes, the table contains notes on some derived conditions in lissamphibians and mammals.

[b] Numbers in this column correspond to the characters indicated in Figure 11.1 and referenced in the text.

[c] Median fins of fishes (dorsal, anal, and caudal fins) are supported by internal skeletal elements and fin rays. These fins reduce roll and apply force to the water column but have no function on land. In cases where extant tetrapods have median fins, such as lissamphibian larvae and secondarily aquatic tetrapods such as dolphins, the fins lack internal skeletal support and fin rays.

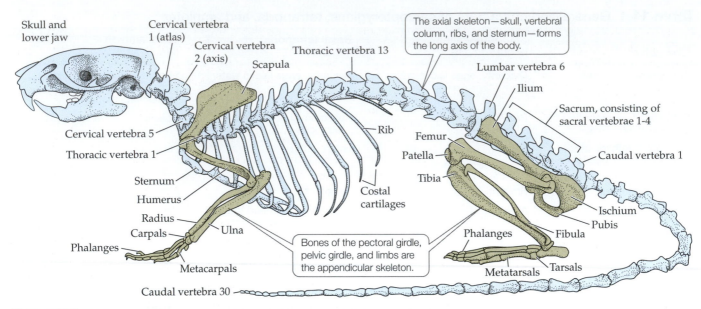

Figure 11.2 Skeleton of a tetrapod. Axial and appendicular skeleton of a classic research animal, the Norway rat, *Rattus norvegicus*. Note the five vertebral regions characteristic of mammals.

retained in tetrapods, including the roofing bones of the human skull (see Section 2.6). In bony fishes, supracleithral bones connect the skull to the pectoral girdle (see Figure 8.14A). As a result, a fish cannot turn its head—it must turn its entire body. The supracleithral bones are reduced or lost in tetrapods, resulting in a flexible neck region that allows movement of the head separately from the rest of the body (see Figure 8.14B).

Vertebrae   In terrestrial tetrapods, the axial skeleton is modified for support on land. Recall from Section 8.5 that zygapophyses on the vertebrae of tetrapods interlock and allow the vertebral column to transfer weight of the body to limbs. (Secondarily aquatic tetrapods, such as cetaceans, have reduced or lost zygapophyses in some regions of the vertebral column). Terrestrial vertebrates also have greater regional differentiation of vertebrae than do fishes. Fishes have two regions: trunk and caudal vertebrae (see Figure 2.10A). Lizards have four regions: cervical, trunk, sacral, and caudal vertebrae. In mammals, trunk vertebrae are differentiated into two distinct types, resulting in five vertebral regions: cervical, thoracic, lumbar, sacral, and caudal (see Figure 11.2).

• **Cervical vertebrae** allow movement of the head relative to the trunk, and muscles that support and move the head attach to processes on cervical vertebrae. In extant tetrapods, the first cervical vertebra is the **atlas**, named for the god in Greek mythology who supported the world on his shoulders. It articulates with the skull and allows the nodding movement (which signifies "yes" in many, though not all,

human cultures). In amniotes, the second cervical vertebra, termed the **axis**, permits rotary movements between the atlas and axis, such as shaking the head "no" (again, the meaning of this gesture differs among human cultures).

• **Trunk vertebrae** often bear ribs, except in lissamphibians, which have almost entirely lost ribs. In mammals, trunk vertebrae are subdivided into **thoracic vertebrae**, which bear ribs, and **lumbar vertebrae**, which lack ribs. Thoracic vertebrae and ribs function in respiration and locomotion; lumbar vertebrae function in body support and locomotion.

• **Sacral vertebrae** articulate tightly with the pelvic girdle in tetrapods and allow the hindlimbs to transfer force to the axial skeleton. Lissamphibians have a single sacral vertebra; lepidosaurs and crocodylians have two sacral vertebrae; and mammals have from three to five. Birds have a synsacrum formed by fusion of sacral vertebrae, other vertebrae, and the pelvic girdle (see Section 21.2).

• **Caudal vertebrae** support the tail and are usually structurally simpler than trunk vertebrae.

Ribs and sternum   Ribs are paired, curved bones that extend from vertebrae and support and enclose the body cavity. In mammals, ribs enclose just the thoracic cavity. The ribs of Paleozoic tetrapods were stouter and more prominent than those of fishes. They played a role in body support as well as locomotion and may have stiffened the trunk in animals that had not yet evolved much postural support from axial musculature (Figure 11.1B #7). In more derived tetrapods, the ribs also function in ventilation. Although ribs on the trunk vertebrae are the most prominent in tetrapods in general, many non-mammalian tetrapods also have small ribs on their cervical vertebrae.

The **sternum**, or breastbone, is a midventral endochondral element found only in tetrapods. It differs in lissamphibians and amniotes and may have evolved independently in the two groups. In amniotes, the distal ends of trunk ribs typically articulate with the sternum, an evolutionary innovation that allowed ribs to function more effectively in inhaling and exhaling air. In both lissamphibians and amniotes, elements of the pectoral girdle articulate with the sternum.

Axial muscles

The axial muscles of tetrapods became highly differentiated compared to those of bony fishes and assumed two new roles: (1) postural support of the axial skeleton (and thus the body), and (2) ventilation of the lungs. Axial muscles are important for maintaining posture on land because the body is not supported by water; without muscular action, the axial skeleton would collapse. Likewise, the method of ventilating lungs differs if the chest is surrounded by air rather than by water. The axial musculature of vertebrates has dorsal **epaxial muscles** and ventral **hypaxial muscles** (see Figure 2.11).

Epaxial muscles The epaxial muscles of lissamphibians form a single longitudinal mass, the dorsalis trunci, and this condition probably characterized the earliest tetrapods (Figure 11.1B #12). A salamander walking on land uses both epaxial and hypaxial muscles to bend the trunk, much as does a swimming fish. Epaxial muscles of a generalized amniote form three major components: the transversospinalis, longissimus, and iliocostalis (Figure 11.1C #12). The transversospinalis connects the neural spines of individual vertebrae along the body, contributing to postural stability. The longissimus extends along the length of the vertebral column and contributes to dorsoventral flexion of the trunk. The iliocostalis of sauropsids extends from the pelvis to the anterior end of the rib cage and contributes to lateral flexion of the trunk.

Hypaxial muscles In the caudal region of a fish's body, hypaxial muscles form a single mass of segmentally organized myomeres. In the trunk region, their hypaxial muscles have two parts: the large external oblique muscle and a smaller wedge of internal oblique muscle adjacent to the ribs (Figure 11.1A #13). In the trunk of basal tetrapods, hypaxial muscles have five parts (Figure 11.1B #13). The subvertebralis acts in conjunction with epaxial muscles to maintain postural stability. The three layers of the body wall surrounding the coelom—an inner layer (transversus abdominis), a middle layer (internal oblique), and an outer layer (external oblique)—function

in breathing and support of viscera. Contraction of the transversus abdominis expels air from the lungs of lissamphibians, which, unlike amniotes, do not use ribs to breathe. A similar function would have been essential for lung ventilation by basal tetrapods. Air-breathing fishes can use surrounding water pressure to force air from the lungs, but tetrapods need muscles to do this.

The rectus abdominis is another hypaxial muscle seen first in tetrapods. It spans the ventral surface of the body from the pectoral to the pelvic girdle and plays a role in posture. (This is the muscle responsible for the "six-pack" abdomen of human bodybuilders.)

Amniotes have further subdivisions of the transversus abdominis, internal oblique, and external oblique in the thorax, where these muscles connect between ribs to form subcostal, internal intercostal, and external intercostal muscles used for inhalation and exhalation (Figure 11.1C #13). The amniote innovation of using elongated ribs and associated muscles for lung ventilation is an important advance in the history of vertebrates because it allowed amniotes to produce a more forceful inhalation. This made it possible to have a longer trachea, which in turn made it possible to have a longer, more mobile neck.

Appendicular skeleton

The components of the **appendicular skeleton** are the pectoral girdle, pelvic girdle, and limbs (see Figure 11.2).

Pectoral and pelvic girdles The pectoral girdle of fishes attaches to the back of the skull; it serves as the origin for muscles that move the gill arches and lower jaw and also supports simple muscle masses that move the pectoral fins. The pelvic girdle of fishes has no connection with the vertebral column, but merely anchors the pelvic fins in the body wall (see Figure 8.14A). Attachment of the pectoral girdle to the skull and lack of attachment of the pelvic girdle to the axial skeleton are unsuited for life on land.

The pectoral girdle of tetrapods is freed from its bony connection with the skull, and support of the body is now its major role. It does not articulate directly with the vertebral column (as the pelvic girdle does), which is why humans can shrug their shoulders but not their hips. The pectoral girdle attaches to the vertebral column by muscles and connective tissue; it attaches to the sternum by muscles, connective tissue, and the clavicle (when present; the clavicle is reduced in size or lost in many mammals). The pelvic girdle of tetrapods consists of three paired bones on each side: the ilium, ischium, and pubis (see Figure 11.2). Blades of the ilium articulate with the axial skeleton via one or more sacral vertebrae.

Tetrapod limbs The skeleton of tetrapod limbs has three regions: stylopodium (humerus or femur); zeugopodium (radius and ulna or tibia and fibula); and autopodium (wrist and hand or ankle and foot; see Figure 8.16). Between the bones of tetrapod limbs are well developed

(A) Pectoral fin muscles of a sturgeon   (B) Superficial (left) and deep (right) forelimb muscles of a lizard

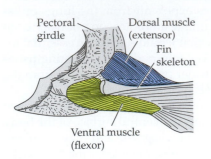

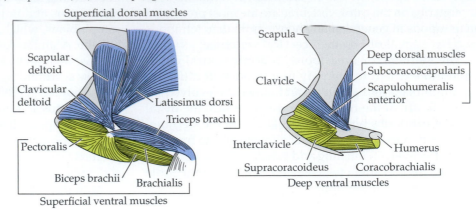

Figure 11.3 Pectoral fin muscles of a fish and forelimb muscles of a tetrapod. (A) Left pectoral fin and associated muscles of a sturgeon. (B) Left forelimb (shoulder region and humerus) and associated superficial and deep muscles of a lizard. Note the greater differentiation of appendicular muscles in the lizard compared with the sturgeon. (After K. Liem et al. 2001. *Functional Anatomy of the Vertebrates*, 3rd ed. Cengage/Harcourt College: Belmont, CA.)

synovial joints (moveable joints with cartilaginous surfaces lubricated by fluid). These synovial joints include shoulder and hip joints, elbow and knee joints, wrist and ankle joints, and hands and feet with joints between the phalanges.

A salamander's ankle joint is between the tibia/fibula and the tarsal bones. In contrast, amniotes have a distinctive new ankle joint, the **mesotarsal joint**, in the middle of the hindfoot between tarsal bones (Figure 11.1C #11). These differences relate to locomotion, described in Section 11.2.

Appendicular muscles   Basal bony fishes such as sturgeons have dorsal and ventral muscle masses originating on the pectoral girdle and inserting at the base of the pectoral fin (**Figure 11.3A**). The dorsal muscle extends the fin, moving it upward and away from the ventral midline. The ventral muscle flexes the fin, pulling it toward the ventral midline. In contrast, tetrapods have differentiated the dorsal and ventral muscle masses into several discrete muscles, as for example in a lizard (**Figure 11.3B**). The highly differentiated appendicular muscles of tetrapods reflect the complexities of moving on land.

Size and scaling

Land-dwelling tetrapods span a vast range of sizes, from frogs less than 2 cm long to the African elephant. Because all structures are subject to laws of physics, the absolute size of an animal profoundly affects its anatomy and physiology. The study of scaling, or how shapes change with size, is known as **allometry** (Greek *allos*, "different"; *metron*, "measure"). Vertebrate biologists use two different types of allometric comparisons: (1) allometric growth during the life of an individual as its body size increases, and (2) allometric comparisons between two or more species.

If a vertebrate's features did not change during growth, all of its components would scale isometrically (Greek *isos*, "similar") and an adult would appear just like a photo enlargement of a juvenile. However, very few body parts scale isometrically. To cite a classic example of allometric growth, the human brain grows rapidly during gestation, infancy, and early childhood before reaching its full size at 6–7 years. Thus, the human skull and facial features become proportionally smaller in relation to the torso and limbs, which continue to grow for many years.

Interspecific allometric comparisons allow us to explore how evolutionary changes in body size relate to fundamental features of the skeleton, muscles, and physiological processes like heart rate. The linear dimensions (height, width, and depth) of an object relate differently to its surface area and its volume. If all three of the object's linear dimensions double isometrically (a twofold change), then the object's surface area increases as the square of the changes in linear dimension, resulting in a fourfold change in surface area. However, volume, and hence weight, increase as the cube of the changes in linear dimension, so a doubling of the object's linear dimensions produces an eightfold change in volume (and weight, assuming that the materials are the same).

Limb bones and muscles support a vertebrate on land. If an animal doubles its size isometrically, its volume (weight) increases eightfold, but the cross-sectional surface area of its limbs increases only fourfold, which poses problems in supporting the greater body weight. Terrestrial tetrapods have two responses to this. The first is to adjust limb posture and flexion of limb joints. For example, relatively small mammals such as many rodents have a crouching posture and flexed limb joints (see Figure 11.2). Larger mammals, such as a domestic cat, have a more upright posture with less flexion of limb joints (**Figure 11.4A**). Still larger mammals (>300 kg) have an even more erect posture and even less flexion of limb joints (**Figure 11.4B**).

(A) Skeleton of a domestic cat

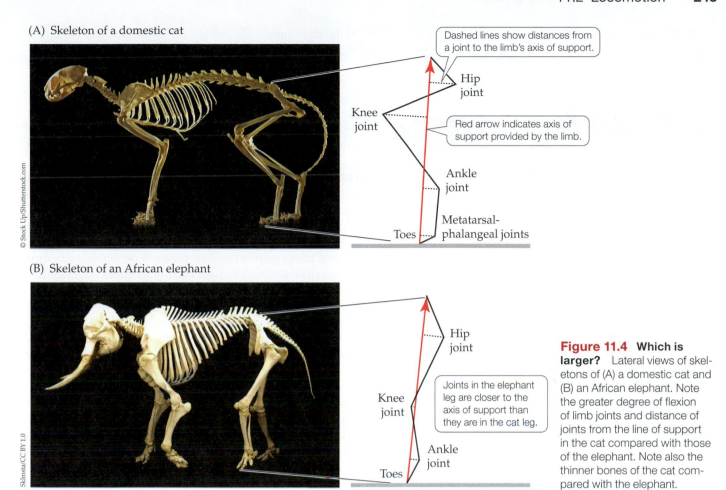

(B) Skeleton of an African elephant

Figure 11.4 **Which is larger?** Lateral views of skeletons of (A) a domestic cat and (B) an African elephant. Note the greater degree of flexion of limb joints and distance of joints from the line of support in the cat compared with those of the elephant. Note also the thinner bones of the cat compared with the elephant.

The second response to supporting greater body weight is to have proportionally thicker limbs with an allometrically greater cross sectional area. For example, elephant bones are much thicker than those of a domestic cat, making it easy to tell from the skeleton which animal is larger even when the two skeletons are depicted at the same size, as in Figure 11.4. The pillarlike weight-bearing stance of elephants and other very large tetrapods, called **graviportal posture**, can support enormous body weights, but reduces the animal's agility.

11.2 Locomotion

LEARNING OBJECTIVES

11.2.1 Explain the difference between a walking gait and a walking-trot gait.

11.2.2 Describe the relationship between stance and locomotion for tetrapods.

11.2.3 Understand how flexion and extension of the vertebral column affect speed and other aspects of terrestrial locomotion.

It requires different forms of locomotion to move through air than through water. Because water is dense, a fish swims by passing a wave along its body—the fins and sides of the body push backward against water and the fish moves forward via lateral undulation (see Section 7.6). Pushing backward against air doesn't move a terrestrial animal forward unless it has wings. Most terrestrial tetrapods (including fliers, when they are on the ground) use legs and feet to transmit a backward force to the substrate. Modes of locomotion of terrestrial tetrapods are more complex than those of fishes.

Axial muscles still participate in locomotion in basal tetrapods, producing lateral bending of the vertebral column during the movements of salamanders, lepidosaurs, and crocodylians. In birds and mammals, however, limb movements largely replace lateral flexion of the trunk. Dorsoventral flexion is an important component of mammalian locomotion. In secondarily aquatic mammals such as cetaceans, axial muscles again assume major roles in powering tail movements; however, in cetaceans, the tail is moved up and down (dorsoventral flexion) rather than side to side (lateral undulation) as in fishes or aquatic reptiles.

Salamanders combine lateral axial movements with diagonal pairs of legs moving together. The left hindlimb and right forelimb move together, followed by movements

(A) Dorsal view of walking-trot gait of a salamander

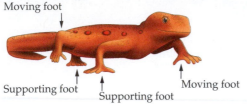

1 Body bends to right; left hindlimb and right forelimb move forward together.

2 Body bends to left; right hindlimb and left forelimb move forward together.

(B) Anterolateral view showing diagonal feet raised from the ground

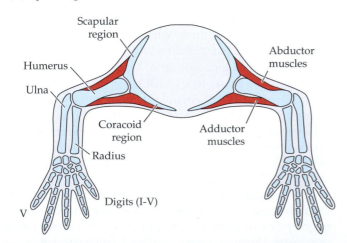

Moving foot

Supporting foot

Supporting foot

Moving foot

Figure 11.5 Walking-trot gait of salamanders. (A) Dorsal view at times 1 and 2 during a walking-trot. (B) Anterolateral view corresponding to time 2. Note that the left hindfoot supports the left side of the body while the right forefoot supports the right side of the body; the right hindfoot and left forefoot are elevated above the substrate and moving forward. (After K. Liem et al. 2001. *Functional Anatomy of the Vertebrates*, 3rd ed. Cengage/Harcourt College: Belmont, CA.)

of the right hindlimb and left forelimb; this gait is known as the walking-trot, illustrated in **Figure 11.5**.

Amniotes, on the other hand, typically employ a walking gait, in which each leg moves independently in succession, usually with three feet on the ground at any given time. Trotting in amniotes is a faster gait in which diagonal pairs of limbs move together (e.g., right front and left rear, as in the walking-trot of salamanders; see Figure 11.5B), but there is a period during which all four feet are off the ground. Humans rely entirely on hindlegs for walking, but we retain the ancestral diagonal coupling of fore- and hindlimbs during walking (and running): we swing our right arm forward when the left leg moves forward, and vice versa. Coupled diagonal movement of paired limbs is probably ancestral for gnathostomes, because epaulette sharks, *Hemiscylium ocellatum*, and African lungfishes, *Protopterus* spp., also move their paired fins in a diagonal fashion when walking on the substrate.

Basal amniotes have a sprawling stance in which the humerus extends laterally away from the body with a nearly 90° bend at the elbow joint (**Figure 11.6A**). The hindlimb is similar. The sprawling stance requires large adductor muscles on the ventral side of the humerus to support the trunk above the limbs. There is a correspondingly large coracoid region of the pectoral girdle to which the adductor muscles attach. Several derived groups of amniotes, such as therian mammals, evolved more upright stances, in which the limbs are placed beneath the

(A) Sprawling stance of a basal amniote

Scapular region

Humerus

Ulna

Coracoid region

Radius

Adductor muscles

Abductor muscles

Digits (I–V)

V

(B) Upright stance of a therian mammal

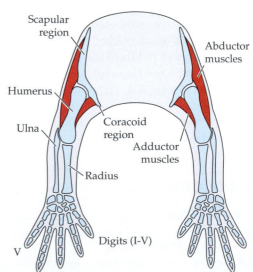

Scapular region

Abductor muscles

Humerus

Ulna

Coracoid region

Adductor muscles

Radius

Digits (I–V)

V

Figure 11.6 Sprawling and upright stances of amniotes. Cross-sections through the pectoral girdle and forelimbs showing sprawling and upright stances of tetrapods. (A) Sprawling stance in a basal amniote such as a lizard. (B) Upright stance of a therian mammal (based on the Virginia opossum, *Didelphis virginiana*). Different stances are associated with differences in sizes of arm muscles and regions of the pectoral girdle, as well as different forms of locomotion (see Figure 11.8).

(A) Thomson's gazelle, *Eudorcas thomsonii*, bounding

(B) Eland, *Taurotragus oryx*, galloping

(C) Cheetah, *Acinonyx jubatus*, legs gathered under body and extended

Figure 11.7 Bounding and galloping. (A) Small antelopes like Thomson's gazelle (15–35 kg) bound. (B) Larger antelopes such as the eland (350–900 kg) gallop. Note that the limbs of the gazelle are more flexed and proportionally longer than those of the eland, and that there is a period of suspension when all four limbs of the gazelle are off the ground and stretched out as it leaps from hindlimbs to forelimbs. (C) In the cheetah—the fastest terrestrial vertebrate—vertebral flexion and extension increase stride length (distance between successive contacts of the same foot) and thereby enhance speed. Positions of the vertebral column are indicated by white dashed lines. (A,B after T.A. McMahon and J.T. Bonner. 1983. *On Size and Life.* New York: Scientific American Books.)

body (**Figure 11.6B**). Therians have reduced the sizes of humeral adductor muscles and the coracoid region of the pectoral girdle (including loss of a separate coracoid bone), because the limbs are underneath the body.

Lateral undulation of the vertebral column plays an important role in the walking-trot gait of salamanders, but for tetrapods with an upright stance, dorsoventral undulation of the vertebral column can play a similarly important role in locomotion. For example, note the vertebral flexion and extension during bounding locomotion (jumping off the hindlegs and landing on the forelegs) of a Thomson's gazelle (**Figure 11.7A**). In contrast to gazelles, larger ungulates such as the eland (**Figure 11.7B**) gallop (as do horses and zebras). The

vertebral column of galloping ungulates is less flexible than that of bounders. Movements of the vertebral column also characterize the fastest terrestrial animal, the cheetah, which can achieve speeds greater than 120 km/h (**Figure 11.7C**).

Many variations in gait, stance, vertebral flexion, and limb movements evolved over the course of tetrapod history (**Figure 11.8**). Bipedalism arose several times, and there were likewise reversals to quadrupedal locomotion. Limblessness and snakelike locomotion also evolved many times in different groups of tetrapods, as did convergent evolution of ricochetal locomotion—bipedal jumping using large hindlimbs and a long tail for balance—within therian mammals (see Figure 1.10).

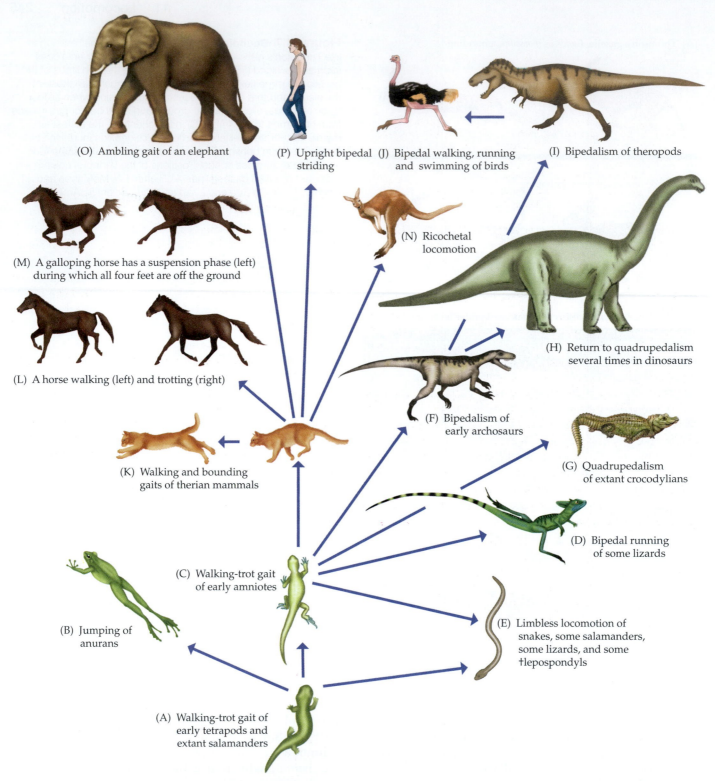

Figure 11.8 **Evolution of terrestrial stance and locomotion.**
(A) The walking-trot gait of basal tetrapods and salamanders is powered by axial movements of the body, with limbs moved in diagonal pairs. (B) Jumping in frogs requires a specialized pelvic girdle and long hindlimbs. (C) Longer limbs are used for the walking-trot gait of basal amniotes and many extant lizards. (D) Some lizards run bipedally using hindlimbs much longer than forelimbs. (E) Limbless locomotion evolved convergently several times within tetrapods. (F) Bipedalism characterized basal archosaurs, which had an upright posture and larger hindlimbs than forelimbs. (G) Extant crocodylians are quadrupedal and have a sprawling stance. (H) Many dinosaurs evolved quadrupedalism. (I) Theropods are bipedal with tiny forelimbs. (J) Birds are bipedal. (K) Therian mammals walk and bound. When bounding, the animal jumps off the two hindlimbs, moves through the air with limbs outstretched, and lands on two forelimbs (or one forelimb and then the other). (L) Horses trot at speeds intermediate between walking and galloping. (M) When galloping, a horse moves one limb at a time, with a period of suspension when all four feet are in the air and the four legs are bunched up, as shown. (N) Ricochetal locomotion is bipedal jumping used by kangaroos and several groups of rodents. It is associated with shorter forelimbs than hindlimbs and a long tail for balance. (O) Elephants and horses use a brisk walking gait, the amble. (P) Humans uniquely use upright bipedal striding.

11.3 Eating

LEARNING OBJECTIVES

11.3.1 Compare the feeding cycles of fishes and terrestrial tetrapods.

11.3.2 Describe differences between the so-called primary tongue of fishes and the tongue of tetrapods.

11.3.3 Understand the relationship between length of hyoid horns and feeding methods in birds.

11.3.4 Explain the lack of salivary glands in fishes and functions of saliva in tetrapods.

Physical differences between water and air profoundly affect the ways in which tetrapods feed. In water, food items are essentially weightless, which means that fishes and aquatic tetrapods such as turtles can suck food into their mouth using buccopharyngeal expansion, a rapid enlargement of the mouth and pharynx. Suction feeding is not an option on land, however, because air is much less viscous than water. (You can suck up noodles along with liquid soup, but you cannot suck up the same noodles if you put them on a plate by themselves.)

Sarcopterygian fishes have a short snout (Figure 11.1A #3), and movements of the jaws and hyoid arch suck water into the mouth for both gill ventilation and feeding. Cranial and trunk muscles power a range of movements for the many bones and joints in a fish's skull. Branchiomeric muscles (muscles that initially develop in the pharyngeal arches; see Section 2.4), such as the adductor mandibulae that closes the jaw, play many roles in feeding movements.

Fishes typically have a two-phase feeding cycle (**Figure 11.9**). Jaw opening is powered by epaxial muscles that raise the head at the craniovertebral joint and hypobranchial muscles that open the jaw and depress the floor of the mouth. Jaw closing is powered by the adductor mandibulae, and a distinct hyoid recovery phase is powered

by sheets of muscle that span the left and right sides of the jaws.

Two modified hyoid arch muscles are important in tetrapod feeding. The first is the depressor mandibulae, which runs from the skull to the back of the lower jaw, where it helps the hypobranchial muscles open the mouth. The second is the sphincter colli, which surrounds the neck and aids in swallowing food.

Basal tetrapods had wide, flat skulls and long snouts, so most of the tooth row was anterior to the eye (Figure 11.1B #3). Many Paleozoic tetrapods lived in or near water and, like lissamphibians today, used buccopharyngeal expansion to suction-feed in water. On land, we infer that they caught prey and swallowed it whole, because they did not have muscles or dentition suitable for chewing. The adductor mandibulae remained the major jaw-closing muscle, and the intermandibularis and interhyoideus muscles functioned to push food backward into the oral cavity.

Amniotes typically have a four-phase feeding cycle of jaw movements: (1) slow opening, (2) fast opening, (3) fast closing, and (4) slow closing. Unlike fishes, there is no hyoid recovery phase (see Figure 11.9). In lissamphibians and sauropsids, jaw opening is powered principally by the depressor mandibulae. In therians, jaw opening results from contraction of the anterior and posterior bellies of the digastric muscle and jaw closing is chiefly powered by the adductor mandibulae or muscles derived from it. The four phases of the amniote feeding cycle allow time to manipulate food in the mouth and crush it between teeth.

Many amniotes use **inertial feeding** after a prey item is captured. The predator jerks its head backward to impart momentum to the prey and simultaneously opens its jaws, loosening the grip on the prey. Then the predator thrusts its head forward over the prey before closing its

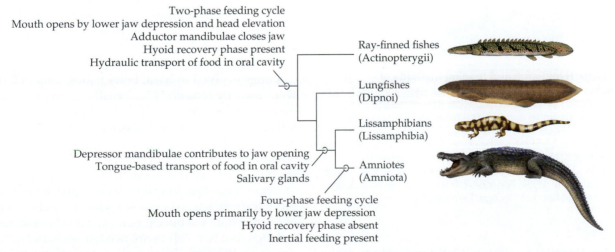

Two-phase feeding cycle
Mouth opens by lower jaw depression and head elevation
Adductor mandibulae closes jaw
Hyoid recovery phase present
Hydraulic transport of food in oral cavity

Ray-finned fishes (Actinopterygii)

Lungfishes (Dipnoi)

Lissamphibians (Lissamphibia)

Depressor mandibulae contributes to jaw opening
Tongue-based transport of food in oral cavity
Salivary glands

Amniotes (Amniota)

Four-phase feeding cycle
Mouth opens primarily by lower jaw depression
Hyoid recovery phase absent
Inertial feeding present

Figure 11.9 Phylogeny of feeding mechanics. Ray-finned fishes have a two-phase feeding cycle consisting of jaw opening and jaw closing. Key changes in tetrapods relate to evolution of tongue-based intraoral transport of food and the four-phase feeding cycle of amniotes, which allow greater processing of food in the mouth. (After K. Liem et al. 2001. *Functional Anatomy of the Vertebrates*, 3rd ed. Cengage/Harcourt College: Belmont, CA.)

(A) Tongue and hyoid horn of a woodpecker

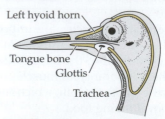

(B) Tongue and hyoid horn of a budgerigar

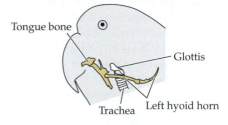

Figure 11.10 Hyoid arch and feeding methods of woodpecker and budgerigar.   (A) Lateral view of the long tongue and hyoid horns of a woodpecker. The photo shows a northern flicker, *Colaptes auratus*. The extended tongue probes for insects in the wood and bark of trees. (B) Short tongue and hyoid horns of a budgerigar, *Melopsittacus undulatus*, in lateral view. Budgerigars, like other members of the parrot family, manipulate food in the mouth using intrinsic muscles of the tongue. (A after K. Liem et al. 2001. *Functional Anatomy of the Vertebrates*, 3rd ed. Cengage/Harcourt College: Belmont, CA.)

jaws again. Inertial feeding allows a predator to make multiple tearing and crushing contacts with prey.

The so-called primary tongue of jawed fishes is small, composed largely of connective tissues, and lacks intrinsic musculature. Fishes use the primary tongue to generate

hydraulic transport of food within the oral cavity (i.e., the food is moved by water currents). In contrast, tetrapod tongues are large, muscular, and mobile and are often interpreted as a key innovation for terrestrial feeding. Tetrapods use the mobile tongue to physically move food in the mouth and transport it to the pharynx.

Many terrestrial salamanders and lizards use sticky tongues to capture prey and transport it into the mouth. Some tetrapods—frogs, salamanders, and chameleons—even project their tongue to capture prey (see Figures 12.4 and 12.6). Mechanisms of tongue projection differ in each group and tongue projection evolved convergently. Some birds, such as woodpeckers and hummingbirds, also protrude their tongue during feeding; the supporting hyoid apparatus has long horns encircling the skull (**Figure 11.10A**). Birds that manipulate food in the mouth using their tongue have short hyoid horns (**Figure 11.10B**).

Salivary glands are present only in tetrapods; saliva is not required to swallow food in water. Salivary gland secretions lubricate food, and in mammals contain enzymes that begin chemical digestion of carbohydrates in the mouth. Some insectivorous mammals, one group of lissamphibians, two species of lizards, and many lineages of snakes evolved venomous salivary secretions to kill prey.

11.4 Breathing Air

LEARNING OBJECTIVES

11.4.1 Describe the steps by which lissamphibians inflate their lungs using a buccal pump.

11.4.2 Explain the negative-pressure aspiration pump used by amniotes.

11.4.3 Distinguish the trachea of lissamphibians from that of amniotes and explain the evolutionary significance of the amniote condition.

The low density and low viscosity of air make tidal ventilation of a saclike lung, such as those found in lissamphibians and most amniotes, energetically feasible. The greater oxygen content of air compared to water reduces the volume of fluid that must be pumped to meet a tetrapod's metabolic requirements.

Lungs evolved in basal bony fishes, long before they were used by tetrapods for breathing on land (see Section 7.1 and **Figure 11.11**). The original mechanism for inflating lungs is called **buccal pumping**. To do this, the floor of the buccal cavity is moved up and down using cranial muscles. A glottal valve separates the buccal cavity from the lung and must be opened to move air into or out of the lungs. In water, fishes can exhale passively, but tetrapods use axial muscles for exhalation. Lissamphibians retain the buccal pumping mechanism for lung inflation, but two valves are needed—a narial valve that closes off the nostrils, as well as the glottal valve. In a typical respiratory cycle of a frog, the narial valve opens, the glottal valve closes, and the floor of the buccal cavity is depressed, sucking air through the nostrils into the buccal cavity. Then the narial valve closes and the glottal valve

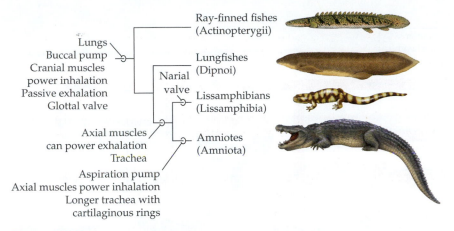

Figure 11.11 Phylogeny of lung ventilation mechanisms.
Some key events in the evolution of lung ventilation and
air-breathing in terrestrial vertebrates. (After K. Liem et al. 2001.
Functional Anatomy of the Vertebrates, 3rd ed. Cengage/
Harcourt College: Belmont, CA.)

opens; muscles raise the floor of the buccal cavity, forcing
air from the buccal cavity into the trachea and lungs. Air
can be expired from the lungs in several ways. In the
first, both narial and glottal valves are opened, allowing
elastic recoil to deflate the lungs, such that air passes from
the lungs into the trachea and buccal cavity, and out the
nostrils. Alternatively, the narial valve can be closed while
the glottal valve is opened; depression of the floor of the
buccal cavity sucks air from the lungs through the trachea
and into the buccal cavity, after which the glottal valve
closes, the narial valve opens, and air is expelled out the
nostrils. Lissamphibians also have an exhalation pump
that functions by contracting body muscles to exhale.

In contrast to lissamphibians, amniotes use a negative-pressure **aspiration pump** (see Figure 11.11). In this
system, there is only a glottal valve. A respiratory cycle
begins by opening the glottal valve and expanding the rib
cage using intercostal muscles (these muscles are derived
from anterior portions of hypaxial musculature; Figure
11.1C #13). Contraction of intercostal muscles creates a
negative pressure (i.e., below atmospheric pressure) within the body cavity, which causes air to be sucked into the
lungs. Air is expelled by compression of the abdominal
cavity, primarily through elastic return of the rib cage
and contraction of the elastic lungs, as well as by contraction of the transversus abdominis musculature. Unlike lissamphibians, axial muscles play a larger role than
cranial muscles in amniote lung ventilation.

Lungs of many lissamphibians are simple sacs with
few internal divisions; the trachea is short, without cartilaginous rings (small cartilaginous nodules may be present), and leads directly to the lungs (Figure 11.1B #6). In
contrast, amniote lungs are subdivided, sometimes in
very complex ways, to increase surface area for gas exchange. The longer trachea of amniotes is strengthened
by cartilaginous rings, and branches into a treelike series
of bronchi leading to each lung (Figure 11.1C #6).

Subdivisions of the lungs differ in sauropsids and synapsids, suggesting independent evolution of complex

lungs. The combination of a trachea with
rings and aspiration pumping allowed
many amniotes to evolve longer necks
than those of lissamphibians or early tetrapods. Amniotes also possess a **larynx**
at the junction of the pharynx and trachea. Derived from skeletal components
of the hyoid arch and posterior visceral-
al arches, the larynx is used for sound
production, most notably in mammals.
Birds also possess a larynx, but it is used
to regulate flow of air into the trachea.
Birds use the **syrinx**, a structure at the
base of the trachea, to vocalize.

11.5 Pumping Blood Uphill

LEARNING OBJECTIVES

11.5.1 Explain why blood pressures are higher in
terrestrial vertebrates than in fishes.

11.5.2 Describe functions of the lymphatic system
and the particular importance of this system
to terrestrial vertebrates.

11.5.3 Compare single circulation in fishes with
double circulation in tetrapods.

11.5.4 Explain why coronary arteries are especially
well-developed in crocodylians, birds, and mammals.

For vertebrates that live in water, the heart needs only to
overcome fluid resistance in order to move blood through
the circulation. Circulation is more difficult for terrestrial vertebrates because blood must be pushed uphill into
portions of the body that are higher off the ground than
the heart. Thus, terrestrial tetrapods have blood pressures
high enough to push blood upward, against the pull of
gravity. Gravity also causes blood to pool in low spots,
such as limbs. Blood flow from the limbs to the heart is
assisted by contractions of adjacent skeletal muscles; veins
in limbs also have valves that resist backflow of blood.

Capillary walls are leaky, and high blood pressure
forces some of the plasma (the liquid part of blood) out
of capillaries and into the intercellular spaces of body
tissues. Lymphatic vessels recover this plasma by a one-
way system of blind-end vessels that parallel veins and
return lymph to the venous system. The lymphatic system
is well developed in some teleost fishes, but it is critically
important on land, where the cardiovascular system is
subject to greater forces of gravity. In tetrapods, valves in
lymphatic vessels prevent backflow while contraction of
adjacent muscles keeps lymph flowing toward the heart,
thereby returning to the venous system blood plasma
forced out of capillaries.

With the loss of gills and evolution of a neck in tetrapods, the heart shifted posteriorly. In fishes, the heart lies
ventral to the gill region anterior to the shoulder girdle,
whereas in tetrapods it lies posterior to the shoulder girdle in the thorax (see Figure 2.5). The sinus venosus and
conus arteriosus, typical of the hearts of chondrichthyans

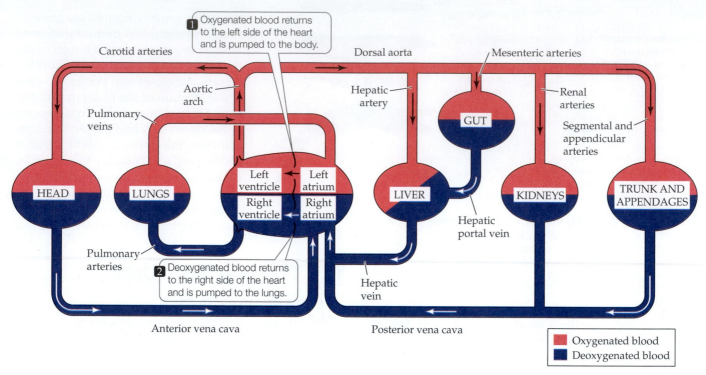

Figure 11.12 **The double-circuit cardiovascular system of a mammal.** Deoxygenated blood returns from the head (via the anterior vena cava), liver (via the hepatic vein), and body (via the posterior vena cava) to the right atrium, which pumps it into the right ventricle. Ventricular contraction pumps blood into the pulmonary arteries to the lungs, where gas exchange occurs. Oxygenated blood returns via pulmonary veins to the left atrium. Atrial contraction pumps blood into the left ventricle and its contraction sends blood into the aortic arch. Paired carotid arteries carry blood from the aortic arch to the head, and the dorsal aorta carries blood to the rest of the body. The liver receives oxygenated blood via the hepatic artery and deoxygenated blood and nutrients absorbed from the gut via the hepatic portal vein. Branches of the dorsal aorta supply the kidneys (renal arteries), trunk, and appendages (segmental and appendicular arteries).

and actinopterygians, are reduced or absent in tetrapod hearts (see Section 2.6).

Tetrapods have a **double circulation** in which the pulmonary circuit supplies the lungs with deoxygenated blood and the systemic circuit supplies the body with oxygenated blood. This is a key difference between circulation in fishes and tetrapods: In fishes, oxygenated blood from the gills flows straight to the body without first returning to the heart, a pattern termed **single circulation**. In contrast, in the double circulation of tetrapods, oxygenated blood from the lungs returns to the heart for additional pumping to the body. Details of heart chambers and blood flow through the heart differ among tetrapods and are described in other chapters, but the familiar double circulation of a mammal serves as an example (**Figure 11.12**). Note the two atria and two ventricles, and that the right side of the heart receives deoxygenated blood returning from the body while the left side receives oxygenated blood returning from the lungs.

Aortic arch 1 is lost in the adults of all extant vertebrates. Correlated with loss of gills in adult tetrapods, aortic arch 2 was lost; adult amniotes also lack aortic arch 5, retaining three aortic arches: arch 3 (carotid arch) serves the head, arch 4 (systemic arch) supplies the body, and arch 6 (pulmonary arch) delivers deoxygenated blood to the lungs.

As in fishes, the aortic arches of lissamphibians branch from a ventral arterial trunk known as the truncus arteriosus, and we hypothesize that a similar arrangement occurred in basal tetrapods (**Figure 11.13A,B**). In amniotes, the pulmonary arteries receive blood from the right ventricle, and the right systemic and carotid arches receive blood from the left ventricle; details of heart anatomy suggest that this condition evolved independently in mammals and other amniotes (**Figure 11.13C**).

A ventricular septum occurs in all extant amniotes, but its form differs in different lineages. For example, a transient ventricular septum forms during ventricular contraction in turtles and lepidosaurs, whereas a permanent ventricular septum is present in crocodylians, birds, and mammals. This phylogenetic pattern indicates that a permanent septum evolved independently in sauropsid (reptiles, including birds) and synapsid (mammal) lineages, an interpretation also supported by developmental evidence.

Gnathostome hearts have a dedicated blood supply from coronary arteries. The coronary arteries are particularly well-developed in crocodylians, birds, and mammals, because their heart muscle requires adequate oxygen delivery to sustain powerful ventricular contractions for generating higher blood pressures than typical of other groups. Also, only the left side of the heart of crocodylians, birds, and mammals contains oxygenated blood, so cardiac muscle of the right atrium and right ventricle cannot obtain oxygen from the blood within them.

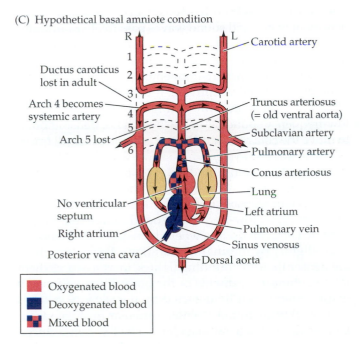

(A) Hypothetical basal tetrapod condition

R — L
— Carotid artery
1
2 — Arch 2 lost
3 — Ventral aorta
Internal gills lost
4 — Subclavian artery
5 — Dorsal part of arch 6 lost in adult (=ductus arteriosus in embryo)
6
Conus arteriosus —
— Pulmonary artery
No ventricular septum
— Lung
Right atrium —
— Left atrium
Sinus venosus —
— Pulmonary vein
Posterior vena cava —
— Dorsal aorta

(B) Hypothetical lissamphibian condition

R — L
— Carotid artery
1
2
Ductus caroticus —
3 — Truncus arteriosus (= old ventral aorta)
4
— Subclavian artery
5
6
Cutaneous artery (to skin) —
— Pulmonary artery
— Conus arteriosus
No ventricular septum
— Lung
Right atrium —
— Left atrium
— Pulmonary vein
Posterior vena cava (with some oxygenated blood from cutaneous vein) —
— Sinus venosus
— Dorsal aorta

(C) Hypothetical basal amniote condition

R — L
— Carotid artery
1
Ductus caroticus lost in adult —
2
3
Arch 4 becomes systemic artery —
4 — Truncus arteriosus (= old ventral aorta)
5 — Subclavian artery
Arch 5 lost —
6 — Pulmonary artery
— Conus arteriosus
— Lung
No ventricular septum
— Left atrium
Right atrium —
— Pulmonary vein
— Sinus venosus
Posterior vena cava —
— Dorsal aorta

■ Oxygenated blood
■ Deoxygenated blood
■ Mixed blood

Figure 11.13 Heart and aortic arches in tetrapods.
(A) Hypothetical basal tetrapod condition. Internal gills and aortic arch 2 have been lost. (B) Hypothetical lissamphibian condition. Patterns of circulation vary for different orders and life stages; this scheme emphasizes features unique to lissamphibians, such as the cutaneous circulation, which returns oxygenated blood to the right side of the heart. (C) Hypothetical basal amniote condition. Ductus caroticus lost in adult. Loss of arch 5 means that arch 4 is the systemic arch, serving postcranial portions of the body. No dedicated circulation for cutaneous gas exchange is present. By long established anatomical convention, the heart and great vessels are drawn in ventral view, which means that the left side of the heart is on the right side of these anatomical schematics.

11.6 Sensory Systems

LEARNING OBJECTIVES

11.6.1 Relate eye structure in the four-eyed fish to vision in both water and air.

11.6.2 Trace the pathway of sound in the ear of a mammal.

11.6.3 Describe how tetrapods couple olfaction and respiration by listing in order the structures through which volatile odorants pass from external nostrils to lungs.

11.6.4 Summarize structure and function of the accessory olfactory system.

Some sensory systems that are exquisitely sensitive in water are useless in air, whereas others work better in air than in water. Air is not dense enough to stimulate the mechanical receptors of the lateral line system and does not conduct electricity well enough to support electroreception. In air, sound waves are too weak to directly set the fluid of the inner ear in motion, so tetrapods must amplify aerial sound waves to detect them. Chemical systems work well on land, however, at least for molecules small enough to be suspended in air (volatile odorants) or stable enough to be deposited on objects in the environment (nonvolatile odorants). Air offers some advantages for vision as well. In this section, we describe how the sensory systems of terrestrial vertebrates differ from those of aquatic vertebrates.

Vision

Vision is more acute in air than in water because light travels through air with less interference than it does through water. Unlike water, air is rarely murky, so vision is more useful as a distance sense in air than water.

The lens shapes of terrestrial and aquatic vertebrates differ, and some terrestrial and aquatic vertebrates focus images on the retina in different ways. The refractive index of the cornea is very similar to that of water, as explained in Section 4.2, so the cornea of fishes has little effect on focusing light rays on the retina. As a result, the job of focusing falls entirely to the lens, which is thick and spherical, and fishes focus light by moving the lens forward and backward within the eye (**Figure 11.14A**). The refractive index of the cornea is greater than that of air, so in air the cornea *does* participate in forming a focused image. Terrestrial vertebrates have thin, elliptical lenses, and mammals and birds

(A) Light paths, cornea, and spherical lens of a fish

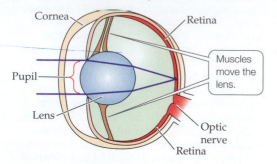

(B) Light paths, cornea, and elliptical lens of a mammal

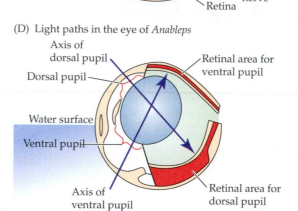

(C) Four-eyed fish, *Anableps anableps*, at water surface

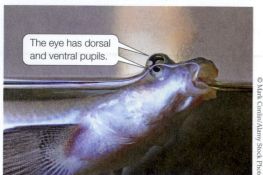

The eye has dorsal and ventral pupils.

© Mark Conlin/Alamy Stock Photo

(D) Light paths in the eye of *Anableps*

Axis of dorsal pupil
Dorsal pupil
Water surface
Ventral pupil
Axis of ventral pupil
Retinal area for ventral pupil
Retinal area for dorsal pupil

Figure 11.14 Vision in water and air. (A) Fishes have thick spherical lenses. The entire lens is moved backward or forward by contraction or relaxation of muscles to adjust for objects at different distances. (B) Terrestrial vertebrates have thin, elliptical lenses. In birds and mammals, focus is adjusted for near or distant objects by muscular contractions that change shape of the lens. (C) The four-eyed fish *Anableps anableps* lives at the water's surface. Each of its eyes has two pupils and two retinas. (D) The ventral pupil admits light from below the water surface, and the dorsal pupil admits light from above. The lens is asymmetric, with a long dimension and a short dimension. The cornea of the ventral pupil is in water and has no effect on focus. The axis of the ventral pupil passes through the long dimension of the lens. The cornea of the dorsal pupil is in air and helps in focusing light on the retina. The axis of the dorsal pupil passes through the short dimension of the lens. (D after J.G. Sivak. 1976. *Vis Res* 16: 531–534.)

focus by changing shape of the lens rather than by moving the lens forward and backward (**Figure 11.14B**).

Anableps anableps, the "four-eyed fish," lives at the water's surface and has eyes designed to see both above and below water (**Figure 11.14C**). Each eye has two pupils, two retinas, and an asymmetric lens with both a long dimension and a short dimension (**Figure 11.14D**). The ventral pupil admits light from below water and, as in other fishes, the cornea of this pupil is in water and has no role in focusing. The axis of the ventral pupil passes through the long dimension of the lens, as you would expect given the thick lenses of other fishes. The dorsal pupil admits light from above the water surface; this cornea is in air and helps focus light. The axis of the dorsal pupil passes through the short dimension of the lens, in a manner reminiscent of terrestrial vertebrates with thinner lenses.

In air, the eye's surface must be protected and kept moist and free of particles. Novel features in terrestrial tetrapods include eyelids, glands that lubricate the eye and keep it moist (including tear-producing lacrimal glands), and a nasolacrimal duct that drains tears from the eyes into the nose (which is why your nose starts to run when you cry).

Hearing and equilibrium

Sound perception is very different in air than in water. The density of animal tissue is nearly the same as the density of water, and sound waves pass freely from water into tissue (for example, sound travels through tissues of the skin and skull to reach the inner ear of fishes). Because water is dense, its movements can directly stimulate the mechanoreceptive hair cells of the lateral line system, which is one way that fishes can detect nearby low-frequency sounds. Air, however, is not dense enough to move the cilia of hair cells, and amniotes lack all traces of the lateral line system.

Information from sound waves is detected and transmitted to the brain by the inner ear. The inner ears of bony fishes contain densely mineralized structures called otoliths that detect low-frequency sounds at relatively short ranges. They can do this when the fish's body is moved even slightly by impinging vibrations. Because otoliths are denser than surrounding tissues, inertia causes them to lag behind movements of the body. Even very small displacements can be detected by the sensory patches (maculae) beneath the otoliths (utricular macula, saccular macula, and lagenar macula). Such vibration-induced sound detection does not work in air, however, because

(A) Middle ear of sauropsid

Sauropsids have one middle ear bone, the stapes.

Tympanum is on surface of head.

Lagena

Auditory tube

Middle ear cavity

(B) Middle ear of mammal

Mammals have three middle ear bones, the malleus, incus, and stapes.

Pinna

Cochlea

Auditory tube

Tympanum is deep in ear canal.

Middle ear cavity

Tensor tympani muscle

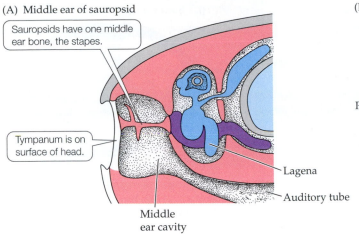

Figure 11.15 Middle ear anatomy of amniotes. Structures of the middle ear amplify and transmit sound waves. (A) The tympanum lies on the surface of the head in some sauropsids and at the end of a short ear canal in others. Vibrations received by the tympanum are transmitted to the oval window by a single middle ear ossicle, the stapes. (B) In mammals, vibrations are transmitted by a chain of three middle-ear ossicles, the malleus, incus, and stapes. Compared with sauropsids, mammals have relatively longer ear canals. Therian mammals have an extensively coiled cochlea in the inner ear (see Figure 11.16E). (See credit details at the end of this chapter.)

air's low viscosity and density cannot easily move the body of a tetrapod.

On land, aerial sound waves are too weak to set the fluid of the inner ear in motion so that sound can be detected. Thus, tetrapods need to amplify aerial sound waves to detect them. Amplification of sound waves and their transfer to the inner ear are functions of the tetrapod middle ear (**Figure 11.15**). On the external surface of the head, many extant tetrapods have a **tympanum** (eardrum), which resonates in synchrony with impinging airborne sound waves. In turn, the tympanum transfers these vibrations to one or more **ossicles**, tiny bones in the air-filled middle ear that transmit the vibrations to the oval window of the fluid-filled inner ear. The area of the tympanum is much larger than that of the oval window, and this difference in area amplifies aerial sound waves.

There must be a physical connection between the tympanum and the oval window to transmit vibrations. This transmission is accomplished in different ways in different groups of tetrapods, but one skeletal element—the stapes, which is derived from the hyomandibula of fishes—is synapomorphic for tetrapods (Figure 11.1 #18) and serves in amplification and transmission of vibrations. Evidence from fossils suggests that the middle ear of lissamphibians evolved independently from the similar system found in basal amniotes. For example, frogs have a single middle ear ossicle homologous to the stapes known as the columella (Latin *columella*, "small column"). The stapes is present in both sauropsids and synapsids. Mammals, which are derived synapsids, incorporate two additional bones—the malleus and incus—to form a chain of three middle ear ossicles connecting the tympanum to the oval window (see Figure 11.15).

The inner ear is a complicated structure that shows both great differences and striking similarities across the vertebrate groups. Unlike fishes (**Figure 11.16A**), extant tetrapods lack large, densely mineralized otoliths in their inner ears. (The much smaller crystals associated with the sensory patches are part of the vestibular system used to detect static equilibrium, the position of the head with respect to gravity; see Section 2.6.)

In terrestrial vertebrates, The footplate of the columella (in frogs) or the stapes (in amniotes) moves the membrane of the oval window in and out in synchrony with vibrations of the tympanum, producing waves of compression in the fluid within the inner ear. Sound waves amplified by the tympanum and oval window stimulate hair cells in the basilar papilla, which lies within the lagena of the inner ear (**Figure 11.16B,C**). The longer lagena of birds forms a straight cochlear duct containing a long basilar papilla (**Figure 11.16D**). In mammals, the lagena has evolved into a long cochlea, which is snail-shaped in therians (**Figure 11.16E**). The greatly elongated basilar papilla within the cochlea is called the organ of Corti; it discriminates frequency and intensity of the vibrations it receives and transmits this information to the central nervous system.

As in all vertebrates, the semicircular canals of tetrapods provide vestibular information about orientation and acceleration of the body.

Olfaction

Volatile odor molecules generally travel faster in air than in water, making olfaction an effective distance sense for terrestrial vertebrates. Nonvolatile molecules do not readily evaporate, and when deposited on conspicuous objects in the environment, can be used to mark

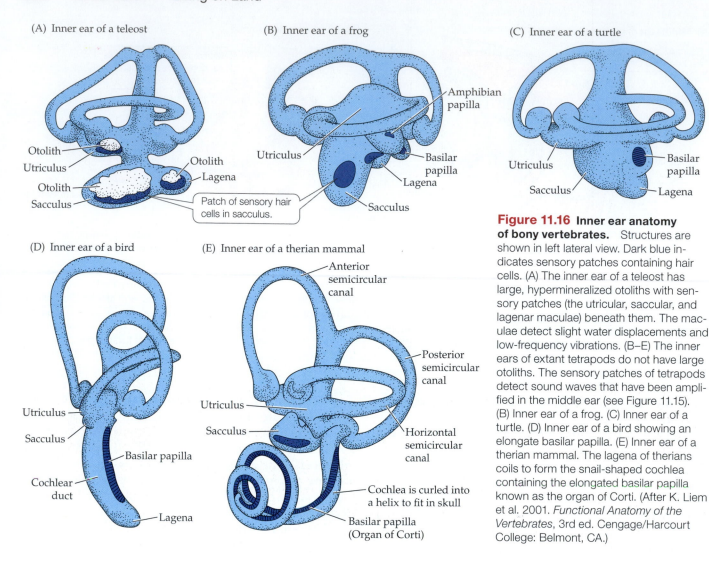

(A) Inner ear of a teleost

Otolith
Utriculus
Otolith
Sacculus
Otolith
Lagena
Patch of sensory hair cells in sacculus.

(B) Inner ear of a frog

Amphibian papilla
Utriculus
Basilar papilla
Lagena
Sacculus

(C) Inner ear of a turtle

Utriculus
Basilar papilla
Sacculus
Lagena

(D) Inner ear of a bird

Utriculus
Sacculus
Basilar papilla
Cochlear duct
Lagena

(E) Inner ear of a therian mammal

Anterior semicircular canal
Posterior semicircular canal
Utriculus
Sacculus
Horizontal semicircular canal
Cochlea is curled into a helix to fit in skull
Basilar papilla (Organ of Corti)

Figure 11.16 Inner ear anatomy of bony vertebrates. Structures are shown in left lateral view. Dark blue indicates sensory patches containing hair cells. (A) The inner ear of a teleost has large, hypermineralized otoliths with sensory patches (the utricular, saccular, and lagenar maculae) beneath them. The maculae detect slight water displacements and low-frequency vibrations. (B–E) The inner ears of extant tetrapods do not have large otoliths. The sensory patches of tetrapods detect sound waves that have been amplified in the middle ear (see Figure 11.15). (B) Inner ear of a frog. (C) Inner ear of a turtle. (D) Inner ear of a bird showing an elongate basilar papilla. (E) Inner ear of a therian mammal. The lagena of therians coils to form the snail-shaped cochlea containing the elongated basilar papilla known as the organ of Corti. (After K. Liem et al. 2001. *Functional Anatomy of the Vertebrates*, 3rd ed. Cengage/Harcourt College: Belmont, CA.)

territories with long-lasting messages. Tetrapods thus have two chemosensory systems: (1) the main olfactory system, used primarily to detect volatile odorants; and (2) the accessory olfactory system, used primarily to detect nonvolatile odorants. The accessory olfactory system was lost in several groups, including birds, bats, and cetaceans. In the case of cetaceans, evidence that they evolved from terrestrial ancestors with an accessory olfactory system is preserved by silenced genes associated with the accessory system.

Receptors for the main olfactory system are located in the olfactory epithelium of the nasal passages. These receptors can be extraordinarily sensitive—some chemicals can be detected at concentrations below 1 part in 1 million trillion (10^{15}) parts of air. With each inhalation, air brought in via the paired external nostrils passes over the olfactory epithelium and enters the back of the mouth via paired **choanae** (internal nostrils; see Section 8.5). From the back of the mouth, air enters the larynx, trachea, bronchi, and lungs. Thus, tetrapods couple olfaction and respiration; except in a few cases, fishes do not couple olfaction and respiration because water drawn in for olfaction does not pass over the gills.

Among tetrapods, mammals probably have the greatest olfactory sensitivity. The area of olfactory epithelium in mammals is increased by curving scrolls of thin, mucosa-coated bone called **turbinates** (**Figure 11.17**). **Ethmoturbinates** are located posteriorly in the nasal cavity, and their name reflects their association with the ethmoid bone, through which pass nerves carrying olfactory signals to the brain. **Nasoturbinates** are located in front of the ethmoturbinates and also may have an olfactory function. Primates, including humans, have a relatively poor sense of smell because our snouts are too short to accommodate large turbinates and an extensive olfactory epithelium found in many other mammals.

Not long ago, mammals were thought to be incapable of smelling when submerged in water because there seemed to be no way they could take in air needed to carry odorant molecules to the olfactory epithelium. We now know this is not the case, at least for some semi-aquatic small mammals. When foraging under water, star-nosed moles and water shrews, exhale air bubbles onto scent trails of potential prey and then re-inhale the bubbles, which carry odorants from the trails to the olfactory epithelium (**Figure 11.18**).

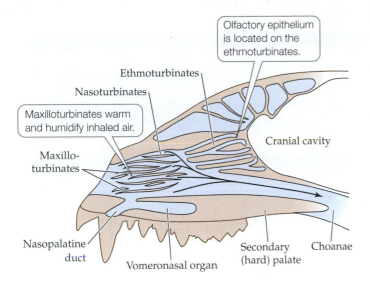

Maxilloturbinates warm and humidify inhaled air.

Olfactory epithelium is located on the ethmoturbinates.

Ethmoturbinates

Nasoturbinates

Cranial cavity

Maxillo-turbinates

Nasopalatine duct

Vomeronasal organ

Secondary (hard) palate

Choanae

Figure 11.17 Turbinates of the skull of a mammal. Seen here in sagittal section through the anterior portion of the skull are three sets of turbinates in the nasal cavity: maxilloturbinates, nasoturbinates, and ethmoturbinates. Odor receptors for the main olfactory system are located on the ethmoturbinates, which are closest to the brain in the cranial cavity. The vomeronasal organ, part of the accessory olfactory system, is a blind pocket in the palate with connections to oral and nasal cavities. (After W.J. Hillenius. 1992. *Paleobiology* 18: 17–29.)

(A) Nose-rays of a star-nosed mole, *Condylura cristata*

From research in Catania Lab; image © Kenneth C. Catania

(B) Nose-rays holding bubbles of air underwater

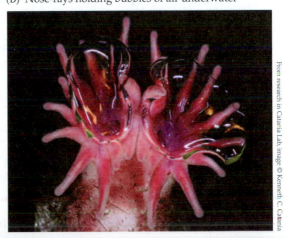

From research in Catania Lab; image © Kenneth C. Catania

(C) Ventral view of an American water shrew, *Sorex palustris*, sniffing underwater

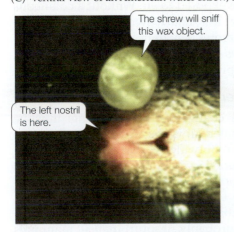

The shrew will sniff this wax object.

The left nostril is here.

It expires a bubble of air from its left nostril, which contacts the wax object.

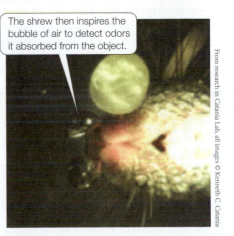

The shrew then inspires the bubble of air to detect odors it absorbed from the object.

From research in Catania Lab; all images © Kenneth C. Catania

Figure 11.18 Star-nosed moles and water shrews can smell underwater. (A) Star-nosed moles have 22 fleshy appendages (rays) surrounding their nostrils. The rays are covered with touch receptors used to detect prey. (B) Star-nosed moles also use olfaction to detect prey, even under water. Close-up of the star of a mole under water showing a bubble of air at each nostril. (C) Water shrew (ventral views, with nose and whiskers to the left) sniffing a green wax object while underwater. The left panel shows the shrew encountering the wax object. The middle panel shows the shrew exhaling bubbles from each nostril, one of which contacts the wax object, picking up odor molecules. The right panel shows the shrew re-inspiring the air bubbles so that odorants from the wax object can travel to the olfactory epithelium.

(A) Ungulate vomeronasal organ and flehmen behavior

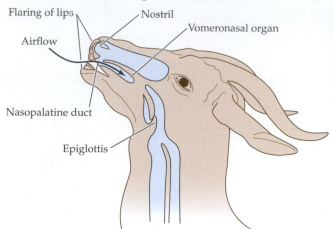

(B) Flehmen behavior of a horse

Figure 11.19 Vomeronasal organ and flehmen behavior of ungulates. (A) Pattern of airflow into the vomeronasal organ of a goat, following direct contact with nonvolatile components of urine. With the lips flared and the head lifted, a sharp inhalation draws air through the nasopalatine duct into the vomeronasal organ where the odorants are detected. (B) Flehmen behavior of a horse. This behavior facilitates transfer of nonvolatile odorants to the vomeronasal organ.

The accessory olfactory system is based on a paired organ unique to tetrapods in the anterior roof of the mouth—the **vomeronasal organ**, sometimes called Jacobson's organ (**Figure 11.19A**). Unlike the main olfactory system, through which air passes as it is inhaled for respiration, the vomeronasal organ is a recess or pouch in different groups of tetrapods. Nonvolatile odorants must be delivered directly to the vomeronasal organ to be detected. For example, when snakes flick their tongues in and out of the mouth, they capture molecules and physically transfer them to the vomeronasal organ. Many male ungulates (hoofed mammals) sniff or taste the urine of a female, a behavior that permits them to determine stage of her reproductive cycle. This sniffing is usually followed by flehmen behavior, in which the male curls his upper lip and strongly inhales pheromone molecules into his vomeronasal organ (**Figure 11.19B**). Primates were thought to have lost the vomeronasal organ system, but some research suggests a remnant may persist in humans for detection of pheromones.

11.7 Conserving Water in a Dry Environment

LEARNING OBJECTIVES

11.7.1 Describe three features of a cutaneous barrier that help limit water loss.

11.7.2 Explain how maxilloturbinates reduce respiratory water loss in mammals.

11.7.3 Compare the opisthonephric kidneys of fishes and lissamphibians with the metanephric kidneys of amniotes.

Aquatic vertebrates are surrounded by water, but for a terrestrial vertebrate water sources are often scattered in time and space. Thus, gaining water and limiting water loss are important for terrestrial tetrapods. In air, water evaporates from the body surface and respiratory system as water vapor; water is also lost through the kidneys and in feces as liquid water. Tetrapods evolved many anatomical specializations to limit water loss, including changes in the skin, respiratory surfaces, and excretory system.

Cutaneous water loss

Skin permeability depends in part on its thickness and degree of keratinization. More important is the role of hydrophobic lipids in limiting water loss through the skin. In rats, the lipid barrier develops shortly before birth, changing the skin from highly permeable to relatively impermeable over a two-day period due to accumulation of lipids. This change correlates with the change in environment: fetal rats develop in amniotic fluid, whereas newborn rats are exposed to air. Without a cutaneous barrier to limit water loss, fatal dehydration would occur. More generally, amniotes benefit from skin impermeability because it allows them to live in habitats that would otherwise be inhospitable.

Respiratory water loss

Tissues in the gas exchange region of the lungs must be kept moist, and inhaled air is warmed to body temperature and saturated with water vapor as it travels to the lungs. This water added as air is inhaled would represent an enormous evaporative loss if it were not recovered as air is exhaled.

In addition to turbinates associated with olfaction (ethmoturbinates and probably nasoturbinates), mammals have **maxilloturbinates** in the anterior part of the nasal cavity that serve a respiratory rather than an olfactory function (see Figure 11.17). Inhaled air is warmed and moistened as it passes over the epithelial covering of the maxilloturbinates, which in turn cools due to evaporative water loss. During exhalation, warm moist air from the lungs passes over the now-cool epithelium of the maxillo-turbinates, cooling the air and condensing water from it. The water is then retained in the nasal passages, keeping them moist.

Birds independently evolved respiratory turbinates that reduce water loss during breathing, and their elongated trachea may also play a role. The turbinates of mammals are complex scrolls of bone. The respiratory turbinates of birds are formed by cartilage and typically consist of a single, sturdy scroll.

Excretory water loss

Tetrapods share the presence of a well-developed urinary bladder (Figure 11.1B,C #15), which receives and stores waste from the kidneys. It is also a site for water recovery in lissamphibians and some sauropsids. Some lepidosaurs, crocodylians, and birds lost the urinary bladder.

Fishes and lissamphibians have elongate **opisthonephric kidneys** lateral to the dorsal aorta and extending the length of the trunk (Figure 11.1A,B #14). Amniotes have compact, often bean-shaped **metanephric kidneys** near the posterior end of the trunk (Figure 11.1C #14). During development, amniote kidney formation recapitulates some aspects of the organ's evolutionary history. For example, early in development, amniote embryos have an elongate kidney drained by the archinephric duct (see Figure 2.15). Later, this portion of the kidney largely degenerates (some portions are retained as parts of the male reproductive system). Paired diverticula of the archinephric ducts grow dorsally into the nephrotome, where they organize the definitive metanephric kidneys, characterized by large numbers of nephrons (each of your kidneys has about 1 million nephrons). The paired diverticula become the ureters that transport waste from the metanephric kidneys to the bladder.

Summary

11.1 Support on Land

Gravity is a fairly minor force in the lives of aquatic vertebrates. In contrast, the skeleton of a terrestrial vertebrate must support the animal's weight against gravity and resist forces exerted during movements.

Whereas fishes have trunk and caudal vertebrae, tetrapods have cervical, trunk, sacral, and caudal vertebrae. The trunk vertebrae of mammals are differentiated into thoracic (rib-bearing) and lumbar (no ribs) vertebrae.

Tetrapods have highly differentiated axial muscles used for locomotion, postural support, and lung ventilation. In contrast to fishes, the pectoral girdle of tetrapods is freed from its connection with the skull and the pelvic girdle articulates with the vertebral column.

The differentiated muscles of the appendicular system allow tetrapod limbs to lift the body off the ground and push against the substrate during locomotion.

Large terrestrial vertebrates have proportionally thicker limb bones and more erect stances than smaller terrestrial vertebrates.

11.2 Locomotion

Salamanders retain the ancestral form of locomotion, the walking-trot gait. Amniotes typically use a walking gait; and mammals may use the faster gait, trotting.

Therian mammals also bound, with flexion of the back increasing stride length. In large therian mammals the bound is modified into the gallop, with less flexion of the back.

Extant salamanders, many lizards, and crocodylians have a sprawling stance and use lateral undulation. Therian

mammals have a more upright stance in which the limbs are beneath the body; upright stance is associated with dorsoventral flexion of the vertebral column.

11.3 Eating

Water is viscous and food items in water are essentially weightless, so fishes and aquatic tetrapods can suck food into their mouth. Suction feeding is not an option on land because air is much less viscous than water.

Tetrapods use their mobile neck and head to seize prey and their large, muscular tongue to manipulate food in the mouth. Derived amniotes typically have a four-phase feeding cycle.

11.4 Breathing Air

Tidal respiration is energetically feasible for air-breathing vertebrates. Lissamphibians use a buccal pump to inflate the lungs.

Amniotes use a negative-pressure aspiration pump to draw air into the lungs.

11.5 Pumping Blood Uphill

In tetrapods, venous blood must be forced upward from regions of the body below the heart. Tetrapods have high blood pressures and valves in the vessels of the limbs to prevent backflow.

The cardiovascular system of fishes has a single circuit, meaning that oxygenated blood does not return to the heart for pumping to body tissues. Tetrapods have a double circuit in which the pulmonary circuit supplies the lungs with deoxygenated blood and the systemic circuit supplies the body with oxygenated blood.

(Continued)

11.6 Sensory Systems

In terrestrial vertebrates, the cornea participates in focusing light on the retina, and the lens is elliptical and not spherical as it is in fishes. Whereas birds and mammals focus an image on the retina by changing shape of the lens, fishes move the entire lens forward and backward.

Because water is dense, its movements can directly stimulate mechanoreceptive hair cells of the lateral line system of fishes. Amniotes lack a lateral line system because air is not dense enough to stimulate hair cells.

Aerial sound waves are relatively weak, so tetrapods must amplify them for detection. Ossicles of the middle ear carry vibrations from the tympanum to the oval window at the entrance to the inner ear. The area of the tympanum is much larger than that of the oval window, and this difference in area, together with vibrations of ossicles, amplifies aerial sound waves, allowing their detection by hair cells in the inner ear.

Unlike most fishes, terrestrial vertebrates couple olfaction and respiration. Tetrapods have a main olfactory system that primarily detects volatile odorants, as well as an accessory olfactory system that primarily detects nonvolatile odorants using the vomeronasal organ.

11.7 Conserving Water in a Dry Environment

Lipids in the skin of amniotes reduce cutaneous water loss. Skin thickness and keratinization also play a role in conserving the body's water.

Respiratory turbinates within the nasal apparatus are thin, bony structures covered with mucosal tissues that reduce water loss when air is inhaled. Other turbinates serve in olfaction. Turbinates evolved independently in mammals and birds.

Fishes and lissamphibians have elongate opisthonephric kidneys. Amniotes have water-conserving, compact, metanephric kidneys with large numbers of nephrons.

Discussion Questions

11.1   Why is bone such a useful structural material for tetrapods on land? Calcified cartilage is lighter than bone; wouldn't that be an advantage for a terrestrial vertebrate?

11.2   Why would we expect secondarily marine tetrapods such as whales to have reduced or lost zygapophyses in some regions of the vertebral column?

11.3   How does the tongue of tetrapods differ from the primary tongue of fishes?

11.4   How does the buccal pump of lissamphibians differ from the negative-pressure aspiration pump of amniotes?

11.5   With respect to the cardiovascular system, differentiate the double circuit of tetrapods from the single circuit of fishes.

11.6   Evaluate the following statement for accuracy: Both fishes and tetrapods couple olfaction with respiration.

Figure credits
Figure 11.15: A after K. Liem et al. 2001. *Functional Anatomy of the Vertebrates*, 3rd ed. Cengage/Harcourt College: Belmont, CA; B after D. Sadava et al. 2017. *Life: The Science of Biology*, 11th ed. Sinauer Associates/Macmillan, New York.

Courtesy of Devin Edmonds

Mimic poison frog, *Ranitomeya imitator*

Lissamphibians

Lissamphibians are non-amniote tetrapods with moist, scaleless skins (Greek *lissos*, "smooth"). The group includes three distinct lineages of extant animals generally referred to simply as "amphibians": Gymnophiona (caecilians), Caudata (salamanders), and Anura (frogs). Common names of lissamphibians can be confusing and often inexact. "Frogs, toads, and treefrogs" and "salamanders and newts" are widely used, and there are any number of regional variations, such as "waterdogs" and "spring lizards." Here we use "frogs" as the general term for all anurans; "toads" are members of the anuran family Bufonidae, and thus are frogs. Similarly, all caudates are salamanders, and "newts" are species in the caudate family Salamandridae, so all newts are salamanders. In contrast to the names "toads" and "newts," which have a phylogenetic basis, "treefrog" is applied to arboreal species of anurans in many different families.

Although the three lineages of Lissamphibia have very different body forms, they are a monophyletic lineage characterized by several shared derived characters, many of which play important roles in their functional biology. In this chapter we will describe the significance of these synapomorphies, including maintaining the ability to detect color at low light levels and balance the inward and outward movement of water and solutes. Lissamphibians are the only tetrapods that metamorphose, and as a result of this dual life history (Greek *amphis*, "double;" *bios*, "life"), lissamphibians play a critical role in the flow of matter and energy in ecosystems. Although most lissamphibians are small and thus are prey for larger tetrapods, some species are toxic (even to predators as large as humans) and others are venomous. Finally, we turn to the worldwide crisis of rapidly declining lissamphibian populations.

12.1 Diversity of Lissamphibians

LEARNING OBJECTIVES

12.1.1 Name and describe the three orders of lissamphibians.

12.1.2 Identify the synapomorphies of lissamphibians.

12.1.3 Describe anuran limbs and locomotion and some of their many specializations.

At first glance, the three lineages of lissamphibians may appear quite different. Frogs lack tails (hence the name Anura, Greek *an*, "without"; *oura*, "tail"), whereas salamanders have long tails (hence the name Caudata, Latin *caudum*, tail). The tails of caecilians are short, as is typical of elongated, burrowing animals. Frogs and most salamanders have four well-developed limbs, whereas all caecilians (and a few salamanders) are limbless. Frogs have long hindlimbs and short, stiff bodies that don't bend when they walk; salamanders have fore- and hindlimbs of equal size and move with lateral undulations; and limbless caecilians burrow with alternating extensions and contractions. Despite these obvious external differences related to locomotor specializations, close examination shows that lissamphibians have many shared derived characters. Anatomical and molecular phylogenetic evidence strongly support the hypothesis shown in **Figure 12.1** that salamanders and frogs are sister taxa. Some anatomical evidence links salamanders and caecilians, but no one has ever formally proposed that frogs and caecilians are sister taxa.

Synapomorphies of Lissamphibia

The most important shared derived character of lissamphibians is a moist, permeable skin. Many Paleozoic non-amniote tetrapods had bony dermal armor; a

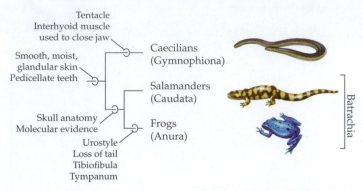

Figure 12.1 **Phylogenetic relationships of lissamphibians.** Salamanders (Caudata) retain the generalized ancestral body form and locomotor modes of tetrapods. Frogs (Anura) are specialized hoppers and jumpers. Most caecilians (Gymnophiona) are fossorial (i.e., they burrow in the earth), although one family is aquatic. Current data support the hypothesis that salamanders and frogs are sister taxa within the monophyletic Lissamphibia. (After K. Liem et al. 2001. *Functional Anatomy of the Vertebrates*, 3rd ed. Cengage/Harcourt College: Belmont, CA.)

permeable, unadorned skin is a derived character shared by lissamphibians. The permeability of lissamphibian skin means that they evaporate water rapidly in dry conditions, but it also allows them to absorb water from wet surfaces and moist soil. Further lissamphibian synapomorphies, some of which are related to their distinctive skin, include:

- **Substantial cutaneous gas exchange** A considerable part of a lissamphibian's exchange of oxygen and carbon dioxide with the environment takes place through its permeable skin (see Section 12.3).

- **Skin glands** Mucus glands embedded in the skin help maintain its moisture (see Section 12.4). In addition, all lissamphibians have granular (poison) glands in the skin (see Section 12.5).

- **Pedicellate, bicuspid teeth** A few stem temnospondyls (the clade that includes lissamphibians) had pedicellate teeth in which the crown and base (pedicel) are composed of dentine and are separated by a narrow zone of uncalcified dentine or fibrous connective tissue. All salamanders and caecilians and most anurans have pedicellate teeth, although teeth have been lost in more than 20 groups of frogs.

- **Carnivory** All extant adult lissamphibians are carnivorous, and relatively little morphological specialization is associated with different dietary habits within groups. Lissamphibians eat almost anything they can catch and swallow. The tongue of aquatic forms is broad, flat, and relatively immobile, although terrestrial lissamphibians can protrude the tongue from the mouth to capture prey.

- **Columella operculum complex** Most lissamphibians use two bones to transmit sounds to the ear. The columella is derived from the hyoid arch and is a shared feature of the lissamphibian clade; it transmits high-frequency sounds to the inner ear. The operculum (not homologous to the operculum of fishes) is a bony or cartilaginous structure attached to the ear capsule of extant salamanders and frogs. The opercularis muscle connects the shoulder girdle to the operculum and transmits low-frequency ground vibrations from the forelimb to the operculum, which transfers the vibrations to the inner ear.

- **Papilla amphibiorum** This special sensory area in the sacculus of the inner ear is present in all lissamphibians and is sensitive to sound frequencies below 1,000 Hz. A second sensory area, the papilla basilaris, detects frequencies above 1,000 Hz; it has been lost in some lineages.

- **Structure of the levator bulbi muscle** This muscle, present in salamanders and frogs and in modified form in caecilians, is a thin sheet in the floor of the eye socket that is innervated by the fifth cranial nerve. It causes the eyes to bulge outward, thereby enlarging the buccal cavity.

- **Green rods** In addition to the "red rods" that are found in all vertebrates and are sensitive to green light, salamanders and frogs have **green rods**, a type of retinal cell that has maximum sensitivity to violet and blue light and is unique to lissamphibians. Caecilians apparently lack green rods; however, the eyes of caecilians are extremely reduced, and the green rods may have been lost.

Although the red and green rods of lissamphibians were described more than a century ago, their significance has only recently been determined. As the sun sets, light shifts toward the blue end of the spectrum—that is, toward the maximum sensitivities of green rods. The dual rod system allows frogs and salamanders to detect colors at light intensities that are 10 to 100 times lower than the lowest perceptible intensities for birds or mammals. Many frogs have bright colors and patterns, often in areas of the body that can be concealed when they are at rest and displayed when they are moving or calling (**Figure 12.2A**). Their retinal green rods probably allow frogs to see these patterns even at night.

In addition, skin fluorescence is widespread among lissamphibians. Blue wavelengths of light are absorbed by the skin and re-emitted as green light (**Figure 12.2B**). The conspicuous colors and fluorescence of many nocturnal frogs and salamanders probably assist them to recognize members of their own species.

Aquatic larvae of lissamphibians and adults of permanently aquatic species have a lateral line system homologous to that of bony fishes (see Figure 4.6). Electroreception based on ampullary organs (see Figure 4.8) is present in the aquatic larvae of caecilians and salamanders, although this sensory system is entirely absent in frogs.

(A) Red-eyed treefrog, *Agalychnis callidryas*, day

Night

(B) Dotted treefrog, *Boana punctata*, day

Night fluorescence

Figure 12.2 Skin coloration in nocturnal frogs.
(A) The Central American red-eyed treefrog is cryptic when it is at rest during the day, but displays colorful markings at night when it is active and calling. Unique retinal green rods probably allow frogs to see these patterns at night, and thus to recognize conspecifics and potential mates. (B) In daylight, the Brazilian treefrog is dull yellow and probably cryptic against vegetation. At night, blue light that penetrates the forest canopy creates a green fluorescence that matches the sensitivity of the frog's retinal red rods.

Salamanders

Salamanders have the most generalized body form and locomotion of the extant lissamphibians. Salamanders are elongated, and all but a few completely aquatic species have four functional limbs. Their walking-trot gait is probably similar to that employed by the earliest tetrapods. It combines the lateral bending characteristic of fish locomotion with leg movements.

The 10 families of Caudata with their approximately 765 species are almost entirely limited to the Northern Hemisphere; their southernmost occurrence is in northern South America (**Table 12.1**). North and Central America have the greatest diversity; more salamander species are found in Tennessee than in all of Europe and Asia combined. Nearly two-thirds of extant salamander species are plethodontids (family Plethodontidae).

Although most salamanders retain the ancestral tetrapod body form, many groups show modifications and adaptations (**Figure 12.3**). Paedomorphosis is widespread, and several families of aquatic salamanders consist solely of paedomorphic forms. Paedomorphs such as the North American mudpuppy (Figure 12.3A) retain larval characters, including larval tooth and bone

patterns, the absence of eyelids, a functional lateral line system, and in some cases external gills.

The largest extant salamanders are the Japanese and Chinese giant salamanders (3 species in the genus *Andrias*), which can reach lengths of more than a meter. The related North American hellbender (*Cryptobranchus alleganiensis*; Figure 12.3B) grows to 60 cm. All are members of Cryptobranchidae (see Table 12.1) and are paedomorphic and permanently aquatic. As their name indicates (Greek *kryptos*, "hidden"; *branchia*, "gills"), they do not retain external gills, although they do have other larval characters.

Many terrestrial salamanders, such as western North American newts (*Taricha* spp., Figure 12.3C) and mole salamanders (*Ambystoma* spp.) have stout bodies and sturdy legs. Others, such as worm salamanders (*Oedipina* spp.; Figure 12.3D), are elongated with tiny legs and use sinusoidal locomotion to move rapidly.

Several lineages of salamanders have adapted to life in caves. The constant temperature and moisture make caves good salamander habitats, and cave-dwelling invertebrates supply food. Plethodontid brook salamanders of the genus *Eurycea* present a continuum, ranging from species with fully metamorphosed adults inhabiting the

(A) North American mudpuppy, *Necturus maculosus*

Courtesy of Todd Pierson

(B) North American hellbender, *Cryptobranchus alleganiensis*

Courtesy of Todd Pierson

(C) California newt, *Taricha torosa*

Steve Jurvetson/CC BY 2.0

(D) Central American worm salamander, *Oedipina sp.*

Courtesy of Wayne Van Devender

(F) Olm, *Proteus anguinus*

© Nature Picture Library/Alamy Stock Photo

(E) Texas blind salamander, *Eurycea rathbuni*

Courtesy of Wayne Van Devender

Figure 12.3 Diversity of salamanders.   The body forms of salamanders reflect differences in their life histories and habitats. (A) The paedomorphic North American mudpuppy, a proteid, retains gills. (B) Other paedomorphic salamanders, such as the North American hellbender, lack gills. (C) The California newt is a terrestrial species with sturdy legs. (D) A Central American worm salamander has extremely small limbs and uses serpentine locomotion. Paedomorphic cave-dwelling salamanders such as the Texas blind salamander (E) and the European olm (F) are unpigmented and eyeless.

twilight zone near cave mouths to fully paedomorphic forms inhabiting the depths of caves or sinkholes. The Texas blind salamander, *Eurycea rathbuni* (Figure 12.3E), is a highly specialized cave dweller—blind, white, with external gills, extremely long legs, and a flattened snout used to probe under pebbles for food. On another continent, a proteid salamander, the European olm, *Proteus anguinus* (Figure 12.3F), has converged on the same body form. Olms may live for as long as a century and are remarkably sedentary; a mark–recapture study in Herzegovina found that many individuals remained within a few square meters for the 8-year duration of the study.

Many plethodontid salamanders have evolved specializations of the hyobranchial apparatus that allow them to project the tongue a considerable distance from the mouth to capture prey (**Figure 12.4**). Muscle-powered tongue projection is ancestral for plethodontids and is retained in some genera in the family, but spring-powered projection systems have evolved independently twice among plethodontids. Spring-powered tongue projection is faster than muscle-powered projection and is far less sensitive to temperature. Relatively minor changes in the form of the tongue and hyoid apparatus are responsible for the switch from muscle-powered to spring-powered tongue projection.

Table 12.1 Extant families of salamanders

Family	Description	Fertilization	Adult size	Number and distribution of species
Hellbenders (Cryptobranchidae)	Very large paedomorphic aquatic salamanders	External	up to 1.8 m	1 species in North America, 3 species in Asia
Asiatic salamanders (Hynobiidae)	Terrestrial or aquatic salamanders with aquatic larvae	External	up to 30 cm	82 species in Asia
Sirens (Sirenidae)	Elongated aquatic salamanders with external gills, lacking the pelvic girdle and hindlimbs	Probably external	up to 75 cm	5 species in southeastern U.S
Salamandrids (Salamandridae)	Terrestrial and aquatic salamanders, most with aquatic larvae; some species of *Salamandra* are viviparous	Internal	up to 20 cm	125 species in Europe, Asia, North America, and extreme northwestern Africa
Mole salamanders (Ambystomatidae)	Small to large terrestrial salamanders with aquatic larvae	Internal	up to 30 cm	32 species in North America
Giant salamanders (Dicamptodontidae)	Terrestrial salamanders with facultative metamorphosis and a single paedomorphic aquatic species	Internal	up to 35 cm	4 species in northwestern U.S., barely extending into Canada
Mudpuppies (Proteidae)	Paedomorphic aquatic salamanders with external gills	Internal	up to 30 cm	7 species of *Necturus* in North America, 1 species of *Proteus* in Europe
Torrent salamanders (Rhyacotritonidae)	Semiaquatic salamanders with aquatic larvae	Internal	up to 10 cm	4 species in coastal coniferous forest streams of the North American Pacific Northwest
Amphiumas (Amphiumidae)	Elongated aquatic salamanders lacking gills	Internal	up to 1 m	3 species in North America
Lungless salamanders (Plethodontidae)	Aquatic or terrestrial salamanders, some with aquatic larvae, others with direct development	Internal	up to 30 cm	487 species in North and Central America, northern South America, Mediterranean Europe, the Korean peninsula

Compiled in part from information on AmphibiaWeb.org (amphibiaweb.org/lists/index.shtml)

Supramonte salamander, *Hydromantes supramontis*, capturing prey

From Deban and Wake, 1997, *Nature* 389: 27–28. Photograph by Stephen M. Deban

Anurans

Whereas salamanders have a limited number of species and a restricted geographic distribution, anurans include more than 50 families containing more than 7,350 species and occur on every continent except Antarctica (Table 12.2). The group is notable not only for its great diversity and wide distribution, but also for the prevalence of complete metamorphosis and the distinctive larval stage (tadpoles), features we will discuss in Section 12.2.

Figure 12.4 Ballistic tongue projection in a plethodontid salamander. Specializations of the hyobranchial apparatus allow some plethodontids, such as the rock- and cave-dwelling supramonte salamander of Sardinia, to project the tongue explosively and for long distances, capturing prey on the sticky tip.

Table 12.2 Major extant ramilies of anurans[a]

Family	Description	Adult size	Number and distribution of species
Tailed frogs (Ascaphidae)	Streambank frogs. The only anurans with an intromittent male organ (the so-called "tail")	up to 50 mm	2 species in the mountains of northwestern North America
Clawed frogs (Pipidae)	Specialized aquatic frogs. Most species of *Xenopus* are polyploids, with some having up to 12 sets of chromosomes. *Xenopus*, *Hymenochirus*, *Pseudohymenochirus*, and some species of *Pipa* have aquatic larvae; other species of *Pipa* have direct development	up to 15 cm	41 species in South America and sub-Saharan Africa
Spadefoot toads (Scaphiopodidae)	Terrestrial frogs with aquatic larvae. They use keratinized structures on the hindfeet to burrow backward into soil, leading to the descriptive common name "spadefoot frogs"	up to 8 cm	7 species from southern Canada to northern Mexico
Horned frogs (Ceratophryidae)	Large, short-legged terrestrial frogs with enormous mouths. They consume insects and even small vertebrates. Tadpoles of *Ceratophrys* and *Lepidobatrachus* are primarily carnivorous	up to 15 cm	12 species in South America
Treefrogs (Hylidae)	The largest anuran clade. Most hylid species are treefrogs, but a few are aquatic or terrestrial	up to 15 cm	1011 species in North, Central, and South America, the West Indies, Europe, Asia, and the Australo-Papuan region
True toads (Bufonidae)	Mainly short-legged and terrestrial. Most have aquatic larvae; some species of *Nectophrynoides* are viviparous	up to 25 cm	633 species worldwide *except for* Australia, Madagascar, and the oceanic islands
Southern frogs (Leptodactylidae)	Semiaquatic and terrestrial frogs with aquatic larvae	up to 25 cm	222 species in southern North America, Central America, northern South America, and the West Indies
Glass frogs (Centrolenidae)	Small arboreal frogs, commonly called glass frogs because the internal organs are visible through the skin of the ventral surface of the body. Oviparous with aquatic larvae; most centrolenids deposit eggs on leaves or rocks above the water	up to 3 cm	159 species in Central America and northern South America
Dart-poison frogs (Dendrobatidae)	Small terrestrial frogs, many brightly colored and toxic. Terrestrial eggs hatch into tadpoles that are transported to water by an adult	up to 6 cm	327 species in Central America and northern South America
New World rain frogs (Eleutherodactylidae)	Small arboreal frogs. Internal fertilization may be widespread, and many species have direct development[b]	up to 9 cm	226 species in Central America, northern South America, and the Greater and Lesser Antilles
Narrow-mouth frogs (Microhylidae)	Terrestrial and arboreal frogs. Many have aquatic larvae, but some species have nonfeeding tadpoles and others lay direct-developing eggs on land	up to 10 cm	688 species in North, Central, and South America, sub-Saharan Africa, India, and Korea to northern Australia
African bullfrogs (Pyxicephalidae)	Terrestrial, semiaquatic, and arboreal frogs. Most have aquatic tadpoles	up to 25 cm	87 species in sub-Saharan Africa
True frogs (Ranidae)	Large clade of aquatic and terrestrial frogs. Most have aquatic tadpoles	up to 30 cm	419 species worldwide *except for* extreme southern South America, South Africa, Madagascar, and most of Australia
Asian treefrogs (Rhacophoridae)	Asian treefrogs that have converged on the body form of hylids, with large heads, large eyes, and enlarged toe pads	up to 2 cm	429 species in sub-Saharan Africa, southern Asia, Japan, and the Philippines
Malagasy poison frogs (Mantellidae)	The major group of frogs on Madagascar. Most are terrestrial, but the family includes arboreal, aquatic, and fossorial species. Species in the genus *Mantella* are notable for resembling Neo-tropical dendrobatids in being brightly colored terrestrial frogs with alkaloid toxins in the skin	up to 10 cm	229 species in Madagascar and the Comoros archipelago

[a]More than 50 families of anurans are recognized with some 7,350 species; only families discussed and illustrated in the text are included here.

[b]A Puerto Rican species, *Eleutherodactylus jasperi*, retains the eggs in the oviducts until the young have passed through metamorphosis and emerge as froglets. This species has not been seen since 1981 and is believed to be extinct

Compiled in part from information on AmphibiaWeb.org (amphibiaweb.org/lists/index.shtml)

(A) †*Triadobatrachus* (B) Extant frog

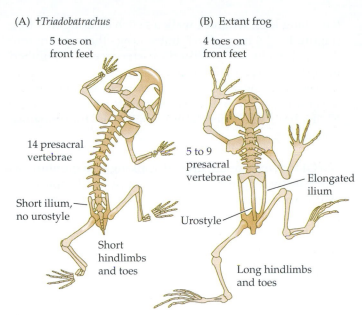

Figure 12.5 †***Triadobatrachus* and a derived anuran.** †*Triadobatrachus* (A), from the Triassic, is the sister group of extant anurans (B). The derived skeletal characters of anurans visible in this comparison include shortening of the body, elongation of the ilia, and fusion of the posterior vertebrae to form a urostyle.

Body form and locomotion Although not all frogs have the ability to jump (specialized aquatic species are swimmers, and some highly terrestrial species are walkers), specialization of the body for jumping is the most conspicuous skeletal feature of anurans (**Figure 12.5**). The hindlimbs and muscles of anurans form a lever system that can catapult the animal into the air. Numerous morphological specializations are associated with this type of locomotion:

- Hindlimbs are elongated, and the tibia and fibula are fused.

- A powerful pelvis strongly fastened to the vertebral column is clearly necessary, as is stiffening of the vertebral column.

- The ilium is elongated and reaches far anteriorly, and the posterior vertebrae are fused into a solid rod, the **urostyle**. The pelvis and urostyle render the posterior half of the trunk rigid.

- The vertebral column is short, with 5–9 presacral vertebrae strongly braced by **zygapophyses**—articulations between adjacent vertebrae (see Figure 8.15)—that restrict lateral bending.

- Strong forelimbs and a flexible pectoral girdle absorb the impact of landing.

- Eyes are large and placed well forward on the head, providing binocular vision.

Anuran jumping ability ranges from hoppers to leapers, arbitrary but useful designations based on the distance an individual can travel in a single jump (**Table 12.3**).

Table 12.3 Locomotor modes of anurans

Category	Maximum distance (body lengths)	Examples
Hoppers	5–10	Bufonids, dendrobatids, scaphiopodids
Jumpers	10–20	Eleutherodactylids, ranids
Leapers	>20	Hylids

The anuran body form probably evolved from a more salamanderlike starting point. Both jumping and swimming have been suggested as the mode of locomotion that made the change advantageous. Salamanders and caecilians swim as fishes do—by passing a sine wave down the body. Anurans have inflexible bodies and swim with simultaneous thrusts of the hind legs. Some paleontologists have proposed that the anuran body form evolved because of the advantages of using the hindlimbs for swimming. An alternative hypothesis traces the anuran body form to the advantage gained by an animal that could rest near the edge of a body of water and escape aquatic or terrestrial predators with a rapid leap followed by locomotion on either land or water.

Several aspects of the natural history of anurans appear to be related to their different modes of locomotion (see Table 12.3). In particular, short-legged species that move by hopping are frequently wide-ranging predators that cover large areas as they search for food. This behavior makes them conspicuous to their own predators, and their short legs prevent them from fleeing rapidly enough to escape. Many of these anurans have potent defensive chemicals that are released from glands in the skin when they are attacked.

Species of frogs that move by jumping or leaping, in contrast to those that hop, are usually sedentary predators that wait in ambush for prey to pass their hiding places. These species are usually cryptically colored, and they often lack chemical defenses. If they are discovered, they rely on a series of rapid leaps to escape. Anurans that forage widely encounter different kinds of prey from those that wait in one spot, and differences in dietary habits may be associated with differences in locomotor mode.

Aquatic anurans use suction to engulf food in water, but most species of semiaquatic and terrestrial anurans have viscoelastic tongues that can be flipped out to trap prey and carry it back to the mouth. Most terrestrial anurans use a catapult-like mechanism to project the tongue (**Figure 12.6**). The tongue flips over as it leaves the mouth, and the portion that makes contact with the prey has a complex microstructure that assists sticky saliva in fastening to the prey and pulling it back into the mouth. The tongues of horned frogs (*Ceratophrys*) have an adhesive strength greater than the frogs' body weight. When the prey is in the mouth, retraction of the eyes assists in swallowing.

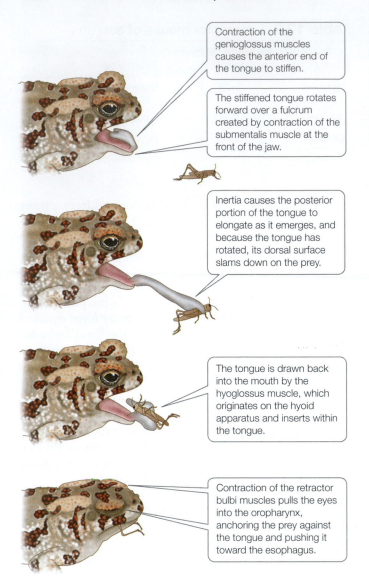

Contraction of the genioglossus muscles causes the anterior end of the tongue to stiffen.

The stiffened tongue rotates forward over a fulcrum created by contraction of the submentalis muscle at the front of the jaw.

Inertia causes the posterior portion of the tongue to elongate as it emerges, and because the tongue has rotated, its dorsal surface slams down on the prey.

The tongue is drawn back into the mouth by the hyoglossus muscle, which originates on the hyoid apparatus and inserts within the tongue.

Contraction of the retractor bulbi muscles pulls the eyes into the oropharynx, anchoring the prey against the tongue and pushing it toward the esophagus.

Figure 12.6 **Terrestrial anurans project their tongues to capture prey.** The tongue is attached at the front of the lower jaw and pivots around the stiffened submentalis muscle as it is flipped out. The tip of the tongue (the portion that is at the rear of the mouth when the tongue is retracted) has glands that secrete sticky mucus that adheres to the prey as the tongue is retracted into the mouth. Retraction of the eyes assists in forcing the prey into the throat.

Diversity   The diversity of anurans exceeds the number of common names that can be used to distinguish the different ecological specialties (**Figure 12.7**). Semi-aquatic frogs like the American bullfrog (Figure 12.7A) have sturdy hind legs and extensive webbing between the toes. Specialized aquatic frogs such as the African clawed frog (Figure 12.7B) have large, webbed hind feet and retain lateral lines as adults. Terrestrial and semiaquatic frogs lose their lateral line systems during metamorphosis.

Terrestrial frogs have little webbing between the toes. Leapers like the Australian rocket frog (Figure 12.7C)

have long legs, whereas walkers such as Budgett's frog (Figure 12.7D) have short limbs. Frogs that squirm into the leaf litter head-first, like the eastern narrow-mouthed toad (Figure 12.7E), have pointed snouts, whereas specialized burrowers that spend most of their lives underground, such as the Australian turtle frog (Figure 12.7F), have blunt heads and stout limbs that they use for digging.

Arboreal frogs like the tiger-striped leaf frog (Figure 12.7G) move by quadrupedal walking and climbing as much as by leaping. They have large heads and eyes, slim waists, and long legs. Flying frogs are gliders, not powered fliers. They have extensive webbing between their toes that forms an air foil allowing them to glide for horizontal distances of 16 m or more, as for example Wallace's flying frog (Figure 12.7H).

Many arboreal species of Hylidae and Rhacophoridae (see Table 12.2) have enlarged toe pads and are called treefrogs. Expanded toe pads are not limited to arboreal frogs, however; some terrestrial species that move across fallen leaves on the forest floor have toe pads, as do some frogs that live on rocks in the spray zone of waterfalls. The surface of the toe pads consists of an epidermal layer of polygonal plates separated by deep channels. Mucus glands distributed over the pads secrete a viscous solution of long-chain, high-molecular-weight polymers. Arboreal frogs use wet adhesion to stick to smooth surfaces. (This is the same mechanism by which a wet scrap of paper sticks to glass.) The watery mucus secreted by the glands on the toe pads forms a layer of fluid between the pad and the surface and establishes a meniscus (curved upper surface) at the interface between air and fluid at the edges of the toes. As long as no air bubble enters the fluid layer, a combination of surface tension and viscosity holds the toe pad and surface together.

Caecilians

The third group of extant lissamphibians, Gymnophiona (Greek *gymnos*, "naked" [i.e., scaleless]; *ophis*, "snake"), is the least known and does not even have an English common name. These are the caecilians, 10 families with 215 species that occur in tropical habitats around the world. Caecilians are terrestrial, with the exception of a few species of Typhlonectidae (**Table 12.4**).

Terrestrial caecilians are fossorial (burrowing), legless lissamphibians that move using alternating contractions and expansions of their segmented bodies (**concertina locomotion**, named after the small, accordionlike musical instrument). The eyes of caecilians are greatly reduced and covered by skin or even by bone. Some species lack eyes entirely, but the retinas of other species have the layered organization that is typical of vertebrates, and these species appear to be able to detect light. Conspicuous grooves in the skin called **annuli** encircle the bodies of caecilians (**Figure 12.8A**). Annuli overlie vertebrae and mark the position of the ribs.

(A) North American bullfrog, *Rana catesbeiana*

© Dee Browning/Shutterstock.com

(B) African clawed frog, *Xenopus laevis*

Lateral line

H. Crisp/CC BY 2.0

(C) Australian rocket frog, *Litoria nasuta*

Froggydarb/CC BY-SA 3.0

(D) Budgett's frog, *Lepidobatrachus laevis*

© Twan Leenders

(E) Eastern narrow-mouthed toad, *Gastrophryne carolinensis*

Courtesy of Todd Pierson

(F) Australian turtle frog, *Myobatrachus gouldii*

Photo by F. Harvey Pough

(G) Tiger-striped leaf frog, *Phyllomedusa tomopterna*

Bernard Dupont/CC BY-SA 3.0

(H) Wallace's flying frog, *Rhacophorus nigropalmatus*

Rushen/CC BY-SA 2.0

Figure 12.7 Diversity of anurans. Body shape, limb length, and foot webbing reflect specializations of anurans for different habitats and different methods of locomotion. (A) A semiaquatic species, the American bullfrog. (B) A specialized aquatic species, the African clawed frog. (C) A terrestrial leaper, the Australian rocket frog. (D) A terrestrial walker and hopper, Budgett's frog. (E) A leaf-litter burrower, the eastern narrow-mouthed toad. (F) A deep burrower, the Australian turtle frog. (G) An arboreal climber, the tiger-striped leaf frog. (H) An arboreal glider, Wallace's flying frog.

Table 12.4 Extant Families of Caecilians

Family	Habitat	Reproduction and development	Adult size	Number and distribution of species
Tailed caecilians (Rhinatrematidae)	Terrestrial, possibly surface dwellers beneath leaf litter	Believed to have aquatic larvae	up to 30 cm	13 species in northern South America
Asian caecilians (Ichthyophiidae)	Subterranean	Oviparous with aquatic larvae	up to 50 cm	57 species in the Philippines, India, Thailand, southern China, mainland Malaysia, Sumatra, and Borneo
Buried-eyed caecilians (Scolecomorphidae)	Subterranean. The eye is attached to the tentacle and moves out from the skull when the tentacle is extended	Both oviparous and viviparous species	up to 45 cm	6 species in tropical Africa
Northeast Indian caecilians (Chikilidae)	Subterranean	Oviparous with direct development	up to 25 cm	4 species in northeastern India
African caecilians (Herpelidae)	Subterranean	Viviparous with direct development; hatchlings of at least one species are dermatophagous	up to 35 cm	10 species in tropical Africa
Common caecilians (Caeciliidae)	Subterranean	Oviparous, possibly also viviparous species; no aquatic larval stage	up to 1.5 m	44 species in Central and South America
Aquatic caecilians (Typhlonectidae)	Aquatic and semiaquatic	Viviparous; embryos scrape lipid-rich cells from the mother's oviduct	up to 1 m	14 species in Central America and northern South America
Indo-African caecilians (Indotyphlidae)	Subterranean	Oviparous and viviparous species, some with aquatic larvae	up to 20 cm	24 species in tropical Africa, Seychelles Islands, and India
Tropical caecilians (Dermophiidae)	Subterranean	Viviparous	up to 60 cm	14 species in Central America, northern South America, West Africa, and Kenya
South American caecilians (Siphonopidae)	Subterranean	Oviparous with direct development; larvae of two species are known to be dermatophagous	up to 0.5 m	28 species in northern and central South America

Compiled in part from information on AmphibiaWeb.org (amphibiaweb.org/lists/index.shtml)

Many species of caecilians have dermal scales in pockets in the annuli; scales are not known in the other extant group of lissamphibians. This may be a character that was lost in Batrachia (frogs and salamanders; see Figure 12.1).

Caecilians uniquely have a pair of protrusible tentacles, one on each side of the snout between the eye and the nostril (**Figure 12.8B**). Some structures that are associated with the eyes of other vertebrates have become associated with the tentacles of caecilians. One of the eye muscles, the retractor bulbi, has become the retractor muscle for the tentacle; the levator bulbi moves the tentacle sheath; and the Harderian gland (which moistens the eye in other tetrapods) lubricates the channel of the tentacle of caecilians. The eyes of caecilians in the African family Scolecomorphidae are attached to the sides of its tentacles. When the tentacle is protruded, the eye is carried along with it, moving out of the tentacular aperture beyond the roofing bones of the skull.

The caecilian tentacle is probably a sensory organ that allows chemical substances to be transported from the animal's surroundings to the vomeronasal organ (an olfactory organ in the roof of the mouth of tetrapods).

(A) A terrestrial caecilian, *Siphonops annulatus*

(B) Detail of the tentacle

Eye

Tentacle

Annuli

(C) An aquatic caecilian, *Typhlonectes natans*

Figure 12.8 Caecilians are elongated, limbless, and nearly eyeless. (A) The ringed caecilian of South America is fossorial. (B) The tentacle, a structure unique to caecilians, lies between the eye and the nostril. (C) Typhlonectids are aquatic caecilians.

Terrestrial caecilians feed on small or elongated prey—termites, earthworms, and larval and adult insects—and the tentacle may allow them to detect the presence of underground prey. Aquatic caecilians capture insect larvae, fishes, and tadpoles (**Figure 12.8C**).

Derived caecilians have solidly roofed skulls. Dual sets of jaw-closing muscles are a unique character of derived caecilians: The adductor mandibulae muscles pull upward on the lower jaw while the interhyoid muscle pulls back and down on the retroarticular process (**Figure 12.9**).

(A) Dorsolateral view of skull, retroarticular process, and interhyoid muscle

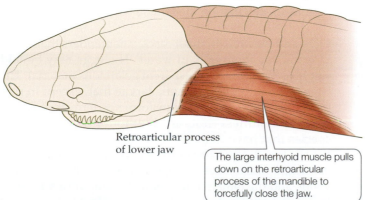

Retroarticular process of lower jaw

The large interhyoid muscle pulls down on the retroarticular process of the mandible to forcefully close the jaw.

(B) Lateral view of skull and mandibular adductor muscles

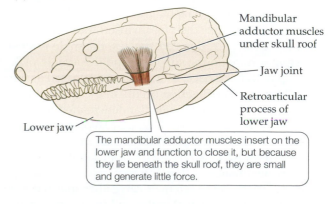

Mandibular adductor muscles under skull roof

Jaw joint

Retroarticular process of lower jaw

Lower jaw

The mandibular adductor muscles insert on the lower jaw and function to close it, but because they lie beneath the skull roof, they are small and generate little force.

Figure 12.9 Derived caecilians have a dual jaw-closing mechanism. The compact heads of these burrowing amphibians limit the force of the adductor mandibulae muscles. Additional force is supplied by the interhyoid muscles acting on the retroarticular process of the mandible. (After W.E. Bemis et al. 1983. *Zool J Linn Soc* 77: 75–96.)

12.2 Life Histories of Lissamphibians

LEARNING OBJECTIVES

12.2.1 Contrast the diversity of reproductive modes in the three clades of lissamphibians.

12.2.2 Explain the benefits of a larval stage in the life history of terrestrial lissamphibians.

12.2.3 Use examples to explain the difference between the specific and the local effects of thyroxine during metamorphosis.

Of all the characters of lissamphibians, none is more remarkable than their range of reproductive modes. Most species of frogs and salamanders lay eggs. The eggs may be deposited in water or on land. The eggs may hatch into aquatic larvae that undergo morphological and physiological changes to become terrestrial adults (metamorphosis), or they may develop into miniatures of the terrestrial adults (direct development). The adults of some species of frogs carry eggs attached to the surface of their bodies. Others carry their eggs in pockets in the skin of the back or flanks, in the vocal sacs, or even in the stomach. In still other species, females retain the eggs in the oviducts and give birth to metamorphosed young. Many frogs and salamanders provide no parental care to their eggs or young, but in other species a parent remains with the eggs and transports tadpoles from the nest to water. The adults of some species of frogs remain with their tadpoles, and in a few species the mother even feeds her tadpoles.

Mating and reproduction in salamanders

Most groups of salamanders have internal fertilization, but Cryptobranchidae, Hynobiidae, and probably Sirenidae retain external fertilization (see Table 12.1). Internal fertilization is accomplished by transferring a packet of sperm (a **spermatophore**) from the male to the female (**Figure 12.10**). The form of the spermatophore differs among species, but all consist of a sperm cap on a gelatinous base.

Females of the hynobiid salamander *Ranodon sibiricus* deposit egg sacs on top of a spermatophore. Males of the Asian salamandrid genus *Euproctus* deposit a spermatophore on the body of a female and then, holding her with their tail or jaws, use their feet to insert the spermatophore into her cloaca. In derived species, the male deposits a spermatophore on the substrate and the female picks up the cap with her cloaca. The sperm are released as the cap dissolves, and fertilization occurs in the oviducts.

A group of all-female salamanders related to *Ambystoma jeffersonianum* is characterized by **kleptogenesis** (Greek *kleptes*, "thief"; *genesis*, "origin"). Females produce eggs that must be activated by sperm from a male of a bisexual species, and males of five different species can be sperm donors. Furthermore, a female can gather spermatophores from different species of males and incorporate variable amounts of DNA from those males into her offspring. The result is an enormously complex genome, varying from triploid to pentaploid, in which mtDNA from the mother

Spermatophore deposited by male axolotl, *Ambystoma mexicanum*

Courtesy of Michael Guglielmelli © A Lot'l Axolotls

Figure 12.10 A spermatophore. Male salamanders deposit spermatophores that contain a capsule of sperm supported on a gelatinous base. The sperm is enclosed in the cream-colored structure; the clear base anchors the spermatophore to the substrate.

is the only consistent component. Identifying critical habitat for conservation of kleptogenic species is complicated by the need to consider the habitat needs of the different species that provide sperm for the kleptogenic species.

Courtship   Courtship patterns are important for species recognition in many vertebrates, and they show great interspecific variation among salamanders. Pheromones probably contribute to species recognition and may stimulate endocrine activity that increases female receptivity.

Pheromone delivery by most salamanders that breed on land involves physical contact between a male and female, during which the male applies secretions of specialized courtship glands (**hedonic glands**) to the nostrils or body of the female. Males of many species of plethodontids have padlike glands under the chin. In some species, the male slaps the female's snout with his chin during courtship. Males of other species spread secretions of the hedonic gland on the female's skin and then abrade the skin with their teeth, inoculating the female with the pheromone.

Eggs and larvae   In most cases, salamanders that breed in water lay their eggs in water. The eggs may be laid singly or in a mass of transparent gelatinous material. The eggs hatch into gilled aquatic larvae that, except in permanently aquatic species, transform into terrestrial adults.

Some families—notably Plethodontidae—include species that have dispensed in part or entirely with an aquatic larval stage. The northern dusky salamander (*Desmognathus fuscus*) lays its eggs beneath a rock or log near water, and the female remains with them until after they have hatched. The larvae have small gills at hatching and may either take up an aquatic existence or move directly to terrestrial life. The red-backed salamander (*Plethodon cinereus*) lays its eggs in a hollow space in a rotten log or beneath a rock. The embryos have gills, but these are re-absorbed before hatching, so the hatchlings are essentially miniature adults.

Parental care is usually provided by the female, as would be expected when fertilization is internal and the male is not present when the eggs are deposited. Fertilization is external in cryptobranchids, hynobiids, and probably sirenids, and the male is the attending parent in cryptobranchids and sirenids; parental care has not been demonstrated in hynobiids.

Viviparity   Only a few species of salamanders, all in the genera *Salamandra* and *Lyciasalamandra*, are viviparous. The European alpine salamander (*S. atra*) gives birth to one or two fully developed young, each about one-third the adult body length, after a gestation period that lasts from 2 to 4 years. Initially the internal clutch contains 20–30 eggs, but only one or two eggs are fertilized and develop into embryos. When the energy in their yolk sacs is exhausted, the embryos consume the unfertilized eggs. When that source of energy is gone, the embryos scrape the reproductive tract of the female with specialized teeth.

Females in some populations of the European fire salamander (*Salamandra salamandra*) produce 20 or more small larvae, each about one-twentieth the length of an adult. The embryos probably get all the energy needed for growth and development from egg yolk. The larvae are released in water and have an aquatic stage that lasts about 3 months. In other populations of this species, the eggs are retained in the oviducts, and when all of the unfertilized eggs have been consumed, some of the embryos cannibalize other embryos. The surviving embryos pass through metamorphosis in the oviducts of the female.

Paedomorphosis   Paedomorphosis, the retention of juvenile characters, has been extensively studied among salamanders. It is the rule in Cryptobranchidae and Proteidae and appears in populations of other species of salamanders in regions where the terrestrial habitat is inhospitable for adults. Paedomorphic individuals do not undergo metamorphosis but retain their larval form, with gills, lateral line and electroreceptive systems, and an aquatic habitat (see Figure 12.3A,B). Some species are facultative paedomorphs that typically metamorphose but under certain environmental conditions will remain in the larval state.

Anuran mating and reproduction

Anurans are the most familiar lissamphibians, in part because of the vocalizations associated with their reproductive behavior. It is not even necessary to be in a rural area to hear them. In springtime, a weed-choked drainage ditch beside a highway or a trash-filled marsh at the edge of a parking lot is likely to attract a few toads or treefrogs that have not yet succumbed to human usurpation of their habitat.

The mating systems of anurans can be divided roughly into **explosive breeding**, in which the breeding season is very short (sometimes only a few days); and **prolonged breeding**, with breeding seasons that may last for several months. Explosive breeders include many species of toads and other anurans that breed in temporary aquatic habitats, such as vernal ponds or pools that form after desert rainstorms. Because these bodies of water do not last long, breeding congregations usually form as soon as the site is available. Males and females arrive at the breeding sites nearly simultaneously, often in large numbers. The numbers of males and females present are approximately equal because the entire population breeds in a short time. Time is the main constraint on how many females a male is able to court, and mating success is usually similar for all the males in a chorus.

In species with prolonged breeding seasons, males usually arrive at the breeding sites first. Males of some species, such as green frogs (*Rana clamitans*), establish territories in which they spend several months, defending the spot against the approach of other males. Males of other species move between daytime retreats and nocturnal calling sites on a daily basis. Females come to the breeding site to breed and leave when they have finished. Only a few females arrive each night, and the number of males at the breeding site is greater than the number of females. Mating success may be skewed, with many males not mating at all and a few males mating several times. Males of anuran species with prolonged breeding seasons compete to attract females, usually by vocalizing. The characteristics of a male frog's vocalization (pitch, length, and repetition rate) may provide information that a female can use to evaluate his quality as a potential mate. This is an active area of study in anuran behavior.

Vocalizations   Anuran calls vary from species to species, and most species have two or three different sorts of calls used in different situations. The most familiar calls are the ones usually referred to as "mating calls," although a less specific term such as **advertisement calls** is preferable. These calls range from the high-pitched *peep* of a spring peeper (*Pseudacris crucifer*) to the bass *jug-o-rum* of an American bullfrog (*Rana catesbeiana*). The characteristics of a call identify the species and sex of the calling individual. Many anuran species are territorial, and males of at least one species, the American bullfrog, can distinguish neighbors from intruders by their calls.

Female frogs are responsive to the advertisement calls of males of their species only for a brief period when their eggs are ready to be laid. Hormones associated with ovulation are thought to sensitize specific cells in the auditory pathway that respond to species-specific characteristics of the male's call. Mixed choruses of anurans are common during the mating season—a dozen species may breed simultaneously in one pond. A female's response to her

(A) Bornean tree-hole frog, *Metaphrynella sundana*

Courtesy of Björn Landers

(B) Indian dancing frog, *Micrixalus kottigeharensis*

Sathyabhama Das Biju/CC BY-SA

Figure 12.11 **Sometimes it takes more than a simple call.** (A) A male Bornean tree-hole frog calling from a hollow in a tree that lowers the frequency of his call, making him sound larger than he really is. (B) A male Indian dancing frog foot-flagging.

own species' mating call is a mechanism for species recognition in that situation.

A female's choice of a mate from among calling males of her own species is based largely on the characteristics of the call, including how loud it is, the dominant frequency, and how long it lasts. Females often prefer loud calls and low frequencies, which indicate that the caller is a large individual. Male Bornean tree-hole frogs (**Figure 12.11A**) make their calls louder by adjusting the dominant frequency by as much as 300 hz to match the resonant frequency of the hollow from which they are calling.

Ambient noise—from a waterfall, for example, or from human activities—can mask the low frequencies in a male's call, making it hard for a female to hear. The calls of some stream-breeding frogs include or are limited to ultrasonic frequencies that are not masked by the sound of flowing water. Several species of torrent frogs supplement or replace calls with visual signals that include foot-flagging—raising one hind leg in the air with the toes extended and then the other, earning them the name dancing frogs (**Figure 12.11B**).

Costs and benefits of vocalizations   The vocalizations of male frogs are costly in at least two senses:

- The amount of energy that goes into call production can be large, and the variations in calling pattern that accompany social interactions among males in a breeding chorus can increase the energy cost per call.
- Vocalization increases the risk of predation for a male frog. A critical function of vocalization is to permit a female to locate a male, but female frogs are not the only animals that can use vocalizations as a cue; predators also find calling males easy to locate.

The túngara frog, *Engystomops pustulatus*, is a small terrestrial species from Central America (**Figure 12.12A**). It breeds in small pools, and breeding assemblies range from a single male to several hundred males. The advertisement call of a male túngara frog is a strange noise, a whine that sounds as if it would be more at home in a video-game arcade than in the tropical night. The whine starts at a frequency of 1,000 Hz and sweeps downward to 500 Hz in about 400 ms. The whine may be produced by itself, or it may be followed by one or several chucks (**Figure 12.12B**).

When a male túngara frog is calling alone in a pond, it usually gives only the whine portion of the call, but as additional males join a chorus, more and more frogs produce calls that include chucks. Male túngara frogs calling in a breeding pond added chucks to their calls when they heard playbacks of recorded calls of other males. That observation suggested that it was the calls of other males that stimulated frogs to make their own calls more complex by adding chucks to the end of the whine.

What advantage would a male frog in a chorus gain from using a whine-chuck call instead of a whine? When female frogs were released individually in the center of an arena where two speakers gave them a choice of one call with and one without chucks, 14 of the 15 females tested moved toward the speaker broadcasting the whine-chuck call.

Female frogs move to whine-chuck calls in preference to whine-only calls, so why do male frogs give whine-chuck calls only when other males are present? The answer lies in another selective force—predation. Túngara frogs in breeding choruses are preyed on by frog-eating bats, and the bats locate the frogs by homing on their vocalizations. In a series of experiments, bats were given a choice of speakers playing recorded calls of túngara frogs. In five experiments at different sites, the bats approached speakers broadcasting whine-chuck calls twice

(A) Air in lungs of túngara frog,
Engystomops pustulatus

Air forced from lungs into vocal sacs

(B) Sonograms of calls

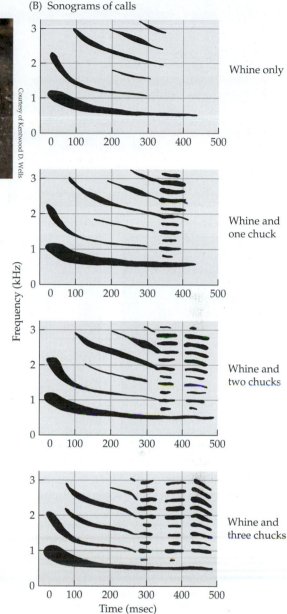

Figure 12.12 **Male túngara frogs change the complexity of their calls.** (A) As the frog calls, air is forced from the lungs into the vocal sac. (B) Time (duration of the sound) is shown on the horizontal axis and frequency (pitch) on the vertical axis. The calls increase in complexity from top (whine only) to bottom (whine followed by three chucks). (B from M.J. Ryan. 1992. *Túngara Frog: A Study in Sexual Selection and Communication.* © University of Chicago Press.)

as frequently as those playing simple whines (168 approaches versus 81). Thus, female frogs are not alone in finding whine-chuck calls more attractive than whine-only calls.

When a male frog shifts from a whine-only to a whine-chuck call, it increases its chances of attracting a female, but simultaneously increases its risk of attracting a predator. In small choruses, competition from other males for females is relatively small and the risk of predation is large. Under those conditions, it is apparently advantageous for a male túngara frog to give simple whines. However, as chorus size increases, competition increases while the risk of predation falls. In that situation, the advantage of giving a complex call apparently outweighs the risks.

Modes of reproduction   The variety of reproductive modes of anurans exceeds that of every other vertebrate lineage except for bony fishes (**Figure 12.13**). External fertilization is ancestral for anurans and is retained in many lineages: the male uses his forelegs to clasp the female in the pectoral region (axillary amplexus) or pelvic region (inguinal amplexus). Internal fertilization has been demonstrated in a few species of anurans and may be more common than we realize. Male tailed frogs of the Pacific Northwest have an extension of the cloaca (the appendage that gives them their name) that they use to introduce sperm into the cloaca of the female (Figure 12.13A). The Puerto Rican coquí, *Eleutherodactylus coqui*, has internal fertilization even though males lack an intromittent organ. Internal fertilization may be widespread among frogs that lay eggs on land, and fertilization must be internal for the few species of viviparous anurans.

Toads retain the ancestral reproductive mode of laying a large number of eggs in water and providing no parental care (Figure 12.13B). Many arboreal frogs lay their eggs on the leaves of trees overhanging water (Figure 12.13C). The eggs undergo embryonic development beyond the reach of aquatic egg predators, and when the tadpoles hatch they drop into the water and take up an aquatic existence. Other species, such as the túngara frog, achieve the same result by constructing foam nests that float on the water surface (Figure 12.13D). The female releases mucus and eggs and the male uses his hind legs to whip the mucus into foam that suspends the eggs. African gray treefrogs construct foam nests in trees overhanding water (Figure 12.13E). When tadpoles in a foam nest hatch, they release an enzyme that dissolves the foam and allows them to drop into the water.

(A) Tailed frog, *Ascaphus truei*

Courtesy of Rob Schell

(B) Puerto Rican crested toads, *Peltophryne lemur*

This egg string contains fertilized eggs that came from the right oviduct.

Jan P. Zegarra/USFWS

(C) Chocolate-foot leaf frogs, *Phasmahyla cochranae*

Courtesy of Mario Sacramento

Photo by Celio Haddad. From Crump, 2015. J. Herp. 49:1-16

(D) Foam nest of túngara frog, *Engystomops pustulosus*

Courtesy of Ryan Taylor

(E) Foam nests of African gray treefrogs, *Chiromantis xerampelina*

Bernard Dupont/CC BY-SA 2.0

(F) Nest excavated by a male blacksmith treefrog, *Boana faber*

From M.L. Crump. 2015. J Herp 49: 1–16. Photo by Celio Haddad

Figure 12.13 **Reproductive modes of anurans.**   (A) The "tail" of a male tailed frog is actually an extension of the cloaca that in these frogs is an intromittent organ. (B) Puerto Rican crested toads in axillary amplexus. Female toads release eggs in two strings, one from each oviduct; only the string from the right oviduct is visible in this photograph. (C) Brazilian chocolate-foot leaf frogs deposit eggs on leaves overhanging a pool. The adult frogs fold the leaf so that the eggs are concealed during their development. (D) Túngara frogs creating a floating foam nest. (E) Foam nests of African gray treefrogs. (F) A male blacksmith frog guarding its eggs.

Although these methods reduce egg mortality, the tadpoles are subjected to predation and competition. Some anurans avoid both problems by finding or constructing breeding sites free from competitors and predators. Male blacksmith frogs excavate a pool on a streambank and call to attract a female. The eggs are deposited in the newly excavated, predator-free pool, and the male remains to guard the clutch (Figure 12.13F).

(A) Savage's glass frog, *Centrolene savagei*, guarding eggs

Mauricio Rivera Correa/CC BY-SA 2.5

(B) Male midwife toad, *Alytes obstetricans*, transporting eggs

Courtesy of Andreas and Christel Nöllert

(C) Male Darwin's frog, *Rhinoderma darwinii*, with eggs in his vocal sac

From M.L. Crump, 2015. J Herp 49:1-16. Courtesy of Martha L. Crump

(D) Female Surinam toad, *Pipa pipa*, with eggs on her back

Photo by F. Harvey Pough

(E) Male sky blue poison frog, *Hyloxalus azureiventris*, transporting tadpoles

Courtesy of Devin Edmonds

(F) Female strawberry poison frog, *Oophaga pumilio*, feeding a tadpole

© Mark Moffett/Minden Pictures

(G) Male African bullfrog, *Pyxicephalus adspersus*, rescuing tadpoles

© Nature Picture Library/Alamy Stock Photo

(H) Male Brazilian rock frog, *Thoropa taophora*, guarding eggs and a tadpole

Courtesy of Celio Haddad

Figure 12.14 Parental care by anurans. Many species of frogs care for their eggs, and some species extend care to the tadpoles. (A) Males of Savage's glass frog remain with their eggs. (B) Male midwife toads carry eggs until they are ready to hatch. (C) The vocal sacs of this male Darwin's frog are stuffed with eggs. (D) A female Surinam toad with eggs on her back. (E) A male sky blue poison frog carrying tadpoles to water. (F) A female strawberry poison frog depositing a trophic egg to feed a tadpole. (G) A male African bullfrog rescuing tadpoles in a drying pool by digging a channel to a deeper pool. (H) A male rock frog guarding eggs and tadpoles simultaneously.

Some frogs have eliminated the tadpole stage entirely. These frogs lay large eggs on land that develop directly into little frogs. This reproductive mode, called direct development, is characteristic of about 20% of anuran species.

Parental care Terrestrial eggs of lissamphibians can be attacked by a variety of predators and parasites, including several species of flies whose maggots consume eggs and embryos. Adults of many frog species guard their eggs and are probably capable of protecting them against these perils (**Figure 12.14**). Males of many frogs that lay their eggs over water remain with them, either sitting beside the eggs or resting on top of them (Figure 12.14A). Most of the terrestrial frogs that lay direct-developing eggs remain with the eggs and will attack an animal that approaches the nest, even when the intruder is larger than the attending parent. Removing the guarding frog frequently results in the eggs being eaten by predators or desiccating and dying before hatching.

Instead of remaining with its eggs, the male European midwife toad carries the eggs with him, gathering the egg strings about his hind legs as the female releases them (Figure 12.14B). A female midwife toad produces about 50 eggs, and males carrying eggs can mate again, eventually carrying as many as three clutches. Males keep the eggs moist by selecting wet habitats or by dunking the egg strings in water. When the embryos have completed development, the male carries the eggs to small pools, where they hatch into aquatic larvae. Male Darwin's frogs carry eggs in their vocal sacs, which extend back to the pelvic region (Figure 12.14C). The embryos pass through metamorphosis in the vocal sacs and hop from their father's mouth as fully developed froglets.

Females of the completely aquatic Surinam toad *Pipa pipa* carry their eggs on their backs (Figure 12.14D). In the breeding season, the skin of a female's back thickens and softens. During egg laying, the male and female in amplexus swim in vertical loops in the water. On the upward part of the loop, the female is above the male and releases a few eggs, which fall onto his ventral surface. He fertilizes them and, on the downward loop, presses them against the female's back. The eggs sink into the female's soft skin, and a cover forms over each one, enclosing it in a small capsule. The eggs of this species complete development and hatch as tiny frogs, whereas other species of *Pipa* hatch as tadpoles.

Tadpoles of many species of tropical frogs grow to metamorphosis in small, water-filled cavities in terrestrial plants like bromeliads; such cavities are called phytotelmata. Dart poison frogs in the family Dendrobatidae (see Table 12.2) deposit eggs in terrestrial nests, and when the tadpoles hatch one of the parents (the male or the female, depending on the dendrobatid species) carries the eggs on its back and releases them in phytotelmata (Figure 12.14E).

Some species of dendrobatids and mantellids feed their tadpoles unfertilized eggs, called trophic eggs (Greek *trophos*, "a nourisher"). These trophic eggs contain toxins that provide chemical defenses to the tadpoles. Female strawberry poison frogs (*Oophaga pumilio*; the generic name means "egg-eater") deposit tadpoles individually in phytotelmata and return at intervals to deposit a trophic egg for each tadpole (Figure 12.14F). That's an impressive feat of navigation for a frog only 2 cm long, because she simultaneously feeds several different tadpoles in plants that are distributed over an area of several square meters. Imitator dart poison frogs (*Ranitomeya imitator*) deposit tadpoles in phytotelmata of bromeliads, and both parents provide care. A territorial male attracts a female by vocalizing and leads her to an egg deposition site where the female deposits eggs. A week later, the male returns and frees each tadpole from its egg membranes, allows it to wiggle onto his back, and carries it to an unoccupied phytotelma. Returning every few days, the male calls from the surface of the water, attracting the tadpole's mother who deposits a trophic egg that the tadpole consumes.

Male African bullfrogs guard their eggs and then continue to guard the tadpoles after they hatch (Figure 12.14G). If the tadpoles become trapped in a drying pool, the male digs a channel to allow the tadpoles to reach deeper water.

Some tadpoles are terrestrial. For example, Brazilian rock frogs deposit eggs in small films of water that seep over exposed rock faces and the terrestrial tadpoles remain on the rocks until they metamorphose. Males engage in vocal duels and wrestling matches to protect a seep, jabbing each other with spines on a thumblike digit called the prepollex. Usually, two females associate with a male, repeatedly mating and depositing clutches of eggs. Because the male and the two females breed frequently, a male is guarding eggs that range in age from newly laid to ready to hatch, as well as tadpoles that have not yet metamorphosed (Figure 12.14H).

Viviparity Species in the African bufonid genus *Nectophrynoides* show a spectrum of reproductive modes. One species deposits eggs that are fertilized externally and hatch into aquatic tadpoles and the remaining species have internal fertilization. Two of them produce young that are nourished by yolk, and the remaining species have embryos that feed on secretions from the walls of the oviduct.

Anuran metamorphosis

For salamanders and caecilians, the transition from larva to adult involves few conspicuous morphological changes: the gills, lateral line, and electroreceptive systems are lost; eyelids and skin glands develop and the adult teeth appear. For anurans, however, the process by which an aquatic tadpole becomes an adult frog involves **complete metamorphosis** in which tadpole structures are broken down and their chemical constituents are completely rebuilt (**Figure 12.15**).

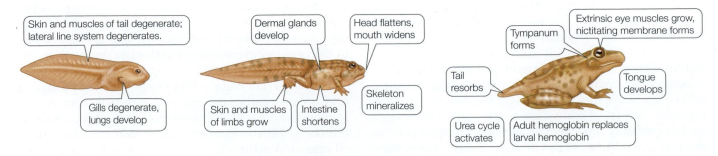

Figure 12.15 Complete metamorphosis of an anuran. The transition from tadpole to frog involves visible and molecular changes in almost every part of the body. (From S.G. Gilbert and M.J.F. Barresi. 2016. *Developmental Biology*, 11th ed. Oxford University Press/Sinauer, Sunderland, MA.)

Table 12.5 Actions of thyroxine during anuran metamorphosis

Structure	Action
Body form and organs	
Skeleton	Calcification
Skin	Formation of dermal glands
Head	Restructuring the mouth
Intestine	Regression and reorganization
Liver	Induction of urea-cycle enzymes
Appendages	
Tail	Reabsorption of skin and muscle
Limbs	Growth and differentiation
Nervous system and sense organs	
Brain	Growth of cerebellum and preoptic nucleus of the hypothalamus
Eyes	Formation of extrinsic eye muscles; formation of nictitating membrane
Respiratory and circulatory systems	
Gills	Degeneration of gill arches and gills
Lungs	Development of lungs
Oxygen transport	Shift from larval to adult hemoglobin

The changes of anuran metamorphosis are stimulated by the actions of thyroxine, and the production and release of thyroxine are controlled by thyroid-stimulating hormone (TSH), a product of the pituitary gland. The action of thyroxine on larval tissues is both specific and local (**Table 12.5**). In other words, it has a different effect in different tissues, and that effect is produced by the presence of thyroxine in the tissue; it does not depend on induction by neighboring tissues.

The final stage of metamorphosis (metamorphic climax) begins with the appearance of the forelimbs and ends with the disappearance of the tail. This is the most rapid part of the process, taking only a few days after a larval period that lasts for weeks or months. One reason for the rapidity of metamorphic climax may be the vulnerability of larvae to predators during this period. A larva with legs and a tail is neither a good tadpole nor a good frog—the legs inhibit swimming, and the tail interferes with jumping. As a result, predators are more successful at catching anurans during metamorphic climax than before hindlegs appear or after the completion of metamorphosis.

The ecology of tadpoles

Although many species of frogs have evolved reproductive modes that bypass an aquatic larval stage and complete metamorphosis, a life history that includes a tadpole has certain advantages. A tadpole is a completely different animal from an adult anuran, both morphologically and ecologically. Tadpoles and frogs are specialized for different habitats and feeding modes and to a variety of feeding niches. Most tadpoles filter food particles from a stream of water, whereas all adult anurans are carnivores that catch prey items individually. Because of these differences, tadpoles can exploit resources that are not available to adult anurans. This advantage may be a factor that has led many species of frogs to retain the ancestral life-history pattern in which an aquatic larva matures into a terrestrial adult. Many aquatic habitats experience annual flushes of primary production when nutrients washed into a pool by rain or melting snow stimulate the rapid growth of algae. The energy and nutrients in this algal bloom are transient resources that are available for a brief time to the organisms able to exploit them.

Tadpoles are eating machines, and filter-feeding and ventilating the gills are related activities. As the stream of water passes through the branchial basket, small food particles are trapped in mucus secreted by epithelial cells. The mucus, along with the particles, is moved from the gill filters to ciliary grooves on the margins of the roof of the pharynx, and then transported posteriorly to the esophagus.

The body forms of tadpoles are finely adapted to their feeding niches (**Figure 12.16**). Surface feeders have mouthparts that unfold into a platter at the water surface, drawing water and floating particles into the mouth (Figure 12.16A). Bottom-feeding tadpoles scrape algae and bacteria from surfaces (Figure 12.16B), and midwater feeders filter particles of food from the water column maintaining their position in the water column with rapid undulations of the end of its thin, nearly transparent tail (Figure 12.16C). Torrent-dwelling tadpoles adhere to rocks in swiftly moving water with sucker-like mouths while scraping algae and bacteria from the rocks (Figure 12.16D). Low fins and powerful tails are characteristic of tadpoles living in swift water. Terrestrial tadpoles reabsorb gills before they hatch and have muscular tails with small fins (Figure 12.16E).

The tadpole of the pancake frog (*Otophryne robusta*) buries itself in sand on the floor of a stream with the tip of its elongated spiracle projecting into the current (Figure 12.16F). The Bernoulli effect draws water into the tadpole's mouth and out the spiracle, creating a passive filter-feeding system. The tadpole's needlelike teeth, which were originally interpreted as indicating a carnivorous diet, prevent sand grains from entering the oral cavity.

Individual tadpoles of species that are normally herbivorous can develop into a cannibalistic morph with a sharp, keratinized beak that allows them to bite off bits of flesh. Eating the freshwater shrimp that occur in some breeding ponds causes tadpoles of the Mexican spadefoot toad (*Spea multiplicata*) to transform into the cannibal morph (**Figure 12.17**). In addition to eating the shrimp, the cannibals prey on other tadpoles, attacking unrelated individuals more often than their own kin.

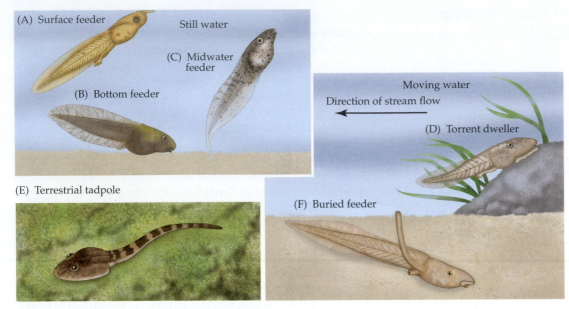

Figure 12.16 **Body forms and mouth structures of tadpoles reflect differences in habitat and diet.** (A) The Kwangshien spadefoot toad (*Panophrys minor*), a surface feeder. (B) The red-legged frog (*Rana aurora*), a bottom scraper. (C) The red-eyed leaf frog (*Agalychnis callidryas*), a midwater feeder. (D) The Australasian waterfall frog (*Litoria nannotis*), a torrent dweller. (E) Tadpoles of the Brazilian rock frog (*Thoropa taophora*) are terrestrial. (F) The pancake frog (*Otophryne robusta*), a buried feeder.

Cannibal morph of Mexican spadefoot toad
(*Spea multiplicata*)

© Wild Horizon/Getty Images

Figure 12.17 **Spadefoot toad tadpoles express alternative phenotypes.** Tadpoles of the Mexican spadefoot toad are typically herbivorous, but those that consume tiny crustaceans (which occur in some breeding ponds) develop into a larger carnivorous morph with a sharp beak that allows them to cannibalize their herbivorous conspecifics.

The feeding mechanisms that make tadpoles such effective collectors of food particles suspended in the water allow them to grow rapidly, but that growth contains the seeds of its own termination. As tadpoles grow bigger, they become less effective at gathering food because of the changing relationship between the size of food-gathering surfaces and the size of their bodies. The branchial surfaces that trap food particles are two-dimensional. Consequently, the food-collecting apparatus of a tadpole increases in size approximately as the square of the linear dimensions of the tadpole. However, the food the tadpole collects must nourish its entire body, and the volume of the body increases in proportion to the cube of the linear dimensions of the tadpole. The result of that relationship is a decreasing effectiveness of food collection as a tadpole grows: the body it must nourish increases in size faster than does its food-collecting apparatus. This mathematical stumbling block might explain why there are no paedomorphic anurans.

Caecilian reproduction and development

The reproductive modes of caecilians are as specialized as their body form and ecology. Fertilization is internal via a male intromittent organ that is protruded from the cloaca. Some species of caecilians lay eggs, and the female coils around the eggs, remaining with them until they hatch (**Figure 12.18A**). Some oviparous species of caecilians have direct development; others, such as Asian tailed caecilians (*Ichthyophis*), lay eggs in underground chambers near water. The eggs hatch into gilled larvae that move into water to complete development.

(A) Female Asian tailed caecilian, *Ichthyophis* sp., with eggs

David Raju/CC BY-SA 4.0

(B) Female ringed caecilian, *Siphonops annulatus*, with young

Courtesy of Carlos Jared

Figure 12.18 Caecilian reproduction. (A) Asian tailed caecilians are oviparous. Here a female coils around her eggs, which will hatch into aquatic larvae. (B) The South American ringed caecilian is a viviparous, direct-developing species notable for dermatophagy (the young feed on the mother's outermost layer of skin).

Viviparity is widespread, however; about 75% of caecilian species are viviparous and **matrotrophic** (receiving nutrition from the mother).

The fetuses of some viviparous species scrape lipid-rich cells from the walls of the oviducts with specialized embryonic teeth. The epithelium of the oviduct proliferates and forms thick beds surrounded by ramifications of connective tissue and capillaries. As the fetuses exhaust their yolk supply, these beds begin to secrete a thick, white, creamy substance that has been called uterine milk. When their yolk supply has been exhausted, the fetuses emerge from their egg membranes, uncurl, and align themselves lengthwise in the oviducts. The fetuses apparently bite the walls of the oviduct, stimulating secretion and stripping some epithelial cells and muscle fibers that they swallow with the uterine milk. Small fetuses are regularly spaced along the oviducts. Large fetuses have their heads spaced at intervals, and the body of one fetus overlaps the head of the next. This spacing probably gives all the fetuses access to the secretory areas on the walls of the oviducts.

Dermatophagy (Greek *derma*, "skin"; *phago*, "to eat") is a remarkable method of feeding the young employed by *Boulengerula taitana* from Kenya and *Siphonops annulatus* and *Microcaecilia dermatophaga* from South America (**Figure 12.18B**). Females give birth to young that are in a relatively undeveloped stage, and the young remain with the mother in subterranean nests. The cells in the stratum corneum (the outermost layer of the skin) of brooding females are thickened and contain vesicles filled with lipids. The young feed by seizing the lipid-rich skin on the mother's body and spinning about their body axis to tear off pieces. The mother remains calm during the feeding bout, which lasts only a few minutes and peels off the outer layer of her skin.

12.3 Respiration and Circulation

LEARNING OBJECTIVES

12.3.1 Explain the importance of a moist skin for gas exchange by lissamphibians.

12.3.2 Explain how oxygenated and deoxygenated blood streams are kept separate in the undivided ventricle of the heart of an anuran.

12.3.3 Explain the differences in the circulatory systems of tadpoles and adult anurans.

Both water and gases pass readily through lissamphibian skin, and all three lineages depend on cutaneous respiration for a significant part of their gas exchange. Under some circumstances, venous blood returning from the skin has a higher oxygen content than blood returning in the pulmonary veins from the lungs. Although the skin permits the passive movement of water and gases, it controls the movement of other compounds. Sodium is actively transported from the outer surface to the inner, and urea is retained by the skin. These characteristics are important in the regulation of osmotic concentration and in facilitating the uptake of water by terrestrial species.

Cutaneous respiration and blood flow

The balance between cutaneous and pulmonary uptake of oxygen varies among species, and within a species it depends on body temperature and the animal's rate of activity. Lissamphibians show increasing reliance on the lungs for oxygen uptake as temperature and activity increase.

The patterns of blood flow within the hearts of adult lissamphibians reflect the use of two respiratory surfaces. The following description is based on the anuran heart (**Figure 12.19**). A septum divides the atrium of the heart into left and right chambers. Oxygenated blood from the

(A) Blood flow when lungs are being ventilated

(B) Blood flow when lungs are not being ventilated

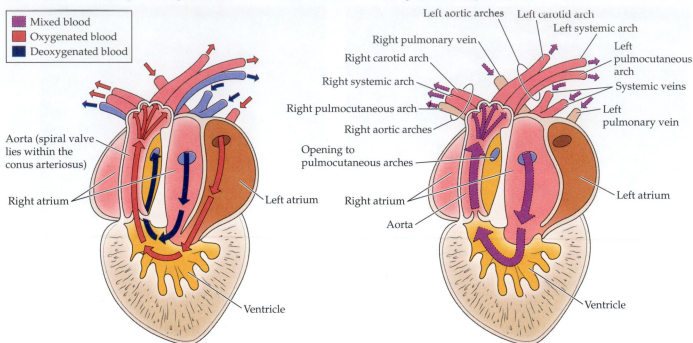

Figure 12.19 Oxygenated and deoxygenated blood flow in the anuran heart. A spiral valve in the conus arteriosus guides blood from the left and right sides of the ventricle to the arches of the left and right aortae. (A) Arrows indicate the pattern of blood flow when lungs are ventilated. Oxygenated blood from the lungs is distributed to all tissues of the body. (B) Blood flow when only cutaneous respiration is taking place (e.g., when the frog is underwater). Oxygenated and deoxygenated blood are mixed. The highest oxygen content is in the systemic veins that drain the skin, and little blood passes through the pulmonary circuit.

lungs flows into the left side of the heart, and deoxygenated blood from the systemic circulation flows into the right side.

The spongy muscular interior of the undivided ventricle minimizes mixing of oxygenated and deoxygenated blood, and the position within the ventricle of a particular parcel of blood appears to determine its fate on leaving the contracting ventricle. The short conus arteriosus contains a spiral valve of tissue that differentially guides blood from the left and right sides of the ventricle to the various arches.

The anatomical relationships within the heart direct oxygenated blood from the pulmonary veins to left atrium, which injects it into the left side of the common ventricle. Contraction of the ventricle ejects blood in streams that spiral out of the pumping chamber, carrying the left-side blood into the ventral portion of the spirally divided conus. This half of the conus is the one from which the carotid and systemic aortic arches arise.

When the lungs are actively ventilated, oxygenated blood returning from them to the heart is selectively distributed to the tissues of the head and body. Deoxygenated venous blood entering the right atrium is directed into the dorsal half of the spiral valve in the conus (see Figure 12.19A). It goes to the pulmocutaneous arches, destined for oxygenation in the lungs. However, when the skin is the primary site of gas exchange (as when a frog

is underwater), the highest oxygen content is in the systemic veins that drain the skin (see Figure 12.19B). In this situation, the lungs may actually be net users of oxygen, and because of vascular constriction little blood passes through the pulmonary circuit.

Because the ventricle is undivided and the majority of the blood arrives from the systemic circuit, the ventral section of the conus receives blood from an overflow of the right side of the ventricle. The scant left atrial supply to the ventricle also flows through the ventral conus. Thus, the most oxygenated blood coming from the heart flows to the tissues of the head and body during this shift in primary respiratory surface, a phenomenon possible only because of the undivided ventricle. Variability of the cardiovascular output in lissamphibians is an essential part of their ability to use alternate respiratory surfaces effectively.

Blood flow in larvae and adults

Larval lissamphibians rely on their gills and skin for gas exchange, whereas adults of species that complete full metamorphosis lose their gills and develop lungs. As the lungs develop, they are increasingly used for respiration. Late in development, tadpoles and partly metamorphosed froglets can be seen swimming to the surface to gulp air. As the gills lose their respiratory function, the carotid arches also change their roles (**Figure 12.20**).

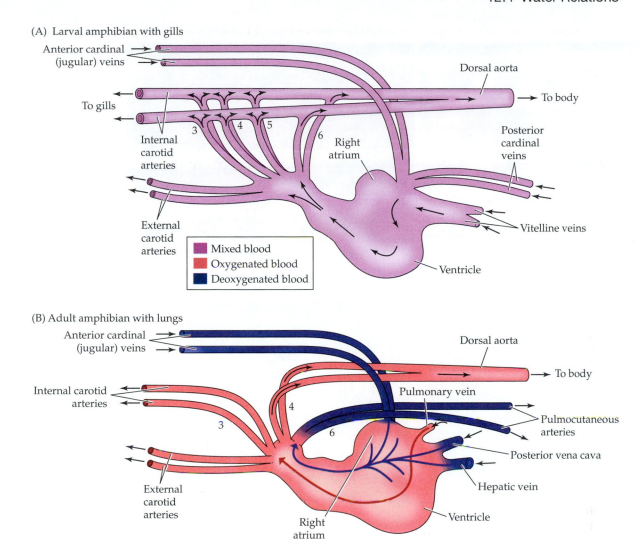

(A) Larval amphibian with gills

(B) Adult amphibian with lungs

Figure 12.20 Changes in the circulatory system at metamorphosis. Blood flow through the aortic arches of a larval lissamphibian (with gills) and an adult (without gills). The head is to the left. Arches 1 and 2 are lost early in embryonic development. The posterior cardinal veins return blood from the posterior body, the vitelline veins carry blood from the intestine, and the hepatic vein brings blood from the liver. (A) In tadpoles, blood is oxygenated in the gills and skin and the heart contains a mixture of oxygenated and deoxygenated blood. Arches 3 through 5 supply blood to the gills and thence to the internal carotid arteries that carry the blood to the head. Arch 6 carries blood to the dorsal aorta.

(B) After metamorphosis, blood is oxygenated in the lungs and skin. The pulmocutaneous veins return oxygenated blood which is distributed to the systemic circulation. Deoxygenated blood returning through the jugular veins, vena cava, and hepatic veins is directed to the lungs and skin via the pulmocutaneous arteries. At metamorphosis, arch 3 becomes the supply vessel for the internal carotid arteries. Initially, arches 4 and 5 supply blood to the dorsal aorta; however, arch 5 is usually lost in anurans, so arch 4 becomes the main route by which blood from the heart enters the aorta. Arch 6 primarily supplies blood to the lungs and skin via the pulmocutaneous arteries.

12.4 Water Relations

LEARNING OBJECTIVES

12.4.1 Describe the role of the skin in the water economy of lissamphibians.

12.4.2 Describe how skin structure and behavioral adjustments allow lissamphibians to reduce cutaneous water loss.

The internal osmotic pressure of lissamphibians is approximately two-thirds that of most other vertebrates—about 200 milliosmoles (mOsm) per kilogram. The primary cause of the dilute body fluids of lissamphibians is low sodium content—approximately 100 millimoles (mM) per kilogram compared with 150 mM/kg in other vertebrates. Lissamphibians can tolerate a doubling of normal sodium concentration, whereas an increase from 150 mM/kg to 170 mM/kg is the maximum humans can tolerate.

Uptake and storage of water

Lissamphibians do not drink water. Because of the permeability of their skins, species that live in aquatic habitats face a continuous osmotic influx of water that they must balance by producing urine. The impressive adaptations of terrestrial lissamphibians are ones that facilitate rehydration from limited sources of water. One such

(A) Pelvic patch of a western toad, *Anaxyrus boreas*, can be pressed against moist substrate to absorb water

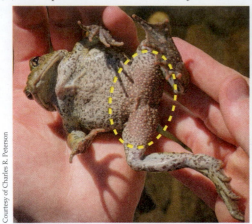

(B) Water beads up on skin of waxy monkey frog, *Phyllomedusa sauvagii*

(C) Chalky white skin of African gray treefrog, *Chiromantis xerampelina*, resists water loss

Figure 12.21 Water uptake and loss in anurans.
(A) A dense network of capillaries in the skin gives the pelvic patch of the western toad a reddish color. A toad pressing its pelvic patch against moist substrate gains water nearly as rapidly as a toad immersed in water. (B,C) A few species of frogs have extremely low rates of cutaneous water loss. (B) The waterproofing mixture of waxes and lipids that this South American waxy monkey frog has spread over its skin has caused water sprayed onto the skin to form beads. (C) The waterproofing mechanism of the African gray treefrog may be related to the iridophores that impart a chalky white color to its skin.

specialization is the **pelvic patch** (**Figure 12.21A**). This area of highly vascularized skin in the pelvic region is responsible for a large portion of an anuran's cutaneous water absorption. Toads that are dehydrated and completely immersed in water rehydrate only slightly faster than those placed in water just deep enough to wet the pelvic area. In arid regions, water may be available only as a thin layer of moisture on a rock or as wet soil, and the pelvic patch allows an anuran to absorb this water.

The urinary bladder plays an important role in the water relations of terrestrial lissamphibians, especially anurans. Amphibian kidneys produce urine that is hyposmolal to the blood, so the urine in the bladder is dilute. Lissamphibians can reabsorb water from urine to replace water they lose by evaporation, and terrestrial lissamphibians have larger bladders than aquatic species. Storage capacities of 20–30% of the body weight of an animal are common for terrestrial anurans, and some species have still larger bladders: the Australian desert frogs *Notaden nichollsi* and *Neobatrachus wilsmorei* can store urine equivalent to about 50% of their body weight, and a bladder volume of 78.9% of body weight has been reported for the Australian moaning frog, *Heleioporus eyrei*.

Cutaneous water loss

Evaporation from the skin of terrestrial and aquatic lissamphibians occurs nearly as rapidly as evaporation from a drop of water. Mucus glands secrete a watery solution of polysaccharides onto the surface of the skin, ensuring that evaporative water loss occurs from the skin surface, rather than from deeper layers.

Many species of arboreal frogs have skin that is less permeable to water than the skin of terrestrial frogs, and a remarkable specialization is seen in a few treefrogs. The South American hylid *Phyllomedusa sauvagii* and the African rhacophorid *Chiromantis xerampelina* lose water through the skin at a rate that is lower than most birds and mammals. Several species of *Phyllomedusa* use their legs to spread lipid-containing secretions of dermal glands over the body surface in a complex sequence of wiping

(A) Water-conserving   (B) Chin up   (C) Low alert   (D) High alert   (E) Calling

Dry nights ⟵————————————————————————⟶ Wet nights

Figure 12.22 Male coquís change their posture to control cutaneous water loss. On dry nights male coquís (*Eleutherodactylus coqui*) adopt a water-conserving posture and on wet nights they vocalize. On nights intermediate between those extremes, the frogs modify their postures to minimize cutaneous water loss while remaining alert to the approach of predators or prey.

movements, forming a waxy coating that conserves water (**Figure 12.21B**). (Similar wiping behavior has been described for an Indian rhacophorid, *Polypedates maculatus*, and an Australian hylid, *Litoria caerulea*.)

The basis for the impermeability of *Chiromantis* is not yet understood but may involve a layer of specialized pigment cells (iridophores) in the skin that contain crystals of guanine. These flat, shiny crystals give the frogs a chalky white appearance during the dry season (**Figure 12.21C**), increasing reflectance and thereby reducing the heat load from solar radiation. The iridophores might form a barrier to water loss, but this function has not been demonstrated.

Chiromantis and *Phyllomedusa* are also unusual because they excrete nitrogenous wastes as precipitated salts of uric acid (as do lizards and birds) rather than as urea. This method of disposing of nitrogen provides still more water conservation for those species.

Behavioral control of cutaneous water loss

For animals with skins as permeable as those of most lissamphibians, cutaneous water loss can limit both the microhabitats they are able to occupy and their periods of activity. For example, red-backed salamanders (*Plethodon cinereus*) forage in the leaf litter on the forest floor and on plant stems. They find more food when they climb plants, but they also lose more water by evaporation, so they climb only on rainy nights.

Some lissamphibians use nuanced behaviors to control evaporative water loss. The Puerto Rican coquí modifies its behavior depending on environmental conditions (**Figure 12.22**). Each male coquí has a calling site on a leaf in the understory vegetation from which it calls to attract a female. Male coquís emerge from their daytime retreats at dusk and move to their calling sites, remaining there until shortly before dawn, when they return to their daytime retreats.

The activities of the frogs vary from night to night, depending on rainfall. On dry nights, males move to their calling sites, but they call only sporadically. Most of the time they rest in a water-conserving posture in which the body and chin are flattened against the leaf surface, the eyes are closed, and the limbs are pressed against the body (Figure 12.22A). A frog in this posture exposes only half its body surface to the air, thereby reducing its rate of evaporative water loss, but because its eyes are closed it cannot see approaching predators or prey.

On nights after an afternoon rainstorm, when the forest is wet, the coquís begin to vocalize soon after dusk. They fall silent for several hours after midnight and resume calling briefly just before dawn. Calling coquís extend their legs and raise themselves off the surface of the leaf (Figure 12.22E). In this position, they lose water by evaporation from the entire body surface. On nights that fall between the extremes of dry and wet, coquís use intermediate postures in which their eyes are open but the amount of exposed body surface is reduced (Figure 12.22B–D).

12.5 Crypsis, Warning Colors, Toxins, and Venoms

LEARNING OBJECTIVES

12.5.1 Describe the characteristics of cryptic and aposematic colors and patterns of lissamphibians.

12.5.2 Distinguish the functions of mucus and granular glands and describe where each type of gland is usually found on the body of a lissamphibian.

12.5.3 Explain the difference between poisonous and venomous animals and give examples of each from Lissamphibia.

Amphibian skin has several important functions in addition to exchanging gases and water with the environment. The colors and patterns of most lissamphibians are cryptic and help to conceal the animals from predators, or fool predators into ignoring them (**Figure 12.23**). Most ground dwelling lissamphibians like the Cameroon toad (Figure 12.23A) have dorsal patterns that are hard to distinguish from dead leaves, and rock-dwelling species such as the Santa Barbara treefrog of Brazil (Figure 12.23B) may blend with lichens. The Taiwan flying frog (Figure 12.23C), like many other arboreal frogs, is green with light-colored linear markings that resemble twigs and stems of plants. The Chinese bird poop frog (Figure 12.23D) avoids predation by looking like something that is of no interest to a predator. Central American glass frogs, such as the Chiriquí glass frog (Figure 12.23E), employ a form of crypsis called edge diffusion. The frogs are nearly transparent when

(A) Cameroon toad, *Sclerophrys camerunensis*

(B) Santa Barbara treefrog, *Bokermannohyla alvarengai*

(C) Taiwan flying frog, *Zhangixalus prasinatus*

(D) Guangxi bird poop frog, *Theloderma albopunctatum*

(E) Chiriquí glass frog, *Hyalinobatrachium petersi*

Circles indicate where the leaf is visible through the tissues of the frog.

Figure 12.23 Crypsis. (A) The colors and pattern of the Cameroon toad match a background of dead leaves. (B) The Santa Barbara treefrog of Brazil rests on lichen covered rocks. (C) Green frogs such as the Taiwan flying frog are hard to detect among leaves and branches. (D) As its name indicates, the Chinese bird poop frog resembles a bird dropping. (E) The transparency of the limbs of glass frogs obscures the outline of the body.

seen from below. Their dorsal surfaces are green, providing camouflage on vegetation, and the transparency of the limbs obscures the outline of the body, making the frog even more difficult for a predator to detect.

Skin glands and toxins

The skin of an amphibian contains mucus glands and granular (poison) glands. Mucus glands secrete mucopolysaccharides and proteoglycans that keep the skin moist, minimize abrasion, and may have antimicrobial properties. Some species of lissamphibians have sticky mucus, while other species have slippery mucus.

Although secretions of the mucus glands of some amphibian species are irritating, an amphibian's primary chemical defense system is its granular glands, which secrete biogenic amines, toxins, alkaloids, and a variety of proteins (**Figure 12.24A**). These glands are concentrated on the dorsal surfaces of the animal (where a predator is most likely to attack), and defense postures of both anurans and salamanders present the glandular areas to potential predators.

A terrestrial caecilian, *Siphonops annulatus*, has a concentration of mucus glands in the head and midbody regions and poison glands in the tail (**Figure 12.24B**). It seems likely that the mucopolysaccharide- and lipid-rich secretions of the glands in the head and midbody provide lubrication as the caecilian burrows through the soil, and the poison glands in the tail deter a predator that attempts to follow the caecilian in its burrow.

Toxicity and diet

A diversity of pharmacologically active substances has been found in amphibian skins. Some of these substances are extremely toxic; others are less toxic but capable of producing unpleasant sensations when a predator bites an amphibian. Biogenic amines such as serotonin and histamine, peptides such as bradykinin, and hemolytic proteins have been found in frogs and salamanders belonging to many families. Many of these substances are named for the animals in which they were discovered—bufotoxin, epibatidine, leptodactyline, and physalaemin are examples.

Cutaneous alkaloids are abundant and diverse among the Neotropical poison frogs, Dendrobatidae (see Table 12.2). Most dendrobatids are brightly colored and move about on the ground in daylight, making no attempt at concealment (**Figure 12.25A**). More than 200 alkaloids have been described from dendrobatids. Most of the alkaloids found in the skins of poison frogs are similar to those found in ants, beetles, and millipedes that live in

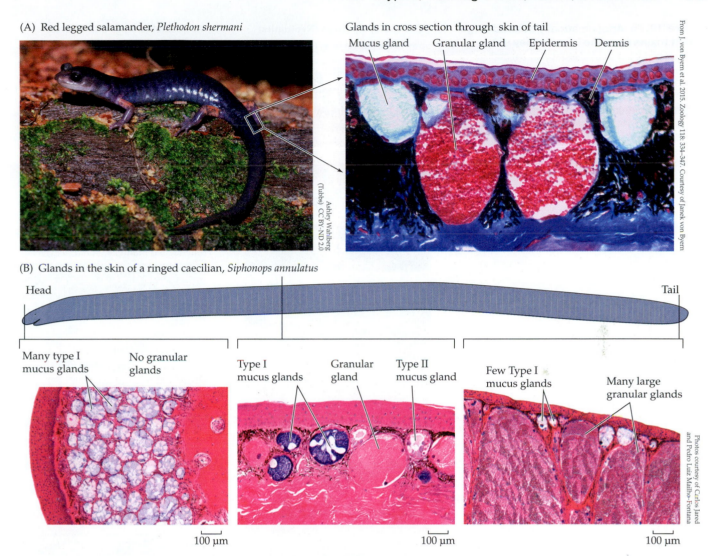

(A) Red legged salamander, *Plethodon shermani*

Glands in cross section through skin of tail

Mucus gland   Granular gland   Epidermis   Dermis

From J. von Byern et al. 2015. Zoology 118: 334–347. Courtesy of Janek von Byern

Ashley Wahlberg (Tubbs) CC BY-ND 2.0

(B) Glands in the skin of a ringed caecilian, *Siphonops annulatus*

Head   Tail

Many type I mucus glands   No granular glands

Type I mucus glands   Granular gland   Type II mucus gland

Few Type I mucus glands   Many large granular glands

100 μm   100 μm   100 μm

Photos courtesy of Carlos Jared and Pedro Luiz Mailho-Fontana

Figure 12.24 Mucus and granular glands. (A) Mucus and granular glands are concentrated on the dorsal surface of the tail of the red-legged salamander. (B) The distribution of mucus and granular glands changes along the length of the body of the ringed caecilian. Type I glands produce lipid-rich mucus that acts as a lubricant. The function of the type II glands has not been determined.

the leaf litter with the frogs, suggesting that frogs obtain alkaloids from their prey.

Mantella frogs endemic to Madagascar provide a striking example of evolutionary convergence with New World dendrobatids (**Figure 12.25B**). Mantellids are small and brightly colored, with colors and patterns that are sometimes very similar to those of dendrobatids. Like dendrobatids, mantellids contain defensive alkaloids obtained from eating ants and millipedes. The ants *Mantella* eat on Madagascar are not closely related to the ants dendrobatids eat in Central and South America, but the alkaloids that protect the ants are similar. Thus, convergent evolution of defensive alkaloids by ants on opposite sides of the world appears to have been a prerequisite for the convergence between the frogs.

Many lissamphibians, including dendrobatid and mantellid frogs, advertise their distasteful properties with conspicuous **aposematic** (warning) colors and behaviors. A predator that makes the mistake of seizing one is likely to spit it out because it is distasteful. The toxins in the skin may also induce vomiting, which reinforces the unpleasant taste for the predator. The predator will remember its unpleasant experience and avoid the distinctly marked animal that produced it.

Some toxic frogs pass toxins on to their tadpoles. The trophic eggs that dendrobatids and mantellids feed to their tadpoles contain alkaloids. By the time the tadpoles metamorphose, they have toxic defenses in place.

Venomous lissamphibians

A toxic animal makes a predator sick after it has been eaten, whereas a venomous animal injects a poisonous substance into predators or prey. A few species of lissamphibians have morphological specializations that qualify them as venomous animals.

About 75 species of West Indian and South American hylids loosely grouped as "casque-headed frogs" are characterized by varying degrees of co-ossification

Figure 12.25 Aposematically colored frogs obtain toxins from the insects they eat. Dendrobatid frogs in the Neotropics (A) and mantellid frogs in Madagascar (B) are protected by toxic alkaloids they ingest from their prey. Frogs in both lineages are aposematically colored, and some of the patterns and colors are convergent.

(A) Dendrobatid frogs (Neotropics)

Strawberry poison frog, *Oophaga pumilio*

Pavel Kirillov/CC BY-SA 2.0

Dyeing poison frog, *Dendrobates tinctorius*

H. Zell/CC BY-SA 3.0

Harlequin poison frog, *Oophaga histrionica*

Courtesy of Frank Steinmann

(B) Mantellid frogs (Madagascar)

Golden mantella, *Mantella aurantiaca*

Frank Vassen/CC BY 2.0

Blue-legged mantella, *Mantella expectata*

Franco Andreone/CC BY-SA 2.5

Harlequin mantella, *Mantella cowanii*

Courtesy of Devin Edmonds

of the skull and skin of the head (**Figure 12.26A**). Many casque-headed frogs retreat into holes and block the openings with their heads, a behavior know as **phragmosis** (Greek *phragma*, "fence"). This behavior reduces cutaneous water loss by about 90% compared to being outside a hole, and also blocks predators from entering. Casque-headed treefrogs produce a wide variety of noxious and toxic skin secretions. Two South American casque-headed frogs, Greening's frog (*Corythomantis greeningi*) and Bruno's casque-headed frog (*Aparasphenodon brunoi*), have bony spines on the skull that pierce the predator's skin, carrying toxins from the granular glands into the wound (**Figure 12.26B,C**).

Three genera of newts, the Asian emperor newts (*Tylototriton*) and crocodile newts (*Echinotriton*) and the European ribbed newts (*Pleurodeles*) are able to inject products of granular glands into the mouth of a predator. Most of these newts have brightly colored warts along the trunk (**Figure 12.26D**). The warts are collections of granular glands, and a sharply pointed rib extends to just beneath the skin in the center of each wart (**Figure 12.26E,F**). When a newt is seized by a predator, the ribs penetrate the body wall, passing through granular glands and carrying toxin into the wounds they create in the predator's mouth.

Caecilians have recently been shown to have glands in the oral mucosa that produce mucus and proteins with ducts that carry these products to the teeth (**Figure 12.26G,H**). These glands, which form from the dental lamina, secrete a viscous fluid that coats the teeth when a caecilian opens its mouth to attack prey. The gland secretions contain proteins found in snake venoms, including enzymes that promote the spread of venom by breaking down the gelatinous material between cells as well as enzymes that attack cell membranes, destroy muscle tissues, and prevent blood from clotting. These glands are present in basal families of the lineage, suggesting that an oral venom system is an ancestral character of the lineage, and the genes coding venom are widely conserved in vertebrates.

(A) Bony maxillary spikes of
   a casque-headed frog

From Jared et al. 2015. Current Biol 25: 2166–2170

(B) Skin covers the spikes of
   Greening's frog, *Corythomantis greeningi*

From Jared et al. 2015. Current Biol 25: 2166–2170

(C) Spikes penetrates skin, punctures granular
   glands, and carries venom into wound

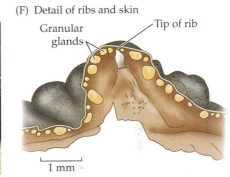

— Granular gland
— Bony spike
— Granular gland

From Jared et al. 2015. Current Biol 25: 2166–2170

(D) Anderson's spiny crocodile newt, *Echinotriton andersoni*

Courtesy of Axel Hernandez

(E) Sharp ribs penetrate skin

Courtesy of Edmund D. Brodie, Jr.

(F) Detail of ribs and skin

Granular glands — — Tip of rib

1 mm

(G) Ringed caecilian, *Siphonops annulatus*

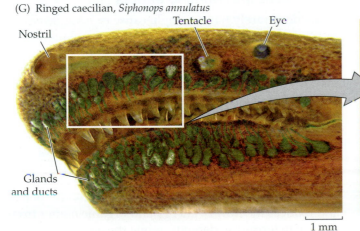

Nostril — Tentacle — Eye

Glands and ducts

1 mm

(H) Detail of area outlined in (G)

1 mm

iScience 23. DOI 10.1016/j.isci.2020.101234/CC BY NC-ND 4.0. G and H from Mailho-Fontana et al. 2020.

Figure 12.26 **Some amphibians are venomous.**
(A) The maxillae of casque-headed frogs such as Greening's frog are studded with bony spikes. Normally the spikes are covered by skin (B), but when a frog is attacked it butts the predator with its head so that the spikes penetrate granular glands (C) and inject venom into the predator. (D) The brightly colored glandular warts along the sides of the body may be an aposematic signal that warns predators to avoid Anderson's crocodile newts. (E, F) When the newt is attacked, the sharp ribs penetrate the body wall, passing through granular glands and carrying toxin into the wounds they create. (G,H) Ducts from glands in the oral mucosa of the ringed caecilian release the products of the glands at the bases of the teeth of the upper and lower jaws.

12.6 Why Are Lissamphibians Vanishing?

LEARNING OBJECTIVES

12.6.1 Describe the plight of lissamphibians worldwide.

12.6.2 Describe the major pathogens that are responsible for continuing declines in populations of lissamphibians.

12.6.3 Describe some ecological and public health consequences of the decline of lissamphibians.

Lissamphibians are vanishing on a worldwide basis. More than 40% of amphibian species are classified as being at some level of risk by the IUCN Red List—a higher proportion than any other major group of vertebrates. The geographic distribution of declining populations is particularly distressing because the highest proportions of species with declining populations are located in Central America, sub-Saharan Africa, and Southeast Asia, which are the global hotspots for amphibian diversity.

In the past 50 years, disease-causing organisms have emerged as a major cause of these extinctions. This phenomenon is recent; most lissamphibian species long ago established stable relationships with local pathogens. But when humans move animals carrying these pathogens to new regions, lethal epidemics spread, as they did when Europeans carried previously unknown pathogens to the New World. One such well-documented epidemic disease is transmitted by ranaviruses, which infect frogs and salamanders and can lead to the extinction of local populations. Fishes are also vulnerable to ranaviruses, and the introduction of sport fishes to new bodies of water has accelerated the spread of ranaviruses among lissamphibians. The most dangerous pathogens currently known to be affecting lissamphibians, however, are chytrid fungi of the genus *Batrachochytrium*.

Chytrid fungi

Batrachochytrium dendrobatidis, referred to as Bd, is responsible for recent disappearances of lissamphibians, especially frogs, in the Americas, Europe, Australia, and New Zealand. The motile reproductive zoospores of the fungus live in water and can penetrate the skin of a frog, causing a disease called chytridiomycosis. When a zoospore enters the skin, it matures into a spherical reproductive body, the zoosporangium, which has branching structures that extend through the skin. These structures interfere with respiration and control of water movement and at high densities kill adult frogs.

Chytridiomycosis has played a large role in the worldwide decline of lissamphibians. The epidemic has been studied especially well in Central and South America. In the late 1980s, two iconic frog species, the golden toad (*Incilius periglenes*) and the harlequin frog (*Atelopus varius*), abruptly vanished from the Monteverde Cloud Forest Reserve in the mountains of northern Costa Rica. The disappearance of those two species was followed by others, and 20 of the 49 frog species that once lived in the reserve have vanished. The reason remained a mystery for several years until Bd was detected and its pathogenic effect was recognized.

The role of chytrid fungi in lissamphibian declines took a horrifying turn in 2013 when a new chytrid, *B. salamandrivorans* ("salamander eater") was identified in a population of fire salamanders (*Salamandra salamandra*) in the Netherlands. Within six months, 90% of the population had died, and in two years it was extinct. Bsal, as the new fungus is known, is substantially more threatening than Bd in several ways:

- Bsal has three times more genes that code for skin-destroying enzymes (metaloproteases) than Bd. These genes are expressed during infection, creating skin ulcers that speed fungal colonization.
- Salamanders do not develop an immune response to Bsal, suggesting that this fungus has an immune-suppressing property that is absent from Bd. Thus, it is unlikely that wild populations of lissamphibians will acquire immunity or that a vaccine can be developed.
- Unlike Bd, Bsal makes two kinds of zoospores, one of which is hardier than the spores of Bd. These spores adhere to the feet of aquatic birds, which can transport them across long distances.
- A single spore of Bsal is sufficient to cause a lethal infection, whereas approximately 10,000 Bd zoosporangia are needed to cause chytridiomycosis.

In addition, Bsal spreads rapidly, even at very low population densities of salamanders, and it is lethal. No treatment or mitigation procedures are effective; rigorous containment is the only method that offers hope of limiting its spread. Genetic analyses indicate that Bsal arose in Asia and was transported to Europe via importation of salamanders for the pet trade. In 2016 the U.S. Fish and Wildlife Service banned importation of more than 200 species of salamanders, and scientists are urging a ban on all importation of amphibians for the pet trade.

Synergisms and domino effects

Extinctions rarely have a single cause or a single effect. Multiple factors simultaneously stress populations and species, and unexpected consequences appear. For lissamphibians, agricultural chemicals, salt used to de-ice highways, and UV-B radiation all have been found to increase the vulnerability of lissamphibians to epidemic diseases.

Because tadpoles and frogs are such different animals (see Section 12.2), the loss of a single anuran species is functionally equivalent to the loss of two species, and decline of amphibian populations has domino effects that extend up and down the food web. The absence of tadpoles alters algae communities in streams and ponds and reduces the flow of energy and organic matter from aquatic to terrestrial habitats, while the absence of adult frogs affects their predators. For example, the Parque Nacional G. D. Omar Torríjos Herrera in Panamá experienced a devastating loss of lissamphibians to Bd in 2004. The abundance of frogs fell by more than 75%, and at least 30 species disappeared. The diets of many species of tropical snakes include frogs and their eggs and tadpoles, and in the years following the Bd epidemic nine species of snakes disappeared from the park, population sizes of the remaining species decreased, and the body condition of individual snakes declined.

Loss of amphibians affects humans as well: tracking the incidence of mosquito-borne diseases in Costa Rica and Panama revealed a doubling in the number of cases of malaria and increases of 23% and 61% in cases of leishmaniasis and dengue in the 8 years following decimation of frog populations by Bd. The cost of malaria prevention measures, such as spraying houses, rose to 3 times pre-Bd levels in Costa Rica and 4 times those levels in Panama.

Summary

12.1 Diversity of Lissamphibians

Lissamphibia is monophyletic, although the three major lineages are morphologically distinct.

- Salamanders (Caudata) have the most generalized body form—elongated, with four limbs in most species.

- Frogs (Anura) have short bodies with nearly inflexible spines. All anurans have four well-developed legs and lack tails as adults.

- Caecilians (Gymnophiona) are elongated, legless, burrowing or aquatic lissamphibians.

The monophyly of lissamphibians is supported by a large number of shared derived characters.

- A moist, permeable skin plays a major role in gas exchange.

- Mucus and granular glands in the skin are important in shaping the ecology and behavior of lissamphibians.

- Pedicillate teeth are shared with a few basal temnospondyls.

- Green rods in the retina are unique to lissamphibians and are sensitive to the violet and blue light that suffuses a forest at night. The red rods found in all vertebrates are sensitive to green wavelengths, and the dual-rod retinas of lissamphibians allow them to detect colors at extremely low light levels.

- The unique papilla amphibiorum in the inner ear is sensitive to low-frequency sounds.

12.2 Life Histories

Aquatic eggs that hatch into free-living aquatic larvae and metamorphose into aquatic or terrestrial adults represent the ancestral reproductive mode of lissamphibians. Extant species have evolved many modifications of that pattern, including terrestrial eggs with or without aquatic larvae, direct development, and viviparity.

Parental care by the male, female, or both parents occurs in many clades. In some species the mother feeds her tadpoles.

All but a few primitive lineages of salamanders have internal fertilization achieved via a spermatophore.

Vocalizations are a prominent part of the mating behavior of most anurans. Males give species-specific advertisement calls that may contain information that females use when selecting mates. Advertisement calls are energetically expensive and may attract predators as well as females.

Nearly all anurans that breed in water have external fertilization; internal fertilization is probably widespread among species that breed on land.

Many anuran species have aquatic larvae (tadpoles), allowing a species to exploit two different resource bases during a lifetime, but loss of the larval stage has occurred in many lineages.

Anuran tadpoles undergo complete metamorphosis in which the tadpole's structures are broken down and the constituents recycled into adult structures. Release of thyroxine, stimulated by thyroid-stimulating hormone, plays a key role in anuran metamorphosis.

Anurans are especially vulnerable to predation in the final stage of complete metamorphosis, when they are unable either to swim or to leap effectively. This stage is usually brief.

Tadpoles are morphologically and ecologically distinct from adult anurans. They filter food particles from a stream of water drawn through the mouth. Many tadpoles use keratinized beaks to rasp algae and bacteria from surfaces, thereby placing suspended particles into the water.

Most caecilians are viviparous and matrotrophic. Males have an intromittent organ, and fertilization is internal. Viviparous species are matrotrophic; embryos of *Typhlonectes* scrape lipid-rich cells from the walls of their mother's oviducts.

The young of three oviparous species of caecilians are known to employ dermatophagy: they tear off and consume pieces of their mother's lipid-rich epidermis.

12.3 Respiration and Circulation

The skin of lissamphibians plays a major role in gas exchange and water uptake and loss.

The spongy interior of the undivided ventricle separates oxygenated and deoxygenated blood streams in the heart while accommodating shifting reliance on cutaneous versus pulmonary gas exchange.

Patterns of blood flow in the circulatory systems of lissamphibians change during metamorphosis as the sites of gas exchange switch from skin and gills to skin and lungs.

12.4 Water Relations

The internal fluid concentrations of lissamphibians are about two-thirds those of most other vertebrates, primarily because of the body fluids' low sodium content.

Lissamphibians do not drink; they absorb water through the skin.

Cutaneous water loss can be a problem even in humid environments, and lissamphibians use sheltered microhabitats and postural adjustments to control evaporation.

(Continued)

Summary *(continued)*

12.5 Crypsis, Warning Colors, Toxins, and Venoms

Granular glands contain toxins that protect lissamphibians. Some of these toxins are synthesized by the amphibian itself, whereas others are derived from toxins in prey.

A few salamanders and frogs are venomous; that is, they can inject the toxic products of their granular glands into a predator.

12.6 Why are Lissamphibians Vanishing?

More than 40% of amphibian species are at risk of extinction.

Virus and fungal diseases have decimated populations of lissamphibians and driven some species to extinction.

Two fungi, *Batrachochytrium dendrobatidis* (Bd) and *B. salamandrivorans* (Bsal), have emerged as major threats to amphibian populations. Bd has been linked to declines of amphibian populations on all continents, most dramatically in North America, Central America, and Australia.

Multiple stresses acting on lissamphibians have a synergistic impact that is greater than the sum of their individual impacts.

Disappearance of anurans causes a domino effect of changes in energy and nutrient flow in aquatic and terrestrial communities, in populations of the anurans predators and prey, and in the incidence of mosquito-borne diseases.

Discussion Questions

12.1 An aquatic egg and larva are ancestral characters of anurans, but terrestrial eggs that hatch into tadpoles or bypass the larval stage entirely have evolved independently in many lineages of anurans. Why didn't all lineages evolve that way? That is, why might it be advantageous for a species of frog to have retained an aquatic egg and a larval stage?

12.2 Four species of salamanders in the genus *Desmognathus* form a streamside salamander guild in the mountains of North Carolina. The four species are found at different distances from the stream, as listed in the table. Suggest one or more hypotheses to explain the relationship between the body sizes of these species and the distance each is found from the water. What prediction can you make for each hypothesis, and how could you test that prediction?

12.3 There are about 10 times as many species of anurans as there are species of salamanders. Can you think of a reason why anurans might be a more successful form of amphibian than salamanders are?

12.4 Describe the ways in which dendrobatid and mantellid frogs have converged.

Species	D. quadramaculatus	D. monticola	D. ochrophaeus	D. anaeus
Adult body length	100–175 mm	83–125 mm	70–100 mm	44–57 mm
Distance from stream	Less than 10 cm	Less than 1 m	1–4 m	4–7 m

13

Synapsids and Sauropsids

Two Ways of Living on Land

Killdeer, *Charadrius vociferus*, defecating in flight

"Amniotes are known by the holes in their heads" is a time-honored phrase in biology that has the merit of being correct. As discussed in Section 9.3, basal tetrapods have a solid bony covering over the roof and sides of their skulls; this is the anapsid condition (Greek *an*, "without"; Greek *aps* "arch"). A solid skull is protective but constrains the size and architecture of jaw muscles, and in the Late Carboniferous, two lineages of amniotes developed fenestrae (Latin *fenestra*, "window")—openings in the skull that allowed jaw muscles to differentiate. The mammalian lineage is characterized by synapsid skulls (Greek *syn*, "together") that have a single fenestra. Members of the reptilian lineages have diapsid skulls (Greek *di*, "two") with two fenestrae (see Figure 9.15). Historically, the classification of amniotes was based largely on these two patterns of skull fenestration. Molecular phylogenetic studies support many core findings based on skull fenestration patterns, but these patterns have changed more than originally recognized.

Extant synapsids—that is to say, all mammals—have synapsid skulls. Extant reptiles (including birds) belong to a group known as Sauropsida that originally had diapsid skulls, although the fenestrae of turtles, snakes, and birds are modified from that original condition. Thus, we can assign extant amniotes to one of two lineages, either Synapsida or Sauropsida. The split between synapsids and sauropsids had occurred by the Late Carboniferous. Mammals are the only extant synapsids (~5,400 species). Extant sauropsids include turtles, lepidosaurs (tuatara, lizards, and snakes), crocodylians, and birds (~22,000 species).

Skull structure is only one way in which the evolutionary histories of Synapsida and Sauropsida diverged as members of both lineages adapted to the challenges of life on land, notably terrestrial locomotion, air-breathing, and conserving water. In this chapter, we describe differences in the ways the lungs and cardiovascular systems of synapsids and sauropsids supply oxygen to the cells and tissues of the body, as well as differences in how their excretory systems conserve water. These differences in structural and functional characters show that there is more than one way to succeed as a terrestrial amniote.

13.1 Conflicts between Locomotion and Respiration

LEARNING OBJECTIVES

13.1.1 Describe differences between lung ventilation with costal movements and with a diaphragm.

13.1.2 Explain conflicts between lung ventilation and locomotion via axial bending and how evolution of the diaphragm eliminated that conflict.

Fast-moving predators that could pursue fleeing prey and fleet-footed prey that could run to escape predators evolved in both the sauropsid and synapsid lineages, as did species capable of powered flight. But running or flying involves much more than just moving limbs rapidly. If an animal is to run very far, or fly, the muscles that move limbs require a steady supply of oxygen. Maintaining oxygen levels that can sustain high levels of activity requires methods of locomotion and respiration that do not conflict with each other. Derived synapsids (mammals) and derived sauropsids (birds) both solved this problem, but in different ways.

Early tetrapods moved with lateral undulations of the trunk, as do most extant salamanders and lizards. Axial muscles power this form of locomotion by bending the trunk from side to side. Limbs and feet are used in alternate pairs (i.e., left front and right rear, right front and left rear) to provide purchase on the substrate as contractions of trunk muscles move the animal (see Figure 11.5).

(A) Lateral axial bending and lung ventilation of a lizard in dorsal view

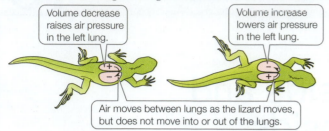

Volume decrease raises air pressure in the left lung.

Volume increase lowers air pressure in the left lung.

Air moves between lungs as the lizard moves, but does not move into or out of the lungs.

(B) Dorsoventral axial bending and lung ventilation of a dog in lateral view

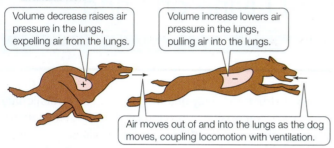

Volume decrease raises air pressure in the lungs, expelling air from the lungs.

Volume increase lowers air pressure in the lungs, pulling air into the lungs.

Air moves out of and into the lungs as the dog moves, coupling locomotion with ventilation.

Figure 13.1 Demands of ventilation and locomotion can conflict. (A) The bending axis of the lizard's thorax is between the two lungs. When the animal moves by bending laterally, the lung on the concave side of the body is compressed and air pressure in that lung increases, while air pressure on the convex side decreases. Air moves between lungs, but little or no air moves into or out of the lungs. (B) The bending axis of a running mammal's thorax is dorsal to the lungs. As the vertebral column bends, volume of the thoracic cavity decreases and pressure in both lungs rises, pushing air out of the lungs. When the vertebral column straightens, volume of the thoracic cavity increases, pressure in the lungs falls and air is pulled into them. (After D.R. Carrier. 1987. *Paleobiology* 13: 326–341.)

The problem with this ancestral mode of locomotion is that axial muscles provide another essential function: basal amniotes ventilated their lungs (that is, took in and expelled air) via **costal ventilation** (Latin *costa*, rib), employing intercostal muscles of the rib cage to create negative pressure within the thorax, drawing air into the

(A) A Permian synapsid, †*Mycterosaurus*

(B) A Triassic synapsid, †*Massetognathus*

Lumbar vertebrae lack ribs

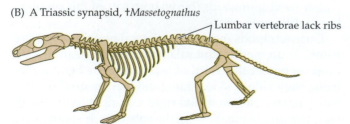

lungs (see Section 11.4). In the ancestral condition, intercostal muscles are asked to (1) bend the trunk unilaterally from side to side for locomotion, and (2) compress the rib cage bilaterally to ventilate the lungs. These two activities cannot happen simultaneously. **Figure 13.1A** illustrates the problem: side-to-side bending of the lizard's rib cage compresses one lung as it expands the other, so air flows from one lung into the other, interfering with airflow in and out via the trachea. In short, the animal cannot run and ventilate its lungs at the same time.

In contrast, dorsoventral flexion of the spine as seen in mammals does not conflict with lung ventilation—quite the contrary, because the inertial forward and backward movements of the viscera alternately compress and expand the lungs as a mammal runs, assisting the intercostal muscles in changing the volume of the lungs (**Figure 13.1B**).

Locomotion based on axial muscles works only for short dashes because any sprint is supported initially by a reservoir of high-energy phosphate compounds present in muscle cells, such as adenosine triphosphate (ATP) and creatine phosphate. When these compounds are used up, muscles switch to anaerobic metabolism, which draws on glycogen stored in the cells and does not require oxygen. When rapid locomotion is sustained beyond a minute or two, supplies of high-energy phosphate compounds and glycogen are used up, and these substances cannot be replenished without a steady supply of oxygen to muscle cells. Because of this conflict between running and breathing, lizards that retain ancestral modes of locomotion and ventilation are limited to short bursts of activity. In more derived sauropsids, including birds, the evolution of lungs with one-way airflow (through-flow) largely solved this problem, as we will explain in Section 13.2.

Synapsids retain the ancestral quadrupedal condition of basal amniotes but with changes to the orientation of the limbs, adopting a more upright posture with limbs held underneath the trunk. Limbs in this position can move more easily fore and aft, and axial bending becomes dorsoventral rather than lateral (see Figure 13.1B), so there is less interference with airflow in the lungs. Early synapsids retained the short limbs, sprawling posture, and long tail that are ancestral characters of amniotes (**Figure 13.2A**). Later synapsids had more upright posture, although not as fully upright as extant mammals (**Figure 13.2B**).

Unlike through-flow ventilation in derived sauropsids, synapsids retain the basal tetrapod condition of in-and-out (bidirectional, or tidal) airflow (see Section 13.2). In

Figure 13.2 Changes in skeletal anatomy of synapsids. (A) Early synapsids retained ribs on all thoracic vertebrae, short legs, and a long tail. (B) Later synapsids lacked ribs on posterior vertebrae and had longer legs and a shorter tail. These derived changes probably coincided with evolution of the diaphragm for respiration, and fore-and-aft movement of the legs during locomotion. (After T.S. Kemp. 2007. *Acta Zool (Stockholm)* 88: 3–22.)

the lineage that led to extant mammals, a second innovation, perhaps evolving in conjunction with changes in skeletal structure and stance, contributed to resolving the conflict between locomotion and respiration. The generalized amniote condition of lung ventilation via costal movement was modified by evolution of a **diaphragm**.

All amniotes share the presence of a **postpulmonary septum**, a sheet of connective tissue in the coelom that separates lungs from digestive organs and forms separate pleural (lungs) and peritoneal (abdominal) cavities (see Figure 2.4C). The mammalian diaphragm represents muscularization of this postpulmonary septum. The position of the diaphragm marks the transition between thoracic and lumbar vertebrae, and the rib cage of mammals defines the extent of the pleural cavity.

The diaphragm is convex anteriorly (i.e., it bulges toward the head) when it is relaxed and flattens when it contracts. Flattening increases the volume of the pulmonary cavity, creating a negative pressure that draws air into the lungs. Simultaneous contraction of the intercostal muscles pulls the ribs forward and outward, expanding the rib cage; you can feel this change when you take a deep breath. Relaxation of the diaphragm permits it to resume its domed shape, and relaxation of the hypaxial muscles allows elastic recoil of the rib cage. These changes raise the pressure in the pulmonary cavity, causing air to be exhaled from the lungs.

Movements of the diaphragm do not conflict with locomotion, and the evolution of the bounding gait of therian mammals (marsupials and placentals; see Figure 11.8) further resolved the conflicting demands of locomotion and respiration. The inertial backward and forward movements of the viscera (especially the liver) with each bounding stride work with the diaphragm to force air into and out of the lungs (see Figure 13.1B). Thus, bounding or galloping mammals inhale and exhale in synchrony with limb movements, and respiration and locomotion work together rather than conflicting. Lung and locomotor synchrony also applies to a hopping kangaroo because the torso is held in a horizontal position. Such synchrony does not occur in humans, however, because we walk erect.

13.2 Lungs and Lung Ventilation: Supplying Oxygen to the Blood

LEARNING OBJECTIVES

13.2.1 Explain the anatomical difference between alveolar and faveolar lungs.

13.2.2 Explain the difference between bidirectional airflow and unidirectional airflow in lung ventilation.

13.2.3 Diagram the passage of air through two respiratory cycles in the lungs of a mammal and a bird.

13.2.4 Identify major stages in the evolution of the avian lung and their independent origins.

The life of an active terrestrial vertebrate requires transferring oxygen molecules (O_2) carried by the circulatory system to cells and tissues. Transfer of O_2 from air to blood is accomplished in the vertebrate lung. Complex lungs evolved in both the synapsid and sauropsid lineages, but the gas-exchange surfaces and networks of air-conducting tubes became organized in very different ways.

Synapsid lungs

Synapsids have compliant lungs and ventilation is tidal and bidirectional. That is, the portion of the lung where gases are exchanged changes shape (i.e., it is compliant), expanding during inhalation and contracting during exhalation; and air flows in and out of the gas-exchange region in a tidal pattern (bidirectional).

The structure of the synapsid respiratory system is an elaboration of the saclike lungs of the earliest tetrapods and of lissamphibians (**Figure 13.3A**). The conducting airways of mammals have a treelike dichotomous pattern of branching (**Figure 13.3B**). Air passes from the trachea through a series of progressively smaller passages, beginning with primary bronchi and extending through many branch points (there are 23 levels of branching in the human respiratory system). The final branches end in cuplike chambers—**alveoli**—that are the sites of gas exchange. This type of lung is called an **alveolar lung**.

Alveoli are tiny (~0.2 mm in diameter, although they vary with body size) and have dense networks of capillaries in their thin walls. Despite their small size, the total surface area of an individual's alveoli is enormous—in humans it is 70 m^2, equal to the floor space of a large room.

Blood in the capillaries of the thin alveolar walls is separated from air in the lumen of the alveolus by 0.6–0.9 µm of tissue. This very short diffusion distance is critically important because a red blood cell passes through an alveolar capillary in less than a second, and in that time it must release CO_2 and take up O_2.

Alveoli expand and contract as lungs are ventilated, and this elastic recoil helps expel air. The alveoli are so tiny that once they are emptied on exhalation, they could not be reinflated were it not for the presence of a surfactant substance secreted by alveolar cells that reduces the surface tension of water (like a film of soap). This substance is far more ancient than the alveolar lung; it has been detected in the lungs of all vertebrates, as well as in gas bladders of bony fishes.

Sauropsid lungs

The lungs of sauropsids are ventilated by unidirectional flow of air over gas-exchange structures called **faveoli** (**Figure 13.3C**). The transition from the compliant alveolar lungs of early amniotes to rigid **faveolar lungs** occurred in a mosaic fashion, with parallel acquisitions and losses of characters along the way.

Evolution of sauropsid lungs   Unidirectional airflow in the lungs is an ancestral character for Reptilia. The ancestral mode of costal lung ventilation was probably retained, and lungs were probably compliant without a

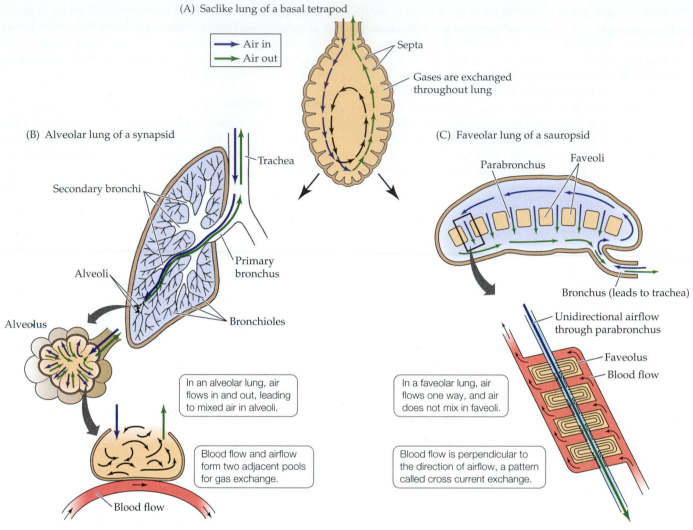

Figure 13.3 Lung evolution in tetrapods. (A) Lissamphibian lungs are simple sacs with peripheral outgrowths (septa) that increase surface area. Airflow is tidal (in-and-out), and the lung contains a mixture of freshly inhaled air (high O_2, low CO_2) and air retained from previous breaths (low O_2, high CO_2). (B) Alveolar lungs of mammals have multiple branching airways (bronchi and bronchioles) with closed-ended alveoli that are the sites of gas exchange. An alveolus contains both fresh and retained air, and blood in capillaries exchanges O_2 and CO_2 with this mixed gas. (C) Faveolar lungs of sauropsids have peripheral air passages connected by thousands of parabronchi with faveoli in their walls. Gas exchange occurs in faveoli, and airflow is unidirectional. Fresh air flows in one direction through the parabronchi and faveoli; blood flows in the opposite direction, crossing the direction of airflow at an angle (i.e., cross-current exchange). (A after F. Overton. 1897. *Applied Physiology*. American Book Co.: New York, Cincinnati, Chicago.)

complex internal structure. Costal ventilation was apparently retained in Lepidosauromorpha, but was replaced by other methods of lung ventilation in Archosauromorpha.

Derived mechanisms of lung ventilation as well as immobilization of the lungs evolved independently in turtles, †pterosaurs, and birds, and reversals and loss of derived characters occurred in some lineages. The mosaic nature of sauropsid lung evolution is described below and diagrammed in **Figure 13.4**.

- *Gastralia* are bones in the ventral abdominal wall (sometimes called ventral ribs) and are a basal amniote character. Gastralia in basal sauropsids were probably immobile. They were firmly embedded in the abdominal muscles, and in early tetrapods may have supported the trunk while a short-legged

animal rested on the ground. Gastralia were retained in most clades, but were lost independently in birds and in squamates (lizards and snakes).

- *Mobile gastralia* were an innovation of archosauria and are retained in extant crocodylians. The ischiotruncus muscle originates on the pelvis and inserts on the gastralia. Its contraction pulls the gastralia caudally, increasing the volume of the abdominal and pulmonary cavities and producing inhalation. Relaxation of the ischiotruncus muscle allows those cavities to contract during exhalation. Mobile gastralia were lost in derived sauropodomorph dinosaurs.

- *Air sacs* in bones anterior and posterior to the lungs evolved independently in †pterosaurs and

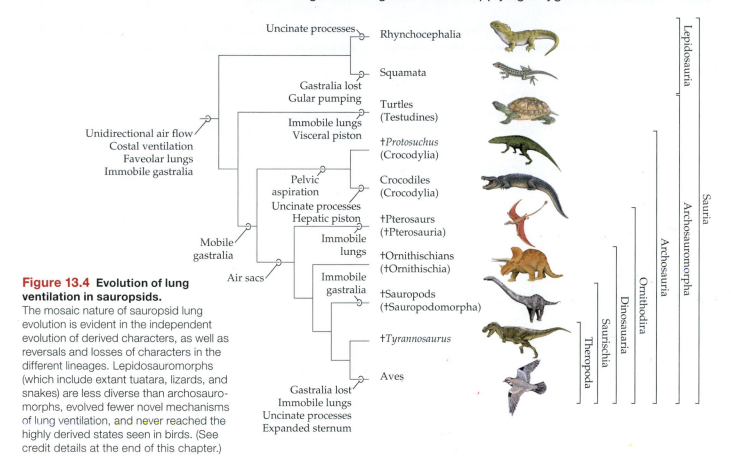

Figure 13.4 Evolution of lung ventilation in sauropsids.
The mosaic nature of sauropsid lung evolution is evident in the independent evolution of derived characters, as well as reversals and losses of characters in the different lineages. Lepidosauromorphs (which include extant tuatara, lizards, and snakes) are less diverse than archosauromorphs, evolved fewer novel mechanisms of lung ventilation, and never reached the highly derived states seen in birds. (See credit details at the end of this chapter.)

saurischian dinosaurs, and are retained in birds, as described below.

- In birds, an *extended sternum* functionally replaced the lost gastralia as a method of ventilating the lungs. The sternum became extended caudally (tailward, or posteriorly), and the caudal end pivots up and down as air is drawn into the posterior air sacs and expelled via the anterior air sacs.

- *Immobile lungs* are characteristic of some derived sauropsids. The lungs of †pterosaurs were firmly anchored to the dorsal ribs and vertebral column, as are the lungs of turtles and birds (see Figure 13.4). Only the ventral portions of these lungs are compliant. The volume of a bird's lung changes by less that 1% during a respiratory cycle.

- *Uncinate processes* are projections on the ribs seen in maniraptoran dinosaurs such as †*Velociraptor* and birds. Muscles that contribute to forceful movements of the ribs and sternum during inhalation and exhalation insert on these processes. Uncinate processes also evolved independently in crocodylians and Rhynchocephalia (tuatara).

In some sauropsid groups, pumping actions facilitate lung ventilation.

- *Gular pump of monitors* Monitor lizards (*Varanus*) ventilate their lungs while they are running by drawing air into the gular region through the nares, then closing the narial valves and forcing the air into the lungs by contraction of gular muscles. This process is like the buccal ventilation of lissamphibians, but it involves expanding and contracting the entire throat rather than only the mouth.

- *Muscular sling of turtles* Encased in rigid shells, turtles cannot move either their ribs or the ventral surface of their bodies to change the volume of the lungs. Instead, they use a sling created by the abdominal muscles. The sling forces the viscera upward and compresses the lungs between the viscera and carapace (dorsal shell) for exhalation; relaxing the muscular sling allows the lungs to expand for inhalation. (see Figure 16.4).

- *Pelvic ventilation of crocodylomorphs* This is a derived character of crocodylomorphs. The pubic bones project anteriorly from the pelvis and are not fused to the ilium and ischium. The ischiopubic muscles originate on the ischium and insert on the pubic bones; contraction of these muscles causes the pubic bones to rotate ventrally and posteriorly, and that movement increases the volume of the abdominal cavity, allowing the pulmonary cavity and lungs to expand.

- *Hepatic piston of crocodylians* Extant crocodylians use movement of the liver to ventilate the lungs. The abdominal muscles contract during exhalation, forcing the liver to move toward the head, thus reducing the volume of the pulmonary cavity. During inhalation, the diaphragmaticus muscle draws the liver posteriorly, increasing the volume of the pulmonary cavity.

Lung ventilation by birds The structure and through-flow ventilation of sauropsid faveolar lungs are perhaps best understood by a more detailed look at the highly derived lungs of birds. The gas-exchange structures of bird lungs are not simple cups, but millions of interconnected small tubules known as **air capillaries** that radiate from the parabronchi (thus the lungs of birds are sometimes called parabronchial lungs; **Figure 13.5A**). Air capillaries intertwine closely with vascular capillaries that carry blood. Air- and blood flow pass in opposite directions in a **cross-current exchange system** (see Figure 13.3C), although they are not exactly parallel because the air and blood capillaries follow winding paths.

The presence of **air sacs** is a second character of the respiratory system of birds. Two groups of air sacs, anterior and posterior, occupy much of the dorsal part of the body and extend into cavities in many of the bones (see Figure 13.5A). The air sacs are large, about nine times the volume of the lungs, poorly vascularized, and do not participate in gas exchange; they are bellows and reservoirs that store air during parts of the respiratory cycle to create a through-flow lung. Two respiratory cycles are required to move a unit of air through the lung (**Figure 13.5B**).

The cross-current flow of air in the air capillaries and blood in the blood capillaries allows efficient exchange of gases, like the countercurrent flow of blood and water in

(A) Positions of air sacs in relation to the lung

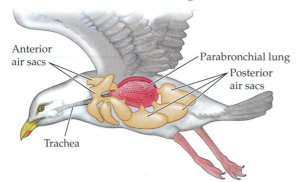

Anterior air sacs

Parabronchial lung

Posterior air sacs

Trachea

(B) The two-breath model for lung ventilation

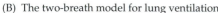

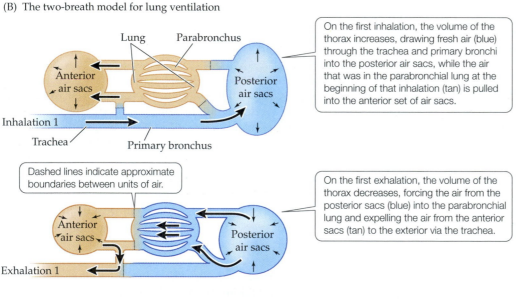

Lung Parabronchus

Anterior air sacs

Posterior air sacs

Inhalation 1

Trachea Primary bronchus

On the first inhalation, the volume of the thorax increases, drawing fresh air (blue) through the trachea and primary bronchi into the posterior air sacs, while the air that was in the parabronchial lung at the beginning of that inhalation (tan) is pulled into the anterior set of air sacs.

Dashed lines indicate approximate boundaries between units of air.

Anterior air sacs

Posterior air sacs

Exhalation 1

On the first exhalation, the volume of the thorax decreases, forcing the air from the posterior sacs (blue) into the parabronchial lung and expelling the air from the anterior sacs (tan) to the exterior via the trachea.

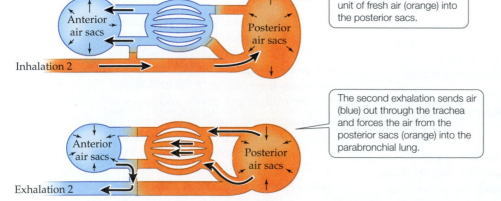

Anterior air sacs

Posterior air sacs

Inhalation 2

The second inhalation draws that unit of air (blue) into the anterior air sacs, and a new unit of fresh air (orange) into the posterior sacs.

The second exhalation sends air (blue) out through the trachea and forces the air from the posterior sacs (orange) into the parabronchial lung.

Anterior air sacs

Posterior air sacs

Exhalation 2

Figure 13.5 Pattern of airflow during inhalation and exhalation by a bird. Note that air flows in only one direction through a parabronchial lung. Two respiratory cycles are required to move a unit of air through the lung. (After D.S. Sadava et al. 2016. *Life: The Science of Biology*, 11th ed. Sinauer Associates & Macmillan Publishing, New York.)

the gills of fishes (see Figure 4.2). In addition, the volumes of the secondary and tertiary bronchi and the air capillaries change very little during ventilation, so the blood vessels are not stretched at each respiratory cycle. As a result, the walls of the air capillaries and blood capillaries of birds are thinner than 0.5 μm, reducing the distance O_2 and CO_2 must diffuse. Rapid diffusion of gas between blood and air is probably one of the mechanisms that allows birds to breathe at very high elevations.

13.3 Circulatory Systems: Supplying Oxygen to Tissues

LEARNING OBJECTIVES

13.3.1 Diagram the path of a red blood cell as it makes a complete transit through the systemic and pulmonary circuits.

13.3.2 Explain how separation between the oxygenated and deoxygenated bloodstreams is maintained in a heart (1) with and (2) without an anatomical septum between the left and right ventricles.

13.3.3 Explain the relationship between blood pressure in the pulmonary and systemic circuits and the presence or absence of a ventricular septum.

Changes in the mechanics of lung ventilation resolved the conflict between locomotion and breathing, and internal divisions of the lungs increased the surface area for gas exchange. These features were essential steps toward occupying adaptive zones that require sustained locomotion, but another element is necessary. To sustain high levels of cellular metabolism, O_2 must be transported rapidly from the lungs to the muscles, and CO_2 must move out of the muscles. The heart and blood vessels of the circulatory system are pipelines through which oxygen is distributed to the tissues. **Figure 13.6** provides schematic diagrams of several synapsid and sauropsid circulatory systems.

Systemic arches of mammals and birds

Developmental studies show that the early embryos of both mammals (synapsids) and birds (sauropsids) have two systemic arches but that one is subsequently lost: mammals retain the left arch as the aorta (see Figure 13.6C) and birds retain the right arch (see Figure 13.6F). However, mammals have no evidence of ever having two systemic arches emerging separately from the heart, as seen in sauropsids, and the left-arched mammalian condition cannot be derived from the sauropsid double-arched condition. Instead, a single vessel emerges (the truncus arteriosus), which then splits into two systemic arches. The carotid arteries of mammals branch from the brachiocephalic artery (a remnant of the right systemic arch). Birds display the opposite condition, in which the carotid arteries branch from the right systemic arch (the aorta) and the left systemic arch has been lost entirely.

The independent reduction to a single arch in both the avian and mammalian lineages suggests that one arch is somehow advantageous in these lineages, although two arches appear to be entirely functional in less derived sauropsids. Perhaps the advantage of a single arch is related to the high blood pressures and high rates of blood flow in the aortic arches of mammals and birds. One vessel with a large diameter creates less friction between flowing blood and the wall of the vessel than would two smaller vessels carrying the same volume of blood. In addition, turbulence may develop where the arches meet, and that would reduce flow.

A powerful heart can produce enough pressure to move blood rapidly through the vessels, but there is a complication: although high blood pressure is needed in the systemic circulation to drive blood from the heart to the limbs and back to the heart, high blood pressure would be bad for the lungs. Lungs are delicate structures; the tissue separating air and blood is extremely thin, and high blood pressure in the lungs would force plasma out of the capillaries into the alveoli or faveoli. If these spaces are partly filled with fluid plasma instead of air—as happens in pneumonia, for example—gas exchange is reduced. Thus, amniotes must maintain different blood pressures in the systemic (body) and pulmonary (lungs) circuits while they are pumping blood at high speed.

Hearts with a ventricular septum: Mammals and birds

Mammals and birds have solved the problem of maintaining different blood pressures in the systemic and pulmonary circuits by separating the ventricle into systemic and pulmonary sides with a permanent septum. This arrangement can be visualized as a figure-eight with the heart in the center, the left loop representing the systemic circuit and the right loop representing the pulmonary circuit (see Figure 13.6).[1] Both derived synapsids and derived sauropsids reached this stage, and they must have done it independently because the relationship of the systemic arches to the left ventricle and the development of the septum are different in the two lineages.

Blood flow in mammalian and avian hearts Blood follows the same paths in the hearts of birds and mammals (**Figure 13.7**). Deoxygenated blood from the posterior part of the body returns to the heart via the inferior vena cava while blood from the head, neck, and upper limbs returns via the paired anterior cardinal veins (which combine near the heart to form the superior vena cava in most placental mammals, incuding humans). This bloodstream enters the right atrium, passes through the right atrioventricular valve into the right ventricle, and then through the pulmonary semilunar valve into the pulmonary artery (which divides into right and left branches leading to the lungs). Oxygenated blood returns from the

---

[1] The systemic side of the ventricle is represented on the left side of Figure 13.6 and the pulmonary side on the right because that is where they appear in conventional depictions of a heart in medical illustrations.

(A) Hypothetical early amniote

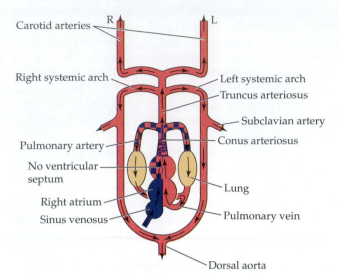

Carotid arteries
R   L

Right systemic arch
Left systemic arch
Truncus arteriosus
Subclavian artery
Pulmonary artery
Conus arteriosus
No ventricular septum
Right atrium
Lung
Sinus venosus
Pulmonary vein
Dorsal aorta

(B) Hypothetical basal synapsid

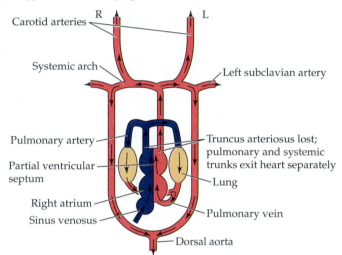

Carotid arteries
R   L

Systemic arch
Left subclavian artery
Pulmonary artery
Truncus arteriosus lost; pulmonary and systemic trunks exit heart separately
Partial ventricular septum
Lung
Right atrium
Sinus venosus
Pulmonary vein
Dorsal aorta

(C) Therian mammal

Carotid arteries
R   L

Right systemic artery partially lost in adults
Left subclavian artery
Left systemic artery becomes the aorta
Pulmonary artery
Complete ventricular septum
Right atrium
Lung
Sinus venosus subsumed into right atrium
Pulmonary vein
Dorsal aorta

	Oxygenated blood
	Deoxygenated blood
	Mixed blood

(D) Generalized sauropsid (e.g., lizard)

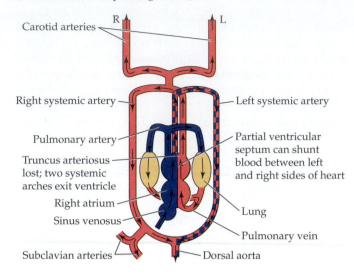

Carotid arteries
R   L

Right systemic artery
Left systemic artery
Pulmonary artery
Partial ventricular septum can shunt blood between left and right sides of heart
Truncus arteriosus lost; two systemic arches exit ventricle
Right atrium
Sinus venosus
Lung
Subclavian arteries
Pulmonary vein
Dorsal aorta

(E) Crocodile

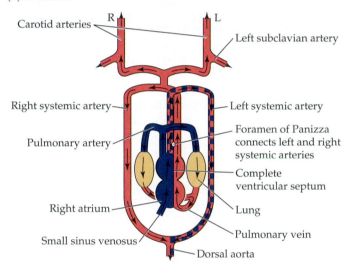

Carotid arteries
R   L

Left subclavian artery
Right systemic artery
Left systemic artery
Pulmonary artery
Foramen of Panizza connects left and right systemic arteries
Complete ventricular septum
Right atrium
Lung
Small sinus venosus
Pulmonary vein
Dorsal aorta

(F) Bird

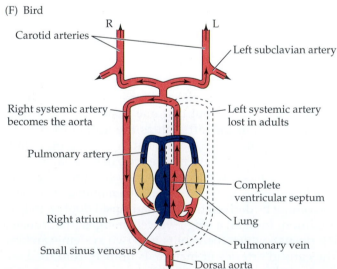

Carotid arteries
R   L

Left subclavian artery
Right systemic artery becomes the aorta
Left systemic artery lost in adults
Pulmonary artery
Complete ventricular septum
Right atrium
Lung
Small sinus venosus
Pulmonary vein
Dorsal aorta

◀ **Figure 13.6** **Amniote hearts and aortic arches.** (A) Hypothetical early amniote condition. A conus arteriosus with a spiral valve and a truncus arteriosus are retained, and a ventricular septum is lacking. This condition, basically like that of lissamphibians, is proposed here to account for differences in these structures between sauropsids and synapsids. (B) Hypothetical early synapsid condition. Mammalian ancestors cannot have had the sauropsid pattern of dual systemic arches. Here a separation of the truncus arteriosus into separate pulmonary and (single) systemic trunks is proposed. Some degree of shunting between pulmonary and systemic circuits and within the heart may have been possible with an incomplete ventricular septum. The sinus venosus is shown as retained because a small sinus venosus is present in monotremes.

(C) Therian mammal. The ventricular septum is complete, and the lower portion of the right systemic arch has been lost. (D) The generalized sauropsid condition is similar to that seen in turtles and lepidosaurs. The truncus arteriosus is divided into a pulmonary arch and two separate systemic arches. The ventricular septum is incomplete, although the ventricle may be complexly subdivided. This system allows blood to be shunted within the heart and between pulmonary and systemic circuits. (E) The ventricular septum of a crocodile is complete, but shunting between left and right systemic arches is still possible via the foramen of Panizza. (F) In birds, the entire left systemic arch is lost, and the sinus venosus is small.

lungs via the right and left pulmonary veins, entering the left atrium and passing through the left atrioventricular valve into the left ventricle from where it exits via the aorta and is distributed via branching arteries.

The ventricular septum (see Figure 13.7) allows mammals and birds to maintain blood pressures in the systemic (left) circuit that are 6–8 times the pressure in the pulmonary (right) circuit (**Figure 13.8**). Thus, an anatomical ventricular septum allows very large pressure differences in the systemic and pulmonary circuits, but it also requires that the volume of blood flowing in the two circuits to be the same.

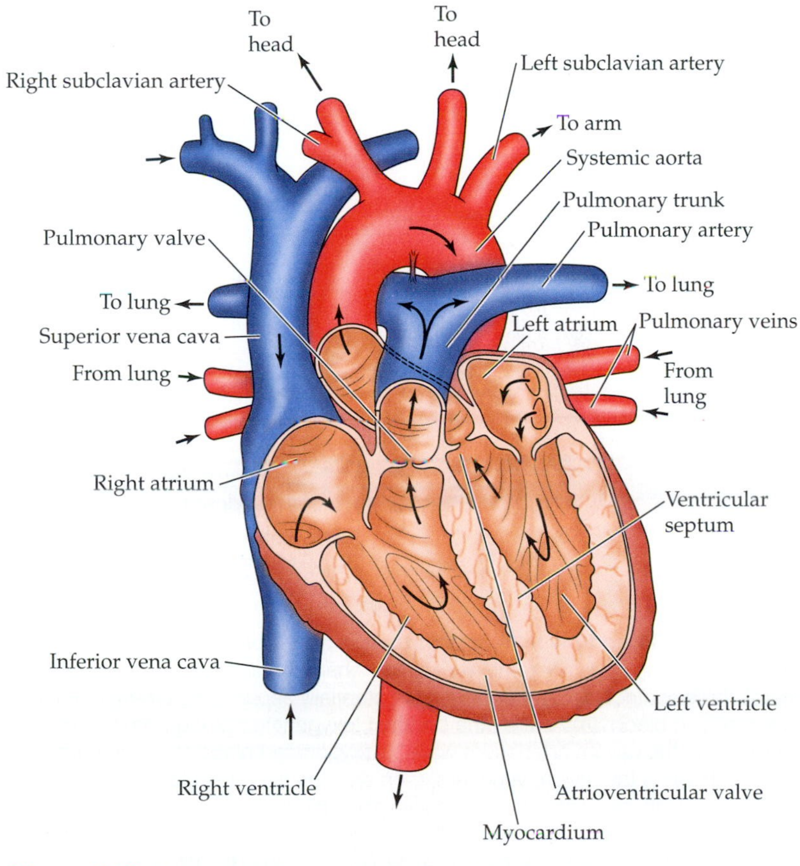

Hearts without a ventricular septum: Turtles and lepidosaurs

The ventricle of a turtle or lepidosaur heart is not divided by a septum, yet oxygenated and deoxygenated blood do not mix. This separation is controlled partly by the relative resistance to flow in the pulmonary and systemic circuits.

The pattern of blood flow can best be explained by considering the morphology of the heart and how intracardiac pressure changes during a heartbeat. **Figure 13.9** is a schematic view of the heart of a turtle. The left and right atria are separate, and a muscular ridge partially divides the ventricle into two spaces, the cavum pulmonale and the cavum venosum. The muscular ridge is not fused to the wall of the ventricle, and thus the cavum pulmonale and the cavum venosum are only partly separated. A third subcompartment of the ventricle, the cavum arteriosum, is located dorsal to the cavum pulmonale and cavum venosum. The cavum arteriosum communicates with the cavum venosum through an intraventricular canal. The pulmonary artery opens from the cavum pulmonale, and the left and right aortic arches open from the cavum venosum.

The right atrium receives deoxygenated blood from the body via the sinus venosus and empties into the cavum venosum, and the left atrium receives oxygenated blood from the lungs and empties into the cavum arteriosum. The atria are separated from the ventricle by flaplike atrioventricular valves that open as the atria contract and then close as the ventricle contracts, preventing blood from being forced back into the atria. The anatomical arrangement of the

Figure 13.7 **Blood flow in a mammalian heart.** Deoxygenated blood from the tissues of the body enters the right atrium and flows through the right atrioventricular valve into the right ventricle. The right ventricle pumps the blood through the pulmonary valve into the pulmonary arteries. Oxygenated blood from the lungs returns via the pulmonary veins to the left atrium and flows through the left atrioventricular valve into the left ventricle. The left ventricle pumps oxygenated blood through the aorta into the systemic circuit. (After R.W. Hill et al. 2016. *Animal Physiology*, 4th ed. Oxford University Press/Sinauer: Sunderland, MA.)

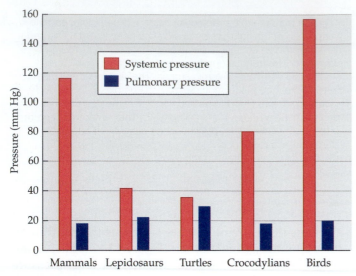

Figure 13.8 Blood pressure in systemic and pulmonary circuits of amniotes. Pressure in the systemic circuit of mammals and birds is 6 or 8 times higher, respectively, than pressure in the pulmonary circuit. Pressures in systemic circuits of lepidosaurs and turtles are much lower than those of mammals and birds, although pulmonary pressures are similar. Crocodylians occupy an intermediate position, with systemic pressure greater than those of turtles and lepidosaurs, but lower than those of mammals and birds.

connections among the atria, their valves, and the three subcompartments of the ventricle is crucial because it is those connections that allow pressure differentials to direct the flow of blood and to prevent mixing of oxygenated and deoxygenated blood. Blood can be shifted from the pulmonary to the systemic circuit (a right-to-left shunt) or in the opposite direction (a left-to-right shunt). The key to these adjustments is changing pressures in the systemic and pulmonary circuits.

The absence of an anatomical septum in the ventricles of turtles and lepidosaurs is not an evolutionary oversight—they were not in the restroom when septa were handed out. Instead, the absence of a septum allows blood to be shifted between the systemic and pulmonary circuits, and the ability to shunt blood is used to control body temperature on a daily and even minute-to-minute basis.

Because they are ectotherms, turtles and lepidosaurs rely on the sun to raise their body temperatures to the levels they maintain during activity (see Chapter 14). Body temperature falls at night and they must bask in the sun to rewarm in the morning. The remarkable attribute of this situation is that most species of lizards and turtles heat more than 40% faster than they cool.

(A) Contraction of atria

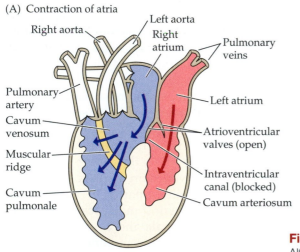

(B) Contraction of ventricle

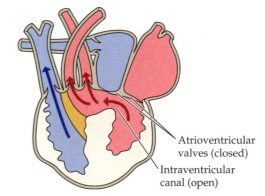

(C) Summary of blood flow through heart

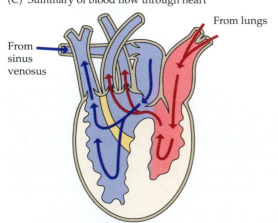

Figure 13.9 Blood flow in hearts of turtles and lepidosaurs. Although the ventricle is anatomically a single chamber, contraction transiently creates three ventricular chambers and substantially separates oxygenated and deoxygenated blood. (A) As the atria contract, oxygenated blood from the left atrium enters the cavum arteriosum while deoxygenated blood from the right atrium first enters the cavum venosum, then crosses the muscular ridge to enter the cavum pulmonale. The atrioventricular valves block the intraventricular canal and prevent mixing of oxygenated and deoxygenated blood. (B) As the ventricle contracts, deoxygenated blood in the cavum pulmonale is expelled through the pulmonary artery and travels to the lungs. The atrioventricular valves close and no longer obstruct the intraventricular canal, forcing oxygenated blood in the cavum arteriosum into the cavum venosum and out through the left and right aortae. Contact between the wall of the ventricle and the muscular ridge prevents mixing of oxygenated and deoxygenated blood. (C) In summary, deoxygenated blood passes from the right atrium through the cavum venosum, across the muscular ridge, into the cavum pulmonale, and out the pulmonary artery into the pulmonary circuit. Oxygenated blood passes from the left atrium into the cavum arteriosum, through the intraventricular canal to the cavum venosum, and out the left and right aortae into the systemic circuit. (After F.H. Pough et al. 2016. *Herpetology*, 4th ed. Oxford University Press/Sinauer, Sunderland, MA.)

These higher heating rates result from two physiological adjustments:

1. Heart rate is higher during heating than during cooling. A high rate of blood flow during warming carries heat from the skin to the core of the body, whereas a low heart rate during cooling delays transport of heat from the core to the skin.

2. The volume of blood flowing to the surface of the body during heating is increased by a pulmonary-to-systemic (right-to-left) intracardiac shunt of blood within the ventricle.

Turtles have a second use for intracardiac shunting: Aquatic turtles cannot breathe when they are diving, and no turtle can breathe when it has withdrawn into its shell for protection from a predator. In those circumstances the lungs cannot oxygenate the blood, and turtles employ a right-to-left (pulmonary-to-systemic) shunt to reduce blood flow to the temporarily nonfunctional lungs.

Shunting blood when the heart has a ventricular septum: Crocodylians

The importance of the intracardiac shunts of turtles and lepidosaurs is demonstrated by crocodylians. Like turtles and lepidosaurs, crocodylians are ectotherms (see Chapter 14). Unlike turtles and lepidosaurs, however, crocodylians have an anatomical septum that divides the ventricle into systemic and pulmonary sides. This septum eliminates the possibility of an *intra*cardiac shunt. Instead, crocodylians have an *extra*cardiac shunt

controlled by differences in blood pressure in the pulmonary and systemic circuits.

Crocodylians increase blood flow through the limbs when they are basking in sunlight to warm themselves. (At least small individuals do—no one has tested this with an adult alligator.) The legs have a large surface area relative to their volume, and blood flowing through the legs rapidly transfers heat to the core of the body. A right-to-left shunt of blood allows crocodylians to increase blood flow through the limbs.

In the crocodylian heart, the right aorta opens from the left ventricle and receives oxygenated blood. The left aorta and the pulmonary artery both exit from the right ventricle (**Figure 13.10**). Blood flow is controlled by the relative resistance to flow in the systemic versus pulmonary circuits, and the pressures change depending on what the animal is doing. Small rounded nodules project into the outflow tract of the pulmonary artery just below the pulmonary valve. The nodules on one side match to nodules on the other side like opposing knuckles and form a valve that is controlled by hormones. Closing the valve increases resistance in the pulmonary circuit and creates a right-to-left shunt of deoxygenated blood into the systemic circuit.

When a crocodylian is active, blood pressure in the left ventricle rises above that of the right ventricle. The left and right aortas of crocodylians are connected via the foramen of Panizza. When pressure in the right aorta exceeds pressure in the left aorta, blood flows through the foramen from the right aorta into the left aorta (see

(A) Typical blood flow through a crocodylian heart

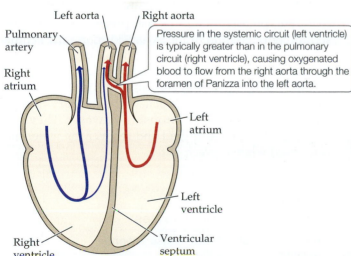

(B) Blood flow through the heart of a basking crocodylian

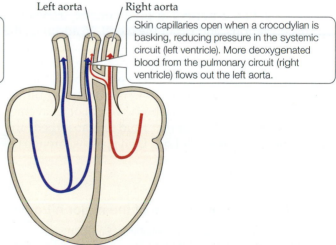

Figure 13.10 Blood flow in a crocodylian heart. A septum completely divides the ventricle of the crocodylian heart into right (pulmonary) and left (systemic) chambers. As in turtles and squamates, the path that blood follows is determined by differences in resistance in the pulmonary and systemic circuits. (A) Pressure in the left ventricle normally is higher than in the right ventricle, and oxygen-rich blood flows from the right aorta into the left aorta through the foramen of Panizza, mixing with some oxygen-poor blood from the right ventricle. (B) When capillary beds in the skin open, resistance (blood pressure) in the systemic circuit falls, allowing more oxygen-poor blood to enter the left aorta while the right aorta continues to receive oxygen-rich blood from the left ventricle. Because the left aorta carries blood to the posterior part of the body and the right aorta carries blood to the brain, the brain always receives oxygen-rich blood. (After F.H. Pough et al. 2016. *Herpetology*, 4th ed. Oxford University Press/Sinauer, Sunderland, MA.)

Figure 13.10A). The increased pressure in the right aorta holds the ventricular valve closed, preventing entry of deoxygenated blood from the right ventricle. Thus, both the right and left aortas receive oxygenated blood during activity.

When a crocodylian is at rest, blood pressure is approximately the same in the right and left ventricles. In this situation deoxygenated blood does flow from the right ventricle into the left aorta and then posteriorly to the viscera (see Figure 13.10B). Note that the *right* aorta supplies blood to the head, so even in this situation the brain receives only oxygenated blood.

The presence of an anatomical ventricular septum and a way to bypass it to shunt blood between the systemic and pulmonary circuits supports the hypothesis that some Mesozoic crocodylomorphs were active predators and may have had high metabolic rates and high blood pressures.

13.4 Getting Rid of Wastes: The Kidneys

LEARNING OBJECTIVES

13.4.1 Identify the primary nitrogenous waste products of amniotes and describe their chemical properties.

13.4.2 Contrast the physiological characters of excretory systems for highly soluble versus highly insoluble nitrogenous wastes.

13.4.3 Describe the processes by which a mammalian kidney adjusts the volume and osmolality of urine.

13.4.4 Describe the locations and significance of extrarenal processing of wastes by sauropsids.

A metanephric kidney drained by a duct called the ureter is a derived character of amniotes. The kidneys are largely responsible for maintaining homeostasis of body fluids, regulating extracellular fluid volume, osmolality, ion concentrations, and pH of the blood. In addition, the kidneys selectively remove water-soluble toxins and produce hormones that control blood pressure. About 25% of the cardiac output of a human passes through the kidneys, which produce about 170 L of glomerular filtrate per day—more than twice the body weight of an average human. More than 99% of that fluid is reabsorbed, and daily urine output is about 1.5 L. Urine passes through the ureters to the cloaca or to a bladder. The synapsid bladder is merely a storage chamber (it has been described as "an organ of social convenience"), but sauropsids continue to process urine after it leaves the kidney.

Nitrogenous waste products

Nitrogenous wastes and ions are excreted primarily in urine, but urine is mostly water, and for a terrestrial animal water is too valuable to waste. Amniotes have three primary nitrogenous waste products (**Table 13.1**).

Ammonia   Metabolism of protein produces ammonia, NH_3. Ammonia is quite toxic, but it is very soluble in water and diffuses rapidly because it is a small molecule. Aquatic non-amniotes (bony fishes and aquatic lissamphibians) excrete a large proportion of their nitrogenous waste as ammonia, and ammonia is a minor nitrogenous waste product of terrestrial amniotes. Human urine and sweat contain small amounts of ammonia.

Urea   Ammonia can be converted to urea, CH_4N_2O, which is less toxic and even more soluble than ammonia. Because it is both soluble and relatively nontoxic, urea can be accumulated within the body and released in a concentrated solution in urine, thereby conserving water. Urea synthesis is an ancestral character of amniotes, and probably of all gnathostomes. Synapsids retained the ancestral pattern of excreting urea, and the mammalian kidney has evolved to be extraordinarily effective in producing concentrated urine.

Uric acid   A complex metabolic pathway converts several nitrogen-containing compounds into uric acid, $C_5H_4O_3N_4$. Unlike ammonia and urea, uric acid is only slightly soluble, and it readily combines with sodium ions (Na^+) and potassium ions (K^+) to precipitate as a salt—sodium or potassium urate. Synapsids normally synthesize and excrete only small quantities of uric acid; excessive synthesis or failure to excrete uric acid can lead to accumulation of uric acid crystals in the joints, a painful condition called gout. Uric acid is the primary nitrogenous waste product of sauropsids, however. They synthesize and excrete uric acid and recover the water that is released when it precipitates.

Table 13.1 Characteristics of the major nitrogenous waste products of vertebrates

Compound	Chemical formula	Molecular weight	Solubility in water[a] (g/L)	Toxicity	Metabolic cost of synthesis	Water conservation efficiency[b]
Ammonia	NH_3	17	520	High	None	1
Urea	CH_4N_2O	60	1,080	Moderate	Low	2
Uric acid	$C_5H_4O_3N_4$	168	0.06	Low	High	4

[a]Solubility changes with temperature; these values refer to normal body temperatures of vertebrates.

[b]Efficiency of water conservation is expressed as number of nitrogen atoms per osmotically active particle. Higher ratios mean more nitrogen is excreted per osmotic unit.

Nitrogen excretion by synapsids: The mammalian kidney

The mammalian kidney is a highly derived organ composed of millions of **nephrons**, the basic unit of kidney structure that is recognizable in nearly all vertebrates (**Figure 13.11**). Each nephron consists of a **glomerulus** that filters the blood and a long tube in which the chemical composition of the filtrate is altered. A portion of this tube, the **loop of Henle**, is a derived character of mammals and is largely responsible for their ability to produce concentrated urine. The mammalian kidney can produce urine more concentrated than that of any non-amniote—and in most cases, more concentrated than that of sauropsids as well (**Table 13.2**).

Urine formation   Urine is concentrated by removing water from the ultrafiltrate that is produced in the glomerulus when water and small molecules are forced out of the capillaries. Because cells are unable to transport water directly, they manipulate the movement of water molecules by transporting ions to create osmotic gradients. In addition, the cells lining the nephron actively reabsorb substances important to the body's economy from the ultrafiltrate and secrete toxic substances into it.

The nephron's activity is a multistep process, described below and in **Figure 13.12**. Each step is localized in a region that has special cell characteristics and distinctive variations in the osmotic environment.

1. An ultrafiltrate is produced at the *glomerulus*. The ultrafiltrate is isosmolal with blood plasma and resembles whole blood after the removal of (a) cellular elements, (b) substances with a molecular weight of 70,000 or higher (primarily proteins), and (c) substances with molecular weights between 15,000 and 70,000, depending on the shapes of the molecules.

2. Actions of the *proximal convoluted tubule* (PCT) decrease the volume of the ultrafiltrate. PCT cells have greatly enlarged surface areas and actively transport Na^+, glucose, and amino acids from the lumen of the tubules to the exterior of the nephron. Chloride ions (Cl^-) and water move passively through the PCT cells in

response to the removal of Na^+. By this process, about two-thirds of the salt is reabsorbed in the PCT and the volume of the ultrafiltrate is reduced by the same amount. Although the urine is still very nearly isosmolal with blood at this stage, the substances contributing to the osmolality of the urine after it has passed through the PCT are at different concentrations than in the blood.

3. The next alteration occurs in the *descending limb of the loop of Henle*. The thin, smooth-surfaced cells of this region freely permit diffusion. Water is lost from the urine, which becomes more concentrated as the descending limb passes through tissues of

(A) Anatomy of a human kidney

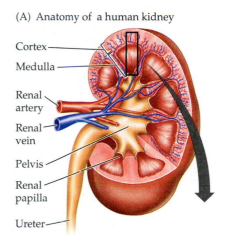

(B) Structure of a nephron

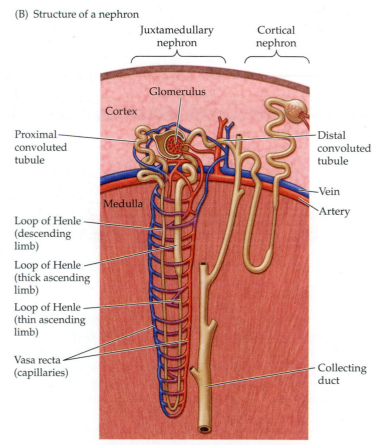

Figure 13.11 Mammalian kidney and nephron. (A) Structural divisions of a human kidney and proximal end of ureter. (B) A mammalian kidney contains millions of nephrons, the functional units of the kidney. General nephron structure, including a glomerulus that filters the blood and tubules in which the blood filtrate is altered into urine, is similar in most vertebrates. The Loop of Henle is a derived character of mammals that allows production of highly concentrated urine. This enlarged diagram is of a section extending from the outer cortical surface to the apex of the renal papilla. (After D.S. Sadava et al. 2016. *Life: The Science of Biology*, 11th ed. Sinauer Associates & Macmillan Publishing, New York.)

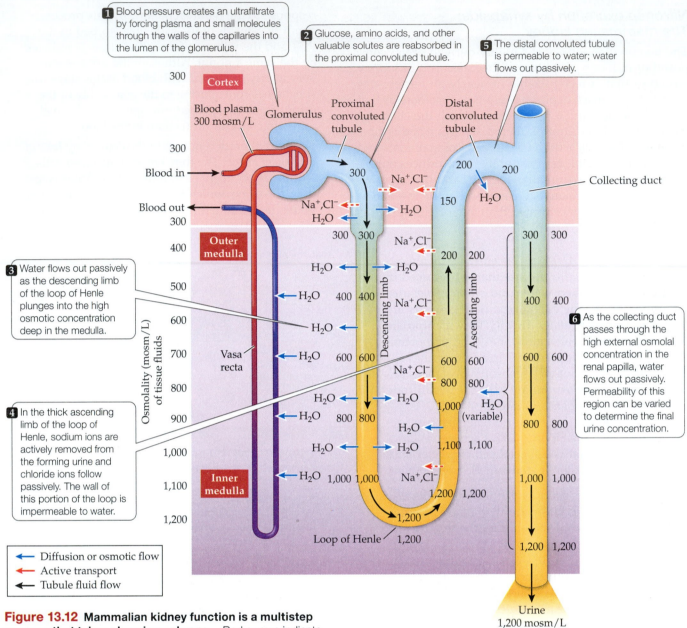

1 Blood pressure creates an ultrafiltrate by forcing plasma and small molecules through the walls of the capillaries into the lumen of the glomerulus.

2 Glucose, amino acids, and other valuable solutes are reabsorbed in the proximal convoluted tubule.

5 The distal convoluted tubule is permeable to water; water flows out passively.

3 Water flows out passively as the descending limb of the loop of Henle plunges into the high osmotic concentration deep in the medulla.

4 In the thick ascending limb of the loop of Henle, sodium ions are actively removed from the forming urine and chloride ions follow passively. The wall of this portion of the loop is impermeable to water.

6 As the collecting duct passes through the high external osmolal concentration in the renal papilla, water flows out passively. Permeability of this region can be varied to determine the final urine concentration.

← Diffusion or osmotic flow
← Active transport
← Tubule fluid flow

Figure 13.12 Mammalian kidney function is a multistep process that takes place in nephrons. Red arrows indicate active transport, blue arrows indicate passive flow. Numbers represent the approximate osmolality of the fluids in indicated regions. (From D.S. Sadava et al. 2016. *Life: The Science of Biology*, 11th ed. Sinauer Associates & Macmillan Publishing, New York.)

increasing osmolality. In humans, the osmolality of the fluid in the descending limb may reach 1,200 millimoles per kilogram (mmol/kg), and other mammals can achieve considerably higher concentrations.

4. The fourth step takes place in the thick *ascending limb of the loop of Henle*, which has cells with numerous large, densely packed mitochondria. ATP produced by mitochondria is used to remove Na+ from the forming urine. Because cells of the ascending limb are impermeable to water, the volume of urine does not decrease as Na+ is removed, and because Na+ was removed without loss of water, the urine is hyposmolal with the body

fluids as it enters the next segment of the nephron. Although this ion-pumping, water-impermeable ascending limb does not concentrate or reduce the volume of the forming urine, it helps set the stage for these important processes.

5. The last portion of the nephron changes in physiological character, but the cells closely resemble those of the ascending loop of Henle. This region, the *terminal portion of the distal convoluted tubule (DCT)*, is permeable to water. Osmolality surrounding the DCT is the same as that of the body fluids, and water in the entering hyposmolal fluid flows outward and equilibrates osmotically. This process reduces the fluid volume to as little as 5% of the original ultrafiltrate.

Table 13.2 Maximum urine concentrations of some synapsids and sauropsids

Species	Maximum observed urine concentration (mmol/kg)	Approximate urine:plasma concentration ratio
Synapsids		
Human (*Homo sapiens*)	1,430	4
Dromedary (*Camelus dromedarius*)	2,800	8
Crest-tailed mulgara (*Dasycercus cristicauda*)	3,231	10
Cat (*Felis domesticus*)	3,250	10
Desert woodrat (*Neotoma lepida*)	4,250	12
Kangaroo rat (*Dipodomys merriami*)	6,382	18
Australian hopping mouse (*Notomys alexis*)	9,370	22
Sauropsids		
Desert tortoise (*Gopherus agassizii*)	337	0.7
American alligator (*Alligator mississippiensis*)	312	0.95
Desert iguana (*Dipsosaurus dorsalis*)	300	0.95
Gopher snake (*Pituophis catenifer*)	342	1
Savannah sparrow (*Passerculus sandwichensis*)	601	1.7
House sparrow (*Passer domesticus*)	826	2.4
House finch (*Carpodacus mexicanus*)	850	2.4

Source: See details at the end of this chapter.

6. The final step in producing a small volume of highly concentrated mammalian urine occurs in the *collecting duct*. Like the descending limb of the loop of Henle, the collecting duct passes through tissues of increasing osmolality that withdraw water from the urine. The significant phenomenon associated with the collecting duct and with the terminal portion of the DCT is variable permeability to water. During excess fluid intake, the collecting duct has low water permeability; only half of the water entering it may be reabsorbed and the remainder excreted. In this way, copious dilute urine can be produced.

Antidiuretic hormone (ADH, also known as **vasopressin)** is released into the circulation in response to an increase in blood osmolality or when blood volume drops. ADH is a polypeptide that is produced in the hypothalamus and stored in the posterior pituitary. When ADH is carried by the circulation to the cells of the terminal portion of the distal convoluted tubule and the collecting duct, it causes them to insert proteins called aquaporins into the plasma membranes of the cells lining the collecting duct.

Aquaporins are tubular proteins with nonpolar (hydrophobic) amino acid residues on the outside and polar (hydrophilic) residues on the inside. When aquaporins are inserted into the plasma membranes of cells, the hydrophobic residues on the aquaporins' outer surfaces anchor them in the phospholipid bilayer of the cell membrane. The hydrophilic residues on the inner surface of the aquaporin then form a channel through which water molecules can flow. These channels make the cells of the terminal portion of the DCT and the collecting duct permeable to water, which flows outward following its osmotic gradient (**Figure 13.13**).

Water that is reabsorbed from the DCT and collecting duct enters the blood, reducing the blood's osmolality and increasing its volume. As water is removed, the volume of the urine in the collecting duct decreases and its concentration increases. The final urine volume leaving the collecting duct may be less than 1% of the original ultrafiltrate volume. In some desert rodents, so little water remains that the urine crystallizes almost immediately upon urination. Alcohol inhibits the release of ADH, inducing a copious urine flow (which can result in dehydrated misery the morning after a drinking binge).

Up to this point the walls of the nephron have been impermeable to urea. As a result, the concentration of urea in the fluid that enters the collecting duct is very high. The walls of the terminal portion of the collecting duct are very permeable to urea, and urea diffuses outward from the collecting duct into the extracellular fluid deep in the kidney. This urea contributes to the high osmolal concentration of the tissues at the lower end of the loop of Henle.

Structure of the nephron   The structural characteristics of the cells that form the walls of the nephron are directly related to the processes that take place in that portion of the nephron. The cells of the PCT contain many mitochondria and have an enormous surface area produced by long, closely spaced microvilli. These structural features reflect the function of the PCT

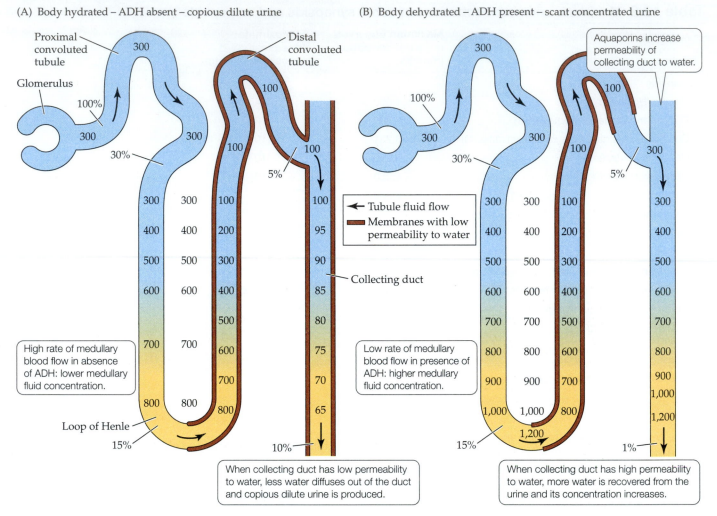

(A) Body hydrated – ADH absent – copious dilute urine

(B) Body dehydrated – ADH present – scant concentrated urine

Proximal convoluted tubule

Distal convoluted tubule

Glomerulus

Aquaporins increase permeability of collecting duct to water.

High rate of medullary blood flow in absence of ADH: lower medullary fluid concentration.

Loop of Henle

← Tubule fluid flow
▬ Membranes with low permeability to water

Collecting duct

Low rate of medullary blood flow in presence of ADH: higher medullary fluid concentration.

When collecting duct has low permeability to water, less water diffuses out of the duct and copious dilute urine is produced.

When collecting duct has high permeability to water, more water is recovered from the urine and its concentration increases.

Figure 13.13 Urine is concentrated in collecting ducts and the loop of Henle. Mammalian kidneys produce dilute urine when the body is hydrated and concentrated urine when the body is dehydrated. (A) When blood osmolality drops below the normal concentration (~300 mmol/kg of body weight), levels of antidiuretic hormone (ADH, or vasopressin) drop and excess body water is excreted. (B) When osmolality rises, ADH levels rise and water is conserved. Numbers show the approximate osmolality of the fluids in the region. Percentages are the volumes of the forming urine relative to the volume of the initial ultrafiltrate. (After R.W. Hill et al. 2016. *Animal Physiology*, 4th ed. Oxford University Press/Sinauer, Sunderland, MA.)

in actively moving Na$^+$ from the lumen of the tubule to the peritubular space and capillaries. Chloride ions (Cl$^-$) follow the electric charge gradient created by the movement of Na$^+$, and water moves in response to the concentration gradient produced by the movement of the two ions.

Farther along the nephron, the cells of the thin segment of the loop of Henle are waferlike and contain fewer mitochondria. The descending limb of the loop of Henle permits passive flow of water. Cells in the thick ascending limb are similar to those in the PCT and actively remove Na$^+$ from the ultrafiltrate.

Finally, cells of the collecting duct appear to be of two kinds. Most seem to be suited to the relatively impermeable state characteristic of periods of sufficient body water. Other cells are rich in mitochondria and have a larger surface area. These are probably the cells that respond to the presence of ADH from the pituitary gland,

triggered by insufficient body fluid. Under the influence of ADH, the collecting duct becomes permeable to water, which flows from the lumen of the duct into the concentrated peritubular fluids.

The loop of Henle The key to producing concentrated urine is the passage of the loop of Henle and collecting ducts through tissues with increasing osmolality. These osmotic gradients are formed and maintained within the mammalian kidney as a result of its structure.

The arrangement within the kidney medulla of the descending and ascending segments of the loop of Henle and its blood supply, the vasa recta, is especially important. These elements create a series of parallel tubes with flow passing in opposite directions in adjacent vessels (countercurrent flow). As a result, Na$^+$ pumped from the ascending limb of the loop of Henle diffuse into the medullary tissues to increase their osmolality, and this

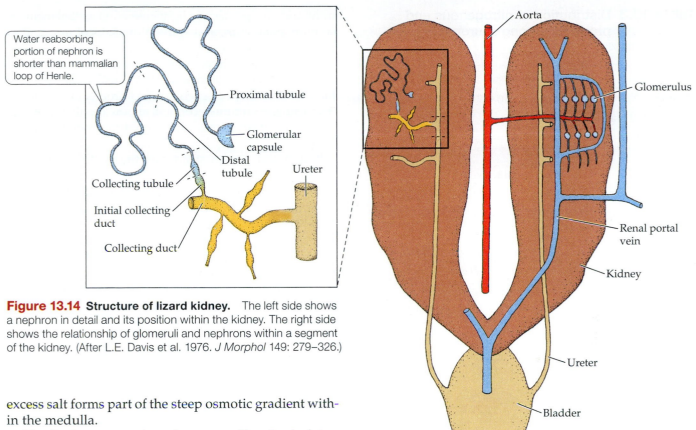

Water reabsorbing portion of nephron is shorter than mammalian loop of Henle.

Proximal tubule

Glomerular capsule

Distal tubule

Ureter

Collecting tubule

Initial collecting duct

Collecting duct

Aorta

Glomerulus

Renal portal vein

Kidney

Ureter

Bladder

Figure 13.14 Structure of lizard kidney. The left side shows a nephron in detail and its position within the kidney. The right side shows the relationship of glomeruli and nephrons within a segment of the kidney. (After L.E. Davis et al. 1976. *J Morphol* 149: 279–326.)

excess salt forms part of the steep osmotic gradient within the medulla.

The final concentration of a mammal's urine is determined by the concentrations of Na^+ and urea accumulated in the fluids of the medulla. Physiological alterations in concentration in the medulla result primarily from the effect of ADH on the rate of blood flushing the medulla. When ADH is present, blood flow into the medulla is retarded and salts accumulate to create a steep osmotic gradient. Another hormone—aldosterone, from the adrenal gland—increases the rate of Na^+ reabsorption into the medulla to promote an increase in medullary salt concentration.

In addition to having these physiological means of concentrating urine, a variety of mammals have morphological alterations of the medulla. Most mammals have two types of nephrons: those with a cortical glomerulus and abbreviated loops of Henle that do not penetrate far into the medulla, and those with juxtamedullary glomeruli, deep within the cortex, with loops that penetrate as far as the papilla (see Figure 13.11B). Obviously, the longer, deeper loops of Henle experience larger osmotic gradients along their lengths. The flow of blood to these two populations of nephrons seems to be independently controlled. Glomeruli in the juxtamedullary nephron are more active in regulating water excretion; cortical glomeruli function in ion regulation.

Nitrogen excretion by sauropsids: Renal and extrarenal routes

All extant representatives of the sauropsid lineage are uricotelic—that is, they can excrete nitrogenous wastes in the form of uric acid. They are not limited to **uricotely,**

however, and many sauropsids also excrete nitrogenous wastes as ammonia or urea (**Table 13.3**).

The strategy for water conservation when uric acid is the primary nitrogenous waste is entirely different from that required when urea is produced. Because urea is soluble, concentrating urine in the kidney can conserve water, but concentrated uric acid would precipitate and block the nephrons. The kidneys of lepidosaurs are elongate and lack the long loops of Henle that allow mammals to reduce the volume of urine and increase its concentration (**Figure 13.14**). The osmolal concentration of urine from the kidneys of lepidosaurs is the same as blood plasma, or even slightly lower. Bird kidneys have two types of nephrons: short-loop nephrons like those of lepidosaurs, and long-loop nephrons that extend downward into the medulla. Long-loop nephrons allow birds to produce urine that is 2–3 times more concentrated than the plasma. These ratios are lower than those of mammals, and even the highest urine-to-plasma ratio recorded for a bird is relatively low compared with mammalian ratios (see Table 13.2).

Precipitation of urate salts If sauropsids depended solely on the urine-concentrating ability of their kidneys, they would excrete all their body water as urine. This is where the low solubility of uric acid becomes advantageous: uric acid precipitates when it enters the cloaca or bladder. The dissolved uric acid combines with ions in the

Table 13.3 Distribution of nitrogenous end products among sauropsids

Group	Total urinary nitrogen (%)		
	Ammonia	Urea	Salts of uric acid
Tuatara	3–4	10–28	65–80
Lizards and snakes	Small	0–8	90–98
Crocodylians	25	0–5	70–75
Turtles			
Aquatic	4–44	45–95	1–24
Desert	3–8	15–50	20–50
Birds			
Omnivores	11–29	1–11	45–72
Carnivores	9–16	1	76–87

Source: After E. H. Larsen. 2014. *Comp Physiol* 4: 405–573.

urine and precipitates as a light-colored mass that includes sodium, potassium, and ammonium salts of uric acid as well as other ions held by complex physical forces. When uric acid and ions precipitate from solution, the urine becomes less concentrated and water is reabsorbed into the blood. In this respect, excretion of nitrogenous wastes as uric acid is even more economical of water than is excretion of urea because the water used to produce urine is reabsorbed and reused, leaving a mixture of whitish urate salts and dark fecal material (see the photograph on the opening page of this chapter).

Water is not the only substance that is reabsorbed from the cloaca, however. Many sauropsids also reabsorb Na$^+$ and K$^+$ and return them to the bloodstream. At first glance, that seems a remarkably inefficient thing to do. After all, it took energy to create the blood pressure that forced ions through the walls of the glomerulus into the urine in the first place, and now more energy is used in the cloaca or bladder to operate the active transport system that returns the ions to the blood. The animal has used two energy-consuming processes, and it is back where it started—with an excess of Na$^+$ and K$^+$ in the blood. Why do that?

Extrarenal salt excretion The solution to this paradox lies in a water-conserving mechanism that is present in many sauropsids: **salt-secreting glands** that provide an extrarenal (i.e., in addition to and separate from the kidney) pathway that disposes of salt with less water than urine. At least four different glands have evolved a salt-secreting function in different sauropsid lineages, and the function evolved independently, and multiple times (**Figure 13.15**). Homologous but non-salt-secreting glands are present in the sister taxa of all lineages with salt glands. Apparently, salt glands evolved by repeated co-option of existing unspecialized glands. Thus, only the *function* of salt glands is novel, not the glands themselves.

Salt glands are characteristic of many species that live in salty habitats (the sea or estuaries) or xeric habitats (deserts), as well as species that eat dry food (such as seeds) or animal flesh, but they are not limited to species in those categories.

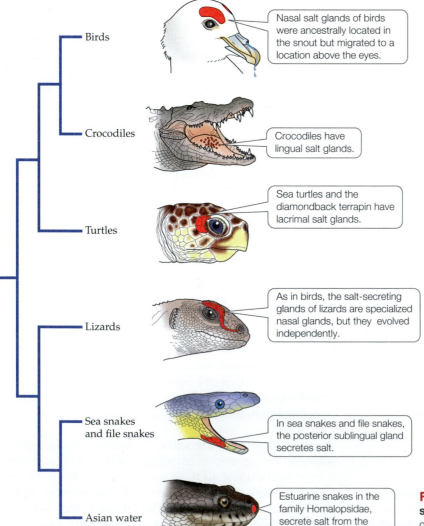

Birds — Nasal salt glands of birds were ancestrally located in the snout but migrated to a location above the eyes.

Crocodiles — Crocodiles have lingual salt glands.

Turtles — Sea turtles and the diamondback terrapin have lacrimal salt glands.

Lizards — As in birds, the salt-secreting glands of lizards are specialized nasal glands, but they evolved independently.

Sea snakes and file snakes — In sea snakes and file snakes, the posterior sublingual gland secretes salt.

Asian water snakes — Estuarine snakes in the family Homalopsidae, secrete salt from the premaxillary gland.

Figure 13.15 **Salt-secreting glands of sauropsids.** Phylogenetic distribution of salt glands reveals that at least four different glands independently evolved salt-secreting properties.

- Salt glands are widespread among marine and shore birds (including gulls, sandpipers, plovers, pelicans, albatrosses, penguins), carnivorous birds (hawks, eagles, vultures, roadrunners), and upland game birds (pheasants, quail, grouse). More surprisingly, salt glands have been identified in freshwater birds (ducks, loons, grebes), three species of doves, and three species of hummingbirds. These are nasal glands that were ancestrally located on the snout, but they have migrated backward and are located above the eyes. Secretions empty into the nasal passages and are expelled with a sneeze or head shake.

- Crocodiles, which have lingual salt glands, are more marine than alligators and gharials, which lack salt-secreting lingual glands.

- Only sea turtles and the diamondback terrapin (*Malaclemys terrapin*, an estuarine species) have salt glands. The lacrimal glands of sea turtles are greatly enlarged (in some species, each gland is larger than the turtle's brain) and secrete a salty fluid. Photographs of nesting sea turtles frequently show clear streaks left by tears flowing through sand adhered to a turtle's head. Those tears are actually salt-gland secretions. Terrestrial turtles, even those that live in deserts, do not have salt glands.

- Salt glands have evolved or been lost many times among lizards; they have been described in 8 families of lizards; 5 of those families also include species without salt glands. The salt-secreting ability of lizards' nasal glands evolved independently of the parallel specialization of the nasal glands of birds,

and the scattered distribution of salt secretion among species within lizard families indicates multiple independent origins of the trait.

- Salt-secreting glands of snakes differ from those of lizards, and two different glands evolved salt-secreting properties. In sea snakes and the marine file snake (*Acrochordus javanicus*), the posterior sublingual gland secretes salt, whereas in estuarine snakes in the Asian family Homalopsidae, the premaxillary gland secretes salt.

Despite their different origins and locations, the structural and functional properties of salt glands are similar. They secrete fluid containing primarily sodium or potassium cations and chloride or bicarbonate (HCO_3^-) anions in high concentrations (**Table 13.4**). Na^+ and Cl^- are the most abundant ions in the salt-gland secretions of marine vertebrates. Insects and plants contain high levels of K^+ and HCO_3^-, and these ions are present in the secretions of herbivorous species.

The total osmolal concentration of the salt-gland secretion may reach 2,000 mmol/kg, more than 6 times the concentration of urine that can be produced by the kidney. This efficiency of excretion is the explanation for the paradox of active uptake of salt from the urine. As ions are actively reabsorbed, water follows passively, so an animal recovers both water and ions from the urine. Ions can then be excreted via the salt gland at much higher concentrations, with a corresponding reduction in the amount of water needed to dispose of the salt. Thus, by investing energy to recover ions from urine, sauropsids with salt glands can conserve water by excreting ions through the more efficient extrarenal route.

Table 13.4 Salt-gland secretions of sauropsids

Species and test conditions	Ion concentration in secretion (mmol/kg)		
	Na⁺	K⁺	Cl⁻
Turtles			
Loggerhead sea turtle (*Caretta caretta*), seawater	732–878	18–31	810–992
Diamondback terrapin (*Malaclemys terrapin*), seawater	322–908	26–40	No data
Lizards			
Desert iguana (*Dipsosaurus dorsalis*), estimated field conditions	180	1,700	1,000
Fringe-toed lizard (*Uma scoparia*), estimated field conditions	639	734	465
Snakes			
Sea snake (*Pelamis platurus*), salt-loaded[a]	620	28	635
Estuarine snake (*Cerberus rynchops*), salt-loaded	414	56	No data
Crocodylian			
Saltwater crocodile (*Crocodylus porosus*), natural diet	663	21	632
Birds			
Black-footed albatross (*Phoebastria nigripes*), salt-loaded[a]	800–900	No data	No data
Herring gull (*Larus argentatus*), salt-loaded	718	24	No data

Source: See details at the end of this chapter.
[a]Extra salt was administered before the test to stimulate the glands.

Summary

13.1 Conflicts Between Locomotion and Respiration

The ancestral mode of terrestrial locomotion that relied on side-to-side bending of the trunk created a conflict between locomotion and lung ventilation. Changes in limb posture and locomotion resolved this conflict independently in the synapsids and sauropsids.

Progressive reduction in the ribs in the posterior region of the trunk of synapsids indicates that they developed a diaphragm that ventilated the lungs with a tidal airflow rather than by costal movements.

13.2 Lungs and Lung Ventilation: Supplying Oxygen to the Blood

Synapsids and sauropsids evolved lungs that rapidly exchange O_2 and CO_2, but they did so in different ways.

Synapsids retained the ancestral saclike lung with tidal inhalation and exhalation. Walls of the gas-exchange surfaces of the lungs (alveoli) are thin, facilitating rapid gas exchange between air and blood.

The lungs of sauropsids are ventilated by unidirectional flow of air over gas-exchange structures called faveoli. At least some portions of sauropsid lungs are immobile and have thinner layers of tissue between air and blood than lungs that must resist stretching and relaxation. Immobilization of the dorsal portion of the lungs evolved independently in different lineages of sauropsids.

Sauropsids, extant and extinct, employed varied mechanisms of lung ventilation, including costal ventilation, a muscular sling, and a hepatic piston.

Unidirectional lung ventilation of immobile lungs is most fully developed in birds, whose immobile lungs do not have to stretch and relax with each breath.

13.3 Circulatory Systems: Supplying Oxygen to Tissues

A heart is a heart if you're a synapsid, but the hearts of sauropsids differ.

- Derived synapsids (mammals) and derived sauropsids (birds) have a ventricular septum that separates the pulmonary and systemic circuits of the circulatory system. This arrangement allows blood pressure to be lower in the pulmonary circuit than in the systemic circuit, but it requires flow to be the same in both circuits.

- Turtles and lepidosaurs lack a permanent ventricular septum. This cardiac anatomy has the advantage of allowing blood to be shifted between circuits.

- Crocodylians have a ventricular septum, but they can shunt blood through the foramen of Panizza to adjust the volume of blood flowing through the pulmonary and systemic circuits.

13.4 Getting Rid of Wastes: The Kidneys

The basis of vertebrate kidney function is the nephron. Each nephron consists of a glomerulus that filters the blood and a long tube in which the chemical composition of the filtrate is altered. A metanephric kidney drained by a ureter is a derived character of amniotes.

Sauropsids and synapsids evolved different methods of eliminating metabolic wastes while minimizing urinary water loss. Nitrogenous wastes are excreted in urine, but because water is a limited resource for terrestrial tetrapods, the excretory system must minimize water loss.

- Synapsids produce a soluble nitrogenous waste product (urea). The nephrons of the synapsid kidney concentrate urea and ions, reabsorb water, and produce a concentrated urine that is stored in a bladder and then excreted.

- The loop of Henle (a portion of the nephron) is a derived character of mammals and is largely responsible for their ability to produce concentrated urine.

- Sauropsids excrete nitrogen in an insoluble form (uric acid). The sauropsid kidney has little capacity to concentrate urine, and uric acid and ions precipitate as urate salts, freeing water that returns to the blood.

Many sauropsids have extrarenal salt-concentrating glands that allow them to excrete salt at very high concentrations.

Discussion Questions

13.1 How does the ancestral locomotor system of tetrapods interfere with respiration?

13.2 Birds regularly reach elevations higher than human mountain climbers can ascend without using auxiliary oxygen. For example, radar tracking of migrating birds shows that they sometimes fly as high as 6,500 m. The alpine chough (*Pyrrhocorax graculus*) lives at elevations of about 8,200 m on Mount Everest and migrating bar-headed geese (*Anser indicus*) pass directly over the summit of the Himalayas at elevations of 9,200 m. Explain the features of the avian respiratory system that allow birds to breathe at such high elevations.

13.3 How would you explain to your roommate that the heart of a turtle is not inferior to the heart of a mammal?

13.4 By the time you feel thirsty, you are already somewhat dehydrated. What is happening to the ADH in your pituitary gland at that point? How does that response contribute to regulating the water content of your body?

13.5 Why is reabsorption of water from the bladder or cloaca essential for sauropsids?

Figure credit
Figure 13.4: Modifed from R. J. Brocklehurst et al. 2020. *Phil Trans R Soc B* 375: 20190140.

Table sources
Table 13.2: G.A. Bartholomew and J.W. Hudson. 1959. *J Mammal* 40: 354–360; T.W. Burns. 1956. *Endocrinology* 58: 243–254; Y. Charnot. 1960. Travaux de l'Institute Scientifique Cérifien, *Série zoologique,* No. 20 Rabat; R.A. Coulson et al. 1950. *Proc Soc Exper Biol Med* 73: 203–206; W.H. Dantzler and B. Schmidt-Nielsen. 1966. *Am J Physiol* 210: 198–210; D.L. Goldstein et al. 1990. *Physiol Zool* 63: 669–682; J.W. Hudson. 1962. *U Calif Pubi Zool*64: 1–56; S. Komadina and S. Solomon. 1970. *Comp Biochem Physiol* 32: 333–343; D.J. La. 1985. *Comp Biochem Physiol Part A: Mol Integr Physiol* 81A: 217–223; E.H. Larsen et al. 2014. *Compr Physiol* 4: 405–573; H. B. Lewis. 1918. *Science* 48: 376; R.E. MacMillen and A.K. Lee. 1969. *Comp Biochem Physiol* 29: 493–514; R.A. McCance and E. Wilkinson. 1947. *J Physiol* 106: 256–263; R.A. McCance and W.F. Young. 1944. *J Physiol* 102: 415–428; R.A. McCance and W.F. Young. 1945. *J Physiol* 104: 296–209; P.G. Prentiss et al. 1959. *Am J Physiol* 196: 625–632; B. Schmidt-Nielsen et al. 1948. *J Cell Comp Physiol* 32: 331–360 and 361–380; K. Schmidt-Nielsen and A.E. Newsome. 1962. *Austral J Biol Sci* 15: 683–689; E. Skadhauge. 1981. Volume 12 of the series *Zoophysiology.* Springer-Verlag: Berlin; E. Skadhauge and S.D. Bradshaw. 1974. *Am J Physiol* 227: 1263–1267; E. Skadhauge. 1974. *J Exper Biol* 61: 269–276; P.C. Withers. 1992. *Comp Animal Physiol.* Table 17.4, p. 863. Saunders College Publishing: New York; A.V. Wolf et al. 1959. *Am J Physiol* 196: 633–641.
Table 13.4: L.S. Babonis and F. Brischoux. 2012. *Integ Comp Biol* 52: 245–256; F. Brischoux et al. 2021. *Proc Royal Soc B* 288: 20203191; Y-C. Cheng et al. 2013. *Zool Studies* 52: 28; W.A. Dunson. 1969. *Am J Physiol* 216: 995–1002; L.C. Hazard. 2001. *Physiol Biochem Zool* 74: 22–31; L.C. Hazard. 2004. In *Iguanas: Biology and Conservation.* A.C. Alberts et al., eds., pp.84–85. University of California Press: Berkeley; L.C. Hazard et al. 2010. *J Exper Zool Part A: Ecol Genet Physiol* 313A: 442–451; H.B. Lillywhite et al. 2019. *PLOS ONE* 14: e0212099; W.N. Holmes and J.G. Phillips. 1985. *Biol Rev* 60: 213–256; J.E. Minnich. 1972. *Comp Biochem Physiol* 41A: 535–549; J.E. Minnich. 1982. In *Biology of the Reptilia,* Vol. 12. Physiology C. C. Gans and F.H. Pough, (Eds.), pp. 325–396. Academic Press: New York; M. Peaker and J.L. Linzell. 1975. *Salt Glands in Birds and Reptiles.* Cambridge University Press, Cambridge; P. Sabat. 2000. *Revista Chilena de Historia Natural* 73: 401–410; K. Schmidt-Nielsen and R. Fange. 1958. *Nature* 182: 783–785; K. Schmidt-Nielsen. 1960. *Circulation* 21: 955–967; J.R. Templeton. 1972. *Symposia of the Zool Soc London* 31: 61–77.

Courtesy of Glenn Tattersall

Heat distribution in a wading flamingo

Ectothermy and Endothermy

Two Ways of Regulating Body Temperature

Earth's terrestrial biomes are vast and varied, and vertebrates can be found in all of them, from steamy equatorial swamps and rainforests to arid hot deserts, to seasonal temperate plains and forests, to dry cold deserts and ice-bound polar regions. Two thermoregulatory strategies, ectothermy and endothermy, allow tetrapods to stabilize their body temperatures in the face of this enormous range of environmental temperatures.

In this chapter we explain the importance of controlling body temperature, describe ectothermy and endothermy, and explain how some vertebrates use elements of both. We then discuss the evolution of endothermy from the ancestral state of ectothermy, a question that has intrigued biologists for decades. Finally, we consider the consequences of ectothermy and endothermy in shaping broad aspects of the lifestyles of different vertebrates and some important consequences of these lifestyles in the context of ecosystems.

14.1 Why Regulate Body Temperature?

LEARNING OBJECTIVES

14.1.1 Explain why thermoregulation is essential for an organism's functioning.

14.1.2 Describe the difference between homeothermy and heterothermy, and explain why these terms are not synonymous with endothermy and ectothermy.

You can think of an organism as a pyramid of processes. For example, biochemical reactions synthesize ATP, that ATP is used by muscles to contract, and those muscular contractions power movements that are the basis of activities. Each of these processes is temperature-sensitive, and some processes are more sensitive to temperature than others; a small change in temperature causes a large change in rate for some reactions and a much smaller change for other reactions. In general, rates increase when temperature rises and decrease when temperature falls.

Metabolic pathways are stepwise biochemical reactions in which the product of each step is the substrate for the next step, and each of these reactions has a different sensitivity to temperature. Matching supply and demand can be vastly simplified by limiting temperature variation, or **thermoregulation**.

Vertebrates employ a variety of behavioral and physiological mechanisms to control body temperature. Before describing these in greater detail, however, we need to clarify two sets of terms used in describing thermoregulation—homeothermy/heterothermy and endothermy/ectothermy—that are often confused and misused. **Homeothermy** describes a stable body temperature, whereas **heterothermy** is a variable (fluctuating) body temperature. A 2°C fluctuation in body temperature is frequently used as an arbitrary dividing line between homeothermy and heterothermy. Endothermy and ectothermy describe the primary source of heat that is used to regulate body temperature. **Endothermy** refers to internal (metabolic) heat production, whereas **ectothermy** indicates that the primary source of heat is outside the body.

The two sets of terms are not synonymous. As we will see in this chapter, there are heterothermal endotherms and homeothermal ectotherms. Indeed, the most homeothermal vertebrates are deep-sea fishes (see Section 7.8), ectotherms that live in an environment where the temperature does not change from one millennium to the next.

14.2 Ectothermal Thermoregulation

LEARNING OBJECTIVES

14.2.1 Describe the pathways of energy exchange for a terrestrial ectotherm.

14.2.2 Describe the behavioral mechanisms employed by terrestrial ectotherms to control heat exchange with the environment.

Ectothermy is the ancestral mode of thermoregulation of vertebrates. Despite its evolutionary antiquity, ectothermy is both complex and an effective way to control body temperature. All objects on the Earth's surface exchange heat with their environment. Ectotherms regulate their body temperatures by balancing the movement of heat inward and outward. During the times they are active, many vertebrate ectotherms maintain body temperatures that are very similar to the body temperatures of birds and mammals.

Energy exchange and mechanisms of ectothermy

A brief discussion of the pathways by which heat moves between an organism and its environment is necessary to understand the thermoregulatory mechanisms employed by terrestrial ectotherms and how thermoregulation interacts with other activities. Both ectotherms and endotherms gain or lose energy by several pathways: solar radiation, infrared (thermal) radiation, convection, conduction, evaporation, and metabolic heat production.

Adjusting the flow of energy through these pathways allows an animal to warm up, cool down, or maintain a stable body temperature.

Figure 14.1 illustrates pathways of thermal energy exchange for an ectotherm. Heat comes primarily from external sources, and the flow through some of the pathways can be either into or out of an organism.

- *Solar radiation* For most ectotherms, the sun is the primary source of heat, and solar energy always results in heat gain. Solar radiation reaches an animal directly when it is in a sunny spot. In addition, solar energy is reflected from clouds and dust particles in the atmosphere and from other objects in the environment. The wavelength distribution of solar radiation is the portion of the solar spectrum that penetrates Earth's atmosphere. About half of this energy is in the visible portion of the spectrum (wavelengths ~390–750 nm); most of the rest is in the infrared (thermal) region of the spectrum (wavelengths >750 nm).

- *Thermal energy* Infrared radiation (IR) is an important part of thermal exchange. All objects, animate or inanimate, radiate energy at wavelengths determined by the absolute temperature of their surface. Objects in the temperature range of animals and Earth's surface (roughly, between −20°C and +50°C) radiate in the infrared portion of the spectrum. Animals continuously radiate energy to the environment and simultaneously receive infrared radiation

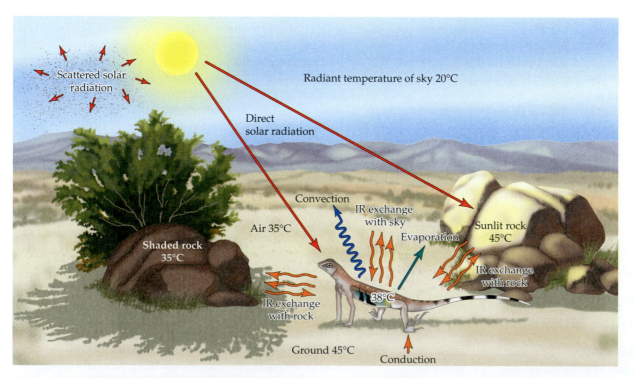

Figure 14.1 Pathways of energy exchange. A terrestrial organism exchanges energy with its environment via several pathways. These are illustrated here in simplified form by a lizard resting on the floor of a desert arroyo. Arrowheads indicate the direction of energy flow. Small adjustments of posture or position can change the magnitude and even the direction of energy exchange in the various pathways, giving the lizard considerable control over its body temperature. IR = infrared radiation. (After F.H. Pough et al. 2016. *Herpetology*, 4th ed. Oxford University Press/Sinauer: Sunderland, MA.)

from the environment. The net heat flow is from the hotter to the cooler surface. Thus, infrared radiation can lead to either gain or loss of heat, depending on the relative temperature of the animal's body surface and the surfaces of the environment. In Figure 14.1, the lizard is cooler than the sunlit rock behind it, and receives more energy from the rock than it loses to the rock. However, the lizard is warmer than the shaded rock in front of it and has a net loss of energy in that exchange. The radiative temperature of a clear sky is about 20°C, so the lizard loses energy by radiation to the sky.

- *Convective exchange* Heat is exchanged between objects and the air via **convection**—the transfer of heat between an animal and a fluid (remember that air is a fluid; see Section 4.1). Convection can result in either gain or loss of heat. If the air temperature is lower than an animal's surface temperature, convection leads to heat loss. If the air is warmer than the animal, convection results in heat gain. In still air, convective currents formed by objects that are warmer than air temperature produce heat loss via natural convection. When the air is moving—that is, when there is a breeze—forced convection replaces natural convection, and the rate of heat exchange is increased. In Figure 14.1, the lizard is warmer than the still air and thus loses heat by natural convection.

- *Conductive exchange* Heat exchange by conduction occurs when objects are in contact. Conductive heat exchange resembles convection in that its direction depends on the relative surface temperatures of the animal and the environment. In Figure 14.1, the lizard gains heat by conduction from the warm ground.

- *Evaporation and condensation* Water evaporates both from the body surface and from the pulmonary system. Each gram of water evaporated represents a loss of about 2,450 joules (J); the exact value changes slightly with the temperature. Evaporation of water transfers heat from the animal to the environment and thus represents a loss of heat. (The reverse situation—condensation of water vapor on an animal—would result in heat gain, but that situation rarely occurs under natural conditions.)

- *Metabolism* Metabolic heat production is the final pathway by which an animal can gain heat. Among ectotherms, metabolic heat gain is usually trivial in relation to the heat gained directly or indirectly from solar energy.

Thermal ecology of ectotherms

Many of the behavioral mechanisms involved in ectothermy are straightforward and are employed not only by ectothermal vertebrates, but also by endothermal insects, birds, and mammals (including humans). On a daily basis, movement back and forth between sunlight and shade is the most obvious thermoregulatory mechanism. This simple approach allows many species of lizards and snakes to keep their body temperature within a range of 2–3°C when they are active; this is the **activity temperature**, a range within which physiological processes and whole-animal performance reach their maxima (**Figure 14.2**).

(A) Western garter snake, *Thamnophis elegans*

Courtesy of Todd Pierson

Figure 14.2 Effect of temperature on performance. An ectotherm's ability to engage in its normal behaviors and physiological processes is temperature-sensitive. Crawling and swimming speeds and speed of digestion by western garter snakes all reach their maximum within or close to the species' activity temperature range. The vertical axis shows the percentage of maximum performance achieved at each temperature. Within the snake's activity range (shaded area), performance level is at least 85% of maximum for all these activities. (B after R.D. Stevenson et al. 1985. *Physiol Biochem Zool* 565: 48–57. © 1985 The University of Chicago Press.)

(B) Behavioral and physiological responses as a function of body temperature

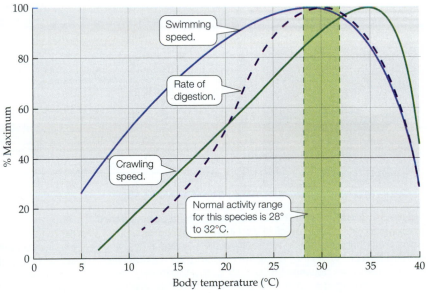

Swimming speed.

Rate of digestion.

Crawling speed.

Normal activity range for this species is 28° to 32°C.

% Maximum

Body temperature (°C)

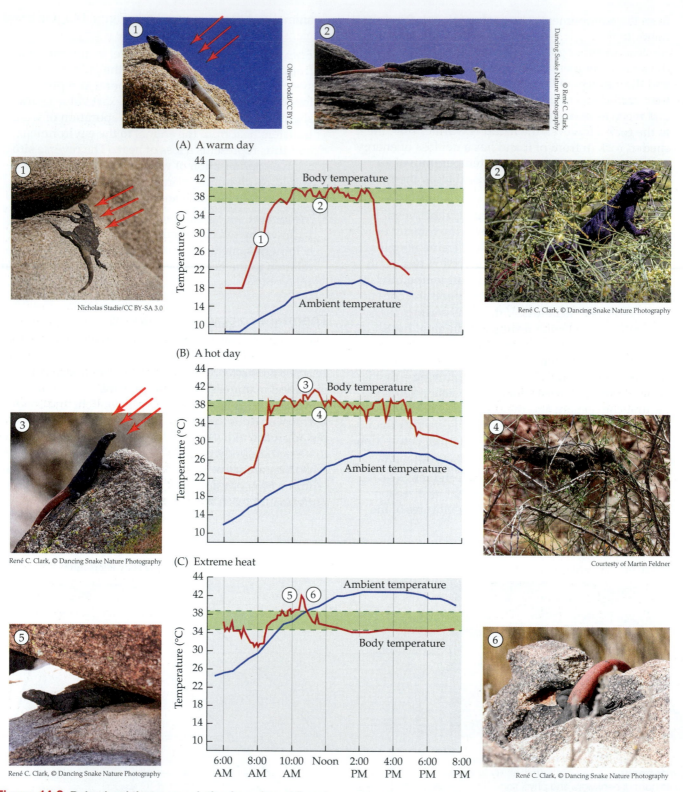

Figure 14.3 **Behavioral thermoregulation by a desert lizard.** Chuckwallas (genus *Sauromalus*) are large, diurnal, herbivorous lizards that inhabit deserts in the southwestern United States and northern Mexico. While they are active, chuckwallas maintain their body temperatures between 36° and 40°C; this thermoregulation is based on changes in posture and selection of microclimates. (A) When they are warming in the morning (①), lizards maximize solar heat absorption by orienting with their backs perpendicular to the rays of the sun and flattening the body to increase the area being heated. When they have reached their activity temperature range they engage in social behavior, feeding, and moving (②). (B) To reduce heating while remaining active, a lizard can (③) face into the sun with its body parallel to the sun's rays so that the sun shines only on the lizard's head and shoulders or (④) climb into a bush where the wind speed and convective cooling are greater than at ground level. (C) When it is extremely hot, a lizard can (⑤) seek shade under an overhanging rock or (⑥) return to the deep crevice where it spent the night. (Graphs after T.J. Case.1976. *J Herpetol* 10: 85–95. Published by Society for the Study of Amphibians and Reptiles.)

Lizards, especially desert species such as the chuckwalla (*Sauromalus ater*, a large, herbivorous lizard that inhabits deserts in the southwestern United States and northern Mexico), are particularly good at behavioral thermoregulation. **Figure 14.3** uses the example of the chuckwalla to illustrate some ways in which desert lizards thermoregulate over the course of a day and in different weathers. (Numbers in the list below correspond to circled numbers 1–6 in the figure.)

1. Early in the morning and on relatively cool days (Figure 14.3 ①), lizards orient perpendicular to the sun's rays and flatten their bodies to expose the largest possible area of their body to solar radiation.

2. When lizards have reached their activity temperature, they move between sun and shade, defending territories and feeding (Figure 14.3 ②).

3. On hotter days (Figure 14.3 ③), a lizard can remain on the surface as the temperature warms by facing directly into the sun and aligning its body parallel to the sun's rays, thus minimizing the portion of its body that receives direct solar radiation.

4. Alternatively, a lizard can climb into a bush, where air temperature is lower than at ground level and air movement is greater (Figure 14.3 ④).

5. When the temperature climbs to extreme levels (Figure 14.3 ⑤), lizards move into shade under overhanging rocks.

6. As a last resort, these lizards retreat into deep crevices in the rocks (Figure 14.3 ⑥).

14.3 Endothermal Thermoregulation

LEARNING OBJECTIVES

14.3.1 Explain the difference in metabolic rates of vertebrate ectotherms and endotherms.

14.3.2 Describe the thermogenic mechanisms of endotherms.

14.3.3 Describe the relationship between metabolic rate and body temperature of an endotherm.

Endothermy, the internal (metabolic) maintenance of body temperature, evolved independently in mammals (synapsids) and birds (sauropsids). The selective factors that led to the evolution of endothermy from the ancestral ectothermic state are unclear (see Section 14.5), but certain costs and benefits of endothermy can be identified. On the benefit side, endothermy allows birds and mammals to maintain their normal body temperature when solar radiation is insufficient or not available—in winter or at night, for example. On the cost side, endothermy is energetically expensive. Endotherms have metabolic rates 7–10 times higher than those of ectotherms of the same body weight. The energy to sustain a high metabolism comes from food, and gram-for-gram endotherms need more food than do ectotherms.

Metabolism is an energetically inefficient process. Only a portion of the energy freed when a chemical bond is broken is captured; the rest is released to the environment as heat. This released energy is the heat endotherms use to maintain their body temperatures at elevated levels. Thus, endotherms regulate their body temperatures by matching the rate of heat produced in the core of the body to the rate of heat lost to the environment across the body surface.

Interplay of heating and cooling allows endotherms to maintain body temperatures at stable levels in the face of fluctuating environmental temperatures. Normal body temperature—**normothermia**—varies for different species of mammals and birds, but most endotherms conform to the generalized diagram in **Figure 14.4**. An endotherm in the **thermoneutral zone** maintains a stable body temperature by adjusting its insulation without changing its metabolic rate. The boundaries of the thermoneutral zone are the **lower critical temperature** and the **upper critical temperature**.

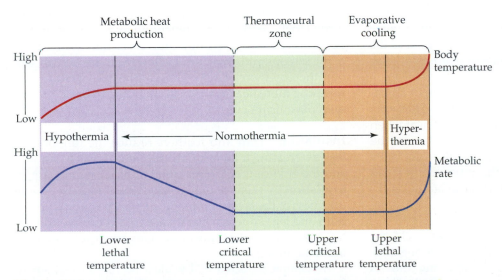

Figure 14.4 Endotherms maintain stable body temperatures over a range of environmental temperatures. Within the *thermoneutral zone*, changing insulation allows a bird or mammal to maintain a stable body temperature. Below *the lower critical temperature* metabolic heat production must increase. Below the *lower lethal temperature*, body temperature falls (hypothermia), reducing metabolic heat production, and death follows. Above the *upper critical temperature*, evaporative cooling maintains normothermy. Above the *upper lethal temperature*, evaporative cooling is no longer sufficient; body temperature rises (hyperthermia), metabolic heat production increases, and an explosive rise in body temperature leads to death.

Below the lower critical temperature, an endotherm must increase its rate of heat production to balance the heat lost from the body. The **lower lethal temperature** is the lowest environmental temperature at which heat production can keep up with heat loss. At temperatures below this, heat loss exceeds heat production, body temperature falls (**hypothermia**), metabolism and heat production decrease, and an uncontrolled drop in body temperature leads to death.

Above the upper critical temperature, an endotherm employs evaporative cooling. The **upper lethal temperature** is the highest environmental temperature at which evaporative cooling can balance heat gain. At higher environmental temperatures, body temperature rises (**hyperthermia**), metabolic heat production increases because tissue temperature is rising, and an explosive increase in temperature leads to death.

Mechanisms of endothermal thermoregulation

As stated above, endotherms generate heat internally and achieve a stable body temperature by matching the rate of metabolic heat production to the rate of heat loss to the environment. Endotherms lose heat to the environment through the same pathways as ectotherms, but the metabolic processes that contribute to internal heat production, as well as the use of insulation to retain heat, are unique to endotherms.

Whole-body metabolism   The metabolic heat production of an organism is the sum of the heat production of all its different tissues and organs. The weight-specific resting metabolic rates of these tissues and organs (expressed in kilojoules per kilogram per day, kJ/kg/day) vary widely. When an animal is at rest, the heart and kidneys are the most metabolically active tissues, consuming twice as much oxygen and releasing twice as much heat as the liver and brain. The metabolism of the remaining organs, of adipose tissue, and of skeletal muscle is even lower.

Thus, the resting metabolic heat production of an endotherm is derived primarily from the heart and kidneys. Muscular activity is not a significant source of heat when an animal is at rest, but intense activity can increase the heat production of skeletal muscle 10–15 times above its resting rate. In cold environments, this muscular heat can compensate for heat loss, but only when an animal is engaged in physical activity.

Shivering and non-shivering thermogenesis   Mammals and birds use both shivering and non-shivering thermogenesis to increase heat production. **Shivering** consists of uncoordinated contraction of postural muscle fibers and transforms chemical energy (ATP) to elastic energy (a contracted muscle fiber) releasing heat in the process. The heat released in muscle tissues is distributed by the circulatory system. Both birds and mammals shiver, although differences in the details of muscle fiber recruitment

during shivering indicate that shivering evolved independently in the two groups.

Non-shivering thermogenesis is an increase in oxidative phosphorylation achieved by short-circuiting a cellular process. One form of this kind of thermogenesis occurs in the mitochondria, where the electron transport chain pumps protons (H^+) from the mitochondrial matrix into the intermembrane space, creating a proton gradient. Most of the protons return to the matrix via ATP synthase embedded in the inner mitochondrial membrane, converting ADP to ATP, but some protons leak back into the matrix without passing through ATP synthase. The effect of this **proton leak** is to increase the rate of oxidative phosphorylation (and thus the rate of heat production) without increasing synthesis of ATP.

Proton leak has two components, basal and inducible. Basal proton leak (which accounts for 20–30% of the resting metabolic rate of laboratory rats) results from the permeability of the inner mitochondrial membrane and is constant. Inducible proton leak is produced by uncoupling proteins (UCPs), which increase the permeability of the inner mitochondrial membrane, allowing additional protons to bypass ATP synthase, thereby increasing heat production. Mammalian UCPs occur in brown fat, a specialized thermogenic tissue; in a subset of cells in subcutaneous white fat; and in skeletal muscle. Birds have an avian-specific UCP. In both mammals and birds, increased proton leakage by release of UCPs is initiated by the nervous system in response to cold.

The sarcoplasmic reticulum (a specialized network of tubules within muscle cells) is a second site of non-shivering thermogenesis. Contraction of skeletal muscle fibers is initiated by neural activity that releases Ca^{2+} from the sarcoplasmic reticulum. These ions are subsequently returned to the sarcoplasmic reticulum by a pump powered by ATP. A protein (sarcolipin) increases the permeability of the sarcoplasmic reticulum, allowing more Ca^{2+} to leak out, thereby increasing oxidative phosphorylation and releasing more heat as the ions are pumped back into the sarcoplasmic reticulum.

Insulation   Hair and feathers reduce the rate of heat loss by creating a layer of air trapped between the skin and the environment. A mammal or bird can increase its insulation by raising its hair or feathers to increase the thickness of the layer of trapped air. (Goose bumps on our arms and legs appear when we are cold because our body hairs rise to a vertical position in an attempt to increase our insulation.)

Insulation is decreased by allowing hair or feathers to lie flat, minimizing the thickness of the layer of trapped air, and by exposing thinly insulated parts of the body (called thermal windows), as a dog does when it lies on its back. Heat loss also can be enhanced by redirecting peripheral blood circulation to the skin, and especially to lightly furred areas such as the ears of jackrabbits and elephants.

Evaporative cooling   The skin of all vertebrates is permeable to water vapor, and this baseline evaporation, called **insensible water loss**, has a cooling effect. Many animals pant, breathing rapidly and shallowly so that evaporation of water from the respiratory system provides a cooling effect. Birds use a rapid fluttering movement of the gular (throat) region to evaporate water for thermoregulation. Humans and some other mammals increase evaporative heat loss by releasing water from sweat glands; water released by sweating evaporates from the surface of the body. Most mammals lack sweat glands and some (kangaroos are an example) increase evaporative cooling by licking their forearms.

Sweating has the advantage of requiring little energy, but because it cools the surface of the skin—the area that is being heated by the environment—it may have only limited effect on the temperature of the body core. Panting and gular fluttering cool internal tissues but require muscular activity, which is an added source of metabolic heat.

14.4 Pure Ectothermy and Pure Endothermy Lie at the Extremes of a Continuum

LEARNING OBJECTIVES

14.4.1 Explain how a species can behave as an ectotherm at some times and as an endotherm at others, and under what conditions this might occur.

14.4.2 Explain the difference between rest-phase torpor and hibernation.

14.4.3 Describe the cycle of body temperature of a male Richardson's ground squirrel from June through March.

Endothermy and ectothermy are not mutually exclusive; many tetrapods combine elements of both modes. A few ectotherms can produce enough metabolic heat to raise their body temperatures, many endotherms are able to allow their body temperatures to fall below normal levels, and some endotherms use the sun to raise their body temperatures.

Endothermal ectotherms

Females of several species of python coil around their eggs and produce heat by rhythmic contraction of their trunk muscles. The rate of muscle contraction increases as air temperature falls, and a female Burmese python (*Python bivittatus*) is able to maintain her body temperature and the temperature of her eggs near 33°C even when air temperatures are as low as 23°C (**Figure 14.5**). Incubation speeds egg development and avoids defects that occur when eggs develop at lower temperatures, but producing that heat requires a substantial expenditure of energy. At 23°C, the metabolic rate of a brooding female python is about 20 times the rate of a python that is not incubating. When a brooding diamond python (*Morelia spilota*) was

given access to a heat lamp in an enclosure adjacent to her eggs, she left the eggs when the heat lamp came on, raised her body temperature by basking under the light, then returned to the eggs, thus combining ectothermal and endothermal mechanisms in an energetically efficient method of brooding her eggs.

Argentinian black-and-white tegu lizards (**Figure 14.6A**) use metabolic heat production to maintain high body temperatures in their burrows during the reproductive season. In late spring and summer, the lizards behave as ectotherms, basking in the sun during the day and retreating to burrows at night where their body temperatures fall to match the burrow temperature. In fall and winter, they hibernate in their burrows, not emerging to bask or feed. Their body temperatures at this time are the same as the temperatures of their burrows (**Figure 14.6B**).

This situation changes during the reproductive season in early spring. Male tegus emerge from their burrows and begin to defend territories and search for mates, while

(A) Female Burmese python, *Python bivittatus*, incubating eggs

© Paul Tessier/123RF

(B) Female's body temperature when incubating

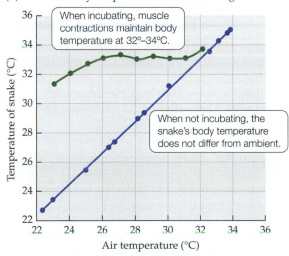

When incubating, muscle contractions maintain body temperature at 32°–34°C.

When not incubating, the snake's body temperature does not differ from ambient.

Figure 14.5 Some female pythons can produce metabolic heat to warm their eggs.   When a female Burmese python is incubating, muscular contractions allow her to maintain the temperature of her body (and of her eggs) between 32° and 34°C. When the python is not incubating, her body temperature does not differ from the ambient temperature. (B from L.H.S. Van Mierop and S. Barnard. 1978. *Copeia* 1978: 615–621. Courtesy of American Society of Ichthyologists and Herpetologists [ASIH].)

(A) Black-and-white tegu, *Salvator merianae*

Bernard Dupont/CC BY 2.0

Figure 14.6 **Tegu lizards can be endothermal during the breeding season.** (A) Black-and-white tegus are large, carnivorous lizards from South America that prey on small mammals and birds as well as frogs, snakes, and smaller lizards. (B) When a tegu is in its burrow during the nonbreeding season, its surface temperature falls to the temperature of the burrow. (C) During the breeding season, tegus can maintain body temperatures as much as 10°C higher than the burrow temperature. (There are two lizards in this image, one warmer than the other.)

(B) Tegu in burrow during nonbreeding season   (C) Two breeding tegus in burrow

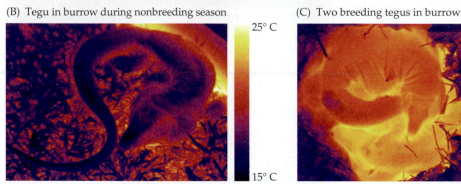

25° C

15° C

B, C from G. Tattersall. 2016. *Zoology* 119: 403–405

females gather nesting material in their burrows. At this time, both males and females display a fivefold increase in metabolic rate that, combined with changes in circulation that retain heat in the core of the body, produces nocturnal body temperatures that are several degrees warmer than the burrow temperatures (**Figure 14.6C**).

Heterothermal endotherms: Torpor and hibernation

Although a warm, stable body temperature allows endotherms to maintain activity even when the environmental temperature is low, the energy cost of maintaining that temperature may be prohibitive. When this is the case, many endotherms allow their body temperatures to fall, enduring periods of hypothermia in a state of **torpor**[1] that lasts from a few hours to 10 months.

Rest-phase torpor   Body temperatures of vertebrates are usually highest during the times they are normally active (the activity phase) and drop during times of inactivity (the rest phase). For example, human body temperature during the day averages about 37.5°C and falls to about 36.5°C at night. Body temperatures of nocturnal vertebrates show the opposite pattern, being higher at night and lower during the day.

[1] The terms "torpor" and "hypothermia" differ in important respects. Torpor is a controlled reduction of body temperature to a new set point and the return to normothermy from torpor is an active process that relies on internal (metabolic) heating. Hypothermia is an uncontrolled drop in body temperature and if return to normothermy occurs, it is a passive process relying on heat from the environment.

Solar heating is an important energy-conserving behavior for at least 5 species of birds and more than 20 species of mammals. On cold nights, roadrunners (ground-dwelling birds of the North American deserts) allow their body temperatures to drop to about 4°C below normal. In the morning they assume a basking posture, drooping the wings and raising the feathers on the back to expose an area of black skin (**Figure 14.7A**). Using heat from the sun to rewarm provides a 41% energy saving compared to using metabolic heat.

Some mammals also bask in the morning to warm after cooling at night. Fat-tailed dunnarts (*Sminthopsis crassicaudata*) are small Australian marsupials that spend fall and winter nights deep in crevices in the soil where they allow their body temperature to fall to that of the surrounding soil—about 15°C. In the morning, a dunnart climbs to the top of the crevice and basks in the sun (**Figure 14.7B**).

Many small diurnal birds and mammals allow their body temperatures to drop at night and use metabolic heat to return to normal temperatures in the morning. Even with the energy cost of rewarming, nocturnal torpor can produce a substantial energy savings compared to remaining normothermal during the night (**Figure 14.8**).

A study of 9 species of hummingbirds in Arizona and Ecuador found that birds became torpid at night in 54.8% of observations. The average duration of torpor was 4.9 hours, and the birds saved an average of 82% of the energy that would have been used to remain normothermal during that period. The duration of torpor was the primary factor influencing the amount of energy saved,

(A) Roadrunner, *Geococcyx californianus*, warming

Raised feathers expose an area of black skin that absorbs solar radiation.

Amy Meredith/CC BY-ND 2.0

(B) Behavioral thermoregulation of fat-tailed dunnart, *Sminthopsis crassicaudata*

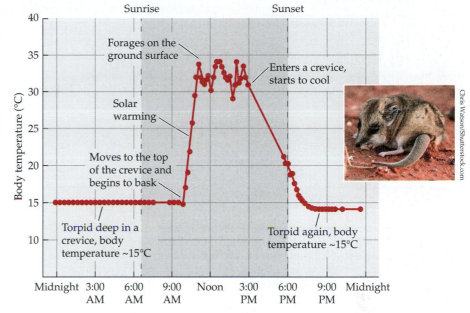

Chris Watson/Shutterstock.com

Sunrise   Sunset

Forages on the ground surface

Enters a crevice, starts to cool

Solar warming

Moves to the top of the crevice and begins to bask

Torpid deep in a crevice, body temperature ~15°C

Torpid again, body temperature ~15°C

Figure 14.7 Solar heating. (A) Facing away from the sun, this roadrunner has drooped its wings and lifted the feathers on its back to expose an area of black skin that absorbs solar radiation. (B) Around 9 AM on the day shown here, a dunnart, its body temperature still near 15°C, climbs to the top of the crevice and basks in the sun; in 2 hours its body temperatures has reached 30°–35°C and it begins to forage. Its body temperature rises and falls as it moves between sun and shade. About 3 PM, the dunnart returns to a crevice and its body temperature cools to 15°C for the night. (B from L. Warnecke et al. 2008. *Naturwissenschaften* 95: 73–78.)

because the cost of warming is the same after a short bout of torpor as after a longer one (**Figure 14.9**).

Hibernation   The deep torpor that occurs during hibernation is a comatose condition, more profound than the deepest sleep. Hibernation is nearly confined to mammals; among birds, hibernation has been demonstrated only for the common poorwill, *Phalaenoptilus nuttallii*.

During hibernation, voluntary motor responses are reduced to sluggish postural changes, although some sensory perception of powerful auditory and tactile stimuli and environmental temperature changes is retained. Perhaps most dramatically, a hibernating animal can arouse spontaneously from this state using heat production by brown fat. Some endotherms can rewarm under their own power from the lowest levels of hibernation, whereas others must warm passively with an increase in environmental temperature until some threshold is reached at which arousal starts.

Deep torpor is more advantageous for a small animal than for a large one. In the first place, the weight-specific energy cost (energy/g) of maintaining a high body temperature is greater for a small animal than for a large one, and as a consequence, a small animal has more to gain from torpor. Second, a small animal cools more rapidly than a large one, so its metabolic rate decreases faster. Furthermore, small animals have less body tissue to rewarm on arousal, and their costs of arousal are correspondingly lower than those of large animals.

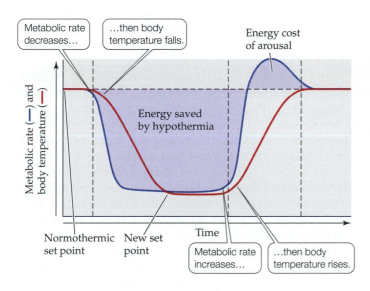

Metabolic rate decreases…

…then body temperature falls.

Energy cost of arousal

Energy saved by hypothermia

Normothermic set point   New set point

Time

Metabolic rate increases…

…then body temperature rises.

Figure 14.8 Changes in body temperature and metabolic rate during rest-phase torpor. As an animal enters rest-phase torpor, a decrease in metabolic rate precedes a fall in body temperature to a new set point. On rewarming, an increase in metabolism precedes the return to normothermia. The metabolic rate during arousal briefly overshoots the resting rate, but the energy cost of arousal is less than the energy saved by the period of torpor.

(A) Male green-crowned brilliant hummingbird, *Heliodoxa jacula*

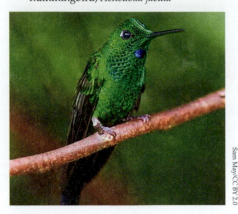

Sam May/CC BY 2.0

(B) Energy use during rest-phase torpor

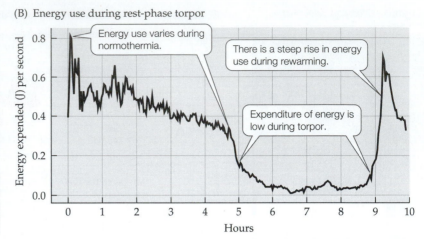

Energy use varies during normothermia.

There is a steep rise in energy use during rewarming.

Expenditure of energy is low during torpor.

Figure 14.9 Hummingbirds can become torpid at night.
(A) Some hummingbirds, such as *Heliodoxa jacula*, enter rest-phase torpor. (B) In the experiment shown here, the bird was active for the first 4.7 hours after being placed in the metabolism chamber, and its temperature was in the normothermal range. Energy expenditure decreased gradually and fell abruptly when the bird entered torpor. The energy used during torpor was only 21% of the cost of remaining normothermal, and in this experiment torpor reduced the energy expenditure during the night by 55%. (B, data from A. Shankar et. al. 2020. *J Avian Biol* 2020: e02305.)

The largest mammals that hibernate are marmots, which weigh about 5 kg. Although bears are popularly believed to hibernate, it is more accurate to say that they enter a state of winter dormancy. Bears in winter dormancy lower their body temperatures only ~5°C below normal levels, and their metabolic rate decreases ~50%. That small reduction in body temperature, combined with large fat stores accumulated before retreating to winter dens, is sufficient to carry bears through the winter, but their torpor is not deep enough to be considered hibernation.

Profound changes occur in a variety of physiological functions of animals in deep torpor. Body temperature drops to within 1°C or less of the surrounding temperature, and in some cases (bats, for example) extended survival is possible at body temperatures just above the freezing point of the tissues. Arctic ground squirrels (*Urocitellus parryii*) allow the temperature of parts of their bodies to fall as low as −2.9°C. Active **metabolic suppression**—temperature-independent changes in gene expression and posttranslational modifications of enzymes in oxidative metabolic pathways—starts before body temperature falls and reduces oxidative metabolism to as little as one-twentieth the rate at normal body temperatures.

Although body temperatures fall dramatically during hibernation, temperature regulation does not entirely cease. Instead, the hypothalamic thermostat is reset. If the body temperature of a hibernating animal falls below the new set point, thermogenic processes bring the hibernating animal back to the regulated level.

Respiration is slow during hibernation, and an animal's overall breathing rate can be less than one breath per minute. Heart rate is drastically reduced, and blood flow to peripheral tissues is virtually shut down, as is blood flow posterior to the diaphragm. Thus, most of the animal's blood is retained in the body's core.

Hibernation is an effective method of conserving energy during long winters, but hibernating animals rewarm at intervals. Heat production by brown fat between the shoulder blades is distributed by the circulatory system. The head, heart, and anterior part of the body warm first, followed several hours later by the posterior regions. Periodic arousals are normal, and these arousals consume a large portion of the total energy used by hibernating mammals. An example of the magnitude of the energy cost of arousal is provided by Richardson's ground squirrels (*Urocitellus richardsonii*) in Alberta, Canada (**Figure 14.10**).

The activity season for ground squirrels in Alberta is short, only about 100 days for adults. Emerging from hibernation in March, males establish territories preparing to court the females that emerge ~2 weeks later. Mature males enter hibernation in June, and females follow ~2 weeks after that. Juveniles of both sexes remain active, increasing their body weight as much as possible before entering hibernation in September. When the squirrels are active, they have body temperatures of 37–38°C; when they are torpid, their temperatures fall as low as 3–4°C. Hibernation consists of periods of torpor and low body temperature alternating with periods of arousal and normothermy throughout the winter.

The periods of arousal account for most of the energy used by a hibernating Richardson's ground squirrel (**Table 14.1**). The energy costs associated with arousal include warming from the hibernation temperature to normothermy (37°C), sustaining normothermy for several hours in a cold burrow, and metabolism as the body temperature slowly decreases during reentry into torpor. For the entire hibernation season, the combined metabolic expenditures for these three phases of the hibernation cycle account for an average of 85% of the total energy used by the squirrel.

Surprisingly, we have no clear understanding of why a ground squirrel undergoes arousals that increase its total winter energy expenditure nearly fivefold. Ground squirrels do not store food in their burrows, so they are

(A) Richardson's ground squirrel, *Urocitellus richardsonii*

Courtesy of Doug Collicutt

(B) Body temperatures of ground squirrel over a 6-month period

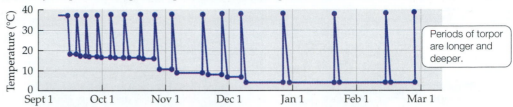

Periods of torpor are longer and deeper.

(C) Body temperatures of ground squirrel during one torpor cycle

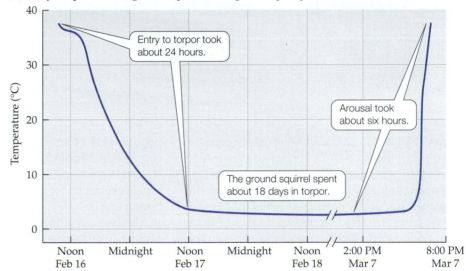

Entry to torpor took about 24 hours.

Arousal took about six hours.

The ground squirrel spent about 18 days in torpor.

Figure 14.10 Hibernation by Richardson's ground squirrel. (A) Adult Richardson's ground squirrel. (B) This squirrel entered hibernation in September, when the temperature in the burrow was about 13°C. As winter progressed and the burrow's temperature fell, the intervals between arousals lengthened and the squirrel's body temperature decreased. In late February, the periods of torpor became shorter, and the squirrel emerged from hibernation in early March. (C) A single torpor cycle. Entry into the cycle began shortly after noon on February 16 and lasted until late afternoon on March 7. In the final 3 hours of the cycle, the squirrel warmed from 3°C to 37°C. (B,C, from L.C.H. Wang and J.W. Hudson 1978. *Strategies in Cold*, pp. 109–145. Academic Press, with permission from Elsevier.)

not using the normothermal periods to eat. They do urinate during normothermy, so eliminating accumulated nitrogenous wastes may be the reason for arousal, and spending some time at a high body temperature may be necessary to carry out other physiological or biochemical activities, such as resynthesizing glycogen, redistributing ions, or synthesizing serotonin. Arousal may also allow a hibernating animal to determine when environmental conditions are suitable for emergence. Whatever their function, the arousals must be important because the squirrel pays a high energetic price for them during a period of extreme energy conservation.

Heterothermal endotherms: Hyperthermia and life in the desert

Hyperthermia Temporarily allowing the body temperature to rise above a species' normal set point allows an animal to conserve water that would otherwise have been used for evaporative cooling. In a desert, conserving water can be more critical than strictly regulating body temperature. Body size determines how hyperthermia can be employed, and small mammals differ from large mammals.

Birds The mobility of birds is an advantage in hot environments, although only soaring birds such as hawks, eagles, and vultures rise high enough to reach cool air. Smaller birds seek cool retreats. In Saudi Arabia, four species of larks have been observed sheltering in burrows of large *Uromastyx* lizards. In North American deserts, horned larks (*Eremophilus alpestris*), poorwills (*Phalaenoptilus nuttallii*), and burrowing owls (*Athene cunicularia*) seek shelter in burrows constructed by rodents and desert

Table 14.1 Energy use during hibernation of a Richardson's ground squirrel.

Month		Percent of total energy per month		
	Entry	Hypothermia	Arousal	Normothermia
July	17.8	8.5	17.2	56.5
September	15.7	19.2	15.2	49.9
November	14.0	20.8	23.1	43.1
January	11.2	24.8	24.1	40.0
March	6.3	3.3	14.1	76.4

Source: From L.C.H. Wang and J.W. Hudson. 1978. *Strategies in Cold: Natural Torpidity and Thermogenesis*. Academic Press. By permission of L.C.H. Wang.

(A) Burrowing owl, *Athene cunicularia*

Mick Thompson/CC BY-SA 3.0

(B) Rosy-faced lovebird, *Agapornis roseicollis*

Courtesy of Kevin McGraw

Figure 14.11 **Desert birds can escape extreme heat.** (A) Burrowing owls are one of several bird species that shelter from the heat in burrows constructed by rodents or desert tortoises. (B) A rosy-faced lovebird in Phoenix has found a vent exhausting cooled air from an air-conditioned building.

tortoises (**Figure 14.11A**). Rosy-faced lovebirds (*Agapornis roseicollis*), native to the Namib Desert of southwestern Africa, are popular pets. Escaped pet lovebirds have established a population in Phoenix, Arizona, where the birds have learned to perch on exhaust vents that are releasing cool air from air-conditioned buildings (**Figure 14.11B**). They do this only during the summer when the air temperature reaches the upper end of their thermoneutral zone.

Rodents White-tailed antelope ground squirrels (*Ammospermophilus leucurus*) are small desert mammals that forage during the day, running across the ground surface. The almost frenetic activity of the squirrels on intensely hot days is a result of the thermoregulatory problems that small animals experience under these conditions. Studies of the white-tailed antelope ground squirrels at Deep Canyon, near Palm Springs in California, illustrate short-term relaxation of body temperature homeostasis.

The heat on summer days at Deep Canyon is intense, and squirrels are exposed to high heat loads for most of the day. They have a bimodal pattern of activity that peaks in the midmorning and again in the early evening (**Figure 14.12**). Relatively few squirrels are active in the middle of the day. Their body temperatures are labile, and the body temperature of an individual squirrel can vary as much as 7.5°C (from 36.1°C to 43.6°C) during a day. The squirrels use this variability of body temperature to store heat during their periods of activity.

(A) White tailed antelope ground squirrel, *Ammospermophilus leucurus*

Marshal Hedin/CC BY 2.0

(B) Losing heat to cool rock

Renee/CC BY 2.0

(C) Body temperature and behavioral thermoregulation of ground squirrel

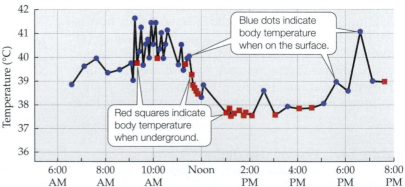

Blue dots indicate body temperature when on the surface.

Red squares indicate body temperature when underground.

Figure 14.12 **Short-term cycles of activity and body temperature.** (A) White-tailed antelope ground squirrels are diurnal, even during the summer when ambient temperatures exceed the species' upper lethal temperature. (B) Early in the day, a squirrel can dump heat by pressing its lightly furred belly against a shaded rock that is still cool from the night. (C) Later in the day, it must seek its burrow to find a cool substrate. (C after M.A. Chappell and G.A. Bartholomew. 1981. *Physiol Zool* 54: 215–223.)

Because the small bodies of antelope ground squirrels heat rapidly, high temperatures limit their bouts of activity to no more than 9–13 minutes. They sprint furiously from one patch of shade to the next, pausing only to seize food or look for predators. The squirrels minimize exposure to the highest temperatures by running across open areas, and they seek shade or their burrows to cool off. Here their small size is an advantage because they cool rapidly, but in the middle of a hot summer day, a squirrel can maintain a body temperature lower than 43°C (the maximum temperature it can tolerate) only by retreating every few minutes to a burrow deeper than 60 cm, where the soil temperature is 30–32°C.

Large mammals A large animal has nowhere to hide from desert conditions. It is too big to burrow underground, and few deserts have vegetation large enough to provide useful shade to an animal much larger than a jackrabbit. However, large body size offers some options not available to smaller animals. Large animals are mobile and can travel long distances to find food or water, whereas small animals may be limited to home ranges only a few meters or tens of meters in diameter.

Large animals have low surface/volume ratios and can be well insulated. Consequently, they absorb heat from the environment slowly. A large body weight gives an animal a large thermal inertia, meaning that it can absorb a large amount of heat energy before its body temperature rises to dangerous levels. Dromedaries are classic large desert animals, with adult body weights of 400–450 kg for females and up to 500 kg for males.

Behavioral mechanisms and the distribution of hair on the body aid camels in reducing their heat load. In summer, dromedaries have hair 5–6 cm long on the neck and back and up to 11 cm long over the hump. On the ventral surface and legs, the hair is only 1.5–2 cm long. Early in the morning, dromedaries lie down on surfaces that have cooled overnight by radiation of heat to the night sky. A dromedary tucks its legs beneath the body, placing its lightly furred legs and ventral surface in contact with cool sand (**Figure 14.13A**). In this position, the camel exposes only its well-protected neck and back to the sun. Sweat cools the skin while the long hair on the camel's back prevents solar heat from reaching the skin.

Dromedaries assemble in small groups and lie pressed closely together throughout the day. Spending a day in the desert sun squashed between two sweaty dromedaries may not be your idea of fun, but in this posture a dromedary reduces its heat gain because it keeps its sides in contact with other dromedaries (at ~40°C) instead of allowing solar radiation to raise the surface temperature of its fur to 70°C or higher.

A dromedary is large enough to use evaporative cooling when water is available, but water is not always available in a desert. The significance of evaporative cooling can be seen by comparing the daily cycle of body temperature in a dromedary that receives water every day with the cycle

(A) Dromedaries, *Camelus dromedarius*

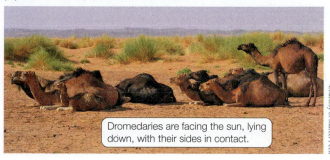

Dromedaries are facing the sun, lying down, with their sides in contact.

Courtesy of Marcel Peek

(B) Dromedary given water daily

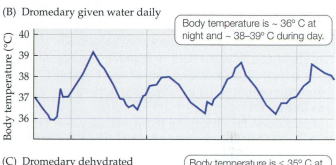

Body temperature is ~ 36° C at night and ~ 38–39° C during day.

(C) Dromedary dehydrated

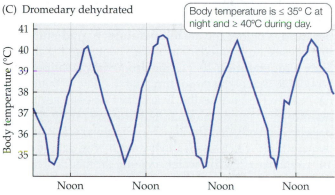

Body temperature is ≤ 35° C at night and ≥ 40°C during day.

Figure 14.13 Daily cycles of body temperature of dromedaries. (A) In the heat of the day, dromedaries face into the sun so that solar radiation falls on their well-insulated necks, shoulders, and backs and press together so solar energy does not strike their sides. Their lightly insulated bellies and legs are in contact with sand that is still cool from the previous night. (B) When a dromedary has daily access to water, its body temperature cools to about 36°C at night and rises to 38–39°C during the day. (C) When the dromedary is dehydrated, its body temperature falls below 35°C at night and rises above 40°C during the day. (B,C after K. Schmidt-Nielsen et al. 1956. *Am J Physiol* 188: 103–112.)

of one that has been deprived of water (**Figure 14.13B,C**). The body temperature of a watered dromedary is 36°C in the early morning and rises to 39°C in midafternoon. When a dromedary is deprived of water, the daily temperature variation doubles. Body temperature falls to 34.5°C at night and climbs to 40.5°C during the day.

The significance of this increased daily fluctuation in body temperature can be assessed in terms of the water that the dromedary would expend in evaporative cooling to prevent the 6°C rise. With a specific heat of 3.5 kJ/kg per 1°C, a 6°C increase in body temperature for a 500-kg dromedary represents storage of 10,500 kJ of heat. Evaporation

Figure 14.14 Countercurrent heat-exchange cools blood going to a dromedary's brain. Venous blood is cooled by evaporation in the nasal passages. When this blood enters the cavernous sinus, it has a temperature of about 37°C. Blood flowing to the brain via the internal carotid arteries has a temperature of about 43°C when it enters the vessels of a rete mirabile in the cavernous sinus. Here it is surrounded by the cool venous blood. This countercurrent mechanism cools the blood to approximately 40°C by the time it reaches the brain.

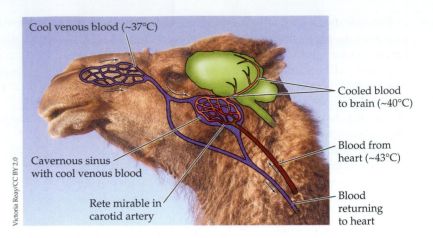

Cool venous blood (~37°C)

Cooled blood to brain (~40°C)

Blood from heart (~43°C)

Blood returning to heart

Cavernous sinus with cool venous blood

Rete mirabile in carotid artery

Victoria Reay/CC BY 2.0

of 1 L of water dissipates approximately 2,400 kJ of heat. Thus, a dromedary would have to evaporate about 4.5 L of water daily to maintain a stable body temperature at the nighttime level, and it can conserve that water by tolerating hyperthermia during the day.

In addition to conserving the water they don't use for evaporative cooling, dromedaries receive an indirect benefit from hyperthermia via a reduction of energy flow from the air to the body. As long as a dromedary's body temperature is below the air temperature, a gradient exists that causes the dromedary to absorb heat from the air. At a body temperature of 40.5°C, the dromedary's temperature is equal to that of the air for much of the day, so there is no net heat exchange; thus the dromedary saves an additional quantity of water by eliminating the temperature gradient between its body and the air. The combined effect of these measures on water loss is illustrated by the data in **Table 14.2**. When deprived of water, a young dromedary reduced its evaporative water loss by 64% and reduced its total daily water loss by more than 50%.

The strategy a dromedary uses is basically the same as that employed by an antelope ground squirrel: saving water by allowing the body temperature to rise until the heat can be dissipated passively. The difference between the two animals is a consequence of the difference in their body sizes. A dromedary weighs 500 kg and can store heat for an entire day and cool off at night, whereas an antelope ground squirrel weighs about 100 g and heats and cools many times during the course of a day.

Keeping a cool head The brain is more sensitive to hyperthermia than other parts of the body, and brain temperatures higher than 43°C rapidly cause damage in most mammals. However, dromedaries and many other large desert mammals can maintain rectal temperatures well above 43°C for hours. They keep their brain temperatures lower than their body temperatures by using a countercurrent heat exchange (see Figure 4.2B) to cool blood before it reaches the brain. The blood supply to the brain passes via the external carotid arteries, and at the base of the brain the arteries break into a rete mirabile (network of capillaries) that lies within a structure called the cavernous sinus (**Figure 14.14**). Blood in the sinus is returning from the walls of the nasal passages where it has been cooled by the evaporation of water. This chilled venous blood cools the warmer arterial blood before it reaches the brain.

A mechanism that produces selective cooling of the brain is widespread among vertebrates, but the anatomical details vary. Many ungulates (antelope, sheep, goats) have a rete mirabile within a cavernous sinus as in Figure 14.14, but horses cool the blood in the internal carotid arteries by passing it through the guttural pouches—outgrowths from the auditory tubes that envelope the internal carotid arteries and are filled with air that is cooler than the blood. Birds use the ophthalmic capillary network to cool blood flowing to the brain. Blood in the ophthalmic capillaries, which lie behind the eyes, is cooled by evaporation of water from the surface of the eyes.

14.5 Evolution of Endothermy

LEARNING OBJECTIVES

14.5.1 Briefly describe four models for the evolution of endothermy.

14.5.2 Give an example of how elements of different models might act synergistically to enhance the value of endothermy.

The question of how endothermy evolved has engaged biologists for decades. The benefits of endothermy as seen in birds and mammals are obvious, but birds and

Table 14.2 Daily water loss of a 250-kg dromedary

Condition	Water loss (L/day) by different routes			
	Urine	Feces	Evaporation	Total
Drinking daily (average of 8 days)	0.9	1.0	10.4	12.3
Not drinking (average of 17 days)	1.4	0.8	3.7	5.9

Source: From K. Schmidt-Nielsen et al. 1956. *Am J Physiol* 188: 103–112.

mammals are highly derived endotherms. Thus, even though comparing extant endotherms with extant ectotherms provides information about how each mode of thermoregulation functions today, it offers little insight into how endothermy evolved. How could so energetically expensive an innovation as endothermy find a place in a world where it was competing with a preexisting, less expensive way to regulate body temperature? How did endothermy increase reproductive success enough to offset its energetic cost?

How did endothermy evolve?

Most explanations of the evolution of endothermy focus either on the direct benefits of having a high and stable body temperature, or on the indirect evolution of endothermy as a by-product of selection for some other character. The four hypotheses discussed below propose proximate benefits that could offset the energetic cost of endothermy. They are not mutually exclusive; more than one of them could apply at different stages during evolutionary change, and selective factors could have been different for mammals and birds.

Direct selection for a high and stable body temperature
The **thermogenic opportunity model** originated with the assumption that basal mammals were nocturnal, which lessened their competition with the predominantly diurnal reptiles of the Mesozoic. This assumption rests partly on characteristics of the visual pigments of extant mammals that suggest such a nocturnal stage in their evolution, and partly on the observation that most extant mammals of the same body size as basal mammals are nocturnal. Even in the comparative warmth of the Mesozoic, nights were sometimes chilly and endothermy might have been advantageous for nocturnal activity.

The **warmer is better model** proposes that an organism benefits from having a high body temperature because biochemical and physiological processes proceed faster or more forcefully as temperature increases. Of course, these rates do not continue to increase indefinitely. At about 45°C many enzymes begin to lose catalytic effectiveness, and many birds and mammals regulate body temperature between 37.5°C and 42.5°C—that is, these animals are close to the maximum temperature for enzyme function but have a margin of safety.

Indirect selection for characters that depend on high and stable body temperature   The **aerobic scope model** focuses on how changes in foraging mode could have affected physiology in the ancestors of birds and mammals. Both ectotherms and endotherms increase their rates of metabolism when they engage in high levels of activity. As an approximation, maximum rates of aerobic metabolism are about 10 times the resting rates. The aerobic scope model proposes that an increase in the maximum metabolic rate must be achieved by an increase in the resting rate and, consequently, by higher rates of internal heat production when an animal is at rest as well as when it is active.

The changes in postcranial skeletons and lung ventilation of sauropsids and synapsids described in the preceding chapter suggest that these animals were developing the changes in the respiratory and circulatory systems that allowed higher aerobic metabolic rates during activity.

The **parental care model** is another hypothesis that attributes the evolution of endothermy to selection for a different character. Endothermy could have several benefits for reproduction, including:

- A warmer temperature increases both the rate of embryonic development and the growth of neonates, shortening the amount of time that neonates are dependent on the parents.

- A stable temperature can increase the viability of embryos by preventing some developmental defects.

- Because some elements of developmental plasticity are temperature-sensitive, the mother can influence the phenotype of her offspring by controlling incubation temperature.

- If some of these mechanisms were acting, selection for reproductive success could have created indirect selection for endothermy.

Evaluating the models

The energetic cost of endothermy is enormous, and that cost must be offset by benefits that are still greater. Incipient endotherms could have allocated the energy used for thermogenesis to other activities, such as faster growth or having more or larger offspring. A model for the evolution of endothermy should present a convincing reason why investing energy in thermogenesis leads to a greater increase in fitness than would have resulted from other uses of that energy.

Although the aerobic scope model has been widely accepted, it has two major weaknesses:

- It rests on the assumption that there is an obligate mechanistic link between resting and maximum metabolic rates, but most comparisons of extant species do not reveal the expected positive correlation between those rates. Failure to find that correlation undermines the hypothesis that selection acting on the maximum metabolic rate would change the resting rate.

- A large proportion of the heat production by extant endotherms occurs in visceral organs, not in the skeletal muscles. Thus, it is not certain that an increase in skeletal muscle metabolism would have much effect on the resting metabolic rate or on thermogenesis when an animal was inactive.

The parental care model also has at least two weaknesses:

- Thermogenesis is energetically expensive, and endotherms incur that expense all year. If the benefit of thermogenesis is realized during reproduction, why not be endothermal for only that part of the year?

- Males play no role in embryogenesis in viviparous species of either synapsids or sauropsids, so why are males endothermal? Participation of males in incubating eggs and neonates appears to be an ancestral character of birds, but there is no evidence of an ancestral role of male mammals in caring for neonates.

Many factors?

As we stated at the start of this section, these four models for the evolution of endothermy are not mutually exclusive. On the contrary, it is likely that an increase in thermogenesis resulting from one model would change the playing field, perhaps making additional heat from other models more beneficial and ultimately leading to a situation in which endothermy is beneficial for both sexes and at all times of the year. For example, selection for the ability to process food and assimilate energy to improve parental care could have favored higher rates of metabolism by the visceral organs and greater food consumption, which in turn would have established a selective advantage for greater ability to search for food via increases in the metabolic capacity of muscles, and expansion of foraging into the nocturnal predator adaptive zone.

14.6 Thermoregulation, Energy Use, and Body Size

LEARNING OBJECTIVES

14.6.1 Explain how a vertebrate's mode of thermoregulation affects its energy requirements.

14.6.2 Explain the difference in the minimum body sizes of ectotherms versus endotherms.

14.6.3 Define gigantothermy and its application to the metabolic rates of dinosaurs.

Life as an animal is costly. In thermodynamic terms, an animal lives by breaking chemical bonds that were formed by plants (if the animal is herbivorous) or by other animals (if it is carnivorous). The animal then uses energy released from those bonds to sustain its own activities. Big animals require more energy (i.e., more food) than small ones, and active animals use more energy than sedentary ones. Thus, vertebrates are particularly expensive because they generally are larger and more mobile than invertebrates.

Energy requirements

In addition to body size and activity, a vertebrate's method of temperature regulation (ectothermy, endothermy, or a combination of the two mechanisms) is a key factor in determining how much energy it uses and therefore how much food it requires. Because ectotherms rely on external sources of energy to raise their body temperatures to the level needed for activity, whereas endotherms use heat generated internally by their metabolism, ectotherms use substantially less energy than

do endotherms. The metabolic rates (rates at which energy is used) of birds and mammals are 7–10 times the metabolic rates of lissamphibians and reptiles of the same body size (**Figure 14.15**).

Body size

Body size is a major difference between ectotherms and endotherms that relates directly to their mode of temperature regulation. The smallest ectothermal tetrapods are much smaller than the smallest endotherms. Few adult birds and mammals have body weights less than 10 g. The smallest species of birds and mammals weigh about 2 g, but many ectotherms are one-tenth that size (0.2 g, or even less; **Figure 14.16**). Amphibians are especially small: 65% of salamanders and 50% of frogs are lighter than 5 g, and 20% of salamander species and 17% of frog species weigh less than 1 g.

Squamates (lizards and snakes) are generally larger than amphibians, but 8% of lizard species and 2% of snake species weigh less than 1 g. At the opposite end of the spectrum, only 3–4% of species of amphibians and squamates weigh more than 100 g, whereas 31% of species of birds and 52% of mammals are heavier than that.

The high energetic cost of endothermy at small body sizes accounts for this difference. As body weight decreases, the weight-specific cost of living (energy expended per gram of body tissue) for an endotherm increases rapidly, becoming nearly infinite at very small body sizes. This is a finite world, and infinite energy requirements are not

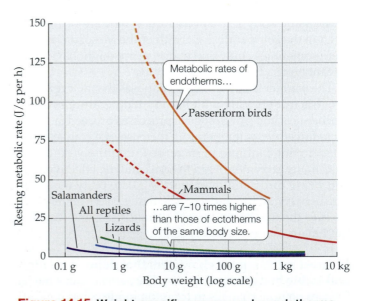

Figure 14.15 Weight-specific energy use by endotherms and ectotherms. Resting metabolic rates of ectothermal and endothermal tetrapods are shown as a function of body size. Metabolic rates of endotherms are 7–10 times higher than those of ectotherms of the same body size. Because weight-specific energy requirements increase at small body sizes, and because rates for ectotherms are substantially lower than those for endotherms, ectotherms can be smaller than endotherms. Dashed portions of the lines for passeriform (perching) birds and mammals show hypothetical extensions into body sizes below the minimum for adults of most species. (After F.H. Pough 1980. *Am Natur* 115: 92–112.)

(A) Smallest mammal, Etruscan shrew, *Suncus etruscus*

Courtesy of Roberto Sindaco

(B) Smallest bird, bumblebee hummingbird, *Selasphorus heloisa*

Ron Knight/CC BY 2.0

(C) Smallest lizard,
    Brookesia nana

From F. Glaw et al.
2021 *Sci Reports* 11: 2522

(D) Smallest frog,
    Paedophryne amauensis

Courtesy of Chris Austin

Figure 14.16 **The smallest species of tetrapod endotherms are enormous compared to the smallest ectotherms.** The smallest mammal, the Etruscan shrew from southern Europe and northern Africa (A) and the smallest bird, the bumblebee hummingbird of Mexico (B), both weigh about 2 g. The smallest lizard, a chameleon from Madagascar (C), weighs 0.2 g. The smallest frog, *Paedophryne amauensis* from Papua New Guinea (D), weighs about 0.1 g.

feasible. Thus, energy apparently sets a lower limit to the body size possible for an endotherm.

The weight-specific energy requirements of ectotherms also increase at small body sizes, but because the energy requirements of ectotherms are about one-tenth those of endotherms of the same body size, an ectotherm can be an order of magnitude smaller than an endotherm before its metabolic rate becomes impossibly large.

Lissamphibians are especially small, and squamates are only slightly larger. These tiny vertebrates can fall prey to spiders, centipedes, beetles, and scorpions (**Figure 14.17**).

Gigantothermy and the body temperatures of dinosaurs

The popularization of "hot-blooded" dinosaurs in the 1970s initiated a controversy about metabolic rates of dinosaurs and their body temperatures that has continued to the present day. Initially, proponents of hot-blooded dinosaurs insisted that dinosaurs must have had metabolic rates similar to those of mammals rather than the much lower metabolic rates characteristic of reptiles.

With a half-century of hindsight, it has become clear that the relationship between high metabolic rates and stable body temperatures that is characteristic of extant endotherms is not applicable to dinosaurs. Body temperatures and thermoregulatory mechanisms are intimately tied to body size, and many dinosaurs were much larger than any extant tetrapods. As a result, the link between stable body temperatures and high metabolic rates seen among extant vertebrates does not apply directly to dinosaurs.

Gigantothermy is a form of thermoregulation characteristic of large animals that have low metabolic rates but nonetheless maintain relatively stable body temperatures that are higher than their surroundings as a result of having low surface/volume ratios. The weight-specific metabolic rates (J/g/h) of vertebrates decrease as body size increases. The weight-specific metabolic rates of elephants are nearly

(A) A spider in its retreat captures and eats a frog

From T.R. Fulgence et al. 2020. *Ecol Evol* 2020;00:1-6/CC BY 4.0

(B) A centipede eating a gecko

Roman Königshofer/CC BY-ND 2.0

(C) A tarantula eating a snake

Courtesy of Gabriela Franzoi Dri

Figure 14.17 **Spiders and centipedes prey on small vertebrates.** (A) This Madagascan spider creates traps for frogs by weaving two leaves together. The spider hides at the back of the enclosure and seizes frogs when they enter the retreat. (B,C) Arthropods such as centipedes and spiders can attack and kill small species of lizards and snakes.

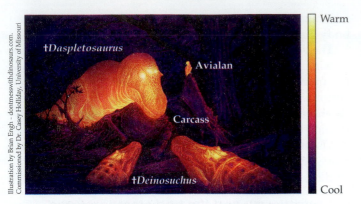

Warm

Cool

Figure 14.18 Thermal windows in the skull? The fronto-parietal fossa (depression) of archosaurs, unlike the fenestrae in the skull, is not associated with jaw muscles. Instead, it contains a dense array of blood vessels that may have provided a cooling system for the brain. This reconstruction of a scene from the Late Cretaceous shows a thermographic view of a theropod (*Daspletosaurus*) and two crocodylians (*Deinosuchus*) facing off at the carcass of a ceratopsian (horned dinosaur) while an avialian (basal bird) watches. The frontoparietal fossae of the theropod and the crocodylians are bright yellow, indicating that they are radiating heat.

as low as those of lizards and crocodylians, even though elephants are homeothermal endotherms. Indeed, during the Pleistocene the geographic range of woolly mammoths extended north of the Arctic Circle.

One biophysical model assumes that dinosaurs had metabolic rates like those of extant crocodylians and predicts that dinosaurs weighing 2,000 kg or more would have had body temperatures above 30°C with a day–night variation of less than 2°C, even in winter or in cold northern and southern latitudes.

Thus, gigantothermy would have allowed large dinosaurs to maintain stable core body temperatures with the low metabolic rates that are characteristic of extant crocodylians. Feathers would not have been necessary for insulation for large dinosaurs; indeed, overheating could have been their major problem, and dinosaurs may have had thermal windows in their skulls to cool their brains (**Figure 14.18**).

14.7 Ectotherms, Endotherms, and Ecosystems

LEARNING OBJECTIVE

14.7.1 Describe how the difference in partitioning of ingested energy by ectotherms and endotherms can affect an entire ecosystem.

The amount of energy required by ectotherms and endotherms is not the only important difference between them; equally significant is what they do with that energy once they have it. Endotherms expend more than 90% of the energy they take in to produce heat to maintain their high body temperatures. Less than 10%—and often as little as 1%—of the energy a bird or mammal assimilates

is available for net conversion (that is, increasing the species' biomass by growth of an individual or production of young). Ectotherms do not rely on metabolic heat; the solar energy they use to warm their bodies is free in the sense that it is not drawn from their food. Thus, most of the energy they ingest is converted into the biomass of their species. On average, ectotherms devote about 50% of the energy in their food to production of new biomass (**Figure 14.19**).

Because of the difference in how energy is used, a given amount of food consumed by an ectotherm produces a much larger increase in biomass than it would in an endotherm. A study at the Hubbard Brook Experimental Forest in New Hampshire showed that although the energy consumption of salamanders was only 20% that of the birds or small mammals in the watershed, the conversion efficiency of salamanders was so high that the annual increment of salamander biomass was equal to that of birds or small mammals.

To say that only 20% of the energy consumption of a terrestrial ecosystem passes through ectotherms makes them sound unimportant relative to birds and mammals, but that conclusion overlooks a critical point. Because many ectotherms are extremely small, the prey species they eat are also tiny—insects the size of fruit flies, for example, and mites that are no larger than poppy seeds. These prey items are too small to be captured by a bird or mammal. But small ectotherms can efficiently convert these minute packages of energy into frog or lizard biomass, and frogs and lizards are large enough for birds and mammals to capture.

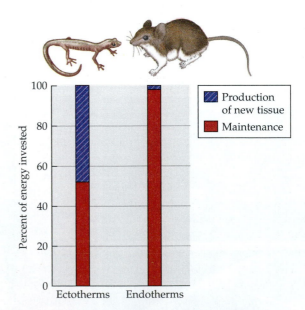

Figure 14.19 Efficiency of biomass production.
On average, ectotherms convert 25 times more of the food they consume to growth and reproduction than endotherms. (Data from F.H. Pough 1980. *Am Natur* 115: 92–112.)

Summary

14.1 Why Regulate Body Temperature?

The rates of biological processes at all levels from metabolic pathways to whole-animal functions are affected by temperature.

Different physiological processes have different temperature sensitivities. Maintaining a stable body temperature (thermoregulation) facilitates coordination of these processes.

The terms "ectothermy" and "endothermy" identify the primary source of the heat energy used to control body temperature: from outside the body (ectothermy) or inside the body (endothermy). "Homeothermy" and "heterothermy" describe the magnitude of variation in body temperature: homeothermy means a stable body temperature, heterothermy means a variable body temperature.

Vertebrates mix and match their thermoregulatory characteristics; endotherms are sometimes heterothermal and ectotherms are sometimes homeothermal.

14.2 Ectothermal Thermoregulation

Ectothermy is the ancestral mode of thermoregulation for tetrapods. It is energetically efficient, and ectotherms are capable of precise control of body temperature, but only when external sources of heat are available.

Solar radiation is a route of heat gain, whereas thermal (infrared) radiation can produce a net gain or net loss of heat. Convective exchange with the air and conductive exchange with the ground can also be routes of gain or loss depending on the relative temperatures of an animal's body surface and the air or ground. Evaporation of water lowers body temperature and metabolic heat production increases body temperature.

Ectotherms balance heat gain from the environment with heat lost to the environment primarily by behavior. Moving between sun and shade is the most obvious thermoregulatory behavior, and small changes in orientation, posture, and location within a habitat fine-tune ectothermal control.

Many ectotherms maintain their body temperature within a range of 2–3°C during activity and metabolic reactions, physiological processes, and whole-animal performance all reach their maxima within this range of temperature.

14.3 Endothermal Thermoregulation

Endothermy allows birds and mammals substantial independence from low environmental temperatures. Endotherms can be active at night and in the winter and can be active year-round in polar and high-mountain environments.

These benefits of endothermy come at an energetic cost. The metabolic rates of endotherms are about ten times higher than those of ectotherms of the same body size, and on a gram-per-gram basis endotherms must find and consume more food than ectotherms of the same body weight.

Endotherms regulate their body temperatures by matching their rate of metabolic heat production to their rate of heat loss. In the normothermal range of environmental temperatures, changes in insulation are sufficient to balance heat loss with heat production. As temperatures cool, endotherms increase metabolic heat production to maintain a stable body temperature. At higher temperatures they use evaporative cooling.

The heart and kidneys are the sources of most metabolic heat production when an endotherm is at rest. Metabolism of skeletal muscle makes little contribution to metabolic heat production by a resting endotherm.

Birds and mammals can use shivering and non-shivering thermogenesis to increase metabolic heat production.

Hair and feathers insulate an endotherm by trapping air. Their insulative value can be increased by raising the hair or feathers to deepen the layer of trapped air and decreased by allowing the hair or feathers to lie flat.

Evaporative cooling is the primary method that endotherms use to increase heat loss. Many mammals pant and birds use gular fluttering; both methods cool the respiratory system. Some mammals sweat, which cools the skin.

14.4 Pure Ectothermy and Pure Endothermy Lie at the Extremes of a Continuum.

Many vertebrates combine ectothermal and endothermal mechanisms to maintain body temperature.

Females of several species of pythons use heat from muscular contractions to warm their eggs, and black-and-white tegu lizards use metabolic heat to raise body temperature during the reproductive season.

Many birds and some mammals reduce energy consumption by engaging in rest-phase torpor, allowing their body temperatures to fall several degrees when they are inactive, and using metabolic heat to warm to normothermia when they resume activity. Nocturnal torpor can save substantial energy compared to remaining normothermal overnight.

Some birds and mammals use solar energy to raise body temperature in the morning.

Some mammals hibernate, allowing body temperatures to fall dramatically and remaining torpid for months at a time. These periods of torpor are interrupted by brief periods of warming, and those periods of normothermy account for most of the energy used by a hibernating animal.

Tolerating hyperthermia allows some desert animals to conserve water. Small diurnal rodents such as ground squirrels allow their body temperatures to rise while they

(Continued)

forage on the ground surface. They then retreat to burrows to cool off before resuming surface activity.

Dromedaries avoid the need for evaporative cooling by tolerating day-long periods of hyperthermia. When they are unable to drink, dromedaries allow their body temperatures to rise during the day and fall at night.

The brain is more sensitive to high temperature than other parts of the body are, and many mammals and birds use countercurrent heat exchange to cool blood before it reaches the brain.

14.5 Evolution of Endothermy

The evolution of endothermy was a major event in vertebrate evolution and endothermy evolved independently in the sauropsid and synapsid lineages.

Four hypotheses about the evolution of endothermy can be loosely assigned to two categories:

- The direct benefits of having a high and constant body temperature (e.g., nocturnal activity or higher rates of physiological processes).

- Selection for a character that secondarily increased metabolic rate and thus metabolic thermogenesis (e.g., a higher aerobic metabolic scope to sustain high levels of activity, or a warmer maternal body temperature that facilitates embryonic development).

These hypotheses are not mutually exclusive, and it is likely that many selective forces were involved.

14.6 Thermoregulation, Energy Use, and Body Size

Because an ectotherm relies on external sources of heat and has no metabolic cost of thermoregulation, it uses only one-seventh to one-tenth as much energy as an endotherm of the same body weight.

Weight-specific metabolic rate increases as body size decreases. That relationship holds for both ectotherms and endotherms, but because the metabolic rate of an ectotherm is an order of magnitude lower than that of an endotherm, ectotherms can maintain smaller body sizes than endotherms without the cost of living becoming prohibitive.

Few endotherms are smaller than 20 g, but most lizards, frogs, and salamanders weigh 20 g or less, and many weigh less than 1 g. Many lissamphibians and lizards are so small that their important predators and competitors are arthropods.

14.7 Ectotherms, Endotherms, and Ecosystems

Endotherms use most of the energy in the food they eat to stay warm, whereas ectotherms devote most of their energy to growth and reproduction.

The combination of small body sizes and efficient conversion of food into their own biomass creates a unique role for ectotherms in terrestrial ecosystems, consuming prey items that are too small for endotherms to capture, and efficiently converting that energy into packages that are large enough for endotherms to eat. As a result, ectothermal vertebrates produce as much new biomass each year as do birds and mammals.

Discussion Questions

14.1 Why does an endotherm's metabolic rate increase above the upper lethal temperature? And why does the metabolic rate fall below the lower lethal temperature?

14.2 The density of sweat glands in human skin is 10 times that of a chimpanzee's as the result of enhanced expression of the *Engrailed 1* gene. Why are sweat glands more important for humans than for chimpanzees?

14.3 Although most fishes are ectotherms, the efficiency of secondary production by fishes is substantially lower than the average of 50% measured for terrestrial ectotherms. What difference between aquatic and terrestrial environments might account for this difference?

14.4 Most mammals have a layer of cutaneous fat beneath the skin, but dromedaries concentrate fat storage in their hump. What is the significance of that difference for the thermoregulation of dromedaries?

14.5 Sea otters, *Enhydra lutris*, are the smallest marine mammals. They live in the north Pacific, where water temperatures range from 0° to ~15°C. Sea otters have a resting metabolic rate about 3 times the rate expected from their body size. What does this observation suggest about their ability to live in cold water?

14.6 To qualify as a scientific hypothesis, a proposed explanation of a phenomenon must be capable of being falsified—that is, disproved by an experiment or by application of data previously gathered. Do any of the proposed explanations of the evolution of endothermy (thermogenic opportunity, warmer is better, indirect selection, parental care) meet that criterion?

© René C. Clark, Dancing Snake Nature Photography

15

Lepidosaurs
Tuatara, Lizards, and Snakes

Male zebra-tailed lizard, *Callisaurus draconoides*

The crown group Lepidosauria is one of the two largest clades of tetrapods, rivaled only by birds. Lepidosaurs include a single species of rhynchocephalian (the tuatara, *Sphenodon punctatus*, the only extant representative of a lineage that thrived during the Mesozoic). Far more familiar, however, are the 7,000+ species of lizards and 4,000+ species of snakes. Although some species are secondarily aquatic, lepidosaurs are predominantly terrestrial tetrapods and occur on every continent except Antarctica (**Table 15.1**).

Within Lepidosauria, Rhynchocephalia (the tuatara) is the sister group of Squamata (lizards and snakes). Snakes (Serpentes) are nested within Squamata (**Figure 15.1**). Although they are distinct in their morphology, lizards and snakes cannot be distinguished phylogenetically, and "lizards" as the word is commonly used is a paraphyletic group that does not include all the descendants of a common ancestor (see Section 1.4). However, lizards and snakes are distinct in many aspects of morphology, ecology, and behavior, and the colloquial separation is useful when discussing them.

In this chapter, we provide a brief overview of the diverse groups of lepidosaurs and their body forms and habitats. We then cover some of the myriad specializations that have arisen in different groups in regard to feeding and foraging, social behavior, and reproduction. We close with a brief discussion of the major threats to the lineage posed by anthropogenic climate change.

15.1 Characters and Diversity of Lepidosaurs

LEARNING OBJECTIVES

15.1.1 Identify the major synapomorphies of lepidosaurs and squam ates.

15.1.2 Describe the ecology and unique features of the tuatara.

15.1.3 Explain the ecological significance of determinate growth for lizards.

15.1.4 Describe the body forms and characteristics of different ecomorphs of lizards and snakes.

15.1.5 Describe the different modes of locomotion of snakes and explain how they are used.

Just as Lissamphibia is characterized by smooth skin (see Chapter 12), skin with overlapping, keratinous scales is synapomorphic for Lepidosauria (Greek *lepisma*, "scale"; *sauros*, "lizard"). Successive generations of epidermal cells are shed at intervals (**Figure 15.2A,B**). Lepidosaurs have a transverse cloacal slit rather than the longitudinal slit that characterizes other tetrapods, and autotomy planes in the caudal vertebrae of many species allow the tail to be shed if it is seized by a predator and then regenerated.

Determinate growth (growth that stops at some genetically determined point rather than continuing throughout the organism's life) is a significant synapomorphy of squamates—which includes all lepidosaurs except the single species of tuatara (see Figure 15.1). Determinate growth is particularly significant for lizards, as we will discuss later in this section. Leglessness is also widely seen; reduction or complete loss of limbs is widespread among some groups of lizards, and all snakes are limbless.

Male squamates have bilateral penises called hemipenes (Greek *hemi*, "half"), but only one is used during copulation (**Figure 15.2C,D**). Behavioral observations have shown that individual snakes and lizards preferentially employ one hemipenis or the other—that is, they are either right- or left-penised.

Figure 15.1 Phylogenetic relationships of Lepidosauria. Lizards and snakes cannot be distinguished phylogenetically; thus "lizards" is a paraphyletic group (i.e., it does not include all the descendants of a common ancestor). The morphologically distinct snakes comprise the monophyletic group Serpentes within Squamata. (After K. Liem et al. 2001. *Functional Anatomy of the Vertebrates*, 3rd ed. Cengage/Harcourt College: Belmont, CA.)

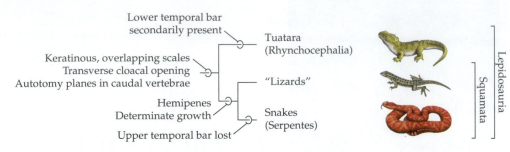

Lower temporal bar secondarily present — Tuatara (Rhynchocephalia)

Keratinous, overlapping scales
Transverse cloacal opening
Autotomy planes in caudal vertebrae

Hemipenes
Determinate growth — Snakes (Serpentes)

Upper temporal bar lost

"Lizards"

Lepidosauria
Squamata

(A) Lizards shed their skin in pieces

(B) Snakes shed their skin intact

(C) Hemipenes of a western diamondback rattlesnake, *Crotalus atrox*

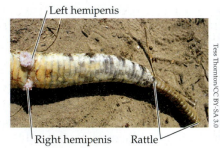

Left hemipenis

Right hemipenis   Rattle

♀

♂

Hemipenis inserted into female's cloaca

(D) Mating house geckos, *Hemidactylus turcicus*

♀

♂

Figure 15.2 Lepidosaurs shed their skins and squamates have two penises. Rough, scaled skin that is shed is a derived character of lepidosaurs. (A) Lizards (and the tuatara) shed the skin in pieces. (B) Snakes shed the skin intact, leaving a "shell" of discarded skin behind. (C) The everted hemipenes of a male rattlesnake show the spines that hold a hemipenis in the cloaca of the female. (D) Intromission by house geckos. The male has curved his tail beneath the female's tail and inserted his left hemipenis into her cloaca.

Modifications of the skull of squamates create flexibility. The frontal and parietal bones meet in a straight line that acts as a hinge, allowing the snout to flex upward. In addition, the quadrates are flexibly attached to the skull and move during feeding. Snakes lost the upper temporal bar, which allows their quadrate to move in three dimensions. Snakes also lost autotomy planes in the tail; a snake cannot regenerate a lost portion of its tail.

Rhynchocephalians and the biology of tuatara

Rhynchocephalians were once a diverse group, with a worldwide distribution in the Mesozoic. Triassic rhynchocephalians were small, with body lengths of only 15–35 cm, and were probably insectivores. During the Jurassic and Cretaceous, rhynchocephalians included herbivorous and marine forms, some of which reached lengths of 1.5 m.

Today, the tuatara (Maori, "spines on the back") is the only extant rhynchocephalian (**Figure 15.3**). Tuatara (no *s* is added to form the plural) once inhabited both the North and South Islands of New Zealand, but the advent of the Maori and the Pacific rats that arrived with them exterminated tuatara on these two main islands about 800 years ago. Today the species is restricted to 32 natural populations on small islands off the coast of New Zealand, 9 translocated populations on small islands, and 5 translocated populations in sanctuaries on the North and South Islands.

Adult tuatara are about 60 cm long. They are nocturnal, and in the cool, foggy nights that characterize their island habitats they cannot raise their body temperatures by basking in

Henry the tuatara, *Sphenodon punctatus*

New Zealand 5¢ coin

Figure 15.3 Henry the tuatara (*Sphenodon punctatus*). Henry came to the Southland Museum in Invercargill, New Zealand, in 1970 and was the model for the New Zealand 5-cent coin.

sunlight as lizards do. Body temperatures ranging from 6°C to 16°C have been reported for active tuatara, and these are low compared with the temperatures of most lizards. During the day, tuatara do bask, raising their body temperatures to 28°C or higher.

Tuatara feed largely on invertebrates but occasionally will eat a frog, lizard, or seabird. The jaws and teeth of tuatara produce a shearing effect during chewing. The upper jaw contains two rows of teeth on each side, one on the maxilla and the other on the palatine. The teeth of the lower jaw fit between the two rows of upper teeth, and the lower jaw closes with an initial vertical movement, followed by an anterior sliding movement. As the lower jaw slides, the food item is bent or sheared between the triangular cusps on the teeth of the upper and lower jaws.

Tuatara live in burrows that they may share with nesting seabirds. The burrows are spaced at intervals of 2–3 m in dense colonies, and both male and female tuatara are territorial. They use vocalizations, behavioral displays, and color change in their social interactions.

The ecology of tuatara rests to a large extent on the exploitation of resources provided by seabird colonies. Tuatara occasionally feed on adult birds, which are most vulnerable to predation at night. More important, the quantities of guano produced by the birds, the scraps of food they bring to their nestlings, and the bodies of dead nestlings attract huge numbers of arthropods that tuatara eat. These arthropods are largely nocturnal and must be hunted when they are active. Thus, the nocturnal activity of tuatara and the low body temperatures resulting from being active at night are probably derived characters that stem from the association of tuatara with colonies of nesting seabirds.

Squamata: Lizards

As mentioned at the start of this section, determinate growth is a derived character for Squamata, and it is especially significant for lizards because it allows adults of many species to remain small enough to eat insects. About 80% of extant species of lizards weigh less than 20 g as adults—smaller than a white mouse (see Section 14.6). Generalized lizards of this size can readily capture insects, whereas large insect-eating vertebrates, such as mammalian anteaters, require morphological specializations to capture tiny prey.

Body forms   Lizards range in size from diminutive geckos and chameleons less than 3 cm long to the Komodo monitor (*Varanus komodoensis*, popularly known as the Komodo dragon), which is 3 m long at maturity and can weigh as much as 70 kg. The Komodo monitor can kill adult water buffalo, but deer and feral goats are its usual prey. Giant varanids were widespread in Eurasia and Australasia during the Neogene. During the Pleistocene, even larger monitors inhabited the islands between Australia and Indonesia and may have preyed on pygmy elephants that also lived on the islands at that time. A reconstruction of the skeleton of a fossil monitor, †*Varanus priscus* from the Pleistocene of Australia, is 5.5 m long, and in life the lizard may have weighed 600 kg. †*Varanus priscus* could well have preyed on humans, as fossil dates indicate that the two species coexisted for at least 10,000 years.

Lizards display an enormous variety of body forms and prey preferences. North and Central American spiny lizards (**Figure 15.4A**) are examples of small, generalized insectivores. Other small lizards have specialized diets: the North American horned lizards (**Figure 15.4B**) and the Australian thorny devil (**Figure 15.4C**) feed on ants. These ant specialists have dorsoventrally flattened bodies, short legs and spines that deter predators.

Most large lizards are herbivores, and many herbivorous lizards, such as the green basilisk (**Figure 15.4D**) are arboreal. Monitor lizards such as goannas (**Figure 15.4E**) and the Komodo dragon mentioned earlier feed on invertebrates and vertebrates, including birds and mammals, and thus are exceptions to the generalization that lizards too large to subsist on insects are herbivores. Another clade of large carnivores, the venomous Gila monster (**Figure 15.4F**) and the four species of beaded lizards are slow-moving, preying on rodents and on eggs and hatchlings in bird nests. Gila monsters are primarily terrestrial, but beaded lizards climb readily and have been found in trees as much as 15 m from the ground.

Many lizard species are arboreal, and true chameleons are the most specialized of these. The toes on a chameleon's **zygodactylous** (Greek *zygos*, "joining"; *daktylos*, "finger or toe") feet are arranged in opposable groups that grasp branches firmly; additional security is provided by a prehensile tail (**Figure 15.4G**). The tongue and hyoid apparatus of chameleons are specialized

(A) Texas spiny lizard *Sceloporus olivaceus*

William L. Farr/CC BY-SA 4.0

(B) Texas horned lizard, *Phrynosoma cornutum*

Courtesy of Todd Pierson

(C) Thorny devil, *Moloch horridus*

Harvey Pough

(D) Green basilisk, *Basiliscus plumifrons*

Bernard DuPont/CC BY-SA 2.0

(E) Racehorse goanna, *Varanus gouldii*

Courtesy of Stephen Zozaya

(F) Gila monster, *Heloderma suspectum*

© René C. Clark, Dancing Snake Nature Photography

(G) Giant chameleon, *Furcifer oustaleti*

Courtesy of Stephen Zozaya

Figure 15.4 Body forms of lizards. (A) Most small lizards, such as the North American spiny lizard, are generalized insectivores. (B,C) North American horned lizards and the Australian thorny devil eat only ants. (D) The green basilisk is an arboreal herbivore. (E) Monitor lizards are elongate, fast-moving predators with long legs. (F) The venomous Gila monster is stout, short-legged, and slow. (G) Chameleons, the most specialized arboreal lizards, have zygodactylous, opposable toes, prehensile tails, and eyes in turrets that can swivel independently.

for ballistic tongue projection (which independently evolved in lissamphibians; see Figure 12.4): the lizard is able to project its tongue forward more than a body's length to capture insects that adhere to its sticky tip. This feeding mechanism requires good eyesight, especially the ability to gauge distances accurately so that the correct trajectory can be employed. Chameleon eyes are elevated in small cones and each eye can move independently. When a chameleon is at rest, its eyes swivel back and forth, viewing its surroundings. When it spots an insect, the lizard fixes both eyes on it and cautiously stalks to within shooting range.

Chameleons have two other distinctions. Labord's chameleon, *Furcifer labordi*, probably has the shortest life span of any tetrapod; and the dwarf chameleon, *Brookesia nana*, is probably the smallest species of amniote. *Furcifer labordi* spends 8 months as an egg, reaches adult size in less than 2 months after hatching, reproduces, and dies. *Brookesia nana* has an adult length of only 22 mm (see Figure 14.16B). Both species are from Madagascar, which has a rich chameleon fauna.

Limblessness Limb reduction has evolved more than 60 times among lizards, and limbless species are characterized by two ecomorphs. Surface-dwelling species have long tails and usually live in dense grass or shrubbery (**Figure 15.5A**). In such a setting, a slim, elongated body can maneuver without legs more easily than a short one, and a long tail contributes to sinusoidal locomotion. In contrast, for **fossorial** (burrowing) species, a tail makes little contribution to locomotion, and in fact creates friction as the animal tunnels. It is not surprising, then, that fossorial lizards have short tails (**Figure 15.5B**).

Amphisbaenians are fossorial lizards with specializations that differ from those of other squamates. Amphisbaenians use their heads for tunneling, and their skulls are rigidly constructed with many of the bones fused.

The head shapes of amphisbaenians correspond to three methods of burrowing.

- Blunt-snouted amphisbaenians burrow by ramming their head into the soil to compact it. Some species have heavily keratinized scales on the snout that are

Table 15.1 Major extant families of lizards[a]

Family	Description	Adult size[b]	Number and distribution of species
Flap-footed lizards (Pygopodidae)	Terrestrial, elongated lizards that move through grass and leaf litter with sinusoidal locomotion. Forelimbs absent, hindlimbs reduced to flaps	~6 to 30 cm	46 species in Australia
Geckos (Gekkonidae)	Terrestrial and arboreal lizards. Many have setae on their toes	~3 to 30 cm	1,433 species worldwide[c]
Skinks (Scincidae)	Terrestrial, arboreal, semifossorial, and semiaquatic lizards. Stout to elongated, many with reduced limbs	~10 to 40 cm	1,730 species worldwide[c]
Amphisbaenians (Amphisbaenia)	Fossorial, elongated lizards. All are legless except for the three species of *Bipes*, which have stout forelimbs	~10 to 80 cm	204 species in the West Indies, South America, Africa, Mediterranean
Lacertids (Lacertidae)	Terrestrial lizards, sometimes referred to as wall lizards. Parthenogenesis occurs in one genus	~10 to 80 cm	358 species in Europe, Africa, and Asia
Whiptails (Teiidae)	Terrestrial lizards, small species are called whiptails or racerunners. About a third of the species of *Aspidoscelis* and *Cnemidophorus* are parthenogenetic	~20 cm to ~1 m	173 species in North America to central South America and the West Indies
Chameleons (Chamaeleonidae)	Primarily arboreal, a few terrestrial species. All have zygodactylous feet, prehensile tails, and ballistic tongues	<3 to 50 cm	219 species in Africa, Madagascar, and southern Spain
Agamids (Agamidae)	Terrestrial or arboreal; a few species are semiaquatic. Australian agamids are called dragons	~10 cm to 1 m	551 species in the Middle East, Asia, Africa, Indoaustralian Archipelago, and Australia
Spiny lizards (Phrynosomatidae)	Terrestrial, rock-dwelling, and arboreal lizards	10 to 30 cm	165 species, North and Central America
Iguanas (Iguanidae)	Herbivorous lizards, terrestrial or arboreal. One semiaquatic marine species in the Galápagos	25 cm to >1 m	45 species, New World, Galápagos Islands, and Fiji
Anoles (Dactyloidae)	Primarily arboreal. Repeated radiations have made *Anolis* a model for studies of evolution and ecomorphology	10 to 50 cm	436 species in the Americas, including the West Indies
Beaded lizards (Helodermatidae)	Terrestrial, heavy-bodied, venomous lizards	50 cm to 1 m	5 species in the southwestern U.S., Mexico, and Central America
Anguids (Anguidae)	Elongate terrestrial lizards. Many are limbless surface-dwellers; *Anniella* are fossorial and *Abronia* are arboreal	20 cm to >1 m	88 species in the Americas, Europe, Middle East, and China
Monitor lizards (Varanidae)	Elongate lizards, primarily terrestrial, some arboreal or semiaquatic. Some are very large (the Komodo dragon, *Varanus komodoensis*, is the world's largest lizard). Most species carnivorous	~20 cm to 3 m	84 species in Australia, East Indies, Asia, and Africa
Snakes (Serpentes)	See Table 15.2		

[a]41 families of lizards are currently recognized. Only families discussed in the text are listed here.
[b]Size is total body length (nose to tail tip).
[c]All continents except Antarctica.

(A) Slender glass lizard, *Ophisaurus attenuatus*

Peter Paplanus/CC BY 2.0

Tail = 70% total length

(B) Checkerboard worm lizard, *Trogonophis wiegmanni*

Courtesy of Ernesto Raya

Tail = 6.5% total length

Figure 15.5 Legless lizards.   (A) The tail of the surface-dwelling slender glass lizard makes up most of its length. (B) In contrast, the tail is less than a tenth of the length of an amphisbaenian, the checkerboard worm lizard.

used with a rotary motion to shave material from the face of the tunnel (**Figure 15.6A**).

- Shovel-snouted species have flat, wedge-shaped heads. The snout is driven into the soil at the front of the tunnel, and the head is lifted to compress loosened soil against the roof of the tunnel (**Figure 15.6B**).

- The heads of keel-snouted amphisbaenians have a vertically oriented sharp-edged scale on the snout. The scale is driven into the soil at the front of the tunnel and anchors the anterior part of the body as it pushes loose soil from side to side in its tunnel (**Figure 15.6C**).

Blunt-snouted amphisbaenians are found near the surface, where the soil is loose. Shovel- and keel-snouted amphisbaenians are able to penetrate denser soil and burrow more deeply.

The skin of amphisbaenians is distinctive. The **annuli** (rings; singular *annulus*) that pass around the circumference of the body are readily apparent from external examination, and dissection shows that the integument is nearly free of connections to the trunk. Thus, the skin forms a tube within which the amphisbaenian's body can slide forward or backward. The separation of trunk and skin is employed during locomotion through tunnels. Integumentary muscles run longitudinally from annulus to annulus. The skin over this area of muscular contraction telescopes and buckles outward, anchoring that part of the amphisbaenian against the wall of its tunnel. Next, contraction of muscles that pass anteriorly from the vertebrae and ribs to the skin slide the trunk forward within the tube of integument. Amphisbaenians use this same mechanism to move backward along their tunnels, contracting muscles that pass posteriorly from the ribs to the skin. In fact, the name "amphisbaenian" is from the Greek *amphis*, "double" and *baino*, "to walk,"

and refers to this ability to move forward and backward with equal facility.

The dental structure of amphisbaenians is also distinctive. They have a single median tooth in the upper jaw, a feature unique to this group of vertebrates (**Figure 15.6D**). The median tooth is part of a specialized dental battery that makes amphisbaenians formidable predators, capable of subduing a wide variety of invertebrates and small vertebrates. The upper tooth fits into the space between two teeth in the lower jaw and forms a set of nippers that can bite out a piece of tissue from a prey item too large for the mouth to engulf.

Squamata: Serpentes

The phylogenetic affinities of Serpentes, the snakes, are nested within squamates (see Figure 15.1), and fossils show that Mesozoic snakes retained legs. All extant snakes are limbless, and they range in size from diminutive burrowing species that feed on termites and grow to only 10 cm, to constrictors that approach 10 m in length (**Table 15.2**).

The fossorial leptotyphlopids (thread snakes; see Figure 15.7D) and typhlopids (blind snakes) are probably similar to basal snakes. These burrowers have shiny scales and reduced eyes. Traces of the pelvic girdle remain in most species of these two groups, but the braincase is snakelike. These anatomical and ecological characters appear consistent with a longstanding hypothesis that snakes evolved from a subterranean lineage of lizards with greatly reduced eyes. The hypothesis that the eyes of extant surface-dwelling snakes secondarily redeveloped after nearly disappearing during a fossorial stage in their evolution explains differences in the visual systems of snakes and lizards, both in their retinal cells and in the way the eye is focused.

Unlike surface-dwelling legless lizards, surface-dwelling snakes have long trunks and relatively short tails.

(A) Angled worm lizard, *Agamodon anguliceps*, blunt-snouted

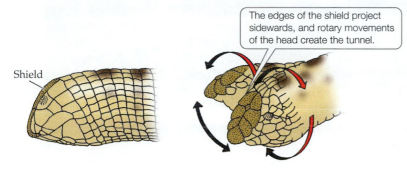

Shield

The edges of the shield project sidewards, and rotary movements of the head create the tunnel.

(B) Florida worm lizard, *Rhineura floridana*, shovel-snouted

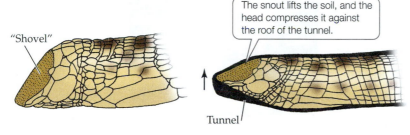

"Shovel"

The snout lifts the soil, and the head compresses it against the roof of the tunnel.

Tunnel

(C) King's worm lizard, *Amphisbaena kingii*, keel-snouted

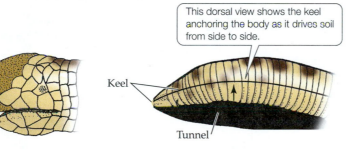

Keel

Keel

This dorsal view shows the keel anchoring the body as it drives soil from side to side.

Tunnel

(D) Median tooth of red worm lizard, *Amphisbaena alba*

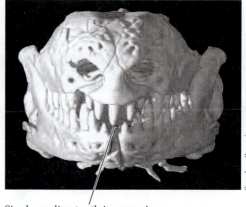

3D HRXCT reconstruction courtesy of DigiMorph.org

Single median tooth in upper jaw

Figure 15.6 Tunneling methods and dentition of amphisbaenians. (A) Some blunt-snouted amphisbaenians rotate the head as they burrow through loosely compacted soil. (B) Shovel-snouted species lift their heads to pack soil into the roof of the tunnel. (C) Keel-snouted species ram the sharp edge of their rostrum into the soil and use it as a fulcrum to pack loosened soil against the sides of the tunnel. (D) All amphisbaenians have a single median tooth in the upper jaw. (A–C adapted from C. Gans. 1974. *Biomechanics*. Philadelphia: Williams & Wilkins.)

This elongation of the trunk results from changed expression of developmental regulatory genes; the changes allow trunk formation to continue for a longer period during development.

Leglessness is also a function of gene expression during development. Among many other functions, the protein Sonic hedgehog (Shh) stimulates limb formation in vertebrates. Shh expression in turn is controlled by the regulatory protein ZRS, and the *ZRS* gene is mutated in snakes. The mutations reduce ZRS activity, and thus expression of Shh, suppressing limb formation. Replacing mouse ZRS with snake ZRS causes mice to be born with stumps instead of legs.

Body forms   The morphology of snakes is highly specialized, and correlated suites of characters identify parallel evolution of ecomorphs in different families. Generalized terrestrial snakes are elongated and have long tails (**Figure 15.7A**). The body length of these snakes is ~10–15 times their maximum circumference, and they feed on slim prey.

In contrast, some vipers, especially African vipers in the genus *Bitis* (**Figure 15.7B**), are exceedingly stout, with body lengths only 5 times their circumference. These snakes can swallow bulky prey, up to and including small antelope. Vipers have long fangs that inject proteolytic (protein-digesting) venom deep into the prey, first killing it and then digesting it from the inside out while the snake's digestive enzymes break it down from the outside in.

Arboreal snakes are exceedingly slim, with body lengths 20–25 times their circumference (**Figure 15.7C**). Their long, thin bodies allow them to spread their weight over many twigs, and they can extend their heads and necks to seize small lizards that sleep on leaves at the tips of branches where other predators cannot reach them.

Fossorial snakes have slim bodies and smooth scales. Like fossorial lizards, they have short tails. The most specialized fossorial snakes have extremely reduced eyes (**Figure 15.7D**).

Sea snakes lack ventral scales and have laterally compressed bodies, flattened tails, and valvular nostrils

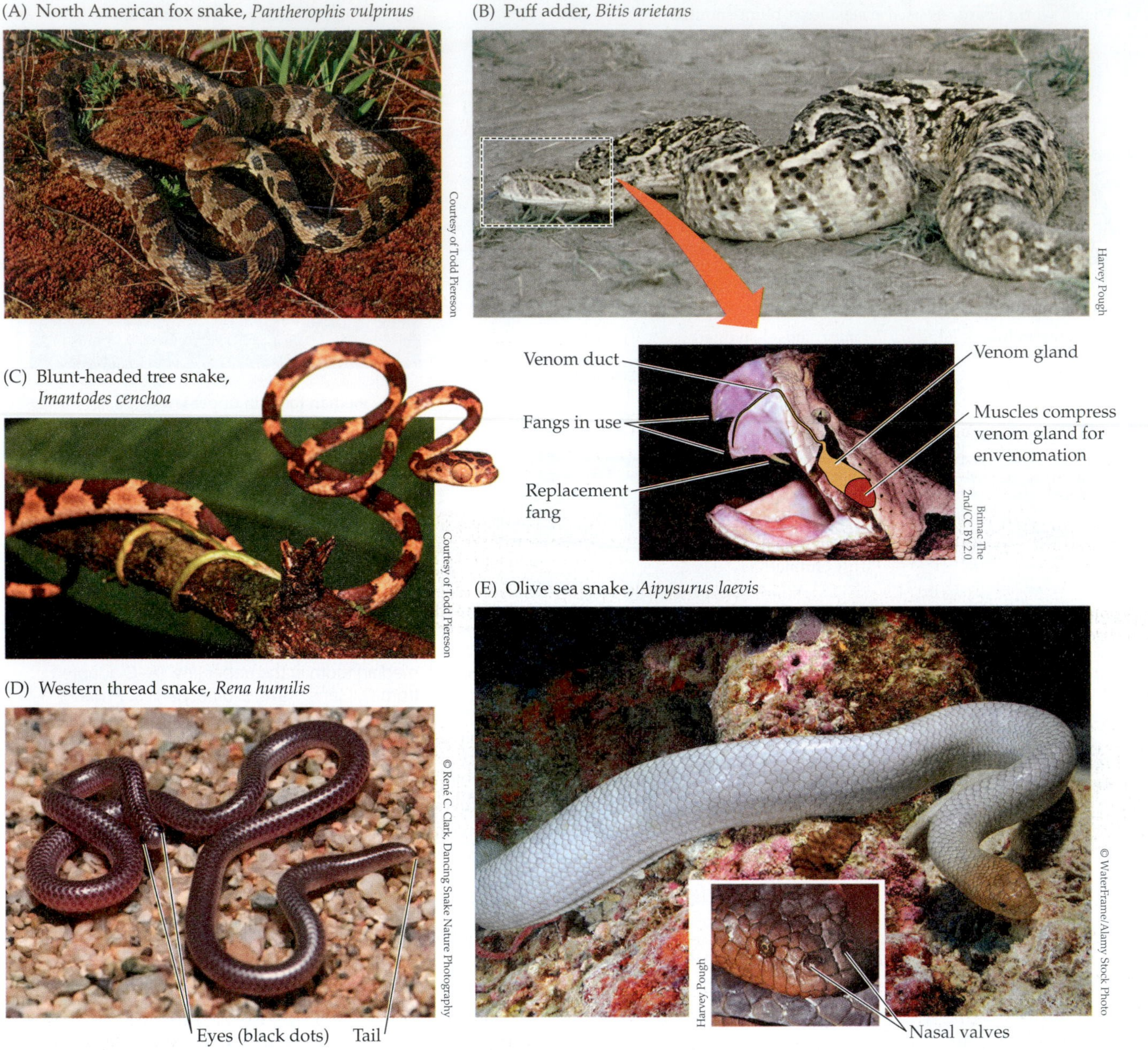
(A) North American fox snake, *Pantherophis vulpinus*
Courtesy of Todd Pierson
(B) Puff adder, *Bitis arietans*
Harvey Pough
Venom duct
Fangs in use
Replacement fang
Venom gland
Muscles compress venom gland for envenomation
Brimac The 2nd/CC BY 2.0
(C) Blunt-headed tree snake, *Imantodes cenchoa*
Courtesy of Todd Pierson
(D) Western thread snake, *Rena humilis*
© René C. Clark, Dancing Snake Nature Photography
Eyes (black dots)
Tail
(E) Olive sea snake, *Aipysurus laevis*
Harvey Pough
Nasal valves
© WaterFrame/Alamy Stock Photo

Figure 15.7 Body forms of snakes. (A) Generalized snakes are slim. (B) Many vipers are stout and have long fangs that deliver venom deep into their prey. (C) Arboreal snakes are slim and can reach prey such as lizards and treefrogs that rest at the tips of twigs. Note the large eyes of this nocturnal species. (D) Fossorial snakes have shiny scales that reduce friction in a tunnel. The blunt tails of many fossorial snakes are hard to distinguish from their heads, and when confronted by a predator the snake waves its tail to deflect attacks from its head. (E) Specialized marine snakes have flattened tails and valvular nostrils.

located on the top of the snout (**Figure 15.7E**). Most sea snakes never leave the water, and the majority of sea snake species are venomous.

Locomotion Snakes use four methods of locomotion, depending on the body form of the species and the substrate over which the snake is moving.

- In **lateral undulation**, the body is thrown into a series of curves, and each curve presses backward against the substrate (**Figure 15.8A**). All snakes can employ this mode of locomotion. "Flying" snakes (**Figure 15.8B**) undulate in the air, and the undulation stabilizes their glide path and increases the distance they can travel.
- **Rectilinear locomotion** is used primarily by heavy-bodied snakes such as large vipers, boas, and pythons. Alternate sections of the body are lifted and pulled forward while the intervening sections of the body rest on the ground, supporting the snake while waves of contraction pass from anterior to posterior (**Figure 15.8C**). Rectilinear locomotion is slow, but it

Table 15.2 Major extant families of snakes[a]

Family	Description	Adult size[b]	Number and distribution of species
Thread snakes (Leptotyphlopidae)	Extremely small, thin, fossorial snakes with reduced eyes, and teeth only on the dentary bones	10 to 30 cm	143 species in the Americas, Africa, and the Middle East
Blind snakes (Typhlopidae)	Small fossorial snakes with reduced eyes, and teeth only on the maxillae	20 to 75 cm	422 species worldwide[c]
Pythons (Pythonidae)	Terrestrial or arboreal constrictors, some are very large (the reticulated python, *Malayopython reticulatus*, is the world's longest snake)	~1 to 10 m	40 species, Africa, Asia, and Australia
Boas (Boidae)	Terrestrial, arboreal, or semiaquatic constrictors, some very large (the green anaconda, *Eunectes marinus*, is the world's heaviest snake and one of the longest)	~50 cm to 9 m	66 species in western North America, Central and South America, and the West Indies
Vipers (Viperidae)	Venomous snakes in which the maxillae rotate, allowing long fangs to rest horizontally when the mouth is closed	~75 cm to 2 m	"True" vipers (Viperinae), 101 species in Eurasia and Africa Pit vipers (Crotalinae), 263 species in Asia and the Americas Fea vipers (Azemiopinae), 2 species in China and Southeast Asia
Elapids (Elapidae)	Venomous terrestrial and marine snakes with short fangs on immobile maxillae	25 cm to 5 m	391 species worldwide[c]
Colubrids (Colubridae)	Mostly elongate terrestrial, arboreal, and semiaquatic snakes. Many have grooved teeth and glands that produce venom	20 cm to ~3 m	2,036 species worldwide[c]

[a]29 families of snakes are currently recognized. Only families discussed in the text are listed here.
[b]Size is total body length (nose to tail tip).
[c]All continents except Antarctica.

is effective even when there are no surface irregularities strong enough to resist the sideward force exerted by lateral undulation. Because the snake moves slowly and in a straight line, it is inconspicuous, and rectilinear locomotion is used by some snakes when stalking prey

- **Concertina locomotion** is used in narrow passages such as rodent burrows that do not provide space for the broad curves of lateral undulation. A snake anchors the posterior part of its body by pressing several loops against the walls of a burrow and extends the front part of its body. When the snake is fully extended, it forms new loops anteriorly and anchors itself with these while it draws the rear end of its body forward (**Figure 15.8D**)

- **Sidewinding** is used primarily by desert-dwelling snakes, where the windblown sand substrate slips away during lateral undulation. The snake raises its body in loops, resting its weight on two or three points that are the only body parts in contact with the ground (**Figure 15.8E**). The loops are swung forward through the air and placed on the ground, with

the points of contact moving smoothly along the body. Force is exerted downward; the lateral component of the force is so small that the snake does not slip sideways. Because the snake's body is extended nearly perpendicular to its line of travel, sidewinding is an effective means of locomotion only for small snakes that live in habitats with few plants or other obstacles.

The ventral scales (scutes) of snakes that travel on solid surfaces have backward-facing projections that allow the snakes to push themselves forward. In contrast, the scutes of sidewinder rattlesnakes (*Crotalus cerastes*) from North American deserts have only a few spikes and the snakes can move in any direction without friction (**Figure 15.8F**). African sidewinding vipers (the horned viper, *Cerastes cerastes*, and the Saharan sand viper, *C. vipera*) have lost the spikes entirely. Their scutes are smooth, with tiny uniform pits scattered across the surface. This difference between African and North American snakes may reflect the ages of the deserts they inhabit—the Sahara is 7–10 million years old, whereas the American desert formed a mere 15–20 thousand years ago.

(A) Lateral undulation

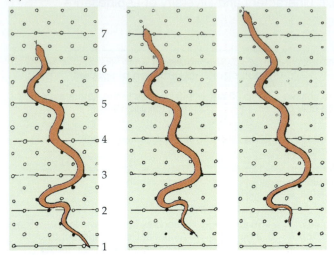

(B) Paradise flying snake, *Chrysopelea paradisi*

Courtesy of J. J. Socha

(C) Rectilinear locomotion

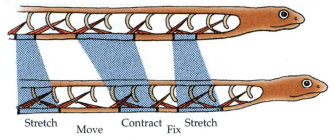

Stretch Move Contract Fix Stretch

(D) Concertina locomotion

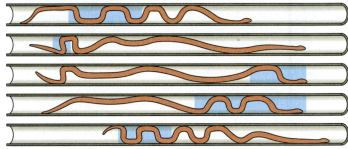

From H.C. Astley and B.C. Jayne 2009 *J Exp Zool* A 311: 207–216

(E) Sidewinding

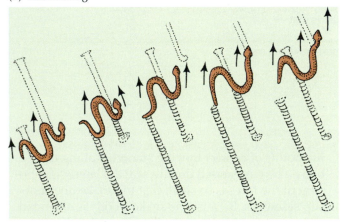

(F) Ventral scutes of rattlesnakes
Crotalus polystictus, rock dweller *Crotalus cerastes*, sand dweller

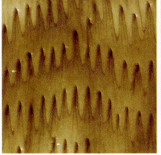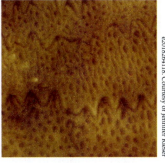

From Rieser et al. *PNAS* 118: e2018264118. Courtesy of Jennifer Rieser

Figure 15.8 Modes of snake locomotion. (A) Lateral undulation. A snake crawling across a board with fixed pegs. The pegs against which the snake is exerting force are shown in solid color. (B) The paradise flying snake uses lateral undulation as it glides through the air. The convexity of the ventral surface provides lift, as it does for a frisbee. (C) Rectilinear locomotion. The upper diagram shows alternate sections of the ventral skin as they are lifted from the substrate. In the lower diagram those sections of skin are pulled forward by muscles that originate on the ribs and insert on the ventral scales. (D) Concertina locomotion. Blue shading indicates the areas at which the snake is anchoring itself by pressing against the sides of the tube. (E) Sidewinding. Force is exerted downward as the snake lifts a loop of its body. The imprints of ventral scales show that the snake is not sliding. (F) The ventral scales of snakes that live on firm substrates have backward-pointing projections that create friction as the snake pushes forward. Sidewinders press downward and the projections on the ventral scales are greatly reduced. The scales are smooth and allow movement in any direction. (See credit details at the end of this chapter.)

15.2 Foraging Modes

LEARNING OBJECTIVES

15.2.1 Describe the difference in foraging behavior of sit-and-wait and widely foraging lizards and snakes.

15.2.2 Explain the significance of the difference in color and pattern of sit-and-wait and widely foraging lizards and snakes.

Squamates use a variety of methods to find, capture, subdue, and swallow prey, and these differences in foraging modes shape the interactions among species in a community. The activity patterns of lizards and snakes range from extremely sedentary ("sit-and-wait") species that spend hours or even days in one place to widely foraging species that are in nearly constant motion (**Figure 15.9**).

(A) Sit-and-wait predators

Sagebrush lizard, *Sceloporus graciosus*

Sit-and-wait lizards perch in elevated sites and use vision to detect moving prey.

Greg Schechter/CC BY 2.0

Western diamondback rattlesnake, *Crotalus atrox*

Tongue (detects scents)

Nostrils

© René C. Clark, Dancing Snake Nature Photography

Sit-and-wait snakes often wait in ambush beside trails that rodents use regularly, sometimes remaining in the same place for days at a time.

Heat-sensitive pit organs allow rattlesnakes to detect and strike prey in the dark.

(B) Widely foraging predators

Giant spotted whiptail lizard, *Aspidoscelis stictogrammus*

Widely foraging lizards hunt at ground level and use scent to detect prey hidden in crevices and under leaves.

© René C. Clark, Dancing Snake Nature Photography

Sonoran whipsnake, *Masticophis bilineatus*

Widely foraging snakes move continuously, exploring places where prey might hide.

Harvey Pough

Figure 15.9 **Two approaches to predation.** (A) Some squamates are sit-and-wait predators that ambush or detect prey from relatively stationary positions. (B) Widely foraging squamates move across hunting territories in search of prey.

Field studies have shown that this spectrum of locomotor behaviors is seen in lizard and snake faunas worldwide. In North America, for example, fence lizards and rattlesnakes are sit-and-wait ambush predators and whiptail lizards and whipsnakes are widely foraging predators. In Australia, dragon lizards (*Ctenophorus*) and death adders (*Acanthophis*) are ambush predators, whereas goannas (*Varanus*) and snakes in the genera *Pseudonaja* and *Pseudechis* forage widely.

Many characters of different squamate species are correlated with their mode of foraging:

- Sit-and-wait lizards make short dashes from a perch to capture prey and then return to the perch, and sit-and-wait snakes wait in ambush for prey to pass. They have patterns of blotches that make them cryptic when they are motionless. High anaerobic metabolic capacities support short dashes to capture prey, but the metabolic substrates stored in muscles are quickly depleted and sit-and-wait species have little ability to sustain locomotion. Some sit-and-wait ambush predators use lures to draw prey within reach (**Figure 15.10**).

- Widely foraging species move nearly continuously, looking under leaves and poking their noses under leaves and into crevices where prey might be hiding. Some widely foraging snakes are trapliners—that is, day after day they follow the same route and check sites where they have captured prey in the past. Widely foraging species have elongated bodies and their solid-colored or striped patterns are cryptic when they are in motion. These snakes and lizards usually have high aerobic metabolic capacities and better endurance but slower sprint speeds than sit-and-wait species.

Territoriality is common among sit-and-wait lizards. Males remain within a limited area that they can survey from their perches (see Figure 15.9A). Widely foraging species are rarely territorial, perhaps because at ground level they have a limited view of their surroundings. Snakes in general are not territorial, although males of some species guard females during the mating season and king cobras guard their nests (see Figure 15.23C).

Although viviparity occurs in species with both foraging modes, it is more common among sit-and-wait than widely foraging species, presumably because retaining bulky embryos in the body interferes with locomotion.

(A) Spider-tailed viper, *Pseudocerastes urarachnoides*

(B) Cantil, *Agkistrodon bilineatus*

(C) Water moccasin, *Agkistrodon piscivorus*

Figure 15.10 Caudal lures. (A) The spider-tailed viper preys on birds that it attracts by wiggling its tail tip, which has elongated scales that look remarkably like the legs of a spider. (B) Juveniles of many species of pit vipers have brightly colored tail tips that they wiggle to attract prey. (C) The yellow tail tips of some pit vipers emit a blue fluorescence that matches the sensitivity of the green rods in the retinas of amphibians (see Section 12.1), probably increasing the effectiveness of the lures at night.

15.3 Skull Kinesis and Feeding

LEARNING OBJECTIVES

15.3.1 Describe the differences between the skulls of tuatara, lizards, and snakes.

15.3.2 Explain the significance of skull kinesis in prey ingestion by snakes.

15.3.3 Explain the roles of snake venom in prey capture, digestion, and defense.

15.3.4 Describe the changes in the internal organs of binge-feeding snakes in the feasting and fasting periods.

Many of the feeding specializations of squamates are related to changes in the structure of the skull and jaws. The ancestral lepidosaur skull was diapsid, with a lower temporal bar formed by a connection between the jugal and quadrate and an upper bar formed by the postorbital and squamosal (**Figure 15.11A**). The tuatara is fully diapsid, albeit secondarily so, having re-evolved the connection forming the lower bar. Lizards lost the lower bar, and snakes lost both the lower and upper bars, increasing the gape and lateral spread of the jaw articulation during feeding (**Figure 15.11B,C**). In addition, lizards and snakes have a hinge between the frontal and parietal bones of the skull that permits movement (skull kinesis). Highly kinetic skulls and jaws with wide gapes accommodate large prey items, but the loss of rigidity at the joints diminishes the crushing force of the jaws.

Feeding specializations of snakes

In popular literature, snakes are sometimes described as "unhinging" their jaws during feeding. That's careless writing and rather silly—unhinged jaws would merely flap back and forth. It is true, however, that snakes have extremely kinetic skulls that allow extensive movement of the jaws. A snake skull contains multiple elements that have flexible joints between them, allowing a staggering degree of complexity in the movements of the skull (**Figure 15.12**). Furthermore, the components are paired, and each side of the head acts independently.

The jaws of derived snakes are much more moveable than the jaws of lizards. Whereas the mandibles of lizards are joined at the front of the mouth by a rigid bony connection; the mandibles of snakes are attached by a stretchy ligament and can spread laterally and move forward or

(A) Tuatara are fully diapsid

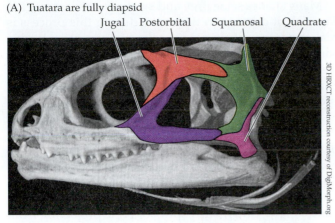

Jugal Postorbital Squamosal Quadrate

3D HRXCT reconstruction courtesy of DigiMorph.org

(B) Lizards have lost the lower bar

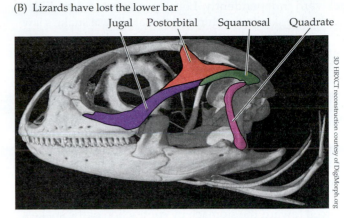

Jugal Postorbital Squamosal Quadrate

3D HRXCT reconstruction courtesy of DigiMorph.org

Figure 15.11 Modifications of the diapsid skull among lepidosaurs. (A) The tuatara has reverted to the ancestral condition for lepidosaurs, with two temporal bars. The upper bar is formed by the postorbital and squamosal, and the lower bar by the jugal and quadrate. (B) Crown group lizards have developed kinesis by eliminating the lower bar; there is a gap between the jugal and quadrate. (C) Skull kinesis is further developed in snakes by loss of both the upper and lower bars.

(C) Snakes have lost both the upper and lower bars

Postorbital Squamosal Quadrate

3D HRXCT reconstruction courtesy of DigiMorph.org

(A) Lateral view of joints and moveable bones of a snake skull and jaw

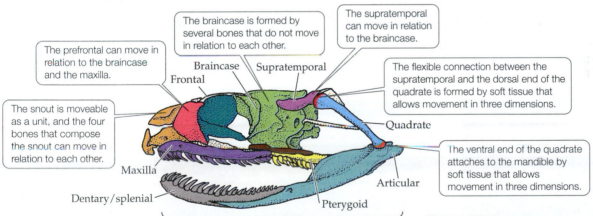

The prefrontal can move in relation to the braincase and the maxilla.

The braincase is formed by several bones that do not move in relation to each other.

The supratemporal can move in relation to the braincase.

The flexible connection between the supratemporal and the dorsal end of the quadrate is formed by soft tissue that allows movement in three dimensions.

The snout is moveable as a unit, and the four bones that compose the snout can move in relation to each other.

Braincase Supratemporal
Frontal

Quadrate

The ventral end of the quadrate attaches to the mandible by soft tissue that allows movement in three dimensions.

Maxilla

Dentary/splenial

Articular

Pterygoid

Mandible

(B) Ventral view of a snake skull and left lower jaw

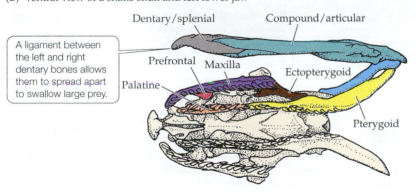

Dentary/splenial Compound/articular

A ligament between the left and right dentary bones allows them to spread apart to swallow large prey.

Prefrontal Maxilla
Palatine

Ectopterygoid

Pterygoid

Figure 15.12 Skull and jawbones of a snake. (A) A snake skull contains multiple elements with moveable connections between them, some of which are shown in this lateral view. Most of the elements are bony, and movement between them occurs in a single plane, but some connections are formed by soft tissue. These connections are rigid when under tension but flexible when they are relaxed, permitting movement in three dimensions. (B) The left and right sides of both the upper and lower jaws have flexible connections, allowing each of those four elements to move independently of the others. All bones of the upper jaws are moveable relative to each other, typically in three dimensions. The palatine and maxilla attach moveably to the prefrontal, which is itself moveable on the frontal, and the pterygoid is movably attached to the quadrate and the compound/articular. In the lower jaws, the left and right mandibles can be protracted and retracted independently of each other and independently of the upper jaws. (After C. Gans 1961. *Am Zool* 1: 217–227.)

backward independently. Loosely connected mandibles and flexible skin in the chin and throat allow a snake's jaw tips to spread, so that the widest part of the prey passes ventral to the articulation of the jaw with the skull.

Snakes usually swallow prey head first, perhaps because that approach presses the prey's limbs against its body, out of the snake's way (**Figure 15.13A**). Swallowing movements take place slowly enough to be easily observed. The mandibular and pterygoid teeth of one side of the head are anchored in the prey, and the head is rotated to advance the opposite jaw as the mandible is protracted and grips the prey ventrally. As this process is repeated, the snake draws the prey item into its mouth. Once the prey has reached the esophagus, it is forced toward the stomach by contraction of the snake's neck muscles.

Many snakes seize prey and swallow it as it struggles. The risk of damage to the snake during this process is real, and various features of snake anatomy seem to give some protection from struggling prey. The frontal and parietal bones of a snake's skull extend downward, entirely enclosing the brain and shielding it from the kicks of the protesting prey as it is swallowed.

Some snakes have highly specialized methods of feeding. Snail-eating snakes use long teeth at the tips of the mandibles to winkle a snail from its shell (**Figure 15.13B**). Egg-eating snakes (several species of *Dasypeltis*; **Figure 15.13C**) have nearly toothless jaws. An egg is swallowed unbroken, then sliced open by hypapophyses—sharp, downward spurs of bone projecting from vertebrae in the neck. Two species of Asian kukri snakes

(A) Ruby-eyed tree viper, *Trimeresurus rubeus*, swallowing a frog

Teeth of the right maxilla and mandible are lifted from the prey…

…then the head rotates to the left, advancing the right maxilla and mandible over the frog's head.

© Wayne Van Devender

(B) Pygmy snail sucker, *Sibon sanniolus*

The mandibles are inserted into opening of the snail's shell.

Maximilian Paradiz/CC BY 2.0

(C) An egg-eating snake, *Dasypeltis scabra*

Vertebral hypapophyses will puncture the egg shell in the esophagus.

Skin stretched between rows of scales allows egg to be swallowed.

Mond76

(D) Kukri snake, *Oligodon fasciolatus*

The snake's head is in the body cavity of the toad and it is swallowing its internal organs.

From Bringsøe et al. 2020. *Herpetozoa* 33: 157–163

Figure 15.13 Feeding methods of snakes. Coordinated head and jaw movements allow snakes to swallow prey. (A) The snake holds the prey with its left maxilla and mandible and disengages the teeth of the right maxilla. Bending its neck to the left, the snake advances its head over the prey. Its backward-curving teeth slide across the surface of the frog like the runners of a sled. Closing the jaws embeds the teeth in the prey. (The fangs, visible in the left-hand photo, are not used during swallowing.) The swallowing process continues with alternating left and right movements until the entire prey has passed through the snake's jaws. (B) Snail-eating snakes use their long mandibles to winkle snails from their shells. (C) Nearly toothless, egg-eating snakes can engulf an egg without breaking it. Here the snake is about to drive projections of the cervical vertebrae (hypapophyses) through the eggshell to expel its contents. (D) A kukri snake forces its head through the flank of a living toad and swallows the internal organs, avoiding the toxic secretions of the toad's granular glands (the white substance that can be seen spreading across the toad's skin).

(*Oligodon*) eviscerate living toads using enlarged teeth in the rear of the maxillae (*oligodon* means "different tooth") to slice through the toad's flank, then inserting the head and swallowing the toad's internal organs (**Figure 15.13D**).

Constriction and venom are predatory specializations that permit a snake to tackle large prey with little risk of injury to itself. Constriction is characteristic of boas and pythons as well as some colubrid snakes (see Table 15.2). Despite travelers' tales of animals crushed to jelly by a python's coils, the process of constriction involves very little pressure. A constrictor seizes prey with its jaws and throws one or more coils of its body around the prey. The loops of the snake's body press against adjacent loops, and friction prevents the prey from forcing the loops open. Each time the prey exhales, the snake takes up the slack by tightening the loops slightly until the increased internal pressure eventually stops the prey's heart.

Venom and fangs

Snakes probably developed venom from an assortment of genes that are widely distributed among amniotes. Basal colubrid snakes may have used venom to immobilize prey. Duvernoy's gland, found in the upper jaw of many extant colubrids, is homologous with the venom glands of viperids and elapids and produces a toxic secretion that immobilizes prey. The presence of Duvernoy's gland appears to be an ancestral character for colubrid snakes, and some extant colubrids have venom that is dangerously toxic; humans have died from bites of the African boomslang (*Dispholidus typus*).

In this context, then, the front-fanged venomous snakes (Elapidae and Viperidae; see Table 15.2) are not a new development, but represent specializations of an ancestral venom-delivery system. A variety of snakes have enlarged teeth (fangs) on the maxillae. Three categories of venomous snakes are recognized:

1. Opisthoglyphous (Greek *opisthen*, "behind"; *glyphe*, "a carving") dentition has evolved in many clades of colubrids. These species have one or more enlarged teeth near the rear of the maxillae, with smaller teeth in front (**Figure 15.14A**). In some forms the fangs are solid; in others, there is a groove on the surface of the fang that may help conduct venom into the wound. The fangs do not engage the prey until it is well into the mouth. Birds and lizards are the primary prey of opisthoglyphs, and the snakes often hold the prey in their mouth until the venom takes effect and the prey has stopped struggling

2. Proteroglyphous (Greek *pro*, "before") snakes include the elapids (cobras, mambas, coral snakes, and sea snakes). The hollow fangs of proteroglyphous snakes are located at the front of the maxillae, and there are often several small, solid teeth behind the fangs (**Figure 15.14B**). The fangs are permanently erect and relatively short.

3. Solenoglyphous (Greek *solen*, "pipe") snakes include the pit vipers of the New World and the true vipers of the Old World. The hollow fangs are the only teeth on the maxillae, which rotate so that the fangs are folded against the roof of the mouth when the jaws are closed (**Figure 15.14C**). This folding mechanism permits solenoglyphous snakes to have long fangs that inject venom deep into the tissues of prey. The venom first kills the prey and then speeds its digestion after it has been swallowed.

Snake venom is a complex mixture of proteins and polypeptides. Hyaluronidase enzymes break down connections between cells and phospholipases destroy cell membranes, allowing venom to spread, while proteolytic enzymes break down muscle tissue.

(A) An opisthoglyph, the African boomslang, *Dispholidus typus*

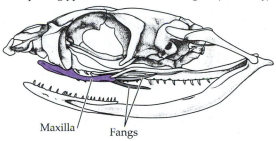

Maxilla Fangs

(B) A proteroglyph, the African mamba, *Dendroaspis jamesoni*

Maxilla

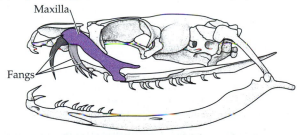

Fangs

(C) A solenoglyph, the African puff adder, *Bitis arietans*

Prefrontal Ectopterygoid

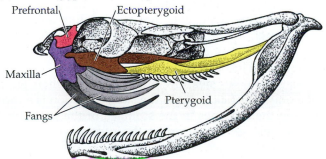

Maxilla

Pterygoid

Fangs

Figure 15.14 Dentition of snakes. (A) Opisthoglyphous snakes have grooved fangs near the rear of the maxillae that conduct venom. (B) Proteroglyphous snakes have permanently erect hollow fangs at the front of the maxillae. (C) Solenoglyphous snakes have hollow fangs on rotating maxillae. When the mouth is closed, the fangs of solenoglyphs lie against the roof of the mouth. When the mouth is opened to strike, the fangs are erected by anterior movement of the pterygoid that is transmitted through the ectopterygoid to the maxilla, causing it to rotate about its articulation with the prefrontal. (See credit details at the end of this chapter.)

Characteristics of snake venom are related to the prey a species of snake eats. Pit vipers with diets that include phylogenetically distant prey have more complex venoms than vipers with phylogenetically narrow diets. That is, a species of viper that eats frogs, lizards, and rodents has more complex venom than a species that eats several different species of rodents.

The toxicities of some venoms are matched to the primary prey of a species of snake. For example, juvenile Australian brown snakes (several species in the genus *Pseudonaja*) feed on lizards and their venom is primarily neurotoxic, whereas the venom of adults, which feed on mammals, is primarily cytotoxic. Each property of venom is maximally effective against the appropriate prey. Northern Pacific rattlesnakes (*Crotalus o. oreganus*) are engaged in an arms race with the California ground squirrels that are their principal prey; as the squirrels evolve greater resistance, the snakes evolve more lethal venom.

Venomous snakes synthesize and store multiple lethal doses of venom in a form that is ready for immediate use but is not toxic to the snake. In rattlesnakes, acidity in the main venom gland inactivates the venom during storage, and secretions from the accessory gland activate the venom when the snake strikes.

Injection of a disabling dose of venom into prey is a very safe prey-catching method. A constrictor must remain in contact with its prey while the prey is dying and thus runs a risk of being injured by the prey's struggles. In contrast, a viper needs only to inject venom and allow the prey to run off to die; the snake then can follow the prey's scent trail and find its corpse. Experiments have shown that a viper can distinguish the scent trail of a mouse it has bitten from the scent trails left by unbitten mice.

Several features of the body form and venom of vipers allow them to eat larger prey in relation to their own body size than can most nonvenomous snakes or elapids. Many vipers are very stout snakes. The triangular head shape usually associated with vipers is produced by the wide-spreading quadrates that allow bulky objects to pass through the mouth, and even a large meal makes little bulge in the stout body of a viper and thus does not interfere with locomotion.

Because snakes swallow prey intact and their teeth make only pinprick holes in the skin, digestion is a race between the enzymes in the stomach of the snake and enzymes released by bacteria in the intestines of the prey. The snake's stomach enzymes attack the prey from the outside, digesting their way through skin and muscle. At the same time, the prey's bacterial enzymes are causing its carcass to rot from the inside out. Bacterial action in the prey's gut releases noxious compounds, and if the snake's enzymes don't reach the gut before decay has progressed too far the snake is forced to regurgitate the putrid mass, also regurgitating any prey it swallowed more recently.

Winning the race with bacteria is especially difficult for snakes that consume bulky prey, and the fangs and venom of vipers have a second important function at this stage. Delivered deep into the body of the prey by the long fangs,

venom speeds digestion. When the venom reaches the intestines, it digests the bacteria, handing victory to the snake.

Hearts and stomachs

Kinetic skulls and safe ways to kill prey enable snakes to consume meals that are large in relation to their own body size. "Binge-feeding" species of snakes that eat large meals at infrequent intervals conserve energy by allowing the intestine to shrink between meals and to hypertrophy rapidly following a meal. The small intestine of a Burmese python (*Python bivittatus*) doubles in size while a meal is being digested and shrinks back when digestion is completed. Digestion requires increased blood flow to the intestine, and the size of the ventricle of the heart increases by 40% within 48 hours after a python swallows a meal. The weights of the liver, pancreas, and kidney also increase and regress, as do the activities of several digestive enzymes.

15.4 Predator Avoidance and Defense

LEARNING OBJECTIVES

15.4.1 Provide examples of cryptic and aposematic patterns and explain how a pattern can be both cryptic and aposematic.

15.4.2 Explain the difference between predator avoidance, predator deterrence, and predator evasion and give examples of each.

15.4.3 Briefly describe the properties of snake venom that function primarily in defense.

15.4.4 Explain the difference between a venomous snake and a poisonous snake.

Squamates are preyed on by birds, mammals, and larger species of squamates, and they have evolved a variety of ways to avoid being attacked, or to evade a predator after they are attacked. Many of these mechanisms are size-specific—small animals can hide more easily than large ones, and large individuals can run faster and fight more effectively than small ones. Temperature also plays a role; many species use one type of defense when they are cold and sluggish and a different type when they are warm and agile. The defensive behavior of the black-and-white tegu lizard (*Salvator merianae*; see Figure 14.6A) combines the effects of both body temperature and ontogenetic increases in size. Juvenile tegus run from predators at all body temperatures, whereas adults (which are large enough to defend themselves) attempt to bite a predator when at low body temperatures, but flee when they are warm and can run rapidly.

Crypsis, aposematism, and mimicry

Avoiding detection by a predator is a sure way to avoid predation; conversely, advertising your presence can be another. Like lissamphibians (see Section 12.5), some species of lizards and snakes are cryptic, which consists of being indistinguishable from one's surroundings or resembling an object of no interest to a predator. Other

(A) Mossy leaf-tail gecko, *Uroplatus sikorae*

Charles James Sharp/CC BY-SA 4.0

(B) Short-tailed horned lizard, *Phrynosoma modestum*

Courtesy of Wade C. Sherbrooke

(C) Casque-headed lizard, *Corytophanes cristatus*

Courtesy of Dale Kalina

Figure 15.15 **Crypsis.** (A) Some arboreal geckos have fringes of skin along the sides of their trunk, tail, and legs that obscure the shape of their body when they rest on a twig. Their dorsal patterns are also cryptic, in this case resembling a patch of lichen. (B) Some small lizards escape detection by looking like pebbles. (C) A casque-headed lizard is usually seen as a silhouette in the dim light of the understory of tropical forest. In this setting, it looks to predators like the stub of a broken tree limb.

species (typically ones that are venomous) are aposematic, with bright colors that advertise danger to potential predators. Some harmless species "cash in" on the protection offered by aposematism by mimicking the warning colorations of dangerous species.

Crypsis is the first line of defense for many squamates. Many arboreal lizards and snakes are green or brown, colors that blend with leaves and twigs. Some lizards have flaps on the flanks of the body and the tail that lie flat against a twig, concealing the rounded contour of the body (**Figure 15.15A**). Some small species of horned lizards resemble pebbles (**Figure 15.15B**). Central American casque-headed lizards have a bony crest on the skull that obscures the shape of the head. These lizards live in the dim light of the forest understory. Resting motionless on a tree trunk and seen in silhouette, they look like the stub of a broken branch (**Figure 15.15C**). When a casque-headed lizard moves, it reveals itself as a lizard, but these lizards are extreme sit-and-wait predators, moving no more than once a day to capture an insect. They are also highly selective predators, attacking only large insects that will provide enough energy to allow a lizard to remain motionless and cryptic until the next day.

Coral snakes are a well known example of snakes whose aposematic colors and patterns render them conspicuous and recognizable by predators (**Figure 15.16A**). In turn, mimicry of coral snakes by rear-fanged and

(A) Eastern coral snake, *Micrurus fulvius*

C.W. Brown, U.S. Geological Survey

(B) Scarlet kingsnake, *Lampropeltis elapsoides*

Courtesy of James H. Harding

Figure 15.16 **Aposematism and mimicry.** (A) Venomous coral snakes are aposematically colored, warning predators to leave them alone. (B) Some nonvenomous snakes mimic the coloration of coral snakes and benefit from a predator's resulting confusion. (The mnemonic "red touch black, OK Jack / Red touch yellow, kills a fellow" works only in the United States; from Mexico southward, some coral snakes have adjacent red and black bands.)

nonvenomous snakes is a widespread phenomenon in the southern United States and in Central and South America (**Figure 15.16B**). Field studies have shown that predatory birds avoid models of snakes with the patterns of coral snakes, even when the mimic's resemblance is imprecise.

Deterrence

Once a snake or lizard has been detected by a predator, deterring attack is a secondary line of defense. Some lizards have markings on the underside of the tail that are not visible when the tail is in its normal position but become conspicuous when the tail is curled over the lizard's back and waved from side to side (**Figure 15.17A**). This behavior, which appears to signal that the lizard is aware of the predator and ready to flee, may cause a predator to move on in search of a less alert meal.

Spines (enlarged, sharp-pointed scales) have two deterrent roles. Many lizards that seek shelter from predators in crevices are spiny; the spines make it difficult for a predator to pull the lizard from its retreat. In addition, spines can be effective deterrents if a predator attacks. Many lizards have spiny tails, and some have spines everywhere (**Figure 15.17B**).

Two clades of Australian geckos deter predators by squirting an evil-smelling, adhesive liquid from hollow spines on the tail (**Figure 15.17C**). The ejected fluid can travel as far as 50 cm, and forms long, sticky filaments that are difficult to remove. Soil particles and bits of plant debris adhere to the filaments, distracting a predator while the gecko makes its escape.

Autotomy

When a squamate has been detected and attacked by a predator, evasion is the remaining option. Running away is usually the first response, but if escape fails many lizards and a few snakes have a unique defense mechanism: they break off a part of their body and run off, leaving the appendage in the predator's jaws. Loss of the tail, or **caudal autotomy**, is the most common form of this defense. Tail muscle has high anaerobic metabolic capacity, and the broken tail writhes and twists for many minutes, distracting the predator and allowing the lizard to escape.

Caudal autotomy occurs in the middle of a tail vertebra, not between vertebrae, and autotomized tails are regenerated with a rod of cartilage replacing the lost vertebrae (**Figure 15.18A**). The cartilage does not break, so subsequent autotomies must be closer to the body than the

(A) Zebra-tailed lizard, *Callisaurus draconoides*

© René C. Clark, Dancing Snake Photography

(B) Sungazer, *Smaug giganteus*

Eric Johnston/CC BY-SA 3.0

(C) Spiny-tailed gecko, *Strophurus krisalys*

Courtesy of Stephen Zozaya

Figure 15.17 Deterrence. (A) Some lizards curl their tails and wave them, signaling to a predator that it has been spotted. (B) Many lizards, especially species that retreat to crevices to escape predators, have spines that make it difficult for a predator to pull them from the rocks. (C) Glands beneath the hollow spines on the tail of some geckos from Australia spray a foul-smelling adhesive substance.

(A) Tail autotomy of Mediterranean house gecko, *Hemidactylus turcicus*

(B) Integumentary autotomy of fish-scaled gecko, *Geckolepis megalepis*

Figure 15.18 Autotomy. (A) Most lizards can autotomize their tail and subsequently regenerate it. The regenerated portion of the tail has a distinctly different pattern from the original tail. (B) Some geckos autotomize portions of their skin. Although the injury looks horrifying (left), the skin and scales will be regenerated within weeks (right).

initial one. Thus, minimizing the amount of its tail that is lost is advantageous. Some lizards automize a large portion of the tail when they are cold and unable to run rapidly, and a smaller portion when they are warm. Caudal autotomy can have at least two costs for a lizard: (1) while its tail is regenerating a lizard's speed is reduced, and (2) a female lizard must redirect energy to regenerating the tail that would otherwise have been used to produce eggs.

In addition to caudal autotomy, some geckos use dermal autotomy as a mode of escape, sacrificing large areas of skin to a predator. The skin splits between the integument (epidermis, connective tissue, and subcutaneous fat) and the underlying dermis. Capillaries in the dermis constrict to prevent blood loss, and the skin regenerates in a few weeks without a scar (**Figure 15.18B**).

Venom and poisons as defense mechanisms

Although venom evolved as a predatory mechanism (see Section 15.3), some specializations enhance its value in defense. Many venomous snakes have distinctive markings or displays, such as the rattles of rattlesnakes and the hoods of cobras that may prevent a predator from attacking a snake (a form of aposematism; **Figure 15.19A,B**; also see Figure 15.23C). Venom is a second line of defense, and many venoms contain substances that produce intense pain without causing tissue damage. Probably these substances make a predator flee before it has injured the snake.

About one-third of the 31 species of cobras have a modification of the fangs that makes venom a defensive weapon. The venom channel in the fangs of these spitting cobras makes a right-angle bend near the tip, so that

(A) Prairie rattlesnake, *Crotalus viridis*

(B) Mozambique spitting cobra, *Naja mossambica*

(C) Green keelback snake, *Rhabdophis plumbicolor*

Figure 15.19 Venomous and poisonous snakes. Although venom's initial function was predation, it can also serve for defense. (A) The buzz produced by rattlesnakes warns potential predators. (B) Spitting cobras can spray venom as far as a meter. (C) Keelback snakes sequester toxins from the toads they eat and can spread the toxins on predators.

venom is directed forward and released as a spray from the front surface of the fang (see Figure 15.19B). Spraying venom is a defensive behavior; a spitting cobra bites like any other cobra when it strikes prey, and it releases more venom when it bites than when it spits. Experimental manipulations using photographs showed that a face with eyes elicited spitting most reliably, while a face without eyes was not as effective. The snake aims at the eyes, and venom causes immediate pain and at least temporary blindness. Spitting behavior has evolved independently three times in cobras, once in Asia and twice in Africa.

A few snakes are poisonous rather than venomous—that is, they contain toxic substances derived from the snakes' prey that sicken a predator. Asian keelback snakes (*Rhabdophis*) that feed on toads (*Bufo*) store neurotoxic bufadienolides in glands beneath the skin (**Figure 15.19C**). When a keelback is attacked, it presses its body against the predator's face and its skin ruptures, spraying the contents of the glands into the predator's mouth and eyes. Species in a derived clade of *Rhabdophis* consume earthworms as their primary prey. Earthworms do not contain bufadienolides, but nonetheless the earthworm-eating species of *Rhabdophis* possess bufadienolides. How can that be? A study of the diets of these snakes revealed that they derive bufadienolides from the larval fireflies they consume.

15.5 Social Behavior

LEARNING OBJECTIVES

15.5.1 Describe behaviors that lizards use in social interactions.

15.5.2 Describe the occurrence of group social behaviors and parental care among squamates.

Squamates employ a variety of visual, auditory, chemical, and tactile signals in behaviors they use to maintain territories and to choose mates. Because humans are primarily visual, we perceive the visual displays of other animals quite readily; thus the extensive repertoires of visual displays of *Anolis* lizards figure heavily in the literature of behavioral ecology. Much less is known about the chemical and tactile signals that are the primary channels of communication for other lizards and for snakes. Many geckos, for example, are nocturnal and use vocalizations during territorial defense and courtship.

Courtship and territoriality

Males of most sit-and-wait species of lizards are territorial and use visual displays for intraspecific communication, whereas widely foraging species are rarely territorial and use olfaction for species and sex identification. A male sit-and-wait lizard spends most of its time perched on a lookout from which it can survey its surroundings and challenge any conspecific that enters its territory. If the intruding lizard is a male, a face-off is likely to follow (**Figure 15.20A**). The two males orient parallel to each other, and each makes itself look as large as possible by extending its legs, flattening its trunk laterally, and tilting toward its opponent.

The colors and patterns of sit-and-wait lizards are typically cryptic, but males of many species have colorful dewlaps—flaps of throat skin—that are extended during territorial disputes and courtship (**Figure 15.20B**). Both males and females of the Pinocchio anole have dewlaps, but they are inconspicuous. In courtship and in territorial displays a male Pinocchio anole raises and lowers his elongate proboscis and wags it from side to side (**Figure 15.20C**).

Other species of lizards signal with their tails. Qinghai toad-headed lizards (*Phrynocephalus vlangalii*) inhabit the Tibetan plateau in western China. Both males and females are territorial, and both use tail displays to defend their burrows against intruders of either sex, although males signal more vigorously to a male intruder than to a

(A) Male zebra-tailed lizards, *Callisaurus draconoides*, facing off

Courtesy of Susan Schalbe

(B) Male fan-throated lizard, *Sitana ponticeriana*, displaying

Yogendra Joshi/CC BY 2.0

(C) Male (left) and female Pinocchio anoles, *Anolis proboscis*

Santiago R. Ron-BioWebEcuador/CC BY 2.0

Figure 15.20 Courtship and territorial behavior of lizards. Sit-and-wait species often use color and movement in territorial encounters (A) and courtship (B). The males in (A) are facing off in whole-body displays; the male in (B) is spreading a colorful dewlap to attract a female. (C) Male Pinocchio anoles wave the elongated proboscis in social interactions.

(A) Male Qinghai toad-headed lizard, *Phrynocephalus vlangalii*

(B) Female tail-coiling display

(C) Male tail-lashing display

Figure 15.21 Tail-curling displays of the Qinghai toad-headed lizard. (A) A male signals to the lizard in the background. (B) Tail displays of females are limited to up and down movements of the curled tail. (C) Males also thrash their tails from side to side. (From R.A. Peters et al. 2016. *Sci Rep* 5: 31573/CC BY.)

female. Details of the display vary with sex (**Figure 15.21**). Females curl the tail and wave it up and down; males curl and wave, and also lash the tail from side to side. Tail displays are fueled by anaerobic metabolism, and the vigor and magnitude of the display provide information about the physical condition of the displaying individual that may play a role in mate selection by both sexes.

Olfaction is the primary sense that male snakes use to identify the sex and species of potential mates and rivals. During the breeding season males of some species of snakes undertake extensive travel, attempting to cross the scent trail of a female and then follow the trail to find her. Males of many species wrestle to determine dominance, especially when a female is present (**Figure 15.22**). Males rear up, pushing against each other as each tries to force its opponent to the ground. The winner (i.e., the snake that pins its rival) mates with the female while the loser flees.

Sociality and parental care

Although in general lizards and snakes are not considered to be social animals and tend not to live in pairs or groups, there are exceptions. Male and female shingleback

(A) Male adder, *Vipera berus*, locates a female

(B) Two males start to wrestle; the gray one has initial advantage

(C) The blue male evades the grasp of the gray male

(D) The blue male pins the gray male

(E) Blue male mating with female

Figure 15.22 Combat and courtship. Males of many species of snakes engage in wrestling bouts to obtain access to a female. Victory is achieved when a male pins his opponent's head to the ground.

(A) A mated pair of shingleback skinks, *Tiliqua rugosa*

P. G. Palmer/C BY-SA 2.0

(B) A family group of Cunningham's skinks,
Egernia cunninghami

Donal Holbern/CC BY 2.0

(C) Female king cobra, *Ophiophagus hannah*,
guarding her nest

Courtesy of K. S. Sajwan

(D) Female five-lined skink, *Plestiodon fasciatus*,
guarding her eggs

Joseph Rogers/CC BY 4.0

(E) Female timber rattlesnake, *Crotalus horridus*,
with her young

Photo by Matthew G. Simon,
courtesy of William Brown

Figure 15.23 Sociality and parental care.   (A) A few species of skinks form stable pair bonds. (B) Most species of the Australian skink genus *Egernia* spend their lives in stable family groups that consist of parents and several generations of offspring. The inset shows a newborn individual between one of its parents and an older sibling. The most common forms of parental care are egg attendance (C, D) and remaining with the young (E).

skinks form long-term pair bonds (**Figure 15.23A**). The members of the pair separate outside the breeding season, but rejoin year after year to mate. Social behavior in the *Egernia* lineage of Australian skinks ranges from solitary species with only transient contacts between individuals to species that form stable, long-term family groups consisting of a monogamous pair of adults and their offspring (**Figure 15.23B**). Family members share a common shelter site, bask close to one another, and defecate on a common scat pile. The parents defend their young, even from dangerous predators such as venomous snakes.

Parental care has been recorded for more than 100 species of squamates. A few species of snakes and a larger number of lizard species protect their nests from predators (**Figure 15.23C,D**). King cobras and some skinks remove dead eggs from the clutch. Some species of pythons brood their eggs. The female coils tightly around the eggs, and in some species, muscular contractions of the female's body

produce sufficient heat to raise the temperature of the eggs to about 30°C, which is substantially warmer than air temperature (see Figure 14.5). Newborn rattlesnakes remain with their mother for 2 weeks or more after birth, not dispersing until they have shed their skins and begun to feed (**Figure 15.23E**), and maternal care of young may be the ancestral condition for pit vipers.

15.6 Reproductive Modes

LEARNING OBJECTIVES

15.6.1 Describe the continuum of reproductive modes of squamates.

15.6.2 Explain the evidence supporting the hybrid origin hypothesis of parthenogenesis.

15.6.3 Defend the proposition that environmental sex determination is (or is not) a plesiomorphic character without a current selective benefit.

Like sharks and bony fishes, squamates have a range of reproductive modes extending from oviparity (development occurs outside the female's body, supported entirely by the yolk—i.e., lecithotrophy) to viviparity (eggs are retained in the oviducts and development is supported by transfer of nutrients from the mother to the fetuses—i.e., matrotrophy). Intermediate conditions include retention of the eggs for a time after they have been fertilized and production of **precocial** young that are reasonably mature and functional at birth, having been nourished primarily by material in the egg's yolk.

Oviparity and viviparity

Oviparity is assumed to be the ancestral reproductive mode for squamates. Viviparity has evolved independently more than 100 times, however, and ~20% of extant squamates are viviparous. A few species of lizards are reproductively bimodal—that is, they include both viviparous and oviparous populations.

Viviparity is not evenly distributed among squamate lineages. Nearly half the origins of viviparity in the group have occurred in skinks, whereas it is unknown in teiid lizards and occurs in only two genera of lacertids. Some viviparous skinks have specialized chorioallantoic placentae; more than 99% of the weight of the fetus of a Brazilian skink (*Brasiliscincus heathi*) results from transport of nutrients across the placenta.

Viviparity has advantages and disadvantages as a reproductive mode. The most commonly cited benefit is the opportunity it provides for a female snake or lizard to use her own thermoregulatory behavior to control the temperature of the embryos during development. This hypothesis is appealing in an ecological context because a relatively short period of retention of the eggs by the female might substantially reduce the total amount of time required for development.

Viviparity potentially lowers reproductive output because a female that is retaining one clutch of eggs cannot produce another. Lizards in warm habitats may produce more than one clutch of eggs in a season, but that is not possible for a viviparous species because internal development takes too long. In a cold climate, however, oviparous lizards are not able to produce more than one clutch of eggs in a breeding season, so viviparity does not necessarily reduce a female's annual reproductive output relative to oviparity. As predicted by that hypothesis, transitions to viviparity among squamates in the past 65 Ma are grouped during long-lasting cold climatic periods.

Viviparity has other costs. The agility of a female lizard is substantially reduced when her embryos are large. Experiments have shown that pregnant female lizards cannot run as fast as non-pregnant females, and snakes capture pregnant lizards more easily than non-pregnant ones. Females of some species become secretive when they are pregnant, perhaps in response to their vulnerability to predation. They reduce their activity and spend more time in hiding. This behavioral adjustment may

contribute to the reduction in body temperature seen in pregnant females of some lizard species, and it probably reduces their rate of prey capture as well.

Parthenogenesis

Parthenogenesis—reproduction by females without fertilization by males—has been identified among squamates in 6 families of lizards and 1 species of snake but is probably more widespread among squamates than this number indicates because parthenogenetic species are not conspicuously different from bisexual species. Parthenogenetic species are typically detected only when a study undertaken for an entirely different purpose reveals that a species contains no males.

Many parthenogenetic species appear to have their origin as interspecific hybrids; these hybrid species are diploid (2*n*), with one set of chromosomes from each parental species. Some parthenogenetic species are triploids (3*n*), usually the result of a backcross of a diploid parthenogenetic individual to a male of one of its bisexual parental species. Less commonly, triploidy can be the result of hybridization of a diploid parthenogenetic species with a male of a bisexual species different from its parental species.

It is common to find the two bisexual parental species and a parthenogenetic species living in overlapping habitats. Parthenogenetic species of whiptail lizards often occur in habitats that are subject to frequent disruption, such as the floodplains of rivers. Habitat disturbance may initially have brought closely related bisexual species together, fostering the hybridization that is the first step in establishing a parthenogenetic species. Once a parthenogenetic species has become established, its reproductive potential is twice that of a bisexual species, because every individual of a parthenogenetic species can produce young. Thus, when a flood or other disaster wipes out most of the lizards in an area, a parthenogenetic species can repopulate a habitat faster than a bisexual species.

Sex determination

Squamates exhibit a spectrum of sex-determining mechanisms. Genotypic sex determination (GSD) characterizes many clades; some are male heterogametic (XY/XX), others female heterogametic (ZZ/ZW). Other taxa have environmental sex determination (ESD), and a few clades have all of those mechanisms (**Table 15.3**). This absence of phylogenetic patterns of sex determination suggests that multiple transitions have occurred between GSD and ESD and between male heterogamety and female heterogamety.

Characterizing species as having GSD or ESD oversimplifies the situation. For example, the sex of veiled chameleons (*Chamaeleo calyptratus*) is determined by the interaction of egg size and incubation temperature: large eggs produce females at low temperatures and males at high temperatures, whereas small eggs have the opposite relationship between incubation temperature and hatchling sex. Bearded dragons (*Pogona vitticeps*) have

Table 15.3 Types of sex determination in squamates

Clade[a]	Modes of sex determination[b]
Pygopodidae	GSD XY/XX
Gekkonidae	ESD; GSD XY/XX, ZZ/ZW
Scincidae	ESD; GSD XY/XX; egg size[c]
Amphisbaenia	GSD ZZ/ZW
Lacertidae	GSD ZZ/ZW
Teiidae	GSD XY/XX
Chamaeleonidae	ESD; GSD XY/XX, ZZ/ZW
Agamidae[d]	GSD ZZ/ZW, ESD
Phrynosomatidae	GSD XY/XX
Iguanidae	GSD XY/XX
Dactyloidae	GSD XY/XX
Helodermatidae	GSD ZZ/ZW
Anguidae	GSD ZZ/ZW
Varanidae	GSD ZZ/ZW
Serpentes (snakes)	Pythonids and boids are GSD XY/XX; others, when known, are GSD ZZ/ZW

[a] Phylogenetic relationships among these groups are shown in Table 15.1.

[b] ESD, environmental sex determination. GSD, genetic sex determination. Male heterogamety, XY/XX. Female heterogamety, ZZ/ZW

[c] One species of skink is known in which ESD, GSD, and egg size interact to determine sex.

dimorphic sex chromosomes—males are ZZ and females are ZW. However, high incubation temperatures can produce male-to-female sex reversal, leading to ZZ females.

Some viviparous species of lizards have ESD, and a pregnant female lizard can determine the sex of her embryos by adjusting her thermoregulatory behavior. Free-ranging female Australian snow skinks, *Niveoscincus ocellatus*) that maintained high body temperatures for many hours a day gave birth to litters that were ~65% female, whereas females that were warm for shorter periods of time had litters that were ~75% male.

The warmer females gave birth earlier in the season than the cooler females, and that observation suggests an advantage for producing females at high incubation temperatures: Birth date influences the adult body size of snow skinks—individuals that are born early in the year are larger when they reproduce than individuals that are born later. Large body size is probably more advantageous for female snow skinks than for males because large females produce larger litters than do small females. A small body size may not be a disadvantage for a male snow skink because males of this species do not compete with other males to hold territories, and even a small male can fertilize the eggs of a large female.

15.7 Climate Change

LEARNING OBJECTIVE

15.7.1 Explain how global climate change affects the daily and seasonal activity of ectotherms such as squamates.

Lepidosaurs are ectotherms, and that characteristic lies at the heart of the effects of global climate change for this group. Ectotherms must use behavior to control their body temperature while they are simultaneously hunting, defending territories, or mating. As global climates warm, lizards and snakes in some populations, particularly in arid regions, are having difficulty integrating thermoregulation with these other activities.

Desert lizards bask in the sunlight to raise their body temperatures to activity levels and move to shaded crevices to avoid overheating (see Figure 14.3). As global temperatures rise, the lizards spend more time in crevices, thereby reducing the time they can spend feeding, defending territories, and searching for mates. The limitation on energy intake may be exacerbated if higher nighttime temperatures increase the lizards' metabolic rates while they are sleeping, causing the lizards to consume energy that could otherwise be devoted to reproduction.

A study of populations of 48 species of spiny lizards (*Sceloporus*) in Mexico revealed the vulnerability of these desert lizards to climate change. Data from Mexican weather stations show that the maximum daily air temperature has increased most during the period from January to May, which corresponds to the breeding season for many *Sceloporus* species. The increase in temperature has been greatest in northern and central Mexico and at high elevations. This information was used to make predictions about the risk of extinction of lizard populations.

In 2009, researchers tested their predictions by returning to 200 sites that had populations of *Sceloporus* in 1975. They found that 12% of the populations had become extinct, and that the change in maximum temperature at the 200 sites was positively correlated with extinctions of local populations (**Figure 15.24A**). A lizard at the sites where populations were extinct would have had to spend most of the daylight hours in a retreat site to escape high temperatures; at the sites where populations were still present, only 4 hours a day were too hot.

Extending the model to 2050 and 2080 expands the area of Mexico in which extinctions are expected to occur and raises the probability of extinction for many populations above 50% (**Figure 15.24B**). Applying the model on a global level and extending it to include 34 families of lizards, researchers predict that extinctions of local populations of lizards will reach 39% by 2080.

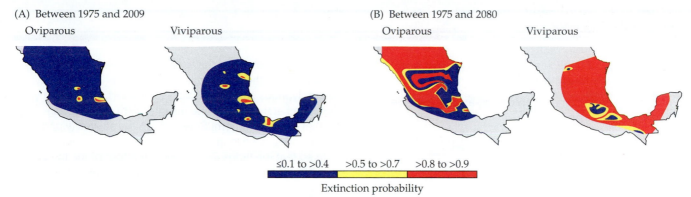

(A) Between 1975 and 2009

Oviparous   Viviparous

(B) Between 1975 and 2080

Oviparous   Viviparous

≤0.1 to >0.4   >0.5 to >0.7   >0.8 to >0.9

Extinction probability

Figure 15.24 Climate change and extinctions of desert lizard populations in Mexico. (A) A survey conducted in 2009 found that 12% of 200 populations of *Sceloporus* lizards documented in 1975 had become extinct. The extinctions were concentrated in areas most affected by rising temperature. Populations at high elevations showed the highest levels of extinction, and viviparous species were more strongly affected than oviparous species. (B) Temperature increases predicted by global climate models are projected to lead to widespread extinctions of *Sceloporus* by 2080. Once again, viviparous species will feel the greatest impact. (After B. Sinervo et al. 2010. *Science* 328: 894–899. © AAAS)

Summary

15.1 Characters and Diversity of Lepidosaurs

Extant lepidosaurs include two lineages: a single species of rhynchocephalian (the tuatara), and more than 11,000 species of squamates (lizards and snakes).

The phylogenetic relationship of snakes is nested within lizards, making lizards in the colloquial sense paraphyletic. Nonetheless, the groups popularly called snakes and lizards are very different in their morphology, ecology, and behavior.

The tuatara has several derived characters, including nocturnal activity, low body temperature, and fully diapsid skull. It is endemic to New Zealand with populations on 32 small islands.

Determinate growth is characteristic of squamates.

More than 80% of lizard species weigh less than 20 g as adults. Lizards of this size can subsist on insects without requiring morphological or ecological specializations to capture them, and most lizards are insectivores.

Most species of lizards that are too large to eat insects are herbivores. Their bulky bodies can accommodate the large intestines needed to digest plants. Monitors and tegus are large and carnivorous; some species prey on other vertebrates.

Many species of lizards are arboreal. Chameleons, the most specialized arboreal lizards, have a prehensile tail, zygodactyl feet, eyes that swivel independently, and a projectile tongue used to capture prey.

Leglessness has evolved more than 60 times among lizards. All extant snakes are legless.

- Surface-dwelling legless lizards have long tails, whereas fossorial species are short-tailed.

- Amphisbaenians, the most specialized legless lizards, are fossorial and can move forward and backward in their tunnels.

- Changes in the expression of regulatory factors during development elongate the trunk and suppress development of the limbs of snakes.

Snakes move in a number of distinctive and sometimes specialized ways.

- Lateral undulation is the generalized form of locomotion employed by all snakes and is used by flying snakes to stabilize their glides.

- Concertina locomotion is used in narrow passages such as burrows; a snake presses the sides of its body against the walls, alternately anchoring and extending the body.

- Heavy-bodied snakes use rectilinear locomotion, successively raising portions of the body from the substrate and moving the skin forward.

- Snakes that move over loose sand use sidewinding, which exerts force downward rather than laterally.

15.2 Foraging Modes

Foraging modes of squamates extend over a spectrum from sit-and-wait ambush predators to widely foraging predators.

- Sit-and-wait species rely primarily on vision to detect prey on the surface; widely foraging species employ chemoreception to detect hidden prey.

(Continued)

Summary *(continued)*

- Most sit-and-wait species of lizards have stout bodies, whereas widely foraging species are slim.

- Viviparity is more common among sit-and-wait species than widely foraging species.

- Many sit-and-wait species are territorial; most widely foraging species are not.

- Sit-and-wait species make infrequent short sprints to capture prey or escape predators and have high anaerobic metabolic capacities but become exhausted quickly. Widely foraging species spend much of their time moving and have high aerobic capacities and greater endurance than sit-and-wait species.

15.3 Skull Kinesis and Feeding

Modifications of the skulls of most lepidosaurs allow kinesis of the skull and jaws during feeding.

- Tuatara have reverted to an akinetic, fully diapsid skull, with two temporal bars.

- Lizards have lost the lower bar, imparting a degree of kinesis to most species.

- Snakes have lost both the lower and upper bar; their highly kinetic skulls have 8 elements that can move independently.

Snakes swallow prey intact, in some cases while it is still struggling, and usually head-first. They draw prey into the mouth with alternating movements of the upper and lower jaws on the left and right sides of the head. The recurved teeth slide forward across the prey as the jaws are advanced and are embedded in the prey as the jaws are retracted.

Many snakes are "binge eaters" that consume large prey items at infrequent intervals. Between meals, the snake's intestine and some internal organs shrink.

Constriction and venom reduce the risk of injury to the snake by killing prey before it is swallowed. Duvernoy's gland, which is present in the upper jaw of many colubrid snakes, is homologous with the venom gland of elapids and vipers.

Opisthoglyphous snakes have one or more enlarged fangs near the rear of the maxillae that conduct venom into prey. Proteroglyphous snakes (elapids) have short, permanently erect fangs at the front of the maxillae with closed channels that conduct venom. Solenoglyphous snakes (vipers and pit vipers) have long fangs on the maxillae, which rotate so that the fangs lie parallel to the roof of the mouth when it is closed and are erected when the snake strikes.

15.4 Predator Avoidance and Defense

Sit-and-wait species, which remain motionless for long periods, often have cryptic colors, patterns, and shapes that conceal them from predators. Many widely foraging species have striped patterns that obscure movement.

Some squamates deter predators by signaling that they are aware of the predator's presence, by threatening a predator, by mimicking the colors and patterns of venomous species, or by spraying a predator with sticky secretions.

Many lizards and a few snakes can autotomize (break off) a portion of the tail. The break occurs within a vertebra, and the missing vertebrae are regenerated with a cartilaginous rod. Some geckos have skin that rips easily, allowing the lizard to escape; the skin subsequently regenerates.

Venom, a predatory mechanism, has been modified to serve a protective function in some clades. Many venoms contain compounds that cause intense pain rather than damaging tissues.

15.5 Social Behavior

Squamates employ visual, auditory, chemical, and tactile signals to maintain territories and choose mates.

Males of sit-and-wait species of lizards are usually territorial and challenge intruding males, often by displaying a colored dewlap combined with push-ups and head bobs.

Snakes and widely foraging lizards are rarely territorial and rely on scent to locate and identify mates.

Summary *(continued)*

The impression that squamates display little sociality may reflect a lack of information.

- Some skinks form monogamous relationships, and others live in family groups.

- More than 100 species of squamates exhibit maternal care.

15.6 Reproductive Modes

Lepidosaurs exhibit both genotypic and environmental sex determination. The distinction between the two is not sharp.

Genotypic sex determination includes male and female heterogamety.

Oviparity is the ancestral reproductive mode for squamates, but viviparity has evolved more than 100 times, primarily in cold climates.

Viviparity allows a female to control the incubation temperature of her embryos, but pregnancy reduces a female's ability to flee from predators and limits reproduction to a single event in a season.

Parthenogenesis is known in six families of lizards and one species of snake, but parthenogenesis is hard to detect and is probably more widespread than realized. Interspecific hybridization of two bisexual species appears to be the starting point for parthenogenesis.

15.7 Lepidosaurs and Climate Change

Because lepidosaurs are ectotherms, climate change threatens their ability to carry out their daily activities. Increasing temperatures limit the time that squamates can be outside their retreat sites, thereby reducing the time they can devote to feeding, defending territories, and seeking mates.

Discussion Questions

15.1 Durophagy (eating hard-shelled prey such as beetles or mollusks) appears as a dietary specialization among lizards, but it is rare among snakes. Why? What specializations of the teeth would you expect to find in durophagous lizards and in snakes?

15.2 North American horned lizards (*Phrynosoma*, Phrynosomatidae) and the Australian thorny devil (*Moloch horridus*, Agamidae) both eat ants almost exclusively. Suggest an explanation for the convergence of ant-eating specialists in these two families on a short-legged, stocky body form and a spine-covered body.

15.3 The text mentions several specializations that allow vipers to eat large prey. Large, heavy-bodied vipers in the genus *Bitis*, such as the African puff adder *Bitis arietans*, provide particularly clear examples of these specializations. List as many of these specializations as you can and explain how they work together.

15.4 Lizards that display sociality are all viviparous. In contrast, oviparous lizards do not appear to form long-term associations with other individuals of their species. What differences between these two groups of lizards might account for the difference in behavior?

Figure credits
Figure 15.8: A,C,E adapted from C. Gans 1974. *Biomechanics*. Philadelphia: Williams & Wilkins; D after H.C. Astley and B.C. Jayne. 2009. *J Exp Zool* A 311: 207–216; F from Rieser et al. 2021. *PNAS* 118: e2018264118.

Turtles

Amazon river turtle, *Podocnemis unifilis*

Toothless and encased in a bony shell with the limb girdles inside the ribs, turtles are remarkable animals. If they had become extinct at the end of the Mesozoic, they would rival dinosaurs and pterosaurs in their novelty. Because they survived, however, turtles are regarded as commonplace and are often used (quite inappropriately) in comparative anatomy courses to represent basal amniotes.

The unique body form of turtles makes them immediately recognizable, but these anatomical rearrangements have obscured morphological characters that are used to determine evolutionary affinities among other vertebrates and it has been difficult to trace turtle evolution. Initial interpretations of the evolutionary relationships of turtles focused on their anapsid skull and placed their origin among stem reptiles. However, the anapsid condition of extant turtles is now recognized as a derived character that is probably a consequence of the development of a rigid trunk and flexible neck. New fossils combined with molecular data and studies of the embryonic development of extant turtles identify turtles as the sister group of Archosauria. The diapsid skulls of two basal turtles, †*Eunotosaurus* and †*Pappochelys*, confirm that relationship.

In this chapter we describe the unique structure of turtles, elements of their behavior and ecology, and explain why many of the characters that have served turtles so well for 200 Ma render them vulnerable in a human-dominated world.

16.1 Form and Function

LEARNING OBJECTIVES

16.1.1 Describe the structure of a turtle shell.

16.1.2 Explain the difference in head retraction of pleurodires and cryptodires.

16.1.3 Explain how a turtle ventilates its lungs.

16.1.4 Trace the steps in the evolution of the turtle body plan.

A turtle shell is unique among tetrapods. Bony body armor was a feature of some †ornithischian dinosaurs as well as extinct and extant mammals (e.g., †glyptodonts and armadillos), but these coverings are anatomically different from the shells of turtles. In addition to the shell, two functional properties set turtles apart from all other amniotes: (1) most species can retract their heads into their shells, and (2) nearly all species must breathe inside a rigid box.

Shell and skeleton

The shell is the most distinctive feature of a turtle. Beta-keratin chemically similar to the beta-keratin in the scales of crocodylians and the feathers of birds (but different from the keratin in lizard scales) forms a layer of horny scutes on the exterior of the **carapace** (upper shell) and **plastron** (lower shell). These scutes do not coincide in number or position with the underlying bones (**Figure 16.1**). The carapace has a row of 5 central scutes, bordered on each side by 4 lateral scutes. The 10–12 marginal scutes on each side turn under the edge of the carapace. The plastron is covered by a series of 6 paired scutes.

The bony inner layer of the carapace is formed primarily by lateral expansion of the ribs (endochondral bone), with the incorporation of some dermal elements. The shell grows faster than the rest of the embryo, and both girdles are enclosed by the rib cage. The plastron is formed largely from dermal ossifications, but the entoplastron is derived from the interclavicle, and the paired epiplastra anterior to it are derived from the clavicles. Processes from the hyoplastron and hypoplastron fuse with the first and fifth pleurals, forming a rigid connection between the plastron and the carapace called the bridge.

(A) External surface of carapace (left) and plastron showing scutes

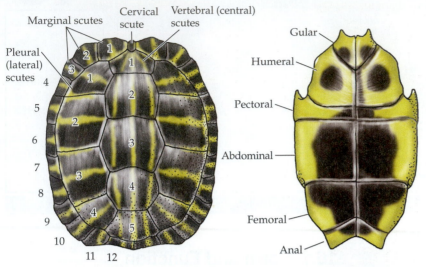

(B) External surface of skeleton of carapace (left) and plastron showing bones

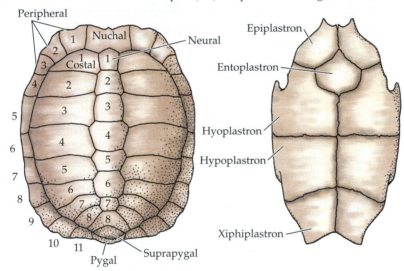

(C) Internal surface of skeleton of carapace showing vertebrae and ribs

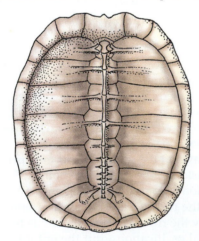

Figure 16.1 **Shell and vertebral column of a turtle.** (A) Epidermal scutes of the carapace and plastron. The carapace has a central (vertebral) row of 5 scutes with 4 lateral (pleural) scutes on each side and 10–12 marginal scutes. The plastron has 6 paired scutes. (B) Dermal bones of the carapace and plastron. (C) Vertebral column, seen from inside the carapace. Note that the anterior ribs articulate with two vertebral centra. (After R. Zangerl. 1969. In C. Gans, [Ed.], *Biology of the Reptilia 1: Morphology*, pp. 311–320. New York: Academic Press.)

Head retraction

Extant turtles have 10 vertebrae in the trunk and 8 in the neck. The centra of the trunk vertebrae are elongated and lie beneath the dermal bones in the dorsal midline of the shell. The centra are constricted in their centers and fused to each other. The neural arches in the anterior two-thirds of the trunk lie between the centra as a result of anterior displacement, and the spinal nerves exit near the middle of the preceding centrum. The ribs are also shifted anteriorly; they articulate with the anterior part of the neurocentral boundary, and in the anterior part of the trunk, where the shift is most pronounced, the ribs extend onto the preceding vertebra.

The two extant lineages of turtles, Cryptodira (Greek *kryptos*, "hidden"; *deire*, "neck") and Pleurodira (Greek *pleura*, "side") can be traced through fossils to the Mesozoic. **Cryptodires** retract the head into the shell by bending the neck in a vertical S shape (**Figure 16.2A,B**). The cervical vertebrae of cryptodires have specialized articulating surfaces between vertebrae, called ginglymi, that permit formation of an S-shaped bend. In most families, the bend is formed by two successive ginglymoidal joints, between the 6th and 7th and the 7th and 8th cervical vertebrae. **Pleurodires** retract the head by bending the neck to the side (horizontally; **Figure 16.2C,D**). This lateral bending is accomplished by simple ball-and-socket or cylindrical joints between adjacent cervical vertebrae.

Lung ventilation

Basal amniotes used movements of the rib cage to draw air into the lungs and to force it out, and most amniotes still employ modifications of that basic mechanism (see Section 13.1). Turtles can't do that; fusion of the ribs with a rigid shell requires a unique method of ventilating the lungs.

The lungs of a turtle are attached to the carapace both dorsally and laterally. Ventrally, the lungs attach to a sheet of connective tissue that is attached to the viscera, and the viscera are underlain by a sling formed by the abdominal muscles (**Figure 16.3**). Turtles draw air into the lungs by contracting muscles that increase the volume of the visceral cavity, allowing the viscera to settle downward. Downward movement of the viscera pulls downward on the lungs,

(A) Cryptodire, map turtle, *Graptemys pseudogeographica*

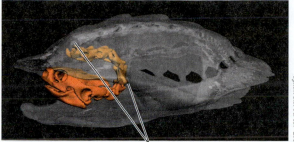

Neck bends in vertical plane

From I. Werneburg et al. 2015
System Biol 64: 187–204

(B) Cryptodire, loggerhead musk turtle, *Sternotherus minor*

Courtesy of Todd Pierson

(C) Pleurodire, Hilaire's side-necked turtle, *Phrynops hilarii*

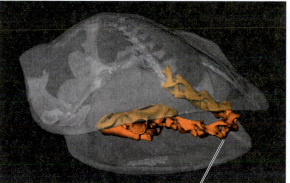

Neck bends in horizontal plane

From I. Werneburg et al. 2015
System Biol 64: 187–204

(D) Pleurodire, long-necked turtle, *Chelodina steindachneri*

Courtesy of Stephen Zozaya

Figure 16.2 Neck retraction by cryptodires and pleurodires.
(A) Cryptodires retract the head into the shell by bending the neck in a vertical S shape. In most families, the bend is formed by successive ginglymoidal joints, one between the 6th and 7th and a second between the 7th and 8th cervical vertebrae. (B) A loggerhead musk turtle shows how this bend allows the head to be pulled directly back into the shell. (C) Pleurodires retract the head by bending the neck laterally. (D) An Australian long-necked turtle shows how the head and neck are folded against the shell opening, leaving part of the neck exposed.

drawing in air. Expiration is accomplished by contracting muscles that force the viscera upward, compressing the lungs and expelling air.

The problems of respiring within a rigid shell are the same for most turtles, but the mechanisms show some variation. Aquatic turtles use the hydrostatic pressure of water to help move air into and out of the lungs. In addition, many aquatic turtles are able to absorb oxygen from and release carbon dioxide into the water. The pharynx and cloaca appear to be the major sites of this aquatic gas exchange. In 1860, the Swiss-American naturalist Louis Agassiz pointed out that the pharynx of softshell turtles (*Apalone*) contains fringelike processes and suggested that these structures are used for underwater

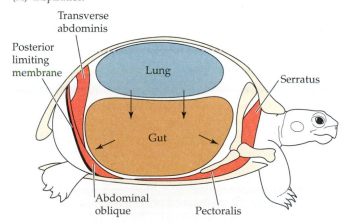

(A) Inspiration

(B) Expiration

Figure 16.3 Schematic view of the lungs and respiratory movements of a tortoise. The abdominal muscles form a sling that moves the viscera to change the volume of the lungs. (A) On inspiration, the serratus and its associated muscles pull the pectoral girdle forward while the abdominal oblique pulls the posterior limiting membrane back. These movements allow the viscera to drop downward, increasing the volume of the lungs. (B) On expiration, the pectoralis pulls the pectoral girdle backward while the transverse abdominis pulls the posterior limiting membrane forward. These movements force the viscera upward, forcing air out of the lungs. The in-and-out movements of the forelimbs and of the soft tissue at the rear of the shell during breathing are conspicuous.

respiration. Subsequent study has shown that softshell turtles use movements of the hyoid apparatus to draw water into and out of the mouth and pharynx (oropharynx) when they are confined underwater. Musk turtles (*Sternotherus*) are also capable of aquatic gas exchange, and histological examination shows that their oropharyngeal region is lined by flat-topped papillae. These structures are highly vascularized and are probably sites of gas exchange.

Australia's Fitzroy River turtle (*Rheodytes leukops*) uses cloacal respiration; its cloacal orifice may be as large as 30 mm in diameter. Large bursae (sacs) open from the wall of the cloaca, and the bursae have a well vascularized lining with numerous projections (villi). The turtle pumps water into and out of the bursae at rates of 15–60 times per minute. Gas exchange through the cloacal bursae meets the turtle's needs, and *Rheodytes* rarely surfaces to breathe.

Evolution of the turtle body plan

In basal diapsids the dorsal ribs, intercostal muscles, and abdominal muscles work together to perform two functions: (1) they control torsion (bending) of the trunk during locomotion, and (2) create the volume changes that ventilate the lungs. As we discussed in Chapter 13, these two functions cannot occur simultaneously—when the trunk bends during locomotion, the lungs cannot be ventilated (see Figure 13.2). Turtles have separated the roles of the ribs and muscles.

- Co-ossification of the ribs with the bony shell makes the trunk rigid and the intercostal muscles have been lost.
- A sling formed by the transverse and oblique abdominal muscles ventilates the lungs, assisted by movements of the limb girdles.

Accompanying changes include reducing the number of dorsal vertebrae and lengthening the vertebrae that remain, shifting the articulation of ribs from on the vertebrae to between the vertebrae, incorporation of the dermal portion of the shoulder girdle into the shell (interclavicle = entoplastron; clavicles = epiplastra; cleithra = nuchal), loss of teeth and formation of a keratinous beak, modifications of the neck associated with head retraction, and loss of both temporal fenestrae.

These anatomical rearrangements can be seen in a series of fossils dating from the Late Permian to the Late Triassic, described below and in **Table 16.1**.

- †*Petrolacosaurus* was a basal diapsid. The body was slim and elongate, with about 25 dorsal vertebrae and ribs. The length and width of the vertebral centra were approximately equal. Intercostal muscles played a major role in ventilating the lungs and stabilizing the trunk during locomotion. Numerous delicate gastralia (see Section 13.2) were present ventrally, and these also played a role in ventilating the lungs and stabilizing the trunk during locomotion.

- Between †*Petrolacosaurus* and †*Eunotosaurus*, the number of dorsal vertebrae and ribs was reduced to 9, the length of the vertebral centra increased to four times their width, and the number of gastralia segments was reduced to a pair of gastralia. †*Eunotosaurus* lies close in time to the separation of the turtle and archosaur lineages, yet division of mechanisms for trunk stabilization and lung ventilation had already begun. The ribs were broadened, intercostal muscles had been lost, and modification of the abdominal muscles as a sling used for lung ventilation had already started.

- Between †*Eunotosaurus* and †*Pappochelys*, the plastron had started to form as the gastralia broadened and lengthened to cover more of the ventral surface. The sternum was lost and the size of the upper fenestra of the skull was reduced.

- Between †*Pappochelys* and †*Eorhynchochelys*, the ribs became shorter and straighter and the upper temporal fenestra became covered so that the skull was anapsid. The anteriormost marginal teeth were lost, producing a partially toothless beak.

- Between †*Eorhynchochelys* and †*Odontochelys*, a complete plastron was formed by expansion and ossification of the gastralia. Keratinous scutes were present, at least on the plastron. Scutes are not preserved on the dorsal surface of the body, but were probably present because no fossil or extant turtle has scales confined to the plastron. Unlike †*Eorhynchochelys*, †*Odontochelys* retained marginal teeth.

- Between †*Odontochelys* and †*Proganochelys*, the carapace became fully formed and a bridge connected the carapace and plastron, forming a complete shell that enclosed the limb girdles. Many features of extant turtles appeared, including emargination of the upper temporal and cheek regions of the skull and changes in the cervical vertebrae related to head retraction. Marginal teeth were lost and a keratinous beak covered the jaws. (This is a second loss of marginal teeth, independent of the reduction of marginal teeth seen in †*Eorhynchochelys*.)

These fossils record a remarkable change from the ancestral diapsid body plan. Increasing the rigidity of the body impaired the ability to increase stride length and speed by bending the trunk, as extant lizards and crocodylians do, and required an entirely new mechanism to ventilate the lungs.

What advantage might have offset this reduction in locomotor ability and change in lung ventilation? A stiff body increases the power that can be applied by the forelimbs during digging, and the ability to construct burrows may account for the initial changes in the evolution of the turtle body plan. Supporting this interpretation, the earliest stem turtles, †*Eunotosaurus* and †*Pappochelys*, are terrestrial. Radiation into aquatic habits came later, seen in †*Eorhynchochelys* and †*Odontochelys*.

Table 16.1 Evolution of the turtle body plan

	Taxon, inferred habitat	Age	Characters
	†*Petrolacosaurus* (basal diapsid) Terrestrial	Late Carboniferous (~300 Ma)	Upper temporal fenestra present Dorsal ribs, intercostal and abdominal muscles stabilize trunk and ventilate lungs Numerous gastralia with a gap between the left and right sides
	†*Eunotosaurus* (transitional form) Terrestrial	Late Permian (~260 Ma)	Upper temporal fenestra present Thoracic vertebrae reduced to 9 and lengthened Separation of locomotion and lung ventilation: • Broadened dorsal ribs stabilize thorax • Intercostal muscles lost • Early form of abdominal muscle sling ventilates lungs Paired gastralia with a gap between the left and right sides
	†*Pappochelys* (transitional form) Terrestrial	Middle Triassic (~240 Ma)	Small upper temporal fenestra present Gastralia broadened, lengthened, and thickened, reducing the gap between the left and right sides
	†*Eorhynchochelys* (oldest known stem turtle) Aquatic	Late Triassic (~228 Ma)	Upper temporal fenestra entirely lost Short, straight dorsal ribs Partial loss of marginal teeth Keratinous beak present
	†*Odontochelys* (stem turtle) Aquatic	Late Triassic (~220 Ma)	Plastron complete Keratinous scales on shell
	†*Proganochelys* (stem turtle) Terrestrial	Late Triassic (~210 Ma)	Carapace complete Shoulder girdle inside shell Ribs articulate between vertebrae Complete loss of marginal teeth

Sources: Compiled from T.R. Lyson et al. 2010. *Biol Ltr* 6: 830–833; T.R. Lyson et al. 2013. *Curr Biol* 23: 1113–1119; T.R. Lyson and G.S. Bever. 2020. *Annu Rev Ecol Evol Syst* 51: 110218–024746; R.R. Schoch and H.D. Sues 2020. *Palaeontology* 63: 375–393/CC BY 4.0; F.E. Peabody. 1952. University of Kansas *Paleont Contrib Vertebrata* 1: 1–41.

16.2 Diversity

LEARNING OBJECTIVES

16.2.1 Explain how the shells of different turtle species reveal their habitat and lifestyle.

16.2.2 Describe some modifications of turtle anatomy and behavior that enhance protection from predators.

Turtles have radiated into terrestrial, freshwater and marine environments. There are, however, no arboreal or aerial species. The 361 extant species of turtles are distributed among 14 families (**Table 16.2**). Nearly three-quarters of extant turtles are cryptodires, and these are the only turtles now found in most of the Northern Hemisphere.

Aquatic and terrestrial species of cryptodires occur in North and South America, Asia, and Africa.

Pleurodires today occur only in freshwater habitats in the Southern Hemisphere, although during the late Mesozoic and early Cenozoic they were distributed worldwide and included both freshwater and marine species. All extant pleurodires are at least semiaquatic, but the shells of some fossil pleurodires suggest that they may have been terrestrial.

Turtle shells are distinctive Turtles are easy to recognize: If it has four legs and a shell, it's a turtle; no other extant tetrapod has a bony external shell. You can tell a lot

Table 16.2 Extant families of turtles

Family	Habitat	Adult size	Number and distribution of species
Pleurodira: 99 species			
Pelomedusidae	Freshwater	Small to medium size (up to 50 cm)	27 species in Africa, Madagascar, and the Seychelles Islands
Podocnemididae	Freshwater	Medium to large size (up to 90 cm)	8 species in northern South America and Madagascar
Chelidae	Freshwater	Small to medium size (up to 50 cm)	64 species in South America, Australia, and New Guinea
Cryptodira: 262 species			
Carettochelyidae	Freshwater	Large (up to 70 cm)	1 species (the pig-nosed turtle) in southern New Guinea and northern Australia
Trionychidae	Freshwater	Small to very large (up to 1.3 m)	33 species in North America, Africa, and Asia
Dermochelyidae	Marine	Extremely large (up to 2.4 m)	1 species (the leatherback sea turtle); Atlantic Ocean from Nova Scotia to Argentina, Pacific Ocean from Japan to Tasmania
Cheloniidae	Marine	Large to very large (up to 1.5 m)	6 species, worldwide in tropical and temperate oceans
Chelydridae	Freshwater	Large (up to 80 cm)	5 species in North and Central America
Dermatemydidae	Freshwater	Medium size (up to 50 cm)	1 species (the Central American river turtle) from Mexico through Nicaragua
Kinosternidae	Freshwater	Small to medium size (up to 40 cm)	30 species in North, Central, and South America
Platysternidae	Freshwater	Small (~18 cm)	1 species (the big-headed turtle); Southeast Asia
Emydidae	Terrestrial, freshwater, coastal	Small to medium size (up to 60 cm)	53 species in North and Central America, 2 species in Europe
Testudinidae[a]	Terrestrial	Small to very large (up to 1 m)	60 species worldwide in temperate and tropical continental regions and on oceanic islands
Geoemydidae	Terrestrial, freshwater, coastal	Small to large (up to 75 cm)	71 species in tropical and subtropical regions of Asia, southern Europe, and North Africa; one genus of Neotropical wood turtles in Central and South America

Source: Based on data in R.C. Thomson et al. 2021. *Proc Natl Acad Sci* 118: e2012215118 and P. Uetz et al. 2021. The Reptile Database, http://www.reptile-database.org, accessed 5/22/2021.

[a] Testudinidae should not be confused with Testudines, a designation (not used in this chapter) that refers to all turtles. The family Testudinidae comprises those species commonly called the tortoises; it is the only family of turtles whose members are exclusively terrestrial.

about the ecology of a turtle's species by looking at its shell. Most tortoises, which are the most terrestrial turtles, have domed carapaces (**Figure 16.4A**). Aquatic swimmers, such as painted turtles, have low carapaces that offer little resistance to movement through water (**Figure 16.4B**). Ossification of the shell of softshell turtles is greatly reduced. The distal ends of the broadened ribs are embedded in flexible connective tissue, and the carapace and plastron are covered with skin. Softshell turtles lie in ambush partly buried in debris on the bottom of a pond (**Figure 16.4C**). Their long necks allow them to reach considerable distances to seize the invertebrates and small fishes on which they feed.

Sea turtles, Cheloniidae and Dermochelyidae, show more extensive specialization for aquatic life than any freshwater turtle; for example, the forelimbs are modified as flippers (**Figure 16.4D**).

Shells provide significant protection A turtle's shell protects it from most predators. Turtles withdraw their head

and legs when threatened by a predator, and the adults of most species have shells that few predators can crush. Some turtles have an additional protective device in the form of flexible hinges that allow the front or rear of the shell to close. The most familiar examples are North American and Asian box turtles (*Terrapene* and *Cuora*, respectively), which have a hinge between the hyoplastral and hypoplastral bones (**Figure 16.5A**). African forest tortoises (*Kinixys*) have a hinge on the posterior part of the carapace (**Figure 16.5B**). The margins of the epidermal shields and the dermal bones of the carapace align, and the hinge runs between the second and third pleural scutes and the fourth and fifth costals. The erratic phylogenetic occurrence of hinged shells and differences among species indicate that shell kinesis has evolved independently many times in turtles.

Many tortoises use their limbs to close gaps in the shell (**Figure 16.5C**). Front legs are folded in front of the head, exposing only the heavily scaled forelimbs; the rear of the shell is sealed off by the soles of the hind feet.

(A) Leopard tortoise, *Stigmochelys pardalis*

(B) Eastern painted turtle, *Chrysemys picta*

Bernard Dupont/CC BY-SA 2.0

Danielle Brigida/CC BY 2.0

(C) Spiny shoftshell, *Apalone spinifera*

(D) Green sea turtle, *Chelonia mydas*

Courtsey of Todd Pierson

James St. John/CC BY 2.0

Figure 16.4 Body forms of turtles.   (A) Most tortoises and many other terrestrial turtles have domed shells that the jaws of mammalian predators cannot easily crush. (B) Aquatic swimmers are streamlined and have large, webbed feet. (C) Softshell turtles lack peripheral ossifications and epidermal scutes. They are ambush predators, and their flat, flexible shells allow them to bury into the substrate, leaving only the head exposed (inset). (D) The forelimbs of sea turtles are modified as flippers and the hindlimbs are used as rudders.

(A) Box turtle, *Terrapene carolina*

(B) African forest tortoise, *Kinixys belliana*

(C) Sonoran desert tortoise, *Gopherus morafkai*

Digom3/CC BY-SA 3.0

Diotime75/CC BY-SA 2.5

Bettina Arrigoni/CC BY 2.0

Hinged plastron

Hinged carapace

Forelimbs close the shell opening

Ineta McParland/CC BY 2.0

Luxanaherandreto/CC BY-SA 4.0

Shelly R. Puffer, USGS

Figure 16.5 Shells provide protection.   (A) Box turtles have a hinge in the plastron that closes both the front and rear openings of the shell. (B) The carapace of the African forest tortoise is hinged and closes the rear opening of the shell. (C) Many tortoises use their forelegs to close the front of the shell.

(A) Hatchling turtle killed by a crow

Fritz Flohr Reynolds/CC BY-SA 3.0

(B) Adult turtle, roadkill

Thomas R. Machnitzki/CC BY 3.0

Figure 16.6 **Shells cannot always provide protection.**
(A) The thin shells of hatchling turtles offer little protection against predators. (B) No shell can protect a turtle from becoming roadkill.

Except when they don't Although hard shells protect adult turtles, hatchlings are in a very different position; they are bite-size and their shells are not rigid enough to resist crushing. Baby turtles are Oreo cookies from a predator's perspective (**Figure 16.6A**). Automobiles also are turtle killers (**Figure 16.6B**); studies have found that the annual mortality of turtle populations from automobiles lies in the range of 1–25%. Deaths of turtles on North American roads are especially high in the eastern United States and the Great Lakes region, where dense networks of roads criss-cross turtle habitat. This is a serious issue because estimates of long-term population viability predict that 2–3% added annual mortality is more than many turtle populations can withstand.

16.3 Social Behavior, Communication, and Courtship

LEARNING OBJECTIVES

16.3.1 Describe the behaviors that turtles employ during courtship and mating.

16.3.2 Describe some social interactions among turtles.

Figure 16.7 **Courtship of pond turtles.**   (A) Male pond turtles follow females, then hover in front of them during courtship. (B–I) Where multiple species of pond turtles occur sympatrically, species-specific markings on the heads and necks play a role in species identification.

Although turtles are popularly viewed as solitary and not very interactive, this impression is probably erroneous. Turtles have a variety of social interactions during which they use tactile, visual, auditory, and olfactory signals, and social behavior is probably more common than has been realized.

Many pond turtles have distinctive stripes of color on their heads, necks, and forelimbs and on their hindlimbs and tail (**Figure 16.7**). Herpetologists use these patterns to distinguish among species, and they are probably species-identification mechanisms for the turtles as well. During the mating season, male pond turtles swim in pursuit of other turtles, and the color and pattern on the posterior limbs may enable males to identify females of their own species. At a later stage of pond turtle courtship, when the male swims backward in front of the female and vibrates his claws against the sides of her head, both sexes can see the patterns on their partner's head, neck, and forelimbs (see Figure 16.7A).

Among terrestrial turtles, the mating behavior of tortoises such as *Gopherus* is best known (**Figure 16.8**). A male must mount a female to mate, balancing on the rear of her carapace. A female tortoise can dislodge an unwanted suitor simply by moving, causing him to topple from his precarious perch (Figure 16.8A). Male tortoises fight during the breeding season, biting and ramming each other (Figure 16.8B–E). Males follow females, sniffing at the cloacal region of the female (Figure 16.8F). Males of some species trail females for days during the breeding season. Head bobs are used during encounters between two males and between a male and a female (Figure 16.8C,G). Insemination requires the male to hook his tail beneath hers to bring their cloacae together (Figure 16.8I). Many tortoises vocalize while they are mating; the sounds they produce have been described as grunts, moans, and bellows.

Figure 16.8 **Combat and courtship.**   This series of events spanned several hours during an afternoon. (A) A small male gopher tortoise (*Gopherus polyphemus*) attempted to mount a female, but was rejected. (B) As the female walked away, a large male arrived. (C–E) After a period of head-bobbing, the males fought and the larger male chased the smaller male away. (F) The large male then returned and followed the female, nudging her and nipping at the rear of her shell. (G–I) After the female turned to confront the male, they engaged in a prolonged bout of head bobbing and the male bit the female's head. Ultimately the female allowed the male to mate.

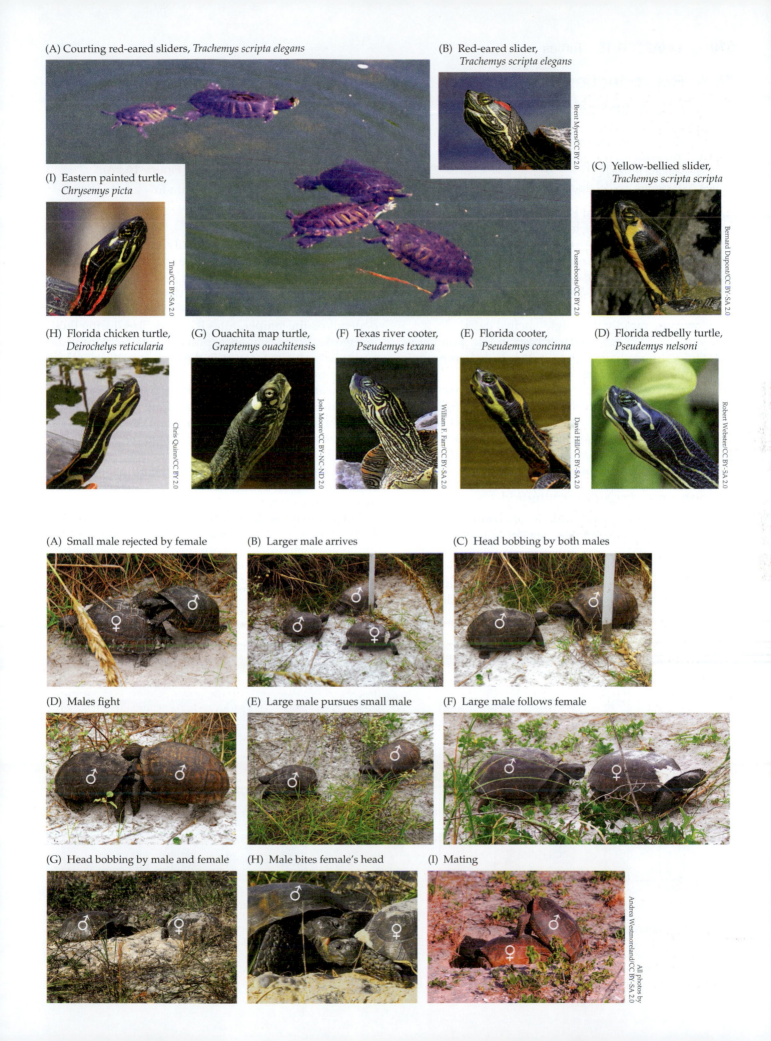

(A) Courting red-eared sliders, *Trachemys scripta elegans*

(B) Red-eared slider, *Trachemys scripta elegans*

Brent Myers/CC BY 2.0

(C) Yellow-bellied slider, *Trachemys scripta scripta*

Bernard Dupont/CC BY-SA 2.0

(I) Eastern painted turtle, *Chrysemys picta*

Tina/CC BY-SA 2.0

Pussrreboots/CC BY 2.0

(H) Florida chicken turtle, *Deirochelys reticularia*

Chris Quinn/CC BY 2.0

(G) Ouachita map turtle, *Graptemys ouachitensis*

Josh Moore/CC BY-NC-ND 2.0

(F) Texas river cooter, *Pseudemys texana*

William F. Farr/CC BY-SA 2.0

(E) Florida cooter, *Pseudemys concinna*

David Hill/CC BY-SA 2.0

(D) Florida redbelly turtle, *Pseudemys nelsoni*

Robert Webster/CC BY-SA 2.0

(A) Small male rejected by female

(B) Larger male arrives

(C) Head bobbing by both males

(D) Males fight

(E) Large male pursues small male

(F) Large male follows female

(G) Head bobbing by male and female

(H) Male bites female's head

(I) Mating

Andrea Westmoreland/CC BY-SA 2.0

All photos by

16.4 Reproduction

LEARNING OBJECTIVES

16.4.1 Describe the patterns of environmental sex determination of turtles.

16.4.2 Describe the occurrence of parental care among turtles.

16.4.3 Describe the behavior of hatchling and juvenile turtles.

All turtles are oviparous. Female turtles use their hind-limbs to excavate a nest in sand or soil and deposit a clutch that ranges from 4–5 eggs for small species to more than 100 eggs for the largest sea turtles. Eggs of most turtle species have soft, flexible shells, although those of some species have rigid shells. Embryonic development typically requires 40–60 days. In general, soft-shelled eggs develop more rapidly than rigid-shelled eggs.

Some turtle species lay their eggs in late summer or fall, and the eggs have a period of arrested embryonic development—**diapause**—during the winter, resuming development when temperatures rise in the spring. The northern snake-necked turtle (*Chelodina rugosa*) from Australia lays its eggs underwater during the wet season. The eggs remain in diapause until the floodwater recedes, begin development when they are exposed to air, and hatch at the start of the next wet season.

Environmental sex determination

The sex of mammals and birds is determined by genes on sex chromosomes (**genetic sex determination, GSD**). Beyond those two groups, however, sex determination is more varied. For some reptiles, sex is determined by the temperature an embryo experiences during development. Such **environmental sex determination (ESD)** is widespread among turtles, apparently universal among crocodylians

(see Section 17.5), and is known for the tuatara and a few species of lizards (see Table 15.2). The switch from one sex to the other usually occurs within a span of 1–2°C.

Two patterns of ESD have been described in turtles:[1]

- Type Ia produces males at low temperatures and females at high temperatures (**Figure 16.9A**).
- Type II produces females at both low and high temperatures and males at intermediate temperatures (**Figure 16.9B**).

Some families of turtles have only one pattern, others have two patterns, and a few have species with both ESD and GSD (**Table 16.3**).

The middle third of embryonic development is the critical period for sex determination—that is, an embryo's sex depends on the temperatures it experiences during those few weeks. When eggs are exposed to a daily temperature cycle, the high temperature in the cycle is most critical for sex determination.

The water content of nest soil interacts with temperature to determine the sex of a hatchling—wet nests produce a higher proportion of males than dry nests, probably because evaporative cooling lowers the temperature of the eggs. And of course, temperatures of natural nests are not completely stable. Daily temperature variation is superimposed on a seasonal cycle of changing environmental temperatures.

Subject to external sources of temperature variation within a nest, a turtle embryo may nevertheless have a limited capacity for thermoregulation while in the egg.

---

[1] A third pattern, type Ib, produces females at low temperatures and males at high temperatures. This pattern occurs in tuatara and a few lizards, but not in turtles.

(A) Type Ia: Males at low temperatures

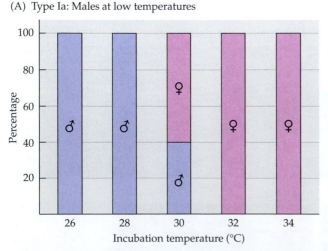

(B) Type II: Females at low and high temperatures

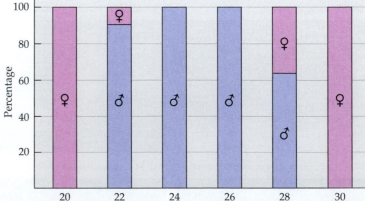

Figure 16.9 Environmental sex determination. (A) Type Ia: Eggs of the loggerhead sea turtle (*Caretta caretta*) hatch into males when they are incubated at 28°C or lower and into females at 32°C and above. Incubation at 30°C (center) produces a mixture of males and females. (B) Type II: Eggs of the North American snapping turtle (*Chelydra serpentina*) produce females at low and high temperatures and males at intermediate temperatures. (A, data from J.J. Bull. 1980. *Q Rev Biol* 55: 3–21; B, data from C.L Yntema. 1976. *J Morphol* 150: 453–462, and J.J. Bull and R.C. Vogt. 1979. *Science* 206: 1186–1188.)

Table 16.3 Sex-determining mechanisms among turtles[a]

Family[a]	Mechanisms[b]
Pleurodira	
Pelomedusidae	ESD Ia, II
Podocnemidae	ESD II
Chelidae	GSD XY
Cryptodira	
Carettochelyidae	ESD Ia
Trionychidae	GSD ZW
Dermochelyidae	ESD Ia
Cheloniidae	ESD Ia
Chelydridae	ESD II
Dermatemydidae	ESD Ia
Kinosternidae	GSD XY; ESD Ia,II
Platysternidae	Unknown
Emydidae	GSD XY; ESD Ia,II
Testudinidae	ESD Ia
Geoemydidae	GSD XY, ZW; ESD Ia,II

[a]The phylogeny of the families is shown in Table 16.2. Data are often available for only a few species in each family, and variation within families is probably understated.

[b]Genetic sex determination (GSD) can be male heterogametic (XY) or female heterogametic (ZW). Environmental sex determination in turtles is type Ia (males at low temperatures, females at high temperatures) or type II (females at low and high temperatures, males at intermediate temperatures).

Embryos of the Chinese softshell turtle (*Pelodiscus sinensis*) rotate within their eggshells so that their backs point toward the source of heat. Temperature measurements revealed a 1°C difference between the sides of the embryo facing toward and away from the source of heat.

Because an embryo's sex changes over a narrow temperature range, and because temperature and moisture differ among nests, both sexes are produced under field conditions, but not always in the same nest. A study of painted turtles (*Chrysemys picta*) found that two-thirds of the nests produced only males or only females while the remaining one-third of the nests produced hatchlings of both sexes. A similar study of Ouachita map turtles (*Graptemys ouachitensis*) produced even more skewed results: only one-seventh of the nests produced both sexes.

All in all, allowing environmental factors to determine the sex of one's offspring sounds like a risky proposition, and several hypotheses have been proposed to explain the benefits of ESD. For example, ESD might be correlated with sexual size dimorphism of adults—that is, for a given species, high incubation temperatures produce the sex that benefits from being larger as an adult. Another perspective suggests that the sex-determining effects of temperature are secondary to the effects of temperature

during development on some important functional character of the hatchlings, such as size at hatching, crawling or swimming speed, or the type of escape behavior employed by an individual, and that selection acts on those characters.

Parental care

Because fertilization is internal and egg deposition occurs days to months after insemination, male turtles are not present when the eggs are laid. Most species of turtles apparently provide no parental care; a female simply leaves after she fills the egg chamber with soil. Female hawksbill (*Eretmochelys imbricata*) and leatherback (*Dermochelys coriacea*) sea turtles move away from the nest in erratic patterns, pausing to scatter sand in several places; they may be creating decoy nests that could make it harder for predators to find the real nest.

Some female tortoises guard their nests and attack egg predators. This behavior usually ends within a few days after nesting, but for two consecutive years a female desert tortoise was still guarding a nest that contained newly hatched young. *Podocnemis expansa*, a South American river turtle, is the only species known to provide posthatching care (**Figure 16.10**). Females nest communally on mid-river sandbars and remain just off the nesting beach while the eggs develop. As the embryos approach hatching, they vocalize within the eggs and continue to vocalize as they enter the river. Females respond by vocalizing in the water, attracting the hatchlings to them. Only after the hatchlings have entered the water do the females move away from the nesting beach, migrating with the hatchlings to flooded forests. Sonar tracking and genetic typing of females and hatchlings showed that some hatchlings migrate with their mother.

Hatching and the behavior of baby turtles

Internal and external cues appear to be important in synchronizing the hatching of turtle embryos. Temperature variation within the nest causes some embryos to develop more rapidly than others. Vocalizations by these embryos may stimulate slower embryos to increase their rates of development in a brief catch-up period shortly before hatching.

Emergence from the nest   The first challenge a turtle faces after hatching is escaping from the nest, and in some instances interactions among all the hatchlings in a nest may be essential. Sea turtle nests are quite deep; the eggs may be buried 50 cm beneath the sand, and the hatchling turtles must struggle upward through the sand to the surface. After several weeks of incubation, the eggs all hatch within a period of a few hours, and 100 or so baby turtles find themselves in a small chamber at the bottom of the nest hole. Spontaneous activity by a few individuals sets the whole group in motion, crawling over and under one another. The turtles at the top of the pile loosen sand from the roof of the chamber as they scramble about, and the

(A) Female river turtles, *Podocnemis expansa*, arrive at nesting beach

(B) Adult female and hatchling with radio transmitters for tracking

(C) Hatchlings vocalize as they enter the water

(D) Hatchlings and adult females depart

Figure 16.10 **Parental care by Amazon river turtles.**
(A) Females arrive at the nesting beaches in groups. Such groups once included thousands of females, but today have only hundreds. (B) Adult females and hatchlings in this study were equipped with radio transmitters allowing them to be tracked. (C) Hatchlings vocalize as they enter the water, and females respond by vocalizing. (D) Hatchlings and adult females depart from the nesting beaches in groups containing hundreds of females and thousands of hatchlings. Maintaining contact with vocalizations, the turtles travel 100 km or more into the flooded forests where they feed.

sand filters down through the mass of baby turtles to the bottom of the chamber.

Periods of a few minutes of frantic activity are interspersed with periods of rest, possibly because the turtles' exertions reduce the concentration of oxygen in the nest and they must wait for more oxygen to diffuse into the nest from the surrounding sand. Gradually, the entire group of hatchlings moves upward through the sand as a unit until it reaches the surface. As the baby turtles approach the surface, high sand temperatures probably inhibit further activity, and the turtles wait a few centimeters below the surface until night falls, when a decrease in temperature triggers their emergence.

All the hatchlings emerge from a nest within a very brief time, and those in different nests that are ready to emerge on a given night all leave their nests at almost the same time. The result is the sudden appearance of hundreds or even thousands of baby turtles on the beach, each one crawling toward the ocean as fast as it can.

Simultaneous emergence is an important feature of the hatching of sea turtles because hatchlings suffer high mortality crossing even the few meters of beach and surf.

Terrestrial predators—crabs, foxes, coatis, peccaries, and other predators—gather at the turtles' breeding beaches at hatching time and await their appearance. Jaguars make long journeys to prey on the baby turtles. In the surf, sharks and predatory bony fishes patrol the beach. Few, if any, baby turtles would get past that gauntlet were it not for the simultaneous emergence that brings them all out at once and temporarily swamps their predators.

The early years Turtles are self-sufficient at hatching and, with the single known exception of the South American river turtle *Podocnemis expansa*, adults provide no parental care. Baby turtles are secretive and rarely encountered in the field. It seems probable that small turtles spend most of their time in concealment because they are so vulnerable to predators.

We know even less about the biology of baby sea turtles than we do about the hatchlings of terrestrial and freshwater species. Where the turtles go in the period following hatching has been a longstanding puzzle in the study of sea turtle life cycles. For example, green sea turtles that hatch on Caribbean beaches in the late summer disappear from sight as soon as they are at sea and are not seen again until they weigh 4–5 kg, some 3 years later. Recent studies have identified nearshore seagrass beds as rookeries for young turtles hatched on the Turks and Caicos Islands, and ocean currents carry hatchlings from Florida's east coast to the relatively calm, seaweed-rich region of the central Atlantic known as the Sargasso Sea.

16.5 Navigation and Migration

LEARNING OBJECTIVES

16.5.1 Describe the types of movements turtles make.

16.5.2 Describe the mechanisms of navigation used by sea turtles.

Pond turtles and terrestrial turtles usually lay their eggs in nests that they construct within their home ranges, often near the shoreline for pond turtles, although female snapping turtles (*Chelydra serpentina*) make overland excursions of 50 m or more to nest sites. The mechanisms of orientation that turtles use to find nest sites are probably the same ones they use to find their way among foraging and resting areas. Familiarity with local landmarks is an effective method of navigation for these turtles, and they may also use the sun for orientation.

Navigation by adult sea turtles

Navigation is a more difficult task for sea turtles, partly because the open ocean lacks conspicuous landmarks and also because feeding and nesting areas are often separated by hundreds or thousands of kilometers. Areas that provide food for sea turtles often lack the characteristics needed for successful nesting, and many sea turtles move long distances between their feeding grounds and their breeding areas. The populations of green sea turtles that breed at Tortuguero and Aves Island feed at sites throughout the Caribbean, and the Ascension Island population feeds on the coast of Brazil.

The ability of sea turtles to navigate over thousands of kilometers of ocean and find their way to nesting beaches that may be no more than tiny coves on a small island is astonishing. Perhaps the most striking example of this homing ability is provided by a green sea turtle colony that has its feeding grounds on the coast of Brazil and nests on Ascension Island, a small volcanic peak 2,200 km east of Brazil and less than 20 km in diameter—a tiny target in the vastness of the South Atlantic.

Magnetic navigation appears to be the primary homing mechanism of sea turtles, which appear to imprint on the magnetic field of the beaches on which they hatch. Chemosensory information may be another component of navigation. For example, the South Atlantic equatorial current flows westward, washing past Ascension Island and continuing toward Brazil, and the odor plume from the island may help guide female turtles back to the island to nest. That is, a female turtle leaving the coast of Brazil may swim upstream in the South Atlantic equatorial current (i.e., up the odor gradient) to locate Ascension Island.

Navigation by hatchling and juvenile sea turtles

Several studies of navigation by hatchling loggerhead sea turtles have shown that they use at least three cues for orientation: light, wave direction, and magnetism. These stimuli play sequential roles in the turtles' behavior.

- When they emerge from their nests, hatchlings crawl toward the brightest light they see. The sky at night is lighter over the ocean than over land, so this behavior brings them to the water's edge. Lights at shopping centers, streetlights, and even porch lights on beachfront houses can confuse loggerheads and other species of sea turtles and lead them inland, where they are crushed on roads or die of overheating the next day.

- In the ocean, loggerhead hatchlings swim into the waves. This response moves them away from shore and ultimately into the Gulf Stream and from there into the North Atlantic Gyre, the current that sweeps around the Atlantic Ocean in a clockwise direction, northward along the coast of the United States, then eastward across the Atlantic (see Figure 16.11).

- Off the coast of Portugal, the North Atlantic Gyre divides into two branches. One turns north toward England, the other swings south past the bulge of Africa and eventually back west across the Atlantic. It is essential for the baby turtles to turn right at Portugal; if they fail to make that turn, they are swept past England into the cold North Atlantic, where they perish. Those that turn southward are eventually carried back to the coast of tropical America— a round-trip that takes 5–7 years. Magnetic orientation appears to keep the turtles within the gyre, and also to tell them when to turn right to catch the southbound current.

We usually think of Earth's magnetic field as providing two-dimensional information (north–south and east–west), but it's more complicated than that. The field loops out of Earth's north and south magnetic poles. At the Equator, the field is essentially parallel to Earth's surface (in other words, an angle of 0 degrees), whereas at the poles it intersects the surface at a 90-degree angle. Thus, the three-dimensional orientation of Earth's magnetic field provides both directional information (which way is magnetic north?) and information about latitude (what is the angle at which the magnetic field intersects Earth's surface?).

When loggerheads in a laboratory wading pool were exposed to an artificial magnetic field with a 57-degree angle of intersection with the surface (an angle characteristic of Florida) they swam toward artificial east—even when the magnetic field was changed 180 degrees, so that the direction they thought was east was actually

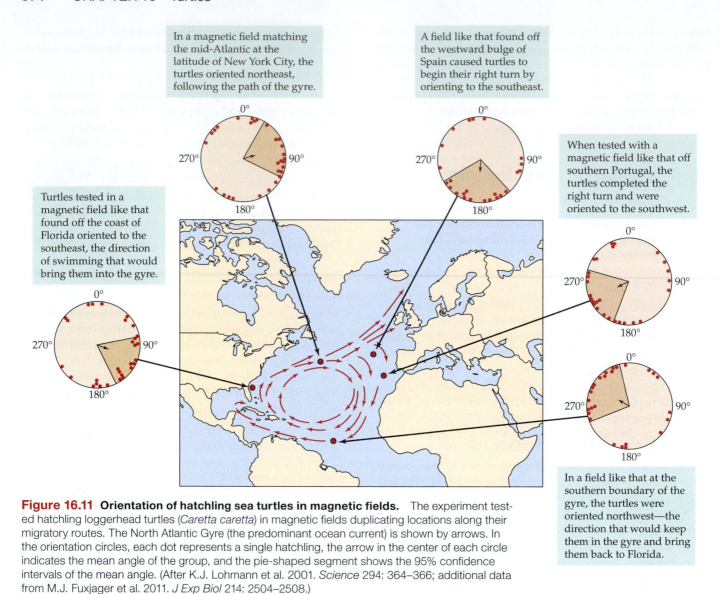

In a magnetic field matching the mid-Atlantic at the latitude of New York City, the turtles oriented northeast, following the path of the gyre.

A field like that found off the westward bulge of Spain caused turtles to begin their right turn by orienting to the southeast.

Turtles tested in a magnetic field like that found off the coast of Florida oriented to the southeast, the direction of swimming that would bring them into the gyre.

When tested with a magnetic field like that off southern Portugal, the turtles completed the right turn and were oriented to the southwest.

In a field like that at the southern boundary of the gyre, the turtles were oriented northwest—the direction that would keep them in the gyre and bring them back to Florida.

Figure 16.11 Orientation of hatchling sea turtles in magnetic fields. The experiment tested hatchling loggerhead turtles (*Caretta caretta*) in magnetic fields duplicating locations along their migratory routes. The North Atlantic Gyre (the predominant ocean current) is shown by arrows. In the orientation circles, each dot represents a single hatchling, the arrow in the center of each circle indicates the mean angle of the group, and the pie-shaped segment shows the 95% confidence intervals of the mean angle. (After K.J. Lohmann et al. 2001. *Science* 294: 364–366; additional data from M.J. Fuxjager et al. 2011. *J Exp Biol* 214: 2504–2508.)

west (**Figure 16.11**). That is, they were able to use a compass sense to determine direction. But that wasn't all they could do: subsequent studies showed that as the three-dimensional magnetic field was changed to match points around the North Atlantic Gyre, the hatchlings turned in the appropriate direction at each location to remain in the gyre and ultimately to return to the coast of Florida. Thus it appears that young loggerheads can use magnetic sensitivity to recognize both direction and latitude.

16.6 Turtles in Trouble

LEARNING OBJECTIVES

16.6.1 Explain why the life history characteristics of turtles place them at risk in a human-dominated world.

16.6.2 Explain why sea turtles are particularly at risk.

16.6.3 Describe the effects of global climate change on turtles.

Turtles have survived with little change since the Triassic, but today more than half the species of turtles face extinction. Habitat loss, pollution, and consumption by humans as food or as traditional medicine are tragically common threats that affect many kinds of vertebrates; other threats are linked to the life history characteristics of turtles. Commercial exploitation is a threat to many species of vertebrates, but the combination of high juvenile mortality, late maturity, and long reproductive lifespans renders turtles especially vulnerable to removal of adults from a population.

Life history

Turtles are long-lived animals, and their life history is based on adult survival. Maturity comes late, and many turtles live as long or longer than humans. Even small species such as the painted turtle do not mature until they are 7 or 8 years old, and females can live for more than 60 years. Many turtles have indeterminate growth—that is, growth continues after adult size is reached—and clutch size increases with body size. Reproductive senescence is delayed; female painted turtles more than 50 years old are still reproducing, although their eggs have lower hatching success than the eggs of younger females.

Long lifetimes are associated with low replacement rates of individuals in a population. Juvenile mortality of turtles is high, but once a turtle reaches adult size it is safe from many natural predators. This life history pattern has been successful for 200 Ma, but the situation changed with the advent of humans because we prey on adult turtles.

Turtles are both delicious and considered medicinal

Turtles are consumed as food and for their supposed medicinal value (**Figure 16.12**). Many Asian cultures revere turtles for their longevity, and turtles have had a prominent role in traditional Chinese medicine for millennia. Turtle meat is said to enrich yin and curb exuberant yang, as well as cure bone diseases, coughs, malaria, and hemorrhoids. Turtle bile mixed with wine is consumed to relieve swelling of the eyes, irregular menstruation, and asthma; turtle skin is used as a remedy for blood diseases; and turtle heads treat concussion, dizziness, and headache.

As China has prospered economically, the demand for turtles has increased. This demand, combined with habitat destruction, has depleted China's turtle populations. As a result, China is a "black hole" for turtles, sucking in the animals from across the world. A 35-month survey recorded nearly 1 million turtles passing through markets in just three cities in China. In 3 years, records show that China legally imported nearly 32 million turtles from the United States, and 700,000 of these were wild-caught adults. Added to that total are the thousands of individuals illegally captured by turtle poachers in the U.S. and shipped to China.

In Latin America, sea turtle eggs are reputed to enhance male virility, creating a demand based on an urban myth. Law enforcement officers from the southeastern U.S. and throughout Central and South America wage a constant battle with egg poachers, who target breeding beaches of sea turtles. Although taking sea turtle eggs is illegal and despite public information campaigns, eggs are sold openly in bars and markets and by mobile vendors (**Figure 16.13**).

In Costa Rica, GPS-enabled decoy sea turtle eggs were used to discover trade routes of trafficked sea turtle eggs. Five decoys were successfully tracked; the shortest trip was 28 m from the nest to a beachside house and another egg travelled only 2 km to a bar. The longest trip, 137 km, went from the beach to the loading bay of a supermarket in the Central Valley. The trafficker probably handed the eggs over to a mobile vendor at the supermarket, because the decoy egg transmitted its final signal from a residence the following day.

Turtles are in demand as pets

More than one-third of reptile species, including endangered species, can be bought and sold on the internet,

(A) Red-footed tortoises, *Chelonoidis carbonarius*

Courtesy of Kate Bruder

(B) Snapping turtle, *Chelydra serpentina*

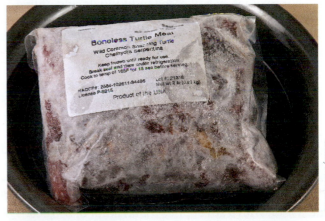

Courtesy of M. H. Siddall

(C) Yellow-headed tortoises, *Indotestudo elongata*, and red-eared turtles, *Trachemys scripta elegans*

David Boté Estrada/CC BY-SA 2.0

Figure 16.12 Turtles in food markets.   (A) Red-footed tortoise meat and eggs for sale in Belen Market, Peru. (B) Frozen snapping turtle meat for sale in markets in New Orleans, U.S.A. (C) Live Asian yellow-headed tortoises (front) and North American red-eared turtles (rear) are sold in the Zhuhai Animal Market, China.

(A) Sea turtle eggs in Mercado de Juchitán Zaragoza, Oaxaca, Mexico

Gzzz/CC BY-SA 4.0

Figure 16.13 Sea turtle eggs and machismo. (A) Sea turtle eggs are sold in markets in Latin America where they are believed to bolster male virility. (B) A poster campaign by WILDCOAST is attempting to counteract that false perception.

(B) Conservation poster

Courtesy of WILDCOAST/www.wildcoast.org

and approximately half of the individuals sold online are collected from the wild. The United States imports and exports more live reptiles for the pet trade than any other country, and the most endangered species command the highest prices.

Madagascar is home to many endangered species, including two critically endangered tortoises that are valued in the pet trade: the radiated tortoise and the ploughshare tortoise (**Figure 16.14**). Both species are protected under Malagasy and international regulations, in spite of which both are targets of illegal wildlife dealers and are offered for sale on the internet. The plowshare tortoise has the unfortunate distinction of being the most valuable reptile in the world, with a street price for an adult of $60,000.

Sea turtles face extra risks

The plight of sea turtles is especially dire, not only because these are among the largest and slowest growing of all turtles, but also because aspects of their biology expose them to additional risks. Adult sea turtles are bycatch (non-target species) in gill net, trawl, and long-line fisheries. In 2000, more than 200,000 loggerhead sea turtles and

(A) Radiated tortoise, *Astrochelys radiata*

Frank Vassen/CC BY 2.0

(B) Plowshare tortoise, *Astrochelys yniphora*

Andrew Routh/CC BY-SA 2.0

Figure 16.14 Turtles in the pet trade. (A) Radiated tortoises and (B) plowshare tortoises, two Critically Endangered species from Madagascar, share the misfortune of being highly valued in the pet trade.

Figure 16.15 Sea turtles, shrimp trawlers, and TEDs.
(A) Shrimp trawlers tow funnel-shaped nets that entrap turtles
and dolphins along with the desired catch. (B) Diagram of a
turtle excluder device, or TED. (C) The TED allows a sea turtle
to escape from a shrimp trawl.

(A) A shrimp trawler

(B) Turtle excluder device (TED)

(C) Loggerhead escaping a shrimp net via a TED

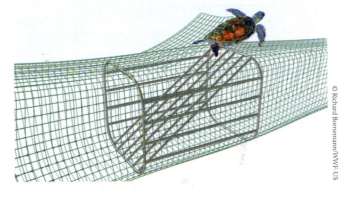

50,000 leatherbacks were captured by longline fisheries,
and most of those turtles died.

Shrimp trawlers   A 1990 report from the National Research Council identified shrimp trawling as the primary
source of mortality for sea turtles in the Gulf of Mexico,
with an estimated 50,000 loggerhead turtles and 5,000
Kemp's ridleys dying annually. The toll attributed to
shrimp trawlers is 10 times the estimated mortality of
sea turtles from all other methods of fishing combined.

Trawl boats tow a pair of large-mouthed, funnel-shaped
nets that engulf everything in their path (**Figure 16.15A**).
Turtles that are swept into the nets are trapped underwater and cannot breathe until the net is raised to the surface
at the end of a tow. Sea turtles normally dive and remain
submerged for substantial periods, and tow times less
than an hour cause little mortality, but mortality rises to
50% after a 2-hour tow and to 75% after 3 hours.

The impact of shrimp trawling on sea turtles was recognized in the 1970s, and turtle excluder devices (TEDs) were
developed to address the problem. A TED is an array of bars
inserted into a net that allows shrimp to pass while divert-
ing turtles out of an opening in the net (**Figure 16.15B,C**).
The trawling industry resisted the use of TEDS for many
years, and it was not until 1992 that TEDs were required
on all U.S. shrimp trawlers from Texas to the North Carolina–Virginia border. A plan formulated in 2016 by the National Oceanic and Atmospheric Administration (NOAA)
extended the use of TEDs to inshore waters where young
sea turtles live and reduced the spacing between bars on
a TED from 4 inches (10 cm) to 3 inches (7.6 cm) to deflect
smaller turtles. That plan was vastly scaled back in December 2019, however, and in 2021 implementation of even the
scaled-back plan was postponed.

Plastic pollution   An estimated 5.25 trillion pieces
of plastic weighing ~270,000 tons foul oceans across the
globe. Between 19 and 23 million metric tons entered aquatic systems in 2016, and that may increase to 53 million
metric tons by 2030. This rampant plastic pollution is a
hazard for all sea life, and sea turtles are especially affected.

Sea turtles consume plastic that looks like their food.
Green sea turtles, which eat sea grass, consume narrow
lengths of green or black plastic, whereas leatherback
turtles, which eat jellyfish, consume plastic bags. Odor
may be an additional stimulus that attracts sea turtles to
plastic. Microorganisms that grow on plastic in the ocean
release dimethyl sulfide, the same odor that sea turtles
(and fishes and sea birds) use to identify their normal
food items.

A diver in the Red Sea described an encounter with a
female hawksbill turtle that had a plastic bag lodged in
her throat:

*I got close to the female and took a few photos of her
eating. Then I realized she had a plastic bag in her
mouth and in fact wasn't able to eat, she was just
nudging the jellyfish and clearly in some distress.
The plastic bag was not only in her mouth but also
8 inches down her throat. I had to hold onto her shell
to wrestle it from her. Almost as soon as I pulled the
bag out, she tried to eat another bag that floated by.
(Bournemouth University lecturer Saeed Rashid in
article by Lauren Fruen. Oct. 4, 2017. http://www.
storytrender.com/26708/26708/)*

Consuming plastic has multiple deleterious consequences for sea turtles (**Figure 16.16**). At a minimum,
plastic occupies space in the digestive tract, apparently
inducing a feeling of satiety and reducing energy intake.
Turtles with plastic comprising 3–25% of the contents of

(A) Hawksbill sea turtle, *Eretmochelys imbricata*

(B) Green sea turtle, *Chelonia mydas*

Stomach | Lg Intestine | Rectum

Courtesy of Kathy Townsend, CSIRO

© Pally/Alamy Stock Photo

Figure 16.16 **Plastic waste is a major cause of sea turtle decline.** (A) A hawksbill turtle ingests a plastic bag, which resembles the jellyfish that are its natural prey. Plastic bags get stuck in the throat; if a turtle does manage to swallow plastic, it clogs the digestive tract. (B) Plastic removed at necropsy from a dead green sea turtle.

their digestive tract weigh ~10–15% less in proportion to their shell length than turtles with less plastic, and their reproductive output is reduced by as much as 88%. Ingesting plastic can also be lethal; on average, 14 pieces of plastic in the gut create a 50% probability of mortality.

Plastic pollution also has indirect effects on sea turtles. As plastic in the sea ages and disintegrates, it releases chemicals that have biological effects, including toxic organochlorine pesticides (such as DDT), carcinogenic polycyclic aromatic hydrocarbons (PAHs), and estrogen-mimicking polychlorinated biphenyls (PCBs) that interfere with reproduction. These persistent organic pollutants, or POPs, are ubiquitous in the sea and are present at every level of the food web, from plankton to sharks and whales. Thus, a sea turtle ingests POPs from its food as well as from the plastic it consumes. Increasing levels of POPs in the blood are associated with increased measures of stress, and have been linked to kidney and liver damage in free-living loggerhead sea turtles.

Exposure to POPs starts early in the life of a sea turtle because POPs from the mother's blood are deposited in the yolks of her eggs. Furthermore, nesting beaches are contaminated with particles of plastic that have washed up from the sea, and these release POPs that can be absorbed by developing embryos. Hatching success and the size of hatchlings of several species of sea turtles decrease as the concentration of POPs increases.

Climate change Sea turtles nest on open beaches where their nests are exposed to full sun. As global temperatures rise, nest temperatures also increase. Low incubation temperatures produce male sea turtles and high temperatures

produce females (Type 1a sex determination; see Table 16.3), so increasing temperatures will skew the sex ratios of populations of sea turtles toward females. This change has already been observed in populations of green sea turtles nesting on beaches of the Australian Great Barrier Reef. Populations of turtles at the southern (cooler) end of the reef are 65–69% female, whereas adult turtles in populations from the warmer northern end of the reef are 87% female. Worse still, more than 99% of juveniles and subadults in the northern populations are female, indicating that these nesting beaches have been producing primarily females for more than two decades.

Paradoxically, global warming also increases the risk of lethally cold water temperatures. As ice covering in the Arctic Ocean has shrunk, the frequency and intensity of polar vortices (southward eruptions of Arctic air) has increased, dropping sea temperatures in the Gulf of Mexico and along the Atlantic coast of North America. When water temperatures drop below 10°C, sea turtles become lethargic and unable to swim, a condition called cold-stunning. NOAA reports that the frequency of cold-stunning events and the number of turtles affected have increased in the past decade. In February 2021, polar air extending south to the Gulf Coast wreaked havoc with residents of Texas and with turtles in the Gulf of Mexico. As dead and dying turtles washed up on beaches, volunteers gathered more than 13,000 cold-stunned turtles and brought them to improvised warming centers. Only about a third of the rescued turtles survived to be released, but without the efforts of the volunteers (whose own homes had no power or water), all of those turtles would have died.

Summary

16.1 Form and Function

The highly specialized morphology of turtles—the trunk and limb girdles enclosed within a shell and the absence of skull fenestrae—has long obscured the phylogenetic affinities of this remarkable lineage. Recent molecular analyses place turtles as the sister group of archosaurs.

The morphology of turtles is unique; no other extant tetrapod has a bony external shell. The turtle shell evolved in stepwise fashion, perhaps initially providing trunk rigidity that enhanced burrowing ability.

- The upper shell (carapace) is formed primarily by expansion of the ribs, which are endochondral bone. The lower shell (plastron) is composed largely of dermal bone with some endochondral elements.

- Beta-keratin, chemically similar to the beta-keratin in the scales of crocodylians and the feathers of birds, forms the horny scutes on the exterior of the carapace and plastron.

The shell incorporates the ribs and encloses the limb girdles of derived turtles.

Because the ribs are incorporated into the carapace, turtles cannot expand and contract the rib cage to ventilate the lungs. Instead, turtles ventilate by moving the viscera to compress and expand the lungs.

Some aquatic turtles employ vascularized tissue in the pharynx or cloaca for gas exchange while they are submerged.

16.2 Diversity

The two extant lineages of turtles, cryptodires and pleurodires, arose during the Mesozoic. The two groups are distinguished by their method of retracting the head into the shell.

- Cryptodires retract the neck and head directly back under the shell using specialized vertebral articulations (ginglymi) that permit formation of an S-shaped bend.

- Pleurodires bend the neck horizontally; their vertebral joints are simpler than those of cryptodires.

About three-quarters of extant turtles are cryptodires. Pleurodires now occur only in the Southern Hemisphere, although fossils show that they once had a broader distribution.

Much can be inferred about the ecology of a turtle from the shape of its shell. Terrestrial turtles often have domed shells that are hard for predators to crush, and aquatic swimmers have streamlined shells.

The forelimbs of sea turtles are modified as paddles.

16.3 Social Behavior, Communication, and Courtship

Although turtles are generally viewed as solitary, social behavior is probably more extensive and phylogenetically widespread among turtles than is yet appreciated.

Turtles engage in social behaviors during the breeding season and communicate with tactile, visual, auditory, and olfactory signals.

Species-specific patterns of stripes and colors on the heads, necks, and tails of many species of pond turtles are probably used to identify conspecifics.

Some turtles vocalize during courtship and mating, emitting sounds that have been described as grunts, moans, and bellows.

16.4 Reproduction

All turtles are oviparous, depositing flexible-shelled or rigid-shelled eggs in nests that females excavate with their hindlimbs.

Most species of turtles exhibit environmental sex determination (ESD), with sex being determined by the temperatures that embryos experience during the middle third of development.

- Type Ia ESD produces males from eggs incubated at low temperatures and females from eggs incubated at high temperatures.

- In type II ESD, both high and low incubation temperatures produce females, and intermediate temperatures produce males.

Spatial variation in nest temperature and interaction between temperature and moisture in sex determination ensure that hatchlings of both sexes are produced.

Few species of turtles provide parental care, and even in those few species care is limited to brief periods of nest attendance. Only one species, a South American river turtle, is known to provide posthatching care.

16.5 Navigation and Migration

Pond turtles and terrestrial turtles may travel considerable distances to find nest sites, probably using landmarks to find their way.

Sea turtles use multiple senses to navigate, including olfaction and magnetic orientation. They can sense both compass directions and their north–south location on Earth's surface, and they make appropriate changes in course to reach their goal.

(Continued)

Summary (continued)

16.6 Turtles in Trouble

More than half of extant species of turtles are facing extinction.

Most turtles require years to mature, and juvenile mortality is high. As result, replacement rates are low, and longevity of adults is a critical factor in the viability of populations.

Adult turtles are relatively safe from natural predators, but not from humans. Anthropogenic mortality of adult turtles strikes at the most vulnerable element of turtle life history—survival of adults.

- Turtles are sold for food in markets, both in the bush and in large cities. Turtle eggs are reputed to have an aphrodisiac effect on human males, and poaching is widespread.

- Turtles are trafficked in the pet trade, with some species selling for thousands of dollars.

- Fishing boats take sea turtles as bycatch, and shrimp trawls and gill nets are major sources of mortality. Methods to reduce bycatch in these fisheries have been developed, but adoption of these methods has been slow.

- Plastic in the sea kills turtles by obstructing the gut of sea turtles that ingest the plastic. In addition, persistent organic pesticides leach from plastic into seawater, affecting the health and reproduction of sea turtles.

Global warming creates two paradoxical threats for turtles, and especially for sea turtles: (1) nest temperatures that produce overwhelmingly female hatchlings, and (2) increased occurrence of extremely low temperatures that cause cold-stunning and death.

Discussion Questions

16.1 Considering the life-history characteristics of most turtles, which of these approaches to conservation of a species of turtle is likely to be more effective, and why: Protecting eggs and nests so that hatchling turtles can disperse safely from the nest or protecting adult female turtles?

16.2 Why is it so difficult to determine the phylogenetic relationships of turtles to other sauropsids?

16.3 Does environmental sex determination offer any benefits to turtles, or is it better regarded as a possibly ancestral system of sex determination that has persisted in most lineages of turtles because it works well enough?

16.4 Describe a simple experiment you could conduct to test the hypothesis that hatchling sea turtles use orientation to Earth's magnetic field to swim in the correct direction when they reach the ocean.

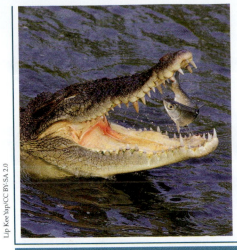

Crocodylians

Saltwater crocodile, *Crocodylus porosus*

Crocodylians include the largest extant species of reptiles, and the only ones that sometimes regard humans as prey. The largest crocodylian, the saltwater crocodile (*Crocodylus porosus*), grows to 6 m or longer and weighs more than 1,000 kg. Not all crocodylians are large, however. Both Africa and South America have forest-dwelling species that are less than 2 m long. All extant crocodylians are semiaquatic predators, but the Mesozoic precursors of extant crocodylians included marine, terrestrial, and herbivorous lineages, as well as enormous species that may have preyed on dinosaurs.

In this chapter we give a brief overview of extant crocodylians followed by a discussion of the evolution of the crocodylian lineage from its origins in the Late Triassic. We then describe the biology of extant crocodylians. Contrary to the impression that comes from seeing well fed captive alligators and crocodiles resting inert for hours, free-living crocodylians have extensive and complex repertoires of predatory, social, and parental behaviors. Finally, we discuss the threats crocodiles and alligators pose to humans, and the even greater threats humans pose to these often very large apex predators as well as to their smaller relatives.

17.1 Diversity of Extant Crocodylians

LEARNING OBJECTIVES

17.1.1 Describe the three extant families of crocodylians and their geographic distributions.

17.1.2 Describe the locomotion of extant crocodylians.

Only 26 extant species of crocodylians are recognized, but molecular studies indicate the existence of several **cryptic species** of crocodiles—that is, species that are indistinguishable morphologically, but are genetically distinct. For example, the Nile crocodile (*Crocodylus niloticus*) was long regarded as a pan-African species, but in 2011 a molecular study of samples from living animals, museum specimens, and mummified crocodiles (some of which were 2,200 years old) identified a cryptic extant sister species, the West African crocodile *Crocodylus suchus*. Cryptic species are an important issue for conservation programs that seek to maintain the genetic diversity of crocodylians.

Distribution of extant crocodylians

Most of the extant species of crocodylians are found in the tropics or subtropics, but three species—the American alligator (*Alligator mississippiensis*), Chinese alligator (*A. sinensis*), and American crocodile (*Crocodylus acutus*)—have ranges that extend into the northern temperate zone, and the range of the Yacare caiman (*Caiman yacare*) extends into the southern temperate zone of Paraguay and Argentina. The two species of *Alligator* (there are only two) are more cold-tolerant than crocodile species. Chinese alligators hibernate in burrows during the winter and American alligators sometimes remain active even in below-freezing conditions, maintaining a breathing hole through thin ice covering the water surface.

Alligators and caimans (*Alligator, Caiman, Melanosuchus,* and *Paleosuchus*)
   Only upper teeth are visible when mouth is closed
   Integumentary sensory organs restricted to the head

Crocodiles (*Crocodylus, Mecistops,* and *Osteolaemus*)
   Upper and lower teeth visible when mouth is closed
   Integumentary sensory organs on head and body
   Salt secreting glands on tongue

Gharial (*Gavialis gangeticus*)
   Elongated snout
   Mandibular symphysis at 20th tooth row
   Upper and lower teeth visible when mouth is closed
   Integumentary sensory organs on head and body

False gharial (*Tomistoma schlegelii*)
   Elongated snout
   Mandibular symphysis at 16th tooth row
   Upper and lower teeth visible when mouth is closed
   Integumentary sensory organs on head and body

Figure 17.1 Some distinguishing features of extant Crocodylia. The three clades of extant crocodylians have distinguishable snouts and dentition. Described species include nine genera (listed in italics). The clade Gavialidae contains two species, the gharial and false gharial. (After H. Wermuth. 1953. *Mitt. Mus. Nat.kd. Berl., Zool. Reihe* 29: 375–511.)

Molecular and morphological characters group extant crocodylians (Crocodylia) in three clades: Alligatoridae, Crocodylidae, and Gavialidae (**Figure 17.1**).

- Alligatoridae includes the two alligator species and six species of caimans. Except for the Chinese alligator, alligatorids are found only in the New World. However, an extensive fossil record reveals their presence in Europe and, to a lesser extent, in Asia, through the Oligocene; their restriction to the Americas dates from the middle Miocene. The American alligator occurs from the U.S. Gulf Coast states to North Carolina and west to Oklahoma, several species of caimans range from Mexico to Paraguay and northern Argentina, and one caiman species is found on the island of Trinidad and Tobago.

- Crocodylidae includes species in the Old and New Worlds. Fossil and molecular evidence indicates that crocodiles migrated from Africa to the Americas within the last 7 million years. The saltwater crocodile (*Crocodylus porosus*) occurs widely in the Indo-Pacific region and penetrates the Indo-Australian Archipelago to northern Australia. In the New World, the American crocodile (*C. acutus*) is quite at home in the sea and occurs in coastal regions from the southern tip of Florida through the Caribbean to northern South America.

- Gavialidae contains two species. The gharial (*Gavialis gangeticus*) once lived in large rivers from northern India to Burma but today is restricted to a small portion of the Ganges River drainage. The gharial has the narrowest snout of any crocodylian; the mandibular symphysis (the point where the mandibles meet at the anterior end of the lower jaw) extends back for half the length of the lower jaw. The false gharial (*Tomistoma schlegelii*) found in the Malay Peninsula, Sumatra, Borneo, and Java has a snout that is nearly as narrow as the gharial's.

Locomotion

Crocodylians swim with lateral sweeps of the tail, steering with their hindfeet (**Figure 17.2A**). A swimming crocodylian can be extremely inconspicuous while it is stalking prey, and it can attack with a frightening surge.

Crocodylians have three modes of terrestrial locomotion:

- *Crawl.* In the belly crawl (**Figure 17.2B**), the legs are extended to the sides of the body and the ventral surface slides across the ground. The belly run is a faster version of the belly crawl. These gaits may look awkward, but they allow crocodylians to attain speeds of 12–14 km/h for short distances.

- *Walk.* For longer terrestrial movements, crocodylians adopt the high walk, with the legs vertically under the body and moving in an anterior-posterior plane rather than extended to the side (**Figure 17.2C**). The high walk is slow, but it allows crocodylians to travel overland for long distances.

(A) Swimming American alligator, *Alligator mississippiensis*

Rajkiran/CC BY-ND 2.0

(B) Belly crawling American alligator

Marianne Serra/CC BY 2.0

(C) High walking American alligator

Gareth Rasberry/CC BY-SA 3.0

(D) Galloping Australian freshwater crocodile, *Crocodylus johnsoni*

From Grigg and Kirshner 2015; photo by Gordon Grigg courtesy of David Kirshner

Figure 17.2 Crocodylians have multiple modes of locomotion. (A) Lateral undulations of the tail propel swimming crocodylians. (B) Sinusoidal movements of the trunk and tail are coordinated with leg movements in the belly crawl and belly run. (C) In the high walk, the legs are held directly beneath the animal. (D) In a gallop, seen only among crocodiles (*Crocodylus*), the vertebral column flexes dorsoventrally.

- *Gallop.* Crocodiles (but not alligators, caimans, or gharials) can gallop, holding the limbs vertically beneath the body and moving the fore- and hindlegs as pairs while dorsoventral flexion of the vertebral column increases stride length (**Figure 17.2D**). Galloping crocodiles can reach speeds of 17 km/h. Extant crocodiles gallop a few tens of meters to water to escape from predators, but predatory terrestrial crocodyliforms of the Mesozoic and Paleocene may have galloped to capture prey.

17.2 The Crocodylomorph Lineage

LEARNING OBJECTIVES

17.2.1 Contrast the ecological and morphological diversity of †notosuchians and neosuchians.

17.2.2 Describe tooth morphologies associated with carnivorous and herbivorous diets of †notosuchians and compare these with the tooth morphology of neosuchians.

The crocodylomorph lineage originated in the Late Triassic and reached its greatest diversity in the Mesozoic. The progression from early crocodylomorphs to the four extant clades is shown in **Figure 17.3**, which illustrates many of the species discussed here.

Unlike extant crocodylians, basal crocodylomorphs were small (about 10 kg) terrestrial predators. †*Terrestrisuchus* (Latin *terra*, "earth"; Greek *souchos*, "crocodile"), a 0.5-m-long predator from the Late Triassic, was lightly built and had hindlegs that were longer than its forelegs, suggesting it could have run bipedally. †*Protosuchus* (1 m) and †*Sphenosuchus* (1.4 m), both from the Early Jurassic, were quadrupedal terrestrial predators.

Herbivory evolved repeatedly in the crocodylomorph lineage, first among †protosuchians and later among †notosuchians and †hylaeochampsids (see Figure 17.3). The Jurassic saw the diversification of two clades of marine crocodyliforms that were distinguished from their terrestrial contemporaries by their large body sizes (500–1,000 kg). †Thalattosuchians appeared in the Early Jurassic and disappeared in the Early Cretaceous (see Section 10.3). The group included two clades: †teleosauroids such as †*Machimosaurus* had body forms similar to those of extant crocodylians, whereas †metriorhynchoids (†*Metriorhynchus*) had limbs modified as paddles and heterocercal tails with a caudal fin (see Figure 17.3). Members of a second marine radiation, †tethysuchians, which appeared in the Late Jurassic and persisted through the Early Eocene, had body forms like those of extant crocodylians (†*Oceanosuchus* in Figure 17.3).

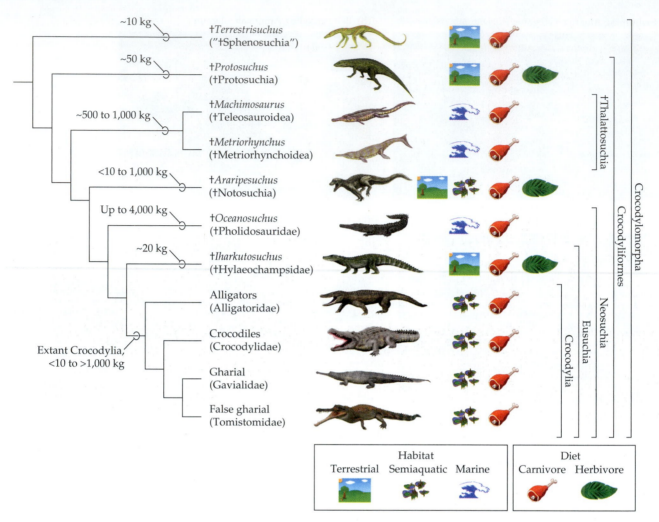

Figure 17.3 The crocodylomorph lineage. Evolution of crocodylomorphs has been marked by repeated changes in habitat (terrestrial, semiaquatic freshwater, marine), diet (carnivorous, omnivorous, herbivorous), and body size. Basal lineages were small, terrestrial carnivores. Transitions to marine and freshwater habitats occurred at least twice, and there were at least three origins of herbivory. Body size increased in some clades and reverted to smaller sizes in others. (See credit details at the end of this chapter.)

The †notosuchian and neosuchian clades separated in the Early Jurassic and diversified during the Jurassic and Cretaceous, with †notosuchians primarily in terrestrial habitats and neosuchians mostly in semiaquatic habitats. That ecological division was not absolute, however—some †notosuchians were semiaquatic and some neosuchians were terrestrial.

†Notosuchia

†Notosuchia was a highly diverse clade in the Southern Hemisphere during the Late Cretaceous, filling adaptive zones that were occupied by small dinosaurs in the Northern Hemisphere. The clade included an enormous diversity of jaw and tooth morphologies indicating a range of ecological niches that far exceeds that of any clade of crocodylomorphs before or since (**Figure 17.4**).

- †*Araripesuchus wegeneri* (Figure 17.4A) was a terrestrial predator. It had long legs and probably pursued its prey.

- †*Kaprosuchus saharicus* (Figure 17.4B), was an aquatic hypercarnivore about 6 m long. It had fanglike teeth in the upper and lower jaws that projected above and below the snout when the mouth was closed.

- †*Anatosuchus minor* (Figure 17.4C), about 1 m long, was probably a semiaquatic generalist; it gets its name from its broad, elongated snout (Latin *anas*, "duck").

- †*Razanandrongobe sakalavae* (Figure 17.4D), a 7-m terrestrial hypercarnivore armed with serrated, laterally flattened teeth, was large enough to prey on dinosaurs.

- †*Simosuchus clarki* (Greek *simos*, "flat-nosed"), only 60 cm long, was a terrestrial herbivore with a blunt snout and a heavy covering of bony plates (Figure 17.4E). Its teeth had multiple cusps like those of extant herbivorous lizards.

- During the Cretaceous a clade of †notosuchians known informally as "mammal-like crocodyliforms" converged on the skull shape and dentition of carnivorous mammals. †*Pakasuchus kapilimai* (Figure 17.4F)

(A) †*Araripesuchus wegneri*, 1 m

Modified from Todd Marshall/CC BY 3.0

(B) Giant aquatic hypercarnivore, †*Kaprosuchus saharicus*, 6 m

Retroarticular process

Carol Abraczinskas/CC BY 3.0

(C) Semiaquatic generalist, †*Anatosuchus minor*, 1 m

Modified from Todd Marshall/CC BY 3.0

(D) †*Razanandrongobe sakalavae* preying on †*Archaeodontosaurus*

Fabio Manucci/CC BY 4.0

(E) Herbivorous †*Simosuchus clarki*, 0.6 m

From Krause D.W., et al. 2010, *J Vert Paleontol* 1: 4–12

(F) Mammal-like heterodont dentition of †*Pakasuchus*, 0.5 m

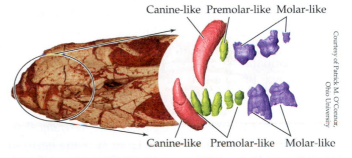

Canine-like Premolar-like Molar-like

Canine-like Premolar-like Molar-like

Courtesy of Patrick M. O'Connor, Ohio University

Figure 17.4 Diversity of †notosuchians. †Notosuchians included (A) terrestrial pursuit predators, (B) semiaquatic hypercarnivores, (C) semiaquatic omnivores, (D) gigantic terrestrial predators, (E) terrestrial herbivores, and (F) terrestrial predators with mammal-like dentition.

had little bony armor, long gracile limbs, and a blunt snout with the nostrils at the tip. Canine-like (caniform) teeth at the front of the jaws were followed by premolar- and molar-like (molariform) teeth like those of carnivorous mammals. The molariform teeth had shearing edges that occluded like those of a cat to slice food. (*Paka* means "cat" in the local Kiswahili language.) These characters all suggest that *Pakasuchus* was a fleet-footed terrestrial predator.

†Notosuchians were a prominent part of the Mesozoic fauna of South America where large species competed with, and probably preyed on, dinosaurs. A fossil site in Argentina that was a subtropical river valley with a meandering stream in the Late Cretaceous records the presence of three clades of dinosaurs (†Sauropoda, †Ankylosauria, and Theropoda) as well as †notosuchian crocodyliforms (**Figure 17.5**).

Neosuchia

Neosuchia was ancestrally a Laurasian group, although it included some Gondwanan lineages. This is the group that gave rise to the extant semiaquatic crocodylians, but during the Mesozoic it included both aquatic and terrestrial species. A clade called †thylaeochampsids had short snouts and heterodont dentition, with large crushing teeth in the rear of the jaw. These characters suggest a herbivorous diet, and the teeth of †*Iharkutosuchus makadii*, an 80-cm terrestrial species from the Late Cretaceous found near the town of Iharkút in western Hungary, had multiple cusps like the teeth of extant herbivorous lizards. The pattern of wear on the teeth indicates that the lower jaw moved from side to side to grind food.

A general increase in body size characterizes Neosuchia, and enormous body size—**gigantism**—is a recurrent theme. †*Sarcosuchus imperator* (Greek *sarkos*, "flesh"; Latin *imperator*, "ruler"), an Early Cretaceous crocodyliform from Africa, had an adult length of 11–12 m and weighed an estimated 8,000 kg. †*Sarcosuchus* appears to have been a generalized predator, probably waiting in ambush at the water's edge, and it was large enough to have preyed on dinosaurs.

Illustration by J. González from A. Paulina-Carabal et al. 2021 *PLOS ONE* 16(9): e0256233/CC BY 4.0

Figure 17.5 The Cerro Fortaleza site in Argentina.
This reconstruction shows dinosaurs and crocodyliforms identified from fossils at this Late Cretaceous site. Three †notosu-
chians appear in the foreground and left side of the figure. In
the center, a small theropod dinosaur (see Chapter 19) faces off
with a †nodosaur while off in the distance two †sauropods (see
Chapter 18) walk away.

†*Deinosuchus* (Greek *deinos*, "terrible"), a Late Cretaceous alligatorid from North America, was up to 10 m
long and weighed 5,000 kg (**Figure 17.6**). †*Deinosuchus*
was semiaquatic, with a body form like that of extant alligators. It was undoubtedly an apex predator; bite marks
matching the teeth of †*Deinosuchus* have been found on
bones of †hadrosaurs (duck-billed dinosaurs).

During the Pliocene and Pleistocene, †*Crocodylus thor-
bjarnarsoni* (7.5 m) inhabited the Lake Turkana basin in
modern-day Kenya, and †*C. anthropophagus* (6 m) (Greek
anthropos, "man"; *phago*, "to eat") lived in the region that
is now Olduvai Gorge in Tanzania. Both of these crocodiles were predators of early humans; fossils of *Homo
habilis* bones with marks made by crocodile teeth have
been found at Olduvai Gorge.

17.3 Predatory Behavior and Diet

LEARNING OBJECTIVES

17.3.1 Explain the relationships among snout shape,
tooth form, and diet of extant crocodylians.

17.3.2 Describe the predatory behaviors of extant
crocodylians and the characters that make stealth
a large component of these behaviors.

The head shapes of extant crocodylians vary from short
and broad to long and slender. Conventional wisdom
equates broad snouts with generalist diets and slender
snouts with piscivory, but field data do not support this
generalization. Of the five extremely slender-snouted
forms—the gharial and false gharial (see Figure 17.1), African slender-snouted crocodile (*Mecistops cataphractus*),
Australian freshwater crocodile (*Crocodylus johnsoni*),
and Orinoco crocodile (*C. intermedius*)—only the gharial
is a specialized piscivore. Even the false gharial is a generalized predator, and there is even a record of a large
false gharial killing and eating a human. Nonetheless,

(A) †*Deinosuchus rugosus*, 12 m compared to a human

1.8 m

Andrey Atuchin/
CC BY-SA 4.0

**Figure 17.6 The Late Cretaceous and early Cenozoic
saw alligatorids of enormous body size.** †*Deinosuchus*
was one of several species displaying gigantism. This species
had the general appearance and ecology of extant alligators
but was 3 times as long and 10 times as heavy as the extant
American alligator.

(B) American alligator, *Alligator mississippiensis*, 4 m

Gareth Rasberry/
CC BY-SA 3.0

broad-snouted species are better able to consume armored prey, such as turtles and the alligator gar.

The teeth of all extant crocodylians are conical and set in sockets along the margins of the jaws, unlike the diverse dentitions seen among †notosuchians. Teeth are replaced during the life of a crocodylian as a new tooth pushes the old tooth out of its socket. A study of Nile crocodiles estimated that each tooth in an adult's jaws had been replaced from 40 to 50 times. At hatching, crocodylians have needle-pointed teeth and the teeth of narrow-snouted species retain this shape throughout life. The teeth of broad-snouted species are also sharp at hatching, but the teeth toward the rear of the jaw of many species become blunter as an individual grows, making them more effective for crushing hard-bodied prey.

The heads of alligatorids are covered with **integumentary sensory organs**, small bulges that are exquisitely sensitive pressure receptors. These organs extend over the entire body of crocodylids and gavialids. Traces of the nerves associated with integumentary sensory organs are present in fossils of semiaquatic Mesozoic crocodyliforms, but not in fossils of terrestrial species.

Sensing water movement is probably the function of the integumentary sensory organs. In complete darkness, American alligators can lunge toward the point of impact of a single drop of water falling on the water surface. Some crocodylians float motionless on the water surface with their mouths open and their forelegs extended and swipe their head sideways to capture fishes detected by the integumentary sensory organs. In addition to detecting prey, crocodylian integumentary sensory organs probably play a role in social interactions.

Some of the predatory stealth of crocodylians results from the presence of a **secondary palate**, a structure that separates the nasal cavity from the oral cavity. The secondary palate of crocodylians allows them to float with just their nostrils and eyes breaking the surface of the water (**Figure 17.7**). Choanae (internal nostrils) lie behind the gular valve, a flap of soft tissue that prevents water from entering the trachea. Gentle movements of the tail can propel a crocodylian through the water with barely detectable ripples, and a powerful tail stroke launches an attack.

After seizing a bird or mammal, a crocodylian drags it under water to drown it. When the prey is dead, the crocodylian bites off large pieces and swallows them whole. Alternatively, crocodylians can use the inertia of a large prey item to pull off pieces: the crocodylian bites the prey, then rotates rapidly around its own long axis (a behavior called the "death roll"), tearing loose the portion it is holding. Sometimes crocodylians wedge a dead animal into a tangle of submerged branches or roots to hold it as the crocodylian pulls chunks of flesh loose. Crocodylians may leave large prey items to decompose for a few days until they can be dismembered easily.

Hunting behavior extends beyond simply waiting in ambush to attack prey. Crocodylians at tourist attractions

(A) Nasal passages and the secondary palate

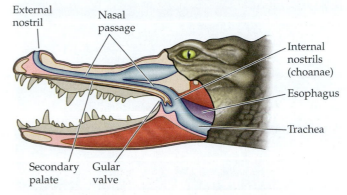

(B) A Nile crocodile, *Crocodylus niloticus*, floating at the surface

Courtesy of Vladimir Dinets

Figure 17.7 A secondary palate allows crocodylians to breathe with only their nostrils and eyes above the water. (A) The secondary palate is a structure that separates the nasal passage from the oral cavity. The internal nostrils (choanae) open into the pharynx behind the gular valve, which forms a seal that prevents water from entering the trachea. (B) Floating with only its nostrils and eyes above the water, a crocodylian is a stealthy predator. (A after G. Grigg. 2015. *Biology and Evolution of Crocodylians*. Illustrations by David Kirshner. Cornell University Press: Ithaca, NY.)

quickly learn what time a tour guide will appear with food, and campers in Northern Australia are asking for trouble if they wash pots and pans on the riverbank every evening. Groups of American alligators and caimans have been observed feeding side by side at the mouths of streams where the current sweeps fish toward them (**Figure 17.8A**). This behavior falls short of true cooperation, but the presence of so many individuals probably increases the capture rate for each participant, because a fish dodging away from one set of jaws is likely to swim into the mouth of an adjacent animal.

Some crocodylians have been reported to augment their ambush hunting strategy with lures. During the breeding season of wading birds, mugger crocodiles (*Crocodylus palustris*) and American alligators lie motionless for hours in shallow water with twigs and sticks balanced on their snouts (**Figure 17.8B**). This behavior is most common near rookeries during the period when birds are gathering sticks to build their nests. A bird that tries to take a stick is at risk of being seized (**Figure 17.8C**).

(A) Yacare caimans, *Caiman yacare*, catching fish at mouth of a stream

Carlos Yamashita, courtesy of F. Wayne King

(B) Mugger crocodile, *Crocodylus palustris*, luring birds with sticks

Courtesy of Vladimir Dinets

(C) American alligator, *Alligator mississippiensis*, consuming an egret

From V. Dinets et al. 2015, *Ethol Ecol* 27: 74–78, photo by Don Specht

(D) American alligator in a roadside ambush

Courtesy of Vladimir Dinets

Figure 17.8 Predatory behaviors of crocodylians.
(A) Yacare caimans at the mouth of a stream in Brazil are facing into the current, capturing fish that must dodge between the closely spaced open mouths. (B) A mugger crocodile with sticks on its head is waiting to ambush nest-building marsh birds. (C) An American alligator consuming a snowy egret that was attracted by a stick on the alligator's head. (D) An American alligator waiting in ambush at the edge of a road at night.

Although crocodylians are semiaquatic, hunting on land appears to be a regular behavior for some species. West African dwarf crocodiles and the two species of dwarf caimans are probably the most terrestrial of the extant crocodylians. American alligators and several species of crocodiles wait in ambush beside trails at night (**Figure 17.8D**). Terrestrial attacks on humans, dogs, and other mammals have been recorded, and prior to the arrival of humans, crocodylians were the apex predators on some Caribbean islands.

17.4 Communication and Social Behavior

LEARNING OBJECTIVES

17.4.1 Describe the sensory modalities used by crocodylians during behavioral interactions.

17.4.2 Explain how the properties of different displays convey information about the signaling individual.

Adult crocodylians use sound extensively in their social behavior. Male crocodylians emit a variety of vocalizations during courtship and territorial displays and also slap their heads and tails against the water (**Figure 17.9**). Vocal displays are especially important for crocodylians such as American alligators that live in dense swamps, because males' territories are often out of sight of other males and of females. The bellow of a male alligator resounds through the swamp, announcing his presence to other alligators. Female alligators also roar, but only males produce subsonic vibrations (~10 Hz, a frequency below the range of human hearing) that cause drops of water to dance on the water surface (**Figure 17.9B**). Subsonic vibrations travel for long distances underwater.

These displays may have different functions: the sharp onset of the sound produced by a head slap facilitates location of its source, but the sound does not travel far. Thus, head slaps may identify the position of a displaying male to nearby individuals. Bellows can be heard at distances of 150 m or more and may stimulate vocalizations from other individuals, creating a crocodylian chorus. Subsonic vibrations travel a kilometer or more underwater, and probably convey information about the size of the crocodylian producing the sound.

(A) Bellowing

(B) Subsonic vibrations

Tristan Loper/CC BY-SA 2.0

Vladimir Dinets

Figure 17.9 **Male American alligators communicate with audible vocalizations and subsonic vibrations.** (A) During the audible portion of its display, a male alligator raises its head and emits a bellow with sound frequencies between 20 and 250 Hz. These frequencies travel through the air. (B) After bellowing, the male sinks until it is partly submerged and produces subsonic vibrations, revealed by the water bubbling around his trunk.

Mandibular glands on the underside of the lower jaw of crocodylians secrete a dark green lipid. To humans the secretion has a musky odor, and it is probably a pheromone. Glandular activity increases during the breeding season. Courting crocodylians employ a variety of signals such as raising their snouts and rubbing their chins on each other while emitting the secretions of their mandibular glands.

As many as 30 American alligators have been reported to assemble in courtship gatherings on nights during the breeding season, swimming within a small area, splashing, hissing, head slapping, and forming pairs (**Figure 17.10**). These gatherings may establish dominance hierarchies and appear to be preludes to mating.

Courtship gathering of *Alligator mississippiensis*

Rusty Clark/CC BY 2.0

Figure 17.10 A courtship gathering. As many as 30 individuals have been observed in these nocturnal groups of American alligators.

17.5 Reproduction and Parental Care

LEARNING OBJECTIVES

17.5.1 Describe the characteristics of environmental sex determination in crocodylians and how these compare with those of squamates and turtles described in Section 15.6 and Section 16.4.

17.5.2 Describe the elements of parental care by crocodylians.

All crocodylians are oviparous, depositing hard-shelled eggs in a nest that is guarded by the female. Alligators and caimans are mound nesters; the female uses her hindlegs to build a pile of soil and vegetation and deposits her eggs in a depression in the center of the mound. About half of the extant species of crocodiles and the false gharial are also mound-nesters. The remaining crocodile species and the gharial deposit eggs in holes that the female excavates with her hindlegs. Fossilized crocodylian nests have been found in Mesozoic deposits, some of them adjacent to the nests of †sauropod dinosaurs.

Environmental sex determination

As is the case for many turtle and squamate species, a crocodylian's sex is determined by the temperature of the environment in which the eggs develop (environmental sex determination, or ESD; see Section 15.6 and Section 16.4). Unlike the varied mechanisms of sex determination seen among turtles and squamates, however, genetic sex determination (GSD) is not known among crocodylians; all species that have been studied display type II ESD, with females developing at low and high temperatures and males at intermediate temperatures (see Figure 16.9). Nests probably have enough internal temperature variation to produce hatchlings of both sexes. For example, temperatures ranged from 33°C to 35°C in the top center of nests of American alligators in marshes, and males

(A) Nest of American alligator, *Alligator mississippiensis*

NPS/Lori Oberhofer

(B) Male mugger crocodile, *Crocodylus palustris*, carrying young

Courtesy of Jeffrey Lang

(C) Crèche of hatchling gharials, *Gavialis gangeticus*

Courtesy of Jeffrey Lang

Figure 17.11 **Crocodylians have extensive parental care.** (A) An adult female American alligator guarding her nest. (B) An adult male mugger crocodile carrying one hatchling from the nest to the water, while two more hatchlings (arrows) await transport. (C) Hatchling gharials assemble in crèches containing hundreds of individuals. (D) Adult male gharials guard the crèches.

(D) Adult male gharial guarding a crèche

Courtesy of Jeffrey Lang

hatched from eggs in this area. At the bottom and sides of the same nests, temperatures ranged from 30°C to 32°C, and eggs from those regions hatched into females.

Parental care

Crocodylians provide parental care that is as extensive as the care provided by birds and lasts longer (**Figure 17.11**). Adult crocodylians have few predators other than humans, but eggs and hatchlings are vulnerable to predation by fishes, lizards, turtles, other species of crocodylians, birds, and mammals. Most often the female is the caregiver, but males of some species participate. In every species that has been studied, the female remains near the nest throughout incubation and defends the clutch.

Baby crocodylians begin to vocalize before they have fully emerged from their eggs, and these vocalizations are loud enough to be heard at some distance. Their calls stimulate one or both parents to open the nest and transport the hatchlings to water. In some species, the parents gently break the eggshells with their teeth to help the

young escape. The sight of a Nile crocodile, with jaws that can crush the leg of a zebra, delicately cracking the shell of an egg little larger than a hen's egg and releasing the hatchling unharmed is truly remarkable.

Young crocodylians remain in a group called a crèche for a considerable period—2 years for the American alligator and 3 years for the spectacled caiman (*Caiman crocodilus*). A crèche contains the young of many parents, and several adults serve as guards. The crèches of gharials can contain up to 1,000 hatchlings, and young males participate in guarding the crèches. Contact calls help juveniles in a crèche maintain cohesion and alert them to the presence of food.

Frightened crocodylian hatchlings emit a distress call that stimulates the parents to come to their defense. In addition to summoning the parents, these vocalizations attract unrelated adults. When staff members at a crocodile farm in Papua New Guinea rescued a hatchling New Guinea crocodile (*Crocodylus novaeguineae*) that had strayed from the pond, the hatchling's distress call brought 20 adult crocodiles surging toward it—and toward the staff members! The dominant male head slapped the water repeatedly and then charged into the chainlink fence where the staff members were standing, while the females swam about, giving deep guttural calls and head slapping the water.

17.6 Threats to and from Crocodylians

LEARNING OBJECTIVES

17.6.1 Explain how the life history characteristics of crocodylians render them vulnerable to direct and indirect anthropogenic threats.

17.6.2 Describe the problems of administering a conservation plan for a species that eats people.

Crocodylians mature late and adult crocodiles have long reproductive lifespans. These life history characteristics make populations vulnerable to human predation on adults. Unregulated exploitation of wild crocodylians for skins has brought many species to the verge of extinction, and in 1975 all species were listed on the Appendices of the Convention on International Trade in Endangered Species of Wild Fauna and Flora (CITES). The listing stopped legal trade (although poaching and smuggling continued), and as populations rebuilt, regulated trade based on sustainable use was reintroduced along with production from captive breeding farms. World trade in legal, regulated crocodylian skins averaged 1.3 million skins per year between 2001 and 2012, and 400–1,000 tons of meat. International trade is now largely legal, sustainable, verifiable, and traceable.

Threats from crocodylians

A few species of crocodylians pose a risk to livestock, pets, and occasionally to people. Warning signs urge caution in shoreline habitats (**Figure 17.12A**). Some attacks on humans are predatory, but many occur when an adult is protecting a nest or crèche of young. Most attacks occur in isolated regions and firm numbers are hard to obtain. The estimated worldwide toll of crocodylian attacks is 300 to 1,000 human fatalities annually. Nile and saltwater crocodiles are most likely to regard humans as prey and these two species account for the majority of these attacks. About 69% of attacks by Nile crocodiles and 49% of attacks by saltwater crocodiles are fatal. Contrary to popular opinion, hippos are the most dangerous animals in Africa (**Figure 17.12B**). More than 80% of attacks by hippos are fatal, and hippos are estimated to cause as many as 3,000 human fatalities annually.

Threats to crocodylians

The major current threats to crocodylian populations are loss and alteration of wetlands. Of course, habitat loss affects almost every species of animal or plant, but its impact is magnified by some of the biological characters of crocodylians.

- Many species of crocodylians are large, and some have extensive home ranges with different places for different activities. For example, the communal nesting sites of gharials on the Chambal River in India are 50–200 km upstream from the feeding areas they use during the monsoon season, and adult gharials annually move between the two locations. A highway crosses the river in two places and dozens of towns and villages, some with populations of

(A) Warning sign in Darwin, Australia

(B) Warning sign in Lake Chamo, Ethiopia

Figure 17.12 Threats from crocodylians. Signs warn of danger from crocodiles in Australia and from crocodiles and hippos in Africa. Although crocodiles are perceived as more fearsome, in Africa there are far more human fatalities from hippo attacks than from crocodiles.

(A) Levee at the edge of the Florida Everglades

From D. Galloway et al. 1999, USGS Circular 1182

(B) Tegu, *Salvator merianae*, leaving American alligator nest with an egg

Courtesy of Frank Mazzotti

(C) Freshwater crocodile, *Crocodylus johnsoni*, eating a cane toad

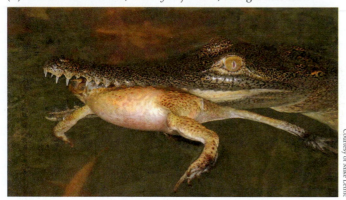

Courtesy of Mike Letnic

Figure 17.13 **Threats to crocodylians.** (A) A network of dikes, canals, and pumping stations on the northern edge of Florida's Everglades National Park provides irrigation and drainage for fields of sugar cane (left side of photograph). (B) A black-and-white tegu, an invasive species in Florida, stealing an egg from an alligator nest. (C) A cane toad will be the last meal this young freshwater crocodile eats.

100,000 or more, line the river. Riverside cultivation, unlicensed fishing, and sand mining make parts of the river uninhabitable for gharials.

- Thousands of miles of levees have been constructed in the southeastern United States to provide water for agriculture (**Figure 17.13A**). The levees are attractive nesting sites for American alligators, but dependence on environmental sex determination makes these structures ecological traps—nests on the unshaded levees are several degrees hotter than nests in the swamps adjacent to the levees, and alligator hatchings from levees are overwhelmingly male.

- Invasive species are a threat to crocodylians in some locations. In Florida, invasive Burmese pythons feed on juvenile and subadult alligators, and introduced tegu and monitor lizards (*Salvator merianae* and *Varanus niloticus*) are nest predators (**Figure 17.13B**).

- Secretions of the paratoid glands of cane toads (*Rhinella marina*) are lethal to predators, even to animals as large as crocodiles (**Figure 17.13C**). Massive die-offs of freshwater crocodiles (*Crocodylus johnsoni*) in the Victoria River of Australia coincided with the 1935 introduction of cane toads; crocodile population densities fell by as much as 77%.

Table 17.1 Response to the Mabuwaya education program[a]

Question	Percentage of "Yes" responses	
	Participating communities	Control communities
"Do you support conservation of the Philippine crocodile?"	62%	21%
"Do other people support conservation of the Philippine crocodile?"	44%	4%
"Do you think that crocodile conservation can benefit the community?"	60%	10%

From M. van Weerd et al. 2012. *The Philippine Crocodile: Ecology, Culture, and Conservation.* Mabuwaya Foundation, Cabagan, Isabela, Philippines.

[a] The survey compared attitudes toward crocodiles in villages that participated in the community education program with attitudes in similar communities (control communities) that did not participate.

Reconciling humans and crocodylians

Conservation of a species that kills people and their livestock faces the challenge of creating a positive public attitude toward the species. Effective conservation programs for crocodylians must reduce damage by promoting safe behaviors among humans and addressing problems while maximizing the ecological, financial, and cultural benefits of protecting wild crocodylians. Above all, these programs must engage their communities at the grassroots level.

The Mabuwaya Foundation in the Philippines started a conservation program for the Critically Endangered Philippine crocodile (*Crocodylus mindorensis*) in 2003. (Mabuwaya is a contraction of the Tagalog words: *Mabuhay*, "long live," and *Buwaya*, "crocodile.") The program combines conservation with extensive problem-solving,

economic, and education elements. The success of the Mabuwaya program is shown by comparing the attitudes toward crocodiles in communities that participate in the program compared to control communities (**Table 17.1**).

A community-based group manages eight local crocodile sanctuaries, protecting nests and increasing hatchling survival by collecting and raising hatchlings and releasing them when they are 2 years old. Crocodile versus livestock conflicts are solved by restoring buffer zones for crocodiles and these buffer zones benefit local economies through forest harvesting and ecotourism. An extensive education program includes puppet shows in elementary schools, visits to crocodile sanctuaries by high school students, community meetings, and distribution of calendars and posters produced each year by students at the Cabagan Campus of the Isabela State University (**Figure 17.14**).

Figure 17.14 Education and crocodile conservation. A poster from the Mabuwaya Foundation's crocodile conservation program, which has the goal of protecting both people and endangered crocodiles.

Summary

17.1 Diversity of Extant Crocodylians

Only 26 species of extant crocodylians are recognized, although additional cryptic species await description.

All extant crocodylians are semiaquatic predators and nearly all live in the tropics or subtropics.

Molecular and morphological phylogenies group crocodylians into three clades.

- Alligatoridae: Alligators and caimans are found in freshwater and, except for the Chinese alligator, are confined to the New World. The teeth on the lower jaws of alligatorids fit into grooves in the upper jaws and are not visible when the mouth is closed.

- Crocodylidae: Crocodiles occur in the Americas, Africa, and Asia. Several species, including the American and Australian saltwater crocodiles, are at home in the sea. The teeth on the lower jaws of crocodylids are visible when the mouth is closed.

- Gavialidae: The gharial (found in a small portion of the Ganges River drainage) and false gharial (from Southeast Asia and Indonesia) are freshwater species with extremely slender snouts.

Crocodylians swim with lateral undulations of the tail and steer with their hindfeet.

On land, crocodylians have three gaits.

- The belly crawl and its speedier version, the belly run, are used for short distances; the limbs extend horizontally, and the ventral surface slides across the ground.

- The high walk, with the legs held vertically under the body, is used for longer terrestrial movements.

- Crocodiles can gallop, moving the fore- and hindlegs as pairs and flexing the vertebral column dorsoventrally to increase the stride length.

17.2 The Crocodylomorph Lineage

The crocodylomorph lineage originated in the late Triassic and reached its greatest diversity in the Mesozoic.

Basal crocodylomorphs were small terrestrial carnivores.

†Notosuchians and neosuchians diverged in the Triassic and diversified during the Jurassic and Cretaceous.

- †Notosuchians radiated in the Southern Hemisphere as terrestrial predators and herbivores and as semiaquatic predators. Large terrestrial †notosuchians survived into the Miocene.

- Neosuchia, an initially Laurasian group that spread to Gondwana, included terrestrial and semiaquatic predators. Gigantism has evolved repeatedly among neosuchians, and some Cretaceous and early Cenozoic species exceeded 10 m in length and weighed as much as 8,000 kg.

17.3 Predatory Behavior and Diet

The shape of the snout is not tightly linked to the dietary habits of extant crocodylians; most species are generalists.

A secondary palate allows crocodylians to breathe with just the nostrils and eyes above the water. A gular valve prevents water from entering the trachea.

Integumentary sensory organs that detect water movement allow crocodylians to locate prey in the dark.

Crocodylians drag terrestrial prey underwater to drown it, then bite off pieces. Sometimes they seize a limb and spin rapidly to pull it off. Large prey items may be wedged underwater and left until they can be dismembered easily.

17.4 Communication and Social Behavior

Crocodylians use sound extensively in social interactions. The bellows of males travel through the air, while vibrations from head slaps and subsonic vibrations are transmitted through the water.

During the mating season, large numbers of individuals of some species assemble in small areas, vocalizing and head-slapping. These aggregations may establish dominance hierarchies and male–female pairings.

17.5 Reproduction and Parental Care

All crocodylians are oviparous, laying hard-shelled eggs in nests constructed by the female.

Alligators and caimans and about half the species of crocodiles build nest mounds of soil and vegetation and deposit the eggs in a hole the female digs in the top of the mound. The remaining species of crocodiles construct nests in sandy substrates.

Crocodylians have type II environmental sex determination, with females produced at low and high temperatures and males at intermediate temperatures.

Parental care includes guarding the nest, assisting hatchlings to emerge, carrying them to the water, and guarding crèches of hatchlings. Females are the primary caregivers, but males of some species help open the nest, transport young to water, and guard crèches.

Contact calls by juveniles maintain cohesion within a crèche, and distress calls bring adults to the rescue.

17.6 Threats to and from Crocodylians

Crocodylians mature late and are long-lived and thus cannot withstand severe adult mortality.

Unregulated exploitation of wild populations for skins has brought many species to the verge of extinction. In many parts of the world sustainable harvest programs have been established and some species have staged dramatic recoveries.

Summary (continued)

Habitat loss and alteration and alien species are threats to crocodylians.

Because crocodylians sometimes prey on pets, livestock, and humans, public education is an essential component of conservation programs.

Discussion Questions

17.1 Some crocodylians construct nest mounds of soil and vegetation, whereas other species dig holes in the ground. The occurrence of mound and hole nests is shown in the phylogeny below. What is the ancestral nest type for crocodylians? What hypotheses can you propose to explain the distribution of those two nest types? What additional information would you gather to test those hypotheses?

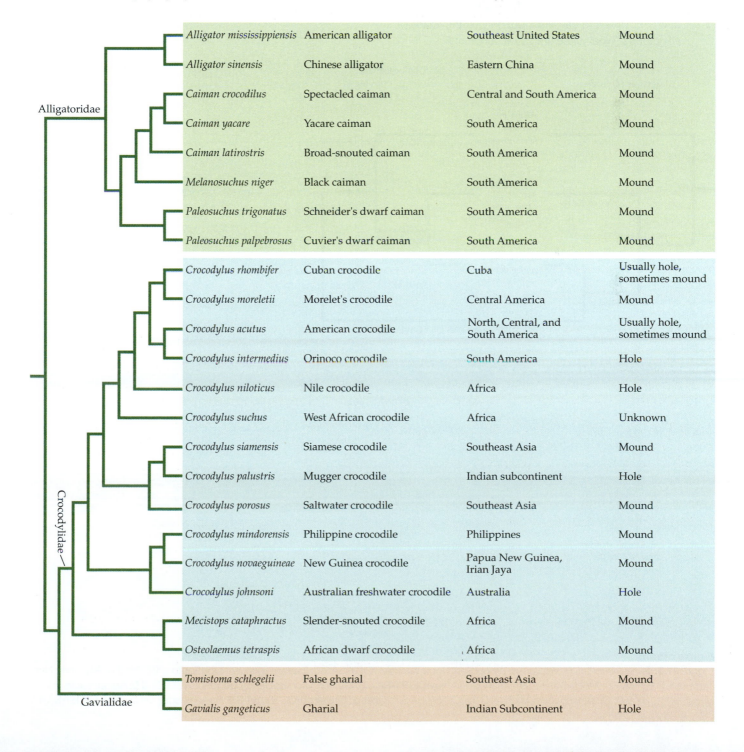

Species	Common name	Distribution	Nest type
Alligator mississippiensis	American alligator	Southeast United States	Mound
Alligator sinensis	Chinese alligator	Eastern China	Mound
Caiman crocodilus	Spectacled caiman	Central and South America	Mound
Caiman yacare	Yacare caiman	South America	Mound
Caiman latirostris	Broad-snouted caiman	South America	Mound
Melanosuchus niger	Black caiman	South America	Mound
Paleosuchus trigonatus	Schneider's dwarf caiman	South America	Mound
Paleosuchus palpebrosus	Cuvier's dwarf caiman	South America	Mound
Crocodylus rhombifer	Cuban crocodile	Cuba	Usually hole, sometimes mound
Crocodylus moreletii	Morelet's crocodile	Central America	Mound
Crocodylus acutus	American crocodile	North, Central, and South America	Usually hole, sometimes mound
Crocodylus intermedius	Orinoco crocodile	South America	Hole
Crocodylus niloticus	Nile crocodile	Africa	Hole
Crocodylus suchus	West African crocodile	Africa	Unknown
Crocodylus siamensis	Siamese crocodile	Southeast Asia	Mound
Crocodylus palustris	Mugger crocodile	Indian subcontinent	Hole
Crocodylus porosus	Saltwater crocodile	Southeast Asia	Mound
Crocodylus mindorensis	Philippine crocodile	Philippines	Mound
Crocodylus novaeguineae	New Guinea crocodile	Papua New Guinea, Irian Jaya	Mound
Crocodylus johnsoni	Australian freshwater crocodile	Australia	Hole
Mecistops cataphractus	Slender-snouted crocodile	Africa	Mound
Osteolaemus tetraspis	African dwarf crocodile	Africa	Mound
Tomistoma schlegelii	False gharial	Southeast Asia	Mound
Gavialis gangeticus	Gharial	Indian Subcontinent	Hole

Alligatoridae

Crocodylidae

Gavialidae

17.2   Phylogenetic hypotheses for crocodylians based on molecular data differ from those based on morphological and paleontological data in the placement of the gharial (*Gavialis gangeticus*) and the false gharial (*Tomistoma schlegelii*). Molecular data place the two species as sister lineages within Gavialidae, whereas morphological and paleontological data place the gharial as the sister lineage of Alligatoridae + Crocodylidae and place the false gharial in Crocodylidae. How does the difference in these phylogenies (shown below) affect our interpretation of the derived versus ancestral condition of crocodylian characters? For example, integumentary sensory organs (ISOs) are present on the heads and bodies of crocodylids and gavialids, but only on the heads of alligatorids. What is the ancestral condition?

17.3   Crèche formation is an uncommon form of parental care, although examples are known among birds and fishes. What conditions might predispose hatchlings of a crocodylian species to form a crèche? What benefits might hatchlings derive from being in a crèche? A guarding adult is related to only a small number of the young in a crèche, so is guarding a crèche an example of altruistic behavior? What costs does an adult incur by guarding a crèche? What benefits might an adult derive from guarding a crèche? What other questions can you pose about crèche formation by crocodylians?

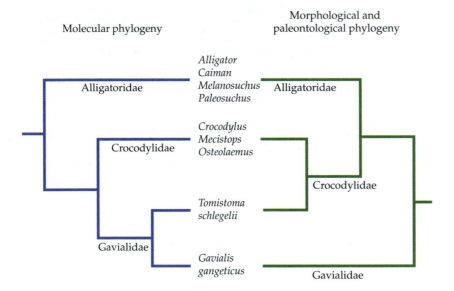

Figure credits
Figure 17.3: *Terrestrisuchus*, *Protosuchus*, crocodiles: Nobu Tamura/CC BY-SA 3.0; *Iharkutosuchus*: Nobu Tamura/CC BY 3.0; Alligators: Nobu Tamura/CC BY-SA 4.0; *Machimosaurus*: Dmitry Bogdanov/CC BY-SA 4.0; *Metriorhynchus*: Dmitry Bogdanov/CC BY 3.0; *Araripesuchus*: Todd Marshall/CC BY 3.0; *Oceanosuchus*: Michael B. H./CC BY-SA 3.0; Gharial: Greverod/CC BY-SA 3.0; False gharial: Drow male/CC BY-SA 4.0.

Avemetatarsalia and the Origin of Dinosauria

A †pterosaur, †*Hatzegopteryx*, an apex predator

The Mesozoic lasted from the close of the Paleozoic 252 Ma to the beginning of the Cenozoic 66 Ma. During these ~186 million years following the end-Permian extinction (the most sweeping mass extinction in Earth's history; see Section 5.5), a diverse fauna of terrestrial vertebrates evolved, only to experience another major mass extinction at the end of the Triassic, followed by recovery and new evolutionary radiations until a third major mass extinction at the end of the Cretaceous ushered in the Cenozoic (see Section 10.4).

Among the newly diversifying lineages of the Mesozoic was Avemetatarsalia ("bird feet"), a group of archosaurs that includes †*Teleocrater*, †pterosaurs, dinosaur precursors, and dinosaurs including birds (**Figure 18.1**). Avemetatarsalians drifted with the moving continents and radiated into many adaptive zones, some of which are occupied today by different vertebrates; other zones, such as those of the enormous herbivorous †sauropods like †*Brontosaurus*, are now unoccupied.

An array of peculiar vertebrates evolved in the Triassic, but after the end-Triassic extinction, the derived avemetatarsalian group Ornithodira took center stage both on land and in the air. The first (†Pterosauria) and second (Aves, the birds) groups of vertebrates to use powered (flapping) flight are ornithodirans, as are the largest terrestrial herbivores and carnivores that ever lived. Some ornithodiran lineages evolved semiaquatic to aquatic forms. However, all ornithodirans other than Aves disappeared at the end of the Cretaceous. Aves persisted and radiated during the Cenozoic into the wide range of extant birds that we will describe in Chapter 21.

In this chapter, we review representative early avemetatarsalians and discuss the biology and evolution of †pterosaurs, †ornithischians, and †sauropodomorphs. Chapter 19 continues the story of dinosaurs, focusing on theropods and the origin of birds. Unlike vertebrate groups covered in previous chapters, these two chapters focus primarily on extinct forms. We include artists' reconstructions of many extinct avemetatarsalians to help you think about what these animals might have looked like, but keep in mind that paleoart is a highly specialized field. The way that artists depict extinct avemetatarsalians—especially certain ever-popular dinosaurs—has changed over time and will continue to change with new discoveries in Mesozoic paleontology.

18.1 Characters and Systematics of Avemetatarsalia

LEARNING OBJECTIVES

18.1.1 Discuss the evolutionary importance of decoupling the functions of forelimbs and hindlimbs in Avemetatarsalia.

18.1.2 Describe phylogenetic relationships among major clades of Avemetatarsalia as presented in Figure 18.1.

18.1.3 Identify synapomorphies for Ornithodira, Dinosauromorpha, and Dinosauria.

18.1.4 Explain why incomplete specimens and problems in diagnosing taxa have made it difficult to develop stable phylogenetic interpretations of Avemetatarsalia.

Avemetatarsalia is one of the two major lineages of archosaurs, the other being Pseudosuchia, the group that includes Crocodylomorpha (see Chapter 9 and Chapter 17). Avemetatarsalia is a large and diverse clade including many familiar extinct vertebrates such as †pterosaurs and non-avian dinosaurs, as well as extant birds. The best

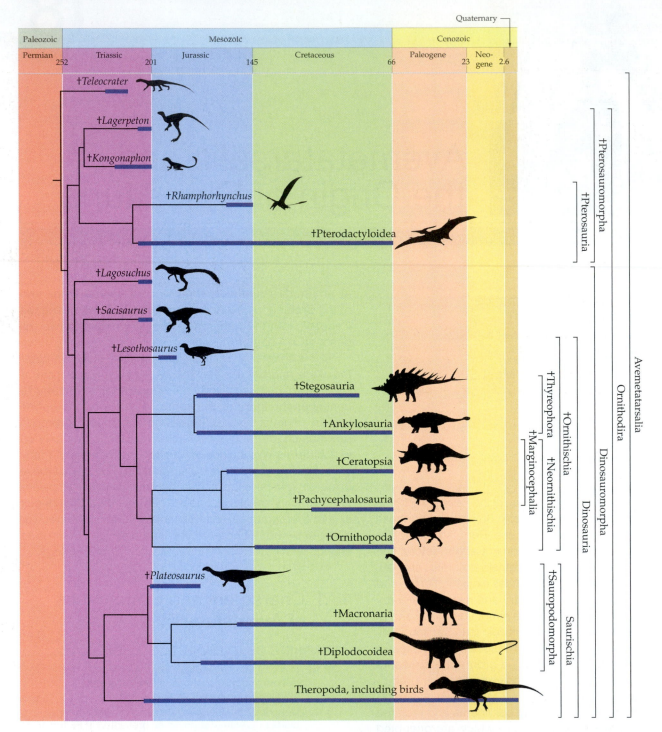

Figure 18.1 Time tree of Avemetatarsalia. The earliest body fossils of avemetatarsalians, such as †*Teleocrater*, occur in rocks from the Middle Triassic. Ornithodira, the clade that includes †pterosaurs and dinosaurs, probably diverged from basal avemetatarsalians about that same time. (See credit details at the end of this chapter.)

defining quality of Avemetatarsalia (although not quite a synapomorphy, because it cannot be crisply defined) is decoupling of the functions of the fore- and hindlimbs, a decoupling that was a prerequisite for the evolution of bipedalism and powered flight within the group.

Because so many groups of avemetatarsalians are known only from fossils, it is important to be cautious about any phylogenetic arrangement of taxa. Hypotheses

of avemetatarsalian relationships change frequently, with seemingly familiar and well established groups being broken up or changing positions on the phylogeny. Such changes often result from discovery of new and better preserved specimens of a previously described taxon, or the powerful influence of a single excellent fossil (the "good fossil effect"; see Section 8.2). Here we adopt a conservative approach, focusing on a few avemetatarsalian

(A) Basal Avemetatarsalia, †*Teleocrater*

(B) Basal †pterosauromorph, †*Kongonaphon*

Liam Elward © Field Museum (2021)

Neil Pezzoni Fanboypiliosopher /CC BY-SA 4.0

Figure 18.2 Basal avemetatarsalians occupied different niches. (A) Forms such as †*Teleocrater* were quadrupedal animals with long necks. (B) Early ornithodirans such as †*Kongonaphon* were arboreal insectivores with long hind limbs. We interpret †*Kongonaphon* and related forms as close relatives of †pterosaurs.

genera as examples leading to the origin of †Pterosauria and Dinosauria (see Figure 18.1), groups that we examine in detail in Sections 18.2, 18.3, and 18.4.

†*Teleocrater*, which lived in the early Triassic of Tanzania, offers clues about the biology of basal members of Avemetatarsalia. A slender, carnivorous quadruped with a long neck, †*Teleocrater* reached total lengths of about 2 m. We think that †*Teleocrater* was quadrupedal because its hindlimbs were not elongated, although its limbs were partially under the body (**Figure 18.2A**).

All remaining avemetatarsalians considered in this chapter belong to Ornithodira (Greek *ornithos*, "bird;" *deire*, "neck"). The name is appropriate because the cervical vertebrae of ornithodirans differ from those of the trunk. The centra of cervical vertebrae are shorter and less tall than those of the trunk, and the cervical ribs of adults typically fuse to the vertebrae. Overall, these features relate to greater neck mobility. Ornithodirans also have long tibiae and metatarsal bones that contribute to overall elongation of the hindlegs, a precursor to the more upright stance and bipedalism typical of more derived forms.

Many groups of ornithodirans departed from this ancestral morphology, and similar ecomorphs evolved in different lineages. Many members of these lineages lived side by side and preyed on or competed with one another. Their diversity is reflected in the many ecological niches they occupied. Flying vertebrates evolved twice within Ornithodira: †pterosaurs in the Late Triassic, and birds in the Jurassic. †Pterosaurs and birds co-occurred from the Jurassic through the Cretaceous, a span of more than 80 million years. Both clades started their evolutionary history as small animals, but some later species were gigantic.

Several lineages of basal ornithodirans radiated in the Triassic. Our understanding of relationships among these lineages is in flux as new fossils are discovered and older specimens are reanalyzed. For example, recent reassessments of †*Lagerpeton*, a 70-cm-tall ornithodiran from the Late Triassic of Argentina, have shown that these slender animals had long hindlimbs. CT scans of †*Lagerpeton* braincases reveal a large anterior semicircular canal in the inner ear, indicating that they had an excellent sense of balance. The strong curvature of the claws on the hand is like that of perchers, climbers, and predators. Fossil trackways in Early Triassic rocks of Poland were made by an animal with †*Lagerpeton*-like feet, meaning that basal ornithodirans were starting to radiate much earlier than previously thought.

The diminutive †*Kongonaphon kely* from the Triassic of Madagascar is related to †*Lagerpeton*, but at only 10 cm it was much smaller (**Figure 18.2B**). Based on its multicusped teeth, it is interpreted as insectivorous (†*Kongonaphon* comes from the Malagasy word *kongona*, "bug," and the Greek *phon*, "slayer"), and it may have been arboreal. Based on these and other features, recent work places †*Lagerpeton* and †*Kongonaphon* in †Pterosauromorpha as the sister group of †Pterosauria (see Figure 18.1).

The sister group of †Pterosauromorpha is Dinosauromorpha (Figure 18.1). Having an offset femoral head—a character associated with rotation of the hindlimbs beneath the body and a more upright stance—is synapomorphic for Dinosauromorpha. Triassic forms thought to be near the base of this group include †*Lagosuchus*, a bipedal carnivore 30–40 cm long (**Figure 18.3A**) and †*Sacisaurus*, a lightly built quadrupedal omnivore or herbivore 1.5–2.5 m long (**Figure 18.3B**). The history of the discovery, study, and naming of †*Lagosuchus* illustrates a frequent problem in working with incomplete fossils. The first specimen described was thought to be undiagnosable (that is, it could not be confidently separated from specimens in other taxa). A later study described new and much more

(A) †*Lagosuchus*

(B) †*Sacisaurus*

Michael B. H. FunkMonk /CC BY-SA 3.0

Nobu Tamura/CC BY-SA 4.0

Figure 18.3 Basal dinosauromorphs. Ornithodirans close to the origin of Dinosauria include lightly built bipedal forms such as (A) †*Lagosuchus* and (B) †*Sacisaurus*

complete material of a similar genus that was named †*Marasuchus*. Subsequent work showed that †*Lagosuchus* and †*Marasuchus* are in fact the same, so †*Marasuchus* is now a synonym, replaced in taxonomic usage by the first-named genus, †*Lagosuchus*.

Dinosauria appeared in the Middle Triassic. Although this group radiated into thousands of fossil and extant species, including the largest terrestrial carnivores and herbivores of all time, dinosaurs were not particularly diverse until the Jurassic (see Figure 18.1). The best synapomorphy for Dinosauria is the presence of a perforated acetabulum, the socket that receives the head of the femur, a character to which we will return in Section 18.3.

Dinosaur diversification occurred in stages. An initial appearance in the Triassic (~235–228 Ma) was followed by a major radiation in the Middle to Late Jurassic (~174–145 Ma), then by several pulses of diversification during the Cretaceous. During their initial appearance, dinosaurs shared the terrestrial environment with many other groups of medium-sized to large amniotes, including basal archosaurs, pseudosuchians, and several synapsid groups. But by the end of the Triassic and throughout the remainder of the Mesozoic, dinosaurs dominated terrestrial ecosystems globally, having evolved diverse body forms, dentitions, and modes of locomotion. Large carnivorous dinosaurs ecologically replaced large pseudosuchian carnivores, and large herbivorous dinosaurs ecologically replaced large herbivorous synapsids.

18.2 †Pterosaurs: Vertebrates Achieve Powered Flight

LEARNING OBJECTIVES _____

18.2.1 Describe the wings and flight of †pterosaurs.

18.2.2 Describe the dietary specializations of †pterosaurs.

18.2.3 Describe the evidence, pro and con, that indicates that †pterosaurs and birds were competitors.

†Pterosaurs (Greek *pteron*, "wing"; *sauros*, "lizard"), one of two independent radiations of ornithodirans with powered (flapping) flight, first appeared in the Late Triassic, some 80 million years before the earliest flying dinosaurs (†*Archaeopteryx*; see Section 19.2). The early evolution of †pterosaurs is somewhat mysterious, but the recent reassessment of †*Lagerpeton* and related forms showed that these early ornithodirans have several traits previously thought to be unique to early †pterosaurs. These include a small body size, tall semicircular canals in the inner ear, and an enlargement of the region of the cerebellum related to balance—all of which contributed to the evolution of powered flight in †Pterosauria.

Basal †pterosaurs retained a long tail stiffened by bony projections that extended anteriorly, overlapping half a dozen vertebrae and preventing the tail from bending either laterally or dorsoventrally. A leaf-shaped expansion at the end of the tail may have acted as a rudder. Derived

†pterodactyloids, which were larger than basal forms and lacked tails and teeth, appeared in the Middle Jurassic and persisted until the end of the Cretaceous (see Figure 18.1).

Structure of †pterosaurs

The stringent mechanical demands of flight are reflected in the structures of flying vertebrates, and it is not surprising that †pterosaurs and birds have many convergently evolved features. Derived †pterosaurs (†pterodactyloids) have reduced or absent teeth; the tail was lost; the sternum developed a keel that served as the origin of strong flight muscles; the thoracic vertebrae became fused into a rigid structure; elements of the pelvic girdle fused with sacral vertebrae to form a synsacrum; the bones were thin-walled; **pneumatization** (air spaces in bones) was extensive; and the eyes and regions of the brain associated with vision and balance were large, whereas olfactory areas were small. †Pterosaurs had a unidirectional flow of air through the lungs (see Figure 13.3), the capacity for high aerobic metabolic rates, and probably bristlelike structures covering the body.

Limbs and locomotion   The limbs of †pterosaurs were large in relation to the trunk and abdomen. The wings were formed by skin stiffened by internal fibers and were completely different from the feathered wings of birds. The primary wing of pterosaurs is formed by the **cheiropatagium** and the **propatagium**. The cheiropatagium contributes the largest surface area of the wing. It is supported anteriorly by the upper and lower arm bones and an extremely elongated fourth finger, and extends posteriorly to the hindlegs (**Figure 18.4A,B**). The propatagium is a small flying membrane extending from the front edge of the wing forward to the neck, supported by the pteroid, a small splintlike bone attached to the wrist.

A posterior membrane, the cruropatagium, provided additional lift at the rear of the body. The cruropatagium of basal †pterosaurs extended between the hindlegs, limiting their independent movement. In †pterodactyloids, the cruropatagium was reduced, extending from the pelvis to the ankle of each leg, allowing the hindlimbs to move independently. Fusion of vertebrae in the thoracic region and presence of a synsacrum (**Figure 18.4C**) strengthened the vertebral column to support up and down movements of the wings. The synsacrum is also important for absorbing stress during takeoffs and landings.

†Pterosaurs walked quadrupedally when on land. However, the rarity of tracks of basal †pterosaurs may indicate that they seldomly walked on land, possibly because they were either arboreal or cliff-dwelling. †Pterodactyloid trackways found in Jurassic and Cretaceous sediments on most continents show that they could stride (**Figure 18.5**).

Aerodynamic tests and modeling of flight efficiency and sinking rate indicate that all †pterosaurs could fly, and that flight efficiency increased throughout the ~150 Ma of †pterosaur evolution. The long, narrow wings of large †pterosaurs (for example, †*Pteranodon*) are like those

(A) Flight membranes of †*Sordes*

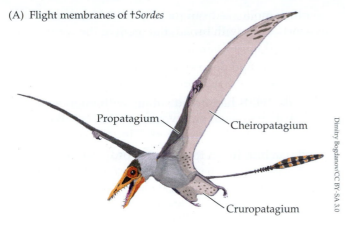

Propatagium

Cheiropatagium

Cruropatagium

Dmitry Bogdanov/CC BY-SA 3.0

(B) Bones of a †pterosaur wing

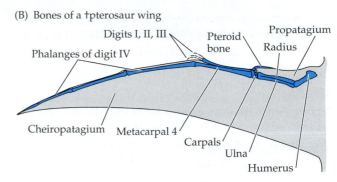

Digits I, II, III

Pteroid bone

Propatagium

Radius

Phalanges of digit IV

Cheiropatagium   Metacarpal 4   Carpals

Ulna

Humerus

Acta Palaeontologica Polonica 59: 109–124/CC BY

Figure 18.4 **Skin formed the flying membranes of** †**pterosaurs.** (A) The cheiropatagium and propatagium were supported by bones of the forelimb; the cruropatagium spanned between the hindlimbs. (B) In the wing, an elongated digit IV supported the cheiropatagium and the unique pteroid bone supported the propatagium. (C) A synsacrum formed by fusion of sacral vertebrae with the pelvic girdle supported the legs during takeoff and landing. (B after K. Liem et al. 2001. *Functional Anatomy of the Vertebrates*, 3rd ed. Cengage/Harcourt College: Belmont, CA; C after E.S. Hyder et al. 2014. *Acta Palaeontologica Polonica* 59: 109–124/CC BY.)

(C) Pelvic girdle and synsacrum of a †pterosaur

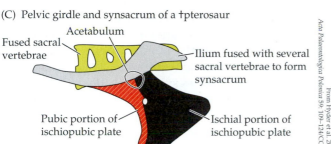

Acetabulum

Fused sacral vertebrae

Ilium fused with several sacral vertebrae to form synsacrum

Pubic portion of ischiopubic plate

Ischial portion of ischiopubic plate

From Hyder et al. 2014. Acta Palaeontologica Polonica 59: 109–124/CC BY

of birds such as albatrosses that soar at high speeds for great distances (soaring flight requires little flapping; see Figure 21.17). The smallest †pterosaur, †*Nemicolopterus crypticus*, was a forest-dweller with a wingspan of only 25 cm. It probably had short, broad elliptical wings like those of birds that fly through woodlands (see Figure 21.16B). The largest species—the †azhdarchids, with wingspans of up to 11 m and body weights of 200 kg or more—had less efficient flight and were probably mostly terrestrial.

Reconstruction of walking †*Pterodactylus* and its trackway

© Mark Witton

How a †pterosaur took flight has long been a subject of debate. Through most of the 20th century, reconstructions of †pterosaurs showed them poised on cliffs, reflecting a belief that launching from a height was the only way a †pterosaur could get into the air. More recently, a bipedal running takeoff was proposed, but both mechanical and aerodynamic arguments have discredited that hypothesis. The most likely method is a quadrupedal leap, which would have allowed the wing and hindlimb muscles to act simultaneously to produce high launch speeds and enough height to prevent the first wingbeats from striking the ground.

Jaws and teeth   †Pterosaurs diversified into the major adaptive zones occupied today by extant birds. Variation in body size and in the jaws and teeth appears to have corresponded to differences in diet, although direct evidence of the diets of †pterosaurs—such as fossilized food in the stomach—is scarce. Recent analyses of tooth wear confirm most earlier interpretations based on jaw and tooth morphology. Generally, early †pterosaurs seem to have had an invertebrate-dominated diet, whereas later forms ate more

Figure 18.5 †**Pterosaurs walked quadrupedally.** A reconstruction of a walking †*Pterodactylus* shows its erect stance and limb movement and the types of trackways that it made. Only the first three fingers of the forelimbs touched the ground; the long fourth finger that supported the cheiropatagium was folded against the body. Because the wrist was rotated outward, impressions of the fingers of the forefoot directed to the side and backward. The toes and soles of the hindlimbs were in contact with the ground.

vertebrates. Some of the morphological specializations of jaws and teeth for different feeding methods and prey are shown in **Figure 18.6** and described here.

- Piscivores had long jaws with pointed teeth (Figure 18.6A).
- Short jaws with large, sharp teeth may have characterized species that preyed on other †pterosaurs as well as on other small tetrapods (Figure 18.6B).
- Small species with short jaws with numerous small, sharp teeth are believed to have been insectivores (Figure 18.6C).
- There were other dietary specializations as well: several †pterodactyloids in the Jurassic and Cretaceous that combined long jaws with extremely long, closely set teeth are believed to have been filter feeders (Figure 18.6D).
- †Dsungaripterids had long, narrow jaws with no anterior teeth; they may have used a pincerlike motion to pluck mollusks from rocks at low tide, then crushed them with broad, flat teeth at the rear of the jaws (Figure 18.6E).
- †Tapejarids had deep skulls with toothless beaks and may have been frugivorous (Figure 18.6F).
- †Istiodactylids had broad snouts with interlocking lancet-shaped teeth and were probably carnivorous, perhaps primarily scavengers (Figure 18.6G).
- †Azhdarchids (from the Uzbek word *azhdarkho*, "a mythical dragon") were terrestrial stalkers that foraged quadrupedally, using their long necks and beaks to seize prey. Large species were apex predators, probably capable of killing dinosaurs (Figure 18.6H).

Described relatively recently, the Cretaceous †azhdarchids have no extant analogues. In behavior and ecology, they probably resembled ground hornbills and marabou storks, except that some were much larger than those birds. The smallest species known was the size of a housecat and had a wingspan of 1.5 m, but most species were

(A) †*Rhamphorhynchus,* piscivore

(B) †*Sericipterus,* aerial predator

(C) †*Anurognathus,* insectivore

(D) †*Pterodaustro,* filter feeder

(E) †*Dsungaripterus,* durophage

(F) †*Tupandactylus,* frugivore

(G) †*Istiodactylus,* scavenger

(H) †*Hatzegopteryx,* apex predator

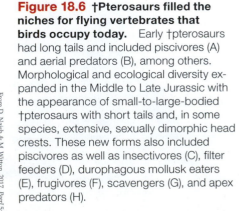

Figure 18.6 †**Pterosaurs filled the niches for flying vertebrates that birds occupy today.** Early †pterosaurs had long tails and included piscivores (A) and aerial predators (B), among others. Morphological and ecological diversity expanded in the Middle to Late Jurassic with the appearance of small-to-large-bodied †pterosaurs with short tails and, in some species, extensive, sexually dimorphic head crests. These new forms also included piscivores as well as insectivores (C), filter feeders (D), durophagous mollusk eaters (E), frugivores (F), scavengers (G), and apex predators (H).

Figure 18.7 The largest †azhdarchids were apex predators. Gracile †azhdarchids such as †*Arambourgiania philadelphiae* (center) could have confronted an adult giraffe eye-to-eye. †*Hatzegopteryx thambema* (right) is an example of a robust †azhdarchid that was not quite as tall as †*A. philadelphiae*. Both were large enough to have preyed on animals the size of humans, and †*H. thambema* may have preyed on dinosaurs.

Giraffe compared to gracile (center) and robust (right) †azhdarchids

© Mark Witton

very large, with wingspans up to 11 m. †*Arambourgiania philadelphiae* could have stood eye-to-eye with a giraffe (**Figure 18.7**).

There are two general types of giant †azhdarchids, gracile and robust. †*Arambourgiania philadelphiae* is a gracile †azhdarchid, with a long neck and relatively light skull. †*Hatzegopteryx thambema*, a robust form, had a shorter neck than †*A. philadelphiae* and a larger, heavier bill (see Figure 18.7). †*Arambourgiania philadelphiae* may have preyed on relatively small animals—up to the size of a human—whereas †*H. thambema* was probably an apex predator, capable of killing medium-sized dinosaurs.

Body covering and head crests Fossils of †*Sordes pilosus* and †*Jeholopterus ningchengensis* from fine-grained sediments show that their skin was covered by fine, hairlike fuzz called **pycnofibers** that probably provided insulation. Pycnofibers were probably colorful, forming patterns used for species and sex recognition. However, it remains debated how widespread such an insulating body covering was among †pterosaurs.

Head crests appeared early in the evolution of †pterosaurs and included a variety of sizes and shapes. Crests could be formed by bones alone, by bones with soft tissue, or by soft tissue alone. Tests of model †pterosaurs in wind tunnels indicate that even large crests had little aerodynamic effect, implying that they had no function in flying. The crests often grew allometrically, becoming larger as an individual grew, but were absent in females. Both characters suggest that the crests were used during intraspecific interactions, such as courtship and territorial disputes.

Reproduction, eggs, and parental care

Like all ornithodirans, †pterosaurs were oviparous. All taxa for which eggs are known retained the flexible eggshells that are believed to be characteristic of the eggs of basal ornithodirans. The eggs were small in relation to the size of the adults. †Pterosaurs appear to have hatched at an advanced stage of development and, like some species of brush turkeys (megapode birds; see Figure 21.32), were probably able to run and fly soon after emerging from the nest (i.e., the young were precocial rather than altricial; see Section 21.8).

Did the evolution of birds doom †pterosaurs?

†Pterosaurs had diversified by the Late Jurassic when the first flying avialans (the lineage that includes birds; see Chapter 19) appeared. †Pterosaurs filled many adaptive

zones now occupied by birds, and thus it is tempting to propose that competition with avialans drove †pterosaurs to extinction, but available evidence does not clearly support that hypothesis. A comparison of the body forms of †pterosaurs and avialans indicates that they occupied statistically different **morphospaces**—that is, they had different combinations of anatomical features, and as a result their lifestyles and ecological niches did not completely overlap.

Soon after avialans appeared, however, †pterosaur wingspans increased from an average of ~1.5 m to ~7 m, whereas the wingspans of avialans remained between 10 cm and 1 m. Character displacement of this sort might result from competition, but the timing of the increase in body size of †pterosaurs also corresponds to the appearance of †pterodactyloids and the gradual disappearance of basal †pterosaurs. If competitive replacement did occur, it was remarkably slow. Avialans appeared in the Late Jurassic and †pterosaurs persisted until the end of the Cretaceous; thus, the two groups lived side by side for nearly 100 million years.

18.3 Dinosaurs: One of the Most Successful Tetrapod Radiations

LEARNING OBJECTIVES

18.3.1 Understand which morphological features allowed dinosaurs to become the ruling vertebrates of terrestrial Mesozoic ecosystems.

18.3.2 Distinguish saurischians and †ornithischians based on morphological features.

18.3.3 Recognize signs of postcranial pneumaticity.

When most people hear the word "dinosaur," they visualize a large animal—either a fearsome predator like †*Tyrannosaurus rex* or an enormous, long-necked herbivore like †*Brontosaurus excelsus*. However, the diversity of sizes and body forms of non-avian dinosaurs extends far beyond those two examples. The huge species were indeed spectacular, but many dinosaurs were about the size of a chicken, and some were even smaller.

After a slow start during the Triassic, when they co-occurred with early avemetatarsalians and ornithodirans, dinosaurs—the saurischian lineage in the Late Triassic, and later the †ornithischian lineage in the Jurassic—rose to dominate terrestrial ecosystems on all continents. An estimated 1,500 to 2,500 species of dinosaurs roamed Earth during the Mesozoic, and in recent years paleontologists have named a new species about every two weeks. In this section we cover the origin and general structure of dinosaurs before giving more specifics for †Ornithischia (Section 18.4) and †Sauropodomorpha (Section 18.5). Theropoda, the lineage that includes Mesozoic avialans, is treated in Chapter 19.

The structure of dinosaurs

The skeletons of large dinosaurs combine the strength required to support animals that weigh thousands or tens of thousands of kilograms with structures that minimized weight of the skeleton. Two key features evolved early within Ornithodira. First, an erect stance with limbs held vertically below the pelvis made weight-bearing and locomotion more efficient, a prerequisite for the significant size increase that occurred during the early radiations of dinosaurs. Second, postcranial pneumatization was widespread in dinosaurs; the open spaces in bones other than those of the cranium are traces of air sacs like those of extant birds (see Figure 21.6) and suggest that these animals had flow-through ventilation of the lungs.

Limbs and locomotion Dinosaurs evolved from small (about 1 m), agile, terrestrial ornithodirans. Features of the hips, limb bones, and ankle acquired early in the dinosaur lineage (or even earlier, in precursors such as †*Lagosuchus*; see Figures 18.1 and 18.3) contributed to their success in the terrestrial realm.

Figure 18.8 shows patterns in the evolution of the pelvic girdle of archosaurs and how the girdle changed within Dinosauria. In early archosaurs such as †*Euparkeria*, the pubis, ischium, and ilium are attached in a triradiate

pattern. The acetabulum, which is the articulation surface for the femur, is in the center. Crocodylomorphs retain this basic arrangement, as do basal avemetatarsalians such as †*Teleocrater* (see Figure 18.2) †Pterosaurs departed from this arrangement. The pubis is fused with the ischium to form a broad ischiopubic plate, which strengthens

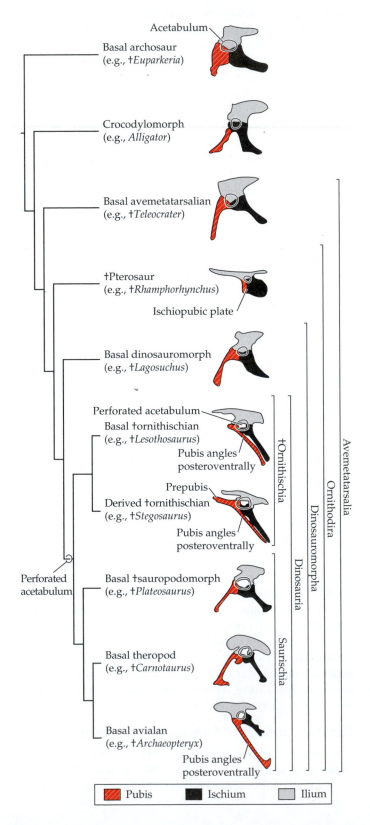

Figure 18.8 Evolution of the pelvis of archosaurs. The pelvis of the Triassic archosaur †*Euparkeria* is composed of the pubis, ischium, and ilium, with the acetabulum (the socket for the head of the femur) at the junction of the three bones. This basic triradiate form of the pelvic girdle was retained in early avemetatarsalians such as †*Teleocrater* and in crocodylomorphs such as alligators. Within Ornithodira, a pelvic girdle in which the ischium and pubis fuse to form an ischiopubic plate evolved in †pterosaurs such as †*Rhamphorhynchus*, while basal dinosauromorphs such as †*Lagosuchus* still had the basic triradiate pelvic girdle. A perforated acetabulum—essentially a hole where the three bones of the pelvis meet—is a synapomorphy of Dinosauria. In †ornithischians such as †*Lesothosaurus* and †*Stegosaurus* the pubis angled posteroventrally. Saurischians such as †*Plateosaurus* and †*Carnotaurus* retained the triradiate pelvic girdle, but it became highly modified in avialans (the theropod lineage that includes †*Archaeopteryx* and birds; see Chapter 19) with the pubis again evolving a posteroventral orientation. (See credit details at the end of this chapter.)

the pelvic girdle for walking upright and landing. Early dinosauromorphs such as †*Lagosuchus* retained the triradiate pelvis and centrally located acetabulum.

The most reliable synapomorphy of Dinosauria is a **perforated acetabulum** (see Figure 18.8). What this means is that the pubis, ischium, and ilium do not fully meet at the acetabulum, leaving an unossified hole—a perforation—in its middle. This character is retained throughout Dinosauria.

One group of dinosaurs evolved modifications of the pelvis that are synapomorphic for the bird-hipped dinosaurs, or †Ornithischia. The name (Greek *ornith*, "bird"; *ischia*, "hip joint") refers to the position of the pubis, for, as in extant birds, the pubis of †ornithischians extends posteroventrally as in the early genus †*Lesothosaurus* (counterintuitively, birds evolved not from †ornithischians but from the other dinosaur clade, Saurischia). Within †Ornithischia are two large clades: †Thyreophora (†stegosaurs and †ankylosaurs) and †Neornithischia (†marginocephalians and †ornithopods). Both have an anterodorsal extension of the pubis, known as a prepubis (see Figure 18.8).

The other large dinosaur group is Saurischia, the "lizard-hipped" dinosaurs. Basal members retained the triradiate pelvis typical of early dinosaurs, both in †Sauropodomorpha (such as †*Plateosaurus*) and early Theropoda (the clade that includes †*Tyrannosaurus*, †*Archaeopteryx*, and birds). Already in †*Archaeopteryx* the typical form of a posteroventrally directed pubis was present (compare the pubis of †*Archaeopteryx* with the pubis of †*Plateosaurus* and †*Carnotaurus* in Figure 18.8)

Changes in stance help to explain why some changes in the anatomy of the pelvis evolved. More basal archosaurs such as †*Euparkeria* had a sprawling posture, with the humerus and femur projecting horizontally and sharp bends at the elbow and knee (**Figure 18.9A**). This posture suffices for small and medium-sized tetrapods, but it does not work for large ones. This is because bones are far more resistant to forces exerted parallel to their long axis (compressive forces) than to forces exerted at an angle

Figure 18.9 Dinosaur pelvic girdles combined upright stance with long strides. The femur was protracted (swung forward) by muscles that originated anterior to the femur. It was retracted (swung backward) by muscles that extended from the base of the tail. (A) The limbs of basal archosaurs such as †*Euparkeria* extended horizontally from the pelvis, like the limbs of extant crocodylians. Femoral protractor muscles originating on the pubis could produce a long stride because the leg swung in a horizontal arc. (B) The limbs of saurischian and †ornithischian dinosaurs were held more vertically beneath the pelvis, and the muscles extending from the pubis to the femur would have been too short to make a long stride, a limitation solved by the evolution of a more anterior origin for the femoral protractors. (C) †Ornithischians rotated the pubis posteriorly and moved the origin of the femoral protractor muscles to the ilium. (D) Derived †ornithischians evolved a forward extension of the pubis (the prepubis) on which the femoral protractors inserted. (E) Saurischian dinosaurs elongated the pubis and ischium and rotated the pubis anteriorly.

(A) Pelvic girdle and hindlimbs showing sprawling and erect stances

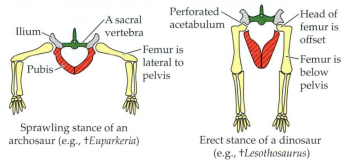

Sprawling stance of an archosaur (e.g., †*Euparkeria*)

Erect stance of a dinosaur (e.g., †*Lesothosaurus*)

(B) Pelvis and femur of a basal archosaur (e.g., †*Euparkeria*)

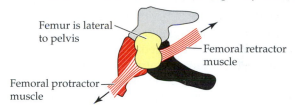

(C) Pelvis, femur and associated muscles of a basal †ornithischian (e.g., †*Lesothosaurus*)

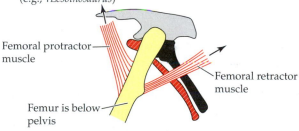

(D) Pelvis, femur and associated muscles of a derived †ornithischian (e.g., †*Stegosaurus*)

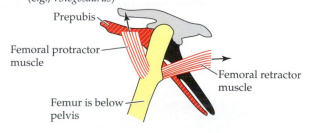

(E) Pelvis, femur and associated muscles of a basal saurischian (e.g., †*Carnotaurus*)

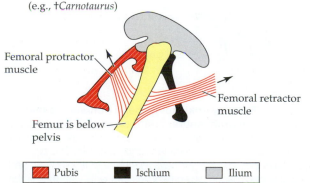

Pubis   Ischium   Ilium

to the long axis (shearing forces). Thus, large mammals such as elephants support their weight on vertical limbs (see Figure 11.4B). Many dinosaurs were far larger than elephants, and their limbs were also oriented vertically.

Changing the angle of the limbs solved the problem of weight-bearing but introduced a new problem. In †Euparkeria, muscles originating on the pubis and inserting on the femur protracted the leg (moved it forward), and muscles originating on the tail retracted the leg (moved it backward; **Figure 18.9B**). The femur of †Euparkeria projected horizontally, so that its pubofemoral muscles extended outward from the pelvis to insert on the femur. Thus, the protractor muscles were long enough to swing the femur through a large arc relative to the ground, generating a long stride length. But moving the legs under the body made the protractor muscles shorter and less effective in moving the femur because a muscle's maximum contraction is about 30% of its resting length. The shorter muscles would have swung the vertical femur through a smaller arc and reduced the stride to a shuffle.

The origins of the femoral protractor muscles of dinosaurs shifted, making the muscles longer and allowing a combination of upright stance with long strides. †Ornithischians and saurischians evolved different ways to do this. In †ornithischians (**Figure 18.9C**), the pubis angles posteroventrally and is close to the ischium; the femoral protractor muscles originated on the ilium. Derived †ornithischians (**Figure 18.9D**) evolved a prepubis that extended beyond the anterior part of the ilium, providing an even more anterior origin for femoral protractor muscles. In saurischians (**Figure 18.9E**), both the pubis and ischium elongated; the pubis became anteroventrally positioned so that the femoral protractor muscles ran back from the pubis to the femur and were long enough to protract it. Although these changes were anatomically different, they were functionally similar. Both changes produced hip articulations that allowed the legs to be

held vertically beneath the pelvis, which supported both heavy bodies and stride lengths that enabled dinosaurs to move rapidly and over long distances.

The ability of dinosaurs to support a heavy body is also related to microstructural changes in limb bones. In both dinosaurs and mammals, limb bones are described as trabeculated because many fine trabeculae within the bone function as reinforcing struts to strengthen the bone and increase its ability to support body weight (see Figure 21.6). There are differences, however, in the trabeculae of dinosaurs and mammals. In dinosaurs, the number of trabeculae and connections between them increase with increasing body weight. In mammals, it is the widths of the trabeculae, not their connectivity, that correlates with heavier body weights. The increased connectivity of many equally thin trabeculae in dinosaurs may explain why dinosaurs could reach larger body masses than terrestrial mammals. The mechanical principle is analogous to load-sharing by trusses in engineering and architecture, and the trabecular structure of bone in dinosaur limbs combined strength with low weight.

A simplified ankle joint, known as a restricted mesotarsal joint, was a third important feature facilitating efficient locomotion (**Figure 18.10**; also see Figure 9.18). This type of joint allows the hindfeet to thrust backward forcefully without twisting. Together, these three evolutionary changes in the pelvic girdles, limb bones, and ankle joints set the stage for increases in body size that are so evident in Mesozoic dinosaurs.

Postcranial pneumatization   Like †pterosaurs, many dinosaurs have postcranial pneumatization in the form of cavities occupied by air sacs. These air sac systems invaded bones, particularly vertebrae and some elements of the limb girdle. Such invasive pneumatization of bones is indicated by distinct foramina piercing the bone surface. For example, there are three systems of air sacs in

(A) Left ankle of an alligator with crocodyloid joint

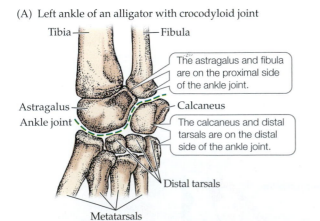

(B) Left ankle of †Lesothosaurus with restricted mesotarsal joint

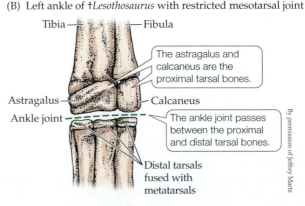

By permission of Jeffrey Martz

Figure 18.10 The way ankle joints bend defines the two extant lineages of archosaurs.   (A) The axis of bending in the ankle joint of crocodylomorphs passes between the astragalus and calcaneus and between the calcaneus and the fibula. This type of joint allows the foot to twist sideways as well as to flex forward and back, and it is one reason why crocodiles can walk with their limbs splayed out or gallop with the legs tucked closer to the body. (B) In contrast to crocodylomorphs, the ankle joint of dinosaurs and birds passes between the proximal (astragalus and calcaneus) and distal tarsal bones, which are fused with the metatarsal bones. This type of ankle joint can flex only backward and forward, which is why it is known as a restricted mesotarsal joint.

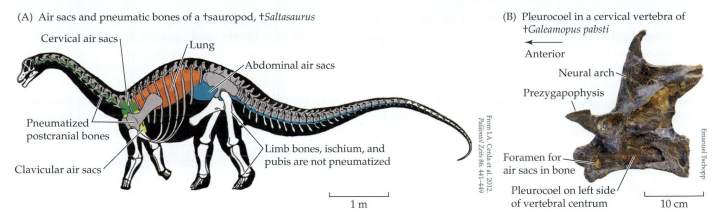

(A) Air sacs and pneumatic bones of a †sauropod, †*Saltasaurus*

Cervical air sacs

Lung

Abdominal air sacs

Pneumatized postcranial bones

Clavicular air sacs

Limb bones, ischium, and pubis are not pneumatized

From I.A. Cerda et al. 2012. *Paläontol Zeits* 86: 441–449

1 m

(B) Pleurocoel in a cervical vertebra of †*Galeamopus pabsti*

Anterior

Neural arch

Prezygapophysis

Foramen for air sacs in bone

Pleurocoel on left side of vertebral centrum

Emanuel Tschopp

10 cm

Figure 18.11 Postcranial pneumatization. (A) Postcranial pneumatization was widespread among ornithodirans and was especially well developed in saurischians such as †*Saltasaurus*. The lung connected with cervical, clavicular, and abdominal air sacs, and many postcranial bones were pneumatized (i.e, air sacs extended into the bones). (B) A cervical vertebra from the †sauropod †*Galeamopus* shows the pleurocoel, a hollow cavity into which cervical air sacs extended; and the foramen, where the air sac system invaded the vertebral centrum.

the †sauropod †*Saltasaurus loricatus*: cervical, clavicular, and abdominal air sacs (**Figure 18.11**). Like the air sacs in extant birds, the air sac systems of non-avian dinosaurs connected to the lungs and allowed highly efficient unidirectional flow of air through the lung (see Figure 13.3C and Figure 13.5). The presence of invasive postcranial pneumatization in both dinosaurs and †pterosaurs suggests that it was present in the common ancestor of Ornithodira and subsequently lost in some lineages, most notably †Ornithischia.

Body covering Fossilized skin impressions consistently show that dinosaurs had nonoverlapping, polygonal scales. Feathers in a non-avian dinosaur were first reported in 1998 in †*Sinosauropteryx prima*, a small theropod. A mid-dorsal crest of feathers extends along the body onto the tail, and they preserve evidence of a banded pattern on the tail (**Figure 18.12**). Since the description of †*Sinosauropteryx*, feathers and featherlike bristles have been discovered in many dinosaur lineages, including †ornithischians.

The presence of feathers in many non-avian dinosaurs does not mean that all of them were feathered. Skin impressions show that large species remained covered with scales. Others had bristlelike structures around the head and body, large imbricated scales on the tail, and a mix of complex featherlike structures and small scales along the limbs.

Other integumentary features known from †sauropod embryos (and from theropods such as †*Juravenator starki*) include modified scales that may have been sense organs like the integumentary sensory structures of crocodylians

(see Section 17.3). †Stegosaurs had rows of polygonal scales and rosettelike structures on their flank and a keratinous sheath covering their dorsal plates. Hence, a combination of different scale patterns and feathered body parts was probably widespread among dinosaurs.

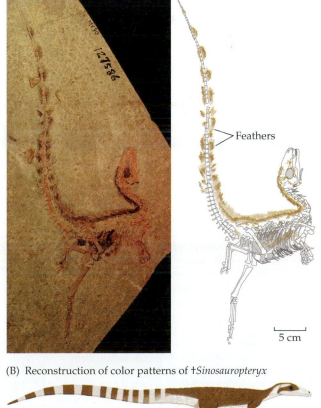

(A) Fossil and interpretive drawing of †*Sinosauropteryx*

Feathers

From F.M. Smithwick et al. 2017. *Curr Biol* 27: 3337–3343.e2. CC BY 4.0

5 cm

(B) Reconstruction of color patterns of †*Sinosauropteryx*

Figure 18.12 The earliest feathers of dinosaurs were keratinaceous filaments. Traces of filaments on the head, spine, and both sides of the tail of the theropod †*Sinosauropteryx* are shown in the sketch and can be seen on the fossil on which it is based. The filaments increase in length from 13 mm on the head to 40 mm on the tail.

Colors and patterns of many feathered dinosaurs were like those of extant birds, with white, brown, red, orange, and black hues being widely distributed. Other taxa, such as †*Archaeopteryx* and †*Microraptor*, seem to have had iridescent plumage like a modern magpie (see Figure 19.12D and Figure 19.18B). Skin with preserved details that indicate color is rare, but some evidence indicates that the †ankylosaur †*Borealopelta* was countershaded, with reddish-to-brown dorsally and a lighter-colored belly.

18.4 †Ornithischia

LEARNING OBJECTIVES

18.4.1 Describe the morphological diversity of †ornithischians.

18.4.2 Explain how we use modern analogs to infer behavior of extinct animals, and the limitations of such an approach.

The earliest unambiguous †ornithischians are known from the Early Jurassic, and most of †Ornithischia's main clades had originated by the end of the Jurassic (see Figure 18.1). Diversity gradually increased throughout the Jurassic and Cretaceous, peaking in the Late Cretaceous.

†Ornithischia includes dinosaurs with many body sizes and forms (**Figure 18.13**). Basal members such as †*Lesothosaurus* were small, bipedal omnivores, but herbivory evolved early (and possibly more than once). All †ornithischians had horny beaks covering the tip of the snout, and two derived lineages independently evolved complex tooth batteries in the back of their jaws. Tooth batteries were formed by several dozen to sometimes thousands of individual teeth, which were tightly packed and grew continuously to form a large grinding surface to process food—that is, they chewed. Mastication (reducing food to a pulp in the mouth before swallowing) is the norm for mammals but is rare in other taxa; most vertebrates either swallow prey whole or tear off and swallow chunks of food. †Ornithischians independently evolved quadrupedal locomotion at least three times: in †Thyreophora, †Marginocephalia, and †Ornithopoda (see Figure 18.13).

- †Thyreophorans were heavily armored animals with either vertically arranged bony plates and spikes (†stegosaurs) or osteoderms (any ossification in the dermis) arranged in a shieldlike body cover (†ankylosaurs).

- †Marginocephalians had heads that were strongly adorned with either bony frills and horns (†ceratopsians) or domelike structures (†pachycephalosaurs).

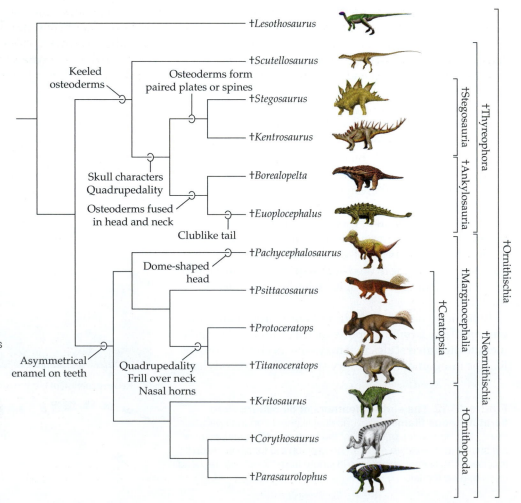

Figure 18.13 Phylogenetic relationships of †Ornithischia. †*Lesothosaurus*, a small Early Jurassic †ornithischian, is shown here as the sister group of †Thyreophora + †Neornithischia. †Thyreophorans have keeled osteoderms that were elaborated differently in its two major subclades, †Stegosauria and †Ankylosauria. †Neornithischia includes †Marginocephalia, such as †ceratopsians, famous for their frills and horns, and †Ornithopoda, which evolved large tooth batteries. Early †ornithischians were bipedal, but quadrupedality evolved independently in both †thyreophorans and †neornithischians. (See credit details at the end of this chapter.)

- †Ornithopods included duck-billed dinosaurs (†hadrosaurs), some of which developed elaborate head crests and reached gigantic body sizes.

†Thyreophora

†Thyreophorans (Greek *thyreos*, "shield"; *phoros*, "bearer") were armored dinosaurs. Their name refers to the parallel rows of large osteoderms extending from the neck to the tail. The two major lineages, †Stegosauria and †Ankylosauria, were quadrupedal, although basal †thyreophorans were bipedal.

†Stegosauria †Stegosaurs (Greek *stegos*, "roof") were most abundant in the Late Jurassic, although some species persisted into the Early Cretaceous (see Figure 18.1). These quadrupedal herbivores were relatively diverse in the Late Jurassic of Laurasia, but only a few species are known from Gondwana; the difference is probably a result of unequal sampling on the two supercontinents.

†Stegosaurs reached up to 9 m in length, with one or several vertical rows of bony plates and spikes along the vertebral spine, short and sturdy forelegs, and long, columnar hindlegs as in †*Kentrosaurus* (**Figure 18.14A**). The skull was small for such a large animal and had a horny beak at the front of the jaws. The teeth were shaped like those of extant herbivorous lizards and show none of the specializations seen in some other †ornithischians.

The distinctive feature of †stegosaurs is a double row of spines or leaf-shaped plates that extends along the vertebral column. In addition, many †stegosaurs had a long spine that projected upward and backward from the shoulder. These spines and plates were hypertrophied keels of the osteoderms that are characteristic of †thyreophorans.

The function of the plates of †*Stegosaurus* has been contested for decades. Initially they were assumed to have provided protection from predators, and some reconstructions show the plates lying flat against the sides of the body as shields. An active defensive function for the plates is not very convincing, however. Even though they would have considerably increased the apparent size of a stegosaur when seen in profile, and their horny sheath may have formed sharp edges, the plates left both the belly and large areas on the sides of the body unprotected.

A role for plates as heat exchangers remains debated. Grooves on the plate surfaces were interpreted as channels for blood vessels that could carry a large flow of blood to be warmed or cooled, according to the needs of the animal. Some of these grooves also connect to large, internally branching canals. However, it is unclear if the vessels in these canals and grooves merely carried nutrients and respiratory gasses to the cells that made an overlying sheath of the plates or if the plates had an active role in thermoregulation.

Combinations of plates and spines were species-specific and potentially dimorphic within a species, suggesting that these characters identified species and possibly sex during behavioral interactions. These different interpretations of plate function are not mutually exclusive, and it is possible that plates initially served one of these functions and that additional roles subsequently evolved.

The tails of †stegosaurs had large spines; in some species, these spines covered the entire tail, whereas in other species

(A) Fossil and interpretive reconstruction of †*Kentrosaurus*

H.Zell/CC BY-SA 3.0

© James Kuether

(B) Fossil and interpretive reconstruction of †*Borealopelta*

Ceratops Yuta/CC BY-SA 4.0

© James Kuether

Figure 18.14 †Thyreophorans had extensive bony armor.
Keeled osteoderms are a derived character of †thyreophorans. In †stegosaurians such as †*Kentrosaurus* (A) the keels of the osteoderms formed spikes or plates. In †ankylosaurs such as †*Borealopelta* (B) some osteoderms fused together, creating platelike armor.

they were restricted to near their tips (see Figure 18.14A). These spines were almost certainly defensive structures, and a biomechanical analysis of the whipping motion of the tail of †*Kentrosaurus* supports that interpretation. Using the tail muscle of an alligator as the basis for calculation, the tail tip of an adult †*Kentrosaurus* could strike a predator at a velocity of 20–40 m/sec. At those speeds, the tip of a spine would have exerted enough force to penetrate skin and muscle to a depth of ~30 cm or to pierce bony armor. A tail vertebra of †*Allosaurus*, a large theropod, shows a healed wound that may have been the result of being punctured by a spike of the contemporaneous †*Stegosaurus*.

†Ankylosauria   †Ankylosaurs (Greek *ankylos*, "stiffening") were heavily armored †ornithischian dinosaurs. Their fossils have been found in Jurassic and Cretaceous deposits on all continents. †Ankylosaurs were quadrupeds that ranged from 2 to 9 m long, with short legs and broad bodies. Fused osteoderms on the neck, back, hips, and tail produced large, shieldlike body coverings (**Figure 18.14B**). Bony plates also covered the skull and jaws (at least one species, †*Euoplocephalus*, even had bony armor on its eyelids).

Soon after the origin of †Ankylosauria in the Early Jurassic, its two main families, †Ankylosauridae and †Nodosauridae, diverged and radiated. Interestingly, †ankylosaurids remained restricted to Laurasia, whereas †nodosaurids seem to have colonized nearly all continents. The disappearance of †nodosaurids from Asia by the Late Cretaceous coincided with the diversification of †ankylosaurids in that region. †Nodosaurids continued to thrive on other continents, however, and in some regions occurred in the same locations as †ankylosaurids. Both clades became extinct at the end of the Cretaceous.

†Ankylosaurs must have been difficult animals to attack, and their evolutionary origin may be linked to increases in the size of carnivorous theropods at that time. Some species of †nodosaurids had spines projecting from the back and sides of the body, while derived species of †ankylosaurids had a lump of bone at the end of the tail that could be swung like a club. All species of †ankylosaurs had broad, flat bodies with armored backs and sides, and the act of merely lying flat on the ground might have been an effective defensive tactic.

†Neornithischia

†Neornithischia includes †Marginocephalia and †Ornithopoda (see Figure 18.13). Both groups were diverse and abundant in the Cretaceous and were important components of the terrestrial fauna.

†Marginocephalia

†Marginocephalians (Latin *margo*, "border") take their name from the bony shelf at the rear of the skull. Two major lineages of †marginocephalians are distinguished: †pachycephalosaurs and †ceratopsians (see Figure 18.13).

†Pachycephalosauria   †Pachycephalosaurs were small to medium-sized (1–5 m) bipedal dinosaurs that lived during the Late Cretaceous. A distinctive feature of this lineage was a pronounced thickening of the skull roof (Greek *pachys*, "thick") in some species, and an array of bony projections on the skull of other species (**Figure 18.15**). The skull dome significantly increased in thickness during ontogeny and is thought to have been used in intraspecific combat.

†Pachycephalosaurs appear relatively late in the dinosaur fossil record and are restricted to the Late Cretaceous of Asia and North America. However, given that their sister taxon †Ceratopsia is known from the Late Jurassic onward, earlier †pachycephalosaurs must have existed. Indeed, a recent reassessment of †ornithischian phylogeny suggests that some species previously attributed to †heterodontosaurids are the basalmost †pachycephalosaurs, but this remains to be confirmed. The †pachycephalosaur lineage remained relatively species-poor until becoming extinct at the end of the Cretaceous.

(A) †*Prenocephale prenes*

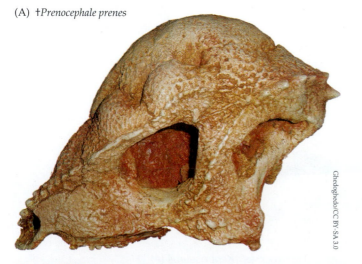

Ghedoghedo/CC BY-SA 3.0

(B) †*Pachycephalosaurus wyomingensis*

Balista/CC BY-SA 3.0

Figure 18.15 **†Pachycephalosaurs had armored heads.**   Both †*Prenocephale* (A) and †*Pachycephalosaurus* (B) had thick, heavily ossified skulls. The skull of †*Pachycephalosaurus* also had thick, bony spikes.

†Ceratopsia †Ceratopsians (Greek *keras*, "horn"; *opsis*, "appearance"), the horned dinosaurs, were the most diverse †marginocephalians. They are mostly known from the Cretaceous, but a few early forms were recently found from the Late Jurassic (see Figure 18.1). The distinctive features of †ceratopsians are the frill over the neck, formed by an enlargement of the parietal and squamosal bones; and a beak similar in shape to that of a parrot. In derived forms, teeth were arranged into batteries in each jaw that allowed them to chew their food, and most †ceratopsians developed distinctive horns on the nose and/or over the eyes.

An early radiation of the small-bodied †Psittacosauridae (Latin *psittaco*, "parrot," in reference to the large beak) occurred during the Early Cretaceous in China and Mongolia. A larger, more famous radiation happened with the evolution of †Ceratopsidae. This radiation led to a diversification of about 35 species in the latest Cretaceous of North America, with different sets of species on the two sides of the Western Interior Seaway that divided North America into two subcontinents at that time. Many of these species were relatively short-lived (a few hundred thousand years), with strong evidence for gradual evolutionary change within lineages.

Although derived †ceratopsians were quadrupedal, they evolved from a bipedal ancestor, and there is evidence that early forms were bipedal when young and quadrupedal as adults. Early †ceratopsians such as †*Psittacosaurus* and †*Protoceratops* were small (cat to sheep size), had a simple frill, and lacked nasal horns (**Figure 18.16A,B**). About 90 Ma, a lineage of larger †ceratopsians (rhinoceros to elephant size) appeared sporting more elaborate frills, long brow horns, and nasal horns. Two groups of these derived †ceratopsians can be distinguished. In short-frilled forms such as †*Triceratops*, the frill extended back over the neck, whereas in the long-frilled forms such as †*Titanoceratops*, the frill extended half the length of the trunk (**Figure 18.16C,D**).

(A) †*Psittacosaurus*

Matt Martyniuk(Dinoguy2)/CC BY-SA 4.0

From J. Vinther et al. 2016. *Curr Bio* 26: 2456–2462. doi.org/10.1016/j.cub.2016.06.065/CC BY 4.0

(B) †*Protoceratops*

Luis Miguel Bugallo Sanchez/CC BY-SA 3.0

(C) †*Triceratops*

Ed T/CC BY-SA 2.0

(D) †*Titanoceratops*

Kurt McKee/CC BY-SA 2.0

LadyofHats/Public Domain

Figure 18.16 Elaboration of †ceratopsian frills and horns.
(A) Basal †ceratopsians such as †*Psittacosaurus* were obligate bipeds as adults. (B) The skulls of early †ceratopsians such as †*Protoceratops* had small frills but lacked horns. Derived †ceratopsians had horns and their frills were more elaborate. (C) The frill of †*Triceratops*, a short-frilled †ceratopsian, extended over its shoulders. (D) Frills of long-frilled †ceratopsians such as †*Titanoceratops* extended half the length of the trunk.

†Ornithopoda

†Ornithopods lived from the Middle Jurassic to the end of the Cretaceous. Most were medium sized, although the largest species rivaled small †sauropods in size. Small species were largely bipedal, while the largest species probably moved mostly on four legs. The group includes several basal clades of relatively low diversity from the Jurassic (including †*Iguanodon*, one of the first dinosaur fossils to be found), and one large family, the †Hadrosauridae.

†Hadrosauridae   †Hadrosaurs (Greek *hadros*, "bulky") were the last group of †ornithopods to evolve, and they were also the most speciose lineage. These large animals reached lengths >15 m and weights >13,000 kg. The anterior portions of their jaws were toothless and sheathed in a horny beak, which in many species was transversely wide, like the bill of a duck (hence their vernacular name of duck-billed dinosaurs). It has been proposed that their radiation during the Cretaceous was linked to the evolution of flowering plants (angiosperms), and there is evidence that toward the end of the Cretaceous, †hadrosaurs replaced †sauropods as the most diverse and abundant herbivores in at least some ecosystems.

There were two subfamilies of †Hadrosauridae: †Saurolophinae and †Lambeosaurinae. †Saurolophines had long nasal regions (**Figure 18.17A–C**). †Lambeosaurines

developed highly distinctive bony crests that were connected to the nasal passage, and which protruded dorsally or posteriorly from their head (**Figure 18.17D–F**). By the Late Cretaceous, †hadrosaurids were present on all continents except Africa, and some forms of both subfamilies were the most abundant and speciose herbivores of their ecosystems, such as the †saurolophine †*Edmontosaurus* in northern North America, and †lambeosaurines in Spain. Species from both †Saurolophinae and †Lambeosaurinae thrived until the end of the Cretaceous.

Like derived †ceratopsians, †hadrosaurs independently evolved tooth batteries that occupied the rear of the jaws. Tooth rows of the upper and lower jaws each contained about 40 teeth packed closely side by side to form a massive chewing surface (**Figure 18.18**). Replacement teeth constantly erupted on the lingual (tongue) side of the tooth battery, growing upward to replace worn teeth. Thus, a †hadrosaur could have several thousand teeth in its mouth, of which several hundred were in use at any given time.

Social behavior of †ornithischian dinosaurs

Although it is impossible to know what a dinosaur was doing in the last moments before it died and embarked on the long trail to becoming a fossil, multiple discoveries of fossil beds that contain many individuals of the same species of †ornithischian support the hypothesis

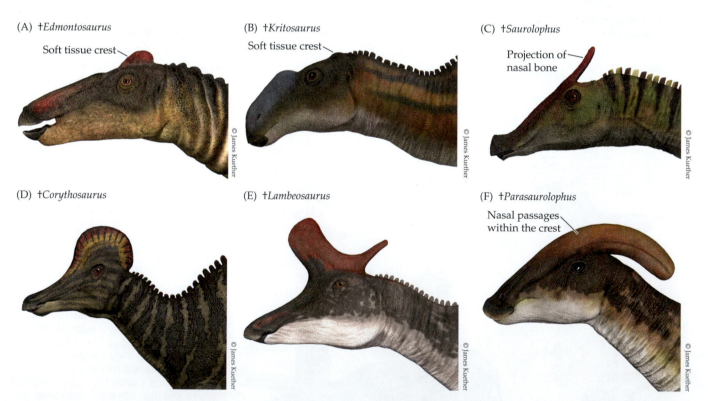

(A) †*Edmontosaurus*   Soft tissue crest

(B) †*Kritosaurus*   Soft tissue crest

(C) †*Saurolophus*   Projection of nasal bone

(D) †*Corythosaurus*

(E) †*Lambeosaurus*

(F) †*Parasaurolophus*   Nasal passages within the crest

© James Kuether

Figure 18.17 Cranial ornamentation of †hadrosaurs. †*Edmontosaurus* (A) and †*Kritosaurus* (B) had crests formed by soft tissue. The closely related †*Saurolophus* (C) had a projection from its nasal bone. †Lambeosaurines such as †*Corythosaurus* (D), †*Lambeosaurus* (E), and †*Parasaurolophus* (F) had elaborate bony crests. The nasal passages of †*Parasaurolophus* ran from the external nares backward through the crests then forward to choanae in the roof of the mouth.

(A) Skull and lower jaw of a †hadrosaur, †*Edmontosaurus*

Left upper tooth battery

The upper and lower tooth batteries met and were worn to form the occlusal surface.

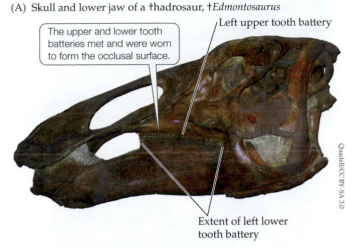

Quadell/CC BY-SA 3.0

Extent of left lower tooth battery

Figure 18.18 Tooth battery of †hadrosaurs. (A) Derived herbivorous †hadrosaurs such as †*Edmontosaurus* had hundreds of teeth in tooth batteries in the upper and lower jaws. Teeth in the upper and lower batteries wore down to form a flat occlusal surface between them, used to grind plant material. (B) Many generations of teeth are exposed in this medial view of the right lower jaw. Wear on the tops of exposed teeth formed the occlusal surface. New teeth erupted constantly at the base of the battery so that the occlusal surface was renewed throughout life.

(B) Tooth battery of right lower jaw viewed from inside

Anterior

Sharp edge of occlusal surface

Museum of the Earth/Courtesy of Maureen Bickley

Tooth 1 is part of the grinding surface

Five teeth in series

Tooth 5 is erupting

Sharp edge of occlusal surface   Worn teeth form grinding surface

that they formed herds and had complex social behaviors. In many cases, the material surrounding the fossils represents a sudden fatal event such as a lahar (volcanic mudflow) or deep deposits of volcanic ash; other spectacular finds include groups of †ornithischians that were trapped in sandstorms. These sudden events imply that the fossils represent a group of individuals that lived and died together. Additional clues about social behavior can be inferred from the skeletons themselves. Still other evidence comes from footprints indicating that hatchlings, juveniles, and adults lived together (**Figure 18.19**).

As mentioned earlier, the skull domes of †pachycephalosaurs increased in thickness during ontogeny and may have been used in male–male combat. Injuries discernable in fossils suggest that male †*Prenocephale* may have engaged in head-on butting contests like those of extant bighorn sheep (**Figures 18.20A,B**), whereas the spikier

†*Hadrosauropodus* and †*Saurexallopus* are known only from footprints

By Karen Carr from A.R. Fiorillo et al. 2018. Sci Rep 8, 11706. doi.org/10.1038/s41598-018-30110-8/CC BY 4.0

heads of †*Pachycephalosaurus* may have been employed in head-to-head shoving matches more like those of bison (**Figure 18.20C,D**).

The frills and horns of †ceratopsians were probably used in social behaviors, including male–male competition. Species of antelope with small horns engage in side-on displays with other males. Competing males swing their heads sideways against the flank of their opponent in a comparison of strength. †*Protoceratops* (see Figure 18.16B) and other early †ceratopsians may have used displays like this. Deer, elk, and moose have large antlers, and males engage with each other head-on, interlocking their antlers and twisting their necks as each individual attempts to knock the other off its feet (**Figure 18.21**). The sturdy horns of †ceratopsians would have been suitable for trials of strength of this sort. Some fossil †*Triceratops* have wounds on the face and frill that match the size and spacing of the large brow horns.

Moose have enormous antlers that consist mostly of flat surfaces. Rival males face each other head-on, twisting and shaking their heads to emphasize the breadth of their antlers. This comparison of antler size is often sufficient to determine dominance, and when moose do engage in trials of strength, they are back-and-forth shoving contests. Long-frilled †ceratopsians, such as †*Titanoceratops*

Figure 18.19 A family group of †hadrosaurs. Footprints at a fossil site in Alaska record the presence of hatchling, juvenile, and adult †hadrosaurs (†*Hadrosauropodus*) as well as a theropod (†*Saurexallopus*). Both are ichnospecies—that is, species named based on footprints rather than body fossils.

(A) Interpretation of head butting by †*Prenocephale*

From J.E. Peterson et al. 2013. *PLOS ONE* 8: e68620/CC BY

(C) Interpretation of head-to-head shoving of †*Pachycephalosaurus*

From J.E. Peterson et al. 2013. *PLOS ONE* 8: e68620/CC BY

(B) Head butting by bighorn sheep, *Ovis canadensis*

© Agnieszka Bacal/Shutterstock.com

(D) Head-to-head shoving of American bison, *Bison bison*

USFWS Midwest/CC BY 2.0

Figure 18.20 †Pachycephalosaurs probably used their heads in male–male combat.   (A,B) The thick, domed skulls of †*Prenocephale prenes* could have protected their brains during head-on impacts of the sort seen in today's bighorn sheep. (C,D) The broad heads and cephalic spines of adult male †*Pachycephalosaurus wyomingensis* may have been used for head-to-head shoving contests like those of extant American bison.

(A) Interpretive reconstruction of fighting †*Arrhinoceratops*

© Universal Images Group North America LLC/DeAgostini/Alamy Stock Photo

(B) Male-male combat in red deer, *Cervus elaphus*

Heather Smithers/CC BY-SA 2.0

Figure 18.21 Male †ceratopsians may have used their horns in combat.   (A) Males of derived species of †ceratopsians such as †*Arrhinoceratops* may have interlocked horns and attempted to twist their opponent's head. (B) Similar behavior is seen in male–male combat among extant deer and elk.

(see Figure 18.16D), may have engaged in similar sorts of displays.

The nasal and brow horns of †ceratopsians would have been formidable weapons and were probably used for defense against carnivorous dinosaurs. A herd of †ceratopsians might have used the same defense as some modern horned mammals, such as muskoxen, forming a defensive ring when a predator approached, with adults on the outside, facing the predator, and juveniles sheltered in the center of the ring.

The cranial ornamentation of †hadrosaurs was unsuited for combat. The enlarged nasal regions of †saurolophines (see Figure 18.17A–C) may have been surrounded by soft tissue that could be inflated by exhaled air, rather like the nasal regions of male elephant seals. The crests of †lambeosaurines were hollow (see Figure 18.17D–F). The curved crest of †*Parasaurolophus* extended back over the shoulders. Air in the hollow crest flowed from the external nares backward through the crest and then forward to the internal nares, located in the palate just anterior to the eyes.

The hollow crests of †lambeosaurines, and possibly also the inflatable soft tissue of †saurolophines, were probably used for visual displays and for vocalizations. Analyses of the inner ears of †lambeosaurines indicate that they would have been able to hear the sound frequencies produced by the resonating columns of air in the nasal passages.

Nesting and parental care by †ornithischians

Inferences about parental care by dinosaurs rest on a firmer basis than inferences about their social behavior. Not only do we have observations of parental care by the extant bracketing groups—crocodylians and birds—but thousands of nests of non-avian dinosaurs have been discovered.

†Ornithischians laid eggs in excavations that might have been filled with rotting vegetation to provide both heat and moisture for the eggs. Similar methods of egg incubation are used by crocodylians and by the brush turkeys of Australia; in contrast, extant birds other than brush turkeys brood their eggs with metabolic heat (see Section 21.9).

Some †ornithischians probably provided an extended period of parental care. A nest colony of the †hadrosaur †*Maiasaura* (Greek *maia*, "good mother") from the Late Cretaceous of Montana contained 15 juveniles about 1 m long (approximately twice the size of hatchlings found in the same area), indicating that the group remained together after they hatched. Teeth of the juvenile †hadrosaurs showed that they had been feeding; some teeth were worn down to one-fourth of their original length. It seems likely that a parent remained with the young. Discovery of a single adult †*Psittacosaurus* with 34 young suggests that juveniles from several clutches may have been assembled in a **crèche**, a group of young animals accompanied by one or more adults. This form of parental care also occurs in many species of crocodylians and birds such as ducks, penguins, and ostriches.

18.5 †Sauropodomorpha

LEARNING OBJECTIVES

18.5.1 Compare the diversity of †ornithischians and †sauropods.

18.5.2 Describe possible functions and physiological problems of having an exceedingly long neck.

18.5.3 Compare and critically assess evidence for herding behavior in †ornithischians and †sauropodomorphs.

†Sauropodomorphs were herbivorous and include gigantic, long-necked, and long-tailed dinosaurs that are centerpieces of paleontology exhibits at many museums. These are the derived forms, however, and small †sauropodomorphs from the Triassic and Early Jurassic were bipedal (see Figure 18.1).

Early †sauropodomorphs had long necks, small heads, and well developed claws on both the fore- and hindfeet. For example, †*Plateosaurus engelhardti* had 10 cervical vertebrae, 15 trunk vertebrae, 3 sacral vertebrae, and about 46 caudal vertebrae. Their long necks suggest that they could browse on plant material up to several meters above the ground, perhaps using their hands and claws to pull branches within reach. The ability to reach tall plants might have been an advantage during the floral shift from low-growing Triassic plants to taller, frond-bearing cycad allies and conifers that occurred in the Late Triassic.

†Sauropods evolved during the Early Jurassic and were highly successful until the end of the Cretaceous. †Sauropods with body lengths >30 m and body masses of ~50,000 kg evolved independently in different lineages, and

Body proportions of a †diplodocoid, †*Barosaurus*

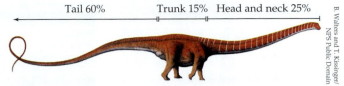

Tail 60%   Trunk 15%   Head and neck 25%

Figure 18.22 †Diplodocoids were mostly neck and tail. An adult †*Barosaurus* was about 23 m long, and 60% of that was the tail.

repeatedly. Among these animals are the largest terrestrial tetrapods that ever existed (for comparison, a large African bush elephant is about 5 m long and weighs 4,000–5,000 kg).

In addition to being enormous, derived †sauropods had extraordinarily long necks and tails (**Figure 18.22**). Their long necks resulted both from lengthening of the cervical vertebrae and from increasing their number, which rose from 10 in early †sauropodomorphs to 12 and eventually 19 in some derived †sauropods. Some of the added cervical vertebrae were modified trunk vertebrae, and shifts in the expression of Hox genes that control the regionalization of the vertebral column were probably the mechanism for this change.

Long necks increased the space in which a †sauropod could feed, by enabling the animal either to reach higher in trees or to swing in a wider arc and gather low-growing vegetation. An alternative (or additional) function of the long necks is based on analogy with giraffes. Male giraffes use their necks in contests with other males, either trying to unbalance each other or swinging their heads to deliver blows (**Figure 18.23A**). Necking duels can last more than 30 minutes and can result in injury or even death. The "necks for sex" hypothesis proposes that male †sauropods engaged in similar behaviors (**Figure 18.23B**). Features of the anatomy of the neck of some †diplodocoids suggest that the neck was crashed down or sideways against an opponent's neck.

The tails of derived †sauropods were even longer than their necks; tails of †diplodocoids contained up to 80 vertebrae and were 10 m long. Multiple functions for such long tails have been suggested. Clearly, the tail acted as a counterbalance when a sauropod kept its long neck stretched out horizontally, which is a pose proposed for most known species. Additionally, the long tail could have been a brace that helped a sauropod to stand bipedally to reach food, and perhaps in courtship displays. The tail might also have been used for defense, as seen in some large extant lizards (**Figure 18.23C,D**). A mechanical analysis of tail-whipping in the †diplodocoid †*Apatosaurus* indicated that it could have produced a sonic boom just as a bullwhip does, except that the dinosaur's boom would have been 200 decibels—loud enough to burst a human eardrum. Another hypothesis suggests that the long tails of †sauropods could have had tactile sensory functions, allowing individuals in a herd to determine the positions of animals behind them.

†Sauropods had remarkably small heads in proportion to their body sizes. Their relatively simple teeth were specialized for cropping vegetation, but not for chewing.

(A) Giraffes necking

Bjørn Christian Tørrissen/CC BY-SA 4.0

(B) †*Brontosaurus* necking

© Mark Witton

(D) †*Barosaurus* fending off two †*Allosaurus*

Fred Wierum/CC BY-SA 4.0

(C) A varanid lizard fending off a leopard

© LatestSightings/Photo by Costa Frangeskides

Figure 18.23 Using long necks and tails. (A,B) The necking combat of male giraffes is the basis for the hypothesis that males of long-necked †sauropods competed for dominance in this manner. (C,D) Some extant lizards whip their tails at predators, and the whiplike tails of †diplodocids might have had the same defensive function.

Early forms had interlocking, spoonlike teeth; more derived forms had peglike teeth with no occlusal contact. It is still a mystery as to how they used their peglike dentition and what kind of plant food they ate, but the fast dental replacement rate (one tooth lasted less than a month in the most extreme cases) suggests that their food was highly abrasive.

Digestion probably proceeded in two steps. **Gastroliths** (stomach stones) pulverized the food mechanically, as do gizzard stones in extant birds. Next, the food was probably digested by enzymes in large gastrointestinal tracts, which may have been multichambered, as in ruminant mammals. The estimated lengths of the gastrointestinal tracts of †sauropods would have allowed slow passage, giving symbiotic microorganisms time to ferment the food.

The relative lengths of the forelimbs and hindlimbs are often used to infer feeding height. For example, †macronarians had longer forelimbs than hindlimbs and are thought to have been high-level browsers (**Figure 18.24A**).

In contrast, †diplodocoids had shorter forelimbs than hindlimbs and are inferred to have been ground-level browsers (**Figure 18.24B**).

Fossil trackways of †sauropods are preserved in many parts of the world and reveal the immense sizes of these animals (**Figure 18.25**). Fossilized tracks show that legs were straight under the body and tails were held horizontally above ground. The limbs were held in an elephantlike pose and moved fore and aft parallel to the midline of the body—the straight-legged locomotion familiar in elephants. In early forms, the left and right feet were only a single foot-width apart. An increase in the width of tracks in the Jurassic and the Cretaceous coincides with the evolution of †Titanosauriformes, a clade characterized by wider hips.

Social behavior of †sauropods

†Sauropods lacked frills and other sexually dimorphic display structures of the sort seen among †ornithischians, but that does not mean that social behavior was

(A) †Macronarian

Large nostril

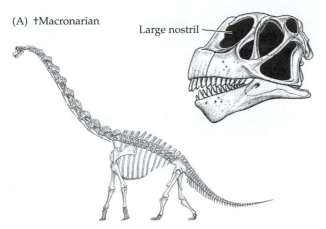

(B) †Diplodocoid

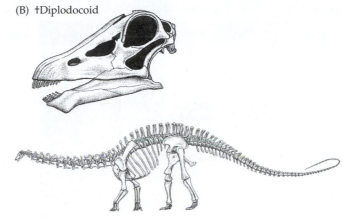

Figure 18.24 †Macronarians and †diplodocoids differed in body proportions and skull shapes. (A) †Macronarians had larger nostrils and relatively longer forelegs than †diplodocoids. The forelimbs of some †macronarians were longer than the hindlimbs, so their backs sloped downward from the shoulders. With their extremely long necks, they may have browsed the treetops. Skulls of early †macronarians were boxy, and spatulate teeth extended the length of the jaws. (B) †Diplodocoids had short forelegs, so their backs sloped down from the hips to the shoulders. They may have fed on large patches of low-growing vegetation while staying in one place by swinging the head in an arc. †Diplodocoids had long, flat snouts, peglike teeth in the front of the jaws, and long tails with a whiplike tip. (From D.E. Fastovsky and D.B. Weishampel. 2005. *Evolution and Extinction of the Dinosaurs*, 2nd ed., John Sibbick, illustrator. © Cambridge University Press 1996, 2005.)

absent; after all, extant crocodylians lack sexually dimorphic ornaments, yet they have an extensive repertoire of social behaviors. Still, the evidence for sociality by †sauropods is so sparse that we can guess that †sauropods had less extensive social interactions than †ornithischians did.

Analyses of oxygen isotope ratios in the teeth of †*Camarasaurus* suggest that these †sauropods made seasonal migrations of several hundred kilometers between upland and lowland environments. We cannot tell whether they traveled as separate individuals moving along the same route at the same time of year, or as a cohesive group containing both adults and juveniles. Identification of the age and species of dinosaurs that made a particular set of tracks can be difficult, and some bonebeds that initially were thought to contain solely juvenile specimens of a single species have since been shown to be composed of a mix of large juveniles and small adults, possibly including two different species.

Trackways from localities around the world show several individual †sauropods moving parallel to each other. The footprints are often of similar size, possibly indicating that these groups consisted of individuals of the same age. Some fossil sites preserve multiple individuals in a restricted size range of a single species of dinosaur, also suggesting segregation by age. Other trackways include prints of large and small individuals, however, and some herds may have included adults and juveniles (**Figure 18.26**).

(A) Early Cretaceous footprints of a †sauropod

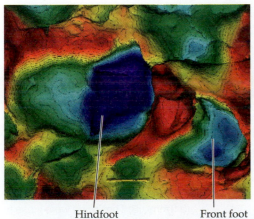

Hindfoot   Front foot

(B) Goolarabooloo Maja Richard Hunter is 1.8 m tall

Figure 18.25 †Sauropod footprints from Early Cretaceous sandstone. Trackways in rocks along the coast of Western Australia, north of the town of Broome, have footprints of 16 different species of dinosaurs, including †thyreophorans, †ornithopods, †sauropods, and theropods. Footprints shown here were made by "Broome †sauropod morphotype A" and are among the largest ever found. (A) Digital analysis uses colors to reveal the depth of the imprint of a hindfoot and a front foot; the depth of the darkest blue areas is 30 cm. (B) Goolarabooloo Maja Richard Hunter, Traditional Custodian of this region, lying beside the print of a hindfoot.

A herd of †macronarians, †*Brachiosaurus*

Figure 18.26 A herd of †*Brachiosaurus*. The hypothesis that juvenile and adult †*Brachiosaurus* formed mixed groups as shown here is supported by some trackways. Other trackways indicate that †sauropods may have formed age-segregated herds.

Nesting and parental care by †sauropodomorphs

Concentrations of nests and eggs ascribed to †sauropodomorphs suggest that these animals had well-defined nesting grounds to which they returned year after year. A site in Argentina yielded fossils of the Early Jurassic †sauropodomorph †*Mussaurus patagonicus* ranging in age from 70-g neonates to 1,500-kg adults, providing the earliest evidence yet found of social cohesion and colonial nesting. Some of the skeletons were in groups of individuals of approximately the same size, suggesting age segregation within the herd.

A nest of the Early Jurassic †sauropodomorph †*Massospondylus carinatus* in South Africa hints at parental care. The embryos in the eggs in this nest appear to be close to hatching, but the ventral portions of the pelvic girdles were poorly developed, the heads were enormous in relation to the bodies, and teeth were virtually absent (**Figure 18.27**). This combination of altricial characters would have made it difficult for the hatchlings to move about or feed themselves and supports the inference that adults of this species might have cared for their young.

With the evolution of gigantism in †sauropods, parental care was probably abandoned. An early juvenile of the †titanosaur †*Rapetosaurus krausei* from Madagascar would have weighed about 3.4 kg, while adults weighed 5,000 kg or more. Unlike †*Massospondylus* hatchlings, the baby †*Rapetosaurus* was precocial, with body proportions like those of an adult, indicating that it was self-sufficient.

Additional evidence suggesting that derived †sauropods did not provide parental care comes from a Late Cretaceous fossil site in Patagonia. Thousands of individuals of an unidentified species of †titanosaur journeyed here to construct nests, and the density of nests reached 11 eggs per m². Considering the size of an adult †titanosaur and the dense spacing of the clutches, any loitering by an adult would have been more likely to crush eggs than to protect them.

(A) A clutch of seven eggs of †*Massospondylus*

(B) Developing †*Massospondylus* in its eggshell

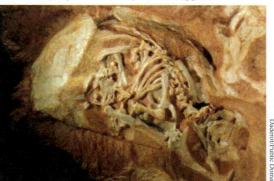

(C) Hatchling and adult †*Massospondylus*

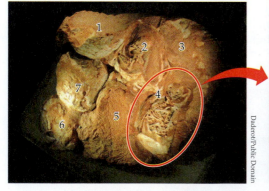

Figure 18.27 Eggs and embryos of †*Massospondylus*. (A) This clutch of seven eggs was laid about 190 Ma and it contains the oldest dinosaur embryos yet found. The eggshells were flexible and very thin (0.1 mm). (B) The young were close to hatching. Their heads were large and the jaws toothless. (C) Forelimb and hindlimb lengths of hatchlings are the same, indicating that they would have been quadrupedal, whereas adult †*Massospondylus* were bipedal.

Summary

18.1 Characters and Systematics of Avemetatarsalia

Early Avemetatarsalia, represented by forms such as †*Teleocrater*, appeared in the middle Triassic. During the late Triassic, and throughout the remainder of the Mesozoic, Avemetatarsalia radiated to become the most abundant, diverse, and ecologically important Mesozoic lineages of tetrapods, including †pterosaurs and dinosaurs. Its included taxon Aves is the most speciose major group of extant tetrapods.

Avemetatarsalians decoupled the functions of the fore and hindlimbs, and this allowed the subsequent evolution of bipedalism and flight.

Ornithodira includes †Pterosauromorpha and Dinosauromorpha. Dinosaurs appeared in the Triassic but did not rise to dominance until the latest Triassic or earliest Jurassic.

During much of the Triassic, avemetatarsalians coexisted with pseudosuchians and synapsids. By the end of the Triassic, carnivorous dinosaurs started to replace pseudosuchian carnivores, and herbivorous dinosaurs had largely replaced herbivorous synapsids.

18.2 †Pterosaurs: Vertebrates Achieve Powered Flight

†Pterosaurs appeared in the Late Triassic, 80 million years before the earliest birds, and persisted until the end of the Cretaceous. They radiated into all adaptive zones occupied by extant birds, as well as into some zones that are not occupied by any extant vertebrates.

Giant †azhdarchids, mostly terrestrial predators, may have been the most remarkable †pterosaurs. Some were as tall as a giraffe, and the largest of them could have preyed on dinosaurs.

All †pterosaurs could fly, probably using a four-legged leap to get into the air. Tracks show that †pterosaurs walked quadrupedally.

The primary wings of †pterosaurs were formed by skin supported by the arm bones and an elongated fourth finger. Some †pterosaurs had a posterior wing membrane between the pelvis and the hind legs.

The skin of at least some †pterosaurs was covered by fine fuzz called pycnofibers.

†Pterosaurs were oviparous, and hatchlings were probably able to run and fly soon after leaving the nest.

Birds and †pterosaurs coexisted for about 100 million years. The origin and diversification of birds coincided with an increase in body size of †pterosaurs, suggesting that birds may have outcompeted †pterosaurs at small body sizes.

18.3 Dinosaurs: One of the Most Successful Tetrapod Radiations

During the Mesozoic, terrestrial archosaurs (pseudosuchians, †pterosaurs, and non-avian dinosaurs) filled the adaptive zones now occupied by mammals and birds. Some dinosaurs, such as the enormous †ornithopods and †sauropods have no extant equivalents.

Increases in body sizes of dinosaurs were possible because of changes in the pelvis and associated musculature.

- Large tetrapods hold their limbs vertically beneath the body because bone is more resistant to compressive forces than to shearing forces.

- As the limbs of dinosaurs moved to a vertical position, the origin of muscles that protract the femur shifted to maintain stride length.

- Microstructural features of the trabeculated limb bones may have allowed dinosaurs to reach large body sizes.

There are two major clades of dinosaurs, †Ornithischia and Saurischia. Saurischia includes †Sauropodomorpha and Theropoda.

18.4 †Ornithischia

†Ornithischians radiated into many body forms. The lineage included bipedal and quadrupedal species; almost all were herbivorous with horny beaks. Two lineages independently evolved tooth batteries to chew their food.

†Thyreophorans (armor-bearing dinosaurs: †stegosaurs and †ankylosaurs) had rows of osteoderms on the back.

- Osteoderms were elaborated into leaf-shaped plates or spines in †stegosaurs.

- Some †ankylosaurs had spines, but they probably relied mostly on thick osteoderms for protection.

†Marginocephalians include the horned dinosaurs (†ceratopsians) and the thick-headed †pachycephalosaurs.

- Although basal †ceratopsians were bipedal, derived forms were quadrupedal. Derived †ceratopsians had a bony frill that extended backward from the skull and horns that, like the horns and antlers of extant mammals, were probably used for defense and in social interactions.

- †Pachycephalosaurs remained bipedal; their distinctive feature was a thick skull roof that was probably used during male-male combat.

†Ornithopods appeared in the Middle Jurassic. Small species were largely bipedal, but the largest species were

(Continued)

Summary *(continued)*

probably mostly quadrupedal, although they may have run bipedally to escape predators.

†Hadrosaurs (duck-billed dinosaurs) were the most speciose group of †ornithopods. The largest species reached lengths of more than 15 m and weighed more than 13,000 kg. They had dental batteries formed by thousands of teeth packed together to form large chewing surfaces.

Many species of †ornithischians probably formed herds. Social interactions probably included male-male competition for access to mates and territories. The plates and spines of †stegosaurs, the frills and horns of †ceratopsians, and the crests of †lambeosaurines were probably visual or auditory signals in these interactions.

Hundreds of fossilized †ornithischian nests have been discovered containing eggs, embryos, hatchlings, or older juveniles. It appears that some †ornithischians attended their nests, and hatchlings of some species probably remained with a parent for extended periods.

18.5 †Sauropodomorpha

Derived †sauropods include the largest tetrapods that ever existed.

Tooth morphology, snout shape, the flexibility of the neck, and relative lengths of the limbs indicate diverging feeding strategies in different clades of †sauropods.

†Diplodocoids with their peglike teeth, wide snout, horizontal neck, and short forelimbs were probably ground-level grazers, whereas some †macronarians had more spatulate teeth, rounded snouts, flexible necks, and long forelimbs and were probably high-level browsers.

Some †sauropods may have formed age-segregated herds, but some trackways include footprints of both large and small individuals.

At least some †sauropods had nesting grounds used by generation after generation. Basal †sauropodomorphs may have provided parental care, but the offspring of ††titanosauriforms were probably precocial and self-sufficient.

Discussion Questions

18.1 Extensive dinosaur faunas inhabited Arctic regions well beyond the polar circle. We know that northern regions were warmer during the Mesozoic than they are now; clearly, they were warm enough for dinosaurs. But Arctic winters would have had long periods of dim light (twilight) in the Mesozoic, as they do now. How might dinosaurs have responded to the seasonal changes in day length in the Arctic?

18.2 Compare the use of inferences about the ecology, behavior, and physiology of extinct animals based on extant phylogenetic brackets with those based on analogies with extant forms—for example, the inferences in this chapter about the social behaviors of †marginocephalians that are based on analogies with modern herbivorous mammals with horns or antlers. How might you test your conclusions in each case?

18.3 The necks of giraffes are the closest extant equivalent to the necks of †sauropod dinosaurs. Two hypotheses have been proposed to explain the functional significance of the necks of giraffes, and those hypotheses can be applied to the necks of †sauropods.

a. Competing browsers hypothesis: The long neck makes more food resources available to a giraffe or †sauropod by allowing it to reach up to browse on vegetation that is not available to short-necked browsers.

b. "Necks for sex" hypothesis: Male giraffes and †sauropods use their necks and heads to batter opposing males during intraspecific combat. Thus, long necks increase the fitness of males by increasing their opportunities to mate.

c. What predictions can you make about the relative lengths of the necks of male and female †sauropods to test these hypotheses?

Figure credits
Figure 18.1: Silhouettes based on *Teleocrater, Kongonaphon*: Fanboyphilosopher (Neil Pezzoni)/CC BY 4.0; *Lagerpeton, Sacisaurus*: Nobu Tamura/CC BY-SA 4.0; Pterodactyloidea, Ceratopsia: Nobu Tamura/CC BY-SA 3.0; *Rhamphorhynchus*: Dmitry Bogdanov/CC BY-SA 3.0; *Lagosuchus*: Michael B.H. FunkMonk/CC BY-SA 3.0; *Lesothosaurus*: Jack Wood/CC BY 4.0; Stegosauria, Ankylosauria: © James Kuether; Pachycephalosauria: Megaraptor-The-Allo/ CC0 1.0; Ornithopoda: Marmelad/CC BY-SA 2.5; *Plateosaurus*: Leandra Walters, Phil Senter, James H. Robins/CC BY 2.5; Macronaria: Michael P. Taylor/CC0 1.0; Diplodocoidea: Fred Wierum/CC BY 4.0; Theropoda: Steve OC/CC BY-SA 4.0.
Figure 18.8: Alligator, *Lagosuchus, Carnotaurus, Archaeopteryx*: J. R. Hutchinson. 2002. *Comp Biochem Physiol Part A Mol Integr Physiol* 133: 1051–1086; *Teleocrater*: S.J. Nesbitt et al. 2017. *Nature* 544: 484-487; *Rhamphorhynchus*: E.S. Hyder et al. 2014. *Acta Palaeontologica Polonica* 59: 109–124/CC BY; *Plateosaurus*: J.R. Hutchinson. 2001. *Zool J Linn Soc* 131: 123–168.
Figure 18.13: *Lesothosaurus*: Nobu Tamura/CC BY-SA 4.0; *Scutellosaurus, Stegosaurus, Kentrosaurus, Borealopelta, Titanoceratops, Corythosaurus*: Nobu Tamura/CC BY-SA 3.0; *Kritosaurus*: Nobu Tamura/CC BY 2.5; *Euoplocephalus*: Nobu Tamura/CC BY 3.0; *Pachycephalosaurus*: Dmitry Bogdanov/CC BY 4.0; *Psittacosaurus*: From J. Vinther et al. 2016. *Curr Biol* 26: P2456-2462/CC BY 4.0; *Protoceratops*: PaleoNeolitic/CC BY 4.0; *Parasaurolophus*: Ryanz720/Public Domain.

Theropods and the Origin of Birds

†*Yi qi*, a basal paravian

This chapter begins with a review of characters and phylogenetic relationships of Mesozoic theropods and descriptions of the major clades. Then, beginning with a description of †*Archaeopteryx*—one of the most famous and important fossil vertebrates ever discovered—we describe the origins of powered flight in Avialae, the theropod lineage that gave rise to birds. We conclude with a discussion of reproduction and parental care among theropods, a subject that has attracted much interest in recent decades.

The earliest members of Theropoda (Greek *therion*, "wild beast"; *pous*, "foot") are known from the Late Triassic of South America, ~230 Ma. Early theropods were small, agile, lightly built, bipedal carnivores with long arms and legs. Carnivory is a theme within Theropoda, and hypercarnivores (animals with a diet of >70% meat) evolved repeatedly within the group. Theropoda includes many iconic Mesozoic dinosaurs, such as †*Tyrannosaurus rex* and †*Velociraptor*, as well as Aves, a group that originated in the Cretaceous and includes about 11,000 extant species of birds.

Theropod evolution has been the subject of intensive research for more than a century. †*Tyrannosaurus* was described in 1905 and †*Velociraptor* in 1924; today the rate of theropod discoveries is at an all-time high, with dozens of new forms found every year. Theropods such as †*Deinonychus*—a 3-m predator from the Cretaceous of Montana, famous for its sickle-shaped claws on the hindfeet described by John Ostrom in 1969—played leading roles in the dinosaur renaissance, a period of renewed scientific and popular interest in dinosaur diversity and biology that continues to this day, with widespread interest in discoveries and new interpretations of how dinosaurs lived.

It is important not to generalize too much based on the anatomy of a single extinct theropod. For example, †*Tyrannosaurus* had small forelimbs and is a familiar theropod, so it is natural to think of theropods as having tiny forelimbs. But this is not true for theropods in general, and we will see many forms with long and greatly enlarged forelimbs as we trace their evolutionary history and the origin of powered flight in avialans.

19.1 Characters and Systematics of Theropods

LEARNING OBJECTIVES

19.1.1 Describe anatomical characters of Late Triassic theropods and how they relate to their biology.

19.1.2 Review the broad evolutionary history of large theropods during the Mesozoic by placing †*Ceratosaurus*, †*Spinosaurus*, †*Allosaurus*, and †*Tyrannosaurus* in their phylogenetic and geological contexts.

Over the last 230 Ma, theropods evolved specialized features and body shapes, and species ranged from tiny bipedal cursors to giant terrestrial hypercarnivores. Therefore, it can be difficult to identify characters that can reliably be seen in every theropod. Bipedality is a general feature, and a condition that predates the origin of Theropoda. Most theropods walk on three-toed hind feet, with the first and fifth digit partially (in the case of digit I) or completely (in many cases of digit V) reduced. Like †pterosaurs and †sauropodomorphs, theropods have an extensive air sac system that invades the vertebral column (mostly cervical and dorsal vertebrae, but in some groups also sacral and caudal vertebrae), as in birds. The hands of non-avian theropods have four or fewer fingers (typically three) held in the same position as human hands, with the palms inward. The hands of many Mesozoic theropods were large and used for grasping, although their size was independently reduced in many groups.

(A) Lightly built skull of †*Velociraptor* has fenestrae and strutlike bones

Nostril

Maxillary fenestra

Antorbital fenestra

Orbit

Upper temporal fenestra

Ziphodont teeth

Mandibular fenestra

Lower temporal fenestra

© William E. Bemis

(B) Using puncture-and-pull feeding to dismember prey

Illustration by Sydney Mohr from A. Torices et al. 2018 *Curr Biol* 28: 1467–1474

Figure 19.1 Skull and lower jaw of a theropod.   (A) †*Velociraptor mongoliensis* had a long skull with large temporal fenestrae to accommodate large jaw closing muscles; its teeth were ziphodont (knifelike), laterally compressed, and serrated. (B) These anatomical features allowed puncture-and-pull feeding, first biting downward and then pulling away so the teeth could tear through the flesh of its victim.

Other characters important to understanding the biology of theropods relate to the skull. For example, †*Velociraptor* has a lightly built skull with strutlike bones and large temporal fenestrae that accommodate large jaw closing muscles (**Figure 19.1A**). Like many other theropods, its skull is narrow, with long jaws suited for fast closing on prey. It has laterally compressed, serrated teeth described as **ziphodont** (knifelike) arranged in a single row along the tooth-bearing bones. Like other archosaurs, its teeth were anchored in sockets in the bones and were replaced throughout life by new teeth that erupted from beneath. †*Velociraptor* used **puncture-and-pull feeding**, first biting downwards to puncture the flesh, then pulling away so that the ziphodont teeth tear through and dismember the prey (**Figure 19.1B**).

Phylogenetic overview of Theropoda

The time tree in **Figure 19.2** traces the evolution of 19 theropod taxa during the Mesozoic. **Figure 19.3** shows their phylogenetic relationships, using selected taxa and characters to explore the history of their diversification. These 19 terminal taxa do not represent the total known diversity of Mesozoic theropods, but all major groups (the 9 clades named on the right of Figure 19.3) are included. The terminal taxon Aves includes the extant birds and will be covered in Chapter 21. The word "bird" was often used in the past for all avialans (as in the statement "†*Archaeopteryx* is a bird"), but we follow current phylogenetic practice and restrict the term "bird" to members of Aves.

Current phylogenetic concepts and classification of theropods owe much to pioneering studies by Jacques Gauthier in the 1980s. In a cladistic analysis of diapsid reptiles, he brought together the first formal phylogenetic evidence that birds are theropod dinosaurs, meaning that any concept of Reptilia that does not include birds is paraphyletic (see Section 1.4). Today, almost any child interested in dinosaurs knows that birds are dinosaurs, but that was not always the case.

At the base of the tree in Figures 19.2 and 19.3 is †*Herrerasaurus* from the Late Triassic of South America,

one of the earliest dinosaurs known. This species has been placed in several different positions within Dinosauria, and the discovery in 1988 of a complete skull set off renewed debates about its phylogenetic placement. †*Herrerasaurus* has ziphodont teeth and grasping hands, and we regard it as a close outgroup of Theropoda.

After an early, restricted radiation of relatively small and swift predators during the Triassic, theropods repeatedly occupied new ecological niches. The Early and Middle Jurassic saw a major radiation and the appearance of two large subclades of Theropoda: †Ceratosauria and Tetanurae (see Figures 19.2 and 19.3).

†Coelophysoids: Early theropods

†Coelophysoids were small, bipedal predators with an elongate skull and many serrated teeth (**Figure 19.4A**). †*Coelophysis bauri* is known from many relatively complete skeletons from the Ghost Ranch locality in New Mexico. Based on its teeth, we infer that its diet changed during ontogeny: individuals specialized on insects and fish while young and hunted terrestrial animals as adults. †*Coelophysis* has a **furcula** (wishbone), and here we interpret the presence of a furcula as a synapomorphy for Theropoda (see Figure 19.3).

The earliest †coelophysoids are from the Late Triassic of northern Pangaea. Later †coelophysoids reached southern Pangaea, but its last members became extinct in the Early Jurassic (see Figure 19.2). After the end-Triassic mass extinction, during which many crocodyliform and synapsid species vanished (see Section 10.4), and a smaller extinction event during the Early Jurassic, Theropoda radiated into two long-lived subclades, †Ceratosauria and Tetanurae (see Figure 19.2).

†Ceratosauria

†Ceratosaurs are generally known for their reduced forearms and short, deep skulls, but the earliest recognized form—†*Saltriovenator*, from the Early Jurassic of Italy—had well developed hands with four fingers. Forms such

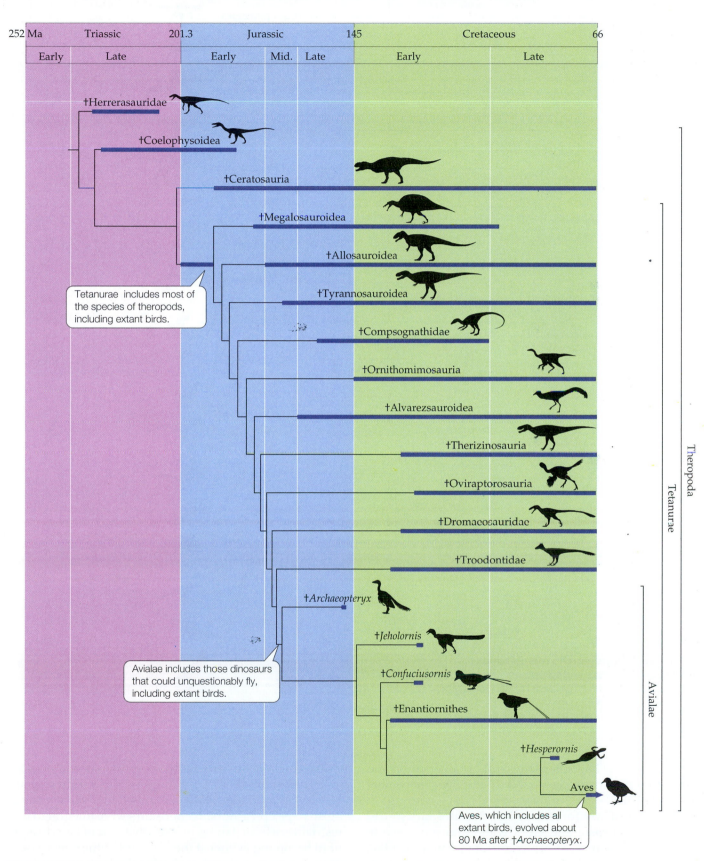

Figure 19.2 Time tree of Theropoda. The theropod lineage originated in the Triassic. Non-avian theropods reached peak diversity during the Cretaceous, with avialans represented today by the many extant species of Aves, the birds. (See credit details at the end of this chapter.)

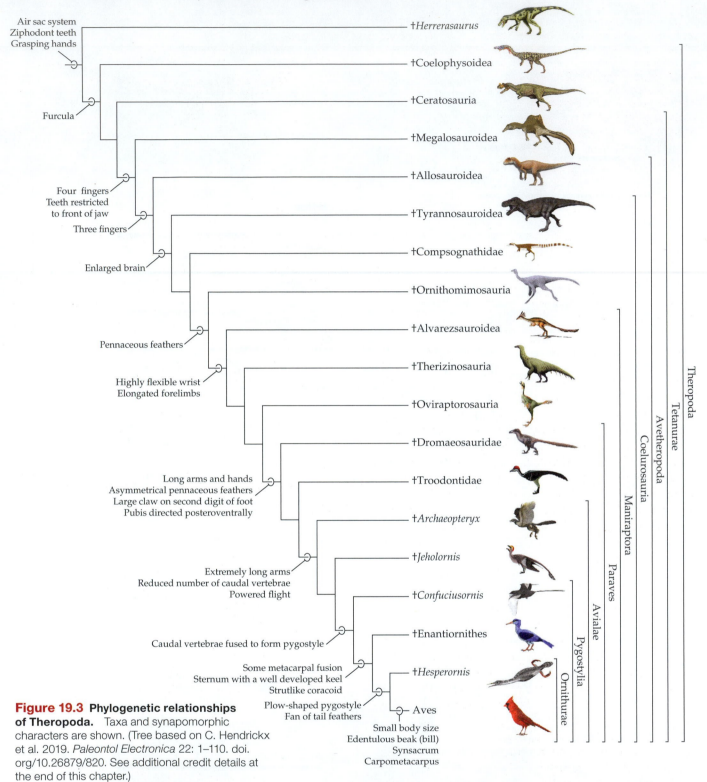

Figure 19.3 Phylogenetic relationships of Theropoda. Taxa and synapomorphic characters are shown. (Tree based on C. Hendrickx et al. 2019. *Paleontol Electronica* 22: 1–110. doi. org/10.26879/820. See additional credit details at the end of this chapter.)

as †*Ceratosaurus* had extensive skull ornamentation and horns that may have been used in intraspecific combat (**Figure 19.4B**).

After their Early Jurassic origin, †ceratosaurs spread over both Laurasia and Gondwana and gave rise to three main lineages, †Ceratosauridae, †Noasauridae, and †Abelisauridae. †Ceratosaurids were large and robust predators that existed until the end of the Early Cretaceous. Small, gracile †noasaurids such as †*Velocisaurus* (**Figure 19.4C**) and huge †abelisaurids such as †*Carnotaurus* (**Figure 19.4D**) continued to diversify. Being mostly apex predators, †ceratosaurs were integral to terrestrial vertebrate faunas (especially those of Gondwana) until becoming extinct at the very end of the Cretaceous. Recent interpretations emphasize the heavily cornified skin and sheaths of the horns (**Figure 19.4E**).

Figure 19.4 Basal theropod groups. (A) †*Coelophysis* was about the size of a turkey. Its furcula (wishbone), a synapomorphy of Theropoda, was about 3 cm wide. (B) †Ceratosauria, an early-branching group of theropods that existed until the end of the Cretaceous, is exemplified by the Late Jurassic †*Ceratosaurus nasicornis*. Note the elaborate skull ornamentation covered with cornified skin. (C) †*Velocisaurus* (not to be confused with †*Velociraptor*) was a small, cursorial †noasaurid. (D) In the Late Cretaceous of Gondwana, †abelisaurids such as †*Carnotaurus* were ecological equivalents of distantly related †tyrannosaurs in the Northern Hemisphere. (E) Restoration of keratinized skin and horns on the head of †*Carnotaurus*.

Tetanurae

Tetanurae (Greek *tetanos*, "tension, stiffness"; *oura*, "tail") includes all theropods that are more closely related to extant birds than to †ceratosaurs. The many lineages within Tetanurae contain imposing predatory species that are icons for dinosaur enthusiasts (for example, †*Tyrannosaurus* and †*Velociraptor*, stars of the novel and movie *Jurassic Park*).

The earliest tetanurans appeared very early in the Jurassic. By the Middle Jurassic the group had become globally distributed, and tetanurans continued to diversify throughout the Cretaceous (see Figure 19.2). Although hypercarnivores have seized the public imagination, some tetanurans ate fish (†*Spinosaurus*), plants (†ornithomimosaurs), or insects (†alvarezsauroids). Gigantism (see Section 14.6) evolved independently in at least five lineages (†megalosauroids, †allosauroids, †tyrannosauroids, †ornithomimosaurs, and †therizinosaurs), while the lineage leading to birds trended toward miniaturization. Thus, Tetanurae includes both the largest and smallest theropods. The largest include extinct forms such as †*Tyrannosaurus*; the smallest is the extant bumblebee hummingbird (see Figure 14.16B).

†Megalosauroids †Megalosauroidea ("big lizards") includes †*Megalosaurus bucklandii* from the Middle Jurassic of England, one of the first dinosaurs to be named (in 1824), and it influenced early concepts of Dinosauria as a group. As its name suggests, †*M. bucklandii* was large, although a complete specimen of this species has never been found.

Perhaps the most famous †megalosauroid is †*Spinosaurus*, which reached truly gigantic body sizes (up to 15 m). †*Spinosaurus* ("spine lizard") is unusual among theropods because it was aquatic. With a snout and teeth like those of extant fish-eating crocodylians, it probably was piscivorous (**Figure 19.5A**). Its limb bones were relatively short with a thick cortex, a feature typical of extant aquatic tetrapods. The tail vertebrae had long, narrow, spinous processes supporting a tailfin that probably contributed the main propulsive force while swimming in Late Cretaceous rivers of North Africa. Additionally, †*Spinosaurus* had elongated dorsal neural spines that formed a dorsal sail of unknown function.

Although †*Spinosaurus* belongs to one of the basalmost tetanuran lineages, its oldest known fossils are from the

(A) Bipedalism, grasping hands, and furcula of a basal theropod, †*Coelophysis*

Shape of furcula

Position of furcula

Grasping hands

1 m

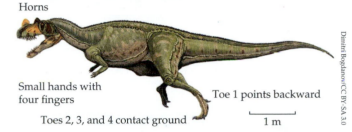

(B) Horns on the head of †*Ceratosaurus*

Horns

Small hands with four fingers

Toes 2, 3, and 4 contact ground

Toe 1 points backward

1 m

(C) †*Velocisaurus* was a small and lightly built †noasaurid.

1 m

(D) The Late Cretaceous †abelisaurid †*Carnotaurus* from South America was ecologically equivalent to †*Tyrannosaurus* in North America

1 m

(E) Detail of the horns and face of †*Carnotaurus*

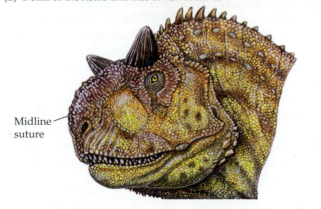

Midline suture

(A) †*Spinosaurus* was a giant aquatic †megalosauroid

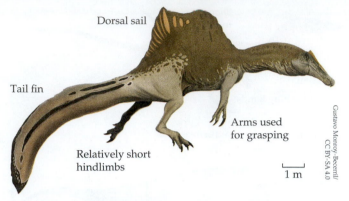

Dorsal sail

Tail fin

Arms used
for grasping

Relatively short
hindlimbs

1 m

Gustavo Monroy-Becerril/
CC BY-SA 4.0

Mike Bowler/CC BY 2.0

†*Spinosaurus* had long jaws
and teeth like a piscivorous
crocodylian.

(B) †*Allosaurus* lived in the Late Jurassic

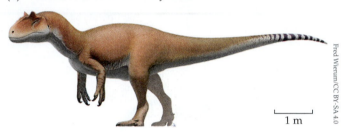

Fred Wierum/CC BY-SA 4.0

1 m

Figure 19.5 **†Megalosauroids and †allosauroids.**
(A) Reflecting its aquatic lifestyle, †*Spinosaurus* had relatively short limbs, a tailfin, and long jaws with teeth like those of extant fish-eating crocodiles. Function of the dorsal sail is unknown. (B) †*Allosaurus* was a large-bodied apex predator with three fingers on each hand, slicing teeth, and two crests running along the dorsolateral edges of the skull.

Late Jurassic, much later than the split between †Spinosauridae and other tetanurans must have occurred. Perhaps their derived morphology and specializations for piscivory make it difficult to identify earlier forms of †Spinosauridae. Although the group was relatively species-poor, its members were widespread throughout Laurasia and Gondwana during the Early Cretaceous before becoming extinct in the earliest Late Cretaceous.

†Allosauroids   †Allosauroidea includes the apex predators †*Allosaurus* (**Figure 19.5B**) and †*Carcharodontosaurus*. These large-bodied animals (the latter reached estimated lengths of almost 13 m and weights of ~8 metric tons) had well developed three-fingered hands, slicing teeth, and crests along the dorsolateral edges of the skull.

There are three main families of †Allosauroidea: †Metriocanthosauridae, †Allosauridae, and †Carcharodontosauridae. †Metriocanthosaurids evolved during the Middle Jurassic in central and eastern Laurasia, and they remained restricted to this geographical area until their extinction in the second half of the Early Cretaceous. †*Allosaurus* became abundant during the Late Jurassic in North America and Europe. †Carcharodontosaurids appeared in the Late Jurassic of Europe, where they co-existed with †allosaurids; by the Early Cretaceous †carcharodontosaurids were the dominant carnivores of the Southern Hemisphere. They diminished in diversity during the Late Cretaceous, and toward the end of the Mesozoic were replaced as apex predators by large-bodied †abelisaurid ceratosaurs.

†Tyrannosauroids   Early †tyrannosauroids were small and lightly built with long arms and legs. Except for

†*Yutyrannus*, a relatively large form from northern China, †tyrannosauroids remained relatively small throughout the Jurassic and Early Cretaceous. Later †tyrannosauroid evolution was marked by great increases in body size, a deepening of the skull, and larger jaw muscles. Much of our knowledge of the †tyrannosaurid skeleton comes from detailed study of a specimen at the Field Museum of Natural History in Chicago. Discovered in 1990 by Susan Hendrickson and nicknamed in her honor, "Sue" is the most complete †*Tyrannosaurus rex* yet discovered. Its sex is unknown, but it is estimated to have been 28 years old at its time of death.

The original mount for Sue was assembled in 2000 and displayed in a huge hall where visitors entering the museum could see it immediately (**Figure 19.6A**). In 2018, based on new discoveries about †*T. rex* biology, Sue's mount was redesigned and relocated to a smaller exhibit space (**Figure 19.6B**). It is interesting to compare the 2000 and 2018 mounts. Gastralia—the series of mobile bones that reinforced the ventral portion of its abdominal cavity and that played a role in breathing (see Section 13.2)—were added. The shape and position of the furcula were corrected; the bone mistakenly installed in the original mount was not the furcula, and it was long and relatively flat (see Figure 19.6A). The correct furcula, which is narrower and more V-shaped, moved the arms lower on the body.

The teeth of †tyrannosauroids were wide, long (up to at least 18 cm), and capable of cutting and crushing bone. The pelvis of a †*Triceratops* found in Montana bears dozens of bite marks from a †*Tyrannosaurus*, and the largest is 1 cm deep and 11.8 cm long, indicating puncture-and-pull feeding (see Figure 19.1B). But that was not the only feeding method †*T. rex* used, for it could also crush bone. We know this because coprolites (fossil feces) left by a †tyrannosaur

(A) Articulated mount of †*Tyrannosaurus rex* at Field Museum, 2000

Old furcula

Shape of old furcula

1 m

© Field Museum of Natural History, GN897.4_2RDc

(B) Articulated mount of †*Tyrannosaurus rex* at Field Museum, 2018

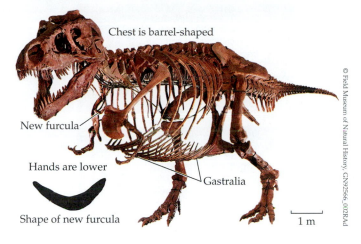

Chest is barrel-shaped

New furcula

Hands are lower

Gastralia

Shape of new furcula

1 m

© Field Museum of Natural History, GN92566_002RAd

Figure 19.6 †*Tyrannosaurus rex*. Two articulated restorations of the famous †*T. rex* specimen in Chicago's Field Museum, nicknamed "Sue" after its discoverer. (A) 2000 mount. (B) 2018 mount. The revised mount shows changes in anatomical interpretations based on new discoveries. (C) Skull of †*T. rex* showing the postorbital boss that may have been used in head-butting combat.

(C) Skull of †*Tyrannosaurus rex*

Orbit

Postorbital boss

Lower temporal fenestra

WehaveaTrex/CC BY-SA 4.0

in Canada contain fragments of crushed bone from a juvenile †ornithischian. Functional anatomical analyses show that large †tyrannosauroids could apply enormous bite forces to crack bones that they then pulverized by repeated bites. Only among synapsids did other amniotes evolve such specialized bone-crushing abilities.

Were †tyrannosauroids primarily predators or mostly scavengers? That distinction is not all-or-nothing; extant mammalian predators consume carrion when it is available, and probably theropods also were opportunistic. Powerful jaws and teeth are as functional for a scavenger as they are for a predator. However, healed tooth marks on the caudal neural spines of a †hadrosaur indicate that it was attacked by a large theropod, probably †*Tyrannosaurus rex*, showing that large theropods sometimes attacked living prey. Such attacks would have been leisurely: models that incorporate the length and strength of limb bones and the sizes of limb muscles indicate that adult †*Tyrannosaurus rex* could not run. It probably walked at ~4.6 km/h, with a maximum speed of 7.5 km/h.

The tiny forelegs of derived †tyrannosaurs are puzzling. Like †ceratosaurs, †tyrannosauroids evolved significantly shorter arms during their radiation, but unlike †ceratosaurs (which had four very short digits), †tyrannosauroids had only two functional fingers. Their arms were too short to reach the mouth, yet the arm bones were robust, and the fingers were tipped with large claws. These characters suggest that the forelegs of †tyrannosaurs had a function, but we do not know what that function was.

The earliest †tyrannosauroids are from the Middle Jurassic (see Figure 19.2). A mostly Asian and European lineage diversified first, but by the Late Jurassic, †tyrannosauroids also inhabited North America. However, they remained relatively low in diversity until the

second half of the Late Cretaceous, when they radiated and attained gigantic body sizes. This radiation marked a major faunal turnover in the Northern Hemisphere, with †tyrannosauroids replacing large †allosauroids (just as, in the Southern Hemisphere, †abelisaurid ceratosaurs replaced large †allosauroids).

†Compsognathids This clade of small Late Jurassic to Early Cretaceous theropods includes, among others, †*Compsognathus*, one of the first small theropods discovered; and the famous feathered specimen of †*Sinosauropteryx* (see Figure 18.12). The specimen of †*Compsognathus* shown in **Figure 19.7** was discovered in Bavaria in the 1850s. There is only one other reasonably complete specimen of †*Compsognathus*, but the two fossils have an outsize importance in the history of theropod paleontology because a 19th-century study drew attention to similarities between †*Compsognathus* and †*Archaeopteryx* (see Section 19.2).

†Ornithomimosaurs †Ornithomimosaurs ("bird-mimic lizards") were cursorial tetanurans of the Cretaceous (**Figure 19.8**). Despite their name, they are not closely related to birds (Aves). †*Ornithomimus* was ostrichlike in size, shape, and perhaps ecology, living in groups in open regions rather than forests. Its long legs had three

†*Compsognathus longipes* was described in 1859

The skull and lower jaw are disarticulated and fragmentary.

Remains of a small lizard, †*Schoenesmahl dyspepsia* were found in the gut.

5 cm

Figure 19.7 †*Compsognathus.* Discovered in the 1850s in Bavarian limestone, this specimen and those of †*Archaeopteryx* (from the same locality) played important roles in the 1860s concept that birds are glorified dinosaurs. That concept did not take hold among scientists until the 1980s, however, at which time phylogenetic systematic methods were used to trace character evolution in theropods.

(A) Skeleton of an †ornithomimosaur, †*Struthiomimus*

1 m

(B) Restoration of †*Ornithomimus*

Figure 19.8 †*Ornithomimosaurs.* (A) Reconstructed skeleton of †*Struthiomimus* from the Late Cretaceous of North America. (B) This restoration of the closely related †*Ornithomimus* is based in part on its convergently evolved similarities to extant ostriches.

weight-bearing toes, the long neck supported a small skull, and its toothless jaws were covered with a horny beak (bill). They retained three digits on the hands; the inner digit was opposable, making it effective for capturing small prey. †Ornithomimosaurs were feathered. The body had a covering of **plumulaceous feathers** (downy feathers lacking a vane), and there were long **pennaceous feathers** (with a central shaft and a vane on each side; see Figure 21.3) on the arms, legs, and tail. Like ostriches, †ornithomimosaurs were omnivores that fed on fruits, insects, small vertebrates, and eggs.

By the earliest Cretaceous, †ornithomimosaurs had spread across the globe, and they continued to diversify during the Middle and Late Cretaceous. By the end of the Mesozoic, several species of †ornithomimosaurs co-occurred in Mongolia and North America.

Maniraptorans   The clade Maniraptora (Latin *manus*, "hand"; *raptor*, "robber") includes the groups †Alvarez-sauroidea to Aves, and thus remains extant (see Figure 19.3). Early maniraptorans had long, slender legs and long arms with grasping fingers. The name Maniraptora refers to the highly flexible wrist joints characteristic of the group. The joint allows the hand to bend inward to rest against the ulna and radius when the front limb is folded (your wrist bends inward only ~90 degrees, but a maniraptoran wrist folds all the way down). This extreme mobility also permitted early maniraptorans to flex the wrists sideways while rotating the hands to seize prey. Birds use the same wrist motion to produce air flow over the primary flight feathers, which are on the portion of the wing corresponding to the hand, to create lift during powered flight (see Section 21.3).

The earliest maniraptorans were probably omnivorous, but herbivory evolved several times, as did insectivory, a reversal to hypercarnivory, and all the dietary specializations we know in birds. The early branching †alvarezsauroids ate wood-nesting termites, an inference based on their uniquely shortened forelimb and the reduction of the hand to a single clawed digit in derived taxa like †*Mononykus* from Mongolia (**Figure 19.9A**). †Therizinosaurs, exemplified by †*Therizinosaurus* (**Figure 19.9B**), were herbivorous, with meter-long claws on their hands, a beak, and a long neck. The enormous claws may have helped them to bend branches to reach leaves or fruits. †Oviraptorosaurs had edentulous (toothless) beaks and a shortened tail; some, such as the specimen of †*Caudipteryx* shown in **Figure 19.9C**, preserve gastroliths (stones in the gastrointestinal tract used to grind food), indicating that they were herbivores (plant material, notably cellulose, is harder to digest than animal matter).

The †dromaeosaurids †*Velociraptor* and †*Deinonychus* (Greek *dein*, "terrible"; *onych*, "claw") were hypercarnivores with ziphodont teeth, puncture-and-pull feeding, and a large claw on the second toes of each hindfoot (**Figure 19.10A**). Early interpretations proposed that †dromaeosaurids hunted in packs and used the claws on the feet to slash prey much larger than themselves (**Figure 19.10B**).

(A) An insectivorous †alvarezsaurid, †*Mononykus*

Each hand had a single claw

10 cm

PaleoNeolithic/CC BY 4.0

(B) The herbivorous †*Therizinosaurus* had huge claws on its hands

ABelov2014/CC BY 3.0

Woudloper/CC BY-SA 3.0

10 cm

(C) Herbivorous †oviraptorosaur, †*Caudipteryx*

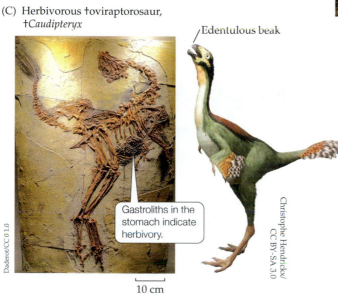

Edentulous beak

Gastroliths in the stomach indicate herbivory.

Daderot/CC 0 1.0

10 cm

Christophe Hendrickx/ CC BY-SA 3.0

Figure 19.9 †Alvarezsauroids, †Therizinosaurs, and †Oviraptorosaurs. (A) †*Mononykus* had a single claw on each hand, which is interpreted as a specialization for breaking into termite mounds. (B) Reaching a length of 10 m and weighing 3,000 kg, two †*Therizinosaurus* dwarf the †pachycephalosaurs (†*Prenocephale*, right front and left rear) and †dromaeosaurid (†*Adasaurus*, left front) also in this scene. The claws of †*Therizinosaurus* are the largest known among vertebrates, reaching lengths of 1 m. These animals were herbivorous and may have used their claws to pull branches and leaves within reach. (C) The herbivorous †oviraptorosaur †*Caudipteryx* had a toothless beak and gastroliths (stones in the gastrointestinal tract that grind food) in the region where extant birds have a gizzard.

Tests using models of the claws show that they are not effective at slicing, however, and probably were not used to slash prey. The current hypothesis draws on behaviors of extant falcons and hawks, which use enlarged talons on the second toe to hold prey as they tear it apart with powerful hooked bills. †Dromaeosaurids may similarly have used their claws and body weight to pin prey to the ground while tearing at it with their jaws. Their feathered arms might have helped them balance on struggling prey (**Figure 19.10C**).

Maniraptorans radiated and dispersed rapidly near the end of the Middle Jurassic. Every maniraptoran lineage shown in Figures 19.2 and 19.3 diversified throughout the Cretaceous, reaching their maximum species diversity towards the end of the Mesozoic. †Troodontidae, the sister group to Avialae, probably adapted to a variety of food sources, with †*Anchiornis* feeding on fish and meat, and others such as †*Jinfengopteryx* being herbivores. However, only one lineage, Aves, survived the end-Cretaceous mass extinction.

Community ecology of theropods

The community structure of most kinds of tetrapods includes carnivorous species with overlapping ranges in adult body size. In South Africa's Kruger National Park, for example, extant carnivorous mammals range in size

from small cats and foxes to large lions and species in between (**Figure 19.11A**). Mesozoic theropods, however, are an exception to that rule. Studies of theropod diversity in the Dinosaur Park Formation of Alberta, Canada have revealed small species (†dromaeosaurids, †troodontids, and †oviraptorosaurs) and very large species of †tyrannosauroids, but very few species that weighed between 100 and 1,000 kg as adults (**Figure 19.11B**). This pattern is consistent in all parts of the world and throughout the Jurassic and Cretaceous. (In contrast to theropod communities, †ornithischian communities included a full range of body sizes.)

This unusual bimodal community structure probably results from three characteristics of giant theropods that are seen most clearly in †tyrannosaurs:

- *Enormous ontogenetic growth.* Oviparity limited the maximum size of hatchling theropods to about 15 kg, even for †*T. rex*, which weighed 6,000–9,000 kg as adults. They reached adult size at about 20 years, having increased the weight at hatching by 500-fold. (Sue, among the largest †*T. rex* yet discovered, is estimated to have weighed ~9,000 kg.)

- *Ontogenetic changes in body proportions.* †Tyrannosaur teenagers had relatively longer legs and slimmer bodies than adults. Adult body proportions were assumed in the late teen years.

- *Ontogenetic changes in the skull and teeth.* The skulls and jaws of juvenile †tyrannosaurs were long, and their teeth were sharp and bladelike. The transition to the deep skulls, shorter jaws, and rounder teeth of adults began in the early teen years.

(A) Skeleton of †*Deinonychus* with inset to show claw

Large claw
on toe II

5 cm

(B) Old interpretation of claw use by †*Deinonychus*

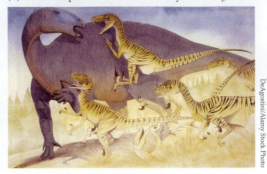

Figure 19.10 Changing interpretations of claw use by †*Deinonychus*. (A) †*Deinonychus* was a fleet-footed bipedal predator. Each foot had a large claw on toe II. (B,C) Initially interpreted as a hypercarnivore that hunted in packs and killed prey by slashing it with its enlarged claws, current interpretations liken its claw use to that of extant large hawks that pin their prey to the ground with enlarged claws and tear off pieces to eat.

(C) Current interpretation of claw use by †*Deinonychus*

These differences suggest that young and adult †tyrannosaurs were functionally very different animals, occupying different feeding niches, using different predatory behaviors, consuming different prey species, and processing food differently in their mouths.

Adult †*Tyrannosaurus rex* could not run, but teenagers, with their longer, slimmer bodies could probably run at ~22 km/h—a speed that would have allowed them to pursue prey as cheetahs do, whereas adults probably stalked their prey and captured it with a short dash in the manner of lions and tigers.

Changes in proportions of the jaws and teeth also would have had important consequences for feeding. Juvenile †*T. rex* probably ate young herbivores and smaller species of theropods, while teenagers captured mid-size prey, and adults fed on large †hadrosaurs. The jaws of a juvenile †*T. rex* could apply a calculated force of 5,500 newtons (N), and juveniles up to their early teens probably used a puncture-and-pull feeding method like that of †*Velociraptor* (see Figure 19.1). Adults used repeated

bites to crush and pulverize bone; the jaws of the largest adults may have been able to apply as much as 34,000 N of force, roughly equivalent to stacking 80 bags of cement on top of a serrated knife directed at a bone.

At the community level, these ontogenetic changes in †*Tyrannosaurus rex* appear to have made them the dominant predators at all body sizes over the course of their growth, dominating the community by monopolizing the feeding niches of theropods in the 100–1,000 kg weight range.

Social behavior of theropods

The fossil record of behaviors is scant at best, and we know less about sociality in Mesozoic theropods than

(A) Carnivorous mammals of Kruger National Park

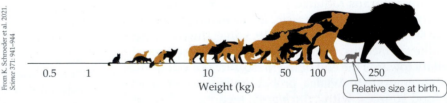

Figure 19.11 Community structure of mammalian carnivores and Mesozoic theropods. (A) The size distribution of adult extant carnivorous mammals at Kruger National Park in South Africa includes species of all sizes. (B) In contrast, the size distribution of adult theropods in the Dinosaur Park Formation in Alberta, Canada, has a conspicuous gap between 100 and 1,000 kg.

(B) Carnivorous dinosaurs of Dinosaur Park Formation with largest carnivore scaled equally to the largest mammalian carnivore at Kruger National Park

we do about the social behaviors of †ornithischians (see Section 18.4). Accumulations of individual skeletons suggest that some large theropods were gregarious, including †carcharodontosaurs and †tyrannosaurs. The remains of twelve large †tyrannosaurs (†*Albertosaurus sarcophagus*) were found amid flood debris in a Late Cretaceous site in Alberta, but there is no way to determine whether this was a social pack swept away as a group or merely individuals feeding on the carcasses of animals killed in an earlier flood. Carcasses carried by river currents often accumulate over time in places where the current slows, and a mass assemblage of †*Allosaurus fragilis* found at the site of a Late Jurassic river in Utah was probably due to this scenario. On the other hand, a tracksite from the Early Cretaceous of Chile shows tracks of large-size theropods walking parallel to each other for long distances. The parallel trails suggest that these animals may have moved as a group.

Evidence of group behavior is a bit stronger for small to medium-size theropods. Parallel trackways are known from several places, most notably Spain and China. Mass accumulations of fossilized bones and skeletons have been found in North and South America and in Asia. For example, more than 20 juveniles of the †ornithomimosaur †*Sinornithomimus dongi* ranging from 1 to 7 years old were trapped in mud in a drying pond at a site in western Mongolia, and at least 18 juvenile and adult individuals of the †oviraptorosaur †*Avimimus* appear to have drowned while crossing a river. †*Sinornithomimus* and †*Avimimus* were herbivorous, making it unlikely that these individuals were drawn together by a temporary abundance of food.

The skulls of many subadult and adult tyrannosaurs bear healed toothmarks that appear to have been inflicted by individuals of the same species, probably during intraspecific combat. That interpretation is supported by the discovery of the tip of a tyrannosaur tooth embedded in one of the skulls. Toothmarks are absent from the skulls of juveniles; they first appear in individuals that are about half of adult size (corresponding to the onset of sexual maturity) and are present in ~60% of adults. The spacing of toothmarks indicates that combatants were size-matched, suggesting that youngsters engaged in practice bouts with one another before they reached adult size. The presence of scars in approximately half of the specimens might indicate that only one sex (probably males) engaged in these combats.

In addition to biting, some †tyrannosauroids probably butted each other. †*Tyrannosaurus rex* had an enlarged postorbital boss that could have made the head a battering ram (see Figure 19.6C), and †abelisaurids had rugose skin on their heads and snouts. Some abelisaurids, such as †*Carnotaurus* ("carnivorous bull"), had horns like those of mammals, consisting of a bony core covered with a cornified sheath (see Figure 19.4E). The snout of †*Carnotaurus* had a horny covering with a midline suture. Cornified snouts like this are present in extant musk oxen (*Ovibos moschatus*) and Cape buffalo (*Syncerus caffer*) and are used in headbutting combats. †Tyrannosauroids may

have butted heads, but their strutlike skulls are less robust than those of today's horned mammals, and †tyrannosauroids may have used their heads for low-impact shoving matches like those of bison (see Figure 18.20D); or they may have swung them against the necks and flanks of opponents like giraffes (see Figure 18.23A).

19.2 †*Archaeopteryx*, Mesozoic Avialans, and the Mosaic Evolution of Avian Characters

LEARNING OBJECTIVES

19.2.1 Place †*Archaeopteryx*, †*Jeholornis*, †*Confuciusornis*, †*Hesperornis*, and †*Asteriornis* in the overall context of Mesozoic history.

19.2.2 Discuss how understanding the mosaic evolution of characters within Theropoda helps us to understand the origin of Aves.

19.2.3 Explain how body size evolution within Tetanurae led to a basic adaptation for flight.

About 80 million years elapsed between the first dinosaurs that could unquestionably fly, a group called Avialae, and the origin of extant birds, Aves (see Figure 19.2). Since the discovery of †*Archaeopteryx* in the 19th century, many other fascinating birdlike fossils have been discovered, and together these show that characters of birds did not evolve all at once, but in a series of intermediate forms. The intermediate forms demonstrate **mosaic evolution**, a concept based on the observation that body parts, sometimes referred to as modules, evolve independently. For example, there are non-flying theropods with feathers but not wings, and avialans that fly but have teeth rather than beaks. In this section we trace these and other changes that occurred during the evolution of Avialae.

Discovery of †Archaeopteryx

Along the shoreline of the Tethys Sea in the Late Jurassic, at a position about where Italy is today, carcasses of †pterosaurs and small dinosaurs such as †*Compsognathus* came to rest on fine sediments accumulating in isolated lagoons behind reefs. Because the lagoons were anoxic and unsuitable for animal life, the carcasses remained virtually untouched by scavengers. Calcium carbonate precipitated from the water column, covering the bodies with fine sediment. The fine particles became cemented together, compressed under centimeters, then meters, then hundreds of meters of overlying sediments to form platelike layers of limestone called plattenkalks.

Since the 1850s, near the southern Bavarian town of Solnhofen, quarries have cut into plattenkalks that formed ~150 Ma. The flat, fine-grained limestone slabs were ideal for lithography, a printing method developed earlier in that century that can produce many copies of highly detailed artwork and text. Because each stone had to be inspected closely to evaluate its lithographic potential, many fossils were discovered.

(A) This feather fossil was the first evidence of †*Archaeopteryx*

The asymmetry indicates it was a primary flight feather.

From R.M. Carney et al. 2012. *Nat Commn* 3: 637

(C) Hindlimb with asymmetrical feathers that form "trousers"

From C. Foth et al. 2014. *Nature* 511: 79–82

(D) Restoration of †*Archaeopteryx* showing feathered "trousers"

Asymmetrical feathers of the "trousers."

© Zhao Chuang PNSO

(B) The Berlin specimen of †*Archaeopteryx*

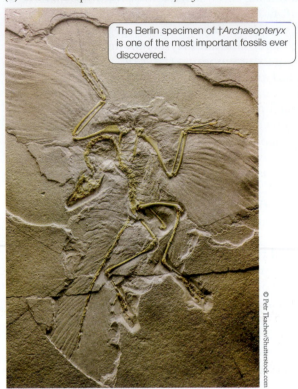

The Berlin specimen of †*Archaeopteryx* is one of the most important fossils ever discovered.

© Petr Tkachev/Shutterstock.com

Figure 19.12 **Fossils of †*Archaeopteryx* from Late Jurassic Solnhofen limestone.** (A) An isolated flight feather found in 1859. Long interpreted as an †*Archaeopteryx* feather, recent research confirms this. (B) The Berlin specimen of †*Archaeopteryx lithographica* found in the 1870s is among the most important and remarkable vertebrate fossils ever discovered because it shows so many features intermediate between theropods such as †*Compsognathus* (see Figure 19.7) and extant birds. (C) Research on a more recent specimen of †*Archaeopteryx* confirms the presence of pennaceous feathers on the legs ("feathered trousers"). (D) A new restoration of †*Archaeopteryx* showing the "feathered trousers."

In 1860, a single isolated feather was discovered in a piece of Solnhofen limestone (**Figure 19.12A**). As you can see, the barbs of this feather are shorter on one side of the shaft than on the other, an asymmetry typical of the primary flight feathers of birds. The ability of the fine-grained stone to preserve barbs provided the first direct evidence that forms like extant birds existed in the middle Mesozoic. In 1861, that specimen was given the name †*Archaeopteryx lithographica*—†*Archaeopteryx* meaning "ancient wing," and *lithographica* refering to the lithographic limestone in which it was found. Soon there was a far more significant find: the skeleton of a crow-size animal with a long tail, surrounded by impressions of feathers just like the one found in 1860.

In 1862, the feathered fossil skeleton was sold to the Natural History Museum in London and formally described in 1863. The London specimen exhibited a mosaic of anatomical features found in dinosaurs and in extant birds. Publication of *On the Origin of Species* in 1859 had prompted scientific searches for evidence of evolution,

and the mosaic of features identifiable in the London specimen made it a centerpiece of that effort.

In 1876, 15 kilometers east of the original locality, workers uncovered an even more complete specimen of †*Archaeopteryx* (**Figure 19.12B**), today housed at the Museum für Naturkunde in Berlin. In addition to its spectacular preservation and near-perfect presentation of the entire body, the Berlin specimen provided new information about the cranial anatomy of †*Archaeopteyx*, including a cranial cavity that could accommodate a large brain like that of extant birds. More recently discovered fossils shed additional light on the anatomy of †*Archaeopteryx*. A specimen described in 2014 shows that there were asymmetrical feathers on the legs, where they formed feathered "trousers," as well as on the wings (**Figure 19.12C,D**).

Figure 19.3 shows †*Archaeopteryx* as the basalmost member of Avialae. Before we can discuss mosaic features of dinosaurs and birds, however, we need to introduce a few more key Cretaceous avialans.

Cretaceous avialans

Early avialans were probably generalized carnivores. Morphological characters such as beak shape indicate that they radiated into many adaptive zones occupied by extant birds, including fish-eaters, insect-eaters, and substrate-probers. Only flying raptors (hawk equivalents) appear to have been rare or absent, possibly because of competition with †pterosaurs and terrestrial paravians.

The timing of avialan diversification is important. †*Archaeopteryx* occurs in the Late Jurassic of Germany, but avialans did not really begin to diversify until the Early Cretaceous (see Figure 19.2). For example, >100 specimens of the turkey-size †*Jeholornis* have been found in Early Cretaceous formations of China (**Figure 19.13A**). †*Jeholornis* could fly but lacked specializations of the shoulder girdle needed for powerful flight. Another abundant early Cretaceous fossil from China is †*Confuciusornis* (**Figure 19.13B**), one of the earliest members of Pygostylia, a group named for reduction and fusion of the tail vertebrae to form a **pygostyle**, a triangular structure that

(A) The Jehol bird, †*Jeholornis*, Early Cretaceous of China

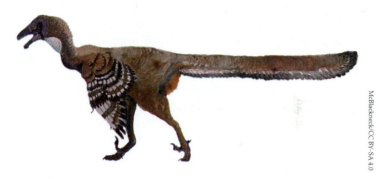

Tiouraren (Y.-C. Tsai)/CC BY-SA 4.0

McBlackneck/CC BY-SA 4.0

(B) Early Cretaceous pygostylian, †*Confuciusornis*

†*Confuciusornis* had a beak.

Long tail feathers in some specimens.

Fused caudal vertebrae formed a pygostyle.

Naturhistorisches Museum Wien/CC BY 2.0

(C) Early Cretaceous †enantiornithine, †*Yuanchuavis*

Luxquine/CC BY-SA 4.0

(D) †*Hesperornis*, a flightless ornithuran from the Late Cretaceous

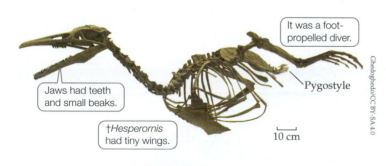

It was a foot-propelled diver.

Jaws had teeth and small beaks.

Pygostyle

†*Hesperornis* had tiny wings.

10 cm

Ghedoghedo/CC BY-SA 4.0

(E) †*Asteriornis*, the "wonder chicken," belongs to Galloanserae, a clade within Aves that includes extant chickens and ducks

The wonder chicken lived about ~66.8 Ma.

Courtesy of Phillip Krzeminski

© Daniel J. Field/University of Cambridge

Figure 19.13 Mesozoic avialans. (A) The Jehol bird was about the size of a turkey. (B) †*Confuciusornis* had a beak and a pygostyle. The long tail feathers of some specimens are probably sexually dimorphic features, like the long tail feathers of many extant birds. (C) †Enantiornithines were very diverse during the Cretaceous and occupied many habitats used by avians but became extinct at the end-Cretaceous extinction. (D) Although it was flightless and had both teeth and a beak, †*Hesperornis* shared skeletal features with Aves, with which it is grouped in Ornithurae. (E) The "wonder chicken" †*Asteriornis maastrichtensis* shows that Aves had already begun to diversify by the end of the Cretaceous. Most avian groups, however, diversified explosively in the early Cenozoic.

supports the tail, as seen today in Aves (see Figure 21.7A). †*Confuciusornis* is also noteworthy for its beak, a feature that evolved independently many times within avialans.

†Enantiornithes (Greek *enantios*, "opposite"; *ornis* "bird") radiated in the Early Cretaceous, an event that coincided with several skeletal adaptations that allowed a more effective flight stroke. These features include a strutlike coracoid, suggesting that sustained powered flight evolved at this stage of avialan evolution. As the dominant group of Cretaceous avialans, †enantiornithines ranged from the size of a sparrow to as large as a turkey vulture. Most species, such as †*Yuanchuavis* (**Figure 19.13C**), were small to medium-size and probably lived in trees, although some had long legs and appear to have been waders, while others had powerful claws like those of extant hawks.

All remaining avialans belong to Ornithurae (Greek *ornis*, "bird"; *oura*, "tail"; see Figure 19.3). Members of this clade have a plow-shaped pygostyle with the tail feathers forming a fan. Ornithurans of the Early Cretaceous were small, finchlike, arboreal species, but by the Late Cretaceous the group included waders, perchers, and secondarily flightless foot-propelled swimmers and divers such as †*Hesperornis* (**Figure 19.13D**).

Estimates based on molecular evidence consistently place the earliest radiation of Aves in the Cretaceous. †*Asteriornis maastrichtensis*, a recently described member of Galloanserae (ducks and chickens), provides fossil evidence that diversification of Aves had started by the latest Cretaceous. Nicknamed the "wonder chicken" because of its mixture of chickenlike and ducklike features, †*A. maastrichtensis* lived less than 1 Ma before the end-Cretaceous mass extinction. However, the sparse fossil record of Aves from the Late Cretaceous suggests that the "big bang" of avian radiation did not happen until the Cenozoic, possibly because the mass extinction left so many niches unoccupied.

The end-Cretaceous mass extinction (see Section 20.4) affected all life on the planet. Widespread forest fires destroyed entire ecosystems and reduced resource availability worldwide. In contrast to the mostly arboreal †enantiornithines, the earliest Aves were mostly ground-dwellers, which made them less dependent on forest ecosystems. In addition, with their small body size, avians required less food, and it is likely that surviving lineages maintained some dietary flexibility, allowing them to capitalize on insects and seeds. These survivors were ready to exploit niches previously occupied by †enantiornithines and †pterosaurs, among others.

Mosaic evolution of some avialan characters

Figure 19.14 and **Table 19.1** illustrate mosaic evolution of some avialan characters by comparing an outgroup, †*Compsognathus*, to three avialans: †*Archaeopteryx*, †*Confuciusornis* (a Cretaceous avialan), and an extant bird,

the northern cardinal (*Cardinalis*). All four taxa have a furcula. Like the basal †sauropodomorph †*Plateosaurus* and the basal theropod †*Carnotaurus* (see Figure 18.8), the pubis of †*Compsognathus* is directed anteroventrally,

(A) Skeleton of †*Compsognathus longipes*

Furcula
Anteroventrally directed pubis

© Scott Hartman

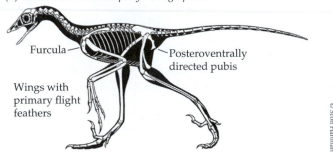

(B) Skeleton of †*Archaeopteryx lithographica*

Furcula
Posteroventrally directed pubis
Wings with primary flight feathers

© Scott Hartman

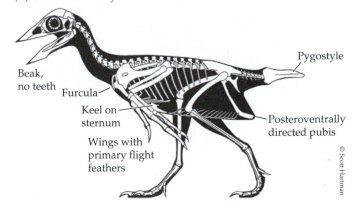

(C) Skeleton of †*Confuciusornis sanctus*

Pygostyle
Beak, no teeth
Furcula
Keel on sternum
Wings with primary flight feathers
Posteroventrally directed pubis

© Scott Hartman

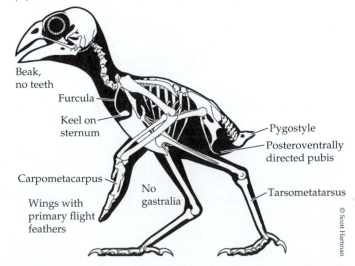

(D) Skeleton of a northern cardinal, *Cardinalis cardinalis*

Beak, no teeth
Furcula
Keel on sternum
Carpometacarpus
No gastralia
Wings with primary flight feathers
Pygostyle
Posteroventrally directed pubis
Tarsometatarsus

© Scott Hartman

Figure 19.14 Mosaic evolution of avialan characters.
Characters shown in this figure are keyed to Table 19.1.

Table 19.1 Mosaic evolution of some avialan characters

Taxon	Furcula	Posteroventrally directed pubis	Wings with primary feathers	Pygostyle	Keel on sternum	Fused bones in arms and legs[a]	Gastralia	Beak
†*Compsognathus*	Yes	No	No	No	No	No	Yes	No
†*Archaeopteryx*	Yes	Yes	Yes	No	No	No	Yes	No
†*Confuciusornis*	Yes	Yes	Yes	Yes	Yes	No	Yes	Yes[b]
Cardinalis	Yes	Yes	Yes	Yes	Yes	Yes	No	Yes[b]

[a] These are the carpometacarpus in the hand and tarsometatarsus in the foot.

[b] Loss of teeth and evolution of a beak occurred independently in †*Confuciusornis* and Aves.

but it is directed progressively more posteroventrally in †*Archaeopteryx*, †*Confuciusornis*, and finally *Cardinalis*. †*Compsognathus* lacks primary flight feathers on its arms, but these are present in the other three genera. †*Compsognathus* and †*Archaeopteryx* have long tails and many caudal vertebrae, but the tail is short in †*Confuciusornis* and *Cardinalis* and they have a pygostyle formed by fusion of several caudal vertebrae. Among the four taxa compared in Figure 19.14, only the cardinal has fully fused **carpometacarpals** (fusions of wrist and finger bones), and **tarsometatarsals** (fusions of ankle and foot bones) as well as loss of gastralia.

†*Compsognathus* and †*Archaeopteryx* have teeth, whereas †*Confuciusornis* and *Cardinalis* have beaks. But beaks evolved independently in the lineages leading to †*Confuciusornis* and *Cardinalis*, and groups phylogenetically much closer to Aves, such as †Enantiornithes and †*Hesperornis*, retained teeth (see Figures 19.2 and 19.13D).

Other avian features

Birds have many other distinctive features including: (1) a paedomorphic skull with an enlarged brain (see Figure 1.7); (2) a short tail and an anteriorly directed femur; (3) highly modified pectoral girdles and forelimbs; (4) a synsacrum composed of many sacral vertebrae fused with the pelvic girdle; and (5) hindlimbs with extensive fusion of bones and digitigrade foot posture (i.e., walking on the toes, with the heels off the ground).

Skull   Bird skulls are paedomorphic theropod skulls, with enlarged orbits and brains. The evolution of an edentulous beak has also been linked with this transition to a paedomorphic skull. However, whereas the enlarged orbits and brains occur relatively late during the evolution of theropods, tooth loss and beaks are widespread among dinosaurs and occur in many other archosaurs and in turtles.

Beaks evolved in several theropod lineages, including †ornithomimosaurs, †therizinosaurs, †oviraptorosaurs, and avialans. In all these groups, early members had teeth; an edentulous beak occurs only in derived taxa. Often, intermediate forms had a beak on the tip of the snout and lower jaw, with teeth further back in the jaws. Some †sauropodomorphs may have had beaks, but

evidence is sparse. Beaks of most †ornithischians were restricted to the rostralmost portion of the skull, with teeth in the rear of the jaws. Some early pseudosuchians, †pterosaurs, and †silesaurids (quadrupedal, herbivorous, Triassic dinosauromorphs) had beaks. Thus, it is likely that the developmental pathway that produces beaks was present in the common archosauromorph ancestor of turtles, crocodylians, †pterosaurs, and dinosaurs.

Girdles and limbs   The avian shoulder girdle has strut-like coracoids, which help the girdle resist pressures exerted on the chest by contractions of the wing muscles. This character is shared by †enantiornithines and Ornithurae (see Figure 19.3), but elongated coracoids occur in Middle Jurassic †tyrannosauroids and more derived theropods. The evolution of strutlike coracoids happened during the Early Cretaceous, coinciding with early radiations of avialans.

In birds, the shape of the furcula has the familiar wishbone shape, and it became an important component of the flight mechanism. The presence of a furcula, however, is a synapomorphy of theropods, which means that the origin of the furcula had nothing to do with flight.

The sternum gradually increased in size in maniraptorans known from the Middle and Late Jurassic. A small keel for the origin of the flight muscles appeared at about the same time as the appearance of the strutlike coracoid in Early Cretaceous avialans, and the co-occurrence of these features may be related to the diversification of avialans at that time.

With the reduction of the tail and the increasing size of the sternum and wings, the center of gravity shifted forward to a position closer to the shoulder girdle. To adapt to this altered weight distribution, the femora (plural of femur) were re-oriented to point forward, bringing the lower limbs below the center of mass. These changes were gradual, and preceded tail truncation: whereas †tyrannosauroids still had vertically oriented femora; those of more derived theropods were increasingly anteriorly rotated. A slightly crouched hindlimb posture is already seen in the derived maniraptoran †*Troodon*, which still had a long, feathered tail. As the tail became truncated during the early evolution of avialans, the femur attained the nearly horizontal position required to keep the feet beneath the center of gravity.

Like †pterosaurs, birds have a synsacrum composed of fused sacral vertebrae and the pelvic girdle (see Figure 21.7). This strengthens the attachment of the girdle to the vertebral column, allowing the column to absorb shocks from landings.

Lung structure and skeletal pneumatization The structure of the avian lung allowed birds to achieve sustained and powerful flight. As discussed in Section 13.2, birds have rigid faveolar lungs with anterior and posterior air sacs that create a unidirectional airflow with a crosscurrent gas exchange (see Figure 13.4). This system, however, was present much earlier within Archosauromorpha—turtles and crocodylians also have faveolar lungs. Theropods inherited from their ancestors a partially rigid lung, probably already with anterior and posterior air sacs. Early in theropod evolution, these air sacs invaded the cervical and anterior trunk vertebrae, pneumatizing them. Pneumatization of the posterior trunk vertebrae, ribs, and long bones evolved independently in as many as 12 theropod lineages as well as in †pterosaurs and †sauropods (see Sections 18.2 and 18.3). Thus, skeletal pneumatization evolved independently of flight.

Associated changes in the skeleton further improved the efficiency of lung ventilation. Bird ribs have bony projections called uncinate processes that extend upward and backward from each rib (see Figure 21.7A). These structures were probably present in all archosaurs but remained mostly cartilaginous, as they are in crocodylians today. In derived maniraptorans like †oviraptorosaurs and †dromaeosaurids, ossified uncinate processes began to occur. These processes act as levers and provide a mechanical advantage for respiratory muscles. Later, in avialans, the bony uncinate processes also fused to the ribs, increasing their mechanical advantage even further. This increase in the efficiency of costal ventilation was accompanied by the loss of gastralia in Aves (see Section 13.2).

Feathers The pennaceous feathers of birds are the most complex integumentary structures of vertebrates, and pennaceous feathers, as well as several other types of feathers, have been identified in many non-avian dinosaurs. The most common are filamentous protofeathers, which are known from nearly all major clades of theropods. The simplest protofeathers, sometimes known as "dinofuzz," were single hollow filaments 1–5 cm long that bear little resemblance to what we think of when we read the word "feather." Among theropods, the most basal known occurrence of protofeathers is in †*Sciurumimus*, an early tetanuran from the Late Jurassic of Germany. The largest taxon known to have had protofeathers is the 9-m †tyrannosauroid †*Yutyrannus* from the Early Cretaceous of China.

The homology of protofeathers with feathers was initially disputed, but fossilized protofeathers of an †alvarezsaurid (†*Shuvuuia deserti*) retain enough organic material to show that the filaments are composed of a type of beta-keratin unique to feathers. This result has been confirmed by studies of protofeathers of many other dinosaurs.

Pennaceous feathers first occur in †ornithomimosaurs, but the greatest diversity of feather types evolved in their sister group, Maniraptora. Different maniraptorans have different feather types along the back, on the forelimbs or hindlimbs or both, and on the tail. Rearrangements of certain feather types and losses of some others are probably associated with powered flight.

Feathers probably evolved in relation to insulation and visual displays. At night, birds keep their feet and eggs warm by crouching and covering them with their body feathers, and by tucking their thinly feathered heads and uninsulated beaks beneath their wings. Feathered theropods, especially smaller individuals, probably did this as well. For example, juvenile †*Ornithomimus* had filamentous feathers on their arms, which may have served as insulation; adults had both filaments and symmetrical pennaceous feathers. This ontogenetic change suggests that feathers of adults had social or reproductive functions, such as display, courtship, or brooding.

A key step in the origin of powered flight was the evolution of asymmetrical pennaceous feathers like those of †*Archaeopteryx* (see Figure 19.12A). Such feathers form the remiges, the set of flight feathers in a bird's wing (see Figure 21.4). Asymmetrical pennaceous feathers are known only in paravians (†dromaeosaurids to birds; see Figure 19.3), but the earliest paravians could not fly. Hence, even asymmetrical pennaceous feathers first had functions such as insulation and display, and were only secondarily adapted for flight.

Color enhances the role of feathers in displays, signaling, and camouflage, and we have some information about the colors of paravians. Scanning electron micrographs of the shapes and distributions of melanosomes (melanin-containing organelles) in the feathers of the paravian †*Anchiornis huxleyi* revealed that it was gray, with a crest of reddish-brown feathers on the top of its head, speckles on the face, and long white and black feathers on the wings and hindlimbs (**Figure 19.15**).

Some paleontologists regard assignment of colors to fossils with skepticism. Bacterial decay can cause postmortem changes in the soft tissues, and it has even been suggested that some of the structures interpreted as pigment organelles are bacteria. However, immunological tests show that both keratin and melanosomes are present in fossil feathers of †*Eoconfuciusornis*, an Early Cretaceous avialan. Analyses of other taxa confirmed the identity of melanosomes in fossils and directly identified different types of melanin in feathers and skin.

Melanosomes are not the only organelles involved in feather coloration, so there will always be some limitations of the method. However, we can tell that the feathers were pigmented and that they had different colors and patterns. Evidence of overall coloration patterns, such as countershading or stripes, does not depend on precise identification of specific hues, but nevertheless yields clues about an animal's biology.

Restoration of feather colors of Late Jurassic †*Anchiornis huxleyi*

© Carl Buell

Figure 19.15 Feather colors in †*Anchiornis*. By studying the distribution of melanosomes in fossils, paleontologists determined the color pattern of this paravian.

There was not a stepwise evolutionary progression in which simple feathers came first and were progressively replaced by more complex feathers. Rather, different types of feathers have different functions. Diversification

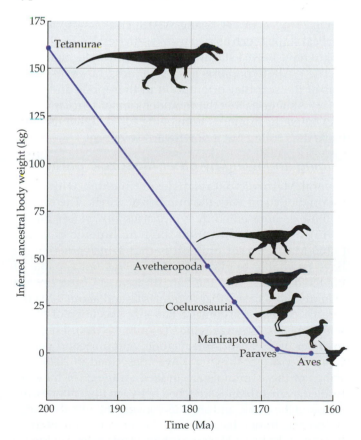

Figure 19.16 Evolution of body size in Tetanurae. During the Jurassic, the inferred ancestral body sizes of tetanuran lineages decreased. (Graph based on M.S.Y. Lee et al. 2014. *Science* 345: 562–566. See additional credit details at the end of this chapter.)

of feather types resulted from extensive duplication and modification of beta-keratin genes. Thus, the widespread distribution of feathers among non-avian dinosaurs emphasizes that the evolution of feathers and the evolution of flight were entirely separate phenomena.

Body size

Many lineages of tetanurans basal to Avialae evolved very large body sizes, such as †tyrannosauroids (6,000–9,000 kg). Even close outgroups of avialans, such as †dromaeosaurids, reached 350 kg. But a substantial reduction in body size characterized avialan evolution (**Figure 19.16**). Starting from basal tetanurans, which averaged about 25 kg, the body sizes of paravians shrank to about 6.5 kg (the size of a large turkey) for †*Caudipteryx*, to about 800 g (a small chicken) for †*Archaeopteryx*, to 300 g (pigeon-size) for †*Epidexipteryx*, and to 30 g (sparrow-size) for †*Scansoriopteryx*.

19.3 Evolution of Powered Flight

LEARNING OBJECTIVES

19.3.1 Discuss adaptations associated with powered flight in birds and how the concept of exaptation contributes to understanding the evolution of these traits.

19.3.2 Describe the several hypotheses for the origin of flight.

19.3.3 Review the evidence for independent evolutionary origins of flight in paravian groups such as †dromaeosaurids.

The origin of powered (flapping) flight in avialans continues to be studied in detail, but there are still many questions about how it happened. One way to start thinking about such questions is to consider what traits are necessary for powered flight. These requirements include: (1) a small and light body; (2) asymmetrical pennaceous feathers arranged to form an effective airfoil; (3) leg bones and legs muscles for takeoffs and landings; and (4) wing bones and muscles for a powerful downstroke. In addition, flight is metabolically expensive, because the brain must process three-dimensional data when in the air and coordinate complex movements for maneuvering.

Some of the traits related to powered flight were present in theropods long before there were any avialans, demonstrating an important concept in evolutionary biology, **exaptation**, which means that a trait has taken on a new or additional function during evolution. Bird feathers are a classic example of exaptation because, as mentioned in the previous section, the earliest feathers served functions such as insulation and display unrelated to flight.

It turns out there are many cases of exaptation in the origin of avialan flight. As we saw in Figure 19.16, small body size—the result of a long-term trend within Tetanurae unrelated to flight—is an exaptation for powered flight in avialans, too. So are asymmetrical pennaceous feathers, which are synapomorphic for paravians.

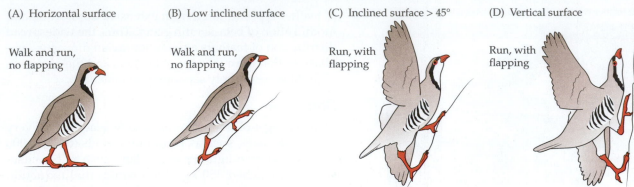

(A) Horizontal surface

Walk and run, no flapping

(B) Low inclined surface

Walk and run, no flapping

(C) Inclined surface > 45°

Run, with flapping

(D) Vertical surface

Run, with flapping

Figure 19.17 Wing-assisted incline running. The chukar partridge walks or runs when moving across level or slightly inclined surfaces (A,B), but flaps its wings when climbing steeper inclines (C,D). (After K.P. Dial. 2003. *Science* 299: 402-404.)

Avialans also inherited long arms and flexible wrists from earlier theropods. And, importantly, early theropods were bipedal cursorial predators, and avialans inherited strong legs and leg muscles suitable for leaping, exaptations for the role of legs in avialan flight.

How—and why—birds got off the ground

Jurassic paravians were cursorial, terrestrial animals. Their long legs allowed them to run to pursue prey and escape predators. The earliest flying paravians did not have to fly very well—even a brief glide on a protowing could have foiled an attack by a terrestrial predator or surprised unsuspecting prey. Enhanced agility is thus a reasonable hypothesis to explain why wings evolved in the first place.

But if wings preceded flight, how did the evolution of powered flight happen? Several scenarios offer ideas about advantages that could be gained by using feathered wings for powered flight. These scenarios fall into two categories, known as "from the trees down" and "from the ground up," depending on the starting point for aerial locomotion.

From the trees down   The ability to glide (a form of passive, or non-flapping, flight in which altitude decreases while horizontal distance is gained; see Section 21.3) evolved many times among vertebrates and is a possible starting point for the evolution of powered flight. The small paravians nearing the transition to flight could probably climb trees, and some may have been arboreal. A paravian ambusher might have used its wings to steer as it swooped down on prey. Alternatively, arboreal paravians that climbed into trees to forage for prey could have glided from one tree to another to avoid descending to the ground, as do extant lizards of the genus *Draco*.

From the ground up   The strongly cursorial (specialized for running) morphology of paravians suggests that the origin of flight should be sought in terrestrial locomotion. One early hypothesis proposed that a cursorial predator might have combined an upward leap with wing-flapping to swat a flying insect to the ground. No extant birds do this, however, and the feathers on the wings of paravians

appear to have been as tightly packed as those on extant birds, making their wings more like a ping-pong paddle than a tennis racquet, so that air resistance to a swing might have rotated a predator away from its prey.

Wing-assisted incline running (WAIR) may provide a more plausible scenario for evolving flight from a cursorial starting point. Juveniles and adults of some species of extant ground-feeding birds like grouse, partridges, and chickens use their wings to ascend steep slopes. On level ground and on slopes <45 degrees, chukar partridges (*Alectoris chukar*) walk or run without flapping their wings. On steeper slopes, however, they run and flap their wings (**Figure 19.17**). Using WAIR, chukars can ascend vertical slopes such as tree trunks, suggesting that WAIR might have been a starting point for powered flight.

The WAIR hypothesis has a weakness, however. Analysis of the scapular orientation in theropods and basal avialans suggests that the shoulder joint of †*Archaeopteryx* prevented the humerus from being lifted higher than the back, thus ruling out a downward wingbeat.

Gliding and flying by other Mesozoic paravians

Several Mesozoic paravians experimented with wing morphology and with the number of wings. The earliest of these is represented by †*Anchiornis huxleyi* (see Figure 19.15), a winged form from the Late Jurassic of China (~160 Ma). Current opinion is that †*Anchiornis* might have been able to glide but that it was not capable of powered flight. From the Late Jurassic of China 159 Ma (slightly more recent than †*Anchiornis* and about 9 Ma before †*Archaeopteryx*) comes a bat- size specimen of †*Yi qi*, which means "strange wing" (**Figure 19.18A**). †*Yi qi* is a member of †Scansoriopterygidae, and a second genus, †*Ambopteryx* ("both wing"), was described in 2020. The wings of these animals are indeed strange, for they are formed by membranes stretched between finger bones, the arm, and the trunk, and are thus convergent with the wings of †pterosaurs and bats. Mostly filamentous feathers cover the rest of the body. However, there is no skeletal evidence of powerful downstroke muscles. Gliding was probably possible but, compared with other gliding vertebrates, the **wing loading** (calculated as the ratio of body weight/wing

(A) Basal paravian, †*Yi qi*   (B) †Dromaeosaurid, †*Microraptor gui*

Figure 19.18 **Paravian wings.**
(A) The †scansoriopterygid †*Yi qi* had batlike membranes of skin that formed a flying surface, probably used for gliding (passive or non-flapping flight). (B) †*Microraptor gui* had flying surfaces on the wing and hindlimb but probably was incapable of extended flight.

area) of †scansoriopterygids suggests that they could travel only short distances.

Hundreds of specimens of several species of †*Microraptor* have been found in Early Cretaceous localities of China, dating from ~20–30 million years after †*Archaeopteryx*. These †dromaeosaurids had feathered wings on all four limbs (**Figure 19.18B**). Whether †*Microraptor* used powered flight has been debated since the discovery of the first specimen in 2000; one 2020 paper concluded that powered flight evolved convergently in †*Microraptor* and avialans. Tests of models of †*Microraptor* wings in wind tunnels suggest that the hindwings increased lift and stability. Hence, although the hindwings were probably not used as flapping devices, they could have supported the animal during flight while the forewings produced the thrust necessary for powered flight.

Because †anchiornithids, †scansoriopterygids, and †dromaeosaurids left no extant descendants, discussions about their capacity for powered flight will continue. Importantly, however, these taxa show that there was widespread experimentation with gliding and flight among early paravians.

19.4 Reproduction and Parental Care by Theropods

LEARNING OBJECTIVES

19.4.1 Review evidence for the assertion that hatching asynchrony evolved before the emergence of Aves.

19.4.2 Explain how we can interpret reproductive behavior of extinct theropods by comparing extinct and extant forms.

Crocodylians and birds form the extant phylogenetic bracket for theropods, and, as discussed in Section 1.4, allow us to form hypotheses about some reproductive characters of theropods. Both crocodylians and birds have "assembly line" oviducts that successively surround a zygote with yolk, albumen, and finally a shell. But there are differences. Crocodylians and turtles have two functional ovaries and produce many small eggs that are deposited in a single laying event; this is probably the ancestral

archosaur condition. Extant birds have a single functional ovary and oviduct and produce fewer and larger eggs that are deposited one at a time over a period of days. Some bird species wait until a full clutch of eggs is laid to begin incubation, which results in synchronous hatching. Others begin incubation immediately after the first egg is laid, leading to asynchronous hatching (see Section 21.9). The eggs of crocodylians are symmetrical, whereas the eggs of extant birds have a narrow end and a blunt end.

Eggs and nests

Fossils of non-avian tetanurans and their nests shed light on the sequence in which the derived reproductive characters of birds appeared. Tetanurans deposited eggs in pairs within a roughly circular depression surrounded by a rim constructed by the adult (**Figure 19.19A**). The pairwise production of eggs, which was initially inferred by their positions in the nest, has been confirmed by the discovery of a female tetanuran with two shelled eggs in the pelvic region. The eggs of tetanurans were asymmetrical, like those of extant birds. Although most tetanuran eggs were laid with the blunt ends facing the center of the clutch, some early avialans (probably †enantiornithines; see Figure 19.3) deposited them in the opposite direction, suggesting that the eggs were at least partly embedded in substrate. Analyses of eggshells of several species of maniraptorans revealed a single origin of egg color before the evolution of Aves, probably somewhere within Maniraptora. The pigments protoporphyrin and biliverdin, which are also present in the eggs of extant birds, were present in most species to different degrees and would have given the eggs a blue-green color (**Figure 19.19B**). Such egg coloration also suggests that the nests of these theropods were at least partially exposed, with the color acting as camouflage or serving for species recognition.

†Oviraptorosaur clutches from China include eggs containing embryos at different stages of development, which indicates hatching asynchrony. These eggs must have been produced and laid at slightly different times. †Oviraptorosaurs produced eggs in pairs, so the reduction to a single ovary and oviduct happened after the evolutionary origin of hatching asynchrony.

Growing evidence also suggests that loss of the right ovary and oviduct, which reduces weight and facilitates take-offs and flying, occurred earlier than the origin of Aves. †*Jeholornis* (see Figure 19.13A) is considered to have

(A) Nest of an †oviraptorosaur, †*Heyuannia huangi*

(B) Texture and blue-green color of eggshells

5 cm

Figure 19.19 **Nests and eggs of a Late Cretaceous †oviraptorosaur.** (A) Eggs are arranged in pairs with their blunt ends pointed to the center of the clutch. This reflects how the eggs were laid. (B) Textured eggshells retained enough pigment to confirm that they were blue-green. The nests of these theropods may have been partially exposed, with the egg color serving as camouflage or for species recognition. (From J. Wiemann et al. 2017. *Peer J* 5: e3706. https://doi.org/10.7717peerj.3706. CC BY 4.0.)

had a single ovary. Its phylogenetic position is at the base of the rapid diversification of avialans in the Cretaceous (see Figure 19.2).

Parental care of hatchlings

Recognition of parental care by theropods lagged discoveries of nests of †ornithischian dinosaurs because of a mistaken identification in 1923. The fossil of a theropod dinosaur that apparently died while attending a nest of eggs was discovered in the Gobi Desert, but its significance was not recognized until 70 years later. The fossilized eggs, which are about 12 cm long and 6 cm in diameter, originally were thought to have been deposited by a small †ceratopsian, †*Protoceratops andrewsi*, because adults of that species were by far the most abundant dinosaurs at the site. The theropod was assumed to have been robbing the nest and was given the name †*Oviraptor philoceratops*, which means "egg-seizer, lover of *Ceratops*."

In 1993, paleontologists from the American Museum of Natural History, the Mongolian Academy of Sciences, and the Mongolian Museum of Natural History discovered a fossilized embryo in an egg identical to the supposed †*Protoceratops* eggs. To their surprise, the embryo was an †*Oviraptor* nearly ready to hatch. With the benefit of hindsight, it is apparent that the adult dinosaur found in 1923 had been resting on its own nest, apparently trying to shelter its eggs from a sandstorm that buried both the adult and its nest. Additional fossils of adult †*Troodon* and †*Deinonychus*, which like †*Oviraptor* were maniraptorans, have subsequently been found sitting on eggs with their legs folded, arms extended, and bellies in contact with the eggs—the same posture that extant ground-nesting birds use when incubating eggs (**Figure 19.20**).

The extensive parental care that characterizes many extant birds originated among non-avian dinosaurs. Male parental care appears to be ancestral for pre-avian tetanurans.

(A) †*Oviraptor philoceratops* brooding eggs

Figure 19.20 **Parental care in theropods.** (A) Restoration of an adult †*Oviraptor philoceratops* brooding a nest of eggs. (B) Male emus brood their eggs in the same way. (C) Parental care by the male continues after the chicks hatch. With only minor changes, this image of an extant male emu with young could be a family group of †*Oviraptor*.

(B) Male emu, *Dromaius novaehollandiae*, brooding eggs

(C) Male emus remain with young post hatching

The relatively large clutch sizes of †*Troodon* and †*Oviraptor* match the clutch sizes of extant birds in which a male incubates the eggs of more than one female. Male incubation and parental care is retained in most species of paleognaths, the most primitive extant birds. Male care is characteristic of cassowaries, emus, rheas, and kiwis; only ostriches sometimes display biparental care. Thus, it is likely that male maniraptorans were the caregivers. Young maniraptorans may have remained with their male parent for extended periods, as do young emus. Biparental care characterizes about 80% of extant bird species but not extant paleognaths.

Summary

19.1 Characters and Systematics of Theropods

†Coelophysoids were among the earliest theropods. These small, swift predators had elongate skulls and were present all over Pangaea during the Late Triassic.

In the Early Jurassic, theropods underwent a significant radiation, resulting in the appearance of the two major lineages †Ceratosauria and Tetanurae.

- †Ceratosaurs were carnivores, and some reached enormous body sizes.

- Tetanurans gave rise to †megalosauroids, including the enigmatic †*Spinosaurus*, a gigantic, fish-eating animal with a sail on its back; †allosauroids such as the widely distributed apex predator †*Allosaurus*; †tyrannosauroids such as the iconic †*T. rex*; and other major lineages, including birds.

†Tyrannosauroidea is probably the most famous clade of theropods because it includes the formidable †*Tyrannosaurus rex*, known from spectacular fossil specimens and star of the movie *Jurassic Park.* However, although the first †tyrannosauroids appeared in the Jurassic, †*T. rex* was from the latest Cretaceous.

- Early †tyrannosauroids were small and lightly built with long arms that they used to seize prey.

- †*Tyrannosaurus* and other derived †tyrannosauroids were enormous, heavily built predators. Strong skulls, powerful jaw muscles, and serrated teeth allowed them to rip flesh from prey and crush bones.

- Forelimbs of derived †tyrannosauroids were greatly reduced, and only two fingers remained in †*Tyrannosaurus*. The arms were too short to lift food to the mouth, but they were robust, and the fingers were clawed, suggesting that the arms and hands were not vestigial. Their function remains a mystery.

†Ornithomimosaurs were ostrichlike in appearance and perhaps also in their ecology. Long hindlimbs in which the femur is longer than the tibia indicate that †ornithomimosaurs were cursorial. The long forelimbs had three grasping fingers. †Ornithomimosaurs had long necks and small skulls, and derived taxa had toothless jaws with beaks. They were probably omnivorous or strictly herbivorous and may have lived in groups, as ostriches do.

Maniraptorans retained the ancestral characters of long legs and long arms with grasping fingers. They include the insectivorous †alvarezsauroids, herbivorous †therizinosauroids, the probably omnivorous †oviraptorosaurs, the hypercarnivorous |dromaeosaurids, †troodontids, and birds. Extensive development of feathers in maniraptorans included downy feathers covering the body as well as pennaceous feathers on the arms, legs, and tail of some species.

The earliest avialans such as †*Archaeopteryx* were probably generalized carnivores. Cretaceous avialans radiated into many adaptive zones now occupied by extant birds, including, among others, fish eaters, insect eaters, and substrate probers. Only flying raptors (hawk equivalents) appear to have been rare or absent, possibly because of competition with †pterosaurs.

An initial radiation of avialans, †enantiornithines, was diverse and successful in the Cretaceous but did not survive the end-Cretaceous extinction. The Late Cretaceous clade Ornithurae contains the ancestor of Aves. Few Cretaceous fossils of basal Aves are known, so we do not know how much of the avian radiation occurred before the end of the Cretaceous.

19.2 †*Archaeopteryx*, Mesozoic Avialans, and the Mosaic Evolution of Avian Characters

Birds provide one of the clearest examples of mosaic evolution and exaptation. Many skeletal characters of birds evolved mosaically and are not directly linked to the evolution of flight. Some were already present in the ancestors of dinosaurs, whereas others evolved within theropods.

- Like other dinosaurs, theropods have a system of air sacs and postcranial pneumatization.

- A furcula (wishbone) is a synapomorphy for Theropoda.

- A wrist joint that allows the hands to be folded against the ulna and radius characterizes maniraptorans.

- The tail became shortened during avialan evolution, culminating in fusion of the last several tail vertebrae into a pygostyle.

Feathers are older than birds.

- Filamentous protofeathers occurred in †ornithischians. †*Sciurumimus*, a basal tetanuran, is the phylogenetically first theropod with filamentous feathers. Such filaments also occurred in a large †tyrannosaur, †*Yutyrannus*.

(Continued)

Summary (continued)

- †Ornithomimosaurs had symmetrical pennaceous feathers.

- Asymmetrical pennaceous feathers characterize Paraves, including forms such as †Microraptor.

A dramatic decrease in body size accompanied the evolution of avialans. Basal tetanurans weighed 1,000 kg or more, the oldest †tyrannosauroids were 100 kg, basal coelurosaurs weighed about 25 kg, †Archaeopteryx 0.5 kg, and early avialans 100 g or less.

19.3 Evolution of Powered Flight

Short glides (passive, non-flapping flight that covers horizontal distance while losing height) may have set the stage for the origin of powered flight.

Asymmetrical pennaceous feathers suggest that many paravians had the potential to fly, but only †Microraptor (a †dromaeosaurid), †Archaeopteryx, and more derived avialans had wing loading ratios that would have allowed powered flight.

During the Cretaceous, †pterosaurs and a diverse array of avialans lived side by side. Several other lineages also experimented with flight or gliding. Some had wings on the hindlimbs, and others had membranous wings. Only a feathered forewing occurs in extant birds.

19.4 Reproduction and Parental Care by Theropods

Extant crocodylians have two functional ovaries and deposit many small, symmetrical eggs in a single egg-laying event, whereas birds have a single functional ovary and deposit one large, asymmetrical egg at a time, completing the clutch over a period of several days.

Early theropods probably buried their eggs, but some maniraptorans built nests with partially exposed eggs. Fossils show that early maniraptorans brooded eggs as birds do, sitting on the clutch with their legs folded and arms extended to cover the eggs.

Uniparental care (probably by the male, as in many palaeognaths) is inferred for non-avian theropods. Biparental care is the derived condition and is characteristic of most birds.

Discussion Questions

19.1 Discuss the evidence for ecological replacement during theropod evolution by comparing the taxa of large carnivorous theropods that occurred in the Jurassic and Cretaceous. What role did the breakup of Pangaea play in the differentiation of theropods in the Northern and Southern Hemispheres?

19.2 How does the concept of mosaic evolution help to explain the mixture of dinosaurian and avian features in †Archaeopteryx?

19.3 Extant megapodes such as the Australian brush turkey Alectura lathami are birds (Aves) that bury their eggs in soil as crocodylians do, rather than brooding them in a nest like most birds. How could you determine whether this mode of incubation is a retained ancestral trait or a derived condition that has reverted to the crocodylian mode?

19.4 Today all large terrestrial predators are mammals, making them obvious examples to compare with large carnivorous theropods. Explain why this comparison may sometimes make sense, but also why it can be misleading.

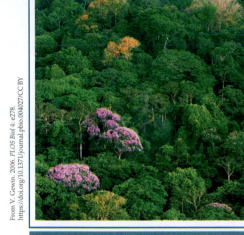

Geography and Ecology of the Cenozoic

Angiosperm-dominated tropical forest, Panama

Throughout the Phanerozoic, changes in continental positions have affected Earth's climates and the ability of vertebrates to disperse between its regions. The geographic continuity of Pangaea in the late Paleozoic and early Mesozoic allowed terrestrial vertebrates to spread across the world, resulting in faunal similarities even among regions far removed from one another. By the late Mesozoic, however, continental separation and epicontinental seas had isolated populations of terrestrial tetrapods and freshwater vertebrates, limiting possibilities for their migration between continents. Over the course of the Cenozoic, the continents reached the positions we see today, and faunas became progressively more different on the different continents.

In this chapter, we describe continental movements of the Cenozoic and the steady cooling from "hothouse Earth" conditions to the unusually stable climate of our own epoch, the Holocene. The episodes of cyclical glacial and interglacial conditions that made the preceding epoch—the Pleistocene—unique are given special attention. After an overview of the aquatic and terrestrial ecosystems of the Cenozoic, we describe the interchange of tetrapod species between North and South America that occurred when the formation of the Isthmus of Panama joined the Laurasian and Gondwanan landmasses of the Western Hemisphere. Finally, we describe several extinction events, including the recent extinctions of many species of mammalian megafauna.

20.1 Continental Geography and Climates

LEARNING OBJECTIVES

20.1.1 Identify and describe the epochs of the Cenozoic.

20.1.2 Describe movements of the continents during the Cenozoic.

20.1.3 Outline the general patterns of global climate change during the Cenozoic and link them to levels of atmospheric CO_2.

The Cenozoic era (Greek *kainos*, "near, recent"; *zoon* "animals") encompasses the most recent 66 Ma of Earth history and includes the Paleogene, Neogene, and Quaternary periods. Because we have so much more data for the three periods of the Cenozoic than we do for many earlier periods, it is typical to refer to the specific named epochs of each period, shown in **Figure 20.1** (compare Figure 20.1 with Figure 5.1 and Figure 10.1). For example, many people are familiar with the Pleistocene epoch, a time characterized by a cycle of glacial advances and retreats popularly called the "Ice Age."

We are living in the Cenozoic, specifically in the Holocene epoch of the Quaternary period of the Cenozoic.

Continental movements

Cenozoic continents were characterized by increasing isolation from one another and by a general trend of movement into higher latitudes, factors leading to new adaptations and increased speciation among organisms. Continental drift in the late Mesozoic and early Cenozoic continued the breakup of Pangaea, which during the Mesozoic split into Laurasia in the north and Gondwana in the south (see Section 10.1).

During the Eocene the landmasses of Laurasia, which had separated into North America and Eurasia, continued to drift northward into higher latitudes. Africa was

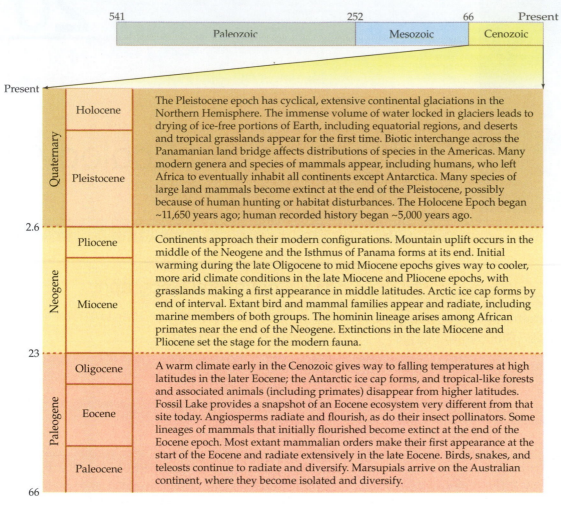

Figure 20.1 Major events of the Cenozoic era. Movement of continental blocks into their current configurations led to the differentiation of distinctive faunas on different continents.

separated from Eurasia and Asia by the Tethys Sea until the Miocene. North and South America were separated by open ocean and would not have a solid land connection until the Pleistocene (**Figure 20.2A**). Meanwhile, rifts formed in Gondwana, pulling southern landmasses farther apart. South America, Antarctica, and Australia separated from Africa during the Cretaceous (although movement of tetrapods between them by island-hopping continued into the early Cenozoic). India, which had broken away from Africa during the Mesozoic, continued to move northward and collided with the Eurasian plate in the middle Cenozoic. This massive collision produced the Himalayas, today's highest mountain range.

By the late Eocene, Australia had separated from Antarctica and was drifting northward. New Zealand, which separated from Australia in the Late Cretaceous, continued its northward drift with a unique indigenous vertebrate fauna. Intermittent land connections between South America and Antarctica existed through most of the Paleogene—Antarctic mammals of the Eocene were basically a subset of South American mammals. Antarctica fully separated from South America by the end of the Eocene, leaving the

ocean gap known today as Drake Passage, allowing the cold Antarctic circumpolar current to develop, and accelerating formation of the Antarctic ice cap.

By the start of the Quaternary, the continents were very close to the positions they occupy today (**Figure 20.2B**).

Cenozoic climates

At the start of the Cenozoic, Earth was still in the "hothouse" state of the late Mesozoic, with temperatures 10–12°C higher than they are today. Over the course of the Paleogene the climate changed from hot and moist to cooler and drier (**Figure 20.3**). Global cooling continued through the Neogene. Cyclical glaciations during the Pleistocene epoch of the Quaternary greatly influenced the ranges of many extant species of vertebrates, and was a further factor in increased isolation and resulting increases in species diversity among many groups.

Cooling down High temperatures continued through the first part of the Cenozoic. At the start of the Eocene, atmospheric CO_2 doubled to 1,000 ppm and the resulting greenhouse effect included a transitory temperature spike

Figure 20.2 Continental positions in the early Eocene and today. Some major events in Earth history—including formation of the Himalayas, separation of South America from Antarctica, and formation of the Isthmus of Panama—influenced ocean currents, global climates, and vertebrate evolution. Dark lines on the continents mark tectonic rift zones. (After C.R. Scotese. 2004. *J Geol* 112: 729–741. Also see C.R. Scotese. 2021. *Annu Rev Earth Planet Sci* 49: 669–718.)

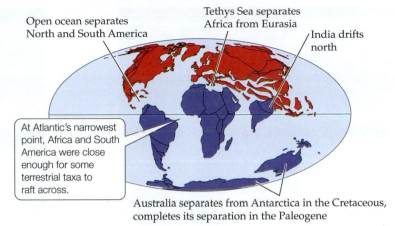

(A) Positions of continents 50 Ma (Eocene)

Open ocean separates North and South America

Tethys Sea separates Africa from Eurasia

India drifts north

At Atlantic's narrowest point, Africa and South America were close enough for some terrestrial taxa to raft across.

Australia separates from Antarctica in the Cretaceous, completes its separation in the Paleogene

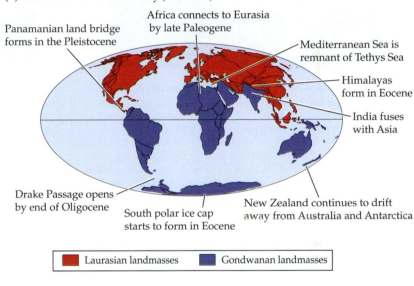

(B) Positions of continents today (Holocene)

Panamanian land bridge forms in the Pleistocene

Africa connects to Eurasia by late Paleogene

Mediterranean Sea is remnant of Tethys Sea

Himalayas form in Eocene

India fuses with Asia

Drake Passage opens by end of Oligocene

South polar ice cap starts to form in Eocene

New Zealand continues to drift away from Australia and Antarctica

■ Laurasian landmasses   ■ Gondwanan landmasses

of around 5–8°C, the Paleocene–Eocene Thermal Maximum (PETM; see Figure 20.3).

By the mid Eocene, the northward drift of North America and Eurasia had increased latitudinal temperature variation within the landmasses. Cooling started in higher latitudes following the maximum Eocene warmth at 50 Ma. There was a brief respite and slight rewarming at 40 Ma, then cooling continued; temperatures plummeted sharply before rising slightly in the Oligocene. Continental movements (such as the separation of South America and Antarctica) changed patterns of ocean currents, resulting in formation of Antarctic ice sheets at the South Pole by ~35 Ma. Temperatures rose to a Neogene high in the mid Miocene, after which further cooling resulted in the formation of Arctic ice sheets at the North Pole around 5 Ma (see Figure 20.3).

The closing of the Isthmus of Panama ~3 Ma connected North and South America. As well as forming a land bridge that species could traverse (see Section 20.3),

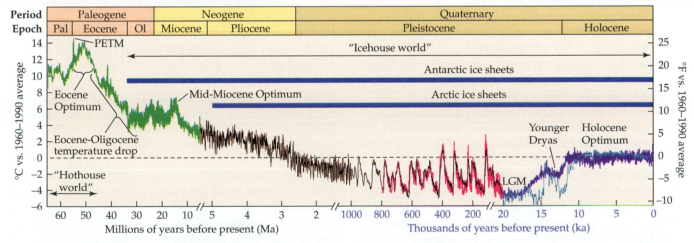

Figure 20.3 Shifting climates of the Cenozoic. Reconstructions of global air temperatures in the Cenozoic are compared with the global average air surface temperature 1960–1990 (dashed horizontal line). Note that major shifts in time scales are indicated by hatches on the x-axis and by changing colors of the traces; for example, a "hothouse world" existed for some 20 million years, half the length of the current "icehouse world." Variegated traces in the Pleistocene and Holocene represent reconstructions drawn from different methodologies. The past 10,000 years have seen relatively stable climatic conditions, referred to as the Holocene Optimum. PETM, Paleocene–Eocene thermal maximum; LGM, last glacial maximum. (Adapted from Glen Fergus/CC BY-SA 3.0.)

(A) Pleistocene (18 ka)

(B) Present

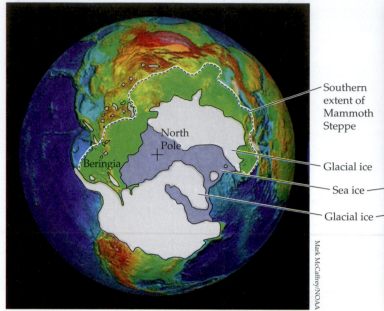

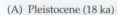

Figure 20.4 **Extent of Northern Hemisphere ice during the Pleistocene and today.** During periods of maximum glaciation, ice sheets 3–4 km deep extended far into the temperate latitudes of the Northern Hemisphere. Mammoth Steppe was a cold, dry biome extending across Eurasia into Canada that was home to large mammals such as †woolly mammoths and †woolly rhinoceroses, as well as extant reindeer and musk ox.

the presence of the Isthmus profoundly affected global climate. Warm waters could no longer circle the Earth; the Gulf Stream formed, carrying warm waters to the high latitudes of western Europe.

Pleistocene glaciations The cycle of continental glaciation and interglacial periods that began in the Pleistocene continues to the present. We still live in a world with abundant polar ice, but at the moment we are in an extended interglacial period—the Holocene Optimum—that began ~10 ka (see Figure 20.3).

The Pleistocene ice ages began ~2.5 Ma and forced many species of vertebrates in North American and Eurasia southward and/or into **glacial refugia**, areas (sometimes quite small) in which they could survive the harsh conditions and from which they subsequently expanded and repopulated as climate conditions changed.

Four major episodes of Pleistocene glaciation were interspersed with 20 or more minor ones. Some 20–16 ka, at the time of the Last Glacial Maximum (LGM; see Figure 20.3), ice 3–4 km thick covered the Northern Hemisphere as far south as 38°N in North America (about the position of New York City) and 50°N in Eurasia (about the position of Frankfurt, Germany; **Figure 20.4**).

Today's climate, the Holocene Optimum, is cool and dry compared with most of the Cenozoic, but warmer than the glacial episodes of the Pleistocene. The diversity of both animals and plants decreases northward and southward from the tropics to the poles in both terrestrial and aquatic habitats. Sea levels today are higher than in glacial episodes because less water is frozen in the polar ice caps.

20.2 Cenozoic Ecosystems

LEARNING OBJECTIVES

20.2.1 Explain the unique value of localities such as the Fossil Lake Lagerstätte for understanding Cenozoic ecosystems.

20.2.2 Describe changing patterns of global vegetation during the Cenozoic.

20.2.3 Provide examples of mammals, birds, and reptiles that radiated during the Cenozoic.

20.2.4 Explain why freshwater teleosts on different continents often provide examples of convergent evolution.

As with earlier eras, we have recovered many Cenozoic fossils from Lagerstätten, particularly in western North America and western Europe. A case in point is the Green River formation in the western United States, an early Eocene fossil bed that extends from southwestern Wyoming into Colorado and Utah, and an area within it called Fossil Lake.

Fossil Lake

Discovered in the 19th century, the Green River locality is among the most productive in the world for early Cenozoic fossils. Within this formation is an especially noteworthy Lagerstätte known as Fossil Lake. Deposition at Fossil Lake lasted for ~2 million years (from ~53 to ~51 Ma), but most of the fossils are from a period of ~20,000 years during which Fossil Lake was ecologically similar to the Florida Everglades of today, with a warm subtropical climate and high biodiversity. The Lagerstätte yields fossil invertebrates like snails, clams, crustaceans, spiders, millipedes, and hundreds of insect species. More than 250 species of plants are known from the locality, ranging from pollen grains to 14-foot-long palm fronds. Some plants even have chew marks, along with the fossilized insects that did the chewing.

Most relevant here, however, Fossil Lake preserves an entire contemporaneous community of vertebrates from the early Eocene, including frogs, salamanders, lizards, snakes, turtles, crocodylians, dog-sized browsing horses, small otterlike mammals related to shrews, and some of the earliest known bats. Dozens of extinct bird species are beautifully preserved as complete skeletons with feathers still attached, making Fossil Lake one of the world's best fossil bird localities. It captures the early rise of passeriform birds, a group that includes more than half of all extant bird species but which was apparently absent prior to the early Eocene.

As you might expect from its name, Fossil Lake is especially renowned for its preservation of aquatic fauna (**Figure 20.5**). An estimated 200,000 specimens are recovered there annually, representing more than 30 types of fishes—from fossil stingrays to paddlefishes, bowfins, and gars, as well as preserved teleosts similar to the arowanas, mooneyes, sandfishes, herring, pike, perches, and basses of today (see Chapter 7). The exquisite preservation of many fossils allows bone-for-bone direct comparisons with extant forms. Some animals died with food in their mouths or stomach regions, so we know what they were eating. There are remarkably complete developmental series for some species, ranging from specimens still coiled up in eggs to specimens several meters long. Some died while swimming in schools, and others provide clues about their reproductive behavior: researchers have uncovered fossils of copulating stingrays, pregnant females, and females surrounded by newborns.

So many fossil fishes have been collected from Fossil Lake that we can estimate the relative abundance of different species present there in the early Cenozoic. Some are superabundant (more than 600,000 specimens of a small fish related to herrings have been excavated), others are exceedingly rare (only one pike fossil has been discovered). Such enormous sample sizes can be difficult to recover even from extant ecosystems, so it is no wonder that generations of paleontologists have used data from Fossil Lake to interpret early Cenozoic paleoecosystems.

Freshwater habitats

Like Fossil Lake, many of the increasingly isolated freshwater habitats of the Cenozoic teemed with diverse fishes. Freshwater teleosts, including minnows, catfishes, and cichlids, radiated in isolation in the interior of the separated continents, increasing their global species diversity and offering many striking examples of convergent evolution. The inability to cross large expanses of salt water explains the occurrence and diversification of freshwater cypriniforms (minnows and allies) in Eurasia, Africa, and North America, which were connected by land and freshwater streams and rivers, but not in South America or Australia which were isolated by salt water. Because of its long isolation, the freshwater fish fauna of Australia is limited. For example, the only native catfishes of Australia belong to a mostly marine family that invaded Australian freshwaters from the sea.

Figure 20.5 Fossil Lake in the early Eocene. Some of the species of vertebrates that lived in the middle of the lake are shown in a reconstruction based on specimens from this unusually rich and comprehensive Lagerstätte.

Mural by Robert Hynes for Fossil Butte National Monument, National Park Service

Marine habitats

The vertebrate fauna of the Cenozoic oceans underwent considerable change in its composition over the course of the era.

Fishes Elasmobranchs and teleosts, particularly the extensive radiation of spiny-finned teleosts (Acanthomorpha, which accounts for about 25% of extant vertebrate species; see Section 7.5), dominated early Cenozoic seas. Coral reefs along the Tethys Sea became hotspots of diversification of marine teleosts in the Eocene. Gone were the giant filter-feeding teleosts of the Mesozoic, their ecological role eventually filled by baleen whales.

A peculiar major extinction of pelagic sharks occurred in the early Miocene (peculiar in the sense that we do not know what caused it), when as much as 90% of shark abundance and 70% of shark diversity disappeared. After this extinction, highly migratory apex predator sharks similar to the pelagic shark fauna of today evolved. This event highlights the importance of events during the Miocene that resulted in modern oceanic ecosystems.

Sauropsids The large †ichthyosaurs, †sauropterygians, †thalattosuchians, and †mosasaurs (see Figure 10.5) were all extinct by the end of the Cretaceous. Over the course of the Cenozoic, the seas were repopulated by new groups, particularly carnivorous mammals, that ecologically replaced these sauropsid apex predators.

Sea turtles and crocodylians survived the Cretaceous–Paleogene (K–Pg) extinctions (see Section 20.4) and prospered during the Cenozoic, which also saw the appearance of two new clades of marine snakes.

Birds Penguins, first known from the Paleocene, cannot fly in air but "fly" very effectively and travel long distances under water. The Southern Hemisphere radiation of penguins during the Cenozoic is somewhat paralleled by a Northern Hemisphere radiation of auks and puffins, a group not closely related to penguins. In the mid-19th century, people hunted the flightless great auk, †*Pinguinus impennis*, to extinction. Other auks and puffins can fly in the air, and some species dive and engage in subaqueous flight, but are not as powerful swimmers as penguins. †Plotopteryids, penguinlike marine birds related to cormorants, radiated in the Pacific and were up to 2 m tall.

Mammals Several groups of mammals entered the marine realm during the Cenozoic. The most diverse and specialized group—Cetacea, the whales—is known from the early Eocene. The radiation of extant toothed and baleen whales occurred in the Oligocene. Sea cows (sirenians) also date from the early Eocene. Pinnipeds (seals, sea lions, and walruses) appeared in the early Miocene.

Terrestrial flora

Modern angiosperm-dominated tropical forests date from the start of the Cenozoic, and the continuing success of angiosperms reduced the diversity and distribution of gymnosperms, especially conifers. During late Eocene cooling, conifers became restricted to high latitudes and high altitudes. Today's boreal forests are mostly conifers.

In the warm world of the Paleocene to middle Eocene, tropical-like forests extended into the Arctic Circle. We use the term "tropical-like" because, although these forests had a multilayered structure like that of modern tropical forests, they were a different assortment of species. Tropical-like Arctic forests were composed of broadleaf plants somewhat similar to those of the swamp-cypress forests and broadleaf floodplain forests of today's southeastern United States.

As global cooling became more profound in the late-middle Eocene (~45 Ma), temperate forests and woodlands, with seasonal climates and winter frosts, spread in the middle and high latitudes as the formerly extensive tropical-like forests became confined to equatorial regions.

Extensive savannas (grasslands with scattered trees) appeared in northern latitudes in the early Miocene, but not until the late Pliocene-early Pleistocene in tropical areas such as East Africa. And, by that time, former savannas at high latitudes had changed into treeless prairie or steppe.

Terrestrial fauna

Most of the extant clades of tetrapods originated in the Late Cretaceous or early Cenozoic. Among mammals, the only remaining synapsid lineage, rodents and bats had extraordinary bursts of diversification throughout the Cenozoic, and members of most other extant orders appeared in the Eocene. While most mammal species were small at the start of the Cenozoic, by the middle Eocene (~40 Ma) mammals larger than 1,000 kg occurred on most continents.

Among the sauropsids, derived clades of snakes (elapids, viperids, and colubrids) diversified in the Miocene and Pliocene, perhaps in part because of the contemporaneous radiation of small mammals, their principal prey. Terrestrial crocodyliforms occurred in the Southern Hemisphere during the Paleocene and Eocene, although only aquatic crocodylians survived into the later Cenozoic. Modern types of birds also diversified, including the passeriforms (perching birds), raptors (eagles, hawks, and vultures), ostriches, and the flightless †moas, †elephant birds, and †terror birds.

During the Last Glacial Maximum, a vast, cold, dry biome known as the Mammoth Steppe stretched from the Iberian Peninsula, across Eurasia, and into present-day Alaska and northwestern Canada (see Figure 20.4). Much of Alaska, Siberia, and Beringia (a land connection between present-day Alaska and northeastern Asia, now under water) was free of ice, allowing dispersal between Asia and North America. The Mammoth Steppe supported a diverse fauna of large mammals, including extinct forms such as †woolly mammoths and †woolly rhinoceroses, extant Arctic mammals such as reindeer and musk ox, and now-tropical mammals such as lions. Images of this extensive ecosystem have become emblematic of the Pleistocene fauna.

20.3 The Great American Biotic Interchange

LEARNING OBJECTIVES

20.3.1 Describe the composition of terrestrial faunas in North and South America before the Great American Biotic Interchange.

20.3.2 Explain how some terrestrial taxa moved north or south before closure of the Central American Seaway.

20.3.3 Summarize the current status of taxa that moved northward and southward between the Americas.

20.3.4 Explain the term "isthmian pair" and its biogeographic significance.

After the breakup of Pangaea in the Mesozoic separated North and South America, the faunas of the two continents developed in isolation for about 100 Ma. Communities of terrestrial vertebrates evolved on both continents, but because they began with different species assemblages and were isolated from each other, they evolved substantial differences.

For most of the Cenozoic, a continuous seaway between North and South America connected what we know today as the Pacific and Atlantic Oceans and allowed marine vertebrates to move barrier free throughout tropical and subtropical ocean waters. The major biogeographic event of the late Cenozoic was the formation of a land bridge—the Isthmus of Panama—between North and South America that allowed the faunas of the two continents to mix and at the same time separated some marine vertebrates, promoting speciation.

Terrestrial vertebrates of North and South America

By the late Cenozoic, both North and South America had full complements of mammalian herbivores and carnivores, but the faunas of the two continents were very different. The North American mammalian fauna had both endemic and Eurasian elements and included burrowing rodents such as voles and prairie dogs; rabbits and hares; weasels; horses; camels; pronghorn antelope; and elephant relatives like †mastodons and †gomphotheres. Apex predators in North America during the Pliocene and Pleistocene included wolves, jaguar, mountain lion, the †cave lion, and both †saber-toothed and †scimitar-toothed cats (†machairodontines). Horses and camelids were present but became extinct on the North American continent at the end of the Pleistocene.

An endemic South American mammalian fauna included both eutherians and metatherians (**Figure 20.6**).

- The diversity of eutherian †meridiungulates in South America from the late Paleocene through the Pleistocene rivaled that of the grazing animals of Africa.

Body forms included the elephantlike †astropotheres, camellike and horselike †litopterns, and tapirlike †xenungulates. †Notoungulates included species ranging 100-fold in size, from †*Toxodon* to †*Pachyrukhos* (see Figure 20.6).

- The eutherian clade Xenarthra includes anteaters, armadillos and †glyptodonts (armored grazers about the size and shape of a Volkswagen), and arboreal sloths and †ground sloths. Both †glyptodonts (†*Doedicurus*) and †ground sloths (†*Megatherium*) increased in size from ~100 kg in the early Miocene to 1,000–2,000 kg in the Pleistocene.

- The metatherian didelphid marsupials (opossums) radiated in the Pliocene, and more than 100 extant species now occur in South and Central America. Didelphidlike metatherians were originally present in North America as well, but were extinct by the end of the Miocene and metatherians did not reappear until the Virginia opossum (*Didelphis virginiana*) moved north during the Pleistocene.

- Apex predators in South American included basal methatherians, the †sparassodonts (see Figure 20.6). Among these were the largest metatherian predator known, †*Proborhyaena gigantea* (600 kg), and a 100-kg saber-toothed catlike scavenger, †*Thylacosmilus atrox*, in addition to a variety of smaller carnivores. These metatherian predators were joined by endemic sauropsid apex predators, including terrestrial †sebecid crocodyliforms, †terror birds (†phorusrhacids), and gigantic boas (†*Titanoboa*). These clades were absent from North America at the time.

In addition to these clades, two groups of eutherian mammals—caviomorph rodents and platyrrhine primates—entered South America from Africa during the early Oligocene, presumably by rafting across the Atlantic at its narrowest point (see Figure 20.2). Although most extant caviomorphs (guinea pigs, chinchillas, agoutis, and capybaras) weigh less than 1 kg, at 1,000 kg the extinct †*Josephoartigasia* is the largest rodent known (see Figure 20.6). Platyrrhines, the New World monkeys, underwent extensive radiation that included the evolution of a prehensile (gripping) tail not found in their Old World counterparts (see Section 24.1).

Faunal interchange

The South American fauna retained its unique nature until the exchange of clades between the two continents known as the **Great American Biotic Interchange**, or **GABI**. An exchange of mammals by island-hopping started before the Isthmus of Panama formed as raccoon relatives moved into South America and †ground sloths entered North America ~9 Ma. The closure of the Isthmus ~3 Ma created a stable land bridge, and by ~2.6 Ma there were major interchanges in both directions (**Figure 20.7**).

More species appear to have migrated from North America to South America than in the opposite direction.

MAMMALIA

Eutheria

Metatheria

†Meridiungulata

†Astrapotheria, †*Granastrapotherium*, 3,000 kg

†Litopterna, †*Macrauchenia* 1,000 kg

†Sparassodonta

†*Proborhyaena*, 600 kg

†Notoungulata, †*Toxodon*, 1,500 kg

†Xenungulata, †*Carodnia*, 175 kg

†*Thylacosmilus*, 100 kg

†Notoungulata, †*Pachyrukhos*, 1.5 kg

Rodentia
Caviomorpha, †*Josephoartigasia*, 1,000 kg

†*Cladosictis*, 6 kg, with †*Pachyrukos* in its jaws

Xenarthra

Cingulata, †*Doedicurus*, 2,000 kg

Pilosa, †*Megatherium*, 4,000 kg

Marsupalia

†Polydolopimorphia, †*Argyrolagus*, 500 g

Didelphimorphia, *Monodelphis*, 100 g

SAUROPSIDA

Crocodyliformes, †*Baurusuchus*, 100 kg

Cariamiformes, †*Paraphysornis*, 180 kg

Serpentes, †*Titanoboa*, 1,100 kg

Figure 20.6 South American endemics.   The endemic mammalian fauna of South America was very different from that of North America before the Isthmus of Panama was complete. Endemic clades of eutherians, ranging in size from 1.5 to 4,000 kg, composed most of the herbivore community. Mammalian carnivores were †sparassodont metatherians. Marsupials were a minor component of the fauna until didelphimorphs radiated in the Miocene into a large number of small to medium size generalized omnivores. The presence of sauropsid apex predators (crocodyliforms, birds, and snakes) was a unique feature of the South American fauna in the middle Cenozoic.

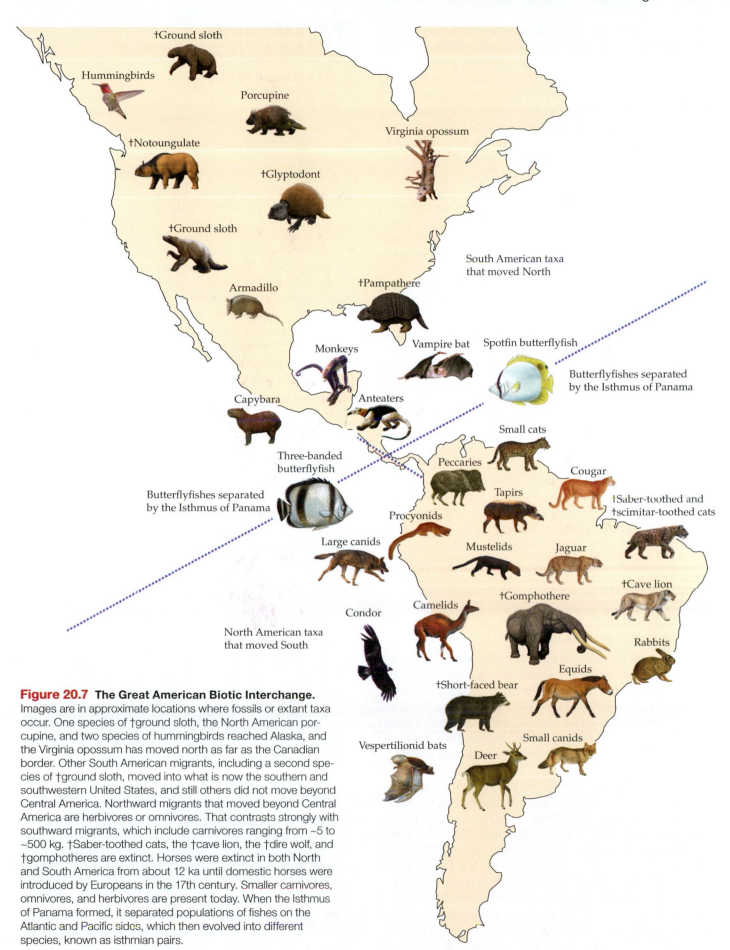

Figure 20.7 The Great American Biotic Interchange.
Images are in approximate locations where fossils or extant taxa occur. One species of †ground sloth, the North American porcupine, and two species of hummingbirds reached Alaska, and the Virginia opossum has moved north as far as the Canadian border. Other South American migrants, including a second species of †ground sloth, moved into what is now the southern and southwestern United States, and still others did not move beyond Central America. Northward migrants that moved beyond Central America are herbivores or omnivores. That contrasts strongly with southward migrants, which include carnivores ranging from ~5 to ~500 kg. †Saber-toothed cats, the †cave lion, the †dire wolf, and †gomphotheres are extinct. Horses were extinct in both North and South America from about 12 ka until domestic horses were introduced by Europeans in the 17th century. Smaller carnivores, omnivores, and herbivores are present today. When the Isthmus of Panama formed, it separated populations of fishes on the Atlantic and Pacific sides, which then evolved into different species, known as isthmian pairs.

Labels within figure:
†Ground sloth
Hummingbirds
Porcupine
Virginia opossum
†Notoungulate
†Glyptodont
†Ground sloth
Armadillo
†Pampathere
South American taxa that moved North
Monkeys
Vampire bat
Spotfin butterflyfish
Butterflyfishes separated by the Isthmus of Panama
Capybara
Anteaters
Small cats
Three-banded butterflyfish
Peccaries
Cougar
Butterflyfishes separated by the Isthmus of Panama
Tapirs
†Saber-toothed and †scimitar-toothed cats
Procyonids
Mustelids
Jaguar
Large canids
†Cave lion
Camelids
†Gomphothere
Condor
Rabbits
North American taxa that moved South
Equids
†Short-faced bear
Vespertilionid bats
Small canids
Deer

Thus, for many years the GABI was viewed as an example of the competitive superiority of Northern Hemisphere mammals, and especially the superiority of placental (eutherian) carnivores over metatherian carnivores. We now consider that interpretation too simplistic. For example, at the peak of their diversity in the early Miocene, South American †sparassodonts (see Figure 20.6) consisted of only 11 species, and these started to decline in the late Miocene and became extinct in the middle Pliocene (~3.5 Ma), well before the final closure of the seaway. Large eutherian carnivores did not appear in South America until the Pliocene, and the replacement of metatherian †sparassodonts by eutherian carnivores looks like opportunistic filling of niches vacated by extinctions of species in a declining clade rather than replacement of one clade by a competitively superior one.

The difference in the current status of mammalian clades that moved north and south during the GABI is more dramatic than the difference in the number of clades that migrated. Representatives of only three South American families of mammals persist in the United States: the Virginia opossum, the nine-banded armadillo, and the North American porcupine. Six additional mammal families of South American origin are present in Central America: anteaters, two- and three-toed tree sloths, capybara, and callitrichid and cebid monkeys.

Compare this with the 16 families of North American mammals that survive in South America. Even more striking, these 16 families comprise nearly half of the extant species of South American mammals. Although it could be argued that this contrast supports the notion of competitive superiority of North American fauna, it's important to note that the South American clades that became extinct in North America were all components of the megafauna that disappeared on both continents, whereas most North American clades that survive in South America are smaller animals.

Marine fauna and isthmian pairs

Formation of the isthmian land bridge also had consequences for marine fishes, which became isolated on the Caribbean and Pacific sides. At one extreme is the tiger shark. It occurs on both sides of the isthmus, but because it has an essentially global population, isthmus formation did not result in speciation. But smaller, less mobile fishes that live in warmer shallow waters diverged soon after the isthmus formed. For example, the spotfin butterflyfish (*Chaetodon ocellatus*) occurs on the Caribbean side of the isthmus; the closely related three-banded butterflyfish (*Chaetodon robustus*) is found on the Pacific side (see Figure 20.7). Molecular clock estimates show that these two species diverged ~3.6 Ma, which agrees with geologic estimates for closure of the Panamanian isthmus. Such **isthmian pairs** are known for other species of shallow-water fishes and invertebrates isolated by the isthmus, and their study provides opportunities to investigate speciation in the context of Earth history.

20.4 Extinctions

LEARNING OBJECTIVES

20.4.1 Compare the four major extinction events of the Cenozoic.

20.4.2 Discuss potential causes of the end-Pleistocene extinctions.

The Cenozoic began in the aftermath of the end-Cretaceous extinctions, the last major mass extinctions to date. Yet recovery at the start of the Cenozoic was remarkably rapid (on a geological time scale).

Four extinctions within the Cenozoic yielded the modern world.

- Late Eocene extinctions were associated with the dramatic drop in temperatures at higher latitudes and resulting environmental changes. Some early groups of mammals disappeared, as did mammals adapted to tropical-like forests, such as early primates. Surviving primate groups became restricted to lower latitudes. The diversity of lissamphibians and reptiles in higher latitudes also decreased dramatically during the late Eocene.

- Late Miocene extinctions were associated with global drying as well as falling temperatures. North America was especially hard hit because animals could not migrate to more tropical areas in South America before formation of the Isthmus of Panama in the Plio-Pleistocene.

- Extinctions at the end of the Pliocene particularly affected marine vertebrates, including cetaceans and giant sharks, and probably were linked to declines in coastal productivity following changes in oceanic circulation.

- The best known extinctions of the Cenozoic occurred at the end of the Pleistocene, when about 30% of mammal genera disappeared. Earlier extinction events in the late Eocene and late Miocene claimed a similar number of mammalian taxa, but they affected mammals of all sizes. In contrast, the Pleistocene extinction affected primarily megafauna (mammals weighing more than 20 kg), which vanished while smaller mammals were unaffected. Other groups, both terrestrial and marine, that had experienced profound extinctions during the Eocene and Miocene were less affected by the Pleistocene extinction.

Many of the megafaunal mammals that became extinct at the end of the Pleistocene are not represented by similarly giant types today. These include †glyptodonts (giant armored mammals related to armadillos) and †giant ground sloths in the Americas and †diprotodontids (marsupials as big as a hippopotamus) in Australia. The extinction also included large and exotic forms of more familiar mammals, such as †saber-toothed cats, †mammoths, †mastodons, the †Irish elk (†*Megaloceros giganteus*), †cave bears (†*Ursus sepleas*), and †woolly rhinoceros (†*Coelodonta antiquitatis*) of Eurasia; and †giant kangaroos (†sthenurines) in Australia. Large terrestrial birds also became extinct; these included herbivores such as †moas (†Dinornithiformes) of New Zealand and †elephant birds (†Aepyornithidae) of Madagascar, and carnivores such as the New World †terror birds (†phorusrhacids).

Much debate surrounds the causes of the Pleistocene extinctions. Many losses occurred at the end of the last glacial period ~12,000 years ago, defining the boundary between the Pleistocene and the Holocene. (Animals appear to be more vulnerable to extinction when the climate changes from glacial to interglacial rather than the other way around, probably because global warming effects occur more rapidly than the effects of cooling.) While the "climate change hypothesis" is an obvious explanation for Pleistocene extinctions, it is only the ending of the *last* glacial period (as opposed to the earlier interglacial periods of the Pleistocene) that brought extinctions of such magnitude.

Megafaunal extinctions of the Pleistocene were concurrent with spread of modern humans and their hunting techniques out of Africa; thus many scientists have concluded that much of the blame should be placed on human activity (the "overkill" hypothesis). In addition, human activities would have resulted in ecosystem disruption even in the absence of excessive hunting.

The climate change and overkill hypotheses are not mutually exclusive, of course; the two forces can act synergistically. In addition, geological and archeological evidence offers a third hypothesis: the Younger Dryas impact event. The impact occurred in Syria ~12.8 ka, spreading debris and igniting wildfires across North and South America, Europe, and Asia. The resulting sudden return to more glacial conditions, known as the Younger Dryas (see Figure 20.3), added to the stress on already vulnerable populations, and the extinction of a variety of megafaunal species appears to coincide with the impact. The end of the Younger Dryas ushered in the Holocene, an epoch of rather remarkable climatic stability that has lasted until the present day.

Summary

20.1 Continental Geography and Climates

Changes in continental positions during the Cenozoic affected ocean currents and created many of the present-day mountain ranges, which in turn affected patterns of rainfall and global climate.

As the breakup of Pangaea continued, continents became increasingly isolated and moved into higher latitudes, with the eventual result of ice forming at the poles. Different continents developed their own faunas that evolved to a greater or lesser extent in isolation.

After a transitory spike in temperatures around 55 Ma (the Paleocene–Eocene Thermal Maximum), global cooling began in the mid Eocene, ~50 Ma, and ice began to form at the South Pole.

Separation of South America and Antarctica allowed the Antarctic circumpolar current to develop, and the Antarctic ice cap formed by ~35 Ma. The Arctic ice cap is considerably more recent; it was established ~5 Ma.

Closure of the Isthmus of Panama ~3 Ma profoundly changed global climate with establishment of the Gulf Stream.

Episodes of extensive glaciation over the past 2.5 Ma resulted in ice sheets covering much of the Northern Hemisphere and a drier world with lower sea levels. We remain in this episodic cycle, although we are in an interglacial period with higher sea levels.

20.2 Cenozoic Ecosystems

Some of our detailed knowledge of Cenozoic ecosystems comes from Lagerstätten such as Fossil Lake, a community of vertebrates, invertebrates, and plants within the early Eocene Green River formation in western North America.

From the start of the Cenozoic, angiosperms dominated tropical forests, reducing the diversity and distribution of gymnosperms, especially conifers.

• During the warm world of the Paleocene, tropical-like forests extended into the Arctic Circle. Global cooling during the Eocene (~4 Ma) resulted in temperate forests and woodlands in the middle and high latitudes and tropical-like forests in equatorial regions.

• Extensive savannas appeared in the early Miocene in northern latitudes, but not until the late Pliocene/early Pleistocene in tropical areas such as East Africa.

Derived clades of snakes diversified in the Miocene and Pliocene, perhaps in response to the contemporaneous radiation of small mammals, their primary prey.

Terrestrial crocodyliforms occurred in the Southern Hemisphere during the Paleocene and Eocene, but only aquatic crocodylians survived into the later Cenozoic.

Modern types of birds diversified in the Cenozoic.

Freshwater teleosts radiated in isolation in the interior lakes and waterways of the separated continents, increasing

(Continued)

Summary *(continued)*

their global species diversity and providing examples of convergent evolution.

Elasmobranchs (sharks and rays) and teleosts (particularly spiny-finned teleosts, Acanthomorpha) dominated early Cenozoic seas. A major, unexplained extinction of pelagic sharks occurred in the early Miocene, after which highly migratory apex predator sharks similar to today's shark fauna evolved, thereby linking Miocene events to modern oceanic ecosystems.

Cenozoic marine tetrapod fauna consisted of mammals (cetaceans, sirenians, and pinnipeds), and birds (penguins in the Southern Hemisphere and auks in the Northern Hemisphere), although sea turtles continued from the Mesozoic.

20.3 The Great American Biotic Interchange

In the early Cenozoic, North and South America, separated by the breakup of Pangaea in the Triassic, were still isolated from each other by the Central American Seaway. The faunas of South and North America were distinctively different for most of the Cenozoic.

Around 3 Ma, formation of the Isthmus of Panama led to exchanges of animals between the two continents, known

as the Great American Biotic Interchange. Many species of endemic South American mammals are now extinct, and around half of its present-day mammalian fauna has North American origins.

Isthmian pairs of species represent speciation that happened in the now-separated Atlantic and Pacific waters after their division by the Isthmus of Panama.

20.4 Extinctions

While the best known Cenozoic extinction took place at the end of the Pleistocene, it mainly affected only larger land animals (megafauna). Extinctions in the late Eocene and late Miocene, both related to declines in higher-latitude temperatures, affected marine and terrestrial animals of all sizes.

Debate continues whether Pleistocene extinctions were mainly caused by climate changes at the end of the last glacial period, or whether the spread of humans across the globe was the major factor, both by their hunting of large animals and by habitat disruption. It is probable that both climatic change and humans played important roles.

Discussion Questions

20.1 How do Earth's ecosystems today differ from those of the early Cenozoic?

20.2 Why did the mammalian faunas of North and South America become so dissimilar during the Cenozoic?

20.3 How are marine vertebrates of the Cenozoic different from those of the Mesozoic?

20.4 Why are the causes of the Pleistocene extinction so controversial?

© Katherine E. Bemis

Extant Birds

Hyacinth macaw, *Anodorhynchus hyacinthinus*

By the end of the Mesozoic, the characters we readily associate with birds—including feathers, flight, and the shelled egg—were well established in the avian lineage, as described in Chapter 19. During the Cenozoic, this "last lineage of dinosaurs" radiated and produced the many and varied species whose beauty and behaviors engage us today.

Flight is a central characteristic of most birds, and, at the structural level, requirements for flight shape many aspects of avian anatomy and physiology. In terms of ecology and behavior, flight provides options not available to terrestrial animals. For example, many birds make long-distance migrations to exploit seasonal changes in resources. Even small species such as hummingbirds travel thousands of kilometers between breeding and non-breeding areas.

A second characteristic of birds is diurnality—most species are active during the day. In general, birds have excellent vision, and color and movement play important roles in their lives. Because humans also are diurnal and alert to colors and movements, birds are popular subjects for behavioral and ecological field studies. The singing behavior of birds has been examined using both ecological approaches and physiological methods to assess neural and hormonal influences on song development, and findings from studies of songbirds inform research on human communication disorders. In short, important areas of biology draw heavily on studies of birds for information that can be generalized to other vertebrates.

In this chapter, we begin with a brief overview of the taxonomic diversity of extant birds. We then turn to the myriad structural specializations for flight, bipedalism, and feeding that have produced this diversity. We describe avian sensory systems, reproduction, and behavior, with special emphasis on communication and navigation, before concluding by briefly enumerating just a few of the many threats faced by birds in our human-dominated postindustrial world.

21.1 Diversity of Aves

LEARNING OBJECTIVES

21.1.1 Explain the functional significance of palatal differences between Paleognathae and Neognathae.

21.1.2 Characterize the major groups within Neognathae.

21.1.3 Identify the character uniting Passeriformes and distinguish among its three extant groups.

Birds are so diverse that ornithologists even debate the number of extant species. The 2021 edition of the International Ornithological Congress's World Bird List (version 11.2) lists 10,912 extant species. Other authorities assert there may be twice that number. A stable, robust phylogenetic tree for extant Aves has eluded generations of ornithologists, and there are still many questions about evolutionary relationships of major groups of living birds. This chapter follows the tree shown in **Figure 21.1**, which incorporates current molecular data.

There is one point of agreement in that two major groups of extant birds, Paleognathae and Neognathae, are distinguished by their palatal anatomy. The names, which mean "old jaw" and "new jaw," highlight a significant evolutionary distinction. The palate of paleognathous

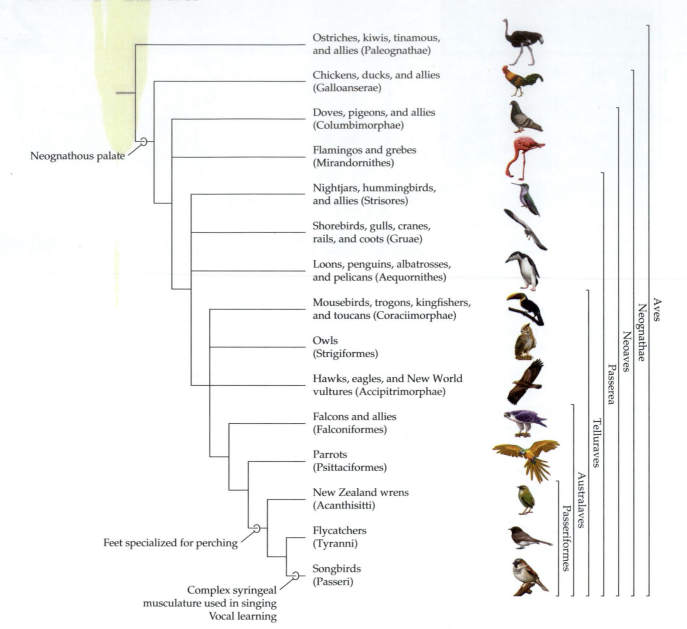

Ostriches, kiwis, tinamous, and allies (Paleognathae)

Chickens, ducks, and allies (Galloanserae)

Doves, pigeons, and allies (Columbimorphae)

Flamingos and grebes (Mirandornithes)

Nightjars, hummingbirds, and allies (Strisores)

Shorebirds, gulls, cranes, rails, and coots (Gruae)

Loons, penguins, albatrosses, and pelicans (Aequornithes)

Mousebirds, trogons, kingfishers, and toucans (Coraciimorphae)

Owls (Strigiformes)

Hawks, eagles, and New World vultures (Accipitrimorphae)

Falcons and allies (Falconiformes)

Parrots (Psittaciformes)

New Zealand wrens (Acanthisitti)

Flycatchers (Tyranni)

Songbirds (Passeri)

Neognathous palate

Feet specialized for perching

Complex syringeal musculature used in singing
Vocal learning

Aves · Neognathae · Neoaves · Passerea · Telluraves · Australaves · Passeriformes

Figure 21.1 Phylogeny of extant birds (Aves). Simplified consensus phylogeny of major groups of birds based on analyses of DNA sequence data. Different studies find different branching patterns, particularly for phylogenetic relationships within Neoaves. (Data from E.L. Braun and R.T. Kimball. 2021. *Birds* 2: 1–22. doi 10.3390/birds2010001. See additional credit details at the end of this chapter.)

birds resembles that of earlier theropods in which bones such as the vomer are relatively large (**Figure 21.2A**). The palatine and pterygoid bones are fused, rendering the palate rigid and inflexible. Extant paleognathous birds include ostriches, emus, rheas, cassowaries, kiwis, and tinamous. Tinamous are reluctant flyers; the others are flightless. The term "ratites" (Latin *ratis*, "raft") is often applied to the flightless forms because they have a smooth, raftlike sternum that lacks a keel, but recent studies conclude that "ratites" is a non-monophyletic group (see Section 1.2) because tinamous are the sister group of cassowaries.

In contrast to Paleognathae, neognathous birds have a lightly built palate with a reduced vomer. To each side, elongated palatine bones have moveable joints with the pterygoid (**Figure 21.2B**). These intercranial joints allow the palate and beak of neognathous birds to move during feeding. This seemingly simple change opened avenues for the evolution of many different bill (beak) shapes and specializations, as we describe in Section 21.5. These diverse bill structures and feeding methods are instrumental in explaining the expansion of neognathous birds into virtually all ecological niches, aquatic and terrestrial, from the Equator to the polar regions (including Antarctica). Some of this diversity is seen in Figure 21.1, described in **Table 21.1**, and detailed in the rest of this chapter.

Galloanserae, the first branch within Neognathae, includes chickens, ducks, and allies. This group diverged from the main stem of avian evolution in the Late Cretaceous. Neoaves includes all other extant birds. There are

(A) Paleognathous palate of a rhea, *Rhea*

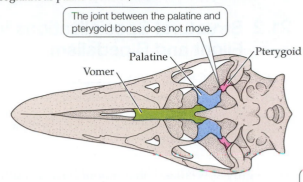

The joint between the palatine and pterygoid bones does not move.

Palatine

Vomer

Pterygoid

(B) Neognathous palate of a wild turkey, *Meleagris*

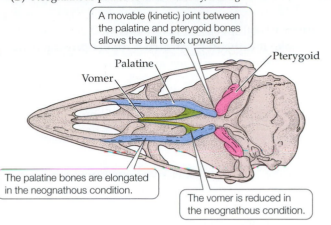

A movable (kinetic) joint between the palatine and pterygoid bones allows the bill to flex upward.

Palatine

Vomer

Pterygoid

The palatine bones are elongated in the neognathous condition.

The vomer is reduced in the neognathous condition.

Figure 21.2 Palatal differences of Paleognathae and Neognathae. The two main groups within extant birds are distinguished by the form of the palate. (A) The paleognathous palate is robust and rigid, resembling that of earlier theropods. (B) The neognathous palate is more lightly built and contributes to cranial kinesis (mobility of the skull bones). The vomer is reduced, and long, thin palatine bones form kinetic joints with the pterygoid the pterygoid bone. These changes allowed the palate and bill to move during feeding, which in turn led to widespread specializations for different food sources and methods of feeding. (See credit details at the end of this chapter.)

Table 21.1 Terminal taxa of extant birds as shown in Figure 21.1

Taxon	Included groups and number of species
Paleognathae	Paleognathous (inflexible) palate, flightless (except for tinamous). Cassowaries and emus (Casuariiformes; 4 species). Kiwis (Apterygiformes; 5 species). Rheas (Rheiformes; 2 or 3 species). Ostriches (Struthioniformes; 2 species). Tinamous (Tinamiformes; 46 species).
Neognathae	Encompasses all nonpaleognathous extant birds, listed below. Flexible palate with intercranial joints. Specializations of the neognathous palate have enabled the radiation of birds into virtually all feeding niches and biomes on the planet.
Galloanserae	Ground-dwelling fowl, including chickens, quail, and megapodes (Galliformes; 290 species). Semiaquatic waterfowl, including ducks, geese, and swans (Anseriformes; 180 species).
Columbimorphae	Flocking seed eaters including pigeons, doves, and allies (Columbiformes; 300 species). Sandgrouses (Pterocliformes; 16 species) and mesites (Mesitornithiformes; 3 species).
Mirandornithes	Flamingos, tropical filter feeders (Phoenicopteriformes; 6 species). Grebes, freshwater diving birds with lobate-webbed toes (Podicipediformes; 20 species).
Strisores	Nighthawks, frogmouths, and poor-wills, nocturnal insectivores, as well as the oilbird, a nocturnal South American frugivore (Caprimulgiformes; 130 species). Hummingbirds and swifts, arboreal birds with specialized flight; swifts feed on insects they capture in flight, and hummingbirds are specialized nectar-feeders that also capture flying insects (Apodiformes; >400 species). Some analyses place arboreal and ground-dwelling cuckoos (Cuculiformes; 150 species) in or near Strisores.
Gruae	Hoatzin, *Opisthocomus hoazin*, a unique herbivorous foregut fermenter of the South American forest (Opisthocomiformes; 1 species). Gulls and auks, carnivorous shorebirds (Charadriiformes; 350 species). Mostly omnivorous cranes (waders), rails (ground-dwellers, some island species flightless), and coots (swimmers/waders) (Gruiformes; 180 species).
Aequornithes	Core water birds. Loons (Gaviiformes; 5 species). Albatrosses and petrels (Procellariiformes; 125 species). Penguins (Sphenisciformes; 17 species). Pelicans (Pelecaniformes; 8 species). Cormorants and boobies (Suliformes; 60 species). Storks, herons, and ibises (Ciconiiformes; 120 species).
Coraciimorphae	Mousebirds (Coliiformes; 6 species). Trogons (Trogoniformes; 46 species). Hornbills (Bucerotiformes; 70 species). Kingfishers (Coraciiformes; 100 species). Woodpeckers and toucans (Piciformes; >300 species).
Strigiformes	Owls. Mostly nocturnal predators with excellent hearing (~200 species).
Accipitrimorphae	New World vultures (Cathartiformes, 7 species). Hawks, eagles, sea eagles, kites, and buzzards (Accipitriformes; 250 species). Scavengers and birds of prey (raptors).
Falconiformes	Falcons, caracaras, and relatives (~60 species). Scavengers and birds of prey.
Psittaciformes	Parrots (~400 species). Arboreal fruit- and seed-eaters with muscled, highly mobile tongues and capacity for vocal learning.
Passeriformes	Perching birds. The three taxa listed below comprise this largest avian clade, with more than half the world's extant species of birds.
Acanthisitti	New Zealand wrens (2 species). Tiny, with weak flight. Endemic to New Zealand.
Tyranni	Also called suboscines. Ovenbirds, antbirds, tyrant flycatchers, and relatives (>1,300 species). Limited control of syringeal muscles, do not produce complex vocalizations.
Passeri	Also called oscines or songbirds. Lyrebirds, crows, sparrows, thrushes, warblers, many other familiar songbirds (>5,000 species). Fine control of syringeal muscles allows complex songs.

more than 10,000 species of neoavians, and several challenges make it difficult to construct robust phylogenetic trees for the many diverse neoavian groups.

- Significant divergences within Neoaves occurred over a remarkably short time in the late Cretaceous and early Paleocene, and it is hard to resolve such rapid divergences even with sophisticated molecular phylogenetic approaches.

- There may have been extensive gene flow by hybridization among early-diverging lineages, further complicating the potential of DNA sequencing to identify clear branching sequences.

- Neoavians exhibit many convergently evolved anatomical specializations, making it difficult to specify robust anatomical characters for many of its subclades. For example, birds of prey (commonly known as raptors) are not a monophyletic group; their similarities in form are due to similar specializations for catching and killing prey, but these specializations arose independently in some groups.

Doves, pigeons, and allies (Columbimorphae), flamingos and grebes (Mirandornithes) group near the base of the Neoaves tree (see Figure 21.1). These placements, based on the most recent phylogenetic analyses, differ from analyses published only a few years ago. More derived birds belong to Passerea. At its base are the nightjars, hummingbirds, and allies (Strisores), shorebirds, gulls, cranes, rails, and coots (Gruae), and the so-called core water birds, which includes loons, penguins, albatrosses, boobies, and pelicans (Aequornithes).

Remaining groups in Figure 21.1 and Table 21.1 belong to Telluraves, the core landbird clade. Recent studies do not resolve phylogenetic relationships among the four clades of Telluraves. The first, Coraciimorphae, includes many cavity-nesting birds such as woodpeckers. Strigiformes, the owls, is distinct from Accipitrimorphae, which includes hawks, eagles, and New World vultures. The fourth and by far the largest group, Australaves, is named for its origin in the Southern Hemisphere and includes falcons (Falconiformes), parrots (Psittaciformes), and perching birds (Passeriformes, also known as passerines).

Passeriformes includes more than 6,500 species—more than half the known extant bird species. Subgroups range in size from the two species of tiny, nearly flightless New Zealand wrens (Acanthisitti) to ~1300 species of flycatchers (Tyranni, sometimes called suboscines) to the more than 5,000 species of songbirds (Passeri, also known as oscines; Latin *oscen*, "songbird"). Passeri includes familiar birds such as crows, warblers, thrushes, sparrows, and many others. Songbirds share derived syringeal muscles used to produce songs, and song production is closely linked to the complex behavioral repertoires of many oscine species.

21.2 Structural Specializations for Flight and Bipedalism

LEARNING OBJECTIVES

21.2.1 Describe structure and function of the five feather types.

21.2.2 Explain key skeletal specializations for flight and bipedalism.

21.2.3 Relate muscle characteristics to type of locomotion.

The highly specialized anatomy of birds is both varied and uniform. Wing shapes reveal different types of flight, feet reflect different forms of locomotion, and bills are specialized for different modes of feeding. Despite these differences, birds have certain uniformities in structure related to specializations for flight and bipedalism.

Body size

Flight imposes a maximum body size on birds. The muscle power needed for takeoff increases by 2.25 for each doubling of body weight. So if species B weighs twice as much as species A, species B will require 2.25 times as much power to fly at its minimum speed. If the proportion of total body weight allocated to flight muscles is constant, the muscles of a large bird must work harder than those of a small bird. In fact it is even more complicated, because power output is a function of both muscular force and wingbeat frequency, and larger birds have lower wingbeat frequencies. Thus, if species B weighs twice as much as species A, species B can develop only 1.59 times as much power from its flight muscles, although it needs 2.25 times as much power to fly. Therefore, larger birds make longer takeoff runs, and a bird could ultimately reach a body weight at which any further increase in size would move it into a realm in which its leg and flight muscles cannot provide enough power to take off.

Flightless birds are spared the problems of producing flapping flight, but even they do not approach the body sizes of large mammals. The heaviest extant birds are the two species of ostriches, which weigh ~150 kg. Extinct †elephant birds weighed an estimated 450 kg. In contrast, the largest extant terrestrial mammal, the African bush elephant, can weigh >5,000 kg.

In addition to body size, body shape is relatively uniform among birds, a fact that relates both to flight and their universal bipedalism. We begin our consideration of flight and bipedalism by exploring structural specializations associated with each.

Feathers

The protein beta-keratin is a synapomorphy of sauropsids and is the main component of feathers. Feathers develop from follicles in the skin and are arranged in tracts called **pterylae**. Pterylae are separated by patches of unfeathered skin, or **apteria**. Some birds, such as penguins, lack pterylae; their short feathers are uniformly distributed over the skin, forming a continuous, insulating layer.

Feather anatomy   Each feather is anchored to its follicle by a tubular base, the **calamus**, which remains firmly implanted within the follicle until molting. A long, tapered, and solid **rachis** extends from the calamus. Closely spaced side branches called **barbs** project from the rachis, and smaller **barbules** branch from opposite sides of the barbs. **Hooklets** extend from barbules and hold adjacent barbs together, forming a flexible, cohesive **vane** (**Figure 21.3A**). A feather is described as pennaceous when a vane is present. Plumulaceous or downy feathers lack a vane (**Figure 21.3B**). Barbules and hooklets maintain the pennaceous character of feather vanes. If a feather's barbs become detached from one another, a bird can draw its slightly open bill along the feather to realign barbules.

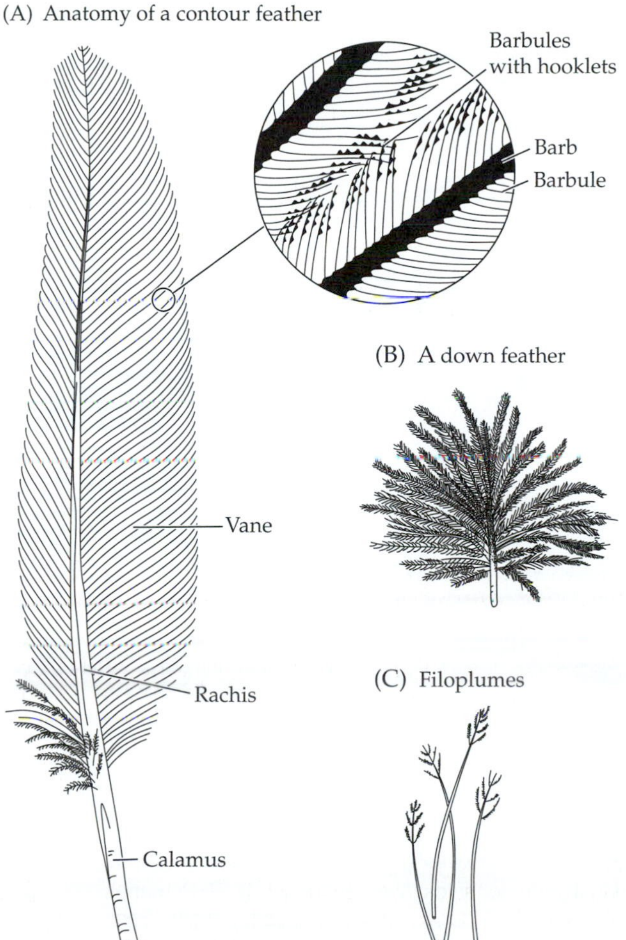

(A) Anatomy of a contour feather

Barbules with hooklets

Barb

Barbule

Vane

Rachis

(B) A down feather

(C) Filoplumes

Calamus

Inferior umbilicus

Figure 21.3 Three types of feathers.   (A) Contour feathers include the outermost feathers on the body, wings (remiges), and tail (rectrices). They have a distinct pennaceous vane as well as a calamus and rachis. Barbs of the vane are held together by barbules and hooklets. (B) Down feathers, which function in thermoregulation, are plumulaceous, lacking a vane. Often they also lack a rachis. (C) Filoplumes are hairlike feathers associated with nerves and receptors near follicles; sensory signals from these feathers help birds correctly position the contour feathers. (See credit details at the end of this chapter.)

A single feather can have both pennaceous and plumulaceous regions (see Figure 21.3A). Near the base of the rachis, barbs and barbules are flexible and barbules lack hooklets, lending a loose, fluffy appearance to this region. This plumulaceous region gives plumage its thermal insulation properties. Farther from the base, the barbs form the tight surface of the vane. The vane is exposed on the exterior surface of the plumage, where it can protect the downy undercoat and shed water.

Ornithologists distinguish five feather types:

- **Contour feathers** are the outermost feathers on a bird's body, wings, and tail. **Remiges** (flight feathers of the wing; singular *remex*) and **rectrices** (flight feathers of the tail; singular *rectrix*) are large, stiff, mostly pennaceous contour feathers modified for flight. Unlike other feathers, remiges are anchored to bones by ligaments. Remiges consist of **primaries** (rooted in the hand, also called the manus) and **secondaries** (rooted in the ulna of the forearm; **Figure 21.4**).

- **Down feathers** are entirely plumulaceous (see Figure 21.3B); the rachis is typically absent. Down feathers provide insulation.

- **Semiplumes** are intermediate in structure between contour and down feathers. They have a rachis, but the barbs do not form a vane because the barbules lack hooklets. Semiplumes are hidden beneath contour feathers, where they provide insulation and help streamline contours of a bird's body.

- **Bristles** are specialized feathers with a stiff rachis, either without barbs or with barbs only on proximal portions. Bristles commonly occur around the eyes and base of the bill, where they keep foreign particles from entering the eyes and nostrils.

- **Filoplumes** are fine, stiff, hairlike feathers with a few short barbs at the tips (see Figure 21.3C). Pressure and vibration receptors in the follicle walls of filoplumes transmit information about position and movement of contour feathers, helping birds adjust their contour feathers for flight, insulation, and display.

Feather pigments   Pigments and structural features produce feather colors. Three types of pigments are widespread among birds: melanins, carotenoids, and porphyrins. Two types of melanins produce dark colors: eumelanin (black, gray, and dark brown), and phaeomelanin (reddish brown and tan; **Figure 21.5A**). Carotenoids are fat-soluble pigments responsible for most red, orange, and yellow colors. Birds obtain carotenoids from their diets, and in some cases color intensity can be used to gauge fitness of a prospective mate. Porphyrins are metal-containing compounds that emit a red fluorescence when exposed to ultraviolet light. Prolonged exposure to sunlight destroys porphyrins, so they are most conspicuous in new plumage.

(A) Dorsal view of wing

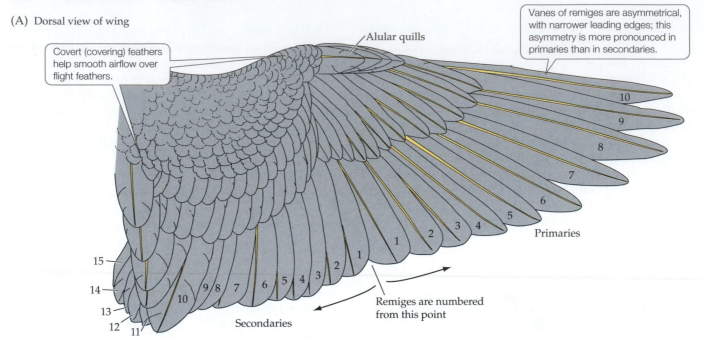

Covert (covering) feathers help smooth airflow over flight feathers.

Alular quills

Vanes of remiges are asymmetrical, with narrower leading edges; this asymmetry is more pronounced in primaries than in secondaries.

Primaries

Remiges are numbered from this point

Secondaries

(B) Ventral view of wing with coverts removed

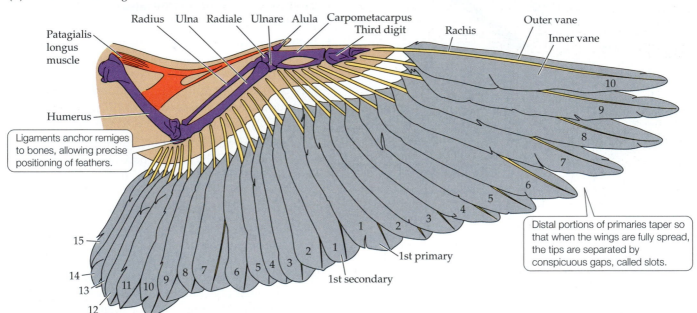

Patagialis longus muscle

Radius   Ulna   Radiale   Ulnare   Alula   Carpometacarpus
Third digit

Rachis

Outer vane

Inner vane

Humerus

Ligaments anchor remiges to bones, allowing precise positioning of feathers.

1st primary

1st secondary

Distal portions of primaries taper so that when the wings are fully spread, the tips are separated by conspicuous gaps, called slots.

Figure 21.4 Wing feathers modified for flight.
Remiges are contour feathers of the wing. Their modifications for flight include primary and secondary feathers; 10 primaries and 15 secondaries are shown in the generalized wing diagrammed here. (A) Alular quills are the 3–5 feathers attached to the alula, the most anterior digit of the wing. This structure is important in flight mechanics (see Section 21.3). (B) Primaries attach to the manus, or hand, which includes the digits. Secondaries attach to the ulna. (See credit details at the end of this chapter.)

Purple, blue, and green are structural colors; they result from tiny, air-filled structures on the surfaces of feather barbs (**Figure 21.5B,C**). When light is scattered by these structures, some wavelengths show constructive interference (intensification of that wavelength), while others show destructive interference (reduced intensity). The spacing of air-filled structures determines which colors are intensified. Structural colors can be combined with pigments; green parakeets, for example, combine a structural blue with a yellow carotenoid.

Streamlining and weight reduction

Birds are the only vertebrates that move fast enough in air for wind resistance and streamlining to be important factors in body shape. Ducks and geese can fly at 80–90 km/h, and peregrine falcons, *Falco peregrinus*, reach speeds of more than 300 km/h when they dive for prey, a behavior known as stooping. Fast-flying birds have many features copied in fast-flying aircraft. For example, contour feathers make smooth junctions between the wings and the body, limiting sources of turbulence that would

(A) Tail feathers of a male wild turkey, *Meleagris gallopavo*

Courtesy of U.S. National Park Service

(B) Tail feathers of a peacock, *Pavo cristatus*

© René C. Clark, Dancing Snake Nature Photography

(C) Two views of the gorget of a male Anna's hummingbird, *Calypte anna*

© René C. Clark, Dancing Snake Nature Photography

Figure 21.5 Pigments and structures combine to produce the colors and patterns of feathers. (A) Eumelanin and phaeomelanin pigments produce the black, brown, tan, and cream regions in the tail feathers of a male wild turkey. (B) Melanins are also responsible for the dramatic colors of a peacock's tail feathers, but this is a structural color caused by interference. (C) The magenta gorget of a male Anna's hummingbird is also produced by interference. The gorget is bright and highly visible from some angles of view (left), but disappears when the bird turns slightly (right).

increase wind resistance. Feet are tucked close to the body during flight (just as an aircraft raises its landing gear) to improving streamlining.

Weight reduction is also key for flight, and reduced body weights evolved as the number and size of some organs decreased; birds lack some organs altogether. For example, almost all birds lack urinary bladders, adult females typically have only the left ovary, and males of most species lack an intromittent organ (see Section 21.8).

Skeleton

Avian skeletons reflect structural modifications for both flight and bipedalism. Many bones of birds are pneumatic (air-filled; **Figure 21.6**), and the skull is especially light. However, the leg bones of birds are proportionally heavier than those of mammals because a bird's two legs have to support the body weight that a quadruped supports with four legs. Here we consider some specializations of the avian skeleton—especially fusion, reduction, and loss of bones—associated with flight and bipedalism.

© The Natural History Museum/Alamy Stock Photo

Cortex   Trabecula   Air pocket

Figure 21.6 Interior of an avian long bone. Air pockets and reinforcing trabeculae in the avian humerus form a hollow core that combines strength and light weight.

Axial skeleton   Birds are diapsids, but the bar between the two temporal fenestrae is lost in extant species (see Section 9.3). The single remaining temporal fenestra merges with the orbit; the eyes and orbits are large, reflecting the importance of vision. The jaws form a bill covered by a horny sheath of keratin called the rhamphotheca. The premaxilla supports the rhamphotheca in the upper bill and the dentary supports it in the lower bill. Extant birds lack teeth, traditionally interpreted as a weight-reducing adaptation for flight. Another hypothesis to explain tooth loss is that tooth development takes a long time and would require longer incubation periods (alligators, which have teeth, hatch about 2 months after the start of incubation, whereas birds hatch 10–30 days after incubation begins). Incubation is a vulnerable time for parents because almost all birds brood eggs using metabolic heat (see Section 21.9), and tooth loss may have been a factor in the evolution of shorter incubation periods.

The vertebral column reflects requirements of flight and bipedal locomotion. Because the forelimbs are specialized for flight, the head and bill are used for feeding and nest building, and to do this, birds need a neck with many highly mobile cervical vertebrae. In contrast, vertebrae in other regions of the body fuse together, creating a rigid body necessary to support up and down strokes of wings and an upright posture when a bird stands (**Figure 21.7A**). Some thoracic vertebrae partially fuse together, and lumbar, sacral, and a few caudal vertebrae fuse with the pelvis to form the **synsacrum** (**Figure 21.7B**). In the caudal region posterior to the synsacrum, several caudal vertebrae remain mobile, but the last few

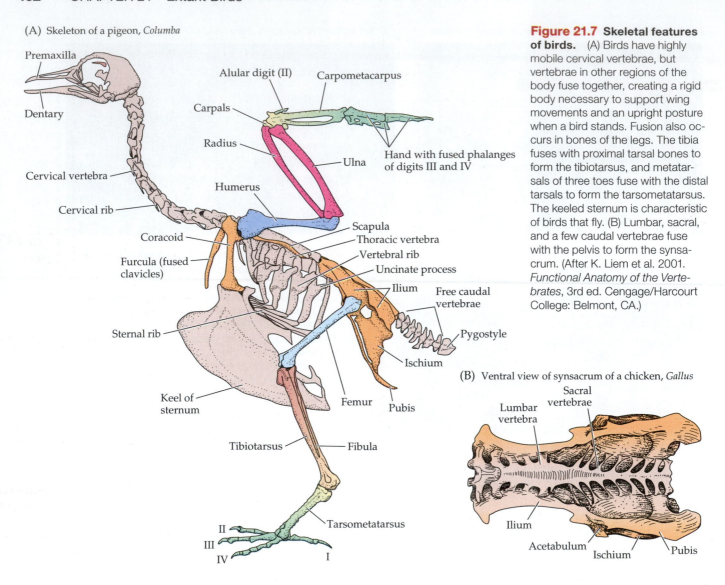

(A) Skeleton of a pigeon, *Columba*

Premaxilla

Dentary

Cervical vertebra

Cervical rib

Coracoid

Furcula (fused clavicles)

Alular digit (II)

Carpals

Radius

Humerus

Carpometacarpus

Hand with fused phalanges of digits III and IV

Ulna

Scapula

Thoracic vertebra

Vertebral rib

Uncinate process

Ilium

Free caudal vertebrae

Pygostyle

Ischium

Sternal rib

Keel of sternum

Femur

Pubis

Tibiotarsus

Fibula

II

III

IV

I

Tarsometatarsus

Figure 21.7 Skeletal features of birds. (A) Birds have highly mobile cervical vertebrae, but vertebrae in other regions of the body fuse together, creating a rigid body necessary to support wing movements and an upright posture when a bird stands. Fusion also occurs in bones of the legs. The tibia fuses with proximal tarsal bones to form the tibiotarsus, and metatarsals of three toes fuse with the distal tarsals to form the tarsometatarsus. The keeled sternum is characteristic of birds that fly. (B) Lumbar, sacral, and a few caudal vertebrae fuse with the pelvis to form the synsacrum. (After K. Liem et al. 2001. *Functional Anatomy of the Vertebrates*, 3rd ed. Cengage/Harcourt College: Belmont, CA.)

(B) Ventral view of synsacrum of a chicken, *Gallus*

Lumbar vertebra

Sacral vertebrae

Ilium

Acetabulum

Ischium

Pubis

fuse to form the **pygostyle** (Greek *pyge*, "rump"; *stylos*, "pillar"). The pygostyle supports the rectrices, and the free caudal vertebrae allow tail movements.

The sternum of flying birds bears a **keel** that is the site for attachment of flight muscles (see Figure 21.7A). Strong fliers have a well-developed keel (including penguins, which have powerful subaqueous flight), but the keel is reduced or absent in flightless species. In most extant birds, the thoracic ribs bear uncinate processes that serve as sites for attachment of respiratory muscles, play an important role in ventilation, and strengthen the rib cage.

Appendicular skeleton   The pectoral girdle consists of the scapula, coracoid, and clavicle on each side. The blade-like scapula extends posteriorly above the ribs and is supported by the stout coracoid, which attaches ventrally to the sternum (see Figure 21.9). The coracoid keeps the chest from collapsing during forceful downstrokes of wings. Additional bracing is provided by the clavicles, which in most birds fuse at their ventral ends to form

the furcula (wishbone). Forelimbs are highly modified for flight and weight reduction. For example, there are only three reduced digits (II, III, and IV) and some carpals (wrist bones) fuse with metacarpals (hand bones) to form the **carpometacarpus** (see Figure 21.7A).

Adaptations for bipedalism include a strong pelvic girdle (hence the synsacrum) and long, strong legs to support weight when standing or landing. The short femur articulates with the tibiotarsus ("drumstick") at the knee joint. The **tibiotarsus**, usually the longest leg bone, forms by fusion of the tibia with proximal tarsal (ankle) bones; the fibula is a thin splint. The ankle joint is between tarsal elements (a mesotarsal joint), as in Mesozoic theropods. All birds lack the fifth toe, and metatarsals associated with three of the remaining toes fuse with the distal tarsals to form the **tarsometatarsus** (see Figure 21.7A). Thus, there are no free tarsal bones in birds because they are part of either the tibiotarsus or the tarsometatarsus. Birds walk with their toes flat on the ground and the tarsometatarsus ("heel") off the ground and projecting upward at an

American flamingo, *Phoenicopterus ruber*

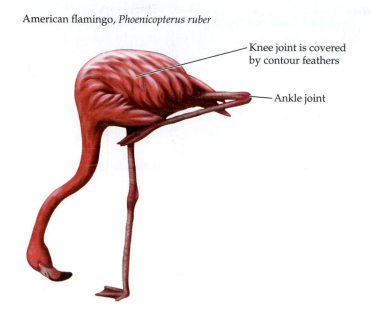

Knee joint is covered by contour feathers

Ankle joint

Figure 21.8 Bird knees only *appear* to point backward. The backward-pointing joint in the left leg of this flamingo is actually an ankle joint, formed where the tibiotarsus meets the tarsometatarsus. The knee is concealed by contour feathers.

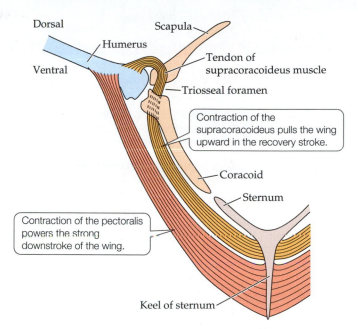

Dorsal

Humerus

Scapula

Ventral

Tendon of supracoracoideus muscle

Triosseal foramen

Contraction of the supracoracoideus pulls the wing upward in the recovery stroke.

Coracoid

Sternum

Contraction of the pectoralis powers the strong downstroke of the wing.

Keel of sternum

Figure 21.9 Major flight muscles of birds. The pectoralis major, responsible for the downstroke, originates on the keel of the sternum and inserts on the ventral side of the humerus. The supra-coracoideus, responsible for the upstroke, originates on the keel and inserts on the dorsal side of the humerus by passing through the triosseal foramen, formed where the coracoid, furcula (not shown), and scapula meet. (See credit details at the end of this chapter.)

angle where it meets the tibiotarsus (digitigrade foot posture). That is why birds appear to have knees that bend backward; the true, forward-bending knee is concealed by contour feathers (**Figure 21.8**).

Muscles

The two major flight muscles of birds are the pectoralis major and supracoracoideus; both are ventral to the vertebral column. The larger **pectoralis major** originates on the keel of the sternum and inserts on the *ventral* side of the humerus (**Figure 21.9**). Its contraction powers the downstroke of the wing. The smaller **supracoracoideus** lies beneath the pectoralis major; it originates on the sternal keel and inserts on the *dorsal* head of the humerus because its tendon passes through the triosseal foramen, where the coracoid, furcula, and scapula meet. Contraction of the supracoracoideus pulls the wing upward (the upstroke, which in most birds is a recovery stroke, meaning that it does not generate thrust; see Section 21.3). Using a chest muscle rather than a back muscle for the upstroke keeps the center of gravity low, stabilizing flight. The supracoracoideus is relatively small in most birds but can be larger in those that use a powered upstroke for hovering, fast aerial pursuit, or underwater flight (as in penguins). The weight ratio of the pectoralis major and supracoracoideus indicates a bird's reliance on a powered upstroke. Such ratios vary from 3:1 (birds with a powered upstroke) to 20:1 (birds in which the upstroke is primarily a recovery stroke).

Major leg muscles are in the upper leg, around the femur, with fewer and smaller muscles around the tibiotarsus, and still fewer, threadlike muscles around the tarsometatarsus. Muscles in upper parts of the leg move its distal regions via long, thin tendons (**Figure 21.10**). This arrangement places most muscle weight near the bird's center of gravity, which aids flight. It also reduces weight distally, enabling swift bipedal locomotion (see Section 21.4).

Relative sizes of flight muscles and leg muscles relate to a bird's primary mode of locomotion. Flight muscles make up 25% of the total body weight of strong fliers such as hummingbirds and swifts; muscles in the tiny legs of these birds account for as little as 2% of body weight. Predatory birds such as eagles, hawks, and owls are strong fliers that use their legs to capture prey; their flight muscles make up 25% of body weight, their leg muscles 10%. Birds that both swim and fly, such as ducks and grebes, have an even division between flight and leg muscles. Finally, primarily terrestrial birds, such as rails, rely on their legs to escape predators and have heavier leg muscles than flight muscles.

Muscle-fiber types and metabolic pathways also distinguish runners from fliers. Differences between dark and white meat of a turkey exemplify these differences. Fowl, especially domestic breeds such as turkeys, rarely fly but can walk and run for long periods. The dark color of their leg muscles comes from myoglobin (the oxygen-binding pigment that facilitates transfer of oxygen from blood to

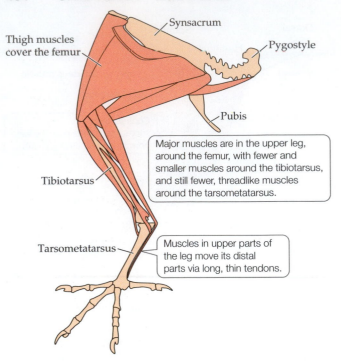

Figure 21.10 Leg muscles of a pigeon. Size and number of muscles decrease from the upper leg to the foot. This arrangement aids flight because muscle weight is concentrated near the bird's center of gravity; it also aids bipedal locomotion by lightening distal parts of the leg, allowing swifter movements. (See credit details at the end of this chapter.)

muscle fibers)—features that indicate a great capacity for aerobic metabolism. Such red muscle fibers contract slowly but have great endurance. In contrast, white muscles of a turkey breast have a relatively poor blood supply, little myoglobin, and little capacity for aerobic metabolism. White muscle fibers contract rapidly and powerfully but have little endurance. The short flights of fowl (including wild species such as pheasants, grouse, and quail) serve primarily to evade predators. An explosive takeoff, fueled by anaerobic metabolic pathways, is followed by a long glide back to the ground. In contrast to fowl, ducks have strong, sustained flight capabilities, and their dark breast muscles rely on aerobic metabolism.

21.3 Wings and Flight

LEARNING OBJECTIVES

21.3.1 Distinguish powered flight from passive flight.

21.3.2 Define four forces acting on bird wings.

21.3.3 Describe relationships between wing proportions and types of flight.

21.3.4 Sketch the four wing shapes recognized by ornithologists.

Avian flight in air is classified as either passive flight or powered flight, and both forms require a light body. **Passive flight**, or non-flapping flight, requires little energy; examples include **soaring** (maintaining or increasing altitude) and **gliding** (decreasing altitude while gaining horizontal

distance). **Powered flight**, perhaps more accurately called flapping flight, requires a lot of energy. In this section, we will consider the mechanics of flight and how wing shape relates to flight characteristics and ecology.

Flight mechanics

A flapping wing is both flexible and porous. The shape, camber (curvature; bird wings have a convex upper surface and concave lower surface), angle relative to the body, and position of individual feathers all change as a wing moves. This is a formidable list of variables, far more complex than those involved in passive flight or the fixed wings of an airplane. Nevertheless, general properties of a bird wing can be described, beginning with the four forces acting on it (**Figure 21.11**).

- **Lift** is the upward force that keeps the bird in the air.
- **Thrust** is the forward force created during flapping flight.
- **Drag** includes backward forces that oppose forward movement through air.
- **Gravity** is the force pulling the bird toward Earth.

Lift and thrust   Wings are airfoils (structures that generate lift) and their curves and shape are critical. Because birds have a cambered wing that tapers at the rear edge, air flows farther and faster over the dorsal surface than the ventral surface (**Figure 21.12A**). Faster airflow over the dorsal surface creates a partial vacuum there (pressure is least where flow is fastest), producing lift. Birds can increase lift by increasing camber of the wing.

A bird can also increase lift by increasing the angle of attack, which is the angle of tilt of the leading edge of the wing (**Figure 21.12B**). However, this works only up to a point; when the angle of attack is too steep, turbulence develops above the wing surface, lift disappears, and the bird stalls. Birds delay or prevent stalling with an **alula**, the moveable, anteriormost digit covered by a few small feathers (alular quills; see Figure 21.4 and Figure 21.7). The alula creates a slot at the front of the wing, thereby increasing the speed of air above the wing and reducing turbulence there (**Figure 21.12C**).

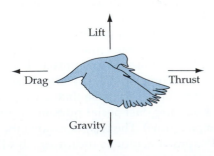

Figure 21.11 Four forces act on a bird's wing. Lift is an upward force that keeps a bird airborne. Thrust is a forward force created during flapping flight. Drag includes backward forces that oppose forward movement. Gravity pulls the bird toward Earth.

(A) Airflow around an airfoil

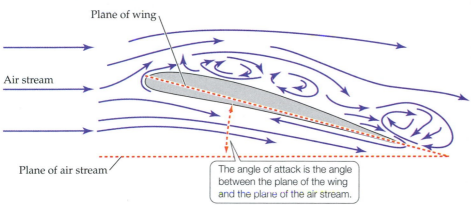

This is air velocity on the dorsal side of the wing.

V_d

Wing

V_l

V_v

V_l

This is the velocity of local air circulation around the wing.

This is air velocity on the ventral side of the wing.

Figure 21.12 **Producing lift and role of the alula in allowing high angles of attack.** (A) Birds have a cambered wing that tapers at the rear edge, so air travels farther and faster over the dorsal surface (V_d) than the ventral surface (V_v). Local circulation of air (V_l) enhances this difference in air speed. Pressure is lowest where speed is highest, resulting in a partial vacuum above the wing, producing lift. (B) Increasing the angle of attack (tilt of the leading edge of the wing relative to the airstream) can increase lift, but causes turbulence above the wing, which leads to stalling. (C) The alula forms a slot at the front of the wing. This slot increases speed of air flow over the wing, reduces turbulence, and delays or prevents stalling. (After K. Liem et al. 2001. *Functional Anatomy of the Vertebrates*, 3rd ed. Cengage/Harcourt College: Belmont, CA.)

(B) A high angle of attack produces greater lift but causes lift-reducing turbulence above wing

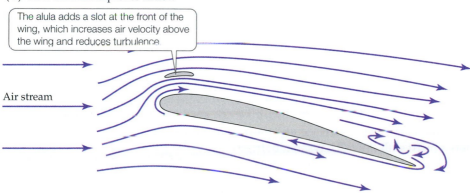

Plane of wing

Air stream

Plane of air stream

The angle of attack is the angle between the plane of the wing and the plane of the air stream.

(C) How an alula improves airflow

The alula adds a slot at the front of the wing, which increases air velocity above the wing and reduces turbulence.

Air stream

Unlike the fixed wings of an airplane, bird wings function not only as airfoils that generate lift but also as propellers that generate forward motion (thrust). A bird cannot continue to fly straight and level unless it develops thrust to balance drag. During the downstroke, primary feathers supported by the hand generate thrust; secondary feathers on the inner wing, which are supported by the forearm, generate lift.

The wings move downward and forward on the downstroke and upward and backward on the upstroke (**Figure 21.13A**). During the downstroke, the trailing edges of the primaries bend upward under air pressure and each feather acts as an individual propeller, biting into the air and generating thrust (**Figure 21.13B**). Thrust is greater than total drag during the downstroke, and the bird accelerates. In small birds, the upstroke generates little or no thrust.

Because the upstroke is mainly a passive recovery stroke, forward speed slows during this part of the wingbeat cycle.

Drag Frictional drag occurs between the air and the surface of a bird's body.[1] In addition, vortices (eddies of swirling air) play a major role in avian flight. For example, vortices formed at the leading edge of the wing increase **induced drag**. Recall that due to the cambered shape of a wing, air pressure is higher beneath the wing than above it. As a result of this pressure difference, local air close to the lower surface swirls upward into the low-pressure area above the wing. This movement of local air creates

---

[1] Recall from Chapter 7 that, in terms of biomechanics, air is a fluid. You may wish to review the discussion of hydrodynamics in Section 7.6.

(A) Wing motions during powered flight

Downstroke · Upstroke · Direction of flight

(B) Primary feathers of bushtit, *Psaltriparus minimus*, act like propellers

Bridget Spencer

Figure 21.13 Generating thrust. (A) During flapping flight, the wings move downward and forward on the downstroke and upward and backward on the upstroke. (B) A bushtit in flight shows the primary feathers twisting and opening during the downstroke, causing these feathers to act as propellers and produce forward thrust. The primaries twist only on the downstroke of small birds, but during both the down- and upstrokes of large birds. (After K. Liem et al. 2001. *Functional Anatomy of the Vertebrates*, 3rd ed. Cengage/Harcourt College: Belmont, CA.)

drag because it opposes movement of the airstream over the ventral surface of the wing (see Figure 21.12A). The circulation is shed as vortices at the tips of the wings (**Figure 21.14**). Induced drag is minimized by longer and more tapered wings, which decrease the surface area that local air must flow around.

Landing Landing is a delicate maneuver. A bird must reduce its velocity during the approach, which increases the risk of stalling prematurely, and then it must stall deliberately as it reaches the perch site. A bird coming in for a landing rotates its body and wings, spreads its tail and extends its feet to create drag, and raises feathers (underwing coverts) on the lower surface of the leading edge of the

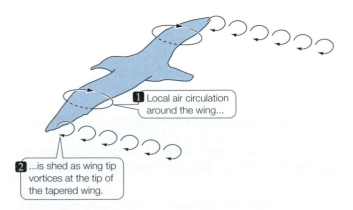

1 Local air circulation around the wing...

2 ...is shed as wing tip vortices at the tip of the tapered wing.

Figure 21.14 Shedding of vortices at wingtips. The local air circulation that produces induced drag is shed as vortices at wing tips. (See credit details at the end of this chapter.)

wings. Underwing covert feathers act like the flaps that rotate out from the lower surface of an airplane's wings during landing: they maintain lift at low speed. Birds also extend the alula as they slow down for landing.

Wing shape and flight

Relative proportions of wing bones reveal much about types of flight. Recall that primary feathers attach to the hand (manus) and generate thrust, whereas secondary feathers attach to the ulna of the forearm and produce lift (**Figure 21.15A**; also see Figure 21.4B). As shown in **Figure 21.15B**, hummingbirds, which engage almost exclusively in powered flight, have a relatively long hand to which numerous primaries attach (for example, 10 primaries in ruby-throated hummingbirds, *Archilochus colubris*). Their forearm is relatively short, and fewer secondaries are rooted in the ulna (6 in ruby-throated hummingbirds). Frigatebirds use mixed flight—powered flight, soaring, and gliding—and have hand and forearm bones of similar length and roughly equal numbers of primaries and secondaries. Albatrosses spend most of their flight time soaring, which requires lift; they have long forearm bones, and many secondary feathers attach to the ulna (32 in the Laysan albatross, *Phoebastria immutabilis*). Their hand is relatively short, with fewer primary feathers (10 in the Laysan albatross).

The **aspect ratio** of a wing is its length divided by its width. Long, narrow wings have high aspect ratios; short, wide wings have low aspect ratios. **Wing loading** is the weight of a bird divided by its total wing area. At any given speed, birds with low wing loading values have more lift. As a general pattern, wing loading values decrease from large to small birds. However, the type of flight also influences wing loading. For example, large birds like vultures that soar, and therefore need lift, have lower wing loading values than do birds of the same size that do not soar.

Ornithologists recognize four types of wings that differ in shape and function.

- *High-speed wings* provide speed and control. They are tapered, lack prominent slotting, and have little camber and a high aspect ratio (**Figure 21.16A**). They lack the elongated forearm found in soaring birds. Aerial foragers such as swallows, as well as birds that migrate long distances such as ducks and terns, have high-speed wings.

(A) General anatomy of wing skeleton and wing

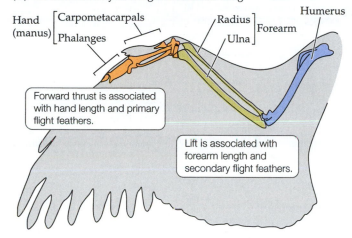

Forward thrust is associated with hand length and primary flight feathers.

Lift is associated with forearm length and secondary flight feathers.

Figure 21.15 Wing proportions and types of flight.
(A) Comparing length of hand (orange) and length of forearm (yellow) predicts type of flight. (B) Hummingbirds engage in powered flight (thrust is key) and have an elongated manus (hand) compared to forearm bones. Frigatebirds engage in mixed flight and their manus and forearm bones are similar in length. Albatrosses engage in dynamic soaring (lift is key) and have a relatively short manus compared to their elongated forearm bones. (A after K. Liem et al. 2001. *Functional Anatomy of the Vertebrates*, 3rd ed. Cengage/Harcourt College: Belmont, CA.)

(B) Wing proportions and types of flight in hummingbird, frigatebird, and albatross

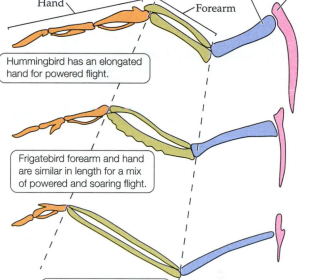

Hummingbird has an elongated hand for powered flight.

Frigatebird forearm and hand are similar in length for a mix of powered and soaring flight.

Albatross has an elongated forearm for soaring flight.

- *Elliptical wings* are short and broad, with low aspect ratios and a high degree of slotting in the outer primaries (**Figure 21.16B**). Birds such as grouse and quail that inhabit areas of dense vegetation have elliptical wings, as do many passeriforms. Elliptical wings allow rapid takeoff and excellent maneuverability in environments with many obstacles.

- *High-aspect-ratio wings* provide the high lift-to-drag ratios needed by oceanic birds such as albatrosses, shearwaters, and petrels that rely on dynamic soaring—a type of soaring in which energy is gained by repeatedly crossing between air masses with different wind speeds. Two notable features of this long and narrow wing are the very elongated forearm to support lift and the absence of slotting at the wing tip (**Figure 21.16C**).

- *Slotted high-lift wings* are broad with an intermediate aspect ratio, deep camber, and marked slotting between primaries (**Figure 21.16D**). Broad wings reduce wing loading and the deep camber and prominent slotting increase lift. This type of wing is found in birds such as hawks, vultures, and condors that engage in static soaring, a type of soaring that occurs over land when birds rely on rising warm air masses (thermals) to obtain lift with little energy expenditure. The highly slotted wings of static soarers enhance maneuverability because the birds can respond to changes in wind currents using individual feathers instead of the entire wing.

Dynamic and static soaring are compared in **Figure 21.17**.

(A) High-speed wings characterize birds that feed in flight or migrate long distances

(B) Elliptical wings enhance maneuverability in complex habitats

(C) High-aspect-ratio wings are used for dynamic soaring

(D) Slotted high-lift wings are used for static soaring

Figure 21.16 Flight characteristics can be deduced from wing shape. (A) High-speed wings have high aspect ratios, low camber, no slotting of primaries, and a swept-back shape. (B) Elliptical wings are short and broad, with a low aspect ratio and slotted primaries. (C) High-aspect-ratio wings are long and narrow and lack slots in outer primaries. The forearm is especially elongated, which is not the case in high-speed wings. (D) Slotted high-lift wings have low wing-loading values, intermediate aspect ratios, and pronounced slotting of primaries. (After R.T. Peterson. 1978. *The Birds*, 2nd ed. New York: Time-Life Books.)

(A) Dynamic soaring over the ocean

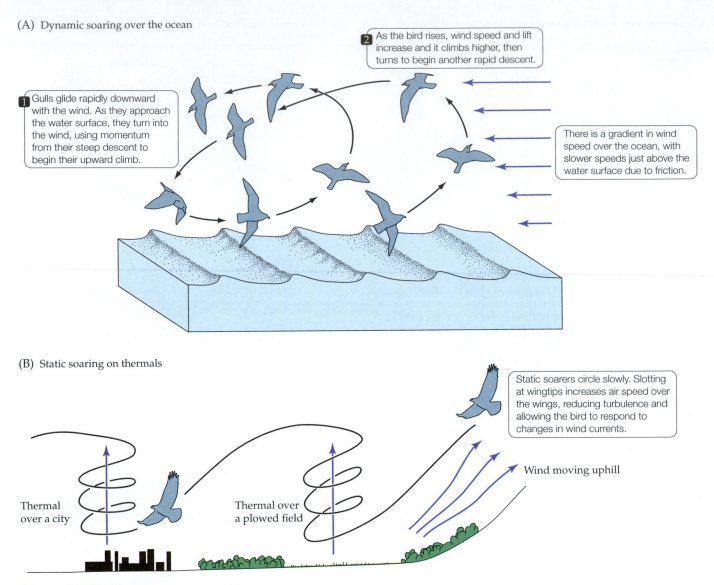

2 As the bird rises, wind speed and lift increase and it climbs higher, then turns to begin another rapid descent.

1 Gulls glide rapidly downward with the wind. As they approach the water surface, they turn into the wind, using momentum from their steep descent to begin their upward climb.

There is a gradient in wind speed over the ocean, with slower speeds just above the water surface due to friction.

(B) Static soaring on thermals

Static soarers circle slowly. Slotting at wingtips increases air speed over the wings, reducing turbulence and allowing the bird to respond to changes in wind currents.

Wind moving uphill

Thermal over a city

Thermal over a plowed field

Figure 21.17 Two types of soaring. (A) Dynamic soarers use the differential in wind speeds over open oceans where winds are persistent. (B) Static soarers take advantage of rising warm air (thermals) over land. Thermals can occur either when wind moves up a cliff face or in areas of differential heating (for example, grasslands are warmer than forests and cities are warmer than surrounding countryside). In such areas, static soaring is an energetically cheap mode of flight. (After K. Liem et al. 2001. *Functional Anatomy of the Vertebrates*, 3rd ed. Cengage/Harcourt College: Belmont, CA.)

21.4 Feet and Locomotion

LEARNING OBJECTIVES

21.4.1 Describe three toe arrangements and their functional significance.

21.4.2 Explain two ways in which swimming underwater evolved in birds.

21.4.3 Compare three types of webbing of aquatic species.

Just as wing proportions and shapes reveal habitat-linked types of flight, foot morphology and toe arrangement tell us something about a bird's ecology, such as whether it spends most of its time in trees, on land, or in water and how it moves in those habitats. We first consider modes of terrestrial locomotion and diversity in number and arrangement of toes. Then, we examine swimming and types of webbing between the toes of aquatic species.

Hopping, walking, and running

When on the ground, birds hop (a succession of jumps in which both legs move together), walk (legs move alternately, with at least one foot in contact with the ground at all times), or run (legs move alternately, with both feet off the ground at some times). Many passeriforms hop, although some larger members of the order, such as ravens and crows, walk. Running birds include large flightless species such as ostriches and emus as well as medium-sized roadrunners (roadrunners can fly but are reluctant to do so).

Toe arrangements   A foot with four toes is the ancestral condition for birds, and most extant species have four toes. The most common toe arrangement is **anisodactyly**:

toes 2, 3, and 4 face forward and toe 1 faces backward (**Figure 21.18A**). Anisodactylous toes are associated with perching, as in passeriforms. (Turkeys and chickens, which also perch, have a similar toe arrangement.)

Zygodactyly is the condition in which toes 2 and 3 face forward and toes 1 and 4 face backward (**Figure 21.18B**). Zygodactylous toes evolved in woodpeckers, owls, the osprey, and parrots. This toe arrangement allows woodpeckers to move on vertical surfaces such as tree trunks, owls and the osprey to hang onto struggling prey, and parrots to manipulate food and climb through vegetation with dexterity.

Didactyly is seen in ostriches. Toes 3 and 4 point forward and toes 1 and 2 have been lost (**Figure 21.18C**). Toe loss contributes to a reduction in weight of distal elements of the leg. Like other vertebrates that are fast runners, ostriches have long, thin legs and relatively small feet. The significance of long legs is obvious—they allow an animal to cover more distance with each stride. The reduction in leg and foot weight involves the physics of momentum. A foot in contact with the ground is motionless, but after pushing off from the ground, the foot and lower part of the leg must accelerate to a speed faster than the trunk of the running animal. Reducing the weight of the foot and leg reduces the energy needed for this change in momentum and allows the animal to run faster.

Swimming

Although no extant birds are fully aquatic, nearly 400 species are specialized for swimming. Almost half of these aquatic species also dive and swim underwater.

Swimming on the surface   Birds that swim on the water's surface use their legs for locomotion. The legs of foot-propelled swimmers such as ducks are at the rear of the body, where the weight of leg muscles interferes least with streamlining, and where the best control of steering can be achieved.

Other specializations include a wide body that increases stability in water, and dense plumage that provides buoyancy and insulation. Also, aquatic birds often have a well-developed **uropygial gland** that produces oil; the birds spread oil onto their feathers during preening to waterproof them (thus it is sometimes called the preen gland). Located dorsally at the base of the tail, the uropygial gland is the only avian sebaceous gland, and may be reduced or lost in nonaquatic birds.

Webbing   Many aquatic birds have webbed feet. **Palmate webbing**—webbing between toes 2, 3, and 4—is most common and occurs in ducks, geese, swans, and gulls (**Figure 21.19A**). **Totipalmate webbing** is webbing between all four toes and is found in pelicans, cormorants, and boobies (**Figure 21.19B**). Finally, coots and grebes have flexible flaps of skin around toes 2, 3, and 4, but the flaps are not connected from one toe to another; this condition is **lobate webbing** (**Figure 21.19C**).

(A) Anisodactylous toes of a perching bird

The first toe points backwards.

Dan Pancamo/CC BY-SA 2.0

(B) Zygodactylous toes of a woodpecker

Toes 2 and 3 point forward.

Toes 1 and 4 point backward.

Ken Thomas/Public Domain

(C) Didactylous toes of an ostrich

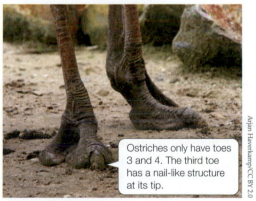

Ostriches only have toes 3 and 4. The third toe has a nail-like structure at its tip.

Arjan Haverkamp/CC BY 2.0

Figure 21.18 Examples of toe arrangements.   (A) Anisodactyly (three toes forward, one backward) is found in perching birds. (B) Zygodactyly (two toes forward, two backward) allows woodpeckers to move on vertical surfaces. (C) Didactyly (two toes of the four are lost) occurs in ostriches and is an adaptation for running.

Diving and swimming underwater   Evolutionary transitions from surface-swimming to subsurface swimming occurred in two independent ways: (1) by further specialization of hindlimbs already adapted for swimming, and (2) by modification of the wings for use as flippers. Highly specialized foot-propelled divers evolved independently among grebes, cormorants, and loons. Wing-propelled divers include auks, diving petrels, shearwaters, and penguins. Penguin wings are an example of

(A) Palmate webbing of Muscovy duck, *Cairina moschata*

Rhododendrites/CC BY-SA 4.0

(B) Totipalmate webbing of blue-footed booby, *Sula nebouxii*

putneymark/CC BY-SA 2.0

(C) Lobate webbing of Eurasian coot, *Fulica atra*

Shyamal/CC BY-SA 3.0

Figure 21.19 Webbed feet characterize swimming birds. (A) Palmate webbing runs between the three forward-facing toes; the fourth toe faces backward and is not included in the webbing. (B) Totipalmate webbing runs between all four toes. (C) Lobate webbing consists of flexible flaps of skin around the three forward-facing toes; the flaps are not connected from one toe to another.

wings that have become modified into structures that function as flippers. Penguin wing bones are short, broad, flat, and fused at the elbow and wrist joints. They also are heavy, lacking the air-filled spaces typical of other birds.

21.5 Bills, Feeding, and Digestion

LEARNING OBJECTIVES

21.5.1 Relate bill shape to diet.

21.5.2 Describe variation in tongue length and its functional significance.

21.5.3 Explain how size and structure of digestive organs reflect dietary specializations.

Birds eat diverse foods, including invertebrates, other vertebrates, and plant parts and products, such as fruits, seeds, flowers, nectar, and less often, leaves. Carnivorous and omnivorous birds capture prey in the air, on the ground, underground, on vegetation, and in water. Herbivores consume plants and parts of plants in both terrestrial and aquatic habitats. In this section, we consider how modifications of the bills, skulls, tongues, and digestive tracts of birds reflect their dietary specializations.

Bills, cranial kinesis, and tongues

Avian bills[2] are diverse in appearance and function, and the cranial kinesis characteristic of neognathous birds allows extensive and independent movements of upper and lower jaws. Tongues also exhibit specializations for different feeding methods.

Bill specializations   Because their forelimbs are specialized as wings, birds rely on bills (and in some cases, feet) for predation. The range of morphological specializations of bills for different feeding methods and prey defies complete description; eight of the many avian feeding modes and bill specializations are shown in **Figure 21.20** and described here.

- Many birds are generalists, with bills that can seize and process both plants and animals (Figure 21.20A).

- The hard coverings of seeds must be removed before the nutritious contents can be digested. Seed-eating birds, or granivores, use different methods to husk seeds before swallowing them (Figure 21.20B).

- Filter-feeding birds use a variety of mechanisms to capture and filter small organisms from water. The broad-tipped bills of roseate spoonbills have a convex upper surface and flat lower surface, which create currents that lift prey from the bottom into the water column, where the spoonbill seizes it (see Figure 21.20C). Flamingos feed in shallow, food-rich waters and have bills specialized for capturing brine shrimp and algae. The bill bends strongly downward and flamingos feed with the bill upside down (see Figure 21.8).

- Many birds that eat animals, such as omnivorous ravens and crows, use heavy, pointed bills to stab prey (Figure 21.20D).

- Probers such as ibises, sandpipers, and kiwis insert their long, narrow bill into a substrate to extract invertebrates (Figure 21.20E).

---

[2] The terms "bill" and "beak" mean the same thing; here we use bill because ornithologists more commonly use it.

(A) American robin, *Turdus migratorius*, a generalist feeder

(B) Blue grosbeak, *Passerina caerulea*, a granivore

(C) Roseate spoonbill, *Platalea ajaja*, a filter feeder

(D) Chihuahuan raven, *Corvus cryptoleucus*, an omnivore

(E) Southern bald ibis, *Geronticus calvus*, a prober

(F) Yellow warbler, *Setophaga petechia*, a gleaner

(G) Peregrine falcon, *Falco peregrinus*, a carnivore

Tomial tooth used to kill prey.

(H) Common merganser, *Mergus merganser*, a piscivore

Figure 21.20 Bills are specialized for feeding method and food source. (A) A familiar generalist, the American robin, feeds on diverse invertebrates as well as fruits and seeds. Its bill can seize both animals and parts of plants. (B) The bill of a granivorous blue grosbeak is strong and deep. The bird slices seeds open using fore-and-aft movements of the lower jaw, then uses the tongue to remove the contents. (C) Roseate spoonbills are notable filter feeders. They immerse their uniquely shaped bill in shallow water and swing it from side to side, feeling for prey. (D) In addition to feeding on berries and grain, omnivorous Chihuahuan ravens prey on small mammals, nestling birds, and lizards, which they stab with their sharp bill. (E) The southern bald ibis is a ground-dwelling bird that probes with its long, narrow bill for invertebrates in dry grasslands. (F) Yellow warblers glean insects from leaf surfaces using their pointed bill. (G) Peregrine falcons are carnivores that sever the spinal cord of stunned prey using tomial "teeth"—sharp projections on each side of the upper part of the bill that fit into corresponding notches on the lower part of the bill. (H) The piscivorous common merganser dives under water to capture fish. Its bill is straight and narrow, with serrated margins to hold slippery prey.

- Gleaners—birds that selectively pick invertebrates from a surface—typically have pointed bills used like forceps to seize small food items (Figure 21.20F).

- Falcons stun prey with the impact of their dive, then quickly kill it with a bite to the neck to disarticulate cervical vertebrae and sever the spinal cord; snapping the neck is made possible by tomial "teeth", sharp projections on each side of the upper bill that fit into corresponding notches of the lower bill (Figure 21.20G). Other birds of prey use talons to kill prey.

- Many fish-eating birds have long bills with specializations for holding slippery prey, including serrated bill edges (Figure 21.20H).

Cranial kinesis As we described in Section 21.1, bones in the skull of neognathous birds can move, although the degree of this cranial kinesis varies among species. Besides the palatine–pterygoid joints shown in Figure 21.2, joints in bird skulls also include a hinge joint where the upper jaw meets the braincase, called the craniofacial hinge (or frontonasal hinge). The craniofacial hinge allows birds to raise and lower the upper jaw (**Figure 21.21A**). This arrangement (1) results in a larger gape, which may assist in swallowing large food items, and (2) allows the upper and lower jaws to move independently, enhancing manipulatory abilities of the bill.

Many birds use their bill to search for hidden food, as when long-billed shorebirds probe in mud and sand to locate worms and crustaceans. These birds display a form of cranial kinesis called **distal rhynchokinesis** (literally, "moving the tip of the bill") in which the flexible zone in the upper jaw has moved toward the tip of the bill, allowing the tip of the upper jaw to be lifted, enabling them to grasp prey under mud and sand (**Figure 21.21B**).

Other functions of the bill Birds use their bills in many activities other than feeding. Such activities include gathering materials and constructing nests; expressing and spreading oil from the uropygial gland; and realigning feather barbs during preening. Some species use their bill when interacting with conspecifics—for example, members of a courting pair may engage in a behavior known as billing, which in some species involves gentle touching of bills and in others forceful clacking together (hence the old-fashioned expression "billing and cooing" to describe romantic vocalizations and other displays of affection between partners).

Bills also are used for thermoregulation. Some toucans regulate blood flow to their large bills as a means of radiating excess body heat, but thermoregulatory function is not limited to birds with unusually large bills; even small birds with unassuming bills radiate heat from their bills. Many birds pant to unload excess heat, but panting results in water loss, whereas radiating heat from a bill does not.

Tongues Bird tongues are almost as variable as bird bills. Some are very short (in groups that swallow large food items whole, such as hornbills or pelicans), while others are very long (in groups with tongue protrusion, such as hummingbirds). In many birds, the dorsal surface and sides of the tongue are covered by many small, keratinized projections (papillae) that help hang onto food and move it toward the esophagus.

Tongue size, structure, musculature, and supporting skeletal structures reveal feeding methods. Woodpeckers drill holes into dead trees and use their long tongues to investigate passageways made by wood-boring insects. The hyoid bones that support the tongue are elongated in woodpeckers and housed in a sheath of muscles that passes around the outside of the skull and rests in the nasal cavity (see Figure 11.10A). When muscles of the sheath contract, the hyoid horns are pushed around the back of the skull, projecting the tongue from the bird's mouth. Barbs on the tip of a woodpecker's tongue impale insects and pull them from their tunnels.

Nectar-feeding birds such as hummingbirds also have long tongues and hyoid horns that wrap around the back of the skull. However, the tip of the tongue of nectar-feeding birds has a spray of hair-thin projections, and capillary forces cause nectar to adhere to the tongue. Other birds, such as parrots, manipulate food in the mouth and have fleshy tongues and short hyoid horns (see Figure 11.10B). Most birds have only extrinsic muscles (muscles outside the tongue) to control tongue movements; parrots,

(A) Bill opening in a domestic chicken, *Gallus gallus domesticus*

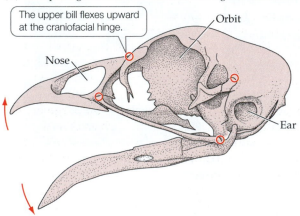

The upper bill flexes upward at the craniofacial hinge.

Orbit

Nose

Ear

(B) Bar-tailed godwit, *Limosa lapponica*

Distal rhynchokinesis

Anne Collins

Figure 21.21 Cranial kinesis. (A) Movement (kinesis) occurs between many bones in the skull (cranium) of a chicken; areas of movement are circled. The craniofacial hinge joint at the junction where the upper jaw meets the braincase allows the upper jaw to move up and down. (B) Long-billed shorebirds such as the bar-tailed godwit probe for worms and crustaceans in soft substrates and can employ distal rhynchokinesis, raising the tip of the upper part of the bill to seize prey. (See credit details at the end of this chapter.)

however, have intrinsic muscles as well, allowing finer control (this is one reason why parrots can be good at mimicking sounds). Finally, a tongue bone extends forward from the hyoid skeleton and lies within the body of birds' tongues (a very different setup than the mammalian tongue, which is devoid of bone). Like hyoid bones, the size and shape of the tongue bone vary in relation to feeding methods.

Digestive tract

Major digestive organs of birds are shown in **Figure 21.22**. The digestive tract of birds differs from those of other vertebrates in some ways, as we describe next. Importantly, the absence of teeth prevents much oral processing of food, and the stomach takes over some of that role.

Esophagus and crop   Birds often gather more food than they can process in a short time, and many species hold the excess in a **crop**, an enlarged portion of the esophagus specialized for temporary food storage (see Figure 21.22). Longitudinal folds allow the crop to expand, and a bird with a crop can gather food rapidly, thus minimizing its exposure time to predators. Additionally, foraging parents of some species gather and store food in the crop, then transport the food back to the nest and regurgitate it to their young.

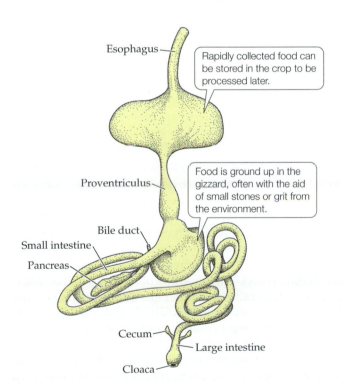

Esophagus

Rapidly collected food can be stored in the crop to be processed later.

Proventriculus

Food is ground up in the gizzard, often with the aid of small stones or grit from the environment.

Bile duct

Small intestine

Pancreas

Cecum

Large intestine

Cloaca

Figure 21.22 Ventral view of the digestive tract of a pigeon. The main organs of the digestive system of a bird include the esophagus, which may contain a crop for temporary storage of food, a two-part stomach (proventriculus and gizzard), small and large intestines, ceca, and cloaca. The gizzard mechanically grinds food, serving the function provided by mammalian teeth. (After K. Liem et al. 2001. *Functional Anatomy of the Vertebrates*, 3rd ed. Cengage/Harcourt College: Belmont, CA.)

Stomach   A typical bird stomach has two distinct chambers: an anterior glandular portion, the **proventriculus**, and a posterior muscular portion, the **gizzard** (see Figure 21.22). Glands in the proventriculus secrete acid and digestive enzymes, and this part of the stomach is especially large in species that swallow large foods such as intact fruits. The gizzard processes food mechanically. Its thick, muscular walls squeeze the contents, and small stones held in the gizzards of many birds help grind food (this is why domestic chickens need to be supplied with grit). In this sense, the gizzard performs the same function as mammalian teeth—that is, the mechanical breakdown of food. Some birds, such as owls and hawks, produce pellets in the gizzard composed of indigestible parts of their prey (fur, teeth, and bones), which they later regurgitate.

Intestines, ceca, and cloaca   As in mammals, the small intestine is the principal site of chemical digestion, where enzymes—some from the pancreas and others from the intestine itself—break down food into molecules that can be absorbed across its walls. The mucosa of the small intestine is modified into a series of folds, layers, and villi (projections) that increase its surface area. The large intestine is relatively short and functions primarily in water reabsorption. Birds generally have a pair of **ceca** (singular, *cecum*), pouches at the junction of small and large intestines (see Figure 21.22). Ceca are small in carnivorous, insectivorous, and seed-eating species, but large in herbivorous and omnivorous species such as cranes, fowl, ducks, geese, and ostriches. Symbiotic microorganisms in the ceca break down plant material during this hindgut fermentation (hindgut in reference to more anterior organs such as the crop and stomach, which are considered the foregut). Ceca also function in water and electrolyte reabsorption and degradation of nitrogenous compounds.

The **cloaca** temporarily stores waste products while water is reabsorbed. Birds precipitate their nitrogenous waste in the form of uric acid, which requires very little water and thus is highly effective at water conservation (see Section 13.4).

Variation in gut morphology between species   Relative sizes of organs of the upper digestive tract vary between species with different diets. Seed-eaters such as the budgerigar have a relatively small crop and a large gizzard for grinding seeds (**Figure 21.23A**). The hoatzin is notable for being a folivore (leaf-eater) that occasionally consumes flowers and fruit and is the only bird known to ferment plant material in the foregut (as mentioned, other birds ferment plants in the ceca, which are part of the hindgut). Symbiotic microorganisms in the hoatzin's large crop break down leaves and other parts of plants, releasing fatty acids for absorption in the intestine. In comparison to its large crop, the proventriculus and gizzard are small because food has been digested before it reaches the stomach (**Figure 21.23B**).

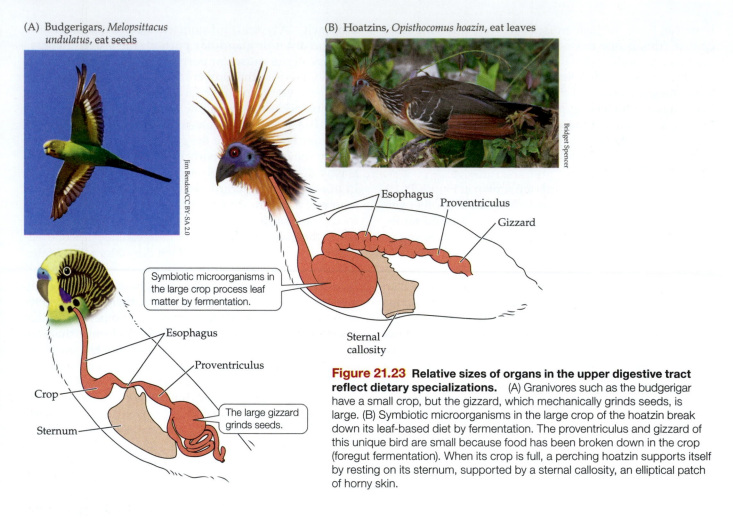

(A) Budgerigars, *Melopsittacus undulatus*, eat seeds

(B) Hoatzins, *Opisthocomus hoazin*, eat leaves

Symbiotic microorganisms in the large crop process leaf matter by fermentation.

Esophagus

Proventriculus

Gizzard

Sternal callosity

Esophagus

Proventriculus

Crop

The large gizzard grinds seeds.

Sternum

Figure 21.23 **Relative sizes of organs in the upper digestive tract reflect dietary specializations.** (A) Granivores such as the budgerigar have a small crop, but the gizzard, which mechanically grinds seeds, is large. (B) Symbiotic microorganisms in the large crop of the hoatzin break down its leaf-based diet by fermentation. The proventriculus and gizzard of this unique bird are small because food has been broken down in the crop (foregut fermentation). When its crop is full, a perching hoatzin supports itself by resting on its sternum, supported by a sternal callosity, an elliptical patch of horny skin.

21.6 Sensory Systems

LEARNING OBJECTIVES

21.6.1 Explain how eye position reflects whether a bird is a predator or prey species.

21.6.2 List four features of owls that contribute to sensitive hearing.

21.6.3 Provide examples of avian behaviors influenced by olfaction.

21.6.4 Describe bill-tip organs and remote touch.

A bird flying rapidly through three-dimensional space requires a continuous flow of sensory information about its position and obstacles. In daylight (and recall that most birds are diurnal), vision is the sense best suited to provide such information, and birds have a well-developed visual system. In fact, the avian visual system remains active even during sleep: when in unihemispheric sleep, a state in which one brain hemisphere is sleeping and the other is awake, the eye neurologically connected to the active hemisphere remains open, allowing birds to remain vigilant while at rest. Hearing is particularly acute in some groups, such as owls, and evidence is increasing that olfaction plays roles in several behaviors of birds. In this section, we consider these familiar senses along with a less familiar sense, remote touch.

Vision

Birds rely on their vision for flying, foraging, choosing a mate, avoiding predators, and scrutinizing the eggs in their nest to make sure they are their own. Brain structure reflects the importance of vision: the optic lobes are large, and the midbrain is an important area for processing visual information. Given their reliance on vision, it is not surprising that birds have large eyes. A bird's eyes can approach the weight of its brain; the weight is displaced caudally (toward the rear of the head) such that the eyeballs meet in the midline of the skull in many species. The eyes of birds are relatively immobile because the six extrinsic eye muscles found in all vertebrates are small in birds—the eye itself takes up most of the space in the large orbit (see Figure 21.21A). To compensate for immobile eyes, birds rely on mobile necks and extreme head-turning abilities. Owls, for example, can turn their head 270 degrees.

Eye position   Because the position of eyes in the head is a tradeoff between breadth of visual field and depth perception, eye position can reveal whether a bird is a primarily a prey or predator species. **Monocular vision**—when the visual fields of the two eyes do not overlap—yields wide fields of view and is associated with eyes positioned on

(A) A woodcock, *Scolopax minor*, can see behind its head

(B) Forward-facing eyes of an osprey, *Pandion haliaetus*, for depth perception

Figure 21.24 **Eye position reflects whether a bird is primarily prey or predator.** (A) The American woodcock feeds by probing for worms in the forest floor. Its laterally placed eyes offer an extensive field of view, such that woodcocks can detect predators approaching from all directions, even from behind. (B) The predatory osprey catches fish and has forward-facing eyes, with good depth perception.

the sides of the head. **Binocular vision** occurs when fields of view of the two eyes overlap and is associated with eyes positioned at the front of the head. **Stereoscopic vision** occurs within the area of overlap of binocular visual fields. The distance between forward-facing eyes means that each retina transmits a slightly different image to the brain. The brain integrates the two images into a single three-dimensional (stereoscopic) image, yielding depth perception. Most birds and mammals have a combination of binocular and monocular vision.

Birds commonly targeted as prey species typically have eyes on the sides of the head (**Figure 21.24A**), a position that yields the extensive fields of view needed to detect predators coming from all directions (including from behind). However, these species have poor depth perception. In contrast, predatory birds have closely spaced, forward-facing eyes (**Figure 21.24B**; also see Figure 21.25A). This eye position yields good depth perception, which is needed to accurately judge distance to prey, but limited fields of view (for example, there is an area directly behind the head that cannot be seen, even with extreme head-turning).

Photoreceptors and color vision   Rods and cones are photoreceptors found in the vertebrate retina (see Section 2.6). Rods record the presence or absence of light and are used in low light conditions; cones are color-sensitive and require higher light levels. Both rods and cones contain opsin photopigments. The world looks different to birds than it does to us because most birds have four photosensitive pigments in their cone cells whereas humans and other primates have only three (see Section 22.3). In other words, birds have tetrachromatic vision, while that of primates is trichromatic.

Three of the avian cone pigments absorb maximally at wavelengths equivalent to those humans perceive—the red, green, and blue regions of the spectrum. The fourth pigment ancestrally responds to wavelengths of about

400 nm, or deep blue. In some species, a mutation in the gene encoding the deep blue pigment shifted its maximum sensitivity into the ultraviolet, thereby extending visual sensitivity of these birds beyond the range of human eyes, and birds perceive colors and patterns that are invisible to humans. Sensitivity to wavelengths in the ultraviolet portion of the spectrum is important in many aspects of bird behavior. For example, ultraviolet reflectance can distinguish males from females, signal the physiological state of prospective mates, and betray the health of nestlings.

As in other sauropsids, most avian cone cells contain oil droplets. The droplets are situated so that light passes through them before reaching the photopigments. Oil droplets range in color from red through orange and yellow to green, and act as filters, absorbing some wavelengths of light and transmitting others.

Hearing

Birds rely on hearing—either alone or in combination with other senses—when communicating and when detecting prey and predators. Most birds have relatively large (compared to head size) tympanic membranes, indicating good auditory sensitivity. The columella (stapes) and its distal cartilaginous extension, the extracolumella, transmit vibrations of the tympanum (eardrum) to the oval window of the inner ear. Sound pressures are amplified during transmission from tympanum to oval window because the area of the tympanum is larger than the area of the oval window. The ratio for birds ranges from 11:1 to 40:1, with high ratios indicating more sensitive hearing (for comparison, the ratio is 21:1 for humans and 36:1 for cats). The cochlea of birds is specialized for fine distinctions of sound frequencies and temporal patterns. Although it is about one-tenth the length of the mammalian cochlea, it has approximately 10 times as many hair cells per unit length.

By all measures, owls are the most acoustically sensitive birds. They have the largest tympani relative to head size and the highest ratios of tympanum area to oval window area. In addition, owls have large cochleae and well-developed auditory centers in the brain. Other

(A) Boreal owl, *Aegolius funereus*

© Greg Schechter CC BY 2.0

(B) Front and dorsal views of skull showing left-right asymmetry

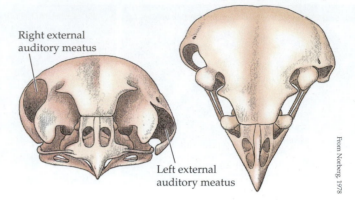

Right external
auditory meatus

Left external
auditory meatus

From Norberg, 1978

Figure 21.25 An owl's ruff and asymmetrical ear openings increase hearing sensitivity.   (A) A boreal owl hunts from perches, typically at night. The ruff of feathers that surrounds the face of many owls focuses and amplifies sound. (B) Frontal and dorsal views of the skull of a boreal owl show asymmetry in the position of the left and right ear openings (the external auditory meatus). Small differences in the time and intensity at which sounds arrive at each ear indicate direction of the source. (B after R.A. Norberg. 1968. *Phil Trans R Soc Lond B* 282: 325–410.)

anatomical features key to the acoustic abilities of owls include a facial ruff and an asymmetrical skull. The ruff is formed by stiff feathers and acts as a parabolic sound reflector, focusing sounds and amplifying them (**Figure 21.25A**). The skulls of many owl species, particularly those that hunt at night, are asymmetrical with respect to placement of external ear openings (**Figure 21.25B**), and these species have the greatest auditory sensitivity. The asymmetry assists with localization of prey: minute differences in the time and intensity at which sounds arrive at the two ears indicate direction of the source, and the brain of an owl integrates time and intensity information with extraordinary sensitivity. Despite asymmetry of the external ear opening, the middle and inner ears of owls are bilaterally symmetrical. The extraordinary hearing ability of owls allows them to locate prey in complete darkness.

Olfaction

For a long time, zoologists thought that birds relied primarily on vision and hearing and had little or no sense of smell. However, evidence of many types—morphological, neuroanatomical, physiological, and behavioral—indicates that birds have a functional olfactory system and that scents play important roles in their lives.

The olfactory bulbs of birds vary from more than 25% of total brain volume in turkey vultures to less than 5% in some passeriforms. In general, this variation reflects the relative importance of olfaction. For example, the olfactory bulbs of turkey vultures, which locate carrion by scent, are four times larger than those of black vultures, which locate carrion using visual cues (**Figure 21.26**). In addition, there are twice as many mitral cells—neurons that carry information from the olfactory bulb to other regions of the brain for integration—in the brains of turkey vultures compared to black vultures. As another example of birds using olfactory cues when foraging, hummingbirds use visual cues to locate flowers with nectar, but avoid nectar sources with odor cues emanating from insects that pose a potential threat.

Birds use olfaction in behaviors other than feeding. Several species, including European starlings and swifts, use smell during navigation. Eurasian blue tits, small songbirds that nest in cavities, use smell to detect the presence of predators at their nests. Finally, there is some evidence that odorant molecules, probably emanating from secretions of the uropygial gland, are detected and can influence species and individual recognition as well as mate choice.

(A) Brain of a turkey vulture, *Cathartes aura*

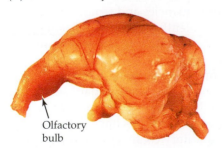

Olfactory
bulb

(B) Brain of a black vulture, *Coragyps atratus*

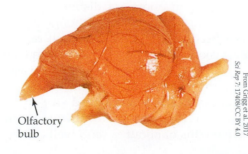

Olfactory
bulb

From Grigg et al. 2017
Sci Rep 7: 17408/CC BY 4.0

Figure 21.26 Size of olfactory bulbs reflects a species' relative reliance on olfaction. (A) Turkey vultures find carrion using olfaction and have large olfactory bulbs. (B) Black vultures find carrion using visual cues and have comparatively smaller olfactory bulbs.

(A) North Island brown kiwi, *Apteryx mantelli*

© Allie Caulfield/CC BY 2.0

(B) Detail of the bill tip

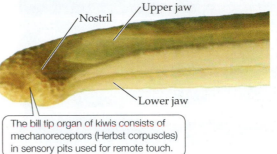

Nostril

Upper jaw

Lower jaw

The bill tip organ of kiwis consists of mechanoreceptors (Herbst corpuscles) in sensory pits used for remote touch.

Figure 21.27 **Kiwis use olfaction and remote touch to locate prey.** (A) Kiwis probe for food in the forest floor, relying on smell and remote touch to find invertebrates. (B) The external nostrils of kiwis are located near the tip of the bill (other birds have nostrils at the base of the bill). Sensory pits in the bone of the bill contain mechanoreceptors for remote touch. (B from G.H. Martin et al. 2007. *PLOS ONE* doi.org/10.1371/journal.pone.0000198 / CC BY 4.0)

Touch

Our coverage of touch focuses on mechanoreceptors of the bill, particularly those called Herbst corpuscles. Herbst corpuscles occur in high concentrations in the bills of five groups of extant birds—kiwis, sandpipers, ibises, waterfowl, and parrots—where they form **bill-tip organs**. Variation in number, arrangement, and precise location of these mechanoreceptors in the bills of these five groups reflects their different foraging modes. For example, the Herbst corpuscles of kiwis, sandpipers, and ibises occur in sensory pits in the bone at the tip of the bill; these birds use their long narrow bills to probe in substrates for invertebrates (**Figure 21.27**). Their bill-tip organs allow detection of prey vibrations in the substrate. We term this sensory modality **remote touch** because it enables the bird to detect prey that is some distance from the bill. Kiwis have enlarged brain regions related to olfaction and touch, and their relative reliance on one sensory modality or the other remains to be determined. Fossil evidence of bill-tip organs in extinct paleognathous birds from the Cretaceous suggests that remote touch is an ancient avian sensory modality.

21.7 Communication

LEARNING OBJECTIVES

21.7.1 Distinguish vocalization from sonation.

21.7.2 Compare vocal learning and auditory learning.

21.7.3 Explain how calls and songs differ in structure and function.

Birds use sounds, colors, and movements when interacting with conspecifics. Birds produce **vocalizations** by passing air through the **syrinx**, a unique avian structure that lies at the junction of the trachea with the primary bronchi. **Sonation**, or nonvocal sound, is typically generated with either the bill or feathers. Many birds combine sounds with visual displays when advertising territories and attracting mates. Here we consider vocalization, sonation, and their use in combination with visual displays.

Vocalization

All vertebrates tested are *auditory* learners, meaning that they can remember sounds. Only a few vertebrates (e.g., songbirds, parrots, hummingbirds, humans, bats, cetaceans, elephants, and pinnipeds) are *vocal* learners. Vocal learners are able to (1) learn and imitate sounds, (2) alter those sounds, and (3) vocalize using the syrinx (birds) or larynx (mammals). Dogs, for example, are auditory but not vocal learners. They can learn human verbal commands such as "sit," "stay," and "down," but (perhaps fortunately) they cannot imitate and vocalize these commands.

Bird vocalizations include calls and songs. **Calls** are simple, brief vocalizations produced by both males and females. Birds use calls, for example, to warn of nearby predators (alarm calls) or to coordinate individuals moving through dense vegetation (contact calls). **Songs** are longer, more complex vocalizations produced in the contexts of reproduction and territoriality. Birds usually have more than one song, and some species have repertoires of several hundred songs or more. Northern mockingbirds imitate the sounds of other bird species, as well as sounds of other vertebrates (cats, dogs, frogs), invertebrates (crickets), and inanimate objects (car alarms, lawn mowers). A typical song consists of a series of phrases, each phrase repeated several times before moving to the next, and a male mockingbird may sing over 200 songs (**Figure 21.28**). A male brown thrasher, *Toxostoma rufum*, may have more than a thousand songs in his repertoire.

Songbirds (Passeri; see Table 21.1) have a complex syrinx and vocal learning capacity. Songbird brains have several song control regions, which are groups of cells involved in hearing, learning, and producing songs.

Historically, only male songbirds were thought to sing. More recently, however, female song has been identified

Northern mockingbird, *Mimus polyglottos*

© Katherine E. Bemis

Figure 21.28 Mockingbirds produce complex songs. Mockingbirds learn sounds throughout life and can imitate the sounds of other birds, as well as sounds of other animals and inanimate objects. Males may have as many as 200 different songs in their repertoire.

in about 70% of songbird species. Early data focused on temperate species, many of which lack female song; in other cases, female songs were present but quieter than those of males, and perhaps simply overlooked.

Mature male songbirds sing primarily during the breeding season. Classic studies showed that for song to develop normally, juveniles must be exposed early in life to songs of adult males of their species in order to learn and memorize appropriate songs. Additionally, males must be able to hear themselves sing, so that they can fine-tune their songs by comparing their own vocal output to the songs they memorized. In some species, the period for learning song occurs early in life and is limited in duration, whereas in others, such as the northern mockingbird, song learning continues throughout life.

Sonation

Examples of nonvocal sounds produced by birds include drumming, whistling, and clapping. However, sometimes sounds with the same name are produced by different species in very different ways. For example, woodpeckers make a drumming sound by pounding their bill on a resonant object, such as a dead tree or metal drainpipe. Both sexes drum, either to claim a territory or attract a mate, and drumming serves as the functional equivalent of song (woodpeckers do not sing, although they do produce calls). Sounds produced when drumming are much louder and more rapid than those made by feeding woodpeckers. Male ruffed grouse, *Bonasa umbellus*, also produce a sound described as drumming, but they do this by moving their wings. A male grouse stands, cups his wings, and rapidly brings them forward and upward toward his chest, without touching the wings to his chest. Despite the lack of contact, the movement is so forceful that some air is compressed between the male's wings

and chest, creating a sound wave; each movement of his wings produces a soft thump, and the thumps occur at ever shorter intervals until they merge into one continuous sound.

Whistling and clapping are often produced by wing feathers. During take-offs and landings, the wings of mourning doves, *Zenaida macroura*, whistle. In other species, the whistling or twittering sounds occur only during very specific display flights. For example, male American woodcocks, *Scolopax minor*, spiral upward during their display flight, and air passing over the outer three primary feathers of each wing produces a twittering sound. Finally, courting male short-eared owls, *Asio flammeus*, rise to about 300 m in the air, hoot many times in succession, enter a shallow dive, and clap their wings below their body in rapid succession, producing a sound reminiscent of applause.

Visual displays

Visual displays are frequently associated with song and sonation. For example, when a male European starling, *Sturnus vulgaris*, advertises his nest cavity to potential mates, he sits near the cavity, singing, and wing-waving. The wing-waving display includes a specific posture (his bill points upward and his tail tucks downward), movements (he simultaneously rotates both wings), and change in the position of certain feathers (he erects his throat feathers).

21.8 Reproduction

LEARNING OBJECTIVES

21.8.1 Trace the pathway of an egg from ovulation to laying.

21.8.2 Define maternal effects.

21.8.3 Compare sex determination in birds and mammals.

21.8.4 Describe how altricial and precocial hatchlings differ.

All birds are oviparous. In this section, we discuss reproductive organs, insemination, egg structure, and embryonic development. We finish with hatching and developmental state of young. Keep in mind that we use the term egg in two ways. We use egg to mean the ovum, or female gamete, which is a single cell, the yolk; technically, this is the correct usage. However, "egg" is commonly used to describe the hard-shelled structure laid by female birds (as well as the hard- or soft-shelled structures laid by lepidosaurs, turtles, crocodylians, and monotremes) that contains not only yolk but also albumen (the white of an egg).

Reproductive organs and insemination

Male birds have two testes that produce sperm. Sperm travel from the testes through deferent ducts to the cloaca. As mentioned in Section 21.3, most male birds lack an intromittent organ, so sperm is transferred by everting the cloaca. At the time of insemination, the female also everts her cloaca and sperm enter during a very brief

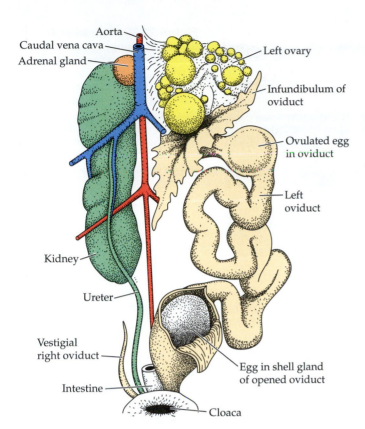

Figure 21.29 Reproductive and excretory organs of a female chicken. Only the left ovary and oviduct are functional in most birds. Ovulated eggs enter the infundibulum, where fertilization occurs, and travel down the left oviduct, spending substantial time in the shell gland, where the egg's shell is added. Eggs are laid through the cloaca. (After K. Liem et al. 2001. *Functional Anatomy of the Vertebrates*, 3rd ed. Cengage/Harcourt College: Belmont, CA.)

Figure labels: Aorta; Caudal vena cava; Adrenal gland; Left ovary; Infundibulum of oviduct; Ovulated egg in oviduct; Left oviduct; Kidney; Ureter; Vestigial right oviduct; Egg in shell gland of opened oviduct; Intestine; Cloaca

moment when the two cloacae come in contact. From the female's cloaca, sperm enter and move up the left oviduct (recall that most female birds have a single functional ovary, usually the left; **Figure 21.29**). During the breeding season, the ovary typically releases an ovum each day until the clutch is complete. A released ovum enters the infundibulum of the oviduct, where fertilization occurs. As the fertilized egg moves along the oviduct, shell membranes are secreted to enclose the yolk and albumen, while carbohydrate and water are added to the albumen. Eggshell formation occurs in the shell gland at the base of the oviduct (see Figure 21.29). At laying, the egg leaves the female's reproductive tract, passing out the cloaca. From ovulation to laying takes about 24 hours, with most of the time (about 20 hours) spent in the shell gland. Female birds can store sperm in the functional oviduct; following release of subsequent ova, these sperm travel to the infundibulum where fertilization can occur. Thus, a single insemination can fertilize several eggs.

Egg structure

Bird eggs are archetypal amniote eggs (see Figure 9.14). The innermost membrane, the amnion, surrounds the embryo while the outermost membrane, the chorion, encloses all embryonic structures and is in contact with the inner side of the shell. The outer portion of the allantois is highly vascularized and presses against the chorion, forming the chorioallantois—the site of embryonic gas exchange. The interior of the allantois is the storage area for nitrogenous wastes in the form of uric acid crystals. Nutrition is provided by the yolk, which is enclosed in the yolk sac and gradually absorbed into the embryo's body. Albumen, which is about 10% protein and 90% water, provides water needed for development.

Because bird embryos have no means of ventilation for gas exchange and depend on a limited store of water, the permeability of the eggshell to oxygen, carbon dioxide, and water vapor is a delicate balance between slowing water loss while permitting outward diffusion of carbon dioxide and inward diffusion of oxygen. The eggshell is penetrated by many tiny pores, which may be blocked to varying degrees with organic or crystalline material, affecting the rate of evaporation of water from the egg. The eggshell grows thinner as calcium is removed from it to form bones of the embryo, increasing rates of diffusion of gases through the shell.

Maternal effects

Female birds deposit diverse substances into the yolk, albumen, and shell. These substances, which include maternal hormones, carotenoids, and vitamins, influence offspring development and phenotype. For example, females of many species deposit androgens in yolk; high levels of yolk androgens are often linked to increased growth rates (preand post-hatching), metabolic rates, nestling begging rates, and aggressiveness. These changes are called **hormone-mediated maternal effects**. (More generally, maternal effects are epigenetic changes—changes that modify expression of genes without changing DNA sequence of the genes—caused by the developmental environment provided by the mother.) Sometimes the amount of androgen deposited by females increases or decreases from the first- to last-laid egg in a clutch. Researchers are examining whether embryos can modify maternally deposited substances, in effect manipulating their own development.

Sex determination

In birds, females are the heterogametic sex, having different sex chromosomes (ZW), and males are the homogametic sex, with matching sex chromosomes (ZZ). (Note that this differs from mammals where males are the heterogametic sex, XY, and females are the homogametic sex, XX.) Presence of the W sex chromosome causes the primordial gonad to secrete estrogen, which stimulates the left gonad to develop as an ovary and the left Müllerian duct system to develop into an oviduct and shell gland. In the absence of estrogen (i.e., when the genotype is ZZ), a male develops. The evolutionary origin of genetic sex determination in birds is unclear because all crocodylians have temperature-dependent sex determination and lack heterogametic sex chromosomes.

(A) Altricial young

Kari Walls/CC BY 2.0

(B) Precocial young

Peter McGowan/USFWS

Figure 21.30 Altricial and precocial young. (A) Altricial young are nearly naked at hatching, unable to thermoregulate, and have closed eyes. They must be fed by their parents. (B) Precocial young are covered with down at hatching, can thermoregulate, and their eyes are open. Soon after hatching, precocial young can walk, swim, and find their own food.

Hatching and developmental state of young

Bird eggs must lose water to hatch because water loss creates an air cell at the blunt end of the egg. The embryo penetrates the membranes of this air cell with its bill 1–2 days before hatching, and ventilation of the lungs starts to replace the chorioallantoic membrane in gas exchange. Shortly before hatching, the chick develops a horny projection on its upper bill (the egg tooth) and an enlarged muscle on the back of the neck (the hatching muscle). The egg tooth is thrust against the shell, forming the first cracks, that widen and allow emergence from the shell. The egg tooth and hatching muscle disappear soon after hatching.

The developmental state of young birds at hatching varies along a continuum from altricial to precocial (**Figure 21.30**). **Altricial** young hatch at a relatively immature state: their eyes are closed; they have little down; they are confined to the nest; and they cannot thermoregulate or feed on their own for some time. Birds with altricial young include passeriforms, owls, woodpeckers, and hawks. In contrast, **precocial** young hatch at a relatively mature state: their eyes are open; they have down; and they can move about, thermoregulate, and feed themselves soon after hatching. Chickens, ducks, and geese have precocial young.

21.9 Parental Care

LEARNING OBJECTIVES

21.9.1 Describe the options for parental care by birds.

21.9.2 Describe categories of parental behavior shown by birds.

21.9.3 Discuss the behaviors of interspecific brood parasites and hosts.

Most birds are monogamous, and in ~80% of bird species both mother and father provide parental care. Because embryos develop in eggs outside the female's body and young birds consume food that both parents can supply, both females and males can perform all categories of parental behavior, including building nests, incubating eggs, and feeding offspring.

In some species, only the mother cares for young (~8%) or more rarely, only the father (~1%). In ~9% of species, three or more individuals provide care; among these **cooperative breeders**, care is typically provided by the mother, father, and one or more offspring from a previous clutch. The remaining ~1% of bird species provide no parental care. These **interspecific brood parasites**, which relinquish all parental duties to adults of another species, are discussed in more detail at the end of this section.

Nest building

Nests protect eggs not only from physical stresses such as heat, cold, and rain, but also from predators. Bird nests range from shallow holes in the ground to enormous structures that represent the combined efforts of hundreds of individuals over many generations (**Figure 21.31**). Nests can be made of woven plant materials, mud, saliva, spiderwebs, and other materials. Given the extensive modifications of avian forelimbs for flight, bills—aided by cranial kinesis and a flexible neck (see Section 21.5)—play an essential role in collecting nest material and constructing nests.

Incubating

Most birds maintain relatively stable egg temperatures via incubation, brooding eggs with metabolic heat to promote embryonic development. To facilitate heat transfer from their own bodies to the eggs, some parents develop brood patches on the breast a few days before the first egg is laid. Prolactin plus estrogen or androgen stimulate formation of brood patches in female and male birds, respectively. The insulating properties of feathers that are so important in thermoregulation become a handicap during brooding, so adult birds either lose or pluck feathers to form one or more patches of bare skin. Blood vessels proliferate and water accumulates in the skin of a brood patch, giving it a spongy texture.

Most species wait to begin incubation until their egg clutch is complete, resulting in synchronous hatching. Other species begin incubation as soon as the first egg

(A) Egg of an oystercatcher, *Charadrius vociferus*

(B) Nest of Anna's hummingbird, *Calypte anna*

(C) Floating nest of American coot, *Fulica americana*

(D) Communal nest of weaverbird, *Philetairus socius*

21.31 Bird nests vary in complexity, size, shape, and building materials. Some nests are no more than shallow depressions; others are elaborate structures. (A) The oystercatcher, like many shorebirds, does not construct a nest, but lays its eggs on bare ground. (B) Anna's hummingbird constructs a tightly woven nest lined on the inside with feathers for insulation and covered on the outside with lichen for concealment. (C) American coots build floating nests using the air-filled stems of aquatic plants. (D) Weaverbirds of western Africa are social and build communal nests that can include hundreds of mated pairs and are occupied for generations. Birds enter the nests through openings at the bottom.

is laid, leading to asynchronous hatching. In the nests of asynchronous hatchers, late-hatched chicks often do poorly and may even be victims of siblicide. In some species, including certain eagles, boobies, and pelicans, siblicide is obligate—the older sibling always kills the younger one(s). In other species, such as the osprey and some egrets, siblicide is facultative—whether older siblings kill the youngest depends on environmental conditions. In obligate species, the younger sibling is insurance for the parents, essentially a potential replacement should the older chick die early on. In facultative species, the youngest sibling serves not only as insurance, but also as extra reproductive value when times are good (i.e., in years when resources are plentiful, parents may be able to raise all chicks).

Megapodes, galliform birds of the Indo-Australian region, do not incubate their eggs. Instead, they bury their eggs in large mound nests and rely on heat from the sun and rotting vegetation to promote embryonic development (**Figure 21.32**). Other megapodes deposit eggs in sand warmed by volcanic hot springs. Megapode chicks are extremely precocial, being fully independent and ready to fly from the time they emerge from the mound.

Australian brush turkey, *Alectura lathami*, and its mound nest

Figure 21.32 Megapodes do not incubate their eggs. Megapodes such as the Australian brush turkey build mounds of soil and vegetation in which eggs are deposited. These birds rely on heat from the sun and rotting vegetation to promote embryonic development. The extremely precocial chicks hatch, emerge from the mound, and require no parental care.

(A) Nestling song thrushes, *Turdus philomelos*

NottsExMiner/CC BY SA 2.0

(B) Gentoo penguin, *Pygoscelis papua*, feeding young

© Andrew Corso

Figure 21.33 Parental care in birds often involves feeding young.
(A) The gaping mouths of nestlings stimulate adults to feed them. The nestling that begs most energetically is most likely to be fed, and arrival of a parent (or almost any other disturbance) stimulates a frenzy of begging. Color contrast between the flanges surrounding the mouth and the interior of the mouth provide a target for adults, as in these song thrush nestlings. (B) An adult gentoo penguin stores crustaceans, squids, and fishes in its digestive tract and once home regurgitates the catch to its offspring.

Feeding young

Methods of food delivery to young birds varies. Adult passeriforms typically carry a single food item (often an insect) in the bill. The parent's arrival at the nest stimulates begging by nestlings, which energetically call and open their mouths wide (**Figure 21.33A**). The parent gives the food item to one nestling and may make several hundred feeding trips a day to ensure that all young in the clutch receive sufficient food. Birds of prey usually deliver a single large prey item to the nest, carrying it with their feet. When their offspring are young, the parent will tear the prey into smaller pieces before feeding it to them. Penguin parents forage at sea for fishes, squids, and crustaceans, store their catch in the digestive tract, and later regurgitate it to offspring (**Figure 21.33B**).

Unlike many seed-eating birds that feed their young insects, which are high in protein and fat, doves and pigeons feed their young crop milk (or "pigeon's milk"). This nutritive mixture is produced by protein- and fat-laden cells that detach from the epithelium of the crop and become suspended in a semi-solid material. Pigeons and doves feed crop milk for a few days after hatching and then begin to mix in seeds. The hormone prolactin—the same hormone that stimulates milk production by mammals—stimulates production of crop milk in both parents as hatching approaches.

Flamingos also feed their young a nutritious mixture after hatching; it is produced by the esophagus and differs in composition from the crop milk of doves and pigeons. This fluid provides convenient nutrition to young flamingos while their complex filter-feeding apparatus develops over the first 2 months posthatching. Production of a nutritious esophageal mixture for offspring also occurs in male emperor penguins, *Aptenodytes forsteri*, when the single egg a male has been incubating hatches before his mate returns from foraging at sea.

Food brought by parents to altricial young also serves as a source of water until young birds can reach locations with water. Precocial young, with their greater mobility, can often, but not always, follow parents to water. Many species of sandgrouse (Pterocliformes), found in dry areas

of Africa, India, and Asia, carry water to their precocial offspring. Parents and young feed primarily on seeds that supply little water. For the first several months after hatching, young do not accompany their parents to watering holes, presumably because the sites are often distant and full of predators. Instead, sandgrouse fathers arrive at watering holes, wade into the water where they rock up and down, soaking specialized belly feathers. Exposure to water prompts barbules of the belly feathers to change orientation, allowing the feathers to soak up and hold water much like a kitchen sponge. Fathers carry the water on the inner surfaces of the belly feathers back to their chicks, which then drink.

Interspecific brood parasitism

As mentioned above, ~1% of bird species, including many species of cuckoos and all species of cowbirds, are interspecific brood parasites. These birds lay their eggs in other species' nests, leaving the foster parents (sometimes called hosts) to raise the parasitic chick. Indeed, the expression "a cuckoo in the nest"—referring to the presence of an unwelcome intruder—derives from this behavior. The biological parents of the parasitic chick benefit because they save time and energy by dispensing with all parental care; this savings allows parasitic females to produce many more eggs in a season than can birds caring for their own young (parasitic brown-headed cowbirds, *Molothrus ater*, can lay up to 50 eggs in a season compared to ~15 eggs laid by non-parasitic passeriforms). Host parents, on the other hand, experience reduced reproductive success, losing all or most of their own young.

Decreases in host reproductive success can result from actions of female brood parasites or their offspring. Female cuckoos remove one or more eggs of the host before laying one of their own in the host nest, and cuckoo chicks methodically push any remaining host eggs or hatchlings out of the nest (**Figure 21.34A**). Young cowbirds simply outcompete host young for parental feedings, causing some or all host young to starve. Parasitic young are favored in begging competitions because they have relatively short incubation periods, which means they hatch

Figure 21.34 Cuckoos are brood parasites. (A) The nest of this reed warbler, *Acrocephalus scirpaceus*, has been parasitized by a European common cuckoo. Soon after hatching, cuckoo nestlings push host eggs or hatchlings out of the nest (even when, as seen here, the host parent is watching). (B) Host parents continue to feed the cuckoo chick, even after it has grown much larger than they are.

(A) Nestling European common cuckoo, *Cuculus canorus*, in host's nest

Host parent watches cuckoo nestling use its back to eject host's egg.

Back of cuckoo nestling with its wings raised

Ejected host egg

blickwinkel/Alamy Stock Photo

(B) Host parent continues to feed nestling cuckoo

Lynn Martin

before host young and often grow more rapidly. Indeed, some parasitic young are larger than the host parents caring for them (**Figure 21.34B**).

Many behaviors characterize brood parasites and their hosts. Female brood parasites simultaneously monitor several potential host nests and lay their egg quickly to avoid detection (cuckoos take only a few seconds to deposit an egg as compared to 20 minutes for most passeriforms). Some brood parasites have eggs or young that resemble those of the host, making it difficult for hosts to identify and reject them. Some host species appear oblivious to the presence of a parasitic egg in their nest, even when it differs in size and color from their own eggs. Other species can detect a foreign egg and either remove it or abandon the nest and start over in a new location. At least for some hosts, recognition and rejection of parasitic eggs and young improve with breeding experience.

21.10 Orientation, Navigation, and Migration

LEARNING OBJECTIVES

21.10.1 Understand how landmark use, compass orientation, and true navigation are experimentally demonstrated.

21.10.2 List cues homing pigeons use to find their way to their home loft.

21.10.3 Discuss costs and benefits of seasonal migration.

Like other animals, birds orient to features of their environment, whether they are making short return trips to the nest to feed young or lengthy seasonal migrations. How do they find their way when moving through the environment, and what cues do they use to reach their goal? **Orientation** refers to an animal determining the direction it is facing. **Navigation** denotes an animal determining how to get from where it is to where it wants to be, and **seasonal migration** refers to long distance movements between breeding and non-breeding areas. In this section, we consider the range of navigational abilities exhibited by birds and the diverse cues they use to reach their destinations. Our discussion of orientation and navigation focuses on how birds respond when experimentally relocated, but the same principles apply to birds that experience natural relocation (for example during severe weather events such as hurricanes).

Navigational abilities

Animals have three levels of navigational abilities: landmark use, compass orientation, and true navigation. Animals may detect landmarks using any sensory modality, including vision, olfaction, and magnetoreception (ability to detect magnetic fields). One way to demonstrate landmark use is to experimentally move or disrupt the landmark(s) and assess whether the animal's orientation is altered. Another is to disrupt the sensory modality with which the animal detects landmarks and see whether orientation is disrupted. When frosted contact lenses were placed on homing pigeons, making it impossible for them to see distant objects and barely able to see nearby ones, the birds arrived in the general vicinity of their home loft (indicating they could determine the direction home without using visual landmarks), but they had difficulty finding the loft itself (indicating visual landmarks might be key in pinpointing the precise location of the home loft).

Compass orientation is the ability to head in a specific direction without using landmarks. This ability is indicated when an experimentally relocated animal does not compensate for its displacement and misses its goal. Birds migrating for the first time to a non-breeding area rely on an inherited program that tells them flight direction and duration. If first-time migrants are experimentally relocated, they miss their population's correct non-breeding area (**Figure 21.35A**). Compasses used by these birds include the sun, stars, and Earth's magnetic field. First-time migrants seem to rely on the magnetic field, although first-timers migrating at night may also use stars.

(A) First-time migrants use compass orientation and cannot find non-breeding area when relocated

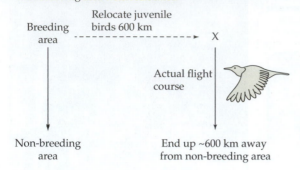

(B) Adult migrants use true navigation to find non-breeding area

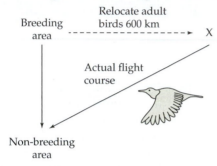

Figure 21.35 Distinguishing compass orientation and true navigation. (A) First-time migrant starlings use compass orientation to find their non-breeding area: when experimentally relocated, they miss their goal by the distance displaced. (B) Adult starlings use true navigation to find their non-breeding area: when experimentally relocated, they compensate for their displacement and reach their goal.

True navigation is the ability to compensate for experimental relocation and reach the goal, without using landmarks and without information about the relocation route. Birds that have migrated to a non-breeding site at least once can then compensate for experimental relocation and still arrive at their correct non-breeding destination (**Figure 21.35B**). This ability requires both a "navigational map," which indicates current location with respect to the goal, and a compass, which indicates direction. Through experience, birds are thought to acquire a navigational map, perhaps based on environmental gradients that increase and decrease in predictable ways, such as the arc of the sun across the sky or the intensity of Earth's magnetic field. Migratory birds are presumed to learn the values of environmental gradients at both their breeding and non-breeding areas (the two goals of their seasonal migratory route), and to be able to extrapolate beyond those areas. If experimentally displaced, a bird could compare gradient values at the unfamiliar location to those learned previously and thus determine its relative displacement. This explains why migrants that have been to the non-breeding site at least once can compensate for relocation. (Note, however, birds relocated great distances, such as from one continent to another, do not always return, suggesting there are limits to true navigational

abilities of experienced migrants.) It also explains why relocated first-time migrants do not reach the correct non-breeding site: because they have not yet learned the gradients at the non-breeding site, they have no basis for comparison to gradients at the site of displacement.

These examples of true navigation have focused on migration between breeding and non-breeding sites. True navigation also is used during movements that are unrelated to migration, for example, during **homing**, when a relocated bird finds its way back to its location prior to being displaced.

Using multiple cues during navigation

Birds typically uses multiple orientation cues, often in a hierarchical manner, and considering distance from home and current conditions (e.g., sunny versus cloudy days), as exemplified by the well-studied homing behavior of racing pigeons. For as long as people have raised and raced pigeons, it has been known that pigeons displaced from their home loft and released in unfamiliar territory typically vanish from sight flying in the direction of their home loft (**Figure 21.36A**). On sunny days, the birds vanish toward home and return rapidly to their lofts. On overcast days, vanishing bearings are less precise, which led to the idea that pigeons use the sun as a compass.

Because the sun's position in the sky changes from dawn to dusk, a bird must know what time of day it is to use the sun to tell direction, and its timekeeping ability requires an internal biological clock. If this timekeeping hypothesis is correct, it should be possible to fool a bird by shifting its internal clock forward or backward. Such experiments have indeed been done, and they support the hypothesis. For example, if the lights are turned on in the pigeon loft 6 hours before sunrise every morning for about 5 days, the birds become accustomed to that artificial sunrise. Then, upon release, they act as if the time is 6 hours later than it actually is—they are directed by the sun, but their internal clocks are wrong by 6 hours. That error should cause the birds to fly off on courses displaced by 90 degrees from the correct course for home, and on sunny days that is exactly what clock-shifted pigeons do (**Figure 21.36B**).

Under cloudy skies, however, clock-shifted pigeons head straight home, despite the 6-hour error in their internal clocks (**Figure 21.36C**). Clearly, then, pigeons have more than one way to navigate. When they can see the sun, they use a sun compass, but when the sun is not visible, they use another mechanism unaffected by the clock-shift.

This second mechanism is probably the ability to sense Earth's magnetic field and use it as a compass. On sunny days, attaching small magnets to a pigeon's head does not affect its ability to navigate, but on cloudy days pigeons with magnets do not return to their home lofts (see Section 16.5 for a discussion of magnetic navigation in sea turtles). Polarized light, ultraviolet light, familiar airborne odors, and visual landmarks provide additional cues.

(A) Homing pigeons, *Columba livia*, departing from release point toward the direction of their home loft

Richard Elzey/CC BY 2.0

Figure 21.36 Orientation by homing pigeons.
(A) When homing pigeons are released, they depart in the direction of their home loft. (B,C) Each dot shows the direction in which a pigeon vanished from sight when it was released in the center of the large circle. The home loft is at the straight up position, and the solid bar in each circle shows the average direction chosen by homing birds. In this experiment, pigeons were clock-shifted by turning on lights in the loft 6 hours before sunrise every morning for 5 days, so that the birds became accustomed to an artificial sunrise. (B) On sunny days, clock-shifted pigeons acted as if the time was 6 hours later than it actually was and flew on courses displaced from the correct course home, illustrating use of a sun compass based on their biological clock. (C) Under cloudy skies, both control and clock-shifted pigeons headed home, illustrating that when the sun compass cannot be used, pigeons use another mechanism unaffected by clock shifting. (B,C modified from W.T. Keeton. 1969. *Science* 165: 922–928.)

(B) Pigeons released on a sunny day use sun compass orientation

Controls

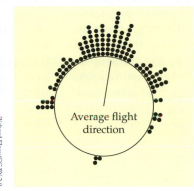

Average flight direction

Experimentally clock shifted 6 hours

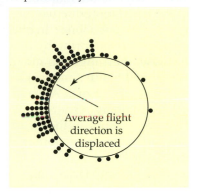

Average flight direction is displaced

(C) Pigeons released on totally overcast day use other information for orientation

Controls

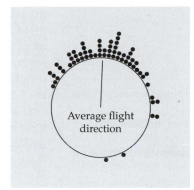

Average flight direction

Experimentally clock shifted 6 hours

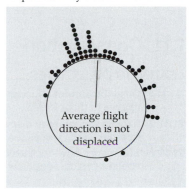

Average flight direction is not displaced

Seasonal migration

About 15% of bird species engage in seasonal migration. Some of these roundtrip journeys are incredible feats of endurance. Arctic terns move between their northern breeding sites in the Arctic and non-breeding sites in Antarctica. For terns nesting in Iceland or Greenland, the round trip can be ~71,000 km. Why do some birds undertake such a grueling journey? What are its costs and benefits?

Costs of migration Migration has significant costs, as evidenced by the number of birds that do not survive it. Indeed, migration is the time of highest mortality during a migratory bird's annual cycle, which consists of four phases: (1) breeding, (2) migration away from breeding grounds, (3) time spent at non-breeding grounds, and (4) migration back to breeding grounds.

As discussed in Section 21.3, powered flight requires a lot of energy. Some birds store fat for weeks preceding migration, but the benefit of this fuel supply must be balanced against the cost of increased weight. Although some birds fly nonstop to their destination, others stop along the migratory route to feed. This layover strategy allows birds to store less fat before beginning their journey.

Birds migrating at night hit obstacles, including skyscrapers. One estimate of the annual mortality of birds in the U.S. indicates that nearly 600 million die from building collisions. In addition, at night urban areas blaze with artificial light, which can disorient migrating birds. One study documented the effects on nocturnally migrating birds of New York City's Tribute in Light at the National September 11 Memorial and Museum, an extreme example of artificial light in a highly urbanized area. Annually since 2002, on the evening of September 11 lights at the Memorial are turned on from dusk to dawn. On this night for each of 7 years, researchers using data from a nearby radar station quantified flock densities, flight speeds, and in-flight vocalizations of migrant birds in the presence and absence of illumination at the Memorial (with the cooperation of the Memorial staff, lights are turned off for 20 minutes when observers estimate more than 1,000 birds are circling in the light beams). Their results showed that when the display was illuminated, birds flew in high-density groups—at slower speeds and in circular flight paths—and frequently vocalized. When the lights were turned off, all signs of these behavioral disruptions disappeared. Such behavioral disruptions drain both time and energy from birds that are already energy-challenged.

Benefits of migration   For migratory behavior to evolve, energetic and mortality costs must be offset by energy gain and reproductive success as a result of migrating to a different habitat. In other words, migrants must produce more offspring by migrating than they would otherwise. For migrants moving from the Neotropics (Central and South America) to summer breeding areas in the Nearctic (North America), one major advantage may be abundant resources, such as rapidly increasing insect populations in the Nearctic. Long summer days at high latitudes are another advantage—longer days provide more time to forage, and longer foraging times may translate into larger clutches and higher survival of offspring. Another benefit of migrating to breeding areas in North America might be to escape competition for food and nest sites in crowded Neotropical habitats.

21.11 Conservation

LEARNING OBJECTIVES

21.11.1 Describe current abundance of North American birds relative to that of 1970.

21.11.2 Explain why grassland species are dramatically declining.

21.11.3 Summarize how declines in abundance can influence species-typical song.

Of the roughly 11,000 extant species of birds, about 1,500 are threatened with extinction. Extinctions, however, are not the only way we assess how a taxon is faring. Some conservation biologists in the continental U.S. and Canada have focused instead on changes in *abundance* of birds—even still-abundant species—to determine trends. Using several long-term, standardized datasets with information on more than 500 species, researchers estimated a net loss in total abundance of 2.9 billion individuals—a reduction of 29%—between 1970 and 2018 (**Figure 21.37A**). Loss in abundance characterized all breeding biomes except wetlands, with grassland species being especially hard hit (**Figure 21.37B**). A separate analysis of 10 years of data from a network of 143 weather radars found a nearly 14% decline in biomass passage of nocturnally migrating birds. Thus, two independent methods of assessing changes in abundance indicate major declines in bird numbers in North America.

Why are grassland species declining so dramatically? Agricultural practices have greatly altered temperate grasslands such as the North American Great Plains. Climate change challenges this already stressed biome with increases in precipitation, temperature, and frequency of severe weather. Dickcissels, *Spiza americana*, are migratory songbirds and obligate grassland specialists that breed in the Great Plains and winter in Central and South America. One long-term capture-recapture study of this species on their breeding grounds revealed that dickcissel abundance declined as precipitation in early summer increased. In contrast, brown-headed cowbirds, which are brood parasites (see Section 21.9), increased in abundance with increasing precipitation at the study area. Thus, at least this one grassland species may be declining due to increasing precipitation associated with climate change combined with increasing cowbird abundance and parasitism.

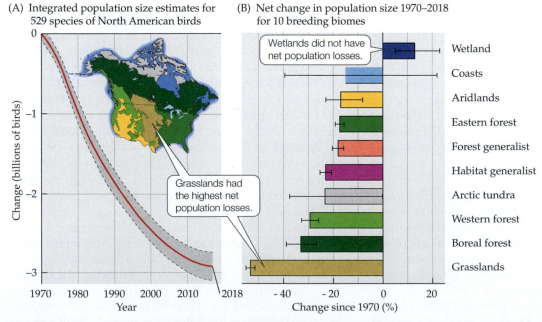

Figure 21.37 Declines in abundance of North American birds, 1970–2018.   (A) Data collected over a 48-year period revealed a net loss of ~2.9 billion birds from the U.S. and Canada. The gray area indicates 95% confidence interval (i.e., the range of values likely containing the actual value). (B) Proportion change since 1970; black bars represent the confidence interval. Loss of abundance characterizes all breeding biomes except wetlands, and was most dramatic in grasslands. (Modified from K.V. Rosenberg et al. 2019. *Science* 366: 120–124.)

Declines in abundance can affect populations in many ways, including disruption of critically important behaviors learned from conspecifics. Recall from Section 21.7 that young male songbirds learn their species-specific song by listening to adult male conspecifics, and that adult song is important for attracting mates. A recent study examined whether decreases in population density could disrupt song development in regent honeyeaters, *Anthochaera phrygia*, critically endangered songbirds of southeastern Australia. From 2015 to 2019 researchers recorded songs (referred to as contemporary songs) and monitored breeding success of males in two regions, one with a relatively high density of regent honeyeaters and the other with a relatively low density. A comparison of contemporary songs with historic recordings (1986–2011) revealed that contemporary songs were less complex and varied dramatically among males. Almost one-third of contemporary males sang songs that differed from their regional norm, and some never sang species-specific songs, instead singing songs of other species. Population density predicted the likelihood of singing the song of another species: males with fewer conspecifics in their area were more likely to sing the song of another species than were males with more conspecifics nearby. When compared with males singing the regional norm, males with atypical songs had reduced breeding success.

Despite these bleak findings, there are reasons to be hopeful. First, birds have many admirers and groups committed to their study and conservation. Research has already shown us ways to help birds, such as turning off artificial lights during times when large numbers of nocturnal migrants are moving through an area. International agreements such as the 1916 Migratory Bird Treaty between the U.S. and Canada, and domestic legislation such as the U.S. Endangered Species Act of 1973, have helped declining species recover, and perhaps new legislation is needed. Absence of declines in wetland species (see Figure 21.37B) probably reflects past efforts at wetland restoration and protection, as well as adaptive harvest management of waterfowl, a management strategy centered on ongoing cycles of monitoring, assessing, and decision-making.

In conclusion, consider a possible change in your own behavior that could help birds. Domestic cats kill some 2.4 billion birds in the U.S. annually, so keeping your cat inside can go a long way to helping birds in your area (and also protects your cat from predators, cars, and diseases).

Summary

21.1 Diversity

Extant birds are divided into Paleognathae and Neognathae.

- Paleognathous birds (tinamous and ratites) have a palate composed of large bones, some of which fuse, forming rigid joints. Neognathous birds have a more lightly built palate with moveable joints.

- Within Neognathae are Galloanserae (chickens, ducks, and allies) and Neoaves (the remaining diversity of extant birds).

21.2 Structural Specializations for Flight and Bipedalism

The aerodynamic and physiological demands of flight constrain many features of birds, including body size and shape.

Birds have five types of feathers: contour, down, semiplumes, bristles, and filoplumes.

Weight-reducing adaptations for flight include decreases in number of some organs (e.g., most adult females have one ovary) and absence of some organs altogether (e.g., most birds lack a urinary bladder).

Specializations of the avian skeleton for flight and bipedalism include strong bones (some pneumatic), fusion of bones (synsacrum, pygostyle, carpometacarpus, tibiotarsus, and tarsometatarsus), and loss of bones (only three digits remain on the hand and the fifth toe is lost).

The keel on the sternum serves as the origin for the pectoralis major (downstroke) and supracoracoideus (upstroke).

Mass of avian leg muscles decreases from the upper leg to the foot, placing most muscle weight near the bird's center of gravity. Lightening the distal part of the leg allows swift bipedal locomotion.

21.3 Wings and Flight

Avian flight is classified as either powered (flapping) or passive (non-flapping). Powered flight requires a lot of energy; passive flight requires little energy. Examples of passive flight include soaring and gliding.

Four forces act on a bird's wings during flight.

- Lift is upward force that keeps the bird in the air.

- Thrust, created during flapping flight, is forward force.

- Drag is backward force that opposes forward movement.

- Gravity pulls the bird toward Earth.

The wings of birds are both airfoils (creating lift) and propellers (creating forward thrust). Primary feathers provide thrust, and secondary feathers provide lift.

(Continued)

Long, narrow wings have high aspect ratios and characterize oceanic birds that soar in strong winds (dynamic soaring). Short, broad wings have low aspect ratios and allow the high maneuverability needed by birds that live in forests.

Wing loading refers to the weight of a bird divided by the area of its wings. Low wing-loading values are characteristic of birds that soar in rising air currents over land (static soaring).

21.4 Feet and Locomotion

Terrestrial locomotion in birds includes walking, hopping, and running.

Most birds have four toes. The most common toe arrangement is anisodactyly, which is associated with perching. Zygodactyly is found in woodpeckers and parrots. Didactyly (two toes) occurs in ostriches, which are flightless runners.

The hind legs that propel aquatic birds are located far back on the body, and the feet are webbed. Webbing may be palmate, totipalmate, or lobate.

21.5 Bills, Feeding, and Digestion

Specialization of the forelimbs as wings and loss of teeth mean that many birds rely on their bills to capture food and all rely on their guts to process it.

Bill shape varies greatly and reflects diet; bird skulls are kinetic.

Birds have a bone in the body of the tongue. Modifications of the tongue are linked to specialized feeding methods.

Many birds have a crop for food storage. The stomach of most birds consists of an anterior glandular portion (proventriculus) and a posterior muscular portion (gizzard).

21.6 Sensory Systems

Avian eyes are large and immobile. Predatory birds have forward-facing eyes and prey species have laterally placed eyes.

Birds have four visual pigments in their retinal cones: red, green, and blue like humans and, for some birds, ultraviolet. Some avian cones have pigmented oil droplets that act as filters.

Owls have the most sensitive hearing among birds and can capture prey in complete darkness. Marked asymmetry of the skull enhances ability to pinpoint source of sound at night.

Contrary to popular belief, many birds have good olfactory sensitivity and use smell when locating food, detecting predators, and recognizing species and individuals.

Some birds have mechanoreceptors in their bill, which they use to detect underground prey via remote touch.

21.7 Communication

Birds produce two kinds of sounds: vocalizations (calls and songs, generated by the syrinx) and sonations (nonvocal sounds, typically generated by bills or feathers). Visual displays frequently accompany both songs and sonations.

Calls are brief, simple vocalizations; songs are longer and more complex.

Songbirds, parrots, and hummingbirds are capable of vocal learning.

21.8 Reproduction

All birds are oviparous, and bird eggs are archetypal shelled amniote eggs.

• The amnion surrounds the developing embryo.

• The chorion is pressed against the eggshell.

• A portion of the allantois contacts the chorion, forming the chorioallantois, the site of gas exchange for the embryo.

• Yolk supplies nutrition and albumen supplies water.

In most birds, insemination occurs when the male's cloaca briefly contacts the female's cloaca. Sperm enter the functional oviduct, and the ovum is fertilized in the infundibulum. The eggshell is added in the shell gland and the egg is laid via the cloaca.

Females deposit substances such as hormones and vitamins in the shell, yolk, and albumen. These substances affect development and phenotype of embryos, epigenetic changes that are called maternal effects.

All birds have genetic sex determination. Females are the heterogametic sex (ZW) and males are homogametic (ZZ).

Developmental state of young at hatching varies on a continuum from altricial (dependence on parental care and feeding) to precocial (high degree of self-sufficiency).

21.9 Parental Care

Parental care in birds is commonly provided by both mother and father, and both sexes can perform all categories of parental behavior.

The bill plays a critical role in collecting nest material and constructing nests, aided by cranial kinesis.

Most species begin incubation once the clutch is complete, leading to synchronous hatching of the eggs. Some species start incubation as soon as the first egg is laid, leading to asynchronous hatching and possibly siblicide.

About 1% of bird species are interspecific brood parasites, laying eggs in the nests of other species to be raised by host parents.

Summary *(continued)*

21.10 Orientation, Navigation, and Migration

Orientation refers to an animal determining the direction it is facing. Navigation denotes an animal determining how to get from where it is to where it wants to be.

Navigational abilities include landmark use, compass orientation, and true navigation. Homing, a form of true navigation, occurs when a relocated bird finds its way back to its original location.

Birds use diverse cues to orient and navigate; these cues include the sun, stars, Earth's magnetic field, odors, landmarks, and polarized and ultraviolet light, in a hierarchical sequence that varies with ambient conditions and distance from goal.

Seasonal migrations are long distance movements between breeding and non-breeding areas.

21.11 Conservation

Almost 3 billion birds have been lost from the U.S. and Canada since 1970. Declines in abundance occur across all biomes except wetlands.

Declines have been most dramatic in grasslands, which have been disrupted by agricultural practices. Increased precipitation associated with climate change also correlates with declines in grassland species and increases in their brood parasites.

Declines in abundance can disrupt the learning of conspecific song by young males. Singing atypical song is associated with reduced fitness, potentially contributing to further declines.

Discussion Questions

21.1 Imagine that you are walking in a penguin colony and you find a bird's humerus. The skin and feathers disappeared long ago, leaving only the bleached bone. You know that the bone must have come from either a penguin or a gull because those are the only two appropriately sized types of birds in the area. How can you determine which bird the bone came from?

21.2 The primary feathers of owls have a comblike leading edge. Given what you learned about airflow and wings, and about how owls hunt, how might this leading edge function?

21.3 Compare digestive tracts, feet, and the placement of eyes in birds of prey and geese.

21.4 Evaluate the following statement for accuracy: All birds are capable of vocal learning.

21.5 Suggest changes that can be made to aid bird conservation. Consider changes from the international level down to the individual level.

Figure credits
Figure 21.1: Ostriches, kiwis, tinamous, and allies: Yathin S Krishnappa/ CC BY-SA 4.0; Chickens, ducks, and allies: Guido Gerding/ CC BY-SA 3.0; Doves, pigeons, and allies: Michal Klajban/ CC BY-SA 4.0; Flamingos and grebes: Lukasz Lukasik/ CC BY-SA 3.0; Nightjars, hummingbirds, and allies: © William E. Bemis; Shorebirds, gulls, cranes, rails, and coots: © Katherine E. Bemis; Looons, penguins, albatrosses, and pelicans: Cristopher Michel/ CC BY 2.0; Mousebirds, trogons, kingfishers,and toucans: Charles J. Sharp, Sharp Photography/ CC BY-SA 3.0; Owls: Abolfazl Fallahzadeh/ Fars Media Corporation/ CC BY 4.0; Hawks, eagles, and New World vultures: Juan lacruz/ CC BY 3.0; Falcons and allies: Charles J. Sharp, Sharp Photography/ CC BY-SA 4.0; Parrots: Luc Viatour/ CC BY 2.0; New Zealand wren: Wynston Cooper/ CC BY-SA 4.0; Flycatcher: © Frank Schulenburg/ CC BY-SA 3.0; Songbirds: H. Zell/ CC BY-SA 3.0
Figure 21.2: After K. Liem et al. 2001. *Functional Anatomy of the Vertebrates*, 3rd ed. Cengage/Harcourt College: Belmont, CA.
Figure 21.3: After K. Liem et al. 2001. *Functional Anatomy of the Vertebrates*, 3rd ed. Cengage/Harcourt College: Belmont, CA.
Figure 21.4: After K. Liem et al. 2001. *Functional Anatomy of the Vertebrates*, 3rd ed. Cengage/Harcourt College: Belmont, CA.
Figure 21.9: After K. Liem et al. 2001. *Functional Anatomy of the Vertebrates*, 3rd ed. Cengage/Harcourt College: Belmont, CA., which is based on T.I. Storer. 1943. *General Zoology*. McGraw-Hill: New York.
Figure 21.10: O.S. Pettingill. 1970. *Ornithology in Laboratory and Field*. Burgess Publishing Company: Minneapolis, MN.
Figure 21.11: After K. Liem et al. 2001. *Functional Anatomy of the Vertebrates*, 3rd ed. Cengage/Harcourt College: Belmont, CA.
Figure 21.14: After K. Liem et al. 2001. *Functional Anatomy of the Vertebrates*, 3rd ed. Cengage/Harcourt College: Belmont, CA.
Figure 21.21A: After K. Liem et al. 2001. *Functional Anatomy of the Vertebrates*, 3rd ed. Cengage/Harcourt College: Belmont, CA.

22

Synapsids and the Origin of Mammals

Short-beaked echidna, *Tachyglossus aculeatus*, a monotreme

Because humans are mammals, we often think of mammals as the dominant vertebrates, but that perspective does not withstand examination. There are barely half as many extant species of mammals as of birds, for example, and as noted in earlier chapters, there are as many species of actinopterygians (ray-finned fishes) as there are all tetrapod species combined (see Figure 1.1).

In some contexts, however, mammals are exceptionally successful. Their size range, from tiny shrews to enormous whales, is impressive, as is the evolution of flight by bats and of echolocation by some rodents, bats, and cetaceans. Mammals have more (and more complex) social systems than most other vertebrates, and these complexities correlate with many features of their biology, from the often-prolonged dependence of young on their mothers to lifelong alliances between individuals that affect social status and reproductive success. Mammals differ from other amniotes in other ways, from reliance on olfaction and hearing as their main sensory systems, to feeding systems based on derived characters affecting chewing and swallowing.

All three extant groups of mammals—monotremes, marsupials, and placentals—were present by the late Mesozoic, along with other groups now extinct. As with the origin of tetrapods (Chapter 8), lissamphibians and amniotes (Chapter 9), and birds (Chapter 19), we rely on analyses of fossilized skeletons of extinct taxa to understand the origin of mammals.

In this chapter we focus on the sequential acquisition of the most important mammalian characters within the synapsid lineage that offer the greatest insight into how and why mammals differ from other vertebrates.

22.1 Synapsid Evolution

LEARNING OBJECTIVES

22.1.1   Compare the skulls of basal eupelycosaurs and therians.

22.1.2   Summarize major postcranial skeletal differences between basal eupelycosaurs and non-mammalian therapsids.

22.1.3   Describe how cynodont therapsids differ from non-cynodont therapsids and what derived features made them more like mammals.

22.1.4   Explain concepts and characters that allow us to recognize Probainognathia and Mammalia.

Synapsida (Greek *syn*, "together"; *apsis*, "loop, arch") includes the extant mammals and their extinct relatives that have lived from the Carboniferous to the present (**Figure 22.1**). The name refers to the single opening, or fenestra, in the temporal region of the skull (**Figure 22.2**). This skull structure distinguishes synapsids from diapsids (two temporal fenestrae; see Figure 9.15).

As mentioned at the start of Chapter 10, synapsids radiated widely in terrestrial habitats during the late Paleozoic, long before the major radiations of diapsids in the Mesozoic. During the Late Carboniferous and throughout the Permian, there were many species of medium to large synapsids (10–200 kg, roughly goat- to pony-size) and a few that weighed more than 500 kg. These Paleozoic forms were the most abundant tetrapods, the apex carnivores, and the largest herbivores in terrestrial food webs. Following the major mass extinction at the end of the Triassic, however, surviving synapsids were much smaller (<1 kg, roughly the size of a rat), and not until the Cenozoic did mammals the size of elephants and whales appear.

In addition to fossils, genetic information has helped us understand key changes in synapsid history, such as the silencing of genes that make yolk proteins in other amniotes, and changes in genes related to color vision. As a result, we can trace the evolutionary history of many of the most interesting features of mammals, including feeding, hearing, locomotion, and breathing. We can

(A) Skull and lower jaw of a basal eupelycosaur, †*Dimetrodon*

(B) Skull and lower jaw of a therian, the dog *Canis*

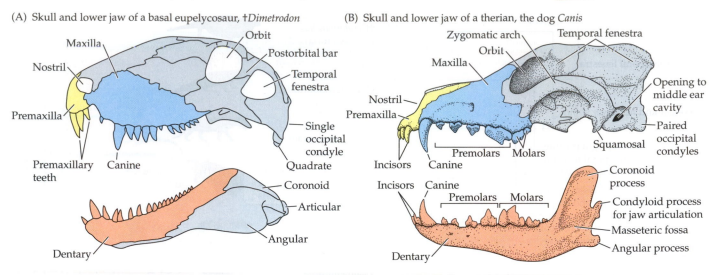

(C) Comparison of occipital region, atlas, and axis vertebrae in basal eupelycosaurs and therians

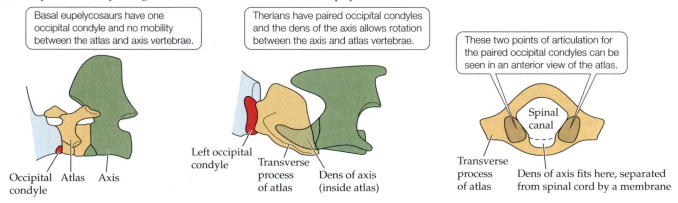

Basal eupelycosaurs have one occipital condyle and no mobility between the atlas and axis vertebrae.

Therians have paired occipital condyles and the dens of the axis allows rotation between the axis and atlas vertebrae.

These two points of articulation for the paired occipital condyles can be seen in an anterior view of the atlas.

Figure 22.3 Cranial bones of a basal eupelycosaur and a therian highlighting major changes over the course of synapsid evolution.   (A) Skull and lower jaw of †*Dimetrodon*, a terrestrial predator. (B) Skull and lower jaw of a domestic dog, *Canis*. The positions and relative sizes of the temporal fenestrae of the two species are important, affecting the jaw-closing musculature, which is more powerful in the dog. (C) Cranioventral joint of a basal eupelycosaur compared to the atlas–axis complex of a therian. Two occipital condyles and the atlas–axis complex give mammals greater mobility of the head on the neck (e.g., nodding "yes" and shaking head "no" are possible). (A after A.S. Romer. 1933. *Vertebrate Paleontology*, 1st ed. University of Chicago Press, Chicago; B after K. Liem et al. 2001. *Functional Anatomy of the Vertebrates*, 3rd ed. Cengage/Harcourt College: Belmont, CA.)

articulation of the lower jaw of †*Dimetrodon* with the skull is between the quadrate bone in the skull and the articular bone in the lower jaw. In contrast, the lower jaw of a dog consists only of the dentary, and its jaw articulation is between the squamosal bone in the skull and the condyloid process of the dentary bone in the lower jaw. The dentary of a dog also has an angular process, which is the insertion site for a portion of the **masseter muscle**, a jaw-closing muscle that originates on the outer face of the zygomatic arch. If you put your hands on your zygomatic arch and clench your teeth, you can feel the masseter muscles bulge as your jaw moves. In Section 22.2 we will trace changes in the jaw joint and bones of the lower jaw and how they relate to the evolution of the series of tiny bones in the mammalian middle ear that transmit vibrations from the eardrum to the inner ear (see Figure 11.15B).

The tooth-bearing bones in the upper jaw of †*Dimetrodon* are the premaxilla and maxilla (see Figure 22.3A). We describe the dentition of †*Dimetrodon* as **homodont** because

all of its teeth have similar shapes, although one tooth is enlarged and is usually interpreted as the homolog of a mammalian canine tooth (teeth posterior to the canines are called cheek teeth or postcanine teeth). Generally, synapsids with homodont dentition do not use their teeth to chew food; rather, they simply cut food into pieces that are small enough to swallow. In contrast, a dog's dentition is **heterodont** because different teeth along its jaw have different shapes (see Figure 22.3B). For example, the teeth attached to the premaxilla are specialized for cutting through food and are called incisors. The maxilla of a mammal typically has three different types of teeth: one enlarged canine tooth, a series of premolar teeth, and a series of molar teeth. The four types of teeth are also present in the lower jaw, with all teeth being attached to the dentary. Synapsids with heterodont dentition typically chew their food into small pieces, making it easier to digest.

Basal eupelycosaurs such as †*Dimetrodon* have a single occipital condyle below the foramen magnum (the

A basal synapsid, †*Cotylorhynchus*

Figure 22.4 Herbivorous basal synapsid. Some species of †*Cotylorhynchus* (†Caseasauria) reached lengths of 6 m and were among the largest synapsids of the Middle Permian.

opening in the braincase for the exit of the spinal cord) that articulates with the atlas vertebra (first cervical vertebra). In contrast, mammals have two occipital condyles—one on each side of the foramen magnum—that articulate with the atlas vertebra (**Figure 22.3C**). The head of a mammal nods up and down, as in the "yes" movements typical of many human cultures. There is little to no movement possible between the atlas vertebra and the axis vertebra (second cervical vertebra) of basal eupelycosaurs. In contrast, mammals have a modified axis vertebra with an upwardly curved projection called the dens (Latin, "tooth") that extends into a space in the atlas vertebra beneath the spinal cord (see Figure 22.3C). This arrangement, referred to as an **atlas–axis complex**, allows mammals to rotate their heads, as humans do when signaling "no" by shaking the head.

Phylogenetic history of synapsids

The most basal clade of Synapsida is †Caseasauria, a group that lived in the Carboniferous and Permian (see Figure 22.1). The earliest forms were lizardlike carnivores

about the size of a cat, but †Caseasauria also includes much larger herbivores such as †*Cotylorhynchus*, which had a distinctively small head and barrel shaped body and reached lengths of 5 to 6 m (**Figure 22.4**).

A basal split within synapsids led to eupelycosaurs, the group that includes mammals. Eupelycosaurs are synapsids with a long, narrow snout. In the remainder of this section, we examine Eupelycosauria and five subclades within it (Therapsida, Cynodontia, Probainognathia, Mammaliaformes, and Mammalia), following the tree and phylogenetic classification in Figure 22.2. We describe representative basal members of each subclade and discuss their characters relevant to the evolution of mammals.

Eupelycosauria   Eupelycosauria is represented by Permian forms such as †*Varanosaurus* (**Figure 22.5A**). These generalized terrestrial and aquatic synapsids superficially resemble extant monitor lizards and reached lengths of 3 m or more. Another Permian eupelycosaur, †*Haptodus*, exhibits many postcranial skeletal features important to understanding synapsid evolution (**Figure 22.5B**). For example, it had a relatively sprawling gait, long feet with a phalangeal formula of 2:3:4:5:3, ribs on all neck and trunk vertebrae, and heavily built limb girdles. †*Haptodus* also retained hind limb retractor muscles that originated on ventral processes of the caudal vertebrae. These caudofemoralis muscles are homologous with the hindlimb retractor muscles of extant sauropsids, such as crocodiles.

Several basal eupelycosaur groups independently evolved dorsal sails supported by elongated neural spines and are commonly known as sailbacks. The best known of these belong to the genus †*Dimetrodon* (discussed above in the comparison of its skull with that of a dog), which encompasses several species spanning a size range from 60 cm to 4.6 m (**Figure 22.6**). Healed spines in some fossils indicate that a web of skin covered them, which also means that there was a blood supply to the skin of the sail. These sails might have been used in temperature regulation and in social signaling.

(A) Basal eupelycosaur, †*Varanosaurus*

(B) Skeleton of a basal eupelycosaur, †*Haptodus*

Figure 22.5 External appearance and skeletal anatomy of basal synapsids. (A) †*Varanosaurus* lived in the Middle Permian and is thought to have been carnivorous. (B) Skeletal features of †*Haptodus* make a good starting point for examining skeletal changes within synapsids. (B modified from P.J. Currie. 1977. *J Paleontol* 51: 927–942.)

Five cervical vertebrae with ribs

Small temporal fenestra

Ribs on all trunk vertebrae

Vertebral interlocking restricts lateral undulation

Two sacral vertebrae

Large pubis and ischium but small ilium

Dentary makes up half of lower jaw

Large clavicle, interclavicle, and coracoid

Sprawling posture

Long feet (phalangeal formula 2:3:4:5:3)

Large processes on caudal vertebrae for caudofemoralis muscle, the major hindlimb retractor

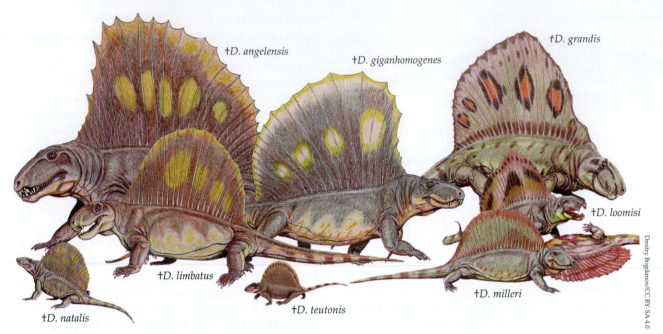

Figure 22.6 The sails of †*Dimetrodon* were probably colorful. These reconstructions accurately represent the sizes and shapes of the dorsal sails of †*Dimetrodon*. The colors and patterns are speculative, but it is likely that they were species-specific, as are the dewlaps of extant lizards (see Figure 15.20B). These species ranged in size from 60 cm (†*D. teutonis*) to 4.6 m (†*D. angelensis*).

Therapsids From their first appearance in the Middle Permian, therapsids occupied drier, more temperate regions than earlier synapsids. The earliest therapsids, such as †*Biarmosuchus*, were carnivores the size of a large dog (**Figure 22.7A**). Other Permian therapsids include †Dinocephalia (herbivores like the cow-sized †*Moschops*, **Figure 22.7B**); †Anomodontia (smaller herbivores like the many species of †dicynodonts, **Figure 22.7C**); †Gorgonopsia (predators like †*Gorgonops*, **Figure 22.7D**); and †Therocephalia.

Found mainly in higher latitudes of the Southern Hemisphere (Gondwana), the non-tropical distribution of early therapsids suggests that they had a degree of endothermy, and some anatomical features support this interpretation. An enlarged temporal fenestra accommodated larger jaw muscles, and the gorgonopsian †*Lycaenops* had a heterodont dentition with incisors, canines, and cheek teeth, which shows that they processed food in the mouth rather than simply swallowing it whole (**Figure 22.7E**). A partial secondary palate separated the nasal passages from the oral cavity.

Therapsids held their limbs underneath the body to a greater extent than did more basal synapsids, providing a more upright stance (see Figure 22.7). A reduced clavicle and interclavicle gave the forelimb a range of motion, increasing stride length and running speed. Therapsids also lack gastralia (present in more basal eupelycosaurs and in outgroups such as crocodylians; see Figure 13.4). Their hindlimb muscles were more mammal-like. Basal synapsids such as †*Haptodus* had large ventral processes on anterior caudal vertebrae because, like sauropsids, they had paired caudofemoralis muscles used to retract the hindlimbs (see Figure 22.5B). Therapsids reduced those processes, the caudofemoralis muscle, and the overall size of the tail (see Figure 22.7E). The expanded ilium in the pelvis served as the origin for gluteal muscles that connected to the greater trochanter of the femur. Gluteal muscles (which give mammals their rounded rear ends) became the major hindlimb extensors of therapsids.

Therapsids have a well developed intertarsal articulation between the astragalus and calcaneus in the ankle, allowing the foot to turn inward for a more upright gait. Their feet are shorter (some phalanges in the third and fourth toes were lost) and serve as levers in limb-based locomotion rather than as holdfasts for axial-based lateral undulation.

†Dicynodonts had tusklike canines and a turtlelike horny beak used to cut vegetation (see Figure 22.7C). They were diverse and abundant in the Late Permian and included long-bodied burrowers and semiaquatic forms such as †*Lystrosaurus* (see Figure 5.10). Most were the size of rabbits or large pigs, although a recently discovered †dicynodont from the Late Triassic of Poland was the size of a large elephant (†*Lisowicia bojani*, ~4,000 kg). Coprolites marking communal defecation spots suggest that †dicynodonts were gregarious.

Medium- to large-size predatory †gorgonopsians such as †*Gorgonops* and †*Lycaenops* occurred in the Late Permian (see Figure 22.7D,E). †*Inostrancevia* was the size of a large bear and had saberlike canines, convergently like

(A) Basal therapsid, †*Biarmosuchus*

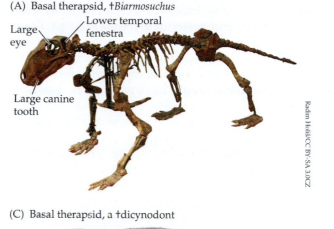

Large eye

Lower temporal fenestra

Large canine tooth

Radim Holiš/CC BY-SA 3.0 CZ

(B) Basal therapsid, a †dinocephalian †*Moschops*

Dmitry Bogdanov/CC BY-SA 3.0

(C) Basal therapsid, a †dicynodont

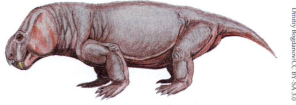

Dmitry Bogdanov/CC BY-SA 3.0

(D) Basal therapsid, a †gorgonopsian, †*Gorgonops*

Dmitry Bogdanov/CC BY 3.0

(E) Skeleton of a †gorgonopsian, †*Lycaenops*

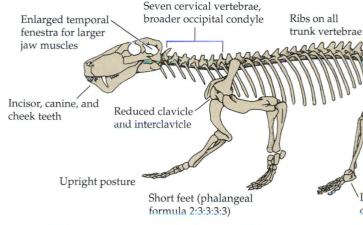

Enlarged temporal fenestra for larger jaw muscles

Seven cervical vertebrae, broader occipital condyle

Ribs on all trunk vertebrae

Vertebral interlocking prevents lateral undulation

Three sacral vertebrae

Expanded ilium for origin of gluteal muscles

Greater trochanter for insertion of gluteal muscles

Reduced processes on caudal vertebrae

Incisor, canine, and cheek teeth

Reduced clavicle and interclavicle

Upright posture

Short feet (phalangeal formula 2:3:3:3:3)

Intertarsal ankle joint between calcaneus and astragalus

Short tail

Figure 22.7 **Anatomy and diversity of basal therapsids.**
(A) Skeleton of a dog-sized Middle Permian basal therapsid, †*Biarmosuchus*, from Russia. It has a more upright posture than in more basal synapsids, but not as upright as in later, more derived therapsids shown in (D). (B) Cow-sized †dinocephalians such as †*Moschops* were herbivorous. The thick skulls led to speculation that males engaged in head-butting. (C) Herbivorous †dicynodonts had horny beaks like a turtle and large tusks. This long-lived group persisted from the Middle Permian to the Late Triassic. (D) †Gorgon-opsians had long canine teeth and a very large gape. They lived from the middle Permian to the end-Permian mass extinction. (E) Skeletal anatomy of a †gorgonopsian showing enlarged temporal fenestra, upright posture, and changes in the pelvic girdle and hindlimb related to evolution of gluteal muscles. (E after E.H. Colbert and R. Broom. 1948. *Bull Am Mus Nat Hist* 89: 353–404.).

those of saber-toothed cats of the late Cenozoic. No †gorgonopsians survived into the Triassic.

Predatory †therocephalians lived in the Late Permian and Early Triassic. The enlarged coronoid process of their dentary provided more area for insertion of jaw-closing muscles, as well as a better lever arm for their action. Based on ridges in their nasal passages, we infer that †therocephalians had cartilaginous maxilloturbinates, scroll-like bones in the nasal passages that humidify and warm incoming air, reducing respiratory water and heat loss (see Figure 11.17). Ethmoturbinate bones, which

house the olfactory epithelium, were probably present in cartilaginous form throughout synapsid evolution. Ossified turbinate bones occur only in mammals.

Some †therocephalians survived the end-Permian extinction, but only †dicynodonts and cynodonts diversified during the Triassic. Simultaneously, archosaurs diversified during the Triassic, giving rise to crocodylomorphs and avemetatarsalians that began to dominate terrestrial ecosystems (see Figure 9.1). Non-mammalian therapsids became an increasingly minor component of the terrestrial fauna during the Triassic and were nearly extinct by its end.

Cynodonts   Cynodont therapsids appeared in the Late Permian and had early success in the Triassic, but most early forms were extinct by the end of that period. Early cynodonts were the size of large dogs. By the Middle Triassic most were about the size of a rabbit, and further miniaturization (to mouse-size or smaller) characterized the cynodont lineage that eventually led to mammals. Several anatomical features show that they were able to process more food and oxygen per day, indicating they had a greater degree of endothermy than the other therapsids (see Section 13.1 and Section 14.5).

Basal cynodonts like †*Thrinaxodon* (**Figure 22.8A**) flourished in the Early Triassic but became extinct in the middle of the period. †*Thrinaxodon* is an important example because it shares several derived features with mammals, such as an enlarged dentary (**Figure 22.8B**). Its postdentary bones were smaller than those of earlier synapsids. The masseteric fossa (a depression on the outer side of the dentary; see Figure 22.3B) indicates that they had a masseter muscle, which closes the jaw and, in mammals, moves the jaw forward and sideways during chewing. Bones on the lower border of the synapsid temporal opening bow out to form the zygomatic arch, which is the origin for the masseter muscle (see Figure 22.3B). Basal cynodonts also had two occipital condyles, enabling more complex movements of the head on the neck.

Cynodonts have multicusped cheek teeth (teeth with cusps both anterior and posterior to the main cusp). Their upper and lower teeth occlude (contact each other during biting) to some extent, allowing a more precise, powerful bite made possible by the new masseter muscle, and the multicusped teeth would have been more resistant to fracture.

†*Thrinaxodon* had a complete bony secondary palate separating its nasal passages from the oral cavity, allowing simultaneous breathing and eating. The secondary palate also allows the tongue to manipulate food against the roof of the mouth and to place it between the upper and lower teeth for processing (see Section 22.3).

Postcranially, changes in the posterior thoracic and lumbar regions of its vertebral column indicate that †*Thrinaxodon* had a diaphragm (see Figure 22.8B and Section 13.1). Its hindlimbs were almost fully mammal-like, with a greater expansion of the iliac blade, further reduction of the pubis and ischium, a mammal-like ankle joint between the astragalus and tibia, and a heel on the calcaneal bone for insertion of the Achilles tendon of the calf muscles. Changes in the shoulder girdle allowed the forelimb to be used to manipulate food.

†Cynognathians were the predominant cynodonts of the Middle Triassic. Dog-size carnivores such as †*Cynognathus* are typical of the group, but there were also herbivores with expanded cheek teeth (the †gomphodonts).

(A) Basal cynodont, †*Thrinaxodon liorhinus*

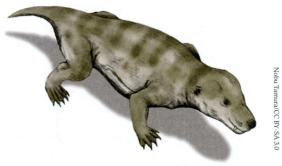

Nobu Tamura/CC BY-SA 3.0

Figure 22.8   A basal cynodont from the Triassic of Gondwana.
(A) †*Thrinaxodon liorhinus* lived in South Africa and Antarctica during the Early Triassic. The species was carnivorous, about the size of large cat. (B) Noteworthy skeletal features of †*Thrinaxodon* include its secondary palate, a depression on the lower jaw for the insertion of the masseter muscle, two occipital condyles, and short, flat ribs in the lumbar region. (B from F.A. Jenkins Jr. 1970. *Evolution* 24: 230–252.)

(B) Skeleton of †*Thrinaxodon*

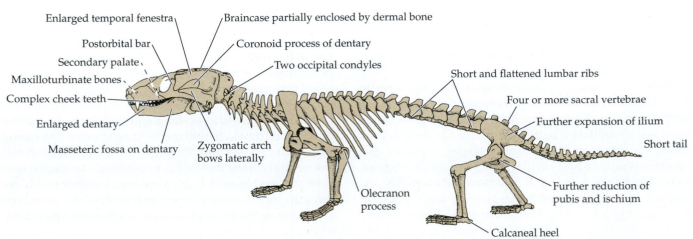

Enlarged temporal fenestra
Braincase partially enclosed by dermal bone
Postorbital bar
Coronoid process of dentary
Secondary palate
Two occipital condyles
Maxilloturbinate bones
Short and flattened lumbar ribs
Complex cheek teeth
Four or more sacral vertebrae
Enlarged dentary
Further expansion of ilium
Masseteric fossa on dentary
Zygomatic arch bows laterally
Short tail
Olecranon process
Further reduction of pubis and ischium
Calcaneal heel

A Late Triassic probainognathian, †*Probainognathus jenseni*

Nobu Tamura/CC BY-SA 4.0

Figure 22.9 **A basal probainognathian.** The clade's name means "progressive jaw," a reference to derived features of the lower jaw (see Section 22.2 and Figure 22.11C). This painting reconstructs †*Probainognathus jenseni* with a long, mobile muzzle, whiskers, and a coat of fur.

Probainognathians The only synapsid clade that survived the end-Triassic extinction is Probainognathia, exemplified by the Late Triassic genus †*Probainognathus*, a rat-size carnivore (**Figure 22.9**). Probainognathians ("progressive jaw") were smaller than cynognathians. They had lost the pineal foramen (the opening in the skull for the pineal eye) that characterized more basal synapsids and had greatly reduced the postdentary bones in the lower jaw (see Section 22.2 about the evolution of the middle ear). Other probainognathians include carnivorous and insectivorous †tritheledontids, herbivorous and rodent-like †tritylodontids, mouse-size carnivorous †brasilodontids, and tiny (shrew-size), insectivorous mammaliaforms. Some †tritheledontids persisted into the Jurassic, and †tritylodontids lived until the Early Cretaceous.

A recent discovery of a †tritylodontid (†*Kayentatherium wellesi*) with about three dozen tiny young illustrates two important reproductive features of early probainognathians. First, the number of young is much greater than in extant mammals, indicating oviparity rather than viviparity. And second, the poorly developed dentition of the hatchlings suggests that the mother provided food.

Derived probainognathians share with mammals an enlarged infraorbital foramen through which the sensory nerves from the snout pass back to the brain. This indicates that they had a highly innervated face, and possibly a mobile, sensitive muzzle with lips and whiskers (see Figure 22.9). They also had an incipient double jaw joint, with the dentary in the lower jaw contacting the squamosal in the skull (see later description of †*Diarthrognathus*).

Mammaliaformes Mammaliaformes includes taxa such as †*Morganucodon* and †*Kuehneotherium* from the latest Triassic–Early Jurassic of Wales; and Mammalia (see Figures 22.1 and 22.2). †*Morganucodon* and †*Kuehneotherium* were the size of shrews, with body weights

<30 g (**Figure 22.10A, B**). However, studies of their teeth showed that they lived up to 10 years, which is much longer than the lifespans of extant shrews. This difference might indicate that they had slower metabolic rates than shrews and were not as fully endothermic as modern mammals.

Compared to earlier probainognathians, mammaliaforms have larger brains, true dentary–squamosal jaw joints with a ball-like articulation on the dentary (although the old quadrate–articular jaw joint remained; see Section 22.2), and teeth that interdigitate with precise occlusion (see Figure 22.13B) and are replaced only once during life (diphyodonty). Evidence from wear on the teeth shows that they moved their jaws sideways in true mastication (chewing).

Mammaliaform teeth have prismatic enamel, which prolongs the life of a tooth's working surface, and the molar teeth have divided roots implanted into deep sockets, enabling them to withstand the force of mastication. Teeth are held in their sockets by the periodontal ligament, which provides sensory feedback, allowing more precise control of occlusion between the teeth and resulting in more effective mastication (**Figure 22.10C**). Mammaliaformes also have an atlas–axis complex (see Figure 22.3C) that allows rotational movements of the head; and 7 cervical vertebrae (first seen in cynodonts such as †*Thrinaxodon*), a number that is retained in all extant mammals except manatees and sloths. In the skeleton, the pubis is shortened, but paired epipubic bones (see Figure 23.1A) are now present (for the insertion of abdominal supportive musculature), and the ilium is rodlike and points forward (see Figure 22.10C).

Mammaliaform diversity in the Mesozoic included carnivores such as †*Sinoconodon*, fish-eaters such as †*Castorocauda* (named for its beaverlike flattened tail), gliders such as †*Volaticotherium*, and arboreal forms such as †*Agilodocodon* (**Figure 22.10D**). However, these animals were small; most were no bigger than a rat, with the largest about the size of a raccoon.

Mammalia Mammalia includes three clades: Prototheria (monotremes and extinct relatives), †Allotheria (†multituberculates and related taxa, now extinct), and Theria (see Figure 22.1). Within Theria are Metatheria (including Marsupialia) and Eutheria (including Placentalia), groups that we consider in Chapter 23.

Many synapomorphies unite extant Mammalia, although it is difficult to know how many of these were present in extinct outgroups of Mammalia. For example, all extant mammals lactate (supply milk from mammary glands) and care for their young. Evidence indicates that lactation evolved only once in mammals, because all extant mammals have similar casein (milk protein) genes. But exactly when this happened in the evolutionary history of Mammaliaformes remains uncertain. In the next section, we trace the evolution of some key skeletal synapomorphies of Mammalia and associated changes in muscles.

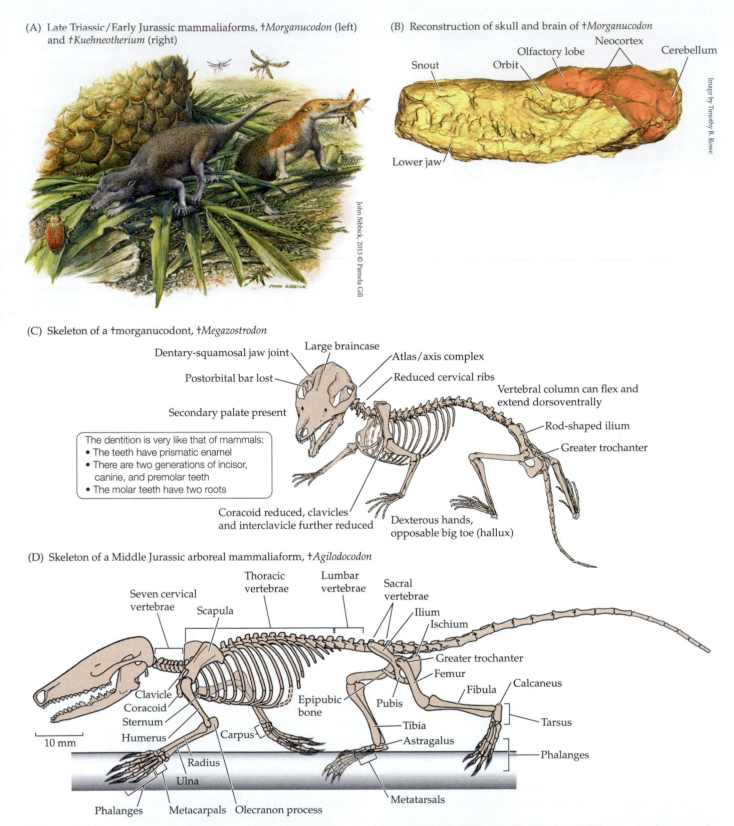

(A) Late Triassic/Early Jurassic mammaliaforms, †*Morganucodon* (left) and †*Kuehneotherium* (right)

John Sibbick, 2013 © Pamela Gill

(B) Reconstruction of skull and brain of †*Morganucodon*

Snout
Orbit
Olfactory lobe
Neocortex
Cerebellum
Lower jaw

Image by Timothy B. Rowe

(C) Skeleton of a †morganucodont, †*Megazostrodon*

Dentary-squamosal jaw joint
Large braincase
Postorbital bar lost
Atlas/axis complex
Reduced cervical ribs
Vertebral column can flex and extend dorsoventrally
Secondary palate present
Rod-shaped ilium
Greater trochanter

The dentition is very like that of mammals:
• The teeth have prismatic enamel
• There are two generations of incisor, canine, and premolar teeth
• The molar teeth have two roots

Coracoid reduced, clavicles and interclavicle further reduced
Dexterous hands, opposable big toe (hallux)

(D) Skeleton of a Middle Jurassic arboreal mammaliaform, †*Agilodocodon*

Seven cervical vertebrae
Thoracic vertebrae
Lumbar vertebrae
Sacral vertebrae
Scapula
Ilium
Ischium
Greater trochanter
Femur
Fibula
Calcaneus
Clavicle
Coracoid
Epipubic bone
Pubis
Tarsus
Sternum
Humerus
Carpus
Tibia
Astragalus
Phalanges
Radius
Ulna
10 mm
Phalanges
Metacarpals
Olecranon process
Metatarsals

Figure 22.10 Basal mammaliaforms. (A) Reconstructions of †*Morganucodon* and †*Kuehneotherium*. Although these two shrew-sized mammaliaforms lived side by side, they ate different foods and had differently shaped molar teeth. The painting shows †*Morganucodon* pursuing a beetle; †*Kuehneotherium* preferred softer-bodied prey. The small pinnae (external ears) are conjectural. (B) Micro-CT reconstruction of †*Morganucodon* shows the large olfactory lobe of the brain, indicative of the importance of olfaction for early mammaliaforms. (C) Skeleton of †*Megazostrodon*, an early mammaliaform related to †*Morganucodon*, highlighting features of the dentition, dentary-squamosal jaw joint, and flexible vertebral column. (D) The earliest known tree climbing mammaliaform, †*Agilodocodon*, was about 13 cm long and had flexible wrist and ankle joints like extant mammals. (C after F.A. Jenkins Jr. and F.R. Parrington. 1976. *Trans R Soc Lond B* 273: 387–431; D after Q.J. Meng et al. 2015. *Science* 374: 764–768.)

22.2 Jaw Joints and Middle Ear Bones

LEARNING OBJECTIVES

22.2.1 Outline two approaches used to interpret evolution of the mammalian jaw joint and middle ear bones.

22.2.2 Trace changes in the jaw joint and middle ear bones from basal synapsids to extant opossums.

22.2.3 Explain why evolutionary changes in synapsid jaws represent a conflict between feeding and hearing.

22.2.4 Evaluate the evidence for convergent evolution of the definitive mammalian middle ear (DMME).

In this section, we examine two of the most important skeletal synapomorphies of Mammalia: (1) loss of the quadrate articular jaw joint and the evolution of the squamosal-dentary jaw joint, and (2) new roles of the quadrate and articular bones in hearing high-frequency sounds. Related to these synapomorphies is a series of evolutionary changes in the position of the middle ear from the lower jaw to the skull—changes that happened convergently at least three times within Mammalia.

The synapsid fossil record and embryological studies of extant mammals document the evolutionary history of the mammalian jaw joint and middle ear bones. Here we examine that transition using examples from basal synapsids such as †*Varanosaurus*, basal cynodonts such as †*Thrinaxodon*, basal probainognathians such as †*Probainognathus*, and developing and adult opossums.

The lower jaw of †*Varanosaurus* consists of several bones, with the tooth-bearing dentary bone forming only the anterior portion of the jaw (**Figure 22.11A**). The jaw articulation is between the articular bone in the lower jaw and the quadrate bone in the skull, which is the original jaw joint of gnathostomes. The large stapes is firmly attached to other bones in the occipital region of the skull. Because of their sizes and relative immobility, neither the quadrate nor the stapes was suited to transmitting airborne sounds to the inner ear. At best, these animals might have been able to hear low-frequency sounds via bone conduction.

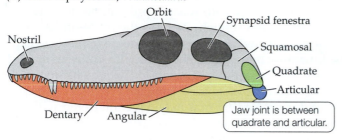

(A) Basal eupelycosaur, †*Varanosaurus*

Nostril · Orbit · Synapsid fenestra · Squamosal · Quadrate · Articular · Dentary · Angular

Jaw joint is between quadrate and articular.

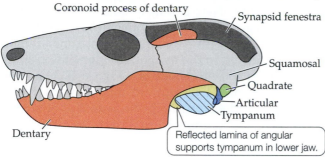

(B) Basal cynodont, †*Thrinaxodon*, with external and middle ear on jaw

Coronoid process of dentary · Synapsid fenestra · Squamosal · Quadrate · Articular · Tympanum · Dentary

Reflected lamina of angular supports tympanum in lower jaw.

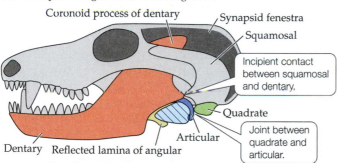

(C) Basal probainognathian, †*Probainognathus*

Coronoid process of dentary · Synapsid fenestra · Squamosal

Incipient contact between squamosal and dentary.

Quadrate · Articular · Dentary · Reflected lamina of angular

Joint between quadrate and articular.

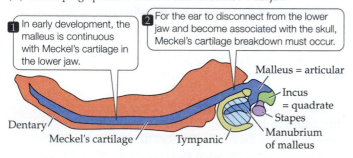

(D) Developing opossum with ear still associated with jaw

1 In early development, the malleus is continuous with Meckel's cartilage in the lower jaw.

2 For the ear to disconnect from the lower jaw and become associated with the skull, Meckel's cartilage breakdown must occur.

Malleus = articular · Incus = quadrate · Stapes · Manubrium of malleus · Dentary · Meckel's cartilage · Tympanic

Figure 22.11 Evolution of mammalian jaw and middle ear ossicles. Skull and lower jaws of (A) †*Varanosaurus*, (B) †*Thrinaxodon*, and (C) †*Probainognathus*. (D) Developing lower jaw of a didelphid opossum. (E) Skull and lower jaw of an adult Virginia opossum. The inset shows detail of middle ear ossicles in the skull. (See credit details at the end of this chapter.)

(E) Adult opossum with ear in skull, disassociated from lower jaw

Nostril · Orbit · Synapsid fenestra · Squamosal

1 Dentary is the only bone in lower jaw.

2 The single jaw joint is between the squamosal and dentary bones.

3 Tympanum and middle ear bones are now associated with skull.

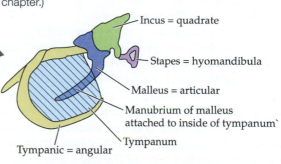

Incus = quadrate · Stapes = hyomandibula · Malleus = articular · Manubrium of malleus attached to inside of tympanum · Tympanum · Tympanic = angular

In contrast, mammalian ears are more sensitive to high-frequency sounds than those of other tetrapods. (To cite a human example, most college students can detect sound vibrations above 10 kHz, although this sensitivity to high frequencies decreases with age.) Unlike basal synapsids, mammals have a single bone in the lower jaw (the dentary), and three bones in the middle ear (the malleus, incus, and stapes, collectively known as middle ear ossicles; see Figure 11.15B) in an enclosed middle ear cavity in the skull.

Embryological studies in the 19th century showed that two bones of the mammalian middle ear are homologs of bones that formed the original jaw joint of gnathostomes. The malleus is homologous with the articular and the incus is homologous with the quadrate (**Figure 22.11B**). Fossils document intermediate stages in the evolutionary change from several bones in the lower jaw to one bone in the lower jaw and three bones in the middle ear. As the postdentary bones decreased in size, the dentary grew a backwards flange (the condylar process) that covered them and eventually contacted the skull, forming a new jaw joint between the dentary and squamosal bones. There was a time when both old and new jaw joints were present side by side. †*Diarthrognathus* (Greek, "two-joint jaw"), a Jurassic †tritheledontid that shows this double-jaw condition, made a splash in vertebrate paleontology when it was described in the 1950s. We now know that a similar jaw arrangement occurred much earlier in Triassic forms such as †*Probainognathus*.

To visualize this transition, compare the skulls of †*Thrinaxodon* and †*Probainognathus* with that of a Virginia opossum (**Figure 22.11B–E**). The lower jaw of the developing opossum shows that the tympanic bone (= angular, which supports the tympanum), malleus (= articular), and incus (= quadrate) develop in the same positions as in †*Thrinaxodon* and †*Probainognathus*, and that the "old jaw joint" persists as the articulation between the malleus and the incus. The malleus and incus of mammals remain tiny in adults.

Anatomical and genomic evidence show that the eardrum and middle ear cavity evolved multiple times within tetrapods. This extraordinary case of convergent evolution was possible because the stapes (which is homologous with the gnathostome hyomandibula; see Figures 3.9B) articulates with the otic capsule that houses the inner ear. If the stapes can vibrate, it can transmit those vibrations to the inner ear. Evolution of a smaller, lighter stapes improves sound conduction. These anatomical features set the stage for convergence.

†*Thrinaxodon* and †*Probainognathus* have a flange on the angular bone of the lower jaw (called the reflected lamina of the angular) that supported a tympanum near the posterior corner of the lower jaw (see Figure 22.11B, C). The stapes of these forms is smaller than in more basal synapsids, and the attachment of the quadrate to the squamosal is looser. The bones could conduct vibrations, although conduction was probably limited because the quadrate and articular bones still formed the jaw joint.

Using the same bones for two functions—as a jaw joint and in hearing—might seem like a clumsy, makeshift arrangement, but recall that in evolution it is only possible to tinker with what is present, not to redesign structures from scratch. Perhaps the trade-off between hearing and jaw function was adequate for early synapsids because they were ectotherms with low rates of food intake. However, derived synapsids with higher metabolic rates required more food, imposing stresses on the jaw joint during feeding that may have interfered with the role of the quadrate and articular in hearing. Thus, the evolution of synapsid jaws addresses a conflict between feeding and hearing.

The evolution of jaw-closing muscles occurred in parallel with changes in the lower jaw. In basal eupelycosaurs such as †*Dimetrodon*, a large adductor mandibulae muscle originated in the temporal region and inserted mainly on the coronoid bone (see Figure 22.3A) of the lower jaw, pulling the jaw upward and backward (**Figure 22.12A**). The adductor mandibulae muscle split to form two parts in cynodonts. The first part, the temporalis, originates in the temporal region, attaches to the coronoid process of the dentary in the lower jaw (a new feature in cynodonts) and pulls the lower jaw upward and backward (**Figure 22.12B**). The second part is the masseter muscle, which originates on the lateral surface of the zygomatic arch, inserts in the masseteric fossa on the main body of the dentary (see Figure 22.3B) and the angular process, and pulls the lower jaw up and forward. Later in the history of synapsids, the masseter of probainognathians (the lineage that includes extant mammals; see Figure 22.2) split to form a deep portion and a superficial portion (**Figure 22.12C**). The deep masseter pulls the lower jaw upward, backward, and sideways, and the superficial masseter pulls the lower jaw upward and forward, allowing even finer control of movements of the lower jaw. Together, these derived jaw muscles form a muscular sling for the lower jaw. This muscular sling may have relieved stresses at the jaw joint, allowing the postdentary bones to become smaller, and to vibrate and thus conduct sound.

Early investigators described cynodont jaw joints as weak because the postdentary bones were loose and wobbly, but a tighter articulation of these bones might have compromised their role as vibrating auditory ossicles. In more derived forms such as †*Probainognathus*, the dentary enlarged as postdentary bones became smaller (see Figure 22.11C). The even-more derived †*Diarthrognathus* is a classic evolutionary intermediate in which the old articular–quadrate joint is next to the new dentary–squamosal articulation. It thus became possible for the wobbly and much smaller quadrate and articular to function as middle ear ossicles, detecting higher-frequency sounds and transmitting vibrations from the eardrum to the stapes and inner ear.

So far in this story, the eardrum and middle ear ossicles are still associated with the posterior end of the lower jaw. In extant mammals, however, the ears are part of the skull, an arrangement termed the **definitive mammalian middle**

(A) Jaw closing muscles in a basal synapsid, †*Dimetrodon*

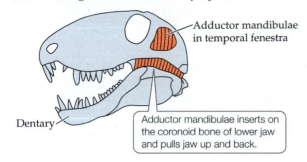

Adductor mandibulae in temporal fenestra

Adductor mandibulae inserts on the coronoid bone of lower jaw and pulls jaw up and back.

Dentary

(B) Jaw closing muscles in a basal cynodont, †*Thrinaxodon*

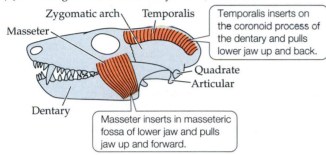

Zygomatic arch   Temporalis

Masseter

Temporalis inserts on the coronoid process of the dentary and pulls lower jaw up and back.

Quadrate
Articular

Dentary

Masseter inserts in masseteric fossa of lower jaw and pulls jaw up and forward.

(C) Jaw closing muscles in basal probainognathian, †*Probainognathus*

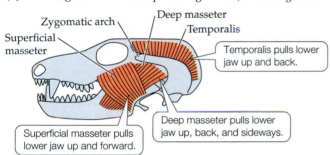

Zygomatic arch   Deep masseter

Superficial masseter   Temporalis

Temporalis pulls lower jaw up and back.

Deep masseter pulls lower jaw up, back, and sideways.

Superficial masseter pulls lower jaw up and forward.

Figure 22.12 Evolution of jaw-closing muscles of synapsids. (A) A single jaw-closing muscle, the adductor mandibulae, was present in basal synapsids such as †*Dimetrodon*. (B) In †*Thrinaxodon*, the adductor mandibulae had differentiated into two muscles, the temporalis (in the position of the original adductor muscle) and the masseter. (C) Probainognathians have further subdivided the masseter into superficial and deep portions. (See credit details at the end of this chapter.)

ear, or **DMME**. In the DMME, the postdentary bones are completely separate from the lower jaw and have become incorporated into the temporal region of the skull. The miniaturized malleus and incus conduct sound to the stapes, and the modified angular bone supports the eardrum. By studying how this transition from the lower jaw to skull happens in developing opossums (see Figure 22.11D), you can see that such a change requires loss of the connection between the malleus and Meckel's cartilage. This happens through a process called Meckel's cartilage breakdown. Amazingly, Meckel's cartilage breakdown evolved independently at least three times—twice among extant mammals (once in the lineage leading to monotremes and again in the lineage leading to therians) and a third time in the lineage leading to †allotherians.

22.3 Other Mammalian Features

LEARNING OBJECTIVES

22.3.1 Compare the teeth, facial musculature, modes of food processing, and metabolic rates of mammals to those of more basal synapsids.

22.3.2 Link the basic features of mammalian skin and skin glands to mammalian physiology.

22.3.3 Explain how oral food processing by mammals differs from that of other tetrapods.

22.3.4 Describe some unique features of mammalian brains and sensory systems.

Having focused largely on skeletal features so far, we now turn to the evolution of other features characteristic of mammals. In this section we discuss anatomical and physiological specializations related to feeding and maintaining body temperature, as well as brains, sensory systems, and physiology of mammals. Within these topics, we provide relevant examples of mammalian behavior.

Teeth

Because hard, enamel-containing teeth are even more likely to fossilize than bones, much of our information about extinct mammals comes from their teeth. Fortunately, mammalian teeth are highly informative about their owners' lifestyles.

Most vertebrates replace teeth throughout life, a pattern termed **polyphyodonty** that cynodonts such as †*Thrinaxodon* retained (**Figure 22.13A**). In contrast, the general mammalian condition is **diphyodonty**, having only two sets of teeth (the so-called milk teeth and a set of permanent, or adult, teeth). Early mammaliaforms such as †*Morganucodon* were diphyodont, as are extant mammals (**Figure 22.13B**). Determinate growth (growth that stops at a genetically specified point in postnatal development) is a prerequisite for diphyodonty, for if our jaws grew continuously, we would need to grow more and larger teeth to keep up with growth of the jaws.

All cheek teeth of cynodonts such as †*Thrinaxodon* have similar shapes, but cheek teeth of mammaliaforms are differentiated into premolars (simple in form and replaced once) and molars (more complex and not replaced; see Figure 22.13B). Molars erupt later in life; our last-erupting molars are called wisdom teeth, so named because they normally erupt at the age—late teens—when we supposedly have gained some wisdom.

Tooth rows in the upper and lower jaws of basal cynodonts like †*Thrinaxodon* are the same distance apart, a condition termed isognathy (**Figure 22.13C**). Animals with an isognathic jaw chew using simple up and down movements with a scissorlike closure. In mammaliaforms, the rows of teeth in the lower jaw are closer together than the upper teeth. Such anisognathous jaws allow chewing to occur only on one side of the jaw at a time, including a grinding side-to-side (lateral-to-medial) movement as well as up-and-down movement, together termed rotary chewing (**Figure 22.13D**). Humans older than about 15 months use rotary chewing to process food.

(A) Cheek teeth of a basal cynodont, †*Thrinaxodon*

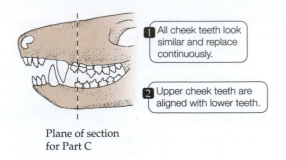

1 All cheek teeth look similar and replace continuously.

2 Upper cheek teeth are aligned with lower teeth.

Plane of section
for Part C

(B) Cheek teeth of a basal mammaliaform, †*Morganucodon*

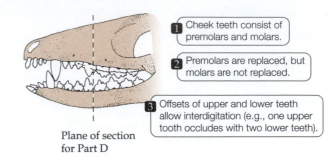

1 Cheek teeth consist of premolars and molars.

2 Premolars are replaced, but molars are not replaced.

3 Offsets of upper and lower teeth allow interdigitation (e.g., one upper tooth occludes with two lower teeth).

Plane of section
for Part D

(C) Isognathous jaws of a basal cynodont, †*Thrinaxodon*

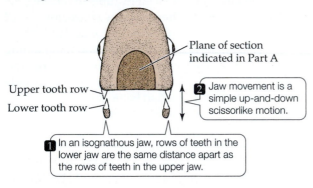

Plane of section indicated in Part A

Upper tooth row
Lower tooth row

2 Jaw movement is a simple up-and-down scissorlike motion.

1 In an isognathous jaw, rows of teeth in the lower jaw are the same distance apart as the rows of teeth in the upper jaw.

(D) Jaws and teeth of a basal mammaliaform, †*Morganucodon*

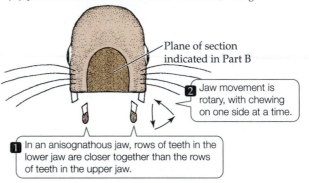

Plane of section indicated in Part B

2 Jaw movement is rotary, with chewing on one side at a time.

1 In an anisognathous jaw, rows of teeth in the lower jaw are closer together than the rows of teeth in the upper jaw.

Figure 22.13 Dentition and jaw movements in basal cynodonts and mammaliaforms. (A,B) Cheek teeth and alignment in †*Thrinaxodon* and †*Morganucodon*. (C) Head-on view of snout of †*Thrinaxodon* shows isognathous jaws and scissorlike jaw movements. (D) Head-on view of snout of †*Morganucodon* shows anisognathous jaws and rotary chewing.

The molars of mammalian upper and lower jaws come together in an interlocking arrangement of the upper and lower teeth in such a way that the cusps of upper teeth interdigitate with those of lower teeth when the jaw closes. This precise **occlusion** enables cusps to cut and crush food thoroughly, creating a larger surface area on which digestive enzymes can act, which promotes rapid digestion.

Molar teeth of the basal mammaliaform †*Morganucodon* have three main cusps in a straight line (**Figure 22.14A**). The tallest of these is the middle cusp. In the slightly more derived mammaliaform †*Kuehneotherium*, the middle cusp has shifted so that the teeth have a triangular form in occlusal ("food's-eye") view and the sides of the triangles increase the shearing ability of the teeth (**Figure 22.14B**). The apices of the triangles of the upper teeth point inward; those of the lower teeth point outward, together forming a reversed triangle occlusion.

Therians have **tribosphenic molars**, a more complex tooth that is interpreted as a key adaptation of the group (**Figure 22.14C,D**). Tribosphenic molars have multiple shearing crests, made possible by the addition of a protocone in the upper molars that now forms a new apex of the triangle; they also have a bowl-shaped surface in the lower molars, the talonid basin (see Figure 22.14C,D), into which the protocone of the upper teeth fits, crushing and pulping food and increasing the amount and diversity of foods that can be processed in a chewing cycle.

Specializations of the palate and tongue for swallowing

Most tetrapods use the mouth and jaws to capture prey or to seize mouthfuls of vegetation. Lissamphibians swallow prey items without using the jaws to process them and lizards hold insects in their mouths, biting down repeatedly to crush the exoskeleton. Carnivorous birds and crocodylians pull pieces from large prey and swallow them whole. In contrast, the thoroughness and complexity of mammalian mastication is unique. Mammals chew on one side of the mouth at a time; the lower jaw moves sideward (from outside to inside), bringing tightly locking teeth into precise occlusion. Food is manipulated with the tongue against the palate and returned to the teeth for further processing until it forms a small bolus that is swallowed. This unique mode of food processing and swallowing, involving muscles of the tongue, pharynx, and esophagus, is known as **deglutition**. Features of the underside of the back of the skull show that the anatomy that facilitates deglutition first evolved in †brasilodontid cynodonts.

Like crocodylians (see Figure 17.7), mammals evolved a secondary palate that separates pathways for air from the oral cavity, an adaptation that allows breathing while eating. The mammalian secondary palate consists of bones that form a hard palate overlain by tissues that form the soft palate. We can trace the evolution of the secondary palate through a series of synapsids. The roof

(A) Molar teeth of a basal mammaliaform, †*Morganucodon*

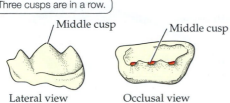

Three cusps are in a row.

Middle cusp   Middle cusp

Lateral view   Occlusal view

(B) Molar teeth of a basal mammaliaform, †*Kuehneotherium*

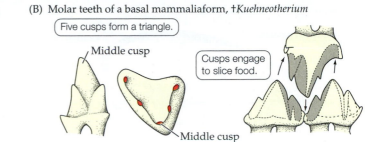

Five cusps form a triangle.

Middle cusp   Cusps engage to slice food.

Middle cusp

Lateral view   Occlusal view   Molars in occlusion

(C) Upper (left) and lower teeth (right) of a therian with tribosphenic molars

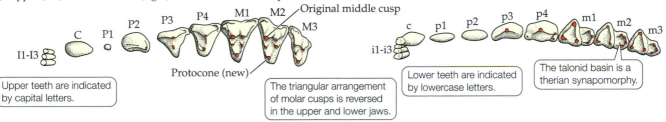

P2   P3   P4   M1   M2   Original middle cusp

C   P1   M3

I1-I3

Protocone (new)

Upper teeth are indicated by capital letters.

The triangular arrangement of molar cusps is reversed in the upper and lower jaws.

c   p1   p2   p3   p4   m1   m2   m3

i1-i3

Lower teeth are indicated by lowercase letters.

The talonid basin is a therian synapomorphy.

(D) Tribosphenic molars of the Virginia opossum, *Didelphis virginiana*

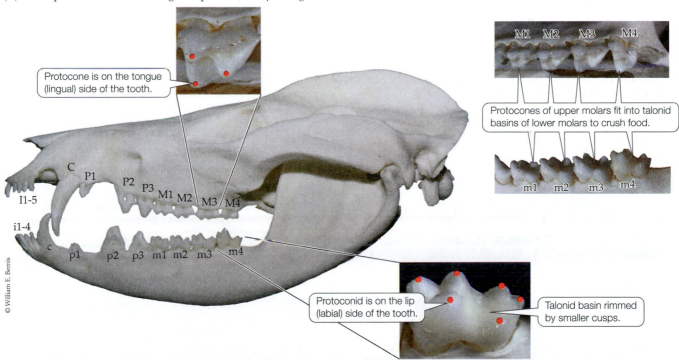

Protocone is on the tongue (lingual) side of the tooth.

M1   M2   M3   M4

Protocones of upper molars fit into talonid basins of lower molars to crush food.

m1   m2   m3   m4

C   P1   P2   P3   M1   M2   M3   M4

I1-5

i1-4

c   p1   p2   p3   m1   m2   m3   m4

© William E. Bemis

Protoconid is on the lip (labial) side of the tooth.

Talonid basin rimmed by smaller cusps.

Figure 22.14 Evolutionary anatomy of molar teeth.
(A) Molar teeth of †*Morganucodon* had a series of three cusps in a row, a condition referred to as triconodont. (B) Cusps in the molar teeth of †*Kuehneotherium* arranged in a triangle with the largest cusp on the lingual (inner) side of the tooth in the upper jaw and the labial (outer) side of the tooth in the lower jaw. This reversed triangle dentition allows efficient slicing of food between the upper and lower teeth. (C) Therians have further modified the molar teeth by the addition of a protocone in the upper molars and a talonid basin in the lower molars; such molars are called tribosphenic molars. (D) Dentition of the Virginia opossum illustrates how the protocones of the upper molars fit into the talonid basins of the lower molars to crush food. (A–C after K. Liem et al. 2001. *Functional Anatomy of the Vertebrates*, 3rd ed. Cengage/Harcourt College: Belmont, CA.)

of the mouth of basal eupelycosaurs such as †*Dimetrodon* is formed by the primary palate (**Figure 22.15A**). Air drawn in through the nose enters the front of the oral cavity, which limits the ability to breathe while eating. Intermediate conditions occur in more derived forms such as †*Probainognathus*, in which extensions of the premaxilla, maxilla, and palatine bones form the hard palate in the anterior half of the oral cavity (**Figure 22.15B**). Mammals have a longer hard palate with a soft palate posterior to it (**Figure 22.15C**).

(A) Sagittal view of synapsid skull based on †*Dimetrodon*

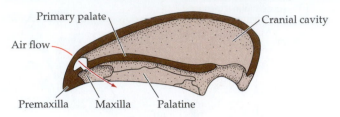

(B) Sagittal view of probainognathian skull based on †*Probainognathus*

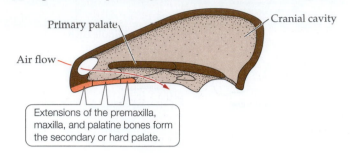

Extensions of the premaxilla, maxilla, and palatine bones form the secondary or hard palate.

(C) Sagittal view of mammal skull and palate based on *Canis*

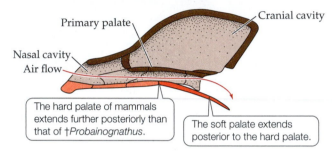

The hard palate of mammals extends further posteriorly than that of †*Probainognathus*.

The soft palate extends posterior to the hard palate.

Figure 22.15 Evolution of the mammalian palate.
(A) Basal synapsids such as †*Dimetrodon* have only a primary palate. (B) Extensions of bones form the secondary (hard) palate of †*Probainognathus*, at least partially separating air flow from the oral cavity. (C) In addition to a larger hard palate, mammals have a soft palate posterior to it. The posterior margin of the soft palate is marked by the uvula, a fleshy extension not shown here. (After K. Liem et al. 2001. *Functional Anatomy of the Vertebrates*, 3rd ed. Cengage/Harcourt College: Belmont, CA, based on A.S. Romer. 1956. *Shorter Edition of the Second Edition of The Vertebrate Life*. Saunders College Publishing: Philadelphia.)

Basal synapsids have only a hard palate, so respiratory gases and food cross pathways in the pharynx (**Figure 22.16A**). In contrast, the nasopharynx (portion of the pharynx behind the nasal cavity and *above* the soft palate) and oropharynx (portion of the pharynx behind the mouth and *below* the soft palate) of mammals are separate. Two seals formed by the tongue, soft palate, and epiglottis can separate the pathways for breathing and swallowing. Seal #1 forms when the back of the tongue presses against the soft palate, which prevents food from entering the pharynx during breathing (**Figure 22.16B**). This is the seal that you use to stop water from entering your throat when you gargle. Seal #2 presses the epiglottis against the back of the soft palate, allowing air to enter the trachea and blocking entrance of food from the oropharynx into the nasopharynx (**Figure**

(A) Basal synapsids could not simultaneously eat and breathe

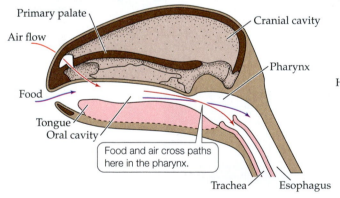

Food and air cross paths here in the pharynx.

(B) In mammals, Seal 1 is formed between the tongue and soft palate

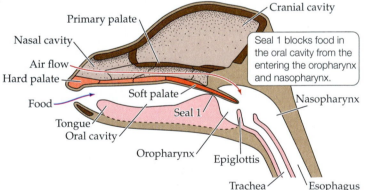

Seal 1 blocks food in the oral cavity from the entering the oropharynx and nasopharynx.

(C) Seal 2 is formed between the epiglottis and soft palate

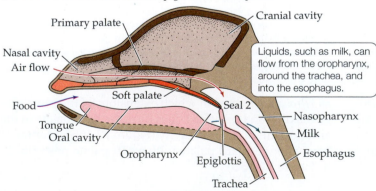

Liquids, such as milk, can flow from the oropharynx, around the trachea, and into the esophagus.

Figure 22.16 Mammals can chew and breathe at the same time. (A) Air flow and food passage from the mouth to the esophagus in basal synapsids required paths of food and air to cross in the pharynx, limiting their ability to process food in the oral cavity. (B,C) In contrast, mammals separate the paths of air and food, allowing them to chew and breathe at the same time. This is possible because of two soft-tissue seals that prevent food from entering the trachea. (B) Seal #1 forms when the back of the tongue presses against the soft palate; this seal allows chewing of food in the mouth and prevents food from entering the pharynx during breathing. Seal #2 presses the epiglottis against the back of the soft palate, allowing air to enter the trachea and preventing solid food in the oropharynx from entering the nasopharynx. Liquids such as milk, however, can pass from the oropharynx, around the trachea, and into the esophagus while this seal is in place.

22.16C). Liquids can pass from the oropharynx, around the trachea, and into the esophagus while this seal is in place. In humans, seal #2 is lost after infancy because the epiglottis becomes separated from contact with the soft palate. This developmental change results in a larger space behind the tongue, which in turn is linked to our increased ability for phonation (sound production and articulation). A trade-off for this enhanced ability to speak is that humans can "swallow the wrong way"—that is, we can inhale food or liquid into the trachea, and even choke to death on food.

Small bones in the base of the throat derived from the hyoid arch (see Figure 2.10C) serve as origins for several muscles, including muscles used in deglutition. In extant sauropsids, these bones are simple and rodlike, associated with a wide gullet where swallowing occurs mainly by pharyngeal constriction. The tiny bones of the hyoid apparatus rarely fossilize, but we know that they were simple in the basal cynodont †*Thrinaxodon*, and that in the early mammaliaform †*Microdocodon* they had the form now seen in extant mammals. The epiglottis was also probably present in early mammaliaforms.

Facial muscles

Facial muscles make our wide variety of facial expressions possible, but they probably evolved in the context of mobile lips and cheeks that enable young mammals to suckle. Our facial muscles are homologs of the neck constrictor muscles of other tetrapods (constrictor colli; **Figure 22.17A**). In sauropsids, which swallow prey with minimal oral processing, constriction of this muscle helps to move food down the esophagus. But because mammals chew food into small pieces and swallow (i.e., deglutition), the constrictor colli muscles were freed to evolve new functions.

Elaboration of the muscles of facial expression occurred differently in mammalian lineages. Facial muscles of monotremes are less complex than those of therians, and the facial muscles of most mammals (including rodents; **Figure 22.17B**) are less complex than those of primates. But although anthropoid primates (humans in particular) have elaborate facial muscles (**Figure 22.17C**), other species are also capable of a diversity of facial expressions, often displaying similar expressions for similar emotions. For example, the snarling expression of an angry human is much like that of an angry dog or an angry horse. Horses use their lips in feeding and can make a wide variety of expressions. Dogs have expressive faces, even more so than their ancestor wolves. Dogs have a new facial muscle (the levator anguli oculi medialis) that raises the inner eyebrow, creating the sad face known as "puppy dog eyes." Early during their domestication, a dog that could produce this expression might have elicited caregiving from humans. Indeed, in a recent study, shelter dogs that raised their inner eyebrow frequently were adopted more readily than those that raised their inner eyebrow less frequently. The 40,000 or so muscles in the elephant's trunk are an extension of the original upper lip muscles.

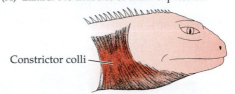

(A) Lizard: No muscles of facial expression

Constrictor colli

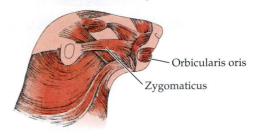

(B) Rodent: Moderate muscles of facial expression

Orbicularis oris

Zygomaticus

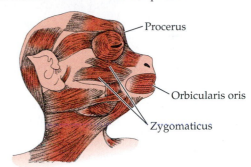

(C) Primate: Extensive muscles of facial expression

Procerus

Orbicularis oris

Zygomaticus

Figure 22.17 Muscles of facial expression. (A) Lizards and other sauropsids lack muscles of facial expression. (B,C) In contrast, mammals have muscles of facial expression, shown here in rodents and primates, that evolved by modification of the constrictor colli musculature. The orbicularis oris purses the lips to form an airtight seal around a nipple during suckling. Primates have a more complex zygomaticus muscle than do rodents. Both rodents and primates can snarl, but primates can also smile, whereas rodents cannot. The procerus muscle allows primates to frown, but rodents lack a procerus and cannot frown. (After K.V. Kardong. 2012. *Vertebrates: Comparative Anatomy*, 1st ed. McGraw-Hill: New York.)

Integument

Mammalian skin and its appendages (hair, whiskers, claws, nails, hooves, and horns) and glands are central to many elements of mammalian biology, including thermoregulation, behavior, locomotion, and communication.

The outer layer of skin, or epidermis, varies in thickness from a few cells in some small rodents to hundreds of cell layers in elephants, rhinoceroses, and hippos; in fact, at one time these animals were grouped together as pachyderms (Greek, "thick skin").

Epidermal cells produce the protein keratin, which forms many important appendages of mammalian skin. The tails of opossums and many rodents have keratinous epidermal scales similar to but softer than those of lizards. Claws, nails, and hooves are accumulations of keratin that protect the terminal phalanx of the digits. Some integumentary appendages are involved in locomotion (e.g.,

(A) Cross section through human skin

(B) Pilosebaceous unit of fur seal

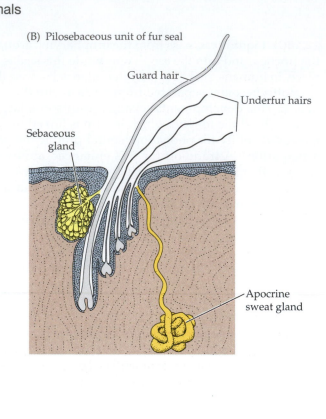

Figure 22.18 Mammalian integument. (A) Most mammals have a barrier layer of flattened, tough cells (the stratum corneum) above layers of cells constantly generated by stem cells in the stratum germinativum. Hair grows from follicles, which are deep invaginations in the epidermis. The skin has sebaceous, eccrine, and apocrine glands that secrete substances such as protective skin oils and (in humans) sweat used in homeostasis and chemical communication. These glands are under hormonal and neuronal control. Receptors for various sensations, including pressure and temperature, are embedded throughout the skin. (B) Pelage (fur) of mammals is organized as pilosebaceous units in which a thicker guard hair stands above finer underfur hairs. (See credit details at the end of this chapter.)

hooves), predation and offense (e.g., claws), defense (e.g., claws, quills), and display (e.g., horns). The most obvious mammalian epidermal structures are hairs.

Hair   Hairs grow from deep invaginations in the germinal layer of the epidermis, called **hair follicles** (**Figure 22.18A**). Hair has many functions in addition to insulation, including camouflage and communication. **Vibrissae** (whiskers) are specialized hairs with sensory properties that grow on the muzzle, around the eyes, and on the lower legs. Touch receptors are associated with vibrissae, and sensation and tactile information may have been the earliest functions of body hair.

Pelage (fur) consists of closely placed hairs, often produced by multiple hair shafts arising from a single complex root (**Figure 22.18B**). Its insulating effect depends on its ability to trap air, and its insulating ability is proportional to the length of the hairs and their density of spacing. Arrector pili muscles that attach midway along the hair shaft pull the hairs erect to deepen the layer of trapped air. Cold stimulates a general contraction of the arrector pili via the sympathetic nerves, as do stressful conditions such as fear and anger. This autonomic reaction occurs even in relatively hairless mammals, including humans, where it produces dimples (goose bumps) on the skin's surface. Hair erection also serves for communication; mammals use it to send a warning of fear or anger, as in the puffed-up fur display of a cat or the raised hackles of a dog, both of which make the animal appear larger.

Hair color depends on the quality and quantity of melanin injected into the forming hair by melanocytes at the base of the hair follicle. Hair replacement occurs either by continuous growth of an individual hair or by periodic **molting**, in which old hairs fall out and are replaced by new hairs. Most mammals have pelage that grows and rests in seasonal phases; molting usually occurs only once or twice a year.

Hair was probably present in early cynodonts, and possibly earlier. We think that insulating fur was present in early mammaliaforms, which were very small and had high surface-area-to-volume ratios, and thus high rates of heat loss. Some direct evidence exists. Dense pelage of the beaverlike Middle Jurassic mammaliaform †*Castorocauda* is preserved as impressions, and fossils of the Early Cretaceous mammal †*Spinolestes* preserve its skin structures (including, as its name suggests, small spines).

Skin glands   Mammals differ from sauropsids in having glandular skin, and the secretory structures of mammalian skin develop from the epidermis. In addition to mammary glands, mammals have three major types of skin glands: sebaceous, eccrine, and apocrine (see Figure 22.18). Sebaceous and apocrine glands are associated with hair follicles; eccrine glands are not. The secretion of all three gland types is under neural and hormonal control.

- **Sebaceous glands** occur over the entire body surface. They produce an oily secretion, sebum, which lubricates and waterproofs the hair and skin. Sheep lanolin, our own greasy hair, and the grease spots that the family dog leaves on the wallpaper are all from sebaceous secretions.

- **Apocrine glands** have restricted distributions in the skin of most mammals, and their secretions appear to be used in chemical communication. In humans, apocrine glands occur in the armpit and pubic regions (these are the secretions we try to mask with deodorants and powders).

- In most mammals, **eccrine glands** occur only on the soles of the feet, prehensile tails, or other areas of skin that contact environmental surfaces, where their secretions improve adhesion or enhance tactile sensation. In some primates (Old World monkeys and apes, including humans), eccrine glands occur over the entire body surface. Human eccrine glands produce sweat, which may be why your hands sweat when you are nervous—the moisture would have given the hands of your primate ancestors a better grip when escaping through forest trees.

It is a common impression that mammals have sweat glands, but in fact very few mammals sweat. Although in humans the eccrine glands produce sweat, hoofed mammals such as pigs and horses produce sweat using apocrine glands. Other mammals induce evaporative cooling by other means; for example, dogs pant and kangaroos lick their forearms (see Section 14.3).

Many mammals have specialized scent glands derived by modification of sebaceous or apocrine glands. Those derived from sebaceous glands secrete a viscous substance, usually employed to mark objects. These scent glands are usually on areas of the body that can be easily applied to objects, such as the face, chin, or feet. Domestic cats often rub their face and chin to mark objects, including their owners. Apocrine glands produce volatile substances that are released into the air. Many members of Carnivora have anal glands that deposit scents along with feces as a form of chemical communication with conspecifics, and skunks can spray foul-smelling fluid from these glands as a defense against predators. Apocrine glands in the ear produce earwax.

Mammary glands have a more complex, branching structure than do other skin glands, but they appear to have evolved by modification of apocrine glands. Their mode of secretion (exocytosis) is similar to that of apocrine glands, and there are similarities in the development of apocrine glands and mammary glands. Mammary glands are, of course, a unique and defining character of mammals.

Lactation, nursing, and suckling

Female mammals **lactate**, feeding their young by producing milk from mammary glands. **Nursing** is the behavior exhibited by lactating females when transferring milk to their young. Nursing rodents crouch over their offspring to transfer milk, nursing domestic pigs lie on their side, and females of other species stand. **Suckling** is the behavior displayed by young when obtaining milk from a nipple. Suckling pups of some species of rodents display tenacious nipple attachment, clinging tightly to the nipple and rarely switching nipples, whereas pups of other species attach less tenaciously, frequently moving from one nipple to another. Species that display tenacious attachment have more pups per nipple, so pups face higher levels of competition for nipples and milk. Suckling is linked to the same changes in the pharyngeal region that relate to the unique mode of mammalian deglutition. The isolation of feeding and respiration via pharyngeal seals (see Figure 22.16B,C) allows an infant to suckle on a nipple while breathing through the nose. (Human infants are obligate nose breathers until they are about 4 months old, a fact that is critical for EMS responders to remember when they respond to a call for an infant in respiratory distress.)

Although monotremes produce milk, their mammary glands lack nipples; their mammary glands are associated with hairs, from which the young lap the milk. The absence of nipples in monotremes could lead us to infer that nipples were absent in earlier mammaliaforms. However, monotremes apparently lost suckling—perhaps related to their specialized bill or beak—so the absence of nipples in monotremes may be a secondary feature.

Lactation allows the production of offspring to be separate from seasonal food supply. Unlike birds, which lay eggs only when there is appropriate food for fledglings, mammals can store food as fat and later convert it into milk. This manner of food provision by the mother alone also means that a female mammal does not need assistance from a male to rear her young. In addition, postnatal nutrition from the mother allows young to be born at a small body size and without erupted teeth.

Male marsupials lack mammary glands, but the glands are present and potentially functional in male monotremes and placentals (indeed, breast cancer can affect human males as well as females). Males of two bat species lactate, and there are records of human males producing milk. It is a mystery why male placental mammals retain mammary glands but normally do not lactate.

Brain and senses

Mammals have the largest brains among vertebrates (see Figure 6.24A). Enlarged brains are a mammalian feature rather than a general synapsid one, and many features of mammalian skulls relate to larger brains. The enlarged part of mammalian cerebral hemispheres—the neocortex or neopallium, which is the area concerned with cognition—forms differently from the enlarged forebrain of birds and other archosaurs. Other unique mammalian features include an infolded cerebellum (the region of the brain concerned with neuromuscular control) and a large representation for cranial nerve VII associated with the facial musculature (a mammalian synapomorphy).

Brains do not usually fossilize, but their shape can sometimes be determined from the endocranial cavity. If the brain fills the endocranial cavity, it leaves impressions on the bones, and a cranial endocast—an internal cast or image of the inside of the cranial cavity—can provide details of brain structure (see Figure 22.10B). Endocasts of the brains of derived cynodonts resemble those of extant sauropsids, but with a larger cerebellum and olfactory lobes. The cerebral hemispheres and regions associated with hearing and vision were still small. Researchers used micro-CT to compare brains of more derived cynodonts with those of early mammaliaforms. Despite the larger brains of early mammaliaforms, including evidence of a small neocortex, expansion of the neocortex did not occur until mammals.

Chemosensation: Smell and taste   As mentioned in Section 2.6, chemosensation—the ability to detect, differentiate, and react to molecules in the external environment—is the most universal of senses, and smell (olfaction) is the primary way that mammals sense their environments (see Section 11.6).

Olfactory genes code for receptor proteins that bind specific chemicals (odorant molecules). The number of olfactory genes is indicative of a vertebrate's sensitivity to smells. Mammals have an average of ~1200 olfactory genes as opposed to ~100 in more basal amniotes. The number of olfactory genes (and hence sense of smell) has become reduced in some mammalian groups, including whales and anthropoid primates. Humans, for example, have small olfactory lobes and only ~350 olfactory genes.

A mobile snout, or rhinarium, is a distinctive feature of mammals. Its elaborated nasal cartilages support muscles used to move the snout, a change correlated with loss of bony processes that previously separated the right and left nostrils, and the nasal septum of mammals is composed primarily of cartilage.

Orthonasal smell is simple sniffing to detect airborne odorants and is used by all tetrapods. Mammals also have retronasal smell, which occurs when air expelled from the lungs picks up odorants released by chewed food as it passes over the olfactory epithelium in the rear of the nose. Orthonasal and retronasal smell, together with sensory information from the lips, tongue, and teeth, generate the sensations of flavors processed in the neocortex.

Basal cynodonts probably could detect flavors, for they had a bony secondary palate (which is correlated with oral food processing) and an enlarged olfactory lobe in the brain. Olfaction and flavor detection were probably not important for more basal synapsids. Only in mammals did ethmoturbinates (see Figure 11.17 and accompanying discussion) become fully ossified, increasing the surface area for the olfactory epithelium tenfold.

Hearing   Unlike monotremes, therians have an external ear, or **pinna** (plural *pinnae*). The earliest evidence of a pinna is in the stem therian †*Spinolestes* from the Early Cretaceous of Spain. The ear canal channels sound collected by the pinna to the eardrum at its base (there is no ear canal in non-mammalian tetrapods and the eardrum is flush with the outside of the head; see Figure 11.15A). Some therians, such as whales and earless seals, secondarily lost the pinna, and the human pinna is unusual in being relatively immobile.

Mammals share a unique modification of the lagena in their inner ear. As shown in Figure 11.16, the basilar papilla, which lies at the entrance to the lagena in all tetrapods, contains the hair cells that detect sounds (see Section 2.6 and Figure 4.6 for a discussion of hair cells). The greatly elongated lagena of mammals accommodates an enlarged basilar papilla known as the organ of Corti that can detect a broad range of frequencies. The long lagena becomes coiled to form the cochlea that fits inside the skull (see Figure 11.15B). The organ of Corti has two types of hair cells, arranged somewhat differently in monotremes and therians, implying an at least partially independent history. The hearing system of therians is extremely acute and allows perception of high-frequency sounds that are inaudible to other vertebrates.

Visual pigments and color vision   Mammals have retinas dominated by **rod cells** that are sensitive to low levels of light but do not perceive color. This is a derived condition compared with the general condition in amniotes, which is to have a retina dominated by **cone cells** containing color-sensitive pigments, or opsins. Most mammals have only two types of opsins for color detection and are thus at best dichromatic (able to perceive only two basic colors, as are red/green color-blind humans).

The mammalian rod-dominated retina came about by the conversion of some cones sensitive to short-wavelength light into rods; cones in mammalian retinas are limited to a small area of acute vision, the fovea, which is the most sensitive area of the retina. Cones require higher light intensity than do rods, and diurnal mammals have a greater concentration of cones in the fovea than do nocturnal ones. Thus, many mammals have good night vision but relatively poor visual acuity or color vision. Night vision is enhanced in some species, such as cats, by a reflective layer called the tapetum lucidum (which is why cats' eyes shine in the dark). The reflected light stimulates additional retinal cells, increasing visual sensitivity in dim light.

Reduction of opsins and decline in color vision appears to have evolved independently in monotremes and therians. Four types of color-photosensitive molecules, each coded by a different gene family, occur in vertebrates, and all four were probably present in early synapsids. The four opsins in cones—RH2 (rodlike opsin), SWS1, SWS2 (short-wavelength-sensitive opsins, sensitive to blue light), and LWS (long-wavelength-sensitive opsin, sensitive to green and red light)—respond to different

(A) Spectral sensitivity of four opsins used in color vision

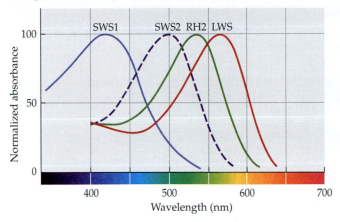

(B) Evolution of four opsin types used for color vision in amniotes

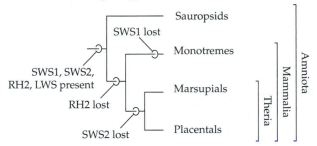

Figure 22.19 Evolution of mammalian visual pigments.
(A) Sauropsid cones have four opsins sensitive to different wavelengths of light used for color vision, and we infer that this was the ancestral condition for synapsids, because this is also the condition in amphibians and bony fishes. Rods contain a fifth visual pigment, RH1; its spectral sensitivity is not shown in the figure. (B) Mammals lost some of these opsins, probably related to the nocturnal habits of basal mammaliaforms. For example, all mammals lack RH2, monotremes lost SWS1, and therians lost SWS2. Human trichromatic vision is based on a duplication and modification of the gene that codes for LWS. (B after G. Jacobs. 2013. *Vis Neurosci* 30: 39–53.)

light wavelengths and thus allow discrimination of colors (**Figure 22.19A**). Most mammals are not sensitive to ultraviolet light, but mutations of the SWS1 opsin gene have produced UV-sensitive opsins in a few rodents and marsupials.

All mammals retain LWS opsin but have lost RH2 opsin. The situation with the two SWS opsins is more complicated. Monotremes retain the SWS2 opsin, whereas therians retain SWS1 (**Figure 22.19B**). This phylogenetic distribution implies convergent loss of color vision in monotremes and therians, suggesting that their common ancestor had both SWS1 and SWS2 opsins and probably was not as nocturnal as many extant mammals.

Old World monkeys and apes are unusual among mammals in having good color vision and a brain specialized for the visual sensory mode. Human trichromatic vision is based on a duplication of the gene for LWS opsin into an L form with maximum sensitivity at 530 nm (more sensitive to green), and an M form with maximum sensitivity at 563 nm (more sensitive to red). This duplication may be why

monkeys and humans are more sensitive to the color red than are most other mammals. Some marsupials (e.g., the honey possum) independently duplicated the LWS opsin gene and are also trichromatic. In contrast, many mammals (primarily nocturnal and aquatic ones, including bats and cetaceans) lost SWS1 opsin and its corresponding gene and thus are monochromatic. A few humans supposedly have an additional opsin gene duplication and are tetrachromatic (but not in the same way that sauropsids are; humans cannot see in the ultraviolet range).

Internal anatomy

Mammalian internal anatomy and physiology differ in many ways from that of other amniotes. Some differences relate to endothermic metabolism, and similar changes evolved convergently in birds. Other differences are uniquely mammalian. We discuss only a few features here.

Mammalian adipose tissue—white fat—is broadly distributed as a subcutaneous layer for insulation. Other vertebrates do not have much subcutaneous fat, although some is present in birds. White fat is also associated with organs such as the heart, intestines, and kidneys, where it has metabolic and cushioning roles. It also occurs in skeletal muscles and surrounds and cushions joint capsules. Adipose tissue is not an inert energy store; fat cells secrete many signaling molecules that coordinate important metabolic processes.

Mammals differ from other vertebrates in the form of their erythrocytes (red blood cells), which lose their nuclei as they develop from stem cells. Mammals are also unique in having platelets, blood cell fragments that form aggregations called thrombi that aid in blood clotting, thus preventing blood loss after injury. Unfortunately, platelets can also form thrombi within coronary and brain vessels, meaning that mammals are the only vertebrates subject to heart attacks and ischemic strokes caused by platelet-induced thrombosis.

22.4 Basal Mammalian Clades

LEARNING OBJECTIVES

22.4.1 Describe the diversity of extant monotremes.

22.4.2 Explain the anatomy of the male and female reproductive tracts of monotremes.

22.4.3 Describe the teeth and jaw movements of †multituberculates.

As shown in Figure 22.2, Mammalia includes four terminal taxa: Prototheria, †Allotheria, Metatheria, and Eutheria. This section describes the diversity of the basal groups Prototheria and †Allotheria. We cover Metatheria and Eutheria—which together form Theria—in Chapter 23.

Prototheria

The only extant clade included in Prototheria is Monotremata (Greek *mono*, "one"; *trema*, "hole"), a name that refers to their single opening of excretory and reproductive

(A) Platypus, *Ornithorhynchus anatinus*

Brisbane City Council/CC BY 2.0

(B) Long-beaked echidna, *Zaglossus bruijni*

© Klaus Rudloff

(C) Short-beaked echidna, *Tachyglossus aculeatus*

Steve Bittinger/CC BY 2.0

(D) Puggle of short-beaked echidna

Courtesy of Stewart Nicol

Figure 22.20 Diversity of monotremes. (A) The semi-aquatic platypus constructs burrows near streams in eastern Australia. It forages for invertebrates underwater, detecting them by using electroreceptors and mechanoreceptors in its duck-like bill. The electroreceptors are not homologous to those of sharks or ray-finned fishes. (B) Long-beaked echidnas live in New Guinea. Two of the three extant species of *Zaglossus* are classified as Critically Endangered by the IUCN. (C) The short-beaked echidna of temperate Australia and New Guinea has fur and stout spines. It uses sturdy forelimbs to break into insect nests and its long tongue to capture them. (D) Newly hatched monotremes, known as puggles, are highly altricial.

tracts via the cloaca. Extant adult monotremes lack teeth, although teeth are present in juvenile platypuses and have been found in adults of fossil species. There are two extant families of monotremes.

- Ornithorhynchidae (Greek *ornis*, "bird"; *rhynchos*, "beak") includes a single species, the platypus *Ornithorhynchus anatinus*, a semiaquatic mammal that eats aquatic invertebrates in the streams of eastern Australia and Tasmania (**Figure 22.20A**).

- Tachyglossidae (Greek *tachy*, "swift"; *glossa*, "tongue") includes three species of long-beaked echidnas (genus *Zaglossus*), which eat earthworms, ants, and termites and occur only in New Guinea; and the short-beaked echidna (*Tachyglossus aculeatus*) of Australia and New Guinea, which eats ants and termites (**Figure 22.20B,C**).

Monotremes have lower metabolic rates than eutherians, and they retain the ancestral amniote condition of oviparity. Their eggs have parchmentlike shells but do not contain enough yolk to sustain the embryo until hatching. Eggs are retained in the uterus where they are nourished by uterine secretions before they are deposited in the nest. The highly altricial young, called puggles, hatch a mere 2–3 weeks after the eggs are laid (**Figure 22.20D**).

A modest Mesozoic diversity of monotremes is known from the Cretaceous of Australia. These were about half the size of extant monotremes, weighing 1–2 kg at most. Some teeth and femur fragments from the Paleocene of Patagonia are the only evidence of monotremes outside Australia and New Guinea. Molecular data show a mid-Cenozoic split of echidnas from platypuses. Echidnas apparently evolved from a semiaquatic lineage, a hypothesis supported by studies of echidna myoglobin, which resembles myoglobin of aquatic animals, and the presence of electroreceptors on the beak.

Extant monotremes have unique specializations, including a leathery bill or beak with electroreceptors and mechanoreceptors used to detect prey underwater or underground. Instead of teeth, adult platypuses have keratinized pads on the jaws (**Figure 22.21A**). Male platypuses have a spur on each hindleg attached to a venom gland, a weapon used in defense and intraspecific fighting. Echidnas have a similar spur, but the secretion of the gland is used for scent marking rather than defense.

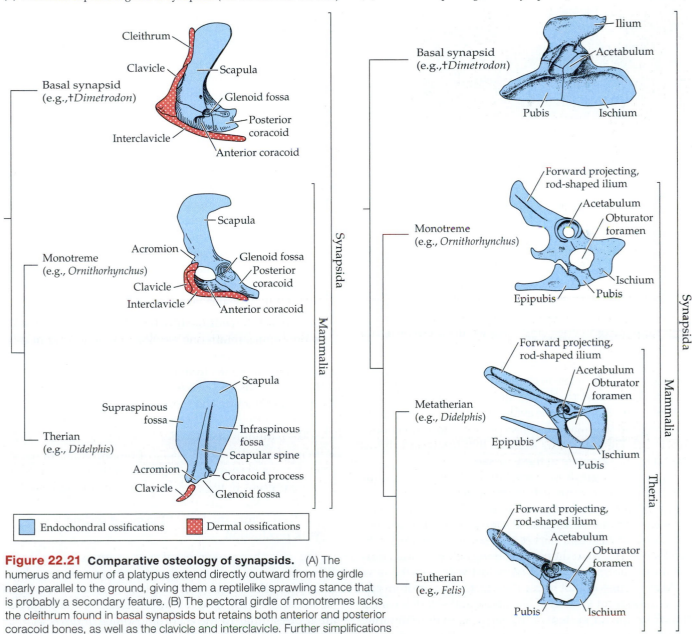

(A) Skeleton of a monotreme, the platypus, *Ornithorhynchus anatinus*

Ilium
Scapula
Femur
Spur
Left epipubic bone
Keratinized pads
Humerus
Clavicle
Interclavicle
Olecranon process of ulna
Broad feet for swimming
Flattened snout and lower jaw
Broad hands for swimming

(B) Evolution of pectoral girdle of synapsids (shown from the left side)

Cleithrum
Clavicle
Scapula
Basal synapsid (e.g., †*Dimetrodon*)
Glenoid fossa
Posterior coracoid
Interclavicle
Anterior coracoid

Scapula
Acromion
Monotreme (e.g., *Ornithorhynchus*)
Glenoid fossa
Posterior coracoid
Clavicle
Interclavicle
Anterior coracoid

Scapula
Supraspinous fossa
Infraspinous fossa
Scapular spine
Therian (e.g., *Didelphis*)
Acromion
Coracoid process
Clavicle
Glenoid fossa

Synapsida
Mammalia

Endochondral ossifications
Dermal ossifications

(C) Evolution of pelvic girdle of synapsids (shown from the left side)

Ilium
Acetabulum
Basal synapsid (e.g.,†*Dimetrodon*)
Pubis
Ischium

Forward projecting, rod-shaped ilium
Acetabulum
Obturator foramen
Monotreme (e.g., *Ornithorhynchus*)
Ischium
Epipubis
Pubis

Forward projecting, rod-shaped ilium
Acetabulum
Obturator foramen
Metatherian (e.g., *Didelphis*)
Epipubis
Ischium
Pubis

Forward projecting, rod-shaped ilium
Acetabulum
Obturator foramen
Eutherian (e.g., *Felis*)
Pubis
Ischium

Synapsida
Mammalia
Theria

Figure 22.21 Comparative osteology of synapsids. (A) The humerus and femur of a platypus extend directly outward from the girdle nearly parallel to the ground, giving them a reptilelike sprawling stance that is probably a secondary feature. (B) The pectoral girdle of monotremes lacks the cleithrum found in basal synapsids but retains both anterior and posterior coracoid bones, as well as the clavicle and interclavicle. Further simplifications of the pectoral girdle occurred in therians, evolutionary changes related to a more upright stance. (C) Pelvic girdle evolution in synapsids also correlates with changes in stance. (See credit details at the end of this chapter.)

(A) Male monotreme (similar to general amniote condition)

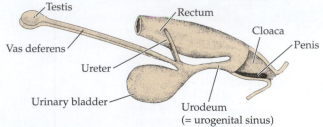

(C) Four-headed penis of an echidna

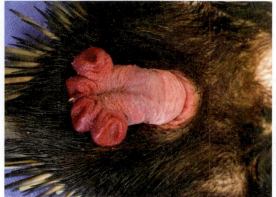

From Fenelon, et al. 2021. Sexual Development 15: 1–10. © 2021 Karger Publishers, Basel, Switzerland

(B) Female monotreme (platypus)

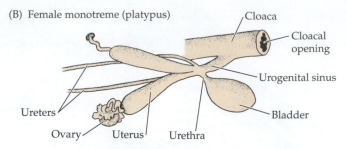

Figure 22.22 Reproductive system of monotremes. (A,B) As in sauropsids, male and female monotremes retain a single opening—the cloaca—for passage of fecal matter, urine, and genital products. (C) Male echidnas have a penis with four heads, or glans. (A after M. Hildebrand and G. Goslow. 1998. *Analysis of Vertebrate Structure*, 5th ed. Wiley: New York; B after M.B. Renfree. 1993. In *Mammal Phylogeny*. F.S. Szalay et al. [eds.], pp. 4–20. Springer-Verlag: New York.)

Like the pectoral girdle of basal synapsids such as †*Dimetrodon*, monotremes have a flat scapula, a clavicle, an interclavicle, and anterior and posterior coracoid bones (**Figure 22.21B**). The glenoid fossa for the articulation of the humerus is between the scapula and posterior coracoid. The cleithrum, a dermal bone that composes much of the pectoral gridle of fishes (see Figure 3.9C) has been lost in monotremes, as well as in therians such as the Virginia opossum. Further evolutionary simplifications characterize the pectoral girdle of therians, which lack interclavicle and anterior coracoid bones, and have reduced the posterior coracoid bone to a small coracoid process of the scapula. These changes and simplifications in ventral elements of the pectoral girdle relate to changes in stance between basal synapsids and mammals (see Figure 11.6). Another important change in therians is the evolution of the scapular spine, dividing the scapula into supraspinous and infraspinous fossae (see Figure 22.21B). Muscles that rotate and abduct the humerus originate in these fossae.

In the pelvic girdle of †*Dimetrodon*, the ilium, ischium, and pubis radiate outwards from the acetabulum (**Figure 22.21C**). In contrast, the rod-shaped ilium of mammals projects forward. Also, there is a large obturator foramen for blood vessels and nerves that serve a portion of the hindlimb. The paired epipubic bones extend anteriorly from the pubis; these bones were lost within eutherians.

Monotremes do not house their testes in a scrotum and have a different system of chromosomal sex determination from the XY system of therians. As in non-amniotes and sauropsids, the intestines, urinary tract, and reproductive tracts of monotremes exit through a common opening, the cloaca (**Figure 22.22A,B**). The two oviducts

of females stay separate and do not fuse in development except at their bases, where they join with the urethra to form the urogenital sinus. When fertilized eggs are present, the oviducts swell to form two uteri, in which the eggs stay for a time. Only the left oviduct is functional in the platypus.

Remarkably, the penis of male echidnas has four heads, or glans (**Figure 12.22C**). A male echidna uses two glans at a time, alternating between the pairs. One captive echidna was able to ejaculate packets of sperm 10 times in succession, and it may be that such rapid-fire ejaculations evolved because of male–male competition to deliver more sperm into a female's reproductive tract.

The genomes of monotremes have a mixture of mammalian and sauropsid characters. The vitellogenin genes that code for yolk formation and the casein genes for milk formation provide a perspective on the evolution of mammalian viviparity and lactation.

- Sauropsids have three functional vitellogenin genes, monotremes retain one vitellogenin gene, and marsupials and placentals have none. The transition of vitellogenin genes to nonfunctional pseudogenes in marsupials and placentals occurred sequentially without affecting the reproductive fitness of those clades, with the last loss of function occurring ~30–70 million years ago.

- In contrast, the casein genes of monotremes, marsupials, and placentals are similar, indicating that they were present in the common ancestor of these clades ~200–310 Ma. Thus, the ability to produce milk evolved before the loss of the genes coding for yolk production.

The sex chromosomes of monotremes are unique. Whereas other amniotes have one pair of sex chromosomes (X/Y or Z/W), monotremes have five pairs (X 1–5 and Y 1–5). The X chromosomes of the platypus are more similar to the Z chromosomes of birds than to the X chromosomes of therians, suggesting that therian X and Y chromosomes evolved after the divergence of monotremes and therians ~166 Ma.

†Allotheria

†Allotheria is more closely related to Theria than to Prototheria and is grouped with Theria in Theriiformes (see Figure 22.2). Its largest subgroup, †Multituberculata, includes dozens of families that lived from the Late Jurassic to the early Cenozoic, but they were a mostly Mesozoic radiation and fared poorly in the end-Cretaceous extinction. These mostly small mammals—rabbit size or smaller, with a few the size of a medium-sized dog (**Figure 22.23**)—had a narrow pelvis, interpreted as evidence that they were viviparous and gave birth to altricial young. Some had spurs like those of extant monotremes.

†Multituberculates are named for their molar teeth, which are broad, multicusped (multituberculed), specialized for grinding rather than shearing, and indicate an omnivorous or herbivorous diet. Many had large, bladelike premolars, probably used to open hard seeds. Most were terrestrial, but some arboreal forms had prehensile tails and ankles that allowed their feet to rotate backward.

Like extant rodents, †multituberculates moved their lower jaw in the fore-and-aft plane rather than from side-to-side. And, like some rodents, some species of †multituberculates were social, living in groups in burrows. Competition with rodents, which first appeared in the late Paleocene, may have been instrumental in the extinction of †multituberculates.

(A) †*Catopsbaatar,* a mouse-sized Late Cretaceous †multituberculate

From Kielan-Jaworowska and Hurum. 2006. Acta Palaeontologica Polonica 51: 393–406. Artwork by Bogusław Waksmundzki/CC BY 4.0

(B) †*Taeniolabis,* a beaver-sized early Cenozoic †multituberculate

Nobu Tamura/CC BY-SA 4.0

Figure 22.23 †Multituberculates. †Multituberculates that lived in the Cretaceous (A) and early Cenozoic (B) had some ecological similarities with extant rodents.

Theria

All remaining species of extant mammals are members of Theria. Like †allotherians, therians are viviparous, but the form of viviparity differs in Metatheria (marsupials) and Eutheria (placentals). We take up the evolutionary history and diversity of Metatheria and Eutheria including their reproductive specializations in Chapter 23.

Summary

22.1 Synapsid Evolution

Synapsids (mammals and their extinct relatives) are distinguished from other amniotes by a single temporal fenestra on each side of the skull.

Paleozoic synapsids were medium to large-size animals (10–200 kg), but by the Jurassic most synapsids weighed 1 kg or less.

Basal synapsids like †*Dimetrodon* lacked a middle ear, and their skulls and teeth differed from modern mammals, indicating that they processed less food and had low metabolic rates.

- The skull had a small temporal fenestra on each side, showing that the volume of jaw-closing muscles was relatively small.

- The teeth were all similar (homodont; apart from a large canine tooth), simple, pointed, and not suitable for chewing food.

Early Permian eupelycosaurs such as †*Varanosaurus,* †*Haptodus,* and †*Dimetrodon* had a sprawling stance. †*Dimetrodon* had a large dorsal sail.

More derived therapsids from the Middle Permian through Triassic may have acquired a degree of endothermy. †Dinocephalians, †dicynodonts, and †gorgonopsians are examples of therapsids.

Compared to earlier synapsids, therapsids had larger temporal fenestrae, heterodont dentition, and a partial secondary palate. They also had a more upright posture and reduction in size of some bones of the pectoral girdle, which allowed a greater range of motion in the forelimb and increased stride length. Their hindlimb muscles were more like those of mammals. Non-mammalian therapsids were nearly extinct by the end of the Triassic.

†*Thrinaxodon,* a basal cynodont that flourished in the Early Triassic, shares several derived features with mammals, including an enlarged dentary, smaller postdentary bones,

(Continued)

Summary *(continued)*

masseter muscle, two occipital condyles, multicusped cheek teeth, complete secondary palate, diaphragm, and ankle joint between the astragalus and tibia. Cynodonts may have been more endothermic than earlier therapsids.

Probainognathians are derived cynodonts and include the Late Triassic genus †*Probainognathus*. More derived forms had extensive innervation of the face and a contact between the dentary and the squamosal, forming an incipient mammal-like jaw joint.

Basal Mammaliaformes include shrew-sized taxa such as †*Morganucodon* and †*Kuehneotherium* from the late Triassic-early Jurassic and Mammalia: these animals were still not as fully endothermic as modern mammals. Compared to probainognathians, mammaliaforms had larger brains, dentary-squamosal jaw joints, diphyodonty, teeth with prismatic enamel, an atlas-axis complex, and seven cervical vertebrae.

Mammalia includes three clades: Prototheria (monotremes and extinct relatives), †Allotheria (†multituberculates and related taxa), and Theria.

22.2 Jaw Joints and Middle Ear Bones

Mammalian ears are more sensitive to high frequency sounds than are those of other tetrapods. Evolution of the mammalian middle ear illustrates a conflict between hearing and feeding.

Mammals have a single dentary bone in the lower jaw that articulates with the squamosal of the skull to form a new jaw joint. Three middle ear bones—the malleus, incus, and stapes—convey vibrations from the tympanum to the inner ear. The malleus and incus are homologues of bones that formed the original jaw joint of gnathostomes, specifically the articular in the lower jaw and the quadrate in the skull.

A change in position of the middle ear from the lower jaw to the skull, a condition called the definitive mammalian middle ear (DMME), accompanied evolution of a new jaw joint. This change happened three times within Mammalia by convergent evolution of Meckel's cartilage breakdown during development.

22.3 Other Mammalian Features

The classic features defining mammals are hair and mammary glands. While these rarely fossilize directly, their presence can be inferred in the earliest mammals.

- The earliest mammals had determinate growth.

- Mammals are diphyodont.

- Mammalian molars are triangular, with a reversed-triangle occlusion between upper and lower teeth. Therians have tribosphenic molars, which are characterized by the presence of a protocone and talonid basin.

- Mammals differ from other tetrapods in chewing food into fine particles and swallowing a discrete bolus of food (deglutition), using the tongue to aid in food processing and swallowing. Pharyngeal anatomy related to deglutition also allows suckling. Presence of the secondary palate, composed of a hard (bony) portion and a soft portion posterior to it, allows simultaneous breathing and feeding.

- Only mammals have muscles of facial expression, which probably evolved in the context of mobile cheeks and lips needed for suckling.

- Mammalian epidermis forms keratin-filled skin cells and derivatives such as hair, claws, nails, and hooves.

- Sebaceous glands lubricate hairs, apocrine glands secrete scents, and eccrine glands act as sweat glands in humans. Mammary glands probably evolved from apocrine glands.

- All female mammals lactate: they have mammary glands that produce milk. Nursing is the behavior by which lactating females transfer milk to young. Only therian mammals have nipples. Suckling is the behavior used by young to obtain milk from a nipple.

- Compared to other tetrapods, mammals have larger brains and rely more on olfaction and hearing than vision (most mammals are nocturnal).

- Basal mammaliaforms had a large cerebellum and large olfactory lobes, but only Mammalia have a large neocortex.

- Monotremes lack the pinna (external ear) and auditory tube of therians. Therian mammals have a unique inner ear structure, including a coiled cochlea, that enables them to hear higher sound frequencies than other tetrapods.

- Mammals use adipose tissue for insulation, metabolic functions, and protection of internal organs and joints.

- The mature red blood cells of mammals are unique in lacking nuclei. Mammals also uniquely have blood platelets that aid in clotting.

22.4 Basal Mammalian Clades

Prototheria includes Monotremata, the platypus of Australia; and four species of echidnas, found today in Australia and New Guinea.

Monotremes have a leathery bill or beak with electroreceptors and mechanoreceptors used to detect prey underwater (platypus) or underground (echidnas). A single opening, the cloaca, serves both the reproductive and excretory tracts. Males possess spurs on the hind legs.

Monotremes retain the general amniote condition of egg-laying. Their eggs have parchmentlike shells and

Summary *(continued)*

contain less yolk than the eggs of sauropsids. Eggs are retained in the uterus and nourished by uterine secretions. The extremely altricial young (puggles) hatch soon after the eggs are laid.

Monotreme genomes retain one vitellogenin gene. The casein genes of monotremes are similar to those of marsupials and placentals, indicating that these genes were present in the common ancestor of these clades.

Monotremes have 5 pairs of sex chromosomes, and the X chromosomes of the platypus more closely resemble the Z chromosomes of birds than the X chromosomes of therians.

†Allotheria is an extinct clade of rodentlike mammals that flourished in the late Mesozoic and early Cenozoic. Its largest clade, †Multituberculata, is named for specialized multicusped molar teeth.

Discussion Questions

22.1 Derived probainognathians have an extensive bony palate, indicating a muscularized and constricted pharynx like extant mammals. What would this have enabled them to do?

22.2 Transitional fossils document evolution of the mammalian ear. How did we know, before these fossils were found, that the malleus and incus evolved from the articular and quadrate bones that formed the original gnathostome jaw joint?

22.3 Explain and defend the assertion that, among tetrapods, only mammals suck and chew.

Figure credits

Figure 22.1: †Caseasauria, †Ohiacodontidae, †Edaphosauridae, †Cynognathia, *Probainognathus*, †Tritheleodontidae, †Allotheria: Nobu Tamura/CC BY-SA 4.0; †Biarmosuchia: Nobu Tamura/CC BY-SA 3.0; †Thrinaxodontidae: Nobu Tamura/CC BY 2.5; †Tritylodontidae: Nobu Tamura/CC BY 3.0; †Varanopidae, †Dinocephalia, †Gorogonopsia: Dmitry Bogdanov/CC BY 3.0; †Sphenacodontidae: Dmitry Bogdanov/CC BY 2.5; †Anomodontia, †Therocephalia: Dmitry Bogdanov/CC BY-SA 3.0; †Brasilitherium: Smokeybjb/CC BY-SA 3.0; †Sinoconodon: Sir Beluga/CC0; †*Morganucodon*: John Sibbick, 2013 © Pamela Gill; Prototheria: Brisbane City Council/CC BY 2.0.

Figure 22.2: *Cotylorhynchus, Varanosaurus, Riograndia,* Multituberculate: Nobu Tamura/CC BY-SA 4.0; *Thrinaxodon*: Nobu Tamura/CC BY 2.5; *Gorgonops*: Dmitry Bogdanov/CC BY 3.0; *Morganucodon*: John Sibbick, 2013 © Pamela Gill; Monotremes: K. Liem et al. 2001. *Functional Anatomy of the Vertebrates*, 3rd ed. Cengage/Harcourt College: Belmont, CA.

Figure 22.11: A,B,E after K. Liem et al. 2001. *Functional Anatomy of the Vertebrates*, 3rd ed. Cengage/Harcourt College: Belmont, CA; C after J.A. Hopson. 1987. *Am Biol Teach* 49: 16–26; D,E inset after A.W. Crompton and F.A. Jenkins Jr. 1979. In *Mesozoic Mammals: The First Two-Thirds of Mammalian History*, JA Lillegraven et al., eds. Berkeley, Univ. California Press.

Figure 22.12: A,B after K. Liem et al. 2001. *Functional Anatomy of the Vertebrates*, 3rd ed. Cengage/Harcourt College: Belmont, CA, based on R.C. Fox. 1964. *University of Kansas Publications, Museum of Natural History* 12: 657–680; C after J.A. Hopson. 1987. *Am Biol Teach* 49: 16–26.

Figure 22.18: A after K. Liem et al. 2001. *Functional Anatomy of the Vertebrates*, 3rd ed. Cengage/Harcourt College: Belmont, CA; B after N. Erdsack et al. 2015. *J R Soc Interface* 12: 20141206, based on D.A. Pabst et al. 1999. *Biology of Marine Mammals*. J.E. Reynolds & S.A. Rommel (Eds.), pp. 15–72. Smithsonian Institution Press: Washington, DC.

Figure 22.21: B after K. Liem et al. 2001. *Functional Anatomy of the Vertebrates*, 3rd ed. Cengage/Harcourt College: Belmont, CA, based on A.S. Romer. 1956. *Shorter Edition of the Second Edition of The Vertebrate Life*. Saunders College Publishing: Philadelphia; C after K. Liem et al. 2001. *Functional Anatomy of the Vertebrates*, 3rd ed. Cengage/Harcourt College: Belmont, CA, based on A.S. Romer. 1956. *Shorter Edition of the Second Edition of The Vertebrate Life*. Saunders College Publishing: Philadelphia and G. Li and Z.X. Luo. 2006. *Nature* 439: 195–200.

23

Therians

Spotted-tail quoll, *Dasyurus maculatus*, a carnivorous marsupial

Chapter 22 traced the evolutionary history of synapsids up to the origin of Mammalia and Theria during the Jurassic period. In this chapter, we turn to Theria and its two clades: Metatheria, which includes extant Marsupialia; and Eutheria, which includes extant Placentalia. Although most therians are terrestrial, there are also burrowing, aquatic, and flying forms. Evolution of these different ways of life led to corresponding diversity in anatomy, body size, and ecology, as well as many striking cases of convergent evolution.

The largest terrestrial and aquatic vertebrates are therians. A large male African bush elephant (*Loxodonta africana*) weighs ~5,500 kg, about the same as a medium-size dinosaur like †*Triceratops*. The blue whale (*Balaenoptera musculus*) weighs ~110,000 kg, which is 1.5 times the estimated weight of the largest dinosaurs (such as the †sauropod †*Argentinosaurus*), making the blue whale the largest animal ever known. However, most therians are small (between 10 g and 100 g), as they were during the Mesozoic. Large therian herbivores and carnivores impact nutrient cycles and energy flow, while smaller therians are important for pollination and seed dispersal, vital roles in Earth's ecosystems.

In this chapter we begin with an overview of therian features, evolution, and the diversity of Metatheria and Eutheria before examining key elements of therian reproductive biology and specializations for feeding and locomotion. We also describe how these features differ between the two major clades.

23.1 Therian Features and Origins of Marsupialia and Placentalia

LEARNING OBJECTIVES

23.1.1 Relate skeletal features of therians to their roles in bounding locomotion.

23.1.2 Describe skeletal and dental differences between marsupials and placentals.

23.1.3 Explain the Mesozoic history of Metatheria, Marsupialia, Eutheria, and Placentalia.

23.1.4 Outline two possible explanations for the loss of epipubic bones in Placentalia.

Therians are viviparous amniotes that give birth to young after a period of internal gestation. During gestation, the embryo (referring to early stages of development, when the major organ systems form) and resulting fetus (referring to later stages of development, when growth of body parts occurs) exchange nutrients, metabolic wastes, respiratory gases, and hormones with the mother via a **placenta**. The placenta forms by modifications of the typical extraembryonic membranes of amniotes (see Figure 9.14A and Section 23.4). Embryos of both marsupials and placentals form placentas, although the two groups differ markedly in their relative lengths of gestation and lactation.

Eggs of therians lack yolk, but therian genomes retain vitellogenin pseudogenes, nonfunctional remnants of genes that code for vitellogenins (yolk proteins) in non-therian amniotes. A membranous shell coat encases marsupial embryos for up to 80% of their short gestation. A marsupial embryo absorbs nutrients and gases through the membrane, but placentation does not start until the embryo "hatches" out of its shell coat. Placentals lack any trace of the shell coat.

Therian skeletons

Therians share several derived skeletal characters related to the evolution of bounding, a locomotor mode unique to therians. During a bound, the animal leaps using its hindlimbs and lands on its forelimbs, so that there is a phase when all four feet are off the ground and the animal is in midair (see Figure 11.7A). Bounding is possible because the coracoid and interclavicle bones are absent in adult therians, and the clavicle is thus the only bony connection with the axial skeleton (**Figure 23.1A**). The clavicle becomes further reduced in therians specialized for cursorial locomotion so that there is no bony connection between the scapula and the axial skeleton, leaving the scapula held by a sling of four muscles (**Figure 23.1B**). These changes allow the scapula of mammalian bounders and cursors to function as an additional limb segment, lengthening the forelimb stride and helping to cushion the impact of landing on the forefeet at the end of a bound. The hingelike elbow joint means that during locomotion the elbow can remain tucked in close to the body while forearm pronation positions the hand face-down (**Figure 23.1C**). The ankle joint is also hingelike, allowing better push-off from the

hind legs, with the astragalus now superimposed (positioned on top of) the calcaneus (**Figure 23.1D**). The prominent calcaneal heel is the attachment site for the Achilles tendon of the calf muscles (the gastrocnemius and soleus muscles), which are powerful extensors of the foot. Finally, the flexible lumbar region of therians allows dorsoventral flexion during locomotion as we saw in cheetahs (see Figure 11.7C, D). Collectively, these postcranial features enable the familiar therian locomotor mode of bounding, easily observed in squirrels. In larger mammals, this bounding locomotion is modified into a gallop.

Almost all marsupials retain epipubic bones that project forward from the pelvis (see Figure 23.1A). Thigh and abdominal muscles that stiffen the torso and resist vertical bending of the trunk during locomotion insert on the epipubic bones. The epipubic bones may limit independent

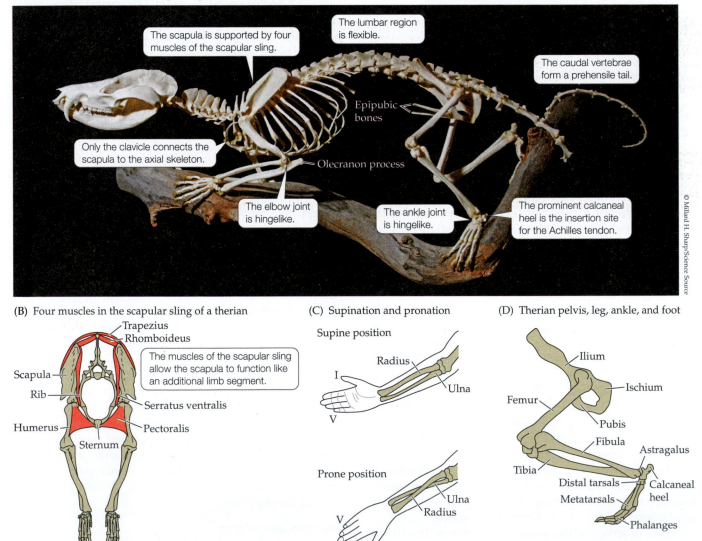

(A) Skeleton of a therian, the Virginia opossum *Didelphis virginiana*

Figure 23.1 Postcranial features of therians related to locomotion. (A) The skeleton of an adult Virginia opossum shows specializations for the unique bounding gait of therians. (B) The highly mobile scapula of therians is supported by a sling of four muscles. (C) Supination refers to turning your hand so that your palm faces upward; pronation is when you turn your hand so that the palm faces downward. The radius rotates to produce these movements. Basal therians can supinate, but cursorial specialists like horses lost this ability. (D) Ankle specializations allow therians to use the foot as a powerful lever. (See credit details at the end of this chapter.)

movement of the hindlimbs in gaits such as the gallop, and the few quadrupedal cursorial marsupials (such the †thylacine; see Figure 23.6B) lost the epipubic bones. Placentals lack the epipubic bones found in metatherians, and we now regard loss of these bones, along with characters of dentition and certain other skeletal characters, as reliable synapomorphies for Placentalia.

Two hypotheses (which are not mutually exclusive) may explain the loss of epipubic bones in placentals. The first is that epipubic bones constrained dorsoventral flexion of the vertebral column when running (this would also explain

their loss in the †thylacine). The second hypothesis relates to differences in the gestation periods of marsupials and placentals. Because the gestation period of placentals is longer than that of marsupials, the fetus reaches a larger size before birth, and rigid epipubic bones might interfere with expansion of the abdomen during pregnancy.

The earliest fossil therians are from the Late Jurassic (**Figure 23.2**). The divergence between Metatheria and Eutheria occurred relatively soon after that, ~160 Ma. Some workers consider the Early Cretaceous †*Sinodelphys* to be an early metatherian, but it may be a basal therian or even

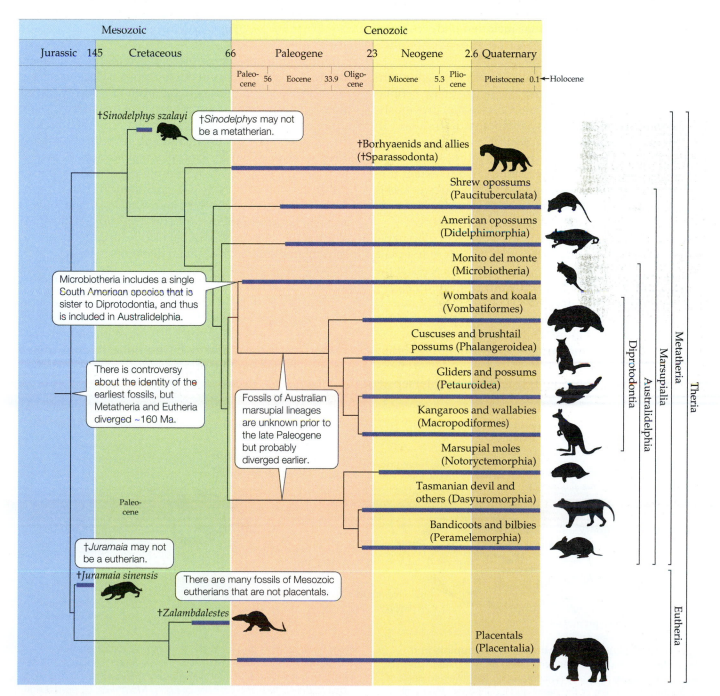

Figure 23.2 Time tree of Theria. The early fossil record of Theria is incomplete but molecular clock estimates place the divergence of Metatheria and Eutheria in the Late Jurassic. Marsupial groups are arranged based on recent molecular phylogenetic interpretations. (Sparassodonta: Nobu Tamura/CC BY-SA 4.0; *Juramaia sinensis, Zalambdalestes*: Nobu Tamura/CC BY-SA 3.0.)

a basal eutherian. The earliest definitive metatherians come from much younger rocks in North America, ~100 Ma after the split between metatherians and eutherians. A possible contender for the earliest eutherian is †*Juramaia sinensis*, from the Late Jurassic of China; however, this may be a basal therian rather than a eutherian. †*Eomaia* is one of several uncontested eutherians known from entire skeletons from the Early Cretaceous of China.

Marsupials originated in the late Cretaceous or early Paleocene (see Figure 23.2) and are the only extant metatherians, but several extinct groups lived in South America during the Cenozoic. Among the most interesting were †sparassodonts of the Eocene–Pliocene, a radiation of carnivorous forms that included badger-, dog-, and bearlike forms, and even a Pliocene saber-toothed form, †*Thylacosmilus*.

Better fossils and improvements in phylogenetic methods have helped paleontologists recognize many nonplacental eutherians from the Late Cretaceous. Previously, paleontologists placed many extinct eutherian taxa in Placentalia; an example is †*Zalambdalestes*, a mouse-size hopping mammal from the Upper Cretaceous of Mongolia, once thought to be a placental related to extant rabbits and rodents. Reassessments of †*Zalambdalestes* fossils show that it has epipubic bones and that its similarities to extant placental taxa are the result of convergence.

In addition to the loss of epipubic bones in placentals, many other skeletal characters distinguish marsupials from placentals. For example, the angular process of the dentary in marsupials bends inward (inflected), allowing insertion of the pterygoideus, a jaw-closing muscle (**Figure 23.3A**) In placentals, the angular process of the dentary is straight (**Figure 23.3B**).

The dentition of marsupials is unusual in that they replace only the last premolar, making them almost monophyodont (having only one set of adult teeth). Marsupials also have distinctive arrangements of teeth. Basal marsupials such as the Virginia opossum have 5 incisors on each side of the upper jaw and 4 on each side of the lower jaw; and 1 canine, 3 premolars, and 4 molars on each side of both the upper and lower jaws, for a total of 50 teeth (see Figure 23.3A). A **dental formula** is a shorthand method of presenting this information, with the number of teeth of each type in one side of the upper jaw in the numerator and the number of teeth of each type in one side of the lower jaw in the denominator. Thus, the total number of teeth is the sum of all the teeth in the upper and lower jaws of one side multiplied by 2. For the opossum, this formula is

$$I\ 5/4, C\ 1/1, P\ 3/3, M\ 4/4 \times 2 = 50\ teeth$$

During ontogeny, most placentals replace the incisors, canines, and premolars but not the molars. Placentals have fewer incisors and molars (maximum of three each) but more premolars (maximum of four) than marsupials. Many adult placentals have fewer teeth than the raccoon's 40 (see Figure 23.3B), but only a few species with specialized diets have more than 40. Examples include armadillos (insect specialists; all teeth are simple pegs) and porpoises (fish-eaters; all teeth have a similar conical shape, a condition called homodonty). Humans, for example, have 2 incisors, 1 canine, 2 premolars, and 2 or 3 molars (the third molar is the "wisdom tooth," which is usually but not always present). Thus the human dental formula is:

$$I\ 2/2, C\ 1/1, P\ 2/2, M\ 2–3/2–3 \times 2 = 28–32\ total\ teeth$$

Many placentals have an elaboration of bone around the inner ear, the auditory bulla, which probably increases auditory acuity (see Figure 23.3B). Most marsupials lack a bulla, and in those that do have a bulla, it forms from a different bone than that of placentals.

(A) Skull, lower jaw, and dental formula of a marsupial, the Virginia opossum, *Didelphis virginiana*

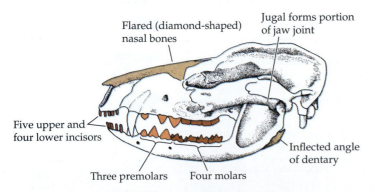

(B) Skull, lower jaw, and dental formula of a placental, the raccoon, *Procyon lotor*

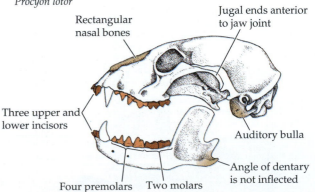

Figure 23.3 Skulls and dentitions of marsupials and placentals. Several key features of the skull and teeth differ between marsupials (A) and placentals (B). The inflected angle of the dentary, the shape of the nasal and jugal bones, the number of teeth, and presence of an auditory bullae are some of the many differences between marsupial and placental skulls. Placentals have fewer incisors and molars but more premolars, as shown in the dental formulae (see text) below the drawings (I, incisors; C, canines; P, premolars; M, molars). (Based on T.E. Lawlor. 1979. *Handbook to the Orders and Families of Living Mammals*, by permission of Barbara Lawlor.)

23.2 Diversity of Marsupials

LEARNING OBJECTIVES

23.2.1 Differentiate the seven clades of marsupials by biogeography (New World versus Australia/New Guinea).

23.2.2 Explain the pathway and timing by which marsupials reached Australia, starting with their likely origin in Asia.

Here we recognize seven extant clades of Marsupialia. Three of these (Paucituberculata, Didelphimorphia, and Microbiotheria) occur in the New World. The other four (Diprotodontia, Notoryctemorphia, Peramelemorphia, and Dasyuromorphia) occur in Australia and New Guinea (**Table 23.1**).

New World marsupial clades Older works recognized "Ameridelphia" for the three groups of marsupials that live in North and South America. However, "Ameridelphia" is a paraphyletic group (see Section 1.4) because the single species of Microbiotheria, which only occurs in South America, is the sister taxon of Diprotodontia, and thus included in Australidelphia (see Figure 23.2).

- Paucituberculata includes the shrew opossums, also known as caenolestids (**Figure 23.4A**). These are small, mainly terrestrial, shrewlike carnivores or invertivores (i.e., they feed on invertebrates other than insects, such as earthworms).

- Didelphimorphia includes more than 100 species of New World opossums in the family Didelphidae. These are small to medium-size marsupials, and most are arboreal or semiarboreal omnivores like the Virginia opossum (**Figure 23.4B**). Some are semiaquatic, like the water opossum (*Chironectes minimus*), which ranges from Mexico to Argentina.

Table 23.1 Terminal Taxa of Marsupials As Shown in Figure 23.2

Taxon	Included families	Number and distribution of species; description
Paucituberculata	Caenolestidae (shrew or rat opossums)	7 species in South America; carnivores and invertivores, 15 to 40 g
Didelphimorphia	Didelphidae (opossums)	111 species in Central and South America, 1 species in North America; semiarboreal omnivores, 20 g to 6 kg
Microbiotheria	Microbiotheriidae (monito del monte)	1 species in South America; arboreal omnivore, 25 g
Vombatiformes	Vombatidae (wombats)	3 species in Australia; herbivorous digging specialists, 20 to 35 kg
Phalangeroidea	Phalangeridae (cuscuses and brushtail possums)	30 species in Australia, New Guinea, and eastern Indonesia; most are nocturnal folivores, up to 5 kg
	Phascolarctidae (koala)	1 species in Australia; arboreal folivore, 4 to 15 kg
Petauroidea	Petauridae (possums, some of them gliders)	11 species in Australia and/or New Guinea; omnivores or insectivores, 95 to 720 g
	Acrobatidae (feather-tail glider, feather-tail possum)	2 species, one in Australia (glider) one in New Guinea (possum); omnivores, 12 to 50 g
	Burramyidae (pygmy possums)	5 species in Australia, New Guinea, and Indonesia; nocturnal omnivores, 10 to 50 g
	Pseudocheiridae (ringtail possums)	17 species in Australia and New Guinea; nocturnal folivores, 200 g to 2 kg
	Tarsipedidae (honey possum)	1 species in Australia; nectarivore, 5 to 10 g
Macropodiformes	Hypsiprymnodontidae (musky rat-kangaroo)	1 species in Australia; omnivore, 360 to 680 g
	Potoroidae (bettongs, potoroos, rat-kangaroos)	8 species in Australia, hopping herbivores, 1 to 3 kg
	Macropodidae (kangaroos and wallabies)	67 species in Australia, New Guinea, and neighboring islands; hopping herbivores, 1 to 90 kg
Notoryctemorphia	Notoryctidae (marsupial moles)	2 species in Australia; sand-swimming insectivores and carnivores, 40 to 70 g
Dasyuromorphia	Dasyuridae (marsupial "mice," quolls, Tasmanian devil)	69 species in Australia and New Guinea; insectivores or carnivores, 2 g to 8 kg
	†Thylacinidae (thylacine)	1 recently extinct species in Australia and New Guinea; nocturnal carnivore, 8 to 30 kg
	Myrmecobiidae (numbat)	1 endangered species in Australia; insectivore, 280 to 700 g
Peramelemorphia	Peramelidae (bandicoots)	20 species in Australia and New Guinea; omnivores, 100 g to 5 kg
	†Chaeropodidae (pig-footed bandicoots)	2 recently extinct species in Australia; nocturnal herbivores, 200 to 600 g
	Thylacomyidae (bilbies)	1 recently extinct and 1 extant species in Australia; nocturnal omnivores, 300 g to 2.5 kg

(A) Dusky shrew opossum, *Caenolestes fuliginosus*

(B) The Virginia opossum, *Didelphis virginiana*, is common across much of its range in North America

(C) Monito del monte, *Dromiciops gliroides*

Figure 23.4 **Extant South and North American marsupials.**   The three extant groups of marsupials in the New World are (A) Paucituberculata, (B) Didelphimorphia, and (C) Microbiotheria. They do not comprise a monophyletic group because Microbiotheria is the extant sister group of Diprotodontia, marsupials of Australia and New Guinea (see Figure 23.5).

- The single species of Microbiotheria, the monito del monte, is a tiny, arboreal marsupial from montane forests of Chile and Argentina (**Figure 23.4C**).

Marsupials of Australia and New Guinea   The four marsupial clades from Australia and New Guinea together with their sister taxon Microbiotheria comprise Australidelphia ("south-lovers"). Most species are in Diprotodontia (Greek *di*, "two"; *pro*, "first"; *odous*, "tooth"), a group named for their procumbent (forward-projecting) lower incisors. Diprotodonts also have syndactylous toes in which the slender phalanges of the second and third digits of the foot are bound together by skin (they retain separate claws). Most of the ~150 species of diprotodonts are herbivores or omnivores. We recognize four clades within Diprotodontia (see Table 23.1).

- Vombatiformes includes wombats and the koala. The three species of burrowing wombats are specialist grazers with continually growing molars (**Figure 23.5A**). Extant wombats are about the size of small pigs (they also look a little like pigs and squeal like pigs),

(A) Common wombat, *Vombatus ursinus*

(B) †Marsupial lion, †*Thylacoleo*, attacks †*Diprotodon*

(C) Koala, *Phascolarctos cinereus*

(D) Australian cuscus, *Phalanger mimicus*

(E) Sugar glider, *Petaurus breviceps*

(F) Long-nosed potoroo, *Potorus tridactylus*

(G) Red kangaroo, *Osphranter rufus*

Figure 23.5 **Diprotodont australidelphians.**   Most extant marsupials are diprotodonts native to Australia and New Guinea. Diprotodontia includes (A–C) Vombatiformes, (D) Phalangeroidea, (E) Petauroidea, and (F,G) Macropodiformes.

but rhinoceros-size wombat relatives (†diprotodontids) roamed Plio-Pleistocene Australian savannas. Carnivory evolved in marsupial lions, wombat relatives such as the Pleistocene †*Thylacoleo* that preyed on other large †diprotodontids (**Figure 23.5B**). The koala is an arboreal relative of wombats (**Figure 23.5C**).

- Phalangeroidea is a group of arboreal, primatelike animals, including cuscuses and brushtail possums[1] (**Figure 23.5D**). Many phalangeroids are nocturnal and arboreal, with grasping hands and digits (phalanges) that inspired the group's name.

- Petauroidea includes the sugar glider (**Figure 23.5E**), which glides between trees using a web of skin stretched between the front and hindlimbs and a feathery tail. The diminutive honey possum or noolbenger, *Tarsipes rostratus*, is the only nectarivorous marsupial (and one of the few nectar-eating therians apart from some bats).

- Macropodiformes has three families of (mostly) hopping marsupials. These include the small, omnivorous rat-kangaroos (potoroos; **Figure 23.5F**) and larger, herbivorous true kangaroos (including wallabies and secondarily arboreal tree-kangaroos). The largest extant kangaroo is the red kangaroo (**Figure 23.5G**), which weighs about 90 kg (about the size of a large

[1] "Possum" refers to arboreal marsupials from Australia, whereas "opossum" is the common name for New World didelphimorphians.

man). Kangaroos three times that size (†Sthenurinae) existed in the Pleistocene; they were too big to hop and probably walked on their hind legs.

Of the three remaining australidelphian clades, there is strong molecular evidence for a sister group relationship between Dasyuromorphia and Peramelemorphia (see Figure 23.2). The third clade includes two species of marsupial moles.

- Notoryctemorphia, the marsupial moles, are small (~40–60 g), desert-dwelling sand swimmers (**Figure 23.6A**). Unlike earth-digging moles of the Northern Hemisphere, sand swimmers cannot build open tunnels because sand collapses behind them. Instead, using both their front and back feet, they "swim" beneath a shallow cover of loose sand. Marsupial moles have reduced eyes, two claws on the forelimb, and eat both invertebrate and vertebrate prey.

- Dasyuromorphia includes the †thylacine †*Thylacinus cynocephalus* (**Figure 23.6B**), which in recent times occurred only on the island of Tasmania off Australia's south coast. It has been declared extinct, although people hope for evidence of its continued existence. Dasyuromorphia also includes many marsupial mice that are more like shrews because they are carnivorous and insectivorous rather than omnivorous; an example is the fat-tailed dunnart (**Figure 23.6C**). There are a few larger carnivorous dasyuromorphs, such as the cat-size quolls (see

(A) Marsupial mole, *Notoryctes typhlops* eating a gecko, *Heteronotia binoei*

(B) A pair of †thylacines, †*Thylacinus cynocephalus*, at the National Zoo in Washington, D.C. in 1902

(C) Fat-tailed dunnart, *Sminthopsis crassicaudata*

(D) Tasmanian devil, *Sarcophilus harrisii*

(E) Numbat, *Myrmecobius fasciatus*

(F) Bilby, *Macrotus lagotis*

Figure 23.6 Other australidelphians. The three non-diprotodont clades of australidelphian marsupials are (A) Notoryctemorphia; (B–E) Dasyuromorphia (including the now-extinct thylacine [B]); and (F) Peramelemorphia.

chapter opener photo) and badger-size Tasmanian devil (**Figure 23.6D**). The numbat mainly eats termites, but it looks more like a miniature fox than a placental anteater (**Figure 23.6E**).

- Peramelemorphia includes bandicoots and bilbies. The pig-footed bandicoot †*Chaeropus ecaudatus*, extinct since the 1950s, was a tiny, spindly-legged, more herbivorous form that resembled the placental elephant shrews of Africa. The bilby (**Figure 23.6F**) looks a little like a rabbit (Australians celebrate an Easter Bilby rather than an Easter Bunny), but is omnivorous rather than herbivorous. Bilbies have syndactylous feet convergently like the syndactylous condition of diprotodonts.

Marsupials and the Australian fauna

Although people consider marsupials quintessential Australian mammals, marsupials did not reach that continent until the early Cenozoic. Although metatherians probably originated in Asia, the earliest marsupials are known from the early Cenozoic of North America, where they enjoyed a modest radiation of small forms. The initial Cenozoic diversification of marsupials occurred primarily in South America, where the more basal types of marsupials persist.

Marsupials probably dispersed to Australia from South America across Antarctica, which was warm and ice-free until ~45 Ma. The earliest known Australian marsupials are from the early Eocene (~55 Ma), and marsupial fossils occur in the Eocene of western Antarctica. The major barrier to dispersal to Australia probably was the midcontinental mountain ridge between western and eastern Antarctica.

Australian marsupials evolved to fill a variety of adaptive zones, with feeding habits ranging from specialized carnivory to complete herbivory, convergently with the evolution of terrestrial placentals elsewhere in the world. When humans reached Australia ~65 ka, they encountered marsupial lions, remarkable predators convergently like placental cats such as jaguars. These carnivores had enormous flesh-shearing teeth, a powerful bite, and greater forearm maneuverability than any extant carnivore. †*Thylacoleo* probably preyed on animals much larger than itself using a large retractable claw on a semiopposable thumb (see Figure 23.5B).

Large macropodiform marsupials such as the red kangaroo (see Figure 23.5G) are primarily grazers, eating green grass, and sometimes browsing on the leaves of shrubs like many placental herbivores, such as impalas. Kangaroos are "ruminantlike," but although they ferment food in an enlarged stomach, they do not chew cud as ruminants do (see Section 23.5).

The Australasian landmasses (Australia, New Zealand, and New Guinea) have been moving closer to Asia throughout the Cenozoic, but there have been few natural migrations of Asian mammals into Australia. For example, monkeys never reached Australasia. In their absence,

tree-kangaroos radiated in tropical forests of northern Australia and New Guinea.

Rodents arrived in Australia by the early Pliocene, probably via dispersal along the island chain between Southeast Asia and New Guinea. Australian rodents are related to the mouse/rat group of Eurasian rodents but evolved into unique Australian forms such as the small, jerboa-like hopping mice and the much larger (otter-size) water rats. True mice and rats arrived later, in the Pleistocene. However, these rodents had surprisingly little overall effect on Australian marsupials. Far more serious threats came from the original invasion by humans from Asia ~65 ka, the arrival of dogs (dingoes) about 4 ka, and the introduction of domestic mammals such as foxes, rabbits, and cats by European colonists during the past few centuries. As a result, some species of Australian marsupials are extinct, and many species are threatened or endangered.

23.3 Diversity of Placentals

LEARNING OBJECTIVES

23.3.1 Recognize the major clades of placental mammals and whether they belong to Atlantogenata or Boreoeutheria.

23.3.2 Describe conservation concerns surrounding pangolins.

Extant placentals belong to approximately 20 distinct orders (**Figure 23.7** and **Table 23.2**). However, placental diversification occurred rapidly and left few reliable morphological clues about phylogenetic relationships among the orders. Here we follow molecular phylogenetic analyses that find two major groupings, a southern Atlantogenata and a northern Boreoeutheria (see Figure 23.7).

Atlantogenata

Clades belonging to Atlantogenata radiated on the continents of Africa (Afrotheria) and South America (Xenarthra), although several clades migrated out of Africa ~17 Ma.

Afrotheria   As its name suggests, Afrotheria is a grouping of endemic African placentals, the product of an extended period of independent, isolated evolution on the African continent.[2] There are two clades within Afrotheria, Afroinsectiphilia and Paenungulata. Afroinsectiphilia includes three groups (see Figure 23.7).

- Afrosoricida includes tenrecs and golden moles. Ancestral tenrecs rafted from mainland Africa to Madagascar, where they underwent extensive radiation.

---

[2]Many groups that we think of as African, such as lions, hyenas, wildebeest, giraffes, and antelope, are boreoeutherians that did not reach Africa until it collided with Eurasia around 17 Ma. Primates (covered in more detail in Chapter 24), are the only long-term African mammals that are not afrotheres.

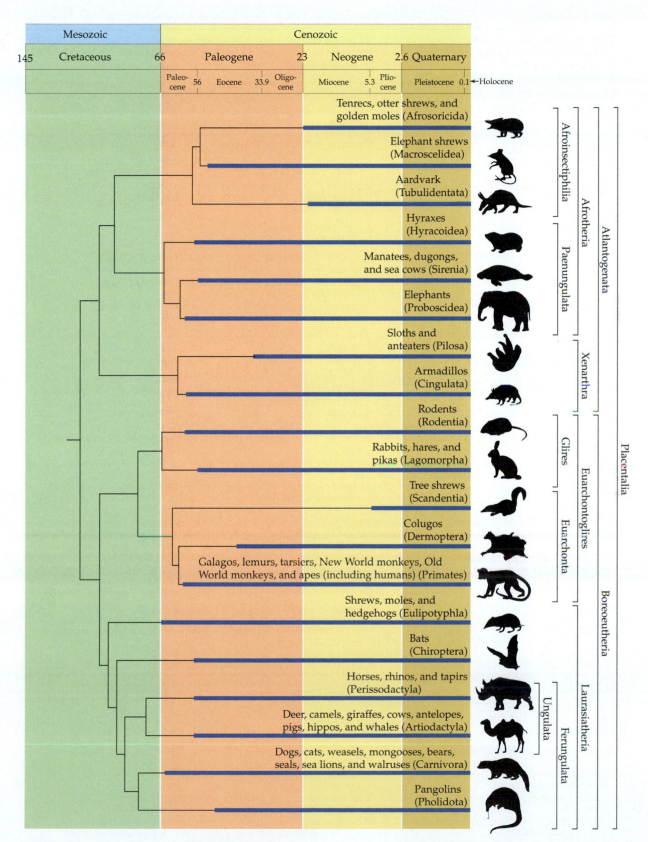

Figure 23.7 Phylogenetic relationships of placentals.
Based on current molecular phylogenetic interpretations, we classify placentals in 19 orders belonging to two major clades—Atlantogenata and Boreoeutheria—with strong biogeographic bases. Atlantogenata includes two groups that radiated in the Southern Hemisphere, Afrotheria of Africa, and Xenarthra of South America. Boreoeutheria comprises Euarchontoglires and Laurasiatheria, two groups that radiated in the Northern Hemisphere.

Table 23.2 Terminal Taxa of Placentals Shown in Figure 23.7

Taxon	Included groups[a]	Number and distribution of species; description
Afrosoricida	Chrysochloridae (golden moles)	21 species in Africa; burrowing insectivores, 20 to 500 g
	Potamogalidae (otter shrews)	3 species in Africa; semiaquatic carnivores, 75 to 950 g
	Tenrecidae (tenrecs)	31 species endemic to Madagascar; primarily invertivores, 5 g to 1 kg
Macroscelidea	Macroscelididae (elephant shrews)	20 species in Africa; primarily insectivores, 50 to 700 g
Tubulidentata	Orycteropidae (aardvark)	1 species in Africa; nocturnal, burrowing myrmecophage, 40 to 60 kg
Hyracoidea	Procaviidae (hyraxes)	5 species in Africa and the Middle East; herbivores, 2 to 5 kg
Sirenia	Dugonidae (dugong, †Steller's sea cow)	2 species, dugong in Indo-West Pacific waters and the recently extinct †Steller's sea cow in Bering Sea; herbivores, 230 to 8,000 kg
	Trichechidae (manatees)	3 species in coastal regions of Florida to northern South America, Amazon Basin, or western coast of Africa; herbivores, 120 to >1,000 kg
Proboscidea	Elephantidae (elephants)	3 species, 1 in tropical Asia, 2 in tropical Africa; enormous, terrestrial herbivores, 4,500 to 7,000 kg.
Pilosa	Bradypodidae (three-toed sloths)	4 species in Central and South America; herbivores, 3.5 to 4.5 kg
	Megalonychidae (two-toed sloths)	2 species in Central and South America; herbivores, 4 to 8 kg
	Cyclopedidae (silky anteater)	1 species from Mexico to South America; myrmecophage, 175 to 400 g
	Myrmecophagidae (anteaters)	3 species in Central and South America; myrmecophages, 3 to 50 kg
Cingulata	Chlamyphoridae (fairy armadillos, 3-banded armadillos, others)	14 species in South America; most are omnivores, 120 g to more than 50 kg
	Dasypodidae (common armadillos)	7 species from North to South America; most are omnivores, 3 to 8 kg
Rodentia	Sciuromorpha (mountain beaver, squirrels, dormice)	3 families, ~300 species worldwide except Australasia and Antarctica; primarily herbivores
	Anomaluromorpha (African flying squirrels, springhares)	3 families, 9 species in sub-Saharan Africa; herbivores
	Castorimorpha (beavers, kangaroo rats, gophers)	3 families, ~109 species in the Northern Hemisphere; herbivores and seed-eaters
	Hystricomorpha (African and American porcupines, capybara, guinea pig)	18 families, ~300 species in Africa and the Americas; herbivores
	Myomorpha (rats, mice, voles, jerboas)	12 families, >1,500 species (about one-fourth of all mammal species), worldwide except Antarctica; omnivores and herbivores
Lagomorpha	Leporidae (rabbits, hares)	67 species worldwide; quadrupedal jumping herbivores, 300 g to 5 kg
	Ochotonidae (pikas)	30 species in North America and Asia; herbivores, 120 to 350 g
Scandentia	Ptilocercidae (pen-tailed tree shrew)	1 species in South/Southeast Asia; omnivore, 40 to 62 g
	Tupaiidae (tree shrews)	19 species in South/Southeast Asia; omnivores, 40 to 340 g
Dermoptera	Cynocephalidae (colugos)	2 species in Southeast Asia; gliding herbivores, 1 to 2 kg
Primates	See Chapter 24	14 families and ~500 species of terrestrial and arboreal frugivores, folivores, insectivores, and omnivores; non-human primates are found primarily in tropical and subtropical areas except Australasia; 85 g to >275 kg
Eulipotyphla	Erinaceidae (hedgehogs, moonrats)	24 species in Eurasia, Africa, and Southeast Asia; omnivores, 40 g to 1.5 kg
	Talpidae (moles, desmans)	59 species in North America, Eurasia, and Southeast Asia; digging invertivores, 10 to 550 g
	Solenodontidae (solenodons)	2 species in Cuba and Hispaniola; burrowing, venomous, insectivores, 0.7 to 1 kg
	Soricidae (shrews)	385 species nearly worldwide; voracious omnivores, 2 to 100 g
Chiroptera	Pteropodoidea (Old World fruit bats)	1 family, 201 species in Eurasia, Africa, and Oceania; primarily frugivores, 40 g to 1.5 kg
	Rhinolophoidea (Old World leaf-nosed bats, horseshoe bats, others)	6 families, 219 species in Africa, Eurasia, Oceania; insectivores, 2 to 30 g
	Emballonuroidea (slit-faced and sheath-tailed bats)	2 families, 69 species in tropics and subtropics worldwide; primarily insectivores but some carnivores, 5 to 105 g
	Noctilionoidea (New World leaf-nosed bats, disk-winged bats, others)	7 families, 248 species from North to South America, Madagascar, and New Zealand; primarily insectivores, but include frugivores, carnivores, and sanguivores, 7 to 200 g
	Vespertilionoidea (vesper bats, free-tailed bats, others)	5 families, 674 species nearly worldwide; insectivores, carnivores, and piscivores, 8 to 220 g

Table 23.2 (*Continued*)

Taxon	Included groups[a]	Number and distribution of species; description
Perissodactyla	Equidae (horses, asses, zebras)	12 species worldwide; herbivores, 200 to 500 kg
	Rhinoceratidae (rhinoceroses)	5 species in Africa and South/Southeast Asia; herbivores, 800 to 2,300 kg
	Tapiridae (tapirs)	4 species in the tropical Americas and Southeast Asia; herbivores, 150 to 400 kg
Artiodactyla	Suidae (pigs)	18 species in Europe, Africa, Asia, East Indies, introduced elsewhere; omnivores or herbivores, 6 to 275 kg
	Tayassuidae (peccaries)	5 species in North to South America; omnivores, 20 to 40 kg
	Hippopotamidae (hippos)	2 species in Africa; herbivores, 180 to 1,300 kg
	Camelidae (camels and llamas)	7 species in Asia, Middle East, and South America; herbivores, 35 to 1,000 kg
	Antilocapridae (pronghorn)	1 species endemic to North America; herbivore, 35 to 70 kg
	Bovidae (antelope, sheep, cattle)	143 species in Africa, Europe, Asia, North America, introduced elsewhere; herbivores, 3 to >1,000 kg
	Cervidae (deer, elk, moose)	~50 species in North and South America, Eurasia, Africa, introduced elsewhere; herbivores, 3 to >700 kg
	Giraffidae (giraffes, okapi)	5 species in Africa; herbivores, 200 to >1,000 kg
	Moschidae (musk deer)	7 species in South Asia; herbivores, up to 17 kg
	Tragulidae (mouse deer)	10 species in Africa and South/Southeast Asia; herbivores, 0.7 to 16 kg
	Mysticeti (baleen whales including right, blue, humpback, and gray whales)	3 families, 16 species; suspension feeders (zooplankton, small crustaceans, fishes), 3,000 to >180,000 kg
	Odontoceti (toothed whales including oceanic and river dolphins, orca, porpoises, pilot and sperm whales, beluga, narwhal)	10 families, 73 species; carnivores (invertebrates and vertebrates), 20 to >40,000 kg
Carnivora	Ailuridae (red panda)	1 species in Nepal and China; eats mostly bamboo, 3 to 6 kg
	Canidae (wolves, foxes, dogs, others)	34 species nearly worldwide; primarily carnivores, 6 to 79 kg
	Mephitidae (skunks, stink badgers)	12 species in the Americas and Southeast Asia; omnivores, 0.5 to 8 kg
	Mustelidae (weasels, badgers, otters)	66 species nearly worldwide; carnivores, 36 g to 45 kg
	Odobenidae (walrus)	1 species in Arctic Ocean and subarctic seas; primarily invertivores, 500 to 2,000 kg
	Otariidae (sea lions)	15 species in Pacific, Atlantic, and Indian oceans; carnivores, 70 to 1,000 kg
	Phocidae (seals)	19 species mostly in Arctic and Antarctic oceans; carnivores, 55 to >2,000 kg
	Procyonidae (raccoons, coatis, kinkajous)	13 species in Americas; omnivores, 1 to 20 kg
	Ursidae (bears)	8 species in Americas, Europe, Asia; primarily omnivores, 25 to 800 kg
	Eupleridae (fossa, striped civet)	10 species endemic to Madagascar; carnivores, 600 g to 9 kg
	Felidae (cats, including bobcat, cheetah, ocelot, lion, leopard, others)	42 species nearly worldwide; carnivores, 1 to 325 kg
	Herpestidae (mongooses)	34 species in Africa, India, and South Asia; carnivores, 300 g to 6 kg
	Hyaenidae (hyenas)	4 species in Africa, Middle East, Central/South Asia; omnivores and a myrmecophage, 7 to 64 kg
	Nandiniidae (African palm civet)	1 species in Africa; arboreal frugivore, 1 to 3 kg
	Prionodontidae (linsangs)	2 species in Asia; arboreal carnivores, 450 to 700 g
	Viverridae (civets and genets)	~36 species in Africa, Europe, South/Southeast Asia; carnivores, 1 to 14 kg
Pholidota	Manidae (pangolins)	8 species in tropical Africa and Asia; myrmecophages, 2 to 33 kg

[a]In most cases included groups are families, identifiable by the suffix -idae. When higher taxa are used (e.g., suborders in Rodentia and superfamilies in Chiroptera), number of families in the group is provided.

There are terrestrial, arboreal, fossorial, and even aquatic forms convergently like mammals that belong to other groups, such as shrews, opossums, hedgehogs, and mice (**Figure 23.8A**). Golden moles of southern Africa are tiny insectivores with large hands and claws that resemble the marsupial moles of Australia (Notoryctemorphia; see Section 23.2).

- Macroscelidea, the elephant shrews, are diurnal and mostly insectivorous. Their common name "elephant" refers not to their size (they are tiny, 50–700 g) but to their long, mobile snout (**Figure 23.8B**).

- Tubulidentata includes a single extant species, *Orycteropus afer*, the aardvark (Afrikaans, "earth pig"). The aardvark is semifossorial, with powerful forelimbs and long claws it uses for digging (**Figure 23.8C**). It uses its long snout and sticky tongue to catch ants and termites that it crushes with flat-crowned, columnlike cheek teeth.

The second group of extant afrotherians, Paenungulata, also has three groups:

- Hyracoidea, the rodentlike hyraxes, are 2–5 kg herbivores, some of which are arboreal (tree hyraxes) and

(A) Lowland streaked tenrec, *Hemicentetes semispinosus*, from Madagascar

Frank Vassen/CC BY 2.0

(B) Golden-rumped elephant shrew, *Rhynchocyon chrysopygus*, from east Africa

Kim/CC BY-SA 2.0

(C) Aardvark, *Orycteropus afer*, a nocturnal burrower

Louise Joubert/CC BY-SA 3.0

(D) Rock hyrax, *Procavia capensis*, from Africa and western Asia

Bernard Dupont/CC BY-SA 2.0

(E) West Indian manatee, *Trichechus manatus*, lives in freshwater and shallow marine habitats

Keith Ramos/NFWS/CC BY 2.0

Figure 23.8 Diversity of afrotherians. The two extant clades of afrotherians are (A–C) Afroinsectiphilia and (D–F) Paenungulata.

(F) African bush elephants, *Loxodonta africana*, in Kruger National Park, South Africa

Derek Keats/CC BY 2.0

others terrestrial (rock hyraxes; **Figure 23.8D**). All are good climbers, with specialized pads on the soles of their feet for traction on surfaces.

- Sirenia includes the fully aquatic, herbivorous dugongs and manatees (**Figure 23.8E**). They grind plant food using keratinized pads on the palate, lower jaw, and tongue. †Steller's sea cow, extinct since the late 18th century, was described from specimens collected in the far northern Bering Sea. Extant species occur in river systems and nearshore marine environments in the Americas and west Africa (three

species of manatees) and along tropical and subtropical coastlines of the Indian and western Pacific Oceans (the single extant species of dugong).

- Proboscidea includes the Asian elephant, *Elephas maximus*, and two species of African elephants (*Loxodonta*, the largest extant land mammals; **Figure 23.8F**). In the Pleistocene, proboscideans also lived in North and South America. They have massive, pillarlike limbs and huge skulls lightened by air spaces (pneumatization) in the bones. An elephant's trunk contains modified muscles of the nose and upper

(A) Brown-throated three-toed sloth, *Bradypus mexicana*

(B) Northern tamandua, *Tamandua mexicana*, foraging in a tree

(C) Mouse-size burrowing pink fairy armadillo, *Chlamyphorus truncatus*, from Argentina

Figure 23.9 Xenarthrans. The two extant clades of xenarthrans are (A,B) Pilosa and (C) Cingulata.

lip and is used in feeding, drinking, smelling, and cooling, such as when water or mud are sucked into the trunk and sprayed on the body. All three species are classified by the IUCN as Endangered.

Xenarthra The second group within Atlantogenata includes two orders (Pilosa and Cingulata) that originated in South America and were confined to this continent until the Great American Biotic Interchange (GABI; see Section 20.3) some 2–3 Ma. Today xenarthrans also occur in North America.

Xenarthrans have relatively low metabolic rates and low body temperatures. They also have unique extra articular surfaces on some of their vertebrae (the Greek roots of Xenarthra translate to "strange joints"). These extra articulations strengthen the axial skeleton, enhancing digging in terrestrial species and reaching in arboreal species. Xenarthrans have simplified dentition and teeth that lack enamel.

- Pilosa includes the sloths and anteaters. Diverse ground sloths, some the size of elephants, inhabited South America during the Cenozoic and dispersed as far north as Alaska during the GABI (see Figure 20.7). The two extant genera of sloths are arboreal and range from ~60 to ~80 cm in length; they are not closely related to each other, but rather related to two different extinct groups of ground sloths. Two-toed sloths are omnivorous, eating plant matter as well as insects and small lizards. Three-toed sloths (**Figure 23.9A**) are folivores (leaf-eaters) with exceptionally slow metabolism. Anteaters are specialized insectivores; completely toothless, they use their long proboscis and sticky, protrusible tongue to feed on ants and termites (**Figure 23.9B**). Sloths and anteaters occur in Central and South America.

- Cingulata, the armadillos, migrated from South America (where they still occur) to North America during the GABI. Prodigious diggers, armadillos use their strong forelimbs to unearth grubs and other invertebrate prey (**Figure 23.9C**).

Boreoeutheria

The remaining extant placental orders belong to Boreoeutheria. Unlike the clades of atlantogenatans, which even today occur predominantly in Africa and South America, boreoeutherians radiated and dispersed globally. The number of orders of Atlantogenata and Boreoeutheria are similar (see Figure 23.7), but there are many more species in the two extant clades of boreoeutherians: Euarchontoglires and Laurasiatheria.

Euarchontoglires This group consists of Glires and Euarchonta (see Figure 23.7). Glires includes only two orders, rodents and lagomorphs, but these orders are large and found more or less worldwide.

- The more than 2,250 species of Rodentia account for about 42% of extant mammal species (**Figure 23.10A–C**). Rodents are known for gnawing on just about anything (Latin *rodens*, "gnawing"); they have a single pair of upper and lower incisors that are ever-growing (see Figure 23.18G). Iron-containing pigments strengthen the enamel on the anterior surfaces of incisors but the posterior surfaces are softer dentine. This combination of a hard anterior surface and softer posterior surface makes rodent incisors like self-sharpening chisels.

- Lagomorpha includes pikas, hares, and rabbits (**Figure 23.10D**). A synapomorphy of lagomorphs is a dental formula that begins with I 2/1 (i.e., each side of the upper jaw has two incisors and each side of the lower jaw has only one incisor). What is unusual about this arrangement is that the much smaller second upper incisor, known as a "peg tooth," lies directly *behind* the first. Peg teeth are used in combination with the lower incisors to cut food. Whereas rabbits and hares are quadrupedal jumpers, pikas scamper around rock crevices.

Three orders of Euarchonta comprise the "other half" of Euarchontoglires.

- Scandentia, tree shrews, are small Asian mammals with long, bushy tails (**Figure 23.10E**). Their

(A) Prairie voles, *Microtus ochrogaster*

© Todd H. Ahern

(B) North American beaver, *Castor canadensis*

Minette Layne/CC BY-SA 2.0

(C) Naked mole rat, *Heterocephalus glaber*

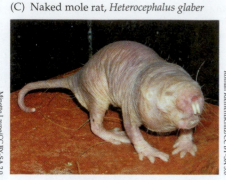

Roman Klementschitz/CC BY-SA 3.0

(D) Eastern cottontail, *Sylvilagus floridanus*

Gareth Rasberry/CC BY-SA 3.0

(E) Pygmy tree shrew, *Tupaia minor*

Paul J. Morris/CC BY-SA 2.0

(F) Sunda colugo, *Galeopterus variegatus*

© iStock.com/thawats

(G) Brown mouse lemur, *Microcebus rufus*

gailhampshire/CC BY 2.0

Figure 23.10 Euarchontoglires. Euarchontoglires is a large clade of boreoeutherians that contains Glires and Euarchonta (see Figure 23.7). Glires includes Rodentia (the largest group of therians; A–C) and Lagomorpha (D). Euarchonta includes (E) Scandentia, (F) Dermoptera (note the baby clinging to the mother's abdomen), and (G) Primates.

locomotion is scansorial—that is, a mix of terrestrial and arboreal—like that of squirrels. Their middle four lower incisors form a tooth comb used in grooming and perhaps feeding.

- Dermoptera, the two extant species of colugos, are also from Asia and have a tooth comb formed by grooved lower incisors. Colugos glide using flaps of skin that extend from the neck to the tips of the fingers, toes, and tail (**Figure 23.10F**). Molecular evidence indicates they are the closest extant relatives of Primates.

- Primates have large brains relative to their body sizes, binocular vision, and greater reliance on vision than olfaction (**Figure 23.10G**). Lemurs and lorises also have a tooth comb, that incorporates the lower canines as well as the incisors. We discuss primates and some of its ~500 extant species, including *Homo sapiens*, in Chapter 24.

Laurasiatheria Currently accepted phylogenetic interpretations of Laurasiatheria ("Laurasian beasts"; see Figure 23.7) are based largely on molecular rather than morphological data.

- Eulipotyphla includes shrews, hedgehogs, moles, and solenodons. These small therians superficially resemble Mesozoic mammaliaforms such as †*Morganucodon* (see Figure 22.10) but are not closely related to them. Shrews occur in most of the world, and are missing only from Australia and polar areas. Most are small (~2–100 g), with high metabolic rates and short lives. Hedgehogs are famous for their defensive behavior of rolling into a ball and erecting their spines, which are stiff hollow hairs (**Figure 23.11A**). There are about 40 species of moles, some of which live on the surface amid leaf litter while others are fossorial, and still others are semiaquatic (see Figure 11.18). The two extant species of solenodons are endemic to the islands of Hispaniola and Cuba, and both species are Critically Endangered. They are large (for a eulipotyphlan; ~1 kg), fossorial insectivores. An extra bone in the nose supports their long cartilaginous snout, and they produce a neurotoxin in their saliva used for defense and food acquisition.

(A) European hedgehog, *Erinaceus europaeus*

Tony Wills/CC BY-SA 3.0

(B) Bechstein's bat, *Myotis bechsteinii*

Dietmar Nill/CC BY 2.5

(C) Indian rhinoceros, *Rhinoceros unicornis*

Aditya Pal/CC BY-SA 4.0

(D) Lesser mouse deer, *Tragulus kanchil*

Sakurai Midori/CC BY-SA 3.0

Photo by Kristian Bell

Canine

(E) Pacific spotted dolphin, *Stenella attenuatus*

NOAA/CC BY 2.0

(F) Aardwolf, *Proteles cristata*

Greg Hume/CC BY-SA 3.0

(G) Indian pangolin, *Manis crassicaudata*

A. J. T. Johnsingh, WWF-India and NCF/CC BY-SA 4.0

Figure 23.11 Laurasiatherians. The basal clades of Laurasiatheria are (A) Eulipotyphla and (B) Chiroptera. The remaining taxa belong to Ferungulata. (C–E) Ungulates comprise the sister groups Perissodactyla (odd-toed ungulates such as rhinoceroses) and Artiodactyla (even-toed ungulates, including deer and whales). (F) Carnivora includes familiar species, such as dogs, and less familiar ones such as the aardwolf. (G) The eight species of pangolins, Pholidota, are illegally poached and traded throughout Asia and are among the most endangered species in the world.

- Chiroptera, the bats (**Figure 23.11B**), are the only mammals capable of powered flight (see Section 23.6). The more than 1,400 bat species represent about 25% of extant mammalian species. They have a nearly worldwide distribution except for polar regions, extreme deserts, and some oceanic islands. Bats perform important ecosystem functions such as controlling flying insects and pollinating more than 300 species of plants, including banana, avocado, mango, and cashew trees.

The remaining orders of Laurasiatheria are in Ferungulata (sometimes spelled Fereuungulata), a group that includes ungulates (hoofed placentals, including whales), carnivores, and pangolins. We interpret Perissodactyla (odd-toed ungulates) and Artiodactyla (even-toed ungulates) as sister taxa within Ungulata (see Figure 23.7), although some researchers do not consider Ungulata to be monophyletic.

- Perissodactyls have either three toes (e.g., rhinoceroses, **Figure 23.11C**) or one toe (horses; see Figure 23.23A).

- Artiodactyls are the even-toed ungulates, including pigs, hippopotamuses, camels, giraffes, antelopes, pronghorn, cattle, sheep, and deer (**Figure 23.11D**). This group also includes Cetacea, the whales (**Figure 23.11E**). Terrestrial artiodactyls can have four toes, as in pigs or hippos, but more cursorial artiodactyls such as cows and antelopes have two toes, reducing weight in distal parts of the limbs. Section 23.6 describes the many specializations of cetaceans for fully aquatic life and swimming.

- Carnivora includes dog- and catlike forms that possess a carnassial apparatus on each side of the jaw, which in adults consists of the last upper premolar and first lower molar (P4/m1; see Figure 23.18H), used to shear meat (carnivorans that consume little or no meat, such as raccoons, have reduced or lost the carnassial apparatus; see Figure 23.3B). Dogs, cats, and hyenas are specialized carnivores, but Carnivora also includes an ant-eating specialist, the aardwolf (**Figure 23.11F**); small, generalized carnivores such as mustelids (e.g., weasels and otters) and viverrids (e.g., meercats and mongooses); omnivorous bears;

and secondarily herbivorous forms such as the giant panda (a bear) and the much smaller red panda (Ailuridae). Also included in Carnivora are the pinnipeds (seals, sea lions, and walrus), secondarily semiaquatic carnivores related to bears (see Section 23.6).

- Pholidota (Greek *pholidotos*, "clad in scales") includes eight extant species of pangolins (**Figure 23.11G**). This final clade in our review of placental diversity has a tragic story. These unusual terrestrial and arboreal mammals of Africa and Asia eat ants and termites, using large claws on their forefeet to tear open the insects' nests and then gathering them on a long, sticky tongue retracted by a muscle that extends from the pangolin's pelvis. Despite legal protection by CITES Appendix 1, which limits international trade and is reserved for the world's most endangered species, pangolins are the most poached animals in the world and continue to be illegally traded internationally. Pangolin scales are used in Chinese traditional medicine, and their meat is considered a delicacy in some countries of Southeast Asia. In 2020, the Chinese government stiffened its internal protections for pangolins and banned their use in traditional medicine, but egregious flouting of laws, regulations, and international treaties designed to protect and conserve irreplaceable wildlife continues and exemplifies the problems in conserving biodiversity.

23.4 Reproduction

LEARNING OBJECTIVES

23.4.1 Describe male and female genitalia of marsupials and placentals.

23.4.2 Compare female reproductive tracts of marsupials and placentals.

23.4.3 Explain differences in placentation between marsupials and placentals.

23.4.4 Describe the innovation in placentals that allowed longer gestation periods.

Theria is the only large group of vertebrates for which viviparity is universal; for vertebrates generally, it is an uncommon reproductive mode. Viviparity places extraordinary demands on the physiology of the mother, and immune systems must be modified so that developing young are not rejected as foreign tissue. Viviparity seems to have evolved only once in mammals, in contrast to multiple origins in Chondrichthyes, Osteichthyes, Lissamphibia, and Lepidosauria.

Neonatal marsupials are extremely altricial (**Figure 23.12A**), although they emerge with well developed shoulders and a precociously developed head—features that allow them to crawl from the birth canal to a nipple and then to latch on to it. Some placentals, including most rodents, are born in a highly altricial state, with closed eyes and ears and no hair; they stay in the nest for several weeks (**Figure 23.12B**). Other placentals, especially ungulates, are born in a precocial state, with open eyes and ears and covered in hair; they are able to follow their mothers within a few hours–sometimes minutes–of birth (**Figure 23.12C**). No matter how precocial, however, all placentals require a period of lactation, both for nutrition and for transfer of antibodies from the mother.

In this section, we describe and compare features of marsupial and placental genitalia, reproductive tracts, placentation, and gestation and consider how the reproductive differences between marsupials and placentals relate to their evolution and diversity.

Genitalia

Therians are the only vertebrates to house the male testicles in a scrotum. During development, testes typically descend into a scrotum outside of the abdominal cavity, which may provide a cooler environment for sperm production than the abdomen. An alternative hypothesis proposes that testicular descent avoids intraabdominal pressures on the testes that might be exerted during bounding or galloping locomotion. The scrotum is anterior to the penis in marsupials but posterior to the penis in most placentals (rabbits are an exception). This difference in

(A) Altricial marsupial (kangaroo)

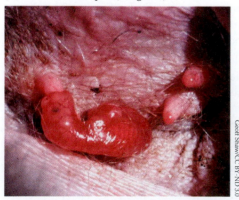

(B) Altricial placental (mice)

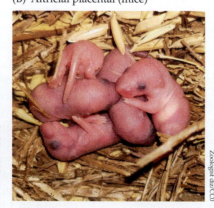

(C) Precocial placental (deer)

Figure 23.12 Altricial and precocial newborn therians. (A) A 50-day-old kangaroo infant suckling on a nipple in its mother's pouch. (B) Altricial placentals such as mice are blind and helpless at birth. (C) Precocial placentals like the white-tailed deer can stand and follow their mothers soon after birth.

location, and differences in physiological control of testicular descent, suggest that the scrotum evolved convergently in marsupials and placentals.

Not all male placentals have a scrotum. For example, testes remain in the abdominal cavity of male whales and pinnipeds. This secondary loss of the scrotum is related to drag reduction and streamlining, key factors for swimming vertebrates (see Section 7.6; cetaceans and some pinnipeds lack ear pinnae, also related to drag reduction). Male elephants and other afrotheres do not have a scrotum, but they retain remnants of genes responsible for testicular descent in other placentals. Thus, the lack of a scrotum in afrotheres is also a secondary loss, and it is probable that the common ancestor of extant placentals had a scrotum.

Male marsupials have a penis with a bifid (forked) glans, whereas placentals have a single glans. Some male placentals, including non-human primates, rodents, insectivores, carnivorans, and bats, have a bone in the penis (the os penis, or baculum). Most male mammals extend the penis from an external sheath (otherwise the only visible portion) only during urination or copulation. The clitoris is the female homolog of the penis but does not pass urine. In the usual therian condition, it lies within the urogenital sinus (the combined vagina and urethra),

where it receives direct stimulation from the penis during copulation. Females of some species have a bone in the clitoris (the os clitoridis, or baubellum).

Urogenital tracts

The digestive, urinary, and reproductive tracts of vertebrates are developmentally and anatomically associated (**Figure 23.13**). Most adult amniotes, including monotremes, have a cloaca that serves as the common exit for all three tracts (see Figure 22.22). Only in therians do the digestive and urogenital tracts have separate openings divided by a wall of connective tissue, the **perineum** (which may be little more than a membrane in marsupials and some small placentals). Modified muscles in the perineal area control the separate urogenital and digestive openings, and among vertebrates only therians have precise control over urination and defecation.

In male therians, both urine (from the bladder) and sperm (from the testes) exit via the urethra of the penis; other male amniotes use the penis only for sperm transmission, and the bladder simply opens into the cloaca (see Figure 22.22). Female therians also have a urethra leading from the bladder; in most, the urethra opens into the vagina (see Figure 23.13B,D), but in primates and some rodents it opens separately to the outside, in front of the vagina.

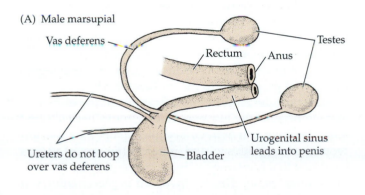
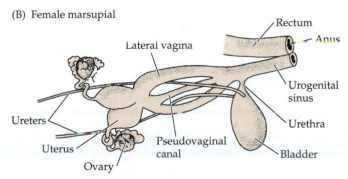
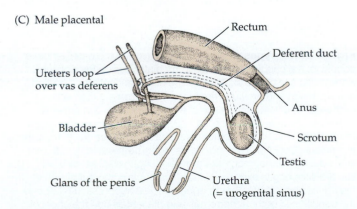
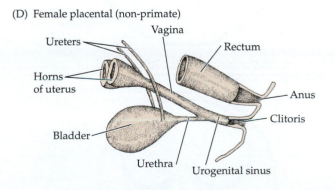

Figure 23.13 Anatomy of therian urogenital tracts. Most adult amniotes have a cloaca, the common exit for products of the digestive, urinary, and reproductive tracts. Therians differ in having separate openings for the digestive tract (anus) and urogenital tract (urogenital sinus). Ureters draining therian kidneys enter the base of the bladder. In male (A) and female (B) marsupials, the ureters pass medial to the reproductive ducts to enter the bladder. Note that female marsupials have paired lateral vaginas and a central pseudovaginal canal. In male (C) and female (D) placentals, the ureters pass laterally around the reproductive ducts to enter the bladder. Female placentals have a single midline vagina.

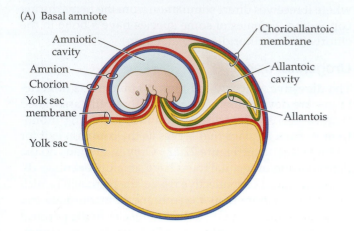

(A) Basal amniote

- Amniotic cavity
- Amnion
- Chorion
- Yolk sac membrane
- Yolk sac
- Chorioallantoic membrane
- Allantoic cavity
- Allantois

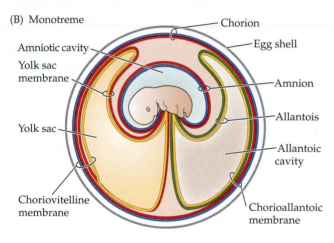

(B) Monotreme

- Chorion
- Egg shell
- Amniotic cavity
- Yolk sac membrane
- Yolk sac
- Amnion
- Allantois
- Allantoic cavity
- Choriovitelline membrane
- Chorioallantoic membrane

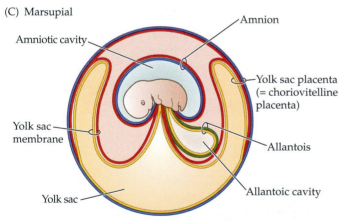

(C) Marsupial

- Amniotic cavity
- Amnion
- Yolk sac placenta (= choriovitelline placenta)
- Yolk sac membrane
- Yolk sac
- Allantois
- Allantoic cavity

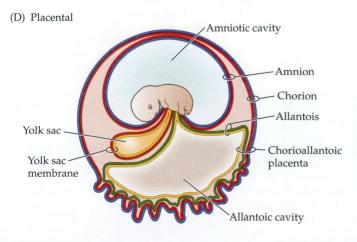

(D) Placental

- Amniotic cavity
- Amnion
- Chorion
- Allantois
- Yolk sac
- Yolk sac membrane
- Chorioallantoic placenta
- Allantoic cavity

Ureters draining therian kidneys enter the base of the bladder rather than the cloaca or urogenital sinus, a condition that differs from other vertebrates, including monotremes. In marsupials, the ureters pass medial to the developing reproductive ducts to enter the bladder, which in females prevents the oviducts from fusing in the midline (see Figure 23.13A,B). The paired lateral vaginas unite anteriorly, and two separate uteri diverge. The lateral vaginas are for passage of sperm (an arrangement complemented by the bifid penis of male marsupials). Young are born through a midline structure, the pseudovaginal canal, which develops the first time a female marsupial gives birth.

In placentals, the ureters pass laterally around the developing reproductive ducts to enter the bladder. The deferent ducts (for passage of sperm) loop around the ureters in their passage from the testes to the urethra (see Figure 23.13C). In female placentals, the paired uteri partially fuse in the midline anterior to the urogenital sinus (see Figure 23.13D). All female placentals have a single midline vagina, but only a few (including humans) have a single median uterus. Most placentals have a bipartite uterus divided lengthwise into left and right sides for some or all its length.

Placentation

Developing therians receive some nutrition from uterine secretions, especially during early stages, but a placenta soon develops from fetal tissues and is the means of exchange for most of the nutrients, metabolic wastes, respiratory gases, and hormones between the mother and fetus. The placenta also produces hormones that signal the state of pregnancy to the mother's body.

Placentas develop by modifications of the embryo's extraembryonic membranes. Recall from Chapter 9 that amniotes have four extraembryonic membranes, the amnion, chorion, yolk sac, and allantois (**Figure 23.14A**), a condition that is retained in monotremes (**Figure 23.14B**). The amnion is not directly involved in placentation but

Figure 23.14 Types of placentation. (A) In the basal amniote condition, the chorion, which lies immediately inside the eggshell, is the major site of gas exchange. The allantois contacts a portion of the chorion and provides the blood vessels needed for gas exchange. The yolk sac is large at the start of development and shrinks as the yolk is metabolized. (B) The eggs of monotremes and the quantity of yolk they contain are smaller compared with the basal amniote condition. The eggshell is highly permeable. At different times during development, the allantois, chorioallantois, and yolk sac are sites of gas exchange. (C) In marsupials, the outgrowth of the chorioallantoic membrane appears to be suppressed relative to the condition in both monotremes and placentals and the choriovitelline placenta is the primary site of gas and nutrient exchange. (D) In placentals, the choriovitelline placenta is usually short-lived, and a chorioallantoic placenta is functional for most of gestation. Only the endoderm can form blood vessels. (Blue, ectoderm; red and green, mesoderm; yellow, endoderm.) (After K. Ferner and A. Mess. 2011. *Resp Physiol Neurobiol* 178: 3950.)

surrounds and protects the embryo. The other three extraembryonic membranes contribute to two main types of therian placentas.

- The the **yolk sac placenta**, also known as a **chorio-vitelline placenta** (**Figure 23.14C**) develops when the chorion fuses with the yolk sac to form a choriovitelline membrane (Latin *vitellus*, "egg yolk"). The choriovitelline membrane develops an exchange surface with the mother's endometrium (uterine lining). In therians, a yolk sac placenta is the first exchange surface to form when a fertilized egg implants in the wall of the uterus. It is typically the only type of placenta formed during the short gestation period of marsupials.

- In placental mammals, the yolk sac placenta is transitory and functions only during preliminary stages of implantation. Subsequently, a **chorioallantoic placenta** forms by fusion of the chorion with the allantois, creating a chorioallantoic membrane that makes up the exchange surface with the endometrium (**Figure 23.14D**). A chorioallantoic placenta is typical for placentals and a transitory chorioallantoic placenta develops near the end of gestation in a few marsupials, primarily bandicoots and bilbies (Peramelemorphia).

The chorioallantoic placenta is more invasive (i.e., burrows more deeply into the uterine wall) in some clades of placentals than others, but there is never a direct connection between the maternal and fetal blood vessels. For example, ungulates such as pigs and horses have an epitheliochorial placenta, in which the epithelium of the mother's endometrium contacts the fetal chorion (note that when naming types of placentas, the maternal tissue always comes first). With an epitheliochorial placenta there is minimal penetration of the mother's endometrium and the uterine lining is not shed at birth.

At the other extreme, anthropoid primates and rodents have a highly invasive hemochorial placenta, in which maternal blood directly contacts the fetal chorion. With hemochorial placentas, the developing embryo becomes partly or completely embedded in the mother's endometrium and the uterine lining sheds at birth. This type of placenta probably never could have evolved if mammals did not have the blood clotting ability bestowed by platelets, ensuring that the mother did not fatally hemorrhage when giving birth.

Gestation

A good way to think about reproductive differences between marsupials and placentals is to remember that marsupials have short gestations and long periods of lactation, whereas placentals have long gestations and short periods of lactation (usually shorter than gestation). In comparison to species with small adult body sizes, larger placentals tend to have longer periods of gestation and larger marsupials have longer periods of lactation.

In marsupials, gestation never exceeds a single estrus cycle (i.e., the cycle of development, maturation, and release of an egg from the ovary; the estrus cycle of humans is known as the menstrual cycle) and is usually considerably shorter. Marsupials are born at an early developmental stage and fasten onto a nipple to complete their development by suckling, receiving nutrition from milk rather than via the placenta. In contrast, hormonal feedback from the placenta suppresses the estrus cycle of placental mammals. As a result, gestation continues for longer (sometimes much longer) than a single estrus cycle, and young are born at a more mature stage than are marsupial young.

The gestation period of kangaroos is relatively long for a marsupial, and production of nutritive uterine secretions is prolonged. As in most marsupials, a kangaroo mother licks a path from the urogenital sinus to the pouch but does not otherwise aid the young in its journey to the nipple. A kangaroo's pouch opens anteriorly, but the pouches of most marsupials open posteriorly, with a shorter distance that the young must travel to reach the nipples within the pouch. (Not all marsupials have pouches; many small South American marsupials and the Australian dasyuromorphs lack them.)

Newborn marsupials have well developed forelimbs and a well developed mouth; the posterior part of their body is less developed than the anterior part. Guided by smell and using their front claws as holdfasts, marsupial neonates crawl from their mother's urogenital sinus to the nipples on her belly, a distance many times their own body length. Upon reaching a nipple they latch on tightly with well developed jaws. The structure of a neonate's pectoral girdle reflects this early journey: it retains the interclavicle and complete coracoid bones typical of monotremes and other amniotes (**Figure 23.15A**), which enable the crawl to the nipple. The metacoracoid later becomes reduced to the coracoid process in adults (**Figure 23.15B**).

When food is plentiful, a female kangaroo can simultaneously have three young at different stages of development: a juvenile out of the pouch (sometimes described as "at foot"), a neonate in the pouch, and an embryo in the mother's reproductive tract (**Figure 23.16**). Female kangaroos have two nipples and can provide milk with different qualities to a neonate in the pouch and to a juvenile that has left the pouch. Neonates receive milk with high protein content and low fat content, whereas juveniles receive low-protein, high-fat milk. The composition of milk produced by the mammary glands is probably determined by how long the young spend suckling per day.

The presence of a suckling neonate in the pouch stimulates the mother's pituitary gland to release prolactin, which in turn inhibits production of progesterone by the corpus luteum, holding the early-stage embryo in **embryonic diapause**, a period of arrested development prior to implantation in the uterus. Diapause breaks and the embryo starts to develop when the neonate reduces its rate of suckling as it begins to leave the pouch. Embryonic diapause enables the mother to space successive births and to separate the timing of mating and fertilization from the start of gestation. Embryonic diapause is

(A) Ventral view of the pectoral girdle of a newborn marsupial

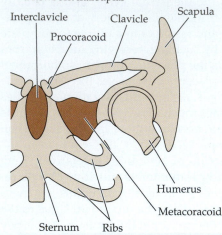

Interclavicle
Clavicle
Scapula
Procoracoid
Humerus
Metacoracoid
Sternum
Ribs

(B) Ventral view of the pectoral girdle of an adult marsupial

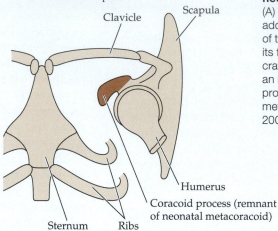

Clavicle
Scapula
Humerus
Coracoid process (remnant of neonatal metacoracoid)
Sternum
Ribs

Figure 23.15 Shoulder girdle of neonatal and adult marsupials. (A) Support from the metacoracoid adds rigidity to the shoulder girdle of the neonate, which helps it use its front claws as holdfasts while it crawls to the pouch. (B) Adults lack an interclavicle and the coracoid process is all that remains of the metacoracoid. (After K.E. Sears. 2004. *Evolution* 58: 2353–2370.)

well developed in kangaroos and often perceived as a derived feature of marsupials, but the phenomenon also occurs in many placentals such as bears.

Evolution of therian viviparity

The first Mesozoic therians were small, most weighing <1 kg. Their viviparous reproduction probably resembled that of small marsupials today—a short gestation period with yolk-sac placenta formation only in the last few days before birth. Because a developing mammal has a genetic makeup different from that of the mother, it causes inflammation during implantation; in marsupials this eventually results in the rejection of the developing young and premature birth. Placentals suppress such inflammation, however, so that gestation can continue for longer than a single estrus cycle. This happens because some connective tissue cells of the maternal endometrium form cells (decidual stromal cells) that dampen the immune response and stabilize implantation. The evolution of this system

for suppressing inflammation allowed longer gestation times to evolve.

The physiology of pregnancy maintenance differs among placentals. Larger placentals tend to have longer gestation periods, and we speculate that longer gestation evolved independently in the four major placental lineages (Afrotheria, Xenarthra, Euarchontoglires, and Laurasiatheria) in response to independent evolutionary increases in body size within each lineage

The marsupial mode of reproduction—short gestation and long lactation—was once thought to be inferior to that of placentals. There is little evidence to support this interpretation, but the marsupial reproductive pattern may have placed some limitations on ecomorphological diversity of marsupials.

• The specialized forelimbs and pectoral girdle of neonates (see Figure 23.15A) needed for the journey from the mother's urogenital sinus to the nipple

A neonate in the pouch (not visible) is attached to a second nipple, from which it receives a constant supply of high-protein, low-fat milk.

A juvenile that has left the pouch and returns to suckle receives high-fat, low-protein milk.

The presence of a suckling neonate stimulates the mother's pituitary gland to release prolactin, initiating a hormonal cascade that maintains an early embryo in diapause.

Toby Hudson/CC BY-SA 3.0

Figure 23.16 A female kangaroo can have offspring in three different stages of development. The mother can nurse both a juvenile that has left the pouch and a neonate in the pouch while maintaining a fertilized embryo in diapause in the reproductive tract.

may have limited possibilities for shoulder evolution and locomotion in marsupials. For example, no marsupials evolved powered flight, which involves substantial modifications of the shoulder region and hand in bats. Also, there are no hoofed marsupials. This might be because if marsupial forefeet were reduced to nongrasping limbs, newborns could not climb to the pouch. This constraint may explain why in large kangaroos—the marsupial equivalents of deer and antelope—only the hindlimbs are specialized for locomotion.

- Marsupial skulls are less variable than those of placentals (even discounting extreme placental types such as cetaceans). This may relate to early ossification of the bones around the mouth, which allows neonates to latch on to the nipple but constrains developmental trajectories of the rest of the skull.

- There are no fully aquatic marsupials. This might be because a marsupial born underwater might be swept away by water currents during its climb to the pouch, and the pouch would not contain air for it to breathe. The only semiaquatic marsupial is the water opossum (*Chironectes minimus*). The young are born on land and a sphincter muscle seals the pouch during short underwater forays.

23.5 Teeth and Feeding Specializations

LEARNING OBJECTIVES

23.5.1 Describe the different types of postcanine teeth of therians.

23.5.2 Explain how different teeth relate to different diets.

23.5.3 Describe how carnivorous and herbivorous therians differ in their teeth, jaw morphology, and jaw muscles.

Therians need to eat and process large quantities of food to fuel their high metabolic rates and starting to process food in the mouth rather than waiting until it reaches the stomach speeds digestion. Chewing reduces food to small particles, increasing the surface area available for digestive enzymes to act. In this section we consider therian teeth and dentitions and examine some functional anatomical implications of different types of jaws and feeding systems.

As shown in Figure 23.3, therians have four types of teeth: incisors, canines, premolars, and molars. Incisors are used for cutting and clipping food and canines for puncturing, holding, and tearing prey. Premolars and molars, sometimes collectively called postcanine or cheek teeth, reduce the food to a bolus for swallowing. Variations in the shapes, numbers, and arrangements of teeth in therians reflect differences in feeding and diet. For example, most herbivores lack canines, which carnivores use to hold and tear prey, although some male herbivores have large canines for display (see Figure 23.11D). Upper canines are generally larger in male primates (including humans) than

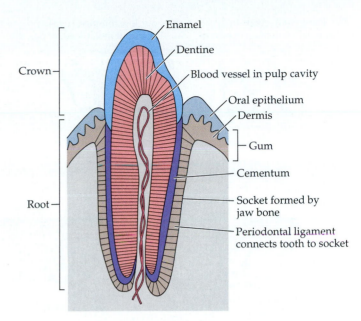

Figure 23.17 Tissues of a mammalian tooth. Enamel on the surface of a tooth is much harder than the mineralized dentine beneath it. The pulp cavity contains dentine-forming cells, blood vessels, and nerves. Cementum forms the outer surface of the tooth below the gum line. The periodontal ligament anchors the tooth in its socket and provides sensory feedback (such as when teeth come into contact with each other). (After K. Liem et al. 2001. *Functional Anatomy of the Vertebrates*, 3rd ed. Cengage/Harcourt College: Belmont, CA.)

in females. The tusks of pigs and walruses are modified canines, whereas elephant tusks are modified incisors.

Cusps and lophs

A generalized therian tooth consists of a crown and root composed of three mineralized tissues, **enamel**, **dentine**, and **cementum**. The enamel-covered crown extends above the gum; the root extends deeply into a socket in the jawbone (**Figure 23.17**). During development, enamel forms by mineralization of an extracellular matrix; the cells that make that matrix are lost once enamel deposition for a tooth is completed, which means that enamel cannot regrow if it is damaged. Enamel is the hardest mineralized tissue, but much of the tissue in a typical tooth is dentine, which is softer than enamel (see Table 2.4). The pulp cavity contains the cells that form dentine, as well as blood vessels and nerves. The periodontal ligament attaches a tooth in its socket in the jaw and allows teeth to move as a mammal grows (which is why braces can bring teeth into new alignments). Rich innervation of the periodontal ligament provides sensory feedback about the teeth and the mechanical forces experienced while chewing.

Cusps and lophs are specialized for different diets

The crowns of the molar teeth of carnivores and insectivores usually have three sharp and pointed cusps (**Figure 23.18A**) as in the original tribosphenic molar typical of therians (see Figure 22.14D). In omnivores and frugivores, the three cusps form rounded, flattened surfaces suitable for crushing and pulping, and addition of a fourth cusp

(A) Three sharp cusps in upper molar teeth of an opossum

(B) Four low rounded cusps in bunodont molar teeth of peccary

(C) Selenodont molar teeth of a deer

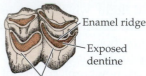

Enamel ridge

Exposed dentine

Crescent-shaped lophs in anterior-posterior directions

(D) Lophodont molar of a rhinoceros

Straight loph along outer edge

Enamel ridge

Exposed dentine

Straight loph in lateral to medial direction

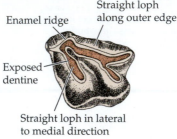

(E) Lamellar molars of a rodent

Exposed dentine

Enamel ridges

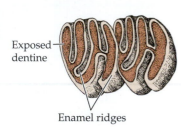

(F) Sections through unworn and worn hypsodont teeth of a horse

Cementum covers a newly erupted tooth

Exposed dentine

Enamel ridges

Worn

Unworn

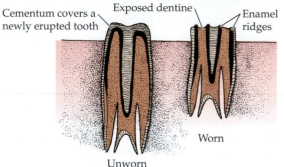

(G) Ever-growing hypselodont teeth of rodents

Self sharpening, chisellike tip

Molars 1 2 3

Premolar 4

Dentine

Diastema

Ever-growing incisors

Enamel is only present on the anterior surfaces of incisors

Diastema

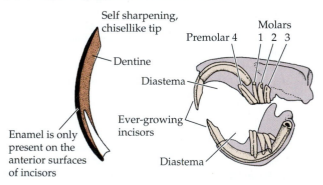

(H) Occlusal views of the carnassial teeth of a carnivore, the coyote *Canis latrans* (anterior to left)

Upper jaw

Lower jaw

Fourth premolar (upper carnassial)

First molar

First molar (lower carnassial)

Second molar

Third molar

Blades of upper and lower carnassial teeth occlude to slice through meat.

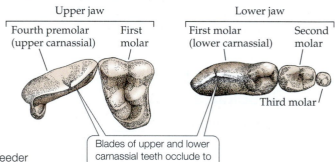

Figure 23.18 Mammalian teeth. (A) Upper molars of a generalist feeder (also see Figure 22.14D). (B) Bunodont molars of an omnivore. (C–F) Lophed (ridged) molars of herbivores. (G) Ever-growing incisors and cheek teeth of a rodent. (H) Occlusal (meshing) view of carnassial teeth of a carnivore, the coyote.

to the upper molars, together with increases in size of the talonid basin in the lower molars, make the teeth appear square rather than triangular. These **bunodont teeth** (Greek *bounos*, "hill") evolved independently in primates and other omnivores such as raccoons and suids (pigs and peccaries; **Figure 23.18B**).

In herbivores that eat more fibrous foods, the simple cusps of bunodont teeth run together into ridges, or **lophs**. Lophed teeth work best once the enamel wears off the top of the loph to expose the underlying dentine. Each loph then consists of a pair of sharp enamel ridges lying on either side of the softer, intervening dentine. When these teeth occlude as the jaws move laterally, they grate the food between multiple sets of flat, shearing blades. Lophed teeth evolved convergently many times among therians, and take different forms: selenodont, lophodont, or lamellar (**Figure 23.18C–E**).

Herbivorous therians face problems of tooth durability because a single set of adult teeth must last a lifetime. Some herbivores eat highly abrasive plant fibers

and leaves that can include silica crystals (phytoliths); they also may ingest grit or soil when feeding close to the ground. To combat this, the cheek teeth of many grazing therians are high-crowned **hypsodont teeth** (Greek *hypsos*, "high"; **Figure 23.18F**). Species with hypsodont teeth have very deep dentary and maxillary bones, and tooth crowns extend deeply into the bones. Cementum covers the entire tooth rather than just the root and base of the crown as in most therians, and cementum fills in the spaces between lophs. As the chewing surface wears away, hypsodont teeth continue to erupt, providing a continuously renewed occlusal surface much as you push out the lead in a mechanical pencil as it wears away. Ultimately, the entire tooth wears away and the animal can no longer eat (most therians in the wild die of natural causes long before their teeth wear out, but domestic horses surviving into their late 20s and 30s often require soft food because they have virtually no cheek teeth left).

Some small therians, such as wombats, rabbits, and rodents, have ever-growing (hypseledont) incisors and molars

(**Figure 23.18G**) which, unlike hypsodont teeth, never completely wear away. Carnivorans use incisors to seize food, canines to puncture and hold prey, and the teeth of the carnassial apparatus (**Figure 23.18H**) to slice through flesh.

Adaptations for herbivory   Some herbivores have lost their incisors. Ruminant artiodactyls have only lower incisors, and African rhinos lack both upper and lower incisors. Herbivores often use all cheek teeth for mastication, and the premolars are molarized (i.e., they have the same shape as molar teeth). Herbivores usually have long snouts, so there is a gap between the incisors (plus the canines, in those species that retain them) and cheek teeth called the **diastema**.

The diastema may allow extra space for the tongue to manipulate food, or it may merely reflect snout elongation that allows the animal to reach into narrow spaces to eat leaves or fruit. Most rodents are herbivorous, but lack especially long snouts, probably because they can manipulate food with their front paws, which ungulates never do; nevertheless, the diastema is present (see Figure 23.18G).

Carnivores and herbivores: Differences in jaw muscles

Therians use a combination of temporalis, masseter, and pterygoideus muscles to close the jaws and the digastric muscle to open the jaws. The relative sizes of the muscles and shapes of the skulls reflect the different demands of processing flesh versus vegetation. The temporalis has its greatest mechanical advantage at the front of the jaw, where the incisors and canines cut into prey. Thus, a large temporalis, reflected in the large temporal fossa in the skull and the tall coronoid process of the jaw, is typical of carnivores, which use a forceful bite with their canines to kill and subdue prey. The occiput (posterior part of the skull) is large in carnivores, and strong muscles used to help restrain struggling prey connect it to the cervical vertebrae (**Figure 23.19A**).

Modifications of the skulls of herbivorous therians allow them to grind tough, resistant food. The protein content of leaves and stems is low, and tough cell walls formed by cellulose (a complex carbohydrate) enclose plant cells. Because of the relatively low caloric value of plant material, herbivores must process large amounts of food per day. To do this, the masseter exerts a crushing force at the back of the tooth row; it also moves the jaw from side to side, helping the lophed molars to rupture cell walls. Herbivorous therians thus have a larger masseter

(A) Carnivore (dog)

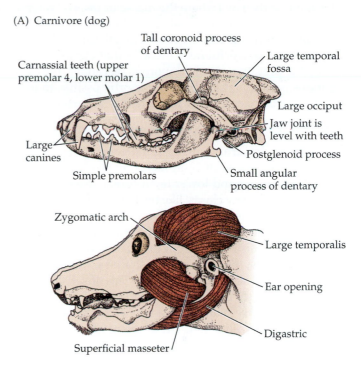

(B) Herbivore (deer)

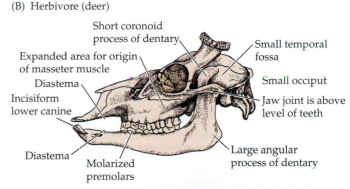

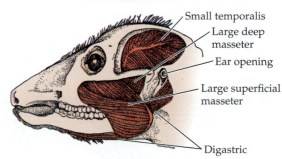

(C) Rodent (beaver)

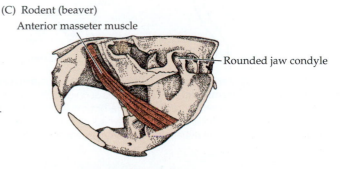

Figure 23.19 Cranial anatomy of carnivorous and herbivorous mammals.   (A) In carnivores, the powerful temporalis muscle and large coronoid process power shearing actions of the carnassial teeth; these teeth also help hold and process struggling prey, using the incisors and canines. (B) Herbivores chew with a sideways grinding movement, and the masseter muscle is more powerful than the temporalis. (C) A rodent skull shows the origin of a portion of the masseter far anterior on the skull. The masseter allows the upper and lower incisors to occlude during gnawing, and also pulls the lower jaw forward (rather than sideways) during mastication.

but a smaller temporalis than carnivores (**Figure 23.19B**). The angle of the jaw (where the masseter inserts) is large, but the coronoid process is short and the temporal fossa is small. Herbivores usually have a small occiput because they lack the powerful neck muscles that carnivores use to subdue prey. In many herbivorous placentals, the cartilaginous partition at the back of the orbit ossifies to form a bony bar that is probably important in absorbing stress from the jaws and protecting the braincase.

Jaw joints In carnivores, as in the general mammalian condition, the jaw joint is on the same level as the tooth row, so the upper and lower teeth contact each other sequentially as the jaw closes, like the blades of a pair of scissors—a morphology well suited for cutting and shearing. The postglenoid process behind the jaw joint prevents the strong temporalis muscle from dislocating the lower jaw.

In herbivores, the jaw joint has shifted so that it is high on the skull and above the tooth row. This offset allows the teeth in the upper and lower jaws to occlude simultaneously, like an offset handle on a kitchen knife that allows the blade to be applied flat. The jaw joint allows sideways motion of the lower jaws to allow left and right grinding action powered by the masseter muscles to shred plant material between the lophs of the upper and lower teeth.

Rodents: Specialized feeders Many rodents have a highly specialized type of food processing. Teeth in their upper and lower jaws are the same distance apart, unlike the condition in most mammals (in which the teeth in the lower jaw are closer to each other than are the teeth in the upper jaw; see Figure 22.13). A rounded jaw condyle (the ball part of the ball and socket joint) allows forward and backward jaw movements, and in derived rodents a new portion of the masseter muscle, the anterior masseter, originates far forward on the skull and pulls the lower jaw

forward into occlusion (**Figure 23.19C**). This arrangement allows chewing on both sides of the jaw at once, a highly efficient mode of food processing.

Digestive tracts

Once food has been mechanically broken down by chewing, it passes down the muscular esophagus to enter the stomach (**Figure 23.20A**). The stomach is the primary food storage organ of mammals, and it typically has elastic, muscular walls. Protein breakdown in the stomach occurs in an acidic environment, which is necessary for the function of the stomach enzyme pepsin. The pylorus, or last section of the stomach, has a valve that allows food to pass into the small intestine.

The small intestine is the longest region of the gastrointestinal tract, and most nutrient absorption takes place there; enzymatic digestion occurs in its relatively basic (alkaline) environment. The first portion, the duodenum, receives secretions via the pancreatic duct, which delivers the enzyme trypsin made by the pancreas; and via the bile duct, which delivers bile from the liver. Trypsin hydrolyzes (breaks the bonds of) protein molecules, while bile emulsifies (breaks into fragments and disperses) fats. The small intestine continues as the jejunum and ileum. The ileum may have a blind sac or cecum at its junction with the large intestine; the cecum absorbs water and electrolytes. The large intestine, or colon, also absorbs water and electrolytes; feces form before the tract ends at the rectum. In some mammals, the cecum and large intestine are packed with microorganisms that ferment cellulose, and nutrients are absorbed in both locations.

The sizes and shapes of the stomach, length of the small intestine and colon, and size and shape of the cecum vary markedly across species. One generality is that species feeding at lower trophic levels (e.g., herbivores) have longer intestines and larger cecae than those that feed

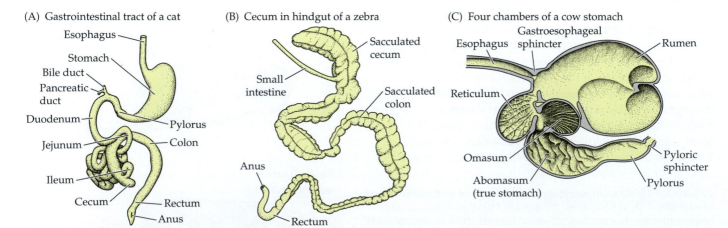

Figure 23.20 Differences in the digestive tracts of carnivorous and herbivorous mammals.   (A) Carnivores such as cats have short and relatively simple digestive tracts. (B,C) Herbivores such as zebras or cows have specializations of the digestive tract in which symbiotic bacteria break down plant materials. (B) Zebras are hindgut fermenters, with bacteria in sacculated (sac-filled) regions of the cecum and colon. (C) Cows have multichambered foreguts (stomachs). Bacterial fermentation takes place in the rumen. (After K. Liem et al. 2001. *Functional Anatomy of the Vertebrates*, 3rd ed. Cengage/Harcourt College: Belmont, CA.)

at higher trophic levels (e.g., carnivores). Mammals that primarily eat meat, such as cats and wolves, have relatively simple gastrointestinal tracts (see Figure 23.20A). Their stomachs are little more than muscular sacs, and the small intestine is relatively short. The cecum is either small or missing. This simplicity is possible because the high nutritional value and relative digestibility of meat makes long, complex intestines unnecessary for extracting adequate nutrition. In contrast, many herbivores have long, convoluted guts because their food is relatively nutrient-poor and difficult to break down.

Fermentation by herbivores Plant cell walls contain complex carbohydrates such as cellulose. No enzymes of the mammalian digestive system can directly break down cellulose, but symbiotic relationships with microorganisms have evolved in many leaf-eating (folivorous) and grass-eating (graminivorous) herbivores. The symbionts live in the animal's gut and release nutrients from cellulose by fermentation, transforming them into absorbable volatile fatty acids. We distinguish hindgut and foregut fermenters.

Hindgut fermenters such as horses, asses, and zebras (**Figure 23.20B**) pass vegetation through the stomach and small intestine into an enormous, sacculated (made up of saclike segments) cecum and colon that are environments for bacterial symbionts. (They are hindgut fermenters because these portions of the digestive tract are after the stomach). Similar specializations of the cecum and colon occur in rhinos and elephants. Rabbits and some rodents just have a greatly enlarged cecum. In larger hindgut fermenters, the products of fermentation are absorbed through the colon wall, but many species of rabbits and rodents are coprophagous (feces-eating). Two types of feces are produced—hard feces with low energy content, and soft feces with high energy content. The soft feces are eaten, allowing the animal to harvest the energy remaining in fermented products that was not absorbed the first time around.

Foregut fermenters such as cows have a more efficient system for cellulose digestion in which plant material is fermented in the stomach before reaching absorptive surfaces of the small intestine. This stomach is multichambered, with its various chambers serving different functions (**Figure 23.20C**). Such animals ruminate, meaning that they regurgitate and chew partially digested material, called the cud. Among artiodactyls, rumination and a multichambered stomach occur both in the ruminants (cattle, antelope, deer, giraffes, and others) and in camelids (camels and llamas). However, stomach anatomy is somewhat different (four chambers in ruminants and three chambers in camelids), and the two multichambered systems probably evolved independently from a simpler one. In cows, coarsely chewed vegetation passes down the esophagus and is shunted into the large rumen, where microbial fermentation begins. Fermented material then passes to the reticulum, where it is packed into cud and sent back up the esophagus for rumination by the teeth. Food that has been reduced to small particles continues into the third chamber, the omasum, which is lined with long folds. Water is reabsorbed in the omasum before the food finally enters the abomasum, sometimes called the true stomach because this is where pepsin digestion of proteins takes place in an acidic environment.

23.6 Locomotion

LEARNING OBJECTIVES

23.6.1 Describe differences in limb anatomy and function between cursorial and fossorial mammals.

23.6.2 Outline anatomical adaptations for swimming in semiaquatic and fully aquatic mammals.

23.6.3 Compare powered flight in bats and birds.

23.6.4 Explain how anatomical and genomic evidence supports the hypothesis that whales are artiodactyls.

Large mammals experience the world differently than smaller ones because of physical size and scaling effects, and often have specialized forms of locomotion, such as galloping (see Sections 11.1 and 11.2). In this section, we explore how therians run, dig, fly, and swim.

Limbs: Speed versus power

Many principles of mammalian locomotion can be understood by comparing the limbs of a cursorial (running) mammal, such as a horse, with those of a mammal that digs, such as an armadillo. In essence, this is a comparison between speed and power and the tradeoffs between them—tradeoffs that are a direct result of differences in the lengths of levers.

Limbs as a whole function as levers when moving an animal, but there are also lever systems within limbs related to the amount of speed or power that a similar muscle mass can produce. To understand how levers relate to therian locomotion, compare the lengths of the in-levers and out-levers of the cursorial horse with those of the fossorial (burrowing) armadillo (**Figure 23.21A,B**). The in-lever for a powerful limb retractor muscle (the latissimus dorsi) is the same length for both animals, but the out-lever for the horse is longer than the out-lever for the armadillo. This means that the limb of the horse will move faster for the same amount of muscular work than will the limb of the armadillo, which is good for speedy running. Conversely, the higher mechanical advantage of the armadillo latissimus dorsi means that the same amount of muscular work generates more power for digging.

Skeletal structure of limbs With these concepts in mind, we can interpret the structure of the limb skeleton and its muscles as they relate to tradeoffs between speed and power. For example, evolutionary modifications that lengthen the legs can contribute to improvements in speed; thus, the horse's front leg is disproportionately longer than the armadillo's (**Figure 23.21C,D**). Most of the elongation occurs in distal (farthest from the torso) portions of the limb—the radius and ulna in the forelimb, the

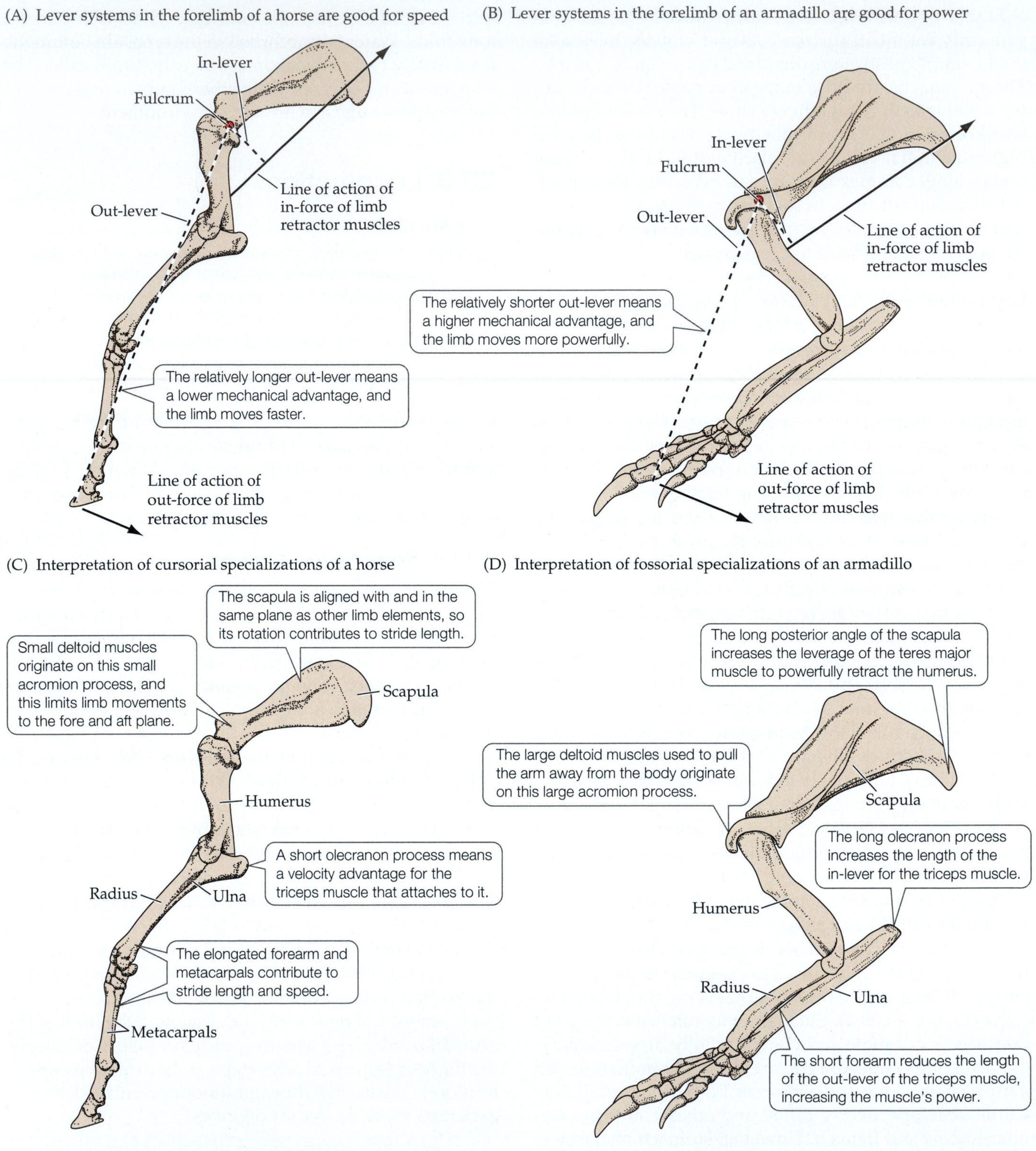

(A) Lever systems in the forelimb of a horse are good for speed

In-lever

Fulcrum

Out-lever

Line of action of
in-force of limb
retractor muscles

The relatively longer out-lever means
a lower mechanical advantage, and
the limb moves faster.

Line of action of
out-force of limb
retractor muscles

(B) Lever systems in the forelimb of an armadillo are good for power

In-lever

Fulcrum

Out-lever

Line of action of
in-force of limb
retractor muscles

The relatively shorter out-lever means
a higher mechanical advantage, and
the limb moves more powerfully.

Line of action of
out-force of limb
retractor muscles

(C) Interpretation of cursorial specializations of a horse

The scapula is aligned with and in the
same plane as other limb elements, so
its rotation contributes to stride length.

Small deltoid muscles
originate on this small
acromion process, and
this limits limb movements
to the fore and aft plane.

Scapula

Humerus

A short olecranon process means
a velocity advantage for the
triceps muscle that attaches to it.

Radius

Ulna

The elongated forearm and
metacarpals contribute to
stride length and speed.

Metacarpals

(D) Interpretation of fossorial specializations of an armadillo

The long posterior angle of the scapula
increases the leverage of the teres major
muscle to powerfully retract the humerus.

The large deltoid muscles used to pull
the arm away from the body originate
on this large acromion process.

Scapula

The long olecranon process
increases the length of the
in-lever for the triceps muscle.

Humerus

Radius

Ulna

The short forearm reduces the length
of the out-lever of the triceps muscle,
increasing the muscle's power.

Figure 23.21 Limbs for speed and power.   Speed and power represent opposite extremes in lever optimizations. (A) Relatively longer out-levers, as in horses, are good for speed. (B) Relatively shorter out-levers, as in the armadillo, are good for power. Understanding how lever systems operate makes it possible to interpret skeletomuscular differences in the limb skeletons of cursorial specialists where speed is key (C) versus digging specialists where power is key (D). (See credit details at the end of this chapter.)

tibia and fibula in the hindlimb, and the metapodials (a collective term used to describe the metacarpal and metatarsal bones in the fore- and hindlimbs). Evolutionary modifications that increase the length of in-lever for key muscles such as the triceps, which extends the forearm, can increase power. For example, note the much longer olecranon process of the ulna in the armadillo compared to that of the horse.

We can interpret other skeletal features, such as processes for the origins of the muscles, for insight into the

Figure 23.22 Foot posture of terrestrial therians. (A) Plantigrade: the entire foot (phalanges to heel) contacts the ground. Humans are plantigrade. (B) Digitigrade: the phalanges contact the ground, but the metatarsals and heel are held off the ground. Dogs and cats are digitigrade. (C) Unguligrade: only the distal ends of the phalanges contact with the ground. (D) Unguligrade mammals like horses and cows protect the tips of their toes with hooves. (A–C after T.A. Vaughan. 1978. *Mammology*, 2nd ed. Saunders College Publishing: Philadelphia.)

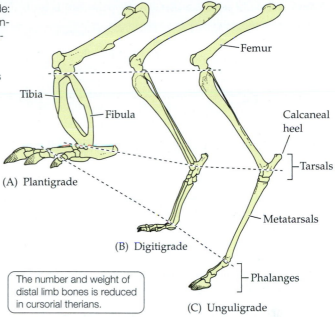

(A) Plantigrade

(B) Digitigrade

The number and weight of distal limb bones is reduced in cursorial therians.

(C) Unguligrade

(D) External anatomy (left) and longitudinal section through hoof of a horse, *Equus*

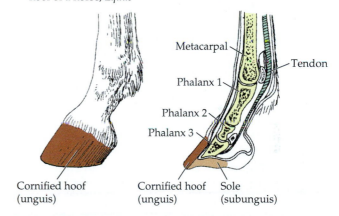

speed–power tradeoff. For example, the large scapula of the armadillo has prominent processes for the origin of the teres major (a muscle that pulls the humerus toward the body) and the deltoid muscles (which pull the humerus up and away from the body; see Figure 23.21D). Neither of these processes is well developed in the horse (see Figure 23.21C).

Distal portions of limbs   These functional anatomical interpretations also help explain different foot postures. Fossorial mammals have a plantigrade foot posture in which the foot and heel contact the ground (**Figure 23.22A**). A plantigrade foot posture is the basal mammalian condition and occurs in mammals that have not become specialized for faster locomotion (your foot posture is plantigrade). The foot of a cursorial mammal has less contact with the ground and the bony elements of the metapodials become elongated to disproportionately contribute to overall limb elongation. Cursorial carnivores have a digitigrade posture in which the phalanges are in contact with the ground, but the metapodials are not (**Figure 23.22B**). The most derived cursors, all of them ungulates, have an unguligrade posture in which only the tips of the digits contact the ground (**Figures 23.22C**). Ungulates protect the tips of the toes with hooves (**Figure 23.22D**). The heavily cornified part, or unguis, protects the sides and very tip of the toe. Behind it is a softer sole, or subunguis.

Cursorial adaptations of ungulate limbs

Another biomechanical theme for cursorial mammals is lightening of the distal portions of the limbs by: (1) reducing or losing digits and (2) by positioning the major locomotor muscles proximally with connections to distal limb elements made by long tendons. Both evolved independently within the two clades of ungulates (odd-toed perissodactyls and even-toed artiodactyls; see Table 23.2). For example, rhinoceroses are heavily built perissodactyls that retain toes II, III and IV, each with a hoof. In contrast, horses are speedy perissodactyls that retain only toe III and have a single hoof (**Figure 23.23A**). Within artiodactyls, forms such as pigs retain toes II–V and have four hooves; more cursorial forms, including cattle, walk on the tips of toes III and IV and have two hooves (**Figure 23.23B**).

There is also a major difference in the ankles of perissodactyls and artiodactyls. The astragalus of perissodactyls has a single trochlea (a pulley-shaped articular surface) for its joint with the lower leg bone (the tibia). In contrast, the astragalus of artiodactyls has two trochleae, one for its joint with the tibia, and one for a new joint with the more

distal bones of the ankle. This double-pulley astragalus allows more movement in the lower leg and may have played a role in the evolution of the fastest artiodactyls, such as pronghorns.

Most of the leg muscle mass of cursors is proximal (near the body), and long tendons transmit forces to the lower limb. The tendons stretch with each stride, storing and then releasing elastic energy, which contributes to locomotor efficiency. Leg tendons of a hopping kangaroo, especially the Achilles tendon that attaches the calf muscle to the calcaneal heel, exemplify elastic storage because the animal bounces on landing as if using a pogo stick. However, all cursorial animals rely on energy storage in tendons for gaits faster than a walk. Humans also rely on elastic energy storage in the Achilles tendon: people with a damaged Achilles tendon—a common sports injury—find running difficult or impossible. Lengthening tendons to increase the amount of stretch and recoil may help explain the evolution of longer limbs and changes in foot posture.

Cursorial specialists restrict motion of the limbs to the fore-and-aft plane so that most of the thrust on the

(A) Ankle and toes of perissodactyls: rhino, *Rhinoceros* (left), horse, *Equus* (right)   (B) Ankle and toes of artiodactyls: pig, *Sus* (left), cow, *Bos* (right)

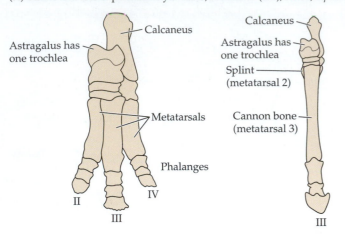

Figure 23.23 Reduction of distal skeletal elements in the limbs of ungulates.   (A) The hindfoot of a non-cursorial perissodactyl (an odd-toed ungulate) such as a rhinoceros has three toes; cursorial perissodactyls such as horses have one toe. (B) The hindfoot of a non-cursorial artiodactyl (an even-toed ungulate) such as a pig has four toes; cursorial artiodactyls such as cows have two toes. Left feet are shown. (After K. Liem et al. 2001. *Functional Anatomy of the Vertebrates*, 3rd ed. Cengage/Harcourt College: Belmont, CA.)

ground contributes to forward movement. To do this, they have reduced or lost the clavicle, and the ribcage becomes more slab-sided, so that the scapula becomes aligned with the axis of limb (see Figure 23.21C). The scapula now swings extensively with the motion of the forelimb, contributing to stride length. The wrist and ankle bones allow motion only in a forward-and-backward plane, and the forelimb cannot be supinated. (Supination is what you do when you turn your hand so that your palm faces upward—a movement that comes from the elbow rather than the wrist, as shown in Figure 23.1C. (A dog can't turn its forepaw very much, and a horse has almost no ability to rotate its hoof.)

Cursorial specializations evolved convergently among many lineages. Early Cenozoic ungulates and carnivorans were not highly cursorial, and at one time we assumed that longer legs evolved in the context of an evolutionary "arms race" between predators and prey: longer legs would give a carnivore more speed to pursue ungulates, resulting in selection for those ungulates with longer legs that could make faster escapes. But the fossil record shows that longer legs evolved in herbivores millions of years before long-legged carnivores appeared. Limb modifications that make a mammal a faster runner also make it more efficient at slower gaits, such as a walk or trot, and the initial evolutionary advantage of longer legs may be that they were more efficient for walking in search of forage across the open plains and savannas of the mid Cenozoic (see Chapter 20).

Digging

Fossorial mammals do several types of digging. Scratch digging at the ground's surface, as seen in armadillos (and in dogs), is most common. Species specialized for scratch digging have a stout pelvis with many sacral vertebrae

that brace their hindlimbs while they dig with forelimbs. Armadillos have extra articular surfaces on some of their vertebrae, a characteristic of xenarthrans that we noted in Section 23.3. These surfaces strengthen the axial skeleton. Fossorial forms that burrow beneath the surface have several types of anatomical specializations. Golden moles and marsupial moles scratch dig with fore- and hindfeet, true moles use a rotation thrust of their greatly expanded hands, and chisel-toothed rodents such as gophers dig with their incisor teeth.

Limbs specialized for digging are almost exactly the opposite of limbs specialized for running; limbs for digging are short and stout, maximizing power at the expense of speed (see Figure 23.21). Digging mammals achieve mechanical advantages in the forelimb by having a long olecranon process (allowing the triceps muscle to retract the hand) and a relatively short forearm. Digging mammals retain all five digits, tipped with stout claws that can break the substrate. They also have large projections on their limb bones that attach to large muscles, as well an enlarged acromion process on the scapula, where strong muscles that retract the limb originate.

Powered flight of bats

Bats are the only mammals capable of powered (= flapping) flight. The order name Chiroptera comes from the Greek words *cheir*, "hand" and *pteron*, "wing"; indeed, bat wings are composed of the phalanges of the hand, which support a web of elastic skin (patagium) between them (**Figure 23.24**). Key anatomical features related to powered flight include elongated forelimb bones (radius, metacarpals, phalanges), a unique calcar bone extending from the heel to support the flight membrane, and a shoulder-locking mechanism that prevents overextension of the wings during the upstroke.

(A) Fruit bat, *Rousettus aegyptiacus*

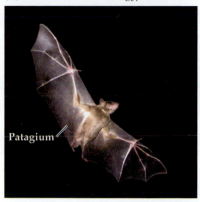

Patagium

(B) Bones of a bat's wing

Thumb   Carpals
Metacarpals   Radius   Humerus
Phalanges
Ulna

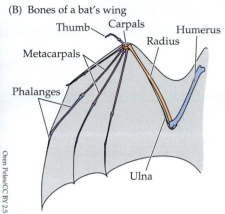

Oren Peles/CC BY 2.5

Figure 23.24 **Bat wings and flight.** (A) Bat wings consist of a web of skin called the patagium. (B) The patagium is supported by elongated forelimb bones (radius, metacarpals, and phalanges). (B after K. Liem et al. 2001. *Functional Anatomy of the Vertebrates*, 3rd ed. Cengage/Harcourt College: Belmont, CA.)

Flight adaptations of bats differ in several ways from those of †pterosaurs and birds. An elongated digit IV supports much of the wing membrane of †pterosaurs (see Figure 18.4B), but birds fuse and lose bones in the hand (see Figure 21.7A). In contrast, bats have elongated bones of the forearm, hand, and digits. The flight membranes of †pterosaurs and bats are skin; birds use feathers attached to the forearm and hand. Birds use chest muscles for both the downstroke (pectoralis muscle) and upstroke (supracoracoideus; see Figure 21.9). Bats, on the other hand, use chest muscles (pectoralis and others) only for the downstroke, and use back muscles (deltoideus and trapezius groups) for the upstroke. Due to extremely flexible joints in their forelimbs, bat flight is highly maneuverable, although generally slower than that of birds.

Swimming

Swimming therians include semiaquatic species that feed and sometimes migrate in water but return to land to give birth. Fully aquatic forms not only feed and migrate in water but also give birth underwater. Semiaquatic and fully aquatic mammals differ in swimming modes and degree of specializations for life in water.

Some semiaquatic mammals are not that different from terrestrial forms except for more paddlelike limbs and denser pelage, features that have evolved many times in groups such as the water opossum (*Chironectes*), water shrew (*Neomys fodiens*), and beavers (*Castor*). Carnivora includes many semiaquatic forms, such as otters (Lutrinae), the sea otter (*Enhydra lutris*), and the polar bear (*Ursus maritimus*). Also within Carnivora are the semiaquatic pinnipeds (seals, sea lions, and walrus), which have many more specializations for swimming and spend comparatively more time in the water. Pinnipeds have thick pelage and a thick layer of subcutaneous fat (blubber) for insulation.

Like humans, many semiaquatic mammals swim using their limbs. For example, beavers use their hindlimbs for propulsion and their tail as a rudder. Using limbs to swim can be inefficient because it creates a lot of drag, but some mammals that swim with their limbs are much

better than others. This is particularly true for pinnipeds (the name means "fin feet"). What may appear on casual inspection to be the tail of a pinniped are its hindlimbs, and the tail is quite short. Hindlimbs of earless seals (no ear pinnae; Phocidae), such as the harbor seal, are permanently set in a turned-backward position, where they power swimming, but this makes them clumsy on land (**Figure 23.25A,B**). Eared seals (ear pinnae present; Otariidae), such as sea lions, use synchronous movements of the forelimbs together with dorsoventral movements of the back and hindlegs to swim. Muscles that power swimming attach to robust neck, shoulder, and forelimb bones (**Figure 23.25C**). Sea lions can turn their hindlegs forward and are much more agile on land than phocids (**Figure 23.25D**). Because both phocids and otariids spend some time on land, they retain zygapophyses on their trunk vertebrae.

Fully aquatic mammals—most of which live in marine habitats—evolved twice, once in Sirenia (dugong and manatees) and once within Artiodactyla (porpoises, dolphins, and whales, the cetaceans). Sirenians are herbivorous; cetaceans are carnivorous. Both sirenians and cetaceans lack pelage and can have thick blubber. They swim using dorsoventral flexion of the vertebral column, which differs from the lateral flexion used by fishes, salamanders, and aquatic sauropsids. They have horizontal tail flukes, but unlike the vertical caudal fins of sharks and teleosts, there is no bony support within the flukes. Sirenians and extant cetaceans use their short forelimbs for braking and steering, not for propulsion. (An exception is the humpback whale, *Megaptera novaeangliae*, which uses its long pectoral fins—Greek *mega*, "giant"; *ptera*, "wing"—to power swimming.)

Cetacean evolution

Extant cetaceans have so many fascinating anatomical, physiological, and behavioral specializations for living in water that is worth exploring how these features evolved. The sequence of anatomical changes is documented by an outstanding fossil record starting in the Eocene of southern Asia, along the shores of the ancient Tethys Sea. It

(A) Skeleton of an earless seal, Phocidae

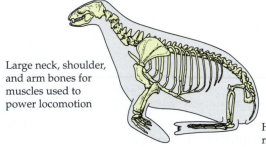

Large lumbar vertebrae for axial muscles used to power locomotion

Forelimbs used for steering

Hindlimbs cannot be rotated beneath the body

(B) Harbor seal, *Phoca vitulina*

Tim Sackton/CC BY-SA 2.0

(C) Skeleton of an eared seal, Otariidae

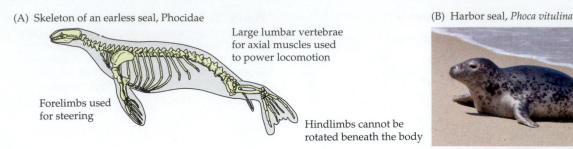

Large neck, shoulder, and arm bones for muscles used to power locomotion

Hindlimbs can be rotated beneath the body

(D) California sea lion, *Zalophus californianus*

Vasek Vinklár/CC BY 2.0

Figure 23.25 Comparative anatomy of swimming of pinnipeds. (A,B) Phocids are earless seals and have highly modified hindlimbs that serve as flippers and use powerful back muscles to power swimming. Graceful in the water, they struggle to move on land. (C,D) Sea lions and other eared seals have large forelimb flippers, use powerful shoulder muscles to power swimming, and are more agile on land. Walruses (not shown) swim with their hindlimbs like eared seals but can turn their feet forward. (See credit details at the end of this chapter.)

shows a progression from semiaquatic forms with four legs to fully aquatic forms with reduced and modified limbs (**Figure 23.26**).

Until the late 20th century, most paleontologists thought that cetaceans were related to carnivorans. In the 1980s, however, discovery of new fossils in Pakistan showed that cetaceans are artiodactyls. This was a surprising finding (partly because of long-held opinions, and also because artiodactyls are herbivorous whereas whales are carnivorous), and it met with some resistance when it was first proposed. The fossils speak clearly, however, and all current anatomical and molecular evidence supports this conclusion. The closest extant terrestrial relatives of cetaceans are hippopotamuses. This finding does not mean that cetaceans evolved from hippos, only that whales and hippos shared a common ancestor some 66 Ma (see Figure 23.26).

Cetacea includes many Eocene fossils that document very rapid evolution of the basic body form of whales. The earliest of the Eocene taxa, the 1.4-m-long †*Pakicetus*, shows several features intermediate between terrestrial artiodactyls and whales (**Figure 23.27**). It has a double-pulley astragalus like terrestrial artiodactyls, but features of its skull, ear region, and caudal vertebrae are whalelike. For example, †*Pakicetus* has dense bones, a long, whalelike rostrum, eyes near the dorsal side of the head (like a hippopotamus, allowing them to see their surroundings as they floated at the surface), robust caudal vertebrae, and a thickened auditory bulla around the ear region like a whale. Its heterodont teeth resemble those of extant piscivorous therians. We interpret †*Pakicetus*

as a nearshore, semiaquatic form that used its limbs for swimming, possibly like an extant otter.

Among the Eocene cetaceans, †*Ambulocetus* (Latin, "walking whale") is the first form known to have space for a fat pad in the lower jaw that extant cetaceans use to conduct aquatic sounds to the middle ear (see Figure 23.26). Others include †*Dalanistes*, which has a long narrow skull and small eyes, and forms that first used dorsoventral flexion for swimming in offshore marine habitats. The isotopic composition of the bones of these derived cetaceans shows that they did not drink fresh water, but obtained their water needs through their food, as do extant cetaceans. However, we know that forms such as †*Maiacetus* still gave birth on land because of a fossil preserved with a fetus about to be born head first, not tail first as in extant whales.

The evolution of other features of extant cetaceans can be traced through the Eocene record, including the evolution of a short neck (achieved by compression and fusion of cervical vertebrae), reduction and eventual loss of the hindlimbs, and changes in the middle and inner ear for underwater hearing, all showing the transition to fully aquatic life.

Neoceti Neoceti has two extant clades (see Figure 23.26): the small to very large toothed whales (Odontoceti: porpoises, dolphins, orcas, and the sperm whale *Physeter macrocephalus*); and the medium-size to giant baleen whales (Mysticeti, including the blue whale *Balaenoptera musculus*, by weight, the largest animal ever known). That

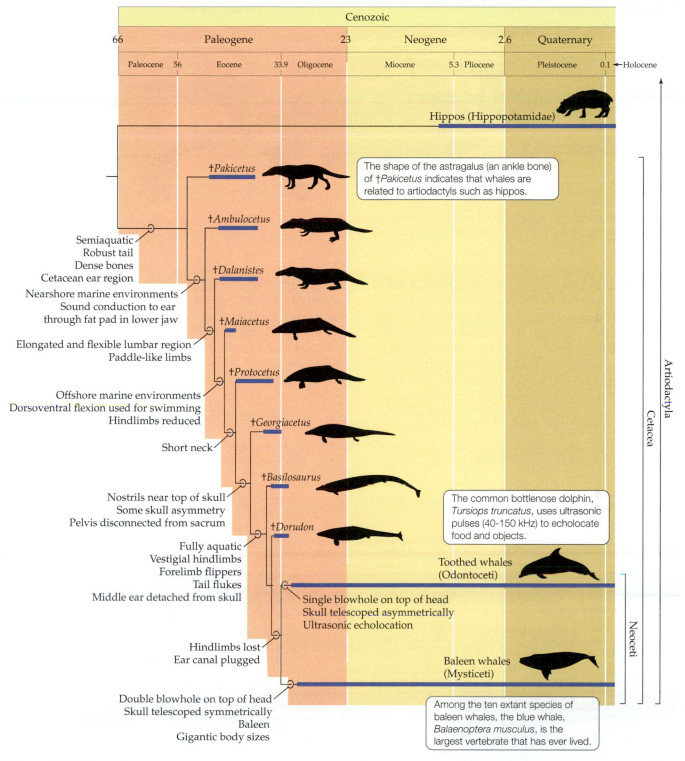

The shape of the astragalus (an ankle bone) of †*Pakicetus* indicates that whales are related to artiodactyls such as hippos.

The common bottlenose dolphin, *Tursiops truncatus*, uses ultrasonic pulses (40–150 kHz) to echolocate food and objects.

Among the ten extant species of baleen whales, the blue whale, *Balaenoptera musculus*, is the largest vertebrate that has ever lived.

Figure 23.26 Evolution of cetaceans. This tree is based on the excellent fossil record of cetacean evolution from semiaquatic limbed forms to fully aquatic forms with highly modified or vestigial limbs. Specimens excavated in Pakistan and other regions that were once the shores of the ancient Tethys Sea document the changes shown here. (See credit details at the end of this chapter.)

the two groups are different from each other is not surprising given that they have been evolving separately for more than 30 Ma.

- Toothed whales eat large prey items such as fishes, squids, and other marine mammals, and use a system of underwater echolocation, or sonar, to find prey.

- Baleen whales lost teeth but have baleen, keratinous sheets that hang from the upper jaw used to strain small items such as zooplankton from the water. Mysticetes cannot echolocate but have exceptional ability to hear low-frequency sounds, which they produce and use for communication.

Restoration of semiaquatic Eocene whale, †*Pakicetus attocki*, from Pakistan showing bones recovered in the field

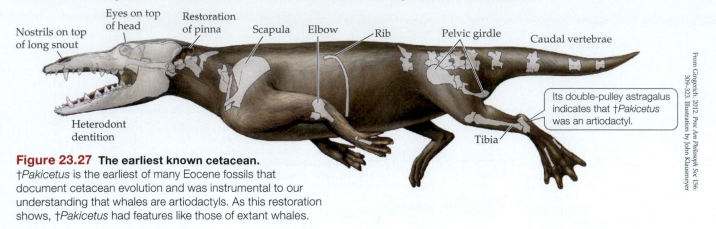

Figure 23.27 The earliest known cetacean.
†*Pakicetus* is the earliest of many Eocene fossils that document cetacean evolution and was instrumental to our understanding that whales are artiodactyls. As this restoration shows, †*Pakicetus* had features like those of extant whales.

From Gingerich. 2012. *Proc Am Philosoph Soc* 156: 309–323. Illustration by John Klausmeyer

A process called telescoping reshaped the facial region of neocetes. Bones of the rostrum (premaxilla and maxilla) become elongated, overlap those of the face (frontals, parietals, and nasals), and compress together to form a shortened braincase at the rear of the skull. During development, the nostrils move posteriorly to form single (in odontocetes) or paired (in mysticetes) blowholes on the top of the head used for breathing at the surface (**Figure 23.28**). Thus, the nasal passages of cetaceans orient vertically rather than horizontally as in other mammals.

Anatomical details and fossils show that telescoping occurred independently in odontocetes and mysticetes. The bones that form the front of the skull of odontocetes are asymmetrical, and we think this happened together with the evolution of a large fatty tissue, known as the melon (Figure 23.28A shows the position of the melon in a sperm whale), used as an acoustic lens for focusing outgoing chirps used in echolocation. In contrast, the skulls of mysticetes are as symmetrical as those of terrestrial artiodactyls such as cows. The earliest baleen whales were small (porpoise size) and retained teeth. They probably

(A) External anatomy of sperm whale, *Physeter macrocephalus*

(B) Skeleton of sperm whale

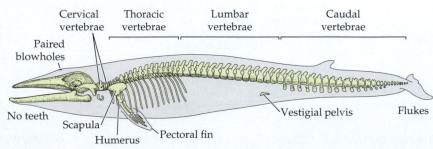

(C) Mother sei whale, *Balaenoptera borealis*, with calf

Note paired blowholes.

Christin Khan/NOAA

(D) Skeleton of fin whale, *Balaenoptera physalus*

Figure 23.28 Neoceti. The two extant clades of Neoceti are toothed whales or odontocetes (A,B) and baleen whales, or mysticetes (C,D). Dorsoventral flexion of the vertebral column powers swimming, and there has been a corresponding increase in the number of vertebrae in the lumbar and caudal regions. (B,D after R. Lydekker. 1894. *The Royal Natural History*, Vol 3. Frederick Warne & Co.: London.)

lost teeth in association with the evolution of suction feeding before baleen evolved.

The initial radiation of Neoceti into echolocation-assisted predation (toothed whales) and suspension feeding (baleen whales) coincided with the Eocene–Oligocene boundary. At about that time, the establishment of the Antarctic Circumpolar Current led to profound changes in ocean productivity and dramatic decreases in temperatures at higher latitudes. With declining global temperatures in the late Miocene, many cetacean families became extinct, while dolphins and porpoises like extant forms appeared. Gigantic baleen whales appeared in the Pliocene, concurrent with increases in seasonal ocean upwelling that provided more nutrients for their prey.

Prior to the start of commercial whaling in the 19th century, North Atlantic baleen whales were 20 times more abundant than they are today. Severe ecological consequences of this slaughter include disruption of oceanic carbon cycles and decreases in marine productivity. Whales play key roles in nutrient flow in marine ecosystems because they feed below the photic zone (see Figure 7.38) but defecate near the surface. This behavior releases iron in the surface waters where it promotes the growth of diatoms, which are food for the krill that whales eat. That cycle (whale poop → diatoms → krill → whales) explains the paradoxical observation that krill biomass has decreased as the abundance of whales decreased. Other vertebrates also eat krill, and a decrease in krill biomass affects them as well. This is one of many reasons why it is imperative to end any and all commercial harvesting of baleen whales.

Cetacean genomics Many features that evolved as cetaceans adapted to life in marine environments do not preserve in fossils. These include adaptations for diving, accumulation of massive fat reserves that serve both metabolic and thermoregulatory functions, and changes in sensation and perception. Genomic studies identified some changes related to the gain or loss of features:

- Mutations in cetacean myoglobin genes enhance that protein's ability to store oxygen in muscles—an essential feature for diving mammals, and one that convergently evolved in pinnipeds and beavers. Even greater changes occurred in the genomes of diving specialists such as the sperm whale.

- Many toothed whales generate sounds at ultrasonic frequencies ("ultrasonic" simply refers to frequencies that humans cannot detect). Echolocation "chirps" range from ~40,000 to ~130,000 Hz, with higher frequencies used to detect smaller prey. Changes related to the reception of high-frequency sound are known in genes that code for prestin, a protein that affects hair cells in the cochlea; echolocating bats, which also use ultrasound, show similar genomic changes.

- The land-based senses of smell and taste do not work well underwater, and many genes that code

for olfaction and taste in terrestrial mammals are pseudogenes in cetaceans.

- Although extant baleen whales lack teeth, they retain pseudogenes for enamel proteins.

- Cetaceans lack hair, but they retain pseudogenes for hair production and growth. And, like other hairless mammals, cetaceans lost the sebaceous glands along with the genes involved in their formation.

23.7 Trophy Hunting and Extinction Risk

LEARNING OBJECTIVES

23.7.1 Explain how trophy hunting can lead to genetic changes in populations.

23.7.2 Explain the Allee effect and how it contributes to an extinction vortex.

Mammals experience the same detrimental human influences that we described in discussions of other groups of vertebrates: habitat loss, harvesting for food, introduction of alien species, and spread of disease. Nearly 25% of terrestrial mammal species are at risk of extinction due to human activities. However, being killed solely to become a hunter's trophy is a risk largely restricted to large mammals and a few species of gamefishes.

Bighorn sheep: A case study

Trophy hunters who want to hang a mounted head on a wall target the individuals that make dramatic trophies, often the ones with the largest horns or antlers. A genetic component accounts for 30–40% of the variance in the size of horns and antlers, and horn size should show rapid evolutionary change under strong selection pressure.

Horns of sheep and other members of the family Bovidae, such as cows and antelope, form from a bony core covered by a keratinous sheath (**Figure 23.29A,B**). As in the case of other hoofed mammals with cranial appendages, they use the horns for display and fighting in male–male interactions. Unlike the antlers of deer, elk, and moose (**Figure 23.29C**), which are shed annually, bovid horns are permanent structures that increase in size throughout the animal's life. Horn size is a key determinant of reproductive success for bighorn rams, and sexual selection favors males with large horns.

A long-term study of rams in a population of bighorn sheep (*Ovis canadensis*) on Ram Mountain in Alberta, Canada, documented a decrease in average horn size that represents genetic change due to selection pressure by trophy hunters (**Figure 23.30**). The situation can be summarized as follows:

- To ensure that only adult rams could be shot, Alberta set the minimum horn length for "legal" rams at 80% of a full circle of curl.

(A) Impala, *Aepyceros melampus*

(B) Sable antelope, *Hippotragus niger*

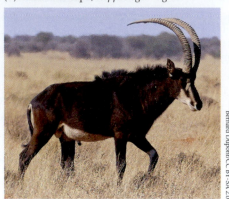

(C) Moose, *Alces alces*

Figure 23.29 Examples of selection resulting from trophy hunting. (A) The average horn lengths of male impalas killed by trophy hunters in Africa declined by 4% between 1974 and 2008.

(B) The horn lengths of sable antelope decreased by 6%.
(C) Similarly, in areas of heavy trophy hunting, the average antler size of Alaskan moose decreased by 15%.

- Rams with fast-growing horns reach that size in about 4 years, and 91% of these rams were killed before they reached 7 years, the age at which large horn size provides an advantage in mating.

- Trophy hunting thus created an unnatural selective advantage for rams with slow-growing horns that reach a smaller maximum size, and in turn these were the individuals whose genes were passed on to succeeding generations. As a result, the average horn size of rams killed by trophy hunters decreased by 14 cm between 1974 and 2011.

Similar results have been reported for other species of wild sheep, wild goats, and African antelopes, as well as for moose (antlers) and elephants (tusks). Recovery from unnatural selection can be slow. Horn size began to recover following relaxation in hunting pressure at Ram Mountain in 1996, but had increased by only 13% as of 2010. Thus, recovery from anthropogenic selection is likely to be much slower than the original response.

Endangering the endangered: The effect of perceived rarity

Sellers commonly attach value to rarity to make products desirable, with phrases like "one of a kind" or "only five left in stock." Permits to hunt big-horn sheep in areas with a reputation for producing trophy heads can draw astonishing bids: in most jurisdictions, they fetch US $20,000–$40,000, but permits for Alberta regularly attract bids exceeding US $400,000.

Endangered status is also a draw for trophy hunters. In 2015 the Namibian government auctioned a permit to kill a male black rhinoceros for US $350,000. This is a critically endangered species, which increased its desirability as a trophy. Between 2004 and 2010, the fees paid by trophy hunters for species that had been moved into more endangered categories on the IUCN Red List nearly doubled, whereas permits for species that remained in the same category or moved to a lower risk category increased by only one-third.

In 2012, trophy hunting generated more than US $217 million for seven sub-Saharan African countries. Most of this revenue derived from the "big five" African game species: lion, leopard, elephant, Cape buffalo, and rhinoceros. Proponents of trophy hunting argue that the revenue generated from hunting supports a broad range of conservation activities, from establishing and maintaining game reserves to fighting poachers. This view maintains that elimination of trophy hunting would increase the loss of biodiversity because habitat protection in game reserves benefits all species in an ecological community, not just those stalked by trophy hunters.

An opposing perspective focuses on the trophy species themselves rather than on the whole community. Many species benefit from the presence of conspecifics—that is, a population's survival rate increases when individuals form groups, a phenomenon known as the **Allee effect**. Familiar examples include the group defenses against predators employed by many ungulates; modifications of the environment that create a suitable environment for the species, such as construction and maintenance of dams by beaver colonies; and cooperative breeding, such as the simultaneous birth of wildebeest calves in a large herd, which swamps the capacity of predators to attack them. For such species, declines in population sizes reduce the Darwinian fitness of each individual in the population. Below a threshold size, a population may crash to extinction.

The extinction vortex

Economic theory holds that exploitation is unlikely to result in the extinction of a species: As the cost of capturing the last few individuals becomes prohibitively high, says the theory, hunters are expected switch to species that are easier to find. Unfortunately for trophy species, increasing

(A) Mature ram with horns describing a full circle of curl

slashvee/CC BY-ND 2.0

Figure 23.30 **Changes in horn length of bighorn sheep.**
(A) Killing rams with full-circle horns is the goal of trophy hunters.
(B) The Ram Mountain population of bighorn sheep in Alberta, Canada, contains approximately 500 rams. Average horn sizes for each year from 1973 to 2010 were standardized (represented by 0) by correcting for the age of the individuals in the population each year (the cohort). During a period of intense trophy hunting (1973–1995), 52 rams were killed and horn length decreased from +5 to –9 cm against standard as fast-growing males were removed from the population before they reached their full reproductive potential at age ~7 years. From 1996 onward, trophy hunting was reduced, with 4 rams killed in 15 years. Horn size slowly increased as fast-growing rams entered the breeding population, but remained well below that of the initial population. (B after G. Pigeon et al. 2016. *Evol Applic* 9: 521–530/CC BY 4.0.)

(B) Average horn size of rams, 1973–2010

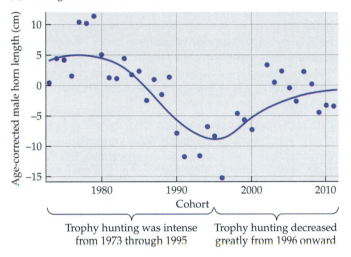

Trophy hunting was intense
from 1973 through 1995

Trophy hunting decreased
greatly from 1996 onward

rarity usually seems to make the remaining individuals even more desirable to some trophy hunters, and the resulting positive feedback loop maintains or increases hunting pressure as a species' numbers dwindle. When an already small population is subject to multiple pressures that feed on one another, the feedback loop quickly becomes what conservation biologists call an **extinction vortex**. Once an extinction vortex is established, the population enters an escalating downward spiral (vortex) and its chances of long-term survival are slim.

A hypothesis known as the anthropogenic Allee effect combines the benefits of conspecifics with the impact of rarity on trophy species. This view holds that the increased incentive provided by rarity ensures that hunting will continue to reduce the size of a population until essential Allee interactions between individuals no longer occur, and an extinction vortex will ensue.

Several life history and ecological characteristics increase the extinction risk of a species, and unfortunately several of them apply to species that are the targets of trophy hunters.

- *Large herbivores often mature late and have low reproductive rates*, which prevents rapid repopulation

when the number of individuals in a population is reduced. Female African elephants, for example, are not sexually mature until they are 11 years old, and males do not reproduce until they are more than 20 years old. Females give birth to a single calf, and the interbirth interval is about 5 years. Female black rhinoceroses reach sexual maturity at about 5 years and males at 6 years; single calves are born at intervals of 3 years or more.

- *Large carnivores need a lot of space.* Trophy hunting may be a factor in the decline of leopard populations in Tanzania. The home ranges of African leopards are large—an average of 229 km^2 for males and 179 km^2 for females. Only 14% of each male's home range overlaps the home ranges of females, and a reduction in the number of leopards in a population reduces the chance of encounters while a female is sexually receptive.

- *A high degree of sociality makes a species vulnerable to the Allee effect.* African wild dogs (*Lycaon pictus*) hunt in packs and expend substantial energy pursuing and bringing down prey. After the prey is killed, spotted hyenas (*Crocuta crocuta*) attempt to chase the dogs away. Packs containing 6 or more dogs can fend off hyenas for 10–15 minutes, which is long enough for the pack to consume the best sections of a carcass, but smaller packs of dogs lose substantial portions of their kills to hyenas. Because hunting is so energetically costly, losing 25% of their kills to hyenas is an unsustainable energy loss, and reduction of pack size below 6 leads to an extinction vortex.

An analysis of the paths to extinction of 10 populations of vertebrates supported the reality of extinction vortices. The final declines to extinction of these populations started from an average population of 70 individuals, and 3 of these populations declined from more than 100 individuals. Complex interactions of multiple factors appear to have been responsible for these extinctions, including the Allee effect.

Summary

23.1 Therian Features and Origins of Marsupialia and Placentalia

Theria has two extant clades: Metatheria (marsupials and their extinct relatives) and Eutheria (placentals and their extinct relatives).

Therians share several unique features related to locomotion: in the shoulder girdle, the coracoid and interclavicle bones have been lost and the scapula is mobile with a scapular spine; the elbow is hingelike; the ankle joint is also hingelike with the superposition of the astragalus on the calcaneus; and in the lumbar region there is greater flexibility between the vertebrae. These features enable the unique bounding locomotion of therians.

Marsupials differ from placentals in having an inflected angular process of the dentary (the angular process of the dentary is straight in placentals), replacement of only the last premolar (placentals replace incisors, canines, and premolars), and retention of epipubic bones (extant placentals lack epipubic bones).

23.2 Diversity of Marsupials

There are seven extant clades of marsupials.

- Three clades (Paucituberculata, Didelphimorphia, and Microbiotheria) occur in the New World.

- Four clades (Diprotodontia, Notoryctemorphia, Peramelemorphia, and Dasyuromorphia) occur in Australia and New Guinea.

23.3 Diversity of Placentals

Most extant species of mammals are placentals. Placentals diversified rapidly, leaving few morphological cues about relationships. Molecular phylogenetic analyses reveal two main groups, Atlantogenata and Boreoeutheria. Atlantogenata includes two groups of species found mainly on southern continents.

- Afrotheria is a large group of African species, including tenrecs, elephant shrews, the aardvark, hyraxes, sirenians, and elephants.

- Xenarthra includes sloths, anteaters, and armadillos native to South America and forms that crossed the Panamanian land bridge during the Great American Biotic Interchange to colonize North America.

Boreoeutheria includes two placental lineages found primarily on the northern continents of Laurasia. Many boreoeutherian clades radiated extensively and today occur worldwide.

- Euarchontoglires includes rodents, rabbits, tree shrews, colugos, and primates.

- Laurasiatheria includes eulipotyphlans (shrews, moles, hedgehogs), bats, ungulates (including cetaceans, the whales), carnivorans (dogs, cats, weasels, bears, and aquatic pinnipeds), and pangolins.

23.4 Reproduction

A scrotum housing the testes outside the body is present in most male marsupials and placentals. It evolved independently in the two groups, as it differs in position (anterior to penis in marsupials, posterior to penis in placentals). The glans of the penis is bifid in marsupials and single in placentals. Many placentals have a bone in the penis (the baculum).

Female marsupials have two separate uteri and two lateral vaginas (for passage of sperm) and a single midline pseudovaginal canal (for passage of young at birth). All female placentals have a single midline vagina; most have a bipartite uterus divided lengthwise into left and right sides for some or all its length. A few species (including humans) have a single median uterus.

All therians have an initial yolk sac (choriovitelline) placenta, which is transitory in placentals but forms the only placenta in most marsupials. In placentals, the yolk sac placenta is subsequently replaced by the chorioallantoic placenta.

Placentals and marsupials differ in the relative lengths of gestation (longer in placentals) and lactation (longer in marsupials). All marsupial young complete their development fastened onto a nipple, and some marsupials house these developing young in a pouch.

A key innovation of placentals is the acquisition of decidual stromal cells in the maternal endometrium. These dampen the maternal inflammatory immune response and enabled the evolution of longer periods of gestation, but longer gestation probably evolved independently in different placental lineages.

23.5 Teeth and Feeding Specializations

Mammals must process large quantities of food per day to fuel their high metabolic rates, and oral breakdown of food to reduce the particle size is an important starting point.

Almost all therian mammals have a dentition consisting of incisors, canines, premolars, and molars, with the form of the molars especially varying in mammals with different diets.

Omnivores have low, blunt molars for crushing (bunodont dentition). Herbivores have molars with ridges (lophs) that shred fibrous food, and cheek teeth (molars + premolars) that may be made more durable by being high-crowned (hypsodont) or ever-growing (hypselodont). Carnivorans have specialized shearing teeth formed by the fourth upper premolar and first lower molar, together called the carnassial apparatus.

Carnivorous and herbivorous mammals differ in skull features, jaw musculature, and digestive tracts.

Summary *(continued)*

- Carnivorous forms have a large temporal fossa in the skull, reflecting a large temporalis muscle best suited for exerting force at the front of the jaw. In contrast, herbivores have a large angular process of the dentary for the attachment of a large masseter muscle, which moves the jaw sideways in a grinding action.

- The jaw joint of carnivorous therians is at the level of the tooth row, but it is above the level of the tooth row in herbivorous species, enabling occlusion of all of the cheek teeth simultaneously when the jaw moves sideways.

- Carnivorous therians have short, simple digestive tracts because adequate nutrition is quickly absorbed from their protein-rich food. Grazers and browsers (which eat grasses and leaves) rely on bacterial symbionts to break down complex carbohydrates such as cellulose; these herbivores have longer digestive tracts with regional specializations for foregut or hindgut fermentation.

23.6 Locomotion

Cursorial (running) mammals have limbs modified for speed at the expense of power. Foot postures are digitigrade or unguligrade and distal limb segments are elongated. Toe loss is common to reduce limb weight. Movements of limbs are restricted to the fore-and-aft plane.

Fossorial (digging) mammals have limbs modified for power at the expense of speed. Forelimb diggers have short, stout arm bones with projections for the attachment of muscles used in digging. Hands are large, with five digits and stout claws.

Bats are the only mammals with powered flight. Their elongated, flexible fingers support a web of elastic skin that yields highly maneuverable flight.

Swimming therians include semiaquatic forms that spend time in water and on land, such as beavers and pinnipeds. Fully aquatic sirenians and cetaceans never leave the water. Most semiaquatic mammals swim using their limbs.

Most fully aquatic mammals swim using dorsoventral flexion of the body and tail.

Cetaceans (whales) are fully aquatic mammals now known to be derived artiodactyls. Transitional forms such as †*Pakicetus* and other Eocene species document the transition from semiaquatic to fully aquatic cetaceans. Molecular phylogenetic evidence places hippos as the living sister group of cetaceans.

Neoceti, the extant whales, contains two major groups

- Odontoceti, toothed whales, includes dolphins, porpoises, and sperm whales. This group uses echolocation to find their prey (typically squids, fishes, or other marine mammals).

- Mysticeti are the baleen whales, which feed on plankton that they filter from seawater using specialized baleen plates. This group includes the largest animal that has ever lived.

Cetacean genomes contain many pseudogenes, relics of once-functional genes no longer of use in aquatic habitats. These include genes for olfaction, taste, hair production, sebaceous gland formation, and in the case of baleen whales, tooth enamel.

23.7 Trophy Hunting and Extinction Risk

Trophy hunting exerts an unnatural selective pressure on some species of mammals, especially the males of large, horned species such as bighorn sheep. Trophy hunters target healthy, mature males; these are the individuals that would normally have long reproductive lives and make the largest genetic contribution to succeeding generations.

Rarity increases the perceived value of a species to trophy hunters, resulting in increased hunting pressure as populations decline—the "anthropogenic Allee effect."

Interaction of the anthropogenic Allee effect with ecological and life history characteristics can drive hunted species into an extinction vortex.

Discussion Questions

23.1 What genetic evidence is there that marsupials and placentals evolved from egg-laying ancestors?

23.2 Australian marsupials are convergently like many placentals (e.g., gliding squirrels and gliding possums). However, there are no marsupials equivalent to a placental cetacean or even to a seal. Why?

23.3 How and why is the mobile scapula of therians essential for their form of locomotion?

23.4 Why do seals and sea lions retain zygapophyses on their vertebrae, whereas extant cetaceans and sirenians do not? How could we use this anatomy to determine the behavior of extinct cetaceans (and what additional evidence do we have in at least one fossil cetacean to back up behavioral predictions)?

Figure credits

Figure 23.1: B after K. Liem et al. 2001. *Functional Anatomy of the Vertebrates*, 3rd ed. Cengage/Harcourt College: Belmont, CA, and W.F. Walker. 1980. *Vertebrate Dissection*, 6th ed. Saunders College Publishing: Philadelphia, PA; D after F.A. Jenkins, Jr. 1971. *J Zool, London* 165: 303-315.

Figure 23.13: A,B after M.B. Renfree. 1993. In *Mammal Phylogeny*. F.S. Szalay et al. (Eds.). New York: Springer; C,D after M. Hildebrand and G. Goslow. 1998. *Analysis of Vertebrate Structures*, 5th ed. New York: Wiley.

Figure 23.21: A,C after K. Liem et al. 2001. *Functional Anatomy of the Vertebrates*, 3rd ed. Cengage/Harcourt College: Belmont, CA, based on J.M. Smith and R.J.G. Savage. 1954. *Zool J Linn Soc London* 42: 603–622; B,D after K. Liem et al. 2001. *Functional Anatomy of the Vertebrates*, 3rd ed. Cengage/Harcourt College: Belmont, CA, based M. Hildebrand and G. Goslow. 1998. *Analysis of Vertebrate Structures*, 5th ed. New York: Wiley, and J.M. Smith and R.J.G. Savage. 1954. *Zool J Linn Soc London* 42: 603-622.

Figure 23.25: A,C after J.E. King. 1964. *Seals of the World*. British Museum (Natural History): London.

Figure 23.26: Tree after C. Zimmer. 2001. *Evolution: The Triumph of an Idea*. Harper Collins, New York.

Courtesy of Lwiro Primate Rehabilitation Centre

Primate Evolution and the Emergence of Humans

Red-tailed monkey, *Cercopithecus ascanius*

Except for humans, primates have largely lived in tropical latitudes since the end of the Eocene, where they have been a moderately successful group in terms of species diversity. Today there are about 514 extant species of primates in 14 families. The so-called wet-nosed primate clade (Strepsirrhini) includes lorises, bushbabies, lemurs, and allies. Other extant primates belong to the dry-nosed primate clade (Haplorhini), which includes Anthropoidea, the clade of monkeys and apes to which humans belong (**Figure 24.1**).

Molecular techniques show that chimpanzees and bonobos are the closest extant relatives of humans, and both molecular dating methods and the fossil record indicate that the divergence of humans from the African great apes occurred about 6.7 Ma. Humans evolved from an arboreal ancestor that did not knuckle-walk, a mode of terrestrial quadrupedal locomotion that evolved independently in gorillas and chimpanzees. Human bipedal walking evolved before the evolution of a large brain. Diverse new fossils show that early human evolution was much more complex than previously thought, with repeated introgression (hybridization) between different lineages of archaic humans.

In this chapter, we review the origins and diversity of extant primates, followed by a discussion of the origins and evolutionary history of Hominini, the humans. We close with a brief description of the effects of *Homo sapiens*, the only surviving hominin species, on other vertebrates and, indeed, on the course of life on Earth.

24.1 Primate Origins and Diversification

LEARNING OBJECTIVES

24.1.1 Explain the significance of the defining features of primates.
24.1.2 Describe the characters of early primate groups.
24.1.3 Distinguish between platyrrhines and catarrhines.
24.1.4 Describe the phylogenetic relationships of the major clades of non-hominoid primates.

An important character shared by all extant species of primates (Latin *primus*, "first") is the presence of opposable thumbs and big toes that can grasp the substrate, especially during arboreal locomotion. Opposable digits also increase dexterity and allow precise manipulation of objects. This precision is associated with the modification of claws into flattened nails, and the presence of fleshy, sensitive pads on the fingertips. This finger morphology is fundamental for climbing and for behaviors such as grooming, which is a component of social bonding in many primates.

Primates have relatively large brains in relation to their body sizes. They are visually driven animals; olfaction is less important than vision, and many genes responsible for odor detection in other mammals are pseudogenes in primates. The evolutionary increase in brain size may have been associated with shifts to a more frugivorous diet and changes in the visual system, including increases in eye size and relocation of the eyes to the front of the face. In this position, the eyes have stereoscopic vision and good depth perception. Binocular vision is critical for gauging distances when jumping from branch to branch and is also important for navigating through three-dimensional environments.

Basal primates

Divergence dates based on molecular phylogenetic information differ from the first appearances of primates in the fossil record. For example, based on molecules the divergence date between the two large extant clades of

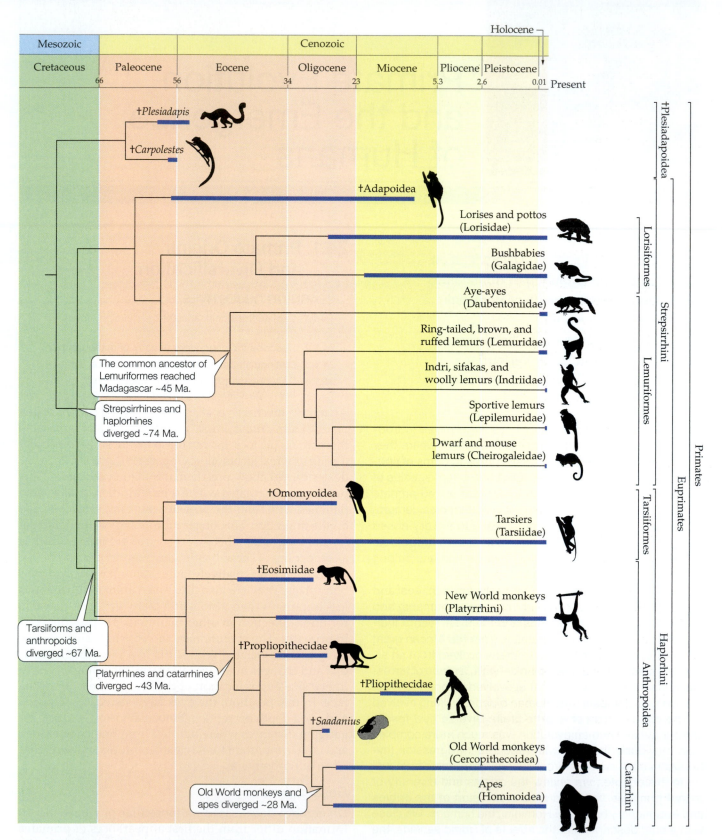

Figure 24.1 Phylogenetic relationships of Primates. The two extant clades of Primates are Strepsirrhini (lorises and relatives as well as the lemurs) and Haplorhini (tarsiers, monkeys, and apes, including humans). Molecular evidence indicates these clades diverged in the late Cretaceous. However, the oldest known fossil primates date only from the late Paleocene of the Cenozoic. (Based on J.I. Bloch and D.M. Boyer. 2002. *Science* 298: 1606–1610.)

primates (Strepsirrhini and Haplorhini) occurred in the Late Cretaceous, ~74 Ma, yet the oldest fossil primates are from the Cenozoic (see Figure 24.1). Explanations for this discrepancy could include problems with estimates based on molecular dating or that we have yet to discover fossils of earlier primates.

Fossils of early primates are known from the early Paleocene in the Northern Hemisphere. One of the best-known examples is †*Plesiadapis cookei*, a generalized arboreal form larger than extant tree squirrels (**Figure 24.2A,B**). †*Plesiadapis* shares some derived features of the teeth and skeleton with more derived primates, but its large, forward-pointing incisors superficially resemble those of rodents. It also retained claws on its digits. Related lineages had teeth indicating diets ranging from frugivory to folivory and insectivory.

The early primate †*Carpolestes simpsoni*, was an arboreal frugivore that had nails on its opposable big toes and claws on the rest of the digits (**Figure 24.2C**). It may have coevolved with angiosperms that provided nectar and fruit and relied on animals like †*Carpolestes* for seed dispersal.

(A) Rodentlike incisor teeth of †*Plesiadapis*

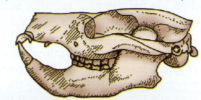

(B) Restoration of †*Plesiadapis cookei*

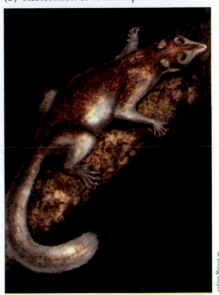

© Doug Boyer

(C) Restoration of †*Carpolestes simpsoni*

© Doug Boyer

Figure 24.2 †*Plesiadapis* and †*Carpolestes*. (A) The earliest primates lacked some features typical of extant primates, such a front-facing orbits, and had enlarged, sharp incisors. (B) †*Carpolestes*, an arboreal frugivore, had an opposable big toe with a nail. (A after E.L. Simons. 1964. *Sci Am* 211: 50–65.)

Euprimates

Euprimates includes all extant primate species as well as extinct lineages (see Figure 24.1). They have flat nails instead of claws on all digits. Claws enable small animals to cling to large-diameter objects such as tree trunks, but nails allowed the evolution of sensitive fingertips that are more useful for locomotion on small-diameter supports such as the terminal branches of a tree, as well as increasing the ability for fine-motor manipulation of objects.

The earliest definitive euprimates are known from the early Eocene of North America and Eurasia. Their diversification in the Northern Hemisphere reflects the tropical-like climates of higher latitudes at that time. These groups declined as climates cooled in the late Eocene and were virtually extinct in northern latitudes by the end of the Eocene.

Figure 24.1 shows the division of Euprimates into two extant clades: Strepsirrhini (Greek *strepsis*, "twisted"; *rhynos*, "nose") and Haplorhini (Greek *haploos*, "simple"). We interpret the lemurlike †Adapoidea as the sister taxon of other strepsirrhines. Exemplified by a Middle Eocene form, †*Darwinius masillae* (**Figure 24.3**), †adapoids share features of the wrist and ankle with extant forms.

Strepsirrhini Most extant strepsirrhines are small and nocturnal, with a tapetum lucidum in the eye that enables them to see in low-light environments. They have long snouts with a rhinarium, a patch of moist, furless skin associated with the olfactory system (similar to the nose of a dog or cat) that is the source of the common name, wet-nosed primates. Synapomorphies of extant strepsirrhines include a toothcomb formed by incisor and canine teeth in the lower jaw and a grooming claw on the second toe; both structures are used to groom the fur. Within Strepsirrhini, both molecular characters and the sparse

Eocene †adapoid, †*Darwinius masillae*

From Franzen et al. 2009, courtesy of the Natural History Museum, Oslo, Norway/CC BY

10 cm

Figure 24.3 The most complete primate fossil known. Even the fur and gut contents of the †adapoid †*Darwinius masillae* are preserved in this fossil from Messel, a middle Eocene (~47 Ma) World Heritage Site in Germany. Here, we interpret lemurlike †adapoids as an extinct sister group of strepsirrhines.

fossil record support a deep split about 58 Ma between Lorisiformes and Lemuriformes.

Lorisiformes are nocturnal arboreal primates, including the bushbabies and pottos of continental Africa and the lorises of Asia. Lorises have large eyes and hands and use their specialized prehensile feet to grip small branches. They make slow, deliberate movements, which is the source of the common name slow loris for the eight species of *Nycticebus* (**Figure 24.4A**). All species of slow lorises are threatened or endangered.

About 100 extant species of Lemuriformes belong to five families endemic to Madagascar. Their common ancestor probably reached Madagascar by rafting from mainland Africa during the Eocene, about 45 Ma (the island of Madagascar separated from mainland Africa during the Mesozoic). During their long, isolated evolutionary history on Madagascar, lemuriforms evolved their own version of primate diversity, including many unusual and distinctive forms (**Figure 24.4B–E**). As

recently as 2 ka, lemurs were more diverse and included bear-size arboreal species resembling koalas and sloths. Much of this diversity is now gone, probably the result of the arrival of humans on Madagascar from Asia about 4 ka. Accounts from European colonizers suggest that some of the giant lemurs may have been alive as recently as 500 years ago. The decline of lemurs continues, and the IUCN (International Union for the Conservation of Nature) classifies 33 extant species as Critically Endangered—one small step away from extinction in the wild—due to deforestation and hunting.

Haplorhini Haplorhines have a short snout with a dry nose and mobile upper lip. They include diurnal species as well as secondarily nocturnal species that lack a tapetum lucidum. Represented today by tarsiers and anthropoids (see Figure 24.1), the earliest haplorhine fossils are tarsiiforms like †*Omomys* and †*Tetonius*, known from the earliest Eocene (**Figure 24.5A**). The tarsiiform brain

(A) Javan slow loris, *Nycticebus javanicus*

Courtesy of Wawan Tarniwan and Little Fireface Project

(B) Aye-aye, *Daubentonia madagascariensis*

David Haring, Duke Lemur Center

(C) Ring-tailed lemur, *Lemur catta*

Courtesy of Amber Walker-Bolton

(D) Coquerel's sifaka, *Propithecus coquereli*

Courtesy of Sergi López-Torres

(E) Golden-brown mouse lemur, *Microcebus ravelobensis*

Courtesy of Malcolm Ramsay

Figure 24.4 Extant strepsirrhines. Like primate outgroups, strepsirrhines have a wet nose. They uniquely share a toothcomb and a grooming claw. (A) Like other lorisiforms, the lorises of Asia are nocturnal and arboreal. Slow lorises have notably slow and deliberate locomotion. (B–E) Lemuriformes radiated on and are endemic to Madagascar. (B) Aye-ayes are arboreal insectivores that use the hooked nail of their thin, elongated middle finger to extract insects from bark and wood. (C) A familiar lemurid is the ring-tailed lemur, a highly social omnivore that lives in groups of 20 or more. (D) Among the 19 species of indriids is Coquerel's sifaka. Skillful climbers with long legs, diurnal sifakas move on the ground using bipedal hopping. (E) Tiny forest dwellers, omnivorous mouse lemurs are the smallest extant primates. They live in social group of up to 15 individuals.

Figure 24.5 Tarsiiforms. Tarsiers have large eyes and elongated tarsal bones. Eocene forms such as †*Tetonius* (A) already showed these features, which are even more obvious in extant tarsiers (B,C). (After A.S. Romer. 1933. *Vertebrate Paleontology.* University of Chicago Press: Chicago.)

(A) Skull of an Eocene tarsiiform, †*Tetonius*   (B) Large orbits of a tarsier skull

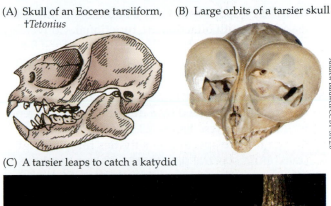

Andrew Bardwell/CC BY-SA 2.0

(C) A tarsier leaps to catch a katydid

© JuergenFreund Photography/juergenfreund.com

was small relative to the brains of extant primates. Forward-facing orbits and long, slender hindlimbs suggest that these early haplorhines were arboreal.

There are about 13 extant species of tarsiers in the islands of Southeast Asia, including Indonesia and the Philippines. They are small primates with large eyes (**Figure 24.5B**). Named for their elongated tarsal bones, tarsiers are adept at climbing and leaping through vegetation (**Figure 24.5C**). The remaining Haplorhini are anthropoids: the monkeys and apes (including humans).

Anthropoids

The origin of anthropoids was related to a shift from nocturnal to diurnal foraging, a change accompanied by a reduction in body size. The earliest known anthropoids were tiny forms such as †*Eosimias sinensis* (<140 g) from the Middle Eocene that ate a mixed diet of fruit and insects. †*Eosimias* and related forms are stem anthropoids, predating the catarrhine–platyrrhine split ~43 Ma (see Figure 24.1). Although early fossil forms were small, most extant species are much larger than strepsirrhines or tarsiers.

Anthropoids have larger brains and smaller olfactory bulbs than other primates. Most species are frugivorous or folivorous, although some small species eat insects. All anthropoids except night monkeys (Aotidae, also known as owl monkeys) are diurnal, and most have complex social systems. Locomotor modes include terrestrial and arboreal quadrupedalism by midsize anthropoids and vertical clinging and leaping by small New World monkeys. Larger arboreal anthropoids typically employ **suspensory locomotion**, moving with the body hanging below the branches. A specialized form of suspensory locomotion is **brachiation**, in which the animal swings from the underside of one branch to the next using its hands. Extant gibbons and the siamang, as well as spider monkeys, brachiate.

In addition to a larger braincase, anthropoids share other derived features of the skull, such as fusion of the paired frontal bones and the presence of a postorbital septum that prevents mechanical disturbance of the eye when the temporalis muscle contracts during chewing (**Figure 24.6**). Unlike strepsirrhines or tarsiers, anthropoids lack specialized grooming claws; they also lack the toothcomb that characterizes extant strepsirrhines.

Extant anthropoids belong to two large clades: Platyrrhini (Greek *platus*, "flat, broad"; *rhynos*, "nose") or New World monkeys; and Catarrhini (Greek *kata*, "downward"), which includes the Old World monkeys and apes (see Figure 24.1). The two groups differ in several important characters.

- *Olfaction* All anthropoids have small olfactory bulbs relative to outgroups, but platyrrhines have a

(A) Skull of a strepsirrhine   (B) Skull of an anthropoid

Unfused frontal bones

Fused frontal bones

Unfused mandibular symphysis

Fused mandibular symphysis

Back of orbit is open

Back of orbit is closed by postorbital septum

Larger brain

Figure 24.6 Cranial differences between strepsirrhines and anthropoids. Anthropoids have a bony septum behind the orbit and the mandibular symphysis joining the two halves of the lower jaws is fused, as are the paired frontal bones. (After J.G. Fleagle. 2013. *Primate Adaptation and Evolution*, 3rd ed. Elsevier: Waltham, MA.)

relatively good sense of smell and most of the genes associated with olfaction are functional. Conversely, only ~50% of catarrhine olfactory genes code for functional receptor proteins, and catarrhines have lost the vomeronasal organ.

- *Color vision* All catarrhines have trichromatic color vision, which evolved by a duplication of the gene for LWS opsin (see Figure 22.19). One gene copy drifted toward spectral sensitivities for green light (maximum spectral sensitivity ~530 nm) and the other toward orange wavelengths (maximum spectral sensitivity ~560 nm). Trichromatic vision is particularly important for detecting ripe fruit and distinguishing it from unripe fruit or green leaves, and it evolved independently by different mechanisms in some platyrrhines, such as howler monkeys.

- *Skulls* Platyrrhines and catarrhines differ in details of their skulls, especially in the ear region. Additionally, while all extant anthropoids have relatively large brains, fossils show that evolutionary enlargement of the brain occurred independently in platyrrhines and catarrhines.

- *Teeth* Platyrrhines retain the plesiomorphic condition of three premolars on each side of the jaw, whereas catarrhines have only two premolars.

New World monkeys

Radiations of platyrrhines, the New World monkeys, paralleled radiations of catarrhine monkeys in Africa and Asia except that there was no radiation of terrestrial platyrrhines equivalent to that of baboons and macaques. Perhaps the extensive radiation of ground sloths in South America inhibited such a primate radiation. However, a striking parallel exists between platyrrhine spider monkeys and catarrhine gibbons: as mentioned above, both groups are specialized arboreal brachiators with extremely long arms by which they swing through the branches. In addition, wrist joint modifications allowing exceptional hand rotation evolved independently in these two groups. However, spider monkeys use their prehensile tail as a "fifth limb" during arboreal locomotion (see Figure 24.7F), whereas gibbons, like all apes, lack a tail.

Platyrrhine diversity   Platyrrhines rafted from Africa and reached South America in the early Oligocene (see Figure 20.2 and Section 20.3). They are an exclusively neotropical radiation, today found primarily in South America although some species live as far north as Mexico. In the past, their range extended to Caribbean islands. As noted above, platyrrhines retain a good sense of smell, and they share the presence of three premolars on each side of the jaw as well as prehensile tails in some members.

There are about 100 extant species of platyrrhines, which we group into three families, Pitheciidae, Cebidae, and Atelidae.

- Pitheciidae include the pitheciines (sakis and uakaris) and callicebines (titis). The white-faced saki (**Figure 24.7A**) is a medium-size arboreal monkey that eats hard foods, particularly fruits and seeds with tough outer coats. The brown titi monkey (**Figure 24.7B**) and related species are unusual among New World monkeys in having complex vocalizations, including dawn duets sung by breeding pairs.

- Cebidae include the cebines (organ grinder and squirrel monkeys), callitrichines (marmosets and tamarins), and aotines (night monkeys). Cebines such as the black-horned capuchin (**Figure 24.7C**) are arboreal quadrupeds that mainly eat fruit and insects and live in groups of several males and females. Capuchins are the only cebids that retain a prehensile tail as adults. Marmosets and tamarins are small and have secondarily evolved clawlike nails on all digits except the big toe. They have simplified molars, and many species eat gum exuded from holes that the marmosets make through the bark of trees. The pygmy marmoset (**Figure 24.7D**) is the smallest monkey, topping out at a body weight around 100 g. Aotines such as the Panamanian night monkey (**Figure 24.7E**) are the only nocturnal anthropoids.

- Atelidae include two subfamilies: atelines (woolly monkeys, spider monkeys, and muriquis), and alouattines (howler monkeys). Spider monkeys have prehensile tails, which they use when brachiating (**Figure 24.7F**). Howler monkeys are among the best known and most intensively studied New World monkeys. Males can make very loud calls due to the enlarged hyoid bone in their throat that houses a resonating chamber (**Figure 24.7G**). Howlers are sexually dimorphic in size and color. They are primarily folivorous and live in mixed groups of several males and females.

Old World monkeys and apes

The nostrils of catarrhines open forward and downward and have smaller nasal openings than those of platyrrhines, and prehensile tails never evolved in catarrhines. Stem catarrhines include †Propliopithecidae, small forms from the late Eocene and early Oligocene of Africa; and †Pliopithecidae, larger Miocene forms that ranged from Africa to Eurasia (see Figure 24.1). The two extant clades of catarrhines are the Old World monkeys (Cercopithecoidea; Greek *kerkos*, "tail"; *pithekos*, "monkey") and the apes (Hominoidea; Latin *homo*, "man"), a group that includes humans.

Molecular estimates date the split between cercopithecoids and hominoids at ~28 Ma, which matches a catarrhine fossil from the late Oligocene of Arabia, †*Saadanius*. The earliest cercopithecoids appear in the late Oligocene (~25 Ma) of Africa around the same time as the first apes. Here we will briefly cover cercopithecoid diversity before the remainder of the chapter turns to Hominoidea, including *Homo sapiens* and our closest relatives.

Extant cercopithecoids include colobines and cercopithecines. Both groups occurred in Eurasia in the late Miocene and Pliocene. The radiation of cercopithecoids in these periods coincided with a reduction in diversity of the earlier radiation of generalized apelike forms. The extant radiation of cercopithecoids is in many respects more derived than apes, as well as being much more diverse.

(A) White-faced saki, *Pithecia pithecia*

(B) Brown titi monkey, *Plecturocebus brunneus*

Figure 24.7 **Extant New World monkeys.** Platyrrhines radiated exclusively in the Western Hemisphere and today occur primarily in South America. Here, we group them into three families, Pitheciidae (A,B), Cebidae (C–E) and Atelidae (F,G). Cebids and atelids are the only non-human primates that occur north of Panama today.

(C) Black-horned capuchin, *Sapajus nigritus*

(D) Pygmy marmoset, *Cebuella pygmaea*

(E) Panamanian night monkey, *Aotus zonalis*

(F) Red-faced spider monkeys, *Ateles paniscus*

(G) Male mantled howler, *Alouatta palliata*

(A) Nepal gray langurs, *Semnopithecus schistaceus*

Aditya Pal/CC BY-SA 4.0

(B) Male proboscis monkey, *Nasalis larvatus*

Colobines have long tails and hindlimbs longer than their forelimbs.

Charles J. Sharp/Sharp Photography/CC BY-SA 4.0

(C) Grooming session, Japanese macaques, *Macaca fuscata*

The fore- and hindlimbs of cercopithecines are similar in length.

Leyo/CC BY-SA 2.5 CH

(D) Blue monkeys, *Cercopithecus mitis stuhlmanni* live in forest canopy

Charles J. Sharp/Sharp Photography/CC BY-SA 4.0

(E) The Hamadryas baboon, *Papio hamadryas*, lives in dry, open habitats

Cercopithecoids walk quadrupedally with the palm contacting the ground.

Cristoph Anton Mitterer/CC BY-SA 2.0

(F) Mandrill, *Mandrillus sphinx*

Malene Thyssen/CC BY-SA 3.0

Figure 24.8 Extant Old World monkeys. Cercopithecoids, the Old World monkeys, are grouped with hominoids (apes, including humans) in Catarrhini (see Figure 24.1). Found throughout Africa and Asia, cercopithecoids include folivorous colobines (A,B), and the more frugivorous cercopithecines (C–E).

- There are ~60 extant species of colobines in Africa and Asia, including colobus monkeys, langurs, and the proboscis monkey (**Figure 24.8A,B**). Commonly called leaf-eating monkeys, they are folivorous, have high-cusped molars, and ferment plant fiber in a complex foregut. Primarily arboreal, colobines have long tails and hindlimbs longer than their forelimbs.

- Most of the ~70 extant species of cercopithecines occur in Africa, although macaques occur in Asia and a small population of the Barbary macaque (*Macaca sylvanus*) lives on Gibraltar, in southwestern Europe. Along with macaques, the group includes guenons (the common name for monkeys in the genus *Cercopithecus*), baboons, and the mandrill, among others (**Figure 24.8C–F**). Cercopithecines are more terrestrial than colobines; their forelimbs and hindlimbs are of similar lengths, and they often have shorter tails. Their hands have a longer thumb and shorter fingers than the hands of colobines. They carry food in cheek pouches and are generally more frugivorous than colobines, a dietary difference reflected in their broad incisors and their bunodont (low-cusped) molars.

24.2 Origin and Evolution of Hominoidea

LEARNING OBJECTIVES
24.2.1 Understand basic differences in behavior among hominoids.
24.2.2 Describe the phylogenetic relationships of hominoids.

As shown in **Figure 24.9**, Hominoidea includes gibbons and the siamang (Hylobatidae) and the great apes, including humans (Hominidae).

The earliest fossil evidence of a hominoid is a single tooth of †*Rukwapithecus* from the late Oligocene (~25 Ma) of eastern Africa, but Miocene hominoids such as †*Ekembo nyanzae* are known from more complete specimens. Most workers interpret these early hominoids as arboreal inhabitants of forested habitats, and their bunodont molars suggest they were frugivorous. Hominoids spread broadly into Eurasia, following the general middle Miocene warming and the connection of Africa to the Eurasian mainland between 16 and 14 Ma.

Unlike other anthropoids, hominoids lack tails. Hominoids also differ from other anthropoids in having a wider, dorsoventrally flattened trunk relative to body length, so that the shoulders, thorax, and hips are

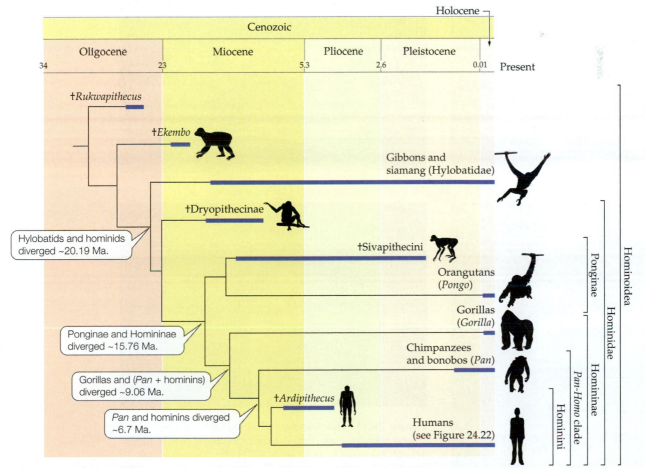

Figure 24.9 Phylogenetic relationships of hominoids. Of the extant hominoids, gibbons and the siamang (Hylobatidae) are sister to the hominids (Hominidae, the apes and humans).

(A) Scapula, ribs, and humerus of an Old World monkey

(B) Scapula, ribs, and humerus of an ape

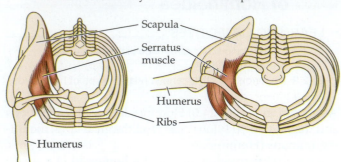

Scapula

Serratus muscle

Humerus

Ribs

Humerus

Figure 24.10 Differences in the position of the scapula in monkeys and apes. (A) The scapula of a monkey lies lateral to a relatively narrow chest. (B) The scapula of an ape lies over a broad, flattened chest. The more curved ribs of apes place the vertebral column closer to the center of gravity, making it easier for apes to balance in an upright position.

proportionately broader than in monkeys (**Figure 24.10**). Hominoids have distinctive dentition: the lower molars of hominoids have five cusps whereas other anthropoids have only four. Additionally, the molar cusps of hominoids are lower, and grooves between the posterior cusps form a distinct Y pattern.

Hylobatidae

Nineteen species of gibbons (*Hylobates*, *Hoolock*, and *Nomascus*) and one species of siamang (*Symphalangus*)

occur in Southeast Asia (**Figure 24.11**). Hylobatids move through trees by brachiation and are bipedal when moving on the ground, holding their arms outstretched for balance like a tightrope walker using a pole. These are the smallest extant apes and differ from other apes (and from most other mammals) in having monogamous social systems. Pairs produce simultaneous calls referred to as duets that strengthen pair bonds and announce that a territory is inhabited by a pair of gibbons. Gibbons were unknown in the fossil record until the recent descriptions of †*Kapi* from the Miocene of India, ~13 Ma.

Hominidae

The best support for monophyly of Hominidae—the clade containing orangutans, gorillas, chimps, and humans —comes from molecular phylogenetic research that

(A) Limb proportions of a white-handed gibbon, *Hylobates lar*

Ladislav Král/CC BY 3.0

(B) White-handed gibbon in Kaeng Krachan National Park, Thailand

JJ Harrison/CC BY 3.0

(C) Siamangs, *Symphalangus syndactylus*

Hendy Wicaksono/CC BY-SA 4.0

(D) Inflated throat pouch of a calling siamang

suneko/CC BY 2.0

Figure 24.11 Hylobatids. The hylobatids of Southeast Asia are the smallest apes. They use their long arms (A,B) to move through the forest by brachiation (C). Both male and female siamangs have throat pouches (D) they use to make loud calls.

shows the divergence between hylobatids and hominids in the Miocene, ~20 Ma. Fossils relevant to the origin of hominids include †dryopithecines, which diversified in Eurasia during the Miocene. Figure 24.9 shows †dryopithecines as the sister group of the other members of Hominidae, a placement based on details of their skulls and postcranial skeletons.

Taxonomic terminology of apes and humans can be confusing. Until some 50 years ago, scientists used the term "hominid" to refer only to the human lineage (*Homo* and extinct relatives). Current taxonomic usage, however, includes all four extant genera of great apes—*Pongo*, *Gorilla*, *Pan*, and *Homo*—in Hominidae. Recent usage (followed in this chapter and illustrated in Figure 24.9) groups the African apes *Gorilla* and *Pan* with *Homo* in the subfamily Homininae, with *Homo* + *Pan* as a subclade. The human lineage alone is the tribe Hominini, and here that is what we mean when we refer to hominins.

Ponginae Ponginae includes the extant orangutans (*Pongo*), which diverged from Homininae about 15.76 Ma (see Figure 24.9). Ponginae also includes †Sivapithecini (named for the Hindu god Shiva), an Asian radiation of apes that lived in forested habitats and had some specializations for suspensory locomotion. †Sivapithecins flourished from about 16 Ma (the Mid Miocene Optimum; see Figure 20.3) to 9 Ma (the start of the late Miocene), when the climate became cooler and drier across the middle latitudes of the Northern Hemisphere. A Pleistocene †sivapithecin, †*Gigantopithecus blacki*, was the largest primate that has ever lived. With an estimated body mass of ~300 kg, it was half again the size of a male gorilla.

Orangutans (Malay, *orang*, "people"; *hutan*, "forest") have extreme sexual dimorphism. Males can weigh twice as much as females and may develop broad cheek pads and a pendulous throat with age (**Figure 24.12A**). When walking quadrupedally, orangutans close their hands into

(A) Male Bornean orangutan, *Pongo pygmaeus*,
 in Tanjung Puting National Park, Indonesia

George Arif Kurniawan/CC BY-SA 4.0

(B) Fist-walking

© iStock.com/Andrey Gudkov

(C) Sumatran orangutan, *Pongo abelii*

©Anatoliy Alekseev/123RF

(D) In the shade

Hendy Wicaksono/CC BY-SA 4.0

Figure 24.12 Orangutan behaviors. (A) The faces of some older male orangutans have large cheek pads. Such males have greater reproductive success than those without cheek pads. (B) Orangutans are highly arboreal but also walk using a unique mode of quadrupedal locomotion termed fist-walking (see Figure 24.13A). (C) Like the other two species of orangutans, the Sumatran orangutan (C) is classified as Critically Endangered by the IUCN. (D) Orangutans demonstrate many behaviors, including the use of leaves to shelter from the sun or rain.

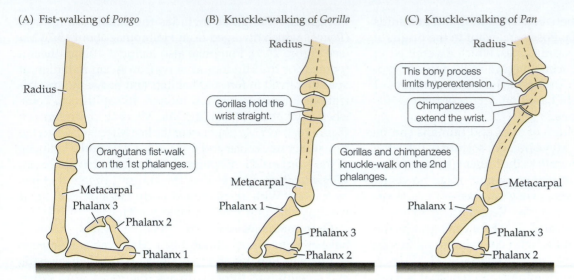

(A) Fist-walking of *Pongo*

(B) Knuckle-walking of *Gorilla*

(C) Knuckle-walking of *Pan*

Radius

Orangutans fist-walk on the 1st phalanges.

Metacarpal
Phalanx 3
Phalanx 2
Phalanx 1

Radius

Gorillas hold the wrist straight.

Metacarpal
Phalanx 1
Gorillas and chimpanzees knuckle-walk on the 2nd phalanges.
Phalanx 3
Phalanx 2

Radius

This bony process limits hyperextension.
Chimpanzees extend the wrist.

Metacarpal
Phalanx 1
Phalanx 3
Phalanx 2

Figure 24.13 Fist- and knuckle-walking. (A) Orangutans fist-walk. (B,C) The knuckle-walks of gorillas and chimpanzees are different, evidence that knuckle-walking evolved independently in the two groups. (B,C after T.L. Kivell and D. Schmitt. 2009. *Proc Natl Acad Sci USA* 106: 14241–14246.)

fists and support themselves on the proximal phalanges, a mode of locomotion known as **fist-walking** (**Figure 24.12B**; **Figure 24.13A**). They also can walk by placing their weight flat on the palms of the hands, a method termed palmigrade walking (which is what humans do when we crawl on all fours). Palmigrade walking occurs in other primates, and we presume it to be plesiomorphic for Hominidae.

Orangutans are more arboreal than other hominids, and usually move slowly through the forest using all four limbs to grasp branches (**Figure 24.12C**). They sometimes walk bipedally on top of branches, supporting themselves by grasping tree limbs overhead. Generally solitary, groups of orangutans usually include only a female and her offspring. Individual orangutans use calls to maintain social contact, and groups of females form in some populations when food resources permit.

Orangutans use tools, including crumpled leaves as sponges to carry water to the mouth, sticks for scratching and removing seeds from fruit pods, and leaf umbrellas as shelters from the sun or heavy downpours (**Figure 24.12D**). At least 24 cultural variants are known based on different patterns of tool use and other behaviors.

Illegal logging, fire, conversion of lowland forest to oil palm plantations, and the pet trade threaten the three extant species of orangutans, all listed as Critically Endangered by the IUCN. An estimated 100,000 wild Bornean orangutans (*Pongo pygmaeus*) survive on the island of Borneo. The other two species—the Sumatran orangutan (*P. abelii*) and the Tapanuli orangutan (*P. tapanuliensis*)—occur on Sumatra, and populations of both are much smaller than those of the Bornean orangutan.

Homininae

Homininae includes gorillas (*Gorilla*) and the *Pan–Homo* clade (see Figure 24.9). *Gorilla* diverged from *Pan–Homo* ~9.06 Ma. The fossil record for early Homininae is scant because conditions for fossilization are poor in forest habitats.

Possible basal gorillas include †*Chororapithecus* (9 isolated teeth) and †*Nakalipithecus* (a partial lower jaw and 11 isolated teeth) known from the mid to late Miocene of Kenya (10–8 Ma). Some teeth from the middle Pleistocene (~1.2 Ma) of Kenya are the only evidence of fossil chimpanzees.

Gorillas Gorillas live in tropical moist forests of sub-Saharan Africa. They are more terrestrial than orangutans or chimpanzees. More folivorous than other Homininae, some populations also eat fruit. Like orangutans, gorillas are sexually dimorphic in body size—males may weigh up to 200 kg, twice the weight of females and have larger canine teeth (**Figure 24.14A**). Unlike orangutans, however, gorillas are social, living in troops led by silverbacks (the largest, most dominant males, named for the silver fur that develops in adult males). A gorilla troop usually consists of a single silverback and several females, with younger males living peripherally. Both sexes leave their troop when they reach adulthood. Gorillas use tools for activities such as food acquisition and social displays (**Figure 24.14B**) but are not known to have population-specific cultures like those of orangutans and chimpanzees. Male gorillas engage in chest-thumping displays (**Figure 24.14C**).

On the ground, gorillas move quadrupedally by knuckle-walking, a derived mode of locomotion in which they support themselves on the dorsal surface of the intermediate phalanges of digits 3 and 4, rather than closing their hands into fists, as orangutans do. The hand and wrist are aligned when a gorilla knuckle-walks (see Figure 24.13B).

The two extant species of gorillas diverged ~3 Ma. The range of the western gorilla, *Gorilla gorilla*, spans several African nations mostly west and north of the Congo River. The species is Critically Endangered. In addition to hunting and habitat loss due to mechanized logging, during the early 2000s the deadly Ebola virus decimated some western gorilla populations, and only ~100,000 individuals remain in the wild.

(A) Mountain gorilla, *Gorilla beringei beringei*, in its Rwandan habitat, yawning

(B) Western gorilla, *Gorilla gorilla*, using twig to extract insects from tree

(C) Chest-thumping display by two male eastern gorillas, *Gorilla beringei*

Figure 24.14 Gorilla behaviors. Male mountain gorillas (A) have large canine teeth. Gorillas use tools for food acquisition (B) as well as for social displays. They are also highly territorial, which can lead to agonistic displays between males (C).

Park in central Africa. Habitat destruction, armed conflict, and poaching threaten the remaining gorillas, and there is heightened concern that Ebola, which is now in the region, could devastate the remaining eastern gorillas.

Chimpanzees   There are two species in the genus *Pan*. The larger and more widely distributed common chimpanzee, *Pan troglodytes*, lives primarily in central and western Africa (**Figure 24.15A**). The bonobo, *Pan paniscus*, is smaller and occurs in central Africa, south of the Congo River (**Figure 24.15B**). The two groups diverged ~2.8 Ma.

Chimpanzees are more arboreal than gorillas, climbing nimbly and moving through forest canopy using walking, leaping, and suspensory locomotion. On the ground, chimpanzees knuckle-walk with a flexed wrist (see Figure 24.14C). This is unlike gorillas and is regarded

Both subspecies of the eastern gorilla, *Gorilla beringei*, also are Critically Endangered. There may be as few as 3,000 individuals of the eastern lowland gorilla (*G. b. graueri*) and only 1,000 mountain gorillas (*G. b. beringei*) living in mountain rainforest habitats such as Virunga National

(A) Common chimpanzee, *Pan troglodytes*

(B) Bonobos, *Pan paniscus*

Figure 24.15 Genus *Pan*: The common chimpanzee and the bonobo. Chimpanzees are larger and more robust than bonobos. (A) A male chimpanzee. (B) Bonobos mate face-to-face, a behavior not seen in any other nonhuman primate.

as evidence that knuckle-walking evolved independently in *Gorilla* and *Pan*.

In the past, the common chimpanzee had a continuous distribution across most of sub-Saharan Africa, but its current distribution is patchy, with four distinct populations recognized as subspecies (**Figure 24.16A**). The best studied population is the eastern chimpanzee, *Pan troglodytes schweinfurthii*, which lives north and east of the Congo River. Jane Goodall began studying this group in Gombe National Park, Tanzania in the early 1960s. During her many decades of field research, she documented chimpanzee tool use, hunting, and conflict, and showed that

chimpanzees have a far more complex behavioral ecology than people had suspected. Sixty years later, Goodall remains a tireless advocate for conservation.

Different wild populations of common chimpanzees have different tool cultures. For example, all six of the cultures shown in **Figure 24.16B** shake branches to attract attention or make aggressive displays and members of several populations open food items by smashing them, but in only two populations do chimpanzees insert a twig into a termite tunnel, withdraw the twig, and then sweep off and eat the attached insects by licking the twig.

(A) Ranges of the four extant subspecies of the common chimpanzee, *Pan troglodytes*, and the bonobo, *Pan paniscus* in sub-Saharan Africa

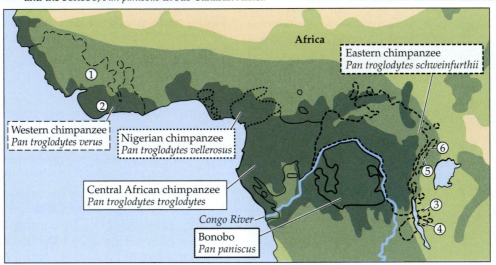

(B) Cultures of six populations of the common chimpanzee, *Pan troglodytes*

	Bossou, Guinea ①	Taï Forest, Côte d'Ivoire ②	Gombe, Tanzania ③	Mahale, Tanzania ④	Kibale, Uganda ⑤	Budongo, Uganda ⑥
Shake a branch to attract attention	+	+	+	+	+	+
Drag a large branch as an aggressive display	+	+	+	+	+	+
Squash a biting insect with a forefinger		+				
Open a food item by smashing it against a log or stone	+	+	+			+
Open a food item by placing it on a hard surface and cracking it with a stone or stick	+	+	+			
Use a twig to extract termites from tunnels				+	+	
Use a twig to capture safari ants	+	+	+			
Use a wad of leaves as a sponge to obtain water from a puddle	+	+	+		+	+
Strike forcefully with a stick		+	+			
Throw a stick or stone at a target	+	+	+	+		
Clasp hands with another chimp during grooming		+			+	+

Figure 24.16 Chimpanzee cultures vary among populations.
(A) There are four extant populations of the common chimpanzee *Pan troglodytes* in forested and savanna habitats north of the Congo River in equatorial Africa. These four isolated populations are recognized as subspecies. The geographic range of the bonobo (*P. paniscus*) lies south of the Congo River. (B) Chimpanzee populations have different cultures (defined here as a repertoire of behaviors). Shown here are behaviors observed in two cultures of the western chimpanzee (*P. t. verus*; cultures 1 and 2) and four cultures of the eastern chimpanzee (*P. t. schweinfurthii*; cultures 3–6). Locations of the sites at which these cultures were studied are indicated in (A). Some behaviors are common to all cultures; squashing an insect with a forefinger, on the other hand, is observed in only one. In four of the cultures, chimpanzees open a hard food item by beating it against a rock or log, but only three of the four cultures have taken the "next step" of placing the food item on hard surface and cracking it with a rock. (Data from A. Whiten et al. 1999. *Nature* 399: 682–685.)

Common chimpanzees are only moderately sexually dimorphic. Like gorillas, they live in groups, but with multiple males instead of a single male. They are very territorial (especially against members of other groups), with a rigid male-dominance structure and frequent aggressive behaviors. Their social organization follows a fission–fusion pattern, meaning that small groups coalesce into a larger group and then split into smaller groups in response to changes in the abundance and spatial distribution of food sources. (You may recall that we described a similar type of behavior among sharks; see Figure 6.25.)

There is evidence that common chimpanzees can plan for future events, such as the availability of ephemeral fruits like ripe figs. Figs are a preferred food of chimpanzees, but trees with ripe figs have a patchy distribution and many animals compete for them. Researchers observing the position of sleeping nests and schedules of a group of five adult female western chimpanzees found that during fig season chimpanzees moved in the evening toward places where they could expect to find ripe figs for breakfast and built their sleeping nests close to the figs. They left their sleeping nests unusually early the next morning (before dawn in some cases) to breakfast on figs. The ability to plan for future events—especially chances to obtain high-quality food—may have played a role in evolutionary increases in brain size and cognition. Researchers hypothesize that such superior cognitive abilities for finding food can offset the higher energetic costs of maintaining a large brain. Referred to as the ecological intelligence hypothesis, similar cognitive abilities may have played a role in the evolution of the human lineage.

Meat is a significant component of the diet of common chimpanzees, which are major predators of red colobus monkeys (*Piliocolobus*). Groups of adult males engage in cooperative hunts, pursuing monkeys through the treetops while females and juveniles follow the hunt at ground level. Males that participate in a hunt and females allied with them share successful kills.

Bonobos live in forested habitats in central Africa south of the Congo River (see Figure 24.16A) and differ from common chimpanzees in many respects.

- Bonobos live in female-centered groups and do not engage in aggressive behaviors as often as common chimpanzees.

- Bonobos are less sexually dimorphic than chimpanzees.

- Genital swelling signals estrus (the time of fertility and short-term sexual receptivity) in female chimpanzees, but female bonobos (like humans) have concealed ovulation and continuous sexual receptivity.

- Bonobos use sex for bonding and for conflict-solving. In addition to heterosexual copulation, sexual behaviors of bonobos include same-sex genital rubbing.

- Bonobos are not known to use tools.

- Meat is not a significant component of the diet of bonobos.

The second group of the *Pan–Homo* clade of Homininae comprises the hominins, which we will discuss in the rest of this chapter.

24.3 Origin and Evolution of Hominini

LEARNING OBJECTIVES

24.3.1 Describe the major evolutionary trends in hominins.

24.3.2 Identify the earliest candidates for potential hominins.

24.3.3 Compare the skulls and diets of †*Australopithecus* and †*Paranthropus*.

24.3.4 Understand the ecological context in which bipedalism evolved.

As stated in the previous section, the *Pan–Homo* clade includes chimpanzees, bonobos, and humans (see Figure 24.9). Humans and their immediate ancestors (i.e., extinct taxa more closely related to *Homo sapiens* than to *Pan*) belong to the tribe Hominini, commonly known as hominins. *Pan* and Hominini diverged in Africa ~6.7 Ma and most of the early diversification of hominins occurred in Africa. †*Ardipithecus* is the sister group of more derived hominins, which include both modern humans and extinct species of *Homo*, among others. All humans alive today belong to one species, *Homo sapiens*.

Distinctive features of hominins

Although humans differ from other great apes in anatomy, culture, and language, molecular studies confirm that we are remarkably close genetically to common chimpanzees and bonobos, having ~98.8% similarity at the genomic level. Given that we are so similar genetically, what explains the great differences between chimpanzees and humans? In broad terms, the sheer size of the genomes (~3.1 billion base pairs) means that there are >37 million base-pair differences between *Pan* and *Homo*. But base-pair differences only partly explain the observed phenotypic differences between chimps and humans because changes in gene regulation and expression patterns during development can produce strikingly different outcomes (see Section 1.5).

Many fossil hominins are known. We can study and compare their anatomy, measure volumes of their braincases, study the external shapes of their brains, predict what they ate based on their dentition, reconstruct their faces and body forms, and in some cases even extract ancient DNA. This material evidence of human evolution is fascinating, but it is impossible to draw a firm line between non-human hominins and ourselves. Becoming human over the last 6.7 Ma was a process, not an on-off switch. For example, both *Pan* and *Homo* are self-aware (conscious) and can predict future events, so by phylogenetic inference our hominin ancestors must also have been self-aware. How these self-aware hominins interacted with their environments and each other over the last 6.7 Ma are central questions with very incomplete, essentially

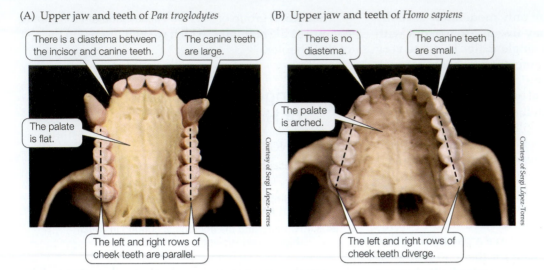

(A) Upper jaw and teeth of *Pan troglodytes*

(B) Upper jaw and teeth of *Homo sapiens*

There is a diastema between the incisor and canine teeth.

The canine teeth are large.

The palate is flat.

The left and right rows of cheek teeth are parallel.

There is no diastema.

The canine teeth are small.

The palate is arched.

The left and right rows of cheek teeth diverge.

Courtesy of Sergi López-Torres

Courtesy of Sergi López-Torres

Figure 24.17 Comparison of the upper jaws and teeth of chimpanzees and humans. (A) Chimpanzees have a long, U-shaped upper jaw with parallel rows of cheek teeth. The palate is flat, and there is a diastema (gap) between the large, pointed canines and the incisors. (B) Humans have a shorter upper jaw, correlated with a shorter muzzle. The upper jaw is V- or bow-shaped, with diverging rows of cheek teeth and a prominently arched palate. The canines are small and blunt, and the teeth are relatively uniform in size with no gaps between them.

unknowable answers. Here our goal is to interpret broad evolutionary patterns and trends.

Many well documented anatomical trends differentiate hominins from chimpanzees. For example, as shown in **Figure 24.17**, the jaws of hominins became shorter, the canine teeth became smaller, and the diastema (gap) between the incisors and canine teeth disappeared. The enamel layer of the teeth became thicker. Other skeletal changes are highlighted in **Figure 24.18**.

- The braincase became greatly enlarged in association with an increase in forebrain size. A prominent vertical forehead evolved.

- Brow ridges and crests for muscle origins on the skull became smaller in association with the reduction in size of the muscles that attach to them. The nose became a more prominent feature of the face, with a distinct bridge and tip.

- The points of articulation of the skull with the vertebral column (the occipital condyles) and the foramen magnum (the hole for the passage of the spinal cord through the skull) shifted from the ancestral position at the rear of the braincase to a position under the braincase. This change balances the skull on top of the vertebral column and signals the appearance of an upright posture.

- The iliac blade became shorter and the pelvis basinlike.

- The rib cage became shorter, resulting in a waist between the ribs and the pelvis.

- In conjunction with bipedalism, the vertebral column evolved an S-shape. Changes in foot anatomy included the evolution of a relatively larger and less divergent big toe that is no longer opposable (**Figure 24.19**).

For many years, scientists linked hominin bipedalism to the appearance of east African savannas, which are grasslands with widely spaced bushes and trees. These savanna environments developed after the formation of the Isthmus of Panama 2.5 Ma blocked the flow of water between the Atlantic and the Pacific Oceans at low latitudes and led to profound global climate changes (see Section 20.1). We now know that hominin bipedalism extends back at least 4.4 Ma, and possibly as far back as 7 Ma, long before the development of East African savannas. Thus, it seems that bipedalism evolved in forested environments and not in relation to savannas.

Early hominins

There is great uncertainty about evolutionary relationships among early hominins, and many of the named taxa are not demonstrably monophyletic. Here we examine some of the best preserved forms to explore the evolution of bipedalism and other features of anatomically modern humans.

†*Ardipithecus* †*Ardipithecus* (Amharic *ardi*, "ground floor") is known from the Middle Awash area of Ethiopia, approximately 5.8–4.4 Ma. Based on the position of its foramen magnum, †*Ardipithecus* was probably an upright biped (**Figure 24.20**). It had small canines and less sexual dimorphism in dentition than gorillas and chimpanzees. Decreased sexual dimorphism in dentition, as in modern humans, is associated with a pair-bonding type of mating system rather than one in which males compete for access to females as in most extant African apes.

The skeleton of †*Ardipithecus* has similarities both to chimpanzees and to more derived hominins. For example, the pelvis indicates capacity for bipedal locomotion.

(A) Skeleton of *Pan troglodytes*

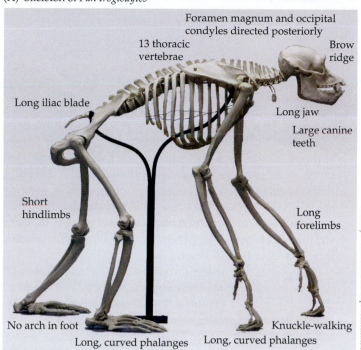

Foramen magnum and occipital condyles directed posteriorly

13 thoracic vertebrae

Brow ridge

Long iliac blade

Long jaw

Large canine teeth

Short hindlimbs

Long forelimbs

No arch in foot

Knuckle-walking

Long, curved phalanges

Long, curved phalanges

Courtesy of Bone Clones Inc./https://boneclones.com

Figure 24.18 Skeletons of a chimpanzee and a human.
Note the shorter iliac blade, less funnel-shaped rib cage, longer legs, and shorter fingers and toes of the human. The elongate pelvis of the chimpanzee forms part of a lever system of the hindlimbs in this quadrupedal primate. The bowl-shaped pelvis of the human supports the abdominal organs during bipedal locomotion.

(B) Skeleton of *Homo sapiens*

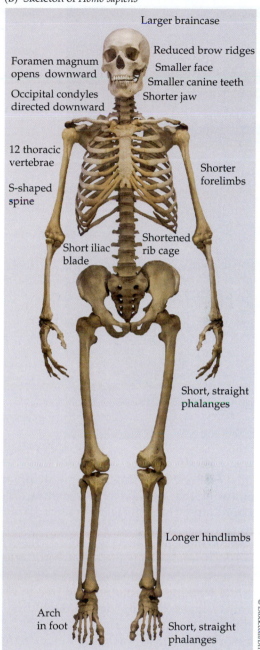

Larger braincase

Reduced brow ridges
Smaller face
Smaller canine teeth
Shorter jaw

Foramen magnum opens downward

Occipital condyles directed downward

12 thoracic vertebrae

S-shaped spine

Shorter forelimbs

Short iliac blade

Shortened rib cage

Short, straight phalanges

Longer hindlimbs

Arch in foot

Short, straight phalanges

© iStock.com/DrPAS

Figure 24.19 Feet of extant hominoids. Hylobatids (A) and orangutans (B) have very long toes and a divergent hallux ("big toe"). Gorillas (C) and chimpanzees (D) have shorter toes and a less divergent hallux. Humans (E) have extremely modified feet in which the large hallux aligns with the other toes.

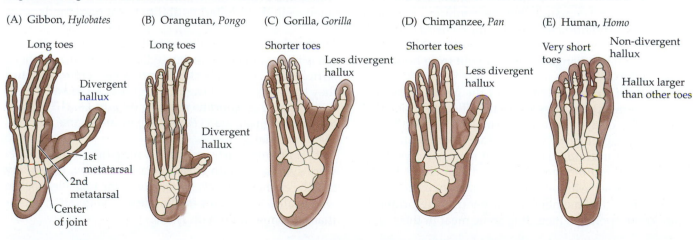

(A) Gibbon, *Hylobates*

Long toes

Divergent hallux

1st metatarsal
2nd metatarsal
Center of joint

(B) Orangutan, *Pongo*

Long toes

Divergent hallux

(C) Gorilla, *Gorilla*

Shorter toes

Less divergent hallux

(D) Chimpanzee, *Pan*

Shorter toes

Less divergent hallux

(E) Human, *Homo*

Very short toes

Non-divergent hallux

Hallux larger than other toes

Restoration of †*Ardipithecus ramidus*

©Julius T. Csotonyi/Science Source

Figure 24.20 †*Ardipithecus ramidus*.   The location of the foramen magnum on the base of the skull and the structure of the pelvis and hindlimbs indicate that †*Ardipithecus* was bipedal. Arboreal characteristics shown in this restoration include long arms, hands with long fingers, and a foot with long toes and a divergent big toe.

However, the foot had long toes and a highly divergent big toe, indicating that †*Ardipithecus* was more arboreal than later hominins. The arms and hands of †*Ardipithecus* were longer than those of later hominins but did not closely resemble those of gorillas or chimpanzees. There is no evidence of knuckle-walking. Instead, the wrist and finger joints suggest that †*Ardipithecus* used its hands for support while walking bipedally along tree branches. These features suggest that bipedalism might first have evolved in trees.

The brain of †*Ardipithecus* was smaller in relation to its body size than the brain of more advanced hominins. Its face sloped less than does the face of a chimpanzee, but more than the faces of more derived hominins. From †*Ardipithecus*, we can conclude that the common ancestor of chimpanzees and humans was not simply like a chimpanzee, and that chimpanzees, too, have undergone substantial evolutionary change since their split from hominins.

†*Australopithecus*   Members of the genus †*Australopithecus* were bipedal apes with a modified dentition that includes the derived features of small canines and thick enamel typical of *Homo sapiens*. Males were larger than females, and individuals grew and matured rapidly, in contrast to the prolonged childhood that characterizes our own species. Microwear analyses of their teeth suggest that the early members of the genus were primarily frugivorous, perhaps including some meat in their diet.

Mary Leakey and her associates discovered tracks made by a foot with a humanlike arch in volcanic ash beds dated between 3.8 and 3.6 Ma at Laetoli in Tanzania. †*Australopithecus afarensis*, which is known from fossils at the same site, probably made these tracks. Another set of tracks nearby has different characteristics, suggesting that a second species of bipedal hominin was also present. Both sets of footprints resemble those of anatomically modern humans preserved on similar substrates, demonstrating that bipedalism in hominins evolved much earlier than did an enlarged brain.

The skeleton of a single young adult female †*Australopithecus afarensis*, popularly known as Lucy, was found in the Afar region of Ethiopia in a deposit dated at 3.2 Ma. Young but fully grown when she died, Lucy was ~1 m tall and weighed ~30 kg. Other finds indicate that males of her species were larger, 40–60 kg. Lucy's teeth and lower jaw are clearly humanlike, but the cranial volume (380–522 cm^3) is closer to that of extant chimpanzees and gorillas. †*Australopithecus afarensis* had a steep forehead (**Figure 24.21A**).

Despite modifications for bipedalism, †*Australopithecus afarensis* retained features that suggest it was semiarboreal. Evidence comes from bones of the fingers and toes, which were longer and more curved than those of anatomically modern humans. †*Australopithecus afarensis* could stand and walk bipedally as we do, but probably spent much time in the trees. A related species, †*Australopithecus africanus*, which lived 3.7–2.1 Ma, had heavy arm bones, suggesting that it may have spent more time in trees than †*A. afarensis*. †*Australopithecus amensis* lived in east Africa 4.2–3.9 Ma. Its fossils are associated with those of forest mammals, reinforcing the notion that early hominin evolution took place in woodlands rather than on the savanna as was once assumed. †*Australopithecus sediba* is known from a single fossil site in South Africa dating to ~1.8 Ma. This species appears to be close to *Homo* in cranial and postcranial features. Like *Homo*, †*A. sediba* had small teeth that limited its ability to eat hard foods.

†*Paranthropus*   Another early diverging genus of hominin is †*Paranthropus*, which lived in eastern and southern Africa 2.7–1 Ma. Their skulls and jaws were robust and they had a prominent sagittal crest that served as the origin for large temporalis muscles (**Figure 24.21B**). These features suggest that †*Paranthropus* ate coarse, fibrous plant material.

†*Paranthropus aethiopicus*, †*P. robustus*, and †*P. boisei* were sympatric with early species of *Homo*. It is likely that the extinction of †*Paranthropus* in the mid-Pleistocene was related to changes in climate and vegetation rather than to competition with *Homo*.

(A) Cranium of †*Australopithecus afarensis*

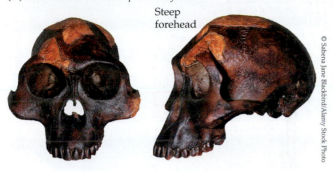

Steep forehead

© Sabena Jane Blackbird/Alamy Stock Photo

(B) Cranium of †*Paranthropus boisei*

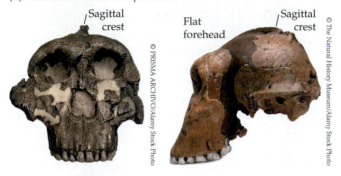

Sagittal crest

Flat forehead

Sagittal crest

© PRISMA ARCHIVO/Alamy Stock Photo

© The Natural History Museum/Alamy Stock Photo

Figure 24.21 Crania of early hominins. (A) The lightly built cranium of †*Australopithecus afarensis* has small teeth. (B) In contrast, the cranium of †*Paranthropus boisei* is massive with a prominent sagittal crest for the origin of a large temporalis muscle, allowing it to crush tough vegetation with its enormous teeth.

24.4 The Genus *Homo*

LEARNING OBJECTIVES

24.4.1 Describe the different species within the genus *Homo*.

24.4.2 Summarize the major migrations of anatomically modern humans around the globe.

24.4.3 Draw a timeline to show when species of *Homo* coexisted.

24.4.4 Describe some examples of hybridization between species of *Homo* and explain how we know this occurred.

Throughout this book we have presented evolutionary relationships as cladograms that show a branching sequence of lineages split into dichotomous clades, as described in Chapter 1. But it is hard to detect, interpret, and depict dichotomous branching patterns for groups that diverged over short periods of geologic time, changed gradually, and hybridized extensively. This is what happened during the Pleistocene evolution of hominins. Based on both anatomical and molecular information, we now know that multiple lineages of hominins spread across the Earth, evolving as local populations with different combinations of ancestral and derived characters, and in many cases interbred with each other **Figure 24.22**.

†Homo habilis

Most researchers regard †*Homo habilis* (Latin *habilis*, "able") as the earliest member of the genus *Homo*. It lived in eastern Africa ~2.8–1.65 Ma and has a larger cranial volume (500–750 cm³) than †*Australopithecus afarensis* (380–522 cm³). †*Homo habilis* also has a smaller face, a smaller jaw, smaller cheek teeth, and larger incisors than †*A. afarensis*. Like †*A. afarensis*, †*H. habilis* was relatively small and retained specializations for climbing. Fossils of †*H. habilis* occur in association with stone artifacts and fossil bones with cut marks, suggesting that they used tools and hunted or scavenged. †*Homo habilis* co-occurred with the more derived †*H. erectus* in eastern Africa for nearly 500 ka.

†Homo erectus

Fossils of †*Homo erectus* (Latin *erectus*, "upright") occur in Africa and Asia, making it the first intercontinentally distributed hominin, and show that this species was able to make ocean voyages. †*Homo erectus* coexisted with other hominin species in southern and eastern Africa and in Asia for at least several hundred thousand years.

Originally described in the late 19th century as †*Pithecanthropus erectus* and known at that time as Java Man or Peking Man, †*Homo erectus* had a large body, relatively small teeth and jaws, and lacked specializations for climbing. The species shared several features with anatomically modern humans.

- *Large size* Substantially larger than earlier hominins (up to 1.85 m tall and weighing at least 65 kg), †*Homo erectus* was about the same size as modern humans.

- *Increased female size and reduced sexual dimorphism* Males were 20–30% larger than females, as in modern humans. Reduced sexual dimorphism in these and later species of *Homo* implies a change from a polygynous mating system (in which males compete for access to females) to monogamous pair-bonding.

- *Body proportions* These resemble the proportions of anatomically modern humans, with short arms, long lower legs, and a narrow pelvis.

- *Larger brains* Cranial volume was larger than in earlier hominins, ranging from 775 to 1,100 cm³ (close to the average of 1,200 cm³ for anatomically modern humans).

- *Tool-making* Hand axes appeared in Africa 1.75 Ma as part of a tool-making industry also known in Europe and western Asia.

- *Humanlike nose* Home erectus had a broad, flat nose with downward-facing nostrils. Other facial features that were less like those of anatomically modern humans include jaw projection beyond the plane of the upper face (a prognathous jaw); no chin; forehead flat and sloping; and prominent bony brow ridges.

- *Relatively small teeth* This feature suggests that food was cooked, because cooked food is easier to chew. Fossils of †*Homo erectus* co-occur with burnt bones at a Pleistocene site (~0.7–0.4 Ma) near Beijing.

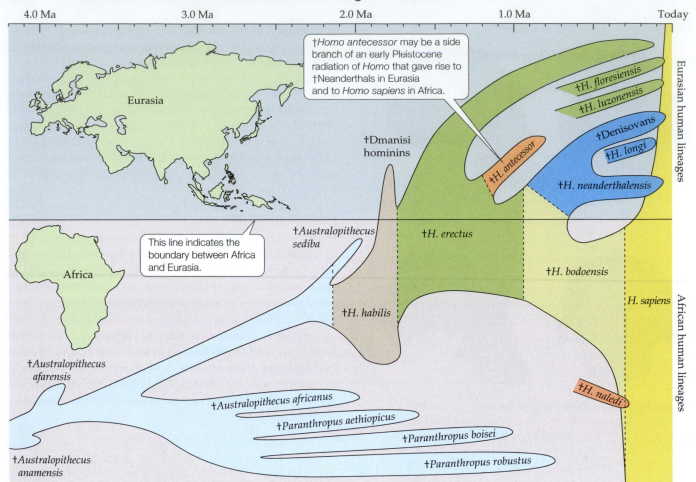

Figure 24.22 Hominin radiations. Hominins originated and diversified in Africa. Separate radiations of hominins migrated into Europe and Asia during the Pleistocene. Local populations of some species overlapped in time and space, allowing interbreeding and the contribution of other genomes to modern *Homo sapiens* (see Figure 24.25). (Inspired by data in J.A. Lindal et al. 2021. *What's in a name? When it comes to human fossils, it's complicated.* theconversation.com/us.)

- *Delayed tooth eruption* Delayed tooth eruption suggests a humanlike extended childhood, which also implies an extended lifespan and humanlike passage of learned information from one generation to the next.

Paleoanthropologists long thought that †*Homo erectus* disappeared between 300 and 200 ka. However, remains of †*H. erectus* from Java date between 53 and 27 ka by two methods, and between 60 and 40 ka by a third. The complex sediments in which these fossils occur are hard to decipher, making these dates controversial, but if they are correct, then late-surviving †*H. erectus* would have been contemporaries of †Neanderthals, †Denisovans, modern *Homo sapiens*, and potentially †*H. floresiensis*, hominins we describe in the remainder of this section.

†Dmanisi hominins

Fossils found at Dmanisi, Georgia (between Turkey and Russia) are the earliest hominins from Eurasia, dating to ~1.8 Ma. A recently described skull combines a small braincase of only 546 cm³ with a very large lower jaw. This cranial volume is much smaller than the average for †*H. erectus* (~900 cm³), and the brain was not only smaller but also simpler than that of *H. erectus*. Although the †Dmanisi hominins resembled †*H. erectus*, their smaller brain, smaller stature (~1.5 m), and prognathous jaw suggest that they may have evolved from a less derived hominin such as †*H. habilis*. Aspects of the elbow and shoulder joints support this interpretation.

†Neanderthals

The first fossils of †Neanderthals (†*Homo neanderthalensis*) were found in the Neander Valley in western Germany in 1856. Since then, †Neanderthal remains have been recovered from a broad geographic range in western Eurasia. The oldest fossil of a †Neanderthal comes from Krapina, Croatia, (130 ka). However, older remains from Sima de los Huesos, Spain (430 ka) might be early members of the Neanderthal lineage. The last populations of †Neanderthals lived in Gibraltar until 28 ka. DNA analyses suggest that †Neanderthals were not directly ancestral to *Homo sapiens*; instead, the two lineages diverged from early archaic forms ~588 ka.

Neanderthals were generally short and stocky compared with modern humans (**Figure 24.23**) but some

†Homo neanderthalensis

Figure 24.23 †Neanderthals had robust, muscular bodies. This restoration shows the barrel-shaped chest, large joints, short limbs, receding forehead, large protruding nose, prominent brow ridges, and receding chin typical of †Neanderthals.

have been estimated to be up to 188 cm—close to 2 m—in height. Their stocky build may have been an adaptation for the cold conditions of Ice Age Europe. The †Neanderthal brain was as large or larger than that of modern *Homo sapiens* and the relative sizes of the brain regions differed. †Neanderthals had a larger occipital area (the region at the back of the head where visual information is processed) than anatomically modern humans, whereas we have a larger middle temporal region (the area in which auditory information, speech, and vision are processed).

†Neanderthals appear to have been much stronger than extant humans. They were stone toolmakers, and there is evidence of symbolic behavior, such as circular constructions of stalagmites—the oldest known constructions made by humans—as well as personal ornaments and cave art. Whether the †Neanderthals had complex speech remains controversial, but they are the first humans known to bury their dead, apparently with considerable ritual. Of special importance are burials at Shanidar Cave in Iraq that include plants known in modern times for their medicinal properties.

†Neanderthals probably hunted large mammals using weapons that required close contact with their prey (they did not have distance weapons such as throwing spears or bows and arrows). Many skeletal remains show evidence of serious injury during life, and the patterns of injury resemble those of present-day rodeo bull riders. Despite the high incidence of injuries, 20% of †Neanderthals were more than 50 years old at the time of death; it was not until after the Middle Ages that human populations again achieved this longevity.

†Denisovan hominins

Fragmentary remains of hominins who lived 200–30 ka have been found in Denisova Cave in the Altai Mountains of southern Siberia. Very little skeletal material is known—only three isolated teeth and a finger bone. †Denisovan molars are distinctive: they are large and lack traits typical of †Neanderthal and human molars. Unlike most cases in paleoanthropology, the main evidence that †Denisovans are an independent hominin lineage comes from molecular studies. They were genetically distinct from †Neanderthals, but contemporaries of †Neanderthals and anatomically modern humans (see Figure 24.22). A fossil of a female hominin found in Siberia (~90 ka) has been determined to be the child of a †Neanderthal and a †Denisovan.

†Homo longi

A newly described species of hominin from China, *†Homo longi* (Mandarin *long*, "dragon"), may be a Denisovan. This "dragon man" is known only from a cranium with one molar dated to ~146 ka. Many features distinguish this specimen from modern humans, including larger, almost square eye sockets, thick brow ridges, and a wide mouth, but the most striking feature is the massive cranial volume (1,420 cm^3). A significantly larger brain size may be one of the key features of Denisovans. Thus, if a DNA analysis of *†H. longi* shows it to be a Denisovan, then we will at last be able to put a face to this mysterious lineage of hominins.

Island species and miniaturization

The "island rule" in evolutionary biology refers to strange phenomena that can occur when animal lineages become isolated on an island: small species evolve to larger body sizes, and large species become miniaturized. The giant land tortoises of Aldabra Island and the Galápagos are familiar examples of island gigantism, and four islands in eastern Indonesia are home to the largest extant species of lizard, the Komodo dragon. During the Pleistocene, giant rats and miniature elephants occurred on islands in Indonesia.

†Homo floresiensis and *†H. luzonensis*, fossil hominins from the islands of Flores in eastern Indonesia and Luzon in the Philippines, respectively, demonstrate miniaturization. Both Flores and Luzon are volcanic islands never connected to continental Asia, which means that the ancestors of these two species would have had to make sea voyages to reach the islands.

Figure 24.24 **Comparison of *Homo sapiens* and †*Homo floresiensis*.** (A) Relative sizes of *Homo sapiens* and †*H. floresiensis*. (B) The cranial volume of †*H. floresiensis* was about 400 cm³, compared with 1,200 cm³ for extant *H. sapiens*. Anthropologists regard †*Homo floresiensis* as a distinct species of human with a relatively derived (though miniature) brain.

(A) Appearance of *Homo sapiens* and †*H. floresiensis*

(B) Skulls of *Homo sapiens* and †*H. floresiensis*

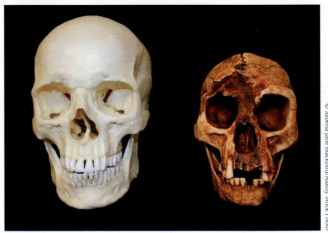

†*Homo floresiensis* During the late Pleistocene Flores Island was inhabited by †*Homo floresiensis*, a species which was only about 1 m tall—much smaller than any other known species of *Homo* (**Figure 24.24**). Skeletal remains of †*H. floresiensis* occur along with stone tools in deposits from ~700–50 ka. †*Homo floresiensis* had primitive canines and premolars but advanced molars—a combination of dental traits not seen in any other hominin. Because of its small size, it was nicknamed "the Hobbit" by the popular press. Although its origin is debated, most paleoanthropologists think †*Homo floresiensis* evolved from a form like †*H. erectus*.

Shortly after the disappearance of †*H. floresiensis*, fossils of modern humans appear at the same sites, dated to ~46 ka. This raises the question of whether modern humans played a role in the extinction of †*H. floresiensis*, as well as whether the two species interbred.

†*Homo luzonensis* Fossils of hominins from Callao Cave on the Philippine island of Luzon dated to before 50 ka and (possibly as old as 700 ka) display a combination of primitive and derived dental traits that differ from those of †*Homo floresiensis*. This species was small, although the fragmentary nature of the remains does not allow an estimate of adult height. Bones of deer and hogs in the cave were probably brought in by hominins because there is no evidence of other large predators on Luzon during the Pleistocene. Cut marks on some of the bones indicate that the carcasses were butchered, although no stone tools have been found.

†Homo naledi

In 2015, †*Homo naledi* (Sotho *naledi*, "star") was described based on the remains of at least 15 individuals recovered from the Dinaledi Chamber, deep in the Rising Star cave system of South Africa. Recent studies suggest that †*H. naledi* lived ~335– 236 ka. †*Homo naledi* stood ~1.5 m tall and had humanlike shoulders, although its pelvis more closely resembled those of earlier hominins. Its foot was much like that of a modern human, but it had longer and more curved finger bones, indicating some ability to climb. It had a small cranial volume, ~465–610 cm³.

How did the bones of †*Homo naledi* end up in the Dinaledi Chamber, which is the deepest (and barely accessible) chamber of the cave system? No other animal remains were found in the chamber, and the hominin bones had not been damaged by scavengers or predators, all of which suggests that the bodies may have been deliberately placed deep in the cave by other individuals of the species.

†Homo bodoensis

†*Homo bodoensis* lived in Africa (and possibly Europe) ~600–300 ka and may have given rise to the earliest *Homo sapiens* (see Figure 24.22). Although this species shows many primitive characters similar to those of †*H. erectus*, its cranial volume is large, ~1,300 cm³. A brain this size places †*H. bodoensis* well within the range found in modern humans.

Origin and radiation of Homo sapiens

We can trace the maternal lineage of an individual by analyzing mitochondrial DNA, and such analyses of people from all over the world show that all living humans trace their mitochondria to a woman who lived in southern Africa ~200 ka. This means that this woman has had an unbroken series of daughters in every generation since then.

A similar approach using the Y chromosome, which is passed only from fathers to sons, indicates that all human males descended from a single individual who lived in Africa ~59 ka. The difference between these estimates—200 ka versus 59 ka—results from uncertainty about the rates of mutation in mitochondrial DNA and the Y chromosome. Both studies indicate that the common ancestor of modern humans lived in Africa, a conclusion reinforced by other genetic information. For example, there is more variation in the genomes of humans in Africa than in the rest of the world combined, which is exactly what we would expect if humans originated in Africa. Furthermore, humans have only about one-tenth the genetic variation of chimpanzees, an observation indicating that human populations were once very small, passing through a genetic bottleneck ~200 ka.

From Africa, anatomically modern humans spread across the world.

- Anatomically modern humans crossed into the Levant region of Asia (the area now called the Middle East) between 194 and 177 ka. About 75 ka, a single dispersal of non-Africans spread across Eurasia and Australasia, reaching the islands of Southeast Asia by 73 ka and Australia by 65 ka. Humans arrived in Siberia between 48 and 44 ka, central Asia by 40 ka, and eastern Asia between 39 and 36 ka.

- Anatomically modern humans arrived in Europe from the Levant region about 45 ka. European populations went through a genetic bottleneck during the Last Glacial Maximum (~25 ka; see Figure 20.3), with evidence of a major population influx from the Levant ~14 ka.

- Human footprints at White Sands National Park in New Mexico confirm the presence of modern humans in the Americas between 23 and 21 ka. Archaeological evidence from Chiquihuite Cave in central Mexico could push back this date back to 26.5 ka. Genetic evidence links the First Peoples of the Americas to Siberians who crossed Beringia during the Pleistocene (see Figure 20.4A).

What happened to the humans who were already there?

As the preceding discussion described, as recently as 30 ka *Homo sapiens* shared the planet with at least three other species—†Neanderthals, †Denisovans, and †*H. floresiensis*; †*H. erectus* may have been present as well. The present-day situation of *Homo sapiens* as the sole existing species of hominin is a phenomenon of the last 27 ka or less. Thus, many areas invaded by anatomically modern humans were already home to other species of *Homo*. What happened to them?

Until recently, most paleoanthropologists supported the "replacement hypothesis," an interpretation of human evolution in which anatomically modern humans completely replaced the populations of other species of *Homo* they encountered, with older populations vanishing without a trace. But genetic studies reveal interbreeding

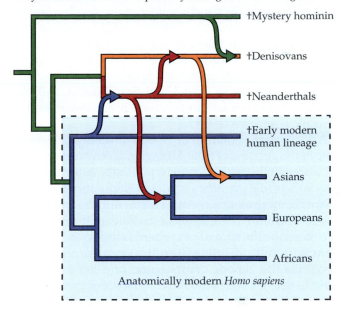

Figure 24.25 Genetic introgression among hominin lineages. Studies of ancient DNA reveal that modern Eurasian populations carry †Neanderthal DNA, reflecting an interbreeding event that occurred before European and Asian populations separated, but after the Eurasian lineage separated from the African lineage. Some modern Asian populations carry †Denisovan DNA as well. †Neanderthals also interbred with †Denisovans, and there is evidence of DNA from a "mystery hominin" lineage present in †Denisovan populations. (After E. Callaway. 2016. *Nature* 10.1038/nature.2016.19394.)

among anatomically modern humans and populations of other *Homo* that they encountered as they spread from Africa. DNA analyses show that both early modern and modern humans interbred with †Neanderthals, modern humans interbred with †Denisovans, †Neanderthals interbred with †Denisovans, and †Denisovans interbred with an unknown population of hominins whose skeletal remains are yet to be discovered (**Figure 24.25**); this "mystery hominin" could potentially be an offshoot of Asian †*H. erectus*.

Comparison of the †Neanderthal genome with the genome of living humans reveals that 2–3% of the nuclear DNA in present-day human populations outside of sub-Saharan Africa came from †Neanderthals, not from our most recent African ancestors. A higher percentage of †Neanderthal DNA is found in east Asian populations. The most plausible interpretation is that modern humans interbred with †Neanderthals in the Middle East between 60 and 50 ka, or even more recently. Remains of an anatomically modern human from Peştera cu Oase, Romania, dated to ~40 ka, had 6–9%, a percentage suggesting that this individual had a †Neanderthal ancestor as recently as 4–6 generations back. Three other modern humans from Bacho Kiro Cave, Bulgaria, dated between 46 and 43 ka, belonged to a westward migration from Asia, and all had †Neanderthal ancestors only a few generations back in their family histories.

Sometime after the admixture of †Neanderthals and modern humans in southwestern Asia, a subgroup of modern *Homo sapiens* moved eastward into Asia, where they interbred with †Denisovans ~45–55 ka. As much as 3.5% of the DNA of descendants of this group of modern humans now living on islands in Southeast Asia and Oceania is Denisovan. East Asian, South Asian, and Native American populations possess a much lower level of Denisovan ancestry (~0.1%).

A 4–8% representation of †Neanderthals or †Denisovans in the genome of non-African modern humans does not discredit the replacement hypothesis—after all, more than 90% of the genome of human populations outside Africa *does* represent a recent African origin—but it calls for modification of the hypothesis. Paleoanthropologists now describe the spread of modern humans as "replacement with hybridization" (or "leaky replacement") to recognize the genetic contribution of other species of *Homo* to the genome of modern humans. For example, the archaic inheritance modern Eurasians have received from †Neanderthals includes genes that boost immune response, define cranial and brain morphology, and promote fertility. On the downside, other †Neanderthal genes are linked to diseases such as depression, skin lesions, blood clots, and urinary tract disorders.

24.5 Evolution of Human Characters

LEARNING OBJECTIVES

24.5.1 Understand the consequences of adopting a bipedal posture.

24.5.2 Understand the consequences of the increase in brain size to human development.

24.5.3 Explain the origins of speech.

In comparison to other hominids, humans share three major derived characters: (1) a bipedal stance and mode of locomotion; (2) a greatly enlarged brain; and (3) the capacity for speech and language. Here we examine possible steps in the evolution of each of these key features.

Bipedalism

Although all modern hominoids can stand erect and walk to some degree on their hindlegs, only hominins display an erect, bipedal, striding mode of locomotion involving specializations of the pelvis and hindlimbs. Such bipedality frees the forelimbs from obligatory functions of support, balance, and locomotion. The most radical changes in the hominin postcranial skeleton are associated with the assumption of a fully erect, bipedal stance in the genus *Homo* (**Figure 24.26**). Anatomical modifications include the S-shaped curvature of the vertebral column, modifications of the pelvis and position of the acetabulum (hip socket), lengthening of the leg bones, and shifting the knee joint directly beneath the midline of the body. Humans also differ from apes in having a longer trunk with a more barrel-shaped (versus funnel-shaped) rib cage, resulting in a distinct waist. The waist allows the pelvis

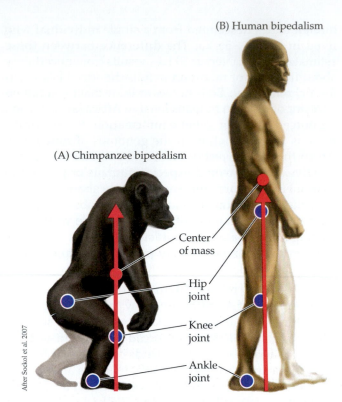

Figure 24.26 Bipedal locomotion in hominids. The trunk, pelvis, and vertebral column of chimpanzees and humans differ in relation to bipedal locomotion. (A) When a chimpanzee walks bipedally, its center of gravity (red dot) is anterior to its hip joint and the chimpanzee must bend its legs at the hip and knee, using muscles to hold its center of gravity over the foot that is its base of support. (B) The S-shaped curve of the human vertebral column places the center of gravity over the supporting foot and most of the weight of a human is transmitted directly through the bones of the supporting leg to the ground. (After M.D. Sockol et al. 2007. *Proc Natl Acad Sci USA* 104: 12265–12269.)

to rotate during bipedal striding without also rotating the upper body (see Figure 24.18).

The S-shaped curvature of the vertebral column is a consequence of bipedal locomotion and develops only when an infant learns to walk. We do not have perfect spines to accommodate the stresses of bipedal locomotion, which are quite different from those encountered by quadrupedal hominids. Consequences of these stresses include high incidences of lower-back problems in modern humans, as well as other problems in the lower trunk, such as hemorrhoids, constipation, and incontinence.

The normal human stance is slightly knock-kneed (that is, our left and right knees are closer to each other than our left and right feet are), allowing us to walk with our feet placed on the midline and reducing rolling of the hips from side to side. This limb position leaves some telltale signatures at the articulation of the femur with the hip and at the knee joint and helps paleoanthropologists deduce whether a fossil species was fully bipedal. An unfortunate consequence of this limb position is that modern humans, especially athletes, are prone to knee dislocations and torn knee ligaments.

Human feet are highly modified for bipedal striding locomotion. Our feet are more arched than those of other hominids and there are corresponding changes in the shapes and positions of the tarsals, metatarsals, and digits (see Figure 24.19). The big toe is no longer opposable as in other primates, and is much larger than the other toes, enabling it to bear the body's weight at the end of a stride.

Large brains

The human brain increased in size threefold over a period of about 2.5 Ma. Our brains are not simply larger versions of ape brains; they have several key differences, such as a relatively much larger prefrontal cortex and a relatively smaller olfactory bulb. We do not know what selective pressures led to large brains. Speculations include increasing ability for social interactions, conceptual complexity, tool use, dealing with rapidly changing ecological conditions, language, or a mixture of these elements.

Brain tissue is metabolically expensive to grow and to maintain (on a gram-for-gram basis, brain tissue has a resting metabolic rate 16 times that of muscle). The basic structure of the human brain is established early in human embryonic development, but human brains continue to grow disproportionately after birth, and this requires continued energy input from the mother. This means that selective pressures for larger brains can be satisfied only in environments that provide sufficient energy for pregnant and lactating females. Thus, the evolution of larger brains may have required increased foraging efficiency (partially achieved through larger female size and mobility) and high-quality foods in substantial quantities (partially achieved by use of tools and fire).

Larger brains also required changes in life-history patterns that probably exaggerated the ancestral primate character of slow rates of pre- and postnatal development, lowering daily energy demands but also reducing a female's lifetime reproductive output. Our prolonged period of childhood may allow offspring to stay with the family long enough to acquire knowledge needed for survival.

Speech and language

Many animals produce sounds for communication, but the use of a symbolic language is unique to humans. Some traits of the vocal apparatus necessary for speech evolved in earlier stages of primate and ape evolution. For example, anatomical structures needed to produce differentiated vowel qualities evolved early in anthropoid evolution (~20 Ma), and voluntary control over movements of the vocal cords occurs in orangutans as well as in humans, which suggests that we inherited this trait from our great ape ancestors.

Controlled speech, however, might not have been possible until a later stage than †*Homo erectus*. The small spinal cord in the thorax of †*H. erectus* suggests that it lacked the complex neural control of intercostal muscles needed to control breathing to talk coherently. The hypoglossal canal (the exit from the skull for cranial nerve XII, which innervates the tongue muscles used in vocalization) is smaller in other hominins than in modern humans and †Neanderthals. Additionally, *FOXP2*, a gene specifically involved in the development of human speech and language, has the same two molecular sequence differences from the chimpanzee condition in both modern humans and †Neanderthals.

Traditional explanations for the limited vocalization of other great apes focused on differences in laryngeal anatomy. We have a descended larynx, which is important to human speech, but recent studies show that this is not unique to anatomically modern humans. Also, speech does not depend only on laryngeal anatomy; there must also be neural changes to quickly process and decode rapidly transmitted sounds. Such processing may be related to increases in volume of the temporal areas of the brain that are associated with speech recognition.

24.6 Humans and Other Vertebrates

LEARNING OBJECTIVES

24.6.1 Explain why the role of *Homo sapiens* in the extinction of Pleistocene megafauna is so contentious.

24.6.2 Describe the arguments pro and con for recognizing the Anthropocene as a geological period.

Ever since the start of the Holocene ~11,650 years ago, humans have disproportionately impacted other vertebrates. Indeed, never in Earth's history has a single animal species so profoundly affected the abundance and prospects for survival of so many other species.

Our closest primate relatives provide a disturbing a case in point. According to the latest Red List assessments by the IUCN, 5 of the 7 species of great apes are Critically Endangered—one step away from extinction. The situation is almost as dire for other primate species: 66% are threatened with extinction. On top of that, 75% of primate species also have declining populations. Illegal hunting poses the greatest threat, and it has led to a decline of more than 70% in the number of eastern gorillas in the past 20 years and to significant losses among other species of primates.

Humans as superpredators and environmental disruptors

Humans are qualitatively and quantitatively different from any other predator. Most apex predators target either young or aged individuals in a prey population, because these are the easiest to capture and kill. In contrast, weapons from spears to automatic rifles allow humans to kill the largest and fittest adults in a population. And that's what humans do, killing mature adult prey 14 times more frequently than other predators do, and not only for purposes of survival (see Section 23.7).

In addition to their direct effect as predators, humans impact other vertebrates by changing the environment. Tens of thousands of years ago, humans were already modifying their environments with fire. Agriculture originated independently on every continent except Antarctica between 12 and 5 ka, and the accompanying transition from nomadic bands to settled populations increased the pace and scope of anthropogenic environmental change. Urbanization, pollution, and accelerating climate change are the most recent manifestations of human impacts on other vertebrates.

Humans also modify the "landscape of fear," which is the spatial distribution of risk perceived by an individual. For example, a long-term study of mountain lions (*Felis concolor*) in the Santa Cruz Mountains of California found that human-induced risk has become the primary determinant of patterns of space use by male mountain lions. Mountain lions travel over more rugged terrain, move faster, and follow longer, more meandering routes to reach their destinations in areas where houses and wildland intermingle (boundary areas) than in wilderness areas. Even sounds associated with humans cause mountain lions to flee from prey they have killed, reducing time they spend feeding by more than half and requiring them to make additional kills. These anthropogenic changes increase the animals' energy expenditure, reduce their food consumption, and interfere with locating prey and finding mates.

The dominant role of humans in shaping this landscape of fear was revealed by the COVID-19 pandemic. Shelter-in-place orders during the spring of 2020 reduced human mobility by ~50%, a phenomenon that has been dubbed the "anthropause." During this time the Santa Cruz mountain lions increased their activity in boundary areas.

Megafaunal extinctions

Despite five large mass extinctions during the Phanerozoic (see Chapters 5, 10, and 20), the average overall species diversity of vertebrates increased rapidly during the past 100 Ma. Like species formation, species extinction is a normal evolutionary process, and the duration of most vertebrate species in the fossil record appears to be from 1 to 10 Ma. But the pattern of long-lived species changed at just about the same time humans became the dominant species in many parts of the world. Today we are in the midst of the "sixth mass extinction," and extinction is progressing at a rate more than 10,000 times that of the end-Cretaceous extinction.

Extinctions began soon after humans arrived The most striking evidence of the impact of humans on extinctions of other vertebrates is the correspondence in the arrival times of humans on continents and islands and the extinction of native megafauna (**Figure 24.27**). Humans reached northern Australia ~65 ka, occupied the dry center of the continent by 49 ka, and had moved south to Tasmania by 39 ka. On a continent-wide basis, the Australian megafauna disappeared about 9 ka after the arrival of humans. Similar connections between the dates of human arrival and megafaunal extinctions exist for other continents.

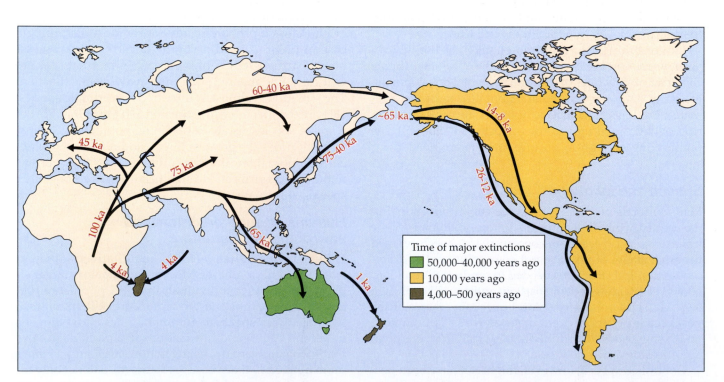

Figure 24.27 Human populations and extinction of native vertebrates.   Modern humans began to leave Africa ~194–177 ka, spreading across Eurasia and into the Americas. Dates in red indicate human arrivals. Extinctions of Pleistocene megafauna correspond with estimates of the dates humans arrived in Australia, the Americas, and on major islands such as New Zealand and Madagascar.

Humans colonized islands later than continents, and extinctions occurred later on islands—between 10 and 4 ka on islands in the Mediterranean, 4 ka on islands in the Arctic Ocean north of Russia, 2–1.5 ka on Madagascar, and ~700 years ago on New Zealand. In each case, dates of megafaunal extinctions closely follow the dates when humans arrived.

Were humans solely responsible?   The multiple coincidences between the arrival of humans and extinctions of native species shown in Figure 24.27 provide strong circumstantial evidence, but not proof, that humans were solely responsible for megafaunal extinctions. For example, two opposing views of the role of humans in the extinction of woolly mammoths were published nearly simultaneously in 2021. One asserts that climate change was the primary factor and that humans played little or no role, whereas the second concludes that humans hastened climate-driven population declines of woolly mammoths by many millennia. A third study contends that woolly mammoths survived until 5,000 years ago.

Few authorities argue that humans were the sole cause of the Pleistocene megafaunal extinctions. Other factors were undoubtedly involved, including climate change, changes in vegetation, the arrival of other alien species, and perhaps new pathogens and parasites introduced by these alien species. Interactions of multiple factors undoubtedly stressed populations of large mammals and birds and in some cases humans tipped them into an extinction vortex (see Section 23.7).

The species that disappeared soon after humans arrived were phylogenetically, ecologically, and geographically diverse. What they had in common was large body size: extinctions overwhelmingly impacted vertebrates with adult weights >44 kg. Life histories of large mammals are characterized by late maturation and late first reproduction, production of few offspring at intervals from 1 to 3 years, long lifespans, and low population densities. Such species cannot rebuild populations rapidly and are vulnerable to overkill (just as we described for sharks; see Figure 6.26).

Is this the Anthropocene?

The overwhelming global impact of humans has led to a proposal that we are now in a new epoch of the Cenozoic, the Anthropocene (Greek *anthropos*, "human"). The suggestion has popular appeal, but reaction from the scientific community has been mixed.

Worldwide impacts of humans can be traced back hundreds or thousands of years. These impacts include

Plastic pollution

© iStock.com/jacus

Figure 24.28 **Plastic makes up nearly 70% of all ocean litter.** Large pieces of plastic are broken down by ultraviolet light and ocean turbulence into microplastic particles that are deposited in marine sediments.

megafaunal extinctions, phenotypic changes in species, effects of agriculture on current plant communities, and reductions in genetic diversity of animal populations. However, geological time periods are based on stratigraphic markers—evidence in geological deposits and ice cores—and many geologists who study these markers are not convinced that we are truly in a new epoch. However, there are several potential stratigraphic markers for the Anthropocene:

- More than 200 minerals and mineral-like materials result from human activities. Portland cement is the most familiar of these, and others include new minerals produced by alteration of existing minerals in the tunnels and dumps of mines.

- Radioactive debris from atmospheric nuclear bomb tests in the mid-20th century may provide a particularly clear signal for at least several million years.

- Scraps of plastic (**Figure 24.28**) are widespread, and are already becoming incorporated into sedimentary deposits; microplastics (particles smaller than 1 mm in diameter) are ubiquitous in ocean sediments, animal tissues, and even in human placentas.

In 2019, the Anthropocene Working Group, composed of geologists, archeologists, and climate and atmospheric scientists, stated that humans' enormous environmental footprint should receive formal stratigraphic recognition as a new geological epoch.

Summary

24.1 Primate Origins and Diversification

The first primates, which were small and arboreal, are known from the early Paleocene of the Northern Hemisphere.

There are two clades of Euprimates: Strepsirrhini and Haplorhini. Strepsirrhini includes the extant Lorisiformes (lorises, pottos, and bushbabies) and Lemuriformes (aye-aye and four families of lemurs).

Lemuriforms reached Madagascar by ~45 Ma and diversified in isolation to occupy many niches filled by other groups of mammals in other regions of the world. They share features such as a tooth comb, grooming claw, and a tapetum lucidum in the eye that enhances nocturnal vision.

By the earliest Eocene, the first haplorhines appeared. These small forms were closely related to extant tarsiers, a group specialized for nocturnal foraging and leaping through the forest canopy.

Haplorhini includes Platyrrhini (New World monkeys) and Catarrhini. Catarrhini contains two groups: Cercopithecoidea (Old World monkeys) and Hominoidea (apes and humans).

• New World monkeys (platyrrhines) are primarily arboreal herbivores ranging in size from tiny marmosets to larger forms such as howler monkeys. Prehensile tails evolved in a subgroup of platyrrhines, and spider monkeys use their tail together with brachiation to move through the forest.

• There are two groups of Old World monkeys (cercopithecoids). Colobines are folivores and largely arboreal. They have a specialized foregut and can break down plant material by fermentation, convergently like the foregut of artiodactyls such as cows. Cercopithecines include the macaques and baboons that live close to or on the ground. Although generally more terrestrial than colobines, some species of cercopithecines, such as the blue monkey, live in the forest canopy.

Molecular estimates date the split between Cercopithecoidea and Hominoidea at ~28 Ma.

24.2 Origin and Evolution of Hominoidea

Apes—including humans—belong to Hominoidea (hominoids). Morphological features such as a broader thoracic cavity and more posterior location of the scapula distinguish hominoids from other catarrhines. Evolutionary increases in brain sizes molded the shape of the hominoid skull, especially in the later part of human evolution.

The first known hominoids occurred in the late Oligocene, ~25 Ma. By the late Miocene, hominoids had diversified and spread throughout Africa, Europe, and Asia.

As well as humans, Hominidae includes orangutans, gorillas, chimpanzees and bonobos, and fossil relatives, collectively referred to as hominids.

24.3 Origin and Evolution of Hominini

Humans and their fossil relatives belong to a tribe within Hominidae called Hominini; members of this tribe are referred to as hominins.

Many hominin fossils have been found in late Pliocene and early Pleistocene deposits of Africa.

• The earliest well known hominin is †*Ardipithecus*, which lived in Ethiopia ~5.8–4.4 Ma.

• †*Australopithecus* (~4.2–2 Ma) and †*Paranthropus* (~2.7–1 Ma) were bipedal. †*Australopithecus* retained many characters associated with arborealism, whereas †*Paranthropus* appear to have been primarily terrestrial.

24.4 The Genus *Homo*

The earliest identified member of the genus *Homo*, †*Homo habilis*, occurred in eastern and southern African from ~2.8–1.6 Ma.

†*Homo erectus*, the first hominin with an intercontinental distribution, lived in Africa and Eurasia from ~2 Ma to at least as recently as 150 ka. †*Homo erectus* had a cranial volume approaching the lower range of *Homo sapiens*, made stone tools, and may have used fire.

†Neanderthals were Eurasian hominins that coexisted with modern humans. They appeared in Europe ~130 ka. The last remaining populations of †Neanderthals are known from Gibraltar, where they lived as recently as 28 ka.

†Denisovans lived in the Altai Mountains of Siberia between 200 and 30 ka. A fossil from Siberia dated to ~90 ka is the child of a †Neanderthal and a †Denisovan, and the cranium of a recently identified hominin species from China, †*Homo longi*, dated to ~146 ka, may be a †Denisovan.

†*Homo floresiensis* and †*H. luzonensis* were island populations of small *Homo* that lived between 700 and 50 ka and are thought to be derived from †*Homo erectus*.

Homo sapiens, the only surviving species of the tribe Hominini, originated in Africa about 300 ka. Humans emigrated from Africa in a series of waves, with populations of humans colonizing all continents by the end of the 20th century.

At least three archaic species of *Homo* were in existence 30 ka, and they interbred with anatomically modern humans.

Summary (continued)

Modern human populations outside of sub-Saharan Africa carry a small percentage (2–3%) of †Neanderthal DNA. †Denisovan DNA is also found among present-day human populations, some of which carry 0.1–3.5% †Denisovan DNA.

The present-day situation, with *Homo sapiens* as the sole existing species of hominin, is a phenomenon of the last 27 ka or less.

24.5 Evolution of Human Characters

Bipedalism appeared in the earliest stages of hominin evolution, long before the evolution of a large brain. †*Ardipithecus* showed evidence of bipedal posture, including an anteriorly positioned foramen magnum and a short, broad pelvis with laterally oriented iliac blades.

Other anatomical modifications for bipedal locomotion include an S-shaped curvature of the vertebral column and position of the knee joint beneath the center of mass of the body.

A cranial volume comparable to that of modern humans was not seen until †*Homo bodoensis* and other archaic forms of the genus *Homo*.

The increase in brain volume in anatomically modern humans may have been caused by social and conceptual complexity, tool use, rapid climate change, language, or the combined action of all these factors.

24.6 Humans and Other Vertebrates

Humans are superpredators, able to kill large animals (megafauna) that otherwise have no natural predators as adults. Dates of extinction of large vertebrates correspond closely with dates when humans arrived on the different continents and islands.

Megafauna species have low population densities, late maturation, low birth rates, and small litter sizes; consequently, they are vulnerable to overkill by human predation.

Humans changed the environment, initially with fire and later with agriculture and urbanization, to the detriment of other vertebrate species. The magnitude of human impact has led to definition of a new geologically and chemically identifiable epoch, the Anthropocene.

Plastic, concrete, and other anthropogenic minerals have been proposed as stratigraphic markers of the Anthropocene.

Discussion Questions

24.1 There are no non-human primates in North America today. What elements of Earth's climate during the Eocene allowed primates to live in North America, and what climatic changes accompanied their disappearance from this continent?

24.2 During the Miocene, when temperatures became warm at higher latitudes, monkeys and apes radiated from Africa into Eurasia. What prevented primates from reappearing in North America during this time?

24.3 What is the functional significance of the S-shaped bend in the vertebral column of *Homo sapiens*?

24.4 Rafting is the most accepted explanation for the arrival of primates from Africa to South America across the South Atlantic Ocean. How could primates raft?

24.5 Suppose that Bigfoot (Sasquatch, an imaginary primate alleged to live in the Pacific Northwest) or the Yeti (an equally imaginary primate that is supposed to live in the Himalayas) were found to be real and to be a derived descendant of †*Homo erectus*. What moral responsibility would we (*Homo sapiens*) have to a species that was as closely related to us as a wolf is to a dog? Would we feel less responsible if it was more closely related to another species of great ape?

The Geologic Timescale

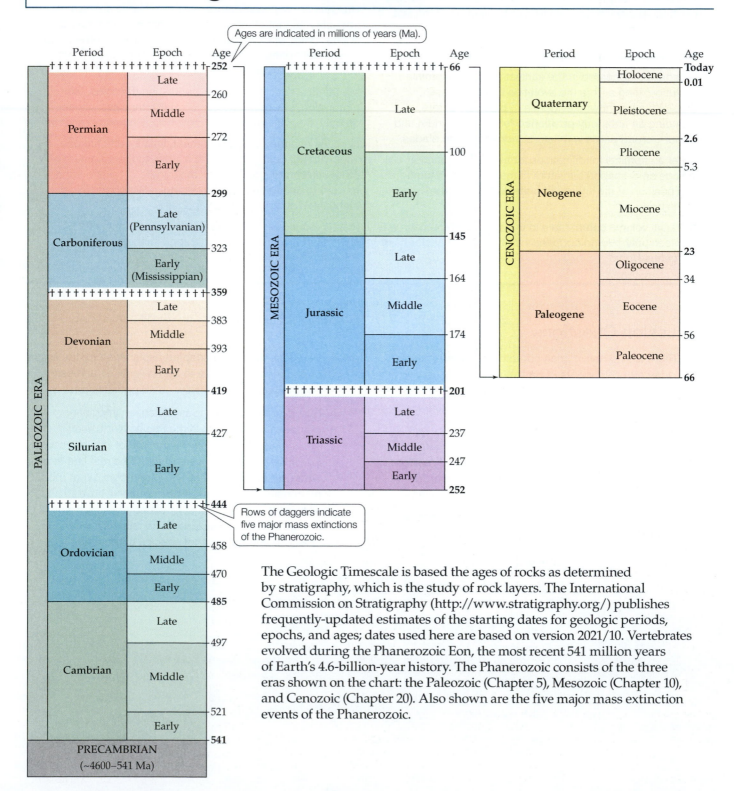

Ages are indicated in millions of years (Ma).

Rows of daggers indicate five major mass extinctions of the Phanerozoic.

The Geologic Timescale is based the ages of rocks as determined by stratigraphy, which is the study of rock layers. The International Commission on Stratigraphy (http://www.stratigraphy.org/) publishes frequently-updated estimates of the starting dates for geologic periods, epochs, and ages; dates used here are based on version 2021/10. Vertebrates evolved during the Phanerozoic Eon, the most recent 541 million years of Earth's 4.6-billion-year history. The Phanerozoic consists of the three eras shown on the chart: the Paleozoic (Chapter 5), Mesozoic (Chapter 10), and Cenozoic (Chapter 20). Also shown are the five major mass extinction events of the Phanerozoic.

Glossary

A

acanthodians Stem chondrichthyans of the Paleozoic known as spiny sharks.

activity temperature The range of body temperatures within which an animal's (especially an ectotherm's) physiological processes and whole-animal performance reach their maxima.

advertisement calls Vocalizations used by individuals to announce their presence, such as when male anurans call to attract mates or declare territory.

aerobic scope model Hypothesis proposing that endothermy is a consequence of selection for high rates of aerobic metabolism that support high levels of physical activity. See *parental care model, thermogenic opportunity model, warmer is better model.*

air capillaries Interconnected small tubes that radiate from the parabronchi of birds; gas exchange occurs between air capillaries and surrounding blood capillaries.

air sacs Spaces in the respiratory passages of sauropsids with faveolar lungs in which air is briefly stored during through-flow lung ventilation.

allantois One of the extraembryonic membranes of amniotes; stores embryonic wastes and fuses with the chorion to form the chorioallantoic membrane, which functions in gas exchange.

Allee effect The phenomenon by which individuals in a population benefit (i.e., have higher survival rates) when they form groups with conspecifics; examples include cooperative breeding and group defenses against predators.

alleles Forms of a gene that differ in their DNA base sequence.

allometry Condition in which the relative sizes of parts of the body change during ontogenetic growth.

altricial Developmental state of neonates that are helpless and dependent on care from parents for some time after birth or hatching. Contrast with *precocial.*

alula Moveable, anteriormost digit of a bird's wing.

alveolar lung A compliant lung, such as that found in mammals, in which gas exchange occurs in saclike chambers (alveoli) at the ends of a series of treelike dichotomous branches of the airways. Airflow in alveolar lungs is tidal. Contrast with *faveolar lungs.*

alveoli Saclike chambers in alveolar lungs that are the sites of gas exchange. Contrast with *faveoli.*

Ambulacraria Group that includes echinoderms and hemichordates.

ammonotely Excretion of nitrogenous wastes primarily as ammonia. Contrast with *ureotely* and *uricotely.*

amnion One of the extraembryonic membranes of amniotes; the inner membrane that encloses the embryo in amniotic fluid, providing protection from mechanical shock.

amniotic egg A shelled egg with four extraembryonic membranes (yolk sac, amnion, chorion, and allantois). This key feature of amniotes requires internal fertilization before the shell is formed and allows laying eggs on land. Modifications of the amniotic egg associated with viviparity characterize therian mammals (e.g., loss of the shell).

amphistylic jaw suspension A type of jaw suspension in early elasmobranchs in which the anterior end of the upper jaw articulated directly with the chondrocranium and the hyomandibula supported the posterior end of the upper jaw.

ampullae of Lorenzini Ampullary organs in the skin on the snout of elasmobranchs; also called electroreceptors.

ampullae See *ampullary organs.*

ampullary organs Flask-shaped organs connected to the surface of the skin by a canal filled with an electrically conductive gel; used to detect electric fields.

anadromous Relating to fishes that move from seawater to freshwater to reproduce. Contrast with *catadromous.*

anadromy Pattern of migration of fishes that move from seawater to freshwater rivers to reproduce. Contrast with *catadromy.*

anapsid Condition in which a skull has no temporal fenestrae (openings in the temporal region); characteristic of tetrapods and retained in the earliest amniotes.

angiosperms Flowering, seed-producing, vascular plants, including grasses and most trees today; about 300,000 extant species. Contrast with *gymnosperms.*

anisodactyly Toe arrangement in birds in which toes 2, 3, and 4 point forward and toe 1 faces backwards; associated with perching.

annuli (singular *annulus*) Grooves in the skin that encircle or partly encircle the body; in caecilians, annuli overlie vertebrae and mark the position of ribs.

antidiuretic hormone (ADH) Hypothalamic hormone released in response to increased blood osmolality or reduced blood volume that promotes recovery of water by the kidney. Also known as *vasopressin.*

apex predator A predator at the top of a food web that often has few or no natural predators. Contrast with *mesopredator.*

apocrine glands Skin glands found in restricted areas (under armpits and in the pubic region of humans); produce an oily secretion used in chemical communication.

apomorphy A character that has changed from its ancestral state—i.e., a derived character.

aposematic Having a character, such as color, sound, odor, or behavior, that advertises an organism's noxious properties to potential predators.

appendicular skeleton Pectoral and pelvic girdles and associated fin or limb bones.

appendicular skeleton Pectoral and pelvic girdles and associated fin or limb bones.

apteria Regions of skin without feathers.

aquaporins Tubular proteins that form water channels through plasma membranes.

arteries Blood vessels that carry blood away from the heart.

aspect ratio Ratio of the length of a fin or wing to its width. A long and narrow wing has a high aspect ratio.

aspiration pump Mechanism for inflating the lungs by contracting intercostal muscles to expand the rib cage, creating a vacuum that sucks air into the lungs. Elastic recoil of the rib cage, which compresses the abdominal cavity, expels air; used by amniotes. Contrast with *buccal pumping.*

atlas–axis complex Arrangement in mammals in which the dens, a projection of the axis vertebra, extends into a space in the atlas vertebra allowing rotation of the head (e.g., shaking head "no" in humans).

atlas First cervical vertebra in tetrapods; articulates with the skull, allowing the nodding movement meaning "yes" (in many human cultures). Contrast with *axis.*

atrium (plural *atria*) 1. A chamber of the heart of vertebrates. 2. A chamber surrounding the gill slits of urochordates and cephalochordates. The atrium exits to the exterior through a pore. 3. In a general sense, an empty space within a structure.

autodiastylic jaw suspension A type of jaw suspension in early chondrichthyans in which processes from the upper jaw attached it rigidly to the chondrocranium.

autopodium Third part of a paired appendage of a tetrapod, the wrist, metacarpals and phalanges or the ankle, metatarsals, and phalanges.

axial skeleton Cranial skeleton, notochord, vertebrae, ribs, sternum, and median fins.

axis Second cervical vertebra in amniotes; permits rotary movements between itself and the atlas such as shaking the head "no" (in many human cultures). Contrast with *atlas*.

B

barbs The primary branches from the rachis of a feather that compose the vane.

barbules Extensions from the barbs of a feather that hook to barbules from an adjacent barb, providing the rigidity that makes a vane an aerodynamic surface.

basal (1) Taxa closer to the root of a particular phylogenetic tree. (2) An endochondral skeletal element that articulates directly with a fin or limb girdle.

base Root of a phylogenetic tree. See *root*.

bathypelagic Portion of the water column extending from 1000 m to 4000 m; perpetually cold and dark. Relatively few species of fishes live below 4000 m and those that do are highly specialized.

Bilateria Animals that at some point in their lives have a body plan with sides that are mirror images of each other.

bill-tip organs Mechanoreceptors (e.g., Herbst corpuscles) in high concentrations at the tip of the bill of some birds; allow remote touch, detection of prey some distance from the bill.

binocular vision Stereoscopic vision by two eyes that have overlapping fields of view.

binomial nomenclature Linnaean system of naming species with two names: a genus name and a species name (formally known as a *specific epithet*)—e.g., *Homo sapiens*.

biofluorescence Absorption of light of one wavelength and reemission at a different wavelength.

bioluminescent Relating to the production of light by a chemical reaction within an organism; in vertebrates, usually occurs in specialized organs known as *photophores*.

biting Method of feeding in which the mouth is opened and closed on large prey, tearing off a chunk of flesh.

blastopore Original opening of the embryonic gut, which forms the anus in deuterostomes.

brachiation A specialized form of arboreal (suspensory) locomotion in which animals swing from the underside of one branch to another using their hands; employed, for example, by extant gibbons, the siamang, and spider monkeys.

branchiostegal rays A fanlike series of dermal bones on the underside of the skull, forming the floor of the gill chamber.

bristles Specialized feathers with a stiff rachis and without barbs, or with barbs on only a portion of the rachis; found around eyes and base of the bill, where they prevent foreign particles from entering the eyes and nostrils.

buccal pumping Mechanism for inflating the lungs by moving the floor of the buccal cavity up and down; used, for example, by lissamphibians. Contrast with *aspiration pump*.

buccopharyngeal pumping Drawing air or water into the mouth and pharynx and expelling it with movements of the floor of the mouth.

bunodont teeth Molar teeth that are square in shape, with four low rounded cusps that crush and pulp food. Typical of omnivorous and frugivorous mammals, including humans.

C

calamus The short, tubular base of the rachis of a feather that implants the feather in its follicle.

calcified cartilage Formed by the deposition of calcium hydroxyapatite in the cartilaginous skeleton of chondrichthyans.

calls Simple, brief vocalizations produced by both male and female birds in contexts such as presence of a predator or staying in contact with conspecifics. Contrast with *songs*.

capillaries Smallest vessels of the circulatory system that are sites of exchange of respiratory gases, nutrients, and waste products between blood and tissues.

carapace The dorsal or upper shell of a turtle. Contrast with *plastron*.

Carboniferous rainforest collapse Change in terrestrial flora that occurred in the late Carboniferous linked to low levels of atmospheric carbon dioxide, aridification, glaciation, and low sea levels.

Carnian Pluvial Episode Transformative episode in the Late Triassic during which previously arid environments experienced wet and rainy conditions only to return to aridity; after the Carnian Pluvial Episode, dinosaurs began to rapidly diversify.

carpometacarpals Bones formed in extant birds by fusion of wrist and finger bones; singular is carpometacarpus.

carpometacarpus In birds, formed by the fusion of some carpals (wrist bones) and metacarpals (hand bones).

catadromous Relating to fishes that move from freshwater to seawater to reproduce. Contrast with *anadromous*.

caudal autotomy Defense mechanism used by many lizards in which the tail is intentionally broken off and left with an attacking predator; the lizard escapes and regenerates the tail.

caudal vertebrae Vertebrae that support the tail and vary in number among tetrapods; typically simpler in structure than trunk vertebrae. Contrast with *cervical vertebrae*, *trunk vertebrae*, and *sacral vertebrae*.

ceca Pouches at the junction of small and large intestines that function in water and electrolyte reabsorption; in birds, also function in hindgut fermentation of plant material.

cementum A mineralized tissue found in mammalian teeth; it covers the outer surface of a tooth below the gum line where it helps fasten the tooth in its socket.

Cenophytic Flora Flora dominated by angiosperms that diversified in the Cretaceous and continues to the present.

Cenozoic era The third and current era of the Phanerozoic lasting from 66 Ma to the present and including the Paleogene, Neogene, and Quaternary periods.

centra (singular *centrum*) Portions of a vertebra that surround the notochord.

centrum (plural *centra*) The portion of a vertebra that surrounds the notochord.

cephalic clasper Spine-encrusted clasper on the top of the head of males of some chimaera species; used during courtship and mating.

Cephalochordata One of two groups of extant non-vertebrate chordates, the lancelets. The other group is Urochordata, the tunicates.

ceratotrichia Flexible fin rays that support the fin webs of chondrichthyans; unjointed and formed from keratin.

cerebellum Region in the hindbrain associated with coordination of motor activity.

cervical vertebrae Vertebrae in the neck region that allow movement of the head relative to the trunk; muscles that support and move the head attach to processes on these vertebrae. Contrast with *trunk vertebrae*, *sacral vertebrae*, and *caudal vertebrae*.

cheiropatagium The primary wing of †pterosaurs consisting of a skin membrane supported by upper and lower arm bones and a very long fourth finger and extending to the hindlegs.

choana (plural *choanae*) Internal opening of a nostril in the oral cavity.

chondrocranium A structure that surrounds the brain. Initially formed by cartilage that is replaced by endochondral bone in most bony fishes and tetrapods.

Chordata Phylum that includes Cephalochordata (lancelets), Urochordata (tunicates), and Vertebrata (vertebrates).

chordate Any member of the phylum Chordata, which includes Cephalochordata (lancelets), Urochordata (tunicates), and Vertebrata (vertebrates); chordates are characterized by presence of a notochord, dorsal hollow neural tube, postanal tail, and endostyle.

chorioallantoic placenta The late-forming placenta of therian mammals, formed by fusion of the allantois to the chorion. In placentals this is the major placenta, while only a few marsupials (bandicoots, bilbies, koala, wombat) form a transitory chorioallantoic placenta towards the end of gestation. Contrast with *yolk sac (choriovitelline) placenta*.

chorion One of the extraembryonic membranes of amniotes; the outer membrane surrounding the embryo and the other extraembryonic membranes that eventually contacts the inside of the eggshell; functions in protection and gas exchange.

choriovitelline placenta The early-forming placenta of all therian mammals, produced by apposition of the yolk sac to the chorion. In marsupials this is the major placenta, in placentals it is transitory; also known as a *yolk sac*. Also found in some sharks. Contrast with *chorioallantoic placenta*.

circulation Movement of blood throughout the body to supply nutrients and oxygen to the tissues and to carry away carbon dioxide and waste products from the tissues.

clade A monophyletic group, which includes an ancestor and all of its descendants.

cladistics Method of investigating evolutionary relationships based on analyses of the distribution of characters.

classification bracket Bracket indicating the taxa included in a group (shown to the right side of the phylogenetic trees in this book). See *extant phylogenetic bracket*.

cleithrum Large dermal bone in the pectoral girdle connected to the skull by bones of the supracleithral series in osteichthyans.

cloaca The common opening of the reproductive and excretory tracts.

closed circulatory system Type of circulatory system in which blood always stays within blood vessels.

coelom Body cavity lined with mesodermal tissue.

common ancestor Most recent individual from which all taxa in a monophyletic group are descended.

compass orientation Ability to head in a specific direction without using landmarks.

complete metamorphosis In anurans, the process by which an aquatic tadpole becomes a frog through the breakdown of larval structures and the building of adult structures.

concertina locomotion Form of locomotion used by snakes in narrow passages where lateral undulation is not possible. The snake anchors the posterior part of its body against the passage walls, extends the front part of its body, anchors the front part of its body, and draws the posterior part of its body forward.

cone cells Photoreceptor cells in the retina that are sensitive to a narrow range of wavelengths (different types of cones are sensitive to different wavelengths); used to perceive colors and for visual acuity.

conodont elements Tooth microfossils of extinct Paleozoic jawless vertebrates, whose relationship to other vertebrates is uncertain.

continental drift Movement of continents across Earth's surface, explained by the theory of plate tectonics.

contour feathers Feathers that streamline a bird, reducing turbulence by smoothing transitions between parts of the body; outermost feathers on a bird's body, wings, and tail.

conus arteriosus An elastic chamber anterior to the ventricle of some gnathostomes.

convection The transfer of heat between an animal and a fluid, such as air.

convergent evolution Independent evolution of similar traits or types of organisms often in relation to biogeographic isolation.

cooperative breeders Breeding arrangement in which three or more individuals (often the mother, father, and an offspring from a previous clutch or litter) provide care for the young.

coronoid process Process on the dentary bone of the lower jaw that serves as a site of attachment for jaw closing muscles; found, for example, in †*Thrinaxodon*, †*Probainognathus*, and extant mammals.

cosmine Shiny mineralized tissue found on bones and scales of sarcopterygians composed of dentine and enamel.

cosmoid scales Thick, rhomboid-shaped scales composed of dentine-like tissue with a shiny layer of true enamel found in basal sarcopterygians. Contrast with *ganoid scales, cycloid scales,* and *ctenoid scales*.

costal ventilation Method of taking in and expelling air that relies primarily on ribs and their associated muscles.

countercurrent exchange System in which flow of two adjacent media occurs in opposite directions; this maximizes the amount of exchange between the two media, such as between respiratory gases in the water and blood in the gills of fishes.

crèche A collection of young from different clutches or litters accompanied by one or more adults; found, for example, in some birds and bats.

Cretaceous Terrestrial Revolution Diversification of terrestrial organisms in the mid-Cretaceous, marking the point at which diversity on land outstripped that in the oceans.

crop An enlarged portion of the esophagus of birds that is specialized for temporary storage of food.

cross-current exchange system An anatomical arrangement of air passages and blood vessels in bird lungs in which air and blood pass in opposite directions, but do not follow parallel pathways as they do in a counter-current exchange system.

crown group Members of a crown group have all of the derived characters of the extant members of a clade. Contrast with *stem group*.

cruropatagium Posterior skin membrane of †pterosaurs that provided additional lift at the rear of the body; extended between the hindlimbs in basal forms but was reduced in derived forms.

cryptic species Species that are indistinguishable morphologically, but are genetically distinct; seen, for example, in crocodiles.

cryptodires Turtles that bend the neck in a vertical S to retract the head into the shell. Contrast with *pleurodires*.

ctenoid scales Thin, bony scales with fine projections known as cteni on the exposed posterior surface of the scale. Contrast with *cosmoid scales, ganoid scales,* and *cycloid scales*.

cupula A cup-shaped gelatinous secretion of the cells of a neuromast organ in which the kinocilium and stereocilia are embedded.

cycloid scales Thin, round, bony scales that lack ornamentation such as ganoine, typical of basal teleosts. Contrast with *cosmoid scales, ganoid scales,* and *ctenoid scales*.

Cyclostomata Clade that includes hagfishes and lampreys. Contrast with *Osteognathostomata*.

cyclostomes Clade that includes hagfishes and lampreys; see *Cyclostomata*.

D

Darwinian fitness Genetic contribution of an individual to succeeding generations relative to the contributions of other members of its population.

deep time Term coined by John McPhee to describe inconceivably long spans of geological time.

definitive mammalian middle ear (DMME) Arrangement in which the ears are part of the temporal region of the skull rather than at the posterior end of the lower jaw; found in extant mammals.

deglutition Unique mode of food processing and swallowing in which food is manipulated with the tongue against the palate, returned to the teeth for additional processing until it forms a small bolus that is swallowed; found in mammals.

demersal eggs Eggs that are deposited on a substrate (e.g., a gravel bottom, underside of a log, or on aquatic vegetation). Contrast with *pelagic eggs*.

demersal   On or near the bottom of a body of water.

dental formula   A shorthand method of presenting the number of teeth of each type—incisors, canines, premolars, and molars—in one side of the upper jaw in the numerator and the number of teeth of each type in one side of the lower jaw in the denominator.

dentary   Tooth-bearing dermal bone forming the anterior part of the lower jaw in osteichthyans.

dentine   A mineralized tissue of vertebrates found in gnathostome teeth, some fish scales, and armor of early vertebrates. In therian mammals, dentine forms much of the tissue of a tooth; it is surrounded by enamel (on the crown) and cementum (on the root).

derived   (1) In reference to a phylogenetic tree, taxa that branch above the first node. (2) A character or character state different from outgroup taxa. See *apomorphy*.

dermal bone   Bone that forms in the dermis of the skin and lacks a cartilaginous precursor.

dermatocranium   Dermal bones that cover a portion of the skull.

dermatophagy   Feeding method used by young of some caecilian species that involves peeling and consuming the outer layer of their mother's lipid-rich skin; feeding on parental skin also occurs in some fish species.

dermis   Dense connective tissue layer of the skin beneath the epidermis; formed from mesoderm.

determinate growth   Growth that does not continue throughout an organism's life but instead ceases at some genetically determined point during the lifespan.

Deuterostomia   Animals in which the blastopore becomes the anus. Contrast with *Protostomia*.

developmental domains   Portions of an embryo specified by expression of patterning genes, such as Hox genes.

developmental regulatory genes   Genes that control developmental processes by activating or repressing expression of other genes.

diapause   A period of arrested embryonic development. For example, some turtles lay their eggs in late summer or early fall, diapause occurs in winter, and development resumes in spring when temperatures rise.

diaphragm   In mammals, the muscular partition between the thoracic cavity, which houses the lungs, and the peritoneal (abdominal) cavity, which houses the viscera.

diapsid   Condition in which there are two fenestrae (upper and lower) in the temporal region on each side of the skull; seen most clearly in the extant tuatara (*Sphenodon*).

diastema   A gap in the row of teeth, usually seen in herbivores between the incisors (plus canines if present) and the cheek teeth (premolars and molars).

didactyly   Condition of having two toes on each foot; found, for example, in ostriches, flightless running birds.

digestion   Breakdown of complex compounds into small molecules that can be absorbed across the wall of the gut and transported to tissues.

diphyodonty   The condition of having only two sets of teeth in a lifetime: in extant mammals, the milk teeth and the permanent teeth. Contrast with *polyphyodonty*.

diploblastic   A metazoan with only two embryonic germ layers, ectoderm and endoderm.

diplorhiny   Condition of having two nostrils and nasal sacs, formed from paired nasal placodes. Contrast with *monorhiny*.

dispersal   Movement of organisms over geological time resulting from the movement of the organisms themselves.

distal rhynchokinesis   Ability of some long-billed birds to raise the tip of the upper part of the bill to grab prey under mud or sand.

double circulation   Type of circulation found in tetrapods in which a pulmonary circuit supplies the lungs with deoxygenated blood and a systemic circuit supplies the body with oxygenated blood. In double circulation, oxygenated blood from the lungs returns to the heart for additional pumping out to the body. Contrast with *single circulation*.

double-pump gill ventilation   Mechanism of ventilating gills that relies on two chambers in the head, the mouth and pharyngeal cavity and the opercular cavity in most vertebrates with gills; sharks have multiple parabranchial cavities, each exiting by a gill slit instead of an opercular cavity.

down feathers   Plumulaceous feathers typically lacking a rachis that provide insulation.

drag   Backward force opposed to forward motion.

durophagy   Eating hard food items such as mollusks.

E

eccrine glands   Skin glands found on the surfaces that contact the substrate in most mammals, the soles of the hands and feet and underside of the tail (if prehensile); produce a watery secretion that aids adhesion. In humans and some other primates, eccrine glands are found over the body surface and produce sweat for evaporative cooling.

ecosystems   Communities of interacting organisms and their abiotic environment.

ectoderm   Outermost embryonic germ layer.

ectothermy   The end of the continuum of thermoregulatory modes at which most of the energy used to raise body temperature comes from external sources (e.g., solar radiation, warm substrate).

embryonic diapause   In mammals, a period of arrested embryonic development prior to implantation in the uterus; found, for example, in some marsupials (e.g., kangaroos) and some placentals (e.g., bears).

enamel   The hardest mineralized tissue of vertebrates found in gnathostome teeth, some fish scales, and armor of early vertebrates. In therian mammals, enamel covers the crown of a tooth.

endochondral bone   Endochondral bone is made by osteocytes that ossify a cartilaginous precursor deep within the body.

endocrine glands   Ductless glands that secrete hormones that travel via the circulatory system.

endoderm   Innermost embryonic germ layer.

endostyle   A ciliated glandular groove on the floor of the pharynx of non-vertebrate chordates and larval lampreys that is homologous with the thyroid gland of vertebrates.

endothermal heterothermy   Condition when one part of an organism's body is warmer than other parts of the body; also called regional endothermy.

endothermy   The end of the continuum of thermoregulatory modes at which most of the energy used to raise body temperature comes from internal sources (i.e., metabolism).

environmental sex determination (ESD)   The situation in which the sex of an individual is determined by a cue that is not genetic; for example, in some reptiles, sex is determined by the temperature the embryo experiences during development. Contrast with *genetic sex determination*.

eon   One of three major divisions of time in Earth history extending from ~ 0.5 to ~2.0 billion years, which are the Archean, Proterozoic, and Phanerozoic; we live in the Phanerozoic eon.

epaxial muscles   Dorsal portion of axial musculature of vertebrates. Contrast with *hypaxial muscles*.

epicontinental seas   Shallow seas that extend into the interior of a continent.

epidermis   Epithelial layer that forms the surface of the skin; formed from ectoderm.

epigenetic effects   Modifications of gene expression during development by non-genetic factors, such as temperature.

epipelagic   Top 200 m of water in the ocean in which light is sufficient to support photosynthesis; much biomass is concentrated here.

epochs   Subdivisions of periods during the Phanerozoic extending thousands to tens of millions of years; we live in the Holocene epoch.

eras One of three major divisions of time in Earth history during the Phanerozoic, extending up to several hundred million years, which are the Paleozoic, Mesozoic, and Cenozoic; we live in the Cenozoic era.

ethmoturbinates Odor receptors of the main olfactory system are located on these scroll-like bones of mammals located in the posterior portion of the nasal cavity where they are associated with the ethmoid bone. Nerves pass through the ethmoid bone to carry olfactory signals to the brain. Contrast with *maxilloturbinates*.

Eugnathostomata Clade containing Acanthodii (including Chondrichthyes) and Osteichthyes.

Euramerica Continent formed from three landmasses that were precursors to Europe, Greenland, and North America, near the Ordovician-Silurian boundary.

euryhaline Capable of living in a wide range of salinities. Contrast with *stenohaline*.

evolutionary developmental biology (evo-devo) An approach combining data from developmental research with phylogenetics.

evolutionary lineage An ancestor and all of its descendants; equivalent to *clade* and *monophyletic group.*

exaptation A term describing a trait that takes on a new or additional function during evolution.

excretion Removal of waste products of metabolism.

exocrine glands Glands with ducts that secrete products such as sweat onto the skin or digestive enzymes into the digestive tract.

explosive breeding Mating pattern of anurans in which the breeding season is very short, sometimes a few days. Contrast with *prolonged breeding*.

extant phylogenetic bracket Using the character states of two extant lineages to infer the character states of extinct taxa that lie between the extant taxa in a phylogeny. See *classification bracket*.

extended specimen Associating additional information such as photographs or DNA sequences with a specimen held in a permanent museum collection.

extinction vortex The situation in which feedback among multiple pressures drives an already small population into a spiral towards extinction.

F

fast-start performance Specialized locomotion used by fishes to escape or dart out and seize prey; associated with a long arrow-shaped body and placement of large median fins at the posterior end of the body.

faveolar lungs Rigid lungs in which gas exchange occurs in tubelike structures (faveoli) lining the walls of the airways. Air flows in one direction in faveolar lungs—from posterior to anterior. Contrast with *alveolar lung*.

faveoli Tubelike structures in faveolar lungs that are the sites of gas exchange. Contrast with *alveoli*.

feeding Taking food into the mouth.

filoplumes Hairlike feathers with few barbs that sense the position of contour feathers.

filter feeding Method of feeding in which very small prey are engulfed with an open mouth. Contrast with *ram feeding* and *suction feeding*.

fist-walking One mode of locomotion employed by extant orangutans when walking quadrupedally; they close their hands into fists and support themselves on proximal phalanges.

flow-through ventilation One-way flow of respiratory media (water or air) past a respiratory surface in gills or lungs. Contrast with *tidal ventilation*.

forebrain Anteriormost portion of vertebrate brain; associated with olfaction.

foregut fermenters Herbivorous mammals, such as cows, in which an enormous, multichambered stomach houses symbiotic microorganisms that break down cellulose, releasing nutrients from consumed plant material. Contrast with *hindgut fermenters*.

fossorial Specialized for burrowing.

furcula The theropod wishbone, formed by fusion of the two clavicles at their central ends.

G

ganoid scales Thick, bony, rhomboid-shaped scales covered with a shiny layer of ganoine found in basal actinopterygians. Contrast with *cosmoid scales, cycloid scales*, and *ctenoid scales*.

ganoine Shiny enamel-like tissue on scales and dermal bones of actinopterygians.

gas bladder Air-filled organ used to regulate buoyancy in bony fishes; also known as the swim bladder.

gastroliths Rocks held in the gastrointestinal tract that help grind food; also called stomach stones and found, for example, in extant crocodylians and birds and extinct †sauropods.

gastrulation Movements of undifferentiated cells of the early embryo resulting in a three-layered embryo, with ectoderm, mesoderm, and endoderm.

gene families Genes that produce structurally related forms of the same protein.

genetic barcoding Using DNA sequencing of short lengths of easily collected DNA, such as portions of the mitogenome, as a tool to identify species.

genetic sex determination (GSD) The situation in which the sex of an individual is determined by genes on sex chromosomes; this occurs, for example, in all birds and mammals. Contrast with *environmental sex determination*.

gigantism A life history pattern characterized by growth to an extremely large body size; seen, for example, in Late Cretaceous and early Cenozoic alligatorids.

gigantothermy Ability of a very large animal with a low metabolic rate to maintain a stable body temperature warmer than its environment due to its low surface area/volume ratio.

gill arches Components of the splanchnocranium that support gill tissue.

gill filaments Synonymous with primary lamellae, which are filamentous structures of internal gills that support secondary lamellae.

gills Vascularized structures supported by gill arches where respiratory gases and other substances are exchanged between the body and surrounding water.

gizzard The posterior muscular stomach of birds and other archosaurs. See *proventriculus*.

glacial refugia Areas into which animals move to survive harsh conditions and from which they later expand and repopulate as climate improves.

gliding Form of passive (non-flapping) flight in which altitude decreases and horizontal distance is gained.

glomerular capsule Portion of a nephron that encloses a tuft of capillaries, the glomerulus, where blood is filtered; from the glomerular capsule, the ultrafiltrate moves through the renal tubule.

glomerulus Capillary tuft associated with a kidney nephron that produces an ultrafiltrate of the blood.

gnathostomes Vertebrates with jaws: among extant vertebrates all except lampreys and hagfishes.

Gondwana Southern supercontinent of the Paleozoic and early Mesozoic, formed from South America, Antarctica, and Australia.

gonochorism Pattern of life history in which an individual begins life as one sex and stays that sex throughout life.

"good fossil effect" Tendency for very well-preserved fossils to drive new ideas and interpretations.

graviportal posture Pillar-like weight-bearing stance of very large tetrapods, such as elephants; associated with reduced agility.

gravity Force pulling an object to Earth.

Great American Biotic Interchange Exchanges of animals between North and South America when the Isthmus of Panama formed around 3 Ma.

green rods Type of retinal cell sensitive to violet and blue light; synapomorphy of lissamphibians (present in salamanders and frogs; possibly lost in caecilians).

gymnosperms Plants that reproduce by seeds (rather than by spores) but do not produce flowers, including extant conifers and cycads. Contrast with *angiosperms*.

H

hair cells Sensory receptors found in the inner ear of vertebrates and in the lateral lines of fishes and aquatic amphibians.

hair follicles Invaginations of the germinal layer of the epidermis that contain individual hairs.

hedonic glands Specialized courtship glands of male salamanders used to deliver pheromones to the female.

hermaphroditic Condition in which an individual animal has functioning gonads of both sexes.

heterocercal caudal fin Caudal fin characteristic of †osteostracans and gnathostomes in which the vertebral column bends upward, making the upper lobe of the fin larger than the ventral lobe; viewed externally the caudal fin is distinctly asymmetrical reflecting the many vertebrae within it (up to 200 or more in some sharks). Contrast with *hypocercal caudal fin* and *homocercal caudal fin*.

heterochrony Changes in the timing of gene expression during development.

heterodont dentition Dentition in which the teeth are regionalized into forms with different shapes and sizes. Contrast with *homodont dentition*.

heterodont Description of a dentition in which the teeth have different shapes. Contrast with *homodont*.

heterometry A change in the intensity of a gene's expression during development resulting in changes in the amount of the gene product.

heterothermy A variable body temperature.

heterotopy A change in the location of a gene's expression during development.

hindbrain Posterior portion of vertebrate brain; associated with sensory input from hearing, lateral line sense, electroreception, taste, and coordination of motor activities, respiration, and circulation.

hindgut fermenters Herbivorous mammals, such as horses, in which an enormous, sacculated cecum and colon house symbiotic microorganisms that break down cellulose, releasing nutrients from consumed plant material. Contrast with *foregut fermenters*.

holostylic jaw articulation A type of jaw suspension in holocephalans in which the upper jaw is fused to the chondrocranium. Contrast with *hyostylic jaw suspension*.

homeothermy A stable body temperature.

homing Form of true navigation in which a relocated animal finds its way back to its location prior to being displaced.

homocercal caudal fin Caudal fin characteristic of teleosteans in which fewer than eight vertebrae turn upward into the dorsal lobe of the caudal fin; although the skeleton of the caudal fin is asymmetrical, when viewed externally the caudal fin appears symmetrical, i.e., the caudal fin

looks the same whether the fish is right side up or upside down.

homodont Description of a dentition in which all the teeth have similar shapes. Contrast with *heterodont*.

hooklets Tiny hooks that extend from barbules and hold adjacent barbs together.

horizontal septum Plane of connective tissue that divides the trunk muscles into epaxial (upper) and hypaxial (lower) portions.

hormone-mediated maternal effects Epigenetic changes caused by the developmental environment provided by the mother.

Hox genes Group of related genes that specify the regional identity of cells along the anterior-posterior axis of the developing embryo. Duplications of Hox genes are thought to have led to the evolution of distinctive structures of vertebrates.

hyostylic jaw suspension A type of jaw suspension in which the anterior end of the upper jaw attaches to the chondrocranium via ligaments and the posterior end articulates with the chondrocranium via a large and mobile hyomandibula. Contrast with *holostylic jaw suspension*.

hypaxial muscles Ventral portion of axial musculature of vertebrates. Contrast with *epaxial muscles*.

hyperosmolal Term used to describe a solution with a lower water potential (i.e., higher solute concentration) than the comparison solution.

hyperthermia Increases in body temperature that occur when environmental temperatures rise above the temperature at which evaporative cooling can keep up with heat gain.

hypocercal caudal fin Caudal fin in which the vertebral column bends downward, making the ventral lobe of the fin larger than the dorsal lobe.

hypodermis Connective tissue layer beneath the dermis; formed from mesoderm.

hyposmolal Term used to describe a solution with a higher water potential (i.e., a lower solute concentration) than the comparison solution.

hypothermia Declines in body temperature that occur when environmental temperatures fall below the temperature at which heat production can keep up with heat loss.

hypsodont teeth Lophed molar teeth that have tall crowns (but retain the tooth root); on tooth eruption much of the crown is contained within the jaw bone, but it emerges throughout life as the exposed crown wears down. Enables mammalian teeth to resist wear; typical of grazing mammals such as horses and many bovids.

I

ichthyoplankton General term to describe eggs and larvae of fishes that occur in pelagic habitats.

induced drag Backward force opposed to forward motion created when local air close to the ventral surface of a bird's wing, for example, opposes movement of the airstream over the ventral surface of the wing.

inertial feeding Method of feeding used by many amniotes once a prey item is captured. First, the predator jerks its head backward to impart momentum to the prey while simultaneously opening its jaws, which loosens its grip on the prey. Then the predator thrusts its head forward over the prey before closing its jaws again.

ingroup Evolutionary group under study. Contrast with *outgroup*.

insensible water loss Amount of body fluid lost daily; routes include diffusion through the skin and evaporation from the respiratory tract.

integumentary sensory organs Extremely sensitive pressure receptors on the surface of crocodylian skin; these receptors probably detect water movement associated with prey and may play a role in social interactions as well.

intergenerational epigenetic effect Epigenetic effects that affect the next generation ($F_0 \rightarrow F_1$).

intermediate mesoderm Segmented mesoderm that forms between somitic and lateral plate mesoderm and gives rise to kidneys, portions of the gonads, and urogenital ducts.

interspecific brood parasites Birds that relinquish all parental care to adults of another species.

intracranial joint Joint between the anterior and posterior portions of the braincase and dermatocranium of sarcopterygians; present in Devonian forms and extant coelacanths but lost in all other extant sarcopterygians.

intragenerational epigenetic effect Epigenetic effects limited to one generation ($F_0 \rightarrow F_0$).

intramembranous ossification Process of ossification by osteocytes associated with a connective tissue membrane.

isosmolal Term used to describe solutions with the same water potential (i.e., with the same solute concentrations).

isthmian pairs Marine species that diverged around the time of closure of the Panamanian Isthmus (~3 Ma); one species lives on the Atlantic side of the isthmus and the other species lives on the Pacific side.

iteroparity Life history pattern in which adults breed more than once.

K

keel Protrusion on the sternum of birds that fly (reduced or absent in flightless species); serves as the origin for flight muscles.

kleptogenesis Reproductive mode found in unisexual (all-female) salamanders; females produce eggs that must be activated by sperm from a male of a bisexual species.

L

labyrinthodont teeth Teeth with infoldings near their base that function to strengthen the attachment to jaw bones.

lactate Feeding young by producing milk from the mammary glands.

Lagerstätte Locality in which fossils exhibit extraordinary preservation and taxonomic diversity.

larynx Structure in amniotes at the junction of the pharynx and trachea used for sound production, especially in mammals. In birds, it is not used for vocalization, serving instead to regulate flow of air into the trachea. Contrast with *syrinx*.

lateral line system Sensory system on the body surface of fishes and aquatic amphibians that detects water movements.

lateral plate mesoderm Unsegmented ventral part of the mesoderm surrounding the gut that gives rise to portions of the viscera and heart.

lateral somitic frontier Boundary in gnathostome embryos between the primaxial region (formed from somites alone and giving rise to such structures as the vertebral column and associated muscles) and the abaxial region (formed by contributions from both somites and lateral plate mesoderm and giving rise to structures such as the paired fins and limbs).

lateral undulation Form of locomotion in which undulatory waves pass along the body to produce thrust.

Laurasia Northern supercontinent of the later Paleozoic, formed from Euramerica and Siberia.

Laurentia Northern supercontinent of the early Paleozoic, formed from North America, Greenland, Scotland, and part of Asia.

lecithotrophy Mode of nutrition in which a developing embryo or fetus receives nourishment from the egg yolk. Contrast with *matrotrophy*.

lepidotrichia Signet-shaped dermal bones in fin rays of osteichthyans that give the fin rays a jointed appearance.

lift Upward force that keeps a bird or bat in the air.

lobate webbing In aquatic birds such as coots and grebes, flexible flaps of skin occur on toes 2, 3, and 4, but the flaps are not connected from one toe to another.

loop of Henle Portion of the renal tubule in nephrons of mammals that dips into the medulla of the kidney; largely responsible for the production of concentrated urine.

lophs Molar teeth in which the cusps seen in bunodont teeth are run together into ridges, which grate the food rather than pulping it. Found in herbivorous mammals such as deer, where they are described as selenodont, and rodents, where they are described as lamellar.

lower critical temperature The temperature representing the lower boundary of the thermoneutral zone; below this temperature, metabolic heat production must increase.

lower lethal temperature The lowest environmental temperature at which heat production can keep up with heat loss; below this temperature, body temperature falls and metabolic heat production decreases, leading to death.

lumbar vertebrae Trunk vertebrae of mammals that lack ribs. Contrast with *thoracic vertebrae*.

M

mammary glands Skin glands that produce milk; a unique and defining feature of mammals.

masseter muscle Jaw-closing muscle of cynodonts and mammals.

matrotrophic Receiving nutrition from the mother. See *matrotrophy*.

matrotrophy A mode of nutrition in which the reproductive tract of the mother supplies the energy for a developing embryo. Contrast with *lecithotrophy*.

maxilla Posterior tooth-bearing dermal bone in the upper jaw of osteichthyans.

maxilloturbinates Scroll-like bones in the nasal cavity of mammals covered by moist epithelium that warms and humidifies air on inspiration and recovers water and heat on expiration. Birds independently evolved cartilaginous respiratory turbinates.

mesenteries Folds of peritoneum that suspend organs in body cavities.

mesoderm Middle embryonic germ layer.

mesopelagic Twilight zone of the ocean extending from 200 m to 1000 m; light penetrates to the bottom of the mesopelagic in clear ocean waters but is insufficient to support photosynthesis. Much biomass is concentrated in the mesopelagic.

Mesophytic Flora Terrestrial flora that existed from the late Carboniferous to the early Cretaceous dominated by seed plants.

mesopredator A predator in the middle of a food web that preys on smaller species and is eaten by other predators. Contrast with *apex predator*.

mesotarsal joint A distinctive joint in the middle of the hindfoot, between proximal and distal rows of tarsal bones; the axis of bending in the foot is a simple hinge.

Mesozoic era The second era of the Phanerozoic lasting from 252 Ma to 66 Ma and including the Triassic, Jurassic, and Cretaceous periods.

Mesozoic Marine Revolution Evolution of a diversified fauna of marine invertebrates resulting from an evolutionary arms race between vertebrate predators with crushing dentitions and their hard-shelled invertebrate prey.

metabolic suppression Changes in gene expression and modifications of enzymes in oxidative metabolic pathways that start before body temperature falls in a mammal entering deep torpor; dramatically reduces oxidative metabolism.

metanephric kidneys Compact, often bean-shaped kidneys located toward the posterior end of the trunk; found in amniotes. Contrast with *opisthonephric kidneys*.

Metazoa The animal kingdom, which includes more than 30 phyla, including Phylum Chordata.

midbrain Middle portion of vertebrate brain; associated with vision.

Modern Synthesis A view of evolution combining information from genetics, natural selection, and population biology that emerged in the mid twentieth century.

molecular scaffolding Using molecular data to determine the branching pattern of phylogenies and morphological data to date the branching.

molting Loss of old hairs and replacement by new ones.

monobasic fin Fin type characteristic of sarcopterygians in which one basal element articulates with the fin girdle. Contrast with *tribasic fin*.

monobasic paired fins See *monobasic fin*.

monocular vision Type of vision in which the visual fields of the two eyes do not overlap.

monophyletic Term describing a group that includes an ancestor and all of its descendants. Contrast with *paraphyletic*.

monorhiny Condition of having a single nostril and nasal sac, formed from a single midline nasal placode. Contrast with *diplorhiny*.

morphospaces Graphical representations of an organism's anatomical characteristics that can be statistically compared with those of other organisms; also called morphological spaces.

mosaic evolution Concept based on the observation that body parts (sometimes called modules) evolve independently.

most parsimonious phylogeny In phylogenetic systematics, the branching sequence that requires the fewest changes to go from the ancestral condition to the derived condition. Also called *parsimony*.

myelin Insulating, lipid-rich sheaths that surround axons and promote rapid conduction of action potentials.

myomeres Blocks of striated muscle fibers arranged along both sides of the body; most conspicuous in fishes.

N

nasoturbinates Located in front of the ethmoturbinates in the nasal cavity of mammals, these scroll-like bones may function in olfaction. Contrast with *maxilloturbinates*.

navigation An animal determines how to get from where it is to where it wants to be.

nephrons Basic functional units of the kidney.

neural crest Embryonic tissue that forms next to neural folds and gives rise to many structures unique to vertebrates; one of three neurogenic tissues (i.e., tissues giving rise to neurons).

neural plate Early embryonic structure that gives rise to the neural tube.

neural tube Precursor to much of the central nervous system, including the brain and spinal cord of vertebrates.

neurogenic placodes Thickenings in the embryonic ectoderm that give rise to sensory organs and nerves in the head unique to vertebrates; one of three neurogenic tissues (i.e., tissues giving rise to neurons).

neuromast organs Clusters of sensory hair cells and associated structures usually enclosed in the lateral lines of fishes and aquatic amphibians.

node Branching point that leads to two or more terminal taxa in a phylogenetic tree.

non-shivering thermogenesis Mechanism of endothermal thermoregulation in which oxidative phosphorylation is increased by short-circuiting a cellular process; occurs, for example, in brown adipose tissue of mammals.

nonplacental viviparity Mode of viviparity in which a developing embryo or fetus grows to term without a placental exchange surface with the mother. Contrast with *placental viviparity*.

normothermia Normal body temperature for an endotherm.

notochord A dorsal stiffening rod that gives the phylum Chordata its name; used as the axial support for swimming of nonvertebrate chordates and present in vertebrate embryos and larvae, but greatly reduced in most adult vertebrates.

nursing Behavior displayed by lactating female mammals when transferring milk to young.

O

occlusion When teeth of the upper jaw contact those of the lower jaw. Mammals have very precise occlusion because the cusps of their upper molars interdigitate with those of lower molars enabling efficient chewing.

odontodes Toothlike elements in the skin composed of dentine and enamel.

olfactores Phylogenetic grouping that contains urochordates (tunicates) and vertebrates.

opercular bone Dermal bone in the head that covers the gills and aids ventilation. Also called *opercle*.

operculum In Osteichthyes, covering of the gill chamber composed of bones and associated soft tissues; in Chimaeriformes composed mostly of soft tissues and some cartilage.

opisthonephric kidneys Elongate kidneys extending the length of the trunk; found in fishes and lissamphibians. Contrast with *metanephric kidneys*.

oral jaws Bony elements that surround the mouth opening. Contrast with *pharyngeal jaws.*

orientation An animal determines the direction it is facing.

osmoregulation Process of maintaining both water and salt balance so that body fluids do not become either too dilute or too concentrated.

osmosis Movement of water across a semipermeable membrane from a solution of higher water potential to a solution of lower water potential (i.e., from a dilute solution to a more concentrated solution).

ossicles One or more tiny bones in the middle ear of tetrapods that transmit vibrations from the tympanum to the oval window of the inner ear.

Osteognathostomata Clade containing all vertebrates with bone. Contrast with *Cyclostomata.*

outgroup A reference group that is less closely related to the group under study (which is known as the ingroup) than the members of the ingroup are related to each other. Contrast with *ingroup.*

oviparity A mode of reproduction in which a female deposits eggs that develop outside her body.

P

paedomorphosis Retention of larval or juvenile characteristics into adult life.

Paleophytic Flora Terrestrial flora that existed from the mid Silurian until the Carboniferous rainforest collapse consisting of mosses, liverworts, hornworts, and tracheophytes, including early seed plants.

Paleozoic era The first era of the Phanerozoic, lasting from 541 Ma to 252 Ma and including the Cambrian, Ordovician, Silurian, Devonian, Carboniferous, and Permian periods.

palmate webbing In aquatic birds such as ducks, geese, and gulls, webbing between toes 2, 3, and 4; the backward-facing toe is not included.

Pangaea Supercontinent that formed in the Carboniferous and persisted until the Middle Jurassic. Vertebrates were able to spread widely across the Earth's surface because there were no intervening oceans.

paraphyletic Referring to a named group of organisms that does not include an ancestor and all of its descendants. Phylogenetic classifications attempt to eliminate naming paraphyletic groups. If such groups must be named, the name is enclosed within quotation marks. Contrast with *monophyletic.*

parental care model Hypothesis proposing that endothermy is a consequence of selection for warmer maternal body temperatures that facilitate embryonic development and growth of neonates, which could shorten the period of parental care. See *aerobic scope model, thermogenic opportunity model, warmer is better model.*

parthenogenesis Reproduction by females without fertilization of the ova by males.

parthenogenetic Reproduction by females without fertilization of the ova by males.

passive flight Non-flapping flight, such as soaring and gliding. Contrast with *powered flight.*

pectoralis major A major flight muscle in birds that originates on the keel of the sternum and inserts on the ventral side of the humerus; its contraction pulls the wing downward (downstroke). Contrast with *supracoracoideus.*

pelage A coat of fur, made up of closely spaced individual hairs.

pelagic eggs Eggs spawned directly into the water column that float because of an enclosed oil droplet. Contrast with *demersal eggs.*

pelvic claspers Paired intromittent organs of male chondrichthyans (one on each pelvic fin) that make internal fertilization universal in this group.

pelvic patch Highly vascularized area in the pelvic region of terrestrial anurans where most of the cutaneous uptake of water occurs.

pennaceous feathers Feathers having a vane on each side of the rachis.

perforated acetabulum Condition in which the bones of the pelvis—pubis, ischium, and ilium—do not fully meet at the acetabulum (socket receiving head of the femur), leaving an unossified hole; synapomorphy of Dinosauria.

pericardial cavity Portion of the body cavity enclosing the heart.

pericardium Lining layer of the pericardial cavity formed by mesoderm.

perichondral bone Bone formed by osteocytes in the membrane surrounding cartilage or bone.

perineum In therian mammals, the wall of connective tissue between openings of the digestive and urogenital tracts.

periods Subdivisions of eras during the Phanerozoic extending millions to ~ 100 million years; we live in the Quaternary period.

peritoneal cavity Portion of the body cavity enclosing the viscera.

peritoneum Lining layer of the peritoneal cavity formed by mesoderm.

Phanerozoic eon The last 541 million years of Earth's 4.6-billion-year history, including the Paleozoic, Mesozoic, and Cenozoic eras. The period during which multicellular life burgeoned and radiated. Contrast with *Precambrian.*

pharyngeal arch Segmentally organized tissues between the pharyngeal gill slits, including skeletal, muscular, nervous, and vascular tissues.

pharyngeal jaws Dermal tooth plates in the pharynx of neopterygian fishes fused to mobile gill arch elements; function in food manipulation and processing. Contrast with *oral jaws.*

pharyngeal slits   Openings in the pharynx used in filter feeding.

pharyngula   Developmental stage in vertebrate embryos characterized by the presence of pharyngeal arches and slits.

pharynx   Throat region.

phenotypic plasticity   Ability of a genotype to produce different phenotypes as a result of epigenetic modification of developmental processes.

photic zone   Top 1000 m of water through which light can penetrate; based on clearest ocean waters.

photophores   Organs specialized to produce light by bioluminescence.

phragmosis   Behavior in which an animal defends itself in a restricted space, such as a burrow or hole, by creating a barrier with its own body; for example, casque-headed frogs retreat into holes and use their head to block the entrance.

phylogenetic systematics   Grouping of taxa according to their evolutionary relationships.

phylogenetic tree   A branching diagram depicting hypotheses of evolutionary relationships among taxa.

phylogeny   Evolutionary relationships based on the branching sequence of lineages.

physoclistous   Condition in teleost fishes in which adults lack a connection between the gas bladder and gut.

physostomous   Condition in bony fishes in which there is a connection between the gas bladder and gut.

pineal eye   A light-sensitive structure homologous to the pineal gland of mammals.

pinna   The external ear, seen in therian mammals.

pitch   Body movements involving tilts up or down.

placenta   An organ formed during pregnancy in some species of vertebrates where the embryo and later fetus exchange nutrients, metabolic wastes, respiratory gases, and hormones with the mother; found, for example, in sharks and therian mammals.

placental viviparity   Mode of viviparity in which a developing embryo or fetus establishes a placental exchange surface with the lining of the mother's uterus. Contrast with *nonplacental viviparity.*

placoid scales   Scales of chondrichthyans, formed from an outer layer of enamel and an inner layer of dentine. Also known as dermal denticles because they resemble teeth.

plastron   The ventral or lower shell of a turtle. Contrast with *carapace.*

plate tectonics   Mechanism by which continental drift occurs by the movement of the underlying tectonic plates.

plesiomorphies   Characters that are unchanged from their ancestral conditions (i.e., ancestral characters).

pleurodires   Turtles that bend the neck to the side (horizontally) to retract the head into the shell. Contrast with *cryptodires.*

plumulaceous feathers   Feathers lacking vanes; also called downy feathers.

pneumatization   Air spaces in bones; found, for example, in †pterosaurs and extant birds.

polyphyodonty   Pattern of tooth replacement in which teeth are replaced throughout life; found in most vertebrates, except most mammals. Contrast with *diphyodonty.*

portal vessels   Blood vessels that connect two capillary beds without passing through the heart.

positive selection   An increase in the frequency of a genetically-based trait in successive generations.

postanal tail   Tail that extends posterior to the anus; characteristic of Phylum Chordata.

postpulmonary septum   Sheet of connective tissue separating the thoracic cavity, which houses the lungs, and the peritoneal (abdominal) cavity, which houses the viscera; its muscularization produced the diaphragm of mammals.

potamodromy   Pattern of migration of fishes that move from large lakes into streams to reproduce.

powered flight   Flapping flight found, for example, in birds and bats.

Precambrian   The more than 4 billion years preceding the Phanerozoic. Covering more than 85% of Earth's planetary history, the Precambrian encompasses the lifeless Hadean and the advent of the earliest life in the Archean and Proterozoic. Contrast with *Phanerozoic.*

precocial   Developmental state of neonates that can locomote and thermoregulate soon after birth or hatching. Contrast with *altricial.*

premaxilla   Anterior tooth-bearing dermal bone in the upper jaw of osteichthyans.

pressure drag   Drag caused by pressure differences as fluids pass around a moving object. Also known as inertial drag, this type of drag is more important at high speeds and large sizes.

primaries   Flight feathers rooted in the hand (manus). Contrast with *secondaries.*

primary lamellae   Filamentous structures of internal gills that support secondary lamellae.

prolonged breeding   Mating pattern of anurans in which the breeding season can last for months. Contrast with *explosive breeding.*

propatagium   Small flying membrane of †pterosaurs extending from the front edge of the wing forward to the neck; supported by the pteroid bone attached to the wrist.

protandry   Pattern of life history in which an individual begins life as a male and later transforms to a female; one form of sequential hermaphroditism. Contrast with *protogyny.*

protogyny   Pattern of life history in which an individual begins life as a female and later transforms to a male; one form of sequential hermaphroditism. Contrast with *protandry.*

proton leak   Movement of protons back into the mitochondrial matrix without passing through ATP synthase; results in an increased rate of oxidative phosphorylation and heat production, without increasing ATP synthesis.

Protostomia   Animals in which the blastopore becomes the mouth. Contrast with *Deuterostomia.*

proventriculus   The anterior glandular stomach of birds and other archosaurs. See *gizzard.*

pterylae   Tracts of follicles in the skin of birds from which feathers grow.

puncture-and-pull feeding   Method of feeding that involves first biting downwards to puncture the flesh and then pulling away so knifelike teeth tear through and dismember the prey. Found, for example, in theropods such as †*Velociraptor.*

pycnofibers   Fine, hairlike fuzz on the skin of some †pterosaurs that probably provided insulation.

pygostyle   Triangular structure formed by fusion of the last few caudal vertebrae of a bird; supports tail feathers. Found within Pygostylia, which includes †*Confuciusornis,* †*Enantiornithes,* †*Hesperornis,* and extant birds.

R

rachis   The central shaft of a feather.

ram feeding   Method of feeding in which small prey are engulfed with an open mouth as the fish swims forward. Contrast with *suction feeding* and *filter feeding.*

ram ventilation   Form of ventilation in which a respiratory current across the gills is generated by a fish swimming with its mouth open.

rectilinear locomotion   Form of locomotion used by heavy-bodied snakes in which waves of contraction pass from anterior to posterior areas of the body; alternate sections of the body are lifted and pulled forward while intervening sections rest on the ground, as the snake moves slowly in a straight line.

rectrices   (singular *rectrix*) Flight feathers of the tail.

remiges   (singular *remex*) Flight feathers of the wing.

remote touch   Sensory ability of some birds to detect prey at some distance from the tip of the bill when probing in mud or sand; based on mechanoreceptors such as Herbst corpuscles that detect prey vibrations.

renal tubule   Portion of a nephron where the ultrafiltrate produced at the glomerulus is processed to recover essential metabolites and water, and remove wastes.

rete mirabile Intertwined capillaries that exchange heat or dissolved substances including respiratory gases between countercurrent flows.

ribs Paired curved bones extending from the trunk vertebrae to support and enclose the body cavity (or the thoracic cavity in mammals); also function in ventilation in derived tetrapods. Present, but smaller, on cervical vertebrae in some non-mammalian tetrapods.

rod cells Photoreceptors in the retina sensitive to a wide range of wavelengths; used to perceive low levels of light but not for high visual acuity.

roll Body movement involving rotation around the body axis.

root Base of a phylogenetic tree; shown near the upper left hand corner of a phylogenetic tree in this book.

ruminate The action of regurgitating and chewing partially digested plant material (the cud); performed by foregut fermenters, such as cows.

runaway greenhouse effect Caused by rapid increase in greenhouse gases and responsible for the end-Permian extinction.

S

sacral vertebrae Vertebrae that tightly articulate with the pelvic girdle in tetrapods, to transfer force from the hindlimbs to the axial skeleton; number varies.

salt-secreting glands Present in many sauropsids, these glands provide an extrarenal pathway (i.e., pathway in addition to the kidney and not involving the kidney) to dispose of salt while conserving water.

Sauropsida The lineage of amniotes represented by extant reptiles (including birds) and their extinct relatives. See *Synapsida*.

seasonal migration Long distance movements between breeding and non-breeding areas.

sebaceous glands Skin glands found over the surface of the body; sebaceous glands produce an oily secretion that lubricates hairs.

secondaries Flight feathers rooted in the ulna of the forearm. Contrast with *primaries*.

secondary lamellae Microscopic projections from the primary lamellae (gill filaments) where gas exchange occurs. See *primary lamellae*.

secondary palate Structure that separates the nasal cavity from the oral cavity; found, for example, in extant crocodylians and mammals.

semelparity Life history pattern in which adults die after breeding once.

semiplumes Feathers intermediate in structure between contour and down feathers; rachis is present, but barbs do not form a vane because the barbules lack hooklets. These feathers function in insulation and streamlining.

shared derived characters A derived character (i.e., one that has changed from its ancestral state) shared by two or more taxa and postulated to have been inherited from their common ancestor; equivalent to *synapomorphy*.

shivering Mechanism of endothermal thermoregulation in which muscle fibers contract in an uncoordinated manner. Shivering converts the chemical energy of ATP into the elastic energy of a contracted muscle fiber, releasing heat.

sidewinding Form of locomotion used by snakes moving across loose substrates, such as sand, in open habitats. The snake raises its body in loops, resting on two or three points where its body contacts the ground, and swings the loops forward.

signal cascades Intracellular pathways that can be triggered by extracellular molecules to either activate or repress gene expression.

simultaneous hermaphroditism Pattern of life history in which an individual has functional ovaries and testes and can function as a male or female in reproduction.

single circulation Type of circulation in fishes in which oxygenated blood from the gills flows straight out to the body, without first returning to the heart. Contrast with *double circulation*.

sinoatrial node Portion of the right atrium that controls the heartbeat of therian mammals.

sinus venosus Posteriormost chamber of the heart in nontherian vertebrates; receives blood from the systemic veins.

sister group (or sister taxon) Monophyletic lineage most closely related to the monophyletic lineage being discussed.

slime glands Paired glands that open via pores to secrete slime composed of mucus and long proteinaceous threads; synapomorphy of hagfishes.

soaring Form of passive (non-flapping) flight in which altitude is maintained or increased.

somites Series of paired segments of the embryonic dorsal mesoderm of vertebrates.

somitic mesoderm Segmented mesoderm derived from somites that gives rise to vertebrae, dermis, and skeletal muscles.

sonation A nonvocal sound, typically generated with either the bill or feathers, used for communication.

songs Long, complex vocalizations produced by birds in the contexts of reproduction and territoriality. Contrast with *calls*.

spermatophore A packet of sperm transferred from male to female during mating of most salamanders; consists of a sperm cap on a gelatinous base and allows internal fertilization without copulation.

splanchnocranium Skeleton associated with the jaws, jaw supports, and gills.

stem group Extinct forms in a clade that lack some of the derived characters that define the crown group. Contrast with *crown group*.

stenohaline Capable of living only within a narrow range of salinities. Contrast with *euryhaline*.

stereoscopic vision Type of vision within the area of overlap of binocular visual fields; yields depth perception.

sternum Midventral endochondral element found only in tetrapods that articulates with elements of the pectoral girdle; often called the breastbone.

stylopodium Upper part of a paired appendage of tetrapod, the humerus or femur.

suckling Behavior displayed by young mammals when obtaining milk from a nipple.

suction feeding Method of feeding in which rapid expansion of the mouth and pharyngeal cavity draws prey into the mouth. Contrast with *ram feeding* and *filter feeding*.

supracleithral series Dermal bones extending from the posterior corners of an osteichthyan skull connecting it to the cleithrum, a large dermal bone in the pectoral girdle.

supracoracoideus A major flight muscle in birds that originates on the keel of the sternum and inserts on the dorsal head of the humerus; its contraction pulls the wing upward (upstroke, often a recovery stroke). Contrast with *supracoracoideus*.

suspensory locomotion Mode of locomotion employed by larger arboreal anthropoids (New World monkeys, Old World monkeys, apes) when moving with the body hanging below the branches.

synapomorphy A derived character (i.e., one that has changed from its ancestral state) shared by two or more taxa and postulated to have been inherited from their common ancestor; equivalent to *shared derived character*.

synapsid Condition in which there is a single lower fenestra in the temporal region on each side of the skull; found, for example, in extant mammals.

Synapsida The lineage of amniotes that includes extant mammals and their extinct relatives. See *Sauropsida*.

synsacrum Structure in birds formed by fusion of lumbar, sacral, and some caudal vertebrae with the pelvis; supports upright, standing posture.

syrinx Structure in birds located at the junction of the trachea with the two bronchi; functions in producing vocalizations. Contrast with *larynx*.

T

tapetum lucidum Shiny crystals of guanine behind the retina that reflect light back through the retina.

tarsometatarsals Bones formed by fusion of the distal tarsal elements with the metatarsals in birds and some nonavian dinosaurs.

tarsometatarsus In birds, the bone formed by fusion of the distal tarsal (ankle) elements with the metatarsals of three toes. See *tibiotarsus*.

taxa (singular *taxon*) Groups of organisms of any rank—e.g., species, genera, families, etc.

taxonomy Discipline that assigns names to organisms.

temporal fenestration Openings (fenestrae) in the temporal region of the skull of amniotes associated with jaw closing muscles. Number and arrangement of fenestrae diagnose the two extant radiations of amniotes, Synapsida (one lower temporal fenestra on each side of the skull) and Sauropsida (two temporal fenestrae—one upper and one lower—on each side of the skull).

terminal taxon A group named at the tip of a branch of a phylogenetic tree; a terminal taxon can be at any taxonomic level, e.g., a species, a genus, a family, etc.

tesserae Small prisms of mineralization that form on the surfaces of some skeletal elements of chondrichthyans, such as the jaws.

thermogenic opportunity model Hypothesis proposing that the adaptive value of endothermy lies in permitting nocturnal activity. See *aerobic scope model, parental care model, warmer is better model.*

thermoneutral zone Range of temperatures within which an endotherm maintains a stable body temperature by adjusting its insulation (for example, position of feathers or hair) without changing its metabolic rate.

thermoregulation Limiting variation in body temperature.

thoracic vertebrae Trunk vertebrae of mammals that bear ribs. Contrast with *lumbar vertebrae.*

thrust Forward force generated during flapping flight of birds and bats.

tibiotarsus In birds, the bone formed by fusion of the tibia and proximal tarsal (ankle) elements. See *tarsometatarsus.*

tidal ventilation Passage of water or air in through the mouth, nose, or gill slits for respiration that then passes out through the same structure(s). Contrast with *flow-through ventilation.*

tooth file Developing teeth aligned in a series behind the functional teeth of sharks; sometimes called a tooth family.

tooth whorls Replacement mechanism for teeth in sharks. See *tooth file.*

topology Arrangement of branches and taxa in a phylogenetic tree.

torpor In endotherms, a controlled reduction in body temperature to a new set point; may last a few hours or many months.

totipalmate webbing In aquatic birds such as pelicans and boobies, webbing between all four toes.

tracheophytes Vascular plants; land plants with internal channels (xylem) for the conduction of water.

transgenerational epigenetic effect Epigenetic effects that affect subsequent generations ($F_0 \rightarrow F_n$).

tribasic In reference to paired fins, the presence of three skeletal elements known as basals articulating with the pectoral or pelvic girdle.

tribasic fin Fin type characteristic of gnathostomes retained in chondrichthyans and actinopterygians in which three basal elements articulate with the fin girdle. Contrast with *monobasic fin.*

tribosphenic molars Type of molar in which a new cusp (protocone) on the upper molars pounds into a new basin (talonid basin) of the lower molars allowing efficient crushing and pulping of food; found in therians.

triploblastic A metazoan with three embryonic germ layers, ectoderm, mesoderm, and endoderm.

true navigation Ability to compensate for experimental relocation and reach the goal without using landmarks and without information about the relocation route.

trunk vertebrae Vertebrae that bear ribs in many tetrapods, except lissamphibians, which have almost entirely lost ribs. Trunk vertebrae are subdivided into thoracic and lumbar vertebrae in mammals. Contrast with *cervical vertebrae, sacral vertebrae,* and *caudal vertebrae.*

turbinates Scroll-like bones in the nasal passages of mammals covered by moist tissue that functions either in olfaction (ethmoturbinates and possibly nasoturbinates) or respiration and water conservation (maxilloturbinates). Birds independently evolved respiratory turbinates, but their turbinates are cartilaginous and consist of a single, sturdy scroll.

tympanum Also called the eardrum, this structure resonates in synchrony with impinging airborne sound waves and transfers vibrations to one or more ossicles in the middle ear of tetrapods.

U

upper critical temperature The temperature representing the upper boundary of the thermoneutral zone; above this temperature, evaporative cooling maintains normal body temperature.

upper lethal temperature The highest environmental temperature at which evaporative cooling can keep up with heat gain; above this temperature, body temperature rises, metabolic heat production rises, and dramatic increases in body temperature result in death.

ureotely Excretion of nitrogenous wastes primarily as urea. Contrast with *ammonotely* and *uricotely.*

uricotely Excretion of nitrogenous wastes primarily as salts of uric acid. Contrast with *ammonotely, ureotely.*

Urochordata One of two groups of extant non-vertebrate chordates, the tunicates. The other group is Cephalochordata, the lancelets.

uropygial gland Sebaceous gland at the base of the tail in birds that produces oil for waterproofing feathers; may be reduced or lost in nonaquatic birds.

urostyle Solid rod of bone formed by fused posterior vertebrae in anurans; this morphological specialization for jumping contributes to a rigid trunk region.

V

vane The aerodynamic surface of a feather formed by interlocking barbules on the barbs that extend from the rachis.

vasopressin Hypothalamic hormone released in response to increased blood osmolality or reduced blood volume that promotes recovery of water by the kidney. Also known as *antidiuretic hormone (ADH).*

veins Blood vessels that return blood to the heart.

ventilation In respiration, movement of a respiratory medium (water or air) across a respiratory surface for gas exchange.

ventilation hypothesis Idea that the origin of jaws is linked to improved ventilation.

ventricle A chamber; the ventricle of the heart is the portion that applies force to eject blood.

vertebrae Serially arranged skeletal elements that form during development in association with the notochord of vertebrates to provide axial support, protect the spinal cord, and serve as the origin for muscles used in locomotion; collectively equivalent to the vertebral column.

vertebral column Term describing the complete series of vertebrae.

vestibular system Semicircular canals and associated sense organs within the inner ear used to detect balance and orientation.

vibrissae Sensory hairs around the snout, also known as whiskers.

viscous drag Drag caused by friction between a moving object and the surrounding fluid at the boundary layer. Viscous drag is important for slowly moving, small objects such as tiny larval fishes.

viviparity Reproductive mode in which the mother gives birth to fully formed young.

vocalizations Sounds produced by birds passing air through the syrinx.

vomeronasal organ (also knowns as *Jacobson's organ*) Paired chemosensory organ in the roof of the mouth of lungfishes and tetrapods (lost in some); part of the accessory olfactory system. Nonvolatile odorants must be delivered directly to the vomeronasal organ to be detected; this contrasts with the main olfactory system through which air carrying volatile odorants is inhaled for olfaction and respiration.

W

warmer is better model Hypothesis proposing that the adaptive value of endothermy lies in the faster rates of biochemical and physiological processes at high temperature. See *aerobic scope model, parental care model, thermogenic opportunity model.*

water balance State when equal amounts of water are entering and leaving an animal's body.

Weberian apparatus In otophysan fishes, small bones derived from vertebrae that connect the gas bladder to the ear.

wing loading The weight of a bird divided by its total wing area.

Y

yaw Body movements involving swings to the right or left.

yolk sac placenta The early-forming placenta of all therian mammals, produced by apposition of the yolk sac to the chorion. In marsupials this is the major placenta, in placentals it is transitory; also known as a choriovitelline placenta. Also found in some sharks. Contrast with *chorioallantoic placenta.*

yolk sac An extraembryonic membrane present in all vertebrates that encloses yolk, the source of nutrition for developing embryos.

Z

zeugopodium Second part of a paired appendage of a tetrapod, the radius and ulna or the tibia and fibula.

ziphodont Term used to describe teeth that are laterally compressed and serrated (i.e., knifelike); found, for example, in theropods such as †*Velociraptor.*

zygapophyses Processes on the neural arches of vertebrae that interlock with adjacent zygapophyses to provide strength and resist torsion of the vertebral column.

zygodactylous A type of foot in which the toes are arranged in two opposable groups; found, for example, in chameleons and some groups of birds, including parrots, owls, and woodpeckers.

zygodactyly A type of foot in which the toes are arranged in two opposable groups; for example, in birds such as woodpeckers and parrots, toes 2 and 3 face forward and toes 1 and 4 face backward.

zygomatic arch The bowed-out lower border of the synapsid temporal opening seen in cynodonts and mammals. In humans this is the cheek bone.

Index

Numbers in *italic* refer to information in an illustration or table.

phylogeny of placentals, *527*
powered flight, 546–547
Baubellum, 535
Baurusuchus, 450
Bd. *See* Batrachochytrium
  dendrobatidis
Beaded lizards, 335, *337*
Beagle (HMS), 11
Beaks
  echidna electroreception, 75
  evolution in birds, 435
  heterometry in birds, 14–15
  mosaic evolution of birds and,
    434, 435
  See also Bills
Bearded dragon (*Pogona vitticeps*),
  355–356
Beardfishes, *142*, *143*, 149
Bears
  number of species and distri-
    bution, 529
  phylogeny of placentals, *527*
  winter dormancy, 322
Beavers (*Castor*), *528*, 547
Bechstein's bat (*Myotis bechsteinii*),
  533
Behavioral thermoregulation,
  315–317
Bellowing, crocodylians, 388, *389*
Belly crawling, crocodylians,
  382, *383*
Beloniformes, *151*, *153*, 156
Beluga sturgeon (*Huso huso*), 137
Beluga whale, *529*
Beta-keratin, 216, 361
Bettas, 154
Bettongs, 523
Biarmosuchia, 492
Biarmosuchus, 496, *497*
Biceps brachii muscle, 242
Bichirs (*Polypterus*), 53, 68, *131*,
  136, 137, 188
Bifid glans, 535
Big toe
  bipedalism in humans, 581
  differences between humans
    and chimpanzees, 572, *573*
  feet of extant hominoids, *573*
  opposable digits of primates,
    557
Bighorn sheep (*Ovis canadensis*),
  414, 551–552, *553*
Bigtoothed pomfret (*Brama
  orcini*), *168*
Bilateria, 21–22
Bilbies, *521*, 523, *525*, 526, *537*
Bile, 542
Bile duct, 542
Biliverdin, 439–440
Billfishes, *151*, *153*, *154*, 155
Billing, 472
Bills
  platypus electroreception, *75*
  specializations in birds,
    470–472
  touch in birds and, 477
  See also Beaks
Bill-tip organs, 477
Binocular vision, 265, 475
Binominal nomenclature, 2–3
Biofluorescence, sharks, 117
Bioluminescence, sharks, 117
Biomass production, 330

Biparental care, 441
Bipedalism
  birds, 461, 462–463
  evolution of, 245, *246*
  hominins, 572, 574
  humans, 580–581
  theropods, 421, 425
Bipes, 337
Bird eggs
  development and structure, 479
  fertilization and development,
    479
  hatching, 439, 480
  incubation, 439, 480–481
  laying of, 439
  maternal effects, 479
Bird flight
  body size and, 458
  compared to bat flight, 547
  evolution of powered flight,
    437–439
  feathers, 458–460
  flight muscles, 463, 464
  a key characteristic of birds, 455
  skeleton, 461–463
  streamlining and weight re-
    duction, 460–461
  wings and flight mechanics,
    464–467, *468*
Bird wings
  feather anatomy and types,
    459, *460*
  flight mechanics, 464–466
  skeletal structure, *460*
  sonation and, 478
  types of, 466–467
  wing shape and flight,
    466–467, *468*
  wing-propelled divers, 469–470
  wing-waving display, 478
Birds (Aves)
  bills, feeding, and digestion,
    470–473, *474*
  bipedalism, 461, 462–463
  blood osmolality, 78
  blood pressure, 297, 299, *300*
  body size, 437
  Cenozoic era, 448
  communication, 477–478
  conservation, 486–487
  cooling of the brain, 326
  defined, 422
  different morphospace from
    pterosaurs, 403
  diversity of, 455–458
  eggs (*see* Bird eggs)
  end-Cretaceous extinctions, 233
  evolution of lung ventilation in
    sauropsids, 295
  evolution of the pelvis, *404*
  extended sternum, 295
  features shared with ptero-
    saurs, 400
  feet and locomotion, 468–470
  flight (*see* Bird flight; Bird
    wings)
  Fossil Lake fossils, 447
  girdles and limbs, 435–436
  hearts and circulation,
    297, *298*, *299*
  heterochrony, 14
  heterometry and beak
    morphology, 14–15

heterothermal endothermy,
  323–324
heterotopy and interdigital
  webbing, 14
hibernation, 321
inner ear, *254*
key features, 6, 435–437, 455
kidneys, 307
leg muscles, 463–464
lung structure and skeletal
  pneumatization, 410, 436
lungs and lung ventilation,
  296–297
maniraptorans and, 428, 429
Mesozoic era, 231, 233
mosaic evolution and,
  431, 434–435
nitrogenous end products, 308
number of extant species, 2
orientation, navigation, and
  migration, 483–486
origin, 399
parental care, 10, 441, 480–483
pennaceous feathers, 436–437
phylogeny of, *456*
phylogeny of amniotes, *209*
phylogeny of extant
  vertebrates, *4*
radiation in the Cretaceous, 434
reproduction, 439, 478–480
respiratory turbinates, 257
rest-phase torpor, 320–321, 322
salt-secreting glands, 308, *309*
sensory systems, 474–477
skull, 435
smallest bird, *329*
syrinx, 249
systemic arches, 297, *298*
time tree and phylogeny of
  theropods, 423, *424*
time tree of Avemetatarsalia,
  398
tongues and feeding, 248
wings (*see* Bird flight; Bird
  wings)
Birth position, ichthyosaurs,
  228, *230*
Bison bison (American bison), *414*
Biting, 115
Bitis, 339, *340*
Bitis arietans (puff adder), *340*, 347
Black rhinoceros, 552, *553*
Black vulture (*Coragyps atratus*), 476
Black-and-white tegu (*Salvator
  merianae*), 320, 348, 392
Black-footed albatross
  (*Phoebastria nigripes*), 309
Black-horned capuchin (*Sapajus
  nigritus*), 562, *563*
Blacksmith treefrog (*Boana faber*),
  274
Blacktip shark (*Carcharhinus
  limbatus*), 116
Black-water diving, 167–169
Bladder. *See* Gas bladders;
  Urinary bladder
Blastopore, 22
Blind snakes, 338, *341*
Blood
  circulatory system, 36–38
  general characteristics, 36–37
  osmolality of vertebrates, *78*
  urochordates, 24

vertebrates and chordates
  compared, 25, *26*
See also Circulation
Blood pressure
  closed circulatory systems, 37
  crocodylians, *300*, 301–302
  mammals and birds, 297, 299,
    300
  terrestrial tetrapods, 249
  turtles and lepidosaurs, 300
Blood vessels
  closed circulatory systems, 37
  countercurrent exchange, 67
  gas glands and buoyancy in
    teleosts, 69
  internal gills, *66*, 67
  *See also individual arteries and
    veins*
Blowholes, 550
Blubber, 547
Blue grosbeak (*Passerina caerulea*),
  471
Blue monkey (*Cercopithecus mitis
  stuhlmanni*), 564
Blue shiner minnow (*Cyprinella
  caerulea*), 147
Blue whale (*Balaenoptera musculus*),
  519, 529, 548, 549
Bluefin tuna, 153
Blue-footed booby (*Sula nebouxii*),
  470
Bluegill sunfish (*Lepomis macro-
  chirus*), 164
Bluehead wrasse (*Thalassoma
  bifasciatum*), 166, 167
Blue-legged mantella (*Mantella
  expectata*), 286
Bluespotted mudskipper (*Boleop
  hthalmus boddarti*), 152
Blunt-headed tree snake
  (*Imantodes cenchoa*), 340
Bluntnose sixgill shark
  (*Hexanchus griseus*), 113
BMPs. *See* Bone morphogenetic
  proteins
Boana faber (blacksmith treefrog),
  274
Boana punctata (dotted treefrog), 261
Boas, 341, 449
Bobcat, 529
Bocaccio rockfish (*Sebastes
  paucispinis*), 165
Body development, 27–28
Body fluids. *See* Osmoregulation
Body forms
  anurans, 265
  convergence in predatory
    marine tetrapods, 232, *233*
  heterochrony and, 13–14
  lizards, 334–336
  locomotion in salamanders, 261
  snakes, 339–340
Body plans
  adaptation to life on land, 238
  evolution in turtles, 364, *365*
  gnathostomes, 52–55
  size and scaling of terrestrial
    tetrapods, 242–243
Body size
  birds, 437, 458
  members of the genus *Homo*,
    575, 576–578